area A	width w	base b	volume V
perimeter P	surface area S	circumference C	area of base B
length l	altitude (height) h	radius r	slant height s

Rectangle

$A = lw \qquad P = 2l + 2w$

Triangle

$A = \dfrac{1}{2}bh$

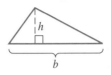

Square

$A = s^2 \qquad P = 4s$

Parallelogram

$A = bh$

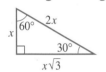

Trapezoid

$A = \dfrac{1}{2}h(b_1 + b_2)$

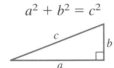

Circle

$A = \pi r^2 \qquad C = 2\pi r$

30°–60° Right Triangle

Right Triangle

$a^2 + b^2 = c^2$

Isosceles Right Triangle

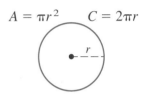

Right Circular Cylinder

$V = \pi r^2 h \qquad S = 2\pi r^2 + 2\pi rh$

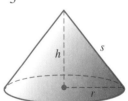

Sphere

$S = 4\pi r^2 \qquad V = \dfrac{4}{3}\pi r^3$

Right Circular Cone

$V = \dfrac{1}{3}\pi r^2 h \qquad S = \pi r^2 + \pi rs$

Pyramid

$V = \dfrac{1}{3}Bh$

Prism

$V = Bh$

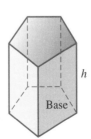

FOURTH EDITION

Elementary and Intermediate Algebra

A Combined Approach

JEROME E. KAUFMANN

KAREN L. SCHWITTERS
Seminole Community College

THOMSON

BROOKS/COLE

Australia • Canada • Mexico • Singapore • Spain
United Kingdom • United States

THOMSON
BROOKS/COLE

Elementary and Intermediate Algebra: A Combined Approach, Fourth Edition
Jerome E. Kaufmann, Karen L. Schwitters

Executive Editor: *Jennifer Huber Laugier*
Development Editor: *Kirsten Markson*
Assistant Editor: *Rebecca Subity*
Editorial Assistant: *Sarah Woicicki*
Technology Project Manager: *Rachael Sturgeon*
Marketing Manager: *Greta Kleinart*
Marketing Assistant: *Jessica Bothwell*
Advertising Project Manager: *Bryan Vann*
Project Manager, Editorial Production: *Hal Humphrey*
Print/Media Buyer: *Barbara Britton*
Production Service: *Susan Graham*
Art Director: *Vernon Boes*

Permissions Editor: *Chelsea Junget*
Text Designer: *Carolyn Deacy*
Photo Researcher: *Sarah Evertson*
Copy Editor: *Susan Graham*
Illustrators: *Network Graphics and Scientific Illustrators*
Cover Designer: *Roger Knox*
Cover Image: *Rowan Moore*
Cover Printer: *Phoenix Color Corp*
Compositor: *G & S Typesetting, Inc.*
Printer: *Quebecor World/Taunton*

For more information about our products, contact us at:
Thomson Learning Academic Resource Center
1-800-423-0563
For permission to use material from this text or product, submit a request online at http://www.thomsonrights.com.
Any additional questions about permissions can be submitted by email to thomsonrights@thomson.com.

Student Edition: ISBN 0-534-49024-7
Annotated Instructor's Edition: ISBN 0-534-49026-3

Thomson Higher Education
10 Davis Drive
Belmont, CA 94002–3098
USA

Asia (including India)
Thomson Learning
5 Shenton Way #01-01
UIC Building
Singapore 068808

Australia/New Zealand
Thomson Learning Australia
102 Dodds Street
Southbank, Victoria 3006
Australia

Canada
Thomson Nelson
1120 Birchmount Road
Toronto, Ontario M1K 5G4
Canada

UK/Europe/Middle East/Africa
Thomson Learning
High Holborn House
50/51 Bedford Row
London WC1R 4LR
United Kingdom

Latin America
Thomson Learning
Seneca, 53
Colonia Polanco
11560 Mexico
D.F. Mexico

Spain (including Portugal)
Thomson Paraninfo
Calle Magallanes, 25
28015 Madrid, Spain

Contents

CHAPTER 7

Factoring, Solving Equations, and Problem Solving   296

CHAPTER 8

A Transition from Elementary Algebra to Intermediate Algebra   338

CHAPTER 9

Rational Expressions   403

CHAPTER 16

Preface

Elementary and Intermediate Algebra: A Combined Approach, Fourth Edition, presents the basic topics of both elementary and intermediate algebra. By combining these topics in one text, we present an organizational format that allows for frequent reinforcement of concepts but eliminates the need to completely reintroduce topics as when two separate texts are used. At any time, if necessary, one can return to the original introduction of a particular topic.

The basic concepts of elementary and intermediate algebra are developed in a logical sequence, but in an easy-to-read manner without excessive technical vocabulary and formalism. *Whenever possible, the algebraic concepts are allowed to develop from their arithmetic counterparts.* The following are two specific examples of this development.

1. Manipulation with simple algebraic fractions begins early (Sections 2.1 and 2.2) when reviewing operations with rational numbers.
2. Manipulation with monomials, without any of the formal vocabulary, is introduced in Section 2.4 when working with exponents.

In the preparation of this edition, special effort was made to incorporate improvements suggested by reviewers and users of the previous editions while preserving the book's many successful features.

New in This Edition

- Every chapter opens with a project problem that uses the Infotrac® College Edition database to research the problem.
- Every section has a Concept Quiz that immediately precedes the problem set. These questions are predominately true/false questions that allow students to check their understanding of the mathematical concepts and definitions introduced in the section. Answers for these questions are located at the end of the respective problem sets.
- Every section now opens with learning objectives. Problem sets contain ample problems so that students can master these objectives.

- Section 3.4 now includes material on solving equations that are identities or contradictions.
- The solutions to inequalities in Section 3.5 are expressed in both set notation and interval notation. Both notations will be used until Chapter 8 when interval notation becomes the typical format. This allows the instructors to choose their preferred format for beginning algebra.
- Section 5.1 now includes material on plotting points in the Cartesian coordinate system.
- The slope of a line has been moved from Section 12.3 to Section 5.3. This allows slope to be introduced immediately after graphing linear equations.
- Section 12.1 now covers a review of slope, distance, and graphing techniques.
- Problems in Chapter 15 have been revised to include more solutions in fraction form, null solution sets, and infinite solution sets.

Other Special Features

- Icons found throughout the text point students to material contained on the Brooks/Cole Website and the Interactive Video Skillbuilder CD-ROM.
- There is a common thread throughout the text, which is "learn a skill, use the skill to solve equations and inequalities, and then use equations and inequalities as problem solving tools." This thread influenced some other decisions.

  1. Approximately 800 word problems are scattered throughout the text. These problems deal with a large variety of applications and constantly show the connections between mathematics and the world around us.
  2. *Many problem-solving suggestions are offered throughout, with special discussions in several sections.* The problem-solving suggestions are demonstrated in more than 100 worked-out examples.
  3. Newly acquired skills are used as soon as possible to solve equations and inequalities. Therefore, equations and inequalities are introduced early in the text and then used throughout in a large variety of problem solving situations.

- As recommended by the American Mathematical Association of Two-Year Colleges, many basic geometric concepts are integrated in problem-solving settings. *Approximately 20 worked-out examples and 200 problems are contained in this text that connect algebra, geometry, and our world.* (For examples, see Problems 17 and 26 on page 170.) The following geometric concepts are presented in problem-solving situations: complementary and supplementary angles, sum of measures of angles of a triangle equals 180°, area and volume formulas, perimeter and circumference formulas, ratio, proportion, Pythagorean theorem, isosceles right triangle, and 30°–60° right triangle relationships.

- The graphing calculator is introduced in Chapter 9 and is used primarily as a teaching tool. The students do not need a graphing calculator to study from this text. Starting in Chapter 9, graphing calculators are incorporated at times to enhance the learning of specific algebraic concepts. These examples are written so that students without a graphing calculator can read and benefit from them. Beginning with Problem Set 9.1, a group of problems called "Graphing Calculator Activities" is included in many of the problem sets. *These activities, which are especially good for small group work, are designed to reinforce concepts (see, for example, Problem Set 12.1), as well as lay groundwork for concepts about to be discussed (see, for example, Problem Set 12.2).* Some of these activities ask students to predict shapes and locations of graphs based on previous graphing experiences and then use the graphing calculator to check their predictions (see, for example, Problem Set 12.6). The graphing calculator is also used as a problem solving tool (see, for example, Problem Set 9.6).
- *Problems called "Thoughts into Words"* are included in every problem set except the review exercises. These problems are designed to encourage students to express in written form their thoughts about various mathematical ideas. For examples, see Problem Sets 2.1, 3.2, 4.4, and 8.1.
- *Problems called "Further Investigations" appear in many of the problem sets. These are "extras" that lend themselves to individual or small group work.* These problems encompass a variety of ideas: some exhibit different approaches to topics covered in the text, some are proofs, some bring in supplementary topics and relationships, and some are more challenging problems. These problems add variety and flexibility to the problem sets, but they could be omitted entirely without disrupting the continuity pattern of the text. For examples, see Problem Sets 1.2, 4.2, 6.3, 7.3, and 8.3.
- Problem solving is a key issue throughout the text. Not only are equations, systems of equations, and inequalities used as problem solving tools, but many other concepts as well. Functions, the distance formula, slope, arithmetic and geometric sequences, the fundamental principle of counting, permutations, combinations, and some basic probability concepts are all developed and then used to solve problems.
- Specific graphing ideas (intercepts, symmetry, restrictions, asymptotes, and transformations) are introduced and used in Chapters 5, 12, 13, and 14. In Section 13.3 the work with parabolas from Chapter 12 is used to motivate definitions for translations, reflections, stretchings, and shrinkings. These transformations are then applied to the graphs of

$$f(x) = x^3, \qquad f(x) = \frac{1}{x}, \qquad f(x) = \sqrt{x}, \qquad \text{and} \qquad f(x) = |x|$$

- All answers for Chapter Review Problems Sets, Chapter Tests, Cumulative Review Problem Sets, and Cumulative Practice Tests appear in the back of the text, along with answers to the odd-numbered problems.
- Please note the exceptionally pleasing design features of this text, including the functional use of color. The open format makes for a continuous

and easy reading flow of material instead of working through a maze of flags, caution symbols, reminder symbols, etc.

Ancillaries

For the Instructor

Annotated Instructor's Edition. This special version of the text contains a Resource Integration Guide that provides chapter-by-chapter correlations to the student and instructor resources available with this textbook. In the AIE, answers are printed next to all respective exercises. Graphs, tables, and other answers appear in an answer section at the back of the text.

Test Bank. The *Test Bank* includes eight tests per chapter as well as three final exams. The tests are made up of a combination of multiple-choice, free-response, true/false, and fill-in-the-blank questions.

Complete Solutions Manual. The *Complete Solutions Manual* provides worked-out solutions to all of the problems in the text.

iLrn Instructor Version. With a balance of efficiency and high performance, simplicity and versatility, *iLrn* gives instructors the power to transform the learning and teaching experience. *iLrn Instructor Version* is made up of two components, *iLrn Testing* and *iLrn Tutorial*. *iLrn Testing* is a revolutionary, Internet-ready, text-specific testing suite that allows instructors to customize exams and track student progress in an accessible, browser-based format. *iLrn* offers full algorithmic generation of problems as well as free-response problems using intuitive mathematical notation. *iLrn Tutorial* is a text-specific, interactive tutorial software program delivered via the Web (at *http://www.iLrn.com*) and is offered in both student and instructor versions. Like *iLrn Testing*, it is browser-based, making it an intuitive mathematical guide even for students with little technological proficiency. So sophisticated, it's simple, *iLrn Tutorial* allows students to work with real math notation in real time, providing instant analysis and feedback. The tracking program built into the instructor version of the software enables instructors to carefully monitor student progress, and the complete integration of the testing, tutorial, and course management components simplifies routine tasks. Results flow automatically to the grade book and instructors can easily communicate to individuals, sections, or entire courses. Visit our newest service for registered users of *iLrn*— the *iLrn* Service and Support Web site called Thomson Learning Connections (TLC). To access this site, go to *http://ilrn.thomsonlearningconnections.com* or click on the Instructor's Support/Technical Support links within *iLrn*.

Text-Specific Videotapes. These text-specific videotape sets, available at no charge to qualified adopters of the text, feature 10- to 20-minute problem-solving lessons that cover each section of every chapter.

For the Student

Student Solutions Manual. The *Student Solutions Manual* provides worked-out solutions to the odd-numbered problems in the text.

Web site (http://mathematics.brookscole.com). Instructors and students have access to a variety of teaching and learning resources. This Web site features everything from book-specific resources to newsgroups.

InfoTrac College Edition. With the adoption of this text, instructors and students will gain anytime, anywhere access to reliable resources with *Infotrac College Edition*, the online library! This fully searchable database offers more than 20 years' worth of full-text articles (not abstracts) from almost 5,000 diverse sources, such as top academic journals, newsletters, and up-to-the-minute periodicals including *Time, Newsweek, Science, Forbes,* and *USA Today*.

iLrn Tutorial Student Version. This text-specific, interactive tutorial software is delivered via the Web (at http://brookscole.com). It is browser-based, making it an intuitive mathematical guide, even for students with little technological proficiency. So sophisticated, it's simple, *iLrn Tutorial* allows students to work with real math notation in real time, providing instant analysis and feedback.

Interactive Video Skillbuilder CD-ROM. The *Interactive Video Skillbuilder CD-ROM* contains video instruction covering each chapter of the text. The problems worked during each video lesson are shown first so that students can try working them before watching the solution. To help students evaluate their progress, each section contains a 10-question Web quiz (the results of which can be emailed to the instructor), and each chapter contains a chapter test with answers to each problem on each test. This dual-platform CD-ROM also includes MathCue tutorial and quizzing software, featuring a Skill Builder that presents problems to solve and evaluates answers with step-by-step explanations; a Quiz function that enables students to generate quiz problems keyed to problem types from each section of the book; a Chapter Test that provides many problems keyed to problem types from each chapter; and a Solution Finder that allows students to enter their own basic problems and receive step-by-step help as if they were working with a tutor. Also new, English/Spanish closed caption translations can be selected to display along with the video instruction. Examples in the book that are taught on video are identified by this logo in the margin.

WebTutor™ Toolbox. Preloaded with content and available via a free access code when packaged with this text, *WebTutor Toolbox* pairs all the content of this text's rich Book Companion Web site with sophisticated course management functionality. Instructors can assign materials (including online quizzes) and have the results flow automatically to the grade book. *WebTutor Toolbox* is ready to use as soon as instructors log on — or instructors can customize its preloaded content by uploading images and other resources, adding weblinks, or creating their own practice materials. Students only have access to student resources on the website. Instructors can enter an access code for password-protected Instructor Resources.

Explorations in Beginning and Intermediate Algebra Using the TI-82/83 with Integrated Appendix Notes for the TI-85/86, Second Edition (0-534-36149-8)
Deborah J. Cochener and Bonnie M. Hodge, both of Austin Peay State University
This user-friendly workbook improves student understanding and retention of algebra concepts through a series of activities and guided explorations using the

graphing calculator. By clearly and succinctly teaching keystrokes, class time is devoted to investigations instead of how to use a graphing calculator.

The Math Student's Guide to the TI-83 Graphing Calculator (0-534-37802-1)
The Math Student's Guide to the TI-86 Graphing Calculator (0-534-37801-3)
The Math Student's Guide to the TI-83 Plus Graphing Calculator (0-534-42021-4)
The Math Student's Guide to the TI-89 Graphing Calculator (0-534-42022-2)
Trish Cabral of Butte College
These videos are designed for students who are new to the graphing calculator or for those who would like to brush up on their skills. Each instructional graphing calculator videotape covers basic calculations, the custom menu, graphing, advanced graphing, matrix operations, trigonometry, parametric equations, polar coordinates, calculus, Statistics I and one-variable data, and Statistics II with linear regression. These wonderful tools are each 105 minutes in length and cover all of the important functions of a graphing calculator.

Mastering Mathematics: How to Be a Great Math Student, Third Edition (0-534-34947-1)
Richard Manning Smith, Bryant College
Providing solid tips for every stage of study, *Mastering Mathematics* stresses the importance of a positive attitude and gives students the tools to succeed in their math course.

Activities Manual for Beginning and Intermediate Algebra, Second Edition
Instructor Edition (0-534-99874-7); Student Edition (0-534-99873-9)
Debbie Garrison, Judy Jones, and Jolene Rhodes, all of Valencia Community College
Designed as a stand-alone supplement for any beginning or intermediate algebra text, *Activities Manual for Beginning and Intermediate Algebra* is a collection of activities written to incorporate the recommendations from the NCTM and from AMATYC's Crossroads. Activities can be used during class or in a laboratory setting to introduce, teach, or reinforce a topic.

Conquering Math Anxiety: A Self-Help Workbook, Second Edition (0-534-38634-2)
Cynthia Arem, Pima Community College
A comprehensive workbook that provides a variety of exercises and worksheets along with detailed explanations of methods to help "math-anxious" students deal with and overcome math fears. This edition now comes with a free relaxation CD-ROM and a detailed list of Internet resources.

Active Arithmetic and Algebra: Activities for Prealgebra and Beginning Algebra (0-534-36771-2)
Judy Jones, Valencia Community College
This activities manual includes a variety of approaches to learning mathematical concepts. Sixteen activities, including puzzles, games, data collection, graphing, and writing activities are included.

Math Facts: Survival Guide to Basic Mathematics, Second Edition (0-534-94734-4)
Algebra Facts: Survival Guide to Basic Algebra (0-534-19986-0)

Theodore John Szymanski, Tompkins-Cortland Community College
This booklet gives easy access to the most crucial concepts and formulas in algebra. Although it is bound, this booklet is structured to work like flash cards.

Acknowledgments

We would like to take this opportunity to thank the following people who served as reviewers for the third and fourth editions of *Elementary and Intermediate Algebra: A Combined Approach* and for the other books in the fourth edition series:

Jan Archibald
Ventura College

Helen Banes
Kirkwood Community College

Paul Raymond Bedard
St. Clair Community College

Julie Reif Bonds
Sonoma State University

Tracy M. Boone
Pennsylvania State University–Altoona

Bethany Chandler
Butler County Community College

Joseph F. Cleary
Massasoit Community College

Delanie Cochran
Indiana University SE

Ronald Conley
Tusculum College

Daniel J. Cronin
New Hampshire Technical Institute

Cynthia Fleck
Wright State University

Corinna M. Goehring
Jackson State Community College

Rockham Goodlet
Howard University

Kay Haralson
Austin Peay State University

Gary Hart
California State University–Dominguez Hills

Lester A. Hemenway
Los Angeles Mission College

Eric Hofelich
Kaskaskia College

Linda Horner
Broward Community College

Laurence C. Huddy, Jr.
Horry-Georgetown Technical College

Josephine Johansen
Rutgers University

Bill Keating
Pikes Peak Community College

Ronnie Kreis
Tusculum College

Gayle L. Krzemien
Pikes Peak Community College

Susan Kutryb
Hudson Valley Community College

Thomas Lawson
Howard University

Alisa Carter Lewis
Tyler Junior College

Sandra Mayo
Los Angeles Mission College

Kathryn T. McClellan
Tarrant County Junior College–Northeast

Jamie McGill
East Tennessee State University

Reed Parr
Salt Lake Community College

C. L. Pinchback
University of Central Arkansas

William Radulovich
Florida Community College at Jacksonville

Rose S. Rheim
Indiana University SE

Karen Rogers Rollins
State University of West Georgia

Martha W. Scarbrough
Motlow State Community College

Mary Lee Seit
Erie Community College

Karen Sharp
Mott Community College

Lawrence Small
Los Angeles Pierce College

Erin Spicer
West Montana College

Katherine R. Struve
Columbus State Community College

Molly Sumner
Pikes Peak Community College

Richard D. Townsend
North Carolina Central University

Mary Voxman
University of Idaho

Dennis W. Watson
Clark College

We are very grateful to the staff of Brooks/Cole, especially Jennifer Huber, Kirsten Markson, Rebecca Subity, and Sarah Woicicki, for their continuous cooperation and assistance throughout this project. We would also like to express our sincere gratitude to Susan Graham and to Hal Humphrey. They continue to make life as an author so much easier by carrying out the details of production in a dedicated and caring way. Additional thanks are due to Arlene Kaufmann who spends numerous hours reading page proofs.

Jerome E. Kaufmann
Karen L. Schwitters

AP/Wide World Photos

Positive and negative integers are used to represent golf scores. A positive integer means over par and a negative integer means under par.

Some Basic Concepts of Arithmetic and Algebra

Karla started 2005 with $500 in her savings account, and she planned to save an additional $50 per month for all of 2005. If we disregard any accumulated interest, the numerical expression $500 + 12(50)$ represents the amount in her savings account at the end of 2005.

The numbers $+2$, -1, -3, $+1$, and -4 are Woody's scores relative to par for five rounds of golf. The numerical expression $2 + (-1) + (-3) + 1 + (-4)$ can be used to determine where Woody stands relative to par at the end of the five rounds.

The temperature at 4 A.M. was $-14°$F. By noon the temperature had increased by $23°$F. The numerical expression $-14 + 23$ gives the temperature at noon.

In the first two chapters of this text, the concept of a **numerical expression** is used as a basis for reviewing addition, subtraction, multiplication, and division of various kinds of numbers. Then the concept of a **variable** enables us to move from numerical

1

expressions to **algebraic expressions** — that is, to start the transition from arithmetic to algebra. Keep in mind that algebra is simply a generalized approach to arithmetic. Many algebraic concepts are extensions of arithmetic ideas. Therefore, we will be using your knowledge of arithmetic to help you with the study of algebra.

InfoTrac Project

Do a subject guide search for **prime numbers**. Find a newspaper article identifying the largest prime number found to date. What are some of the reasons given for finding large prime numbers, and what are some of the modern day uses for them? Find 5 prime numbers greater than 100. Now factor the following number into primes: 3042.

1.1   Numerical and Algebraic Expressions

Objectives

- Recognize basic vocabulary and symbols associated with sets.
- Simplify numerical expressions according to the order of operations.
- Evaluate algebraic expressions.

In arithmetic, we use symbols such as 4, 8, 17, and π to represent numbers. We indicate the basic operations of addition, subtraction, multiplication, and division by the symbols $+$, $-$, $\cdot$, and $\div$, respectively. Thus we can formulate specific **numerical expressions.** For example, we can write the indicated sum of eight and four as $8 + 4$.

In algebra, the concept of a **variable** provides the basis for generalizing. By using x and y to represent *any* number, we can use the expression $x + y$ to represent the indicated sum of *any two* numbers. The x and y in such an expression are called *variables,* and the phrase $x + y$ is called an **algebraic expression.** We commonly use letters of the alphabet such as x, y, z, and w as variables. The key idea is that they represent numbers; therefore, as we review various operations and properties pertaining to numbers, we are building the foundation for our study of algebra.

Many of the notational agreements made in arithmetic are extended to algebra with a few slight modifications. The following chart summarizes the notational agreements pertaining to the four basic operations. Note the variety of ways to write a product by using parentheses to indicate multiplication. Actually, the *ab* form is the simplest and the form probably used most often; expressions such as *abc*, 6*x*, and 7*xyz* all indicate multiplication. Also note the various forms for indicating division; the fractional form, $\dfrac{c}{d}$, is generally used in algebra, although the other forms do serve a purpose at times.

Operation	Arithmetic	Algebra	Vocabulary
Addition	$4 + 6$	$x + y$	The *sum* of x and y
Subtraction	$7 - 2$	$w - z$	The *difference* of w and z
Multiplication	$9 \cdot 8$	$a \cdot b, a(b), (a)b,$ $(a)(b),$ or ab	The *product* of a and b
Division	$8 \div 2, \dfrac{8}{2}, 2\overline{)8}$	$c \div d, \dfrac{c}{d},$ or $d\overline{)c}$	The *quotient* of c and d

As we review arithmetic ideas and introduce algebraic concepts, it is convenient to use some of the basic vocabulary and symbols associated with sets. A **set** is a collection of objects, and the objects are called **elements** or **members** of the set. In arithmetic and algebra, the elements of a set are often numbers. To communicate about sets, we use set braces, { }, to enclose the elements (or a description of the elements), and we use capital letters to name sets. For example, we can represent a set A, which consists of the vowels of the alphabet, in these ways:

$A = \{$vowels of the alphabet$\}$ Word description, or

$A = \{a, e, i, o, u\}$ List or roster description

We can modify the listing approach if the number of elements is quite large. For example, all the letters of the alphabet can be listed as

$\{a, b, c, \ldots, z\}$

We simply begin by writing enough elements to establish a pattern, and then the three dots indicate that the set continues in that pattern. The final entry indicates the last element of the pattern. If we write

$\{1, 2, 3, \ldots\}$

the set begins with the counting numbers, 1, 2, and 3. The three dots indicate that it continues in a like manner forever; there is no last element. A set that consists of no elements is called the **null set** (written $\varnothing$).

Two sets are said to be *equal* if they contain exactly the same elements. For example,

$\{1, 2, 3\} = \{2, 1, 3\}$

because both sets contain the same elements; the order in which the elements are written doesn't matter. The slash mark through the equality symbol denotes "not equal to." Thus if $A = \{1, 2, 3\}$ and $B = \{1, 2, 3, 4\}$, we can write $A \neq B$, which we read as "set A is not equal to set B."

Simplifying Numerical Expressions

Now let's simplify some numerical expressions that involve the set of **whole numbers** — that is, the set $\{0, 1, 2, 3, \ldots\}$.

EXAMPLE 1

Simplify $8 + 7 - 4 + 12 - 7 + 14$.

Solution

The additions and subtractions should be performed from left to right in the order in which they appear. Thus $8 + 7 - 4 + 12 - 7 + 14$ simplifies to 30. ■

EXAMPLE 2

Simplify $7(9 + 5)$.

Solution

The parentheses indicate the product of 7 and the quantity $9 + 5$. Perform the addition inside the parentheses first, and then multiply; $7(9 + 5)$ thus simplifies to $7(14)$, which is 98. ■

EXAMPLE 3

Simplify $(7 + 8) \div (4 - 1)$.

Solution

First, we perform the operations inside the parentheses; $(7 + 8) \div (4 - 1)$ thus becomes $15 \div 3$, which is 5. ■

We frequently express a problem such as Example 3 in the form $\dfrac{7 + 8}{4 - 1}$. We don't need parentheses in this case because the fraction bar indicates that the sum of 7 and 8 is to be divided by the difference $4 - 1$. A problem may, however, contain both parentheses and fraction bars, as the next example illustrates.

EXAMPLE 4

Simplify $\dfrac{(4 + 2)(7 - 1)}{9} + \dfrac{7}{10 - 3}$.

Solution

First simplify above and below the fraction bars, and then proceed to evaluate as follows:

$$\frac{(4 + 2)(7 - 1)}{9} + \frac{7}{10 - 3} = \frac{(6)(6)}{9} + \frac{7}{7}$$

$$= \frac{36}{9} + 1 = 4 + 1 = 5$$

■

EXAMPLE 5

Simplify $7 \cdot 9 + 5$.

Solution

If there are no parentheses to indicate otherwise, multiplication takes precedence over addition. First perform the multiplication, and then do the addition; $7 \cdot 9 + 5$ therefore simplifies to $63 + 5$, which is 68. ■

(Compare Example 2 and Example 5, and note the difference in meaning.)

EXAMPLE 6   Simplify $8 + 4 \cdot 3 - 14 \div 2$.

Solution

The multiplication and division should be done first in the order in which they appear, from left to right. Thus $8 + 4 \cdot 3 - 14 \div 2$ simplifies to $8 + 12 - 7$. We then perform the addition and subtraction in the order in which they appear, which simplifies $8 + 12 - 7$ to 13. ■

EXAMPLE 7   Simplify $18 \div 3 \cdot 2 + 8 \cdot 10 \div 2$.

Solution

If we perform the multiplications and divisions first in the order in which they appear, and then do the additions and subtractions, our work takes on the following format:

$$18 \div 3 \cdot 2 + 8 \cdot 10 \div 2 = 6 \cdot 2 + 80 \div 2 = 12 + 40 = 52$$ ■

EXAMPLE 8   Simplify $5 + 6[2(3 + 9)]$.

Solution

We use brackets for the same purpose as parentheses. In such a problem, we need to simplify *from the inside out,* performing the operations in the innermost parentheses first.

$$\begin{aligned} 5 + 6[2(3 + 9)] &= 5 + 6[2(12)] \\ &= 5 + 6[24] \\ &= 5 + 144 = 149 \end{aligned}$$ ■

Let us now summarize the ideas presented in the preceding examples on **simplifying numerical expressions.** When we simplify a numerical expression, the operations should be performed in the order listed below:

Order of Operations

1. Perform the operations inside the symbols of inclusion (parentheses and brackets) and above and below each fraction bar. Start with the innermost inclusion symbol.
2. Perform all multiplications and divisions in the order in which they appear from left to right.
3. Perform all additions and subtractions in the order in which they appear from left to right.

Evaluating Algebraic Expressions

We can use the concept of a variable to generalize from numerical expressions to algebraic expressions. Each of the following is an example of an algebraic expression:

$$3x + 2y \qquad 5a - 2b + c \qquad 7(w + z)$$

$$\frac{5d + 3e}{2c - d} \qquad 2xy + 5yz \qquad (x + y)(x - y)$$

An algebraic expression takes on a numerical value whenever each variable in the expression is replaced by a specific number. For example, if x is replaced by 9 and z by 4, the algebraic expression $x - z$ becomes the numerical expression $9 - 4$, which simplifies to 5. We say that $x - z$ **has a value** of 5 when x equals 9 and z equals 4. The value of $x - z$, when x equals 25 and z equals 12, is 13. The general algebraic expression $x - z$ has a specific value each time x and z are replaced by numbers.

Consider the next examples, which illustrate the process of finding a value of an algebraic expression. The process is often referred to as **evaluating algebraic expressions.**

E X A M P L E   9   Find the value of $3x + 2y$ when x is replaced by 5 and y by 17.

Solution

The following format is convenient for such problems:

$$
\begin{aligned}
3x + 2y &= 3(5) + 2(17) \quad \text{when } x = 5 \text{ and } y = 17 \\
&= 15 + 34 \\
&= 49
\end{aligned}
$$

■

In Example 9, for the algebraic expression, $3x + 2y$, note that the multiplications "3 times x" and "2 times y" are implied without the use of parentheses. The algebraic expression switches to a numerical expression when numbers are substituted for variables; in that case, parentheses are used to indicate the multiplication. We could also use the raised dot to indicate multiplication; that is, $3(5) + 2(17)$ could be written as $3 \cdot 5 + 2 \cdot 17$. Furthermore, note that once we have a numerical expression, our previous agreements for simplifying numerical expressions are in effect.

E X A M P L E   1 0   Find the value of $12a - 3b$ when $a = 5$ and $b = 9$.

Solution

$$
\begin{aligned}
12a - 3b &= 12(5) - 3(9) \quad \text{when } a = 5 \text{ and } b = 9 \\
&= 60 - 27 \\
&= 33
\end{aligned}
$$

■

EXAMPLE   11   Evaluate $4xy + 2xz - 3yz$ when $x = 8$, $y = 6$, and $z = 2$.

Solution

$$4xy + 2xz - 3yz = 4(8)(6) + 2(8)(2) - 3(6)(2) \quad \text{when } x = 8, y = 6, \text{ and}$$
$$z = 2$$
$$= 192 + 32 - 36$$
$$= 188$$

EXAMPLE   12   Evaluate $\dfrac{5c + d}{3c - d}$ for $c = 12$ and $d = 4$.

Solution

$$\frac{5c + d}{3c - d} = \frac{5(12) + 4}{3(12) - 4} \quad \text{for } c = 12 \text{ and } d = 4$$
$$= \frac{60 + 4}{36 - 4}$$
$$= \frac{64}{32}$$
$$= 2$$

EXAMPLE   13   Evaluate $(2x + 5y)(3x - 2y)$ when $x = 6$ and $y = 3$.

Solution

$$(2x + 5y)(3x - 2y) = (2 \cdot 6 + 5 \cdot 3)(3 \cdot 6 - 2 \cdot 3) \quad \text{when } x = 6 \text{ and}$$
$$y = 3$$
$$= (12 + 15)(18 - 6)$$
$$= (27)(12)$$
$$= 324$$

CONCEPT QUIZ

For Problems 1–10, answer true or false.

1. The expression ab indicates the sum of a and b.

2. Any of the following notations, $(a)b, a \cdot b, a(b)$, can be used to indicate the product of a and b.

3. The phrase $2x + y - 4z$ is called an algebraic expression.

4. A set is a collection of objects, and the objects are called terms.

5. The sets $\{2, 4, 6, 8\}$ and $\{6, 4, 8, 2\}$ are equal.

6. The set $\{1, 3, 5, 7, \ldots\}$ has a last element of 99.

7. The null set has one element.

8. To evaluate $24 \div 6 \cdot 2$, the first operation that should be performed is to multiply 6 times 2.

9. To evaluate $6 + 8 \cdot 3$, the first operation that should be performed is to multiply 8 times 3.

10. The algebraic expression $2(x + y)$ simplifies to 24, if x is replaced by 10 and y is replaced by 0.

PROBLEM SET 1.1

For Problems 1–34, simplify each of the numerical expressions.

1. $9 + 14 - 7$

2. $32 - 14 + 6$

3. $7(14 - 9)$

4. $8(6 + 12)$

5. $16 + 5 \cdot 7$

6. $18 - 3(5)$

7. $4(12 + 9) - 3(8 - 4)$

8. $7(13 - 4) - 2(19 - 11)$

9. $4(7) + 6(9)$

10. $8(7) - 4(8)$

11. $6 \cdot 7 + 5 \cdot 8 - 3 \cdot 9$

12. $8(13) - 4(9) + 2(7)$

13. $(6 + 9)(8 - 4)$

14. $(15 - 6)(13 - 4)$

15. $6 + 4[3(9 - 4)]$

16. $92 - 3[2(5 - 2)]$

17. $16 \div 8 \cdot 4 + 36 \div 4 \cdot 2$

18. $7 \cdot 8 \div 4 - 72 \div 12$

19. $\dfrac{8 + 12}{4} - \dfrac{9 + 15}{8}$

20. $\dfrac{19 - 7}{6} + \dfrac{38 - 14}{3}$

21. $56 - [3(9 - 6)]$

22. $17 + 2[3(4 - 2)]$

23. $7 \cdot 4 \cdot 2 \div 8 + 14$

24. $14 \div 7 \cdot 8 - 35 \div 7 \cdot 2$

25. $32 \div 8 \cdot 2 + 24 \div 6 - 1$

26. $48 \div 12 + 7 \cdot 2 \div 2 - 1$

27. $4 \cdot 9 \div 12 + 18 \div 2 + 3$

28. $5 \cdot 8 \div 4 - 8 \div 4 \cdot 3 + 6$

29. $\dfrac{6(8 - 3)}{3} + \dfrac{12(7 - 4)}{9}$

30. $\dfrac{3(17 - 9)}{4} + \dfrac{9(16 - 7)}{3}$

31. $83 - \dfrac{4(12 - 7)}{5}$

32. $78 - \dfrac{6(21 - 9)}{4}$

33. $\dfrac{4 \cdot 6 + 5 \cdot 3}{7 + 2 \cdot 3} + \dfrac{7 \cdot 9 + 6 \cdot 5}{3 \cdot 5 + 8 \cdot 2}$

34. $\dfrac{7 \cdot 8 + 4}{5 \cdot 8 - 10} + \dfrac{9 \cdot 6 - 4}{6 \cdot 5 - 20}$

For Problems 35–54, evaluate each algebraic expression for the given values of the variables.

35. $7x + 4y$ for $x = 6$ and $y = 8$

36. $8x + 6y$ for $x = 9$ and $y = 5$

37. $16a - 9b$ for $a = 3$ and $b = 4$

38. $14a - 5b$ for $a = 7$ and $b = 9$

39. $4x + 7y + 3xy$ for $x = 4$ and $y = 9$

40. $x + 8y + 5xy$ for $x = 12$ and $y = 3$

41. $14xz + 6xy - 4yz$ for $x = 8$, $y = 5$, and $z = 7$

42. $9xy - 4xz + 3yz$ for $x = 7$, $y = 3$, and $z = 2$

43. $\dfrac{54}{n} + \dfrac{n}{3}$ for $n = 9$

44. $\dfrac{n}{4} + \dfrac{60}{n} - \dfrac{n}{6}$ for $n = 12$

45. $\dfrac{y + 16}{6} + \dfrac{50 - y}{3}$ for $y = 8$

46. $\dfrac{w + 57}{9} + \dfrac{90 - w}{7}$ for $w = 6$

47. $(x + y)(x - y)$ for $x = 8$ and $y = 3$

48. $(x + 2y)(2x - y)$ for $x = 7$ and $y = 4$

49. $(5x - 2y)(3x + 4y)$ for $x = 3$ and $y = 6$

50. $(3a + b)(7a - 2b)$ for $a = 5$ and $b = 7$

51. $6 + 3[2(x + 4)]$ for $x = 7$

52. $9 + 4[3(x + 3)]$ for $x = 6$

53. $81 - 2[5(n + 4)]$ for $n = 3$

54. $78 - 3[4(n - 2)]$ for $n = 4$

For Problems 55–60, find the value of $\dfrac{bh}{2}$ for each set of values for the variables b and h.

55. $b = 8$ and $h = 12$ **56.** $b = 6$ and $h = 14$

57. $b = 7$ and $h = 6$ **58.** $b = 9$ and $h = 4$

59. $b = 16$ and $h = 5$ **60.** $b = 18$ and $h = 13$

For Problems 61–66, find the value of $\dfrac{h(b_1 + b_2)}{2}$ for each each set of values for the variables h, b_1, and b_2. (Subscripts are used to indicate that b_1 and b_2 are different variables.)

61. $h = 17$, $b_1 = 14$, and $b_2 = 6$

62. $h = 9$, $b_1 = 12$, and $b_2 = 16$

63. $h = 8$, $b_1 = 17$, and $b_2 = 24$

64. $h = 12$, $b_1 = 14$, and $b_2 = 5$

65. $h = 18$, $b_1 = 6$, and $b_2 = 11$

66. $h = 14$, $b_1 = 9$, and $b_2 = 7$

67. You should be able to do calculations like those in Problems 1–34 both *with* and *without* a calculator. Be sure that you can do Problems 1–34 with your calculator, and be sure to use the parentheses key when appropriate.

■ ■ ■ Thoughts into words

68. Explain the difference between a numerical expression and an algebraic expression.

69. Your friend keeps getting an answer of 45 when simplifying $3 + 2(9)$. What mistake is he making, and how would you help him?

■ ■ ■ Further investigations

Grouping symbols can affect the order in which the arithmetic operations are performed. For the following problems, insert parentheses so that the expression is equal to the given value.

70. Insert parentheses so that $36 + 12 \div 3 + 3 + 6 \cdot 2$ is equal to 20.

71. Insert parentheses so that $36 + 12 \div 3 + 3 + 6 \cdot 2$ is equal to 50.

72. Insert parentheses so that $36 + 12 \div 3 + 3 + 6 \cdot 2$ is equal to 38.

73. Insert parentheses so that $36 + 12 \div 3 + 3 + 6 \cdot 2$ is equal to 55.

Answers to Concept Quiz

1. False   **2.** True   **3.** True   **4.** False   **5.** True   **6.** False   **7.** False   **8.** False   **9.** True   **10.** False

1.2   Prime and Composite Numbers

Objectives

■   Identify whole numbers greater than one as prime or composite.

■   Factor a whole number into a product of prime numbers.

■   Find the greatest common factor of two or more whole numbers.

■   Find the least common multiple of two or more whole numbers.

Occasionally terms in mathematics have a special meaning in the discussion of a particular topic. Such is the case with the term "divides" as it is used in this section.

We say that 6 *divides* 18 because 6 times the whole number 3 produces 18, but 6 *does not divide* 19 because there is no whole number such that 6 times the number produces 19. Likewise 5 *divides* 35 because 5 times the whole number 7 produces 35, but 5 *does not divide* 42 because there is no whole number such that 5 times the number produces 42. We can use this general definition:

DEFINITION 1.1

Given that a and b are whole numbers, with a not equal to zero, a *divides* b if and only if there exists a whole number k such that $a \cdot k = b$.

REMARK: Note the use of the variables a, b, and k in the statement of a general definition. Also note that the definition merely generalizes the concept of "divides," which we introduced in the specific examples that precede the definition.

The following statements further clarify Definition 1.1. Pay special attention to the italicized words because they indicate some of the terminology used for this topic.

1. 8 *divides* 56 because $8 \cdot 7 = 56$.
2. 7 *does not divide* 38 because there is no whole number k such that $7 \cdot k = 38$.
3. 3 is a *factor* of 27 because $3 \cdot 9 = 27$.
4. 4 is *not a factor* of 38 because there is no whole number k such that $4 \cdot k = 38$.
5. 35 is a *multiple* of 5 because $5 \cdot 7 = 35$.
6. 29 is *not a multiple* of 7 because there is no whole number k such that $7 \cdot k = 29$.

The **factor** terminology is used extensively. We say that 7 and 8 are factors of 56 because $7 \cdot 8 = 56$; 4 and 14 are also factors of 56 because $4 \cdot 14 = 56$. The factors of a number are also the divisors of the number.

Now consider two special kinds of whole numbers called **prime numbers** and **composite numbers** according to the following definition.

DEFINITION 1.2

A **prime number** is a whole number, greater than 1, that has no factors (divisors) other than itself and 1. Whole numbers, greater than 1, that are not prime numbers are called **composite numbers.**

The prime numbers less than 50 are 2, 3, 5, 7, 11, 13, 17, 19, 23, 29, 31, 37, 41, 43, and 47. Note that each number has no factors other than itself and 1. An interesting point is that the set of prime numbers is an infinite set; that is, the prime numbers go on forever, and there is no *largest* prime number.

We can express every composite number as the indicated product of prime numbers. Consider these examples:

$$4 = 2 \cdot 2 \qquad 6 = 2 \cdot 3 \qquad 8 = 2 \cdot 2 \cdot 2 \qquad 10 = 2 \cdot 5 \qquad 12 = 2 \cdot 2 \cdot 3$$

The indicated product of prime numbers is sometimes called the **prime factored form** of the number.

We can use various procedures to find the prime factors of a given composite number. For our purposes, the simplest technique is to factor the composite number into any two easily recognized factors and then to continue to factor each of these until we obtain only prime factors. Consider these examples:

$$18 = 2 \cdot 9 = 2 \cdot 3 \cdot 3 \qquad\qquad 27 = 3 \cdot 9 = 3 \cdot 3 \cdot 3$$
$$24 = 4 \cdot 6 = 2 \cdot 2 \cdot 2 \cdot 3 \qquad 150 = 10 \cdot 15 = 2 \cdot 5 \cdot 3 \cdot 5$$

It does not matter which two factors we choose first. For example, we might start by expressing 18 as $3 \cdot 6$ and then factor 6 into $2 \cdot 3$, which produces a final result of $18 = 3 \cdot 2 \cdot 3$. Either way, 18 contains two prime factors of 3 and one prime factor of 2. The order in which we write the prime factors is not important.

Greatest Common Factor

We can use the prime factorization form of two composite numbers to conveniently find their **greatest common factor.** Consider this example:

$$42 = 2 \cdot 3 \cdot 7$$
$$70 = 2 \cdot 5 \cdot 7$$

Note that 2 is a factor of both, as is 7. Therefore, 14 (the product of 2 and 7) is the greatest common factor of 42 and 70. In other words, 14 is the largest whole number that divides both 42 and 70. The following examples should further clarify the process of finding the greatest common factor of two or more numbers.

EXAMPLE 1 Find the greatest common factor of 48 and 60.

Solution

$$48 = 2 \cdot 2 \cdot 2 \cdot 2 \cdot 3$$
$$60 = 2 \cdot 2 \cdot 3 \cdot 5$$

Because two 2s and a 3 are common to both, the greatest common factor of 48 and 60 is $2 \cdot 2 \cdot 3 = 12$. ∎

EXAMPLE 2 Find the greatest common factor of 21 and 75.

Solution

$$21 = 3 \cdot 7$$
$$75 = 3 \cdot 5 \cdot 5$$

Because only a 3 is common to both, the greatest common factor is 3. ∎

EXAMPLE 3

Find the greatest common factor of 24 and 35.

Solution

$$24 = 2 \cdot 2 \cdot 2 \cdot 3$$
$$35 = 5 \cdot 7$$

Because there are no common prime factors, the greatest common factor is 1. ■

The concept of greatest common factor can be extended to more than two numbers, as the next example demonstrates.

EXAMPLE 4

Find the greatest common factor of 24, 56, and 120.

Solution

$$24 = 2 \cdot 2 \cdot 2 \cdot 3$$
$$56 = 2 \cdot 2 \cdot 2 \cdot 7$$
$$120 = 2 \cdot 2 \cdot 2 \cdot 3 \cdot 5$$

Because three 2s are common to the numbers, the greatest common factor of 24, 56, and 120 is $2 \cdot 2 \cdot 2 = 8$. ■

Least Common Multiple

We stated that 35 is a *multiple of* 5 because $5 \cdot 7 = 35$. The set of all whole numbers that are multiples of 5 consists of 0, 5, 10, 15, 20, 25, and so on. In other words, 5 times each successive whole number ($5 \cdot 0 = \mathbf{0}, 5 \cdot 1 = \mathbf{5}, 5 \cdot 2 = \mathbf{10}, 5 \cdot 3 = \mathbf{15}$, etc.) produces the multiples of 5. In a like manner, the set of multiples of 4 consists of 0, 4, 8, 12, 16, and so on.

It is sometimes necessary to determine the smallest common *nonzero* multiple of two or more whole numbers. We use the phrase **least common multiple** to designate this nonzero number. For example, the least common multiple of 3 and 4 is 12, which means that 12 is the smallest nonzero multiple of both 3 and 4. Stated another way, 12 is the smallest nonzero whole number that is divisible by both 3 and 4. Likewise, we say that the least common multiple of 6 and 8 is 24.

If we cannot determine the least common multiple by inspection, then using the prime factorization form of composite numbers is helpful. Study the solutions to the following examples very carefully as we develop a systematic technique for finding the least common multiple of two or more numbers.

EXAMPLE 5

Find the least common multiple of 24 and 36.

Solution

Let's first express each number as a product of prime factors:

$$24 = 2 \cdot 2 \cdot 2 \cdot 3$$
$$36 = 2 \cdot 2 \cdot 3 \cdot 3$$

Because 24 contains three 2s, the least common multiple must have three 2s. Also, because 36 contains two 3s, we need to put two 3s in the least common multiple. The least common multiple of 24 and 36 is therefore $2 \cdot 2 \cdot 2 \cdot 3 \cdot 3 = 72$. ■

If the least common multiple is not obvious by inspection, then we can proceed as follows:

STEP 1   Express each number as a product of prime factors.

STEP 2   The least common multiple contains each different prime factor as many times as the *most* times it appears in any one of the factorizations from step 1.

E X A M P L E   6   Find the least common multiple of 48 and 84.

Solution

$$48 = 2 \cdot 2 \cdot 2 \cdot 2 \cdot 3$$
$$84 = 2 \cdot 2 \cdot 3 \cdot 7$$

We need four 2s in the least common multiple because of the four 2s in 48. We need one 3 because of the 3 in each of the numbers, and one 7 is needed because of the 7 in 84. The least common multiple of 48 and 84 is $2 \cdot 2 \cdot 2 \cdot 2 \cdot 3 \cdot 7 = 336$. ■

E X A M P L E   7   Find the least common multiple of 12, 18, and 28.

Solution

$$12 = 2 \cdot 2 \cdot 3$$
$$18 = 2 \cdot 3 \cdot 3$$
$$28 = 2 \cdot 2 \cdot 7$$

The least common multiple is $2 \cdot 2 \cdot 3 \cdot 3 \cdot 7 = 252$. ■

E X A M P L E   8   Find the least common multiple of 8 and 9.

Solution

$$8 = 2 \cdot 2 \cdot 2$$
$$9 = 3 \cdot 3$$

The least common multiple is $2 \cdot 2 \cdot 2 \cdot 3 \cdot 3 = 72$. ■

C O N C E P T   Q U I Z

For Problems 1–5, answer true or false.

1. Every even whole number greater than 2 is a composite number.

2. Two is the only even prime number.

3. One is a prime number.

4. The prime factored form of 24 is $2 \cdot 2 \cdot 6$.

5. Some whole numbers are both prime and composite numbers.

PROBLEM SET 1.2

For Problems 1–20, classify each statement as true or false.

1. 8 divides 56

2. 9 divides 54

3. 6 does not divide 54

4. 7 does not divide 42

5. 96 is a multiple of 8

6. 78 is a multiple of 6

7. 54 is not a multiple of 4

8. 64 is not a multiple of 6

9. 144 is divisible by 4

10. 261 is divisible by 9

11. 173 is divisible by 3

12. 149 is divisible by 7

13. 11 is a factor of 143

14. 11 is a factor of 187

15. 9 is a factor of 119

16. 8 is a factor of 98

17. 3 is a prime factor of 57

18. 7 is a prime factor of 91

19. 4 is a prime factor of 48

20. 6 is a prime factor of 72

For Problems 21–30, classify each number as prime or composite.

21. 53   **22.** 57   **23.** 59

24. 61   **25.** 91   **26.** 81

27. 89   **28.** 97   **29.** 111

30. 101

For Problems 31–42, factor each composite number into a product of prime numbers. For example, $18 = 2 \cdot 3 \cdot 3$.

31. 26   **32.** 16   **33.** 36

34. 80   **35.** 49   **36.** 92

37. 56   **38.** 144   **39.** 120

40. 84   **41.** 135   **42.** 98

For Problems 43–54, find the greatest common factor of the given numbers.

43. 12 and 16   **44.** 30 and 36

45. 56 and 64   **46.** 72 and 96

47. 63 and 81   **48.** 60 and 72

49. 84 and 96   **50.** 48 and 52

51. 36, 72, and 90   **52.** 27, 54, and 63

53. 48, 60, and 84   **54.** 32, 80, and 96

For Problems 55–66, find the least common multiple of the given numbers.

55. 6 and 8   **56.** 8 and 12

57. 12 and 16   **58.** 9 and 12

59. 28 and 35   **60.** 42 and 66

61. 49 and 56   **62.** 18 and 24

63. 8, 12, and 28   **64.** 6, 10, and 12

65. 9, 15, and 18   **66.** 8, 14, and 24

■ ■ ■ **Thoughts into words**

67. How would you explain the concepts "greatest common factor" and "least common multiple" to a friend who missed class during that discussion?

68. Is it always true that the greatest common factor of two numbers is less than the least common multiple of those same two numbers? Explain your answer.

■ ■ ■ **Further investigations**

69. The numbers 0, 2, 4, 6, 8, and so on are multiples of 2. They are also called *even* numbers. Why is 2 the only even prime number?

70. Find the smallest nonzero whole number that is divisible by 2, 3, 4, 5, 6, 7, and 8.

71. Find the smallest whole number greater than 1 that produces a remainder of 1 when divided by 2, 3, 4, 5, or 6.

72. What is the greatest common factor of x and y if x and y are both prime numbers and x does not equal y? Explain your answer.

73. What is the greatest common factor of x and y if x and y are nonzero whole numbers and y is a multiple of x? Explain your answer.

74. What is the least common multiple of x and y if they are both prime numbers and x does not equal y? Explain your answer.

75. What is the least common multiple of x and y if the greatest common factor of x and y is 1? Explain your answer.

Familiarity with a few basic divisibility rules will be helpful for determining the prime factors of some numbers. For example, if you can quickly recognize that 51 is divisible by 3, then you can divide 51 by 3 to find another factor of 17. Because 3 and 17 are both prime numbers, we have $51 = 3 \cdot 17$. The divisibility rules for 2, 3, 5, and 9 are given below.

Rule for 2

A whole number is divisible by 2 if and only if the units digit of its base-10 numeral is divisible by 2. (In other words, the units digit must be 0, 2, 4, 6, or 8.)

EXAMPLES   68 is divisible by 2 because 8 is divisible by 2.
57 is not divisible by 2 because 7 is not divisible by 2.

Rule for 3

A whole number is divisible by 3 if and only if the sum of the digits of its base-10 numeral is divisible by 3.

EXAMPLES   51 is divisible by 3 because $5 + 1 = 6$ and 6 is divisible by 3.
144 is divisible by 3 because $1 + 4 + 4 = 9$ and 9 is divisible by 3.
133 is not divisible by 3 because $1 + 3 + 3 = 7$ and 7 is not divisible by 3.

Rule for 5

A whole number is divisible by 5 if and only if the units digit of its base-10 numeral is divisible by 5. (In other words, the units digit must be 0 or 5.)

EXAMPLES   115 is divisible by 5 because 5 is divisible by 5.
172 is not divisible by 5 because 2 is not divisible by 5.

Rule for 9

A whole number is divisible by 9 if and only if the sum of the digits of its base-10 numeral is divisible by 9.

EXAMPLES   765 is divisible by 9 because $7 + 6 + 5 = 18$ and 18 is divisible by 9.
147 is not divisible by 9 because $1 + 4 + 7 = 12$ and 12 is not divisible by 9.

For Problems 76–85, use the previous divisibility rules to help determine the prime factorization of each number.

76. 118	**77.** 76	**78.** 201
79. 123	**80.** 85	**81.** 115
82. 117	**83.** 441	**84.** 129
85. 153		

Answers to Concept Quiz

1. True   **2.** True   **3.** False   **4.** False   **5.** False

1.3   Integers: Addition and Subtraction

Objectives

■   Know the terminology associated with sets of integers.

■   Add and subtract integers.

■   Evaluate algebraic expressions for integer values.

■   Apply the concepts of adding and subtracting integers to model problems.

"A record temperature of 35° *below* zero was recorded on this date in 1904." "The IMDigital stock closed *down* 3 points yesterday." "On a first-down sweep around left end, Faulk *lost* 7 yards." "The West Coast Manufacturing Company reported *assets* of 50 million dollars and *liabilities* of 53 million dollars for 2004." Such examples illustrate our need for negative numbers.

The number line is a helpful visual device for our work at this time. We can associate the set of whole numbers with evenly spaced points on a line as indicated in Figure 1.1. For each nonzero whole number, we can associate its *negative* to the

Figure 1.1

left of zero; with 1 we associate −1, with 2 we associate −2, and so on, as indicated in Figure 1.2. The set of whole numbers, along with −1, −2, −3, and so on, is called the set of **integers.**

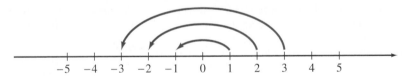

Figure 1.2

The following terminology is used with reference to the integers:

$\{\ldots, -3, -2, -1, 0, 1, 2, 3, \ldots\}$	Integers
$\{1, 2, 3, 4, \ldots\}$	Positive integers
$\{0, 1, 2, 3, 4, \ldots\}$	Nonnegative integers
$\{\ldots, -3, -2, -1\}$	Negative integers
$\{\ldots, -3, -2, -1, 0\}$	Nonpositive integers

The symbol −1 can be read as "negative one," "opposite of one," or "additive inverse of one." The "opposite of" and "additive inverse of" terminology is very helpful when working with variables. For example, reading the symbol $-x$ as "opposite of x" or "additive inverse of x" emphasizes an important issue. Because x can be any integer, $-x$ (the opposite of x) can be zero, positive, or negative. If x is a positive integer, then $-x$ is negative. If x is a negative integer, then $-x$ is positive. If x is zero, then $-x$ is zero. These statements can be written and illustrated on the number lines as in Figure 1.3.

If $x = 3$,
then $-x = -(3) = -3$.

If $x = -3$,
then $-x = -(-3) = 3$.

If $x = 0$,
then $-x = -(0) = 0$.

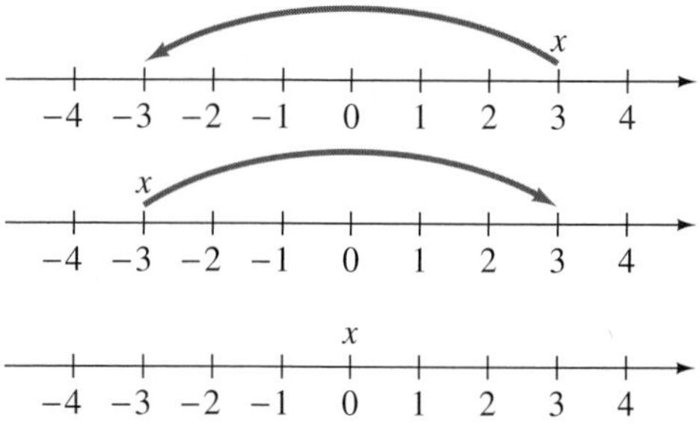

Figure 1.3

From this discussion we can recognize the following general property:

PROPERTY 1.1

If a is any integer, then

$$-(-a) = a$$

(The opposite of the opposite of any integer is the integer itself.)

Addition of Integers

The number line is also a convenient visual aid for interpreting the *addition of integers*. Consider the following examples and their number-line interpretations as shown in Figure 1.4.

Problem	Number line interpretation	Sum
$3 + 2$	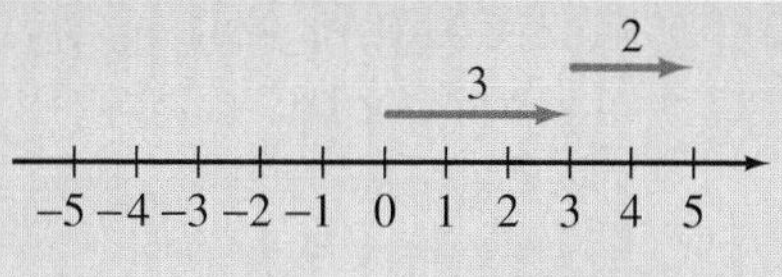	$3 + 2 = 5$
$3 + (-2)$	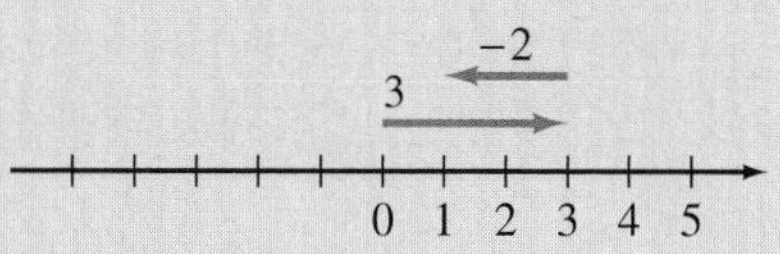	$3 + (-2) = 1$
$-3 + 2$	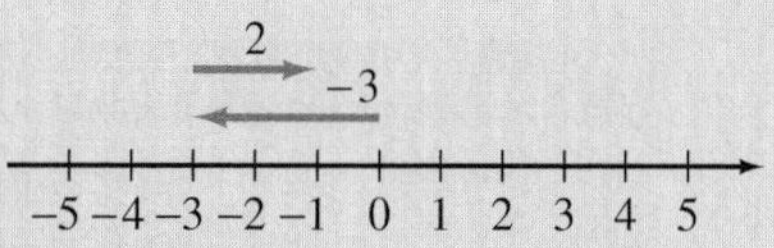	$-3 + 2 = -1$
$-3 + (-2)$	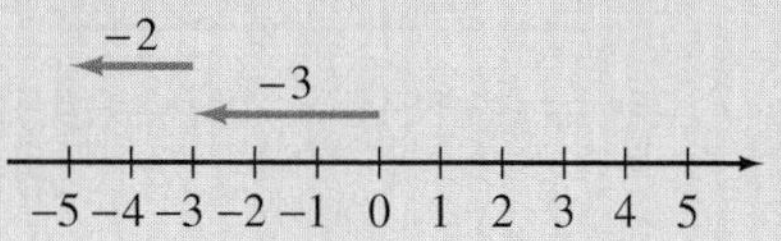	$-3 + (-2) = -5$

Figure 1.4

Once you get a feel for movement on the number line, simply forming a mental image of this movement is sufficient. Consider the next addition problems, and mentally picture the number-line interpretation. Be sure that you agree with all of our answers.

$$5 + (-2) = 3 \qquad -6 + 4 = -2 \qquad -8 + 11 = 3$$
$$-7 + (-4) = -11 \qquad -5 + 9 = 4 \qquad 9 + (-2) = 7$$
$$14 + (-17) = -3 \qquad 0 + (-4) = -4 \qquad 6 + (-6) = 0$$

The last example illustrates a general property that you should note: **Any integer plus its opposite equals zero.**

REMARK: Profits and losses pertaining to investments also provide a good physical model for interpreting the addition of integers. A loss of $25 on one investment, along with a profit of $60 on a second investment, produces an overall profit of $35. We can express this as $-25 + 60 = 35$. You may want to check the preceding examples using a profit and loss interpretation.

Even though all problems that involve the addition of integers could be done by using the number-line interpretation, it is sometimes convenient to give a more precise description of the addition process. For this purpose, we need to consider briefly the concept of absolute value. The **absolute value** of a number is the distance between the number and zero on the number line. For example, the absolute value of 6 is 6. The absolute value of -6 is also 6. The absolute value of 0 is 0. Vertical bars on either side of a number denote absolute value. Thus we write

$$|6| = 6 \qquad |-6| = 6 \qquad |0| = 0$$

Note that the absolute value of a positive number is the number itself, but the absolute value of a negative number is its opposite. Thus the absolute value of any number except 0 is positive, and the absolute value of 0 is 0.

We can describe the process of **adding integers** precisely by using the concept of absolute value.

Two Positive Integers

The sum of two positive integers is the sum of their absolute values. (The sum of two positive integers is a positive integer.)

$$43 + 54 = |43| + |54| = 43 + 54 = 97$$

Two Negative Integers

The sum of two negative integers is the opposite of the sum of their absolute values. (The sum of two negative integers is a negative integer.)

$$(-67) + (-93) = -(|-67| + |-93|)$$
$$= -(67 + 93)$$
$$= -160$$

One Positive and One Negative Integer

The sum of a positive and a negative integer can be found by subtracting the smaller absolute value from the larger absolute value and giving the result the sign of the original number that has the larger absolute value. If the integers have the same absolute value, then their sum is zero.

$$82 + (-40) = |82| - |-40|$$
$$= 82 - 40$$
$$= 42$$

$$74 + (-90) = -(|-90| - |74|)$$
$$= -(90 - 74)$$
$$= -16$$

$$(-17) + 17 = |-17| - |17|$$
$$= 17 - 17$$
$$= 0$$

Zero and Another Integer

The sum of zero and any integer is the integer itself.

$$0 + (-46) = -46$$
$$72 + 0 = 72$$

The following examples further demonstrate how to add integers. Be sure that you agree with each of the results.

$$-18 + (-56) = -(|-18| + |-56|) = -(18 + 56) = -74$$
$$-71 + (-32) = -(|-71| + |-32|) = -(71 + 32) = -103$$
$$64 + (-49) = |64| - |-49| = 64 - 49 = 15$$
$$-56 + 93 = |93| - |-56| = 93 - 56 = 37$$
$$-114 + 48 = -(|-114| - |48|) = -(114 - 48) = -66$$
$$45 + (-73) = -(|-73| - |45|) = -(73 - 45) = -28$$
$$46 + (-46) = 0 \qquad -48 + 0 = -48$$
$$(-73) + 73 = 0 \qquad 0 + (-81) = -81$$

It is true that this *absolute value approach* does precisely describe the process of adding integers, but don't forget about the number-line interpretation. Included in the next problem set are other physical models for interpreting the addition of integers. Some people find these models very helpful.

Subtraction of Integers

The following examples illustrate a relationship between addition and subtraction of *whole numbers:*

$$7 - 2 = 5 \quad \text{because } 2 + 5 = 7$$
$$9 - 6 = 3 \quad \text{because } 6 + 3 = 9$$
$$5 - 1 = 4 \quad \text{because } 1 + 4 = 5$$

This same relationship between addition and subtraction holds for *all integers:*

$$5 - 6 = -1 \quad \text{because } 6 + (-1) = 5$$
$$-4 - 9 = -13 \quad \text{because } 9 + (-13) = -4$$
$$-3 - (-7) = 4 \quad \text{because } -7 + 4 = -3$$
$$8 - (-3) = 11 \quad \text{because } -3 + 11 = 8$$

Now consider a further observation:

$$5 - 6 = -1 \qquad \text{and} \qquad 5 + (-6) = -1$$
$$-4 - 9 = -13 \qquad \text{and} \qquad -4 + (-9) = -13$$
$$-3 - (-7) = 4 \qquad \text{and} \qquad -3 + 7 = 4$$
$$8 - (-3) = 11 \qquad \text{and} \qquad 8 + 3 = 11$$

The previous examples help us realize that we can state the subtraction of integers in terms of the addition of integers. More precisely, a general description for the **subtraction of integers** follows:

Subtraction of Integers

If a and b are integers, then $a - b = a + (-b)$.

It may be helpful for you to read $a - b = a + (-b)$ as "*a* minus *b* is equal to *a* plus the opposite of *b*." Every subtraction problem can be changed into an equivalent addition problem, as illustrated by these examples.

$$6 - 13 = 6 + (-13) = -7$$
$$9 - (-12) = 9 + 12 = 21$$
$$-8 - 13 = -8 + (-13) = -21$$
$$-7 - (-8) = -7 + 8 = 1$$

It should be apparent that addition of integers is a key operation. Being able to add integers effectively is indispensable for further work in algebra.

Evaluating Algebraic Expressions

Let's conclude this section by evaluating some algebraic expressions using negative and positive integers.

E X A M P L E   1   Evaluate each algebraic expression for the given values of the variables.

(a) $x - y$ for $x = -12$ and $y = 20$

(b) $-a + b$ for $a = -8$ and $b = -6$

(c) $-x - y$ for $x = 14$ and $y = -7$

Solution

(a) $x - y = -12 - 20$ when $x = -12$ and $y = 20$

$\qquad\qquad = -12 + (-20)$

$\qquad\qquad = -32$

(b) $-a + b = -(-8) + (-6)$ when $a = -8$ and $b = -6$

$\qquad\qquad = 8 + (-6)$

$\qquad\qquad = 2$

(c) $-x - y = -(14) - (-7)$ when $x = 14$ and $y = -7$

$\qquad\qquad = -14 + 7$

$\qquad\qquad = -7$

CONCEPT QUIZ

For Problems 1–4, match the description with the set of numbers.

1. $\{\ldots, -3, -2, -1\}$ **A.** Positive integers

2. $\{1, 2, 3, \ldots\}$ **B.** Negative integers

3. $\{0, 1, 2, 3, \ldots\}$ **C.** Nonnegative integers

4. $\{\ldots, -3, -2, -1, 0\}$ **D.** Nonpositive integers

For Problems 5–10, answer true or false.

5. The number zero is considered to be a positive integer.

6. The number zero is considered to be a negative integer.

7. The absolute value of a number is the distance between the number and one on the number line.

8. The $|-4|$ is -4.

9. The opposite of -5 is 5.

10. a minus b is equivalent to a plus the opposite of b.

PROBLEM SET 1.3

For Problems 1–10, use the number-line interpretation to find each sum.

1. $5 + (-3)$

2. $7 + (-4)$

3. $-6 + 2$

4. $-9 + 4$

5. $-3 + (-4)$

6. $-5 + (-6)$

7. $8 + (-2)$

8. $12 + (-7)$

9. $5 + (-11)$

10. $4 + (-13)$

For Problems 11–30, find each sum.

11. $17 + (-9)$ **12.** $16 + (-5)$

13. $8 + (-19)$ **14.** $9 + (-14)$

15. $-7 + (-8)$ **16.** $-6 + (-9)$

17. $-15 + 8$ **18.** $-22 + 14$

19. $-13 + (-18)$ **20.** $-15 + (-19)$

21. $-27 + 8$ **22.** $-29 + 12$

23. $32 + (-23)$ **24.** $27 + (-14)$

25. $-25 + (-36)$ **26.** $-34 + (-49)$

27. $54 + (-72)$ **28.** $48 + (-76)$

29. $-34 + (-58)$ **30.** $-27 + (-36)$

For Problems 31–50, subtract as indicated.

31. $3 - 8$ **32.** $5 - 11$

33. $-4 - 9$ **34.** $-7 - 8$

35. $5 - (-7)$ **36.** $9 - (-4)$

37. $-6 - (-12)$ **38.** $-7 - (-15)$

39. $-11 - (-10)$ **40.** $-14 - (-19)$

41. $-18 - 27$ **42.** $-16 - 25$

43. $34 - 63$ **44.** $25 - 58$

45. $45 - 18$ **46.** $52 - 38$

47. $-21 - 44$ **48.** $-26 - 54$

49. $-53 - (-24)$ **50.** $-76 - (-39)$

For Problems 51–66, add or subtract as indicated.

51. $6 - 8 - 9$ **52.** $5 - 9 - 4$

53. $-4 - (-6) + 5 - 8$ **54.** $-3 - 8 + 9 - (-6)$

55. $5 + 7 - 8 - 12$ **56.** $-7 + 9 - 4 - 12$

57. $-6 - 4 - (-2) + (-5)$

58. $-8 - 11 - (-6) + (-4)$

59. $-6 - 5 - 9 - 8 - 7$ **60.** $-4 - 3 - 7 - 8 - 6$

61. $7 - 12 + 14 - 15 - 9$ **62.** $8 - 13 + 17 - 15 - 19$

63. $-11 - (-14) + (-17) - 18$

64. $-15 + 20 - 14 - 18 + 9$

65. $16 - 21 + (-15) - (-22)$

66. $17 - 23 - 14 - (-18)$

The horizontal format is used extensively in algebra, but occasionally the vertical format shows up. You should therefore have some exposure to the vertical format. Find the sums for Problems 67–78.

67. $\begin{array}{r} 5 \\ -9 \\ \hline \end{array}$ **68.** $\begin{array}{r} 8 \\ -13 \\ \hline \end{array}$ **69.** $\begin{array}{r} -13 \\ -18 \\ \hline \end{array}$

70. $\begin{array}{r} -14 \\ -28 \\ \hline \end{array}$ **71.** $\begin{array}{r} -18 \\ 9 \\ \hline \end{array}$ **72.** $\begin{array}{r} -17 \\ 9 \\ \hline \end{array}$

73. $\begin{array}{r} -21 \\ 39 \\ \hline \end{array}$ **74.** $\begin{array}{r} -15 \\ 32 \\ \hline \end{array}$ **75.** $\begin{array}{r} 27 \\ -19 \\ \hline \end{array}$

76. $\begin{array}{r} 31 \\ -18 \\ \hline \end{array}$ **77.** $\begin{array}{r} -53 \\ 24 \\ \hline \end{array}$ **78.** $\begin{array}{r} 47 \\ -28 \\ \hline \end{array}$

For Problems 79–90, do the *subtraction* problems in vertical format.

79. $\begin{array}{r} 5 \\ 12 \\ \hline \end{array}$ **80.** $\begin{array}{r} 8 \\ 19 \\ \hline \end{array}$ **81.** $\begin{array}{r} 6 \\ -9 \\ \hline \end{array}$

82. $\begin{array}{r} 13 \\ -7 \\ \hline \end{array}$ **83.** $\begin{array}{r} -7 \\ -8 \\ \hline \end{array}$ **84.** $\begin{array}{r} -6 \\ -5 \\ \hline \end{array}$

85. $\begin{array}{r} 17 \\ -19 \\ \hline \end{array}$ **86.** $\begin{array}{r} 18 \\ -14 \\ \hline \end{array}$ **87.** $\begin{array}{r} -23 \\ 16 \\ \hline \end{array}$

88. $\begin{array}{r} -27 \\ 15 \\ \hline \end{array}$ **89.** $\begin{array}{r} -12 \\ 12 \\ \hline \end{array}$ **90.** $\begin{array}{r} -13 \\ -13 \\ \hline \end{array}$

For Problems 91–100, evaluate each algebraic expression for the given values of the variables.

91. $x - y$ for $x = -6$ and $y = -13$

92. $-x - y$ for $x = -7$ and $y = -9$

93. $-x + y - z$ for $x = 3$, $y = -4$, and $z = -6$

94. $x - y + z$ for $x = 5$, $y = 6$, and $z = -9$

95. $-x - y - z$ for $x = -2$, $y = 3$, and $z = -11$

96. $-x - y + z$ for $x = -8$, $y = -7$, and $z = -14$

97. $-x + y + z$ for $x = -11$, $y = 7$, and $z = -9$

98. $-x - y - z$ for $x = 12$, $y = -6$, and $z = -14$

99. $x - y - z$ for $x = -15$, $y = 12$, and $z = -10$

100. $x + y - z$ for $x = -18$, $y = 13$, and $z = 8$

A game such as football can also be used to interpret addition of integers. A gain of 3 yards on one play followed by a loss of 5 yards on the next play places the ball 2 yards *behind* the initial line of scrimmage, and this may be expressed as $3 + (-5) = -2$. Use this "football interpretation" to find the sums for Problems 101–110.

101. $4 + (-7)$ **102.** $3 + (-5)$

103. $-4 + (-6)$ **104.** $-2 + (-5)$

105. $-5 + 2$ **106.** $-10 + 6$

107. $-4 + 15$ **108.** $-3 + 22$

109. $-12 + 17$ **110.** $-9 + 21$

For Problems 111–120, use the "profit and loss interpretation" for the addition of integers that was illustrated in the remark on page 18.

111. $60 + (-125)$ **112.** $50 + (-85)$

113. $-55 + (-45)$ **114.** $-120 + (-220)$

115. $-70 + 45$ **116.** $-125 + 45$

117. $-120 + 250$ **118.** $-75 + 165$

119. $145 + (-65)$ **120.** $275 + (-195)$

121. The temperature at 5 A.M. was $-17°$F. By noon the temperature had increased by $14°$F. Use the addition of integers to describe this situation, and determine the temperature at noon.

122. The temperature at 6 P.M. was $-6°$F, and by 11 P.M. the temperature had dropped another $5°$F. Use the subtraction of integers to describe this situation, and determine the temperature at 11 P.M.

123. Megan shot rounds of 3 over par, 2 under par, 3 under par, and 5 under par for a four-day golf tournament. Use the addition of integers to describe this situation, and determine how much over or under par she was for the tournament.

124. The annual report of a company contained these figures: a loss of $615,000 for 2000, a loss of $275,000 for 2001, a loss of $70,000 for 2002, and a profit of $115,000 for 2003. Use the addition of integers to describe this situation, and determine the company's total loss or profit for the four-year period.

125. Suppose that during a five-day period a share of Dell stock recorded the following gains and losses:

Monday	Tuesday	Wednesday
Lost $2	Gained $1	Gained $3

Thursday	Friday
Gained $1	Lost $2

Use the addition of integers to describe this situation and to determine the amount of gain or loss for the five-day period.

126. The Dead Sea is approximately thirteen hundred ten feet below sea level. Suppose that you are standing eight hundred five feet above the Dead Sea. Use the addition of integers to describe this situation and to determine your elevation.

127. Use your calculator to check your answers for Problems 51–66.

■ ■ ▨ Thoughts into words

128. The statement $-6 - (-2) = -6 + 2 = -4$ can be read as "negative six minus negative two equals negative six plus two, which equals negative four." Express each equation in words.

  (a) $8 + (-10) = -2$

  (b) $-7 - 4 = -7 + (-4) = 11$

  (c) $9 - (-12) = 9 + 12 = 21$

  (d) $-5 + (-6) = -11$

129. The algebraic expression $-x - y$ can be read as "the opposite of x minus y." Give each expression in words.

  (a) $-x + y$

  (b) $x - y$

  (c) $-x - y + z$

Answers to Concept Quiz

  1. B   **2.** A   **3.** C   **4.** D   **5.** False   **6.** False   **7.** False   **8.** False   **9.** True   **10.** True

1.4   **Integers: Multiplication and Division**

Objectives

■   Multiply and divide integers.

■   Evaluate algebraic expressions involving the multiplication and division of integers.

■   Apply the concepts of multiplying and dividing integers to model problems.

Multiplication of Integers

Multiplication of whole numbers may be interpreted as repeated addition. For example, $3 \cdot 4$ means the sum of three 4s; thus $3 \cdot 4 = 4 + 4 + 4 = 12$. Consider the following examples, which use the repeated-addition idea to find the product of a positive integer and a negative integer.

$$3(-2) = -2 + (-2) + (-2) = -6$$
$$2(-4) = -4 + (-4) = -8$$
$$4(-1) = -1 + (-1) + (-1) + (-1) = -4$$

Note the use of parentheses to indicate multiplication. Sometimes both numbers are enclosed in parentheses; in this case we would have $(3)(-2)$.

When multiplying whole numbers, we realize that the order in which we multiply two factors does not change the product; in other words, $2(3) = 6$ and $3(2) = 6$. Using this idea, we can now handle a negative integer times a positive integer:

$$(-2)(3) = (3)(-2) = (-2) + (-2) + (-2) = -6$$
$$(-3)(2) = (2)(-3) = (-3) + (-3) = -6$$
$$(-4)(3) = (3)(-4) = (-4) + (-4) + (-4) = -12$$

Finally, let's consider the product of two negative integers. The following pattern helps us with the reasoning for this situation.

$$4(-3) = -12$$
$$3(-3) = -9$$
$$2(-3) = -6$$
$$1(-3) = -3$$
$$0(-3) = 0 \qquad \text{The product of zero and any integer is zero.}$$
$$(-1)(-3) = ?$$

Certainly, to continue this pattern, the product of -1 and -3 has to be 3. This type of reasoning helps us to realize that the product of any two negative integers is a positive integer.

Using the concept of absolute value, we can now precisely describe the **multiplication of integers:**

Multiplying Integers

1. The product of two positive integers or two negative integers is the product of their absolute values.
2. The product of a positive and a negative integer (either order) is the opposite of the product of their absolute values.
3. The product of zero and any integer is zero.

The next examples illustrate this description of multiplication:

$$(-5)(-2) = |-5| \cdot |-2| = 5 \cdot 2 = 10$$
$$(7)(-6) = -(|7| \cdot |-6|) = -(7 \cdot 6) = -42$$
$$(-8)(9) = -(|-8| \cdot |9|) = -(8 \cdot 9) = -72$$
$$(-14)(0) = 0$$
$$(0)(-28) = 0$$

These examples show a step-by-step process for multiplying integers. In reality, however, the key issue is to remember whether the product is positive or negative. In other words, we need to remember that *the product of two positive integers or two negative integers is a positive integer,* and that *the product of a positive integer and a negative integer (in either order) is a negative integer.* Then we can avoid the step-by-step analysis and simply write the results as follows:

$$(7)(-9) = -63$$
$$(8)(7) = 56$$
$$(-5)(-6) = 30$$
$$(-4)(12) = -48$$

Division of Integers

By looking back at our knowledge of whole numbers, we can get some guidance for our work with integers. We know, for example, that $\frac{8}{2} = 4$ because $2 \cdot 4 = 8$. In other words, we find the quotient of two whole numbers by looking at a related multiplication problem. In the following examples, we have used this same link between multiplication and division to determine the quotients:

$$\frac{8}{-2} = -4 \quad \text{because } (-2)(-4) = 8$$

$$\frac{-10}{5} = -2 \quad \text{because } (5)(-2) = -10$$

$$\frac{-12}{-4} = 3 \quad \text{because } (-4)(3) = -12$$

$$\frac{0}{-6} = 0 \quad \text{because } (-6)(0) = 0$$

$\dfrac{-9}{0}$ is undefined   because no number times 0 produces -9

$\dfrac{0}{0}$ is undefined   because any number times 0 equals 0.

                           Remember that division by zero is undefined!

Dividing Integers

1. The quotient of two positive or two negative integers is the quotient of their absolute values.
2. The quotient of a positive integer and a negative integer (or a negative and a positive) is the opposite of the quotient of their absolute values.
3. The quotient of zero and any nonzero integer (zero divided by any nonzero integer) is zero.

The next examples illustrate this description of division:

$$\frac{-8}{-4} = \frac{|-8|}{|-4|} = \frac{8}{4} = 2$$

$$\frac{-14}{2} = -\left(\frac{|-14|}{|2|}\right) = -\left(\frac{14}{2}\right) = -7$$

$$\frac{15}{-3} = -\left(\frac{|15|}{|-3|}\right) = -\left(\frac{15}{3}\right) = -5$$

$$\frac{0}{-4} = 0$$

For practical purposes, the objective is to determine whether the quotient is positive or negative. *The quotient of two positive integers or two negative integers is positive,* and *the quotient of a positive integer and a negative integer or of a negative integer and a positive integer is negative.* We then can simply write the quotients as follows without showing all of the steps:

$$\frac{-18}{-6} = 3 \qquad\qquad \frac{-24}{12} = -2 \qquad\qquad \frac{36}{-9} = -4$$

REMARK: Occasionally people use the phrase "two negatives make a positive." We hope they realize that the reference is to multiplication and division only; in addition, the sum of two negative integers is still a negative integer. It is probably best to avoid such imprecise statements.

Simplifying Numerical Expressions

Now we can simplify numerical expressions involving any or all of the four basic operations with integers. Keep in mind the agreements on the order of operations we stated in Section 1.1.

EXAMPLE 1

Simplify $-4(-3) - 7(-8) + 3(-9)$.

Solution

$$
\begin{aligned}
-4(-3) - 7(-8) + 3(-9) &= 12 - (-56) + (-27) \\
&= 12 + 56 + (-27) \\
&= 41
\end{aligned}
$$

EXAMPLE 2

Simplify $\dfrac{-8 - 4(5)}{-4}$.

Solution

$$
\begin{aligned}
\frac{-8 - 4(5)}{-4} &= \frac{-8 - 20}{-4} \\
&= \frac{-28}{-4} \\
&= 7
\end{aligned}
$$

Evaluating Algebraic Expressions

Evaluating algebraic expressions often involves using two or more operations with integers. The final examples of this section illustrate such situations.

EXAMPLE 3

Find the value of $3x + 2y$ when $x = 5$ and $y = -9$.

Solution

$$
\begin{aligned}
3x + 2y &= 3(5) + 2(-9) \quad \text{when } x = 5 \text{ and } y = -9 \\
&= 15 + (-18) \\
&= -3
\end{aligned}
$$

EXAMPLE 4

Evaluate $-2a + 9b$ for $a = 4$ and $b = -3$.

Solution

$$
\begin{aligned}
-2a + 9b &= -2(4) + 9(-3) \quad \text{when } a = 4 \text{ and } b = -3 \\
&= -8 + (-27) \\
&= -35
\end{aligned}
$$

| EXAMPLE 5 | Find the value of $\dfrac{x - 2y}{4}$ when $x = -6$ and $y = 5$. |

Solution

$$\frac{x - 2y}{4} = \frac{-6 - 2(5)}{4} \quad \text{when } x = -6 \text{ and } y = 5$$

$$= \frac{-6 - 10}{4}$$

$$= \frac{-16}{4}$$

$$= -4$$

CONCEPT QUIZ

For Problems 1–8, answer true or false.

1. The product of two negative integers is a positive integer.

2. The product of a positive integer and a negative integer is a positive integer.

3. When multiplying three negative integers the product is negative.

4. The rules for adding integers and the rules for multiplying integers are the same.

5. The quotient of two negative integers is negative.

6. The quotient of a positive integer and zero is a positive integer.

7. The quotient of a negative integer and zero is zero.

8. The product of zero and any integer is zero.

PROBLEM SET 1.4

For Problems 1–40, find the product or quotient (that is, multiply or divide) as indicated.

1. $5(-6)$

2. $7(-9)$

3. $\dfrac{-27}{3}$

4. $\dfrac{-35}{5}$

5. $\dfrac{-42}{-6}$

6. $\dfrac{-72}{-8}$

7. $(-7)(8)$

8. $(-6)(9)$

9. $(-5)(-12)$

10. $(-7)(-14)$

11. $\dfrac{96}{-8}$

12. $\dfrac{-91}{7}$

13. $14(-9)$

14. $17(-7)$

15. $(-11)(-14)$

16. $(-13)(-17)$

17. $\dfrac{135}{-15}$

18. $\dfrac{-144}{12}$

19. $\dfrac{-121}{-11}$

20. $\dfrac{-169}{-13}$

21. $(-15)(-15)$

22. $(-18)(-18)$

23. $\dfrac{112}{-8}$

24. $\dfrac{112}{-7}$

25. $\dfrac{0}{-8}$

26. $\dfrac{-8}{0}$

27. $\dfrac{-138}{-6}$

28. $\dfrac{-105}{-5}$

29. $\dfrac{76}{-4}$

30. $\dfrac{-114}{6}$

31. $(-6)(-15)$

32. $\dfrac{0}{-14}$

33. $(-56) \div (-4)$

34. $(-78) \div (-6)$

35. $(-19) \div 0$

36. $(-90) \div 15$

37. $(-72) \div 18$

38. $(-70) \div 5$

39. $(-36)(27)$

40. $(42)(-29)$

For Problems 41–60, simplify each numerical expression.

41. $3(-4) + 5(-7)$

42. $6(-3) + 5(-9)$

43. $7(-2) - 4(-8)$

44. $9(-3) - 8(-6)$

45. $(-3)(-8) + (-9)(-5)$

46. $(-7)(-6) + (-4)(-3)$

47. $5(-6) - 4(-7) + 3(2)$

48. $7(-4) - 8(-7) + 5(-8)$

49. $\dfrac{13 + (-25)}{-3}$

50. $\dfrac{15 + (-36)}{-7}$

51. $\dfrac{12 - 48}{6}$

52. $\dfrac{16 - 40}{8}$

53. $\dfrac{-7(10) + 6(-9)}{-4}$

54. $\dfrac{-6(8) + 4(-14)}{-8}$

55. $\dfrac{4(-7) - 8(-9)}{11}$

56. $\dfrac{5(-9) - 6(-7)}{3}$

57. $-2(3) - 3(-4) + 4(-5) - 6(-7)$

58. $2(-4) + 4(-5) - 7(-6) - 3(9)$

59. $-1(-6) - 4 + 6(-2) - 7(-3) - 18$

60. $-9(-2) + 16 - 4(-7) - 12 + 3(-8)$

For Problems 61–76, evaluate each algebraic expression for the given values of the variables.

61. $7x + 5y$ for $x = -5$ and $y = 9$

62. $4a + 6b$ for $a = -6$ and $b = -8$

63. $9a - 2b$ for $a = -5$ and $b = 7$

64. $8a - 3b$ for $a = -7$ and $b = 9$

65. $-6x - 7y$ for $x = -4$ and $y = -6$

66. $-5x - 12y$ for $x = -5$ and $y = -7$

67. $\dfrac{5x - 3y}{-6}$ for $x = -6$ and $y = 4$

68. $\dfrac{-7x + 4y}{-8}$ for $x = 8$ and $y = 6$

69. $3(2a - 5b)$ for $a = -1$ and $b = -5$

70. $4(3a - 7b)$ for $a = -2$ and $b = -4$

71. $-2x + 6y - xy$ for $x = 7$ and $y = -7$

72. $-3x + 7y - 2xy$ for $x = -6$ and $y = 4$

73. $-4ab - b$ for $a = 2$ and $b = -14$

74. $-5ab + b$ for $a = -1$ and $b = -13$

75. $(ab + c)(b - c)$ for $a = -2, b = -3,$ and $c = 4$

76. $(ab - c)(a + c)$ for $a = -3, b = 2,$ and $c = 5$

For Problems 77–82, find the value of $\dfrac{5(F - 32)}{9}$ for each of the given values for F.

77. $F = 59$

78. $F = 68$

79. $F = 14$

80. $F = -4$

81. $F = -13$

82. $F = -22$

For Problems 83–88, find the value of $\dfrac{9C}{5} + 32$ for each of the given values for C.

83. $C = 25$

84. $C = 35$

85. $C = 40$

86. $C = 0$

87. $C = -10$

88. $C = -30$

89. Monday morning Thad bought 800 shares of a stock at $19 per share. During that week the stock went up $2 per share on one day and dropped $1 per share on each of the other four days. Use multiplication and addition of integers to describe this situation, and determine the value of the 800 shares at closing time on Friday.

90. In one week a small company showed a profit of $475 for one day and a loss of $65 for each of the other four days. Use multiplication and addition of integers to describe this situation, and determine the company's profit or loss for the week.

91. At 6 P.M. the temperature was 5°F. For the next four hours the temperature dropped 3° per hour. Use multiplication and addition of integers to describe this situation and to find the temperature at 10 P.M.

92. For each of the first three days of a golf tournament, Jason shot 2 strokes under par. Then for each of the last two days of the tournament, he shot 4 strokes over par. Use multiplication and addition of integers to describe this situation and to determine Jason's score relative to par for the five-day tournament.

93. Use a calculator to check your answers for Problems 41–60.

■ ■ ■ Thoughts into words

94. Your friend keeps getting an answer of -7 when simplifying the expression $-6 + (-8) \div 2$. What mistake is she making and how would you help her?

95. Make up a problem that you can solve using $6(-4) = -24$.

96. Make up a problem that could be solved using $(-4)(-3) = 12$.

97. Explain why $\dfrac{0}{4} = 0$ but $\dfrac{4}{0}$ is undefined.

Answers to Concept Quiz

1. True   **2.** False   **3.** True   **4.** False   **5.** False   **6.** False   **7.** False   **8.** True

1.5   Use of Properties

Objectives

■   Recognize the properties of integers.

■   Apply the properties of integers to simplify numerical expressions.

■   Simplify algebraic expressions.

We will begin this section by listing and briefly commenting on some of the basic properties of integers. We will then show how these properties facilitate manipulation with integers and also serve as a basis for some algebraic computation.

Commutative Property of Addition

If a and b are integers, then

$$a + b = b + a$$

Commutative Property of Multiplication

If a and b are integers, then

$$ab = ba$$

Addition and multiplication are said to be commutative operations. This means that the order in which you add or multiply two integers does not affect the result. For example, $3 + 5 = 5 + 3$ and $7(8) = 8(7)$. It is also important to realize that subtraction and division *are not* commutative operations; order does make a difference. For example, $8 - 7 \neq 7 - 8$ and $16 \div 4 \neq 4 \div 16$.

Associative Property of Addition

If a, b, and c are integers, then

$$(a + b) + c = a + (b + c)$$

Associative Property of Multiplication

If a, b, and c are integers, then

$$(ab)c = a(bc)$$

Our arithmetic operations are binary operations. We only operate (add, subtract, multiply, or divide) on two numbers at a time. Therefore when we need to operate on three or more numbers, the numbers must be grouped.

The associative properties can be thought of as grouping properties. For example, $(-8 + 3) + 9 = -8 + (3 + 9)$. Changing the grouping of the numbers for addition does not affect the result. This is also true for multiplication as $[(-6)(5)](-4) = (-6)[(5)(-4)]$ illustrates. Addition and multiplication are associative operations. Subtraction and division *are not* associative operations. For example, $(8 - 4) - 7 = -3$, whereas $8 - (4 - 7) = 11$. An example showing that division is not associative is $(8 \div 4) \div 2 = 1$, whereas $8 \div (4 \div 2) = 4$.

Identity Property of Addition

If a is an integer, then

$$a + 0 = 0 + a = a$$

We refer to 0 as the *identity element* for addition. This simply means that the sum of any integer and 0 is exactly the same integer. For example, $-197 + 0 = 0 + (-197) = -197$.

Identity Property of Multiplication

If a is an integer, then

$$a(1) = 1(a) = a$$

We call 1 the *identity element* for multiplication. The product of any integer and 1 is exactly the same integer. For example, $(-573)(1) = (1)(-573) = -573$.

Additive Inverse Property

For every integer a, there exists an integer $-a$ such that

$$a + (-a) = (-a) + a = 0$$

The integer $-a$ is called the *additive inverse* of a or the *opposite* of a. Thus 6 and -6 are additive inverses, and their sum is 0. The additive inverse of 0 is 0.

Multiplication Property of Zero

If a is an integer, then

$$a(0) = (0)(a) = 0$$

In other words, the product of 0 and any integer is 0. For example, $(-873)(0) = (0)(-873) = 0$.

Multiplicative Property of Negative One

If a is an integer, then

$$(a)(-1) = (-1)(a) = -a$$

The product of any integer and -1 is the opposite of the integer. For example, $(-1)(48) = (48)(-1) = -48$.

Distributive Property

If a, b, and c are integers, then

$$a(b + c) = ab + ac$$

The distributive property involves both addition and multiplication. We say that **multiplication distributes over addition.** For example, $3(4 + 7) = 3(4) + 3(7)$. Because $b - c = b + (-c)$, it follows that **multiplication also distributes over subtraction.** This could be stated as $a(b - c) = ab - ac$. For example, $7(8 - 2) = 7(8) - 7(2)$.

Let's now consider some examples that use these properties to help with certain types of manipulations.

E X A M P L E   1   Find the sum $[43 + (-24)] + 24$.

Solution

In such a problem, it is much more advantageous to group -24 and 24. Thus

$$[43 + (-24)] + 24 = 43 + [(-24) + 24] \quad \text{Associative property for addition}$$
$$= 43 + 0$$
$$= 43$$

E X A M P L E   2   Find the product $[(-17)(25)](4)$.

Solution

In this problem, it is easier to group 25 and 4. Thus

$$[(-17)(25)](4) = (-17)[(25)(4)] \quad \text{Associative property for multiplication}$$
$$= (-17)(100)$$
$$= -1700 \qquad \blacksquare$$

E X A M P L E   3   Find the sum $17 + (-24) + (-31) + 19 + (-14) + 29 + 43$.

Solution

Certainly we could add in the order in which the numbers appear. However, because addition is *commutative* and *associative*, we can change the order, and group in any convenient way. For example, we can add all the positive integers, add all of the negative integers, and then add these two results. It might be convenient to use the vertical format:

$$
\begin{array}{rrr}
17 & & \\
19 & -24 & \\
29 & -31 & 108 \\
\underline{43} & \underline{-14} & \underline{-69} \\
108 & -69 & 39
\end{array}
$$

$\blacksquare$

For a problem such as Example 3, it might be advisable first to work out the problem by adding in the order in which the numbers appear and then to use the rearranging and regrouping idea as a check. Don't forget the link between addition and subtraction: A problem such as $18 - 43 + 52 - 17 - 23$ can be changed to $18 + (-43) + 52 + (-17) + (-23)$.

E X A M P L E   4   Simplify $(-75)(-4 + 100)$.

Solution

For such a problem, it might be convenient to apply the *distributive property* and then to simplify. Thus

$$(-75)(-4 + 100) = (-75)(-4) + (-75)(100)$$
$$= 300 + (-7500)$$
$$= -7200 \qquad \blacksquare$$

E X A M P L E   5   Simplify $19(-26 + 25)$.

Solution

For this problem, we are better off *not* applying the distributive property but simply adding the numbers inside the parentheses and then finding the indicated product. Thus

$$19(-26 + 25) = 19(-1) = -19 \qquad \blacksquare$$

EXAMPLE 6   Simplify $27(104) + 27(-4)$.

Solution

Keep in mind that the *distributive property* enables us to change from the form $a(b + c)$ to $ab + ac$, or from $ab + ac$ to $a(b + c)$. In this problem we want to use the latter change:

$$27(104) + 27(-4) = 27[104 + (-4)]$$
$$= 27(100)$$
$$= 2700 \qquad \blacksquare$$

Examples 4, 5, and 6 demonstrate an important issue. Sometimes the form $a(b + c)$ is the most convenient, but at other times the form $ab + ac$ is better. A suggestion regarding this issue — a suggestion that also applies to the use of the other properties — is to *think first* and then decide whether or not you can use the properties to make the manipulations easier.

Combining Similar Terms

Algebraic expressions such as these:

$$3x \qquad 5y \qquad 7xy \qquad -4abc \qquad z$$

are called **terms.** A term is an indicated product that may have any number of factors. We call the variables in a term **literal factors,** and we call the numerical factor the **numerical coefficient.** Thus in $7xy$, the x and y are literal factors, and 7 is the numerical coefficient. The numerical coefficient of the term $-4abc$ is -4. Because $z = 1(z)$, the numerical coefficient of the term z is 1. Terms that have the same literal factors are called **like terms** or **similar terms.** Some examples of similar terms are

$$3x \text{ and } 9x \qquad\qquad 14abc \text{ and } 29abc$$
$$7xy \text{ and } -15xy \qquad 4z, 9z, \text{ and } -14z$$

We can simplify algebraic expressions that contain similar terms by using a form of the distributive property. Consider these examples:

$$3x + 5x = (3 + 5)x$$
$$= 8x$$

$$-9xy + 7xy = (-9 + 7)xy$$
$$= -2xy$$

$$18abc - 27abc = (18 - 27)abc$$
$$= [18 + (-27)]abc$$
$$= -9abc$$

$$4x + x = (4 + 1)x \qquad \text{Don't forget that } x = 1(x).$$
$$= 5x$$

More complicated expressions might first require that we rearrange terms by using the commutative property.

$$7x + 3y + 9x + 5y = 7x + 9x + 3y + 5y \qquad \text{Commutative property for addition}$$
$$= (7 + 9)x + (3 + 5)y \qquad \text{Distributive property}$$
$$= 16x + 8y$$

$$9a - 4 - 13a + 6 = 9a + (-4) + (-13a) + 6$$
$$= 9a + (-13a) + (-4) + 6 \qquad \text{Commutative property for addition}$$
$$= [9 + (-13)]a + 2 \qquad \text{Distributive property}$$
$$= -4a + 2$$

As you become more adept at handling the various simplifying steps, you may want to do the steps mentally and thereby go directly from the given expression to the simplified form.

$$19x - 14y + 12x + 16y = 31x + 2y$$
$$17ab + 13c - 19ab - 30c = -2ab - 17c$$
$$9x + 5 - 11x + 4 + x - 6 = -x + 3$$

Simplifying some algebraic expressions requires repeated applications of the distributive property, as the next examples demonstrate:

$$5(x - 2) + 3(x + 4) = 5(x) - 5(2) + 3(x) + 3(4) \qquad \text{Distributive property}$$
$$= 5x - 10 + 3x + 12$$
$$= 5x + 3x - 10 + 12 \qquad \text{Commutative property}$$
$$= 8x + 2$$

$$-7(y + 1) - 4(y - 3) = -7(y) - 7(1) - 4(y) - 4(-3)$$
$$= -7y - 7 - 4y + 12 \qquad \text{Be careful with this sign.}$$
$$= -7y - 4y - 7 + 12$$
$$= -11y + 5$$

$$5(x + 2) - (x + 3) = 5(x + 2) - 1(x + 3) \qquad \text{Remember that } -a = -1a.$$
$$= 5(x) + 5(2) - 1(x) - 1(3)$$
$$= 5x + 10 - x - 3$$
$$= 5x - x + 10 - 3$$
$$= 4x + 7$$

After you are sure of each step, you can use a more simplified format.

$$5(a + 4) - 7(a - 2) = 5a + 20 - 7a + 14$$
$$= -2a + 34$$

$$9(z - 7) + 11(z + 6) = 9z - 63 + 11z + 66$$
$$= 20z + 3$$

$$-(x - 2) + (x + 6) = -x + 2 + x + 6$$
$$= 8$$

Back to Evaluating Algebraic Expressions

Simplifying by combining similar terms aids in the process of evaluating some algebraic expressions. The last examples of this section illustrate this idea.

EXAMPLE 7   Evaluate $8x - 2y + 3x + 5y$ for $x = 3$ and $y = -4$.

Solution

Let's first simplify the given expression.

$$8x - 2y + 3x + 5y = 11x + 3y$$

Now we can evaluate for $x = 3$ and $y = -4$.

$$11x + 3y = 11(3) + 3(-4)$$
$$= 33 + (-12)$$
$$= 21$$

EXAMPLE 8   Evaluate $2ab + 5c - 6ab + 12c$ for $a = 2$, $b = -3$, and $c = 7$.

Solution

$$2ab + 5c - 6ab + 12c = -4ab + 17c$$
$$= -4(2)(-3) + 17(7) \quad \text{When } a = 2, b = -3, \text{ and } c = 7$$
$$= 24 + 119$$
$$= 143$$

EXAMPLE 9   Evaluate $8(x - 4) + 7(x + 3)$ for $x = 6$.

Solution

$$8(x - 4) + 7(x + 3) = 8x - 32 + 7x + 21 \quad \text{Distributive property}$$
$$= 15x - 11$$
$$= 15(6) - 11 \quad \text{When } x = 6$$
$$= 79$$

CONCEPT QUIZ

For Problems 1–10, answer true or false.

1. Addition is a commutative operation.

2. Subtraction is a commutative operation.

3. $[(2)(-3)](7) = (2)[(-3)(7)]$ is an example of the associative property for multiplication.

4. $[(8)(5)](-2) = (-2)[(8)(5)]$ is an example of the associative property for multiplication.

5. Zero is the identity element for addition.

6. The integer $-a$ is the additive inverse of a.

7. The additive inverse of 0 is 0.

8. The numerical coefficient of the term $-8xy$ is 8.

9. The numerical coefficient of the term ab is 1.

10. $6xy$ and $-2xyz$ are similar terms.

PROBLEM SET 1.5

For Problems 1–12, state the property that justifies each statement. For example, $3 + (-4) = (-4) + 3$ because of the commutative property for addition.

1. $3(7 + 8) = 3(7) + 3(8)$

2. $(-9)(17) = 17(-9)$

3. $-2 + (5 + 7) = (-2 + 5) + 7$

4. $-19 + 0 = -19$

5. $143(-7) = -7(143)$

6. $5[9 + (-4)] = 5(9) + 5(-4)$

7. $-119 + 119 = 0$

8. $-4 + (6 + 9) = (-4 + 6) + 9$

9. $-56 + 0 = -56$

10. $5 + (-12) = -12 + 5$

11. $[5(-8)]4 = 5[-8(4)]$

12. $[6(-4)]8 = 6[-4(8)]$

For Problems 13–30, simplify each numerical expression. Don't forget to take advantage of the properties if they can be used to simplify the computation.

13. $(-18 + 56) + 18$

14. $-72 + [72 + (-14)]$

15. $36 - 48 - 22 + 41$

16. $-24 + 18 + 19 - 30$

17. $(25)(-18)(-4)$

18. $(2)(-71)(50)$

19. $(4)(-16)(-9)(-25)$

20. $(-2)(18)(-12)(-5)$

21. $37(-42 - 58)$

22. $-46(-73 - 27)$

23. $59(36) + 59(64)$

24. $-49(72) - 49(28)$

25. $15(-14) + 16(-8)$

26. $-9(14) - 7(-16)$

27. $17 + (-18) - 19 - 14 + 13 - 17$

28. $-16 - 14 + 18 + 21 + 14 - 17$

29. $-21 + 22 - 23 + 27 + 21 - 19$

30. $24 - 26 - 29 + 26 + 18 + 29 - 17 - 10$

For Problems 31–62, simplify each algebraic expression by combining similar terms.

31. $9x - 14x$

32. $12x - 14x + x$

33. $4m + m - 8m$

34. $-6m - m + 17m$

35. $-9y + 5y - 7y$

36. $14y - 17y - 19y$

37. $4x - 3y - 7x + y$

38. $9x + 5y - 4x - 8y$

39. $-7a - 7b - 9a + 3b$

40. $-12a + 14b - 3a - 9b$

41. $6xy - x - 13xy + 4x$

42. $-7xy - 2x - xy + x$

43. $5x - 4 + 7x - 2x + 9$

44. $8x + 9 + 14x - 3x - 14$

45. $-2xy + 12 + 8xy - 16$

46. $14xy - 7 - 19xy - 6$

47. $-2a + 3b - 7b - b + 5a - 9a$

48. $-9a - a + 6b - 3a - 4b - b + a$

49. $13ab + 2a - 7a - 9ab + ab - 6a$

50. $-ab - a + 4ab + 7ab - 3a - 11ab$

51. $3(x + 2) + 5(x + 6)$

52. $7(x + 8) + 9(x + 1)$

53. $5(x - 4) + 6(x + 8)$

54. $-3(x + 2) - 4(x - 10)$

55. $9(x + 4) - (x - 8)$

56. $-(x - 6) + 5(x - 9)$

57. $3(a - 1) - 2(a - 6) + 4(a + 5)$

58. $-4(a + 2) + 6(a + 8) - 3(a - 6)$

59. $-2(m + 3) - 3(m - 1) + 8(m + 4)$

60. $5(m - 10) + 6(m - 11) - 9(m - 12)$

61. $(y + 3) - (y - 2) - (y + 6) - 7(y - 1)$

62. $-(y - 2) - (y + 4) - (y + 7) - 2(y + 3)$

For Problems 63–80, simplify each algebraic expression and then evaluate the resulting expression for the given values of the variables.

63. $3x + 5y + 4x - 2y$ for $x = -2$ and $y = 3$

64. $5x - 7y - 9x - 3y$ for $x = -1$ and $y = -4$

65. $5(x - 2) + 8(x + 6)$ for $x = -6$

66. $4(x - 6) + 9(x + 2)$ for $x = 7$

67. $8(x + 4) - 10(x - 3)$ for $x = -5$

68. $-(n + 2) - 3(n - 6)$ for $n = 10$

69. $(x - 6) - (x + 12)$ for $x = -3$

70. $(x + 12) - (x - 14)$ for $x = -11$

71. $2(x + y) - 3(x - y)$ for $x = -2$ and $y = 7$

72. $5(x - y) - 9(x + y)$ for $x = 4$ and $y = -4$

73. $2xy + 6 + 7xy - 8$ for $x = 2$ and $y = -4$

74. $4xy - 5 - 8xy + 9$ for $x = -3$ and $y = -3$

75. $5x - 9xy + 3x + 2xy$ for $x = 12$ and $y = -1$

76. $-9x + xy - 4xy - x$ for $x = 10$ and $y = -11$

77. $(a - b) - (a + b)$ for $a = 19$ and $b = -17$

78. $(a + b) - (a - b)$ for $a = -16$ and $b = 14$

79. $-3x + 7x + 4x - 2x - x$ for $x = -13$

80. $5x - 6x + x - 7x - x - 2x$ for $x = -15$

81. Use a calculator to check your answers for Problems 13–30.

■ ■ ■ **Thoughts into words**

82. State in your own words the associative property for addition of integers.

83. State in your own words the distributive property for multiplication over addition.

84. Is $2 \cdot 3 \cdot 5 \cdot 7 \cdot 11 + 7$ a prime or a composite number? Defend your answer.

Answers to Concept Quiz

1. True   **2.** False   **3.** True   **4.** False   **5.** True   **6.** True   **7.** True   **8.** False   **9.** True   **10.** False

CHAPTER 1

SUMMARY

(1.1) To **simplify a numerical expression,** perform the operations in the following order:

1. Perform the operations inside the symbols of inclusion (parentheses and brackets) and above and below each fraction bar. Start with the innermost inclusion symbol.
2. Perform all multiplications and divisions in the order in which they appear from left to right.
3. Perform all additions and subtractions in the order in which they appear from left to right.

To **evaluate an algebraic expression,** substitute the given values for the variables into the algebraic expression, and simplify the resulting numerical expression.

(1.2) A **prime number** is a whole number greater than 1 that has no factors (divisors) other than itself and 1. Whole numbers greater than 1 that are not prime numbers are called **composite numbers.** Every composite number has one and only one prime factorization.

The **greatest common factor** of 12 and 18 is 6, which means that 6 is the largest whole number divisor of both 12 and 18.

The **least common multiple** of 12 and 18 is 36, which means that 36 is the smallest nonzero multiple of both 12 and 18.

(1.3) The number line is a convenient visual aid for interpreting **addition of integers.**

Subtraction of integers is defined in terms of addition: $a - b$ means $a + (-b)$.

(1.4) To **multiply integers** we must remember that *the product of two positives or two negatives is positive* and

that *the product of a positive and a negative (in either order) is negative.*

To **divide integers** we must remember that *the quotient of two positives or two negatives is positive* and that *the quotient of a positive and a negative (or a negative and a positive) is negative.*

(1.5) The following basic properties help with numerical manipulations and serve as a basis for algebraic computations:

Commutative Properties

For addition: $a + b = b + a$
For multiplication: $ab = ba$

Associative Properties

For addition: $(a + b) + c = a + (b + c)$
For multiplication: $(ab)c = a(bc)$

Identity Properties

For addition: $a + 0 = 0 + a = a$
For multiplication: $a(1) = 1(a) = a$

Additive Inverse Property

$a + (-a) = (-a) + a = 0$

Multiplication Property of Zero

$a(0) = 0(a) = 0$

Multiplication Property of Negative One

$-1(a) = a(-1) = -a$

Distributive Properties

$a(b + c) = ab + ac$
$a(b - c) = ab - ac$

CHAPTER 1 REVIEW PROBLEM SET

In Problems 1–10, perform the indicated operations.

1. $7 + (-10)$

2. $(-12) + (-13)$

3. $8 - 13$

4. $-6 - 9$

5. $-12 - (-11)$

6. $-17 - (-19)$

7. $(13)(-12)$

8. $(-14)(-18)$

9. $(-72) \div (-12)$

10. $117 \div (-9)$

For Problems 11–15, classify each number as prime or composite.

11. 73   **12.** 87

13. 63   **14.** 81

15. 91

For Problems 16–20, express each number as the product of prime factors.

16. 24   **17.** 63

18. 57   **19.** 64

20. 84

21. Find the greatest common factor of 36 and 54.

22. Find the greatest common factor of 48, 60, and 84.

23. Find the least common multiple of 18 and 20.

24. Find the least common multiple of 15, 27, and 35.

For Problems 25–38, simplify each numerical expression.

25. $(19 + 56) + (-9)$

26. $43 - 62 + 12$

27. $8 + (-9) + (-16) + (-14) + 17 + 12$

28. $19 - 23 - 14 + 21 + 14 - 13$

29. $3(-4) - 6$ **30.** $(-5)(-4) - 8$

31. $(5)(-2) + (6)(-4)$

32. $(-6)(8) + (-7)(-3)$

33. $(-6)(3) - (-4)(-5)$

34. $(-7)(9) - (6)(5)$

35. $\dfrac{4(-7) - (3)(-2)}{-11}$ **36.** $\dfrac{(-4)(9) + (5)(-3)}{1 - 18}$

37. $3 - 2[4(-3 - 1)]$ **38.** $-6 - [3(-4 - 7)]$

39. A record high temperature of 125°F occurred in Laughlin, Nevada on June 29, 1994. A record low temperature of $-50°F$ occurred in San Jacinto, Nevada on January 8, 1937. Find the difference between the record high and low temperatures.

40. In North America, the highest elevation — Mt. McKinley, Alaska — is 20,320 feet above sea level. The lowest elevation in North America, at Death Valley, California, is 282 feet below sea level. Find the absolute value of the difference in elevation between Mt. McKinley and Death Valley.

41. As a running back in a football game, Marquette carried the ball 7 times. On two plays he gained 6 yards each play, on another play he lost 4 yards, on the next three plays he gained 8 yards per play, and on the last play he lost 1 yard. Write a numerical expression that gives Marquette's overall yardage for the game, and simplify that expression.

42. Shelley started the month with $3278 in her checking account. During the month she deposited $175 each week for 4 weeks and had debit charges of $50, $189, $160, $20, and $115. What is the balance in her checking account after these deposits and debits?

For Problems 43–54, simplify each algebraic expression by combining similar terms.

43. $12x + 3x - 7x$

44. $9y + 3 - 14y - 12$

45. $8x + 5y - 13x - y$

46. $9a + 11b + 4a - 17b$

47. $3ab - 4ab - 2a$

48. $5xy - 9xy + xy - y$

49. $3(x + 6) + 7(x + 8)$

50. $5(x - 4) - 3(x - 9)$

51. $-3(x - 2) - 4(x + 6)$

52. $-2x - 3(x - 4) + 2x$

53. $2(a - 1) - a - 3(a - 2)$

54. $-(a - 1) + 3(a - 2) - 4a + 1$

For Problems 55–68, evaluate each algebraic expression for the given values of the variables.

55. $5x + 8y$ for $x = -7$ and $y = -3$

56. $7x - 9y$ for $x = -3$ and $y = 4$

57. $\dfrac{-5x - 2y}{-2x - 7}$ for $x = 6$ and $y = 4$

58. $\dfrac{-3x + 4y}{3x}$ for $x = -4$ and $y = -6$

59. $-2a + \dfrac{a - b}{a - 2}$ for $a = -5$ and $b = 9$

60. $\dfrac{2a + b}{b + 6} - 3b$ for $a = 3$ and $b = -4$

61. $5a + 6b - 7a - 2b$ for $a = -1$ and $b = 5$

62. $3x + 7y - 5x + y$ for $x = -4$ and $y = 3$

63. $2xy + 6 + 5xy - 8$ for $x = -1$ and $y = 1$

64. $7(x + 6) - 9(x + 1)$ for $x = -2$

65. $-3(x - 4) - 2(x + 8)$ for $x = 7$

66. $2(x - 1) - (x + 2) + 3(x - 4)$ for $x = -4$

67. $(a - b) - (a + b) - b$ for $a = -1$ and $b = -3$

68. $2ab - 3(a - b) + b + a$ for $a = 2$ and $b = -5$

CHAPTER 1

TEST

  applies to all problems in this Chapter Test.

For Problems 1–10, simplify each numerical expression.

1. $6 + (-7) - 4 + 12$

2. $7 + 4(9) + 2$

3. $-4(2 - 8) + 14$

4. $5(-7) - (-3)(8)$

5. $8 \div (-4) + (-6)(9) - 2$

6. $(-8)(-7) + (-6) - (9)(12)$

7. $\dfrac{6(-4) - (-8)(-5)}{-16}$

8. $-14 + 23 - 17 - 19 + 26$

9. $(-14)(4) \div 4 + (-6)$

10. $6(-9) - (-8) - (-7)(4) + 11$

11. It was reported on the 5 o'clock weather show that the current temperature was 7°F. The forecast was for the temperature to drop 13 degrees by 6:00 A.M. If the forecast is correct, what will the temperature be at 6:00 A.M.?

For Problems 12–17, evaluate each algebraic expression for the given values of the variables.

12. $7x - 9y$ for $x = -4$ and $y = -6$

13. $-4a - 6b$ for $a = -9$ and $b = 12$

14. $3xy - 8y + 5x$ for $x = 7$ and $y = -2$

15. $5(x - 4) - 6(x + 7)$ for $x = -5$

16. $3x - 2y - 4x - x + 7y$ for $x = 6$ and $y = -7$

17. $\dfrac{-x - y}{y - x}$ for $x = -9$ and $y = -6$

18. Classify 79 as a prime or a composite number.

19. Express 360 as a product of prime factors.

20. Find the greatest common factor of 36, 60, and 84.

21. Find the least common multiple of 9 and 24.

22. State the property of integers demonstrated by $[-3 + (-4)] + (-6) = -3 + [(-4) + (-6)]$.

23. State the property of integers demonstrated by $8(25 + 37) = 8(25) + 8(37)$.

24. Simplify $-7x + 9y - y + x - 2y - 7x$ by combining similar terms.

25. Simplify $-2(x - 4) - 5(x + 7) - 6(x - 1)$ by applying the distributive property and combining similar terms.

People that watch the stock market are familiar with rational numbers expressed in decimal form.

AP/Wide World Photos

Real Numbers

aleb left an estate valued at $750,000. His will states that three-fourths of the estate is to be divided equally among his three children. The numerical expression $\left(\dfrac{1}{3}\right)\left(\dfrac{3}{4}\right)(750,000)$ can be used to determine how much each of his three children should receive.

When the market opened on Monday morning, Garth bought some shares of a stock at $13.25 per share. The rational numbers $0.75, -1.50, 2.25, -0.25$, and -0.50 represent the daily changes in the market for that stock for the week. We use the numerical expression $13.25 + 0.75 + (-1.50) + 2.25 + (-0.25) + (-0.50)$ to determine the value of one share of Garth's stock when the market closed on Friday.

The width of a rectangle is w feet, and its length is 4 feet more than three times its width. The algebraic expression $2w + 2(3w + 4)$ represents the perimeter of the rectangle.

In this chapter we use the concepts of **numerical** and **algebraic expressions** to review some computational skills from arithmetic and to continue the transition from arithmetic to algebra. However, the set of **rational numbers** now becomes the primary

43

focal point. We urge you to use this chapter to fine-tune your arithmetic skills so that the algebraic concepts in subsequent chapters can be built upon a solid foundation.

InfoTrac Project

●Do a subject guide search on **decimal fractions.** Find a periodical article on the Cincinnati Stock Exchange. Write a sentence summarizing the article. If you bought 100 shares of stock at $\$27\frac{5}{8}$ and sold them at $\$36\frac{3}{4}$, what was your profit? Now convert the two stock prices into decimal form and calculate the profit using decimals. Which was easier? Which fractions converted into decimals would render your answer less accurate than doing the calculations with fractions?

2.1   Rational Numbers: Multiplication and Division

Objectives

- Reduce fractions to lowest terms.
- Multiply and divide fractions.
- Solve application problems involving multiplication and division of fractions.

Any number that can be written in the form $\dfrac{a}{b}$, where a and b are integers and b is not zero, is called a **rational number.** (The form $\dfrac{a}{b}$ is called a fraction or sometimes a common fraction.) Here are some examples of rational numbers:

$$\frac{1}{2} \qquad \frac{7}{9} \qquad \frac{15}{7} \qquad \frac{-3}{4} \qquad \frac{5}{-7} \qquad \frac{-11}{-13}$$

All integers are rational numbers because every integer can be expressed as the indicated quotient of two integers. For example,

$$6 = \frac{6}{1} = \frac{12}{2} = \frac{18}{3}, \text{ etc.}$$

$$27 = \frac{27}{1} = \frac{54}{2} = \frac{81}{3}, \text{ etc.}$$

$$0 = \frac{0}{1} = \frac{0}{2} = \frac{0}{3}, \text{ etc.}$$

Our work in Chapter 1 with the division of negative integers helps with the next three examples:

$$-4 = \frac{-4}{1} = \frac{-8}{2} = \frac{-12}{3}, \text{ etc.}$$

$$-6 = \frac{6}{-1} = \frac{12}{-2} = \frac{18}{-3}, \text{ etc.}$$

$$10 = \frac{10}{1} = \frac{-10}{-1} = \frac{-20}{-2}, \text{ etc.}$$

Observe the following general property:

PROPERTY 2.1

$$\frac{-a}{b} = \frac{a}{-b} = -\frac{a}{b} \quad \text{and} \quad \frac{-a}{-b} = \frac{a}{b}$$

Therefore, we can write the rational number $\dfrac{-2}{3}$, for example, as $\dfrac{2}{-3}$ or $-\dfrac{2}{3}$. (However, we seldom express rational numbers with negative denominators.)

Multiplying Rational Numbers

We define multiplication of rational numbers in common fractional form as follows:

DEFINITION 2.1

If a, b, c, and d are integers, with b and d not equal to zero, then

$$\frac{a}{b} \cdot \frac{c}{d} = \frac{a \cdot c}{b \cdot d}$$

To multiply rational numbers in common fractional form, we simply *multiply numerators and multiply denominators*. Because the numerators and denominators are integers, our previous agreements pertaining to multiplication of integers hold for the rationals. That is, *the product of two positive rational numbers or of two negative rational numbers is a positive rational number. The product of a positive rational number and a negative rational number (in either order) is a negative rational number.* Furthermore, we see from the definition that the commutative and associative properties hold for multiplication of rational numbers. We are free to rearrange and regroup factors as we do with integers.

The following examples illustrate the definition for multiplying rational numbers.

$$\frac{1}{3} \cdot \frac{2}{5} = \frac{1 \cdot 2}{3 \cdot 5} = \frac{2}{15}$$

$$\frac{3}{4} \cdot \frac{5}{7} = \frac{3 \cdot 5}{4 \cdot 7} = \frac{15}{28}$$

$$\frac{-2}{3} \cdot \frac{7}{9} = \frac{-2 \cdot 7}{3 \cdot 9} = \frac{-14}{27} \quad \text{or} \quad -\frac{14}{27}$$

$$\frac{1}{5} \cdot \frac{9}{-11} = \frac{1 \cdot 9}{5(-11)} = \frac{9}{-55} \quad \text{or} \quad -\frac{9}{55}$$

$$-\frac{3}{4} \cdot \frac{7}{13} = \frac{-3}{4} \cdot \frac{7}{13} = \frac{-3 \cdot 7}{4 \cdot 13} = \frac{-21}{52} \quad \text{or} \quad -\frac{21}{52}$$

$$\frac{3}{5} \cdot \frac{5}{3} = \frac{3 \cdot 5}{5 \cdot 3} = \frac{15}{15} = 1$$

The last example is a very special case. *If the product of two numbers is 1, the numbers are said to be **reciprocals** of each other.*

Using Definition 2.1 and applying the multiplication property of one, the fraction $\dfrac{a \cdot k}{b \cdot k}$, where b and k are nonzero integers, simplifies as shown:

$$\frac{a \cdot k}{b \cdot k} = \frac{a}{b} \cdot \frac{k}{k} = \frac{a}{b} \cdot 1 = \frac{a}{b}$$

This result is stated as Property 2.2.

PROPERTY 2.2

If b and k are nonzero integers, and a is any integer, then

$$\frac{a \cdot k}{b \cdot k} = \frac{a}{b}$$

We often use Property 2.2 when we work with rational numbers. It is called the **fundamental principle of fractions** and provides the basis for equivalent fractions. In the following examples, we will use this property to *reduce fractions to lowest terms* or *express fractions in simplest or reduced form.*

EXAMPLE 1   Reduce $\dfrac{12}{18}$ to lowest terms.

Solution

$$\frac{12}{18} = \frac{2 \cdot 6}{3 \cdot 6} = \frac{2}{3} \cdot \frac{6}{6} = \frac{2}{3} \cdot 1 = \frac{2}{3}$$

EXAMPLE 2   Change $\dfrac{14}{35}$ to simplest form.

Solution

$$\frac{14}{35} = \frac{2 \cdot \cancel{7}}{5 \cdot \cancel{7}} = \frac{2}{5}$$ Divide a common factor of 7 out of both the numerator and denominator.

EXAMPLE 3   Express $\dfrac{-24}{32}$ in reduced form.

Solution

$$\frac{-24}{32} = -\frac{24}{32} = -\frac{3 \cdot \cancel{8}}{4 \cdot \cancel{8}} = -\frac{3}{4} \qquad \frac{-a}{b} = -\frac{a}{b}$$

EXAMPLE 4

Reduce $-\dfrac{72}{90}$.

Solution

$$-\frac{72}{90} = -\frac{2 \cdot 2 \cdot 2 \cdot 3 \cdot 3}{2 \cdot 3 \cdot 3 \cdot 5} = -\frac{4}{5}$$

Use the prime factored forms of the numerator and denominator to help recognize common factors. ■

The fractions may contain variables in the numerator or denominator (or both), but this creates no great difficulty. Our thought processes remain the same, as these next examples illustrate. Variables that appear in denominators represent *nonzero* integers.

EXAMPLE 5

Reduce $\dfrac{9x}{17x}$.

Solution

$$\frac{9x}{17x} = \frac{9 \cdot x}{17 \cdot x} = \frac{9}{17}$$

■

EXAMPLE 6

Simplify $\dfrac{8x}{36y}$.

Solution

$$\frac{8x}{36y} = \frac{2 \cdot 2 \cdot 2 \cdot x}{2 \cdot 2 \cdot 3 \cdot 3 \cdot y} = \frac{2x}{9y}$$

■

EXAMPLE 7

Express $\dfrac{-9xy}{30y}$ in reduced form.

Solution

$$\frac{-9xy}{30y} = -\frac{9xy}{30y} = -\frac{3 \cdot 3 \cdot x \cdot y}{2 \cdot 3 \cdot 5 \cdot y} = -\frac{3x}{10}$$

■

EXAMPLE 8

Reduce $\dfrac{-7abc}{-9ac}$.

Solution

$$\frac{-7abc}{-9ac} = \frac{7abc}{9ac} = \frac{7abc}{9ac} = \frac{7b}{9} \qquad \frac{-a}{-b} = \frac{a}{b}$$

■

We are now ready to consider multiplication problems with the understanding that *the final answer should be expressed in reduced form.* Study the following examples carefully, because we have used different formats to handle such problems.

EXAMPLE 9   Multiply $\dfrac{7}{9} \cdot \dfrac{5}{14}$.

Solution

$$\frac{7}{9} \cdot \frac{5}{14} = \frac{7 \cdot 5}{9 \cdot 14} = \frac{\cancel{7} \cdot 5}{3 \cdot 3 \cdot 2 \cdot \cancel{7}} = \frac{5}{18}$$
■

EXAMPLE 10   Find the product of $\dfrac{8}{9}$ and $\dfrac{18}{24}$.

Solution

$$\frac{\overset{1}{\cancel{8}}}{\underset{1}{\cancel{9}}} \cdot \frac{\overset{2}{\cancel{18}}}{\underset{3}{\cancel{24}}} = \frac{2}{3}$$ Divide a common factor of 8 out of 8 and 24 and a common factor of 9 out of 9 and 18.
■

EXAMPLE 11   Multiply $\left(-\dfrac{6}{8}\right)\left(\dfrac{14}{32}\right)$.

Solution

$$\left(-\frac{6}{8}\right)\left(\frac{14}{32}\right) = -\frac{\overset{3}{\cancel{6}} \cdot \overset{7}{\cancel{14}}}{\underset{4}{\cancel{8}} \cdot \underset{16}{\cancel{32}}} = -\frac{21}{64}$$ Divide a common factor of 2 out of 6 and 8 and a common factor of 2 out of 14 and 32.
■

EXAMPLE 12   Multiply $\left(-\dfrac{9}{4}\right)\left(-\dfrac{14}{15}\right)$.

Solution

$$\left(-\frac{9}{4}\right)\left(-\frac{14}{15}\right) = \frac{\cancel{3} \cdot 3 \cdot 2 \cdot 7}{2 \cdot 2 \cdot \cancel{3} \cdot 5} = \frac{21}{10}$$ Immediately we recognize that *a negative times a negative is positive.*
■

EXAMPLE 13   Multiply $\dfrac{9x}{7y} \cdot \dfrac{14y}{45}$.

Solution

$$\frac{9x}{7y} \cdot \frac{14y}{45} = \frac{\cancel{9} \cdot x \cdot \overset{2}{\cancel{14}} \cdot \cancel{y}}{\cancel{7} \cdot \cancel{y} \cdot \underset{5}{\cancel{45}}} = \frac{2x}{5}$$
■

EXAMPLE   14   Multiply $\dfrac{-6c}{7ab} \cdot \dfrac{14b}{5c}$.

Solution

$$\frac{-6c}{7ab} \cdot \frac{14b}{5c} = -\frac{2 \cdot 3 \cdot \cancel{c} \cdot 2 \cdot \cancel{7} \cdot \cancel{b}}{\cancel{7} \cdot a \cdot \cancel{b} \cdot 5 \cdot \cancel{c}} = -\frac{12}{5a}$$

Dividing Rational Numbers

The following example motivates a definition for division of rational numbers in fractional form.

$$\frac{\dfrac{3}{4}}{\dfrac{2}{3}} = \left(\frac{\dfrac{3}{4}}{\dfrac{2}{3}}\right)\left(\frac{\dfrac{3}{2}}{\dfrac{3}{2}}\right) = \frac{\left(\dfrac{3}{4}\right)\left(\dfrac{3}{2}\right)}{\left(\dfrac{2}{3}\right)\left(\dfrac{3}{2}\right)} = \frac{\left(\dfrac{3}{4}\right)\left(\dfrac{3}{2}\right)}{1} = \left(\dfrac{3}{4}\right)\left(\dfrac{3}{2}\right) = \frac{9}{8}$$

↑

Notice that this is a form of 1 and $\dfrac{3}{2}$ is the reciprocal of $\dfrac{2}{3}$.

In other words, $\dfrac{3}{4}$ divided by $\dfrac{2}{3}$ is equivalent to $\dfrac{3}{4}$ times $\dfrac{3}{2}$. The following definition for division should seem reasonable:

DEFINITION 2.2

If b, c, and d are nonzero integers, and a is any integer, then

$$\frac{a}{b} \div \frac{c}{d} = \frac{a}{b} \cdot \frac{d}{c}$$

Note that to divide $\dfrac{a}{b}$ by $\dfrac{c}{d}$, we multiply $\dfrac{a}{b}$ times the reciprocal of $\dfrac{c}{d}$, which is $\dfrac{d}{c}$. The following examples demonstrate the important steps of a division problem.

$$\frac{2}{3} \div \frac{1}{2} = \frac{2}{3} \cdot \frac{2}{1} = \frac{4}{3}$$

$$\frac{5}{6} \div \frac{3}{4} = \frac{5}{6} \cdot \frac{4}{3} = \frac{5 \cdot 4}{6 \cdot 3} = \frac{5 \cdot 2 \cdot 2}{2 \cdot 3 \cdot 3} = \frac{10}{9}$$

$$-\frac{9}{12} \div \frac{3}{6} = -\frac{\overset{3}{\cancel{9}}}{\underset{2}{\cancel{12}}} \cdot \frac{\overset{1}{\cancel{6}}}{\underset{1}{\cancel{3}}} = -\frac{3}{2}$$

$$\left(-\frac{27}{56}\right) \div \left(-\frac{33}{72}\right) = \left(-\frac{27}{56}\right)\left(-\frac{72}{33}\right) = \frac{\overset{9}{\cancel{27}} \cdot \overset{9}{\cancel{72}}}{\underset{7}{\cancel{56}} \cdot \underset{11}{\cancel{33}}} = \frac{81}{77}$$

$$\frac{6}{7} \div 2 = \frac{6}{7} \cdot \frac{1}{2} = \frac{\overset{3}{\cancel{6}}}{7} \cdot \frac{1}{\underset{1}{\cancel{2}}} = \frac{3}{7}$$

$$\frac{5x}{7y} \div \frac{10}{28y} = \frac{5x}{7y} \cdot \frac{28y}{10} = \frac{5 \cdot x \cdot \overset{4}{\cancel{28}} \cdot \cancel{y}}{7 \cdot \cancel{y} \cdot \underset{2}{\cancel{10}}} = 2x$$

PROBLEM 1

Frank has purchased 50 candy bars to make s'mores for the Boy Scout troop. If he uses $\frac{2}{3}$ of a candy bar for each s'more, how many s'mores will he be able to make?

Solution

To find how many s'mores can be made, we need to divide 50 by $\frac{2}{3}$.

$$50 \div \frac{2}{3} = 50 \cdot \frac{3}{2} = \frac{\overset{25}{\cancel{50}}}{1} \cdot \frac{3}{\underset{1}{2}} = \frac{75}{1} = 75$$

So Frank can make 75 s'mores.   ■

CONCEPT QUIZ

For Problems 1–10, answer true or false.

1. 6 is a rational number.

2. $\frac{1}{8}$ is a rational number.

3. $\frac{-2}{-3} = \frac{2}{3}$

4. $\frac{-5}{3} = \frac{5}{-3}$

5. The product of a negative rational number and a positive rational number is a positive rational number.

6. If the product of two rational numbers is 1, the numbers are said to be reciprocals.

7. The reciprocal of $\frac{-3}{7}$ is $\frac{7}{3}$.

8. $\frac{10}{25}$ is reduced to lowest terms.

9. $\frac{4ab}{7c}$ is reduced to lowest terms.

10. To divide $\frac{m}{n}$ by $\frac{p}{q}$, we multiply $\frac{m}{n}$ by $\frac{q}{p}$.

PROBLEM SET 2.1

For Problems 1–24, reduce each fraction to lowest terms.

1. $\dfrac{8}{12}$

2. $\dfrac{12}{16}$

3. $\dfrac{16}{24}$

4. $\dfrac{18}{32}$

5. $\dfrac{15}{9}$

6. $\dfrac{48}{36}$

7. $\dfrac{-8}{48}$

8. $\dfrac{-3}{15}$

9. $\dfrac{27}{-36}$

10. $\dfrac{9}{-51}$

11. $\dfrac{-54}{-56}$

12. $\dfrac{-24}{-80}$

13. $\dfrac{24x}{44x}$

14. $\dfrac{15y}{25y}$

15. $\dfrac{9x}{21y}$

16. $\dfrac{4y}{30x}$

17. $\dfrac{14xy}{35y}$

18. $\dfrac{55xy}{77x}$

19. $\dfrac{-20ab}{52bc}$

20. $\dfrac{-23ac}{41c}$

21. $\dfrac{-56yz}{-49xy}$

22. $\dfrac{-21xy}{-14ab}$

23. $\dfrac{65abc}{91ac}$

24. $\dfrac{68xyz}{85yz}$

For Problems 25–58, multiply or divide as indicated, and express answers in reduced form.

25. $\dfrac{3}{4} \cdot \dfrac{5}{7}$

26. $\dfrac{4}{5} \cdot \dfrac{3}{11}$

27. $\dfrac{2}{7} \div \dfrac{3}{5}$

28. $\dfrac{5}{6} \div \dfrac{11}{13}$

29. $\dfrac{3}{8} \cdot \dfrac{12}{15}$

30. $\dfrac{4}{9} \cdot \dfrac{3}{2}$

31. $\dfrac{-6}{13} \cdot \dfrac{26}{9}$

32. $\dfrac{3}{4} \cdot \dfrac{-14}{12}$

33. $\dfrac{7}{9} \div \dfrac{5}{9}$

34. $\dfrac{3}{11} \div \dfrac{7}{11}$

35. $\dfrac{1}{4} \div \dfrac{-5}{6}$

36. $\dfrac{7}{8} \div \dfrac{14}{-16}$

37. $\left(-\dfrac{8}{10}\right)\left(-\dfrac{10}{32}\right)$

38. $\left(-\dfrac{6}{7}\right)\left(-\dfrac{21}{24}\right)$

39. $-9 \div \dfrac{1}{3}$

40. $-10 \div \dfrac{1}{4}$

41. $\dfrac{5x}{9y} \cdot \dfrac{7y}{3x}$

42. $\dfrac{4a}{11b} \cdot \dfrac{6b}{7a}$

43. $\dfrac{6a}{14b} \cdot \dfrac{16b}{18a}$

44. $\dfrac{5y}{8x} \cdot \dfrac{14z}{15y}$

45. $\dfrac{10x}{-9y} \cdot \dfrac{15}{20x}$

46. $\dfrac{3x}{4y} \cdot \dfrac{-8w}{9z}$

47. $ab \cdot \dfrac{2}{b}$

48. $3xy \cdot \dfrac{4}{x}$

49. $\left(-\dfrac{7x}{12y}\right)\left(-\dfrac{24y}{35x}\right)$

50. $\left(-\dfrac{10a}{15b}\right)\left(-\dfrac{45b}{65a}\right)$

51. $\dfrac{3}{x} \div \dfrac{6}{y}$

52. $\dfrac{6}{x} \div \dfrac{14}{y}$

53. $\dfrac{5x}{9y} \div \dfrac{13x}{36y}$

54. $\dfrac{3x}{5y} \div \dfrac{7x}{10y}$

55. $\dfrac{-7}{x} \div \dfrac{9}{x}$

56. $\dfrac{8}{y} \div \dfrac{28}{-y}$

57. $\dfrac{-4}{n} \div \dfrac{-18}{n}$

58. $\dfrac{-34}{n} \div \dfrac{-51}{n}$

For Problems 59–74, perform the operations as indicated, and express answers in lowest terms.

59. $\dfrac{3}{4} \cdot \dfrac{8}{9} \cdot \dfrac{12}{20}$

60. $\dfrac{5}{6} \cdot \dfrac{9}{10} \cdot \dfrac{8}{7}$

61. $\left(-\dfrac{3}{8}\right)\left(\dfrac{13}{14}\right)\left(-\dfrac{12}{9}\right)$

62. $\left(-\dfrac{7}{9}\right)\left(\dfrac{5}{11}\right)\left(-\dfrac{18}{14}\right)$

63. $\left(\dfrac{3x}{4y}\right)\left(\dfrac{8}{9x}\right)\left(\dfrac{12y}{5}\right)$

64. $\left(\dfrac{2x}{3y}\right)\left(\dfrac{5y}{x}\right)\left(\dfrac{9}{4x}\right)$

65. $\left(-\dfrac{2}{3}\right)\left(\dfrac{3}{4}\right) \div \dfrac{1}{8}$

66. $\dfrac{3}{4} \cdot \dfrac{4}{5} \div \dfrac{1}{6}$

67. $\dfrac{5}{7} \div \left(-\dfrac{5}{6}\right)\left(-\dfrac{6}{7}\right)$

68. $\left(-\dfrac{3}{8}\right) \div \left(-\dfrac{4}{5}\right)\left(\dfrac{1}{2}\right)$

69. $\left(-\dfrac{6}{7}\right) \div \left(\dfrac{5}{7}\right)\left(-\dfrac{5}{6}\right)$

70. $\left(-\dfrac{4}{3}\right) \div \left(\dfrac{4}{5}\right)\left(\dfrac{3}{5}\right)$

71. $\left(\dfrac{4}{9}\right)\left(-\dfrac{9}{8}\right) \div \left(-\dfrac{3}{4}\right)$

72. $\left(-\dfrac{7}{8}\right)\left(\dfrac{4}{7}\right) \div \left(-\dfrac{3}{2}\right)$

73. $\left(\dfrac{5}{2}\right)\left(\dfrac{2}{3}\right) \div \left(-\dfrac{1}{4}\right) \div (-3)$

74. $\dfrac{1}{3} \div \left(\dfrac{3}{4}\right)\left(\dfrac{1}{2}\right) \div 2$

75. Maria's department has $\dfrac{3}{4}$ of all of the accounts within the ABC Advertising Agency. Maria is personally responsible for $\dfrac{1}{3}$ of all accounts in her department. For what portion of all of the accounts at ABC is Maria personally responsible?

76. Pablo has a board that is $4\frac{1}{2}$ feet long, and he wants to to cut it into three pieces all of the same length (see Figure 2.1). Find the length of each of the three pieces.

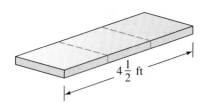

$4\frac{1}{2}$ ft

Figure 2.1

77. A recipe for birthday cake calls for $\frac{3}{4}$ cup of sugar. How much sugar is needed to make 3 cakes?

78. Caleb left an estate valued at $750,000. His will states that three-fourths of the estate is to be divided equally among his three children. How much should each receive?

79. One of Arlene's recipes calls for $3\frac{1}{2}$ cups of milk. If she wants to make one-half of the recipe, how much milk should she use?

80. The total length of the four sides of a square is $8\frac{2}{3}$ yards. How long is each side of the square?

81. If it takes $3\frac{1}{4}$ yards of material to make drapes for 1 window, how much material is needed for 5 windows?

82. If your calculator is equipped to handle rational numbers in $\frac{a}{b}$ form, use it to check your answers for Problems 1–12 and 59–74.

■ ■ ■ **Thoughts into words**

83. State in your own words the property

$$-\frac{a}{b} = \frac{-a}{b} = \frac{a}{-b}$$

84. Find the mistake in this simplification process:

$$\frac{1}{2} \div \left(\frac{2}{3}\right)\left(\frac{3}{4}\right) \div 3 = \frac{1}{2} \div \frac{1}{2} \div 3 = \frac{1}{2} \cdot 2 \cdot \frac{1}{3} = \frac{1}{3}$$

How would you correct the error?

■ ■ ■ **Further investigations**

85. The division problem $35 \div 7$ can be interpreted as "how many 7s are there in 35?" Likewise, a division problem such as $3 \div \frac{1}{2}$ can be interpreted as "how many halves are there in 3?" Use this "how many" interpretation to do each division problem.

a. $4 \div \frac{1}{2}$ **b.** $3 \div \frac{1}{4}$

c. $5 \div \frac{1}{8}$ **d.** $6 \div \frac{1}{7}$

e. $\frac{5}{6} \div \frac{1}{6}$ **f.** $\frac{7}{8} \div \frac{1}{8}$

86. Estimation is important in mathematics. In each of the following, estimate whether the answer is greater than 1 or less than 1 by using the "how many" idea from Problem 85.

a. $\frac{3}{4} \div \frac{1}{2}$ **b.** $1 \div \frac{7}{8}$

c. $\frac{1}{2} \div \frac{3}{4}$ **d.** $\frac{8}{7} \div \frac{7}{8}$

e. $\frac{2}{3} \div \frac{1}{4}$ **f.** $\frac{3}{5} \div \frac{3}{4}$

87. Reduce each fraction to lowest terms. Don't forget that we presented some divisibility rules in Problem Set 1.2.

a. $\frac{99}{117}$ **b.** $\frac{175}{225}$

c. $\frac{-111}{123}$ **d.** $\frac{-234}{270}$

e. $\frac{270}{495}$ **f.** $\frac{324}{459}$

g. $\frac{91}{143}$ **h.** $\frac{187}{221}$

2.2 Rational Numbers: Addition and Subtraction

Objectives

- ▪ Add and subtract rational numbers in fractional form.
- ▪ Combine similar terms whose coefficients are rational numbers in fractional form.
- ▪ Solve application problems that involve the addition and subtraction of rational numbers in fractional form.

Suppose that it is one-fifth of a mile between your dorm and the student center, and two-fifths of a mile between the student center and the library along a straight line as indicated in Figure 2.2. The total distance between your dorm and the library is three-fifths of a mile, and we write $\frac{1}{5} + \frac{2}{5} = \frac{3}{5}$.

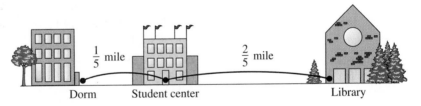

Dorm   Student center   Library

Figure 2.2

A pizza is cut into seven equal pieces and you eat two of the pieces (see Figure 2.3). How much of the pizza remains? We represent the whole pizza by $\frac{7}{7}$ and then conclude that $\frac{7}{7} - \frac{2}{7} = \frac{5}{7}$ of the pizza remains.

Figure 2.3

These examples motivate the following definition for addition and subtraction of rational numbers in $\dfrac{a}{b}$ form.

DEFINITION 2.3

If a, b, and c are integers, and b is not zero, then

$$\frac{a}{b} + \frac{c}{b} = \frac{a + c}{b} \qquad \text{Addition}$$

$$\frac{a}{b} - \frac{c}{b} = \frac{a - c}{b} \qquad \text{Subtraction}$$

We say that *rational numbers* **with common denominators** *can be added or subtracted by adding or subtracting the numerators and placing the results over the common denominator.* Consider these examples:

$$\frac{3}{7} + \frac{2}{7} = \frac{3 + 2}{7} = \frac{5}{7}$$

$$\frac{7}{8} - \frac{2}{8} = \frac{7 - 2}{8} = \frac{5}{8}$$

$$\frac{2}{6} + \frac{1}{6} = \frac{2 + 1}{6} = \frac{3}{6} = \frac{1}{2} \qquad \text{We agree to reduce the final answer.}$$

$$\frac{3}{11} - \frac{5}{11} = \frac{3 - 5}{11} = \frac{-2}{11} \qquad \text{or} \qquad -\frac{2}{11}$$

$$\frac{5}{x} + \frac{7}{x} = \frac{5 + 7}{x} = \frac{12}{x}$$

$$\frac{9}{y} - \frac{3}{y} = \frac{9 - 3}{y} = \frac{6}{y}$$

In the last two examples, we must specify that the variables x and y cannot be equal to zero in order to exclude division by zero. It is always necessary to restrict denominators to nonzero values, although we will not take the time or space to list such restrictions for every problem.

How do we add or subtract if the fractions do not have a common denominator? We use the fundamental principle of fractions, $\dfrac{a}{b} = \dfrac{a \cdot k}{b \cdot k}$, and obtain equivalent fractions that have a common denominator. **Equivalent fractions** are fractions that name the same number. Consider the following example, which shows the details.

EXAMPLE 1   Add $\dfrac{1}{2} + \dfrac{1}{3}$.

Solution

$$\frac{1}{2} = \frac{1 \cdot 3}{2 \cdot 3} = \frac{3}{6} \qquad \frac{1}{2} \text{ and } \frac{3}{6} \text{ are equivalent fractions that name the same number.}$$

$$\frac{1}{3} = \frac{1 \cdot 2}{3 \cdot 2} = \frac{2}{6} \qquad \frac{1}{3} \text{ and } \frac{2}{6} \text{ are equivalent fractions that name the same number.}$$

$$\frac{1}{2} + \frac{1}{3} = \frac{3}{6} + \frac{2}{6} = \frac{3 + 2}{6} = \frac{5}{6} \qquad \blacksquare$$

Note that in Example 1 we chose 6 as our common denominator, and 6 is the least common multiple of the original denominators 2 and 3. (Recall that the least common multiple is the smallest nonzero whole number divisible by the given numbers.) In general, we use the least common multiple of the denominators of the fractions to be added or subtracted as a **least common denominator** (LCD).

Recall from Section 1.2 that the least common multiple is found either by inspection or by using the prime factorization forms of the numbers. Let's consider some examples involving these procedures.

EXAMPLE 2   Add $\dfrac{1}{4} + \dfrac{2}{5}$.

Solution

By inspection we see that the LCD is 20. Thus both fractions can be changed to equivalent fractions that have a denominator of 20.

$$\frac{1}{4} + \frac{2}{5} = \frac{1 \cdot 5}{4 \cdot 5} + \frac{2 \cdot 4}{5 \cdot 4} = \frac{5}{20} + \frac{8}{20} = \frac{13}{20}$$

Use of fundamental
principle of fractions

$\blacksquare$

EXAMPLE 3   Subtract $\dfrac{5}{8} - \dfrac{7}{12}$.

Solution

By inspection the LCD is 24. Thus

$$\frac{5}{8} - \frac{7}{12} = \frac{5 \cdot 3}{8 \cdot 3} - \frac{7 \cdot 2}{12 \cdot 2} = \frac{15}{24} - \frac{14}{24} = \frac{1}{24} \qquad \blacksquare$$

If the LCD is not obvious by inspection, then we can use the technique from Chapter 1 to find the least common multiple. We proceed as follows:

STEP 1   Express each denominator as a product of prime factors.

STEP 2   The LCD contains each different prime factor as many times as the *most* times it appears in any one of the factorizations from step 1.

EXAMPLE 4 Add $\dfrac{5}{18} + \dfrac{7}{24}$.

Solution

If we cannot find the LCD by inspection, then we can use the prime factorization forms.

$$\left.\begin{array}{l} 18 = 2\cdot 3\cdot 3 \\ 24 = 2\cdot 2\cdot 2\cdot 3 \end{array}\right\} \;\rightarrow\; \text{LCD} = 2\cdot 2\cdot 2\cdot 3\cdot 3 = 72$$

$$\frac{5}{18} + \frac{7}{24} = \frac{5\cdot 4}{18\cdot 4} + \frac{7\cdot 3}{24\cdot 3} = \frac{20}{72} + \frac{21}{72} = \frac{41}{72}$$ ∎

EXAMPLE 5 Subtract $\dfrac{3}{14} - \dfrac{8}{35}$.

Solution

$$\left.\begin{array}{l} 14 = 2\cdot 7 \\ 35 = 5\cdot 7 \end{array}\right\} \;\rightarrow\; \text{LCD} = 2\cdot 5\cdot 7 = 70$$

$$\frac{3}{14} - \frac{8}{35} = \frac{3\cdot 5}{14\cdot 5} - \frac{8\cdot 2}{35\cdot 2} = \frac{15}{70} - \frac{16}{70} = \frac{-1}{70} \quad \text{or} \quad -\frac{1}{70}$$ ∎

EXAMPLE 6 Add $\dfrac{-5}{8} + \dfrac{3}{14}$.

Solution

$$\left.\begin{array}{l} 8 = 2\cdot 2\cdot 2 \\ 14 = 2\cdot 7 \end{array}\right\} \;\rightarrow\; \text{LCD} = 2\cdot 2\cdot 2\cdot 7 = 56$$

$$\frac{-5}{8} + \frac{3}{14} = \frac{-5\cdot 7}{8\cdot 7} + \frac{3\cdot 4}{14\cdot 4} = \frac{-35}{56} + \frac{12}{56} = \frac{-23}{56} \quad \text{or} \quad -\frac{23}{56}$$ ∎

EXAMPLE 7 Add $-3 + \dfrac{2}{5}$.

Solution

$$-3 + \frac{2}{5} = \frac{-3\cdot 5}{1\cdot 5} + \frac{2}{5} = \frac{-15}{5} + \frac{2}{5} = \frac{-15 + 2}{5} = \frac{-13}{5} \quad \text{or} \quad -\frac{13}{5}$$ ∎

Denominators that contain variables do not complicate the situation very much, as the next examples illustrate.

EXAMPLE 8   Add $\dfrac{2}{x} + \dfrac{3}{y}$.

Solution

By inspection, the LCD is xy.

$$\frac{2}{x} + \frac{3}{y} = \frac{2 \cdot y}{x \cdot y} + \frac{3 \cdot x}{y \cdot x} = \frac{2y}{xy} + \frac{3x}{xy} = \frac{2y + 3x}{xy}$$

Commutative property

EXAMPLE 9   Subtract $\dfrac{3}{8x} - \dfrac{5}{12y}$.

Solution

$$\left.\begin{array}{l} 8x = 2 \cdot 2 \cdot 2 \cdot x \\ 12y = 2 \cdot 2 \cdot 3 \cdot y \end{array}\right\} \longrightarrow \text{LCD} = 2 \cdot 2 \cdot 2 \cdot 3 \cdot x \cdot y = 24xy$$

$$\frac{3}{8x} - \frac{5}{12y} = \frac{3 \cdot 3y}{8x \cdot 3y} - \frac{5 \cdot 2x}{12y \cdot 2x} = \frac{9y}{24xy} - \frac{10x}{24xy} = \frac{9y - 10x}{24xy}$$

EXAMPLE 10   Add $\dfrac{7}{4a} + \dfrac{-5}{6bc}$.

Solution

$$\left.\begin{array}{l} 4a = 2 \cdot 2 \cdot a \\ 6bc = 2 \cdot 3 \cdot b \cdot c \end{array}\right\} \longrightarrow \text{LCD} = 2 \cdot 2 \cdot 3 \cdot a \cdot b \cdot c = 12abc$$

$$\frac{7}{4a} + \frac{-5}{6bc} = \frac{7 \cdot 3bc}{4a \cdot 3bc} + \frac{-5 \cdot 2a}{6bc \cdot 2a} = \frac{21bc}{12abc} + \frac{-10a}{12abc} = \frac{21bc - 10a}{12abc}$$

Simplifying Numerical Expressions

Let's now consider simplifying numerical expressions that contain rational numbers. As with integers, we first do multiplications and divisions and then perform the additions and subtractions. In these next examples, only the major steps are shown, so be sure that you can fill in all of the details.

EXAMPLE 11   Simplify $\dfrac{3}{4} + \dfrac{2}{3} \cdot \dfrac{3}{5} - \dfrac{1}{2} \cdot \dfrac{1}{5}$.

Solution

$$\frac{3}{4} + \frac{2}{3} \cdot \frac{3}{5} - \frac{1}{2} \cdot \frac{1}{5} = \frac{3}{4} + \frac{2}{5} - \frac{1}{10} \qquad \text{Perform the multiplications.}$$

$$= \frac{15}{20} + \frac{8}{20} - \frac{2}{20} = \frac{15 + 8 - 2}{20} = \frac{21}{20} \qquad \begin{array}{l} \text{Change to} \\ \text{equivalent} \\ \text{fractions and} \\ \text{combine the} \\ \text{numerators.} \end{array}$$

EXAMPLE 12   Simplify $\dfrac{3}{5} \div \dfrac{8}{5} + \left(-\dfrac{1}{2}\right)\left(\dfrac{1}{3}\right) + \dfrac{5}{12}$.

Solution

$$\dfrac{3}{5} \div \dfrac{8}{5} + \left(-\dfrac{1}{2}\right)\left(\dfrac{1}{3}\right) + \dfrac{5}{12} = \dfrac{3}{5} \cdot \dfrac{5}{8} + \left(-\dfrac{1}{2}\right)\left(\dfrac{1}{3}\right) + \dfrac{5}{12}$$

Change division to multiply by the reciprocal.

$$= \dfrac{3}{8} + \dfrac{-1}{6} + \dfrac{5}{12}$$

$$= \dfrac{9}{24} + \dfrac{-4}{24} + \dfrac{10}{24}$$

Change to equivalent fractions.

$$= \dfrac{9 + (-4) + 10}{24}$$

$$= \dfrac{15}{24} = \dfrac{5}{8}$$

Reduce! ■

The distributive property, $a(b + c) = ab + ac$, holds true for rational numbers and (as with integers) can be used to facilitate manipulation.

EXAMPLE 13   Simplify $12\left(\dfrac{1}{3} + \dfrac{1}{4}\right)$.

Solution

For help in this situation, let's change the form by applying the distributive property.

$$12\left(\dfrac{1}{3} + \dfrac{1}{4}\right) = 12\left(\dfrac{1}{3}\right) + 12\left(\dfrac{1}{4}\right)$$

$$= 4 + 3$$

$$= 7$$ ■

EXAMPLE 14   Simplify $\dfrac{5}{8}\left(\dfrac{1}{2} + \dfrac{1}{3}\right)$.

Solution

In this case it may be easier not to apply the distributive property but to work with the expression in its given form.

$$\dfrac{5}{8}\left(\dfrac{1}{2} + \dfrac{1}{3}\right) = \dfrac{5}{8}\left(\dfrac{3}{6} + \dfrac{2}{6}\right)$$

$$= \dfrac{5}{8}\left(\dfrac{5}{6}\right)$$

$$= \dfrac{25}{48}$$ ■

Examples 13 and 14 emphasize a point made in Chapter 1. *Think first* and decide whether or not you can use the properties to make the manipulations easier. The next example illustrates the combining of similar terms that have fractional coefficients.

EXAMPLE 15   Simplify $\frac{1}{2}x + \frac{2}{3}x - \frac{3}{4}x$ by combining similar terms.

Solution

The distributive property, along with our knowledge of adding and subtracting rational numbers, provides the basis for working this type of problem.

$$\frac{1}{2}x + \frac{2}{3}x - \frac{3}{4}x = \left(\frac{1}{2} + \frac{2}{3} - \frac{3}{4}\right)x$$

$$= \left(\frac{6}{12} + \frac{8}{12} - \frac{9}{12}\right)x$$

$$= \frac{5}{12}x \qquad \blacksquare$$

PROBLEM 1   Brian brought 5 cups of flour with him on a camping trip. He wants to make biscuits and cake for tonight's supper. It takes $\frac{3}{4}$ of a cup of flour for the biscuits and $2\frac{3}{4}$ cups of flour for the cake. How much flour will be left over for the rest of his camping trip?

Solution

Let's do this problem in two steps. First, add the amounts of flour needed for the biscuits and cake.

$$\frac{3}{4} + 2\frac{3}{4} = \frac{3}{4} + \frac{11}{4} = \frac{14}{4} = \frac{7}{2}$$

Then to find the amount of flour left over, we will subtract $\frac{7}{2}$ from 5.

$$5 - \frac{7}{2} = \frac{10}{2} - \frac{7}{2} = \frac{3}{2} = 1\frac{1}{2}$$

So $1\frac{1}{2}$ cups of flour remain. $\qquad \blacksquare$

CONCEPT QUIZ

For Problems 1–8, answer true or false.

1. To add rational numbers with common denominators, add the numerators and place the result over the common denominator.

2. When adding $\dfrac{2}{c} + \dfrac{6}{c}$, c can be equal to zero.

3. Fractions that name the same number are called equivalent fractions.

4. The least common multiple of the denominators can always be used as a common denominator when adding or subtracting fractions.

5. To subtract $\dfrac{3}{8}$ and $\dfrac{1}{5}$, we need to find equivalent fractions with a common denominator.

6. To multiply $\dfrac{5}{7}$ and $\dfrac{2}{3}$, we need to find equivalent fractions with a common denominator.

7. Either 20, 40, or 60 can be used as a common denominator when adding $\dfrac{1}{4}$ and $\dfrac{3}{5}$, but 20 is the least common denominator.

8. When adding $\dfrac{2x}{ab}$ and $\dfrac{3y}{bc}$, the least common denominator is ac.

PROBLEM SET 2.2

For Problems 1–64, add or subtract as indicated, and express answers in lowest terms.

1. $\dfrac{2}{7} + \dfrac{3}{7}$

2. $\dfrac{3}{11} + \dfrac{5}{11}$

3. $\dfrac{7}{9} - \dfrac{2}{9}$

4. $\dfrac{11}{13} - \dfrac{6}{13}$

5. $\dfrac{3}{4} + \dfrac{9}{4}$

6. $\dfrac{5}{6} + \dfrac{7}{6}$

7. $\dfrac{11}{12} - \dfrac{3}{12}$

8. $\dfrac{13}{16} - \dfrac{7}{16}$

9. $\dfrac{1}{8} - \dfrac{5}{8}$

10. $\dfrac{2}{9} - \dfrac{5}{9}$

11. $\dfrac{5}{24} + \dfrac{11}{24}$

12. $\dfrac{7}{36} + \dfrac{13}{36}$

13. $\dfrac{8}{x} + \dfrac{7}{x}$

14. $\dfrac{17}{y} + \dfrac{12}{y}$

15. $\dfrac{5}{3y} + \dfrac{1}{3y}$

16. $\dfrac{3}{8x} + \dfrac{1}{8x}$

17. $\dfrac{1}{3} + \dfrac{1}{5}$

18. $\dfrac{1}{6} + \dfrac{1}{8}$

19. $\dfrac{15}{16} - \dfrac{3}{8}$

20. $\dfrac{13}{12} - \dfrac{1}{6}$

21. $\dfrac{7}{10} + \dfrac{8}{15}$

22. $\dfrac{7}{12} + \dfrac{5}{8}$

23. $\dfrac{11}{24} + \dfrac{5}{32}$

24. $\dfrac{5}{18} + \dfrac{8}{27}$

25. $\dfrac{5}{18} - \dfrac{13}{24}$

26. $\dfrac{1}{24} - \dfrac{7}{36}$

27. $\dfrac{5}{8} - \dfrac{2}{3}$

28. $\dfrac{3}{4} - \dfrac{5}{6}$

29. $-\dfrac{2}{13} - \dfrac{7}{39}$

30. $-\dfrac{3}{11} - \dfrac{13}{33}$

31. $-\dfrac{3}{14} + \dfrac{1}{21}$

32. $-\dfrac{3}{20} + \dfrac{14}{25}$

33. $-4 - \dfrac{3}{7}$

34. $-2 - \dfrac{5}{6}$

35. $\dfrac{3}{4} - 6$ **36.** $\dfrac{5}{8} - 7$

37. $\dfrac{3}{x} + \dfrac{4}{y}$ **38.** $\dfrac{5}{x} + \dfrac{8}{y}$

39. $\dfrac{7}{a} - \dfrac{2}{b}$ **40.** $\dfrac{13}{a} - \dfrac{4}{b}$

41. $\dfrac{2}{x} + \dfrac{7}{2x}$ **42.** $\dfrac{5}{2x} + \dfrac{7}{x}$

43. $\dfrac{10}{3x} - \dfrac{2}{x}$ **44.** $\dfrac{13}{4x} - \dfrac{3}{x}$

45. $\dfrac{1}{x} - \dfrac{7}{5x}$ **46.** $\dfrac{2}{x} - \dfrac{17}{6x}$

47. $\dfrac{3}{2y} + \dfrac{5}{3y}$ **48.** $\dfrac{7}{3y} + \dfrac{9}{4y}$

49. $\dfrac{5}{12y} - \dfrac{3}{8y}$ **50.** $\dfrac{9}{4y} - \dfrac{5}{9y}$

51. $\dfrac{1}{6n} - \dfrac{7}{8n}$ **52.** $\dfrac{3}{10n} - \dfrac{11}{15n}$

53. $\dfrac{5}{3x} + \dfrac{7}{3y}$ **54.** $\dfrac{3}{2x} + \dfrac{7}{2y}$

55. $\dfrac{8}{5x} + \dfrac{3}{4y}$ **56.** $\dfrac{1}{5x} + \dfrac{5}{6y}$

57. $\dfrac{7}{4x} - \dfrac{5}{9y}$ **58.** $\dfrac{2}{7x} - \dfrac{11}{14y}$

59. $-\dfrac{3}{2x} - \dfrac{5}{4y}$ **60.** $-\dfrac{13}{8a} - \dfrac{11}{10b}$

61. $3 + \dfrac{2}{x}$ **62.** $\dfrac{5}{x} + 4$

63. $2 - \dfrac{3}{2x}$ **64.** $-1 - \dfrac{1}{3x}$

For Problems 65–80, simplify each numerical expression, and express answers in reduced form.

65. $\dfrac{1}{4} - \dfrac{3}{8} + \dfrac{5}{12} - \dfrac{1}{24}$ **66.** $\dfrac{3}{4} + \dfrac{2}{3} - \dfrac{1}{6} + \dfrac{5}{12}$

67. $\dfrac{5}{6} + \dfrac{2}{3} \cdot \dfrac{3}{4} - \dfrac{1}{4} \cdot \dfrac{2}{5}$ **68.** $\dfrac{2}{3} + \dfrac{1}{2} \cdot \dfrac{2}{5} - \dfrac{1}{3} \cdot \dfrac{1}{5}$

69. $\dfrac{3}{4} \cdot \dfrac{6}{9} - \dfrac{5}{6} \cdot \dfrac{8}{10} + \dfrac{2}{3} \cdot \dfrac{6}{8}$

70. $\dfrac{3}{5} \cdot \dfrac{5}{7} + \dfrac{2}{3} \cdot \dfrac{3}{5} - \dfrac{1}{7} \cdot \dfrac{2}{5}$

71. $4 - \dfrac{2}{3} \cdot \dfrac{3}{5} - 6$ **72.** $3 + \dfrac{1}{2} \cdot \dfrac{1}{3} - 2$

73. $\dfrac{4}{5} - \dfrac{10}{12} - \dfrac{5}{6} \div \dfrac{14}{8} + \dfrac{10}{21}$

74. $\dfrac{3}{4} \div \dfrac{6}{5} + \dfrac{8}{12} \cdot \dfrac{6}{9} - \dfrac{5}{12}$

75. $24\left(\dfrac{3}{4} - \dfrac{1}{6}\right)$ Don't forget the distributive property!

76. $18\left(\dfrac{2}{3} + \dfrac{1}{9}\right)$

77. $64\left(\dfrac{3}{16} + \dfrac{5}{8} - \dfrac{1}{4} + \dfrac{1}{2}\right)$

78. $48\left(\dfrac{5}{12} - \dfrac{1}{6} + \dfrac{3}{8}\right)$

79. $\dfrac{7}{13}\left(\dfrac{2}{3} - \dfrac{1}{6}\right)$ **80.** $\dfrac{5}{9}\left(\dfrac{1}{2} + \dfrac{1}{4}\right)$

For Problems 81–96, simplify each algebraic expression by combining similar terms.

81. $\dfrac{1}{3}x + \dfrac{2}{5}x$ **82.** $\dfrac{1}{4}x + \dfrac{2}{3}x$

83. $\dfrac{1}{3}a - \dfrac{1}{8}a$ **84.** $\dfrac{2}{5}a - \dfrac{2}{7}a$

85. $\dfrac{1}{2}x + \dfrac{2}{3}x + \dfrac{1}{6}x$ **86.** $\dfrac{1}{3}x + \dfrac{2}{5}x + \dfrac{5}{6}x$

87. $\dfrac{3}{5}n - \dfrac{1}{4}n + \dfrac{3}{10}n$ **88.** $\dfrac{2}{5}n - \dfrac{7}{10}n + \dfrac{8}{15}n$

89. $n + \dfrac{4}{3}n - \dfrac{1}{9}n$ **90.** $2n - \dfrac{6}{7}n + \dfrac{5}{14}n$

91. $-n - \dfrac{7}{9}n - \dfrac{5}{12}n$ **92.** $-\dfrac{3}{8}n - n - \dfrac{3}{14}n$

93. $\dfrac{3}{7}x + \dfrac{1}{4}y + \dfrac{1}{2}x + \dfrac{7}{8}y$

94. $\frac{5}{6}x + \frac{3}{4}y + \frac{4}{9}x + \frac{7}{10}y$

95. $\frac{2}{9}x + \frac{5}{12}y - \frac{7}{15}x - \frac{13}{15}y$

96. $-\frac{9}{10}x - \frac{3}{14}y + \frac{2}{25}x + \frac{5}{21}y$

97. Beth wants to make three sofa pillows for her new sofa. After consulting the chart provided by the fabric shop, she decides to make a 12″ round pillow, an 18″ square pillow, and a 12″ × 16″ rectangular pillow. According to the chart, how much fabric will Beth need to purchase?

Fabric Shop Chart

10″ round	$\frac{3}{8}$ yard
12″ round	$\frac{1}{2}$ yard
12″ square	$\frac{5}{8}$ yard
18″ square	$\frac{3}{4}$ yard
12″ × 16″ rectangular	$\frac{7}{8}$ yard

98. Vinay has a board that is $6\frac{1}{2}$ feet long. If he cuts off a piece $2\frac{3}{4}$ feet long, how long is the remaining piece of board?

99. Mindy takes a daily walk of $2\frac{1}{2}$ miles. One day a thunderstorm forced her to stop her walk after $\frac{3}{4}$ of a mile. By how much was her walk shortened that day?

100. Blake Scott leaves $\frac{1}{4}$ of his estate to the Boy Scouts, $\frac{2}{5}$ to the local cancer fund, and the rest to his church. What fractional part of the estate does the church receive?

101. A triangular plot of ground measures $14\frac{1}{2}$ yards by by $12\frac{1}{3}$ yards by $9\frac{5}{6}$ yards. How many yards of fencing is needed to enclose the plot?

102. For her exercise program, Lian jogs for $2\frac{1}{2}$ miles, then walks for $\frac{3}{4}$ of a mile, and finally jogs for another $1\frac{1}{4}$ miles. Find the total distance that Lian covers.

103. If your calculator handles rational numbers in $\frac{a}{b}$ form, use it to check your answers for Problems 65–80.

■ ■ □ **Thoughts into words**

104. Give a step-by-step description of how to add the rational numbers $\frac{3}{8}$ and $\frac{5}{18}$.

105. Give a step-by-step description of how to add the fractions $\frac{5}{4x}$ and $\frac{7}{6x}$.

106. The will of a deceased collector of antique automobiles specified that his cars be left to his three children. Half were to go to his elder son, $\frac{1}{3}$ to his daughter, and $\frac{1}{9}$ to his younger son. At the time of his death,

17 cars were in the collection. The administrator of his estate borrowed a car to make 18. Then he distributed the cars as follows:

Elder son: $\frac{1}{2}(18) = 9$

Daughter: $\frac{1}{3}(18) = 6$

Younger son: $\frac{1}{9}(18) = 2$

This totaled 17 cars, so he then returned the borrowed car. Where is the error in this problem?

2.3   Real Numbers and Algebraic Expressions

Objectives

- Classify real numbers.

- Add, subtract, multiply, and divide rational numbers in decimal form.

- Combine similar terms whose coefficients are rational numbers in decimal form.

- Evaluate algebraic expressions when the variables are rational numbers.

- Solve application problems that involve the operations of rational numbers in decimal form.

We classify decimals — also called decimal fractions — as **terminating, repeating,** or **nonrepeating.** Here are examples of these classifications:

Terminating decimals	Repeating decimals	Nonrepeating decimals
0.3	0.333333 . . .	0.5918654279 . . .
0.26	0.5466666 . . .	0.26224222722229 . . .
0.347	0.14141414 . . .	0.145117211193111148 . . .
0.9865	0.237237237 . . .	0.645751311 . . .

A repeating decimal has a block of digits that repeats indefinitely. This repeating block of digits may contain any number of digits, and it may or may not begin repeating immediately after the decimal point. Technically, a terminating decimal can be thought of as repeating zeros after the last digit. For example, $0.3 = 0.30 = 0.300 = 0.3000$, and so on.

In Section 2.1 we defined a rational number to be any number that can be written in the form $\frac{a}{b}$, where a and b are integers and b is not zero. *A rational number can also be defined as any number that has a terminating or repeating decimal representation.* Thus we can express rational numbers in either common-fraction form or decimal-fraction form, as the next examples illustrate. A repeating decimal can also be written by using a bar over the digits that repeat, for example, $0.\overline{14}$.

Terminating decimals	Repeating decimals
$\frac{3}{4} = 0.75$	$\frac{1}{3} = 0.3333 . . .$
$\frac{1}{8} = 0.125$	$\frac{2}{3} = 0.66666 . . .$

Terminating decimals	Repeating decimals
$\dfrac{5}{16} = 0.3125$	$\dfrac{1}{6} = 0.166666\ldots$
$\dfrac{7}{25} = 0.28$	$\dfrac{1}{12} = 0.08333\ldots$
$\dfrac{2}{5} = 0.4$	$\dfrac{14}{99} = 0.14141414\ldots$

The nonrepeating decimals are called **irrational numbers,** and they do appear in forms other than decimal form. For example, $\sqrt{2}$, $\sqrt{3}$, and π are irrational numbers. An approximate decimal representation for each of these follows.

$$\left. \begin{array}{l} \sqrt{2} = 1.414213562373\ldots \\ \sqrt{3} = 1.73205080756887\ldots \\ \pi = 3.14159265358979\ldots \end{array} \right\} \quad \text{Nonrepeating decimals}$$

(We will do more work with irrational numbers in Chapter 10.)

The rational numbers together with the irrationals form the set of **real numbers.** This tree diagram of the real-number system is helpful for summarizing some basic ideas.

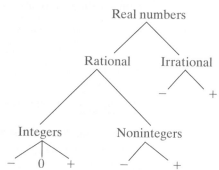

Any real number can be traced down through the diagram as follows:

5 is real, rational, an integer, and positive.

-4 is real, rational, an integer, and negative.

$\dfrac{3}{4}$ is real, rational, a noninteger, and positive.

0.23 is real, rational, a noninteger, and positive.

$-0.161616\ldots$ is real, rational, a noninteger, and negative.

$\sqrt{7}$ is real, irrational, and positive.

$-\sqrt{2}$ is real, irrational, and negative.

The properties of integers that we discussed in Section 1.5 are true for all real numbers. We restate them here for your convenience. The multiplicative inverse property has been added to the list. A discussion follows.

Commutative Property of Addition

If a and b are real numbers, then

$$a + b = b + a$$

Commutative Property of Multiplication

If a and b are real numbers, then

$$ab = ba$$

Associative Property of Addition

If a, b, and c are real numbers, then

$$(a + b) + c = a + (b + c)$$

Associative Property of Multiplication

If a, b, and c are real numbers then

$$(ab)c = a(bc)$$

Identity Property of Addition

If a is any real number, then

$$a + 0 = 0 + a = a$$

Identity Property of Multiplication

If a is any real number, then

$$a(1) = 1(a) = a$$

Additive Inverse Property

For every real number a, there exists an integer $-a$ such that

$$a + (-a) = (-a) + a = 0$$

Multiplication Property of Zero

If a is any real number, then

$$a(0) = (0)(a) = 0$$

Multiplicative Property of Negative One

If a is any real number, then

$$(a)(-1) = (-1)(a) = -a$$

Multiplicative Inverse Property

For every nonzero real number a, there exists a real number $\dfrac{1}{a}$, such that

$$a\left(\frac{1}{a}\right) = \frac{1}{a}(a) = 1$$

Distributive Property

If a, b, and c are real numbers, then

$$a(b + c) = ab + ac$$

The number $\dfrac{1}{a}$ is called the **multiplicative inverse of a** or the **reciprocal of a.**

For example, the reciprocal of 2 is $\dfrac{1}{2}$ and $2\left(\dfrac{1}{2}\right) = \dfrac{1}{2}(2) = 1$. Likewise, the reciprocal of $\dfrac{1}{2}$ is $\dfrac{1}{\frac{1}{2}}$. Therefore, 2 and $\dfrac{1}{2}$ are said to be reciprocals (or multiplicative inverses) of each other. Also, $\dfrac{2}{5}$ and $\dfrac{5}{2}$ are multiplicative inverses and $\left(\dfrac{2}{5}\right)\left(\dfrac{5}{2}\right) = 1$. Because division by zero is undefined, zero does not have a reciprocal.

Basic Operations with Decimals

The basic operations with decimals may be related to the corresponding operations with common fractions. For example, $0.3 + 0.4 = 0.7$ because $\dfrac{3}{10} + \dfrac{4}{10} = \dfrac{7}{10}$, and $0.37 - 0.24 = 0.13$ because $\dfrac{37}{100} - \dfrac{24}{100} = \dfrac{13}{100}$. In general, to add or subtract decimals, we add or subtract the hundredths, the tenths, the ones, the tens, and so on. To keep place values aligned, we line up the decimal points.

	Addition		Subtraction	
	1	1 11	6 16	8 11 13
	2.14	5.214	7.6̸	9.23̸5̸
	3.12	3.162	4.9	6.781
	5.16	7.218	2.7	2.454
	10.42	8.914		
		24.508		

We can use examples such as the following to help formulate a general rule for multiplying decimals.

Because $\dfrac{7}{10} \cdot \dfrac{3}{10} = \dfrac{21}{100}$, then $(0.7)(0.3) = 0.21$.

Because $\dfrac{9}{10} \cdot \dfrac{23}{100} = \dfrac{207}{1000}$, then $(0.9)(0.23) = 0.207$.

Because $\dfrac{11}{100} \cdot \dfrac{13}{100} = \dfrac{143}{10,000}$, then $(0.11)(0.13) = 0.0143$.

In general, to multiply decimals, we (1) multiply the numbers and ignore the decimal points, and then (2) insert the decimal point in the product so that the number of digits to the right of the decimal point in the product is equal to the sum of the numbers of digits to the right of the decimal point in each factor.

(0.7)	$\times$	(0.3)	$=$	0.21
↑		↑		↑
One digit to right	$+$	One digit to right	$=$	Two digits to right

(0.9)		(0.23)	$=$	0.207
↑		↑		↑
One digit to right	$+$	Two digits to right	$=$	Three digits to right

(0.11)		(0.13)	$=$	0.0143
↑		↑		↑
Two digits to right	$+$	Two digits to right	$=$	Four digits to right

We frequently use the vertical format when multiplying decimals.

41.2	One digit to right
0.13	Two digits to right
1236	
412	
5.356	Three digits to right

0.021	Three digits to right
0.03	Two digits to right
0.00063	Five digits to right

Note that in the last example, we actually multiplied $3 \cdot 21$ and then inserted three 0s to the left so that there would be five digits to the right of the decimal point.

Once again let's look at some links between common fractions and decimals.

Because $\dfrac{6}{10} \div 2 = \dfrac{\overset{3}{\cancel{6}}}{10} \cdot \dfrac{1}{2} = \dfrac{3}{10}$, we have $2\overline{)0.6}$ with quotient 0.3.

Because $\dfrac{39}{100} \div 13 = \dfrac{\overset{3}{\cancel{39}}}{100} \cdot \dfrac{1}{\cancel{13}} = \dfrac{3}{100}$, we have $13\overline{)0.39}$ with quotient 0.03.

Because $\dfrac{85}{100} \div 5 = \dfrac{\overset{17}{\cancel{85}}}{100} \cdot \dfrac{1}{\cancel{5}} = \dfrac{17}{100}$, we have $5\overline{)0.85}$ with quotient 0.17.

In general, to divide a decimal by a nonzero whole number, we (1) place the decimal point in the quotient directly above the decimal point in the dividend

$$\text{Divisor}\overline{)\,\text{Dividend}} \quad \text{with Quotient}$$

and then (2) divide as with whole numbers, except that in the division process, 0s are placed in the quotient immediately to the right of the decimal point in order to show the correct place value.

$$\begin{array}{r} 0.121 \\ 4\overline{)0.484} \end{array} \qquad \begin{array}{r} 0.24 \\ 32\overline{)7.68} \\ \underline{6\,4} \\ 1\,28 \\ \underline{1\,28} \end{array} \qquad \begin{array}{r} 0.019 \\ 12\overline{)0.228} \\ \underline{12} \\ 108 \\ \underline{108} \end{array} \qquad \text{Zero needed to show the correct place value}$$

Don't forget that *you can check division by multiplication.* For example, because $(12)(0.019) = 0.228$, we know that our last division example is correct.

We can easily handle problems involving division by a decimal by changing to an equivalent problem that has a whole-number divisor. Consider the following examples, in which we have changed the original division problem to fractional form to show the reasoning involved in the procedure.

$$0.6\overline{)0.24} \;\rightarrow\; \frac{0.24}{0.6} = \left(\frac{0.24}{0.6}\right)\left(\frac{10}{10}\right) = \frac{2.4}{6} \;\rightarrow\; \begin{array}{r} 0.4 \\ 6\overline{)2.4} \end{array}$$

$$0.12\overline{)0.156} \;\rightarrow\; \frac{0.156}{0.12} = \left(\frac{0.156}{0.12}\right)\left(\frac{100}{100}\right) = \frac{15.6}{12} \;\rightarrow\; \begin{array}{r} 1.3 \\ 12\overline{)15.6} \\ \underline{12} \\ 36 \\ \underline{36} \end{array}$$

$$1.3\overline{)0.026} \;\rightarrow\; \frac{0.026}{1.3} = \left(\frac{0.026}{1.3}\right)\left(\frac{10}{10}\right) = \frac{0.26}{13} \;\rightarrow\; \begin{array}{r} 0.02 \\ 13\overline{)0.26} \\ \underline{26} \end{array}$$

The following format is commonly used with such problems:

$$\begin{array}{r} 5.6 \\ 0.21.\overline{)1.17.6} \\ \underline{1\,05} \\ 12\,6 \\ \underline{12\,6} \end{array} \qquad \text{The arrows indicate that both the divisor and dividend were multiplied by 100, which changes the divisor to a whole number.}$$

$$\begin{array}{r} 0.04 \\ 3.7.\overline{)0.1.48} \\ \underline{1\,48} \end{array} \qquad \text{The divisor and dividend were multiplied by 10.}$$

Our agreements for operating with positive and negative integers extend to all real numbers. For example, the product of two negative real numbers is a positive real number. Make sure that you agree with the following results. (You may need to do some work on scratch paper, because the steps are not shown.)

$$0.24 + (-0.18) = 0.06 \qquad\qquad (-0.4)(0.8) = -0.32$$
$$-7.2 + 5.1 = -2.1 \qquad\qquad (-0.5)(-0.13) = 0.065$$
$$-0.6 + (-0.8) = -1.4 \qquad\qquad (1.4) \div (-0.2) = -7$$
$$2.4 - 6.1 = -3.7 \qquad\qquad (-0.18) \div (0.3) = -0.6$$

$$0.31 - (-0.52) = 0.83 \qquad\qquad (-0.24) \div (-4) = 0.06$$
$$(0.2)(-0.3) = -0.06$$

Numerical and algebraic expressions may contain the decimal form as well as the fractional form of rational numbers. We continue to follow the rule of doing the multiplications and divisions *first* and then the additions and subtractions, unless parentheses indicate otherwise. The following examples illustrate a variety of situations that involve both the decimal and the fractional forms of rational numbers.

EXAMPLE 1

Simplify $6.3 \div 7 + (4)(2.1) - (0.24) \div (-0.4)$.

Solution

$$
\begin{aligned}
6.3 \div 7 + (4)(2.1) - (0.24) \div (-0.4) &= 0.9 + 8.4 - (-0.6) \\
&= 0.9 + 8.4 + 0.6 \\
&= 9.9
\end{aligned}
$$

EXAMPLE 2

Evaluate $\dfrac{3}{5}a - \dfrac{1}{7}b$ for $a = \dfrac{5}{2}$ and $b = -1$.

Solution

$$
\begin{aligned}
\frac{3}{5}a - \frac{1}{7}b &= \frac{3}{5}\left(\frac{5}{2}\right) - \frac{1}{7}(-1) \quad \text{for } a = \frac{5}{2} \text{ and } b = -1 \\
&= \frac{3}{2} + \frac{1}{7} \\
&= \frac{21}{14} + \frac{2}{14} \\
&= \frac{23}{14}
\end{aligned}
$$

EXAMPLE 3

Evaluate $\dfrac{1}{2}x + \dfrac{2}{3}x - \dfrac{1}{5}x$ for $x = -\dfrac{3}{4}$.

Solution

First, let's *combine similar terms* by using the distributive property.

$$
\begin{aligned}
\frac{1}{2}x + \frac{2}{3}x - \frac{1}{5}x &= \left(\frac{1}{2} + \frac{2}{3} - \frac{1}{5}\right)x \\
&= \left(\frac{15}{30} + \frac{20}{30} - \frac{6}{30}\right)x \\
&= \frac{29}{30}x
\end{aligned}
$$

Now we can evaluate.

$$\frac{29}{30}x = \frac{29}{30}\left(-\frac{3}{4}\right) \quad \text{when } x = -\frac{3}{4}$$

$$= \frac{29}{\underset{10}{30}}\left(-\frac{\overset{1}{3}}{4}\right) = -\frac{29}{40}$$

E X A M P L E 4   Evaluate $2x + 3y$ for $x = 1.6$ and $y = 2.7$.

Solution

$$2x + 3y = 2(1.6) + 3(2.7) \quad \text{when } x = 1.6 \text{ and } y = 2.7$$
$$= 3.2 + 8.1$$
$$= 11.3$$

E X A M P L E 5   Evaluate $0.9x + 0.7x - 0.4x + 1.3x$ for $x = 0.2$.

Solution

First, let's *combine similar terms* by using the distributive property.

$$0.9x + 0.7x - 0.4x + 1.3x = (0.9 + 0.7 - 0.4 + 1.3)x = 2.5x$$

Now we can evaluate.

$$2.5x = (2.5)(0.2) \quad \text{for } x = 0.2$$
$$= 0.5$$

P R O B L E M 1   A layout artist is putting together a group of images. She has four images with widths of 1.35 centimeters, 2.6 centimeters, 5.45 centimeters, and 3.2 centimeters, respectively. If the images are set side by side, what will be their combined width?

Solution

To find the combined width, we need to add the four widths.

$$
\begin{array}{r}
{\scriptstyle 1\ 1} \\
1.35 \\
2.6 \\
5.45 \\
+\ \ 3.2 \\
\hline
12.60
\end{array}
$$

So the combined width would be 12.6 centimeters.

CONCEPT QUIZ

For Problems 1–10, answer true or false.

1. A rational number can be defined as any number that has a terminating or repeating decimal representation.

2. A repeating decimal has a block of digits that repeat only once.

3. Every irrational number is also classified as a real number.

4. The rational numbers along with the irrational numbers form the set of natural numbers.

5. 0.141414. . . is a rational number.

6. $-\sqrt{5}$ is real, irrational, and negative.

7. 0.35 is real, rational, integer, and positive.

8. The reciprocal of c, where $c \neq 0$, is also the multiplicative inverse of c.

9. Any number multiplied by its multiplicative inverse gives a result of 0.

10. Zero does not have a multiplicative inverse.

PROBLEM SET 2.3

For Problems 1–8, classify the real numbers by tracing down the diagram on page 64.

1. -2
2. $1/3$
3. $\sqrt{5}$
4. $-0.09090909\ldots$
5. 0.16
6. $-\sqrt{3}$
7. $-8/7$
8. 0.125

For Problems 9–34, perform each indicated operation.

9. $0.37 + 0.25$
10. $7.2 + 4.9$
11. $2.93 - 1.48$
12. $14.36 - 5.89$
13. $(-4.7) + 1.4$
14. $(-14.1) + 9.5$
15. $-3.8 + 11.3$
16. $-2.5 + 14.8$
17. $6.6 - (-1.2)$
18. $18.3 - (-7.4)$
19. $-11.5 - (-10.6)$
20. $-14.6 - (-8.3)$
21. $(0.4)(2.9)$
22. $(0.3)(3.6)$
23. $(-0.8)(0.34)$
24. $(-0.7)(0.67)$
25. $(9)(-2.7)$
26. $(8)(-7.6)$
27. $(-0.7)(-64)$
28. $(-0.9)(-56)$
29. $1.56 \div 1.3$
30. $7.14 \div 2.1$
31. $5.92 \div (-0.8)$
32. $-2.94 \div 0.6$
33. $-0.266 \div (-0.7)$
34. $-0.126 \div (-0.9)$

For Problems 35–48, simplify each numerical expression.

35. $16.5 - 18.7 + 9.4$
36. $17.7 + 21.2 - 14.6$
37. $0.34 - 0.21 - 0.74 + 0.19$
38. $-5.2 + 6.8 - 4.7 - 3.9 + 1.3$
39. $0.76(0.2 + 0.8)$
40. $9.8(1.8 - 0.8)$
41. $0.6(4.1) + 0.7(3.2)$
42. $0.5(74) - 0.9(87)$
43. $7(0.6) + 0.9 - 3(0.4) + 0.4$
44. $-5(0.9) - 0.6 + 4.1(6) - 0.9$

45. $(0.96) \div (-0.8) + 6(-1.4) - 5.2$

46. $(-2.98) \div 0.4 - 5(-2.3) + 1.6$

47. $5(2.3) - 1.2 - 7.36 \div 0.8 + 0.2$

48. $0.9(12) \div 0.4 - 1.36 \div 17 + 9.2$

For Problems 49–60, simplify each algebraic expression by combining similar terms.

49. $x - 0.4x - 1.8x$

50. $-2x + 1.7x - 4.6x$

51. $5.4n - 0.8n - 1.6n$

52. $6.2n - 7.8n - 1.3n$

53. $-3t + 4.2t - 0.9t + 0.2t$

54. $7.4t - 3.9t - 0.6t + 4.7t$

55. $3.6x - 7.4y - 9.4x + 10.2y$

56. $5.7x + 9.4y - 6.2x - 4.4y$

57. $0.3(x - 4) + 0.4(x + 6) - 0.6x$

58. $0.7(x + 7) - 0.9(x - 2) + 0.5x$

59. $6(x - 1.1) - 5(x - 2.3) - 4(x + 1.8)$

60. $4(x + 0.7) - 9(x + 0.2) - 3(x - 0.6)$

For Problems 61–74, evaluate each algebraic expression for the given values of the variables. Don't forget that for some problems, it might be helpful to combine similar terms first and then to evaluate.

61. $x + 2y + 3z$ for $x = \dfrac{3}{4}$, $y = \dfrac{1}{3}$, and $z = -\dfrac{1}{6}$

62. $2x - y - 3z$ for $x = -\dfrac{2}{5}$, $y = -\dfrac{3}{4}$, and $z = \dfrac{1}{2}$

63. $\dfrac{3}{5}y - \dfrac{2}{3}y - \dfrac{7}{15}y$ for $y = -\dfrac{5}{2}$

64. $\dfrac{1}{2}x + \dfrac{2}{3}x - \dfrac{3}{4}x$ for $x = \dfrac{7}{8}$

65. $-x - 2y + 4z$ for $x = 1.7$, $y = -2.3$, and $z = 3.6$

66. $-2x + y - 5z$ for $x = -2.9$, $y = 7.4$, and $z = -6.7$

67. $5x - 7y$ for $x = -7.8$ and $y = 8.4$

68. $8x - 9y$ for $x = -4.3$ and $y = 5.2$

69. $0.7x + 0.6y$ for $x = -2$ and $y = 6$

70. $0.8x + 2.1y$ for $x = 5$ and $y = -9$

71. $1.2x + 2.3x - 1.4x - 7.6x$ for $x = -2.5$

72. $3.4x - 1.9x + 5.2x$ for $x = 0.3$

73. $-3a - 1 + 7a - 2$ for $a = 0.9$

74. $5x - 2 + 6x + 4$ for $x = -1.1$

75. Tanya bought 400 shares of one stock at \$14.35 per share and 250 shares of another stock at \$16.68 per share. How much did she pay for the 650 shares?

76. On a trip Brent bought these amounts of gasoline: 9.7 gallons, 12.3 gallons, 14.6 gallons, 12.2 gallons, 13.8 gallons, and 15.5 gallons. How many gallons of gasoline did he purchase on the trip?

77. Kathrin has a piece of copper tubing that is 76.4 centimeters long. She needs to cut it into four pieces all of equal length. Find the length of each piece.

78. On a trip Biance filled the gas tank and noted that the odometer read 24876.2 miles. After the next filling the odometer read 25170.5 miles. It took 13.5 gallons of gasoline to fill the tank. How many miles per gallon did she get on that tank of gas?

79. The total length of the four sides of a square is 18.8 centimeters. How long is each side of the square?

80. When the market opened on Monday morning, Garth bought some shares of a stock at \$13.25 per share. The daily changes in the market for that stock for the week were 0.75, −1.50, 2.25, −0.25, and −0.50. What was the value of one share of that stock when the market closed on Friday afternoon?

81. Victoria bought two pounds of Gala apples at \$1.79 per pound and three pounds of Fuji apples at \$.99 per pound. How much did she spend for the apples?

82. In 2003 the average speed of the winner of the Daytona 500 was 133.87 miles per hour. In 2000 the average speed of the winner was 155.669 miles per hour. How much faster was the average speed of the winner in 2000 compared to the winner in 2003?

83. Use a calculator to check your answers for Problems 35–48.

■ ■ ■ **Thoughts into words**

84. At this time how would you describe the difference between arithmetic and algebra?

85. How have the properties of the real numbers been used thus far in your study of arithmetic and algebra?

86. Do you think that $2\sqrt{2}$ is a rational or an irrational number? Defend your answer.

■ ■ ■ **Further investigations**

87. Without doing the actual dividing, defend the statement, "$\frac{1}{7}$ produces a repeating decimal." [*Hint:* Think about the possible remainders when dividing by 7.]

88. Express each of the following in repeating decimal form.

  a. $\frac{1}{7}$ **b.** $\frac{2}{7}$ **c.** $\frac{4}{9}$

  d. $\frac{5}{6}$ **e.** $\frac{3}{11}$ **f.** $\frac{1}{12}$

89. a. How can we tell that $\frac{5}{16}$ will produce a terminating decimal?

  b. How can we tell that $\frac{7}{15}$ will not produce a terminating decimal?

  c. Determine which of the following will produce a terminating decimal: $\frac{7}{8}, \frac{11}{16}, \frac{5}{12}, \frac{7}{24}, \frac{11}{75}, \frac{13}{32}, \frac{17}{40}, \frac{11}{30}, \frac{9}{20}, \frac{3}{64}$.

Answers to Concept Quiz

1. True **2.** False **3.** True **4.** False **5.** True **6.** True **7.** False **8.** True **9.** False **10.** True

2.4 Exponents

Objectives

■ Know the definition and terminology for exponential notation.

■ Simplify numerical expressions that involve exponents.

■ Simplify algebraic expressions by combining similar terms.

■ Evaluate algebraic expressions that involve exponents.

We use exponents to indicate repeated multiplication. For example, we can write $5 \cdot 5 \cdot 5$ as 5^3, where the 3 indicates that 5 is to be used as a factor three times. The following general definition is helpful.

> **DEFINITION 2.4**
>
> If n is a positive integer, and b is any real number, then
>
> $$b^n = \underbrace{bbb \cdots b}_{n \text{ factors of } b}$$

We refer to the b as the **base** and to n as the **exponent.** The expression b^n can be read as "b to the nth **power.**" We frequently associate the terms **squared** and **cubed** with exponents of 2 and 3, respectively. For example, b^2 is read as "b squared" and b^3 as "b cubed." An exponent of 1 is usually not written, so b^1 is written as b.

The following examples further clarify the concept of an exponent:

$$2^3 = 2 \cdot 2 \cdot 2 = 8 \qquad\qquad (0.6)^2 = (0.6)(0.6) = 0.36$$

$$3^5 = 3 \cdot 3 \cdot 3 \cdot 3 \cdot 3 = 243 \qquad \left(\frac{1}{2}\right)^4 = \frac{1}{2} \cdot \frac{1}{2} \cdot \frac{1}{2} \cdot \frac{1}{2} = \frac{1}{16}$$

$$(-5)^2 = (-5)(-5) = 25 \qquad -5^2 = -(5 \cdot 5) = -25$$

We especially want to call your attention to the last two examples. Note that $(-5)^2$ means that -5 is the base and that it is to be used as a factor twice. However, -5^2 means that 5 is the base and that after 5 is squared, we take the opposite of that result.

Exponents provide a way of writing algebraic expressions in compact form. Sometimes we need to change from the compact form to an expanded form, as these next examples demonstrate.

$$x^4 = x \cdot x \cdot x \cdot x \qquad\qquad (2x)^3 = (2x)(2x)(2x)$$

$$2y^3 = 2 \cdot y \cdot y \cdot y \qquad\qquad (-2x)^3 = (-2x)(-2x)(-2x)$$

$$-3x^5 = -3 \cdot x \cdot x \cdot x \cdot x \cdot x \qquad -x^2 = -(x \cdot x)$$

$$a^2 + b^2 = a \cdot a + b \cdot b$$

At other times we need to change from an expanded form to a more compact form using the exponent notation:

$$3 \cdot x \cdot x = 3x^2$$

$$2 \cdot 5 \cdot x \cdot x \cdot x = 10x^3$$

$$3 \cdot 4 \cdot x \cdot x \cdot y = 12x^2 y$$

$$7 \cdot a \cdot a \cdot a \cdot b \cdot b = 7a^3 b^2$$

$$(2x)(3y) = 2 \cdot x \cdot 3 \cdot y = 2 \cdot 3 \cdot x \cdot y = 6xy$$

$$(3a^2)(4a) = 3 \cdot a \cdot a \cdot 4 \cdot a = 3 \cdot 4 \cdot a \cdot a \cdot a = 12a^3$$

$$(-2x)(3x) = -2 \cdot x \cdot 3 \cdot x = -2 \cdot 3 \cdot x \cdot x = -6x^2$$

The commutative and associative properties for multiplication enable us to rearrange and regroup factors in the last three examples above.

We can use the concept of *exponent* to extend our work with combining similar terms, operating with fractions, and evaluating algebraic expressions. Study the following examples very carefully; they should help you pull together many ideas.

E X A M P L E   1   Simplify $4x^2 + 7x^2 - 2x^2$ by combining similar terms.

Solution

By applying the distributive property, we obtain

$$4x^2 + 7x^2 - 2x^2 = (4 + 7 - 2)x^2$$

$$= 9x^2$$

E X A M P L E   2   Simplify $-8x^3 + 9y^2 + 4x^3 - 11y^2$ by combining similar terms.

Solution

By rearranging terms and then applying the distributive property, we get

$$-8x^3 + 9y^2 + 4x^3 - 11y^2 = -8x^3 + 4x^3 + 9y^2 - 11y^2$$
$$= (-8 + 4)x^3 + (9 - 11)y^2$$
$$= -4x^3 - 2y^2$$

E X A M P L E   3   Simplify $-7x^2 + 4x + 3x^2 - 9x$.

Solution

$$-7x^2 + 4x + 3x^2 - 9x = -7x^2 + 3x^2 + 4x - 9x$$
$$= (-7 + 3)x^2 + (4 - 9)x$$
$$= -4x^2 - 5x$$

As soon as you feel comfortable with this process of combining similar terms, you may want to do some of the steps mentally. Then your work may appear as follows:

$$9a^2 + 6a^2 - 12a^2 = 3a^2$$
$$6x^2 + 7y^2 - 3x^2 - 11y^2 = 3x^2 - 4y^2$$
$$7x^2y + 5xy^2 - 9x^2y + 10xy^2 = -2x^2y + 15xy^2$$
$$2x^3 - 5x^2 - 10x - 7x^3 + 9x^2 - 4x = -5x^3 + 4x^2 - 14x$$

The next two examples illustrate how we handle exponents when reducing fractions.

E X A M P L E   4   Reduce $\dfrac{8x^2y}{12xy}$.

Solution

$$\frac{8x^2y}{12xy} = \frac{2 \cdot 2 \cdot 2 \cdot \cancel{x} \cdot x \cdot \cancel{y}}{2 \cdot 2 \cdot 3 \cdot \cancel{x} \cdot \cancel{y}} = \frac{2x}{3}$$

E X A M P L E   5   Reduce $\dfrac{15a^2b^3}{25a^3b}$.

Solution

$$\frac{15a^2b^3}{25a^3b} = \frac{3 \cdot \cancel{5} \cdot \cancel{a} \cdot \cancel{a} \cdot \cancel{b} \cdot b \cdot b}{\cancel{5} \cdot 5 \cdot \cancel{a} \cdot \cancel{a} \cdot a \cdot \cancel{b}} = \frac{3b^2}{5a}$$

The next three examples show how you may use exponents when multiplying and dividing fractions.

E X A M P L E   6   Multiply $\left(\dfrac{4x}{6y}\right)\left(\dfrac{12y^2}{7x^2}\right)$ and express the answer in reduced form.

Solution

$$\left(\frac{4x}{6y}\right)\left(\frac{12y^2}{7x^2}\right) = \frac{4 \cdot \overset{2}{\cancel{12}} \cdot \cancel{x} \cdot \cancel{y} \cdot y}{\cancel{6} \cdot 7 \cdot \cancel{y} \cdot \cancel{x} \cdot x} = \frac{8y}{7x}$$

E X A M P L E   7   Multiply and simplify $\left(\dfrac{8a^3}{9b}\right)\left(\dfrac{12b^2}{16a}\right)$.

Solution

$$\left(\frac{8a^3}{9b}\right)\left(\frac{12b^2}{16a}\right) = \frac{8 \cdot \overset{\overset{2}{4}}{\cancel{12}} \cdot \cancel{a} \cdot a \cdot a \cdot \cancel{b} \cdot b}{\underset{3}{\cancel{9}} \cdot \underset{2}{\cancel{16}} \cdot \cancel{b} \cdot \cancel{a}} = \frac{2a^2b}{3}$$

E X A M P L E   8   Divide and express in reduced form $\dfrac{-2x^3}{3y^2} \div \dfrac{4}{9xy}$.

Solution

$$\frac{-2x^3}{3y^2} \div \frac{4}{9xy} = -\frac{2x^3}{3y^2} \cdot \frac{9xy}{4} = -\frac{\overset{1}{\cancel{2}} \cdot \overset{3}{\cancel{9}} \cdot x \cdot x \cdot x \cdot x \cdot \cancel{y}}{\cancel{3} \cdot \underset{2}{\cancel{4}} \cdot \cancel{y} \cdot y} = -\frac{3x^4}{2y}$$

The next two examples illustrate how we handle exponents in the denominator when adding and subtracting fractions.

E X A M P L E   9   Add $\dfrac{4}{x^2} + \dfrac{7}{x}$.

Solution
The LCD is x^2. Thus

$$\frac{4}{x^2} + \frac{7}{x} = \frac{4}{x^2} + \frac{7 \cdot x}{x \cdot x} = \frac{4}{x^2} + \frac{7x}{x^2} = \frac{4 + 7x}{x^2}$$

E X A M P L E   1 0   Subtract $\dfrac{3}{xy} - \dfrac{4}{y^2}$.

Solution

$$\left.\begin{array}{l} xy = x \cdot y \\ y^2 = y \cdot y \end{array}\right\} \;\longrightarrow\; \text{The LCD is } xy^2.$$

$$\frac{3}{xy} - \frac{4}{y^2} = \frac{3 \cdot y}{xy \cdot y} - \frac{4 \cdot x}{y^2 \cdot x} = \frac{3y}{xy^2} - \frac{4x}{xy^2}$$
$$= \frac{3y - 4x}{xy^2}$$

Remember that exponents indicate repeated multiplication. Therefore, to simplify numerical expressions containing exponents, we proceed as follows:

1. Perform the operations inside the symbols of inclusion (parentheses and brackets) and above and below each fraction bar. Start with the innermost inclusion symbol.
2. Compute all indicated powers.
3. Perform all multiplications and divisions in the order in which they appear from left to right.
4. Perform all additions and subtractions in the order in which they appear from left to right.

Keep these steps in mind as we evaluate some algebraic expressions containing exponents.

EXAMPLE 11 Evaluate $3x^2 - 4y^2$ for $x = -2$ and $y = 5$.

Solution

$$3x^2 - 4y^2 = 3(-2)^2 - 4(5)^2 \quad \text{when } x = -2 \text{ and } y = 5$$
$$= 3(-2)(-2) - 4(5)(5)$$
$$= 12 - 100$$
$$= -88$$

EXAMPLE 12 Find the value of $a^2 - b^2$ when $a = \frac{1}{2}$ and $b = -\frac{1}{3}$.

Solution

$$a^2 - b^2 = \left(\frac{1}{2}\right)^2 - \left(-\frac{1}{3}\right)^2 \quad \text{when } a = \frac{1}{2} \text{ and } b = -\frac{1}{3}$$
$$= \frac{1}{4} - \frac{1}{9}$$
$$= \frac{9}{36} - \frac{4}{36}$$
$$= \frac{5}{36}$$

EXAMPLE 13   Evaluate $5x^2 + 4xy$ for $x = 0.4$ and $y = -0.3$.

Solution

$$5x^2 + 4xy = 5(0.4)^2 + 4(0.4)(-0.3) \quad \text{when } x = 0.4 \text{ and } y = -0.3$$
$$= 5(0.16) + 4(-0.12)$$
$$= 0.80 + (-0.48)$$
$$= 0.32$$

CONCEPT QUIZ

For Problems 1–10, answer true or false.

1. Exponents are used to indicate repeated additions.

2. In the expression b^n, b is called the base and n is called the number.

3. The term *cubed* is associated with an exponent of three.

4. For the term $3x$, the exponent on the x is one.

5. In the expression $(-4)^3$, the base is 4.

6. In the expression -4^3, the base is 4.

7. Changing from an expanded notation to an exponential notation, $5 \cdot 5 \cdot 5 \cdot a \cdot b \cdot b = 5^3ab$.

8. When simplifying $2x^3 + 5x^3$, the result would be $7x^6$.

9. The least common multiple for xy^2 and x^2y^3 is x^2y^3.

10. The term *squared* is associated with an exponent of two.

PROBLEM SET 2.4

For Problems 1–20, find the value of each numerical expression. For example, $2^4 = 2 \cdot 2 \cdot 2 \cdot 2 = 16$.

1. 2^6

2. 2^7

3. 3^4

4. 4^3

5. $(-2)^3$

6. $(-2)^5$

7. -3^2

8. -3^4

9. $(-4)^2$

10. $(-5)^4$

11. $\left(\dfrac{2}{3}\right)^4$

12. $\left(\dfrac{3}{4}\right)^3$

13. $-\left(\dfrac{1}{2}\right)^3$

14. $-\left(\dfrac{3}{2}\right)^3$

15. $\left(-\dfrac{3}{2}\right)^2$

16. $\left(-\dfrac{4}{3}\right)^2$

17. $(0.3)^3$

18. $(0.2)^4$

19. $-(1.2)^2$

20. $-(1.1)^2$

For Problems 21–34, simplify each numerical expression.

21. $3^2 + 2^3 - 4^3$

22. $2^4 - 3^3 + 5^2$

23. $(-2)^3 - 2^4 - 3^2$

24. $(-3)^3 - 3^2 - 6^2$

25. $5(2)^2 - 4(2) - 1$

26. $7(-2)^2 - 6(-2) - 8$

27. $-2(3)^3 - 3(3)^2 + 4(3) - 6$

28. $5(-3)^3 - 4(-3)^2 + 6(-3) + 1$

29. $-7^2 - 6^2 + 5^2$

30. $-8^2 + 3^4 - 4^3$

31. $-3(-4)^2 - 2(-3)^3 + (-5)^2$

32. $-4(-3)^3 + 5(-2)^3 - (4)^2$

33. $\dfrac{-3(2)^4}{12} + \dfrac{5(-3)^3}{15}$

34. $\dfrac{4(2)^3}{16} - \dfrac{2(3)^2}{6}$

For Problems 35–46, use exponents to express each algebraic expression in a more compact form. For example, $3 \cdot 5 \cdot x \cdot x \cdot y = 15x^2y$ and $(3x)(2x^2) = 6x^3$.

35. $9 \cdot x \cdot x$

36. $8 \cdot x \cdot x \cdot x \cdot y$

37. $3 \cdot 4 \cdot x \cdot y \cdot y$

38. $7 \cdot 2 \cdot a \cdot a \cdot b \cdot b \cdot b$

39. $-2 \cdot 9 \cdot x \cdot x \cdot x \cdot x \cdot y$

40. $-3 \cdot 4 \cdot x \cdot y \cdot z \cdot z$

41. $(5x)(3y)$

42. $(3x^2)(2y)$

43. $(6x^2)(2x^2)$

44. $(-3xy)(6xy)$

45. $(-4a^2)(-2a^3)$

46. $(-7a^3)(-3a)$

For Problems 47–58, simplify each expression by combining similar terms.

47. $3x^2 - 7x^2 - 4x^2$

48. $-2x^3 + 7x^3 - 4x^3$

49. $-12y^3 + 17y^3 - y^3$

50. $-y^3 + 8y^3 - 13y^3$

51. $7x^2 - 2y^2 - 9x^2 + 8y^2$

52. $5x^3 + 9y^3 - 8x^3 - 14y^3$

53. $\dfrac{2}{3}n^2 - \dfrac{1}{4}n^2 - \dfrac{3}{5}n^2$

54. $-\dfrac{1}{2}n^2 + \dfrac{5}{6}n^2 - \dfrac{4}{9}n^2$

55. $5x^2 - 8x - 7x^2 + 2x$

56. $-10x^2 + 4x + 4x^2 - 8x$

57. $x^2 - 2x - 4 + 6x^2 - x + 12$

58. $-3x^3 - x^2 + 7x - 2x^3 + 7x^2 - 4x$

For Problems 59–68, reduce each fraction to simplest form.

59. $\dfrac{9xy}{15x}$

60. $\dfrac{8x^2y}{14x}$

61. $\dfrac{22xy^2}{6xy^3}$

62. $\dfrac{18x^3y}{12xy^4}$

63. $\dfrac{7a^2b^3}{17a^3b}$

64. $\dfrac{9a^3b^3}{22a^4b^2}$

65. $\dfrac{-24abc^2}{32bc}$

66. $\dfrac{4a^2c^3}{-22b^2c^4}$

67. $\dfrac{-5x^4y^3}{-20x^2y}$

68. $\dfrac{-32xy^2z^4}{-48x^3y^3z}$

For Problems 69–86, perform the indicated operations and express answers in reduced form.

69. $\left(\dfrac{7x^2}{9y}\right)\left(\dfrac{12y}{21x}\right)$

70. $\left(\dfrac{3x}{8y^2}\right)\left(\dfrac{14xy}{9y}\right)$

71. $\left(\dfrac{5c}{a^2b^2}\right) \div \left(\dfrac{12c}{ab}\right)$

72. $\left(\dfrac{13ab^2}{12c}\right) \div \left(\dfrac{26b}{14c}\right)$

73. $\dfrac{6}{x} + \dfrac{5}{y^2}$

74. $\dfrac{8}{y} - \dfrac{6}{x^2}$

75. $\dfrac{5}{x^4} - \dfrac{7}{x^2}$

76. $\dfrac{9}{x} - \dfrac{11}{x^3}$

77. $\dfrac{3}{2x^3} + \dfrac{6}{x}$

78. $\dfrac{5}{3x^2} + \dfrac{6}{x}$

79. $\dfrac{-5}{4x^2} + \dfrac{7}{3x^2}$

80. $\dfrac{-8}{5x^3} + \dfrac{10}{3x^3}$

81. $\dfrac{11}{a^2} - \dfrac{14}{b^2}$

82. $\dfrac{9}{x^2} + \dfrac{8}{y^2}$

83. $\dfrac{1}{2x^3} - \dfrac{4}{3x^2}$

84. $\dfrac{2}{3x^3} - \dfrac{5}{4x}$

85. $\dfrac{3}{x} - \dfrac{4}{y} - \dfrac{5}{xy}$

86. $\dfrac{5}{x} + \dfrac{7}{y} - \dfrac{1}{xy}$

For Problems 87–100, evaluate each algebraic expression for the given values of the variables.

87. $4x^2 + 7y^2$ for $x = -2$ and $y = -3$

88. $5x^2 + 2y^3$ for $x = -4$ and $y = -1$

89. $3x^2 - y^2$ for $x = \dfrac{1}{2}$ and $y = -\dfrac{1}{3}$

90. $x^2 - 2y^2$ for $x = -\dfrac{2}{3}$ and $y = \dfrac{3}{2}$

91. $x^2 - 2xy + y^2$ for $x = -\dfrac{1}{2}$ and $y = 2$

92. $x^2 + 2xy + y^2$ for $x = -\dfrac{3}{2}$ and $y = -2$

93. $-x^2$ for $x = -8$

94. $-x^3$ for $x = 5$

95. $-x^2 - y^2$ for $x = -3$ and $y = -4$

96. $-x^2 + y^2$ for $x = -2$ and $y = 6$

97. $-a^2 - 3b^3$ for $a = -6$ and $b = -1$

98. $-a^3 + 3b^2$ for $a = -3$ and $b = -5$

99. $y^2 - 3xy$ for $x = 0.4$ and $y = -0.3$

100. $x^2 + 5xy$ for $x = -0.2$ and $y = -0.6$

101. Use a calculator to check your answers for Problems 1–34.

■ ■ ■ **Thoughts into words**

102. Your friend keeps getting an answer of 16 when simplifying -2^4. What mistake is he making and how would you help him?

103. Explain how you would simplify $\dfrac{12x^2y}{18xy}$.

Answers to Concept Quiz

1. False **2.** False **3.** True **4.** True **5.** False **6.** True **7.** False **8.** False **9.** True **10.** True

2.5 Translating from English to Algebra

Objectives

■ Translate algebraic expressions into English phrases.

■ Translate English phrases into algebraic expressions.

■ Write algebraic expressions for converting units of measure within a measurement system.

In order to use the tools of algebra for solving problems, we must be able to translate back and forth between the English language and the language of algebra. In this section we will translate algebraic expressions into English phrases (word phrases) and English phrases into algebraic expressions.

Let's begin by considering the following translations from algebraic expressions to word phrases.

Algebraic expression	Word phrase
$x + y$	The sum of x and y
$x - y$	The difference of x and y
$y - x$	The difference of y and x
xy	The product of x and y
$\dfrac{x}{y}$	The quotient of x and y
$3x$	The product of 3 and x
$x^2 + y^2$	The sum of x squared and y squared
$2xy$	The product of 2, x, and y
$2(x + y)$	Two times the quantity x plus y
$x - 3$	Three less than x

Now let's consider the reverse process: translating from word phrases to algebraic expressions. Part of the difficulty in translating from English to algebra is that different word phrases translate into the same algebraic expression. Thus we need to become familiar with *different ways of saying the same thing,* especially when referring to the four fundamental operations. The next examples should acquaint you with some of the phrases used in the basic operations.

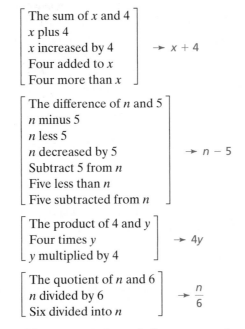

Often a word phrase indicates more than one operation. Furthermore, the standard vocabulary of *sum, difference, product,* and *quotient* may be replaced by other terminology. Study the following translations very carefully. Also remember that the commutative property holds for addition and multiplication but not for subtraction and division. Therefore the phrase "x plus y" can be written as $x + y$ or $y + x$. However the phrase "x minus y" means that y must be subtracted from x, and the phrase is written as $x - y$. So be very careful of phrases that involve subtraction or division.

Word phrase	Algebraic expression
The sum of two times x and three times y	$2x + 3y$
The sum of the squares of a and b	$a^2 + b^2$
Five times x divided by y	$\dfrac{5x}{y}$
Two more than the square of x	$x^2 + 2$
Three less than the cube of b	$b^3 - 3$
Five less than the product of x and y	$xy - 5$
Nine minus the product of x and y	$9 - xy$
Four times the sum of x and 2	$4(x + 2)$
Six times the quantity w minus 4	$6(w - 4)$

Suppose you are told that the sum of two numbers is 12, and one of the numbers is 8. What is the other number? The other number is $12 - 8$, which equals 4. Now suppose you are told that the product of two numbers is 56, and one of the numbers is 7. What is the other number? The other number is $56 \div 7$, which equals 8. The following examples illustrate the use of these addition–subtraction and multiplication–division relationships.

E X A M P L E   1

The sum of two numbers is 83, and one of the numbers is x. What is the other number?

Solution

Using the addition–subtraction relationship, we can represent the other number by $83 - x$. ∎

E X A M P L E   2

The difference of two numbers is 14. The smaller number is n. What is the larger number?

Solution

Because the smaller number plus the difference must equal the larger number, we can represent the larger number by $n + 14$. ∎

EXAMPLE 3

The product of two numbers is 39, and one of the numbers is y. Represent the other number.

Solution

Using the multiplication–division relationship, we can represent the other number by $\dfrac{39}{y}$. ∎

The English statement may not contain key words such as *sum, difference, product,* or *quotient.* Instead, the statement may describe a physical situation; from this description you need to deduce the operations involved. We now make some suggestions for handling such situations.

EXAMPLE 4

Arlene can type 70 words per minute. How many words can she type in m minutes?

Solution

In 10 minutes she would type $70(10) = 700$ words. In 50 minutes she would type $70(50) = 3500$ words. Thus in m minutes she would type $70m$ words. ∎

Note the use of some specific examples [$70(10) = 700$ and $70(50) = 3500$] to help formulate the general expression. This technique of first formulating some specific examples and then generalizing can be very effective.

EXAMPLE 5

Lynn has n nickels and d dimes. Express, in cents, this amount of money.

Solution

Three nickels and 8 dimes would be $5(3) + 10(8) = 95$ cents. Thus n nickels and d dimes would be $5n + 10d$ cents. ∎

EXAMPLE 6

A train travels at the rate of r miles per hour. How far will it travel in 8 hours?

Solution

Suppose that a train travels at 50 miles per hour. Using the formula *distance equals rate times time,* we find that it would travel $50 \cdot 8 = 400$ miles. Therefore, at r miles per hour, it would travel $r \cdot 8$ miles. We usually write the expression $r \cdot 8$ as $8r$. ∎

EXAMPLE 7

The cost of a 5-pound box of candy is d dollars. How much is the cost per pound for the candy?

Solution

The price per pound is figured by dividing the total cost by the number of pounds. Therefore, we represent the price per pound by $\dfrac{d}{5}$. ∎

The English statement to be translated into algebra may contain some geometric ideas. For example, suppose that we want to express in inches the length of a line segment that is f feet long. Because 1 foot = 12 inches, we can represent f feet by 12 times f, written as $12f$ inches.

Table 2.1 lists some of the basic relationships pertaining to linear measurements in the English and metric systems. (Additional listings of both systems are located inside the back cover of this book.)

Table 2.1

English system	Metric system
12 inches = 1 foot	1 kilometer = 1000 meters
3 feet = 36 inches = 1 yard	1 hectometer = 100 meters
5280 feet = 1760 yards = 1 mile	1 dekameter = 10 meters
	1 decimeter = 0.1 meter
	1 centimeter = 0.01 meter
	1 millimeter = 0.001 meter

EXAMPLE 8

The distance between two cities is k kilometers. Express this distance in meters.

Solution

Because 1 kilometer equals 1000 meters, we need to multiply k by 1000. Therefore, $1000k$ represents the distance in meters. ■

EXAMPLE 9

The length of a line segment is i inches. Express that length in yards.

Solution

To change from inches to yards, we must divide by 36. Therefore $\dfrac{i}{36}$ represents in yards the length of the line segment. ■

EXAMPLE 10

The width of a rectangle is w centimeters, and the length is 5 centimeters less than twice the width. What is the length of the rectangle? What is the perimeter of the rectangle? What is the area of the rectangle?

Solution

We can represent the length of the rectangle by $2w - 5$. Now we can sketch a rectangle as in Figure 2.4 and record the given information. The perimeter of a rectangle is the sum of the lengths of the four sides. Therefore, the perimeter, in centimeters, is given by $2w + 2(2w - 5)$, which can be written as $2w + 4w - 10$ and then simplified to $6w - 10$. The area of a rectangle is the product of the length and width.

Therefore, the area in square centimeters is given by $w(2w - 5) = w \cdot 2w + w(-5) = 2w^2 - 5w$.

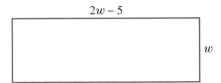

Figure 2.4

EXAMPLE   11   The length of a side of a square is x feet. Express the length of a side in inches. What is the area of the square in square inches?

Solution

Because 1 foot equals 12 inches, we need to multiply x by 12. Therefore, $12x$ represents the length of a side in inches. The area of a square is the length of a side squared. So the area in square inches is given by $(12x)^2 = (12x)(12x) = 12 \cdot 12 \cdot x \cdot x = 144x^2$.

CONCEPT QUIZ

For Problems 1–10, match the English phrase with its algebraic expression.

1. The product of x and y A. $x - y$

2. Two less than x B. $x + y$

3. x subtracted from 2   C. $\dfrac{x}{y}$

4. The difference of x and y D. $x - 2$

5. The quotient of x and y E. xy

6. The sum of x and y F. $x^2 - y$

7. Two times the sum of x and y G. $2(x + y)$

8. Two times x plus y H. $2 - x$

9. x squared minus y I. $x + 2$

10. Two more than x J. $2x + y$

PROBLEM SET 2.5

For Problems 1–12, write a word phrase for each of the algebraic expressions. For example, lw can be expressed as "the product of l and w."

1. $a - b$ **2.** $x + y$

3. $\frac{1}{3}Bh$ **4.** $\frac{1}{2}bh$

5. $2(l + w)$ **6.** πr^2

7. $\frac{A}{w}$ **8.** $\frac{C}{\pi}$

9. $\frac{a + b}{2}$ **10.** $\frac{a - b}{4}$

11. $3y + 2$ **12.** $3(x - y)$

For Problems 13–36, translate each word phrase into an algebraic expression. For example, "the sum of x and 14" translates into $x + 14$.

13. The sum of l and w

14. The difference of x and y

15. The product of a and b

16. The product of $\frac{1}{3}$, B, and h

17. The quotient of d and t

18. r divided into d

19. The product of l, w, and h

20. The product of π and the square of r

21. x subtracted from y

22. The difference of x and y

23. Two larger than the product of x and y

24. Six plus the cube of x

25. Seven minus the square of y

26. The quantity, x minus 2, cubed

27. The quantity, x minus y, divided by 4

28. Eight less than x

29. Ten less x

30. Nine times the quantity, n minus 4

31. Ten times the quantity, n plus 2

32. The sum of four times x and five times y

33. Seven subtracted from the product of x and y

34. Three times the sum of n and 2

35. Twelve less than the product of x and y

36. Twelve less the product of x and y

For Problems 37–72, answer the question with an algebraic expression.

37. The sum of two numbers is 35, and one of the numbers is n. What is the other number?

38. The sum of two numbers is 100, and one of the numbers is x. What is the other number?

39. The difference of two numbers is 45, and the smaller number is n. What is the other number?

40. The product of two numbers is 25, and one of the numbers is x. What is the other number?

41. Janet is y years old. How old will she be in 10 years?

42. Hector is y years old. How old was he 5 years ago?

43. Debra is x years old, and her mother is 3 years less than twice as old as Debra. How old is Debra's mother?

44. Jack is x years old, and Dudley is 1 year more than three times as old as Jack. How old is Dudley?

45. Donna has d dimes and q quarters in her bank. How much money, in cents, does she have?

46. Andy has c cents that is all in dimes. How many dimes does he have?

47. A car travels d miles in t hours. What is the rate of the car?

48. If g gallons of gasoline cost d dollars, what is the price per gallon?

49. If p pounds of candy cost d dollars, what is the price per pound?

50. Sue can type x words per minute. How many words can she type in 1 hour?

51. Larry's annual salary is *d* dollars. What is his monthly salary?

52. Nancy's monthly salary is *d* dollars. What is her annual salary?

53. If *n* represents a whole number, what represents the next larger whole number?

54. If *n* represents an even number, what represents the next larger even number?

55. If *n* represents an odd number, what represents the next larger odd number?

56. Maria is *y* years old, and her sister is twice as old. What represents the sum of their ages?

57. Willie is *y* years old and his father is 2 years less than twice Willie's age. What represents the sum of their ages?

58. Harriet has *p* pennies, *n* nickels, and *d* dimes. How much money, in cents, does she have?

59. The perimeter of a rectangle is *y* yards and *f* feet. What is the perimeter expressed in inches?

60. The perimeter of a triangle is *m* meters and *c* centimeters. What is the perimeter in centimeters?

61. A rectangular plot of ground is *f* feet long. What is its length in yards?

62. The height of a telephone pole is *f* feet. What is the height in yards?

63. The width of a rectangle is *w* feet, and its length is three times the width. What is the perimeter of the rectangle in feet?

64. The width of a rectangle is *w* feet, and its length is 1 foot more than twice its width. What is the perimeter of the rectangle in feet?

65. The length of a rectangle is *l* inches and its width is 2 inches less than one-half of its length. What is the perimeter of the rectangle in inches?

66. The length of a rectangle is *l* inches and its width is 3 inches more than one-third of its length. What is the perimeter of the rectangle in inches?

67. The first side of a triangle is *f* feet long. The second side is 2 feet longer than the first side. The third side is twice as long as the second side. What is the perimeter of the triangle in inches?

68. The first side of a triangle is *y* yards long. The second side is 3 yards shorter than the first side. The third side is three times as long as the second side. What is the perimeter of the triangle in feet?

69. The width of a rectangle is *w* yards, and the length is twice the width. What is the area of the rectangle in square yards?

70. The width of a rectangle is *w* yards, and the length is 4 yards more than the width. What is the area of the rectangle in square yards?

71. The length of a side of a square is *s* yards. What is the area of the square in square feet?

72. The length of a side of a square is *y* centimeters. What is the area of the square in square millimeters?

■ ■ ■ **Thoughts into words**

73. What does the phrase "translating from English to algebra" mean to you?

74. Your friend is having trouble with Problems 61 and 62. For example, for Problem 61 she doesn't know if the answer should be $3f$ or $\dfrac{f}{3}$. What can you do to help her?

CHAPTER 2

SUMMARY

(2.1) The property $\dfrac{a \cdot k}{b \cdot k} = \dfrac{a}{b}$ is used to express fractions in reduced form.

To **multiply** rational numbers in common fractional form, we multiply numerators, multiply denominators, and express the result in reduced form.

To **divide** rational numbers in common fractional form, we multiply by the reciprocal of the divisor.

(2.2) Addition and **subtraction** of rational numbers in common fractional form are based on these equations:

$$\frac{a}{b} + \frac{c}{b} = \frac{a+c}{b} \qquad \text{Addition}$$

$$\frac{a}{b} - \frac{c}{b} = \frac{a-c}{b} \qquad \text{Subtraction}$$

To add or subtract fractions that do not have a common denominator, we use the **fundamental principle of fractions,** $\dfrac{a}{b} = \dfrac{a \cdot k}{b \cdot k}$, and obtain equivalent fractions that have a common denominator.

(2.3) To **add** or **subtract decimals,** we write the numbers in a column so that the decimal points are lined up, and then we add or subtract just as we do with integers.

To **multiply decimals,** we (1) multiply the numbers, ignoring the decimal points, and then (2) insert the decimal point in the product so that the number of digits to the right of the decimal point in the product is equal to the sum of the number of digits to the right of the decimal point in each factor.

To **divide a decimal by a nonzero whole number,** we (1) place the decimal point in the quotient directly above the decimal point in the dividend, and then (2) divide as with whole numbers; in the division process, we place zeros in the quotient immediately to the right of the decimal point (if necessary) to show the correct place value.

To **divide by a decimal,** we change to an equivalent problem that has a whole-number divisor.

(2.4) Expressions of the form b^n, where

$$b^n = bbb \cdots b \qquad \text{n factors of } b$$

are read as "b to the nth power"; b is the *base,* and n is the *exponent.*

(2.5) To translate from English phrases to algebraic expressions, we must be familiar with the standard vocabulary of *sum, difference, product,* and *quotient,* as well as with other terms used to express the same ideas.

CHAPTER 2 REVIEW PROBLEM SET

For Problems 1–15, evaluate each numerical expression.

1. 2^6

2. $(-3)^3$

3. -4^2

4. 5^3

5. $\left(\dfrac{3}{4}\right)^2$

6. $-\left(\dfrac{1}{2}\right)^2$

7. $\left(\dfrac{1}{2} + \dfrac{2}{3}\right)^2$

8. $(0.6)^3$

9. $(0.12)^2$

10. $(0.06)^2$

11. $\left(-\dfrac{2}{3}\right)^3$

12. $\left(-\dfrac{1}{2}\right)^4$

13. $\left(\dfrac{1}{4} - \dfrac{1}{2}\right)^3$

14. $(0.5)^2$

15. $\left(\dfrac{1}{2} + \dfrac{1}{3} - \dfrac{1}{6}\right)^2$

For Problems 16–25, perform the indicated operations and express answers in reduced form.

16. $\dfrac{3}{8} + \dfrac{5}{12}$

17. $\dfrac{9}{14} - \dfrac{3}{35}$

18. $\dfrac{2}{3} + \dfrac{-3}{5}$

19. $\dfrac{7}{x} + \dfrac{9}{2y}$

20. $\dfrac{5}{xy} - \dfrac{8}{x^2}$

21. $\left(\dfrac{7y}{8x}\right)\left(\dfrac{14x}{35}\right)$

22. $\left(\dfrac{6xy}{9y^2}\right) \div \left(\dfrac{15y}{18x^2}\right)$ **23.** $\left(\dfrac{-3x}{12y}\right)\left(\dfrac{8y}{-7x}\right)$

24. $\left(-\dfrac{4y}{3x}\right)\left(-\dfrac{3x}{4y}\right)$ **25.** $\left(\dfrac{6n}{7}\right)\left(\dfrac{9n}{8}\right)$

For Problems 26–37, simplify each numerical expression.

26. $\dfrac{1}{6} + \dfrac{2}{3} \cdot \dfrac{3}{4} - \dfrac{5}{6} \div \dfrac{8}{6}$ **27.** $\dfrac{3}{4} \cdot \dfrac{1}{2} - \dfrac{4}{3} \cdot \dfrac{3}{2}$

28. $\dfrac{7}{9} \cdot \dfrac{3}{5} + \dfrac{7}{9} \cdot \dfrac{2}{5}$ **29.** $\dfrac{4}{5} \div \dfrac{1}{5} \cdot \dfrac{2}{3} - \dfrac{1}{4}$

30. $\dfrac{2}{3} \cdot \dfrac{1}{4} \div \dfrac{1}{2} + \dfrac{2}{3} \cdot \dfrac{1}{4}$

31. $0.48 + 0.72 - 0.35 - 0.18$

32. $0.81 + (0.6)(0.4) - (0.7)(0.8)$

33. $1.28 \div 0.8 - 0.81 \div 0.9 + 1.7$

34. $(0.3)^2 + (0.4)^2 - (0.6)^2$

35. $(1.76)(0.8) + (1.76)(0.2)$

36. $(2^2 - 2 - 2^3)^2$ **37.** $1.92(0.9 + 0.1)$

For Problems 38–43, simplify each algebraic expression by combining similar terms. Express answers in reduced form when they involve common fractions.

38. $\dfrac{3}{8}x^2 - \dfrac{2}{5}y^2 - \dfrac{2}{7}x^2 + \dfrac{3}{4}y^2$

39. $0.24ab + 0.73bc - 0.82ab - 0.37bc$

40. $\dfrac{1}{2}x + \dfrac{3}{4}x - \dfrac{5}{6}x + \dfrac{1}{24}x$

41. $1.4a - 1.9b + 0.8a + 3.6b$

42. $\dfrac{2}{5}n + \dfrac{1}{3}n - \dfrac{5}{6}n$ **43.** $n - \dfrac{3}{4}n + 2n - \dfrac{1}{5}n$

For Problems 44–49, evaluate each algebraic expression for the given values of the variables.

44. $\dfrac{1}{4}x - \dfrac{2}{5}y$ for $x = \dfrac{2}{3}$ and $y = -\dfrac{5}{7}$

45. $a^3 + b^2$ for $a = -\dfrac{1}{2}$ and $b = \dfrac{1}{3}$

46. $2x^2 - 3y^2$ for $x = 0.6$ and $y = 0.7$

47. $0.7w + 0.9z$ for $w = 0.4$ and $z = -0.7$

48. $\dfrac{3}{5}x - \dfrac{1}{3}x + \dfrac{7}{15}x - \dfrac{2}{3}x$ for $x = \dfrac{15}{17}$

49. $\dfrac{1}{3}n + \dfrac{2}{7}n - n$ for $n = 21$

For Problems 50–57, answer each question with an algebraic expression.

50. The sum of two numbers is 72, and one of the numbers is n. What is the other number?

51. Joan has p pennies and d dimes. How much money, in cents, does she have?

52. Ellen types x words in an hour. What is her typing rate per minute?

53. Harry is y years old. His brother is 3 years less than twice as old as Harry. How old is Harry's brother?

54. Larry chose a number n. Cindy chose a number 3 more than five times the number chosen by Larry. What number did Cindy choose?

55. The height of a file cabinet is y yards and f feet. How tall is the file cabinet in inches?

56. The length of a rectangular room is m meters. How long in centimeters is the room?

57. Corrine has n nickels, d dimes, and q quarters. How much money, in cents, does she have?

For Problems 58–67, translate each word phrase into an algebraic expression.

58. Five less than n

59. Five less n

60. Ten times the quantity, x minus 2

61. Ten times x minus 2

62. x minus 3

63. d divided by r

64. x squared plus 9

65. x plus 9, the quantity squared

66. The sum of the cubes of x and y

67. Four less than the product of x and y

CHAPTER 2

TEST

🔲 **applies to all problems in this Chapter Test.**

1. Find the value of each expression.

   a. $(-3)^4$ **b.** -2^6 **c.** $(0.2)^3$

2. Express $\dfrac{42}{54}$ in reduced form.

3. Simplify $\dfrac{18xy^2}{32y}$.

For Problems 4–7, simplify each numerical expression.

4. $5.7 - 3.8 + 4.6 - 9.1$

5. $0.2(0.4) - 0.6(0.9) + 0.5(7)$

6. $-0.4^2 + 0.3^2 - 0.7^2$

7. $\left(\dfrac{1}{3} - \dfrac{1}{4} + \dfrac{1}{6}\right)^4$

For Problems 8–11, perform the following indicated operations and express answers in reduced form.

8. $\dfrac{5}{12} \div \dfrac{15}{8}$

9. $-\dfrac{2}{3} - \dfrac{1}{2}\left(\dfrac{3}{4}\right) + \dfrac{5}{6}$

10. $3\left(\dfrac{2}{5}\right) - 4\left(\dfrac{5}{6}\right) + 6\left(\dfrac{7}{8}\right)$

11. $4\left(\dfrac{1}{2}\right)^3 - 3\left(\dfrac{2}{3}\right)^2 + 9\left(\dfrac{1}{4}\right)^2$

For Problems 12–17, perform the indicated operations, and express answers in simplest form.

12. $\dfrac{8x}{15y} \cdot \dfrac{9y^2}{6x}$

13. $\dfrac{6xy}{9} \div \dfrac{y}{3x}$

14. $\dfrac{4}{x} - \dfrac{5}{y^2}$

15. $\dfrac{3}{2x} + \dfrac{7}{6x}$

16. $\dfrac{5}{3y} + \dfrac{9}{7y^2}$

17. $\left(\dfrac{15a^2b}{12a}\right)\left(\dfrac{8ab}{9b}\right)$

For Problems 18 and 19, simplify each algebraic expression by combining similar terms.

18. $3x - 2xy - 4x + 7xy$

19. $-2a^2 + 3b^2 - 5b^2 - a^2$

For Problems 20–23, evaluate each algebraic expression for the given values of the variables.

20. $x^2 - xy + y^2$ for $x = \dfrac{1}{2}$ and $y = -\dfrac{2}{3}$

21. $0.2x - 0.3y - xy$ for $x = 0.4$ and $y = 0.8$

22. $\dfrac{3}{4}x - \dfrac{2}{3}y$ for $x = -\dfrac{1}{2}$ and $y = \dfrac{3}{5}$

23. $3x - 2y + xy$ for $x = 0.5$ and $y = -0.9$

24. David has n nickels, d dimes, and q quarters. How much money, in cents, does he have?

25. Hal chose a number n. Sheila chose a number 3 less than four times the number that Hal chose. Express the number that Sheila chose in terms of n.

CUMULATIVE REVIEW PROBLEM SET *Chapters 1 and 2*

For Problems 1–12, simply each numerical expression.

1. $16 - 18 - 14 + 21 - 14 + 19$

2. $7(-6) - 8(-6) + 4(-9)$

3. $6 - [3 - (10 - 12)]$

4. $-9 - 2[4 - (-10 + 6)] - 1$

5. $\dfrac{-7(-4) - 5(-6)}{-2}$

6. $\dfrac{5(-3) + (-4)(6) - 3(4)}{-3}$

7. $\dfrac{3}{4} + \dfrac{1}{3} \div \dfrac{4}{3} - \dfrac{1}{2}$

8. $\left(\dfrac{2}{3}\right)\left(-\dfrac{3}{4}\right) - \left(\dfrac{5}{6}\right)\left(\dfrac{4}{5}\right)$

9. $\left(\dfrac{1}{2} - \dfrac{2}{3}\right)^2$

10. -4^3

11. $\dfrac{0.0046}{0.000023}$

12. $(0.2)^2 - (0.3)^3 + (0.4)^2$

For Problems 13–20, evaluate each algebraic expression for the given values of the variables.

13. $3xy - 2x - 4y$ for $x = -6$ and $y = 7$

14. $-4x^2y - 2xy^2 + xy$ for $x = -2$ and $y = -4$

15. $\dfrac{5x - 2y}{3x}$ for $x = \dfrac{1}{2}$ and $y = -\dfrac{1}{3}$

16. $0.2x - 0.3y + 2xy$ for $x = 0.1$ and $y = 0.3$

17. $-7x + 4y + 6x - 9y + x - y$ for $x = -0.2$ and $y = 0.4$

18. $\dfrac{2}{3}x - \dfrac{3}{5}y + \dfrac{3}{4}x - \dfrac{1}{2}y$ for $x = \dfrac{6}{5}$ and $y = -\dfrac{1}{4}$

19. $\dfrac{2}{n} - \dfrac{3}{n^2}$ for $n = 2$

20. $-ab + \dfrac{1}{5}a - \dfrac{2}{3}b$ for $a = -2$ and $b = \dfrac{3}{4}$

For Problems 21–24, express each number as a product of prime factors.

21. 54

22. 78

23. 91

24. 153

For Problems 25–28, find the greatest common factor of the given numbers.

25. 42 and 70

26. 63 and 81

27. 28, 36, and 52

28. 48, 66, and 78

For Problems 29–32, find the least common multiple of the given numbers.

29. 20 and 28

30. 40 and 100

31. 12, 18, and 27

32. 16, 20, and 80

For Problems 33–38, simplify each algebraic expression by combining similar terms.

33. $\dfrac{2}{3}x - \dfrac{1}{4}y - \dfrac{3}{4}x - \dfrac{2}{3}y$

34. $-n - \dfrac{1}{2}n + \dfrac{3}{5}n + \dfrac{5}{6}n$

35. $3.2a - 1.4b - 6.2a + 3.3b$

36. $-(n - 1) + 2(n - 2) - 3(n - 3)$

37. $-x + 4(x - 1) - 3(x + 2) - (x + 5)$

38. $2a - 5(a + 3) - 2(a - 1) - 4a$

For Problems 39–46, perform the indicated operations and express answers in reduced form.

39. $\dfrac{5}{12} - \dfrac{3}{16}$

40. $\dfrac{3}{4} - \dfrac{5}{6} - \dfrac{7}{9}$

41. $\dfrac{5}{xy} - \dfrac{2}{x} + \dfrac{3}{y}$

42. $-\dfrac{7}{x^2} + \dfrac{9}{xy}$

43. $\left(\dfrac{7x}{9y}\right)\left(\dfrac{12y}{14}\right)$

44. $\left(-\dfrac{5a}{7b^2}\right)\left(-\dfrac{8ab}{15}\right)$

45. $\left(\dfrac{6x^2y}{11}\right) \div \left(\dfrac{9y^2}{22}\right)$

46. $\left(-\dfrac{9a}{8b}\right) \div \left(\dfrac{12a}{18b}\right)$

For Problems 47–50, answer the question with an algebraic expression.

47. Hector has p pennies, n nickels, and d dimes. How much money in cents does he have?

48. Ginny chose a number n. Penny chose a number 5 less than 4 times the number chosen by Ginny. What number did Penny choose?

49. The height of a flagpole is y yards, f feet, and i inches. How tall is the flagpole in inches?

50. A rectangular room is x meters by y meters. What is its perimeter in centimeters?

© Richard Radstone/Getty Images/Taxi

The inequality $\dfrac{95 + 82 + 93 + 84 + s}{5} \geq 90$ can be used to determine that Ashley needs a 96 or higher on her fifth exam to have an average of 90 or higher for the five exams if she got 95, 82, 93, and 84 on her first four exams.

Equations, Inequalities, and Problem Solving

Carlos paid $645 to have his air conditioner repaired. Included in the bill were $360 for parts and a charge for 3 hours of labor. How much was Carlos charged for each hour of labor? If we let h represent the hourly charge, then the equation $360 + 3h = 645$ can be used to determine that the per-hour charge is $95.

Tracy received a cell phone bill for $136.74. Included in the $136.74 were a monthly-plan charge of $39.99 and a charge for 215 extra minutes. How much is Tracy being charged for each extra minute? If we let c represent the charge per minute, then the equation $39.99 + 215c = 136.74$ can be used to determine that the charge for each extra minute is $0.45.

Chris had scores of 93, 86, and 89 on her first three algebra tests. What score must she get on the fourth test to have an average of 90 or higher for the four tests? If we let x represent Chris's fourth test grade, then we can use the inequality $\dfrac{93 + 86 + 89 + x}{4} \geq 90$ to determine that Chris needs to score 92 or higher.

Throughout this book we follow a common theme: Develop some new skills, use the skills to help solve equations and inequalities, and finally, use the equations and inequalities to solve applied problems. In this chapter we want to use the skills we developed in the first two chapters to solve equations and inequalities and begin our work with applied problems.

InfoTrac Project

Do a subject guide search on **per capita debt,** and find an article on central Canada. Write a brief summary of the article. Suppose the revenue received from taxes in Ontario can be represented by the expression $1,500,000x + 10,000,000$, where x represents the average amount of taxes paid by each citizen of Ontario. If the projected expenditures for the next year are expected to be $495,250,000, what would each citizen of Ontario have to pay in order to have a balanced budget?

3.1 Solving First-Degree Equations

Objectives

■ Solve first-degree equations using the addition-subtraction property of equality.

■ Solve first-degree equations using the multiplication-division property of equality.

These are examples of **numerical statements:**

$$3 + 4 = 7 \qquad 5 - 2 = 3 \qquad 7 + 1 = 12$$

The first two are true statements, and the third is a false statement.

When you use x as a variable, statements like these

$$x + 3 = 4 \qquad 2x - 1 = 7 \qquad x^2 = 4$$

are called **algebraic equations** in x. We call a number a a **solution** or **root** of an equation if a true numerical statement is formed when we substitute a for x. (We also say that a satisfies the equation.) For example, 1 is a solution of $x + 3 = 4$ because substituting 1 for x produces the true numerical statement $1 + 3 = 4$. We call the set of all solutions of an equation its **solution set.** Thus the solution set of $x + 3 = 4$ is $\{1\}$. Likewise, the solution set of $2x - 1 = 7$ is $\{4\}$ and the solution set of $x^2 = 4$ is $\{-2, 2\}$. **Solving an equation** refers to the process of determining the solution set. Remember that a set that consists of no elements is called the **empty** or **null set** and is denoted by $\varnothing$. Thus we say that the solution set of $x = x + 1$ is $\varnothing$; that is, there are no real numbers that satisfy $x = x + 1$.

In this chapter we will consider techniques for solving **first-degree equations of one variable.** This means that the equations contain only one variable, and this variable has an exponent of 1. Here are some examples of first-degree equations of one variable:

$$3x + 4 = 7 \qquad 0.8w + 7.1 = 5.2w - 4.8$$

$$\frac{1}{2}y + 2 = 9 \qquad 7x + 2x - 1 = 4x - 1$$

Equivalent equations are equations that have the same solution set. For example,

$$5x - 4 = 3x + 8$$
$$2x = 12$$
$$x = 6$$

are all equivalent equations; this can be verified by showing that 6 is the solution for all three equations.

As we work with equations, we can use the following properties of equality:

PROPERTY 3.1   *Properties of Equality*

For all real numbers, a, b, and c,

1. $a = a$ Reflexive property
2. if $a = b$, then $b = a$ Symmetric property
3. if $a = b$ and $b = c$, then $a = c$ Transitive property
4. if $a = b$, then a may be replaced by b, or b may be replaced by a, in any statement, without changing the meaning of the statement   Substitution property

---

The general procedure for solving an equation is to continue replacing the given equation with equivalent but simpler equations until we obtain an equation of the form **variable = constant** or **constant = variable.** Thus in the preceding example, $5x - 4 = 3x + 8$ was simplified to $2x = 12$, which was further simplified to $x = 6$, from which the solution of 6 is obvious. The exact procedure for simplifying equations is our next concern.

Two properties of equality play an important role in the process of solving equations. The first of these is the **addition-subtraction property of equality,** which we state as follows:

PROPERTY 3.2   *Addition-Subtraction Property of Equality*

For all real numbers a, b, and c,

1. $a = b$ if and only if $a + c = b + c$.
2. $a = b$ if and only if $a - c = b - c$.

Property 3.2 states that *any number can be added to or subtracted from both sides of an equation, and the result is an equivalent equation.* Consider the use of this property in the next four examples.

EXAMPLE 1   Solve $x - 8 = 3$.

Solution

$$x - 8 = 3$$
$$x - 8 + 8 = 3 + 8 \qquad \text{Add 8 to both sides.}$$
$$x = 11$$

The solution set is $\{11\}$. ■

REMARK: It is true that a simple equation like Example 1 can be solved *by inspection*. That is to say, we could think, "some number minus 8 produces 3," and obviously, the number is 11. However, as the equations become more complex, the technique of solving by inspection becomes ineffective. This is why it is necessary to develop more formal techniques for solving equations. Therefore we will begin developing such techniques even with very simple equations.

EXAMPLE 2   Solve $x + 14 = -8$.

Solution

$$x + 14 = -8$$
$$x + 14 - 14 = -8 - 14 \qquad \text{Subtract 14 from both sides.}$$
$$x = -22$$

The solution set is $\{-22\}$. ■

EXAMPLE 3   Solve $n - \dfrac{1}{3} = \dfrac{1}{4}$.

Solution

$$n - \frac{1}{3} = \frac{1}{4}$$
$$n - \frac{1}{3} + \frac{1}{3} = \frac{1}{4} + \frac{1}{3} \qquad \text{Add } \frac{1}{3} \text{ to both sides.}$$
$$n = \frac{3}{12} + \frac{4}{12}$$
$$n = \frac{7}{12}$$

The solution set is $\left\{\dfrac{7}{12}\right\}$. ■

EXAMPLE 4

Solve $0.72 = y + 0.35$.

Solution

$$0.72 = y + 0.35$$
$$0.72 - 0.35 = y + 0.35 - 0.35 \qquad \text{Subtract 0.35 from both sides.}$$
$$0.37 = y$$

The solution set is $\{0.37\}$.

Note in Example 4 that the final equation is $0.37 = y$ instead of $y = 0.37$. Technically, the **symmetric property of equality** (if $a = b$, then $b = a$) would permit us to change from $0.37 = y$ to $y = 0.37$, but such a change is not necessary to determine that the solution is 0.37. You should also realize that you could apply the symmetric property to the original equation. Thus $0.72 = y + 0.35$ becomes $y + 0.35 = 0.72$, and subtracting 0.35 from both sides would produce $y = 0.37$.

We should make at this time one other comment that pertains to Property 3.2. Because subtracting a number is equivalent to adding its opposite, Property 3.2 could be stated only in terms of addition. Thus to solve an equation such as Example 4, we could add -0.35 to both sides rather than subtracting 0.35 from both sides.

The other important property for solving equations is the **multiplication-division property of equality.**

PROPERTY 3.3 *Multiplication-Division Property of Equality*

For all real numbers, a, b, and c, where $c \neq 0$,

1. $a = b$ if and only if $ac = bc$.

2. $a = b$ if and only if $\dfrac{a}{c} = \dfrac{b}{c}$.

Property 3.3 states that *we get an equivalent equation whenever both sides of a given equation are multiplied or divided by the same nonzero real number.* The following examples illustrate the use of this property.

EXAMPLE 5

Solve $\dfrac{3}{4}x = 6$.

Solution

$$\frac{3}{4}x = 6$$
$$\frac{4}{3}\left(\frac{3}{4}x\right) = \frac{4}{3}(6) \qquad \text{Multiply both sides by } \frac{4}{3} \text{ because } \left(\frac{4}{3}\right)\left(\frac{3}{4}\right) = 1.$$
$$x = 8$$

The solution set is $\{8\}$.

EXAMPLE 6

Solve $5x = 27$.

Solution

$$5x = 27$$

$$\frac{5x}{5} = \frac{27}{5} \qquad \text{Divide both sides by 5.}$$

$$x = \frac{27}{5} \qquad \frac{27}{5} \text{ could be expressed as } 5\frac{2}{5} \text{ or } 5.4.$$

The solution set is $\left\{ \dfrac{27}{5} \right\}$.

EXAMPLE 7

Solve $-\dfrac{2}{3}p = \dfrac{1}{2}$.

Solution

$$-\frac{2}{3}p = \frac{1}{2}$$

$$\left(-\frac{3}{2}\right)\left(-\frac{2}{3}p\right) = \left(-\frac{3}{2}\right)\left(\frac{1}{2}\right) \qquad \begin{array}{l} \text{Multiply both sides by } -\dfrac{3}{2} \\[6pt] \text{because } \left(-\dfrac{3}{2}\right)\left(-\dfrac{2}{3}\right) = 1. \end{array}$$

$$p = -\frac{3}{4}$$

The solution set is $\left\{ -\dfrac{3}{4} \right\}$.

EXAMPLE 8

Solve $26 = -6x$.

Solution

$$26 = -6x$$

$$\frac{26}{-6} = \frac{-6x}{-6} \qquad \text{Divide both sides by } -6.$$

$$-\frac{26}{6} = x \qquad \frac{26}{-6} = -\frac{26}{6}$$

$$-\frac{13}{3} = x \qquad \text{Don't forget to reduce!}$$

The solution set is $\left\{ -\dfrac{13}{3} \right\}$.

Look back at Examples 5–8 and you will notice that we divided both sides of the equation by the coefficient of the variable whenever the coefficient was an integer; otherwise, we used the multiplication part of Property 3.3. Technically, because dividing by a number is equivalent to multiplying by its reciprocal, Property 3.3 could be stated only in terms of multiplication. Thus to solve an equation such as $5x = 27$, we could multiply both sides by $\dfrac{1}{5}$ instead of dividing both sides by 5.

EXAMPLE 9

Solve $0.2n = 15$.

Solution

$$0.2n = 15$$

$$\frac{0.2n}{0.2} = \frac{15}{0.2} \qquad \text{Divide both sides by 0.2.}$$

$$n = 75$$

The solution set is $\{75\}$.

CONCEPT QUIZ

For Problems 1–10, answer true or false.

1. Equivalent equations have the same solution set.

2. $x^2 = 9$ is a first-degree equation.

3. The set of all solutions is called a solution set.

4. If the solution set is the null set, then the equation has at least one solution.

5. Solving an equation refers to obtaining any other equivalent equation.

6. If 5 is a solution, then a true numerical statement is formed when 5 is substituted for the variable in the equation.

7. Any number can be subtracted from both sides of an equation, and the result is an equivalent equation.

8. Any number can divide both sides of an equation to obtain an equivalent equation.

9. By the reflexive property, if $y = 2$ then $2 = y$.

10. By the transitive property, if $x = y$ and $y = 4$, then $x = 4$.

PROBLEM SET 3.1

Use the properties of equality to help solve each equation.

1. $x + 9 = 17$

2. $x + 7 = 21$

3. $x + 11 = 5$

4. $x + 13 = 2$

5. $-7 = x + 2$

6. $-12 = x + 4$

7. $8 = n + 14$

8. $6 = n + 19$

9. $21 + y = 34$

10. $17 + y = 26$

11. $x - 17 = 31$

12. $x - 22 = 14$

13. $14 = x - 9$

14. $17 = x - 28$

15. $-26 = n - 19$

16. $-34 = n - 15$

17. $y - \dfrac{2}{3} = \dfrac{3}{4}$

18. $y - \dfrac{2}{5} = \dfrac{1}{6}$

19. $x + \dfrac{3}{5} = \dfrac{1}{3}$

20. $x + \dfrac{5}{8} = \dfrac{2}{5}$

21. $b + 0.19 = 0.46$

22. $b + 0.27 = 0.74$

23. $n - 1.7 = -5.2$

24. $n - 3.6 = -7.3$

25. $15 - x = 32$

26. $13 - x = 47$

27. $-14 - n = 21$

28. $-9 - n = 61$

29. $7x = -56$

30. $9x = -108$

31. $-6x = 102$

32. $-5x = 90$

33. $5x = 37$

34. $7x = 62$

35. $-18 = 6n$

36. $-52 = 13n$

37. $-26 = -4n$

38. $-56 = -6n$

39. $\dfrac{t}{9} = 16$

40. $\dfrac{t}{12} = 8$

41. $\dfrac{n}{-8} = -3$

42. $\dfrac{n}{-9} = -5$

43. $-x = 15$

44. $-x = -17$

45. $\dfrac{3}{4}x = 18$

46. $\dfrac{2}{3}x = 32$

47. $-\dfrac{2}{5}n = 14$

48. $-\dfrac{3}{8}n = 33$

49. $\dfrac{2}{3}n = \dfrac{1}{5}$

50. $\dfrac{3}{4}n = \dfrac{1}{8}$

51. $\dfrac{5}{6}n = -\dfrac{3}{4}$

52. $\dfrac{6}{7}n = -\dfrac{3}{8}$

53. $\dfrac{3x}{10} = \dfrac{3}{20}$

54. $\dfrac{5x}{12} = \dfrac{5}{36}$

55. $\dfrac{-y}{2} = \dfrac{1}{6}$

56. $\dfrac{-y}{4} = \dfrac{1}{9}$

57. $-\dfrac{4}{3}x = -\dfrac{9}{8}$

58. $-\dfrac{6}{5}x = -\dfrac{10}{14}$

59. $-\dfrac{5}{12} = \dfrac{7}{6}x$

60. $-\dfrac{7}{24} = \dfrac{3}{8}x$

61. $-\dfrac{5}{7}x = 1$

62. $-\dfrac{11}{12}x = -1$

63. $-4n = \dfrac{1}{3}$

64. $-6n = \dfrac{3}{4}$

65. $-8n = \dfrac{6}{5}$

66. $-12n = \dfrac{8}{3}$

67. $1.2x = 0.36$

68. $2.5x = 17.5$

69. $30.6 = 3.4n$

70. $2.1 = 4.2n$

71. $-3.4x = 17$

72. $-4.2x = 50.4$

■ ■ ▨ Thoughts into words

73. Describe the difference between a numerical statement and an algebraic equation.

74. Are the equations $6 = 3x + 1$ and $1 + 3x = 6$ equivalent equations? Defend your answer.

Answers to Concept Quiz

1. True   **2.** False   **3.** True   **4.** False   **5.** False   **6.** True   **7.** True   **8.** False   **9.** False   **10.** True

3.2   **Equations and Problem Solving**

Objectives

■   Solve first-degree equations using both the addition-subtraction property of equality and the multiplication-division property of equality.

■   Declare variables and write equations to solve word problems.

We often need to use more than one property of equality to help find the solution of an equation. Consider the next examples.

E X A M P L E   1

Solve $3x + 1 = 7$.

Solution

$$3x + 1 = 7$$
$$3x + 1 - 1 = 7 - 1 \qquad \text{Subtract 1 from both sides.}$$
$$3x = 6$$
$$\frac{3x}{3} = \frac{6}{3} \qquad \text{Divide both sides by 3.}$$
$$x = 2$$

We can *check* the potential solution by substituting it into the original equation to see whether we get a true numerical statement.

   Check

$$3x + 1 = 7$$
$$3(2) + 1 \stackrel{?}{=} 7$$
$$6 + 1 \stackrel{?}{=} 7$$
$$7 = 7$$

Now we know that the solution set is $\{2\}$.   ■

E X A M P L E   2

Solve $5x - 6 = 14$.

Solution

$$5x - 6 = 14$$
$$5x - 6 + 6 = 14 + 6 \qquad \text{Add 6 to both sides.}$$
$$5x = 20$$
$$\frac{5x}{5} = \frac{20}{5} \qquad \text{Divide both sides by 5.}$$
$$x = 4$$

Check

$$5x - 6 = 14$$
$$5(4) - 6 \overset{?}{=} 14$$
$$20 - 6 \overset{?}{=} 14$$
$$14 = 14$$

The solution set is $\{4\}$.

EXAMPLE 3

Solve $4 - 3a = 22$.

Solution

$$4 - 3a = 22$$
$$4 - 3a - 4 = 22 - 4 \qquad \text{Subtract 4 from both sides.}$$
$$-3a = 18$$
$$\frac{-3a}{-3} = \frac{18}{-3} \qquad \text{Divide both sides by } -3.$$
$$a = -6$$

 Check

$$4 - 3a = 22$$
$$4 - 3(-6) \overset{?}{=} 22$$
$$4 + 18 \overset{?}{=} 22$$
$$22 = 22$$

The solution set is $\{-6\}$.

Note that in Examples 1, 2, and 3, we first used the addition-subtraction property and then used the multiplication-division property. In general, this sequence of steps provides the easiest format for solving such equations. Perhaps you should convince yourself of that fact by doing Example 1 again, this time using the multiplication-division property first and then the addition-subtraction property.

EXAMPLE 4

Solve $19 = 2n + 4$.

Solution

$$19 = 2n + 4$$
$$19 - 4 = 2n + 4 - 4 \qquad \text{Subtract 4 from both sides.}$$
$$15 = 2n$$
$$\frac{15}{2} = \frac{2n}{2} \qquad \text{Divide both sides by 2.}$$
$$\frac{15}{2} = n$$

Check

$$19 = 2n + 4$$

$$19 \stackrel{?}{=} 2\left(\frac{15}{2}\right) + 4$$

$$19 \stackrel{?}{=} 15 + 4$$

$$19 = 19$$

The solution set is $\left\{\dfrac{15}{2}\right\}$. ■

Word Problems

In the last section of Chapter 2 we translated English phrases into algebraic expressions. We are now ready to extend that idea to the translation of English *sentences* into algebraic *equations*. Such translations enable us to use the concepts of algebra to solve word problems. Let's consider some examples.

PROBLEM 1

A certain number added to 17 yields a sum of 29. What is the number?

Solution

Let n represent the number to be found. The sentence "a certain number added to 17 yields a sum of 29" translates to the algebraic equation $17 + n = 29$. To solve this equation, we use these steps:

$$17 + n = 29$$
$$17 + n - 17 = 29 - 17$$
$$n = 12$$

The solution is 12, which is the number asked for in the problem. ■

We often refer to the statement "let n represent the number to be found" as **declaring the variable.** We need to choose a letter to use as a variable and indicate what it represents for a specific problem — which may seem like an insignificant idea, but as the problems become more complex, the process of declaring the variable becomes even more important. We could solve a problem such as Problem 1 without setting up an algebraic equation; however, as problems increase in difficulty, the translation from English to an algebraic equation becomes a key issue. Therefore, even with these relatively simple problems, we need to concentrate on the translation process.

PROBLEM 2

Six years ago Bill was 13 years old. How old is he now?

Solution

Let y represent Bill's age now; therefore, $y - 6$ represents his age six years ago. Thus

$$y - 6 = 13$$
$$y - 6 + 6 = 13 + 6$$
$$y = 19$$

Bill is presently 19 years old. ■

PROBLEM 3   Betty worked 8 hours Saturday and earned $66. How much did she earn per hour?

Solution A

Let x represent the amount Betty earned per hour. The number of hours worked times the wage per hour yields the total earnings. Thus

$$8x = 66$$

$$\frac{8x}{8} = \frac{66}{8}$$

$$x = 8.25$$

Betty earned $8.25 per hour.

Solution B

Let y represent the amount Betty earned per hour. The wage per hour equals the total wage divided by the number of hours. Thus

$$y = \frac{66}{8}$$

$$y = 8.25$$

Betty earned $8.25 per hour.   ■

Sometimes we can use more than one equation to solve a problem. In Solution A we set up the equation in terms of multiplication, whereas in Solution B we were thinking in terms of division.

PROBLEM 4   If 2 is subtracted from five times a certain number, the result is 28. Find the number.

Solution

Let n represent the number to be found. Translating the first sentence in the problem into an algebraic equation, we obtain

$$5n - 2 = 28$$

To solve this equation we proceed as follows:

$$5n - 2 + 2 = 28 + 2$$

$$5n = 30$$

$$\frac{5n}{5} = \frac{30}{5}$$

$$n = 6$$

The number to be found is 6.   ■

PROBLEM 5   The cost of a five-day vacation cruise package was $534. This cost included $339 for the cruise and an amount for 2 nights of lodging on shore. Find the cost per night of the on-shore lodging.

Solution

Let n represent the cost for one night of lodging; then $2n$ represents the total cost of lodging. Thus the cost for the cruise and lodging is the total cost of $534. We can proceed as follows:

$$\text{Cost of cruise} + \text{Cost of lodging} = \$534$$
$$339 \qquad + \qquad 2n \qquad = 534$$

To solve this equation we proceed as follows:

$$339 + 2n = 534$$
$$2n = 195$$
$$\frac{2n}{2} = \frac{195}{2}$$
$$n = 97.50$$

The cost of lodging per night is $97.50.

CONCEPT QUIZ

For Problems 1–5, answer true or false.

1. Only one property of equality is necessary to solve any equation.

2. Substituting the solution into the original equation to obtain a true numerical statement can be used to check potential solutions.

3. The statement "let x represent the number" is referred to as checking the variable.

4. Sometimes there can be two approaches to solving a word problem.

5. To solve the equation, $\frac{1}{3}x - 2 = 7$, you could begin by either adding 2 to both sides of the equation or by multiplying both sides of the equation by 3.

For Problems 6–10, match the English sentence with its algebraic equation.

6. Three added to a number is 24.

7. The product of 3 and a number is 24.

8. Three less than a number is 24.

9. The quotient of a number and three is 24.

10. A number subtracted from 3 is 24.

A. $3x = 24$

B. $3 - x = 24$

C. $x + 3 = 24$

D. $x - 3 = 24$

E. $\frac{x}{3} = 24$

PROBLEM SET 3.2

For Problems 1–40, solve each equation.

1. $2x + 5 = 13$

2. $3x + 4 = 19$

3. $5x + 2 = 32$

4. $7x + 3 = 24$

5. $3x - 1 = 23$

6. $2x - 5 = 21$

7. $4n - 3 = 41$

8. $5n - 6 = 19$

9. $6y - 1 = 16$

10. $4y - 3 = 14$

11. $2x + 3 = 22$

12. $3x + 1 = 21$

13. $10 = 3t - 8$

14. $17 = 2t + 5$

15. $5x + 14 = 9$

16. $4x + 17 = 9$

17. $18 - n = 23$

18. $17 - n = 29$

19. $-3x + 2 = 20$

20. $-6x + 1 = 43$

21. $7 + 4x = 29$

22. $9 + 6x = 23$

23. $16 = -2 - 9a$

24. $18 = -10 - 7a$

25. $-7x + 3 = -7$

26. $-9x + 5 = -18$

27. $17 - 2x = -19$

28. $18 - 3x = -24$

29. $-16 - 4x = 9$

30. $-14 - 6x = 7$

31. $-12t + 4 = 88$

32. $-16t + 3 = 67$

33. $14y + 15 = -33$

34. $12y + 13 = -15$

35. $32 - 16n = -8$

36. $-41 = 12n - 19$

37. $17x - 41 = -37$

38. $19y - 53 = -47$

39. $29 = -7 - 15x$

40. $49 = -5 - 14x$

For each of the following problems, (a) choose a variable and indicate what it represents in the problem, (b) set up an equation that represents the situation described, and (c) solve the equation.

41. Twelve added to a certain number is 21. What is the number?

42. A certain number added to 14 is 25. Find the number.

43. Nine subtracted from a certain number is 13. Find the number.

44. A certain number subtracted from 32 is 15. What is the number?

45. Suppose that two items cost $43. If one of the items costs $25, what is the cost of the other item?

46. Eight years ago Rosa was 22 years old. Find Rosa's present age.

47. Six years from now, Nora will be 41 years old. What is her present age?

48. Chris bought eight pizzas for a total of $83.60. What was the price per pizza?

49. Chad worked 6 hours Saturday for a total of $43.50. How much per hour did he earn?

50. Jill worked 8 hours Saturday at $7.50 per hour. How much did she earn?

51. If 6 is added to three times a certain number, the result is 24. Find the number.

52. If 2 is subtracted from five times a certain number, the result is 38. Find the number.

53. Nineteen is 4 larger than three times a certain number. Find the number.

54. If nine times a certain number is subtracted from 7, the result is 52. Find the number.

55. Forty-nine is equal to 6 less than five times a certain number. Find the number.

56. Seventy-one is equal to 2 more than three times a certain number. Find the number.

57. If 1 is subtracted from six times a certain number, the result is 47. Find the number.

58. Five less than four times a number equals 31. Find the number.

59. If eight times a certain number is subtracted from 27, the result is 3. Find the number.

60. Twenty is 22 less than six times a certain number. Find the number.

61. A jeweler has priced a diamond ring at $550 (see Figure 3.1). This price represents $50 less than twice the cost of the ring to the jeweler. Find the cost of the ring to the jeweler.

Figure 3.1

62. Todd is on a 1750-calorie-per-day diet plan. This plan permits 650 calories less than twice the number of calories permitted by Lerae's diet plan. How many calories are permitted by Lerae's plan?

63. The length of a rectangular floor is 18 meters (see Figure 3.2). This represents 2 meters less than five times the width of the floor. Find the width of the floor.

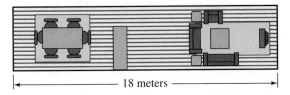

|← 18 meters →|

Figure 3.2

64. An executive is earning $85,000 per year. This represents $15,000 less than twice her salary 4 years ago. Find her salary 4 years ago.

65. In the year 2000, it was estimated that there were 874 million speakers of Mandarin Chinese. This was 149 million less than three times the speakers of the English language. By this estimate how many million speakers of the English language were there in the year 2000?

66. A bill from a limousine company was $510. This included $150 for the service and $80 for each hour of use. Find the number of hours that the limousine was used.

67. Robin paid $454 for a car DVD system. This included $379 for the DVD player and $60 an hour for installation. Find the number of hours it took to install the DVD system.

68. Tracy received a cell phone bill for $136.74. Included in the $136.74 were a charge of $39.99 for the monthly plan and a charge for 215 extra minutes. How much is Tracy being charged for each extra minute?

■ ■ ■ Thoughts into words

69. Give a step-by-step description of how you would solve the equation $17 = -3x + 2$.

70. What does the phrase "declare a variable" mean when it refers to solving a word problem?

71. Suppose that you are helping a friend with his homework and he solves the equation $19 = 14 - x$ like this:

$$19 = 14 - x$$
$$19 + x = 14 - x + x$$
$$19 + x = 14$$
$$19 + x - 19 = 14 - 19$$
$$x = -5$$

The solution set is $\{-5\}$.

Does he have a correct solution set? What would you tell him about his method of solving the equation?

3.3   More on Solving Equations and Problem Solving

Objectives

■   Solve first-degree equations by simplifying both sides and then applying properties of equality.

■   Solve word problems representing several quantities in terms of the same variable.

As equations become more complex, we need additional steps to solve them, so we must organize our work carefully to minimize the chances for error. Let's begin this section with some suggestions for solving equations, and then we will illustrate a *solution format* that is effective.

We can summarize the process of solving first-degree equations of one variable as follows:

STEP 1 Simplify both sides of the equation as much as possible.

STEP 2 Use the addition-subtraction property of equality to isolate a term that contains the variable on one side of the equation and a constant on the other.

STEP 3 Use the multiplication-division property of equality to make the coefficient of the variable 1.

The following examples illustrate this step-by-step process for solving equations. Study them carefully and be sure that you understand each step.

E X A M P L E   1   Solve $5y - 4 + 3y = 12$.

Solution

$$5y - 4 + 3y = 12$$
$$8y - 4 = 12 \qquad \text{Combine similar terms on the left side.}$$
$$8y - 4 + 4 = 12 + 4 \qquad \text{Add 4 to both sides.}$$
$$8y = 16$$
$$\frac{8y}{8} = \frac{16}{8} \qquad \text{Divide both sides by 8.}$$
$$y = 2$$

The solution set is $\{2\}$. You can do the check alone now!   ∎

E X A M P L E   2   Solve $7x - 2 = 3x + 9$.

Solution

Note that both sides of the equation are in simplified form; thus we can begin by using the subtraction property of equality.

$$7x - 2 = 3x + 9$$
$$7x - 2 - 3x = 3x + 9 - 3x \qquad \text{Subtract } 3x \text{ from both sides.}$$
$$4x - 2 = 9$$
$$4x - 2 + 2 = 9 + 2 \qquad \text{Add 2 to both sides.}$$
$$4x = 11$$
$$\frac{4x}{4} = \frac{11}{4} \qquad \text{Divide both sides by 4.}$$
$$x = \frac{11}{4}$$

The solution set is $\left\{\dfrac{11}{4}\right\}$.   ∎

EXAMPLE 3   Solve $5n + 12 = 9n - 16$.

Solution

$$5n + 12 = 9n - 16$$
$$5n + 12 - 9n = 9n - 16 - 9n \qquad \text{Subtract } 9n \text{ from both sides.}$$
$$-4n + 12 = -16$$
$$-4n + 12 - 12 = -16 - 12 \qquad \text{Subtract } 12 \text{ from both sides.}$$
$$-4n = -28$$
$$\frac{-4n}{-4} = \frac{-28}{-4} \qquad \text{Divide both sides by } -4.$$
$$n = 7$$

The solution set is $\{7\}$.   ■

Word Problems

As we expand our skill in solving equations, we also expand our ability to solve word problems. No one definite procedure will ensure success at solving word problems, but the following suggestions can be helpful.

Suggestions for Solving Word Problems

1. Read the problem carefully and make sure that you understand the meanings of all the words. Be especially alert for any technical terms in the statement of the problem.
2. Read the problem a second time (perhaps even a third time) to get an overview of the situation described and to determine the known facts as well as what is to be found.
3. Sketch any figure, diagram, or chart that might be helpful in analyzing the problem.
4. Choose a meaningful variable to represent an unknown quantity in the problem (perhaps *t* if time is an unknown quantity); represent any other unknowns in terms of that variable.
5. Look for a **guideline** that you can use to set up an equation. A guideline might be a formula such as *distance equals rate times time*, or a statement of a relationship, such as *the sum of the two numbers is 28.* A guideline may also be indicated by a figure or diagram that you sketch for a particular problem.
6. Form an equation that contains the variable and that translates the conditions of the guideline from English to algebra.
7. Solve the equation and use the solution to determine all facts requested in the problem.
8. **Check all answers against the original statement of the problem.**

If you decide not to check an answer, at least use the *reasonableness of answer* idea as a partial check. That is, ask yourself, "is this answer reasonable?" For example, if the problem involves two investments that total $10,000, then an answer of $12,000 for one investment is certainly *not reasonable.*

Now let's consider some problems and use these suggestions.

PROBLEM 1   Find two consecutive even numbers whose sum is 74.

Solution

To solve this problem, we must know the meaning of the technical phrase "two consecutive even numbers." Two consecutive even numbers are two even numbers that have one and only one whole number between them. For example, 2 and 4 are consecutive even numbers. Now we can proceed as follows: Let n represent the first even number; then $n + 2$ represents the next even number. Because their sum is 74, we can set up and solve the following equation:

$$n + (n + 2) = 74$$
$$2n + 2 = 74$$
$$2n + 2 - 2 = 74 - 2$$
$$2n = 72$$
$$\frac{2n}{2} = \frac{72}{2}$$
$$n = 36$$

If $n = 36$, then $n + 2 = 38$; thus the numbers are 36 and 38.

Check

To check your answers for Problem 1, determine whether they satisfy the conditions stated in the original problem. Because 36 and 38 are two consecutive even numbers, and $36 + 38 = 74$ (their sum is 74), we know that the answers are correct. ∎

Suggestion 5 in our list of problem-solving suggestions was to "look for a *guideline* that can be used to set up an equation." The guideline may not be explicitly stated in the problem but may instead be implied by the nature of the problem. Consider the next example.

PROBLEM 2   Barry sells bicycles on a salary-plus-commission basis. He receives a monthly salary of $300 and a commission of $15 for each bicycle that he sells. How many bicycles must he sell in a month to have a total monthly salary of $750?

Solution

Let b represent the number of bicycles to be sold in a month. Then $15b$ represents Barry's commission for those bicycles. The *guideline* "fixed salary plus commission equals total monthly salary" generates the following equation:

Fixed salary + Commission = Total monthly salary

$$\$300 \quad + \quad 15b \quad = \quad \$750$$

Let's solve this equation.

$$300 + 15b - 300 = 750 - 300$$
$$15b = 450$$
$$\frac{15b}{15} = \frac{450}{15}$$
$$b = 30$$

He must sell 30 bicycles per month. (Does this number check?)   ■

Geometric Problems

Sometimes the guideline for setting up an equation to solve a problem is based on a geometric relationship. Several basic geometric relationships pertain to angle measure. Let's state three of these relationships and then consider some problems.

1. Two angles for which the sum of their measures is 90° (the symbol ° indicates degrees) are called **complementary angles.**
2. Two angles for which the sum of their measures is 180° are called **supplementary angles.**
3. The sum of the measures of the three angles of a triangle is 180°.

PROBLEM 3

One of two complementary angles is 14° larger than the other. Find the measure of each of the angles.

Solution
If we let a represent the measure of the smaller angle, then $a + 14$ represents the measure of the larger angle. Because they are complementary angles, their sum is 90° and we can proceed as follows:

$$a + a + 14 = 90$$
$$2a + 14 = 90$$
$$2a + 14 - 14 = 90 - 14$$
$$2a = 76$$
$$\frac{2a}{2} = \frac{76}{2}$$
$$a = 38$$

If $a = 38$, then $a + 14 = 52$, and the angles have measures of 38° and 52°.   ■

PROBLEM 4

Find the measures of the three angles of a triangle if the second is three times the first and the third is twice the second.

Solution
If we let a represent the measure of the smallest angle, then $3a$ and $2(3a)$ represent the measures of the other two angles. Therefore we can set up and solve the

following equation:

$$a + 3a + 2(3a) = 180$$
$$a + 3a + 6a = 180$$
$$10a = 180$$
$$\frac{10a}{10} = \frac{180}{10}$$
$$a = 18$$

If $a = 18$, then $3a = 54$ and $2(3a) = 108$, so the angles have measures of $18°$, $54°$, and $108°$. ∎

CONCEPT QUIZ

For Problems 1–8, answer true or false.

1. If n represents a whole number, then $n + 1$ would represent the next consecutive whole number.

2. If n represents an odd whole number, then $n + 1$ would represent the next consecutive odd whole number.

3. If n represents an even whole number, then $n + 2$ would represent the next consecutive even whole number.

4. The sum of the measures of two complementary angles is $90°$.

5. The sum of the measures of two supplementary angles is $360°$.

6. The measure of the three angles in a triangle is $120°$.

7. In checking word problems it is sufficient to check the solution in the equation.

8. For a word problem, the reasonableness of an answer is appropriate as a partial check.

PROBLEM SET 3.3

For Problems 1–32, solve each equation.

1. $2x + 7 + 3x = 32$

2. $3x + 9 + 4x = 30$

3. $7x - 4 - 3x = -36$

4. $8x - 3 - 2x = -45$

5. $3y - 1 + 2y - 3 = 4$

6. $y + 3 + 2y - 4 = 6$

7. $5n - 2 - 8n = 31$

8. $6n - 1 - 10n = 51$

9. $-2n + 1 - 3n + n - 4 = 7$

10. $-n + 7 - 2n + 5n - 3 = -6$

11. $3x + 4 = 2x - 5$

12. $5x - 2 = 4x + 6$

13. $5x - 7 = 6x - 9$

14. $7x - 3 = 8x - 13$

15. $6x + 1 = 3x - 8$

16. $4x - 10 = x + 17$

17. $7y - 3 = 5y + 10$

18. $8y + 4 = 5y - 4$

19. $8n - 2 = 11n - 7$

20. $7n - 10 = 9n - 13$

21. $-2x - 7 = -3x + 10$

22. $-4x + 6 = -5x - 9$

23. $-3x + 5 = -5x - 8$

24. $-4x + 7 = -6x + 4$

25. $-7 - 6x = 9 - 9x$

26. $-10 - 7x = 14 - 12x$

27. $2x - 1 - x = 3x - 5$

28. $3x - 4 - 4x = 5 - 5x + 3x$

29. $5n - 4 - n = -3n - 6 + n$

30. $4x - 3 + 2x = 8x - 3 - x$

31. $-7 - 2n - 6n = 7n - 5n + 12$

32. $-3n + 6 + 5n = 7n - 8n - 9$

Solve each of the following problems by setting up and solving an algebraic equation.

33. The sum of a number plus four times the number is 85. What is the number?

34. A number subtracted from three times the number yields 68. Find the number.

35. Find two consecutive odd numbers whose sum is 72.

36. Find two consecutive even numbers whose sum is 94.

37. Find three consecutive even numbers whose sum is 114.

38. Find three consecutive odd numbers whose sum is 159.

39. Two more than three times a certain number is the same as 4 less than seven times the number. Find the number.

40. One more than five times a certain number is equal to 11 less than nine times the number. What is the number?

41. The sum of a number and five times the number equals 18 less than three times the number. Find the number.

42. One of two supplementary angles is five times as large as the other. Find the measure of each angle.

43. One of two complementary angles is 6 less than twice the other angle. Find the measure of each angle.

44. If two angles are complementary, and the difference of their measures is 62°, find the measure of each angle.

45. If two angles are supplementary, and the larger angle is 20° less than three times the smaller angle, find the measure of each angle.

46. Find the measures of the three angles of a triangle if the largest is 14° less than three times the smallest, and the other angle is 4° more than the smallest.

47. One of the angles of a triangle has a measure of 40°. Find the measures of the other two angles if the difference of their measures is 10°.

48. Jesstan worked as a telemarketer on a salary-plus-commission basis. He was paid a salary of $300 a week and $12 commission for each sale. If his earnings for the week were $960, how many sales did he make?

49. Marci sold an antique vase in an online auction for $69.00. This was $15 less than twice what she paid for it. What price did she pay for the vase?

50. A set of wheels sold in an online auction for $560. This was $35 more than three times the opening bid. How much was the opening bid?

51. Suppose that Bob is paid two times his normal hourly rate for each hour worked in excess of 40 hours in a week. Last week he earned $504 for 48 hours of work. What is his hourly wage?

52. Last week on an algebra test, the highest grade was 9 points less than three times the lowest grade. The sum of the two grades was 135. Find the lowest and highest grades on the test.

53. At a university-sponsored concert, there were three times as many women as men. A total of 600 people attended the concert. How many men and how many women attended?

54. Suppose that a triangular lot is enclosed with 135 yards of fencing (see Figure 3.3). The longest side of the lot is 5 yards more than twice the length of the shortest side. The other side is 10 yards longer than the shortest side. Find the lengths of the three sides of the lot.

Figure 3.3

55. The textbook for a biology class cost $15 more than twice the cost of a used textbook for college algebra. If the cost of the two books together is $129, find the cost of the biology book.

56. A nutrition plan counts grams of fat, carbohydrates, and fiber. The grams of carbohydrates are to be 15 more than twice the grams of fat. The grams of fiber are to be three less than the grams of fat. If the grams of carbohydrate, fat, and fiber must total 48 grams for a dinner meal, how many grams of each would be in the meal?

57. At a local restaurant, $275 in tips is to be shared between the server, bartender, and busboy. The server gets $25 more than three times the amount the busboy receives. The bartender gets $50 more than the amount the busboy receives. How much will the server receive?

■ ■ ■ **Thoughts into words**

58. Give a step-by-step description of how you would solve the equation $3x + 4 = 5x - 2$.

59. Suppose your friend solved the problem "find two consecutive odd integers whose sum is 28" like this:

$$x + x + 1 = 28$$
$$2x = 27$$
$$x = \frac{27}{2} = 13\frac{1}{2}$$

She claims that $13\frac{1}{2}$ will check in the equation.

Where has she gone wrong, and how would you help her?

■ ■ ■ **Further investigations**

60. Solve each of these equations.

a. $7x - 3 = 4x - 3$

b. $-x - 4 + 3x = 2x - 7$

c. $-3x + 9 - 2x = -5x + 9$

d. $5x - 3 = 6x - 7 - x$

e. $7x + 4 = -x + 4 + 8x$

f. $3x - 2 - 5x = 7x - 2 - 5x$

g. $-6x - 8 = 6x + 4$

h. $-8x + 9 = -8x + 5$

Answers to Concept Quiz

1. True **2.** False **3.** True **4.** True **5.** False **6.** False **7.** False **8.** True

3.4   **Equations Involving Parentheses and Fractional Forms**

Objectives

■   Solve first-degree equations that involve the use of the distributive property.

■   Solve first-degree equations that involve fractional forms.

■   Solve first-degree equations that are contradictions.

■   Solve first-degree equations that are identities.

■   Solve word problems where variable quantities are multiplied by a rate.

We will use the distributive property frequently in this section as we expand our techniques for solving equations. Recall that in symbolic form, the distributive property states that $a(b + c) = ab + ac$. The following examples illustrate the use of this property to *remove parentheses*. Pay special attention to the last two examples, which involve a negative number in front of the parentheses.

$$3(x + 2) = \quad 3 \cdot x + 3 \cdot 2 \quad = 3x + 6$$

$$5(y - 3) = \quad 5 \cdot y - 5 \cdot 3 \quad = 5y - 15 \quad [a(b - c) = ab - ac]$$

$$2(4x + 7) = \quad 2(4x) + 2(7) \quad = 8x + 14$$

$$-1(n + 4) = \quad (-1)(n) + (-1)(4) \quad = -n - 4$$

$$-6(x - 2) = \quad (-6)(x) - (-6)(2) \quad = -6x + 12$$

$\downarrow$

Do this step mentally!

It is often necessary to solve equations in which the variable is part of an expression enclosed in parentheses. The distributive property is used to remove the parentheses, and then we proceed in the usual way. Consider the following examples. (Note that when solving an equation, we are beginning to show *only the major steps.*)

EXAMPLE 1   Solve $4(x + 3) = 2(x - 6)$.

Solution

$$4(x + 3) = 2(x - 6)$$
$$4x + 12 = 2x - 12 \qquad \text{Applied distributive property on each side}$$
$$2x + 12 = -12 \qquad \text{Subtracted } 2x \text{ from both sides}$$
$$2x = -24 \qquad \text{Subtracted 12 from both sides}$$
$$x = -12 \qquad \text{Divided both sides by 2}$$

The solution set is $\{-12\}$.   ■

It may be necessary to use the distributive property to remove more than one set of parentheses and then to combine similar terms. Consider the next two examples.

EXAMPLE 2 Solve $6(x - 7) - 2(x - 4) = 13$.

Solution

$$6(x - 7) - 2(x - 4) = 13 \qquad \text{Be careful with this sign!}$$
$$6x - 42 - 2x + 8 = 13 \qquad \text{Distributive property}$$
$$4x - 34 = 13 \qquad \text{Combined similar terms}$$
$$4x = 47 \qquad \text{Added 34 to both sides}$$
$$x = \frac{47}{4} \qquad \text{Divided both sides by 4}$$

The solution set is $\left\{ \dfrac{47}{4} \right\}$.

In a previous section we solved equations such as $x - \dfrac{2}{3} = \dfrac{3}{4}$ by adding $\dfrac{2}{3}$ to both sides. If an equation contains several fractions, then it is usually easier to *clear the equation of all fractions* by multiplying both sides by the least common denominator of all the denominators. Perhaps several examples will clarify this idea.

EXAMPLE 3 Solve $\dfrac{1}{2}x + \dfrac{2}{3} = \dfrac{5}{6}$.

Solution

$$\frac{1}{2}x + \frac{2}{3} = \frac{5}{6}$$

$$6\left(\frac{1}{2}x + \frac{2}{3}\right) = 6\left(\frac{5}{6}\right) \qquad \text{6 is the LCD of 2, 3, and 6.}$$

$$6\left(\frac{1}{2}x\right) + 6\left(\frac{2}{3}\right) = 6\left(\frac{5}{6}\right) \qquad \text{Distributive property}$$

$$3x + 4 = 5 \qquad \begin{array}{l}\text{Note how the equation has been } \textit{cleared} \\ \textit{of all fractions.}\end{array}$$

$$3x = 1$$

$$x = \frac{1}{3}$$

The solution set is $\left\{ \dfrac{1}{3} \right\}$.

EXAMPLE 4   Solve $\dfrac{5n}{6} - \dfrac{1}{4} = \dfrac{3}{8}$.

Solution

$$\frac{5n}{6} - \frac{1}{4} = \frac{3}{8}$$ Remember $\dfrac{5n}{6} = \dfrac{5}{6}n$.

$$24\left(\frac{5n}{6} - \frac{1}{4}\right) = 24\left(\frac{3}{8}\right)$$ 24 is the LCD of 6, 4, and 8.

$$24\left(\frac{5n}{6}\right) - 24\left(\frac{1}{4}\right) = 24\left(\frac{3}{8}\right)$$ Distributive property

$$20n - 6 = 9$$

$$20n = 15$$

$$n = \frac{15}{20} = \frac{3}{4}$$

The solution set is $\left\{\dfrac{3}{4}\right\}$. ■

We use many of the ideas presented in this section to help solve the equations in the next two examples. Study the solutions carefully and be sure that you can supply reasons for each step. It might be helpful to cover up the solutions and try to solve the equations on your own.

EXAMPLE 5   Solve $\dfrac{x + 3}{2} + \dfrac{x + 4}{5} = \dfrac{3}{10}$.

Solution

$$\frac{x + 3}{2} + \frac{x + 4}{5} = \frac{3}{10}$$

$$10\left(\frac{x + 3}{2} + \frac{x + 4}{5}\right) = 10\left(\frac{3}{10}\right)$$ 10 is the LCD of 2, 5, and 10.

$$10\left(\frac{x + 3}{2}\right) + 10\left(\frac{x + 4}{5}\right) = 10\left(\frac{3}{10}\right)$$ Distributive property

$$5(x + 3) + 2(x + 4) = 3$$

$$5x + 15 + 2x + 8 = 3$$

$$7x + 23 = 3$$

$$7x = -20$$

$$x = -\frac{20}{7}$$

The solution set is $\left\{-\dfrac{20}{7}\right\}$. ■

EXAMPLE 6

Solve $\dfrac{x-1}{4} - \dfrac{x-2}{6} = \dfrac{2}{3}$.

Solution

$$\frac{x-1}{4} - \frac{x-2}{6} = \frac{2}{3}$$

$$12\left(\frac{x-1}{4} - \frac{x-2}{6}\right) = 12\left(\frac{2}{3}\right) \qquad \text{12 is the LCD of 4, 6, and 3.}$$

$$12\left(\frac{x-1}{4}\right) - 12\left(\frac{x-2}{6}\right) = 12\left(\frac{2}{3}\right) \qquad \text{Distributive property}$$

$$3(x-1) - 2(x-2) = 8$$

$$3x - 3 - 2x + 4 = 8 \qquad \text{Be careful with this sign!}$$

$$x + 1 = 8$$

$$x = 7$$

The solution set is $\{7\}$. ∎

Contradictions and Identities

All of the equations we have solved thus far are conditional equations. For instance, the equation $3x = 12$ is a true statement under the condition that $x = 4$. Now we will consider two other types of equations — contradictions and identities. When the equation is not true under any condition, then the equation is called a contradiction. The solution set for a contradiction is the empty or null set and is denoted by $\varnothing$. When an equation is true for any permissible value of the variable for which the equation is defined, the equation is called an identity and the solution set for an identity is the set of all real numbers for which the equation is defined. We will denote the set of all real numbers as {all reals}. The following examples show the solutions for these types of equations.

EXAMPLE 7

Solve $4x + 5 = 2(2x - 8)$

Solution

$$4x + 5 = 2(2x - 8)$$

$$4x + 5 = 4x - 16 \qquad \text{Distributive property}$$

$$5 = -16 \qquad \text{Subtracted } 4x \text{ from both sides}$$

The result is a false statement. Therefore the equation is a contradiction. There is no value of x that will make the equation a true statement, and hence the solution set is the empty set, $\varnothing$. ∎

EXAMPLE 8

Solve $5(x + 3) + 2x - 4 = 7x + 11$

Solution

$$5(x + 3) + 2x - 4 = 7x + 11 \qquad \text{Distributive property}$$
$$5x + 15 + 2x - 4 = 7x + 11$$
$$7x + 11 = 7x + 11 \qquad \text{Combined similar terms}$$
$$11 = 11 \qquad \text{Subtracted } 7x \text{ from both sides}$$

The last step gives an equation with no variable terms, but the equation is a true statement. This equation is an *identity* and any real number is a solution. The solution set would be written as {all reals}. ■

Word Problems

We are now ready to solve some word problems using equations of the different types presented in this section. Again, it might be helpful for you to attempt to solve the problems on your own before looking at the book's approach.

PROBLEM 1

Loretta has 19 coins (quarters and nickels) that amount to $2.35. How many coins of each kind does she have?

Solution

Let q represent the number of quarters. Then $19 - q$ represents the number of nickels. We can use the following guideline to help set up an equation:

Value of quarters in cents + Value of nickels in cents = Total value in cents

$$25q \qquad + \qquad 5(19 - q) \qquad = \qquad 235$$

We can solve the equation in this way:

$$25q + 95 - 5q = 235$$
$$20q + 95 = 235$$
$$20q = 140$$
$$q = 7$$

If $q = 7$, then $19 - q = 12$, so she has 7 quarters and 12 nickels. ■

PROBLEM 2

Find a number such that 4 less than two-thirds the number is equal to one-sixth the number.

Solution

Let n represent the number. Then $\frac{2}{3}n - 4$ represents 4 less than two-thirds the number, and $\frac{1}{6}n$ represents one-sixth the number.

$$\frac{2}{3}n - 4 = \frac{1}{6}n$$

$$6\left(\frac{2}{3}n - 4\right) = 6\left(\frac{1}{6}n\right)$$

$$4n - 24 = n$$

$$3n - 24 = 0$$

$$3n = 24$$

$$n = 8$$

The number is 8.

PROBLEM 3

Lance is paid $1\frac{1}{2}$ times his normal hourly rate for each hour he works in excess of 40 hours in a week. Last week he worked 50 hours and earned \$462. What is his normal hourly rate?

Solution

Let x represent his normal hourly rate. Then $\frac{3}{2}x$ represents $1\frac{1}{2}$ times his normal hourly rate. We can use the following guideline to help set up the equation:

Regular wages for first 40 hours + Wages for 10 hours of overtime = Total wages

$$40x \qquad + \qquad 10\left(\frac{3}{2}x\right) \qquad = \qquad 462$$

We get

$$40x + 15x = 462$$

$$55x = 462$$

$$x = 8.40$$

His normal hourly rate is \$8.40.

PROBLEM 4

Find three consecutive whole numbers such that the sum of the first plus twice the second plus three times the third is 134.

Solution

Let n represent the first whole number. Then $n + 1$ represents the second whole number, and $n + 2$ represents the third whole number. We have

$$n + 2(n + 1) + 3(n + 2) = 134$$

$$n + 2n + 2 + 3n + 6 = 134$$

$$6n + 8 = 134$$

$$6n = 126$$

$$n = 21$$

The numbers are 21, 22, and 23.

Keep in mind that the problem-solving suggestions we offered in Section 3.3 simply outline a general algebraic approach to solving problems. You will add to this list throughout this course and in any subsequent mathematics courses that you take. Furthermore, you will be able to pick up additional problem-solving ideas from your instructor and from fellow classmates as you discuss problems in class. Always be on the alert for any ideas that might help you become a better problem solver.

CONCEPT QUIZ

For Problems 1–10, answer true or false.

1. To solve an equation of the form $a(x + b) = 14$, the associative property would be applied to remove the parentheses.

2. Multiplying both sides of an equation by the common denominator of all fractions in the equation clears the equation of all fractions.

3. If Jack has 15 coins (dimes and quarters), and x represents the number of dimes, then $x - 15$ represents the number of quarters.

4. The equation $3(x + 1) = 3x + 3$ has an infinite number of solutions.

5. The equation $2x = 0$ has no solution.

6. The equation $4x + 5 = 4x + 3$ has no solution.

7. The solution set for an equation that is a contradiction is the null set.

8. For a conditional equation, the solution set is the set of all real numbers.

9. When an equation is true for any permissible value of the variable, then the equation is called an identity.

10. When an equation is true for only certain values of the variable, then the equation is called a contradiction.

PROBLEM SET 3.4

For Problems 1–60, solve each equation.

1. $7(x + 2) = 21$

2. $4(x + 4) = 24$

3. $5(x - 3) = 35$

4. $6(x - 2) = 18$

5. $-3(x + 5) = 12$

6. $-5(x - 6) = -15$

7. $4(n - 6) = 5$

8. $3(n + 4) = 7$

9. $6(n + 7) = 8$

10. $8(n - 3) = 12$

11. $-10 = -5(t - 8)$

12. $-16 = -4(t + 7)$

13. $5(x - 4) = 4(x + 6)$

14. $6(x - 4) = 3(2x + 5)$

15. $8(x + 1) = 9(x - 2)$

16. $4(x - 7) = 5(x + 2)$

17. $8(t + 5) = 4(2t + 10)$

18. $7(t - 5) = 5(t + 3)$

19. $2(6t + 1) = 4(3t - 1)$

20. $6(t + 5) = 2(3t + 15)$

21. $-2(x - 6) = -(x - 9)$

22. $-(x + 7) = -2(x + 10)$

23. $-3(t - 4) - 2(t + 4) = 9$

24. $5(t - 4) - 3(t - 2) = 12$

25. $3(n - 10) - 5(n + 12) = -86$

26. $4(n + 9) - 7(n - 8) = 83$

27. $3(x + 1) + 4(2x - 1) = 5(2x + 3)$

28. $4(x - 1) + 5(x + 2) = 3(x - 8)$

29. $-(x + 2) + 2(x - 3) = -2(x - 7)$

30. $-2(x + 6) + 3(3x - 2) = -3(x - 4)$

31. $5(2x - 1) - (2x + 4) = 4(2x + 3) - 21$

32. $3(4x + 1) - 2(2x + 1) = -2(x - 1) - 1$

33. $-(a - 1) - (3a - 2) = 6 + 2(a - 1)$

34. $3(2a - 1) - 2(5a + 1) = 4(3a + 4)$

35. $3(x - 1) + 2(x - 3) = -4(x - 2) + 10(x + 4)$

36. $-2(x - 4) - (3x - 2) = -2 + (-5x + 2)$

37. $3 - 7(x - 1) = 9 - 6(2x + 1)$

38. $8 - 5(2x + 1) = 2 - 6(x - 3)$

39. $\dfrac{3}{4}x - \dfrac{2}{3} = \dfrac{5}{6}$

40. $\dfrac{1}{2}x - \dfrac{4}{3} = -\dfrac{5}{6}$

41. $\dfrac{5}{6}x + \dfrac{1}{4} = -\dfrac{9}{4}$

42. $\dfrac{3}{8}x + \dfrac{1}{6} = -\dfrac{7}{12}$

43. $\dfrac{1}{2}x - \dfrac{3}{5} = \dfrac{3}{4}$

44. $\dfrac{1}{4}x - \dfrac{2}{5} = \dfrac{5}{6}$

45. $\dfrac{n}{3} + \dfrac{5n}{6} = \dfrac{1}{8}$

46. $\dfrac{n}{6} + \dfrac{3n}{8} = \dfrac{5}{12}$

47. $\dfrac{5y}{6} - \dfrac{3}{5} = \dfrac{2y}{3}$

48. $\dfrac{3y}{7} + \dfrac{1}{2} = \dfrac{y}{4}$

49. $\dfrac{h}{6} + \dfrac{h}{8} = 1$

50. $\dfrac{h}{4} + \dfrac{h}{3} = 1$

51. $\dfrac{x + 2}{3} + \dfrac{x + 3}{4} = \dfrac{13}{3}$

52. $\dfrac{x - 1}{4} + \dfrac{x + 2}{5} = \dfrac{39}{20}$

53. $\dfrac{x - 1}{5} - \dfrac{x + 4}{6} = -\dfrac{13}{15}$

54. $\dfrac{x + 1}{7} - \dfrac{x - 3}{5} = \dfrac{4}{5}$

55. $\dfrac{x + 8}{2} - \dfrac{x + 10}{7} = \dfrac{3}{4}$

56. $\dfrac{x + 7}{3} - \dfrac{x + 9}{6} = \dfrac{5}{9}$

57. $\dfrac{x - 2}{8} - 1 = \dfrac{x + 1}{4}$

58. $\dfrac{x - 4}{2} + 3 = \dfrac{x - 2}{4}$

59. $\dfrac{x + 1}{4} = \dfrac{x - 3}{6} + 2$

60. $\dfrac{x + 3}{5} = \dfrac{x - 6}{2} + 1$

Solve each of the following problems by setting up and solving an appropriate algebraic equation.

61. Find two consecutive whole numbers such that the smaller plus four times the larger equals 39.

62. Find two consecutive whole numbers such that the smaller subtracted from five times the larger equals 57.

63. Find three consecutive whole numbers such that twice the sum of the two smallest numbers is 10 more than three times the largest number.

64. Find four consecutive whole numbers such that the sum of the first three equals the fourth number.

65. The sum of two numbers is 17. If twice the smaller is 1 more than the larger, find the numbers.

66. The sum of two numbers is 53. If three times the smaller is 1 less than the larger, find the numbers.

67. Find a number such that 20 more than one-third of the number equals three-fourths of the number.

68. The sum of three-eighths of a number and five-sixths of the same number is 29. Find the number.

69. The difference of two numbers is 6. One-half of the larger number is 5 larger than one-third of the smaller. Find the numbers.

70. The difference of two numbers is 16. Three-fourths of the larger number is 14 larger than one-half of the smaller number. Find the numbers.

71. Suppose that a board 20 feet long is cut into two pieces. Four times the length of the shorter piece is 4 feet less than three times the length of the longer piece. Find the length of each piece.

72. Ellen is paid "time and a half" for each hour over 40 hours worked in a week. Last week she worked 44 hours and earned $391. What is her normal hourly rate?

73. Lucy has 35 coins consisting of nickels and quarters amounting to $5.75. How many coins of each kind does she have?

74. Suppose that Julian has 44 coins consisting of pennies and nickels. If the number of nickels is 2 more than twice the number of pennies, find the number of coins of each kind.

75. Max has a collection of 210 coins consisting of nickels, dimes, and quarters. He has twice as many dimes as nickels and 10 more quarters than dimes. How many coins of each kind does he have?

76. Ginny has a collection of 425 coins consisting of pennies, nickels, and dimes. She has 50 more nickels than pennies and 25 more dimes than nickels. How many coins of each kind does she have?

77. Maida has 18 coins consisting of dimes and quarters amounting to $3.30. How many coins of each kind does she have?

78. Ike has some nickels and dimes amounting to $2.90. The number of dimes is 1 less than twice the number of nickels. How many coins of each kind does he have?

79. Mario has a collection of 22 specimens in his aquarium consisting of crabs, fish, and plants. There are three times as many fish as crabs. There are two more plants than crabs. How many specimens of each kind are in the collection?

80. Tickets for a concert were priced at $8 for students and $10 for nonstudents (see Figure 3.4). There are 1500 tickets sold for a total of $12,500. How many student tickets were sold?

Figure 3.4

81. The supplement of an angle is 30° more than twice its complement. Find the measure of the angle.

82. The sum of the measure of an angle and three times its complement is 202°. Find the measure of the angle.

83. In triangle ABC, the measure of angle A is 2° less than one-fifth the measure of angle C. The measure of angle B is 5° less than one-half the measure of angle C. Find the measures of the three angles of the triangle.

84. If one-fourth of the complement of an angle plus one-fifth of the supplement of the angle equals 36°, find the measure of the angle.

85. The supplement of an angle is 10° less than three times its complement. Find the size of the angle.

86. In triangle ABC, the measure of angle C is eight times the measure of angle A, and the measure of angle B is 10° more than the measure of angle C. Find the measure of each angle of the triangle.

■ ■ ■ **Thoughts into words**

87. Discuss how you would solve the equation

$3(x - 2) - 5(x + 3) = -4(x + 9)$

88. Why must potential answers to word problems be checked back in the original statement of the problem?

89. Consider these two solutions:

$$3(x + 2) = 9 \qquad\qquad 3(x - 4) = 7$$

$$\frac{3(x + 2)}{3} = \frac{9}{3} \qquad\qquad \frac{3(x - 4)}{3} = \frac{7}{3}$$

$$x + 2 = 3 \qquad\qquad x - 4 = \frac{7}{3}$$

$$x = 1 \qquad\qquad x = \frac{19}{3}$$

Are both of these solutions correct? Comment on the effectiveness of the approaches.

90. Make up an equation whose solution set is the null set. Explain why the solution set is null.

91. Make up an equation whose solution set is the set of all real numbers. Explain why the solution set is all real numbers.

■ ■ ■ **Further investigations**

92. Solve each of the following equations.

a. $-2(x - 1) = -2x + 2$

b. $3(x + 4) = 3x - 4$

c. $5(x - 1) = -5x - 5$

d. $\dfrac{x - 3}{3} + 4 = 3$

e. $\dfrac{x + 2}{3} + 1 = \dfrac{x - 2}{3}$

f. $\dfrac{x - 1}{5} - 2 = \dfrac{x - 11}{5}$

g. $4(x - 2) - 2(x + 3) = 2(x + 6)$

h. $5(x + 3) - 3(x - 5) = 2(x + 15)$

i. $7(x - 1) + 4(x - 2) = 15(x - 1)$

93. Find three consecutive integers such that the sum of the smallest integer and the largest integer is equal to twice the middle integer.

3.5   Inequalities

Objectives

■   Solve first-degree inequalities.

■   Write the solution set of an inequality in set-builder notation or interval notation.

■   Graph the solution set of an inequality.

Just as we use the symbol = to represent *is equal to*, we also use the symbols < and > to represent *is less than* and *is greater than,* respectively. The following are examples of **statements of inequality.** Note that the first four are true statements, and the last two are false.

$$6 + 4 > 7$$

$$8 - 2 < 14$$

$4 \cdot 8 > 4 \cdot 6$

$5 \cdot 2 < 5 \cdot 7$

$5 + 8 > 19$

$9 - 2 < 3$

Algebraic inequalities contain one or more variables. Here are some examples of algebraic inequalities:

$x + 3 > 4$

$2x - 1 < 6$

$x^2 + 2x - 1 > 0$

$2x + 3y < 7$

$7ab < 9$

An algebraic inequality such as $x + 1 > 2$ is neither true nor false as it stands; it is called an **open sentence.** Each time a number is substituted for x, the algebraic inequality $x + 1 > 2$ becomes a numerical statement that is either true or false. For example, if $x = 0$, then $x + 1 > 2$ becomes $0 + 1 > 2$, which is false. If $x = 2$, then $x + 1 > 2$ becomes $2 + 1 > 2$, which is true. **Solving an inequality** refers to the process of finding the numbers that make an algebraic inequality a true numerical statement. We say that such numbers, which are called the **solutions of the inequality,** *satisfy* the inequality. The set of all solutions of an inequality is called its **solution set.** We often state solution sets for inequalities with **set builder notation.** For example, the solution set for $x + 1 > 2$ is the set of real numbers greater than 1, expressed as $\{x | x > 1\}$. The set builder notation $\{x | x > 1\}$ is read as "the set of all x such that x is greater than 1." We sometimes graph solution sets for inequalities on a number line; the solution set for $\{x | x > 1\}$ is pictured in Figure 3.5.

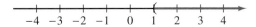

Figure 3.5

The left-hand parenthesis at 1 indicates that 1 is *not* a solution, and the red part of the line to the right of 1 indicates that all real numbers greater than 1 are solutions. We refer to the red portion of the number line as the *graph* of the solution set $\{x | x > 1\}$.

It is also convenient to express solution sets of inequalities using **interval notation.** The solution set $\{x | x > 6\}$ is written as $(6, \infty)$ using interval notation. In interval notation parentheses are used to indicate exclusion of the endpoint. The $>$ and $<$ symbols in inequalities also indicate the exclusion of the endpoint. So when the inequality has a $>$ or $<$ symbol, the interval notation uses a parenthesis. This is consistent with the use of parentheses on the number line.

In this same example, $\{x | x > 6\}$, the solution set has no upper endpoint, so the infinity symbol, ∞, is used to indicate that the interval continues indefinitely. The solution set for $\{x | x < 3\}$ is written as $(-\infty, 3)$ in interval notation. Here the solution set has no lower endpoint, so a negative sign precedes the infinity symbol because the interval is extending indefinitely in the opposite direction. The infinity symbol always has a parenthesis in interval notation because there is no actual endpoint to include.

The solution set $\{x|x \geq 5\}$ is written as $[5, \infty)$ using interval notation. In interval notation square brackets are used to indicate inclusion of the endpoint. The $\geq$ and $\leq$ symbols in inequalities also indicate the inclusion of the endpoint. So when the inequality has a $\geq$ or $\leq$ symbol, the interval notation uses a square bracket. Again the use of a bracket in interval notation is consistent with the use of a bracket on the number line.

The examples in the table (Figure 3.6) below contain some simple algebraic inequalities, their solution sets, graphs of the solution sets, and the solution sets written in interval notation. Look them over very carefully to be sure you understand the symbols.

Algebraic inequality	Solution set	Graph of solution set	Interval notation
$x < 2$	$\{x|x < 2\}$	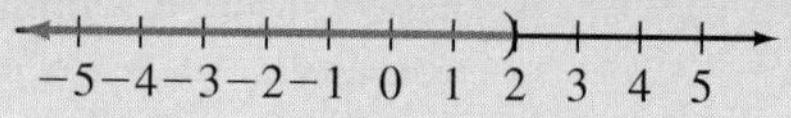 $-5-4-3-2-1\ 0\ 1\ 2\ 3\ 4\ 5$	$(-\infty, 2)$
$x > -1$	$\{x|x > -1\}$	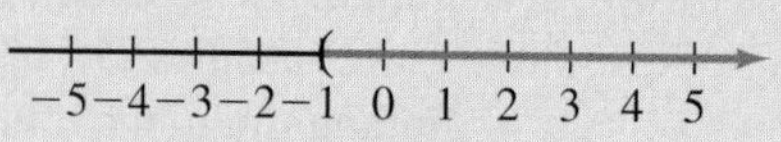 $-5-4-3-2-1\ 0\ 1\ 2\ 3\ 4\ 5$	$(-1, \infty)$
$3 < x$	$\{x|x > 3\}$	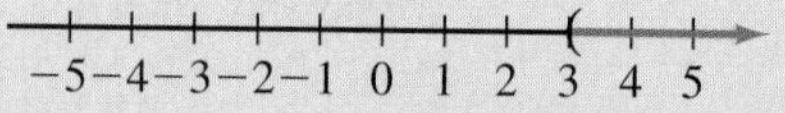 $-5-4-3-2-1\ 0\ 1\ 2\ 3\ 4\ 5$	$(3, \infty)$
$x \geq 1$ ($\geq$ is read "greater than or equal to")	$\{x|x \geq 1\}$	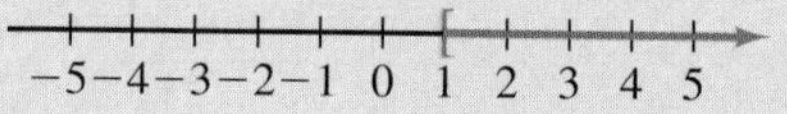 $-5-4-3-2-1\ 0\ 1\ 2\ 3\ 4\ 5$	$[1, \infty)$
$x \leq 2$ ($\leq$ is read "less than or equal to")	$\{x|x \leq 2\}$	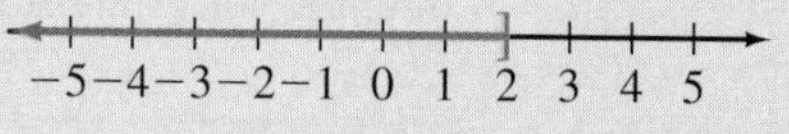 $-5-4-3-2-1\ 0\ 1\ 2\ 3\ 4\ 5$	$(-\infty, 2]$
$1 \geq x$	$\{x|x \leq 1\}$	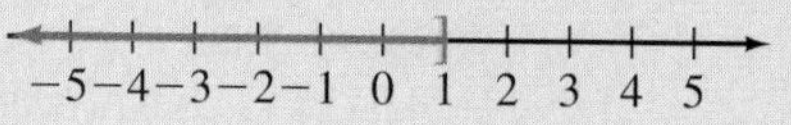 $-5-4-3-2-1\ 0\ 1\ 2\ 3\ 4\ 5$	$(-\infty, 1]$

Figure 3.6

The general process for solving inequalities closely parallels that for solving equations. We continue to replace the given inequality with equivalent but simpler inequalities. For example,

$$2x + 1 > 9 \qquad \textbf{(1)}$$
$$2x > 8 \qquad \textbf{(2)}$$
$$x > 4 \qquad \textbf{(3)}$$

are all equivalent inequalities; that is, they have the same solutions. Thus to solve (1) we can find the solutions of (3), which are obviously all numbers greater than 4. The exact procedure for simplifying inequalities is based primarily on two properties. The first of these is the **addition-subtraction property of inequality.**

> **PROPERTY 3.4**   *Addition-Subtraction Property of Inequality*
>
> For all real numbers a, b, and c,
>
> **1.** $a > b$ if and only if $a + c > b + c$.
> **2.** $a > b$ if and only if $a - c > b - c$.

Property 3.4 states that any number can be added to or subtracted from both sides of an inequality, and the result is an equivalent inequality. The property is stated in terms of $>$, but analogous properties exist for $<$, $\geq$, and $\leq$. Consider the use of this property in the next three examples.

E X A M P L E   1   Solve $x - 3 > -1$ and graph the solutions.

Solution

$$x - 3 > -1$$
$$x - 3 + 3 > -1 + 3 \quad \text{Add 3 to both sides.}$$
$$x > 2$$

The solution set is $\{x \mid x > 2\}$, and it can be graphed as shown in Figure 3.7. The solution written in interval notation is $(2, \infty)$.

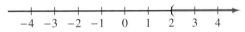

Figure 3.7

E X A M P L E   2   Solve $x + 4 \leq 5$ and graph the solutions.

Solution

$$x + 4 \leq 5$$
$$x + 4 - 4 \leq 5 - 4 \quad \text{Subtract 4 from both sides.}$$
$$x \leq 1$$

The solution set is $\{x \mid x \leq 1\}$, and it can be graphed as shown in Figure 3.8. The solution written in interval notation is $(-\infty, 1]$.

Figure 3.8

E X A M P L E   3   Solve $5 > 6 + x$ and graph the solutions.

Solution

$$5 > 6 + x$$
$$5 - 6 > 6 + x - 6 \quad \text{Subtract 6 from both sides.}$$
$$-1 > x$$

Because $-1 > x$ is equivalent to $x < -1$, the solution set is $\{x|x < -1\}$. It can be graphed as shown in Figure 3.9. The solution written in interval notation is $(-\infty, -1)$.

Figure 3.9

Now let's look at some numerical examples to see what happens when both sides of an inequality are multiplied or divided by some number.

$$4 > 3 \quad \rightarrow \quad 5(4) > 5(3) \quad \rightarrow \quad 20 > 15$$

$$-2 > -3 \quad \rightarrow \quad 4(-2) > 4(-3) \quad \rightarrow \quad -8 > -12$$

$$6 > 4 \quad \rightarrow \quad \frac{6}{2} > \frac{4}{2} \quad \rightarrow \quad 3 > 2$$

$$8 > -2 \quad \rightarrow \quad \frac{8}{4} > \frac{-2}{4} \quad \rightarrow \quad 2 > -\frac{1}{2}$$

Note that multiplying or dividing both sides of an inequality by a positive number produces an inequality of the same sense. This means that if the original inequality is *greater than*, then the new inequality is *greater than*, and if the original is *less than*, then the resulting inequality is *less than*.

Now note what happens when we multiply or divide both sides by a negative number:

$$3 < 5 \quad \rightarrow \quad -2(3) > -2(5) \quad \rightarrow \quad -6 > -10$$

$$-4 < 1 \quad \rightarrow \quad -5(-4) > -5(1) \quad \rightarrow \quad 20 > -5$$

$$14 > 2 \quad \rightarrow \quad \frac{14}{-2} < \frac{2}{-2} \quad \rightarrow \quad -7 < -1$$

$$-3 > -6 \quad \rightarrow \quad \frac{-3}{-3} < \frac{-6}{-3} \quad \rightarrow \quad 1 < 2$$

Multiplying or dividing both sides of an inequality *by a negative number reverses the sense of the inequality.* Property 3.5 summarizes these ideas.

PROPERTY 3.5 *Multiplication-Division Property of Inequality*

(a) For all real numbers, a, b, and c, with $c > 0$,
   1. $a > b$ if and only if $ac > bc$

   2. $a > b$ if and only if $\dfrac{a}{c} > \dfrac{b}{c}$

(b) For all real numbers, a, b, and c, with $c < 0$,
   1. $a > b$ if and only if $ac < bc$

   2. $a > b$ if and only if $\dfrac{a}{c} < \dfrac{b}{c}$

Similar properties hold when each inequality is reversed or when $>$ is replaced with $\geq$ and $<$ is replaced with $\leq$. For example, if $a \leq b$ and $c < 0$, then $ac \geq bc$ and $\dfrac{a}{c} \geq \dfrac{b}{c}$.

Observe the use of Property 3.5 in the next three examples.

EXAMPLE 4

Solve $2x > 4$.

Solution

$$2x > 4$$

$$\frac{2x}{2} > \frac{4}{2} \qquad \text{Divide both sides by 2.}$$

$$x > 2$$

The solution set is $\{x | x > 2\}$ or $(2, \infty)$ in interval notation. ∎

EXAMPLE 5

Solve $\dfrac{3}{4}x \leq \dfrac{1}{5}$.

Solution

$$\frac{3}{4}x \leq \frac{1}{5}$$

$$\frac{4}{3}\left(\frac{3}{4}x\right) \leq \frac{4}{3}\left(\frac{1}{5}\right) \qquad \text{Multiply both sides by } \frac{4}{3}.$$

$$x \leq \frac{4}{15}$$

The solution set is $\left\{x | x \leq \dfrac{4}{15}\right\}$ or $\left(-\infty, \dfrac{4}{15}\right]$ in interval notation. ∎

EXAMPLE 6

Solve $-3x > 9$.

Solution

$$-3x > 9$$

$$\frac{-3x}{-3} < \frac{9}{-3} \qquad \text{Divide both sides by } -3, \text{ which reverses the inequality.}$$

$$x < -3$$

The solution set is $\{x | x < -3\}$ or $(-\infty, -3)$ in interval notation. ∎

As we mentioned earlier, many of the same techniques used to solve equations may be used to solve inequalities. However, we must be extremely careful when we apply Property 3.5. Study the following examples and note the similarities between solving equations and solving inequalities.

E X A M P L E   7

Solve $4x - 3 > 9$.

Solution

$$4x - 3 > 9$$
$$4x - 3 + 3 > 9 + 3 \quad \text{Add 3 to both sides.}$$
$$4x > 12$$
$$\frac{4x}{4} > \frac{12}{4} \quad \text{Divide both sides by 4.}$$
$$x > 3$$

The solution set is $\{x \mid x > 3\}$ or $(3, \infty)$ in interval notation. ■

E X A M P L E   8

Solve $-3n + 5 < 11$.

Solution

$$-3n + 5 < 11$$
$$-3n + 5 - 5 < 11 - 5 \quad \text{Subtract 5 from both sides.}$$
$$-3n < 6$$
$$\frac{-3n}{-3} > \frac{6}{-3} \quad \text{Divide both sides by } -3, \text{ which reverses} \\ \text{the inequality.}$$
$$n > -2$$

The solution set is $\{n \mid n > -2\}$ or $(-2, \infty)$ in interval notation. ■

Checking the solutions for an inequality presents a problem. Obviously we cannot check all of the infinitely many solutions for a particular inequality. However, by checking at least one solution, especially when the multiplication-division property was used, we might catch the common mistake of forgetting to reverse the sense of the inequality. In Example 8 we are claiming that all numbers greater than -2 will satisfy the original inequality. Let's check one such number in the original inequality — say, -1.

$$-3n + 5 < 11$$
$$-3(-1) + 5 \overset{?}{<} 11$$
$$3 + 5 \overset{?}{<} 11$$
$$8 < 11$$

Thus -1 satisfies the original inequality. If we had forgotten to reverse the sense of the inequality when we divided both sides by -3, our answer would have been $n < -2$, and the check would have detected the error.

CONCEPT QUIZ

For Problems 1–10, answer true or false.

1. Numerical statements of inequality are always true.

2. The algebraic statement $x + 4 > 6$ is called an open sentence.

3. The algebraic inequality $2x > 10$ has one solution.

4. The algebraic inequality $x < 3$ has an infinite number of solutions.

5. The set-building notation $\{x|x < -5\}$ is read "the set of variables that are particular to $x < -5$."

6. When graphing the solution set of an inequality, a square bracket is used to include the endpoint.

7. The solution set of the inequality $x \geq 4$ is written $(4, \infty)$.

8. The solution set of the inequality $x < -5$ is written $(-\infty, -5)$.

9. When multiplying both sides of an inequality by a negative number, the sense of the inequality stays the same.

10. When adding a negative number to both sides of an inequality, the sense of the inequality stays the same.

PROBLEM SET 3.5

For Problems 1–10, determine whether each numerical inequality is *true* or *false*.

1. $2(3) - 4(5) < 5(3) - 2(-1) + 4$

2. $5 + 6(-3) - 8(-4) > 17$

3. $\dfrac{2}{3} - \dfrac{3}{4} + \dfrac{1}{6} > \dfrac{1}{5} + \dfrac{3}{4} - \dfrac{7}{10}$

4. $\dfrac{1}{2} + \dfrac{1}{3} < \dfrac{1}{3} + \dfrac{1}{4}$

5. $\left(-\dfrac{1}{2}\right)\left(\dfrac{4}{9}\right) > \left(\dfrac{3}{5}\right)\left(-\dfrac{1}{3}\right)$

6. $\left(\dfrac{5}{6}\right)\left(\dfrac{8}{12}\right) < \left(\dfrac{3}{7}\right)\left(\dfrac{14}{15}\right)$

7. $\dfrac{3}{4} + \dfrac{2}{3} \div \dfrac{1}{5} > \dfrac{2}{3} + \dfrac{1}{2} \div \dfrac{3}{4}$

8. $1.9 - 2.6 - 3.4 < 2.5 - 1.6 - 4.2$

9. $0.16 + 0.34 > 0.23 + 0.17$

10. $(0.6)(1.4) > (0.9)(1.2)$

For Problems 11–22, state the solution set and graph it on a number line.

11. $x > -2$ 12. $x > -4$

13. $x \leq 3$ 14. $x \leq 0$

15. $2 < x$ 16. $-3 \leq x$

17. $-2 \geq x$ 18. $1 > x$

19. $-x > 1$ 20. $-x < 2$

21. $-2 < -x$ 22. $-1 > -x$

For Problems 23–60, solve each inequality.

41. $4x - 3 \leq 21$

42. $5x - 2 \geq 28$

23. $x + 6 < -14$

24. $x + 7 > -15$

43. $-2x - 1 \geq 41$

44. $-3x - 1 \leq 35$

25. $x - 4 \geq -13$

26. $x - 3 \leq -12$

45. $6x + 2 < 18$

46. $8x + 3 > 25$

27. $4x > 36$

28. $3x < 51$

47. $3 > 4x - 2$

48. $7 < 6x - 3$

29. $6x < 20$

30. $8x > 28$

49. $-2 < -3x + 1$

50. $-6 > -2x + 4$

31. $-5x > 40$

32. $-4x < 24$

51. $-38 \geq -9t - 2$

52. $36 \geq -7t + 1$

33. $-7n \leq -56$

34. $-9n \geq -63$

53. $5x - 4 - 3x > 24$

54. $7x - 8 - 5x < 38$

35. $48 > -14n$

36. $36 < -8n$

55. $4x + 2 - 6x < -1$

56. $6x + 3 - 8x > -3$

37. $16 < 9 + n$

38. $19 > 27 + n$

57. $-5 \geq 3t - 4 - 7t$

58. $6 \leq 4t - 7t - 10$

39. $3x + 2 > 17$

40. $2x + 5 < 19$

59. $-x - 4 - 3x > 5$

60. $-3 - x - 3x < 10$

■ ■ ▨ **Thoughts into words**

61. Do the "greater than" and "less than" relations possess the symmetric property? Explain your answer.

62. Is the solution set for $x < 3$ the same as that for $3 > x$? Explain your answer.

63. How would you convince someone that it is necessary to reverse the sense of the inequality when multiplying both sides of an inequality by a negative number?

■ ■ ▨ **Further investigations**

Solve each of the following inequalities.

64. $x + 3 < x - 4$

65. $x - 4 < x + 6$

66. $2x + 4 > 2x - 7$

67. $5x + 2 > 5x + 7$

68. $3x - 4 - 3x > 6$

69. $-2x + 7 + 2x > 1$

70. $-5 \leq -4x - 1 + 4x$

71. $-7 \geq 5x - 2 - 5x$

Answers to Concept Quiz

1. False   **2.** True   **3.** False   **4.** True   **5.** False   **6.** True   **7.** False   **8.** True   **9.** False   **10.** True

3.6 Inequalities, Compound Inequalities, and Problem Solving

Objectives

■ Solve inequalities that involve the use of the distributive property.

■ Solve inequalities that involve fractional forms.

■ Determine the solution set for compound inequality statements.

■ Solve word problems that translate into inequality statements.

Let's begin this section by solving three inequalities with the same basic steps we used with equations. Again, be careful when applying the multiplication-division property of inequality.

EXAMPLE 1   Solve $5x + 8 \leq 3x - 10$.

Solution

$$5x + 8 \leq 3x - 10$$
$$5x + 8 - 3x \leq 3x - 10 - 3x \qquad \text{Subtract } 3x \text{ from both sides.}$$
$$2x + 8 \leq -10$$
$$2x + 8 - 8 \leq -10 - 8 \qquad \text{Subtract 8 from both sides.}$$
$$2x \leq -18$$
$$\frac{2x}{2} \leq \frac{-18}{2} \qquad \text{Divide both sides by 2.}$$
$$x \leq -9$$

The solution set is $\{x | x \leq -9\}$ or $(-\infty, -9]$.   ■

EXAMPLE 2   Solve $4(x + 3) + 3(x - 4) \geq 2(x - 1)$.

Solution

$$4(x + 3) + 3(x - 4) \geq 2(x - 1)$$
$$4x + 12 + 3x - 12 \geq 2x - 2 \qquad \text{Distributive property}$$
$$7x \geq 2x - 2 \qquad \text{Combine similar terms.}$$
$$7x - 2x \geq 2x - 2 - 2x \qquad \text{Subtract } 2x \text{ from both sides.}$$
$$5x \geq -2$$
$$\frac{5x}{5} \geq \frac{-2}{5} \qquad \text{Divide both sides by 5.}$$
$$x \geq -\frac{2}{5}$$

The solution set is $\left\{ x \middle| x \geq -\frac{2}{5} \right\}$ or $\left[-\frac{2}{5}, \infty \right)$.   ■

EXAMPLE 3 Solve $-\dfrac{3}{2}n + \dfrac{1}{6}n < \dfrac{3}{4}$.

Solution

$$-\frac{3}{2}n + \frac{1}{6}n < \frac{3}{4}$$

$$12\left(-\frac{3}{2}n + \frac{1}{6}n\right) < 12\left(\frac{3}{4}\right) \quad \text{Multiply both sides by 12, the LCD of all denominators.}$$

$$12\left(-\frac{3}{2}n\right) + 12\left(\frac{1}{6}n\right) < 12\left(\frac{3}{4}\right) \quad \text{Distributive property}$$

$$-18n + 2n < 9$$

$$-16n < 9$$

$$\frac{-16n}{-16} > \frac{9}{-16} \quad \text{Divide both sides by } -16, \text{ which reverses the inequality.}$$

$$n > -\frac{9}{16}$$

The solution set is $\left\{ n \mid n > -\dfrac{9}{16} \right\}$ or $\left(-\dfrac{9}{16}, \infty\right)$. ■

In Example 3 we are claiming that all numbers greater than $-\dfrac{9}{16}$ will satisfy the original inequality. Let's check one number — say, 0.

$$-\frac{3}{2}n + \frac{1}{6}n < \frac{3}{4}$$

$$-\frac{3}{2}(0) + \frac{1}{6}(0) \overset{?}{<} \frac{3}{4}$$

$$0 < \frac{3}{4}$$

Therefore, 0 satisfies the original inequality. If we had forgotten to reverse the inequality sign when we divided both sides by -16, then our answer would have been $n < -\dfrac{9}{16}$, and the check would have detected the error.

Compound Statements

The words "and" and "or" are used in mathematics to form compound statements. We use "and" and "or" to join two inequalities to form a compound inequality.

Consider the compound inequality

$$x > 2 \quad \text{and} \quad x < 5$$

For the solution set, we must find values of x that make both inequalities true statements. The solution set of a compound inequality formed by the word "and" is the **intersection** of the solution sets of the two inequalities. The intersection of two sets, denoted by ∩, contains the elements that are common to both sets. For example, if $A = \{1, 2, 3, 4, 5, 6\}$ and $B = \{0, 2, 4, 6, 8, 10\}$, then $A \cap B = \{2, 4, 6\}$. So to find the solution set of the compound inequality $x > 2$ and $x < 5$, we find the solution set for each inequality and then determine the solutions that are common to both solution sets.

EXAMPLE 4

Graph the solution set for the compound inequality $x > 2$ and $x < 5$, and write the solution set in interval notation.

Solution

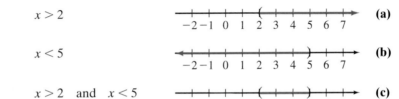

$x > 2$ (a)

$x < 5$ (b)

$x > 2$ and $x < 5$ (c)

Figure 3.10

Thus all numbers greater than 2 and less than 5 are included in the solution set $\{x|2 < x < 5\}$, and the graph is shown in Figure 3.10(c). In interval notation the solution set is $(2, 5)$.

EXAMPLE 5

Graph the solution set for the compound inequality $x \leq 1$ and $x \leq 4$, and write the solution set in interval notation.

Solution

$x \leq 1$ (a)

$x \leq 4$ (b)

$x \leq 1$ and $x \leq 4$ (c)

Figure 3.11

The intersection of the two solution sets is $x \leq 1$. The solution set $\{x|x \leq 1\}$ contains all the numbers that are less than or equal to 1, and the graph is shown in Figure 3.11(c). In interval notation the solution set is $(-\infty, 1]$.

The solution set of a compound inequality formed by the word "or" is the **union** of the solution sets of the two inequalities. The union of two sets, denoted by ∪, contains all the elements in both sets. For example, if $A = \{0, 1, 2\}$ and $B = \{1, 2, 3, 4\}$, then $A \cup B = \{0, 1, 2, 3, 4\}$. Note that even though 1 and 2 are in both set A and set B, there is no need to write them twice in $A \cup B$.

To find the solution set of the compound inequality

$$x > 1 \qquad \text{or} \qquad x > 3$$

we find the solution set for each inequality and then take all the values that satisfy either inequality or both.

E X A M P L E   6

Graph the solution set for $x > 1$ or $x > 3$ and write the solution in interval notation.

Solution

$x > 1$ **(a)**

$x > 3$ **(b)**

$x > 1$ or $x > 3$ **(c)**

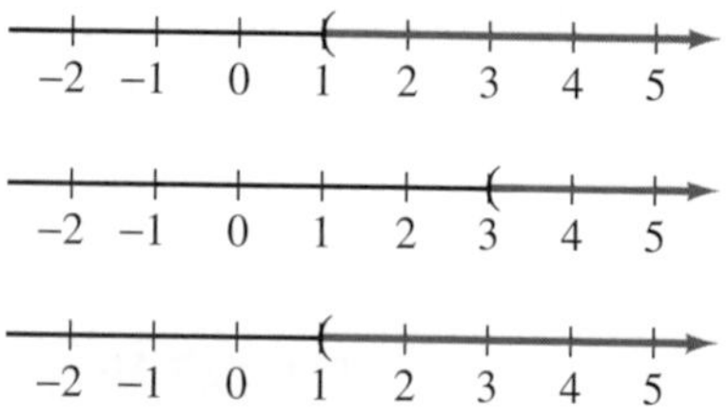

Figure 3.12

Thus all numbers greater than 1 are included in the solution set $\{x \mid x > 1\}$, and the graph is shown in Figure 3.12(c). The solution set is written as $(1, \infty)$ in interval notation. ∎

E X A M P L E   7

Graph the solution set for $x \leq 0$ or $x \geq 2$, and write the solution in interval notation.

Solution

$x \leq 0$ **(a)**

$x \geq 2$ **(b)**

$x \leq 0$ or $x \geq 2$ **(c)**

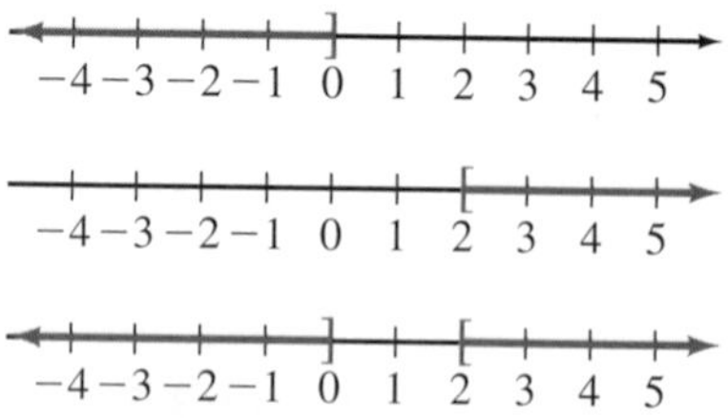

Figure 3.13

Thus all numbers less than or equal to 0 and all numbers greater than or equal to 2 are included in the solution set $\{x \mid x \leq 0 \text{ or } x \geq 2\}$, and the graph is shown in Figure 3.13(c). Since the solution set contains two intervals that are not continuous, a ∪ symbol is used in the interval notation. The solution set is written as $(-\infty, 0] \cup [2, \infty)$ in interval notation. ■

Back to Problem Solving

Let's consider some word problems that translate into inequality statements. The *suggestions for solving word problems* in Section 3.3 apply here, except that the situation described in a problem will translate into an inequality instead of an equation.

PROBLEM 1

Ashley had scores of 95, 82, 93, and 84 on her first four exams of the semester. What score must she get on the fifth exam to have an average of 90 or higher for the five exams?

Solution

Let s represent the score needed on the fifth exam. Because we find the average by adding all five scores and dividing by 5 (the number of exams), we can solve this inequality:

$$\frac{95 + 82 + 93 + 84 + s}{5} \geq 90$$

We use the following steps:

$$\frac{354 + s}{5} \geq 90 \qquad \text{Simplify numerator of left side.}$$

$$5\left(\frac{354 + s}{5}\right) \geq 5(90) \qquad \text{Multiply both sides by 5.}$$

$$354 + s \geq 450$$

$$354 + s - 354 \geq 450 - 354 \qquad \text{Subtract 354 from both sides.}$$

$$s \geq 96$$

She must receive a score of 96 or higher on the fifth exam. ■

PROBLEM 2

The Cubs have won 40 baseball games and have lost 62 games. They have 60 more games to play. To win more than 50% of all their games, how many of the remaining 60 games must they win?

Solution

Let w represent the number of games they must win out of the 60 games remaining. Because they are playing a total of $40 + 62 + 60 = 162$ games, to win more than 50% of their games, they will have to win more than 81 games. Thus we have the inequality

$$w + 40 > 81$$

Solving this yields

$$w > 41$$

They need to win at least 42 of the remaining 60 games.   ■

CONCEPT QUIZ

For Problems 1–5, answer true or false.

1. The solution set of a compound inequality formed by the word "and" is an intersection of the solution sets of the two inequalities.

2. The solution set of a compound inequality formed by the words "and" or "or" is a union of the solution sets of the two inequalities.

3. The intersection of two sets contains the elements that are common to both sets.

4. The union of two sets contains all the elements in both sets.

5. The intersection of set A and set B is denoted by $A \cap B$.

For Problems 6–10, match the compound statement with the graph of its solution set (Figure 3.14).

6. $x > 4$ or $x < -1$

7. $x > 4$ and $x > -1$

8. $x > 4$ or $x > -1$

9. $x \leq 4$ and $x \geq -1$

10. $x > 4$ or $x \geq -1$

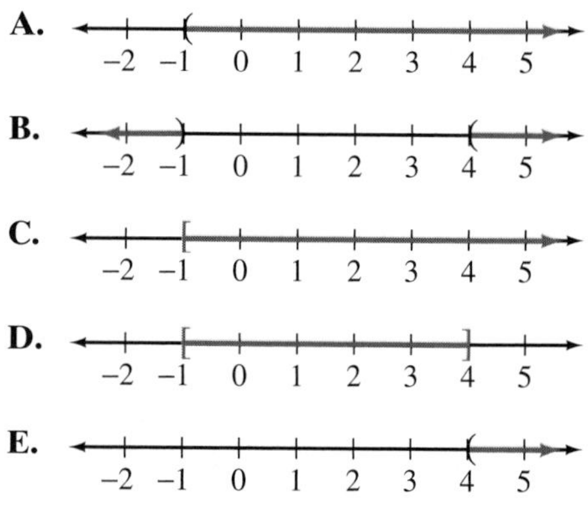

A.

B.

C.

D.

E.

Figure 3.14

PROBLEM SET 3.6

For Problems 1–50, solve each inequality.

1. $3x + 4 > x + 8$

2. $5x + 3 < 3x + 11$

3. $7x - 2 < 3x - 6$

4. $8x - 1 > 4x - 21$

5. $6x + 7 > 3x - 3$

6. $7x + 5 < 4x - 12$

7. $5n - 2 \leq 6n + 9$

8. $4n - 3 \geq 5n + 6$

9. $2t + 9 \geq 4t - 13$

10. $6t + 14 \leq 8t - 16$

11. $-3x - 4 < 2x + 7$

12. $-x - 2 > 3x - 7$

13. $-4x + 6 > -2x + 1$

14. $-6x + 8 < -4x + 5$

15. $5(x - 2) \leq 30$

16. $4(x + 1) \geq 16$

17. $2(n + 3) > 9$

18. $3(n - 2) < 7$

19. $-3(y - 1) < 12$

20. $-2(y + 4) > 18$

21. $-2(x + 6) > -17$

22. $-3(x - 5) < -14$

23. $3(x - 2) < 2(x + 1)$

24. $5(x + 3) > 4(x - 2)$

25. $4(x + 3) > 6(x - 5)$

26. $6(x - 1) < 8(x + 5)$

27. $3(x - 4) + 2(x + 3) < 24$

28. $2(x + 1) + 3(x + 2) > -12$

29. $5(n + 1) - 3(n - 1) > -9$

30. $4(n - 5) - 2(n - 1) < 13$

31. $\dfrac{1}{2}n - \dfrac{2}{3}n \geq -7$

32. $\dfrac{3}{4}n + \dfrac{1}{6}n \leq 1$

33. $\dfrac{3}{4}n - \dfrac{5}{6}n < \dfrac{3}{8}$

34. $\dfrac{2}{3}n - \dfrac{1}{2}n > \dfrac{1}{4}$

35. $\dfrac{3x}{5} - \dfrac{2}{3} > \dfrac{x}{10}$

36. $\dfrac{5x}{4} + \dfrac{3}{8} < \dfrac{7x}{12}$

37. $n \geq 3.4 + 0.15n$

38. $x \geq 2.1 + 0.3x$

39. $0.09t + 0.1(t + 200) > 77$

40. $0.07t + 0.08(t + 100) > 38$

41. $0.06x + 0.08(250 - x) \geq 19$

42. $0.08x + 0.09(2x) \leq 130$

43. $\dfrac{x - 1}{2} + \dfrac{x + 3}{5} > \dfrac{1}{10}$

44. $\dfrac{x + 3}{4} + \dfrac{x - 5}{7} < \dfrac{1}{28}$

45. $\dfrac{x + 2}{6} - \dfrac{x + 1}{5} < -2$

46. $\dfrac{x - 6}{8} - \dfrac{x + 2}{7} > -1$

47. $\dfrac{n + 3}{3} + \dfrac{n - 7}{2} > 3$

48. $\dfrac{n - 4}{4} + \dfrac{n - 2}{3} < 4$

49. $\dfrac{x - 3}{7} - \dfrac{x - 2}{4} \leq \dfrac{9}{14}$

50. $\dfrac{x - 1}{5} - \dfrac{x + 2}{6} \geq \dfrac{7}{15}$

For Problems 51–66, graph the solutions for each compound inequality.

51. $x > -1$ and $x < 2$

52. $x > 1$ and $x < 4$

53. $x < -2$ or $x > 1$

54. $x < 0$ or $x > 3$

55. $x > -2$ and $x \leq 2$

56. $x \geq -1$ and $x < 3$

57. $x > -1$ and $x > 2$

58. $x < 2$ and $x < 3$

59. $x > -4$ or $x > 0$

60. $x < 2$ or $x < 4$

61. $x > 3$ and $x < -1$

62. $x < -3$ and $x > 6$

63. $x \leq 0$ or $x \geq 2$

64. $x \leq -2$ or $x \geq 1$

65. $x > -4$ or $x < 3$

66. $x > -1$ or $x < 2$

Solve each of the following problems by setting up and solving an appropriate inequality.

67. Five more than three times a number is greater than 26. Find the numbers that satisfy this relationship.

68. Fourteen increased by twice a number is less than or equal to three times the number. Find the numbers that satisfy this relationship.

69. Suppose that the perimeter of a rectangle is to be no greater than 70 inches and that the length of the rectangle must be 20 inches. Find the largest possible value for the width of the rectangle.

70. One side of a triangle is three times as long as another side. The third side is 15 centimeters long. If the perimeter of the triangle is to be no greater than 75 centimeters, find the largest lengths that the other two sides can be.

71. Sue bowled 132 and 160 in her first two games. What must she bowl in the third game to have an average of at least 150 for the three games?

72. Mike has scores of 87, 81, and 74 on his first three algebra tests. What score must he get on the fourth test to have an average of 85 or higher for the four tests?

73. This semester Lance has scores of 96, 90, and 94 on his first three algebra exams. What must he average on the last two exams to have an average greater than 92 for all five exams?

74. The Mets have won 45 baseball games and lost 55 games. They have 62 more games to play. To win more than 50% of all their games, how many of the remaining 62 games must they win?

75. An Internet business has costs of $4000 plus $32 per sale. The business receives revenue of $48 per sale. How many sales would insure that the revenues exceed the costs?

76. The average height of the two forwards and the center of a basketball team is 6 feet, 8 inches. What must the average height of the two guards be so that the team average is at least 6 feet, 4 inches?

77. Scott shot rounds of 82, 84, 78, and 79 on the first four days of the golf tournament. What must he shoot on the fifth day of the tournament to average 80 or less for the five days?

78. Sydney earns $2300 a month. To qualify for a mortgage, her monthly payments must be less than 35% of her monthly income. Her monthly payments must be less than what amount to qualify for the mortgage?

■ ■ ■ **Thoughts into words**

79. Give a step-by-step description of how you would solve the inequality $3x - 2 > 4(x + 6)$.

80. Find the solution set for each of the following compound statements, and in each case explain your reasoning.

   a. $x > 2$ and $5 > 4$ **b.** $x > 2$ or $5 > 4$

   c. $x > 2$ and $4 > 10$ **d.** $x > 2$ or $4 > 10$

Answers to Concept Quiz

1. True   **2.** False   **3.** True   **4.** True   **5.** True   **6.** B   **7.** E   **8.** A   **9.** D   **10.** C

CHAPTER 3

SUMMARY

(3.1) Numerical equations may be true or false. **Algebraic equations** (open sentences) contain one or more variables. **Solving an equation** refers to the process of finding the number (or numbers) that make(s) an algebraic equation a true statement. A **first-degree equation of one variable** is an equation that contains only one variable, and this variable has an exponent of 1.

Properties 3.1, 3.2, and 3.3 provide the basis for solving equations. Be sure that you can use these properties to solve the variety of equations presented in this chapter.

(3.2) It is often necessary to use both the addition-subtraction and multiplication-division properties of equality to solve an equation.

Be sure to *declare your variable* as you translate English sentences into algebraic equations.

(3.3) Keep these suggestions in mind as you solve word problems:

1. Read the problem carefully.
2. Sketch any figure, diagram, or chart that might be helpful.
3. Choose a meaningful variable.
4. Look for a *guideline.*
5. Form an equation or inequality.
6. Solve the equation or inequality.
7. Check your answers.

(3.4) The **distributive property** is used to *remove parentheses.*

If an equation contains several fractions, then it is usually advisable to *clear the equation of all fractions* by multiplying both sides by the least common denominator of all the denominators in the equation.

(3.5) Properties 3.4 and 3.5 provide the basis for solving **algebraic inequalities.** Be sure that you can use these properties to solve the variety of inequalities presented in this chapter.

We can use many of the same techniques that we used to solve equations to solve inequalities, *but* we must be very careful when multiplying or dividing both sides of an inequality by the same number. *Don't forget* that when multiplying or dividing both sides of an inequality by a *negative number,* you must **reverse** the sense of the resulting inequality.

(3.6) The words "and" and "or" are used to form compound inequalities.

The solution set of a compound inequality formed by the word "and" is the **intersection** of the solution sets of the two inequalities. The solution set of a compound inequality formed by the word "or" is the **union** of the solution sets of the two inequalities.

To solve inequalities involving "and," we must satisfy all of the conditions. Thus the compound inequality $x > 1$ *and* $x < 3$ is satisfied by all numbers between 1 and 3.

To solve inequalities involving "or," we must satisfy one or more of the conditions. Thus the compound inequality $x < -1$ *or* $x > 2$ is satisfied by (a) all numbers less than -1, or (b) all numbers greater than 2, or (c) both (a) and (b).

CHAPTER 3 REVIEW PROBLEM SET

In Problems 1–20, solve each of the equations.

1. $9x - 2 = -29$

2. $-3 = -4y + 1$

3. $7 - 4x = 10$

4. $6y - 5 = 4y + 13$

5. $4n - 3 = 7n + 9$

6. $7(y - 4) = 4(y + 3)$

7. $2(x + 1) + 5(x - 3) = 11(x - 2)$

8. $-3(x + 6) = 5x - 3$

9. $\dfrac{2}{5}n - \dfrac{1}{2}n = \dfrac{7}{10}$

10. $\dfrac{3n}{4} + \dfrac{5n}{7} = \dfrac{1}{14}$

11. $\dfrac{x - 3}{6} + \dfrac{x + 5}{8} = \dfrac{11}{12}$

12. $\dfrac{n}{2} - \dfrac{n - 1}{4} = \dfrac{3}{8}$

13. $-2(x - 4) = -3(x + 8)$

14. $3x - 4x - 2 = 7x - 14 - 9x$

15. $5(n - 1) - 4(n + 2) = -3(n - 1) + 3n + 5$

16. $\dfrac{x - 3}{9} = \dfrac{x + 4}{8}$

17. $\dfrac{x - 1}{-3} = \dfrac{x + 2}{-4}$

18. $-(t - 3) - (2t + 1) = 3(t + 5) - 2(t + 1)$

19. $\dfrac{2x - 1}{3} = \dfrac{3x + 2}{2}$

20. $3(2t - 4) + 2(3t + 1) = -2(4t + 3) - (t - 1)$

For Problems 21–36, solve each inequality.

21. $3x - 2 > 10$

22. $-2x - 5 < 3$

23. $2x - 9 \geq x + 4$

24. $3x + 1 \leq 5x - 10$

25. $6(x - 3) > 4(x + 13)$

26. $2(x + 3) + 3(x - 6) < 14$

27. $\dfrac{2n}{5} - \dfrac{n}{4} < \dfrac{3}{10}$

28. $\dfrac{n + 4}{5} + \dfrac{n - 3}{6} > \dfrac{7}{15}$

29. $-16 < 8 + 2y - 3y$

30. $-24 > 5x - 4 - 7x$

31. $-3(n - 4) > 5(n + 2) + 3n$

32. $-4(n - 2) - (n - 1) < -4(n + 6)$

33. $\dfrac{3}{4}n - 6 \leq \dfrac{2}{3}n + 4$

34. $\dfrac{1}{2}n - \dfrac{1}{3}n - 4 \geq \dfrac{3}{5}n + 2$

35. $-12 > -4(x - 1) + 2$

36. $36 < -3(x + 2) - 1$

For Problems 37–40, graph the solution set for each of the compound inequalities.

37. $x > -3$ and $x < 2$ **38.** $x < -1$ or $x > 4$

39. $x < 2$ or $x > 0$ **40.** $x > 1$ and $x > 0$

Set up an equation or an inequality and solve each of the following problems.

41. Three-fourths of a number equals 18. Find the number.

42. Nineteen is 2 less than three times a certain number. Find the number.

43. The difference of two numbers is 21. If 12 is the smaller number, find the other number.

44. One subtracted from nine times a certain number is the same as 15 added to seven times the number. Find the number.

45. Monica had scores of 83, 89, 78, and 86 on her first four exams. What score must she get on the fifth exam so that her average for all five exams is 85 or higher?

46. The sum of two numbers is 40. Six times the smaller number equals four times the larger. Find the numbers.

47. Find a number such that two less than two-thirds of the number is one more than one-half of the number.

48. Ameya's average score for her first three psychology exams was 84. What must she get on the fourth exam so that her average for the four exams is 85 or higher?

49. Miriam has 30 coins consisting of nickels and dimes amounting to $2.60. How many coins of each kind does she have?

50. Suppose that Russ has a bunch of nickels, dimes, and quarters amounting to $15.40. The number of dimes is 1 more than three times the number of nickels, and the number of quarters is twice the number of dimes. How many coins of each kind does he have?

51. The supplement of an angle is 14° more than three times the complement of the angle. Find the measure of the angle.

52. Pam rented a car from a rental agency that charges $25 a day and $0.20 per mile. She kept the car for 3 days and her bill was $215. How many miles did she drive during that 3-day period?

CHAPTER 3

TEST

applies to all problems in this Chapter Test.

For Problems 1–12, solve each of the equations.

1. $7x - 3 = 11$

2. $-7 = -3x + 2$

3. $4n + 3 = 2n - 15$

4. $3n - 5 = 8n + 20$

5. $4(x - 2) = 5(x + 9)$

6. $9(x + 4) = 6(x - 3)$

7. $5(y - 2) + 2(y + 1) = 3(y - 6)$

8. $\dfrac{3}{5}x - \dfrac{2}{3} = \dfrac{1}{2}$

9. $\dfrac{x - 2}{4} = \dfrac{x + 3}{6}$

10. $\dfrac{x + 2}{3} + \dfrac{x - 1}{2} = 2$

11. $\dfrac{x - 3}{6} - \dfrac{x - 1}{8} = \dfrac{13}{24}$

12. $-5(n - 2) = -3(n + 7)$

For Problems 13–18, solve each of the inequalities.

13. $3x - 2 < 13$

14. $-2x + 5 \geq 3$

15. $3(x - 1) \leq 5(x + 3)$

16. $-4 > 7(x - 1) + 3$

17. $-2(x - 1) + 5(x - 2) < 5(x + 3)$

18. $\dfrac{1}{2}n + 2 \leq \dfrac{3}{4}n - 1$

For Problems 19 and 20, graph the solution set for each compound inequality.

19. $x \geq -2$ and $x \leq 4$

20. $x < 1$ or $x > 3$

For Problems 21–25, set up an equation or an inequality and solve each problem.

21. A car repair bill without the tax was $441. This included $153 for parts and 4 hours of labor. Find the hourly rate that was charged for labor.

22. Suppose that a triangular plot of ground is enclosed with 70 meters of fencing. The longest side of the lot is two times the length of the shortest side, and the third side is 10 meters longer than the shortest side. Find the length of each side of the plot.

23. Tina had scores of 86, 88, 89, and 91 on her first four history exams. What score must she get on the fifth exam to have an average of 90 or higher for the five exams?

24. Sean has 103 coins consisting of nickels, dimes, and quarters. The number of dimes is 1 less than twice the number of nickels, and the number of quarters is 2 more than three times the number of nickels. How many coins of each kind does he have?

25. In triangle ABC, the measure of angle C is one-half the measure of angle A, and the measure of angle B is 30° more than the measure of angle A. Find the measure of each angle of the triangle.

CUMULATIVE PRACTICE TEST *Chapters 1-3*

For Problems 1–4, simplify each numerical expression.

1. $3(-4) - 2 + (-3)(-6) - 1$

2. $-(2)^7$

3. $6.2 - 7.1 - 3.4 + 1.9$

4. $-\dfrac{2}{3} + \dfrac{1}{2} - \dfrac{1}{4}$

For Problems 5–7, evaluate each algebraic expression for the given values of the variables.

5. $-4x + 2y - xy$ for $x = -2$ and $y = 3$

6. $\dfrac{1}{5}x - \dfrac{2}{3}y$ for $x = -\dfrac{1}{2}$ and $y = \dfrac{1}{6}$

7. $0.2(x - y) - 0.3(x + y)$ for $x = 0.1$ and $y = -0.2$

8. Find the greatest common factor of 48, 60, and 96.

9. Find the least common multiple of 9 and 12.

10. Simplify $\dfrac{3}{8}x + \dfrac{3}{7}y - \dfrac{5}{12}x - \dfrac{3}{4}y$ by combining similar terms.

11. Simplify $-(x - 2) + 6(x + 4) - 2(x - 7)$ by applying the distributive property and combining similar terms.

For Problems 12–15, perform the indicated operations and express answers in simplest form.

12. $\left(\dfrac{5x^2 y}{6xy^2}\right)\left(\dfrac{8x^2 y^2}{30y}\right)$

13. $\dfrac{5}{x} - \dfrac{6}{y^2}$

14. $\left(\dfrac{ab^3}{3a^2}\right) \div \left(\dfrac{ab}{9b^2}\right)$

15. $\dfrac{5}{3y} - \dfrac{2}{5y^2}$

For Problems 16–18, solve each equation.

16. $-3(x + 4) = -4(x - 1)$

17. $\dfrac{x + 1}{4} - \dfrac{x - 2}{3} = -2$

18. $2(2x - 1) + 3(x - 3) = -4(x + 7)$

For Problems 19 and 20, solve the inequality.

19. $2(x - 1) \geq 3(x - 6)$

20. $-2 < -(x - 1) - 4$

21. Express 300 as the product of prime factors.

22. Graph the solutions for the compound inequality $x \geq -1$ and $x < 3$.

For Problems 23–25, use an equation or inequality to help solve each problem.

23. On Friday and Saturday nights, the police made a total of 42 arrests at a DUI checkpoint. On Saturday night they made 6 more than three times the arrests of Friday night. Find the number of arrests for each night.

24. For a wedding reception, the caterer charges a $125 fee plus $35 per person for dinner. If Peter and Rynette must keep the cost of the caterer to less than $2500, how many people can attend the reception?

25. Find two consecutive odd numbers whereby the smaller plus five times the larger equals 76.

© Photodisc

The equation $s = 3 + 0.6s$ can be used to determine how much the owner of a pizza parlor must charge for a pizza if it costs $3 to make the pizza, and he wants to make a profit of 60% based on the selling price.

Formulas and Problem Solving

Kirk starts jogging at the rate of 5 miles per hour. One-half hour later, Consuela starts jogging on the same route at 7 miles per hour. How long will it take Consuela to catch Kirk? If we let t represent the time that Consuela jogs, then $t + \dfrac{1}{2}$ represents Kirk's time. We can use the equation $7t = 5\left(t + \dfrac{1}{2} \right)$ to determine that Consuela should catch Kirk in $1\dfrac{1}{4}$ hours.

We used the **formula** *distance equals rate times time,* which is usually expressed as $d = rt$, to set up the equation $7t = 5\left(t + \dfrac{1}{2} \right)$. Throughout this chapter we will use a variety of formulas in a problem-solving setting to connect algebraic and geometric concepts.

InfoTrac Project   Do a subject guide search on **mathematics,** choose subdivisions and then formulae. Find the article "Fracture formula yields volcanic forecasts." Write a brief summary as to what these scientists hope the formula will enable them to do. In 1826 George Ohm discovered a formula showing the relationship between voltage, amperage, and resistance. Because of his work, the unit-to-measure resistance (ohm) is named after him. According to his formula, electrical resistance in ohms is equal to voltage (E) divided by current (I); $R = \frac{E}{I}$. If E is 5.9 volts and I is 0.35 amps, what is the resistance? Now if $R = 120$ ohms and the current is 0.01 amps, find the voltage.

4.1   Ratio, Proportion, and Percent

Objectives

- Solve proportions.
- Use a proportion to convert a fraction to a percent.
- Solve basic percent problems.
- Solve word problems using proportions.

Ratio

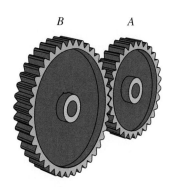

B   A

Figure 4.1

In Figure 4.1, as gear A revolves four times, gear B will revolve three times. We say that the *gear ratio* of A to B is 4 to 3, or the gear ratio of B to A is 3 to 4. Mathematically, a **ratio** *is the comparison of two numbers by division.* We can write the gear ratio of A to B in these equivalent expressions:

$$4 \text{ to } 3 \qquad 4\!:\!3 \qquad \frac{4}{3}$$

We express ratios as fractions in reduced form. For example, if there are 7500 women and 5000 men at a certain university, then the ratio of women to men is $\frac{7500}{5000} = \frac{3}{2}$.

Proportion

A statement of equality between two ratios is called a **proportion.** For example,

$$\frac{2}{3} = \frac{8}{12}$$

is a proportion that states that the ratios $\frac{2}{3}$ and $\frac{8}{12}$ are equal. In the general proportion

$$\frac{a}{b} = \frac{c}{d} \qquad b \neq 0 \text{ and } d \neq 0$$

if we multiply both sides of the equation by the common denominator, bd, we obtain

$$(bd)\left(\frac{a}{b}\right) = (bd)\left(\frac{c}{d}\right)$$
$$ad = bc$$

Let's state this as a property of proportions.

$$\frac{a}{b} = \frac{c}{d} \quad \text{if and only if } ad = bc, \text{ where } b \neq 0 \text{ and } d \neq 0$$

The products ad and bc are commonly called **cross products.** Thus the property states that the cross products in a proportion are equal.

EXAMPLE 1   Solve $\dfrac{x}{20} = \dfrac{3}{4}$.

Solution

$$\frac{x}{20} = \frac{3}{4}$$
$$4x = 60 \qquad \text{Cross products are equal.}$$
$$x = 15$$

The solution set is $\{15\}$.

EXAMPLE 2   Solve $\dfrac{x-3}{5} = \dfrac{x+2}{4}$.

Solution

$$\frac{x-3}{5} = \frac{x+2}{4}$$
$$4(x-3) = 5(x+2) \qquad \text{Cross products are equal.}$$
$$4x - 12 = 5x + 10 \qquad \text{Distributive property}$$
$$-12 = x + 10 \qquad \text{Subtracted } 4x \text{ from both sides}$$
$$-22 = x \qquad \text{Subtracted 10 from both sides}$$

The solution set is $\{-22\}$.

If a variable appears in one or both of the denominators, then certain restrictions must be imposed to avoid division by zero, as the next example illustrates.

EXAMPLE 3   Solve $\dfrac{7}{a-2} = \dfrac{4}{a+3}$.

Solution

$$\dfrac{7}{a-2} = \dfrac{4}{a+3} \qquad a \neq 2 \text{ and } a \neq -3$$

$7(a+3) = 4(a-2)$	Cross products are equal.
$7a + 21 = 4a - 8$	Distributive property
$3a + 21 = -8$	Subtracted 4a from both sides
$3a = -29$	Subtracted 21 from both sides
$a = -\dfrac{29}{3}$	Divided both sides by 3

The solution set is $\left\{-\dfrac{29}{3}\right\}$.   ■

EXAMPLE 4   Solve $\dfrac{x}{4} + 3 = \dfrac{x}{5}$.

Solution

This is *not* a proportion, so let's multiply both sides by 20, the least common denominator, to clear the equation of all fractions.

$$\dfrac{x}{4} + 3 = \dfrac{x}{5}$$

$20\left(\dfrac{x}{4} + 3\right) = 20\left(\dfrac{x}{5}\right)$	Multiply both sides by 20.
$20\left(\dfrac{x}{4}\right) + 20(3) = 20\left(\dfrac{x}{5}\right)$	Distributive property
$5x + 60 = 4x$	
$x + 60 = 0$	Subtracted 4x from both sides
$x = -60$	Subtracted 60 from both sides

The solution set is $\{-60\}$.   ■

REMARK: Example 4 demonstrates the importance of *thinking first before pushing the pencil*. Because the equation was not in the form of a proportion, we needed to revert to a previous technique for solving it.

Problem Solving Using Proportions

Some word problems can be conveniently set up and solved using the concepts of ratio and proportion. Consider the next examples.

PROBLEM 1

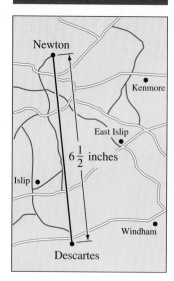

Figure 4.2

On the map in Figure 4.2, 1 inch represents 20 miles. If two cities are $6\frac{1}{2}$ inches apart on the map, find the number of miles between the cities.

Solution

Let m represent the number of miles between the two cities. Now let's set up a proportion where one ratio compares distances in inches on the map, and the other ratio compares *corresponding* distances in miles on land.

$$\frac{1}{6\frac{1}{2}} = \frac{20}{m}$$

To solve this equation, we equate the cross products.

$$m(1) = \left(6\frac{1}{2}\right)(20)$$

$$m = \left(\frac{13}{2}\right)(20) = 130$$

The distance between the two cities is 130 miles.   ■

PROBLEM 2

A sum of $1750 is to be divided between two people in the ratio of 3 to 4. How much does each person receive?

Solution

Let d represent the amount of money to be received by one person. Then $1750 - d$ represents the amount for the other person. We set up this proportion:

$$\frac{d}{1750 - d} = \frac{3}{4}$$

$$4d = 3(1750 - d)$$
$$4d = 5250 - 3d$$
$$7d = 5250$$
$$d = 750$$

If $d = 750$, then $1750 - d = 1000$; therefore, one person receives $750, and the other person receives $1000.   ■

Percent

The word **percent** means *per one hundred,* and we use the symbol % to express it. For example, we write 7 percent as 7%, which means $\frac{7}{100}$, or 0.07. In other words, percent is a special kind of ratio — a ratio in which the denominator is always 100. Proportions provide a convenient basis for changing common fractions to percents. Consider the following examples.

EXAMPLE 5

Express $\dfrac{7}{20}$ as a percent.

Solution

We are asking, "what number compares to 100 as 7 compares to 20?" Therefore, if we let n represent that number, we can set up the proportion like this:

$$\frac{n}{100} = \frac{7}{20}$$

$20n = 700$ Cross products are equal.

$n = 35$

Thus $\dfrac{7}{20} = \dfrac{35}{100} = 35\%$. ∎

EXAMPLE 6

Express $\dfrac{5}{6}$ as a percent.

Solution

$$\frac{n}{100} = \frac{5}{6}$$

$6n = 500$ Cross products are equal.

$$n = \frac{500}{6} = \frac{250}{3} = 83\frac{1}{3}$$

Therefore $\dfrac{5}{6} = 83\dfrac{1}{3}\%$. ∎

Some Basic Percent Problems

What is 8% of 35? Fifteen percent of what number is 24? Twenty-one is what percent of 70? These are the three basic types of percent problems. Each of these problems can be solved easily by translating it into, and solving, a simple algebraic equation.

PROBLEM 3

What is 8% of 35?

Solution

Let n represent the number to be found. The *is* refers to equality, and *of* means multiplication. Thus the question translates into

$$n = (8\%)(35)$$

which can be solved as follows:

$$n = (0.08)(35)$$

$$= 2.8$$

Therefore 2.8 is 8% of 35. ∎

PROBLEM 4   Fifteen percent of what number is 24?

Solution

Let n represent the number to be found.

$$(15\%)(n) = (24)$$
$$0.15n = 24$$
$$15n = 2400 \qquad \text{Multiplied both sides by 100}$$
$$n = 160$$

Therefore 15% of 160 is 24.

PROBLEM 5   Twenty-one is what percent of 70?

Solution

Let r represent the percent to be found.

$$21 = r(70)$$
$$\frac{21}{70} = r$$
$$\frac{3}{10} = r \qquad \text{Reduce!}$$
$$\frac{30}{100} = r \qquad \text{Changed } \frac{3}{10} \text{ to } \frac{30}{100}$$
$$30\% = r$$

Therefore 21 is 30% of 70.

PROBLEM 6   Seventy-two is what percent of 60?

Solution

Let r represent the percent to be found.

$$72 = r(60)$$
$$\frac{72}{60} = r$$
$$\frac{6}{5} = r$$
$$\frac{120}{100} = r \qquad \text{Changed } \frac{6}{5} \text{ to } \frac{120}{100}$$
$$120\% = r$$

Therefore, 72 is 120% of 60.

Again it is helpful to get into the habit of checking answers for *reasonableness*. We also suggest that you alert yourself to a potential computational error by *estimating* the answer before you actually do the problem. For example, prior to doing Problem 6, you may have estimated: "Because 72 is larger than 60, the answer has to be greater than 100%. Furthermore 1.5 (or 150%) times 60 equals 90." Therefore, you can estimate the answer to be somewhere between 100% and 150%. That may seem rather broad, but many times such an estimate will detect a computational error.

CONCEPT QUIZ

For Problems 1–8, answer true or false.

1. A ratio is the comparison of two numbers by division.

2. The ratio of 7 to 3 can be written 3:7.

3. A proportion is a statement of equality between two ratios.

4. For the proportion $\dfrac{x}{3} = \dfrac{y}{5}$, the cross product would be $5x = 3y$.

5. The algebraic statement, $\dfrac{w}{2} = \dfrac{w}{5} + 1$, is a proportion.

6. The word *percent* means parts per one thousand.

7. For the proportion $\dfrac{a+1}{a-2} = \dfrac{5}{7}$, $a \neq -1$ and $a \neq 2$.

8. If the cross product of a proportion is $wx = yz$, then $\dfrac{x}{z} = \dfrac{y}{w}$.

PROBLEM SET 4.1

For Problems 1–32, solve each equation.

1. $\dfrac{x}{6} = \dfrac{3}{2}$

2. $\dfrac{x}{9} = \dfrac{5}{3}$

3. $\dfrac{5}{12} = \dfrac{n}{24}$

4. $\dfrac{7}{8} = \dfrac{n}{16}$

5. $\dfrac{x}{3} = \dfrac{5}{2}$

6. $\dfrac{x}{7} = \dfrac{4}{3}$

7. $\dfrac{x-2}{4} = \dfrac{x+4}{3}$

8. $\dfrac{x-6}{7} = \dfrac{x+9}{8}$

9. $\dfrac{x+1}{6} = \dfrac{x+2}{4}$

10. $\dfrac{x-2}{6} = \dfrac{x-6}{8}$

11. $\dfrac{h}{2} - \dfrac{h}{3} = 1$

12. $\dfrac{h}{5} + \dfrac{h}{4} = 2$

13. $\dfrac{x+1}{3} - \dfrac{x+2}{2} = 4$

14. $\dfrac{x-2}{5} - \dfrac{x+3}{6} = -4$

15. $\dfrac{-4}{x+2} = \dfrac{-3}{x-7}$

16. $\dfrac{-9}{x+1} = \dfrac{-8}{x+5}$

17. $\dfrac{-1}{x-7} = \dfrac{5}{x-1}$

18. $\dfrac{3}{x-10} = \dfrac{-2}{x+6}$

19. $\dfrac{3}{2x-1} = \dfrac{2}{3x+2}$

20. $\dfrac{1}{4x+3} = \dfrac{2}{5x-3}$

21. $\dfrac{n+1}{n} = \dfrac{8}{7}$

22. $\dfrac{5}{6} = \dfrac{n}{n+1}$

23. $\dfrac{x-1}{2} - 1 = \dfrac{3}{4}$

24. $-2 + \dfrac{x+3}{4} = \dfrac{5}{6}$

25. $-3 - \dfrac{x+4}{5} = \dfrac{3}{2}$

26. $\dfrac{x-5}{3} + 2 = \dfrac{5}{9}$

27. $\dfrac{n}{150-n} = \dfrac{1}{2}$

28. $\dfrac{n}{200-n} = \dfrac{3}{5}$

29. $\dfrac{300-n}{n} = \dfrac{3}{2}$

30. $\dfrac{80-n}{n} = \dfrac{7}{9}$

31. $\dfrac{-1}{5x-1} = \dfrac{-2}{3x+7}$

32. $\dfrac{-3}{2x-5} = \dfrac{-4}{x-3}$

For Problems 33–44, use proportions to change each common fraction to a percent.

33. $\dfrac{11}{20}$

34. $\dfrac{17}{20}$

35. $\dfrac{3}{5}$

36. $\dfrac{7}{25}$

37. $\dfrac{1}{6}$

38. $\dfrac{5}{7}$

39. $\dfrac{3}{8}$

40. $\dfrac{1}{16}$

41. $\dfrac{3}{2}$

42. $\dfrac{5}{4}$

43. $\dfrac{12}{5}$

44. $\dfrac{13}{6}$

For Problems 45–56, answer the question by setting up and solving an appropriate equation.

45. What is 7% of 38?   **46.** What is 35% of 52?

47. 15% of what number is 6.3?

48. 55% of what number is 38.5?

49. 76 is what percent of 95?

50. 72 is what percent of 120?

51. What is 120% of 50?

52. What is 160% of 70?

53. 46 is what percent of 40?

54. 26 is what percent of 20?

55. 160% of what number is 144?

56. 220% of what number is 66?

For Problems 57–73, solve each problem using a proportion.

57. A house plan has a scale where 1 inch represents 6 feet. Find the dimensions of a rectangular room that measures $2\dfrac{1}{2}$ inches by $3\dfrac{1}{4}$ inches on the house plan.

58. On a certain map, 1 inch represents 15 miles. If two cities are 7 inches apart on the map, find the number of miles between the cities.

59. Suppose that a car can travel 264 miles using 12 gallons of gasoline. How far will it go on 15 gallons?

60. Jesse used 10 gallons of gasoline to drive 170 miles. How much gasoline will he need to travel 238 miles?

61. If the ratio of the length of a rectangle to its width is $\dfrac{5}{2}$, and the width is 24 centimeters, find its length.

62. If the ratio of the width of a rectangle to its length is $\dfrac{4}{5}$, and the length is 45 centimeters, find the width.

63. A saltwater solution is made by dissolving 3 pounds of salt in 10 gallons of water (see Figure 4.3).

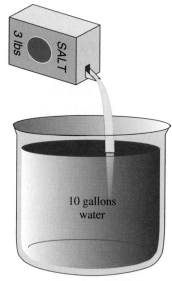

Figure 4.3

At this rate, how many pounds of salt are needed for 25 gallons of water?

64. A home valued at $150,000 is assessed $2700 in real estate taxes. At the same rate, how much are the taxes on a home assessed at $175,000?

65. If 20 pounds of fertilizer will cover 1500 square feet of lawn, how many pounds are needed for 2500 square feet?

66. It was reported that a flu epidemic is affecting six out of every ten college students in a certain part of the country. At this rate, how many students will be affected at a university of 15,000 students?

67. A preelection poll indicated that three out of every seven eligible voters were going to vote in an upcoming election. At this rate, how many people are expected to vote in a city of 210,000 eligible voters?

68. A board 28 feet long is cut into two pieces, and the lengths of the two pieces are in the ratio of 2 to 5. Find the lengths of the two pieces.

69. In a nutrition plan, the ratio of calories to grams of carbohydrates is 16 to 1. According to this ratio, how many grams of carbohydrates would be in a plan that has 2200 calories?

70. The ratio of male students to female students at a certain university is 5 to 4. If there is a total of 6975 students, find the number of male students and the number of female students.

71. An investment of $500 earns $45 in a year. At the same rate, how much additional money must be invested to raise the earnings to $72 per year?

72. A sum of $1250 is to be divided between two people in the ratio of 2 to 3. How much does each person receive?

73. An inheritance of $180,000 is to be divided between a child and the local cancer fund in the ratio of 5 to 1. How much money will the child receive?

■ ■ ■ Thoughts into words

74. Explain the difference between a ratio and a proportion.

75. What is wrong with this solution?

$$\frac{x}{2} + 4 = \frac{x}{6}$$

$$6\left(\frac{x}{2} + 4\right) = 2(x)$$

$$3x + 24 = 2x$$

$$x = -24$$

Explain how it should be solved.

76. Estimate an answer for each of the following problems, and explain how you arrived at your estimate. Then work out the problem to see how well you estimated.

a. The ratio of female students to male students at a small private college is 5 to 3. If there is a total of 1096 students, find the number of male students.

b. If 15 pounds of fertilizer will cover 1200 square feet of lawn, how many pounds are needed for 3000 square feet?

c. An investment of $5000 earns $300 interest in a year. At the same rate, how much money must be invested to earn $450?

d. If the ratio of the length of a rectangle to its width is 5 to 3, and the length is 70 centimeters, find its width.

■ ■ ■ Further investigations

Solve each of the following equations. Don't forget that division by zero is undefined.

77. $\dfrac{3}{x-2} = \dfrac{6}{2x-4}$

78. $\dfrac{8}{2x+1} = \dfrac{4}{x-3}$

79. $\dfrac{5}{x-3} = \dfrac{10}{x-6}$

80. $\dfrac{6}{x-1} = \dfrac{5}{x-1}$

81. $\dfrac{x-2}{2} = \dfrac{x}{2} - 1$

82. $\dfrac{x+3}{x} = 1 + \dfrac{3}{x}$

4.2   More on Percents and Problem Solving

Objectives

- Solve word problems involving discount.
- Solve word problems involving selling price.
- Use the simple interest formula to solve problems.

We can solve the equation $x + 0.35 = 0.72$ by subtracting 0.35 from both sides of the equation. Another technique for solving equations that contain decimals is to *clear the equation of all decimals* by multiplying both sides by an appropriate power of 10. The following examples demonstrate the use of that strategy in a variety of situations.

EXAMPLE 1   Solve $0.5x = 14$.

Solution

$$0.5x = 14$$
$$5x = 140 \qquad \text{Multiplied both sides by 10}$$
$$x = 28 \qquad \text{Divided both sides by 5}$$

The solution set is $\{28\}$.   ■

EXAMPLE 2   Solve $x + 0.07x = 0.13$.

Solution

$$x + 0.07x = 0.13$$
$$100(x + 0.07x) = 100(0.13) \qquad \text{Multiply both sides by 100.}$$
$$100(x) + 100(0.07x) = 100(0.13) \qquad \text{Distributive property}$$
$$100x + 7x = 13$$
$$107x = 13$$
$$x = \frac{13}{107}$$

The solution set is $\left\{\dfrac{13}{107}\right\}$.   ■

EXAMPLE 3   Solve $0.08y + 0.09y = 3.4$.

Solution

$$0.08y + 0.09y = 3.4$$
$$8y + 9y = 340 \qquad \text{Multiplied both sides by 100}$$
$$17y = 340$$
$$y = 20$$

The solution set is $\{20\}$.   ■

EXAMPLE 4

Solve $0.10t = 560 - 0.12(t + 1000)$.

Solution

$$0.10t = 560 - 0.12(t + 1000)$$
$$10t = 56,000 - 12(t + 1000) \qquad \text{Multiplied both sides by 100}$$
$$10t = 56,000 - 12t - 12,000 \qquad \text{Distributive property}$$
$$22t = 44,000$$
$$t = 2000$$

The solution set is $\{2000\}$. ■

Problems Involving Percents

We can solve many consumer problems with an equation approach. For example, here is a general guideline regarding discount sales:

> Original selling price − Discount = Discount sale price

Let's work some examples using our algebraic techniques along with this basic guideline.

PROBLEM 1

Amy bought a dress at a 30% discount sale for $35. What was the original price of the dress?

Solution

Let p represent the original price of the dress. We can use the basic discount guideline to set up an algebraic equation:

Original selling price − Discount = Discount sale price

$$(100\%)(p) \quad - (30\%)(p) = \quad \$35$$

Solving this equation, we get

$$(100\%)(p) - (30\%)(p) = 35$$
$$(70\%)(p) = 35$$
$$0.7p = 35$$
$$7p = 350$$
$$p = 50$$

The original price of the dress was $50. ■

Don't forget that if an item is on sale for 30% off, then you are going to pay $100\% - 30\% = 70\%$ of the original price. Thus at a 30% discount sale, you can buy a $50 dress for $(70\%)(\$50) = \$35$. (Note that we just checked our answer for Problem 1.)

PROBLEM 2

Find the cost of a $60 pair of jogging shoes on sale for 20% off.

Solution

Let x represent the discount sale price. Because the shoes are on sale for 20% off, we must pay 80% of the original price.

$$x = (80\%)(60)$$
$$= (0.8)(60) = 48$$

The sale price is $48. ◼

Here is another equation that is useful in consumer problems:

> Selling price = Cost + Profit

Profit (also called *markup, markon, margin,* and *margin of profit*) may be stated in different ways: as a percent of the selling price, as a percent of the cost, or simply in terms of dollars and cents. Let's consider some problems where the profit is either a percent of the selling price or a percent of the cost.

PROBLEM 3

A retailer has some shirts that cost him $20 each. He wants to sell them at a profit of 60% of the cost. What should the selling price be on the shirts?

Solution

Let s represent the selling price. The basic relationship *selling price equals cost plus profit* can be used as a guideline:

Selling price = Cost + Profit (% of cost)

$$s = \$20 + (60\%)(20)$$

Solving this equation, we obtain

$$s = 20 + (60\%)(20)$$
$$= 20 + (0.6)(20)$$
$$= 20 + 12$$
$$= 32$$

The selling price should be $32. ◼

PROBLEM 4

Kathrin bought a painting for $120 and later decided to resell it. She made a profit of 40% of the selling price. What did she receive for the painting?

Solution

We can use the same basic relationship as a guideline, except this time the profit is a percent of the selling price. Let s represent the selling price.

Selling price = Cost + Profit (% of selling price)

$$s = 120 + (40\%)(s)$$

We can solve this equation:

$$s = 120 + (40\%)(s)$$
$$s = 120 + 0.4s$$
$$0.6s = 120 \qquad\qquad \text{Subtracted 0.4s from both sides}$$
$$s = \frac{120}{0.6} = 200$$

She received $200 for the painting.   ∎

We can also translate certain types of investment problems into algebraic equations. In some of these problems, we use the simple interest formula $i = Prt$, where i represents the amount of interest earned by investing P dollars at a yearly rate of r percent for t years.

P R O B L E M   5

John invested $9300 for 2 years and received $1395 in interest. Find the annual interest rate John received on his investment.

Solution

$$i = Prt$$
$$1395 = 9300r(2)$$
$$1395 = 18600r$$
$$\frac{1395}{18600} = r$$
$$0.075 = r$$

The annual interest rate is 7.5%.   ∎

P R O B L E M   6

How much principal must be invested to receive $1500 in interest when the investment is made for 3 years at an annual interest rate of 6.25%?

Solution

$$i = Prt$$
$$1500 = P(0.0625)(3)$$
$$1500 = P(0.1875)$$
$$\frac{1500}{0.1875} = P$$
$$8000 = P$$

The principal must be $8000.   ∎

PROBLEM 7

How much monthly interest will be charged on a credit card bill with a balance of $754 when the credit card company charges an 18% annual interest rate?

Solution

$$i = Prt$$

$$i = 754(0.18)\left(\frac{1}{12}\right) \quad \text{Remember, 1 month is } \frac{1}{12} \text{ of a year.}$$

$$i = 11.31$$

The interest charge would be $11.31.   ■

CONCEPT QUIZ

For Problems 1–5, answer true or false.

1. To clear the decimals from the equation, $0.5x + 1.24 = 0.07x + 1.8$, you would multiply both sides of the equation by 10.

2. If an item is on sale for 35% off, then you are going to pay 65% of the original price.

3. Profit is always a percent of the selling price.

4. In the formula, $i = Prt$, the r represents the interest return.

5. The basic relationship, *selling price equals cost plus profit,* can be used whether the profit is based on selling price or cost.

PROBLEM SET 4.2

For Problems 1–22, solve each equation.

1. $x - 0.36 = 0.75$

2. $x - 0.15 = 0.42$

3. $x + 7.6 = 14.2$

4. $x + 11.8 = 17.1$

5. $0.62 - y = 0.14$

6. $7.4 - y = 2.2$

7. $0.7t = 56$

8. $1.3t = 39$

9. $x = 3.36 - 0.12x$

10. $x = 5.3 - 0.06x$

11. $s = 35 + 0.3s$

12. $s = 40 + 0.5s$

13. $s = 42 + 0.4s$

14. $s = 24 + 0.6s$

15. $0.07x + 0.08(x + 600) = 78$

16. $0.06x + 0.09(x + 200) = 63$

17. $0.09x + 0.1(2x) = 130.5$

18. $0.11x + 0.12(3x) = 188$

19. $0.08x + 0.11(500 - x) = 50.5$

20. $0.07x + 0.09(2000 - x) = 164$

21. $0.09x = 550 - 0.11(5400 - x)$

22. $0.08x = 580 - 0.1(6000 - x)$

For Problems 23–38, set up an equation and solve each problem.

23. Tom bought an electric drill at a 30% discount sale for $35. What was the original price of the drill?

24. Magda bought a dress for $140, which represents a 20% discount off the original price. What was the original price of the dress?

25. Find the cost of a $4800 wide-screen television that is on sale for 25% off.

26. Byron purchased a computer monitor at a 10% discount sale for $121.50. What was the original price of the monitor?

27. Suppose that Jack bought a $32 putter on sale for 35% off. How much did he pay for the putter?

28. Mindy bought a 13-inch portable color TV for 20% off the list price. The list price was $149.95. What did she pay for the TV?

29. Pierre paid $126 for a coat that was listed for $180. What rate of discount did he receive?

30. Phoebe paid $32 for a pair of sandals that were listed for $40. What rate of discount did she receive?

31. A retailer has some toe rings that cost him $5 each. He wants to sell them at a profit of 70% of the cost. What should be the selling price of the toe rings?

32. A retailer has some video games that cost her $25 each. She wants to sell them at a profit of 80% of the cost. What price should she charge for the games?

33. The owner of a pizza parlor wants to make a profit of 55% of the cost for each pizza sold. If it costs $8 to make a pizza, at what price should it be sold?

34. Produce in a food market usually has a high markup because of loss due to spoilage. If a head of lettuce costs a retailer $0.40, at what price should it be sold to realize a profit of 130% of the cost?

35. Jewelry has a very high markup rate. If a ring costs a jeweler $400, what should its price be to yield a profit of 60% of the selling price?

36. If a box of candy costs a retailer $2.50, and he wants to make a profit of 50% based on the selling price, what price should he charge for the candy?

37. If the cost of a pair of shoes is $32 for a retailer and he sells them for $44.80, what is his rate of profit based on the cost?

38. A retailer has some candle sets that cost her $24. If she sells them for $31.20, find her rate of profit based on the cost.

For Problems 39–46, use the formula $i = Prt$ to reach a solution.

39. Find the annual interest rate if $560 in interest is earned when $3500 is invested for 2 years.

40. How much interest will be charged on a student loan if $8000 is borrowed for 9 months at a 19.2% annual interest rate?

41. How much principal, invested at 8% annual interest for 3 years, is needed to earn $1000?

42. How long will $2400 need to be invested at a 5.5% annual interest rate to earn $330?

43. What will be the interest earned on a $5000 certificate of deposit invested at 6.8% annual interest for 10 years?

44. One month a credit card company charged $38.15 in interest on a balance of $2725. What annual interest rate is the credit card company charging?

45. How much is a month's interest on a mortgage balance of $95,000 at an 8% annual interest rate?

46. For how many years must $2000 be invested at a 5.4% annual interest rate to earn $162?

■ ■ ■ **Thoughts into words**

47. What is wrong with this solution?

$$1.2x + 2 = 3.8$$
$$10(1.2x) + 2 = 10(3.8)$$
$$12x + 2 = 38$$
$$12x = 36$$
$$x = 3$$

How should it be solved?

48. From a consumer's standpoint, would you prefer that a retailer figure his profit on the basis of cost or the selling price of an item? Explain your answer.

■■■ Further investigations

49. A retailer buys an item for $40, resells it for $50, and claims that he is making only a 20% profit. Is his claim correct?

50. A store has a special discount sale of 40% off on all items. It also advertises an additional 10% off on items bought in quantities of a dozen or more. How much will it cost to buy a dozen items of some particular kind that regularly sell for $5 per item? (Be careful, a 40% discount followed by a 10% discount is not equal to a 50% discount.)

51. Is a 10% discount followed by a 40% discount the same as a 40% discount followed by a 10% discount? Justify your answer.

52. Some people use the following formula for determining the selling price of an item when the profit is based on a percent of the selling price.

$$\text{Selling price} = \frac{\text{Cost}}{100\% - \text{Percent of profit}}$$

Show how this formula is developed.

For Problems 55–62, solve each equation, and express the solution in decimal form. Your calculator might be helpful.

53. $2.4x + 5.7 = 9.6$

54. $-3.2x - 1.6 = 5.8$

55. $0.08x + 0.09(800 - x) = 68.5$

56. $0.10x + 0.12(720 - x) = 80$

57. $7x - 0.39 = 0.03$ **58.** $9x - 0.37 = 0.35$

59. $0.2(t + 1.6) = 3.4$ **60.** $0.4(t - 3.8) = 2.2$

Answers to Concept Quiz

1. False   **2.** True   **3.** False   **4.** False   **5.** True

4.3   Formulas: Geometric and Others

Objectives

■   Solve formulas for a specific variable when given the numerical values for the remaining variables.

■   Solve formulas for a specific variable.

■   Apply geometric formulas.

■   Solve an equation for a specific variable.

To find the distance traveled in 3 hours at a rate of 50 miles per hour, we multiply the rate by the time. Thus the distance is $50(3) = 150$ miles. We usually state the rule *distance equals rate times time* as a **formula:** $d = rt$. Formulas are simply rules that we state in symbolic language and express as equations. Thus the formula $d = rt$ is an equation that involves three variables: d, r, and t.

As we work with formulas, it is often necessary to solve for a specific variable when we know numerical values for the remaining variables. Consider the next examples.

EXAMPLE 1

Solve $d = rt$ for r if $d = 330$ and $t = 6$.

Solution

Substitute 330 for d and 6 for t in the given formula:

$$330 = r(6)$$

Solving this equation yields

$$330 = 6r$$
$$55 = r$$ ◼

EXAMPLE 2

Solve $C = \dfrac{5}{9}(F - 32)$ for F if C $= 10$. (This formula expresses the relationship between the Fahrenheit and Celsius temperature scales.)

Solution

Substitute 10 for C to obtain

$$10 = \frac{5}{9}(F - 32)$$

We can solve this equation.

$$\frac{9}{5}(10) = \frac{9}{5}\left(\frac{5}{9}\right)(F - 32) \qquad \text{Multiply both sides by } \frac{9}{5}.$$
$$18 = F - 32$$
$$50 = F$$ ◼

Sometimes it may be convenient to change a formula's form by using the properties of equality. For example, we can change the formula $d = rt$ as follows:

$$d = rt$$
$$\frac{d}{r} = \frac{rt}{r} \qquad \text{Divide both sides by } r.$$
$$\frac{d}{r} = t$$

We say that the formula $d = rt$ **has been solved for the variable t.** The formula can also be **solved for r:**

$$d = rt$$

$$\frac{d}{t} = \frac{rt}{t} \qquad \text{Divide both sides by } t.$$

$$\frac{d}{t} = r$$

Geometric Formulas

There are several formulas in geometry that we use quite often. Let's briefly review them at this time; we will use them periodically throughout the remainder of the text. These formulas (along with some others) and Figures 4.4 through 4.14 are also listed in the inside front cover of this text.

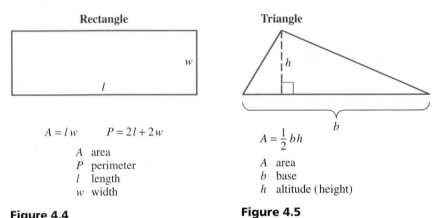

Rectangle

$A = lw \qquad P = 2l + 2w$

A area
P perimeter
l length
w width

Figure 4.4

Triangle

$A = \frac{1}{2}bh$

A area
b base
h altitude (height)

Figure 4.5

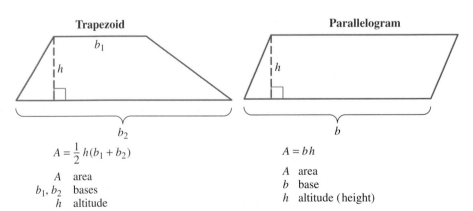

Trapezoid

$A = \frac{1}{2}h(b_1 + b_2)$

A area
b_1, b_2 bases
h altitude

Figure 4.6

Parallelogram

$A = bh$

A area
b base
h altitude (height)

Figure 4.7

Circle

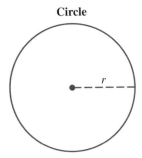

$A = \pi r^2 \qquad C = 2\pi r$

A area
C circumference
r radius

Figure 4.8

Sphere

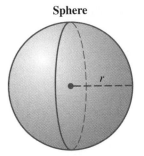

$V = \dfrac{4}{3}\pi r^3 \quad S = 4\pi r^2$

S surface area
V volume
r radius

Figure 4.9

Prism

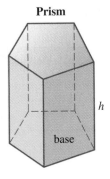

$V = Bh$

V volume
B area of base
h altitude (height)

Figure 4.10

Rectangular Prism

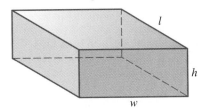

$V = lwh \qquad S = 2hw + 2hl + 2lw$

V volume
S total surface area
w width
l length
h altitude (height)

Figure 4.11

Pyramid

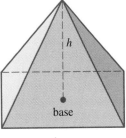

$V = \dfrac{1}{3}Bh$

V volume
B area of base
h altitude (height)

Figure 4.12

Right Circular Cylinder

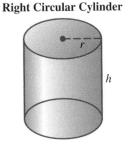

$V = \pi r^2 h \qquad S = 2\pi r^2 + 2\pi rh$

V volume
S total surface area
r radius
h altitude (height)

Figure 4.13

Right Circular Cone

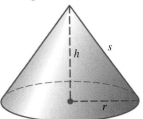

$V = \dfrac{1}{3}\pi r^2 h \qquad S = \pi r^2 + \pi rs$

V volume
S total surface area
r radius
h altitude (height)
s slant height

Figure 4.14

EXAMPLE 3 Solve $C = 2\pi r$ for r.

Solution

$$C = 2\pi r$$

$$\frac{C}{2\pi} = \frac{2\pi r}{2\pi} \qquad \text{Divide both sides by } 2\pi.$$

$$\frac{C}{2\pi} = r$$

EXAMPLE 4 Solve $V = \frac{1}{3}Bh$ for h.

Solution

$$V = \frac{1}{3}Bh$$

$$3(V) = 3\left(\frac{1}{3}Bh\right) \qquad \text{Multiply both sides by 3.}$$

$$3V = Bh$$

$$\frac{3V}{B} = \frac{Bh}{B} \qquad \text{Divide both sides by } B.$$

$$\frac{3V}{B} = h$$

EXAMPLE 5 Solve $P = 2l + 2w$ for w.

Solution

$$P = 2l + 2w$$

$$P - 2l = 2l + 2w - 2l \qquad \text{Subtract } 2l \text{ from both sides.}$$

$$P - 2l = 2w$$

$$\frac{P - 2l}{2} = \frac{2w}{2} \qquad \text{Divide both sides by 2.}$$

$$\frac{P - 2l}{2} = w$$

EXAMPLE 6 Find the total surface area of a right circular cylinder that has a radius of 10 inches and a height of 14 inches.

Solution

Let's sketch a right circular cylinder and record the given information as in Figure 4.15. We substitute 10 for r and 14 for h in the formula for finding the total surface area of a right circular cylinder.

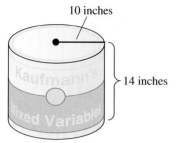

Figure 4.15

$$S = 2\pi r^2 + 2\pi rh$$
$$= 2\pi(10)^2 + 2\pi(10)(14)$$
$$= 200\pi + 280\pi$$
$$= 480\pi$$

The total surface area is 480π square inches.   ■

In Example 6 we used the figure to record the given information, and it also served as a reminder of the geometric figure under consideration. Now let's consider an example where a figure is very useful in the analysis of the problem.

E X A M P L E   7

A sidewalk 3 feet wide surrounds a rectangular plot of ground that measures 75 feet by 100 feet. Find the area of the sidewalk.

Solution

Let's make a sketch and record the given information as in Figure 4.16.

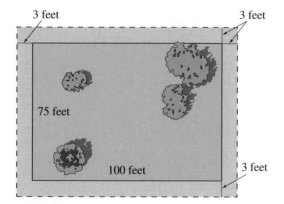

Figure 4.16

We can find the area of the sidewalk by subtracting the area of the rectangular plot from the area of the plot plus the sidewalk (the large dashed rectangle). The width of the large rectangle is $75 + 3 + 3 = 81$ feet, and its length is $100 + 3 + 3 = 106$ feet, so

$$A = (81)(106) - (75)(100)$$
$$= 8586 - 7500$$
$$= 1086$$

The area of the sidewalk is 1086 square feet.   ■

Changing Forms of Equations

In Chapter 5 you will be working with equations that contain two variables. At times you will need to solve for one variable in terms of the other variable — that is, to change the form of the equation as we have been doing with formulas. The next examples illustrate, once again, how we can use the properties of equality for such situations.

EXAMPLE 8 Solve $3x + y = 4$ for x.

Solution

$$3x + y = 4$$
$$3x + y - y = 4 - y \qquad \text{Subtract } y \text{ from both sides.}$$
$$3x = 4 - y$$
$$\frac{3x}{3} = \frac{4 - y}{3} \qquad \text{Divide both sides by 3.}$$
$$x = \frac{4 - y}{3}$$

EXAMPLE 9 Solve $4x - 5y = 7$ for y.

Solution

$$4x - 5y = 7$$
$$4x - 5y - 4x = 7 - 4x \qquad \text{Subtract } 4x \text{ from both sides.}$$
$$-5y = 7 - 4x$$
$$\frac{-5y}{-5} = \frac{7 - 4x}{-5} \qquad \text{Divide both sides by } -5.$$
$$y = \frac{7 - 4x}{-5}\left(\frac{-1}{-1}\right) \qquad \begin{array}{l}\text{Multiply numerator and denominator}\\ \text{of fraction on the right by } -1.\end{array}$$
$$y = \frac{4x - 7}{5} \qquad \begin{array}{l}\text{We commonly do this so that the}\\ \text{denominator is positive.}\end{array}$$

EXAMPLE 10 Solve $y = mx + b$ for m.

Solution

$$y = mx + b$$
$$y - b = mx + b - b \qquad \text{Subtract } b \text{ from both sides.}$$
$$y - b = mx$$
$$\frac{y - b}{x} = \frac{mx}{x} \qquad \text{Divide both sides by } x.$$
$$\frac{y - b}{x} = m$$

CONCEPT QUIZ

For Problems 1–10, match the correct formula for each.

1. Area of a rectangle

2. Circumference of a circle

3. Volume of a rectangular prism

4. Area of a triangle

5. Area of a circle

6. Volume of a right circular cylinder

7. Perimeter of a rectangle

8. Volume of a sphere

9. Area of a parallelogram

10. Area of a trapezoid

A. $A = \pi r^2$

B. $V = lwh$

C. $P = 2l + 2w$

D. $V = \dfrac{4}{3}\pi r^3$

E. $A = lw$

F. $A = bh$

G. $A = \dfrac{1}{2}h(b_1 + b_2)$

H. $A = \dfrac{1}{2}bh$

I. $C = 2\pi r$

J. $V = \pi r^2 h$

PROBLEM SET 4.3

For Problems 1–10, solve for the specified variable using the given facts.

1. Solve $d = rt$ for t if $d = 336$ and $r = 48$.

2. Solve $d = rt$ for r if $d = 486$ and $t = 9$.

3. Solve $i = Prt$ for P if $i = 200$, $r = 0.08$, and $t = 5$.

4. Solve $i = Prt$ for t if $i = 880$, $P = 2750$, and $r = 0.04$.

5. Solve $F = \dfrac{9}{5}C + 32$ for C if F = 68.

6. Solve $C = \dfrac{5}{9}(F - 32)$ for F if C = 15.

7. Solve $V = \dfrac{1}{3}Bh$ for B if $V = 112$ and $h = 7$.

8. Solve $V = \dfrac{1}{3}Bh$ for h if $V = 216$ and $B = 54$.

9. Solve $A = P + Prt$ for t if $A = 5080$, $P = 4000$, and $r = 0.03$.

10. Solve $A = P + Prt$ for P if $A = 1032$, $r = 0.06$, and $t = 12$.

For Problems 11–32, use the geometric formulas given in this section to help find solutions.

11. Find the perimeter of a rectangle that is 14 centimeters long and 9 centimeters wide.

12. If the perimeter of a rectangle is 80 centimeters and its length is 24 centimeters, find its width.

13. If the perimeter of a rectangle is 108 inches, and its length is $3\dfrac{1}{4}$ feet, find its width in inches.

14. How many yards of fencing would it take to enclose a rectangular plot of ground that is 69 feet long and 42 feet wide?

15. A dirt path 4 feet wide surrounds a rectangular garden that is 38 feet long and 17 feet wide. Find the area of the dirt path.

16. Find the area of a cement walk 3 feet wide that surrounds a rectangular plot of ground 86 feet long and 42 feet wide.

17. Suppose that paint costs $6.00 per liter and that 1 liter will cover 9 square meters of surface. We are going to paint (on one side only) 50 rectangular pieces of wood of the same size, which have a length of 60 centimeters and a width of 30 centimeters. What will be the total cost of the paint?

18. A lawn is in the shape of a triangle with one side 130 feet long and the altitude to that side 60 feet long. Will one sack of fertilizer, which covers 4000 square feet, be enough to fertilize the lawn?

19. Find the length of an altitude of a trapezoid with bases of 8 inches and 20 inches and an area of 98 square inches.

20. A flower garden is in the shape of a trapezoid with bases of 6 yards and 10 yards. The distance between the bases is 4 yards. Find the area of the garden.

21. The diameter of a metal washer is 4 centimeters, and the diameter of the hole is 2 centimeters (see Figure 4.17). How many square centimeters of metal are there in 50 washers? Express the answer in terms of π.

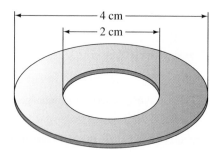

Figure 4.17

22. Find the area of a circular plot of ground that has a radius of length 14 meters. Use $3\frac{1}{7}$ as an approximation for π.

23. Find the area of a circular region that has a diameter of 1 yard. Express the answer in terms of π.

24. Find the area of a circular region if the circumference is 12π units. Express the answer in terms of π.

25. Find the total surface area and the volume of a sphere that has a radius 9 inches long. Express the answers in terms of π.

26. A circular pool is 34 feet in diameter and has a flagstone walk around it that is 3 feet wide (see Figure 4.18). Find the area of the walk. Express the answer in terms of π.

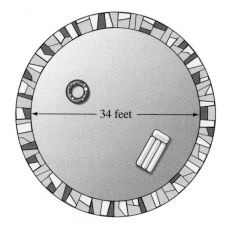

Figure 4.18

27. Find the volume and total surface area of a right circular cylinder that has a radius of 8 feet and a height of 18 feet. Express answers in terms of π.

28. Find the total surface area and volume of a sphere that has a diameter 12 centimeters long. Express the answers in terms of π.

29. If the volume of a right circular cone is 324π cubic inches and a radius of the base is 9 inches long, find the height of the cone.

30. Find the volume and total surface area of a tin can if the radius of the base is 3 centimeters, and the height of the can is 10 centimeters. Express answers in terms of π.

31. If the total surface area of a right circular cone is 65π square feet and a radius of the base is 5 feet long, find the slant height of the cone.

32. If the total surface area of a right circular cylinder is 104π square meters and a radius of the base is 4 meters long, find the height of the cylinder.

For Problems 33–44, solve each formula for the indicated variable. (Before doing these problems, cover the right-hand column and see how many of the formulas you recognize!)

33. $V = Bh$ for h Volume of a prism

34. $A = lw$ for l Area of a rectangle

35. $V = \frac{1}{3}Bh$ for B Volume of a pyramid

36. $A = \frac{1}{2}bh$ for h Area of a triangle

37. $P = 2l + 2w$ for w Perimeter of a rectangle

38. $V = \pi r^2 h$ for h Volume of a cylinder

39. $V = \frac{1}{3}\pi r^2 h$ for h Volume of a cone

40. $i = Prt$ for t Simple interest formula

41. $F = \frac{9}{5}C + 32$ for C   Celsius to Fahrenheit

42. $A = P + Prt$ for t Simple interest formula

43. $A = 2\pi r^2 + 2\pi rh$ for h Surface area of a cylinder

44. $C = \frac{5}{9}(F - 32)$ for F   Fahrenheit to Celsius

For Problems 45–60, solve each equation for the indicated variable.

45. $3x + 7y = 9$ for x

46. $5x + 2y = 12$ for x

47. $9x - 6y = 13$ for y

48. $3x - 5y = 19$ for y

49. $-2x + 11y = 14$ for x

50. $-x + 14y = 17$ for x

51. $y = -3x - 4$ for x

52. $y = -7x + 10$ for x

53. $\frac{x - 2}{4} = \frac{y - 3}{6}$ for y

54. $\frac{x + 1}{3} = \frac{y - 5}{2}$ for y

55. $ax - by - c = 0$ for y

56. $ax + by = c$ for y

57. $\frac{x + 6}{2} = \frac{y + 4}{5}$ for x

58. $\frac{x - 3}{6} = \frac{y - 4}{8}$ for x

59. $m = \frac{y - b}{x}$ for y

60. $y = mx + b$ for x

■ ■ ■ **Thoughts into words**

61. Suppose that both the length and width of a rectangle are doubled. How does this affect the perimeter of the rectangle? Defend your answer.

62. Suppose that the length of the radius of a circle is doubled. How does this affect the area of the circle? Defend your answer.

63. Some people *subtract 32 and then divide by 2* to estimate the change from a Fahrenheit temperature reading to a Celsius reading. Why does this give an estimate and how good is the estimate?

■ ■ ■ **Further investigations**

For each of the following problems, use 3.14 as an approximation for π. Your calculator should be helpful with these problems.

64. Find the area of a circular plot of ground that has a radius 16.3 meters long. Express your answer to the nearest tenth of a square meter.

65. Find, to the nearest tenth of a square centimeter, the area of the shaded ring in Figure 4.19.

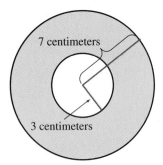

7 centimeters

3 centimeters

Figure 4.19

66. Find, to the nearest square inch, the area of each of these pizzas: 10-inch diameter, 12-inch diameter, and 14-inch diameter.

67. Find, to the nearest square centimeter, the total surface area of the tin can in Figure 4.20.

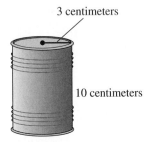

3 centimeters

10 centimeters

Figure 4.20

68. Find, to the nearest square centimeter, the total surface area of a baseball that has a radius of 4 centimeters (see Figure 4.21).

4 centimeters

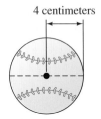

Figure 4.21

69. Find, to the nearest cubic inch, the volume of a softball that has a diameter of 5 inches (see Figure 4.22).

5-inch diameter

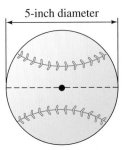

Figure 4.22

70. Find, to the nearest cubic meter, the volume of the figure shown in Figure 4.23.

8 meters

20 meters

12 meters

Figure 4.23

4.4 Problem Solving

Objectives

■ Apply problem solving techniques such as drawing diagrams, sketching figures, and using a guideline to solve word problems.

■ Solve word problems involving simple interest.

■ Solve word problems involving the perimeter of rectangles, triangles, or circles.

■ Solve word problems involving distance, rate, and time.

Let's begin this section by restating the suggestions for solving word problems that we offered in Section 3.3.

Suggestions for Solving Word Problems

1. Read the problem carefully, and make sure that you understand the meanings of all the words. Be especially alert for any technical terms used in the statement of the problem.

2. Read the problem a second time (perhaps even a third time) to get an overview of the situation being described and to determine the known facts as well as what is to be found.

3. Sketch any figure, diagram, or chart that might be helpful in analyzing the problem.

4. Choose a meaningful variable to represent an unknown quantity in the problem (perhaps *t* if time is the unknown quantity); represent any other unknowns in terms of that variable.

5. Look for a guideline that can be used to set up an equation. A guideline might be a formula, such as *selling price equals cost plus profit,* or a relationship, such as *interest earned from a 9% investment plus interest earned from a 10% investment equals total amount of interest earned.* A guideline may also be illustrated by a figure or diagram that you sketch for a particular problem.

6. Form an equation that contains the variable that translates the conditions of the guideline from English into algebra.

7. Solve the equation, and use the solution to determine all the facts requested in the problem.

8. **Check all answers back in the original statement of the problem.**

Again we emphasize the importance of suggestion 5. Determining the guideline to follow when setting up the equation is a vital part of the analysis of a problem. Sometimes the guideline is a formula — such as one of the formulas we presented in the previous section and accompanying problem set. Let's consider a problem of that type.

PROBLEM 1 How long will it take $500 to double itself if it is invested at 8% simple interest?

Solution

Let's use the basic simple interest formula, $i = Prt$, where i represents interest, P is the principal (money invested), r is the rate (percent), and t is the time in years. For $500 to "double itself" means that we want the original $500 to earn another $500 in interest. Thus using $i = Prt$ as a guideline, we can proceed as follows:

$$i = Prt$$
$$500 = 500(8\%)(t)$$

Now let's solve this equation:

$$500 = 500(0.08)(t)$$
$$1 = 0.08t$$
$$100 = 8t$$
$$\frac{100}{8} = t$$
$$12\frac{1}{2} = t$$

It will take $12\frac{1}{2}$ years. ∎

If the problem involves a geometric formula, then a sketch of the figure is helpful for recording the given information and analyzing the problem. The next problem illustrates this idea.

PROBLEM 2 The length of a football field is 40 feet more than twice its width, and the perimeter of the field is 1040 feet. Find the length and width of the field.

Solution

Because the length is stated in terms of the width, we can let w represent the width, and then $2w + 40$ represents the length (see Figure 4.24). A guideline for this problem is the perimeter formula $P = 2l + 2w$. Thus we use the following equation to set up and solve the problem:

$$P = 2l + 2w$$
$$1040 = 2(2w + 40) + 2w$$
$$1040 = 4w + 80 + 2w$$
$$1040 = 6w + 80$$
$$960 = 6w$$
$$160 = w$$

If $w = 160$, then $2w + 40 = 2(160) + 40 = 360$. Thus the football field is 360 feet long and 160 feet wide.

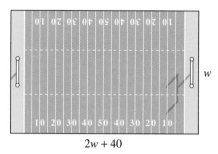

$2w + 40$

Figure 4.24

Sometimes the formulas we use when we are analyzing a problem are different from those we use as a guideline for setting up the equation. For example, uniform-motion problems involve the formula $d = rt$, but the main guideline for setting up an equation for such problems is usually a statement about either *times*, *rates*, or *distances*. Let's consider an example.

PROBLEM 3

Pablo leaves city A on a moped and travels toward city B at 18 miles per hour. At the same time, Cindy leaves city B on a moped and travels toward city A at 23 miles per hour. The distance between the two cities is 123 miles. How long will it take before Pablo and Cindy meet on their mopeds?

Solution

First, sketch a diagram as in Figure 4.25. Then let t represent the time that Pablo travels and also the time that Cindy travels.

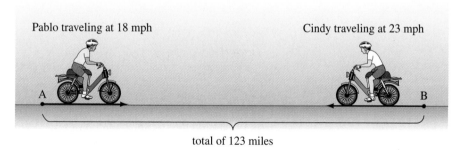

Pablo traveling at 18 mph   Cindy traveling at 23 mph

A   B

total of 123 miles

Figure 4.25

Distance Pablo travels + Distance Cindy travels = Total distance

$$18t \qquad + \qquad 23t \qquad = \qquad 123$$

Solving this equation yields

$$18t + 23t = 123$$
$$41t = 123$$
$$t = 3$$

They both travel for 3 hours.   ■

Some people find it helpful to use a chart to organize the known and unknown facts in a uniform-motion problem. We will illustrate with an example.

P R O B L E M   4   A car leaves a town traveling at 60 kilometers per hour. How long will it take a second car traveling at 75 kilometers per hour to catch the first car if the second car leaves 1 hour later and travels the same route?

    Solution

Let t represent the time of the second car. Then $t + 1$ represents the time of the first car, because it travels 1 hour longer. We can now record the information of the problem in a chart.

	Rate	Time	Distance
First car	60	$t + 1$	$60(t + 1)$
Second car	75	t	$75t$ $d = rt$

Because the second car is to overtake the first car, the distances must be equal.

Distance of second car = Distance of first car
$$75t \qquad = \qquad 60(t + 1)$$

We can solve this equation:

$$75t = 60(t + 1)$$
$$75t = 60t + 60$$
$$15t = 60$$
$$t = 4$$

The second car should overtake the first car in 4 hours. (Check the answer!)   ■

We should like to offer one bit of advice at this time. Don't become discouraged if solving word problems is giving you trouble. Problem solving is not a skill that you can develop overnight. It takes time, patience, hard work, and an open mind. Keep giving it your best shot, and gradually you should become more confident in your approach to such problems. Furthermore we realize that some (perhaps many) of these problems may not seem "practical" to you; however, keep in mind that the real goal here is to develop your skill in applying problem-solving techniques. Finding and using a guideline, sketching a figure to record information and help in the analysis, estimating an answer before attempting to solve the problem, and using a chart to record information are the key issues.

C O N C E P T   Q U I Z

Arrange the following steps for solving word problems into the correct order.

A. Declare a variable and represent any other unknown quantities in terms of that variable.

B. Check the answer back in the original statement of the problem.

C. Write an equation for the problem and remember to look for a formula or guideline that could be used to write the equation.

D. Read the problem carefully and be sure that you understand all the terms in the stated problem.

E. Sketch a diagram or figure that helps you analyze the problem.

F. Solve the equation and determine the answer to the question asked in the problem.

PROBLEM SET 4.4

For Problems 1–12, solve each equation. These equations are of the types you will be using in Problems 13–40.

1. $950(0.12)t = 950$

2. $1200(0.09)t = 1200$

3. $l + \dfrac{1}{4}l - 1 = 19$

4. $l + \dfrac{2}{3}l + 1 = 41$

5. $500(0.08)t = 1000$

6. $800(0.11)t = 1600$

7. $s + (2s - 1) + (3s - 4) = 37$

8. $s + (3s - 2) + (4s - 4) = 42$

9. $\dfrac{5}{2}r + \dfrac{5}{2}(r + 6) = 135$

10. $\dfrac{10}{3}r + \dfrac{10}{3}(r - 3) = 90$

11. $24\left(t - \dfrac{2}{3}\right) = 18t + 8$

12. $16t + 8\left(\dfrac{9}{2} - t\right) = 60$

Set up an equation, and solve each of the following problems. Keep in mind the suggestions we offered in this section.

13. How long will it take $4000 to double itself if it is invested at 8% simple interest?

14. How many years will it take $4000 to double itself if it is invested at 5% simple interest?

15. How long will it take $8000 to triple itself if it is invested at 6% simple interest?

16. How many years will it take $500 to earn $750 in interest if it is invested at 6% simple interest?

17. The length of a rectangle is three times its width. If the perimeter of the rectangle is 112 inches, find its length and width.

18. The width of a rectangle is one-half of its length. If the perimeter of the rectangle is 54 feet, find its length and width.

19. Suppose that the length of a rectangle is 2 centimeters less than three times its width. The perimeter of the rectangle is 92 centimeters. Find the length and width of the rectangle.

20. Suppose that the length of a certain rectangle is 1 meter more than five times its width. The perimeter of the rectangle is 98 meters. Find the length and width of the rectangle.

21. The width of a rectangle is 3 inches less than one-half of its length. If the perimeter of the rectangle is 42 inches, find the area of the rectangle.

22. The width of a rectangle is 1 foot more than one-third of its length. If the perimeter of the rectangle is 74 feet, find the area of the rectangle.

23. The perimeter of a triangle is 100 feet. The longest side is 3 feet less than twice the shortest side, and the third side is 7 feet longer than the shortest side. Find the lengths of the sides of the triangle.

24. A triangular plot of ground has a perimeter of 54 yards. The longest side is twice the shortest side, and the third side is 2 yards longer than the shortest side. Find the lengths of the sides of the triangle.

25. The second side of a triangle is 1 centimeter more than three times the first side. The third side is 2 centimeters longer than the second side. If the perimeter is 46 centimeters, find the length of each side of the triangle.

26. The second side of a triangle is 3 meters less than twice the first side. The third side is 4 meters longer than the second side. If the perimeter is 58 meters, find the length of each side of the triangle.

27. The perimeter of an equilateral triangle is 4 centimeters more than the perimeter of a square, and the length of a side of the triangle is 4 centimeters more than the length of a side of the square. Find the length of a side of the equilateral triangle. (An equilateral triangle has three sides of the same length.)

28. Suppose that a square and an equilateral triangle have the same perimeter. Each side of the equilateral triangle is 6 centimeters longer than each side of the square. Find the length of each side of the square. (An equilateral triangle has three sides of the same length.)

29. Suppose that the length of a radius of a circle is the same as the length of a side of a square. If the circumference of the circle is 15.96 centimeters greater than the perimeter of the square, find the length of a radius of the circle. (Use 3.14 as an approximation for π.)

30. The circumference of a circle is 2.24 centimeters more than six times the length of a radius. Find the radius of the circle. (Use 3.14 as an approximation for π.)

31. Sandy leaves a town traveling in her car at a rate of 45 miles per hour. One hour later, Monica leaves the same town traveling the same route at a rate of 50 miles per hour. How long will it take Monica to overtake Sandy?

32. Two cars start from the same place traveling in opposite directions. One car travels 4 miles per hour faster than the other car. Find their speeds if after 5 hours they are 520 miles apart.

33. The distance between city A and city B is 325 miles. A freight train leaves city A and travels toward city B at 40 miles per hour. At the same time, a passenger train leaves city B and travels toward city A at 90 miles per hour. How long will it take the two trains to meet?

34. Kirk starts jogging at 5 miles per hour. Half an hour later Consuela starts jogging on the same route at 7 miles per hour. How long will it take Consuela to catch Kirk?

35. A car leaves a town at 40 miles per hour. Two hours later, a second car leaves the town traveling the same route and overtakes the first car in 5 hours and 20 minutes. How fast was the second car traveling?

36. Two airplanes leave St. Louis at the same time and fly in opposite directions. If one travels at 500 kilometers per hour and the other at 600 kilometers per hour, how long will it take for them to be 1925 kilometers apart (see Figure 4.26)?

Figure 4.26

37. Two trains leave from the same station at the same time, one traveling east and the other traveling west.

At the end of $9\frac{1}{2}$ hours they are 1292 miles apart. If the rate of the train traveling east is 8 miles per hour faster than the other train, find their rates.

38. Dawn started on a 58-mile trip on her moped at 20 miles per hour. After a while the motor "kicked out," and she pedaled the remainder of the trip at 12 miles per hour. The entire trip took $3\frac{1}{2}$ hours. How far had Dawn traveled when the motor on the moped quit running?

39. Jeff leaves home and rides his bicycle out into the country for 3 hours. On his return trip along the same route, it takes him three-quarters of an hour longer. If his rate on the return trip was 2 miles per hour slower than his rate on the trip out into the country, find the total round-trip distance.

40. In $1\frac{1}{4}$ hours more time, Rita, riding her bicycle at 12 miles per hour, rode 2 miles farther than Sonya, who was riding her bicycle at 16 miles per hour. How long did each girl ride?

■ ■ ▢ **Thoughts into words**

41. Suppose that your friend analyzes Problem 31 as follows: "Sandy has traveled 45 miles before Monica starts. Because Monica travels 5 miles per hour faster than Sandy, it will take her $\frac{45}{5} = 9$ hours to catch Sandy." How would you react to this analysis of the problem?

42. Summarize the new ideas about problem solving that you have acquired thus far in this course.

Answers to Concept Quiz

D E A C F B

4.5   More About Problem Solving

Objectives

■   Solve word problems involving mixture.

■   Solve word problems involving age.

■   Solve word problems involving distance, rate, and time.

Let's begin this section with an important but often overlooked facet of problem solving: the importance of *looking back* over your solution and considering some of the following questions:

   1. Is your answer to the problem a reasonable answer? Does it agree with the answer you estimated before doing the problem?
   2. Have you checked your answer by substituting it back into the conditions stated in the problem?
   3. Do you now see another plan that you could use to solve the problem? Perhaps even another guideline could be used.

4. Do you now see that this problem is closely related to another problem that you have previously solved?

5. Have you "tucked away for future reference" the technique you used to solve this problem?

Looking back over the solution of a newly solved problem can often lay important groundwork for solving problems in the future.

Now let's consider three problems that we often refer to as mixture problems. No basic formula applies for all of these problems, but the suggestion that *you think in terms of a pure substance* is often helpful in setting up a guideline. For example, a phrase such as "a 30% solution of acid" means that 30% of the amount of solution is acid and the remaining 70% is water.

PROBLEM 1   How many milliliters of pure acid must be added to 150 milliliters of a 30% solution of acid to obtain a 40% solution? (See Figure 4.27.)

REMARK: If a guideline is not obvious from reading the problem, you may want to guess an answer and then check that guess. Suppose we guess that 30 milliliters of pure acid need to be added. To check we must determine whether the final solution is 40% acid. Because we started with $0.30(150) = 45$ milliliters of pure acid and added our guess of 30 milliliters, the final solution will have $45 + 30 = 75$ milliliters of pure acid. The final amount of solution is $150 + 30 = 180$ milliliters. Thus the final solution is $\dfrac{75}{180} = 41\dfrac{2}{3}\%$ pure acid.

Solution

We hope that by guessing and checking our guess, we recognize this guideline:

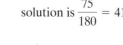

| Amount of pure acid in original solution | $+$ | Amount of pure acid to be added | $=$ | Amount of pure acid in final solution |

150 milliliters
30% solution

Figure 4.27

Let p represent the amount of pure acid to be added. Then, using the guideline, we form this equation:

$$(30\%)(150) + p = 40\%(150 + p)$$

Now let's solve this equation to determine the amount of pure acid to be added.

$$(0.30)(150) + p = 0.40(150 + p)$$
$$45 + p = 60 + 0.4p$$
$$0.6p = 15$$
$$p = \frac{15}{0.6} = 25$$

We must add 25 milliliters of pure acid. (You should check this answer.)   ■

PROBLEM 2

Suppose that you have a supply of a 30% solution of alcohol and a 70% solution of alcohol. How many quarts of each should be mixed to produce a 20-quart solution that is 40% alcohol?

Solution

We can use a guideline similar to the one in Problem 1:

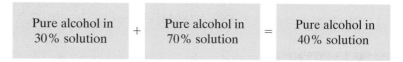

| Pure alcohol in 30% solution | + | Pure alcohol in 70% solution | = | Pure alcohol in 40% solution |

Let x represent the amount of 30% solution. Then $20 - x$ represents the amount of 70% solution. Now, using the guideline, we translate to

$$(30\%)(x) + (70\%)(20 - x) = (40\%)(20)$$

Solving this equation, we obtain

$$0.30x + 0.70(20 - x) = 8$$
$$30x + 70(20 - x) = 800$$
$$30x + 1400 - 70x = 800$$
$$-40x = -600$$
$$x = 15$$

Therefore, $20 - x = 5$. We should mix 15 quarts of the 30% solution with 5 quarts of the 70% solution. ∎

PROBLEM 3

A 4-gallon radiator is full and contains a 40% solution of antifreeze. How much needs to be drained out and replaced with pure antifreeze to obtain a 70% solution?

Solution

We can use this guideline:

| Pure antifreeze in the original solution | − | Pure antifreeze in the solution drained out | + | Pure antifreeze added | = | Pure antifreeze in the final solution |

Let x represent the amount of pure antifreeze to be added. Then x also represents the amount of the 40% solution to be drained out. Thus the guideline translates into the following equation:

$$(40\%)(4) - (40\%)(x) + x = (70\%)(4)$$

We can then solve this equation:

$$0.4(4) - 0.4x + x = 0.7(4)$$
$$1.6 + 0.6x = 2.8$$
$$0.6x = 1.2$$
$$x = 2$$

Therefore, we must drain out 2 gallons of the 40% solution and then add 2 gallons of pure antifreeze. (Checking this answer is a worthwhile exercise!)   ■

P R O B L E M   4   A woman invests a total of $5000. Part of it is invested at 4% and the remainder at 6%. Her total yearly interest from the two investments is $260. How much did she invest at each rate?

Solution

Let x represent the amount invested at 6%. Then $5000 - x$ represents the amount invested at 4%. Use the following guideline:

Interest earned from 6% investment	+	Interest earned from 4% investment	=	Total interest earned
↓		↓		↓
$(6\%)(x)$	+	$(4\%)(\$5000 - x)$	=	$\$260$

Solving this equation yields

$$(6\%)(x) + (4\%)(5000 - x) = 260$$
$$0.06x + 0.04(5000 - x) = 260$$
$$6x + 4(5000 - x) = 26{,}000$$
$$6x + 20{,}000 - 4x = 26{,}000$$
$$2x + 20{,}000 = 26{,}000$$
$$2x = 6000$$
$$x = 3000$$

Therefore, $5000 - x = 2000$.

She invested $3000 at 6% and $2000 at 4%.   ■

P R O B L E M   5   An investor invests a certain amount of money at 3%. Then he finds a better deal and invests $5000 more than that amount at 5%. His yearly income from the two investments is $650. How much did he invest at each rate?

Solution

Let x represent the amount invested at 3%. Then $x + 5000$ represents the amount invested at 5%.

$$(3\%)(x) + (5\%)(x + 5000) = 650$$
$$0.03x + 0.05(x + 5000) = 650$$

$$3x + 5(x + 5000) = 65{,}000$$
$$3x + 5x + 25{,}000 = 65{,}000$$
$$8x + 25{,}000 = 65{,}000$$
$$8x = 40{,}000$$
$$x = 5000$$

Therefore, $x + 5000 = 10{,}000$.

He invested $5000 at 3% and $10,000 at 5%.   ■

We now consider a problem where the key to solving the problem is the process of representing the various unknown quantities in terms of one variable.

PROBLEM 6   Jody is 6 years younger than her sister Cathy, and in 7 years Jody will be three-fourths as old as Cathy is at that time. Find their present ages.

Solution

By letting c represent Cathy's present age, we can represent all of the unknown quantities like this:

c:	Cathy's present age
$c - 6$:	Jody's present age
$c + 7$:	Cathy's age in 7 years
$c - 6 + 7$ or $c + 1$:	Jody's age in 7 years

The statement *in 7 years Jody will be three-fourths as old as Cathy is at that time* serves as the guideline, so we can set up and solve the following equation:

$$c + 1 = \frac{3}{4}(c + 7)$$
$$4c + 4 = 3(c + 7)$$
$$4c + 4 = 3c + 21$$
$$c = 17$$

Therefore, Cathy's present age is 17 and Jody's present age is $17 - 6 = 11$.   ■

CONCEPT QUIZ

For Problems 1–7, answer true or false.

1. The phrase "a 40% solution of alcohol" means that 40% of the amount of solution is alcohol.

2. The amount of pure acid in 300 ml of a 30% solution is 100 ml.

3. If we want to produce 10 quarts by mixing solution A and solution B, the amount of solution A needed could be represented by x and the amount of solution B would then be represented by $10 - x$.

4. The formula $d = rt$ is equivalent to $r = \dfrac{d}{t}$.

5. The formula $d = rt$ is equivalent to $t = \dfrac{r}{d}$.

6. If y represents John's current age, then his age four years ago would be represented by $y - 4$.

7. If Shane's current age is represented by x, then his age in 10 years would be represented by $10x$.

PROBLEM SET 4.5

For Problems 1–12, solve each equation. You will be using these types of equations in Problems 13–37.

1. $0.3x + 0.7(20 - x) = 0.4(20)$

2. $0.4x + 0.6(50 - x) = 0.5(50)$

3. $0.2(20) + x = 0.3(20 + x)$

4. $0.3(32) + x = 0.4(32 + x)$

5. $0.7(15) - x = 0.6(15 - x)$

6. $0.8(25) - x = 0.7(25 - x)$

7. $0.4(10) - 0.4x + x = 0.5(10)$

8. $0.2(15) - 0.2x + x = 0.4(15)$

9. $20x + 12\left(4\dfrac{1}{2} - x\right) = 70$

10. $30x + 14\left(3\dfrac{1}{2} - x\right) = 97$

11. $3t = \dfrac{11}{2}\left(t - \dfrac{3}{2}\right)$ **12.** $5t = \dfrac{7}{3}\left(t + \dfrac{1}{2}\right)$

Set up an equation and solve each of the following problems.

13. How many milliliters of pure acid must be added to 100 milliliters of a 10% acid solution to obtain a 20% solution?

14. How many liters of pure alcohol must be added to 20 liters of a 40% solution to obtain a 60% solution?

15. How many centiliters of distilled water must be added to 10 centiliters of a 50% acid solution to obtain a 20% acid solution?

16. How many milliliters of distilled water must be added to 50 milliliters of a 40% acid solution to reduce it to a 10% acid solution?

17. Suppose that we want to mix some 30% alcohol solution with some 50% alcohol solution to obtain 10 quarts of a 35% solution. How many quarts of each kind should we use?

18. We have a 20% alcohol solution and a 50% solution. How many pints must be used from each to obtain 8 pints of a 30% solution?

19. How much water needs to be removed from 20 gallons of a 30% salt solution to change it to a 40% salt solution?

20. How much water needs to be removed from 30 liters of a 20% salt solution to change it to a 50% salt solution?

21. Suppose that a 12-quart radiator contains a 20% solution of antifreeze. How much solution needs to be drained out and replaced with pure antifreeze to obtain a 40% solution of antifreeze?

22. A tank contains 50 gallons of a 40% solution of antifreeze. How much solution needs to be drained out and replaced with pure antifreeze to obtain a 50% solution?

23. How many gallons of a 15% salt solution must be mixed with 8 gallons of a 20% salt solution to obtain a 17% salt solution?

24. How many liters of a 10% salt solution must be mixed with 15 liters of a 40% salt solution to obtain a 20% salt solution?

25. Thirty ounces of a punch containing 10% grapefruit juice is added to 50 ounces of a punch containing 20%

grapefruit juice. Find the percent of grapefruit juice in the resulting mixture.

26. Suppose that 20 gallons of a 20% salt solution are mixed with 30 gallons of a 25% salt solution. What is the percent of salt in the resulting solution?

27. Suppose that the perimeter of a square equals the perimeter of a rectangle. The width of the rectangle is 9 inches less than twice the side of the square, and the length of the rectangle is 3 inches less than twice the side of the square. Find the dimensions of the square and the rectangle.

28. The perimeter of a triangle is 40 centimeters. The longest side is 1 centimeter longer than twice the shortest side. The other side is 2 centimeters shorter than the longest side. Find the lengths of the three sides.

29. Andy starts walking from point A at 2 miles per hour. Half an hour later, Aaron starts walking from point A at $3\frac{1}{2}$ miles per hour and follows the same route. How long will it take Aaron to catch up with Andy?

30. Suppose that Karen, riding her bicycle at 15 miles per hour, rode 10 miles farther than Michelle, who was riding her bicycle at 14 miles per hour. Karen rode for 30 minutes longer than Michelle. How long did Michelle and Karen each ride their bicycles?

31. Pam is half as old as her brother Bill. Six years ago, Bill was four times older than Pam. How old is each now?

32. Suppose that the sum of the present ages of Tom and his father is 100 years. Ten years ago, Tom's father was three times as old as Tom was at that time. Find their present ages.

33. Suppose that Cory invested a certain amount of money at 3% interest and $750 more than that amount at 5%. His total yearly interest was $157.50. How much did he invest at each rate?

34. Nina received an inheritance of $12,000 from her grandmother. She invested part of it at 6% interest and the remainder at 8%. If the total yearly interest from both investments was $860, how much did she invest at each rate?

35. Udit received $1200 from his parents as a graduation present. He invested part of it at 4% interest and the remainder at 6%. If the total yearly interest amounted to $62, how much did he invest at each rate?

36. Sally invested a certain sum of money at 3%, twice that sum at 4%, and three times that sum at 6%. Her total yearly interest from all three investments was $145. How much did she invest at each rate?

37. If $2000 is invested at 4% interest, how much money must be invested at 7% interest so that the total return for both investments averages 6%?

38. Fawn invested a certain amount of money at 3% interest and $1250 more than that amount at 5%. Her total yearly interest was $134.50. How much did she invest at each rate?

39. A sum of $2300 is invested, part of it at 6% interest and the remainder at 8%. If the interest earned by the 8% investment is $100 more than the interest earned by the 6% investment, find the amount invested at each rate.

40. If $3000 is invested at 9% interest, how much money must be invested at 12% so that the total return for both investments averages 11%?

41. How can $5400 be invested, part of it at 8% and the remainder at 10%, so that the two investments will produce the same amount of interest?

42. A sum of $6000 is invested, part of it at 5% interest and the remainder at 7%. If the interest earned by the 5% investment is $160 less than the interest earned by the 7% investment, find the amount invested at each rate.

Answers to Concept Quiz

1. True   **2.** False   **3.** True   **4.** True   **5.** False   **6.** True   **7.** False

SUMMARY

(4.1) A **ratio** is the comparison of two numbers by division.

A statement of equality between two ratios is a **proportion**. In a proportion, the *cross products* are equal; that is to say,

$$\text{If } \frac{a}{b} = \frac{c}{d}, \quad \text{then } ad = bc, \quad b \neq 0 \text{ and } d \neq 0$$

We can use the cross product property to solve equations that are in the form of a proportion.

A variety of word problems can be set up and solved using proportions.

The concept of **percent** means *per one hundred* and is therefore a special ratio — a ratio that has a denominator of 100. Proportions provide a convenient way to change common fractions to percents.

(4.2) To solve equations that contain decimals, we can clear the equation of all decimals by multiplying both sides by an appropriate power of 10.

Many consumer problems involve the concept of *percent* and can be solved with an equation approach. We frequently use the basic relationships *selling price equals cost plus profit* and *original selling price minus discount equals discount sale price.*

(4.3) Formulas are rules stated in symbolic form. We can solve a formula such as $P = 2l + 2w$ for $l \left(l = \dfrac{P - 2w}{2} \right)$ or for $w \left(w = \dfrac{P - 2l}{2} \right)$ by applying the properties of equality. Many of the formulas used in this section connect algebra, geometry, and the real world.

(4.4) and **(4.5)** Don't forget these suggestions for solving word problems:

1. Read the problem carefully.
2. Sketch a figure, diagram, or chart that might be helpful to organize the facts.
3. Choose a meaningful variable.
4. Look for a guideline.
5. Use the guideline to help set up an equation.
6. Solve the equation and determine the facts requested in the problem.
7. Check your answers back in the original statement of the problem.

Item 4, determining a guideline, is often the key component when solving a problem. Many times we use formulas as guidelines. Reviewing the examples in Sections 4.4 and 4.5 should help you better understand the role of formulas in problem solving.

CHAPTER 4 REVIEW PROBLEM SET

In Problems 1–5, solve each equation.

1. $0.5x + 0.7x = 1.7$

2. $0.07t + 0.12(t - 3) = 0.59$

3. $0.1x + 0.12(1700 - x) = 188$

4. $x - 0.25x = 12$

5. $0.2(x - 3) = 14$

6. Solve $P = 2l + 2w$ for w if $P = 50$ and $l = 19$.

7. Solve $F = \dfrac{9}{5}C + 32$ for C if $F = 77$.

8. Solve $A = P + Prt$ for t.

9. Solve $2x - 3y = 13$ for x.

10. Find the area of a trapezoid that has one base 8 inches long and the other base 14 inches long, if the altitude between the two bases is 7 inches.

11. If the area of a triangle is 27 square centimeters and the length of one side is 9 centimeters, find the length of the altitude to that side.

12. If the total surface area of a right circular cylinder is 152π square feet and a radius of a base is 4 feet long, find the height of the cylinder.

Set up an equation and solve each of the following problems.

13. Eighteen is what percent of 30?

14. The sum of two numbers is 96 and their ratio is 5 to 7. Find the numbers.

15. Fifteen percent of a certain number is 6. Find the number.

16. Suppose that the length of a certain rectangle is 5 meters more than twice the width. The perimeter of the rectangle is 46 meters. Find the length and width of the rectangle.

17. Two airplanes leave from the same airport in Chicago at the same time and fly in opposite directions. If one travels at 350 miles per hour and the other at 400 miles per hour, how long will it be before they are 1125 miles apart?

18. How many liters of pure alcohol must be added to 10 liters of a 70% solution to obtain a 90% solution?

19. A copper wire 110 centimeters long was bent in the shape of a rectangle. The length of the rectangle was 10 centimeters more than twice the width. Find the dimensions of the rectangle.

20. Seventy-eight yards of fencing were purchased to enclose a rectangular garden. The length of the garden is 1 yard less than three times its width. Find the length and width of the garden.

21. The ratio of the complement of an angle to the supplement of the angle is 7 to 16. Find the measure of the angle.

22. If a car uses 18 gallons of gasoline for a 369-mile trip, at the same rate of consumption how many gallons will it use on a 615-mile trip?

23. A sum of $2100 is invested, part of it at 3% interest and the remainder at 5%. If the interest earned by the 5% investment is $51 more than the interest earned by the 3% investment, find the amount invested at each rate.

24. A retailer has some sweaters that cost $28 each. At what price should the sweaters be sold to obtain a profit of 30% of the selling price?

25. Anastasia bought a dress on sale for $39, and the original price of the dress was $60. She received a discount of what percent?

26. One angle of a triangle has a measure of 47°. Of the other two angles, one of them is 3° less than three times the other angle. Find the measures of the two remaining angles.

27. Connie rides out into the country on her bicycle at a speed of 10 miles per hour. An hour later Jay leaves from the same place that Connie did and rides his bicycle along the same route at 12 miles per hour. How long will it take Jay to catch Connie?

28. How many gallons of a 10% salt solution must be mixed with 12 gallons of a 15% salt solution to obtain a 12% salt solution?

29. Suppose that 20 ounces of a punch containing 20% orange juice is added to 30 ounces of a punch containing 30% orange juice. Find the percent of orange juice in the resulting mixture.

30. How much interest is due on a 2-year student loan when $3500 is borrowed at a 5.25% annual interest rate?

CHAPTER 4 *TEST*

📼 **applies to all problems in this Chapter Test.**

For Problems 1–10, solve each equation.

1. $\dfrac{x+2}{4} = \dfrac{x-3}{5}$

2. $\dfrac{-4}{2x-1} = \dfrac{3}{3x+5}$

3. $\dfrac{x-1}{6} - \dfrac{x+2}{5} = 2$

4. $\dfrac{x+8}{7} - 2 = \dfrac{x-4}{4}$

5. $\dfrac{n}{20-n} = \dfrac{7}{3}$

6. $\dfrac{h}{4} + \dfrac{h}{6} = 1$

7. $0.05n + 0.06(400 - n) = 23$

8. $s = 35 + 0.5s$

9. $0.07n = 45.5 - 0.08(600 - n)$

10. $12t + 8\left(\dfrac{7}{2} - t\right) = 50$

11. Solve $F = \dfrac{9C + 160}{5}$ for C.

12. Solve $y = 2(x - 4)$ for x.

13. Solve $\dfrac{x+3}{4} = \dfrac{y-5}{9}$ for y.

For Problems 14–16, use the geometric formulas given in this chapter to help you find the solution.

14. Find the area of a circular region if the circumference is 16π centimeters. Express the answer in terms of π.

15. If the perimeter of a rectangle is 100 inches and its length is 32 inches, find the area of the rectangle.

16. The area of a triangular plot of ground is 133 square yards. If the length of one side of the plot is 19 yards, find the length of the altitude to that side.

For Problems 17–25, set up an equation and solve.

17. Express $\dfrac{5}{4}$ as a percent.

18. Thirty-five percent of what number is 24.5?

19. Cora bought a digital camera for $132.30, which represented a 30% discount off the original price. What was the original price of the camera?

20. A retailer has some lamps that cost her $40 each. She wants to sell them at a profit of 30% of the cost. What price should she charge for the lamps?

21. Hugh paid $48 for a pair of golf shoes that were listed for $80. What rate of discount did he receive?

22. The election results in a certain precinct indicated that the ratio of female voters to male voters was 7 to 5. If a total of 1500 people voted, how many women voted?

23. A car leaves a city traveling at 50 miles per hour. One hour later a second car leaves the same city traveling on the same route at 55 miles per hour. How long will it take the second car to overtake the first car?

24. How many centiliters of pure acid must be added to 6 centiliters of a 50% acid solution to obtain a 70% acid solution?

25. How long will it take $4000 to double itself if it is invested at 9% simple interest?

CUMULATIVE REVIEW PROBLEM SET *Chapters 1-4*

For Problems 1–10, simplify each algebraic expression by combining similar terms.

1. $7x - 9x - 14x$

2. $-10a - 4 + 13a + a - 2$

3. $5(x - 3) + 7(x + 6)$

4. $3(x - 1) - 4(2x - 1)$

5. $-3n - 2(n - 1) + 5(3n - 2) - n$

6. $6n + 3(4n - 2) - 2(2n - 3) - 5$

7. $\frac{1}{2}x - \frac{3}{4}x + \frac{2}{3}x - \frac{1}{6}x$

8. $\frac{1}{3}n - \frac{4}{15}n + \frac{5}{6}n - n$

9. $0.4x - 0.7x - 0.8x + x$

10. $0.5(x - 2) + 0.4(x + 3) - 0.2x$

For Problems 11–20, evaluate each algebraic expression for the given values of the variables.

11. $5x - 7y + 2xy$ for $x = -2$ and $y = 5$

12. $2ab - a + 6b$ for $a = 3$ and $b = -4$

13. $-3(x - 1) + 2(x + 6)$ for $x = -5$

14. $5(n + 3) - (n + 4) - n$ for $n = 7$

15. $\left(\frac{1}{x} + \frac{1}{y}\right)^2$ for $x = \frac{1}{2}$ and $y = \frac{1}{3}$

16. $\frac{3}{4}n - \frac{1}{3}n + \frac{5}{6}n$ for $n = -\frac{2}{3}$

17. $2a^2 - 4b^2$ for $a = 0.2$ and $b = -0.3$

18. $x^2 - 3xy - 2y^2$ for $x = \frac{1}{2}$ and $y = \frac{1}{4}$

19. $5x - 7y - 8x + 3y$ for $x = 9$ and $y = -8$

20. $\frac{3a - b - 4a + 3b}{a - 6b - 4b - 3a}$ for $a = -1$ and $b = 3$

For Problems 21–26, evaluate each expression.

21. 3^4

22. -2^6

23. $(0.4)^3$

24. $\left(-\frac{1}{2}\right)^5$

25. $\left(\frac{1}{2} + \frac{1}{3}\right)^2$

26. $\left(\frac{3}{4} - \frac{7}{8}\right)^3$

For Problems 27–38, solve each equation.

27. $-5x + 2 = 22$

28. $3x - 4 = 7x + 4$

29. $7(n - 3) = 5(n + 7)$

30. $2(x - 1) - 3(x - 2) = 12$

31. $\frac{2}{5}x - \frac{1}{3} = \frac{1}{3}x + \frac{1}{2}$

32. $\frac{t - 2}{4} + \frac{t + 3}{3} = \frac{1}{6}$

33. $\frac{2n - 1}{5} - \frac{n + 2}{4} = 1$

34. $0.09x + 0.12(500 - x) = 54$

35. $-5(n - 1) - (n - 2) = 3(n - 1) - 2n$

36. $\frac{-2}{x - 1} = \frac{-3}{x + 4}$

37. $0.2x + 0.1(x - 4) = 0.7x - 1$

38. $-(t - 2) + (t - 4) = 2\left(t - \frac{1}{2}\right) - 3\left(t + \frac{1}{3}\right)$

For Problems 39–46, solve each inequality.

39. $4x - 6 > 3x + 1$

40. $-3x - 6 < 12$

41. $-2(n - 1) \le 3(n - 2) + 1$

42. $\frac{2}{7}x - \frac{1}{4} \ge \frac{1}{4}x + \frac{1}{2}$

43. $0.08t + 0.1(300 - t) > 28$

44. $-4 > 5x - 2 - 3x$

45. $\dfrac{2}{3}n - 2 \geq \dfrac{1}{2}n + 1$

46. $-3 < -2(x - 1) - x$

For Problems 47–54, set up an equation or an inequality and solve each problem.

47. Erin's salary this year is $32,000. This represents $2000 more than twice her salary 5 years ago. Find her salary 5 years ago.

48. One of two supplementary angles is 45° less than four times the other angle. Find the measure of each angle.

49. Jaamal has 25 coins, consisting of nickels and dimes, amounting to $2.10. How many coins of each kind does he have?

50. Hana bowled 144 and 176 in her first two games. What must she bowl in the third game to have an average of at least 150 for the three games?

51. A board 30 feet long is cut into two pieces whose lengths are in the ratio of 2 to 3. Find the lengths of the two pieces.

52. A retailer has some shoes that cost him $32 per pair. He wants to sell them at a profit of 20% of the selling price. What price should he charge for the shoes?

53. Two cars start from the same place traveling in opposite directions. One car travels 5 miles per hour faster than the other car. Find their speeds if after 6 hours they are 570 miles apart.

54. How many liters of pure alcohol must be added to 15 liters of a 20% solution to obtain a 40% solution?

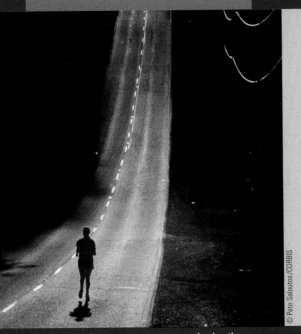

Using the concept of slope, we can set up and solve the proportion $\dfrac{30}{100} = \dfrac{y}{5280}$ to determine how much vertical change a highway having a 30% grade has in a horizontal distance of 1 mile.

© Pete Saloutos/CORBIS

Coordinate Geometry and Linear Systems

A man that weighs 185 pounds burns approximately 10 calories per minute when running. The equation $y = 10x$, where y represents the calories burned, and x represents the time in minutes spent running describes the relationship between time spent running and calories burned. We can use a graph of the equation (shown in Figure 5.1) to answer questions like, "How much time spent running is necessary to burn 820 calories?".

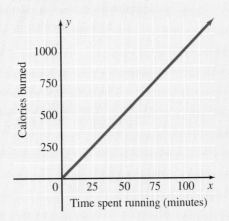

Figure 5.1

In this chapter we will associate pairs of real numbers with points in a geometric plane. This will provide the basis for obtaining pictures of algebraic equations and inequalities in two variables. Finally, we will work with systems of equations that will provide us with even more problem-solving power.

InfoTrac Project

Do a subject guide search on volume. Under "cubic content" find a periodical article on Marketed Production of Natural Gas, by State, 1995–2001. Make a table of values for years (x) and millions of cubic feet marketed (y) for Utah for years 1995 to 1998. Use your table of values to graph the data. Is the number of cubic feet being marketed increasing or decreasing? Now plot the data for Utah in 1999 on the same graph. What change did that make to your graph? Now add the data for Utah in 2000 to your graph. Write a sentence describing what happened to your graph from 1995 to 1999 and from 1999 to 2000. Give a possible explanation for the change. Connect any two of your data points and calculate the slope of the line that would pass through those two points. Have a friend choose two different points. Were your slopes the same? Why or why not?

5.1 Cartesian Coordinate System

Objectives

■   Plot points on a rectangular coordinate system.

■   Solve equations for the specified variable.

■   Draw graphs of equations by plotting points.

Now let's consider two number lines (one vertical and one horizontal) perpendicular to each other at the point we associate with zero on both lines (Figure 5.2). We refer to these number lines as the horizontal and vertical **axes** or, together, as the **coordinate axes.** They partition the plane into four parts called **quadrants.** The quadrants are numbered counterclockwise from I to IV, as indicated in Figure 5.2. The point of intersection of the two axes is called the **origin.**

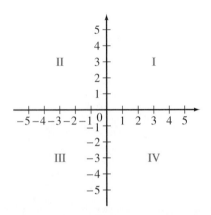

Figure 5.2

It is now possible to set up a one-to-one correspondence between ordered pairs of real numbers and the points in a plane. To each ordered pair of real numbers there corresponds a unique point in the plane, and to each point there corresponds a unique ordered pair of real numbers. We have indicated a part of this correspondence in Figure 5.3. The ordered pair (3, 1) corresponds to a point A, and this means that point A is located 3 units to the right of and 1 unit up from the origin. (The ordered pair (0, 0) corresponds to the origin.) The ordered pair $(-2, 4)$ corresponds to point B, and this means that point B is located 2 units to the left of and 4 units up from the origin. Make sure that you agree with all of the other points plotted in Figure 5.3.

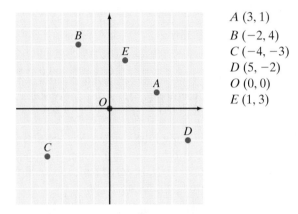

$A\ (3, 1)$
$B\ (-2, 4)$
$C\ (-4, -3)$
$D\ (5, -2)$
$O\ (0, 0)$
$E\ (1, 3)$

Figure 5.3

REMARK: The notation $(-2, 4)$ was used earlier in this text to indicate an interval of the real number line. Now we are using the same notation to indicate an ordered pair of real numbers. This double meaning should not be confusing because the context of the material will always indicate which meaning of the notation is being used. Throughout this chapter we will be using the ordered-pair interpretation.

In general, we refer to the real numbers a and b in an ordered pair (a, b) associated with a point as the **coordinates of the point.** The first number, a, called the **abscissa,** is the directed distance of the point from the vertical axis, measured parallel to the horizontal axis. The second number, b, called the **ordinate,** is the directed distance of the point from the horizontal axis, measured parallel to the vertical axis (see Figure 5.4(a)). Thus in the first quadrant all points have a positive abscissa and a positive ordinate. In the second quadrant all points have a negative abscissa and a positive ordinate. We have indicated the sign situations for all four quadrants in Figure 5.4(b). This system of associating points with ordered pairs of real numbers is called the **Cartesian coordinate system** or the **rectangular coordinate system.**

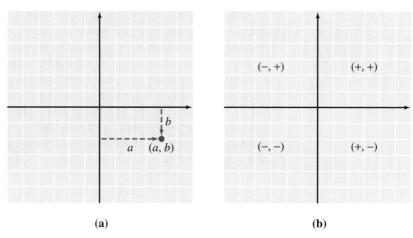

(a)   **(b)**

Figure 5.4

Plotting points on a rectangular coordinate system can be helpful when analyzing data to determine a trend or relationship. The following example shows the plot of some data.

EXAMPLE 1

The chart below shows the Friday and Saturday scores of golfers in terms of par. Plot the charted information on a rectangular coordinate system. Let Friday's score be the first number in the ordered pair, and let Saturday's score be the second number in the ordered pair, for each golfer.

	Mark	Ty	Vinay	Bill	Herb	Rod
Friday's score	1	−2	−1	4	−3	0
Saturday's score	3	−2	0	7	−4	1

Solution

The ordered pairs are as follows:

Mark $(1, 3)$ Ty $(-2, -2)$ Vinay $(-1, 0)$ Bill $(4, 7)$ Herb $(-3, -4)$ Rod $(0, 1)$

The points are plotted on the rectangular coordinate system in Figure 5.5. In the study of statistics, this graph of the charted data would be called a scatterplot. For this plot, the points appear to follow a straight-line path, which suggests that there is a linear correlation between Friday's score and Saturday's score.

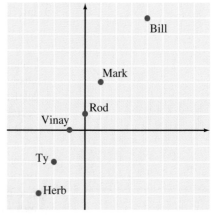

Figure 5.5

In Example 1, we used a rectangular coordinate system to plot data points. Now we will extend our use of the rectangular coordinate system to graph the solutions for equations in two variables. Let's begin by looking at the solutions for the equation $y = x + 3$. A *solution* of an equation in two variables is an ordered pair of real numbers that satisfies the equation. When using the variables x and y, we agree that the first number of an ordered pair is a value for x, and the second number is a value for y. We see that $(1, 4)$ is a solution for $y = x + 3$ because when x is replaced by 1 and y by 4, the result is the true numerical statement $4 = 1 + 3$. Likewise, $(-1, 2)$ is a solution for $y = x + 3$ because $2 = -1 + 3$ is a true statement. We can find infinitely many pairs of real numbers that satisfy $y = x + 3$ by arbitrarily choosing values for x and, for each value of x chosen, determining a corresponding value for y. Let's use a table to record some of the solutions for $y = x + 3$.

Choose x	Determine y from $y = x + 3$	Solution for $y = x + 3$
0	3	$(0, 3)$
1	4	$(1, 4)$
3	6	$(3, 6)$
5	8	$(5, 8)$
-1	2	$(-1, 2)$
-3	0	$(-3, 0)$
-5	-2	$(-5, -2)$

Now we can locate the point associated with each ordered pair on a rectangular coordinate system; we label the horizontal axis the **x axis** and the vertical axis the **y axis,** as shown in Figure 5.6(a). The straight line in Figure 5.6(b), which is drawn through the points represents all of the infinitely many solutions of the equation $y = x + 3$ and is called the graph of the equation.

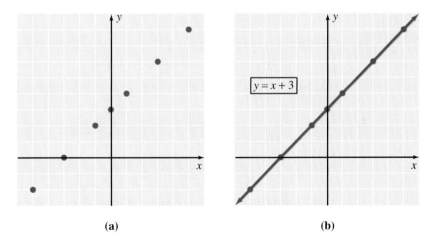

(a)   (b)

Figure 5.6

The following examples further illustrate the process of graphing equations.

E X A M P L E   2   Graph $y = x^2$.

Solution

First we set up a table of some of the solutions.

x	y	Solutions (x, y)
0	0	$(0, 0)$
1	1	$(1, 1)$
2	4	$(2, 4)$
3	9	$(3, 9)$
−1	1	$(-1, 1)$
−2	4	$(-2, 4)$
−3	9	$(-3, 9)$

Then we plot the points associated with the solutions as in Figure 5.7(a). Finally we connect the points with a smooth curve in Figure 5.7(b). This curve is called a **parabola;** we will study parabolas in more detail in Chapter 11.

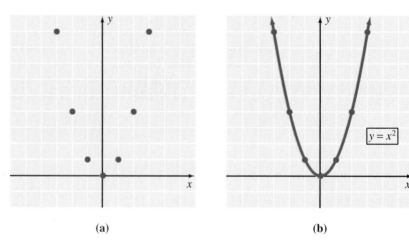

(a)   (b)

Figure 5.7

How many solutions do we need to have in a table of values? There is no definite answer to this question other than a sufficient number for the graph of the equation to be determined. In other words, we need to plot points until we can determine the nature of the curve.

EXAMPLE 3

Graph $y = -(x + 2)^2$.

Solution

First, let's set up a table of values.

x	y
0	−4
1	−9
−1	−1
−2	0
−3	−1
−4	−4
−5	−9

From the table we can plot the ordered pairs $(0, -4)$, $(1, -9)$, $(-1, -1)$, $(-2, 0)$, $(-3, -1)$, $(-4, -4)$, and $(-5, -9)$. Connecting these points with a smooth curve produces Figure 5.8.

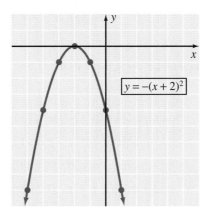

$$y = -(x + 2)^2$$

Figure 5.8

EXAMPLE 4

Graph $2x + 3y = 6$.

Solution

First let's change the form of the equation to make it easier to find solutions. We can solve either for x in terms of y or for y in terms of x. Let's solve for y.

$$2x + 3y = 6$$
$$3y = -2x + 6$$
$$y = \frac{-2x + 6}{3}$$

Now we can set up a table of values. If we look carefully at our equation we recognize that choosing values of x that are multiples of 3 will produce integer values for y. This is not necessary, but it does make computations easier and plotting points associated with pairs of integers is more exact than getting involved with fractions. Plotting these points and connecting them produces Figure 5.9.

x	y
0	2
3	0
6	−2
−3	4
−6	6

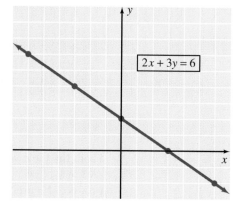

Figure 5.9

To graph an equation in two variables, x and y, use these steps:

STEP 1 Solve the equation for y in terms of x or for x in terms of y, if it is not already in such a form.

STEP 2 Set up a table of ordered pairs that satisfy the equation.

STEP 3 Plot the points associated with the ordered pairs.

STEP 4 Connect the points with a smooth curve.

Let's conclude this section with two more examples that illustrate step 1.

EXAMPLE 5 Solve $4x + 9y = 12$ for y.

Solution

$$4x + 9y = 12$$
$$9y = 12 - 4x \quad \text{Subtracted } 4x \text{ from both sides}$$
$$y = \frac{12 - 4x}{9} \quad \text{Divided both sides by 9}$$

EXAMPLE 6

Solve $4x - 5y = 6$ for y.

Solution

$$4x - 5y = 6$$
$$-5y = 6 - 4x \qquad \text{Subtracted } 4x \text{ from both sides}$$
$$y = \frac{6 - 4x}{-5} \qquad \text{Divided both sides by } -5$$
$$y = \frac{4x - 6}{5} \qquad \frac{6-4x}{-5} \text{ can be changed to } \frac{-6+4x}{5} \text{ by multiplying}$$
$$\text{numerator and denominator by } -1.$$

CONCEPT QUIZ

For Problems 1–5, answer true or false.

1. In a rectangular coordinate system, the coordinate axes partition the plane into four parts called quadrants.

2. Quadrants are named with Roman numerals and numbered clockwise.

3. The real numbers in an ordered pair are referred to as the coordinates of the point.

4. The equation $y = x + 3$ has an infinite number of ordered pairs that satisfy the equation.

5. The point of intersection of the coordinate axes is called the origin.

For Problems 6–10, match the points plotted in Figure 5.10 with their coordinates.

6. $(-3, 1)$

7. $(4, 0)$

8. $(3, -1)$

9. $(0, 4)$

10. $(-1, -3)$

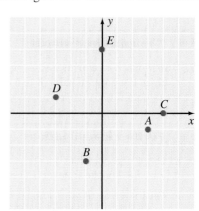

Figure 5.10

PROBLEM SET 5.1

1. Maria, a biology student, designed an experiment to test the effects of changing the amount of light and the amount of water given to selected plants. In the experiment the amounts of water and light given to the plant were randomly changed. The chart shows the amount of light and water above or below the normal amount given to the plant for six days. Plot the charted information on a rectangular coordinate system. Let the change in light be the first number in the ordered pair, and let the change in water be the second number in the ordered pair.

	Mon.	Tue.	Wed.	Thu.	Fri.	Sat.
Change in amount of light	1	−2	−1	4	−3	0
Change in amount of water	−3	4	−1	0	−5	1

2. Chase is studying the monthly percent changes in the stock price for two different companies. Using the data in the table below, plot the points for each month. Let the percent change for XM Inc. be the first number of the ordered pair, and let the percent change for Icom be the second number in the ordered pair.

	Jan.	Feb.	Mar.	Apr.	May	Jun.
XM Inc.	1	−2	−1	4	−3	0
Icom	−3	4	2	−5	−1	1

For Problems 3–12, solve the equation for the variable indicated.

3. $3x + 7y = 13$ for y

4. $5x + 9y = 17$ for y

5. $x − 3y = 9$ for x

6. $2x − 7y = 5$ for x

7. $−x + 5y = 14$ for y

8. $−2x − y = 9$ for y

9. $−3x + y = 7$ for x

10. $−x − y = 9$ for x

11. $−2x + 3y = −5$ for y

12. $3x − 4y = −7$ for y

For Problems 13–36, graph each of the equations.

13. $y = x + 1$

14. $y = x + 4$

15. $y = x − 2$

16. $y = −x − 1$

17. $y = (x − 2)^2$

18. $y = (x + 1)^2$

19. $y = x^2 − 2$

20. $y = x^2 + 1$

21. $y = \frac{1}{2}x + 3$

22. $y = \frac{1}{2}x − 2$

23. $x + 2y = 4$

24. $x + 3y = 6$

25. $2x − 5y = 10$

26. $5x − 2y = 10$

27. $y = x^3$

28. $y = x^4$

29. $y = −x^2$

30. $y = −x^3$

31. $y = x$

32. $y = −x$

33. $y = −3x + 2$

34. $3x − y = 4$

35. $y = 2x^2$

36. $y = −3x^2$

■■■ Thoughts into words

37. How would you convince someone that there are infinitely many ordered pairs of real numbers that satisfy the equation $x + y = 9$?

38. Explain why no points of the graph of the equation $y = x^2 + 1$ will lie below the x axis.

■■■ Further investigations

39. a. Graph the equations $y = x^2 + 2$, $y = x^2 + 4$, and $y = x^2 − 3$ on the same set of axes.

b. On the basis of your graphs in part (a), sketch a graph of $y = x^2 − 1$ without plotting any points.

40. a. Graph the equations $y = (x − 2)^2$, $y = (x − 4)^2$, and $y = (x + 3)^2$ on the same set of axes.

b. On the basis of your graphs in part (a), sketch a graph of $y = (x + 5)^2$ without plotting any points.

41. a. Graph the equations $y = (x − 1)^2 + 2$, $y = (x − 3)^2 − 2$, and $y = (x + 2)^2 + 3$ on the same set of axes.

b. On the basis of your graphs in part (a), sketch a graph of $y = (x + 1)^2 − 4$ without plotting any points.

5.2 Graphing Linear Equations

Objectives

■ Find the *x* and *y* intercepts for linear equations.

■ Graph linear equations.

■ Use linear equations to model problems.

The following table summarizes some of our results from graphing equations in the previous section and its accompanying problem set:

Equation	Type of graph produced
$y = x + 3$	Straight line
$y = x^2$	Parabola
$2x + 3y = 6$	Straight line
$y = -3x + 2$	Straight line
$y = x^2 - 2$	Parabola
$y = (x - 2)^2$	Parabola
$5x - 2y = 10$	Straight line
$3x - y = 4$	Straight line
$y = x^3$	No name will be given at this time, but not a straight line
$y = x$	Straight line
$y = \frac{1}{2}x + 3$	Straight line

In this table pay special attention to the equations that produce a straight-line graph. They are called **linear equations in two variables.** In general, any equation of the form $Ax + By = C$, where A, B, and C are constants (A and B not both zero) and x and y are variables, is a linear equation in two variables, and its graph is a straight line.

We should clarify two points about our description of a linear equation in two variables. First, the choice of x and y for variables is arbitrary. We can choose any two letters to represent the variables. An equation such as $3m + 2n = 7$ can be considered a linear equation in two variables. So that we are not constantly changing the labeling of the coordinate axes when graphing equations, however, it is much easier to use the same two variables in all equations. Thus we will go along with convention and use x and y as our variables. Second, the statement "any equation of the form $Ax + By = C$" technically means *any equation of the form $Ax + By = C$ or equivalent to that form.* For example, the equation $y = x + 3$, which has a straight-line graph, is equivalent to $-x + y = 3$.

The knowledge that any equation of the form $Ax + By = C$ produces a straight-line graph, along with the fact that two points determine a straight line, makes graphing linear equations in two variables a simple process. We merely find two solutions, plot the corresponding points, and connect the points with a straight line. It is probably wise to find a third point as a check point. Let's consider an example.

EXAMPLE 1

Graph $2x - 3y = 6$.

Solution

Let $x = 0$; then

$$2(0) - 3y = 6$$
$$-3y = 6$$
$$y = -2$$

Thus $(0, -2)$ is a solution.

Let $y = 0$; then

$$2x - 3(0) = 6$$
$$2x = 6$$
$$x = 3$$

Thus $(3, 0)$ is a solution.

Let $x = -3$; then

$$2(-3) - 3y = 6$$
$$-6 - 3y = 6$$
$$-3y = 12$$
$$y = -4$$

Thus $(-3, -4)$ is a solution.

We can plot the points associated with these three solutions and connect them with a straight line to produce the graph of $2x - 3y = 6$ in Figure 5.11.

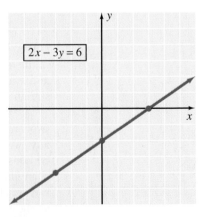

Figure 5.11

Let us briefly review our approach to Example 1. Note that we did not begin by solving either for y in terms of x or for x in terms of y. The reason for this is that we know the graph is a straight line, so there is no need for an extensive table of values. Thus there is no real benefit to changing the form of the original equation. The first two solutions indicate where the line intersects the coordinate axes. The ordinate of the point $(0, -2)$ is called the **y intercept,** and the abscissa of the point $(3, 0)$ is the **x intercept** of this graph. That is, the graph of the equation $2x - 3y = 6$ has a y intercept of -2 and an x intercept of 3. In general the intercepts are easy to find. You can let $x = 0$ and solve for y to find the y intercept, and you can let $y = 0$ and solve for x to find the x intercept. The third solution, $(-3, -4)$, serves as a check point. If $(-3, -4)$ had not been on the line determined by the two intercepts, then we would have known that we had made an error.

EXAMPLE 2

Graph $x + 2y = 4$.

Solution

Without showing all of our work, we present the following table to indicate the intercepts and a check point.

x	y	
0	2	Intercepts
4	0	
2	1	Check point

We plot the points $(0, 2)$, $(4, 0)$, and $(2, 1)$ and connect them with a straight line to produce the graph in Figure 5.12.

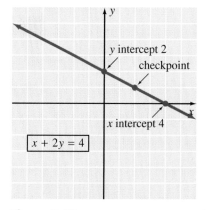

Figure 5.12

EXAMPLE 3

Graph $2x + 3y = 7$.

Solution

The intercepts and a check point are listed in the table. Finding intercepts may involve fractions, but the computation is usually easy. We plot the points from the table and show the graph of $2x + 3y = 7$ in Figure 5.13.

x	y	
0	$\frac{7}{3}$	
$\frac{7}{2}$	0	Intercepts
2	1	Check point

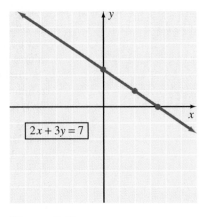

Figure 5.13

EXAMPLE 4 Graph $y = 2x$.

Solution

Note that $(0, 0)$ is a solution; thus this line intersects both axes at the origin. Because both the x intercept and the y intercept are determined by the origin, $(0, 0)$, we need another point to graph the line. Then we should find a third point as a check point. These results are summarized in the following table. The graph of $y = 2x$ is shown in Figure 5.14.

x	y	
0	0	Intercept
2	4	Additional point
−1	−2	Check point

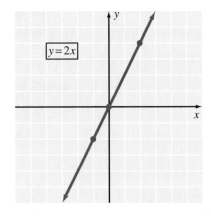

Figure 5.14

EXAMPLE 5 Graph $y = -2$.

Solution

Because we are considering linear equations in two variables, the equation $y = -2$ is equivalent to $0x + y = -2$. Now we can see that y will be equal to -2 for any value of x. Some of the solutions are listed in the table. The graph of all of the solutions is the horizontal line indicated in Figure 5.15.

x	y
−1	−2
0	−2
1	−2
3	−2

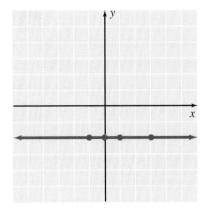

Figure 5.15

EXAMPLE 6

Graph $x = 3$.

Solution

Because we are considering linear equations in *two variables,* the equation $x = 3$ is equivalent to $x + 0(y) = 3$. Now we can see that any value of y can be used, but the x value must always be 3. Therefore, some of the solutions are (3, 0), (3, 1), (3, 2), (3, − 1), and (3, −2). The graph of all of the solutions is the vertical line indicated in Figure 5.16.

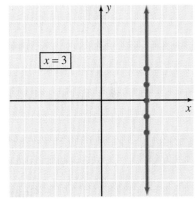

$x = 3$

Figure 5.16

Applications of Linear Equations

Linear equations in two variables can be used to model many different types of real-world problems. For example, suppose that a retailer wants to sell some items at a profit of 30% of the cost of each item. If we let s represent the selling price and c the cost of each item, then we can use the equation

$$s = c + 0.3c = 1.3c$$

to determine the selling price of each item on the basis of the cost of the item. For example, if the cost of an item is \$4.50, then the retailer should sell it for $s = (1.3)(4.5) =$ \$5.85.

By finding values that satisfy the equation $s = 1.3c$, we can create this table:

c	1	5	10	15	20
s	1.3	6.5	13	19.5	26

From the table we see that if the cost of an item is $15, then the retailer should sell it for $19.50 in order to make a profit of 30% of the cost. Furthermore, because this is a linear relationship, we can obtain exact values for costs that fall between the values given in the table. For example, because a c value of 12.5 is halfway between the c values of 10 and 15, the corresponding s value is halfway between the s values of 13 and 19.5. Therefore, a c value of 12.5 produces

$$s = 13 + \frac{1}{2}(19.5 - 13) = 16.25$$

Thus, if the cost of an item is $12.50, the retailer should sell it for $16.25.

Now let's get a picture (graph) of this linear relationship. We can label the horizontal axis c and the vertical axis s and we can use the origin along with one ordered pair from the table to produce the straight-line graph in Figure 5.17. (Because of the type of application, only nonnegative values for c and s are appropriate.)

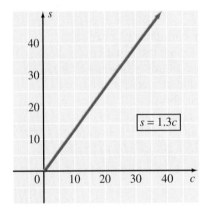

Figure 5.17

From the graph we can approximate s values on the basis of given c values. For example, if $c = 30$, then by reading up from 30 on the c axis to the line and then across to the s axis, we see that s is a little less than 40. (We get an exact s value of 39 by using the equation $s = 1.3c$.)

Many formulas that are used in various applications are linear equations in two variables. For example, the formula $C = \frac{5}{9}(F - 32)$, which converts temperatures from the Fahrenheit scale to the Celsius scale, is a linear relationship. Using this equation, we can determine that 14°F is equivalent to

$$C = \frac{5}{9}(14 - 32) = \frac{5}{9}(-18) = -10°C$$

Let's use the equation $C = \dfrac{5}{9}(F - 32)$ to create a table of values:

F	−22	−13	5	32	50	68	86
C	−30	−25	−15	0	10	20	30

Reading from the table, we see for example that $-13°F = -25°C$ and $68°F = 20°C$.

To graph the equation $C = \dfrac{5}{9}(F - 32)$ we can label the horizontal axis F and the vertical axis C and plot two points that are given in the table. Figure 5.18 shows the graph of the equation.

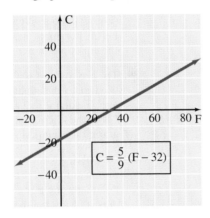

Figure 5.18

From the graph we can approximate C values on the basis of given F values. For example, if $F = 80°$, then by reading up from 80 on the F axis to the line and then across to the C axis, we see that C is approximately 25°. Likewise, we can obtain approximate F values on the basis of given C values. For example, if $C = -25°$, then by reading across from −25 on the C axis to the line and then up to the F axis, we see that F is approximately −15°.

CONCEPT QUIZ

For Problems 1–10, answer true or false.

1. The graph of $y = x^2$ is a straight line.

2. Any equation of the form $Ax + By = C$, where A, B, and C (A and B not both zero) are constants and x and y are variables, has a graph that is a straight line.

3. The equations $2x + y = 4$ and $y = -2x + 4$ are equivalent.

4. The y intercept of the graph of $3x + 4y = -12$ is -4.

5. The x intercept of the graph of $3x + 4y = -12$ is -4.

6. Determining just two points is sufficient to graph a straight line.

7. The graph of $y = 4$ is a vertical line.

8. The graph of $x = 4$ is a vertical line.

9. The graph of $y = -1$ has a y intercept of -1.

10. The graph of every linear equation has a y intercept.

PROBLEM SET 5.2

For Problems 1–36, graph each linear equation.

1. $x + y = 2$
2. $x + y = 4$
3. $x - y = 3$
4. $x - y = 1$
5. $x - y = -4$
6. $-x + y = 5$
7. $x + 2y = 2$
8. $x + 3y = 5$
9. $3x - y = 6$
10. $2x - y = -4$
11. $3x - 2y = 6$
12. $2x - 3y = 4$
13. $x - y = 0$
14. $x + y = 0$
15. $y = 3x$
16. $y = -2x$
17. $x = -2$
18. $y = 3$
19. $y = 0$
20. $x = 0$
21. $y = -2x - 1$
22. $y = 3x - 4$
23. $y = \frac{1}{2}x + 1$
24. $y = \frac{2}{3}x - 2$
25. $y = -\frac{1}{3}x - 2$
26. $y = -\frac{3}{4}x - 1$
27. $4x + 5y = -10$
28. $3x + 5y = -9$
29. $-2x + y = -4$
30. $-3x + y = -5$
31. $3x - 4y = 7$
32. $4x - 3y = 10$
33. $y + 4x = 0$
34. $y - 5x = 0$
35. $x = 2y$
36. $x = -3y$

37. Suppose that the daily profit from an ice cream stand is given by the equation $p = 2n - 4$, where n represents the number of gallons of ice cream mix used in a day, and p represents the number of dollars of profit. Label the horizontal axis n and the vertical axis p, and graph the equation $p = 2n - 4$ for nonnegative values of n.

38. The cost (c) of playing an online computer game for a time (t) in hours is given by the equation $c = 3t + 5$. Label the horizontal axis t and the vertical axis c, and graph the equation for nonnegative values of t.

39. The area of a sidewalk whose width is fixed at 3 feet can be given by the equation $A = 3l$, where A represents the area in square feet and l represents the length in feet. Label the horizontal axis l and the vertical axis A, and graph the equation $A = 3l$ for nonnegative values of l.

40. An online grocery store charges for delivery based on the equation $C = 0.30p$, where C represents the cost in dollars, and p represents the weight of the groceries in pounds. Label the horizontal axis p and the vertical axis C, and graph the equation $C = 0.30p$ for nonnegative values of p.

41. At \$0.06 per kilowatt-hour, the equation $A = 0.06t$ determines the amount, A, of an electric bill for t hours. Complete this table of values:

Hours, t	696	720	740	775	782
Dollars and cents, A					

42. Suppose that a used-car dealer determines the selling price of his cars by using a markup of 60% of the cost. If s represents the selling price and c the cost, this equation applies:

$$s = c + 0.6c = 1.6c$$

Complete this table using the equation $s = 1.6c$.

Dollars, c	250	325	575	895	1095
Dollars, s					

43. a. The equation $F = \dfrac{9}{5}C + 32$ converts temperatures from degrees Celsius to degrees Fahrenheit. Complete this table:

C	0	5	10	15	20	−5	−10	−15	−20	−25
F										

b. Graph the equation $F = \dfrac{9}{5}C + 32$.

c. Use your graph from part (b) to approximate values for F when C = 25°, 30°, −30°, and −40°.

d. Check the accuracy of your readings from the graph in part (c) by using the equation $F = \dfrac{9}{5}C + 32$.

■ ■ ■ **Thoughts into words**

44. Your friend is having trouble understanding why the graph of the equation $y = 3$ is a horizontal line containing the point $(0, 3)$. What might you do to help him?

45. How do we know that the graph of $y = -4x$ is a straight line that contains the origin?

46. Do all graphs of linear equations have x intercepts? Explain your answer.

47. How do we know that the graphs of $x - y = 4$ and $-x + y = -4$ are the same line?

■ ■ ■ **Further investigations**

From our previous work with absolute value, we know that $|x + y| = 2$ is equivalent to $x + y = 2$ and to $x + y = -2$. Therefore, the graph of $|x + y| = 2$ consists of the two lines $x + y = 2$ and $x + y = -2$. Graph each of the following equations.

48. $|x + y| = 1$

49. $|x - y| = 2$

50. $|2x + y| = 4$

51. $|3x - y| = 6$

Answers to Concept Quiz

1. False   **2.** True   **3.** True   **4.** False   **5.** True   **6.** True   **7.** False   **8.** True   **9.** True   **10.** False

5.3   Slope of a Line

Objectives

■   Find the slope of a line between two points.

■   Given the equation of a line, find two points on the line, and use those points to determine the slope of the line.

■   Graph lines, given a point and the slope.

■   Solve word problems that involve slope.

In Figure 5.19, note that the line associated with $4x - y = 4$ is *steeper* than the line associated with $2x - 3y = 6$. Mathematically, we use the concept of **slope** to discuss the steepness of lines. The slope of a line is the ratio of the vertical change to the horizontal change as we move from one point on a line to another point. We indicate this in Figure 5.20 with the points P_1 and P_2.

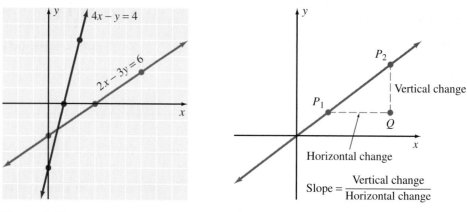

Figure 5.19   **Figure 5.20**

We can give a precise definition for slope by considering the coordinates of the points P_1, P_2, and Q in Figure 5.21. Since P_1 and P_2 represent any two points on the line, we assign the coordinates (x_1, y_1) to P_1 and (x_2, y_2) to P_2. The point Q is the same distance from the y axis as P_2 and the same distance from the x axis as P_1. Thus we assign the coordinates (x_2, y_1) to Q (see Figure 5.21). It should now be apparent that the vertical change is $y_2 - y_1$, and the horizontal change is $x_2 - x_1$. Thus we have the following definition for slope.

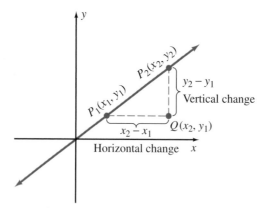

Figure 5.21

DEFINITION 5.1

If points P_1 and P_2 with coordinates (x_1, y_1) and (x_2, y_2), respectively, are any two different points on a line, then the slope of the line (denoted by m) is

$$m = \frac{y_2 - y_1}{x_2 - x_1}, \quad x_1 \neq x_2$$

Using Definition 5.1, we can easily determine the slope of a line if we know the co-ordinates of two points on the line.

EXAMPLE 1   Find the slope of the line determined by each of the following pairs of points.

    (a) $(2, 1)$ and $(4, 6)$

    (b) $(3, 2)$ and $(-4, 5)$

    (c) $(-4, -3)$ and $(-1, -3)$

Solution

    (a) Let $(2, 1)$ be P_1 and $(4, 6)$ be P_2 as in Figure 5.22; then we have

$$m = \frac{y_2 - y_1}{x_2 - x_1} = \frac{6 - 1}{4 - 2} = \frac{5}{2}$$

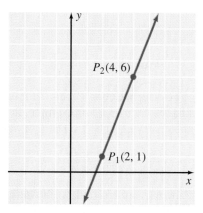

Figure 5.22

    (b) Let $(3, 2)$ be P_1 and $(-4, 5)$ be P_2 as in Figure 5.23.

$$m = \frac{y_2 - y_1}{x_2 - x_1} = \frac{5 - 2}{-4 - 3} = \frac{3}{-7} = -\frac{3}{7}$$

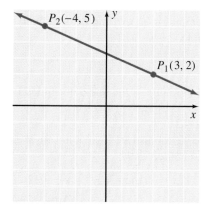

Figure 5.23

(c) Let $(-4, -3)$ be P_1 and $(-1, -3)$ be P_2 as in Figure 5.24.

$$m = \frac{y_2 - y_1}{x_2 - x_1} = \frac{-3 - (-3)}{-1 - (-4)} = \frac{0}{3} = 0$$

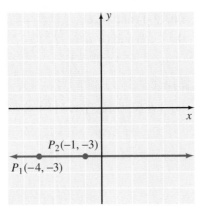

Figure 5.24

The designation of P_1 and P_2 in such problems is arbitrary and does not affect the value of the slope. For example, in part (a) of Example 1 suppose that we let $(4, 6)$ be P_1 and $(2, 1)$ be P_2. Then we obtain

$$m = \frac{y_2 - y_1}{x_2 - x_1} = \frac{1 - 6}{2 - 4} = \frac{-5}{-2} = \frac{5}{2}$$

The parts of Example 1 illustrate the three basic possibilities for slope; that is, the slope of a line can be *positive, negative,* or *zero*. A line that has a positive slope rises as we move from left to right, as in part (a). A line that has a negative slope falls as we move from left to right, as in part (b). A horizontal line, as in part (c), has a slope of 0. Finally, we need to realize that **the concept of slope is undefined for vertical lines.** This is because, for any vertical line, the change in x as we move from one point to another is zero. Thus the ratio $\frac{y_2 - y_1}{x_2 - x_1}$ will have a denominator of zero and be undefined. So in Definition 5.1, the restriction $x_1 \neq x_2$ is made.

EXAMPLE 2   Find the slope of the line determined by the equation $3x + 4y = 12$.

Solution

Since we can use any two points on the line to determine the slope of the line, let's find the intercepts.

$$\text{If } x = 0, \text{ then } 3(0) + 4y = 12$$
$$4y = 12$$
$$y = 3$$

$$\text{If } y = 0, \text{ then } 3x + 4(0) = 12$$

$$3x = 12$$

$$x = 4$$

Using $(0, 3)$ as P_1 and $(4, 0)$ as P_2, we have

$$m = \frac{y_2 - y_1}{x_2 - x_1} = \frac{0 - 3}{4 - 0} = \frac{-3}{4} = -\frac{3}{4}$$ ■

We need to emphasize one final idea pertaining to the concept of slope. The slope of a line is a **ratio** of vertical change to horizontal change. A slope of $\frac{3}{4}$ means that for every 3 units of vertical change, there is a corresponding 4 units of horizontal change. So starting at some point on the line, we could move to other points on the line as follows:

$$\frac{3}{4} = \frac{6}{8} \qquad \text{by moving 6 units } up \text{ and 8 units to the } right$$

$$\frac{3}{4} = \frac{15}{20} \qquad \text{by moving 15 units } up \text{ and 20 units to the } right$$

$$\frac{3}{4} = \frac{\frac{3}{2}}{2} \qquad \text{by moving } 1\frac{1}{2} \text{ units } up \text{ and 2 units to the } right$$

$$\frac{3}{4} = \frac{-3}{-4} \qquad \text{by moving 3 units } down \text{ and 4 units to the } left$$

Likewise, a slope of $-\frac{5}{6}$ indicates that starting at some point on the line, we could move to other points on the line as follows:

$$-\frac{5}{6} = \frac{-5}{6} \qquad \text{by moving 5 units } down \text{ and 6 units to the } right$$

$$-\frac{5}{6} = \frac{5}{-6} \qquad \text{by moving 5 units } up \text{ and 6 units to the } left$$

$$-\frac{5}{6} = \frac{-10}{12} \qquad \text{by moving 10 units } down \text{ and 12 units to the } right$$

$$-\frac{5}{6} = \frac{15}{-18} \qquad \text{by moving 15 units } up \text{ and 18 units to the } left$$

EXAMPLE 3   Graph the line that passes through the point $(0, -2)$ and has a slope of $\frac{1}{3}$.

Solution

To graph, plot the point $(0, -2)$. Furthermore, because the slope $= \frac{\text{vertical change}}{\text{horizontal change}} = \frac{1}{3}$, we can locate another point on the line by starting from the point $(0, -2)$ and moving 1 unit up and 3 units to the right to obtain the point $(3, -1)$. Because two points determine a line, we can draw the line (Figure 5.25).

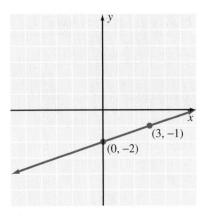

Figure 5.25

REMARK: Because $m = \dfrac{1}{3} = \dfrac{-1}{-3}$, we can locate another point by moving 1 unit down and 3 units to the left from the point $(0, -2)$.   ■

E X A M P L E   4

Graph the line that passes through the point $(1, 3)$ and has a slope of -2.

Solution

To graph the line, plot the point $(1, 3)$. We know that $m = -2 = \dfrac{-2}{1}$. Furthermore, because the slope $= \dfrac{\text{vertical change}}{\text{horizontal change}} = \dfrac{-2}{1}$, we can locate another point on the line by starting from the point $(1, 3)$ and moving 2 units down and 1 unit to the right to obtain the point $(2, 1)$. Because two points determine a line, we can draw the line (Figure 5.26).

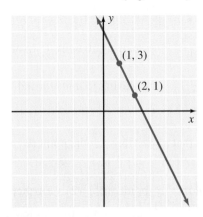

Figure 5.26

REMARK: Because $m = -2 = \dfrac{-2}{1} = \dfrac{2}{-1}$ we can locate another point by moving 2 units up and 1 unit to the left from the point $(1, 3)$. ■

Applications of Slope

The concept of slope has many real-world applications even though the word "slope" is often not used. For example, the highway in Figure 5.27 is said to have a "grade" of 17%. This means that for every horizontal distance of 100 feet, the highway rises or drops 17 feet. In other words, the absolute value of slope of the highway is $\dfrac{17}{100}$.

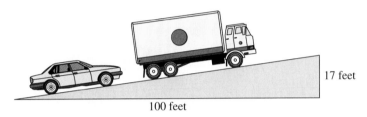

17 feet

100 feet

Figure 5.27

PROBLEM 1

A certain highway has a 3% grade. How many feet does it rise in a horizontal distance of 1 mile?

Solution

A 3% grade means a slope of $\dfrac{3}{100}$. Therefore, if we let y represent the unknown vertical distance and use the fact that 1 mile = 5280 feet, we can set up and solve the following proportion:

$$\frac{3}{100} = \frac{y}{5280}$$

$$100y = 3(5280) = 15{,}840$$

$$y = 158.4$$

The highway rises 158.4 feet in a horizontal distance of 1 mile. ■

A roofer, when making an estimate to replace a roof, is concerned about not only the total area to be covered but also the "pitch" of the roof. (Contractors do not define pitch the same way that mathematicians define slope, but both terms refer to "steepness.") The two roofs in Figure 5.28 might require the same number of shingles, but the roof on the left will take longer to complete because the pitch is so great that scaffolding will be required.

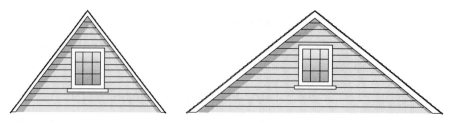

Figure 5.28

The concept of slope is also used in the construction of flights of stairs. The terms "rise" and "run" are commonly used, and the steepness (slope) of the stairs can be expressed as the ratio of rise to run. In Figure 5.29, the stairs on the left with the ratio of $\dfrac{10}{11}$ are steeper than the stairs on the right, which have a ratio of $\dfrac{7}{11}$.

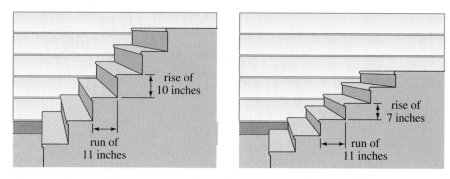

Figure 5.29

Technically, the concept of slope is involved in most situations where the idea of an incline is used. Hospital beds are constructed so that both the head-end and the foot-end can be raised or lowered; that is, the slope of either end of the bed can be changed. Likewise, treadmills are designed so that the incline (slope) of the platform can be raised or lowered as desired. Perhaps you can think of several other applications of the concept of slope.

CONCEPT QUIZ

For Problems 1–8, answer true or false.

1. The concept of slope of a line pertains to the steepness of the line.

2. The slope of a line is the ratio of the horizontal change to the vertical change moving from one point to another point on the line.

3. A line that has a negative slope falls as we move from left to right.

4. The slope of a vertical line is 0.

5. The slope of a horizontal line is 0.

6. A line cannot have a slope of 0.

7. A slope of $\dfrac{-5}{2}$ is the same as a slope of $-\dfrac{5}{-2}$.

8. A slope of 5 means that for every unit of horizontal change there is a corresponding 5 units of vertical change.

PROBLEM SET 5.3

For Problems 1–20, find the slope of the line determined by each pair of points.

1. $(7, 5), (3, 2)$

2. $(9, 10), (6, 2)$

3. $(-1, 3), (-6, -4)$

4. $(-2, 5), (-7, -1)$

5. $(2, 8), (7, 2)$

6. $(3, 9), (8, 4)$

7. $(-2, 5), (1, -5)$

8. $(-3, 4), (2, -6)$

9. $(4, -1), (-4, -7)$

10. $(5, -3), (-5, -9)$

11. $(3, -4), (2, -4)$

12. $(-3, -6), (5, -6)$

13. $(-6, -1), (-2, -7)$

14. $(-8, -3), (-2, -11)$

15. $(-2, 4), (-2, -6)$

16. $(-4, -5), (-4, 9)$

17. $(-1, 10), (-9, 2)$

18. $(-2, 12), (-10, 2)$

19. $(a, b), (c, d)$

20. $(a, 0), (0, b)$

21. Find y if the line through the points $(7, 8)$ and $(2, y)$ has a slope of $\dfrac{4}{5}$.

22. Find y if the line through the points $(12, 14)$ and $(3, y)$ has a slope of $\dfrac{4}{3}$.

23. Find x if the line through the points $(-2, -4)$ and $(x, 2)$ has a slope of $-\dfrac{3}{2}$.

24. Find x if the line through the points $(6, -4)$ and $(x, 6)$ has a slope of $-\dfrac{5}{4}$.

For Problems 25–32, you are given one point on a line and the slope of the line. Find the coordinates of three other points on the line.

25. $(3, 2), m = \dfrac{2}{3}$

26. $(4, 1), m = \dfrac{5}{6}$

27. $(-2, -4), m = \dfrac{1}{2}$

28. $(-6, -2), m = \dfrac{2}{5}$

29. $(-3, 4), m = -\dfrac{3}{4}$

30. $(-2, 6), m = -\dfrac{3}{7}$

31. $(4, -5), m = -2$

32. $(6, -2), m = 4$

For Problems 33–40, sketch the line determined by each pair of points, and decide whether the slope of the line is positive, negative, or zero.

33. $(2, 8), (7, 1)$

34. $(1, -2), (7, -8)$

35. $(-1, 3), (-6, -2)$

36. $(7, 3), (4, -6)$

37. $(-2, 4), (6, 4)$

38. $(-3, -4), (5, -4)$

39. $(-3, 5), (2, -7)$

40. $(-1, -1), (1, -9)$

For Problems 41–48, graph the line that passes through the given point and has the given slope.

41. $(3, 1)$ $m = \dfrac{2}{3}$

42. $(-1, 0)$ $m = \dfrac{3}{4}$

43. $(-2, 3)$ $m = -1$

44. $(1, -4)$ $m = -3$

45. $(0, 5)$ $m = \dfrac{-1}{4}$

46. $(-3, 4)$ $m = \dfrac{-3}{2}$

47. $(2, -2)$ $m = \dfrac{3}{2}$

48. $(3, -4)$ $m = \dfrac{5}{2}$

For Problems 49–68, find the coordinates of two points on the given line, and then use those coordinates to find the slope of the line.

49. $3x + 2y = 6$

50. $4x + 3y = 12$

51. $5x - 4y = 20$

52. $7x - 3y = 21$

53. $x + 5y = 6$

54. $2x + y = 4$

55. $2x - y = -7$

56. $x - 4y = -6$

57. $y = 3$

58. $x = 6$

59. $-2x + 5y = 9$

60. $-3x - 7y = 10$

61. $6x - 5y = -30$

62. $7x - 6y = -42$

63. $y = -3x - 1$

64. $y = -2x + 5$

65. $y = 4x$

66. $y = 6x$

67. $y = \dfrac{2}{3}x - \dfrac{1}{2}$

68. $y = -\dfrac{3}{4}x + \dfrac{1}{5}$

69. Suppose that a highway rises a distance of 135 feet in a horizontal distance of 2640 feet. Express the grade of the highway to the nearest tenth of a percent.

70. The grade of a highway up a hill is 27%. How much change in horizontal distance is there if the vertical height of the hill is 550 feet? Express the answer to the nearest foot.

71. If the ratio of rise to run is to be $\dfrac{3}{5}$ for some stairs, and the measure of the rise is 19 centimeters, find the measure of the run to the nearest centimeter.

72. If the ratio of rise to run is to be $\dfrac{2}{3}$ for some stairs and the measure of the run is 28 centimeters, find the measure of the rise to the nearest centimeter.

73. A county ordinance requires a $2\dfrac{1}{4}$% "fall" for a sewage pipe from the house to the main pipe at the street. How much vertical drop must there be for a horizontal distance of 45 feet? Express the answer to the nearest tenth of a foot.

■ ■ ■ **Thoughts into words**

74. How would you explain the concept of slope to someone who was absent from class the day it was discussed?

75. If one line has a slope of $\dfrac{2}{3}$ and another line has a slope of 2, which line is steeper? Explain your answer.

76. Why do we say that the slope of a vertical line is undefined?

77. Suppose that a line has a slope of $\dfrac{3}{4}$ and contains the point $(5, 2)$. Are the points $(-3, -4)$ and $(14, 9)$ also on the line? Explain your answer.

Answers to Concept Quiz

1. True   **2.** False   **3.** True   **4.** False   **5.** True   **6.** False   **7.** False   **8.** True

5.4   Systems of Two Linear Equations

Objectives

■   Solve linear systems of two equations by graphing.

■   Solve linear systems of two equations by the substitution method.

■   Use a system of equations to solve word problems.

Solving Linear Systems by Graphing

Suppose that we graph $x - 2y = 4$ and $x + 2y = 8$ on the same set of axes, as shown in Figure 5.30. The ordered pair $(6, 1)$, which is associated with the point of intersection of the two lines, satisfies both equations. That is, $(6, 1)$ is the solution for

$x - 2y = 4$ and $x + 2y = 8$. To check this, we can substitute 6 for x and 1 for y in both equations.

$$x - 2y = 4 \quad \text{becomes} \quad 6 - 2(1) = 4$$
$$x + 2y = 8 \quad \text{becomes} \quad 6 + 2(1) = 8$$

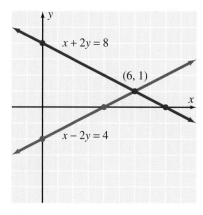

Figure 5.30

Thus we say that $\{(6, 1)\}$ is the solution set of the system

$$\begin{pmatrix} x - 2y = 4 \\ x + 2y = 8 \end{pmatrix}$$

Two or more linear equations in two variables considered together are called a **system of linear equations.** Here are three systems of linear equations:

$$\begin{pmatrix} x - 2y = 4 \\ x + 2y = 8 \end{pmatrix} \quad \begin{pmatrix} 5x - 3y = 9 \\ 3x + 7y = 12 \end{pmatrix} \quad \begin{pmatrix} 4x - y = 5 \\ 2x + y = 9 \\ 7x - 2y = 13 \end{pmatrix}$$

To **solve a system of linear equations** means to find all of the ordered pairs that are solutions of all of the equations in the system. In this chapter, we will consider only systems of *two* linear equations in two variables. There are several techniques for solving systems of linear equations. We will use three of them in this chapter: a graphing method and a substitution method in this section and another method in the following section.

To solve a system of linear equations by **graphing,** we proceed as in the opening discussion of this section. We graph the equations on the same set of axes, and then the ordered pairs associated with any points of intersection are the solutions to the system. Let's consider another example.

EXAMPLE 1

Solve the system $\begin{pmatrix} x + y = 5 \\ x - 2y = -4 \end{pmatrix}$.

Solution

Let's find the intercepts and a check point for each of the lines.

x + y = 5		
x	**y**	
0	5	Intercepts
5	0	
2	3	Check point

x − 2y = −4		
x	**y**	
0	2	Intercepts
−4	0	
−2	1	Check point

Figure 5.31 shows the graphs of the two equations.

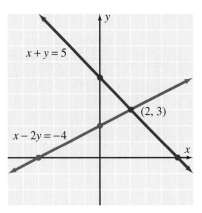

Figure 5.31

It appears that (2, 3) is the solution of the system. To check it, we can substitute 2 for x, and 3 for y in both equations.

$x + y = 5$ becomes $2 + 3 = 5$ A true statement

$x - 2y = -4$ becomes $2 - 2(3) = -4$ A true statement

Therefore, $\{(2, 3)\}$ is the solution set.   ■

It should be evident that solving systems of equations by graphing requires accurate graphs. In fact, unless the solutions are integers, it is really quite difficult to obtain exact solutions from a graph. Checking a solution is particularly important when you use the graphing approach. By checking you can be absolutely sure that you are *reading* the correct solution from the graph.

Figure 5.32 shows the three possible cases for the graph of a system of two linear equations in two variables.

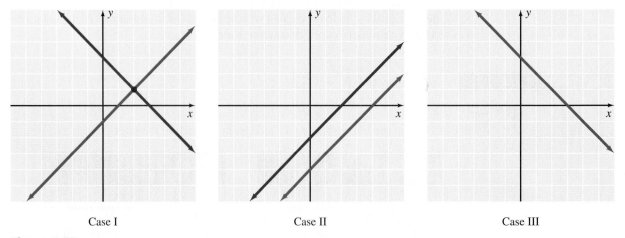

Case I   Case II   Case III

Figure 5.32

CASE I   The graphs of the two equations are two lines that intersect at one point. There is *one solution,* and we call the system a **consistent system.**

CASE II   The graphs of the two equations are parallel lines. There is *no solution,* and we call the system an **inconsistent system.**

CASE III   The graphs of the two equations are the same line. There *are infinitely many* solutions to the system. Any pair of real numbers that satisfies one of the equations will also satisfy the other equation, and we say the equations are **dependent.**

Thus as we solve a system of two linear equations in two variables, we know what to expect. The system will have no solutions, one ordered pair as a solution, or infinitely many ordered pairs as solutions. Most of the systems that we will be working with in this text will have one solution.

Substitution Method

It should be evident that solving systems of equations by graphing requires accurate graphs. In fact, unless the solutions are integers, it is quite difficult to obtain exact solutions from a graph. Thus we will consider some other methods for solving systems of equations.

The **substitution method** works quite well with systems of two linear equations in two unknowns.

STEP 1   Solve one of the equations for one variable in terms of the other variable if neither equation is in such a form. (If possible, make a choice that will avoid fractions.)

STEP 2   Substitute the expression obtained in step 1 into the other equation. This produces an equation in one variable.

STEP 3   Solve the equation obtained in step 2.

STEP 4   Use the solution obtained in step 3, along with the expression obtained in step 1, to determine the solution of the system.

Now let's look at some examples that illustrate the substitution method.

EXAMPLE   2   Solve the system $\left(\begin{array}{c} x + y = 16 \\ y = x + 2 \end{array} \right)$.

Solution

Because the second equation states that y equals $x + 2$, we can substitute $x + 2$ for y in the first equation.

$$x + y = 16 \quad \xrightarrow{\text{Substitute } x + 2 \text{ for } y.} \quad x + (x + 2) = 16$$

Now we have an equation with one variable that we can solve in the usual way.

$$x + (x + 2) = 16$$
$$2x + 2 = 16$$
$$2x = 14$$
$$x = 7$$

Substituting 7 for x in one of the two original equations (let's use the second one) yields

$$y = 7 + 2 = 9$$

 Check

To check, we can substitute 7 for x and 9 for y in both of the original equations.

$$7 + 9 = 16 \qquad \text{A true statement}$$
$$9 = 7 + 2 \qquad \text{A true statement}$$

The solution set is $\{(7, 9)\}$.

E X A M P L E   3   Solve the system $\begin{pmatrix} 3x - 7y = 2 \\ x + 4y = 1 \end{pmatrix}$.

Solution

Let's solve the second equation for x in terms of y.

$$x + 4y = 1$$
$$x = 1 - 4y$$

Now we can substitute $1 - 4y$ for x in the first equation.

$$3x - 7y = 2 \quad \xrightarrow{\text{Substitute } 1 - 4y \text{ for } x.} \quad 3(1 - 4y) - 7y = 2$$

Let's solve this equation for y.

$$3(1 - 4y) - 7y = 2$$
$$3 - 12y - 7y = 2$$
$$-19y = -1$$
$$y = \frac{1}{19}$$

Finally, we can substitute $\frac{1}{19}$ for y in the equation $x = 1 - 4y$.

$$x = 1 - 4\left(\frac{1}{19}\right)$$

$$= 1 - \frac{4}{19}$$

$$= \frac{15}{19}$$

The solution set is $\left\{ \left(\frac{15}{19}, \frac{1}{19} \right) \right\}$.

E X A M P L E   4   Solve the system $\left(\begin{matrix} 5x - 6y = -4 \\ 3x + 2y = -8 \end{matrix} \right)$.

Solution

Note that solving either equation for either variable will produce a fractional form. Let's solve the second equation for y in terms of x.

$$3x + 2y = -8$$

$$2y = -8 - 3x$$

$$y = \frac{-8 - 3x}{2}$$

Now we can substitute $\dfrac{-8 - 3x}{2}$ for y in the first equation.

$$5x - 6y = -4 \xrightarrow{\text{Substitute } \frac{-8 - 3x}{2} \text{ for } y.} 5x - 6\left(\frac{-8 - 3x}{2} \right) = -4$$

Solving this equation yields

$$5x - 6\left(\frac{-8 - 3x}{2} \right) = -4$$

$$5x - 3(-8 - 3x) = -4$$

$$5x + 24 + 9x = -4$$

$$14x = -28$$

$$x = -2$$

Substituting -2 for x in $y = \dfrac{-8 - 3x}{2}$ yields

$$y = \frac{-8 - 3(-2)}{2}$$

$$y = \frac{-8 + 6}{2}$$

$$y = \frac{-2}{2}$$

$$= -1$$

The solution set is $\{(-2, -1)\}$.

EXAMPLE 5   Solve the system $\begin{pmatrix} 2x + y = 4 \\ 4x + 2y = 7 \end{pmatrix}$.

Solution

Let's solve the first equation for y in terms of x.

$$2x + y = 4$$
$$y = 4 - 2x$$

Now we can substitute $4 - 2x$ for y in the second equation.

$$4x + 2y = 7 \quad \xrightarrow{\text{Substitute } 4 - 2x \text{ for } y.} \quad 4x + 2(4 - 2x) = 7$$

Let's solve this equation for x.

$$4x + 2(4 - 2x) = 7$$
$$4x + 8 - 4x = 7$$
$$8 = 7$$

The statement $8 = 7$ is a contradiction, and therefore the original system is inconsistent; it has no solution. The solution set is $\varnothing$.   ■

EXAMPLE 6   Solve the system. $\begin{pmatrix} y = 2x + 1 \\ 4x - 2y = -2 \end{pmatrix}$

Solution

Because the first equation states that y equals $2x + 1$, we can substitute $2x + 1$ for y in the second equation.

$$4x - 2y = -2 \quad \xrightarrow{\text{Substitute } 2x + 1 \text{ for } y.} \quad 4x - 2(2x + 1) = -2$$

Let's solve this equation for x.

$$4x - 2(2x + 1) = -2$$
$$4x - 4x - 2 = -2$$
$$-2 = -2$$

We obtained a true statement, $-2 = -2$, which indicates that the system has an infinite number of solutions. Any ordered pair that satisfies one of the equations will also satisfy the other equation. Thus the solution set is any ordered pair on the line, $y = 2x + 1$, and the solution set can be written as $\{(x, y) | y = 2x + 1\}$.   ■

Problem Solving

Many word problems that we solved earlier in this text by using one variable and one equation can also be solved by using a system of two linear equations in two variables. In fact, in many of these problems you may find it much more natural to use two variables. Let's consider some examples.

P R O B L E M   1

Anita invested some money at 8% and $400 more than that amount at 9%. The yearly interest from the two investments was $87. How much did Anita invest at each rate?

Solution

Let x represent the amount invested at 8%, and let y represent the amount invested at 9%. The problem translates into this system:

The amount invested at 9% was $400 more than at 8%. $\begin{pmatrix} y = x + 400 \\ 0.08x + 0.09y = 87 \end{pmatrix}$
The yearly interest from the two investments was $87.

From the first equation, we can substitute $x + 400$ for y in the second equation and solve for x.

$$0.08x + 0.09(x + 400) = 87$$
$$0.08x + 0.09x + 36 = 87$$
$$0.17x = 51$$
$$x = 300$$

Therefore, Anita invested $300 at 8% and $300 + $400 = $700 at 9%. ■

P R O B L E M   2

The proceeds from a concession stand that sold hamburgers and hot dogs at the baseball game were $575.50. The price of a hot dog was $2.50, and the price of a hamburger was $3.00. If a total of 213 hot dogs and hamburgers were sold, how many of each kind were sold?

Solution

Let x equal the number of hot dogs sold, and let y equal the number of hamburgers sold. The problem translates into this system:

The number sold $\begin{pmatrix} x + y = 213 \\ 2.50x + 3.00y = 575.50 \end{pmatrix}$
The proceeds from the sales

Let's begin by solving the first equation for y.

$$x + y = 213$$
$$y = 213 - x$$

Now we will substitute $213 - x$ for y in the second equation and solve for x.

$$2.50x + 3.00(213 - x) = 575.50$$
$$2.50x + 639.00 - 3.00x = 575.50$$
$$-0.5x + 639.00 = 575.50$$
$$-0.5x = -63.50$$
$$x = 127$$

Therefore, there were 127 hot dogs sold and $213 - 127 = 86$ hamburgers sold. ∎

CONCEPT QUIZ

For Problems 1–8, answer true or false.

1. To *solve a system of equations* means to find all the ordered pairs that satisfy all of the equations in the system.

2. A consistent system of linear equations will have more than one solution.

3. If the graph of a system of two distinct linear equations results in two parallel lines, then the system has no solution.

4. Every system of equations has a solution.

5. If the graphs of the two equations in a system are the same line, then equations in the system are dependent.

6. To solve a system of two equations in variables x and y, it is sufficient to just find a value for x.

7. For the system $\begin{pmatrix} 2x + y = 4 \\ x + 5y = 10 \end{pmatrix}$, the ordered pair $(1, 2)$ is a solution.

8. Graphing a system of equations is the most accurate method to find the solution of the system.

PROBLEM SET 5.4

For Problems 1–20, use the graphing method to solve each system.

1. $\begin{pmatrix} x + y = 1 \\ x - y = 3 \end{pmatrix}$

2. $\begin{pmatrix} x - y = 2 \\ x + y = -4 \end{pmatrix}$

3. $\begin{pmatrix} x + 2y = 4 \\ 2x - y = 3 \end{pmatrix}$

4. $\begin{pmatrix} 2x - y = -8 \\ x + y = 2 \end{pmatrix}$

5. $\begin{pmatrix} x + 3y = 6 \\ x + 3y = 3 \end{pmatrix}$

6. $\begin{pmatrix} y = -2x \\ y - 3x = 0 \end{pmatrix}$

7. $\begin{pmatrix} x + y = 0 \\ x - y = 0 \end{pmatrix}$

8. $\begin{pmatrix} 3x - y = 3 \\ 3x - y = -3 \end{pmatrix}$

9. $\begin{pmatrix} 3x - 2y = 5 \\ 2x + 5y = -3 \end{pmatrix}$

10. $\begin{pmatrix} 2x + 3y = 1 \\ 4x - 3y = -7 \end{pmatrix}$

11. $\begin{pmatrix} y = -2x + 3 \\ 6x + 3y = 9 \end{pmatrix}$

12. $\begin{pmatrix} y = 2x + 5 \\ x + 3y = -6 \end{pmatrix}$

13. $\begin{pmatrix} y = 5x - 2 \\ 4x + 3y = 13 \end{pmatrix}$

14. $\begin{pmatrix} y = x - 2 \\ 2x - 2y = 4 \end{pmatrix}$

15. $\begin{pmatrix} y = 4 - 2x \\ y = 7 - 3x \end{pmatrix}$ **16.** $\begin{pmatrix} y = 3x + 4 \\ y = 5x + 8 \end{pmatrix}$

17. $\begin{pmatrix} y = 2x \\ 3x - 2y = -2 \end{pmatrix}$ **18.** $\begin{pmatrix} y = 3x \\ 4x - 3y = 5 \end{pmatrix}$

19. $\begin{pmatrix} 7x - 2y = -8 \\ x = -2 \end{pmatrix}$ **20.** $\begin{pmatrix} 3x + 8y = -1 \\ y = -2 \end{pmatrix}$

For Problems 21–46, solve each system by using the substitution method.

21. $\begin{pmatrix} x + y = 20 \\ x = y - 4 \end{pmatrix}$ **22.** $\begin{pmatrix} x + y = 23 \\ y = x - 5 \end{pmatrix}$

23. $\begin{pmatrix} y = -3x - 18 \\ 5x - 2y = -8 \end{pmatrix}$ **24.** $\begin{pmatrix} 4x - 3y = 33 \\ x = -4y - 25 \end{pmatrix}$

25. $\begin{pmatrix} x = -3y \\ 7x - 2y = -69 \end{pmatrix}$ **26.** $\begin{pmatrix} 9x - 2y = -38 \\ y = -5x \end{pmatrix}$

27. $\begin{pmatrix} x + 2y = 5 \\ 3x + 6y = -2 \end{pmatrix}$ **28.** $\begin{pmatrix} 4x + 2y = 6 \\ y = -2x + 3 \end{pmatrix}$

29. $\begin{pmatrix} 3x - 4y = 9 \\ x = 4y - 1 \end{pmatrix}$ **30.** $\begin{pmatrix} y = 3x - 5 \\ 2x + 3y = 6 \end{pmatrix}$

31. $\begin{pmatrix} y = \frac{2}{5}x - 1 \\ 3x + 5y = 4 \end{pmatrix}$ **32.** $\begin{pmatrix} y = \frac{3}{4}x - 5 \\ 5x - 4y = 9 \end{pmatrix}$

33. $\begin{pmatrix} 7x - 3y = -2 \\ x = \frac{3}{4}y + 1 \end{pmatrix}$ **34.** $\begin{pmatrix} 5x - y = 9 \\ x = \frac{1}{2}y - 3 \end{pmatrix}$

35. $\begin{pmatrix} 2x + y = 12 \\ 3x - y = 13 \end{pmatrix}$ **36.** $\begin{pmatrix} -x + 4y = -22 \\ x - 7y = 34 \end{pmatrix}$

37. $\begin{pmatrix} 4x + 3y = -40 \\ 5x - y = -12 \end{pmatrix}$ **38.** $\begin{pmatrix} x - 5y = 33 \\ -4x + 7y = -41 \end{pmatrix}$

39. $\begin{pmatrix} 3x + y = 2 \\ 11x - 3y = 5 \end{pmatrix}$ **40.** $\begin{pmatrix} 2x - y = 9 \\ 7x + 4y = 1 \end{pmatrix}$

41. $\begin{pmatrix} 4x - 8y = -12 \\ 3x - 6y = -9 \end{pmatrix}$ **42.** $\begin{pmatrix} 2x - 4y = -6 \\ 3x - 6y = 10 \end{pmatrix}$

43. $\begin{pmatrix} 4x - 5y = 3 \\ 8x + 15y = -24 \end{pmatrix}$ **44.** $\begin{pmatrix} 2x + 3y = 3 \\ 4x - 9y = -4 \end{pmatrix}$

45. $\begin{pmatrix} 6x - 3y = 4 \\ 5x + 2y = -1 \end{pmatrix}$ **46.** $\begin{pmatrix} 7x - 2y = 1 \\ 4x + 5y = 2 \end{pmatrix}$

For Problems 47–58, solve each problem by setting up and solving an appropriate system of linear equations.

47. Doris invested some money at 7% and some money at 8%. She invested $6000 more at 8% than she did at 7%. Her total yearly interest from the two investments was $780. How much did Doris invest at each rate?

48. Suppose that Gus invested a total of $8000, part of it at 4% and the remainder at 6%. His yearly income from the two investments was $380. How much did he invest at each rate?

49. Find two numbers whose sum is 131 such that one number is 5 less than three times the other.

50. The difference of two numbers is 75. The larger number is 3 less than four times the smaller number. Find the numbers.

51. In a class of 50 students, the number of women is 2 more than five times the number of men. How many women are there in the class?

52. In a recent survey, 1000 registered voters were asked about their political preferences. The number of men in the survey was 5 less than one-half the number of women. Find the number of men in the survey.

53. The perimeter of a rectangle is 94 inches. The length of the rectangle is 7 inches more than the width. Find the dimensions of the rectangle.

54. Two angles are supplementary, and the measure of one of them is 20° less than three times the measure of the other angle. Find the measure of each angle.

55. A deposit slip listed $700 in cash to be deposited. There were 100 bills, some of them five-dollar bills and the remainder ten-dollar bills. How many bills of each denomination were deposited?

56. Cindy has 30 coins, consisting of dimes and quarters, that total $5.10. How many coins of each kind does she have?

57. The income from a student production was $27,500. The price of a student ticket was $8, and nonstudent tickets were sold at $15 each. Three thousand tickets were sold. How many tickets of each kind were sold?

58. Sue bought 3 packages of cookies and 2 sacks of potato chips for $7.35. Later she bought 2 packages of cookies and 5 sacks of potato chips for $9.63. Find the price of a package of cookies.

■ ■ ■ Thoughts into words

59. Discuss the strengths and weaknesses of solving a system of linear equations by graphing.

60. Determine a system of two linear equations for which the solution is $(5, 7)$. Are there other systems that have the same solution set? If so, find at least one more system.

61. Give a general description of how to use the substitution method to solve a system of two linear equations in two variables.

62. Is it possible for a system of two linear equations in two variables to have exactly two solutions? Defend your answer.

63. Explain how you would use the substitution method to solve the system

$$\begin{pmatrix} 2x + 5y = 5 \\ 5x - y = 9 \end{pmatrix}$$

Answers to Concept Quiz

1. True **2.** False **3.** True **4.** False **5.** True **6.** False **7.** False **8.** False

---

5.5 Elimination-by-Addition Method

Objectives

■   Solve linear systems of equations by the elimination-by-addition method.

■   Solve word problems using a system of two linear equations.

We found in the previous section that the substitution method for solving a system of two equations and two unknowns works rather well. However, as the number of equations and unknowns increases, the substitution method becomes quite unwieldy. In this section we are going to introduce another method, called the **elimination-by-addition method.** We shall introduce it here, using systems of two linear equations in two unknowns. Later in the text, we shall extend its use to three linear equations in three unknowns.

The elimination-by-addition method involves replacing systems of equations with simpler equivalent systems until we obtain a system from which we can easily extract the solutions. **Equivalent systems of equations are systems that have exactly the same solution set.** We can apply the following operations or transformations to a system of equations to produce an equivalent system.

> **1.** Any two equations of the system can be interchanged.
> **2.** Both sides of any equation of the system can be multiplied by any nonzero real number.
> **3.** Any equation of the system can be replaced by the sum of the equation and a nonzero multiple of another equation.

Now let's see how to apply these operations to solve a system of two linear equations in two unknowns.

E X A M P L E   1

Solve the system $\begin{pmatrix} 3x + 2y = 1 \\ 5x - 2y = 23 \end{pmatrix}$.

(1)
(2)

Solution

Let's replace equation (2) with an equation we form by multiplying equation (1) by 1 and then adding that result to equation (2).

$\begin{pmatrix} 3x + 2y = 1 \\ 8x \quad\;\; = 24 \end{pmatrix}$

(3)
(4)

From equation (4) we can easily obtain the value of x.

$$8x = 24$$
$$x = 3$$

Then we can substitute 3 for x in equation (3).

$$3x + 2y = 1$$
$$3(3) + 2y = 1$$
$$2y = -8$$
$$y = -4$$

The solution set is $\{(3, -4)\}$. Check it! ∎

E X A M P L E   2

Solve the system $\begin{pmatrix} x + 5y = -2 \\ 3x - 4y = -25 \end{pmatrix}$.

(1)
(2)

Solution

Let's replace equation (2) with an equation we form by multiplying equation (1) by -3 and then adding that result to equation (2).

$\begin{pmatrix} x + 5y = -2 \\ -19y = -19 \end{pmatrix}$

(3)
(4)

From equation (4) we can obtain the value of y.

$$-19y = -19$$
$$y = 1$$

Now we can substitute 1 for y in equation (3).

$$x + 5y = -2$$
$$x + 5(1) = -2$$
$$x = -7$$

The solution set is $\{(-7, 1)\}$. ∎

Note that our objective has been to produce an equivalent system of equations such that one of the variables can be *eliminated* from one equation. We accomplish this by multiplying one equation of the system by an appropriate number and then *adding* that result to the other equation. Thus the method is called *elimination by addition*. Let's look at another example.

EXAMPLE 3

Solve the system $\begin{pmatrix} 2x + 5y = 4 \\ 5x - 7y = -29 \end{pmatrix}$. (1)
(2)

Solution

Let's form an equivalent system where the second equation has no x term. First, we can multiply equation (2) by -2.

$$\begin{pmatrix} 2x + 5y = 4 \\ -10x + 14y = 58 \end{pmatrix}$$ (3)
(4)

Now we can replace equation (4) with an equation that we form by multiplying equation (3) by 5 and then adding that result to equation (4).

$$\begin{pmatrix} 2x + 5y = 4 \\ 39y = 78 \end{pmatrix}$$ (5)
(6)

From equation (6) we can find the value of y.

$$39y = 78$$
$$y = 2$$

Now we can substitute 2 for y in equation (5).

$$2x + 5y = 4$$
$$2x + 5(2) = 4$$
$$2x = -6$$
$$x = -3$$

The solution set is $\{(-3, 2)\}$. ∎

EXAMPLE 4

Solve the system $\begin{pmatrix} 3x - 2y = 5 \\ 2x + 7y = 9 \end{pmatrix}$. (1)
(2)

Solution

We can start by multiplying equation (2) by -3.

$$\begin{pmatrix} 3x - 2y = 5 \\ -6x - 21y = -27 \end{pmatrix}$$ (3)
(4)

Now we can replace equation (4) with an equation we form by multiplying equation (3) by 2 and then adding that result to equation (4).

$$\begin{pmatrix} 3x - 2y = 5 \\ -25y = -17 \end{pmatrix}$$ (5)
(6)

From equation (6) we can find the value of y.

$$-25y = -17$$
$$y = \frac{17}{25}$$

Now we can substitute $\dfrac{17}{25}$ for y in equation (5).

$$3x - 2y = 5$$

$$3x - 2\left(\frac{17}{25}\right) = 5$$

$$3x - \frac{34}{25} = 5$$

$$3x = 5 + \frac{34}{25}$$

$$3x = \frac{125}{25} + \frac{34}{25}$$

$$3x = \frac{159}{25}$$

$$x = \left(\frac{159}{25}\right)\left(\frac{1}{3}\right) = \frac{53}{25}$$

The solution set is $\left\{\left(\dfrac{53}{25}, \dfrac{17}{25}\right)\right\}$. (Perhaps you should check this result!) ■

Which Method to Use

We can use both the elimination-by-addition and the substitution methods to obtain exact solutions for any system of two linear equations in two unknowns. Sometimes we need to decide which method to use on a particular system. As we have seen with the examples thus far in this section and with those in the previous section, many systems lend themselves to one or the other method by the original format of the equations. Let's emphasize that point with some more examples.

EXAMPLE 5   Solve the system $\begin{pmatrix} 4x - 3y = 4 \\ 10x + 9y = -1 \end{pmatrix}$. **(1)**
(2)

Solution

Because changing the form of either equation in preparation for the substitution method would produce a fractional form, we are probably better off using the elimination-by-addition method. Let's replace equation (2) with an equation we form by multiplying equation (1) by 3 and then adding that result to equation (2).

$$\begin{pmatrix} 4x - 3y = 4 \\ 22x = 11 \end{pmatrix}$$ **(3)**
(4)

From equation (4) we can determine the value of x.

$$22x = 11$$

$$x = \frac{11}{22} = \frac{1}{2}$$

Now we can substitute $\frac{1}{2}$ for x in equation (3).

$$4x - 3y = 4$$

$$4\left(\frac{1}{2}\right) - 3y = 4$$

$$2 - 3y = 4$$

$$-3y = 2$$

$$y = -\frac{2}{3}$$

The solution set is $\left\{\left(\frac{1}{2}, -\frac{2}{3}\right)\right\}$. ■

EXAMPLE 6   Solve the system $\begin{pmatrix} 6x + 5y = -3 \\ y = -2x - 7 \end{pmatrix}$.   **(1)** **(2)**

Solution

Because the second equation is of the form *y equals,* let's use the substitution method. From the second equation we can substitute $-2x - 7$ for y in the first equation.

$$6x + 5y = -3 \quad \xrightarrow{\text{Substitute } -2x - 7 \text{ for } y.} \quad 6x + 5(-2x - 7) = -3$$

Solving this equation yields

$$6x + 5(-2x - 7) = -3$$

$$6x - 10x - 35 = -3$$

$$-4x - 35 = -3$$

$$-4x = 32$$

$$x = -8$$

We substitute -8 for x in the second equation to get

$$y = -2(-8) - 7$$

$$= 16 - 7 = 9$$

The solution set is $\{(-8, 9)\}$. ■

Sometimes we need to simplify the equations of a system before we can decide which method to use for solving the system. Let's consider an example of that type.

EXAMPLE 7   Solve the system $\begin{pmatrix} \dfrac{x - 2}{4} + \dfrac{y + 1}{3} = 2 \\ \dfrac{x + 1}{7} + \dfrac{y - 3}{2} = \dfrac{1}{2} \end{pmatrix}$.   **(1)** **(2)**

Solution

First, we need to simplify the two equations. Let's multiply both sides of equation (1) by 12 and simplify.

$$12\left(\frac{x-2}{4} + \frac{y+1}{3}\right) = 12(2)$$

$$3(x-2) + 4(y+1) = 24$$

$$3x - 6 + 4y + 4 = 24$$

$$3x + 4y - 2 = 24$$

$$3x + 4y = 26$$

Let's multiply both sides of equation (2) by 14.

$$14\left(\frac{x+1}{7} + \frac{y-3}{2}\right) = 14\left(\frac{1}{2}\right)$$

$$2(x+1) + 7(y-3) = 7$$

$$2x + 2 + 7y - 21 = 7$$

$$2x + 7y - 19 = 7$$

$$2x + 7y = 26$$

Now we have the following system to solve.

$$\begin{pmatrix} 3x + 4y = 26 \\ 2x + 7y = 26 \end{pmatrix} \qquad \textbf{(3)} \\ \textbf{(4)}$$

Probably the easiest approach is to use the elimination-by-addition method. We can start by multiplying equation (4) by -3.

$$\begin{pmatrix} 3x + 4y = 26 \\ -6x - 21y = -78 \end{pmatrix} \qquad \textbf{(5)} \\ \textbf{(6)}$$

Now we can replace equation (6) with an equation we form by multiplying equation (5) by 2 and then adding that result to equation (6).

$$\begin{pmatrix} 3x + 4y = 26 \\ -13y = -26 \end{pmatrix} \qquad \textbf{(7)} \\ \textbf{(8)}$$

From equation (8) we can find the value of y.

$$-13y = -26$$

$$y = 2$$

Now we can substitute 2 for y in equation (7).

$$3x + 4y = 26$$

$$3x + 4(2) = 26$$

$$3x = 18$$

$$x = 6$$

The solution set is $\{(6, 2)\}$.

REMARK: Don't forget that to check a problem like Example 7, you must check the potential solutions back in the *original* equations.

In Section 5.4, we explained that you can tell whether a system of two linear equations in two unknowns has no solution, one solution, or infinitely many solutions by graphing the equations of the system. That is, the two lines may be parallel (no solution), or they may intersect in one point (one solution), or they may coincide (infinitely many solutions).

From a practical viewpoint, the systems that have one solution deserve most of our attention. However, we do need to be able to deal with the other situations because they occur occasionally. The next two examples illustrate the type of thing that happens when we encounter a *no solution* or *infinitely many solutions* situation when using the elimination-by-addition method.

EXAMPLE 8   Solve the system $\begin{pmatrix} 2x + y = 1 \\ 4x + 2y = 3 \end{pmatrix}$.   (1)   (2)

Solution

Use the elimination-by-addition method. Let's replace equation (2) with an equation we form by multiplying equation (1) by -2 and then adding that result to form equation (2).

$$\begin{pmatrix} 2x + y = 1 \\ 0 + 0 = 1 \end{pmatrix}$$ (3)   (4)

The false numerical statement $0 + 0 = 1$ implies that the system has no solution. Thus the solution set is $\varnothing$.   ∎

EXAMPLE 9   Solve the system $\begin{pmatrix} 5x + y = 2 \\ 10x + 2y = 4 \end{pmatrix}$.   (1)   (2)

Solution

Use the elimination-by-addition method. Let's replace equation (2) with an equation we form by multiplying equation (1) by -2 and then adding that result to equation (2).

$$\begin{pmatrix} 5x + y = 2 \\ 0 + 0 = 0 \end{pmatrix}$$ (3)   (4)

The *true numerical statement* $0 + 0 = 0$ implies that the system has *infinitely many solutions.* Any ordered pair that satisfies one of the equations will also satisfy the other equation. Thus the solution set can be expressed as

$$\{(x, y)|5x + y = 2\}$$ ∎

PROBLEM 1   A 25% chlorine solution is to be mixed with a 40% chlorine solution to produce 12 gallons of a 35% chlorine solution. How many gallons of each solution should be mixed?

Solution

Let x represent the gallons of 25% chlorine solution, and let y represent the gallons of 40% chlorine solution.

Then one equation of the system will be $x + y = 12$. For the other equation we need to multiply the number of gallons of each solution by its percentage of chlorine. That gives the equation $0.25x + 0.40y = 0.35(12)$. So we need to solve the following system:

$$\begin{pmatrix} x + y = 12 \\ 0.25x + 0.40y = 4.2 \end{pmatrix} \qquad \begin{matrix} \textbf{(1)} \\ \textbf{(2)} \end{matrix}$$

Use the elimination-by-addition method. Let's replace equation (2) with an equation we form by multiplying equation (1) by -0.25 and then adding that result to equation (2).

$$\begin{pmatrix} x + y = 12 \\ 0.15y = 1.2 \end{pmatrix} \qquad \begin{matrix} \textbf{(3)} \\ \textbf{(4)} \end{matrix}$$

From equation (4) we can find the value of y.

$$0.15y = 1.2$$
$$y = 8$$

Now we can substitute 8 into equation (3).

$$x + 8 = 12$$
$$x = 4$$

Therefore, we need 4 gallons of the 25% chlorine solution and 8 gallons of the 40% solution.

CONCEPT QUIZ

For Problems 1–5, answer true or false.

1. Any two equations of a system can be interchanged to obtain an equivalent system.

2. Any equation of a system can be multiplied on both sides by zero to obtain an equivalent system.

3. The objective of the elimination-by-addition method is to produce an equivalent system with an equation where one of the variables has been eliminated.

4. Either the substitution method or the elimination-by-addition method can be used for any linear system of equations.

5. If an equivalent system for an original system is $\begin{pmatrix} 3x - 5y = 7 \\ 0 + 0 = 0 \end{pmatrix}$, then the original system is inconsistent and has no solution.

PROBLEM SET 5.5

For Problems 1–16, use the elimination-by-addition method to solve each system.

1. $\begin{pmatrix} 2x + 3y = -1 \\ 5x - 3y = 29 \end{pmatrix}$ **2.** $\begin{pmatrix} 3x - 4y = -30 \\ 7x + 4y = 10 \end{pmatrix}$

3. $\begin{pmatrix} 6x - 7y = 15 \\ 6x + 5y = -21 \end{pmatrix}$ **4.** $\begin{pmatrix} 5x + 2y = -4 \\ 5x - 3y = 6 \end{pmatrix}$

5. $\begin{pmatrix} x - 2y = -12 \\ 2x + 9y = 2 \end{pmatrix}$ **6.** $\begin{pmatrix} x - 4y = 29 \\ 3x + 2y = -11 \end{pmatrix}$

7. $\begin{pmatrix} 4x + 7y = -16 \\ 6x - y = -24 \end{pmatrix}$ **8.** $\begin{pmatrix} 6x + 7y = 17 \\ 3x + y = -4 \end{pmatrix}$

9. $\begin{pmatrix} 3x - 2y = 5 \\ 2x + 5y = -3 \end{pmatrix}$ **10.** $\begin{pmatrix} 4x + 3y = -4 \\ 3x - 7y = 34 \end{pmatrix}$

11. $\begin{pmatrix} 7x - 2y = 4 \\ 7x - 2y = 9 \end{pmatrix}$ **12.** $\begin{pmatrix} 5x - y = 6 \\ 10x - 2y = 12 \end{pmatrix}$

13. $\begin{pmatrix} 5x + 4y = 1 \\ 3x - 2y = -1 \end{pmatrix}$ **14.** $\begin{pmatrix} 2x - 7y = -2 \\ 3x + y = 1 \end{pmatrix}$

15. $\begin{pmatrix} 8x - 3y = 13 \\ 4x + 9y = 3 \end{pmatrix}$ **16.** $\begin{pmatrix} 10x - 8y = -11 \\ 8x + 4y = -1 \end{pmatrix}$

For Problems 17–44, solve each system by using either the substitution or the elimination-by-addition method, whichever seems more appropriate.

17. $\begin{pmatrix} 5x + 3y = -7 \\ 7x - 3y = 55 \end{pmatrix}$ **18.** $\begin{pmatrix} 4x - 7y = 21 \\ -4x + 3y = -9 \end{pmatrix}$

19. $\begin{pmatrix} x = 5y + 7 \\ 4x + 9y = 28 \end{pmatrix}$ **20.** $\begin{pmatrix} 11x - 3y = -60 \\ y = -38 - 6x \end{pmatrix}$

21. $\begin{pmatrix} x = -6y + 79 \\ x = 4y - 41 \end{pmatrix}$ **22.** $\begin{pmatrix} y = 3x + 34 \\ y = -8x - 54 \end{pmatrix}$

23. $\begin{pmatrix} 4x - 3y = 2 \\ 5x - y = 3 \end{pmatrix}$ **24.** $\begin{pmatrix} 3x - y = 9 \\ 5x + 7y = 1 \end{pmatrix}$

25. $\begin{pmatrix} 5x - 2y = 1 \\ 10x - 4y = 7 \end{pmatrix}$ **26.** $\begin{pmatrix} 4x + 7y = 2 \\ 9x - 2y = 1 \end{pmatrix}$

27. $\begin{pmatrix} 3x - 2y = 7 \\ 5x + 7y = 1 \end{pmatrix}$ **28.** $\begin{pmatrix} 2x - 3y = 4 \\ y = \dfrac{2}{3}x - \dfrac{4}{3} \end{pmatrix}$

29. $\begin{pmatrix} -2x + 5y = -16 \\ x = \dfrac{3}{4}y + 1 \end{pmatrix}$ **30.** $\begin{pmatrix} y = \dfrac{2}{3}x - \dfrac{3}{4} \\ 2x + 3y = 11 \end{pmatrix}$

31. $\begin{pmatrix} y = \dfrac{2}{3}x - 4 \\ 5x - 3y = 9 \end{pmatrix}$ **32.** $\begin{pmatrix} 5x - 3y = 7 \\ x = \dfrac{3y}{4} - \dfrac{1}{3} \end{pmatrix}$

33. $\begin{pmatrix} \dfrac{x}{6} + \dfrac{y}{3} = 3 \\ \dfrac{5x}{2} - \dfrac{y}{6} = -17 \end{pmatrix}$ **34.** $\begin{pmatrix} \dfrac{3x}{4} - \dfrac{2y}{3} = 31 \\ \dfrac{7x}{5} + \dfrac{y}{4} = 22 \end{pmatrix}$

35. $\begin{pmatrix} -(x - 6) + 6(y + 1) = 58 \\ 3(x + 1) - 4(y - 2) = -15 \end{pmatrix}$

36. $\begin{pmatrix} -2(x + 2) + 4(y - 3) = -34 \\ 3(x + 4) - 5(y + 2) = 23 \end{pmatrix}$

37. $\begin{pmatrix} 5(x + 1) - (y + 3) = -6 \\ 2(x - 2) + 3(y - 1) = 0 \end{pmatrix}$

38. $\begin{pmatrix} 2(x - 1) - 3(y + 2) = 30 \\ 3(x + 2) + 2(y - 1) = -4 \end{pmatrix}$

39. $\begin{pmatrix} \dfrac{1}{2}x - \dfrac{1}{3}y = 12 \\ \dfrac{3}{4}x + \dfrac{2}{3}y = 4 \end{pmatrix}$ **40.** $\begin{pmatrix} \dfrac{2}{3}x + \dfrac{1}{5}y = 0 \\ \dfrac{3}{2}x - \dfrac{3}{10}y = -15 \end{pmatrix}$

41. $\begin{pmatrix} \dfrac{2x}{3} - \dfrac{y}{2} = -\dfrac{5}{4} \\ \dfrac{x}{4} + \dfrac{5y}{6} = \dfrac{17}{16} \end{pmatrix}$ **42.** $\begin{pmatrix} \dfrac{x}{2} + \dfrac{y}{3} = \dfrac{5}{72} \\ \dfrac{x}{4} + \dfrac{5y}{2} = -\dfrac{17}{48} \end{pmatrix}$

43. $\begin{pmatrix} \dfrac{3x + y}{2} + \dfrac{x - 2y}{5} = 8 \\ \dfrac{x - y}{3} - \dfrac{x + y}{6} = \dfrac{10}{3} \end{pmatrix}$

44. $\begin{pmatrix} \dfrac{x - y}{4} - \dfrac{2x - y}{3} = -\dfrac{1}{4} \\ \dfrac{2x + y}{3} + \dfrac{x + y}{2} = \dfrac{17}{6} \end{pmatrix}$

For Problems 45–55, solve each problem by setting up and solving an appropriate system of equations.

45. A 10% salt solution is to be mixed with a 20% salt solution to produce 20 gallons of a 17.5% salt solution. How many gallons of the 10% solution and how many gallons of the 20% solution will be needed?

46. A small town library buys a total of 35 books that cost $1022. Some of the books cost $22 each, and the remainder cost $34 per book. How many books of each price did the library buy?

47. Suppose that on a particular day the cost of 3 tennis balls and 2 golf balls is $12. The cost of 6 tennis balls and 3 golf balls is $21. Find the cost of 1 tennis ball and the cost of 1 golf ball.

48. For moving purposes, the Hendersons bought 25 cardboard boxes for $97.50. There were two kinds of boxes; the large ones cost $7.50 per box, and the small ones cost $3 per box. How many boxes of each kind did they buy?

49. A motel in a suburb of Chicago rents single rooms for $62 per day and double rooms for $82 per day. If a total of 55 rooms were rented for $4210, how many of each kind were rented?

50. Suppose that one solution is 50% alcohol and another solution is 80% alcohol. How many liters of each solution should be mixed to make 10.5 liters of a 70% alcohol solution?

51. If the numerator of a certain fraction is increased by 5 and the denominator is decreased by 1, the resulting fraction is $\frac{8}{3}$. However, if the numerator of the original fraction is doubled and the denominator is increased by 7, the resulting fraction is $\frac{6}{11}$. Find the original fraction.

52. A man bought 2 pounds of coffee and 1 pound of butter for a total of $18.75. A month later the prices had not changed (this makes it a fictitious problem), and he bought 3 pounds of coffee and 2 pounds of butter for $29.50. Find the price per pound of both the coffee and the butter.

53. Suppose that we have a rectangular book cover. If the width is increased by 2 centimeters, and the length is decreased by 1 centimeter, then the area is increased by 28 square centimeters. However, if the width is decreased by 1 centimeter and the length is increased by 2 centimeters, then the area is increased by 10 square centimeters. Find the dimensions of the book cover.

54. A blueprint indicates a master bedroom in the shape of a rectangle. If the width is increased by 2 feet and the length remains the same, then the area is increased by 36 square feet. However, if the width is increased by 1 foot and the length is increased by 2 feet, then the area is increased by 48 square feet. Find the dimensions of the room as indicated on the blueprint.

55. A fulcrum is placed so that weights of 60 pounds and 100 pounds are in balance. If 20 pounds are subtracted from the 100-pound weight, then the 60-pound weight must be moved 1 foot closer to the fulcrum to preserve the balance. Find the original distance between the 60-pound and 100-pound weights.

■ ■ ▨ **Thoughts into words**

56. Give a general description of how to use the elimination-by-addition method to solve a system of two linear equations in two variables.

57. Explain how you would use the elimination-by-addition method to solve the system

$$\begin{pmatrix} 3x - 4y = -1 \\ 2x - 5y = 9 \end{pmatrix}$$

58. How do you decide whether to solve a system of linear equations in two variables by using the substitution method or the elimination-by-addition method?

■ ■ ■ **Further investigations**

59. There is another way of telling whether a system of two linear equations in two unknowns is consistent, inconsistent, or dependent without taking the time to graph each equation. It can be shown that any system of the form

$$a_1 x + b_1 y = c_1$$
$$a_2 x + b_2 y = c_2$$

has one and only one solution if

$$\frac{a_1}{a_2} \neq \frac{b_1}{b_2} \qquad \text{Consistent}$$

that it has no solution if

$$\frac{a_1}{a_2} = \frac{b_1}{b_2} \neq \frac{c_1}{c_2} \qquad \text{Inconsistent}$$

and that it has infinitely many solutions if

$$\frac{a_1}{a_2} = \frac{b_1}{b_2} = \frac{c_1}{c_2} \qquad \text{Dependent}$$

Determine whether each of the following systems is consistent, inconsistent, or dependent.

a. $\begin{pmatrix} 4x - 3y = 7 \\ 9x + 2y = 5 \end{pmatrix}$ **b.** $\begin{pmatrix} 5x - y = 6 \\ 10x - 2y = 19 \end{pmatrix}$

c. $\begin{pmatrix} 5x - 4y = 11 \\ 4x + 5y = 12 \end{pmatrix}$ **d.** $\begin{pmatrix} x + 2y = 5 \\ x - 2y = 9 \end{pmatrix}$

e. $\begin{pmatrix} x - 3y = 5 \\ 3x - 9y = 15 \end{pmatrix}$ **f.** $\begin{pmatrix} 4x + 3y = 7 \\ 2x - y = 10 \end{pmatrix}$

g. $\begin{pmatrix} 3x + 2y = 4 \\ y = -\frac{3}{2}x - 1 \end{pmatrix}$ **h.** $\begin{pmatrix} y = \frac{4}{3}x - 2 \\ 4x - 3y = 6 \end{pmatrix}$

60. A system such as

$$\begin{pmatrix} \dfrac{3}{x} + \dfrac{2}{y} = 2 \\ \dfrac{2}{x} - \dfrac{3}{y} = \dfrac{1}{4} \end{pmatrix}$$

is not a system of linear equations but can be transformed into a linear system by changing variables. For example, when we substitute u for $\dfrac{1}{x}$ and v for $\dfrac{1}{y}$ in the above system we get

$$\begin{pmatrix} 3u + 2v = 2 \\ 2u - 3v = \dfrac{1}{4} \end{pmatrix}$$

We can solve this "new" system either by elimination by addition or by substitution (we will leave the details for you) to produce $u = \dfrac{1}{2}$ and $v = \dfrac{1}{4}$. Therefore, because $u = \dfrac{1}{x}$ and $v = \dfrac{1}{y}$, we have

$$\frac{1}{x} = \frac{1}{2} \qquad \text{and} \qquad \frac{1}{y} = \frac{1}{4}$$

Solving these equations yields

$$x = 2 \qquad \text{and} \qquad y = 4$$

The solution set of the original system is $\{(2, 4)\}$. Solve each of the following systems.

a. $\begin{pmatrix} \dfrac{1}{x} + \dfrac{2}{y} = \dfrac{7}{12} \\ \dfrac{3}{x} - \dfrac{2}{y} = \dfrac{5}{12} \end{pmatrix}$ **b.** $\begin{pmatrix} \dfrac{2}{x} + \dfrac{3}{y} = \dfrac{19}{15} \\ -\dfrac{2}{x} + \dfrac{1}{y} = -\dfrac{7}{15} \end{pmatrix}$

c. $\begin{pmatrix} \dfrac{3}{x} - \dfrac{2}{y} = \dfrac{13}{6} \\ \dfrac{2}{x} + \dfrac{3}{y} = 0 \end{pmatrix}$ **d.** $\begin{pmatrix} \dfrac{4}{x} + \dfrac{1}{y} = 11 \\ \dfrac{3}{x} - \dfrac{5}{y} = -9 \end{pmatrix}$

e. $\begin{pmatrix} \dfrac{5}{x} - \dfrac{2}{y} = 23 \\ \dfrac{4}{x} + \dfrac{3}{y} = \dfrac{23}{2} \end{pmatrix}$ **f.** $\begin{pmatrix} \dfrac{2}{x} - \dfrac{7}{y} = \dfrac{9}{10} \\ \dfrac{5}{x} + \dfrac{4}{y} = -\dfrac{41}{20} \end{pmatrix}$

61. Solve the following system for x and y.

$$\begin{pmatrix} a_1 x + b_1 y = c_1 \\ a_2 x + b_2 y = c_2 \end{pmatrix}$$

Answers to Concept Quiz

1. True **2.** False **3.** True **4.** True **5.** False

5.6   Graphing Linear Inequalities

Objectives

■   Graph linear inequalities.

■   Graph systems of two linear inequalities.

Linear inequalities in two variables are of the form $Ax + By > C$ or $Ax + By < C$, where A, B, and C are real numbers, and A and B not both zero. (Combined linear equality and inequality statements are of the form $Ax + By \geq C$ or $Ax + By \leq C$.) Graphing linear inequalities is almost as easy as graphing linear equations. The following discussion leads to a simple step-by-step process. Let's consider the next equation and related inequalities.

$$x - y = 2$$
$$x - y > 2$$
$$x - y < 2$$

The graph of $x - y = 2$ is shown in Figure 5.33. The line divides the plane into two half-planes, one above the line and one below the line. In Figure 5.34(a) we have indicated coordinates for several points above the line. Note that for each point, the ordered pair of real numbers satisfies the inequality $x - y < 2$. This is true for *all points* in the half-plane above the line. Therefore, the graph of $x - y < 2$ is the half-plane above the line, indicated by the shaded region in Figure 5.34(b). We use a dashed line to indicate that points on the line do not satisfy $x - y < 2$.

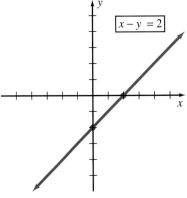

Figure 5.33

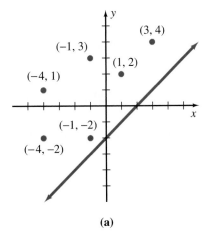

(a)

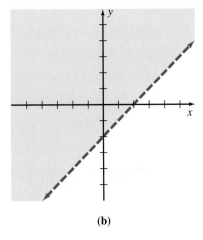

(b)

Figure 5.34

In Figure 5.35(a) the coordinates of several points below the line $x - y = 2$ are indicated. Note that for each point, the ordered pair of real numbers satisfies the inequality $x - y > 2$. This is true for *all points* in the half-plane below the line. Therefore, the graph of $x - y > 2$ is the half-plane below the line, as indicated by the shaded region in Figure 5.35(b).

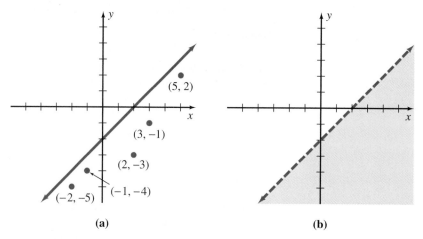

(a)   (b)

Figure 5.35

On the basis of this discussion, we suggest the following steps for graphing linear inequalities.

> **STEP 1** Graph the corresponding equality. Use a solid line if equality is included in the given statement. Use a dashed line if equality is not included.
>
> **STEP 2** Choose a "test point" that is not on the line and substitute its coordinates into the inequality statement. (The origin is a convenient point to use if it is not on the line.)
>
> **STEP 3** The graph of the given inequality is:
>   **(a)** the half-plane that contains the test point if the inequality is satisfied by the coordinates of the point, or
>   **(b)** the half-plane that does not contain the test point if the inequality is not satisfied by the coordinates of the point.

Let's apply these steps to some examples.

EXAMPLE 1   Graph $2x + y > 4$.

Solution

STEP 1   Graph $2x + y = 4$ as a dashed line, because equality is not included in the given statement $2x + y > 4$.

STEP 2 Choose the origin as a test point, and substitute its coordinates into the inequality.

$2x + y > 4$ becomes $2(0) + 0 > 4$ A false statement

STEP 3 Because the test point does not satisfy the given inequality, the graph is the half-plane that does not contain the test point. Thus the graph of $2x + y > 4$ is the half-plane above the line, as indicated in Figure 5.36.

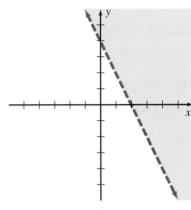

Figure 5.36

E X A M P L E   2   Graph $y \leq 2x$.

Solution

STEP 1 Graph $y = 2x$ as a solid line, because equality is included in the given statement.

STEP 2 Because the origin is on the line, we need to choose another point as a test point. Let's use $(3, 2)$.

$y \leq 2x$ becomes $2 \leq 2(3)$ A true statement

STEP 3 Because the test point satisfies the given inequality, the graph is the half-plane that contains the test point. Thus the graph of $y \leq 2x$ is the line, along with the half-plane below the line, as indicated in Figure 5.37.

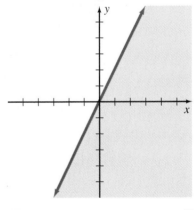

Figure 5.37

Systems of Linear Inequalities

It is now easy to use a graphing approach to solve a system of linear inequalities. For example, the solution set of a system of linear inequalities, such as

$$\begin{pmatrix} x + y < 1 \\ x - y > 1 \end{pmatrix}$$

is the intersection of the solution sets of the individual inequalities. In Figure 5.38(a) we indicated the solution set for $x + y < 1$, and in Figure 5.38(b) we indicated the solution set for $x - y > 1$. Then, in Figure 5.38(c) we shaded the region that represents the intersection of the two shaded regions in parts (a) and (b); thus it is the solution of the given system. The shaded region in Figure 5.38(c) consists of all points that are below the line $x + y = 1$ *and also* are below the line $x - y = 1$.

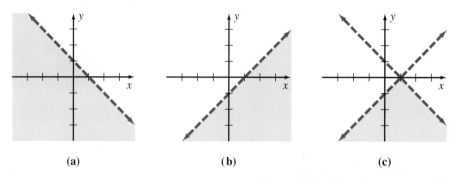

(a)   (b)   (c)

Figure 5.38

Let's solve another system of linear inequalities.

E X A M P L E   3   Solve the system $\begin{pmatrix} 2x + 3y \le 6 \\ x - 4y < 4 \end{pmatrix}$.

Solution

Let's first graph the individual inequalities. The solution set for $2x + 3y \le 6$ is shown in Figure 5.39(a), and the solution set for $x - 4y < 4$ is shown in Figure 5.39(b). (Note the solid line in part (a) and the dashed line in part (b).) Then, in Figure 5.39(c) we shaded the intersection of the graphs in parts (a) and (b). Thus we represented the solution set for the given system by the shaded region in Figure 5.39(c). This region consists of all points that are on or below the line $2x + 3y = 6$ *and also* are above the line $x - 4y = 4$.

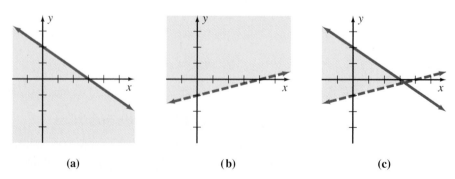

(a)   (b)   (c)

Figure 5.39

REMARK: Remember that the shaded region in Figure 5.39(c) represents the solution set of the given system. Parts (a) and (b) were drawn only to help determine the final shaded region. With some practice, you may be able to go directly to part (c) without actually sketching the graphs of the individual inequalities.

CONCEPT QUIZ

For Problems 1–5, answer true or false.

1. The ordered pair $(2, -3)$ satisfies the linear inequality $2x + y > 1$.

2. A dashed line on the graph indicates that the points on the line do not satisfy the inequality.

3. Any point can be used as a test point to determine the half-plane that is the solution of the inequality.

4. The solution of a system of inequalities is the intersection of the solution sets of the individual inequalities.

5. The ordered pair $(1, 4)$ satisfies the system of linear inequalities $\begin{pmatrix} x + y > 2 \\ 2x + y < 3 \end{pmatrix}$.

PROBLEM SET 5.6

For Problems 1–20, graph each inequality.

1. $x + y > 1$

2. $2x + y > 4$

3. $3x + 2y < 6$

4. $x + 3y < 3$

5. $2x - y \geq 4$

6. $x - 2y \geq 2$

7. $4x - 3y \leq 12$

8. $3x - 4y \leq 12$

9. $y > -x$

10. $y < x$

11. $2x - y \geq 0$

12. $3x - y \leq 0$

13. $-x + 2y < -2$

14. $-2x + y > -2$

15. $y \leq \dfrac{1}{2}x - 2$

16. $y \geq -\dfrac{1}{2}x + 1$

17. $y \geq -x + 4$

18. $y \leq -x - 3$

19. $3x + 4y > -12$

20. $4x + 3y > -12$

For Problems 21–30, indicate the solution set for each system of linear inequalities by shading the appropriate region.

21. $\begin{pmatrix} 2x + 3y > 6 \\ x - y < 2 \end{pmatrix}$

22. $\begin{pmatrix} x - 2y < 4 \\ 3x + y > 3 \end{pmatrix}$

23. $\begin{pmatrix} x - 3y \geq 3 \\ 3x + y \leq 3 \end{pmatrix}$

24. $\begin{pmatrix} 4x + 3y \leq 12 \\ 4x - y \geq 4 \end{pmatrix}$

25. $\begin{pmatrix} y \geq 2x \\ y < x \end{pmatrix}$

26. $\begin{pmatrix} y \leq -x \\ y > -3x \end{pmatrix}$

27. $\begin{pmatrix} y < -x + 1 \\ y > -x - 1 \end{pmatrix}$

28. $\begin{pmatrix} y > x - 2 \\ y < x + 3 \end{pmatrix}$

29. $\begin{pmatrix} y < \dfrac{1}{2}x + 2 \\ y < \dfrac{1}{2}x - 1 \end{pmatrix}$

30. $\begin{pmatrix} y > -\dfrac{1}{2}x - 2 \\ y > -\dfrac{1}{2}x + 1 \end{pmatrix}$

■ ■ ■ Thoughts into words

31. Explain how you would graph the inequality $-x - 2y > 4$.

32. Why is the point $(3, -2)$ not a good test point to use when graphing the inequality $3x - 2y \leq 13$?

■ ■ ■ Further investigations

For Problems 33–36, indicate the solution set for each system of linear inequalities by shading the appropriate region.

33. $\begin{pmatrix} y > x + 1 \\ y < x - 1 \end{pmatrix}$

34. $\begin{pmatrix} x \geq 0 \\ y \geq 0 \\ 3x + 4y \leq 12 \\ 2x + y \leq 4 \end{pmatrix}$

35. $\begin{pmatrix} x \geq 0 \\ y \geq 0 \\ 2x + y \leq 4 \\ 2x - 3y \leq 6 \end{pmatrix}$

36. $\begin{pmatrix} x \geq 0 \\ y \geq 0 \\ 3x + 5y \geq 15 \\ 5x + 3y \geq 15 \end{pmatrix}$

Answers to Concept Quiz

1. False **2.** True **3.** False **4.** True **5.** False

CHAPTER 5

SUMMARY

(5.1) The **Cartesian** or **rectangular coordinate system** involves a one-to-one correspondence between ordered pairs of real numbers and the points of a plane. The system provides the basis for a study of coordinate geometry, which is a link between algebra and geometry. In this section when we are given an algebraic equation, we want to find its geometric graph.

One graphing technique is to plot a sufficient number of points to determine the graph of the equation.

(5.2) Any equation of the form $Ax + By = C$, where A, B, and C are constants (A and B not both zero) and x and y are variables, is a **linear equation in two variables,** and its graph is a **straight line.**

To **graph** a linear equation, we can find two solutions (the intercepts are usually easy to determine), plot the corresponding points, and then connect the points with a straight line.

(5.3) If points P_1 and P_2 with coordinates (x_1, y_1) and (x_2, y_2), respectively, are any two points on a line, then the **slope** of the line (denoted by m) is given by

$$m = \frac{y_2 - y_1}{x_2 - x_1}, \qquad x_1 \neq x_2$$

The slope of a line is a **ratio** of vertical change to horizontal change. The slope of a line can be negative, positive, or zero. The concept of slope is not defined for vertical lines.

(5.4) Solving a system of two linear equations by graphing produces one of these three possibilities:

1. The graphs of the two equations are two intersecting lines, which indicates *one solution* for the system, which is called a **consistent system.**

2. The graphs of the two equations are two parallel lines, which indicates *no solution* for the system, which is called an **inconsistent system.**

3. The graphs of the two equations are the same line, which indicates *infinitely many solutions* for the sys-

tem. We refer to the equations as a set of **dependent** equations.

Here are the steps of the **substitution method** for solving a system of equations:

STEP 1 Solve one of the equations for one variable in terms of the other variable if neither equation is in such a form. (If possible, make a choice that will avoid fractions.)

STEP 2 Substitute the expression obtained in step 1 into the other equation to produce an equation with one variable.

STEP 3 Solve the equation obtained in step 2.

STEP 4 Use the solution obtained in step 3, along with the expression obtained in step 1, to determine the solution of the system.

(5.5) The **elimination-by-addition method** involves replacing systems of equations with equivalent systems until we reach a system for which the solutions can be easily determined. We can perform the following operations or transformations on a system to produce an equivalent system:

1. Any two equations of the system can be interchanged.

2. Both sides of any equation of the system can be multiplied by any nonzero real number.

3. Any equation of the system can be replaced by the *sum* of that equation and a nonzero multiple of another equation.

(5.6) **Linear inequalities** in two variables are of the form $Ax + By > C$ or $Ax + By < C$. To graph a linear inequality, we suggest the following steps:

STEP 1 Graph the corresponding equality. Use a solid line if equality is included in the original statement. Use a dashed line if equality is not included.

STEP 2 Choose a test point not on the line and substitute its coordinates into the inequality.

STEP 3 The graph of the original inequality is

(a) the half-plane that contains the test point if the inequality is satisfied by the coordinates of that point, or

(b) the half-plane that does not contain the test point if the inequality is not satisfied by the coordinates of the point.

The solution set of a system of linear inequalities is the intersection of the solution sets of the individual inequalities.

CHAPTER 5 REVIEW PROBLEM SET

For Problems 1–6, graph each equation.

1. $2x - 5y = 10$

2. $y = -\dfrac{1}{3}x + 1$

3. $y = 2x^2 + 1$

4. $y = -x^2 - 2$

5. $2x - 3y = 0$

6. $y = -x^3$

For Problems 7–10, determine the x and y intercepts for the graph of each equation.

7. $4x + y = -4$

8. $x - 2y = 2$

9. $y = 3x + 6$

10. $y = -4x + 1$

11. Find the slope of the line determined by the points $(2, -5)$ and $(-1, 1)$.

12. Find the slope of the line determined by the points $(4, -3)$ and $(4, 2)$.

13. Solve the system $\begin{pmatrix} 2x + y = 4 \\ x - y = 5 \end{pmatrix}$ by using the graphing method.

For Problems 14–25, solve each system by using either the substitution method or the elimination-by-addition method.

14. $\begin{pmatrix} 2x - y = 1 \\ 3x - 2y = -5 \end{pmatrix}$

15. $\begin{pmatrix} 2x + 5y = 7 \\ x = -3y + 1 \end{pmatrix}$

16. $\begin{pmatrix} 3x + 2y = 7 \\ 4x - 5y = 3 \end{pmatrix}$

17. $\begin{pmatrix} 9x + 2y = 140 \\ x + 5y = 135 \end{pmatrix}$

18. $\begin{pmatrix} \dfrac{1}{2}x + \dfrac{1}{4}y = -5 \\ \dfrac{2}{3}x - \dfrac{1}{2}y = 0 \end{pmatrix}$

19. $\begin{pmatrix} x + y = 1000 \\ 0.07x + 0.09y = 82 \end{pmatrix}$

20. $\begin{pmatrix} y = 5x + 2 \\ 10x - 2y = 1 \end{pmatrix}$

21. $\begin{pmatrix} 5x - 7y = 9 \\ y = 3x - 2 \end{pmatrix}$

22. $\begin{pmatrix} 10t + u = 6u \\ t + u = 12 \end{pmatrix}$

23. $\begin{pmatrix} t = 2u \\ 10t + u - 36 = 10u + t \end{pmatrix}$

24. $\begin{pmatrix} u = 2t + 1 \\ 10t + u + 10u + t = 110 \end{pmatrix}$

25. $\begin{pmatrix} y = -\dfrac{2}{3}x \\ \dfrac{1}{3}x - y = -9 \end{pmatrix}$

For Problems 26–33, solve each problem by setting up and solving a system of two linear equations in two variables.

26. The sum of two numbers is 113. The larger number is 1 less than twice the smaller number. Find the numbers.

27. Last year Mark invested a certain amount of money at 6% annual interest and $50 more than that amount at 8%. He received $39.00 in interest. How much did he invest at each rate?

28. Cindy has 43 coins consisting of nickels and dimes. The total value of the coins is $3.40. How many coins of each kind does she have?

29. The length of a rectangle is 1 inch more than three times the width. If the perimeter of the rectangle is 50 inches, find the length and width.

30. Two angles are complementary, and one of them is 6° less than twice the other one. Find the measure of each angle.

31. Two angles are supplementary, and the larger angle is 20° less than three times the smaller angle. Find the measure of each angle.

32. Four cheeseburgers and five milkshakes cost a total of $25.50. Two milkshakes cost $1.75 more than one cheeseburger. Find the cost of a cheeseburger and also find the cost of a milkshake.

33. Three bottles of orange juice and two bottles of water cost $6.75. On the other hand, two bottles of juice and three bottles of water cost $6.15. Find the cost per bottle of each.

34. Graph the inequality $-2x + y < 4$.

35. Graph the inequality $3x + 2y \geq -6$.

36. Solve, by graphing, the system of inequalities
$$\begin{pmatrix} -x + 2y > 2 \\ 3x - y > 3 \end{pmatrix}.$$

TEST

applies to all problems in this Chapter Test.

1. Is $(-2, -3)$ a solution of $7x - 2y = -8$?

2. Is $(-1, -5)$ a solution of $y = -x^2 - 4$?

3. Find the x intercept(s) of the graph of $y = -x^2 + 16$.

4. Find the x intercept(s) of the graph of $-3x + 4y = -12$.

5. Find the slope of the line determined by the points $(3, -1)$ and $(-1, 1)$.

For Problems 6–9, graph each equation.

6. $5x + 3y = 15$

7. $-2x + y = -4$

8. $y = -\dfrac{1}{2}x - 2$

9. $y = 2x^2 - 3$

10. Is $\{(-2, 4)\}$ the solution set of the system $\begin{pmatrix} 3x - 2y = -14 \\ 5x + y = 14 \end{pmatrix}$?

11. Is $\{(1, -5)\}$ the solution set of the system $\begin{pmatrix} -x - 3y = 14 \\ 2x + 5y = -23 \end{pmatrix}$?

12. Solve the system $\begin{pmatrix} 3x - 2y = -4 \\ 2x + 3y = 19 \end{pmatrix}$ by graphing.

13. Solve the system $\begin{pmatrix} x - 3y = -9 \\ 4x + 7y = 40 \end{pmatrix}$ using the elimination-by-addition method.

14. Solve the system $\begin{pmatrix} 5x + y = -14 \\ 6x - 7y = -66 \end{pmatrix}$ using the substitution method.

15. Solve the system $\begin{pmatrix} 2x - 7y = 26 \\ 3x + 2y = -11 \end{pmatrix}$.

16. Solve the system $\begin{pmatrix} 8x + 5y = -6 \\ 4x - y = 18 \end{pmatrix}$.

17. Is $(3, -2)$ a member of the solution set of the inequality $3x - 2y > 6$?

18. Is $(-3, -5)$ a member of the solution set of the inequality $-2x - y \le 4$?

19. Is $(2, 1)$ a member of the solution set of the system of inequalities $\begin{pmatrix} 5x - y < 10 \\ 3x + 2y > 6 \end{pmatrix}$?

20. Graph the inequality $3x - 2y > -6$.

21. Graph the inequality $y \le -3x$.

22. Solve, by graphing, the system of inequalities $\begin{pmatrix} x - 2y \le 4 \\ 2x + y \le 4 \end{pmatrix}$.

For Problems 23–25, solve each problem by setting up and solving a system of two linear equations in two variables.

23. Kelsey has a collection of 40 coins, consisting of dimes and quarters, worth $7.60. How many coins of each kind does she have?

24. The length of a rectangle is 1 inch less than twice the width of the rectangle. If the perimeter of the rectangle is 40 inches, find the length of the rectangle.

25. One solution contains 30% alcohol and another solution contains 80% alcohol. Some of each of the two solutions are mixed to produce 5 liters of a 60% alcohol solution. How many liters of the 80% alcohol solution are used?

CUMULATIVE PRACTICE TEST *Chapters 1-5*

For Problems 1–3, simplify each numerical expression.

1. $3.9 - 4.6 - 1.2 + 0.4$

2. $\left(\dfrac{1}{2}\right)^3 + \left(\dfrac{1}{4}\right)^2 - \dfrac{1}{8}$

3. $(0.2)^3 - (0.4)^3 + (1.2)^2$

For Problems 4–6, evaluate each algebraic expression for the given values of the variables.

4. $\dfrac{1}{2}x - \dfrac{2}{5}y + xy$ for $x = -3$ and $y = \dfrac{1}{2}$

5. $1.1x - 2.3y + 2.5x + 1.6y$ for $x = 0.4$ and $y = 0.7$

6. $2(x + 6) - 3(x + 9) - 4(x - 5)$ for $x = 17$

For Problems 7–9, perform the indicated operations and express answers in simplest form.

7. $\dfrac{3}{x} + \dfrac{2}{y} - \dfrac{4}{xy}$

8. $\left(\dfrac{7xy}{9xy^2}\right)\left(\dfrac{12x}{14y}\right)$

9. $\left(\dfrac{2ab^2}{7a}\right) \div \left(\dfrac{4b}{21a}\right)$

For Problems 10–14, solve each equation.

10. $2(x - 3) - (x + 4) = 3(x + 10)$

11. $\dfrac{2}{3}x + \dfrac{1}{4} - \dfrac{1}{2}x = -1$

12. $\dfrac{x + 1}{4} - \dfrac{x - 2}{6} = \dfrac{3}{8}$

13. $\dfrac{x + 6}{7} = \dfrac{x - 2}{8}$

14. $0.06x + 0.07(1800 - x) = 120$

15. Solve $2x - 3y = 13$ for y.

16. Solve the system $\begin{pmatrix} 5x - 2y = -20 \\ 2x + 3y = 11 \end{pmatrix}$.

17. Solve the system $\begin{pmatrix} 4x - 9y = 47 \\ 7x + y = 32 \end{pmatrix}$.

18. Solve the inequality $3(x - 2) < 4(x + 6)$

19. Solve the inequality $\dfrac{1}{4}x - 2 \geq \dfrac{2}{3}x + 1$

20. Graph the equation $y = -2x - 3$.

21. Graph the equation $y = -2x^2 - 3$.

22. Graph the inequality $x - 2y > 4$.

For Problems 23–25, use an equation, an inequality, or a system of equations to help solve each problem.

23. Last week on an algebra test, the highest grade was 9 points less then three times the lowest grade. The sum of the two grades was 135. Find the lowest and highest grades on the test.

24. Suppose that Derwin shot rounds of 82, 84, 78, and 79 on the first four days of a golf tournament. What must he shoot on the fifth day of the tournament to average 80 or less for the five days?

25. A 10% salt solution is to be mixed with a 15% salt solution to produce 10 gallons of a 13% salt solution. How many gallons of the 10% salt solution and how many gallons of the 15% salt solution will be needed?

The average distance between the sun and the earth is approximately 93,000,000 miles. Using scientific notation, 93,000,000 can be written as $(9.3)(10^7)$.

© Tom McCarthy/PhotoEdit

Exponents and Polynomials

A strip with a uniform width is shaded along both sides and both ends of a rectangular poster that measures 12 inches by 16 inches. How wide is the strip if one-half of the poster is shaded? If we let x represent the width of the strip, then we can use the equation $16(12) - (16 - 2x)(12 - 2x) = 96$ to determine that the width of the strip is 2 inches.

The equation we used to solve this problem is called a quadratic equation. Quadratic equations belong to a larger classification called polynomial equations. To solve problems involving polynomial equations, we need to develop some basic skills that pertain to polynomials. That is, we need to be able to add, subtract, multiply, divide, and factor polynomials. Chapters 6 and 7 will help you develop those skills as you work through problems that involve quadratic equations.

InfoTrac Project
Do a keyword search on consumer debt and find an article titled "Consumer, Business Debt are in a Fast Rise." Write a summary of the article and include in your summary the total dollars of nonfinancial debt outstanding in the United States. Write this number in scientific notation. According to the 2000 census, the population of the United States is 281,421,906. Round this figure to the nearest million and convert it to scientific notation. Using scientific notation, calculate the amount of nonfinancial debt outstanding per person in the United States.

6.1   **Addition and Subtraction of Polynomials**

Objectives

■ Know the definition of monomial, binomial, trinomial, and polynomial.

■ Determine the degree of a polynomial.

■ Add polynomials.

■ Subtract polynomials using either a vertical or a horizontal format.

In earlier chapters, we called algebraic expressions such as $4x$, $5y$, $-6ab$, $7x^2$, and $-9xy^2z^3$ *terms.* Recall that a term is an indicated product that may contain any number of factors. The variables in a term are called *literal factors,* and the numerical factor is called the *numerical coefficient* of the term. Thus in $-6ab$, a and b are literal factors, and the numerical coefficient is -6. Terms that have the same literal factors are *similar* or *like* terms.

Terms that contain variables with only whole numbers as exponents are called **monomials.** The previously listed terms, $4x$, $5y$, $-6ab$, $7x^2$, and $-9xy^2z^3$, are all monomials. (We will work later with some algebraic expressions, such as $7x^{-1}y^{-1}$ and $4a^{-2}b^{-3}$ that are not monomials.) The **degree of a monomial** is the sum of the exponents of the literal factors. Here are some examples:

$4xy$ is of degree 2.

$5x$ is of degree 1.

$14a^2b$ is of degree 3.

$-17xy^2z^3$ is of degree 6.

$-9y^4$ is of degree 4.

If the monomial contains only one variable, then the exponent of the variable is the degree of the monomial. Any nonzero constant term is said to be of degree zero.

A **polynomial** is a monomial or a finite sum (or difference) of monomials. The **degree of a polynomial** is the degree of the term with the highest degree in the polynomial. Some special classifications of polynomials are made according to the number of terms. We call a one-term polynomial a **monomial,** a two-term polynomial a **binomial,** and a three-term polynomial a **trinomial.** The following examples illustrate some of this terminology:

The polynomial $5x^3y^4$ is a monomial of degree 7.
The polynomial $4x^2y - 3xy$ is a binomial of degree 3.
The polynomial $5x^2 - 6x + 4$ is a trinomial of degree 2.
The polynomial $9x^4 - 7x^3 + 6x^2 + x - 2$ is given no special name but is of degree 4.

Adding Polynomials

In the preceding chapters, you have worked many problems involving the addition and subtraction of polynomials. For example, simplifying $4x^2 + 6x + 7x^2 - 2x$ to $11x^2 + 4x$ by combining similar terms can actually be considered the addition problem $(4x^2 + 6x) + (7x^2 - 2x)$. At this time we will simply review and extend some of those ideas.

EXAMPLE 1 Add $5x^2 + 7x - 2$ and $9x^2 - 12x + 13$.

Solution

We commonly use the horizontal format for such work. Thus

$$(5x^2 + 7x - 2) + (9x^2 - 12x + 13) = (5x^2 + 9x^2) + (7x - 12x) + (-2 + 13)$$
$$= 14x^2 - 5x + 11 \qquad \blacksquare$$

The commutative, associative, and distributive properties provide the basis for rearranging, regrouping, and combining similar terms.

EXAMPLE 2 Add $5x - 1, 3x + 4$, and $9x - 7$.

Solution

$$(5x - 1) + (3x + 4) + (9x - 7) = (5x + 3x + 9x) + [-1 + 4 + (-7)]$$
$$= 17x - 4 \qquad \blacksquare$$

EXAMPLE 3 Add $-x^2 + 2x - 1, 2x^3 - x + 4$, and $-5x + 6$.

Solution

$$(-x^2 + 2x - 1) + (2x^3 - x + 4) + (-5x + 6)$$
$$= (2x^3) + (-x^2) + (2x - x - 5x) + (-1 + 4 + 6)$$
$$= 2x^3 - x^2 - 4x + 9 \qquad \blacksquare$$

Subtracting Polynomials

Recall from Chapter 1 that $a - b = a + (-b)$. We define subtraction as *adding the opposite*. This same idea extends to polynomials in general. The opposite of a polynomial is formed by taking the opposite of each term. For example, the opposite of $(2x^2 - 7x + 3)$ is $-2x^2 + 7x - 3$. Symbolically, we express this as

$$-(2x^2 - 7x + 3) = -2x^2 + 7x - 3$$

Now let's consider some subtraction problems.

EXAMPLE 4   Subtract $2x^2 + 9x - 3$ from $5x^2 - 7x - 1$.

Solution

Use the horizontal format.

$$(5x^2 - 7x - 1) - (2x^2 + 9x - 3) = (5x^2 - 7x - 1) + (-2x^2 - 9x + 3)$$
$$= (5x^2 - 2x^2) + (-7x - 9x) + (-1 + 3)$$
$$= 3x^2 - 16x + 2 \qquad \blacksquare$$

EXAMPLE 5   Subtract $-8y^2 - y + 5$ from $2y^2 + 9$.

Solution

$$(2y^2 + 9) - (-8y^2 - y + 5) = (2y^2 + 9) + (8y^2 + y - 5)$$
$$= (2y^2 + 8y^2) + (y) + (9 - 5)$$
$$= 10y^2 + y + 4 \qquad \blacksquare$$

Later, when dividing polynomials, you will need to use a vertical format to subtract polynomials. Let's consider two such examples.

EXAMPLE 6   Subtract $3x^2 + 5x - 2$ from $9x^2 - 7x - 1$.

Solution

$$\begin{array}{l} 9x^2 - 7x - 1 \\ 3x^2 + 5x - 2 \end{array}$$ Notice which polynomial goes on the bottom and the alignment of similar terms in columns.

Now we can *mentally form the opposite of the bottom polynomial* and add.

$$\begin{array}{l} 9x^2 -\ 7x - 1 \\ \underline{3x^2 +\ 5x - 2} \\ 6x^2 - 12x + 1 \end{array}$$ The opposite of $3x^2 + 5x - 2$ is $-3x^2 - 5x + 2$. $\qquad \blacksquare$

E X A M P L E   7   Subtract $15y^3 + 5y^2 + 3$ from $13y^3 + 7y - 1$.

Solution

$$13y^3 \qquad\quad + 7y - 1 \qquad \text{Similar terms are arranged in columns.}$$
$$\underline{15y^3 + 5y^2 \qquad\quad + 3}$$
$$-2y^3 - 5y^2 + 7y - 4 \qquad \text{We mentally formed the opposite of the bottom polynomial and added.}$$ ■

We can use the distributive property along with the properties $a = 1(a)$ and $-a = -1(a)$ when adding and subtracting polynomials. The next examples illustrate this approach.

E X A M P L E   8   Perform the indicated operations:

$$(3x - 4) + (2x - 5) - (7x - 1)$$

Solution

$$(3x - 4) + (2x - 5) - (7x - 1)$$
$$= 1(3x - 4) + 1(2x - 5) - 1(7x - 1)$$
$$= 1(3x) - 1(4) + 1(2x) - 1(5) - 1(7x) - 1(-1)$$
$$= 3x - 4 + 2x - 5 - 7x + 1$$
$$= 3x + 2x - 7x - 4 - 5 + 1$$
$$= -2x - 8$$ ■

Certainly we can do some of the steps mentally; Example 9 gives a possible format.

E X A M P L E   9   Perform the indicated operations:

$$(-y^2 + 5y - 2) - (-2y^2 + 8y + 6) + (4y^2 - 2y - 5)$$

Solution

$$(-y^2 + 5y - 2) - (-2y^2 + 8y + 6) + (4y^2 - 2y - 5)$$
$$= -y^2 + 5y - 2 + 2y^2 - 8y - 6 + 4y^2 - 2y - 5$$
$$= -y^2 + 2y^2 + 4y^2 + 5y - 8y - 2y - 2 - 6 - 5$$
$$= 5y^2 - 5y - 13$$ ■

When we use the horizontal format, as in Examples 8 and 9, we use parentheses to indicate a quantity. In Example 8 the quantities $(3x - 4)$ and $(2x - 5)$ are to be added; from this result we are to subtract the quantity $(7x - 1)$. Brackets,

[ ], are also sometimes used as grouping symbols, especially if there is a need to indicate quantities within quantities. To remove the grouping symbols, perform the indicated operations, starting with the innermost set of symbols. Let's consider two examples of this type.

EXAMPLE 10

Perform the indicated operations:

$$3x - [2x + (3x - 1)]$$

Solution

First we need to add the quantities $2x$ and $(3x - 1)$.

$$3x - [2x + (3x - 1)] = 3x - [2x + 3x - 1]$$
$$= 3x - [5x - 1]$$

Now we need to subtract the quantity $[5x - 1]$ from $3x$.

$$3x - [5x - 1] = 3x - 5x + 1$$
$$= -2x + 1$$

EXAMPLE 11

Perform the indicated operations.

$$8 - \{7x - [2 + (x - 1)] + 4x\}$$

Solution

Start with the innermost set of grouping symbols (the parentheses) and proceed as follows:

$$8 - \{7x - [2 + (x - 1)] + 4x\} = 8 - \{7x - [2 + x - 1] + 4x\}$$
$$= 8 - \{7x - [x + 1] + 4x\}$$
$$= 8 - (7x - x - 1 + 4x)$$
$$= 8 - (10x - 1)$$
$$= 8 - 10x + 1$$
$$= -10x + 9$$

As a final example in this section, we look at polynomials in a geometric setting.

EXAMPLE 12

Suppose that a parallelogram and a rectangle have dimensions as indicated in Figure 6.1. Find a polynomial that represents the sum of the areas of the two figures.

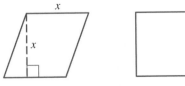

Figure 6.1

Solution

Using the area formulas $A = bh$ and $A = lw$ for parallelograms and rectangles, respectively, we can represent the sum of the areas of the two figures as follows:

Area of the parallelogram $x(x) = x^2$

Area of the rectangle $20(x) = 20x$

We can represent the total area by $x^2 + 20x$. ∎

CONCEPT QUIZ

For Problems 1–5, answer true or false.

1. The degree of the monomial $4x^2y$ is 3.

2. The degree of the polynomial $2x^4 - 5x^3 + 7x^2 - 4x + 6$ is 10.

3. A three-term polynomial is called a binomial.

4. A polynomial is a monomial or a finite sum of monomials.

5. Monomial terms must have whole number exponents for each variable.

For Problems 6–10, match the polynomial with its description.

6. $5xy^2$	**A.** Monomial of degree 5
7. $5xy^2 + 3x^2$	**B.** Binomial of degree 5
8. $5x^2y + 3xy^4$	**C.** Monomial of degree 3
9. $3x^5 + 2x^3 + 5x - 1$	**D.** Binomial of degree 3
10. $3x^2y^3$	**E.** Polynomial of degree 5

PROBLEM SET 6.1

For Problems 1–8, determine the degree of each polynomial.

1. $7x^2y + 6xy$ **2.** $4xy - 7x$

3. $5x^2 - 9$ **4.** $8x^2y^2 - 2xy^2 - x$

5. $5x^3 - x^2 - x + 3$ **6.** $8x^4 - 2x^2 + 6$

7. $5xy$ **8.** $-7x + 4$

For Problems 9–22, add the polynomials.

9. $3x + 4$ and $5x + 7$ **10.** $3x - 5$ and $2x - 9$

11. $-5y - 3$ and $9y + 13$

12. $x^2 - 2x - 1$ and $-2x^2 + x + 4$

13. $-2x^2 + 7x - 9$ and $4x^2 - 9x - 14$

14. $3a^2 + 4a - 7$ and $-3a^2 - 7a + 10$

15. $5x - 2$, $3x - 7$, and $9x - 10$

16. $-x - 4$, $8x + 9$, and $-7x - 6$

17. $2x^2 - x + 4$, $-5x^2 - 7x - 2$, and $9x^2 + 3x - 6$

18. $-3x^2 + 2x - 6$, $6x^2 + 7x + 3$, and $-4x^2 - 9$

19. $-4n^2 - n - 1$ and $4n^2 + 6n - 5$

20. $-5n^2 + 7n - 9$ and $-5n - 4$

21. $2x^2 - 7x - 10$, $-6x - 2$, and $-9x^2 + 5$

22. $7x - 11$, $-x^2 - 5x + 9$, and $-4x + 5$

For Problems 23–34, subtract the polynomials using a horizontal format.

23. $7x + 1$ from $12x + 6$

24. $10x + 3$ from $14x + 13$

25. $5x - 2$ from $3x - 7$

26. $7x - 2$ from $2x + 3$

27. $-x - 1$ from $-4x + 6$

28. $-3x + 2$ from $-x - 9$

29. $x^2 - 7x + 2$ from $3x^2 + 8x - 4$

30. $2x^2 + 6x - 1$ from $8x^2 - 2x + 6$

31. $-2n^2 - 3n + 4$ from $3n^2 - n + 7$

32. $3n^2 - 7n - 9$ from $-4n^2 + 6n + 10$

33. $-4x^3 - x^2 + 6x - 1$ from $-7x^3 + x^2 + 6x - 12$

34. $-4x^2 + 6x - 2$ from $-3x^3 + 2x^2 + 7x - 1$

For Problems 35–44, subtract the polynomials using a vertical format.

35. $3x - 2$ from $12x - 4$

36. $-4x + 6$ from $7x - 3$

37. $-5a - 6$ from $-3a + 9$

38. $7a - 11$ from $-2a - 1$

39. $8x^2 - x + 6$ from $6x^2 - x + 11$

40. $3x^2 - 2$ from $-2x^2 + 6x - 4$

41. $-2x^3 - 6x^2 + 7x - 9$ from $4x^3 + 6x^2 + 7x - 14$

42. $4x^3 + x - 10$ from $3x^2 - 6$

43. $2x^2 - 6x - 14$ from $4x^3 - 6x^2 + 7x - 2$

44. $3x - 7$ from $7x^3 + 6x^2 - 5x - 4$

For Problems 45–64, perform the indicated operations.

45. $(5x + 3) - (7x - 2) + (3x + 6)$

46. $(3x - 4) + (9x - 1) - (14x - 7)$

47. $(-x - 1) - (-2x + 6) + (-4x - 7)$

48. $(-3x + 6) + (-x - 8) - (-7x + 10)$

49. $(x^2 - 7x - 4) + (2x^2 - 8x - 9) - (4x^2 - 2x - 1)$

50. $(3x^2 + x - 6) - (8x^2 - 9x + 1) - (7x^2 + 2x - 6)$

51. $(-x^2 - 3x + 4) + (-2x^2 - x - 2)$
   $- (-4x^2 + 7x + 10)$

52. $(-3x^2 - 2) + (7x^2 - 8) - (9x^2 - 2x - 4)$

53. $(3a - 2b) - (7a + 4b) - (6a - 3b)$

54. $(5a + 7b) + (-8a - 2b) - (5a + 6b)$

55. $(n - 6) - (2n^2 - n + 4) + (n^2 - 7)$

56. $(3n + 4) - (n^2 - 9n + 10) - (-2n + 4)$

57. $7x + [3x - (2x - 1)]$

58. $-6x + [-2x - (5x + 2)]$

59. $-7n - [4n - (6n - 1)]$

60. $9n - [3n - (5n + 4)]$

61. $(5a - 1) - [3a + (4a - 7)]$

62. $(-3a + 4) - [-7a + (9a - 1)]$

63. $13x - \{5x - [4x - (x - 6)]\}$

64. $-10x - \{7x - [3x - (2x - 3)]\}$

65. Subtract $5x - 3$ from the sum of $4x - 2$ and $7x + 6$.

66. Subtract $7x + 5$ from the sum of $9x - 4$ and $-3x - 2$.

67. Subtract the sum of $-2n - 5$ and $-n + 7$ from $-8n + 9$.

68. Subtract the sum of $7n - 11$ and $-4n - 3$ from $13n - 4$.

69. Find a polynomial that represents the perimeter of the rectangle in Figure 6.2.

```
┌────────────────────────┐
│                        │  x − 2
└────────────────────────┘
        3x + 5
```

Figure 6.2

70. Find a polynomial that represents the area of the shaded region in Figure 6.3. The length of a radius of the larger circle is r units, and the length of a radius of the smaller circle is 4 units.

Figure 6.3

71. Find a polynomial that represents the sum of the areas of the rectangles and squares in Figure 6.4.

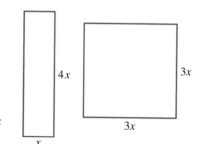

Figure 6.4

72. Find a polynomial that represents the total surface area of the rectangular solid in Figure 6.5.

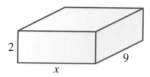

Figure 6.5

■ ■ ■ **Thoughts into words**

73. Explain how to subtract the polynomial

$$3x^2 + 6x - 2 \text{ from } 4x^2 + 7.$$

74. Is the sum of two binomials always another binomial? Defend your answer.

75. Is the sum of two binomials ever a trinomial? Defend your answer.

Answers to Concept Quiz

1. True   **2.** False   **3.** False   **4.** True   **5.** True   **6.** C   **7.** D   **8.** B   **9.** E   **10.** A

6.2 Multiplying Monomials

Objectives

■   Apply the properties of exponents to multiply monomials.

■   Multiply a polynomial by a monomial.

■   Use products of monomials to represent the area or volume of geometric figures.

In Section 2.4, we used exponents and some of the basic properties of real numbers to simplify algebraic expressions into a more compact form; for example,

$$(3x)(4xy) = 3 \cdot 4 \cdot x \cdot x \cdot y = 12x^2y$$

Actually, we were **multiplying monomials,** and it is this topic that we will pursue now. We can make multiplying monomials easier by using some basic properties of exponents. These properties are the direct result of the definition of an exponent.

The following examples lead to the first property:

$$x^2 \cdot x^3 = (x \cdot x)(x \cdot x \cdot x) = x^5$$
$$a^3 \cdot a^4 = (a \cdot a \cdot a)(a \cdot a \cdot a \cdot a) = a^7$$
$$b \cdot b^2 = (b)(b \cdot b) = b^3$$

In general,

$$b^n \cdot b^m = \underbrace{(b \cdot b \cdot b \cdot \cdots \cdot b)}_{n \text{ factors of } b}\underbrace{(b \cdot b \cdot b \cdot \cdots \cdot b)}_{m \text{ factors of } b}$$

$$= \underbrace{b \cdot b \cdot b \cdot \cdots \cdot b}_{(n + m) \text{ factors of } b}$$

$$= b^{n+m}$$

PROPERTY 6.1

If b is any real number and n and m are positive integers, then

$$b^n \cdot b^m = b^{n+m}$$

Property 6.1 states that when multiplying powers with the same base, we add exponents.

E X A M P L E   1

Multiply.

(a) $x^4 \cdot x^3$ **(b)** $a^8 \cdot a^7$

Solution

(a) $x^4 \cdot x^3 = x^{4+3} = x^7$ **(b)** $a^8 \cdot a^7 = a^{8+7} = a^{15}$ ■

Another property of exponents is demonstrated by these examples.

$$(x^2)^3 = x^2 \cdot x^2 \cdot x^2 = x^{2+2+2} = x^6$$
$$(a^3)^2 = a^3 \cdot a^3 = a^{3+3} = a^6$$
$$(b^3)^4 = b^3 \cdot b^3 \cdot b^3 \cdot b^3 = b^{3+3+3+3} = b^{12}$$

In general,

$$(b^n)^m = \underbrace{b^n \cdot b^n \cdot b^n \cdot \cdots \cdot b^n}_{m \text{ factors of } b^n}$$

$$= b^{\overbrace{n+n+n+\cdots+n}^{m \text{ of these } n\text{'s}}}$$

$$= b^{mn}$$

PROPERTY 6.2

If b is any real number and m and n are positive integers, then

$$(b^n)^m = b^{mn}$$

Property 6.2 states that when raising a power to a power, we multiply exponents.

EXAMPLE 2   Raise each to the indicated power.

(a) $(x^4)^3$ (b) $(a^5)^6$

Solution

(a) $(x^4)^3 = x^{3 \cdot 4} = x^{12}$ (b) $(a^5)^6 = a^{6 \cdot 5} = a^{30}$ ■

The third property of exponents that we will use in this section raises a monomial to a power.

$$(2x)^3 = (2x)(2x)(2x) = 2 \cdot 2 \cdot 2 \cdot x \cdot x \cdot x = 2^3 \cdot x^3$$
$$(3a^4)^2 = (3a^4)(3a^4) = 3 \cdot 3 \cdot a^4 \cdot a^4 = (3)^2(a^4)^2$$
$$(-2xy^5)^2 = (-2xy^5)(-2xy^5) = (-2)(-2)(x)(x)(y^5)(y^5) = (-2)^2(x)^2(y^5)^2$$

In general,

$$(ab)^n = \underbrace{ab \cdot ab \cdot ab \cdot \cdots \cdot ab}_{n \text{ factors of } ab}$$

$$= \underbrace{(a \cdot a \cdot a \cdot \cdots \cdot a)}_{n \text{ factors of } a}\underbrace{(b \cdot b \cdot b \cdot \cdots \cdot b)}_{n \text{ factors of } b}$$

$$= a^n b^n$$ ■

PROPERTY 6.3

If a and b are real numbers and n is a positive integer, then

$$(ab)^n = a^n b^n$$

Property 6.3 states that when raising a monomial to a power, we raise each factor to that power.

EXAMPLE 3   Raise each to the indicated power.

(a) $(2x^2y^3)^4$ (b) $(-3ab^5)^3$

Solution

(a) $(2x^2y^3)^4 = (2)^4(x^2)^4(y^3)^4 = 16x^8y^{12}$

(b) $(-3ab^5)^3 = (-3)^3(a^1)^3(b^5)^3 = -27a^3b^{15}$ ■

Consider the following examples in which we use the properties of exponents to help simplify the process of multiplying monomials.

1. $(3x^3)(5x^4) = 3 \cdot 5 \cdot x^3 \cdot x^4$

$\qquad = 15x^7 \qquad x^3 \cdot x^4 = x^{3+4} = x^7$

2. $(-4a^2b^3)(6ab^2) = -4 \cdot 6 \cdot a^2 \cdot a \cdot b^3 \cdot b^2$

$\qquad = -24a^3b^5$

3. $(xy)(7xy^5) = 1 \cdot 7 \cdot x \cdot x \cdot y \cdot y^5 \qquad$ The numerical coefficient of xy is 1.

$\qquad = 7x^2y^6$

4. $\left(\dfrac{3}{4}x^2y^3\right)\left(\dfrac{1}{2}x^3y^5\right) = \dfrac{3}{4} \cdot \dfrac{1}{2} \cdot x^2 \cdot x^3 \cdot y^3 \cdot y^5$

$\qquad\qquad\qquad = \dfrac{3}{8}x^5y^8$

It is a simple process to raise a monomial to a power when using the properties of exponents. Study the next examples.

5. $(2x^3)^4 = (2)^4(x^3)^4 \quad$ by using $(ab)^n = a^nb^n$

$\qquad = (2)^4(x^{12}) \quad$ by using $(b^n)^m = b^{mn}$

$\qquad = 16x^{12}$

6. $(-2a^4)^5 = (-2)^5(a^4)^5$

$\qquad = -32a^{20}$

7. $\left(\dfrac{2}{5}x^2y^3\right)^3 = \left(\dfrac{2}{5}\right)^3(x^2)^3(y^3)^3$

$\qquad\qquad = \dfrac{8}{125}x^6y^9$

8. $(0.2a^6b^7)^2 = (0.2)^2(a^6)^2(b^7)^2$

$\qquad = 0.04a^{12}b^{14}$

Sometimes problems involve first raising monomials to a power and then multiplying the resulting monomials, as in the following examples.

9. $(3x^2)^3(2x^3)^2 = (3)^3(x^2)^3(2)^2(x^3)^2$

$\qquad\qquad = (27)(x^6)(4)(x^6)$

$\qquad\qquad = 108x^{12}$

10. $(-x^2y^3)^5(-2x^2y)^2 = (-1)^5(x^2)^5(y^3)^5(-2)^2(x^2)^2(y)^2$

$\qquad\qquad\qquad = (-1)(x^{10})(y^{15})(4)(x^4)(y^2)$

$\qquad\qquad\qquad = -4x^{14}y^{17}$

The distributive property, along with the properties of exponents, form a basis for finding the product of a monomial and a polynomial. The next examples illustrate these ideas.

11. $(3x)(2x^2 + 6x + 1) = (3x)(2x^2) + (3x)(6x) + (3x)(1)$
$$= 6x^3 + 18x^2 + 3x$$

12. $(5a^2)(a^3 - 2a^2 - 1) = (5a^2)(a^3) - (5a^2)(2a^2) - (5a^2)(1)$
$$= 5a^5 - 10a^4 - 5a^2$$

13. $(-2xy)(6x^2y - 3xy^2 - 4y^3)$
$$= (-2xy)(6x^2y) - (-2xy)(3xy^2) - (-2xy)(4y^3)$$
$$= -12x^3y^2 + 6x^2y^3 + 8xy^4$$

Once you feel comfortable with this process, you may want to perform most of the work mentally and simply write down the final result. See whether you understand the following examples.

14. $3x(2x + 3) = 6x^2 + 9x$

15. $-4x(2x^2 - 3x - 1) = -8x^3 + 12x^2 + 4x$

16. $ab(3a^2b - 2ab^2 - b^3) = 3a^3b^2 - 2a^2b^3 - ab^4$

We conclude this section by making a connection between algebra and geometry.

EXAMPLE 4

Suppose that the dimensions of a rectangular solid are represented by x, $2x$, and $3x$, as shown in Figure 6.6. Express the volume and total surface area of the figure.

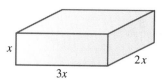

x

$2x$

$3x$

Figure 6.6

Solution

Using the formula $V = lwh$, we can express the volume of the rectangular solid as $(2x)(3x)(x)$, which equals $6x^3$. The total surface area can be described as follows:

Area of front and back rectangles: $2(x)(3x) = 6x^2$
Area of left side and right side: $2(2x)(x) = 4x^2$
Area of top and bottom: $2(2x)(3x) = 12x^2$

We can represent the total surface area by $6x^2 + 4x^2 + 12x^2$ or $22x^2$.   ■

CONCEPT QUIZ

For Problems 1–6, answer true or false.

1. When multiplying factors with the same base, add the exponents.

2. $3^2 \cdot 3^2 = 9^4$

3. $2x^2 \cdot 3x^3 = 6x^6$

4. $(x^2)^3 = x^5$

5. $(-4x^3)^2 = -4x^6$

6. To simplify $(3x^2y)(2x^3y^2)^4$, use the order of operations to first raise $2x^3y^2$ to the fourth power, and then multiply the monomials.

PROBLEM SET 6.2

For Problems 1–30, multiply using the properties of exponents to help with the manipulation.

1. $(5x)(9x)$

2. $(7x)(8x)$

3. $(3x^2)(7x)$

4. $(9x)(4x^3)$

5. $(-3xy)(2xy)$

6. $(6xy)(-3xy)$

7. $(-2x^2y)(-7x)$

8. $(-5xy^2)(-4y)$

9. $(4a^2b^2)(-12ab)$

10. $(-3a^3b)(13ab^2)$

11. $(-xy)(-5x^3)$

12. $(-7y^2)(-x^2y)$

13. $(8ab^2c)(13a^2c)$

14. $(9abc^3)(14bc^2)$

15. $(5x^2)(2x)(3x^3)$

16. $(4x)(2x^2)(6x^4)$

17. $(4xy)(-2x)(7y^2)$

18. $(5y^2)(-3xy)(5x^2)$

19. $(-2ab)(-ab)(-3b)$

20. $(-7ab)(-4a)(-ab)$

21. $(6cd)(-3c^2d)(-4d)$

22. $(2c^3d)(-6d^3)(-5cd)$

23. $\left(\dfrac{2}{3}xy\right)\left(\dfrac{3}{5}x^2y^4\right)$

24. $\left(-\dfrac{5}{6}x\right)\left(\dfrac{8}{3}x^2y\right)$

25. $\left(-\dfrac{7}{12}a^2b\right)\left(\dfrac{8}{21}b^4\right)$

26. $\left(-\dfrac{9}{5}a^3b^4\right)\left(-\dfrac{15}{6}ab^2\right)$

27. $(0.4x^5)(0.7x^3)$

28. $(-1.2x^4)(0.3x^2)$

29. $(-4ab)(1.6a^3b)$

30. $(-6a^2b)(-1.4a^2b^4)$

For Problems 31–46, raise each monomial to the indicated power. Use the properties of exponents to help with the manipulation.

31. $(2x^4)^2$

32. $(3x^3)^2$

33. $(-3a^2b^3)^2$

34. $(-8a^4b^5)^2$

35. $(3x^2)^3$

36. $(2x^4)^3$

37. $(-4x^4)^3$

38. $(-3x^3)^3$

39. $(9x^4y^5)^2$

40. $(8x^6y^4)^2$

41. $(2x^2y)^4$

42. $(2x^2y^3)^5$

43. $(-3a^3b^2)^4$

44. $(-2a^4b^2)^4$

45. $(-x^2y)^6$

46. $(-x^2y^3)^7$

For Problems 47–60, multiply by using the distributive property.

47. $5x(3x + 2)$

48. $7x(2x + 5)$

49. $3x^2(6x - 2)$

50. $4x^2(7x - 2)$

51. $-4x(7x^2 - 4)$

52. $-6x(9x^2 - 5)$

53. $2x(x^2 - 4x + 6)$

54. $3x(2x^2 - x + 5)$

55. $-6a(3a^2 - 5a - 7)$

56. $-8a(4a^2 - 9a - 6)$

57. $7xy(4x^2 - x + 5)$

58. $5x^2y(3x^2 + 7x - 9)$

59. $-xy(9x^2 - 2x - 6)$

60. $xy^2(6x^2 - x - 1)$

For Problems 61–70, remove the parentheses by multiplying and then simplify by combining similar terms; for example,

$$3(x - y) + 2(x - 3y) = 3x - 3y + 2x - 6y$$
$$= 5x - 9y$$

61. $5(x + 2y) + 4(2x + 3y)$

62. $3(2x + 5y) + 2(4x + y)$

63. $4(x - 3y) - 3(2x - y)$

64. $2(5x - 3y) - 5(x + 4y)$

65. $2x(x^2 - 3x - 4) + x(2x^2 + 3x - 6)$

66. $3x(2x^2 - x + 5) - 2x(x^2 + 4x + 7)$

67. $3[2x - (x - 2)] - 4(x - 2)$

68. $2[3x - (2x + 1)] - 2(3x - 4)$

69. $-4(3x + 2) - 5[2x - (3x + 4)]$

70. $-5(2x - 1) - 3[x - (4x - 3)]$

For Problems 71–80, perform the indicated operations and simplify.

71. $(3x)^2(2x^3)$

72. $(-2x)^3(4x^5)$

73. $(-3x)^3(-4x)^2$

74. $(3xy)^2(2x^2y)^4$

75. $(5x^2y)^2(xy^2)^3$

76. $(-x^2y)^3(6xy)^2$

77. $(-a^2bc^3)^3(a^3b)^2$

78. $(ab^2c^3)^4(-a^2b)^3$

79. $(-2x^2y^2)^4(-xy^3)^3$

80. $(-3xy)^3(-x^2y^3)^4$

81. Express in simplified form the sum of the areas of the two rectangles shown in Figure 6.7.

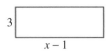 

Figure 6.7

82. Express in simplified form the volume and the total surface area of the rectangular solid in Figure 6.8.

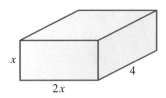

Figure 6.8

83. Represent the area of the shaded region in Figure 6.9. The length of a radius of the smaller circle is x, and the length of a radius of the larger circle is $2x$.

Figure 6.9

84. Represent the area of the shaded region in Figure 6.10.

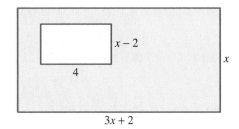

Figure 6.10

85. How would you explain to someone why the product of x^3 and x^4 is x^7 and not x^{12}?

86. Suppose your friend was absent from class the day that this section was discussed. How would you help her understand why the property $(b^n)^m = b^{mn}$ is true?

87. How can Figure 6.11 be used to demonstrate geometrically that $x(x + 2) = x^2 + 2x$?

Figure 6.11

■ ■ ■ **Further investigations**

For Problems 88–97, find each of the indicated products. Assume that the variables in the exponents represent positive integers; for example,

$$(x^{2n})(x^{4n}) = x^{2n+4n} = x^{6n}$$

88. $(x^n)(x^{3n})$

89. $(x^{2n})(x^{5n})$

90. $(x^{2n-1})(x^{3n+2})$

91. $(x^{5n+2})(x^{n-1})$

92. $(x^3)(x^{4n-5})$

93. $(x^{6n-1})(x^4)$

94. $(2x^n)(3x^{2n})$

95. $(4x^{3n})(-5x^{7n})$

96. $(-6x^{2n+4})(5x^{3n-4})$

97. $(-3x^{5n-2})(-4x^{2n+2})$

Answers to Concept Quiz

1. True **2.** False **3.** False **4.** False **5.** False **6.** True

6.3 Multiplying Polynomials

Objectives

■ Use the distributive property to find the product of two binomials.

■ Use the shortcut pattern to find the product of two binomials.

■ Use the pattern to find the square of a binomial.

■ Use the pattern to find the product of $(a + b)(a - b)$.

In general, to go from multiplying a monomial times a polynomial to multiplying two polynomials requires the use of the distributive property twice. Consider some examples.

EXAMPLE 1   Find the product of $(x + 3)$ and $(y + 4)$.

Solution

$$
\begin{aligned}
(x + 3)(y + 4) &= x(y + 4) + 3(y + 4) \\
&= x(y) + x(4) + 3(y) + 3(4) \\
&= xy + 4x + 3y + 12
\end{aligned}
$$
  ■

Note that each term of the first polynomial is multiplied times each term of the second polynomial.

EXAMPLE 2   Find the product of $(x - 2)$ and $(y + z + 5)$.

Solution

$$
\begin{aligned}
(x - 2)(y + z + 5) &= x(y) + x(z) + x(5) - 2(y) - 2(z) - 2(5) \\
&= xy + xz + 5x - 2y - 2z - 10
\end{aligned}
$$
  ■

Frequently, multiplying polynomials will produce similar terms that can be combined to simplify the resulting polynomial.

EXAMPLE 3

Multiply $(x + 3)(x + 2)$.

Solution

$$\begin{aligned} (x + 3)(x + 2) &= x(x + 2) + 3(x + 2) \\ &= x^2 + 2x + 3x + 6 \\ &= x^2 + 5x + 6 \end{aligned}$$

EXAMPLE 4

Multiply $(x - 4)(x + 9)$.

Solution

$$\begin{aligned} (x - 4)(x + 9) &= x(x + 9) - 4(x + 9) \\ &= x^2 + 9x - 4x - 36 \\ &= x^2 + 5x - 36 \end{aligned}$$

EXAMPLE 5

Multiply $(x + 4)(x^2 + 3x + 2)$.

Solution

$$\begin{aligned} (x + 4)(x^2 + 3x + 2) &= x(x^2 + 3x + 2) + 4(x^2 + 3x + 2) \\ &= x^3 + 3x^2 + 2x + 4x^2 + 12x + 8 \\ &= x^3 + 7x^2 + 14x + 8 \end{aligned}$$

EXAMPLE 6

Multiply $(2x - y)(3x^2 - 2xy + 4y^2)$.

Solution

$$\begin{aligned} (2x - y)(3x^2 - 2xy + 4y^2) &= 2x(3x^2 - 2xy + 4y^2) - y(3x^2 - 2xy + 4y^2) \\ &= 6x^3 - 4x^2y + 8xy^2 - 3x^2y + 2xy^2 - 4y^3 \\ &= 6x^3 - 7x^2y + 10xy^2 - 4y^3 \end{aligned}$$

Perhaps the most frequently used type of multiplication problem is the product of two binomials. It will be a big help later if you can become proficient at multiplying binomials without showing all of the intermediate steps. This is quite easy to do if you use a three-step shortcut pattern demonstrated by the following examples.

EXAMPLE 7   Multiply $(x + 5)(x + 7)$.

Solution

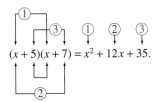

$(x + 5)(x + 7) = x^2 + 12x + 35.$

STEP 1   Multiply $x \cdot x$.

STEP 2   Multiply $5 \cdot x$ and $7 \cdot x$ and combine them.

STEP 3   Multiply $5 \cdot 7$.

Figure 6.12

EXAMPLE 8   Multiply $(x - 8)(x + 3)$.

Solution

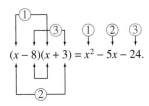

$(x - 8)(x + 3) = x^2 - 5x - 24.$

Figure 6.13

EXAMPLE 9   Multiply $(3x + 2)(2x - 5)$.

Solution

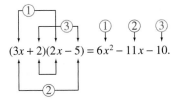

$(3x + 2)(2x - 5) = 6x^2 - 11x - 10.$

Figure 6.14

The mnemonic device FOIL is often used to remember the pattern for multiplying binomials. The letters in FOIL represent First, Outside, Inside, and Last. If you look back at Examples 7 through 9, step 1 is to find the product of the first term in each binomial; step 2 is to find the product of the outside terms and the inside terms; and step 3 is to find the product of the last terms in each binomial. Now see whether *you* can use the pattern to find these products:

$$(x + 3)(x + 7)$$
$$(3x + 1)(2x + 5)$$
$$(x - 2)(x - 3)$$
$$(4x + 5)(x - 2)$$

Your answers should be $x^2 + 10x + 21$, $6x^2 + 17x + 5$, $x^2 - 5x + 6$, and $4x^2 - 3x - 10$.

Keep in mind that the shortcut pattern applies only to finding the product of two binomials. For other situations, such as finding the product of a binomial and a trinomial, we suggest showing the intermediate steps:

$$\begin{aligned}
(x + 3)(x^2 + 6x - 7) &= x(x^2) + x(6x) - x(7) + 3(x^2) + 3(6x) - 3(7) \\
&= x^3 + 6x^2 - 7x + 3x^2 + 18x - 21 \\
&= x^3 + 9x^2 + 11x - 21
\end{aligned}$$

Perhaps you could omit the first step, and shorten the form as follows:

$$\begin{aligned}
(x - 4)(x^2 - 5x - 6) &= x^3 - 5x^2 - 6x - 4x^2 + 20x + 24 \\
&= x^3 - 9x^2 + 14x + 24
\end{aligned}$$

Remember that you are multiplying each term of the first polynomial times each term of the second polynomial and combining similar terms.

Exponents are also used to indicate repeated multiplication of polynomials. For example, we can write $(x + 4)(x + 4)$ as $(x + 4)^2$. Thus to square a binomial, we simply write it as the product of two equal binomials and apply the shortcut pattern.

$$(x + 4)^2 = (x + 4)(x + 4) = x^2 + 8x + 16$$
$$(x - 5)^2 = (x - 5)(x - 5) = x^2 - 10x + 25$$
$$(2x + 3)^2 = (2x + 3)(2x + 3) = 4x^2 + 12x + 9$$

When you square binomials, be careful not to forget the middle term. In other words, $(x + 3)^2 \neq x^2 + 3^2$; instead, $(x + 3)^2 = (x + 3)(x + 3) = x^2 + 6x + 9$.

The next example suggests a format to use when cubing a binomial.

$$\begin{aligned}
(x + 4)^3 &= (x + 4)(x + 4)(x + 4) \\
&= (x + 4)(x^2 + 8x + 16) \\
&= x(x^2 + 8x + 16) + 4(x^2 + 8x + 16) \\
&= x^3 + 8x^2 + 16x + 4x^2 + 32x + 64 \\
&= x^3 + 12x^2 + 48x + 64
\end{aligned}$$

Special Product Patterns

When we multiply binomials, some special patterns occur that you should recognize. We can use these patterns to find products and later to factor polynomials. We will state each of the patterns in general terms, followed by examples to illustrate the use of each pattern.

PATTERN

$$(a + b)^2 = (a + b)(a + b) = a^2 \quad + \quad 2ab \quad + \quad b^2$$

Square of
first term   +   Twice the   +   Square of
of binomial   product of   second term
                     the two terms   of binomial
                     of binomial

Examples

$$(x + 4)^2 = x^2 + 8x + 16$$
$$(2x + 3y)^2 = 4x^2 + 12xy + 9y^2$$
$$(5a + 7b)^2 = 25a^2 + 70ab + 49b^2$$

PATTERN

$$(a - b)^2 = (a - b)(a - b) = a^2 \quad - \quad 2ab \quad + \quad b^2$$

Square of
first term   −   Twice the   +   Square of
of binomial   product of   second term
                     the two terms   of binomial
                     of binomial

Examples

$$(x - 8)^2 = x^2 - 16x + 64$$
$$(3x - 4y)^2 = 9x^2 - 24xy + 16y^2$$
$$(4a - 9b)^2 = 16a^2 - 72ab + 81b^2$$

PATTERN

$$(a + b)(a - b) = a^2 \quad - \quad b^2$$

Square of
first term   −   Square of
of binomial   second term
                     of binomial

Examples

$$(x + 7)(x - 7) = x^2 - 49$$
$$(2x + y)(2x - y) = 4x^2 - y^2$$
$$(3a - 2b)(3a + 2b) = 9a^2 - 4b^2$$

As you might expect, there are geometric interpretations for many of the algebraic concepts presented in this section. We will give you the opportunity to make some of these connections between algebra and geometry in the next problem set. We conclude this section with a problem that allows us to use some algebra and geometry.

EXAMPLE 10

A rectangular piece of tin is 16 inches long and 12 inches wide, as shown in Figure 6.15. From each corner, a square piece x inches on a side is cut out. The flaps are then turned up to form an open box. Find polynomials that represent the volume and outside surface area of the box.

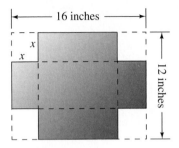

Figure 6.15

Solution

The length of the box is $16 - 2x$, the width is $12 - 2x$, and the height is x. From the volume formula $V = lwh$, the polynomial $(16 - 2x)(12 - 2x)(x)$, which simplifies to $4x^3 - 56x^2 + 192x$, represents the volume.

The outside surface area of the box is the area of the original piece of tin minus the four corners that were cut off. Therefore the polynomial $16(12) - 4x^2$ (or $192 - 4x^2$) represents the outside surface area of the box. ∎

CONCEPT QUIZ

For Problems 1–5, answer true or false.

1. The algebraic expression $(x + y)^2$ is called the square of a binomial.

2. The algebraic expression $(x + y)(x + 2xy + y)$ is called the product of two binomials.

3. The mnemonic device FOIL stands for first, outside, inside, and last.

4. $(a + 2)^2 = a^2 + 4$

5. $(y + 3)(y - 3) = y^2 + 9$

PROBLEM SET 6.3

For Problems 1–10, find the indicated products by applying the distributive property; for example,

$$(x + 1)(y + 5) = x(y) + x(5) + 1(y) + 1(5)$$
$$= xy + 5x + y + 5$$

1. $(x + 2)(y + 3)$

2. $(x + 3)(y + 6)$

3. $(x - 4)(y + 1)$

4. $(x - 5)(y + 7)$

5. $(x - 5)(y - 6)$

6. $(x - 7)(y - 9)$

7. $(x + 2)(y + z + 1)$

8. $(x + 4)(y - z + 4)$

9. $(2x + 3)(3y + 1)$

10. $(3x - 2)(2y - 5)$

For Problems 11–36, find the indicated products by applying the distributive property and combining similar terms. Use the following format to show your work:

$$(x + 3)(x + 8) = x(x) + x(8) + 3(x) + 3(8)$$
$$= x^2 + 8x + 3x + 24$$
$$= x^2 + 11x + 24$$

11. $(x + 3)(x + 7)$

12. $(x + 4)(x + 2)$

13. $(x + 8)(x - 3)$

14. $(x + 9)(x - 6)$

15. $(x - 7)(x + 1)$

16. $(x - 10)(x + 8)$

17. $(n - 4)(n - 6)$

18. $(n - 3)(n - 7)$

19. $(3n + 1)(n + 6)$

20. $(4n + 3)(n + 6)$

21. $(5x - 2)(3x + 7)$

22. $(3x - 4)(7x + 1)$

23. $(x + 3)(x^2 + 4x + 9)$

24. $(x + 2)(x^2 + 6x + 2)$

25. $(x + 4)(x^2 - x - 6)$

26. $(x + 5)(x^2 - 2x - 7)$

27. $(x - 5)(2x^2 + 3x - 7)$

28. $(x - 4)(3x^2 + 4x - 6)$

29. $(2a - 1)(4a^2 - 5a + 9)$

30. $(3a - 2)(2a^2 - 3a - 5)$

31. $(3a + 5)(a^2 - a - 1)$

32. $(5a + 2)(a^2 + a - 3)$

33. $(x^2 + 2x + 3)(x^2 + 5x + 4)$

34. $(x^2 - 3x + 4)(x^2 + 5x - 2)$

35. $(x^2 - 6x - 7)(x^2 + 3x - 9)$

36. $(x^2 - 5x - 4)(x^2 + 7x - 8)$

For Problems 37–80, find the indicated products by using the shortcut pattern for multiplying binomials.

37. $(x + 2)(x + 9)$

38. $(x + 3)(x + 8)$

39. $(x + 6)(x - 2)$

40. $(x + 8)(x - 6)$

41. $(x + 3)(x - 11)$

42. $(x + 4)(x - 10)$

43. $(n - 4)(n - 3)$

44. $(n - 5)(n - 9)$

45. $(n + 6)(n + 12)$

46. $(n + 8)(n + 13)$

47. $(y + 3)(y - 7)$

48. $(y + 2)(y - 12)$

49. $(y - 7)(y - 12)$

50. $(y - 4)(y - 13)$

51. $(x - 5)(x + 7)$

52. $(x - 1)(x + 9)$

53. $(x - 14)(x + 8)$

54. $(x - 15)(x + 6)$

55. $(a + 10)(a - 9)$

56. $(a + 7)(a - 6)$

57. $(2a + 1)(a + 6)$

58. $(3a + 2)(a + 4)$

59. $(5x - 2)(x + 7)$

60. $(2x - 3)(x + 8)$

61. $(3x - 7)(2x + 1)$

62. $(5x - 6)(4x + 3)$

63. $(4a + 3)(3a - 4)$

64. $(5a + 4)(4a - 5)$

65. $(6n - 5)(2n - 3)$

66. $(4n - 3)(6n - 7)$

67. $(7x - 4)(2x + 3)$

68. $(8x - 5)(3x + 7)$

69. $(5 - x)(9 - 2x)$

70. $(4 - 3x)(2 + x)$

71. $(-2x + 3)(4x - 5)$

72. $(-3x + 1)(9x - 2)$

73. $(-3x - 1)(3x - 4)$

74. $(-2x - 5)(4x + 1)$

75. $(8n + 3)(9n - 4)$

76. $(6n + 5)(9n - 7)$

77. $(3 - 2x)(9 - x)$

78. $(5 - 4x)(4 - 5x)$

79. $(-4x + 3)(-5x - 2)$

80. $(-2x + 7)(-7x - 3)$

For Problems 81–110, use the pattern $(a + b)^2 = a^2 + 2ab + b^2$, $(a - b)^2 = a^2 - 2ab + b^2$, or the pattern $(a + b)(a - b) = a^2 - b^2$ to find the indicated products.

81. $(x + 7)^2$

82. $(x + 9)^2$

83. $(5x - 2)(5x + 2)$

84. $(6x + 1)(6x - 1)$

85. $(x - 1)^2$

86. $(x - 4)^2$

87. $(3x + 7)^2$

88. $(2x + 9)^2$

89. $(2x - 3)^2$

90. $(4x - 5)^2$

91. $(2x + 3y)(2x - 3y)$

92. $(3a - b)(3a + b)$

93. $(1 - 5n)^2$

94. $(2 - 3n)^2$

95. $(3x + 4y)^2$

96. $(2x + 5y)^2$

97. $(3 + 4y)^2$

98. $(7 + 6y)^2$

99. $(1 + 7n)(1 - 7n)$

100. $(2 + 9n)(2 - 9n)$

101. $(4a - 7b)^2$

102. $(6a - b)^2$

103. $(x + 8y)^2$

104. $(x + 6y)^2$

105. $(5x - 11y)(5x + 11y)$

106. $(7x - 9y)(7x + 9y)$

107. $x(8x + 1)(8x - 1)$

108. $3x(5x + 7)(5x - 7)$

109. $-2x(4x + y)(4x - y)$

110. $-4x(2 - 3x)(2 + 3x)$

For Problems 111–118, find the indicated products. Don't forget that $(x + 2)^3$ means $(x + 2)(x + 2)(x + 2)$.

111. $(x + 2)^3$

112. $(x + 4)^3$

113. $(x - 3)^3$

114. $(x - 1)^3$

115. $(2n + 1)^3$

116. $(3n + 2)^3$

117. $(3n - 2)^3$

118. $(4n - 3)^3$

119. Explain how Figure 6.16 can be used to demonstrate geometrically that $(x + 3)(x + 5) = x^2 + 8x + 15$.

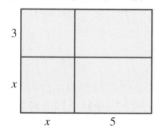

Figure 6.16

120. Explain how Figure 6.17 can be used to demonstrate geometrically that $(x + 5)(x - 3) = x^2 + 2x - 15$.

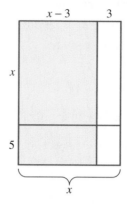

Figure 6.17

121. A square piece of cardboard is 14 inches long on each side. From each corner, a square piece x inches on a side is cut out, as shown in Figure 6.18. The flaps are then turned up to form an open box. Find polynomials that represent the volume and the outside surface area of the box.

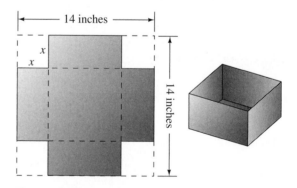

Figure 6.18

■ ■ ■ Thoughts into words

122. Describe the process of multiplying two polynomials.

123. Illustrate as many uses of the distributive property as you can.

124. Determine the number of terms in the product of $(x + y + z)$ and $(a + b + c)$ without doing the multiplication. Explain how you arrived at your answer.

■ ■ ■ Further investigations

125. The following two patterns result from cubing binomials:

$$(a + b)^3 = a^3 + 3a^2b + 3ab^2 + b^3$$
$$(a - b)^3 = a^3 - 3a^2b + 3ab^2 - b^3$$

Use these patterns to redo Problems 111–118.

126. Find a pattern for the expansion of $(a + b)^4$. Then use that pattern to expand $(x + 2)^4$, $(x + 3)^4$, and $(2x + 1)^4$.

127. We can use some of the product patterns to do arithmetic computations mentally. For example, let's use the pattern $(a + b)^2 = a^2 + 2ab + b^2$ to compute 31^2 mentally. Your thought process should be $31^2 = (30 + 1)^2 = 30^2 + 2(30)(1) + 1^2 = 961$. Compute each of the following numbers mentally and then check your answers.

a. 21^2 **b.** 41^2 **c.** 71^2

d. 32^2 **e.** 52^2 **f.** 82^2

128. Use the pattern $(a - b)^2 = a^2 - 2ab + b^2$ to compute each of the following numbers mentally and then check your answers.

a. 19^2 **b.** 29^2 **c.** 49^2

d. 79^2 **e.** 38^2 **f.** 58^2

129. Every whole number with a units digit of 5 can be represented by the expression $10x + 5$, where x is a whole number. For example, $35 = 10(3) + 5$ and $145 = 10(14) + 5$. Now observe the following pattern for squaring such a number:

$$(10x + 5)^2 = 100x^2 + 100x + 25$$
$$= \boxed{100x(x + 1) + 25}$$

The pattern inside the dashed box can be stated as "add 25 to the product of x, $x + 1$, and 100." Thus, to compute 35^2 mentally, we can figure

$$35^2 = 3(4)(100) + 25 = 1225$$

Compute each of the following numbers mentally and then check your answers.

a. 15^2 **b.** 25^2 **c.** 45^2

d. 55^2 **e.** 65^2 **f.** 75^2

g. 85^2 **h.** 95^2 **i.** 105^2

Answers to Concept Quiz

1. True   **2.** False   **3.** True   **4.** False   **5.** False

6.4 Dividing by Monomials

Objectives

■   Apply the properties of exponents to divide monomials.

■   Divide polynomials by monomials.

To develop an effective process for dividing by a monomial, we must rely on yet another property of exponents. This property is also a direct consequence of the definition of exponent and is illustrated by the following examples.

$$\frac{x^5}{x^2} = \frac{x \cdot x \cdot x \cdot x \cdot x}{x \cdot x} = x^3$$

$$\frac{a^4}{a^3} = \frac{a \cdot a \cdot a \cdot a}{a \cdot a \cdot a} = a$$

$$\frac{y^7}{y^3} = \frac{\cancel{y} \cdot \cancel{y} \cdot \cancel{y} \cdot y \cdot y \cdot y \cdot y}{\cancel{y} \cdot \cancel{y} \cdot \cancel{y}} = y^4$$

$$\frac{x^4}{x^4} = \frac{\cancel{x} \cdot \cancel{x} \cdot \cancel{x} \cdot \cancel{x}}{\cancel{x} \cdot \cancel{x} \cdot \cancel{x} \cdot \cancel{x}} = 1$$

$$\frac{y^3}{y^3} = \frac{\cancel{y} \cdot \cancel{y} \cdot \cancel{y}}{\cancel{y} \cdot \cancel{y} \cdot \cancel{y}} = 1$$

PROPERTY 6.4

If b is any nonzero real number, and n and m are positive integers, then

1. $\dfrac{b^n}{b^m} = b^{n-m}$ when $n > m$

2. $\dfrac{b^n}{b^m} = 1$ when $n = m$

(The situation when $n < m$ is discussed in a later section.)

Applying Property 6.4 to the previous examples yields these results:

$$\frac{x^5}{x^2} = x^{5-2} = x^3$$

$$\frac{a^4}{a^3} = a^{4-3} = a^1 \qquad \text{Usually written as } a$$

$$\frac{y^7}{y^3} = y^{7-3} = y^4$$

$$\frac{x^4}{x^4} = 1$$

$$\frac{y^3}{y^3} = 1$$

Property 6.4, along with our knowledge of dividing integers, provides the basis for dividing a monomial by another monomial. Consider the next examples.

$$\frac{16x^5}{2x^3} = 8x^{5-3} = 8x^2 \qquad \qquad \frac{-81a^{12}}{-9a^4} = 9a^{12-4} = 9a^8$$

$$\frac{-35x^9}{5x^4} = -7x^{9-4} = -7x^5 \qquad \frac{45x^4}{9x^4} = 5 \qquad \frac{x^4}{x^4} = 1$$

$$\frac{56y^6}{-7y^2} = -8y^{6-2} = -8y^4 \qquad \frac{54x^3y^7}{-6xy^5} = -9x^{3-1}y^{7-5} = -9x^2y^2$$

Recall that $\dfrac{a}{c} + \dfrac{b}{c} = \dfrac{a+b}{c}$; this same property $\left(\text{except viewed as } \dfrac{a+b}{c} = \dfrac{a}{c} + \dfrac{b}{c}\right)$ serves as the basis for dividing a polynomial by a monomial. Consider these examples:

$$\frac{25x^3 + 10x^2}{5x} = \frac{25x^3}{5x} + \frac{10x^2}{5x} \qquad \frac{a+b}{c} = \frac{a}{c} + \frac{b}{c}$$

$$= 5x^2 + 2x$$

$$\frac{-35x^8 - 28x^6}{7x^3} = \frac{-35x^8}{7x^3} - \frac{28x^6}{7x^3} \qquad \frac{a-b}{c} = \frac{a}{c} - \frac{b}{c}$$

$$= -5x^5 - 4x^3$$

To divide a polynomial by a monomial, we simply divide each term of the polynomial by the monomial. Here are some additional examples:

$$\frac{12x^3y^2 - 14x^2y^5}{-2xy} = \frac{12x^3y^2}{-2xy} - \frac{14x^2y^5}{-2xy} = -6x^2y + 7xy^4$$

$$\frac{48ab^5 + 64a^2b}{-16ab} = \frac{48ab^5}{-16ab} + \frac{64a^2b}{-16ab} = -3b^4 - 4a$$

$$\frac{33x^6 - 24x^5 - 18x^4}{3x} = \frac{33x^6}{3x} - \frac{24x^5}{3x} - \frac{18x^4}{3x}$$

$$= 11x^5 - 8x^4 - 6x^3$$

As with many skills, once you feel comfortable with the process, you may want to perform some of the steps mentally. Your work could take on the following format.

$$\frac{24x^4y^5 - 56x^3y^9}{8x^2y^3} = 3x^2y^2 - 7xy^6$$

$$\frac{13a^2b - 12ab^2}{-ab} = -13a + 12b$$

CONCEPT QUIZ

For Problems 1–5, answer true or false.

1. When dividing factors with the same base, add the exponents.

2. $\dfrac{10a^6}{2a^2} = 8a^4$

3. $\dfrac{y^8}{y^4} = y^2$

4. $\dfrac{6x^5 + 3x}{3x} = 2x^4$

5. $\dfrac{x^3}{x^3} = 0$

PROBLEM SET 6.4

For Problems 1–24, divide the monomials.

1. $\dfrac{x^{10}}{x^2}$

2. $\dfrac{x^{12}}{x^5}$

3. $\dfrac{4x^3}{2x}$

4. $\dfrac{8x^5}{4x^3}$

5. $\dfrac{-16n^6}{2n^2}$

6. $\dfrac{-54n^8}{6n^4}$

7. $\dfrac{72x^3}{-9x^3}$

8. $\dfrac{84x^5}{-7x^5}$

9. $\dfrac{65x^2y^3}{5xy}$

10. $\dfrac{70x^3y^4}{5x^2y}$

11. $\dfrac{-91a^4b^6}{-13a^3b^4}$

12. $\dfrac{-72a^5b^4}{-12ab^2}$

13. $\dfrac{18x^2y^6}{xy^2}$

14. $\dfrac{24x^3y^4}{x^2y^2}$

15. $\dfrac{32x^6y^2}{-x}$

16. $\dfrac{54x^5y^3}{-y^2}$

17. $\dfrac{-96x^5y^7}{12y^3}$

18. $\dfrac{-84x^4y^9}{14x^4}$

19. $\dfrac{-ab}{ab}$

20. $\dfrac{6ab}{-ab}$

21. $\dfrac{56a^2b^3c^5}{4abc}$

22. $\dfrac{60a^3b^2c}{15a^2c}$

23. $\dfrac{-80xy^2z^6}{-5xyz^2}$

24. $\dfrac{-90x^3y^2z^8}{-6xy^2z^4}$

33. $\dfrac{-24n^8 + 48n^5 - 78n^3}{-6n^3}$

34. $\dfrac{-56n^9 + 84n^6 - 91n^2}{-7n^2}$

35. $\dfrac{-60a^7 - 96a^3}{-12a}$

36. $\dfrac{-65a^8 - 78a^4}{-13a^2}$

37. $\dfrac{27x^2y^4 - 45xy^4}{-9xy^3}$

38. $\dfrac{-40x^4y^7 + 64x^5y^8}{-8x^3y^4}$

39. $\dfrac{48a^2b^2 + 60a^3b^4}{-6ab}$

40. $\dfrac{45a^3b^4 - 63a^2b^6}{-9ab^2}$

41. $\dfrac{12a^2b^2c^2 - 52a^2b^3c^5}{-4a^2bc}$

42. $\dfrac{48a^3b^2c + 72a^2b^4c^5}{-12ab^2c}$

43. $\dfrac{9x^2y^3 - 12x^3y^4}{-xy}$

44. $\dfrac{-15x^3y + 27x^2y^4}{xy}$

45. $\dfrac{-42x^6 - 70x^4 + 98x^2}{14x^2}$

46. $\dfrac{-48x^8 - 80x^6 + 96x^4}{16x^4}$

47. $\dfrac{15a^3b - 35a^2b - 65ab^2}{-5ab}$

48. $\dfrac{-24a^4b^2 + 36a^3b - 48a^2b}{-6ab}$

49. $\dfrac{-xy + 5x^2y^3 - 7x^2y^6}{xy}$

50. $\dfrac{-9x^2y^3 - xy + 14xy^4}{-xy}$

For Problems 25–50, perform each division of polynomials by monomials.

25. $\dfrac{8x^4 + 12x^5}{2x^2}$

26. $\dfrac{12x^3 + 16x^6}{4x}$

27. $\dfrac{9x^6 - 24x^4}{3x^3}$

28. $\dfrac{35x^8 - 45x^6}{5x^4}$

29. $\dfrac{-28n^5 + 36n^2}{4n^2}$

30. $\dfrac{-42n^6 + 54n^4}{6n^4}$

31. $\dfrac{35x^6 - 56x^5 - 84x^3}{7x^2}$

32. $\dfrac{27x^7 - 36x^5 - 45x^3}{3x}$

■ ■ ■ **Thoughts into words**

51. How would you explain to someone why the quotient of x^8 and x^2 is x^6 and not x^4?

52. Your friend is having difficulty with problems such as $\dfrac{12x^2y}{xy}$ and $\dfrac{36x^3y^2}{-xy}$ where there appears to be no numerical coefficient in the denominator. What can you tell him that might help?

6.5   Dividing by Binomials

Objective

■   Divide polynomials by binomials.

Perhaps the easiest way to explain the process of dividing a polynomial by a binomial is to work out a few examples and describe the step-by-step procedure as we go along.

EXAMPLE 1   Divide $x^2 + 5x + 6$ by $x + 2$.

Solution

STEP 1   Use the conventional long division format from arithmetic, and arrange both the dividend and the divisor in descending powers of the variable.

$x + 2 \overline{)x^2 + 5x + 6}$

STEP 2   Find the first term of the quotient by dividing the first term of the dividend by the first term of the divisor.

$x + 2 \overline{)x^2 + 5x + 6}$ $\dfrac{x}{}$ $\dfrac{x^2}{x} = x$

STEP 3   Multiply the entire divisor by the term of the quotient found in step 2, and position this product to be subtracted from the dividend.

$x + 2 \overline{)x^2 + 5x + 6}$ $x(x + 2) =$
$\quad\quad x^2 + 2x$ $x^2 + 2x$

STEP 4   Subtract.
Remember to add the opposite!

$x + 2 \overline{)x^2 + 5x + 6}$
$\quad\quad x^2 + 2x$
$\quad\quad\quad 3x + 6$

STEP 5   Repeat the process beginning with step 2; use the polynomial that resulted from the subtraction in step 4 as a new dividend.

$x + 2 \overline{)x^2 + 5x + 6}$ $\dfrac{3x}{x} = 3$
$\quad x + 3$
$\quad\quad x^2 + 2x$
$\quad\quad\quad 3x + 6$ $3(x + 2) =$
$\quad\quad\quad 3x + 6$ $3x + 6$

Thus $(x^2 + 5x + 6) \div (x + 2) = x + 3$, which can be checked by multiplying $(x + 2)$ and $(x + 3)$.

$$(x + 2)(x + 3) = x^2 + 5x + 6$$ ■

A division problem such as $(x^2 + 5x + 6) \div (x + 2)$ can also be written as $\dfrac{x^2 + 5x + 6}{x + 2}$. Using this format, we can express the final result for Example 1 as $\dfrac{x^2 + 5x + 6}{x + 2} = x + 3$. (Technically the restriction $x \neq -2$ should be made to avoid division by zero.)

In general, to check a division problem we can multiply the divisor times the quotient and add the remainder. This can be expressed as

Dividend = (Divisor)(Quotient) + Remainder

Sometimes the remainder is expressed as a fractional part of the divisor. The relationship then becomes

$$\frac{\text{Dividend}}{\text{Divisor}} = \text{Quotient} + \frac{\text{Remainder}}{\text{Divisor}}$$

EXAMPLE 2   Divide $2x^2 - 3x - 20$ by $x - 4$.

Solution

STEP 1 $x - 4 \overline{)2x^2 - 3x - 20}$

STEP 2
$$x - 4 \overline{)\begin{array}{l} 2x \\ 2x^2 - 3x - 20 \end{array}} \qquad \frac{2x^2}{x} = 2x$$

STEP 3
$$x - 4 \overline{)\begin{array}{l} 2x \\ 2x^2 - 3x - 20 \\ 2x^2 - 8x \end{array}} \qquad 2x(x - 4) = 2x^2 - 8x$$

STEP 4
$$x - 4 \overline{)\begin{array}{l} 2x \\ 2x^2 - 3x - 20 \\ 2x^2 - 8x \\ 5x - 20 \end{array}}$$

STEP 5
$$x - 4 \overline{)\begin{array}{l} 2x + 5 \\ 2x^2 - 3x - 20 \\ 2x^2 - 8x \\ 5x - 20 \\ 5x - 20 \end{array}} \qquad \frac{5x}{x} = 5$$
$$5(x - 4) = 5x - 20$$

 Check

$$(x - 4)(2x + 5) = 2x^2 - 3x - 20$$

Therefore, $\dfrac{2x^2 - 3x - 20}{x - 4} = 2x + 5.$

Now let's continue to think in terms of the step-by-step division process but organize our work in the typical long-division format.

E X A M P L E   3   Divide $12x^2 + x - 6$ by $3x - 2$.

Solution

$$
\begin{array}{r}
4x + 3 \\
3x - 2 \overline{) 12x^2 + x - 6} \\
\underline{12x^2 - 8x} \\
9x - 6 \\
\underline{9x - 6}
\end{array}
$$

 Check

$$(3x - 2)(4x + 3) = 12x^2 + x - 6$$

Therefore, $\dfrac{12x^2 + x - 6}{3x - 2} = 4x + 3.$ ■

Each of the next three examples illustrates another aspect of the division process. Study them carefully; then you should be ready to work the exercises in the next problem set.

E X A M P L E   4   Perform the division $(7x^2 - 3x - 4) \div (x - 2)$.

Solution

$$
\begin{array}{r}
7x + 11 \\
x - 2 \overline{) 7x^2 - 3x - 4} \\
\underline{7x^2 - 14x} \\
11x - 4 \\
\underline{11x - 22} \\
18 \quad \longleftarrow \text{A remainder of 18}
\end{array}
$$

Check
Just as in arithmetic, we check by *adding* the remainder to the product of the divisor and quotient.

$$(x - 2)(7x + 11) + 18 \stackrel{?}{=} 7x^2 - 3x - 4$$
$$7x^2 - 3x - 22 + 18 \stackrel{?}{=} 7x^2 - 3x - 4$$
$$7x^2 - 3x - 4 = 7x^2 - 3x - 4$$

Therefore, $\dfrac{7x^2 - 3x - 4}{x - 2} = 7x + 11 + \dfrac{18}{x - 2}.$ ■

E X A M P L E   5   Perform the division $\dfrac{x^3 - 8}{x - 2}$.

Solution

$$
\begin{array}{r}
x^2 + 2x\ + 4 \\
x - 2\overline{)x^3 + 0x^2 + 0x - 8} \\
\underline{x^3 - 2x^2} \\
2x^2 + 0x - 8 \\
\underline{2x^2 - 4x} \\
4x - 8 \\
\underline{4x - 8}
\end{array}
$$

←—— Notice the insertion of x^2 and x terms with zero coefficients.

 Check

$$(x - 2)(x^2 + 2x + 4) \overset{?}{=} x^3 - 8$$
$$x^3 + 2x^2 + 4x - 2x^2 - 4x - 8 \overset{?}{=} x^3 - 8$$
$$x^3 - 8 = x^3 - 8$$

Therefore, $\dfrac{x^3 - 8}{x - 2} = x^2 + 2x + 4.$ ■

E X A M P L E   6   Perform the division $\dfrac{x^3 + 5x^2 - 3x - 4}{x^2 + 2x}$.

Solution

$$
\begin{array}{r}
x\ + 3 \\
x^2 + 2x\overline{)x^3 + 5x^2 - 3x - 4} \\
\underline{x^3 + 2x^2} \\
3x^2 - 3x - 4 \\
\underline{3x^2 + 6x} \\
-9x - 4
\end{array}
$$

←—— A remainder of $-9x - 4$

We stop the division process when the degree of the remainder is less than the degree of the divisor.

 Check

$$(x^2 + 2x)(x + 3) + (-9x - 4) \overset{?}{=} x^3 + 5x^2 - 3x - 4$$
$$x^3 + 3x^2 + 2x^2 + 6x - 9x - 4 \overset{?}{=} x^3 + 5x^2 - 3x - 4$$
$$x^3 + 5x^2 - 3x - 4 = x^3 + 5x^2 - 3x - 4$$

Therefore, $\dfrac{x^3 + 5x^2 - 3x - 4}{x^2 + 2x} = x + 3 + \dfrac{-9x - 4}{x^2 + 2x}.$ ■

CONCEPT QUIZ

For Problems 1–6, answer true or false.

1. A division problem written as $(x^2 - x - 6) \div (x - 1)$ could also be written as $\dfrac{x^2 - x - 6}{x - 1}$.

2. The division of $\dfrac{x^2 + 7x + 12}{x + 3} = x + 4$ could be checked by multiplying $(x + 4)$ by $(x + 3)$.

3. For the division problem $(2x^2 + 5x + 9) \div (2x + 1)$ the remainder is 7. The remainder for the division problem can be expressed as $\dfrac{7}{2x + 1}$.

4. In general, to check a division problem we can multiply the divisor times the quotient and subtract the remainder.

5. If a term is inserted to act as a placeholder, then the coefficient of the term must be zero.

6. When performing division, the process ends when the degree of the remainder is less than the degree of the divisor.

PROBLEM SET 6.5

For Problems 1–40, perform the divisions.

1. $(x^2 + 16x + 48) \div (x + 4)$

2. $(x^2 + 15x + 54) \div (x + 6)$

3. $(x^2 - 5x - 14) \div (x - 7)$

4. $(x^2 + 8x - 65) \div (x - 5)$

5. $(x^2 + 11x + 28) \div (x + 3)$

6. $(x^2 + 11x + 15) \div (x + 2)$

7. $(x^2 - 4x - 39) \div (x - 8)$

8. $(x^2 - 9x - 30) \div (x - 12)$

9. $(5n^2 - n - 4) \div (n - 1)$

10. $(7n^2 - 61n - 90) \div (n - 10)$

11. $(8y^2 + 53y - 19) \div (y + 7)$

12. $(6y^2 + 47y - 72) \div (y + 9)$

13. $(20x^2 - 31x - 7) \div (5x + 1)$

14. $(27x^2 + 21x - 20) \div (3x + 4)$

15. $(6x^2 + 25x + 8) \div (2x + 7)$

16. $(12x^2 + 28x + 27) \div (6x + 5)$

17. $(2x^3 - x^2 - 2x - 8) \div (x - 2)$

18. $(3x^3 - 7x^2 - 26x + 24) \div (x - 4)$

19. $(5n^3 + 11n^2 - 15n - 9) \div (n + 3)$

20. $(6n^3 + 29n^2 - 6n - 5) \div (n + 5)$

21. $(n^3 - 40n + 24) \div (n - 6)$

22. $(n^3 - 67n - 24) \div (n + 8)$

23. $(x^3 - 27) \div (x - 3)$

24. $(x^3 + 8) \div (x + 2)$

25. $\dfrac{27x^3 - 64}{3x - 4}$

26. $\dfrac{8x^3 + 27}{2x + 3}$

27. $\dfrac{1 + 3n^2 - 2n}{n + 2}$

28. $\dfrac{x + 5 + 12x^2}{3x - 2}$

29. $\dfrac{9t^2 + 3t + 4}{-1 + 3t}$

30. $\dfrac{4n^2 + 6n - 1}{4 + 2n}$

31. $\dfrac{6n^3 - 5n^2 - 7n + 4}{2n - 1}$

32. $\dfrac{21n^3 + 23n^2 - 9n - 10}{3n + 2}$

33. $\dfrac{4x^3 + 23x^2 - 30x + 32}{x + 7}$

34. $\dfrac{5x^3 - 12x^2 + 13x - 14}{x - 1}$

35. $(x^3 + 2x^2 - 3x - 1) \div (x^2 - 2x)$

36. $(x^3 - 6x^2 - 2x + 1) \div (x^2 + 3x)$

37. $(2x^3 - 4x^2 + x - 5) \div (x^2 + 4x)$

38. $(2x^3 - x^2 - 3x + 5) \div (x^2 + x)$

39. $(x^4 - 16) \div (x + 2)$

40. $(x^4 - 81) \div (x - 3)$

■ ■ ■ **Thoughts into words**

41. Give a step-by-step description of how you would do the division problem $(2x^3 + 8x^2 - 29x - 30) \div (x + 6)$.

42. How do you know by inspection that the answer to the following division problem is incorrect?

$$(3x^3 - 7x^2 - 22x + 8) \div (x - 4) = 3x^2 + 5x + 1$$

Answers to Concept Quiz

1. True   **2.** True   **3.** True   **4.** False   **5.** True   **6.** True

---

6.6   Zero and Negative Integers as Exponents

Objectives

■   Apply the properties of exponents including negative and zero exponents.

■   Write numbers in scientific notation.

■   Write numbers expressed in scientific notation in standard decimal notation.

■   Use scientific notation to evaluate numerical expressions.

Thus far in this text, we have used only positive integers as exponents. The next definition and properties serve as a basis for our work with exponents.

DEFINITION 6.1

If n is a positive integer and b is any real number, then

$$b^n = \underbrace{bbb \cdots b}_{n \text{ factors of } b}$$

PROPERTY 6.5

If m and n are positive integers and a and b are real numbers, except $b \neq 0$ whenever it appears in a denominator, then

1. $b^n \cdot b^m = b^{n+m}$

2. $(b^n)^m = b^{mn}$

3. $(ab)^n = a^n b^n$

4. $\left(\dfrac{a}{b}\right)^n = \dfrac{a^n}{b^n}$ Part 4 has not been stated previously.

5. $\dfrac{b^n}{b^m} = b^{n-m}$ When $n > m$

   $\dfrac{b^n}{b^m} = 1$ When $n = m$

Property 6.5 pertains to the use of positive integers as exponents. Zero and the negative integers can also be used as exponents. First, let's consider the use of 0 as an exponent. We want to use 0 as an exponent in such a way that the basic properties of exponents will continue to hold. Consider the example $x^4 \cdot x^0$. If part 1 of Property 6.5 is to hold, then

$$x^4 \cdot x^0 = x^{4+0} = x^4$$

Note that x^0 *acts like* 1 because $x^4 \cdot x^0 = x^4$. This suggests the following definition:

DEFINITION 6.2

If b is a nonzero real number, then

$b^0 = 1$

According to Definition 6.2, the following statements are all true.

$4^0 = 1$

$(-628)^0 = 1$

$\left(\dfrac{4}{7}\right)^0 = 1$

$n^0 = 1, \qquad n \neq 0$

$(x^2 y^5)^0 = 1, \qquad x \neq 0 \text{ and } y \neq 0$

A similar line of reasoning indicates how negative integers should be used as exponents. Consider the example $x^3 \cdot x^{-3}$. If part 1 of Property 6.5 is to hold, then

$$x^3 \cdot x^{-3} = x^{3+(-3)} = x^0 = 1$$

Thus x^{-3} must be the reciprocal of x^3 because their product is 1; that is,

$$x^{-3} = \frac{1}{x^3}$$

This process suggests the following definition:

DEFINITION 6.3

If n is a positive integer and b is a nonzero real number, then

$$b^{-n} = \frac{1}{b^n}$$

According to Definition 6.3, the following statements are all true.

$$x^{-6} = \frac{1}{x^6}$$

$$2^{-3} = \frac{1}{2^3} = \frac{1}{8}$$

$$10^{-2} = \frac{1}{10^2} = \frac{1}{100} \qquad \text{or} \qquad 0.01$$

$$\frac{1}{x^{-4}} = \frac{1}{\frac{1}{x^4}} = x^4$$

$$\left(\frac{2}{3}\right)^{-2} = \frac{1}{\left(\frac{2}{3}\right)^2} = \frac{1}{\frac{4}{9}} = \frac{9}{4}$$

REMARK: Note in the last example that $\left(\frac{2}{3}\right)^{-2} = \left(\frac{3}{2}\right)^2$. In other words, to raise a fraction to a negative power, we can invert the fraction and raise it to the corresponding positive power.

We can verify (we will not do so in this text) that all parts of Property 6.5 hold for *all integers*. In fact, we can replace part 5 with this statement.

Replacement for part 5 of Property 6.5

$$\frac{b^n}{b^m} = b^{n-m} \quad \text{for all integers } n \text{ and } m$$

The next examples illustrate the use of this new property. In each example, we have simplified the original expression and used only positive exponents in the final result.

$$\frac{x^2}{x^5} = x^{2-5} = x^{-3} = \frac{1}{x^3}$$

$$\frac{a^{-3}}{a^{-7}} = a^{-3-(-7)} = a^{-3+7} = a^4$$

$$\frac{y^{-5}}{y^{-2}} = y^{-5-(-2)} = y^{-5+2} = y^{-3} = \frac{1}{y^3}$$

$$\frac{x^{-6}}{x^{-6}} = x^{-6-(-6)} = x^{-6+6} = x^0 = 1$$

The properties of exponents provide a basis for simplifying certain types of numerical expressions, as the following examples illustrate.

$$2^{-4} \cdot 2^6 = 2^{-4+6} = 2^2 = 4$$

$$10^5 \cdot 10^{-6} = 10^{5+(-6)} = 10^{-1} = \frac{1}{10} \qquad \text{or} \qquad 0.1$$

$$\frac{10^2}{10^{-2}} = 10^{2-(-2)} = 10^{2+2} = 10^4 = 10,000$$

$$(2^{-3})^{-2} = 2^{-3(-2)} = 2^6 = 64$$

Having the use of all integers as exponents also expands the type of work that we can do with algebraic expressions. In each of the following examples, we have simplified a given expression and used only positive exponents in the final result.

$$x^8 x^{-2} = x^{8+(-2)} = x^6$$

$$a^{-4} a^{-3} = a^{-4+(-3)} = a^{-7} = \frac{1}{a^7}$$

$$(y^{-3})^4 = y^{-3(4)} = y^{-12} = \frac{1}{y^{12}}$$

$$(x^{-2} y^4)^{-3} = (x^{-2})^{-3}(y^4)^{-3} = x^6 y^{-12} = \frac{x^6}{y^{12}}$$

$$\left(\frac{x^{-1}}{y^2}\right)^{-2} = \frac{(x^{-1})^{-2}}{(y^2)^{-2}} = \frac{x^2}{y^{-4}} = x^2 y^4$$

$$(4x^{-2})(3x^{-1}) = 12x^{-2+(-1)} = 12x^{-3} = \frac{12}{x^3}$$

$$\left(\frac{12x^{-6}}{6x^{-2}}\right)^{-2} = (2x^{-6-(-2)})^{-2} = (2x^{-4})^{-2}$$

$$= (2)^{-2}(x^{-4})^{-2}$$

$$= \left(\frac{1}{2^2}\right)(x^8) = \frac{x^8}{4}$$

Scientific Notation

Many scientific applications of mathematics involve the use of very large and very small numbers. For example:

The speed of light is approximately 29,979,200,000 centimeters per second.

A light year — the distance light travels in 1 year — is approximately 5,865,696,000,000 miles.

A gigahertz equals 1,000,000,000 hertz.

The length of a typical virus cell equals 0.000000075 of a meter.

The length of a diameter of a water molecule is 0.0000000003 of a meter.

Working with numbers of this type in standard form is quite cumbersome. It is much more convenient to represent very small and very large numbers in **scientific notation,** sometimes called scientific form. A number is in scientific notation when it is written as the product of a number between 1 and 10 (including 1) and an integral power of 10. Symbolically, a number in scientific notation has the form $(N)(10^k)$, where $1 \leq N < 10$ and k is an integer. For example, 621 can be written as $(6.21)(10^2)$ and 0.0023 can be written as $(2.3)(10^{-3})$.

To switch from ordinary notation to scientific notation, we can use the following procedure.

Write the given number as the product of a number greater than or equal to 1 and less than 10, and an integral power of 10. To determine the exponent of 10, count the number of places that the decimal point moved when going from the original number to the number between 1 and 10. This exponent is (a) negative if the original number is less than 1, (b) positive if the original number is greater than 10, and (c) zero if the original number itself is between 1 and 10.

Thus we can write

$0.000179 = (1.79)(10^{-4})$ According to part (a) of the procedure

$8175 = (8.175)(10^3)$ According to part (b)

$3.14 = (3.14)(10^0)$ According to part (c)

We can express the applications given earlier in scientific notation as follows:

Speed of light: $29,979,200,000 = (2.99792)(10^{10})$ centimeters per second

Light year: $5,865,696,000,000 = (5.865696)(10^{12})$ miles

Gigahertz: $1,000,000,000 = (1)(10^9)$ hertz

Length of a virus cell: $0.000000075 = (7.5)(10^{-8})$ meter

Diameter of a water molecule: $0.0000000003 = (3)(10^{-10})$ meter

To switch from scientific notation to ordinary decimal notation, we can use the following procedure:

Move the decimal point the number of places indicated by the exponent of 10. The decimal point is moved to the right if the exponent is positive and to the left if it is negative.

Thus we can write

$(4.71)(10^4) = 47,100$ Two zeros are needed for place value purposes.

$(1.78)(10^{-2}) = 0.0178$ One zero is needed for place value purposes.

The use of scientific notation along with the properties of exponents can make some arithmetic problems much easier to evaluate. The next examples illustrate this point.

E X A M P L E   1

Evaluate $(4000)(0.000012)$.

Solution

$$
\begin{aligned}
(4000)(0.000012) &= (4)(10^3)(1.2)(10^{-5}) \\
&= (4)(1.2)(10^3)(10^{-5}) \\
&= (4.8)(10^{-2}) \\
&= 0.048
\end{aligned}
$$

■

E X A M P L E   2

Evaluate $\dfrac{960,000}{0.032}$.

Solution

$$
\begin{aligned}
\frac{960,000}{0.032} &= \frac{(9.6)(10^5)}{(3.2)(10^{-2})} \\
&= (3)(10^7) \qquad \frac{10^5}{10^{-2}} = 10^{5-(-2)} = 10^7 \\
&= 30,000,000
\end{aligned}
$$

■

E X A M P L E   3

Evaluate $\dfrac{(6000)(0.00008)}{(40,000)(0.006)}$.

Solution

$$
\begin{aligned}
\frac{(6000)(0.00008)}{(40,000)(0.006)} &= \frac{(6)(10^3)(8)(10^{-5})}{(4)(10^4)(6)(10^{-3})} \\
&= \frac{(48)(10^{-2})}{(24)(10^1)} \\
&= (2)(10^{-3}) \qquad \frac{10^{-2}}{10^1} = 10^{-2-1} = 10^{-3} \\
&= 0.002
\end{aligned}
$$

■

CONCEPT QUIZ

For Problems 1–6, answer true or false.

1. Any nonzero number raised to the zero power is equal to one.

2. The algebraic expression x^{-2} is the reciprocal of x^2 for $x \neq 0$.

3. To raise a fraction to a negative exponent, we can invert the fraction and raise it to the corresponding positive exponent.

4. $\dfrac{1}{y^{-3}} = y^{-3}$

5. A number in scientific notation has the form $(N)(10^k)$ where $1 \le N < 10$, and k is any real number.

6. A number is less than zero if the exponent is negative when the number is written in scientific notation.

PROBLEM SET 6.6

For Problems 1–30, evaluate each numerical expression.

1. 3^{-2}

2. 2^{-5}

3. 4^{-3}

4. 5^{-2}

5. $\left(\dfrac{3}{2}\right)^{-1}$

6. $\left(\dfrac{3}{4}\right)^{-2}$

7. $\dfrac{1}{2^{-4}}$

8. $\dfrac{1}{3^{-1}}$

9. $\left(-\dfrac{4}{3}\right)^{0}$

10. $\left(-\dfrac{1}{2}\right)^{-3}$

11. $\left(-\dfrac{2}{3}\right)^{-3}$

12. $(-16)^0$

13. $(-2)^{-2}$

14. $(-3)^{-2}$

15. $-(3^{-2})$

16. $-(2^{-2})$

17. $\dfrac{1}{\left(\dfrac{3}{4}\right)^{-3}}$

18. $\dfrac{1}{\left(\dfrac{3}{2}\right)^{-4}}$

19. $2^6 \cdot 2^{-9}$

20. $3^5 \cdot 3^{-2}$

21. $3^6 \cdot 3^{-3}$

22. $2^{-7} \cdot 2^2$

23. $\dfrac{10^2}{10^{-1}}$

24. $\dfrac{10^1}{10^{-3}}$

25. $\dfrac{10^{-1}}{10^2}$

26. $\dfrac{10^{-2}}{10^{-2}}$

27. $(2^{-1} \cdot 3^{-2})^{-1}$

28. $(3^{-1} \cdot 4^{-2})^{-1}$

29. $\left(\dfrac{4^{-1}}{3}\right)^{-2}$

30. $\left(\dfrac{3}{2^{-1}}\right)^{-3}$

For Problems 31–84, simplify each algebraic expression and express your answers using positive exponents only.

31. $x^6 x^{-1}$

32. $x^{-2} x^7$

33. $n^{-4} n^2$

34. $n^{-8} n^3$

35. $a^{-2} a^{-3}$

36. $a^{-4} a^{-6}$

37. $(2x^3)(4x^{-2})$

38. $(5x^{-4})(6x^7)$

39. $(3x^{-6})(9x^2)$

40. $(8x^{-8})(4x^2)$

41. $(5y^{-1})(-3y^{-2})$

42. $(-7y^{-3})(9y^{-4})$

43. $(8x^{-4})(12x^4)$

44. $(-3x^{-2})(-6x^2)$

45. $\dfrac{x^7}{x^{-3}}$

46. $\dfrac{x^2}{x^{-4}}$

47. $\dfrac{n^{-1}}{n^3}$

48. $\dfrac{n^{-2}}{n^5}$

49. $\dfrac{4n^{-1}}{2n^{-3}}$

50. $\dfrac{12n^{-2}}{3n^{-5}}$

51. $\dfrac{-24x^{-6}}{8x^{-2}}$

52. $\dfrac{56x^{-5}}{-7x^{-1}}$

53. $\dfrac{-52y^{-2}}{-13y^{-2}}$

54. $\dfrac{-91y^{-3}}{-7y^{-3}}$

55. $(x^{-3})^{-2}$

56. $(x^{-1})^{-5}$

57. $(x^2)^{-2}$

58. $(x^3)^{-1}$

59. $(x^3 y^4)^{-1}$

60. $(x^4 y^{-2})^{-2}$

61. $(x^{-2} y^{-1})^3$

62. $(x^{-3} y^{-4})^2$

63. $(2n^{-2})^3$

64. $(3n^{-1})^4$

65. $(4n^3)^{-2}$

66. $(2n^2)^{-3}$

67. $(3a^{-2})^4$

68. $(5a^{-1})^2$

69. $(5x^{-1})^{-2}$

70. $(4x^{-2})^{-2}$

71. $(2x^{-2} y^{-1})^{-1}$

72. $(3x^2 y^{-3})^{-2}$

73. $\left(\dfrac{x^2}{y}\right)^{-1}$

74. $\left(\dfrac{y^2}{x^3}\right)^{-2}$

75. $\left(\dfrac{a^{-1}}{b^2}\right)^{-4}$

76. $\left(\dfrac{a^3}{b^{-2}}\right)^{-3}$

77. $\left(\dfrac{x^{-1}}{y^{-3}}\right)^{-2}$

78. $\left(\dfrac{x^{-3}}{y^{-4}}\right)^{-1}$

79. $\left(\dfrac{x^2}{x^3}\right)^{-1}$

80. $\left(\dfrac{x^4}{x}\right)^{-2}$

81. $\left(\dfrac{2x^{-1}}{x^{-2}}\right)^{-3}$

82. $\left(\dfrac{3x^{-2}}{x^{-5}}\right)^{-1}$

83. $\left(\dfrac{18x^{-1}}{9x}\right)^{-2}$

84. $\left(\dfrac{35x^2}{7x^{-1}}\right)^{-1}$

For Problems 85–94, write each number in scientific notation; for example, $786 = (7.86)(10^2)$.

85. 321

86. 74

87. 8000

88. 500

89. 0.00246

90. 0.017

91. 0.0000179

92. 0.00000049

93. 87,000,000

94. 623,000,000,000

For Problems 95–106, write each number in standard decimal form; for example, $(1.4)(10^3) = 1400$.

95. $(8)(10^3)$

96. $(6)(10^2)$

97. $(5.21)(10^4)$

98. $(7.2)(10^3)$

99. $(1.14)(10^7)$

100. $(5.64)(10^8)$

101. $(7)(10^{-2})$

102. $(8.14)(10^{-1})$

103. $(9.87)(10^{-4})$

104. $(4.37)(10^{-5})$

105. $(8.64)(10^{-6})$

106. $(3.14)(10^{-7})$

For Problems 107–118, use scientific notation and the properties of exponents to evaluate each numerical expression.

107. $(0.007)(120)$

108. $(0.0004)(13)$

109. $(5,000,000)(0.00009)$

110. $(800,000)(0.0000006)$

111. $\dfrac{6000}{0.0015}$

112. $\dfrac{480}{0.012}$

113. $\dfrac{0.00086}{4300}$

114. $\dfrac{0.0057}{30,000}$

115. $\dfrac{0.00039}{0.0013}$

116. $\dfrac{0.0000082}{0.00041}$

117. $\dfrac{(0.0008)(0.07)}{(20,000)(0.0004)}$

118. $\dfrac{(0.006)(600)}{(0.00004)(30)}$

■■■ Thoughts into words

119. Is the following simplification process correct?

$$(2^{-2})^{-1} = \left(\frac{1}{2^2}\right)^{-1} = \left(\frac{1}{4}\right)^{-1} = \frac{1}{\left(\frac{1}{4}\right)^1} = 4$$

Can you suggest a better way to do the problem?

120. Explain the importance of scientific notation.

■■■ Further investigations

121. Use your calculator to redo Problems 1–16. Be sure that your answers are equivalent to the answers you obtained without the calculator.

122. Use your calculator to evaluate $(140,000)^2$. Your answer should be displayed in scientific notation; the format of the display depends on the particular calculator. For example, it may look like $\boxed{1.96 \quad 10}$ or $\boxed{1.96E + 10}$. Thus in ordinary notation, the answer is 19,600,000,000. Use your calculator to evaluate each expression. Express final answers in ordinary notation.

a. $(9000)^3$

b. $(4000)^3$

c. $(150,000)^2$

d. $(170,000)^2$

e. $(0.012)^5$

f. $(0.0015)^4$

g. $(0.006)^3$

h. $(0.02)^6$

123. Use your calculator to check your answers to Problems 107–118.

Answers to Concept Quiz

1. True   **2.** True   **3.** True   **4.** False   **5.** False   **6.** False

CHAPTER 6

SUMMARY

(6.1) Terms that contain variables with only whole numbers as exponents are called **monomials.** A **polynomial** is a monomial or a finite sum (or difference) of monomials. Polynomials of one term, two terms, and three terms are called **monomials, binomials,** and **trinomials,** respectively.

Addition and subtraction of polynomials are based on using the distributive property and combining similar terms.

(6.2) and **(6.3)** The following properties of exponents serve as a basis for multiplying polynomials.

1. $b^n \cdot b^m = b^{n+m}$

2. $(b^n)^m = b^{mn}$

3. $(ab)^n = a^n b^n$

(6.4) The following properties of exponents serve as a basis for dividing monomials.

1. $\dfrac{b^n}{b^m} = b^{n-m}$ when $n > m$

2. $\dfrac{b^n}{b^m} = 1$ when $n = m$

Dividing a polynomial by a monomial is based on the property

$$\frac{a + b}{c} = \frac{a}{c} + \frac{b}{c}.$$

(6.5) To review the division of a polynomial by a binomial, turn to Section 6.5 and study the examples carefully.

(6.6) We use the following two definitions to expand our work with exponents to include zero and the negative integers.

DEFINITION 6.2

If b is a nonzero real number, then

$$b^0 = 1$$

DEFINITION 6.3

If n is a positive integer, and b is a nonzero real number, then

$$b^{-n} = \frac{1}{b^n}$$

The following properties of exponents are true for all integers.

1. $b^n \cdot b^m = b^{n+m}$

2. $(b^n)^m = b^{mn}$

3. $(ab)^n = a^n b^n$

4. $\left(\dfrac{a}{b}\right)^n = \dfrac{a^n}{b^n}$ $b \neq 0$ whenever it appears in a denominator.

5. $\dfrac{b^n}{b^m} = b^{n-m}$

To represent a number in scientific notation, express it as the product of a number between 1 and 10 (including 1) and an integral power of 10.

CHAPTER 6 REVIEW PROBLEM SET

For Problems 1–4, perform the additions and subtractions.

1. $(5x^2 - 6x + 4) + (3x^2 - 7x - 2)$

2. $(7y^2 + 9y - 3) - (4y^2 - 2y + 6)$

3. $(2x^2 + 3x - 4) + (4x^2 - 3x - 6) - (3x^2 - 2x - 1)$

4. $(-3x^2 - 2x + 4) - (x^2 - 5x - 6) - (4x^2 + 3x - 8)$

For Problems 5–12, remove parentheses and combine similar terms.

5. $5(2x - 1) + 7(x + 3) - 2(3x + 4)$

6. $3(2x^2 - 4x - 5) - 5(3x^2 - 4x + 1)$

7. $6(y^2 - 7y - 3) - 4(y^2 + 3y - 9)$

8. $3(a - 1) - 2(3a - 4) - 5(2a + 7)$

9. $-(a + 4) + 5(-a - 2) - 7(3a - 1)$

10. $-2(3n - 1) - 4(2n + 6) + 5(3n + 4)$

11. $3(n^2 - 2n - 4) - 4(2n^2 - n - 3)$

12. $-5(-n^2 + n - 1) + 3(4n^2 - 3n - 7)$

For Problems 13–20, find the indicated products.

13. $(5x^2)(7x^4)$

14. $(-6x^3)(9x^5)$

15. $(-4xy^2)(-6x^2y^3)$

16. $(2a^3b^4)(-3ab^5)$

17. $(2a^2b^3)^3$

18. $(-3xy^2)^2$

19. $5x(7x + 3)$

20. $(-3x^2)(8x - 1)$

For Problems 21–40, find the indicated products. Be sure to simplify your answers.

21. $(x + 9)(x + 8)$

22. $(3x + 7)(x + 1)$

23. $(x - 5)(x + 2)$

24. $(y - 4)(y - 9)$

25. $(2x - 1)(7x + 3)$

26. $(4a - 7)(5a + 8)$

27. $(3a - 5)^2$

28. $(x + 6)(2x^2 + 5x - 4)$

29. $(5n - 1)(6n + 5)$

30. $(3n + 4)(4n - 1)$

31. $(2n + 1)(2n - 1)$

32. $(4n - 5)(4n + 5)$

33. $(2a + 7)^2$

34. $(3a + 5)^2$

35. $(x - 2)(x^2 - x + 6)$

36. $(2x - 1)(x^2 + 4x + 7)$

37. $(a + 5)^3$

38. $(a - 6)^3$

39. $(x^2 - x - 1)(x^2 + 2x + 5)$

40. $(n^2 + 2n + 4)(n^2 - 7n - 1)$

For Problems 41–48, perform the divisions.

41. $\dfrac{36x^4y^5}{-3xy^2}$

42. $\dfrac{-56a^5b^7}{-8a^2b^3}$

43. $\dfrac{-18x^4y^3 - 54x^6y^2}{6x^2y^2}$

44. $\dfrac{-30a^5b^{10} + 39a^4b^8}{-3ab}$

45. $\dfrac{56x^4 - 40x^3 - 32x^2}{4x^2}$

46. $(x^2 + 9x - 1) \div (x + 5)$

47. $(21x^2 - 4x - 12) \div (3x + 2)$

48. $(2x^3 - 3x^2 + 2x - 4) \div (x - 2)$

For Problems 49–60, evaluate each expression.

49. $3^2 + 2^2$

50. $(3 + 2)^2$

51. 2^{-4}

52. $(-5)^0$

53. -5^0

54. $\dfrac{1}{3^{-2}}$

55. $\left(\dfrac{3}{4}\right)^{-2}$

56. $\dfrac{1}{\left(\dfrac{1}{4}\right)^{-1}}$

57. $\dfrac{1}{(-2)^{-3}}$

58. $2^{-1} + 3^{-2}$

59. $3^0 + 2^{-2}$

60. $(2 + 3)^{-2}$

For Problems 61–72, simplify each of the following and express your answers using positive exponents only.

61. x^5x^{-8}

62. $(3x^5)(4x^{-2})$

63. $\dfrac{x^{-4}}{x^{-6}}$

64. $\dfrac{x^{-6}}{x^{-4}}$

65. $\dfrac{24a^5}{3a^{-1}}$

66. $\dfrac{48n^{-2}}{12n^{-1}}$

67. $(x^{-2}y)^{-1}$

68. $(a^2b^{-3})^{-2}$

69. $(2x)^{-1}$

70. $(3n^2)^{-2}$

71. $(2n^{-1})^{-3}$

72. $(4ab^{-1})(-3a^{-1}b^2)$

For Problems 73–76, write each expression in standard decimal form.

73. $(6.1)(10^2)$

74. $(5.6)(10^4)$

75. $(8)(10^{-2})$

76. $(9.2)(10^{-4})$

For Problems 77–80, write each number in scientific notation.

77. 9000

78. 47

79. 0.047

80. 0.00021

For Problems 81–84, use scientific notation and the properties of exponents to evaluate each expression.

81. $(0.00004)(12,000)$

82. $(0.0021)(2000)$

83. $\dfrac{0.0056}{0.0000028}$

84. $\dfrac{0.00078}{39,000}$

CHAPTER 6

TEST

⬛📟 **applies to all problems in this Chapter Test.**

1. Find the sum of $-7x^2 + 6x - 2$ and $5x^2 - 8x + 7$.

2. Subtract $-x^2 + 9x - 14$ from $-4x^2 + 3x + 6$.

3. Remove parentheses and combine similar terms for the expression $3(2x - 1) - 6(3x - 2) - (x + 7)$.

4. Find the product $(-4xy^2)(7x^2y^3)$.

5. Find the product $(2x^2y)^2(3xy^3)$.

For Problems 6–12, find the indicated products and express answers in simplest form.

6. $(x - 9)(x + 2)$

7. $(n + 14)(n - 7)$

8. $(5a + 3)(8a + 7)$

9. $(3x - 7y)^2$

10. $(x + 3)(2x^2 - 4x - 7)$

11. $(9x - 5y)(9x + 5y)$

12. $(3x - 7)(5x - 11)$

13. Find the indicated quotient: $\dfrac{-96x^4y^5}{-12x^2y}$.

14. Find the indicated quotient: $\dfrac{56x^2y - 72xy^2}{-8xy}$.

15. Find the indicated quotient:
    $(2x^3 + 5x^2 - 22x + 15) \div (2x - 3)$.

16. Find the indicated quotient:
    $(4x^3 + 23x^2 + 36) \div (x + 6)$.

17. Evaluate $\left(\dfrac{2}{3}\right)^{-3}$.

18. Evaluate $4^{-2} + 4^{-1} + 4^0$.

19. Evaluate $\dfrac{1}{2^{-4}}$.

20. Find the product $(-6x^{-4})(4x^2)$ and express the answer using a positive exponent.

21. Simplify $\left(\dfrac{8x^{-1}}{2x^2}\right)^{-1}$ and express the answer using a positive exponent.

22. Simplify $(x^{-3}y^5)^{-2}$ and express the answer using positive exponents.

23. Write 0.00027 in scientific notation.

24. Express $(9.2)(10^6)$ in standard decimal form.

25. Evaluate $(0.000002)(3000)$.

CUMULATIVE REVIEW PROBLEM SET *Chapters 1-6*

For Problems 1–10, evaluate each of the numerical expressions.

1. $5 + 3(2 - 7)^2 \div 3 \cdot 5$

2. $8 \div 2 \cdot (-1) + 3$ **3.** $7 - 2^2 \cdot 5 \div (-1)$

4. $4 + (-2) - 3(6)$ **5.** $(-3)^4$

6. -2^5 **7.** $\left(\dfrac{2}{3}\right)^{-1}$ **8.** $\dfrac{1}{4^{-2}}$

9. $\left(\dfrac{1}{2} - \dfrac{1}{3}\right)^{-2}$ **10.** $2^0 + 2^{-1} + 2^{-2}$

For Problems 11–16, evaluate each algebraic expression for the given values of the variables.

11. $\dfrac{2x + 3y}{x - y}$ for $x = \dfrac{1}{2}$ and $y = -\dfrac{1}{3}$

12. $\dfrac{2}{5}n - \dfrac{1}{3}n - n + \dfrac{1}{2}n$ for $n = -\dfrac{3}{4}$

13. $\dfrac{3a - 2b - 4a + 7b}{-a - 3a + b - 2b}$ for $a = -1$ and $b = -\dfrac{1}{3}$

14. $-2(x - 4) + 3(2x - 1) - (3x - 2)$ for $x = -2$

15. $(x^2 + 2x - 4) - (x^2 - x - 2) + (2x^2 - 3x - 1)$
for $x = -1$

16. $2(n^2 - 3n - 1) - (n^2 + n + 4) - 3(2n - 1)$
for $n = 3$

For Problems 17–29, find the indicated products.

17. $(3x^2y^3)(-5xy^4)$ **18.** $(-6ab^4)(-2b^3)$

19. $(-2x^2y^5)^3$ **20.** $-3xy(2x - 5y)$

21. $(5x - 2)(3x - 1)$ **22.** $(7x - 1)(3x + 4)$

23. $(-x - 2)(2x + 3)$ **24.** $(7 - 2y)(7 + 2y)$

25. $(x - 2)(3x^2 - x - 4)$

26. $(2x - 5)(x^2 + x - 4)$

27. $(2n + 3)^3$ **28.** $(1 - 2n)^3$

29. $(x^2 - 2x + 6)(2x^2 + 5x - 6)$

For Problems 30–34, perform the indicated divisions.

30. $\dfrac{-52x^3y^4}{13xy^2}$ **31.** $\dfrac{-126a^3b^5}{-9a^2b^3}$

32. $\dfrac{56xy^2 - 64x^3y - 72x^4y^4}{8xy}$

33. $(2x^3 + 2x^2 - 19x - 21) \div (x + 3)$

34. $(3x^3 + 17x^2 + 6x - 4) \div (3x - 1)$

For Problems 35–38, simplify each expression and express your answers using positive exponents only.

35. $(-2x^3)(3x^{-4})$ **36.** $\dfrac{4x^{-2}}{2x^{-1}}$

37. $(3x^{-1}y^{-2})^{-1}$ **38.** $(xy^2z^{-1})^{-2}$

For Problems 39–41, use scientific notation and the properties of exponents to help evaluate each numerical expression.

39. $(0.00003)(4000)$ **40.** $(0.0002)(0.003)^2$

41. $\dfrac{0.00034}{0.0000017}$

For Problems 42–49, solve each of the equations.

42. $5x + 8 = 6x - 3$

43. $-2(4x - 1) = -5x + 3 - 2x$

44. $\dfrac{y}{2} - \dfrac{y}{3} = 8$

45. $6x + 8 - 4x = 10(3x + 2)$

46. $1.6 - 2.4x = 5x - 65$

47. $-3(x - 1) + 2(x + 3) = -4$

48. $\dfrac{3n + 1}{5} + \dfrac{n - 2}{3} = \dfrac{2}{15}$

49. $0.06x + 0.08(1500 - x) = 110$

For Problems 50–55, solve each of the inequalities.

50. $2x - 7 \leq -3(x + 4)$

51. $6x + 5 - 3x > 5$

52. $4(x - 5) + 2(3x + 6) < 0$

53. $-5x + 3 > -4x + 5$

54. $\dfrac{3x}{4} - \dfrac{x}{2} \leq \dfrac{5x}{6} - 1$

55. $0.08(700 - x) + 0.11x \geq 65$

For Problems 56–60, graph each equation.

56. $y = x^2 - 1$ **57.** $y = 2x + 3$

58. $y = -5x$ **59.** $x - 2y = 6$

60. $y = -\dfrac{1}{2}x + 2$

For Problems 61–67, set up an equation or a system of equations to help solve each of the problems.

61. The sum of 4 and three times a certain number is the same as the sum of the number and 10. Find the number.

62. Fifteen percent of some number is 6. Find the number.

63. Lou has 18 coins consisting of dimes and quarters. If the total value of the coins is $3.30, how many coins of each denomination does he have?

64. A sum of $1500 is invested, part of it at 8% interest and the remainder at 9%. If the total interest amounts to $128, find the amount invested at each rate.

65. How many gallons of water must be added to 15 gallons of a 12% salt solution to change it to a 10% salt solution?

66. Two airplanes leave Atlanta at the same time and fly in opposite directions. If one travels at 400 miles per hour and the other at 450 miles per hour, how long will it take them to be 2975 miles apart?

67. The length of a rectangle is 1 meter more than twice its width. If the perimeter of the rectangle is 44 meters, find the length and width.

CHAPTER 7

Algebraic equations can be used to solve a large variety of problems involving geometric relationships.

© Journal Courier/The Image Works

Factoring, Solving Equations, and Problem Solving

A flower garden is in the shape of a right triangle with one leg 7 meters longer than the other leg and the hypotenuse 1 meter longer than the longer leg. Find the lengths of all three sides of the right triangle. A popular geometric formula, called the Pythagorean theorem, serves as a guideline for setting up an equation to solve this problem. We can use the equation $x^2 + (x + 7)^2 = (x + 8)^2$ to determine that the sides of the right triangle are 5 meters, 12 meters, and 13 meters long.

The distributive property has allowed us to combine similar terms and multiply polynomials. In this chapter, we will see yet another use of the distributive property as we learn how to **factor polynomials.** Factoring polynomials will allow us to solve other kinds of equations, which will, in turn, help us to solve a greater variety of word problems.

InfoTrac Project

● Do a subject guide search on 'arches' and find a periodical article named "Thorn in my Shrine." Write a summary of the article. In the article the St. Louis Gateway Arch is mentioned. This arch is 630 feet high (shorter than the proposed arch in Buffalo, New York). To find out how far apart the legs of the St. Louis Gateway Arch are, solve the following equation using factoring:

$$-\frac{7}{10}x^2 + 42x = 0.$$

7.1   Factoring by Using the Distributive Property

Objectives

■   Find the greatest common factor.

■   Factor out the greatest common factor.

■   Factor by grouping.

■   Solve equations by factoring.

In Chapter 1, we found the *greatest common factor* of two or more whole numbers by inspection or by using the prime factored form of the numbers. For example, by inspection we see that the greatest common factor of 8 and 12 is 4. This means that 4 is the largest whole number that is a factor of both 8 and 12. If it is difficult to determine the greatest common factor by inspection, then we can use the prime factorization technique as follows:

$$42 = 2 \cdot 3 \cdot 7$$
$$70 = 2 \cdot 5 \cdot 7$$

We see that $2 \cdot 7 = 14$ is the greatest common factor of 42 and 70.

It is meaningful to extend the concept of greatest common factor to monomials. Consider the next example.

EXAMPLE 1   Find the greatest common factor of $8x^2$ and $12x^3$.

Solution

$$8x^2 = 2 \cdot 2 \cdot 2 \cdot x \cdot x$$
$$12x^3 = 2 \cdot 2 \cdot 3 \cdot x \cdot x \cdot x$$

Therefore, the greatest common factor is $2 \cdot 2 \cdot x \cdot x = 4x^2$.   ■

By "the greatest common factor of two or more monomials" we mean the monomial with the largest numerical coefficient and highest power of the variables that is a factor of the given monomials.

E X A M P L E   2

Find the greatest common factor of $16x^2y$, $24x^3y^2$, and $32xy$.

Solution

$$16x^2y = 2 \cdot 2 \cdot 2 \cdot 2 \cdot x \cdot x \cdot y$$
$$24x^3y^2 = 2 \cdot 2 \cdot 2 \cdot 3 \cdot x \cdot x \cdot x \cdot y \cdot y$$
$$32xy = 2 \cdot 2 \cdot 2 \cdot 2 \cdot 2 \cdot x \cdot y$$

Therefore, the greatest common factor is $2 \cdot 2 \cdot 2 \cdot x \cdot y = 8xy$. ∎

We have used the distributive property to multiply a polynomial by a monomial; for example,

$$3x(x + 2) = 3x^2 + 6x$$

Suppose we start with $3x^2 + 6x$ and want to express it in factored form. We use the distributive property in the form $ab + ac = a(b + c)$.

$$3x^2 + 6x = 3x(x) + 3x(2) \qquad \text{3x is the greatest common factor}$$
$$\text{of } 3x^2 \text{ and } 6x.$$
$$= 3x(x + 2) \qquad \text{Use the distributive property.}$$

The next four examples further illustrate this process of **factoring out the greatest common monomial factor.**

E X A M P L E   3

Factor $12x^3 - 8x^2$.

Solution

$$12x^3 - 8x^2 = 4x^2(3x) - 4x^2(2)$$
$$= 4x^2(3x - 2) \qquad ab - ac = a(b - c)$$ ∎

E X A M P L E   4

Factor $12x^2y + 18xy^2$.

Solution

$$12x^2y + 18xy^2 = 6xy(2x) + 6xy(3y)$$
$$= 6xy(2x + 3y)$$ ∎

E X A M P L E   5

Factor $24x^3 + 30x^4 - 42x^5$.

Solution

$$24x^3 + 30x^4 - 42x^5 = 6x^3(4) + 6x^3(5x) - 6x^3(7x^2)$$
$$= 6x^3(4 + 5x - 7x^2)$$ ∎

E X A M P L E   6   Factor $9x^2 + 9x$.

Solution

$$9x^2 + 9x = 9x(x) + 9x(1)$$
$$= 9x(x + 1)$$ ■

We want to emphasize the point made just before Example 3. It is important to realize that we are factoring out the *greatest* common monomial factor. We could factor an expression such as $9x^2 + 9x$ in Example 6 as $9(x^2 + x)$, $3(3x^2 + 3x)$, $3x(3x + 3)$, or even $\frac{1}{2}(18x^2 + 18x)$, but it is the form $9x(x + 1)$ that we want. We can accomplish this by factoring out the greatest common monomial factor; we sometimes refer to this process as **factoring completely.** A polynomial with integral coefficients is in completely factored form if these conditions are met:

1. It is expressed as a product of polynomials with integral coefficients.
2. No polynomial, other than a monomial, within the factored form can be further factored into polynomials with integral coefficients.

Thus $9(x^2 + x)$, $3(3x^2 + 3x)$, and $3x(3x + 2)$ are not completely factored because they violate condition 2. The form $\frac{1}{2}(18x^2 + 18x)$ violates both conditions 1 and 2.

Sometimes there may be a **common binomial factor** rather than a common monomial factor. For example, each of the two terms of $x(y + 2) + z(y + 2)$ has a binomial factor of $(y + 2)$. Thus we can factor $(y + 2)$ from each term and get

$$x(y + 2) + z(y + 2) = (y + 2)(x + z)$$

Consider a few more examples involving a common binomial factor:

$$a(b + c) - d(b + c) = (b + c)(a - d)$$
$$x(x + 2) + 3(x + 2) = (x + 2)(x + 3)$$
$$x(x + 5) - 4(x + 5) = (x + 5)(x - 4)$$

It may be that the original polynomial exhibits no apparent common monomial or binomial factor, which is the case with

$$ab + 3a + bc + 3c$$

However, by factoring a from the first two terms and c from the last two terms, we see that

$$ab + 3a + bc + 3c = a(b + 3) + c(b + 3)$$

Now a common binomial factor of $(b + 3)$ is obvious, and we can proceed as before:

$$a(b + 3) + c(b + 3) = (b + 3)(a + c)$$

This factoring process is called **factoring by grouping.** Let's consider two more examples of factoring by grouping.

$$x^2 - x + 5x - 5 = x(x - 1) + 5(x - 1) \quad \text{Factor } x \text{ from first two terms and}$$
$$5 \text{ from last two terms.}$$
$$= (x - 1)(x + 5) \quad \text{Factor common binomial factor of}$$
$$(x - 1) \text{ from both terms.}$$
$$6x^2 - 4x - 3x + 2 = 2x(3x - 2) - 1(3x - 2) \quad \text{Factor } 2x \text{ from first two terms}$$
$$\text{and } -1 \text{ from last two terms.}$$
$$= (3x - 2)(2x - 1) \quad \text{Factor common binomial factor of}$$
$$(3x - 2) \text{ from both terms.}$$

Back to Solving Equations

Suppose we are told that the product of two numbers is 0. What do we know about the numbers? Do you agree we can conclude that at least one of the numbers must be 0? The next property formalizes this idea.

> **PROPERTY 7.1**
>
> For all real numbers a and b,
>
> $$ab = 0 \quad \text{if and only if } a = 0 \text{ or } b = 0$$

Property 7.1 provides us with another technique for solving equations.

EXAMPLE 7

Solve $x^2 + 6x = 0$.

Solution

To solve equations by applying Property 7.1, one side of the equation must be a product, and the other side of the equation must be zero. This equation already has zero on the right-hand side of the equation, but the left-hand side of this equation is a sum. We will factor the left-hand side, $x^2 + 6x$, to change the sum into a product.

$$x^2 + 6x = 0$$
$$x(x + 6) = 0$$
$$x = 0 \quad \text{or} \quad x + 6 = 0 \quad ab = 0 \text{ if and only if } a = 0 \text{ or } b = 0$$
$$x = 0 \quad \text{or} \quad x = -6$$

The solution set is $\{-6, 0\}$. (Be sure to check both values in the original equation.)

EXAMPLE 8

Solve $x^2 = 12x$.

Solution

In order to solve this equation by Property 7.1, we will first get zero on the right-hand side of the equation by adding $-12x$ to each side. Then we factor the expression on the left-hand side of the equation.

$$x^2 = 12x$$
$$x^2 - 12x = 0 \qquad \text{Added } -12x \text{ to both sides}$$
$$x(x - 12) = 0$$
$$x = 0 \quad \text{or} \quad x - 12 = 0 \qquad ab = 0 \text{ if and only if } a = 0 \text{ or } b = 0$$
$$x = 0 \quad \text{or} \qquad\qquad x = 12$$

The solution set is {0, 12}.

REMARK: Note in Example 8 that we *did not* divide both sides of the original equation by x. Doing so would cause us to lose the solution of 0.

EXAMPLE 9   Solve $4x^2 - 3x = 0$.

Solution

$$4x^2 - 3x = 0$$
$$x(4x - 3) = 0$$
$$x = 0 \quad \text{or} \quad 4x - 3 = 0 \qquad ab = 0 \text{ if and only if } a = 0 \text{ or } b = 0$$
$$x = 0 \quad \text{or} \qquad\quad 4x = 3$$
$$x = 0 \quad \text{or} \qquad\quad x = \frac{3}{4}$$

The solution set is $\left\{ 0, \dfrac{3}{4} \right\}$.

EXAMPLE 1 0   Solve $x(x + 2) + 3(x + 2) = 0$.

Solution

In order to solve this equation by Property 7.1, we will factor the left-hand side of the equation. The greatest common factor of the terms is $(x + 2)$.

$$x(x + 2) + 3(x + 2) = 0$$
$$(x + 2)(x + 3) = 0$$
$$x + 2 = 0 \quad \text{or} \quad x + 3 = 0 \qquad ab = 0 \text{ if and only if } a = 0 \text{ or } b = 0$$
$$x = -2 \quad \text{or} \qquad x = -3$$

The solution set is $\{-3, -2\}$.

Each time we expand our equation-solving capabilities, we also gain more techniques for solving word problems. Let's solve a geometric problem with the ideas we learned in this section.

PROBLEM 1

The area of a square is numerically equal to twice its perimeter. Find the length of a side of the square.

Solution

Sketch a square and let s represent the length of each side (see Figure 7.1). Then the area is represented by s^2 and the perimeter by $4s$. Thus

$$s^2 = 2(4s)$$
$$s^2 = 8s$$
$$s^2 - 8s = 0$$
$$s(s - 8) = 0$$
$$s = 0 \quad \text{or} \quad s - 8 = 0$$
$$s = 0 \quad \text{or} \quad s = 8$$

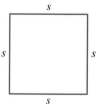

Figure 7.1

Because 0 is not a reasonable answer to the problem, the solution is 8. (Be sure to check this solution in the original statement of the problem!)

CONCEPT QUIZ

For Problems 1–6, answer true or false.

1. The greatest common factor of $6x^2y^3 - 12x^3y^2 + 18x^4y$ is $2x^2y$.

2. If the factored form of a polynomial can be factored further, then it has not met the conditions to be considered "factored completely."

3. Common factors are always monomials.

4. If the product of x and y is zero, then x is zero or y is zero.

5. The factored form $3a(2a^2 + 4)$ is factored completely.

6. The solutions for the equation $x(x + 2) = 7$ are 7 and 5.

PROBLEM SET 7.1

For Problems 1–10, find the greatest common factor of the given expressions.

1. $24y$ and $30xy$

2. $32x$ and $40xy$

3. $60x^2y$ and $84xy^2$

4. $72x^3$ and $63x^2$

5. $42ab^3$ and $70a^2b^2$

6. $48a^2b^2$ and $96ab^4$

7. $6x^3$, $8x$, and $24x^2$

8. $72xy$, $36x^2y$, and $84xy^2$

9. $16a^2b^2$, $40a^2b^3$, and $56a^3b^4$

10. $70a^3b^3$, $42a^2b^4$, and $49ab^5$

For Problems 11–46, factor each polynomial completely.

11. $8x + 12y$

12. $18x + 24y$

13. $14xy - 21y$

14. $24x - 40xy$

15. $18x^2 + 45x$

16. $12x + 28x^3$

17. $12xy^2 - 30x^2y$

18. $28x^2y^2 - 49x^2y$

19. $36a^2b - 60a^3b^4$

20. $65ab^3 - 45a^2b^2$

21. $16xy^3 + 25x^2y^2$

22. $12x^2y^2 + 29x^2y$

23. $64ab - 72cd$

24. $45xy - 72zw$

25. $9a^2b^4 - 27a^2b$

26. $7a^3b^5 - 42a^2b^6$

27. $52x^4y^2 + 60x^6y$

28. $70x^5y^3 - 42x^8y^2$

29. $40x^2y^2 + 8x^2y$

30. $84x^2y^3 + 12xy^3$

31. $12x + 15xy + 21x^2$

32. $30x^2y + 40xy + 55y$

33. $2x^3 - 3x^2 + 4x$

34. $x^4 + x^3 + x^2$

35. $44y^5 - 24y^3 - 20y^2$

36. $14a - 18a^3 - 26a^5$

37. $14a^2b^3 + 35ab^2 - 49a^3b$

38. $24a^3b^2 + 36a^2b^4 - 60a^4b^3$

39. $x(y + 1) + z(y + 1)$

40. $a(c + d) + 2(c + d)$

41. $a(b - 4) - c(b - 4)$

42. $x(y - 6) - 3(y - 6)$

43. $x(x + 3) + 6(x + 3)$

44. $x(x - 7) + 9(x - 7)$

45. $2x(x + 1) - 3(x + 1)$

46. $4x(x + 8) - 5(x + 8)$

For Problems 47–60, use the process of *factoring by grouping* to factor each polynomial.

47. $5x + 5y + bx + by$

48. $7x + 7y + zx + zy$

49. $bx - by - cx + cy$

50. $2x - 2y - ax + ay$

51. $ac + bc + a + b$

52. $x + y + ax + ay$

53. $x^2 + 5x + 12x + 60$

54. $x^2 + 3x + 7x + 21$

55. $x^2 - 2x - 8x + 16$

56. $x^2 - 4x - 9x + 36$

57. $2x^2 + x - 10x - 5$

58. $3x^2 + 2x - 18x - 12$

59. $6n^2 - 3n - 8n + 4$

60. $20n^2 + 8n - 15n - 6$

For Problems 61–84, solve each equation.

61. $x^2 - 8x = 0$

62. $x^2 - 12x = 0$

63. $x^2 + x = 0$

64. $x^2 + 7x = 0$

65. $n^2 = 5n$

66. $n^2 = -2n$

67. $2y^2 - 3y = 0$

68. $4y^2 - 7y = 0$

69. $7x^2 = -3x$

70. $5x^2 = -2x$

71. $3n^2 + 15n = 0$

72. $6n^2 - 24n = 0$

73. $4x^2 = 6x$

74. $12x^2 = 8x$

75. $7x - x^2 = 0$

76. $9x - x^2 = 0$

77. $13x = x^2$

78. $15x = -x^2$

79. $5x = -2x^2$

80. $7x = -5x^2$

81. $x(x + 5) - 4(x + 5) = 0$

82. $x(3x - 2) - 7(3x - 2) = 0$

83. $4(x - 6) - x(x - 6) = 0$

84. $x(x + 9) = 2(x + 9)$

For Problems 85–91, set up an equation and solve each problem.

85. The square of a number equals nine times that number. Find the number.

86. Suppose that four times the square of a number equals 20 times that number. What is the number?

87. The area of a square is numerically equal to five times its perimeter. Find the length of a side of the square.

88. The area of a square is 14 times as large as the area of a triangle. One side of the triangle is 7 inches long, and the altitude to that side is the same length as a side of the square. Find the length of a side of the square. Also find the areas of both figures, and be sure that your answer checks.

89. Suppose that the area of a circle is numerically equal to the perimeter of a square whose length of a side is the same as the length of a radius of the circle. Find the length of a side of the square. Express your answer in terms of π.

90. One side of a parallelogram, an altitude to that side, and one side of a rectangle all have the same measure. If an adjacent side of the rectangle is 20 centimeters long, and the area of the rectangle is twice the area of the parallelogram, find the areas of both figures.

91. The area of a rectangle is twice the area of a square. If the rectangle is 6 inches long, and the width of the rectangle is the same as the length of a side of the square, find the dimensions of both the rectangle and the square.

■ ■ ■ Thoughts into words

92. Suppose that your friend factors $24x^2y + 36xy$ like this:

$$24x^2y + 36xy = 4xy(6x + 9)$$
$$= (4xy)(3)(2x + 3)$$
$$= 12xy(2x + 3)$$

Is this correct? Would you suggest any changes?

93. The following solution is given for the equation $x(x - 10) = 0$.

$$x(x - 10) = 0$$
$$x^2 - 10x = 0$$
$$x(x - 10) = 0$$
$$x = 0 \quad \text{or} \quad x - 10 = 0$$
$$x = 0 \quad \text{or} \quad x = 10$$

The solution set is $\{0, 10\}$. Is this solution correct? Would you suggest any changes?

■ ■ ■ Further investigations

94. The total surface area of a right circular cylinder is given by the formula $A = 2\pi r^2 + 2\pi rh$, where r represents the radius of a base, and h represents the height of the cylinder. For computational purposes, it may be more convenient to change the form of the right side of the formula by factoring it.

$$A = 2\pi r^2 + 2\pi rh$$
$$= 2\pi r(r + h)$$

Use $A = 2\pi r(r + h)$ to find the total surface area of each of the following cylinders. Use $\dfrac{22}{7}$ as an approximation for π.

a. $r = 7$ centimeters and $h = 12$ centimeters

b. $r = 14$ meters and $h = 20$ meters

c. $r = 3$ feet and $h = 4$ feet

d. $r = 5$ yards and $h = 9$ yards

95. The formula $A = P + Prt$ yields the total amount of money accumulated (A) when P dollars is invested at r percent simple interest for t years. For computational purposes it may be convenient to change the right side of the formula by factoring.

$$A = P + Prt$$
$$= P(1 + rt)$$

Use $A = P(1 + rt)$ to find the total amount of money accumulated for each of the following investments.

a. $100 at 8% for 2 years

b. $200 at 9% for 3 years

c. $500 at 10% for 5 years

d. $1000 at 10% for 10 years

For Problems 96–99, solve each equation for the indicated variable.

96. $ax + bx = c$ for x

97. $b^2x^2 - cx = 0$ for x

98. $5ay^2 = by$ for y

99. $y + ay - by - c = 0$ for y

7.2   Factoring the Difference of Two Squares

Objectives

■   Factor the difference of two squares.

■   Solve equations by factoring the difference of two squares.

In Section 6.3, we noted some special multiplication patterns. One of these patterns was

$$(a - b)(a + b) = a^2 - b^2$$

We can view this same pattern as follows:

Difference of Two Squares

$$a^2 - b^2 = (a - b)(a + b)$$

To apply the pattern is a fairly simple process, as these next examples illustrate. The steps inside the box are often performed mentally.

$$x^2 - 36 = (x)^2 - (6)^2 = (x - 6)(x + 6)$$

$$4x^2 - 25 = (2x)^2 - (5)^2 = (2x - 5)(2x + 5)$$

$$9x^2 - 16y^2 = (3x)^2 - (4y)^2 = (3x - 4y)(3x + 4y)$$

$$64 - y^2 = (8)^2 - (y)^2 = (8 - y)(8 + y)$$

Because multiplication is commutative, the order of writing the factors is not important. For example, $(x - 6)(x + 6)$ can also be written as $(x + 6)(x - 6)$.

You must be careful not to assume an analogous factoring pattern for the *sum* of two squares; it does not exist. For example, $x^2 + 4 \neq (x + 2)(x + 2)$ because $(x + 2)(x + 2) = x^2 + 4x + 4$. We say that the **sum of two squares is not factorable using integers.** The phrase "using integers" is necessary because $x^2 + 4$ could be written as $\frac{1}{2}(2x^2 + 8)$, but such *factoring* is of no help. Furthermore, we do not consider $(1)(x^2 + 4)$ as factoring $x^2 + 4$.

It is possible that both the technique of *factoring out a common monomial factor* and the *difference of two squares* pattern can be applied to the same polynomial. In general, it is best to look for a common monomial factor first.

EXAMPLE 1   Factor $2x^2 - 50$.

Solution

$$2x^2 - 50 = 2(x^2 - 25) \qquad \text{Common factor of 2}$$
$$= 2(x - 5)(x + 5) \qquad \text{Difference of squares}$$

In Example 1, by expressing $2x^2 - 50$ as $2(x - 5)(x + 5)$, we say that it has been **factored completely.** That means the factors $2, x - 5$, and $x + 5$ cannot be factored any further using integers.

EXAMPLE 2   Factor completely $18y^3 - 8y$.

Solution

$$18y^3 - 8y = 2y(9y^2 - 4) \qquad \text{Common factor of } 2y$$
$$= 2y(3y - 2)(3y + 2) \qquad \text{Difference of squares}$$

Sometimes it is possible to apply the difference-of-squares pattern more than once. Consider the next example.

EXAMPLE 3   Factor completely $x^4 - 16$.

Solution

$$x^4 - 16 = (x^2 + 4)(x^2 - 4)$$
$$= (x^2 + 4)(x + 2)(x - 2)$$

The following examples should help you to summarize the factoring ideas presented thus far.

$$5x^2 + 20 = 5(x^2 + 4)$$
$$25 - y^2 = (5 - y)(5 + y)$$
$$3 - 3x^2 = 3(1 - x^2) = 3(1 + x)(1 - x)$$
$$36x^2 - 49y^2 = (6x - 7y)(6x + 7y)$$
$$a^2 + 9 \quad \text{is not factorable using integers}$$
$$9x + 17y \quad \text{is not factorable using integers}$$

Solving Equations

Each time we learn a new factoring technique, we also develop more power for solving equations. Let's consider how we can use the difference-of-squares factoring pattern to help solve certain kinds of equations.

EXAMPLE 4

Solve $x^2 = 25$.

Solution

$$x^2 = 25$$
$$x^2 - 25 = 0$$
$$(x + 5)(x - 5) = 0$$

$x + 5 = 0$ or $x - 5 = 0$ Remember: $ab = 0$ if and only if $a = 0$ or $b = 0$.

$x = -5$ or $x = 5$

The solution set is $\{-5, 5\}$. Check these answers! ∎

EXAMPLE 5

Solve $9x^2 = 25$.

Solution

$$9x^2 = 25$$
$$9x^2 - 25 = 0$$
$$(3x + 5)(3x - 5) = 0$$

$3x + 5 = 0$ or $3x - 5 = 0$

$3x = -5$ or $3x = 5$

$x = -\dfrac{5}{3}$ or $x = \dfrac{5}{3}$

The solution set is $\left\{-\dfrac{5}{3}, \dfrac{5}{3}\right\}$. ∎

EXAMPLE 6

Solve $5y^2 = 20$.

Solution

$$5y^2 = 20$$
$$\frac{5y^2}{5} = \frac{20}{5} \qquad \text{Divide both sides by 5.}$$
$$y^2 = 4$$
$$y^2 - 4 = 0$$
$$(y + 2)(y - 2) = 0$$

$y + 2 = 0$ or $y - 2 = 0$

$y = -2$ or $y = 2$

The solution set is $\{-2, 2\}$. Check it! ∎

EXAMPLE 7

Solve $x^3 - 9x = 0$.

Solution

$$x^3 - 9x = 0$$
$$x(x^2 - 9) = 0$$
$$x(x - 3)(x + 3) = 0$$

$x = 0$ or $x - 3 = 0$ or $x + 3 = 0$

$x = 0$ or $x = 3$ or $x = -3$

The solution set is $\{-3, 0, 3\}$.   ∎

The more we know about solving equations, the more easily we can solve word problems.

PROBLEM 1

The combined area of two squares is 20 square centimeters. Each side of one square is twice as long as a side of the other square. Find the lengths of the sides of each square.

Solution

We can sketch two squares and label the sides of the smaller square s (see Figure 7.2). Then the sides of the larger square are $2s$. The sum of the areas of the two squares is 20 square centimeters, so we set up and solve the following equation:

$$s^2 + (2s)^2 = 20$$
$$s^2 + 4s^2 = 20$$
$$5s^2 = 20$$
$$s^2 = 4$$
$$s^2 - 4 = 0$$
$$(s + 2)(s - 2) = 0$$

$s + 2 = 0$ or $s - 2 = 0$

$s = -2$ or $s = 2$

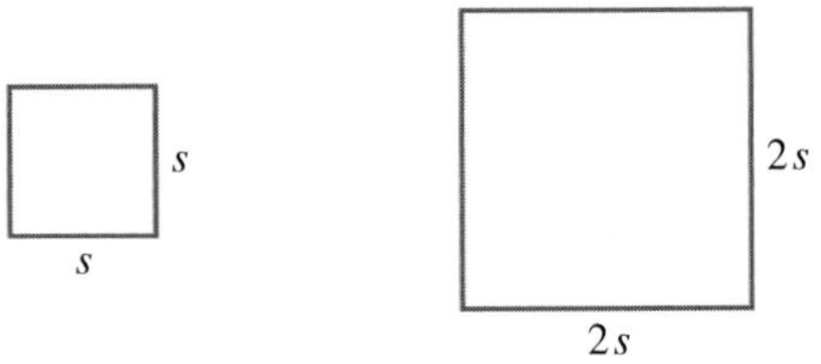

Figure 7.2

Because s represents the length of a side of a square, we must disregard the solution -2. Thus one square has sides of length 2 centimeters, and the other square has sides of length $2(2) = 4$ centimeters.   ∎

CONCEPT QUIZ

For Problems 1–5, answer true or false.

1. A binomial that has two perfect square terms that are subtracted is called the difference of two squares.

2. The sum of two squares is factorable using integers.

    3. When factoring it is usually best to look for a common factor first.

    4. The polynomial $4x^2 + y^2$ factors into $(2x + y)(2x + y)$.

    5. The completely factored form of $y^4 - 81$ is $(y^2 + 9)(y^2 - 9)$.

PROBLEM SET 7.2

For Problems 1–12, use the difference-of-squares pattern to factor each polynomial.

1. $x^2 - 1$

2. $x^2 - 25$

3. $x^2 - 100$

4. $x^2 - 121$

5. $x^2 - 4y^2$

6. $x^2 - 36y^2$

7. $9x^2 - y^2$

8. $49y^2 - 64x^2$

9. $36a^2 - 25b^2$

10. $4a^2 - 81b^2$

11. $1 - 4n^2$

12. $4 - 9n^2$

For Problems 13–40, factor each polynomial completely. Indicate any that are not factorable using integers. Don't forget to look for a common monomial factor first.

13. $5x^2 - 20$

14. $7x^2 - 7$

15. $8x^2 + 32$

16. $12x^2 + 60$

17. $2x^2 - 18y^2$

18. $8x^2 - 32y^2$

19. $x^3 - 25x$

20. $2x^3 - 2x$

21. $x^2 + 9y^2$

22. $18x - 42y$

23. $45x^2 - 36xy$

24. $16x^2 + 25y^2$

25. $36 - 4x^2$

26. $75 - 3x^2$

27. $4a^4 + 16a^2$

28. $9a^4 + 81a^2$

29. $x^4 - 81$

30. $16 - x^4$

31. $x^4 + x^2$

32. $x^5 + 2x^3$

33. $3x^3 + 48x$

34. $6x^3 + 24x$

35. $5x - 20x^3$

36. $4x - 36x^3$

37. $4x^2 - 64$

38. $9x^2 - 9$

39. $75x^3y - 12xy^3$

40. $32x^3y - 18xy^3$

For Problems 41–64, solve each equation.

41. $x^2 = 9$

42. $x^2 = 1$

43. $4 = n^2$

44. $144 = n^2$

45. $9x^2 = 16$

46. $4x^2 = 9$

47. $n^2 - 121 = 0$

48. $n^2 - 81 = 0$

49. $25x^2 = 4$

50. $49x^2 = 36$

51. $3x^2 = 75$

52. $7x^2 = 28$

53. $3x^3 - 48x = 0$

54. $x^3 - x = 0$

55. $n^3 = 16n$

56. $2n^3 = 8n$

57. $5 - 45x^2 = 0$

58. $3 - 12x^2 = 0$

59. $4x^3 - 400x = 0$

60. $2x^3 - 98x = 0$

61. $64x^2 = 81$

62. $81x^2 = 25$

63. $36x^3 = 9x$

64. $64x^3 = 4x$

For Problems 65–76, set up an equation and solve the problem.

65. Forty-nine less than the square of a number equals zero. Find the number.

66. The cube of a number equals nine times the number. Find the number.

67. Suppose that five times the cube of a number equals 80 times the number. Find the number.

68. Ten times the square of a number equals 40. Find the number.

69. The sum of the areas of two squares is 234 square inches. Each side of the larger square is five times the length of a side of the smaller square. Find the length of a side of each square.

70. The difference of the areas of two squares is 75 square feet. Each side of the larger square is twice the length of a side of the smaller square. Find the length of a side of each square.

71. Suppose that the length of a certain rectangle is $2\frac{1}{2}$ times its width, and the area of that same rectangle is 160 square centimeters. Find the length and width of the rectangle.

72. Suppose that the width of a certain rectangle is three-fourths of its length and that the area of this same rectangle is 108 square meters. Find the length and width of the rectangle.

73. The sum of the areas of two circles is 80π square meters. Find the length of a radius of each circle if one of them is twice as long as the other.

74. The area of a triangle is 98 square feet. If one side of the triangle and the altitude to that side are of equal length, find the length.

75. The total surface area of a right circular cylinder is 100π square centimeters. If a radius of the base and the altitude of the cylinder are the same length, find the length of a radius.

76. The total surface area of a right circular cone is 192π square feet. If the slant height of the cone is equal in length to a diameter of the base, find the length of a radius.

■ ■ ■ **Thoughts into words**

77. How do we know that the equation $x^2 + 1 = 0$ has no solutions in the set of real numbers?

78. Why is the following factoring process incomplete?

$16x^2 - 64 = (4x + 8)(4x - 8)$

How should the factoring be done?

79. Consider the following solution:

$$4x^2 - 36 = 0$$
$$4(x^2 - 9) = 0$$
$$4(x + 3)(x - 3) = 0$$
$$4 = 0 \quad \text{or} \quad x + 3 = 0 \quad \text{or} \quad x - 3 = 0$$
$$4 = 0 \quad \text{or} \quad x = -3 \quad \text{or} \quad x = 3$$

The solution set is $\{-3, 3\}$. Is this a correct solution? Do you have any suggestion to offer the person who worked on this problem?

■ ■ ■ **Further investigations**

The following patterns can be used to factor the sum of two cubes and the difference of two cubes, respectively.

$a^3 + b^3 = (a + b)(a^2 - ab + b^2)$
$a^3 - b^3 = (a - b)(a^2 + ab + b^2)$

Consider these examples:

$x^3 + 8 = (x)^3 + (2)^3 = (x + 2)(x^2 - 2x + 4)$
$x^3 - 1 = (x)^3 - (1)^3 = (x - 1)(x^2 + x + 1)$

Use the sum-of-two-cubes and the difference-of-two-cubes patterns to factor each polynomial.

80. $x^3 + 1$ **81.** $x^3 - 8$

82. $n^3 - 27$

83. $n^3 + 64$

84. $8x^3 + 27y^3$

85. $27a^3 - 64b^3$

86. $1 - 8x^3$

87. $1 + 27a^3$

88. $x^3 + 8y^3$

89. $8x^3 - y^3$

90. $a^3b^3 - 1$

91. $27x^3 - 8y^3$

92. $8 + n^3$

93. $125x^3 + 8y^3$

94. $27n^3 - 125$

95. $64 + x^3$

Answers to Concept Quiz

1. True   **2.** False   **3.** True   **4.** False   **5.** False

7.3   Factoring Trinomials of the Form $x^2 + bx + c$

Objectives

■   Factor trinomials of the form $x^2 + bx + c$.

■   Use factoring of trinomials to solve equations.

■   Solve word problems involving consecutive numbers.

■   Use the Pythagorean theorem to solve problems.

One of the most common types of factoring used in algebra is to express a trinomial as the product of two binomials. In this section, we will consider trinomials where the coefficient of the squared term is 1—that is, trinomials of the form $x^2 + bx + c$.

Again, to develop a factoring technique we first look at some multiplication ideas. Consider the product $(x + r)(x + s)$, and use the distributive property to show how each term of the resulting trinomial is formed.

$$(x + r)(x + s) = x(x) + x(s) + r(x) + r(s)$$

$$x^2 \quad + \quad (s + r)x \quad + \quad rs$$

Note that the coefficient of the middle term is the *sum* of r and s and that the last term is the *product* of r and s. These two relationships are used in the next examples.

E X A M P L E   1

Factor $x^2 + 7x + 12$.

Solution

We need to fill in the blanks with two numbers whose sum is 7 and whose product is 12.

$$x^2 + 7x + 12 = (x + \underline{\quad})(x + \underline{\quad})$$

This can be done by setting up a table showing possible numbers.

Product	Sum
$1(12) = 12$	$1 + 12 = 13$
$2(6) = 12$	$2 + 6 = 8$
$3(4) = 12$	$3 + 4 = 7$

The bottom line contains the numbers that we need. Thus

$$x^2 + 7x + 12 = (x + 3)(x + 4)$$ ■

E X A M P L E   2   Factor $x^2 - 11x + 24$.

Solution

We need two numbers whose product is 24 and whose sum is -11.

Product	Sum
$(-1)(-24) = 24$	$-1 + (-24) = -25$
$(-2)(-12) = 24$	$-2 + (-12) = -14$
$(-3)(-8) = 24$	$-3 + (-8) = -11$
$(-4)(-6) = 24$	$-4 + (-6) = -10$

The third line contains the numbers that we want. Thus

$$x^2 - 11x + 24 = (x - 3)(x - 8)$$

E X A M P L E   3   Factor $x^2 + 3x - 10$.

Solution

We need two numbers whose product is -10 and whose sum is 3.

Product	Sum
$1(-10) = -10$	$1 + (-10) = -9$
$-1(10) = -10$	$-1 + 10 = 9$
$2(-5) = -10$	$2 + (-5) = -3$
$-2(5) = -10$	$-2 + 5 = 3$

The bottom line is the key line. Thus

$$x^2 + 3x - 10 = (x + 5)(x - 2)$$

E X A M P L E   4   Factor $x^2 - 2x - 8$.

Solution

We need two numbers whose product is -8 and whose sum is -2.

Product	Sum
$1(-8) = -8$	$1 + (-8) = -7$
$-1(8) = -8$	$-1 + 8 = 7$
$2(-4) = -8$	$2 + (-4) = -2$
$-2(4) = -8$	$-2 + 4 = 2$

The third line has the information we want.

$$x^2 - 2x - 8 = (x - 4)(x + 2)$$

The tables in the last four examples illustrate one way of organizing your thoughts for such problems. We showed complete tables; that is, for Example 4, we included the bottom line even though the desired numbers were obtained in the third line. If you use such tables, keep in mind that as soon as you get the desired numbers, the table need not be continued beyond that point. Furthermore, many times you may be able to find the numbers without using a table. The key ideas are the product and sum relationships.

EXAMPLE 5

Factor $x^2 - 13x + 12$.

Solution

Product	Sum
$(-1)(-12) = 12$	$(-1) + (-12) = -13$

We need not complete the table.

$$x^2 - 13x + 12 = (x - 1)(x - 12)$$

In the next example, we refer to the concept of absolute value. Recall that the absolute value is the number without regard for the sign. For example,

$$|4| = 4 \quad \text{and} \quad |-4| = 4$$

EXAMPLE 6

Factor $x^2 - x - 56$.

Solution

Note that the coefficient of the middle term is -1. Therefore, we are looking for two numbers whose product is -56; their sum is -1, so the absolute value of the negative number must be one larger than the absolute value of the positive number. The numbers are -8 and 7, and we have

$$x^2 - x - 56 = (x - 8)(x + 7)$$

EXAMPLE 7

Factor $x^2 + 10x + 12$.

Solution

Product	Sum
$1(12) = 12$	$1 + 12 = 13$
$2(6) = 12$	$2 + 6 = 8$
$3(4) = 12$	$3 + 4 = 7$

Because the table is complete and no two factors of 12 produce a sum of 10, we conclude that

$$x^2 + 10x + 12$$

is not factorable using integers.

In a problem such as Example 7, we need to be sure that we have tried all possibilities before we conclude that the trinomial is not factorable.

Back to Solving Equations

The property $ab = 0$ if and only if $a = 0$ or $b = 0$ continues to play an important role as we solve equations that involve the factoring ideas of this section. Consider the following examples.

EXAMPLE 8   Solve $x^2 + 8x + 15 = 0$.

Solution

$$x^2 + 8x + 15 = 0$$
$$(x + 3)(x + 5) = 0 \qquad \text{Factor the left side.}$$
$$x + 3 = 0 \qquad \text{or} \qquad x + 5 = 0 \quad \text{Use } ab = 0 \text{ if and only if } a = 0 \text{ or } b = 0.$$
$$x = -3 \qquad \text{or} \qquad x = -5$$

The solution set is $\{-5, -3\}$. ∎

EXAMPLE 9   Solve $x^2 + 5x - 6 = 0$.

Solution

$$x^2 + 5x - 6 = 0$$
$$(x + 6)(x - 1) = 0$$
$$x + 6 = 0 \qquad \text{or} \qquad x - 1 = 0$$
$$x = -6 \qquad \text{or} \qquad x = 1$$

The solution set is $\{-6, 1\}$. ∎

EXAMPLE 10   Solve $y^2 - 4y = 45$.

Solution

$$y^2 - 4y = 45$$
$$y^2 - 4y - 45 = 0$$
$$(y - 9)(y + 5) = 0$$
$$y - 9 = 0 \qquad \text{or} \qquad y + 5 = 0$$
$$y = 9 \qquad \text{or} \qquad y = -5$$

The solution set is $\{-5, 9\}$. ∎

Don't forget that we can always check to be absolutely sure of our solutions. Let's check the solutions for Example 10. If $y = 9$, then $y^2 - 4y = 45$ becomes

$$9^2 - 4(9) \stackrel{?}{=} 45$$
$$81 - 36 \stackrel{?}{=} 45$$
$$45 = 45$$

If $y = -5$, then $y^2 - 4y = 45$ becomes

$$(-5)^2 - 4(-5) \overset{?}{=} 45$$
$$25 + 20 \overset{?}{=} 45$$
$$45 = 45$$

Back to Problem Solving

The more we know about factoring and solving equations, the more easily we can solve word problems.

PROBLEM 1

Find two consecutive integers whose product is 72.

Solution

Let n represent one integer. Then $n + 1$ represents the next integer.

$$n(n + 1) = 72 \qquad \text{The product of the two integers is 72.}$$
$$n^2 + n = 72$$
$$n^2 + n - 72 = 0$$
$$(n + 9)(n - 8) = 0$$
$$n + 9 = 0 \qquad \text{or} \qquad n - 8 = 0$$
$$n = -9 \qquad \text{or} \qquad n = 8$$

If $n = -9$, then $n + 1 = -9 + 1 = -8$. If $n = 8$, then $n + 1 = 8 + 1 = 9$. Thus the consecutive integers are -9 and -8 or 8 and 9. ■

PROBLEM 2

A rectangular plot is 6 meters longer than it is wide. The area of the plot is 16 square meters. Find the length and width of the plot.

Solution

We let w represent the width of the plot, and then $w + 6$ represents the length (see Figure 7.3).

$w + 6$

Figure 7.3

Using the area formula $A = lw$, we obtain

$$w(w + 6) = 16$$
$$w^2 + 6w = 16$$
$$w^2 + 6w - 16 = 0$$
$$(w + 8)(w - 2) = 0$$
$$w + 8 = 0 \qquad \text{or} \qquad w - 2 = 0$$
$$w = -8 \qquad \text{or} \qquad w = 2$$

The solution -8 is not possible for the width of a rectangle, so the plot is 2 meters wide, and its length $(w + 6)$ is 8 meters. ∎

The Pythagorean theorem, an important theorem pertaining to right triangles, can also serve as a guideline for solving certain types of problems. The Pythagorean theorem states that **in any right triangle, the square of the longest side** (*called the hypotenuse*) **is equal to the sum of the squares of the other two sides** (*called legs*); see Figure 7.4. We can use this theorem to help solve a problem.

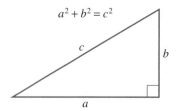

Figure 7.4

PROBLEM 3

Suppose that the lengths of the three sides of a right triangle are consecutive whole numbers. Find the lengths of the three sides.

Solution

Let s represent the length of the shortest leg. Then $s + 1$ represents the length of the other leg, and $s + 2$ represents the length of the hypotenuse. Using the Pythagorean theorem as a guideline, we obtain the following equation:

$$\overbrace{s^2 + (s + 1)^2}^{\text{Sum of squares of two legs}} = \overbrace{(s + 2)^2}^{\text{Square of hypotenuse}}$$

Solving this equation yields

$$s^2 + s^2 + 2s + 1 = s^2 + 4s + 4$$
$$2s^2 + 2s + 1 = s^2 + 4s + 4$$
$$s^2 + 2s + 1 = 4s + 4 \qquad \text{Add } -s^2 \text{ to both sides.}$$
$$s^2 - 2s + 1 = 4 \qquad \text{Add } -4s \text{ to both sides.}$$
$$s^2 - 2s - 3 = 0 \qquad \text{Add } -4 \text{ to both sides.}$$
$$(s - 3)(s + 1) = 0$$

$$s - 3 = 0 \quad \text{or} \quad s + 1 = 0$$
$$s = 3 \quad \text{or} \quad s = -1$$

The solution of -1 is not possible for the length of a side, so the shortest side (s) is of length 3. The other two sides ($s + 1$ and $s + 2$) have lengths of 4 and 5.   ■

CONCEPT QUIZ

For Problems 1–6, answer true or false.

1. Any trinomial of the form $x^2 + bx + c$ can be factored (using integers) into the product of two binomials.

2. To factor $x^2 - 4x - 60$ we look for two numbers whose product is -60 and whose sum is -4.

3. A trinomial of the form $x^2 + bx + c$ will never have a common factor other than 1.

4. If n represents an odd integer, then $n + 1$ represents the next consecutive odd integer.

5. The Pythagorean theorem only applies to right triangles.

6. In a right triangle the longest side is called the hypotenuse.

PROBLEM SET 7.3

For Problems 1–30, factor each trinomial completely. Indicate any that are not factorable using integers.

1. $x^2 + 10x + 24$

2. $x^2 + 9x + 14$

3. $x^2 + 13x + 40$

4. $x^2 + 11x + 24$

5. $x^2 - 11x + 18$

6. $x^2 - 5x + 4$

7. $n^2 - 11n + 28$

8. $n^2 - 7n + 10$

9. $n^2 + 6n - 27$

10. $n^2 + 3n - 18$

11. $n^2 - 6n - 40$

12. $n^2 - 4n - 45$

13. $t^2 + 12t + 24$

14. $t^2 + 20t + 96$

15. $x^2 - 18x + 72$

16. $x^2 - 14x + 32$

17. $x^2 + 5x - 66$

18. $x^2 + 11x - 42$

19. $y^2 - y - 72$

20. $y^2 - y - 30$

21. $x^2 + 21x + 80$

22. $x^2 + 21x + 90$

23. $x^2 + 6x - 72$

24. $x^2 - 8x - 36$

25. $x^2 - 10x - 48$

26. $x^2 - 12x - 64$

27. $x^2 + 3xy - 10y^2$

28. $x^2 - 4xy - 12y^2$

29. $a^2 - 4ab - 32b^2$

30. $a^2 + 3ab - 54b^2$

For Problems 31–50, solve each equation.

31. $x^2 + 10x + 21 = 0$

32. $x^2 + 9x + 20 = 0$

33. $x^2 - 9x + 18 = 0$

34. $x^2 - 9x + 8 = 0$

35. $x^2 - 3x - 10 = 0$

36. $x^2 - x - 12 = 0$

37. $n^2 + 5n - 36 = 0$

38. $n^2 + 3n - 18 = 0$

39. $n^2 - 6n - 40 = 0$

40. $n^2 - 8n - 48 = 0$

41. $t^2 + t - 56 = 0$

42. $t^2 + t - 72 = 0$

43. $x^2 - 16x + 28 = 0$

44. $x^2 - 18x + 45 = 0$

45. $x^2 + 11x = 12$

46. $x^2 + 8x = 20$

47. $x(x - 10) = -16$

48. $x(x - 12) = -35$

49. $-x^2 - 2x + 24 = 0$

50. $-x^2 + 6x + 16 = 0$

For Problems 51–68, set up an equation and solve each problem.

51. Find two consecutive integers whose product is 56.

52. Find two consecutive odd whole numbers whose product is 63.

53. Find two consecutive even whole numbers whose product is 168.

54. One number is 2 larger than another number. The sum of their squares is 100. Find the numbers.

55. Find four consecutive integers such that the product of the two larger integers is 22 less than twice the product of the two smaller integers.

56. Find three consecutive integers such that the product of the two smaller integers is 2 more than ten times the largest integer.

57. One number is 3 smaller than another number. The square of the larger number is 9 larger than ten times the smaller number. Find the numbers.

58. The area of the floor of a rectangular room is 84 square feet. The length of the room is 5 feet more than its width. Find the length and width of the room.

59. Suppose that the width of a certain rectangle is 3 inches less than its length. The area is numerically 6 less than twice the perimeter. Find the length and width of the rectangle.

60. The sum of the areas of a square and a rectangle is 64 square centimeters. The length of the rectangle is 4 centimeters more than a side of the square, and the width of the rectangle is 2 centimeters more than a side

of the square. Find the dimensions of the square and the rectangle.

61. The perimeter of a rectangle is 30 centimeters, and the area is 54 square centimeters. Find the length and width of the rectangle. [*Hint:* Let w represent the width; then $15 - w$ represents the length.]

62. The perimeter of a rectangle is 44 inches, and its area is 120 square inches. Find the length and width of the rectangle.

63. An apple orchard contains 84 trees. The number of trees per row is 5 more than the number of rows. Find the number of rows.

64. A room contains 54 chairs. The number of rows is 3 less than the number of chairs per row. Find the number of rows.

65. Suppose that one leg of a right triangle is 7 feet shorter than the other leg. The hypotenuse is 2 feet longer than the longer leg. Find the lengths of all three sides of the right triangle.

66. Suppose that one leg of a right triangle is 7 meters longer than the other leg. The hypotenuse is 1 meter longer than the longer leg. Find the lengths of all three sides of the right triangle.

67. Suppose that the length of one leg of a right triangle is 2 inches less than the length of the other leg. If the length of the hypotenuse is 10 inches, find the length of each leg.

68. The length of one leg of a right triangle is 3 centimeters more than the length of the other leg. The length of the hypotenuse is 15 centimeters. Find the lengths of the two legs.

■ ■ ■ **Thoughts into words**

69. What does the expression "not factorable using integers" mean to you?

70. Discuss the role that factoring plays in solving equations.

71. Explain how you would solve the equation $(x - 3)(x + 4) = 0$

and also how you would solve $(x - 3)(x + 4) = 8$.

■ ■ ■ **Further investigations**

For Problems 72–75, factor each trinomial and assume that all variables appearing as exponents represent positive integers.

72. $x^{2a} + 10x^a + 24$

73. $x^{2a} + 13x^a + 40$

74. $x^{2a} - 2x^a - 8$

75. $x^{2a} + 6x^a - 27$

76. Suppose that we want to factor $n^2 + 26n + 168$ so that we can solve the equation $n^2 + 26n + 168 = 0$. We need to find two positive integers whose product is 168 and whose sum is 26. Because the constant term, 168, is rather large, let's look at it in prime factored form:

$$168 = 2 \cdot 2 \cdot 2 \cdot 3 \cdot 7$$

Now we can mentally form two numbers by using all of these factors in different combinations. Using two 2s and the 3 in one number and the other 2 and the 7 in another number produces $2 \cdot 2 \cdot 3 = 12$ and $2 \cdot 7 = 14$. Therefore, we can solve the given equation as follows:

$$n^2 + 26n + 168 = 0$$
$$(n + 12)(n + 14) = 0$$

$$n + 12 = 0 \qquad \text{or} \qquad n + 14 = 0$$
$$n = -12 \qquad \text{or} \qquad n = -14$$

The solution set is $\{-14, -12\}$.

Solve each of the following equations.

a. $n^2 + 30n + 216 = 0$

b. $n^2 + 35n + 294 = 0$

c. $n^2 - 40n + 384 = 0$

d. $n^2 - 40n + 375 = 0$

e. $n^2 + 6n - 432 = 0$

f. $n^2 - 16n - 512 = 0$

7.4 Factoring Trinomials of the Form $ax^2 + bx + c$

Objectives

■   Factor trinomials where the leading coefficient is not 1.

■   Solve equations that involve factoring.

Now let's consider factoring trinomials where the coefficient of the squared term is not 1. We first illustrate an informal trial-and-error technique that works quite well for certain types of trinomials. This technique simply relies on our knowledge of multiplication of binomials.

E X A M P L E   1

Factor $2x^2 + 7x + 3$.

Solution

By looking at the first term, $2x^2$, and the positive signs of the other two terms, we know that the binomials are of the form

$$(2x + \underline{\quad})(x + \underline{\quad})$$

Because the factors of the constant term, 3, are 1 and 3, we have only two possibilities to try:

$$(2x + 3)(x + 1) \qquad \text{or} \qquad (2x + 1)(x + 3)$$

By checking the middle term of both of these products, we find that the second one yields the correct middle term of $7x$. Therefore,

$$2x^2 + 7x + 3 = (2x + 1)(x + 3)$$ ■

EXAMPLE 2

Factor $6x^2 - 17x + 5$.

Solution

First, we note that $6x^2$ can be written as $2x \cdot 3x$ or $6x \cdot x$. Second, because the middle term of the trinomial is negative and the last term is positive, we know that the binomials are of the form

$$(2x - \underline{\hspace{1cm}})(3x - \underline{\hspace{1cm}}) \quad \text{or} \quad (6x - \underline{\hspace{1cm}})(x - \underline{\hspace{1cm}})$$

The factors of the constant term, 5, are 1 and 5, so we have the following possibilities:

$$(2x - 5)(3x - 1) \qquad (2x - 1)(3x - 5)$$
$$(6x - 5)(x - 1) \qquad (6x - 1)(x - 5)$$

By checking the middle term for each of these products, we find that the product $(2x - 5)(3x - 1)$ produces the desired term of $-17x$. Therefore,

$$6x^2 - 17x + 5 = (2x - 5)(3x - 1)$$

EXAMPLE 3

Factor $8x^2 - 8x - 30$.

Solution

First, we note that the polynomial $8x^2 - 8x - 30$ has a common factor of 2. Factoring out the common factor gives us $2(4x^2 - 4x - 15)$. Now we need to factor $4x^2 - 4x - 15$.

Now, we note that $4x^2$ can be written as $4x \cdot x$ or $2x \cdot 2x$. Second, the last term, -15, can be written as $(1)(-15)$, $(-1)(15)$, $(3)(-5)$, or $(-3)(5)$. Thus we can generate the possibilities for the binomial factors as follows:

Using 1 and −15	**Using −1 and 15**
$(4x - 15)(x + 1)$	$(4x - 1)(x + 15)$
$(4x + 1)(x - 15)$	$(4x + 15)(x - 1)$
$(2x + 1)(2x - 15)$	$(2x - 1)(2x + 15)$

Using 3 and −5	**Using −3 and 5**
$(4x + 3)(x - 5)$	$(4x - 3)(x + 5)$
$(4x - 5)(x + 3)$	$(4x + 5)(x - 3)$
✓ $(2x - 5)(2x + 3)$	$(2x + 5)(2x - 3)$

By checking the middle term of each of these products, we find that the product indicated with a check mark produces the desired middle term of $-4x$. Therefore,

$$8x^2 - 8x - 30 = 2(2x - 5)(2x + 3)$$

Let's pause for a moment and look back over Examples 1, 2, and 3. Obviously, Example 3 created the most difficulty because we had to consider so many possibilities. We have suggested one possible format for considering the possibilities, but as you practice such problems, you may develop a format of your own that works better for you. Whatever format you use, the key idea is to organize your work so that you consider all possibilities. Let's look at another example.

EXAMPLE 4

Factor $4x^2 + 6x + 9$.

Solution

First, we note that $4x^2$ can be written as $4x \cdot x$ or $2x \cdot 2x$. Second, because the middle term is positive and the last term is positive, we know that the binomials are of the form

$$(4x + \underline{\quad})(x + \underline{\quad}) \quad \text{or} \quad (2x + \underline{\quad})(2x + \underline{\quad})$$

Because 9 can be written as $9 \cdot 1$ or $3 \cdot 3$, we have only the following five possibilities to try:

$$(4x + 9)(x + 1) \quad (4x + 1)(x + 9)$$
$$(4x + 3)(x + 3) \quad (2x + 1)(2x + 9)$$
$$(2x + 3)(2x + 3)$$

When we try all of these possibilities, we find that none of them yields a middle term of $6x$. Therefore, $4x^2 + 6x + 9$ is *not factorable* using integers.   ■

REMARK: Example 4 illustrates the importance of organizing your work so that you try *all* possibilities before you conclude that a particular trinomial is not factorable.

Now We Can Solve More Equations

The ability to factor certain trinomials of the form $ax^2 + bx + c$ provides us with greater equation-solving capabilities. Consider the next examples.

EXAMPLE 5

Solve $3x^2 + 17x + 10 = 0$.

Solution

$$3x^2 + 17x + 10 = 0$$
$$(x + 5)(3x + 2) = 0 \qquad \text{Factoring } 3x^2 + 17x + 10 \text{ as } (x + 5)(3x + 2) \text{ may require some extra work on scratch paper.}$$

$$x + 5 = 0 \qquad \text{or} \qquad 3x + 2 = 0 \quad ab = 0 \text{ if and only if } a = 0 \text{ or } b = 0$$
$$x = -5 \qquad \text{or} \qquad 3x = -2$$
$$x = -5 \qquad \text{or} \qquad x = -\frac{2}{3}$$

The solution set is $\left\{ -5, -\frac{2}{3} \right\}$. Check it!   ■

EXAMPLE 6   Solve $24x^2 + 2x - 15 = 0$.

Solution

$$24x^2 + 2x - 15 = 0$$
$$(4x - 3)(6x + 5) = 0$$

$$4x - 3 = 0 \quad \text{or} \quad 6x + 5 = 0$$
$$4x = 3 \quad \text{or} \quad 6x = -5$$
$$x = \frac{3}{4} \quad \text{or} \quad x = -\frac{5}{6}$$

The solution set is $\left\{-\dfrac{5}{6}, \dfrac{3}{4}\right\}$. ■

CONCEPT QUIZ

For Problems 1–5, answer true or false.

1. Any trinomial of the form $ax^2 + bx + c$ can be factored (using integers) into the product of two binomials.

2. To factor $2x^2 - x - 3$, we look for two numbers whose product is -3 and whose sum is -1.

3. A trinomial of the form $ax^2 + bx + c$ will never have a common factor other than 1.

4. The factored form $(x + 3)(2x + 4)$ is factored completely.

5. The difference-of-squares polynomial $9x^2 - 25$ could be written as the trinomial $9x^2 + 0x - 25$.

PROBLEM SET 7.4

For Problems 1–30, factor each of the trinomials completely. Indicate any that are not factorable using integers.

1. $3x^2 + 7x + 2$

2. $2x^2 + 9x + 4$

3. $6x^2 + 19x + 10$

4. $12x^2 + 19x + 4$

5. $4x^2 - 25x + 6$

6. $5x^2 - 22x + 8$

7. $12x^2 - 31x + 20$

8. $8x^2 - 30x + 7$

9. $5y^2 - 33y - 14$

10. $6y^2 - 4y - 16$

11. $4n^2 + 26n - 48$

12. $4n^2 + 17n - 15$

13. $2x^2 + x + 7$

14. $7x^2 + 19x + 10$

15. $18x^2 + 45x + 7$

16. $10x^2 + x - 5$

17. $21x^2 - 90x + 24$

18. $6x^2 - 17x + 12$

19. $8x^2 + 2x - 21$

20. $9x^2 + 15x - 14$

21. $9t^2 - 15t - 14$

22. $12t^3 - 20t^2 - 25t$

23. $12y^2 + 79y - 35$

24. $9y^2 + 52y - 12$

25. $6n^2 + 2n - 5$

26. $20n^2 - 27n + 9$

27. $14x^2 + 55x + 21$

28. $15x^2 + 34x + 15$

29. $20x^2 - 31x + 12$

30. $8t^2 - 3t - 4$

For Problems 31–50, solve each equation.

31. $2x^2 + 13x + 6 = 0$

32. $3x^2 + 16x + 5 = 0$

33. $12x^2 + 11x + 2 = 0$

34. $15x^2 + 56x + 20 = 0$

35. $3x^2 - 25x + 8 = 0$

36. $4x^2 - 31x + 21 = 0$

37. $15n^2 - 41n + 14 = 0$

38. $6n^2 - 31n + 40 = 0$

39. $6t^2 + 37t - 35 = 0$

40. $2t^2 + 15t - 27 = 0$

41. $16y^2 - 18y - 9 = 0$

42. $9y^2 - 15y - 14 = 0$

43. $9x^2 - 6x - 8 = 0$

44. $12n^2 + 28n - 5 = 0$

45. $10x^2 - 29x + 10 = 0$

46. $4x^2 - 16x + 15 = 0$

47. $6x^2 + 19x = -10$

48. $12x^2 + 17x = -6$

49. $16x(x + 1) = 5$

50. $5x(5x + 2) = 8$

■ ■ ■ **Thoughts into words**

51. Explain your thought process when factoring $24x^2 - 17x - 20$.

52. Your friend factors $8x^2 - 32x + 32$ as follows:

$$8x^2 - 32x + 32 = (4x - 8)(2x - 4)$$
$$= 4(x - 2)(2)(x - 2)$$
$$= 8(x - 2)(x - 2)$$

Is she correct? Do you have any suggestions for her?

53. Your friend solves the equation $8x^2 - 32x + 32 = 0$ as follows:

$$8x^2 - 32x + 32 = 0$$
$$(4x - 8)(2x - 4) = 0$$

$4x - 8 = 0$	or	$2x - 4 = 0$
$4x = 8$	or	$2x = 4$
$x = 2$	or	$x = 2$

The solution set is {2}. Is she correct? Do you have any changes to recommend?

Answers to Concept Quiz

1. False   **2.** False   **3.** False   **4.** False   **5.** True

7.5   Factoring, Solving Equations, and Problem Solving

Objectives

■   Factor perfect-square trinomials.

■   Recognize the different types of factoring patterns.

■   Use factoring to solve equations.

■   Solve word problems that involve factoring.

Factoring

Before we summarize our work with factoring techniques, let's look at two more special factoring patterns. These patterns emerge when multiplying binomials. Consider the following examples.

$$(x + 5)^2 = (x + 5)(x + 5) = x^2 + 10x + 25$$
$$(2x + 3)^2 = (2x + 3)(2x + 3) = 4x^2 + 12x + 9$$
$$(4x + 7)^2 = (4x + 7)(4x + 7) = 16x^2 + 56x + 49$$

In general, $(a + b)^2 = (a + b)(a + b) = a^2 + 2ab + b^2$. Also,

$$(x - 6)^2 = (x - 6)(x - 6) = x^2 - 12x + 36$$
$$(3x - 4)^2 = (3x - 4)(3x - 4) = 9x^2 - 24x + 16$$
$$(5x - 2)^2 = (5x - 2)(5x - 2) = 25x^2 - 20x + 4$$

In general, $(a - b)^2 = (a - b)(a - b) = a^2 - 2ab + b^2$. Thus we have the following patterns.

Perfect-Square Trinomials

$$a^2 + 2ab + b^2 = (a + b)^2$$
$$a^2 - 2ab + b^2 = (a - b)^2$$

Trinomials of the form $a^2 + 2ab + b^2$ or $a^2 - 2ab + b^2$ are called **perfect-square trinomials**. They are easy to recognize because of the nature of their terms. For example, $9x^2 + 30x + 25$ is a perfect-square trinomial for these reasons:

1. The first term is a square: $(3x)^2$
2. The last term is a square: $(5)^2$
3. The middle term is twice the product of the quantities being squared in the first and last terms: $2(3x)(5)$

Likewise, $25x^2 - 40xy + 16y^2$ is a perfect-square trinomial for these reasons:

1. The first term is a square: $(5x)^2$
2. The last term is a square: $(4y)^2$
3. The middle term is twice the product of the quantities being squared in the first and last terms: $2(5x)(4y)$

Once we know that we have a perfect-square trinomial, the factoring process follows immediately from the two basic patterns.

$$9x^2 + 30x + 25 = (3x + 5)^2$$
$$25x^2 - 40xy + 16y^2 = (5x - 4y)^2$$

Here are some additional examples of perfect-square trinomials and their factored forms.

$$x^2 - 16x + 64 = \boxed{(x)^2 - 2(x)(8) + (8)^2} = (x - 8)^2$$
$$16x^2 - 56x + 49 = \boxed{(4x)^2 - 2(4x)(7) + (7)^2} = (4x - 7)^2$$
$$25x^2 + 20xy + 4y^2 = \boxed{(5x)^2 + 2(5x)(2y) + (2y)^2} = (5x + 2y)^2$$
$$1 + 6y + 9y^2 = \boxed{(1)^2 + 2(1)(3y) + (3y)^2} = (1 + 3y)^2$$
$$4m^2 - 4mn + n^2 = \boxed{(2m)^2 - 2(2m)(n) + (n)^2} = (2m - n)^2$$

You may want to do this step mentally after you feel comfortable with the process.

In this chapter, we have considered some basic factoring techniques one at a time, but you must be able to apply them as needed in a variety of situations. Let's first summarize the techniques and then consider some examples.

These are the techniques we have discussed in this chapter:

1. Factoring by using the distributive property to factor out the greatest common monomial or binomial factor
2. Factoring by grouping
3. Factoring by applying the difference-of-squares pattern
4. Factoring by applying the perfect-square-trinomial pattern
5. Factoring trinomials of the form $x^2 + bx + c$ into the product of two binomials
6. Factoring trinomials of the form $ax^2 + bx + c$ into the product of two binomials

As a general guideline, **always look for a greatest common monomial factor first,** and then proceed with the other factoring techniques.

In each of the following examples, we have factored completely whenever possible. Study them carefully and note the factoring techniques we used.

1. $2x^2 + 12x + 10 = 2(x^2 + 6x + 5) = 2(x + 1)(x + 5)$

2. $4x^2 + 36 = 4(x^2 + 9)$ Remember that the *sum* of two squares is not factorable using integers unless there is a common factor.

3. $4t^2 + 20t + 25 = (2t + 5)^2$ If you fail to recognize a perfect-trinomial square, no harm is done. Simply proceed to factor into the product of two binomials, and then you will recognize that the two binomials are the same.

4. $x^2 - 3x - 8$ is not factorable using integers. This becomes obvious from the table.

Product	Sum
$1(-8) = -8$	$1 + (-8) = -7$
$-1(8) = -8$	$-1 + 8 = 7$
$2(-4) = -8$	$2 + (-4) = -2$
$-2(4) = -8$	$-2 + 4 = 2$

No two factors of -8 produce a sum of -3.

5. $6y^2 - 13y - 28 = (2y - 7)(3y + 4)$. We found the binomial factors as follows:

$(y + \underline{\quad})(6y - \underline{\quad})$

    or $1 \cdot 28$ or $28 \cdot 1$

$(y - \underline{\quad})(6y + \underline{\quad})$ $2 \cdot 14$ or $14 \cdot 2$

    or $4 \cdot 7$ or $\boxed{7 \cdot 4}$

$(2y - \underline{\quad})(3y + \underline{\quad})$

    or

$(2y + \underline{\quad})(3y - \underline{\quad})$

6. $32x^2 - 50y^2 = 2(16x^2 - 25y^2) = 2(4x + 5y)(4x - 5y)$

Solving Equations by Factoring

Each time we considered a new factoring technique in this chapter, we used that technique to help solve some equations. It is important that you be able to recognize which technique works for a particular type of equation.

EXAMPLE 1   Solve $x^2 = 25x$.

Solution

$$x^2 = 25x$$
$$x^2 - 25x = 0$$
$$x(x - 25) = 0$$
$$x = 0 \quad \text{or} \quad x - 25 = 0$$
$$x = 0 \quad \text{or} \quad x = 25$$

The solution set is $\{0, 25\}$. Check it! ■

EXAMPLE 2   Solve $x^3 - 36x = 0$.

Solution

$$x^3 - 36x = 0$$
$$x(x^2 - 36) = 0$$
$$x(x + 6)(x - 6) = 0$$
$$x = 0 \quad \text{or} \quad x + 6 = 0 \quad \text{or} \quad x - 6 = 0 \quad \text{If } abc = 0, \text{ then } a = 0$$
$$\text{or } b = 0 \text{ or } c = 0.$$
$$x = 0 \quad \text{or} \quad x = -6 \quad \text{or} \quad x = 6$$

The solution set is $\{-6, 0, 6\}$. Does it check? ■

EXAMPLE 3   Solve $10x^2 - 13x - 3 = 0$.

Solution

$$10x^2 - 13x - 3 = 0$$
$$(5x + 1)(2x - 3) = 0$$

$5x + 1 = 0$	or	$2x - 3 = 0$
$5x = -1$	or	$2x = 3$
$x = -\dfrac{1}{5}$	or	$x = \dfrac{3}{2}$

The solution set is $\left\{-\dfrac{1}{5}, \dfrac{3}{2}\right\}$. Does it check?   ∎

EXAMPLE 4   Solve $4x^2 - 28x + 49 = 0$.

Solution

$$4x^2 - 28x + 49 = 0$$
$$(2x - 7)^2 = 0$$
$$(2x - 7)(2x - 7) = 0$$

$2x - 7 = 0$	or	$2x - 7 = 0$
$2x = 7$	or	$2x = 7$
$x = \dfrac{7}{2}$	or	$x = \dfrac{7}{2}$

The solution set is $\left\{\dfrac{7}{2}\right\}$.   ∎

Pay special attention to the next example. We need to change the form of the original equation before we can apply the property $ab = 0$ if and only if $a = 0$ or $b = 0$. The unique feature of this property is that an indicated product is set equal to zero.

EXAMPLE 5   Solve $(x + 1)(x + 4) = 40$.

Solution

$$(x + 1)(x + 4) = 40$$
$$x^2 + 5x + 4 = 40$$
$$x^2 + 5x - 36 = 0$$
$$(x + 9)(x - 4) = 0$$

$x + 9 = 0$	or	$x - 4 = 0$
$x = -9$	or	$x = 4$

The solution set is $\{-9, 4\}$. Check it!   ∎

EXAMPLE 6   Solve $2n^2 + 16n - 40 = 0$.

Solution

$$2n^2 + 16n - 40 = 0$$
$$2(n^2 + 8n - 20) = 0$$
$$n^2 + 8n - 20 = 0 \qquad \text{Multiplied both sides by } \frac{1}{2}$$
$$(n + 10)(n - 2) = 0$$
$$n + 10 = 0 \qquad \text{or} \qquad n - 2 = 0$$
$$n = -10 \qquad \text{or} \qquad n = 2$$

The solution set is $\{-10, 2\}$. Does it check?   ■

Problem Solving

The preface to this book states that a common thread throughout the book is *to learn a skill, to use that skill to help solve equations,* and then *to use equations to help solve problems.* This approach should be very apparent in this chapter. Our new factoring skills have provided more ways of solving equations, which in turn gives us more power to solve word problems. We conclude the chapter by solving a few more problems.

PROBLEM 1   Find two numbers whose product is 65 if one of the numbers is 3 more than twice the other number.

Solution

Let n represent one of the numbers; then $2n + 3$ represents the other number. Because their product is 65, we can set up and solve the following equation:

$$n(2n + 3) = 65$$
$$2n^2 + 3n - 65 = 0$$
$$(2n + 13)(n - 5) = 0$$
$$2n + 13 = 0 \qquad \text{or} \qquad n - 5 = 0$$
$$2n = -13 \qquad \text{or} \qquad n = 5$$
$$n = -\frac{13}{2} \qquad \text{or} \qquad n = 5$$

If $n = -\dfrac{13}{2}$, then $2n + 3 = 2\left(-\dfrac{13}{2}\right) + 3 = -10$. If $n = 5$, then $2n + 3 = 2(5) + 3 = 13$. Thus the numbers are $-\dfrac{13}{2}$ and -10, or 5 and 13.   ■

P R O B L E M   2

The area of a triangular sheet of paper is 14 square inches. One side of the triangle is 3 inches longer than the altitude to that side. Find the length of the one side and the length of the altitude to that side.

Solution

Let h represent the altitude to the side. Then $h + 3$ represents the length of the side of the triangle (see Figure 7.5). Because the formula for finding the area of a triangle is $A = \dfrac{1}{2}bh$, we have

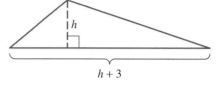

Figure 7.5

$$\frac{1}{2}h(h + 3) = 14$$

$$h(h + 3) = 28 \qquad \text{Multiplied both sides by 2}$$

$$h^2 + 3h = 28$$

$$h^2 + 3h - 28 = 0$$

$$(h + 7)(h - 4) = 0$$

$$h + 7 = 0 \qquad \text{or} \qquad h - 4 = 0$$

$$h = -7 \qquad \text{or} \qquad h = 4$$

The solution -7 is not reasonable. Thus the altitude is 4 inches, and the length of the side to which that altitude is drawn is 7 inches.   ■

P R O B L E M   3

A strip with a uniform width is shaded along both sides and both ends of a rectangular poster with dimensions 12 inches by 16 inches. How wide is the strip if one-half of the poster is shaded?

Solution

Let x represent the width of the shaded strip of the poster in Figure 7.6. The area of the strip is one-half of the area of the poster; therefore, it is $\dfrac{1}{2}(12)(16) = 96$ square inches.

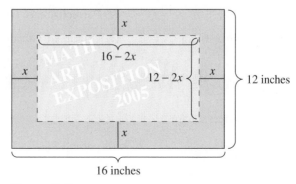

Figure 7.6

Furthermore, we can represent the area of the strip around the poster by *the area of the poster minus the area of the unshaded portion.* Thus we can set up and solve the following equation:

Area of poster $-$ Area of unshaded portion $=$ Area of strip

$$16(12) \quad - \quad (16 - 2x)(12 - 2x) \quad = \quad 96$$

$$192 - (192 - 56x + 4x^2) = 96$$
$$192 - 192 + 56x - 4x^2 = 96$$
$$-4x^2 + 56x - 96 = 0$$
$$x^2 - 14x + 24 = 0$$
$$(x - 12)(x - 2) = 0$$
$$x - 12 = 0 \quad \text{or} \quad x - 2 = 0$$
$$x = 12 \quad \text{or} \quad x = 2$$

Obviously, the strip cannot be 12 inches wide because the total width of the poster is 12 inches. Thus we must disregard the solution 12 and conclude that the strip is 2 inches wide. ∎

CONCEPT QUIZ

For Problems 1–7, match each factoring problem with the name of the type of pattern that would be used to factor the problem.

1. $x^2 + 2xy + y^2$ **A.** Trinomial with an x-squared coefficient of one

2. $x^2 - y^2$ **B.** Common binomial factor

3. $ax + ay + bx + by$ **C.** Difference of two squares

4. $x^2 + bx + c$ **D.** Common factor

5. $ax^2 + bx + c$ **E.** Factor by grouping

6. $ax^2 + ax + a$ **F.** Perfect-square trinomial

7. $(a + b)x + (a + b)y$ **G.** Trinomial with an x-squared coefficient of not one

PROBLEM SET 7.5

For Problems 1–12, factor each of the perfect-square trinomials.

1. $x^2 + 4x + 4$

2. $x^2 + 18x + 81$

3. $x^2 - 10x + 25$

4. $x^2 - 24x + 144$

5. $9n^2 + 12n + 4$

6. $25n^2 + 30n + 9$

7. $16a^2 - 8a + 1$

8. $36a^2 - 84a + 49$

9. $4 + 36x + 81x^2$

10. $1 - 4x + 4x^2$

11. $16x^2 - 24xy + 9y^2$

12. $64x^2 + 16xy + y^2$

For Problems 13–40, factor each polynomial completely. Indicate any that are not factorable using integers.

13. $2x^2 + 17x + 8$

14. $x^2 + 19x$

15. $2x^3 - 72x$

16. $30x^2 - x - 1$

17. $n^2 - 7n - 60$

18. $4n^3 - 100n$

19. $3a^2 - 7a - 4$

20. $a^2 + 7a - 30$

21. $8x^2 + 72$

22. $3y^3 - 36y^2 + 96y$

23. $9x^2 + 30x + 25$

24. $5x^2 - 5x - 6$

25. $15x^2 + 65x + 70$

26. $4x^2 - 20xy + 25y^2$

27. $24x^2 + 2x - 15$

28. $9x^2y - 27xy$

29. $xy + 5y - 8x - 40$

30. $xy - 3y + 9x - 27$

31. $20x^2 + 31xy - 7y^2$

32. $2x^2 - xy - 36y^2$

33. $24x^2 + 18x - 81$

34. $30x^2 + 55x - 50$

35. $12x^2 + 6x + 30$

36. $24x^2 - 8x + 32$

37. $5x^4 - 80$

38. $3x^5 - 3x$

39. $x^2 + 12xy + 36y^2$

40. $4x^2 - 28xy + 49y^2$

For Problems 41–70, solve each equation.

41. $4x^2 - 20x = 0$

42. $-3x^2 - 24x = 0$

43. $x^2 - 9x - 36 = 0$

44. $x^2 + 8x - 20 = 0$

45. $-2x^3 + 8x = 0$

46. $4x^3 - 36x = 0$

47. $6n^2 - 29n - 22 = 0$

48. $30n^2 - n - 1 = 0$

49. $(3n - 1)(4n - 3) = 0$

50. $(2n - 3)(7n + 1) = 0$

51. $(n - 2)(n + 6) = -15$

52. $(n + 3)(n - 7) = -25$

53. $2x^2 = 12x$

54. $-3x^2 = 15x$

55. $t^3 - 2t^2 - 24t = 0$

56. $2t^3 - 16t^2 - 18t = 0$

57. $12 - 40x + 25x^2 = 0$

58. $12 - 7x - 12x^2 = 0$

59. $n^2 - 28n + 192 = 0$

60. $n^2 + 33n + 270 = 0$

61. $(3n + 1)(n + 2) = 12$

62. $(2n + 5)(n + 4) = -1$

63. $x^3 = 6x^2$

64. $x^3 = -4x^2$

65. $9x^2 - 24x + 16 = 0$

66. $25x^2 + 60x + 36 = 0$

67. $x^3 + 10x^2 + 25x = 0$

68. $x^3 - 18x^2 + 81x = 0$

69. $24x^2 + 17x - 20 = 0$

70. $24x^2 + 74x - 35 = 0$

For Problems 71–88, set up an equation and solve each problem.

71. Find two numbers whose product is 15 such that one of the numbers is seven more than four times the other number.

72. Find two numbers whose product is 12 such that one of the numbers is four less than eight times the other number.

73. Find two numbers whose product is -1. One of the numbers is three more than twice the other number.

74. Suppose that the sum of the squares of three consecutive integers is 110. Find the integers.

75. One number is one more than twice another number. The sum of the squares of the two numbers is 97. Find the numbers.

76. One number is one less than three times another number. If the product of the two numbers is 102, find the numbers.

77. In an office building, a room contains 54 chairs. The number of chairs per row is three less than twice the number of rows. Find the number of rows and the number of chairs per row.

78. An apple orchard contains 85 trees. The number of trees in each row is three less than four times the number of rows. Find the number of rows and the number of trees per row.

79. Suppose that the combined area of two squares is 360 square feet. Each side of the larger square is three times as long as a side of the smaller square. How big is each square?

80. The area of a rectangular slab of sidewalk is 45 square feet. Its length is 3 feet more than four times its width. Find the length and width of the slab.

81. The length of a rectangular sheet of paper is 1 centimeter more than twice its width, and the area of the rectangle is 55 square centimeters. Find the length and width of the rectangle.

82. Suppose that the length of a certain rectangle is three times its width. If the length is increased by 2 inches and the width increased by 1 inch, the newly formed rectangle has an area of 70 square inches. Find the length and width of the original rectangle.

83. The area of a triangle is 51 square inches. One side of the triangle is 1 inch less than three times the length of the altitude to that side. Find the length of that side and the length of the altitude to that side.

84. Suppose that a square and a rectangle have equal areas. Furthermore, suppose that the length of the rectangle is twice the length of a side of the square and the width of the rectangle is 4 centimeters less than the length of a side of the square. Find the dimensions of both figures.

85. A strip of uniform width is to be cut off of both sides and both ends of a sheet of paper that is 8 inches by 11 inches in order to reduce the size of the paper to an area of 40 square inches. Find the width of the strip.

86. The sum of the areas of two circles is 100π square centimeters. The length of a radius of the larger circle is 2 centimeters more than the length of a radius of the smaller circle. Find the length of a radius of each circle.

87. The sum of the areas of two circles is 180π square inches. The length of a radius of the smaller circle is 6 inches less than the length of a radius of the larger circle. Find the length of a radius of each circle.

88. A strip of uniform width is shaded along both sides and both ends of a rectangular poster that is 18 inches by 14 inches. How wide is the strip if the unshaded portion of the poster has an area of 165 square inches?

■ ■ ■ **Thoughts into words**

89. When factoring polynomials, why do you think that it is best to look for a greatest common monomial factor first?

90. Explain how you would solve $(4x - 3)(8x + 5) = 0$ and also how you would solve $(4x - 3)(8x + 5) = -9$.

91. Explain how you would solve

$$(x + 2)(x + 3) = (x + 2)(3x - 1)$$

Do you see more than one approach to this problem?

Answers to Concept Quiz

1. F or A   **2.** C   **3.** E   **4.** A   **5.** G   **6.** D   **7.** B

CHAPTER 7

SUMMARY

(7.1) The distributive property in the form $ab + ac = a(b + c)$ provides the basis for **factoring out a greatest common monomial or binomial factor.**

Rewriting an expression such as $ab + 3a + bc + 3c$ as $a(b + 3) + c(b + 3)$ and then factoring out the common binomial factor of $b + 3$ so that $a(b + 3) + c(b + 3)$ becomes $(b + 3)(a + c)$, is called **factoring by grouping.**

The property $ab = 0$ if and only if $a = 0$ or $b = 0$ provides us with another technique for solving equations.

(7.2) This factoring pattern is called the **difference of two squares:**

$$a^2 - b^2 = (a - b)(a + b)$$

(7.3) The following multiplication pattern provides a technique for factoring trinomials of the form $x^2 + bx + c$.

$$(x + r)(x + s) = x^2 + rx + sx + rs$$
$$= x^2 + (r + s)x + rs$$

Sum of r and s Product of r and s

(7.4) To review a technique for factoring trinomials of the form $ax^2 + bx + c$, turn to Section 7.4 and study Examples 1–4.

(7.5) As a general guideline for **factoring completely,** always look for a greatest common monomial or binomial factor *first*, and then proceed with one or more of the following techniques:

1. Apply the difference-of-squares pattern.

2. Apply the perfect-square-trinomial pattern.

3. Factor a trinomial of the form

   $$x^2 + bx + c$$

   into the product of two binomials.

4. Factor a trinomial of the form

   $$ax^2 + bx + c$$

   into the product of two binomials.

CHAPTER 7 REVIEW PROBLEM SET

For Problems 1–24, factor completely. Indicate any polynomials that are not factorable using integers.

1. $x^2 - 9x + 14$

2. $3x^2 + 21x$

3. $9x^2 - 4$

4. $4x^2 + 8x - 5$

5. $25x^2 - 60x + 36$

6. $n^3 + 13n^2 + 40n$

7. $y^2 + 11y - 12$

8. $3xy^2 + 6x^2y$

9. $x^4 - 1$

10. $18n^2 + 9n - 5$

11. $x^2 + 7x + 24$

12. $4x^2 - 3x - 7$

13. $3n^2 + 3n - 90$

14. $x^3 - xy^2$

15. $2x^2 + 3xy - 2y^2$

16. $4n^2 - 6n - 40$

17. $5x + 5y + ax + ay$

18. $21t^2 - 5t - 4$

19. $2x^3 - 2x$

20. $3x^3 - 108x$

21. $16x^2 + 40x + 25$

22. $xy - 3x - 2y + 6$

23. $15x^2 - 7xy - 2y^2$

24. $6n^4 - 5n^3 + n^2$

For Problems 25–44, solve each equation.

25. $x^2 + 4x - 12 = 0$

26. $x^2 = 11x$

27. $2x^2 + 3x - 20 = 0$

28. $9n^2 + 21n - 8 = 0$

29. $6n^2 = 24$

30. $16y^2 + 40y + 25 = 0$

31. $t^3 - t = 0$

32. $28x^2 + 71x + 18 = 0$

33. $x^2 + 3x - 28 = 0$

34. $(x - 2)(x + 2) = 21$

35. $5n^2 + 27n = 18$

36. $4n^2 + 10n = 14$

37. $2x^3 - 8x = 0$

38. $x^2 - 20x + 96 = 0$

39. $4t^2 + 17t - 15 = 0$

40. $3(x + 2) - x(x + 2) = 0$

41. $(2x - 5)(3x + 7) = 0$

42. $(x + 4)(x - 1) = 50$

43. $-7n - 2n^2 = -15$ **44.** $-23x + 6x^2 = -20$

Set up an equation and solve each of the following problems.

45. The larger of two numbers is one less than twice the smaller number. The difference of their squares is 33. Find the numbers.

46. The length of a rectangle is 2 centimeters less than five times the width of the rectangle. The area of the rectangle is 16 square centimeters. Find the length and width of the rectangle.

47. Suppose that the combined area of two squares is 104 square inches. Each side of the larger square is five times as long as a side of the smaller square. Find the size of each square.

48. The longer leg of a right triangle is one unit shorter than twice the length of the shorter leg. The hypotenuse is one unit longer than twice the length of the shorter leg. Find the lengths of the three sides of the triangle.

49. The product of two numbers is 26 and one of the numbers is one larger than six times the other number. Find the numbers.

50. Find three consecutive positive odd whole numbers such that the sum of the squares of the two smaller numbers is nine more than the square of the largest number.

51. The number of books per shelf in a bookcase is one less than nine times the number of shelves. If the bookcase contains 140 books, find the number of shelves.

52. The combined area of a square and a rectangle is 225 square yards. The length of the rectangle is eight times the width of the rectangle, and the length of a side of the square is the same as the width of the rectangle. Find the dimensions of the square and the rectangle.

53. Suppose that we want to find two consecutive integers such that the sum of their squares is 613. What are they?

54. If numerically the volume of a cube equals the total surface area of the cube, find the length of an edge of the cube.

55. The combined area of two circles is 53π square meters. The length of a radius of the larger circle is 1 meter more than three times the length of a radius of the smaller circle. Find the length of a radius of each circle.

56. The product of two consecutive odd whole numbers is one less than five times their sum. Find the numbers.

57. Sandy has a photograph that is 14 centimeters long and 8 centimeters wide. She wants to reduce the length and width by the same amount so that the area is decreased by 40 square centimeters. By what amount should she reduce the length and width?

58. Suppose that a strip of uniform width is plowed along both sides and both ends of a garden that is 120 feet long and 90 feet wide (see Figure 7.7). How wide is the strip if the garden is half plowed?

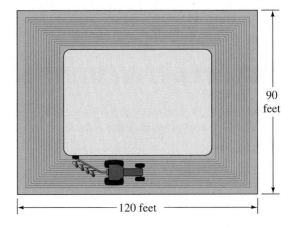

90 feet

120 feet

Figure 7.7

CHAPTER 7

TEST

 applies to all problems in this Chapter Test.

For Problems 1–10, factor each expression completely.

1. $x^2 + 3x - 10$ 2. $x^2 - 5x - 24$

3. $2x^3 - 2x$ 4. $x^2 + 21x + 108$

5. $18n^2 + 21n + 6$ 6. $ax + ay + 2bx + 2by$

7. $4x^2 + 17x - 15$ 8. $6x^2 + 24$

9. $30x^3 - 76x^2 + 48x$ 10. $28 + 13x - 6x^2$

For Problems 11–21, solve each equation.

11. $7x^2 = 63$

12. $x^2 + 5x - 6 = 0$

13. $4n^2 = 32n$

14. $(3x - 2)(2x + 5) = 0$

15. $(x - 3)(x + 7) = -9$

16. $x^3 + 16x^2 + 48x = 0$

17. $9(x - 5) - x(x - 5) = 0$

18. $3t^2 + 35t = 12$

19. $8 - 10x - 3x^2 = 0$

20. $3x^3 = 75x$

21. $25n^2 - 70n + 49 = 0$

For Problems 22–25, set up an equation and solve each problem.

22. The length of a rectangle is 2 inches less than twice its width. If the area of the rectangle is 112 square inches, find the length of the rectangle.

23. The length of one leg of a right triangle is 4 centimeters more than the length of the other leg. The length of the hypotenuse is 8 centimeters more than the length of the shorter leg. Find the length of the shorter leg.

24. A room contains 112 chairs. The number of chairs per row is five less than three times the number of rows. Find the number of chairs per row.

25. If numerically the volume of a cube equals twice the total surface area, find the length of an edge of the cube.

CUMULATIVE REVIEW PROBLEM SET *Chapters 1-7*

For Problems 1–6, evaluate each algebraic expression for the given values of the variables. You may first want to simplify the expression or change its form by factoring.

1. $3x - 2xy - 7x + 5xy$ for $x = \dfrac{1}{2}$ and $y = 3$

2. $7(a - b) - 3(a - b) - (a - b)$ for $a = -3$ and $b = -5$

3. $ab + b^2$ for $a = 0.4$ and $b = 0.6$

4. $x^2 - y^2$ for $x = -6$ and $y = 4$

5. $x^2 + x - 72$ for $x = -10$

6. $3(x - 2) - (x + 3) - 5(x + 6)$ for $x = -6$

For Problems 7–14, evaluate each numerical expression.

7. 3^{-3}

8. $\left(\dfrac{2}{3}\right)^{-1}$

9. $\left(\dfrac{1}{2} + \dfrac{1}{3}\right)^0$

10. $\left(\dfrac{1}{3} + \dfrac{1}{4}\right)^{-1}$

11. -4^{-2}

12. -4^2

13. $\dfrac{1}{\left(\dfrac{2}{5}\right)^2}$

14. $(-3)^{-3}$

For Problems 15–26, perform the indicated operations and express answers in simplest form.

15. $\dfrac{7}{5x} + \dfrac{2}{x} - \dfrac{3}{2x}$

16. $\dfrac{4x}{5y} \div \dfrac{12x^2}{10y^2}$

17. $(-5x^2y)(7x^3y^4)$

18. $(9ab^3)^2$

19. $(-3n^2)(5n^2 + 6n - 2)$

20. $(5x - 1)(3x + 4)$

21. $(2x + 5)^2$

22. $(x + 2)(2x^2 - 3x - 1)$

23. $(x^2 - x - 1)(x^2 + 2x - 3)$

24. $(-2x - 1)(3x - 7)$

25. $\dfrac{24x^2y^3 - 48x^4y^5}{8xy^2}$

26. $(28x^2 - 19x - 20) \div (4x - 5)$

For Problems 27–36, factor each polynomial completely.

27. $3x^3 + 15x^2 + 27x$

28. $x^2 - 100$

29. $5x^2 - 22x + 8$

30. $8x^2 - 22x - 63$

31. $n^2 + 25n + 144$

32. $nx + ny - 2x - 2y$

33. $3x^3 - 3x$

34. $2x^3 - 6x^2 - 108x$

35. $36x^2 - 60x + 25$

36. $3x^2 - 5xy - 2y^2$

For Problems 37–46, solve each equation.

37. $3(x - 2) - 2(x + 6) = -2(x + 1)$

38. $x^2 = -11x$

39. $0.2x - 3(x - 0.4) = 1$

40. $5n^2 - 5 = 0$

41. $x^2 + 5x - 6 = 0$

42. $\dfrac{2x + 1}{2} + \dfrac{3x - 4}{3} = 1$

43. $2(x - 1) - x(x - 1) = 0$

44. $6x^2 + 19x - 7 = 0$

45. $(2x - 1)(x - 8) = 0$

46. $(x + 1)(x + 6) = 24$

For Problems 47–51, solve each inequality.

47. $-3x - 2 \geq 1$

48. $18 < 2(x - 4)$

49. $3(x - 2) - 2(x + 1) \leq -(x + 5)$

50. $\dfrac{2}{3}x - \dfrac{1}{4}x - 1 > 3$

51. $0.08x + 0.09(2x) \leq 130$

For Problems 52–55, graph each equation.

52. $y = -3x + 5$

53. $y = \dfrac{1}{4}x + 2$

54. $3x - y = 3$

55. $y = 2x^2 - 4$

For Problems 56–59, solve each system of equations.

56. $\begin{pmatrix} 7x - 2y = -34 \\ x + 2y = -14 \end{pmatrix}$

57. $\begin{pmatrix} 5x + 3y = -9 \\ 3x - 5y = 15 \end{pmatrix}$

58. $\begin{pmatrix} 6x - 11y = -58 \\ 8x + y = 1 \end{pmatrix}$

59. $\begin{pmatrix} \dfrac{1}{2}x + \dfrac{2}{3}y = -1 \\ \dfrac{2}{5}x - \dfrac{1}{3}y = -6 \end{pmatrix}$

60. Is 91 a prime or composite number?

61. Find the greatest common factor of 18 and 48.

62. Find the least common multiple of 6, 8, and 9.

63. Express $\dfrac{7}{4}$ as a percent.

64. Express 0.0024 in scientific notation.

65. Express $(3.14)(10^3)$ in ordinary decimal notation.

66. Graph on a number line the solutions for the compound inequality $x < 0$ or $x > 3$.

67. Find the area of a circular region if the circumference is 8π centimeters. Express the answer in terms of π.

68. Thirty percent of what number is 5.4?

69. Graph the inequality $3x - 2y < -6$.

For Problems 70–85, use an equation, an inequality, or a system of equations to help solve each problem.

70. One leg of a right triangle is 2 inches longer than the other leg. The hypotenuse is 4 inches longer than the shorter leg. Find the lengths of the three sides of the right triangle.

71. How many milliliters of a 65% solution of hydrochloric acid must be added to 40 milliliters of a 30% solution of hydrochloric acid to obtain a 55% solution?

72. A landscaping border 28 feet long is bent into the shape of a rectangle. The length of the rectangle is 2 feet more than the width. Find the dimensions of the rectangle.

73. Two motorcyclists leave Daytona Beach at the same time and travel in opposite directions. If one travels at 55 miles per hour and the other travels at 65 miles per hour, how long will it take for them to be 300 miles apart?

74. Find the length of an altitude of a trapezoid with bases of 10 centimeters and 22 centimeters and an area of 120 square centimeters.

75. If a car uses 16 gallons of gasoline for a 352-mile trip, at the same rate of consumption, how many gallons will it use on a 594-mile trip?

76. If two angles are supplementary, and the larger angle is 20° less than three times the smaller angle, find the measure of each angle.

77. Find the measures of the three angles of a triangle if the largest angle is 10° more than twice the smallest, and the other angle is 10° larger than the smallest angle.

78. Zorka has 175 coins consisting of pennies, nickels, and dimes. The number of dimes is five more than twice the number of pennies, and the number of nickels is 10 more than the number of pennies. How many coins of each kind does she have?

79. Rashed has some dimes and quarters amounting to $7.65. The number of quarters is three less than twice the number of dimes. How many coins of each kind does he have?

80. Ashley has scores of 85, 87, 90, and 91 on her first four algebra tests. What score must she get on the fifth test to have an average of 90 or better for the five tests?

81. The ratio of girls to boys in a certain school is six to five. If there is a total of 1650 students in the school, find the number of girls and the number of boys.

82. If a ring costs a jeweler $750, at what price should it be sold for the jeweler to make a profit of 70% based on the selling price?

83. Suppose that the jeweler in Problem 82 would be satisfied with a 70% profit based on the cost of the ring. At what price should he sell the ring?

84. How many quarts of pure alcohol must be added to 6 quarts of a 30% solution to obtain a 40% solution?

85. Suppose that the cost of five tennis balls and four golf balls is $17. Furthermore, suppose that at the same prices, the cost of three tennis balls and seven golf balls is $20.55. Find the cost of one tennis ball and the cost of one golf ball.

A quadratic equation can be solved to determine the width of a uniform strip trimmed off both the sides and ends of a sheet of paper to obtain a specified area for the sheet of paper.

© Pete Saloutos /CORBIS

A Transition from Elementary Algebra to Intermediate Algebra

O bserve the following five rows of numbers.

Row 1				1		1
Row 2			1		2		1
Row 3		1		3		3		1
Row 4	1		4		6		4		1
Row 5	1		5		10		10		5		1

This configuration can be extended indefinitely. Do you see a pattern that will create row 6? Of what significance is this configuration of numbers? These questions are answered in Section 8.4.

As the title indicates, our primary objective in this chapter is to review briefly some concepts of elementary algebra and, in some instances, to extend the concepts into intermediate algebra territory. For example, in Sections 8.1 and 8.2, we review some basic techniques for solving equations and inequalities. Then, in Section 8.3, these techniques are extended to solve equations and inequalities that involve absolute value.

In Section 8.4, operations on polynomials are reviewed, and the multiplication of binomials is extended to binomial expansions in general. Likewise, in Section 8.5 the division of polynomials is reviewed and then extended to synthetic division. Various techniques for factoring polynomials are reviewed in Section 8.6, and one or two new techniques are introduced.

This chapter should help you make a smooth transition from elementary algebra to intermediate algebra. We have indicated the new material in this chapter with a "New" symbol.

InfoTrac Project

Do a subject guide search on the metric system. Find a periodical article on why we should learn the metric system. Write a brief summary of the article. If the length of a game room is 5 meters less than twice the width, what is the length if the area is 117 square meters?

8.1   Equations: A Brief Review

Objectives

■   Apply properties of equality to solve linear equations.

■   Set up and solve proportions.

■   Solve systems of two linear equations.

■   Write equations to represent word problems and solve the equations.

An **algebraic equation** such as $3x + 1 = 13$ is neither true nor false as it stands; for this reason, it is sometimes called an open sentence. Each time that a number is substituted for x (the variable), the algebraic equation $3x + 1 = 13$ becomes a **numerical statement** that is either true or false. For example, if $x = 2$, then $3x + 1 = 13$ becomes $3(2) + 1 = 13$, which is a false statement. If $x = 4$, then $3x + 1 = 13$ becomes $3(4) + 1 = 13$, which is a true statement. **Solving an algebraic equation** refers to the process of finding the number (or numbers) that make(s) the algebraic equation a true numerical statement. Such numbers are called the **solutions** or **roots** of the equation and are said to *satisfy* the equation. The set of all solutions of an equation is called its solution set. Thus the solution set of $3x + 1 = 13$ is {4}.

Equivalent equations are equations that have the same solution set. For example,

$$3x + 1 = 13, \qquad 3x = 12, \qquad \text{and} \qquad x = 4$$

are equivalent equations because {4} is the solution set of each. The general procedure for solving an equation is to continue replacing the given equation with equivalent but simpler equations until an equation of the form "variable = constant" or "constant = variable" is obtained. Thus in the previous example, $3x + 1 = 13$ simplifies to $3x = 12$, which simplifies to $x = 4$, from which the solution set {4} is obvious.

Techniques for solving equations revolve around the following basic properties of equality:

> ### PROPERTY 8.1 *Properties of Equality*
>
> For all real numbers, a, b, and c,
>
> 1. $a = a$ Reflexive property
> 2. If $a = b$, then $b = a$.   Symmetric property
> 3. If $a = b$ and $b = c$, then $a = c$.   Transitive property
> 4. If $a = b$, then a may be replaced by b, or b may be replaced by a, in any statement without changing the meaning of the statement.
>   Substitution property
> 5. $a = b$ if and only if $a + c = b + c$.   Addition property
> 6. $a = b$ if and only if $ac = bc$, where $c \neq 0$.   Multiplication property

In Chapter 3 we stated an "addition–subtraction property of equality" but pointed out that because subtraction is defined in terms of "adding the opposite," only an addition property is technically necessary. Likewise, because division can be defined in terms of "multiplying by the reciprocal," only a multiplication property is necessary.

Now let's use some examples to review the process of solving equations.

EXAMPLE 1   Solve the equation $-3x + 1 = -8$.

Solution

$$-3x + 1 = -8$$
$$-3x = -9 \qquad \text{Added } -1 \text{ to both sides.}$$
$$x = 3 \qquad \text{Multiplied both sides by } -\frac{1}{3}.$$

The solution set is {3}.

Don't forget that to be absolutely sure of a solution set, we must check the solution(s) back into the original equation. Thus for Example 1, substituting 3 for x in $-3x + 1 = -8$ produces $-3(3) + 1 = -8$, which is a true statement. Our solution set is indeed {3}. We will not use the space to show all checks, but remember their importance.

EXAMPLE 2   Find the solution set of $-3n + 6 + 5n = 9n - 4 - 6n$.

Solution

$$-3n + 6 + 5n = 9n - 4 - 6n$$
$$2n + 6 = 3n - 4 \qquad \text{Combined similar terms on both sides.}$$
$$6 = n - 4 \qquad \text{Added } -2n \text{ to both sides.}$$
$$10 = n \qquad \text{Added 4 to both sides.}$$

The solution set is {10}.

Equations Containing Parentheses

If an equation contains parentheses, we may need to apply the distributive property, combine similar terms, and then apply the addition and multiplication properties.

EXAMPLE 3

Solve $3(2x - 5) - 2(4x + 3) = -1$.

Solution

$$3(2x - 5) - 2(4x + 3) = -1$$
$$3(2x) - 3(5) - 2(4x) - 2(3) = -1 \quad \text{Apply the distributive property twice.}$$
$$6x - 15 - 8x - 6 = -1$$
$$-2x - 21 = -1 \quad \text{Combine similar terms.}$$
$$-2x = 20$$
$$x = -10$$

The solution set is $\{-10\}$. Perhaps you should check this solution!

Equations Containing Fractional Forms

If an equation contains fractional forms, then it is usually easier to multiply both sides by the least common denominator of all of the denominators.

EXAMPLE 4

Solve $\dfrac{3x + 2}{4} + \dfrac{2x - 5}{6} = \dfrac{3}{8}$.

Solution

$$\frac{3x + 2}{4} + \frac{2x - 5}{6} = \frac{3}{8}$$
$$24\left(\frac{3x + 2}{4} + \frac{2x - 5}{6}\right) = 24\left(\frac{3}{8}\right) \quad \text{24 is the LCD of 4, 6, and 8.}$$
$$24\left(\frac{3x + 2}{4}\right) + 24\left(\frac{2x - 5}{6}\right) = 24\left(\frac{3}{8}\right) \quad \text{Apply the distributive property.}$$
$$6(3x + 2) + 4(2x - 5) = 9$$
$$18x + 12 + 8x - 20 = 9$$
$$26x - 8 = 9$$
$$26x = 17$$
$$x = \frac{17}{26}$$

The solution set is $\left\{\dfrac{17}{26}\right\}$.

If the equation contains some decimal fractions, then multiplying both sides of the equation by an appropriate power of 10 usually works quite well.

E X A M P L E   5

Solve $0.04x + 0.06(1200 - x) = 66$.

Solution

$$0.04x + 0.06(1200 - x) = 66$$
$$100[0.04x + 0.06(1200 - x)] = 100(66)$$
$$100(0.04x) + 100[0.06(1200 - x)] = 100(66)$$
$$4x + 6(1200 - x) = 6600$$
$$4x + 7200 - 6x = 6600$$
$$-2x + 7200 = 6600$$
$$-2x = -600$$
$$x = 300$$

The solution set is {300}.

Proportions

Recall that a statement of equality between two ratios is a **proportion.** For example, $\dfrac{3}{4} = \dfrac{15}{20}$ is a proportion that states that the ratios $\dfrac{3}{4}$ and $\dfrac{15}{20}$ are equal. A general property of proportions states the following:

$$\frac{a}{b} = \frac{c}{d} \quad \text{if and only if } ad = bc, \text{ where } b \neq 0 \text{ and } d \neq 0$$

The products ad and bc are commonly called **cross products.** Thus the property states that the cross products in a proportion are equal. This becomes the basis of another equation-solving process. If a variable appears in one or both denominators, then restrictions need to be imposed to avoid division by zero.

E X A M P L E   6

Solve $\dfrac{3}{5x - 2} = \dfrac{4}{7x + 3}$.

Solution

$$\frac{3}{5x - 2} = \frac{4}{7x + 3} \qquad x \neq \frac{2}{5} \quad \text{and} \quad x \neq -\frac{3}{7}$$
$$3(7x + 3) = 4(5x - 2) \qquad \text{Cross products are equal.}$$
$$21x + 9 = 20x - 8 \qquad \text{Apply the distributive property on both sides.}$$
$$x = -17$$

The solution set is {−17}.

Equations That Are Algebraic Identities

An equation that is satisfied by all numbers for which both sides of the equation are defined is called an **algebraic identity.** For example,

$$5(x + 1) = 5x + 5 \qquad x^2 - 9 = (x + 3)(x - 3)$$

$$\frac{1}{x} + \frac{2}{x} = \frac{3}{x} \qquad\qquad x^2 - x - 12 = (x - 4)(x + 3)$$

are all algebraic identities. In the third identity, x cannot equal zero, so the statement $\frac{1}{x} + \frac{2}{x} = \frac{3}{x}$ is true for all real numbers except 0. The other identities listed are true for all real numbers. Sometimes the original form of the equation does not explicitly indicate that it is an identity. Consider the following example.

EXAMPLE 7

Solve $5(x + 3) - 3(x - 5) = 2(x + 15)$.

Solution

$$5(x + 3) - 3(x - 5) = 2(x + 15)$$
$$5x + 15 - 3x + 15 = 2x + 30$$
$$2x + 30 = 2x + 30$$

At this step it becomes obvious that we have an algebraic identity. No restrictions are necessary, so the solution set, written in set builder notation, is $\{x|x \text{ is a real number}\}$. ∎

Equations That Are Contradictions

By inspection we can tell that the equation $x + 1 = x + 2$ has no solutions, because adding 1 to a number cannot produce the same result as adding 2 to that number. Thus the solution set is ∅ (the null set or empty set). Likewise, we can determine by inspection that the solution set for each of the following equations is ∅.

$$3x - 1 = 3x - 4 \qquad 5(x - 1) = 5x - 1 \qquad \frac{1}{x} + \frac{2}{x} = \frac{5}{x}$$

Now suppose that the solution set is *not* obvious by inspection.

EXAMPLE 8

Solve $4(x - 2) - 2(x + 3) = 2(x + 6)$.

Solution

$$4(x - 2) - 2(x + 3) = 2(x + 6).$$
$$4x - 8 - 2x - 6 = 2x + 12$$
$$2x - 14 = 2x + 12$$

At this step we might recognize that the solution set is $\varnothing$, or we could continue by adding $-2x$ to both sides to produce

$$-14 = 12$$

Because we have logically arrived at a contradiction ($-14 = 12$), the solution set is $\varnothing$.   ■

Problem Solving

Remember that one theme throughout this text is "learn a skill, then use the skill to help solve equations and inequalities, and then use equations and inequalities to help solve problems." Being able to solve problems is the end result of this sequence; it is what we want to achieve. You may want to turn back to Section 3.3 and refresh your memory of the problem-solving suggestions given there.

Suggestion 5 is "look for a **guideline** that can be used to set up an equation." Such a guideline may or may not be explicitly stated in the problem.

PROBLEM 1   A 10-gallon container is full and contains a 40% solution of antifreeze. How much needs to be drained out and replaced with pure antifreeze to obtain a 70% solution?

Solution

We can use the following guideline for this problem.

Pure antifreeze in the original solution	$-$	Pure antifreeze in the solution drained out	$+$	Pure antifreeze added	$=$	Pure antifreeze in the final solution

Let x represent the amount of pure antifreeze to be added. Then x also represents the amount of the 40% solution to be drained out. Thus the guideline translates into the equation

$$40\%(10) - 40\%(x) + x = 70\%(10)$$

We can solve this equation as follows:

$$0.4(10) - 0.4x + x = 0.7(10)$$
$$4 + 0.6x = 7$$
$$0.6x = 3$$
$$6x = 30$$
$$x = 5$$

Therefore, we need to drain out 5 gallons of the 40% solution and replace it with 5 gallons of pure antifreeze. (Be sure you can check this answer!)   ■

Recall that in Chapter 5 we found that sometimes it is easier to solve a problem using two equations and two unknowns than using one equation with one unknown. Also at that time, we used three different techniques for solving a system of two linear equations in two variables: (1) a graphing approach, which was not very efficient when we wanted exact solutions, (2) a substitution method, and (3) an elimination-by-addition method. Now let's use a system of equations to help solve a problem. Furthermore, we will solve the system twice to review both the substitution method and the elimination-by-addition method.

PROBLEM 2

Jose bought 1 pound of bananas and 3 pounds of tomatoes for $5.66. At the same prices, Jessica bought 4 pounds of bananas and 5 pounds of tomatoes for $10.81. Find the price per pound for bananas and for tomatoes.

Solution

Let b represent the price per pound for bananas, and let t represent the price per pound for tomatoes. The problem translates into the following system of equations.

$$\left(\begin{array}{l} b + 3t = 5.66 \\ 4b + 5t = 10.81 \end{array} \right)$$

Let's solve this system using the substitution method. The first equation can be written as $b = 5.66 - 3t$. Now we can substitute $5.66 - 3t$ for b in the second equation.

$$4(5.66 - 3t) + 5t = 10.81$$
$$22.64 - 12t + 5t = 10.81$$
$$22.64 - 7t = 10.81$$
$$-7t = -11.83$$
$$t = 1.69$$

Finally, we can substitute 1.69 for t in $b = 5.66 - 3t$.

$$b = 5.66 - 3(1.69)$$
$$= 5.66 - 5.07$$
$$= 0.59$$

Therefore, the price for bananas is $0.59 per pound, and the price for tomatoes is $1.69 per pound. ∎

For review purposes, let's solve the system of equations in Problem 2 using the elimination-by-addition method.

$$\left(\begin{array}{l} b + 3t = 5.66 \\ 4b + 5t = 10.81 \end{array} \right)$$

Multiply the first equation by -4, and add that result to the second equation to produce a new second equation.

$$\left(\begin{array}{l} b + 3t = 5.66 \\ -7t = -11.83 \end{array} \right)$$

Now we can determine the value of t.

$$-7t = -11.83$$
$$t = 1.69$$

Substitute 1.69 for t in $b + 3t = 5.66$.

$$b + 3(1.69) = 5.66$$
$$b + 5.07 = 5.66$$
$$b = 0.59$$

A t value of 1.59 and a b value of 0.59 agree with our previous work.

CONCEPT QUIZ

For Problems 1–5, solve the equations by inspection, and match the equation with its solution set.

1. $2x + 5 = 2x + 8$

2. $\dfrac{2}{x} = \dfrac{1}{7}$

3. $\dfrac{1}{x} + \dfrac{3}{x} = \dfrac{1}{3}$

4. $4x - 2 = 2(2x - 1)$

5. $5x + 7 = 3x + 7$

A. $\{0\}$

B. $\{x \mid x \text{ is a real number}\}$

C. $\{14\}$

D. $\varnothing$

E. $\{12\}$

PROBLEM SET 8.1

For Problems 1–26, solve each equation.

1. $5x - 4 = 16$

2. $-4x + 3 = -13$

3. $-6 = 7x + 1$

4. $8 = 6x - 4$

5. $-2x + 8 = -3x + 14$

6. $7 - 3x = 6 + 3x$

7. $4x - 6 - 5x = 3x + 1 - x$

8. $x - 2 - 3x = 2x + 1 - 5x$

9. $6(2x + 5) = 5(3x - 4)$

10. $-2(4x - 7) = -(9x + 4)$

11. $-2(3x - 1) - 3(2x + 5) = -5(2x - 8)$

12. $4(5x - 2) + (x - 4) = 6(3x - 10)$

13. $\dfrac{2}{3}x - \dfrac{1}{2} + \dfrac{1}{4}x = \dfrac{5}{8}$

14. $\dfrac{3}{4}x + \dfrac{2}{5} - x = \dfrac{3}{10}$

15. $\dfrac{2x - 1}{3} - \dfrac{3x + 2}{5} = -2$

16. $\dfrac{4x + 1}{6} + \dfrac{2x - 3}{5} = \dfrac{4}{15}$

17. $\dfrac{-3}{2x - 1} = \dfrac{2}{4x + 7}$

18. $\dfrac{4}{5x - 2} = \dfrac{6}{7x + 3}$

19. $\dfrac{x + 2}{3} + 1 = \dfrac{x - 2}{3}$

20. $3(x + 1) - 5(x - 2) = -2(x - 7)$

21. $0.09x + 0.11(x + 125) = 68.75$

22. $0.08(x + 200) = 0.07x + 20$

23. $\dfrac{x - 1}{5} - 2 = \dfrac{x - 11}{5}$

24. $3(x - 1) + 2(x + 4) = 5(x + 1)$

25. $0.05x - 4(x + 0.5) = 1.2$

26. $0.5(3x + 0.7) = 20.6$

For Problems 27–36, solve each system of equations.

27. $\begin{pmatrix} x - 2y = 10 \\ 3x + 7y = -22 \end{pmatrix}$ **28.** $\begin{pmatrix} 4x - 5y = -21 \\ 3x + y = 8 \end{pmatrix}$

29. $\begin{pmatrix} 5x - 2y = -3 \\ 3x - 4y = 15 \end{pmatrix}$ **30.** $\begin{pmatrix} 7x - 5y = 6 \\ 3x + 2y = -14 \end{pmatrix}$

31. $\begin{pmatrix} 2x - y = 6 \\ 4x - 2y = -1 \end{pmatrix}$ **32.** $\begin{pmatrix} y = 4x - 3 \\ y = 3x + 1 \end{pmatrix}$

33. $\begin{pmatrix} 3x + 4y = -18 \\ 5x - 7y = -30 \end{pmatrix}$ **34.** $\begin{pmatrix} 4x - 6y = -4 \\ 2x - 3y = -2 \end{pmatrix}$

35. $\begin{pmatrix} \dfrac{1}{2}x - \dfrac{2}{3}y = -8 \\ \dfrac{3}{2}x + \dfrac{1}{3}y = -10 \end{pmatrix}$ **36.** $\begin{pmatrix} \dfrac{3}{4}x + \dfrac{2}{5}y = 13 \\ \dfrac{1}{3}x - \dfrac{3}{10}y = 1 \end{pmatrix}$

For Problems 37–52, use an equation or a system of equations to help you solve each problem.

37. Find three consecutive integers whose sum is −45.

38. Tina is paid time-and-a-half for each hour worked over 40 hours in a week. Last week she worked 45 hours and earned $380. What is her normal hourly rate?

39. There are 51 students in a certain class. The number of females is 5 less than three times the number of males. Find the number of females and the number of males in the class.

40. The sum of the present ages of Eric and his father is 58 years. In 10 years, his father will be twice as old as Eric will be at that time. Find their present ages.

41. Kaitlin went on a shopping spree, spending a total of $124 on a skirt, a sweater, and a pair of shoes. The cost of the sweater was $\dfrac{8}{7}$ of the cost of the skirt. The shoes cost $8 less than the skirt. Find the cost of each item.

42. The ratio of male students to female students at a certain university is 5 to 7. If there is a total of 16,200 students, find the number of male and the number of female students.

43. If each of two opposite sides of a square is increased by 3 centimeters and each of the other two sides is decreased by 2 centimeters, the area is increased by 8 square centimeters. Find the length of a side of the square.

44. Desa invested a certain amount of money at 4% interest and $1500 more than that amount at 7%. Her total yearly interest was $435. How much did she invest at each rate?

45. Suppose that an item costs a retailer $50. How much more profit could be gained by fixing a 50% profit based on selling price rather than a 50% profit based on the cost?

46. Find three consecutive integers such that the sum of the smallest integer and the largest integer is equal to twice the middle integer.

47. The ratio of the weight of sodium to that of chlorine in common table salt is 5 to 3. Find the amount of each element in a salt block that weighs 200 pounds.

48. Sean bought 5 lemons and 3 limes for $1.70. At the same prices, Kim bought 4 lemons and 7 limes for $2.05. Find the price per lemon and the price per lime.

49. One leg of a right triangle is 7 meters longer than the other leg. If the length of the hypotenuse is 17 meters, find the length of each leg.

50. The perimeter of a rectangle is 44 inches and its area is 112 square inches. Find the length and width of the rectangle.

51. Domenica and Javier start from the same location at the same time and ride their bicycles in opposite directions for 4 hours, at which time they are 140 miles apart. Domenica rides 3 miles per hour faster than Javier. Find the rate of each rider.

52. A container has 6 liters of a 40% alcohol solution in it. How much pure alcohol should be added to raise it to a 60% solution?

53. Explain how a trial-and-error approach could be used to solve Problem 49.

54. Now try a trial-and-error approach to solve Problem 41. What kind of difficulty are you having?

55. Suppose that your friend analyzes Problem 51 as follows: If they are 140 miles apart in 4 hours, then together they would need to average $\frac{140}{4} = 35$ miles per hour. Accordingly, we need two numbers whose sum is 35, and one number must be 3 larger than the other number. Thus Javier rides at 16 miles per hour and Domenica at 19 miles per hour. How would you react to this analysis of the problem?

Answers to Concept Quiz

1. D   **2.** C   **3.** E   **4.** B   **5.** A

8.2   Inequalities: A Brief Review

Objectives

■   Review of properties and techniques for solving inequalities.

■   Express solution sets in interval notation.

■   Review solving compound inequalities.

■   Review solving word problems that involve inequalities.

Just as we use the symbol $=$ to represent "is equal to," we also use the symbols $<$ and $>$ to represent "is less than" and "is greater than," respectively. Thus various **statements of inequality** can be made.

$a < b$ means a is less than b

$a \le b$ means a is less than or equal to b

$a > b$ means a is greater than b

$a \ge b$ means a is greater than or equal to b

An **algebraic inequality** such as $x - 2 < 6$ is neither true nor false as it stands and is called an open sentence. For each numerical value substituted for x, the algebraic inequality $x - 2 < 6$ becomes a numerical statement of inequality that is true or false. For example, if $x = 10$, then $x - 2 < 6$ becomes $10 - 2 < 6$, which is false. If $x = 5$, then $x - 2 < 6$ becomes $5 - 2 < 6$, which is true. **Solving an algebraic inequality** is the process of finding the numbers that make it a true numerical statement. Such numbers are called the **solutions** of the inequality and are said to **satisfy** it.

The general process for solving inequalities closely parallels that for solving equations. We continue to replace the given inequality with equivalent but simpler inequalities until the solution set is obvious. The following property provides the basis for producing equivalent inequalities. (Because subtraction can be defined in terms of addition, and division can be defined in terms of multiplication, we state the property at this time in terms of only addition and multiplication.)

PROPERTY 8.2

1. For all real numbers a, b, and c,

   $a > b$ if and only if $a + c > b + c$.

2. For all real numbers, a, b, and c, with $c > 0$,

   $a > b$ if and only if $ac > bc$

3. For all real numbers, a, b, and c, with $c < 0$,

   $a > b$ if and only if $ac < bc$

Similar properties exist if $>$ is replaced by $<$, $\le$, or $\ge$. Part 1 of Property 8.2 is commonly called the **addition property of inequality.** Parts 2 and 3 together make up the **multiplication property of inequality.** Pay special attention to part 3. If both sides of an inequality are multiplied by a negative number, the inequality symbol must be reversed. For example, if both sides of $-3 < 5$ are multiplied by -2, then the inequality $6 > -10$ is produced. In the following example, note the use of the distributive property, as well as both the addition and multiplication properties of inequality.

EXAMPLE 1   Solve $3(2x - 1) < 8x - 7$.

Solution

$$3(2x - 1) < 8x - 7$$
$$6x - 3 < 8x - 7 \qquad \text{Apply distributive property to left side.}$$
$$-2x - 3 < -7 \qquad \text{Add } -8x \text{ to both sides.}$$
$$-2x < -4 \qquad \text{Add 3 to both sides.}$$
$$-\frac{1}{2}(-2x) > -\frac{1}{2}(-4) \qquad \text{Multiply both sides by } -\frac{1}{2}, \text{ which reverses the inequality.}$$
$$x > 2$$

The solution set is $\{x \mid x > 2\}$.   ■

A graph of the solution set $\{x \mid x > 2\}$ in Example 1 is shown in Figure 8.1. The parenthesis indicates that 2 does not belong to the solution set.

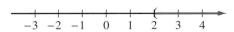

Figure 8.1

Checking the solutions of an inequality presents a problem. Obviously, we cannot check all of the infinitely many solutions for a particular inequality. However, by checking at least one solution, especially when the multiplication property has been used, we might catch a mistake of forgetting to change the type of inequality. In Example 1 we are claiming that all numbers greater than 2 will satisfy the original inequality. Let's check the number 3.

$$3(2x - 1) < 8x - 7$$
$$3[2(3) - 1] \overset{?}{<} 8(3) - 7$$
$$3(5) < 17$$
$$15 < 17 \qquad\qquad \text{It checks!}$$

Interval Notation

It is also convenient to express solution sets of inequalities by using **interval notation.** For example, the notation $(2, \infty)$ refers to the interval of all real numbers greater than 2. As on the graph in Figure 8.1, the left-hand parenthesis indicates that 2 is not to be included. The infinity symbol, ∞, along with the right-hand parenthesis, indicates that there is no right-hand endpoint. Following (Figure 8.2) is a partial list of interval notations, along with the sets and graphs that they represent. Note the use of square brackets to *include* endpoints.

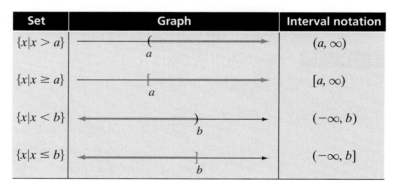

Set	Graph	Interval notation
$\{x\|x > a\}$		(a, ∞)
$\{x\|x \geq a\}$		$[a, \infty)$
$\{x\|x < b\}$		$(-\infty, b)$
$\{x\|x \leq b\}$		$(-\infty, b]$

Figure 8.2

E X A M P L E   2   Solve $-3x + 5x - 2 \geq 8x - 7 - 9x$.

Solution

$$-3x + 5x - 2 \geq 8x - 7 - 9x$$
$$2x - 2 \geq -x - 7 \qquad\qquad \text{Combine similar terms on both sides.}$$
$$3x - 2 \geq -7 \qquad\qquad \text{Add } x \text{ to both sides.}$$
$$3x \geq -5 \qquad\qquad \text{Add 2 to both sides.}$$

$$\frac{1}{3}(3x) \ge \frac{1}{3}(-5) \qquad \text{Multiply both sides by } \frac{1}{3}.$$

$$x \ge -\frac{5}{3}$$

The solution set is $\left[-\frac{5}{3}, \infty\right)$.

E X A M P L E   3   Solve $4(x - 3) > 9(x + 1)$.

Solution

$$4(x - 3) > 9(x + 1)$$
$$4x - 12 > 9x + 9 \qquad \text{Apply the distributive property.}$$
$$-5x - 12 > 9 \qquad \text{Add } -9x \text{ to both sides.}$$
$$-5x > 21 \qquad \text{Add 12 to both sides.}$$
$$-\frac{1}{5}(-5x) < -\frac{1}{5}(21) \qquad \text{Multiply both sides by } -\frac{1}{5}, \text{ which reverses the inequality.}$$
$$x < -\frac{21}{5}$$

The solution set is $\left(-\infty, -\frac{21}{5}\right)$.

The next example will solve the inequality without indicating the justification for each step. Be sure that you can supply the reasons for the steps.

E X A M P L E   4   Solve $3(2x + 1) - 2(2x + 5) < 5(3x - 2)$.

Solution

$$3(2x + 1) - 2(2x + 5) < 5(3x - 2)$$
$$6x + 3 - 4x - 10 < 15x - 10$$
$$2x - 7 < 15x - 10$$
$$-13x - 7 < -10$$
$$-13x < -3$$
$$-\frac{1}{13}(-13x) > -\frac{1}{13}(-3)$$
$$x > \frac{3}{13}$$

The solution set is $\left(\frac{3}{13}, \infty\right)$.

EXAMPLE 5   Solve $4 \leq 2(x - 3) - (x + 4)$

Solution

$$4 \leq 2(x - 3) - (x + 4)$$
$$4 \leq 2x - 6 - x - 4$$
$$4 \leq x - 10$$
$$14 \leq x$$
$$x \geq 14$$

The solution set is $[14, \infty)$.   ■

REMARK: In the solution for Example 2, the solution set could be determined from the statement $14 \leq x$. However, you may find it easier to use the equivalent statement $x \geq 14$ to determine the solution set.

EXAMPLE 6   Solve $\dfrac{x - 4}{6} - \dfrac{x - 2}{9} \leq \dfrac{5}{8}$.

Solution

$$\frac{x - 4}{6} - \frac{x - 2}{9} \leq \frac{5}{18}$$

$$18\left(\frac{x - 4}{6} - \frac{x - 2}{9}\right) \leq 18\left(\frac{5}{18}\right) \quad \text{Multiply both sides by the LCD.}$$

$$18\left(\frac{x - 4}{6}\right) - 18\left(\frac{x - 2}{9}\right) \leq 18\left(\frac{5}{18}\right) \quad \text{Distributive property}$$

$$3(x - 4) - 2(x - 2) \leq 5$$
$$3x - 12 - 2x + 4 \leq 5$$
$$x - 8 \leq 5$$
$$x \leq 13$$

The solution set is $(-\infty, 13]$.   ■

EXAMPLE 7   Solve $0.08x + 0.09(x + 100) \geq 43$.

Solution

$$0.08x + 0.09(x + 100) \geq 43$$

$$100[0.08x + 0.09(x + 100)] \geq 100(43) \quad \text{Multiply both sides by 100.}$$

$$100(0.08x) + 100[0.09(x + 100)] \geq 4300$$

$$8x + 9(x + 100) \geq 4300$$

$$8x + 9x + 900 \geq 4300$$

$$17x + 900 \geq 4300$$

$$17x \geq 3400$$
$$x \geq 200$$

The solution set is $[200, \infty)$.

Compound Statements

We use the words *and* and *or* in mathematics to form **compound statements.** The following are examples of compound numerical statements that use *and.* We call such statements **conjunctions.** We agree to call a conjunction true only if all of its component parts are true. Statements 1 and 2 below are true, but statements 3, 4, and 5 are false.

1. $3 + 4 = 7$ and $-4 < -3$ True
2. $-3 < -2$ and $-6 > -10$ True
3. $6 > 5$ and $-4 < -8$ False
4. $4 < 2$ and $0 < 10$ False
5. $-3 + 2 = 1$ and $5 + 4 = 8$ False

We call compound statements that use *or* **disjunctions.** The following are examples of disjunctions that involve numerical statements.

6. $0.14 > 0.13$ or $0.235 < 0.237$ True
7. $\dfrac{3}{4} > \dfrac{1}{2}$ or $-4 + (-3) = 10$ True
8. $-\dfrac{2}{3} > \dfrac{1}{3}$ or $(0.4)(0.3) = 0.12$ True
9. $\dfrac{2}{5} < -\dfrac{2}{5}$ or $7 + (-9) = 16$ False

A disjunction is true if at least one of its component parts is true. In other words, disjunctions are false only if all of the component parts are false. Thus statements 6, 7, and 8 are true, but statement 9 is false.

Now let's consider finding solutions for some compound statements that involve algebraic inequalities. Keep in mind that our previous agreements for labeling conjunctions and disjunctions true or false form the basis for our reasoning.

EXAMPLE 8

Graph the solution set for the conjunction $x > -1$ and $x < 3$. Also express the solution set in interval notation and in set-builder notation.

Solution

The key word is *and,* so we need to satisfy both inequalities. Thus all numbers between -1 and 3 are solutions, which we indicate on a number line as in Figure 8.3.

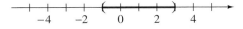

Figure 8.3

Using interval notation, we can represent the interval enclosed in parentheses in Figure 8.3 by $(-1, 3)$. Using set-builder notation, we can express the same interval as $\{x|-1 < x < 3\}$, where the statement $-1 < x < 3$ is read "negative one is less than x, and x is less than three." In other words, x is between -1 and 3. ■

Example 8 represents another concept that pertains to sets. The set of all elements common to two sets is called the **intersection** of the two sets. Thus in Example 8 we found the intersection of the two sets $\{x|x > -1\}$ and $\{x|x < 3\}$ to be the set $\{x|-1 < x < 3\}$. In general, we define the intersection of two sets as follows:

DEFINITION 8.1

The **intersection** of two sets A and B (written $A \cap B$) is the set of all elements that are in both set A and set B. Using set builder notation, we can write

$$A \cap B = \{x|x \in A \quad and \quad x \in B\}$$

E X A M P L E   9

Solve the conjunction $3x + 1 > -5$ *and* $2x + 5 > 7$, and graph its solution set on a number line.

Solution

First, let's simplify both inequalities.

$$
\begin{array}{ccc}
3x + 1 > -5 & \text{and} & 2x + 5 > 7 \\
3x > -6 & \text{and} & 2x > 2 \\
x > -2 & \text{and} & x > 1
\end{array}
$$

Because this is a conjunction, we must satisfy both inequalities. Thus all numbers greater than 1 are solutions, and the solution set is $(1, \infty)$. We show the graph of the solution set in Figure 8.4.

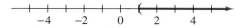

Figure 8.4   ■

We can solve a conjunction such as $3x + 1 > -3$ and $3x + 1 < 7$, in which the same algebraic expression (in this case $3x + 1$) is contained in both inequalities, by using the **compact form** $-3 < 3x + 1 < 7$ as follows:

$$-3 < 3x + 1 < 7$$
$$-4 < 3x < 6 \qquad \text{Add –1 to the left side, middle, and right side.}$$
$$-\frac{4}{3} < x < 2 \qquad \text{Multiply through by } \frac{1}{3}.$$

The solution set is $\left(-\frac{4}{3}, 2\right)$.

The word *and* ties the concept of a conjunction to the set concept of intersection. In a like manner, the word *or* links the idea of a disjunction to the set concept of **union.** We define the union of two sets as follows:

> **DEFINITION 8.2**
>
> The **union** of two sets A and B (written $A \cup B$) is the set of all elements that are in set A or in set B, or in both. Using set builder notation, we can write
>
> $$A \cup B = \{x | x \in A \quad or \quad x \in B\}$$

EXAMPLE 10   Graph the solution set for the disjunction $x < -1 \; or \; x > 2$, and express it using interval notation.

Solution

The key word is *or,* so all numbers that satisfy either inequality (or both) are solutions. Thus all numbers less than -1, along with all numbers greater than 2, are the solutions. The graph of the solution set is shown in Figure 8.5.

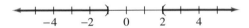

Figure 8.5

Using interval notation and the set concept of union, we can express the solution set as $(-\infty, -1) \cup (2, \infty)$. ■

Example 10 illustrates that in terms of set vocabulary, the solution set of a disjunction is the union of the solution sets of the component parts of the disjunction. Note that there is *no compact form* for writing $x < -1 \; or \; x > 2$.

EXAMPLE 11   Solve the disjunction $2x - 5 < -11 \; or \; 5x + 1 \geq 6$, and graph its solution set on a number line.

Solution

First, let's simplify both inequalities.

$$2x - 5 < -11 \quad or \quad 5x + 1 \geq 6$$
$$2x < -6 \quad or \quad 5x \geq 5$$
$$x < -3 \quad or \quad x \geq 1$$

This is a disjunction, and all numbers less than -3, along with all numbers greater than or equal to 1, will satisfy it. Thus the solution set is $(-\infty, -3) \cup [1, \infty)$. Its graph is shown in Figure 8.6.

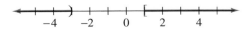

Figure 8.6 ■

In summary, to solve a compound sentence involving an inequality, proceed as follows:

1. Solve separately each inequality in the compound sentence.
2. If it is a *conjunction*, the solution set is the *intersection* of the solution sets of each inequality.
3. If it is a *disjunction*, the solution set is the *union* of the solution sets of each inequality.

The following agreements (Figure 8.7) on the use of interval notation should be added to the list on page 350.

Set	Graph	Interval notation
$\{x \mid a < x < b\}$	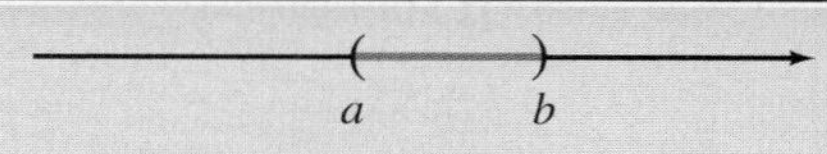	(a, b)
$\{x \mid a \leq x < b\}$	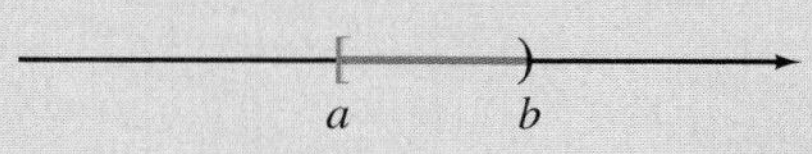	$[a, b)$
$\{x \mid a < x \leq b\}$		$(a, b]$
$\{x \mid a \leq x \leq b\}$	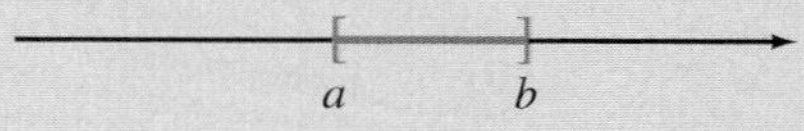	$[a, b]$
$\{x \mid x \text{ is a real number}\}$		$(-\infty, \infty)$

Figure 8.7

Problem Solving

We will conclude this section with some word problems that contain inequality statements.

PROBLEM 1

Sari had scores of 94, 84, 86, and 88 on her first four exams of the semester. What score must she obtain on the fifth exam to have an average of 90 or better for the five exams?

Solution

Let s represent the score Sari needs on the fifth exam. Because the average is computed by adding all scores and dividing by the number of scores, we have the following inequality to solve.

$$\frac{94 + 84 + 86 + 88 + s}{5} \geq 90$$

Solving this inequality, we obtain

$$\frac{352 + s}{5} \geq 90$$

$$5\left(\frac{352 + s}{5}\right) \geq 5(90) \qquad \text{Multiply both sides by 5.}$$

$$352 + s \geq 450$$

$$s \geq 98$$

Sari must receive a score of 98 or better. ■

PROBLEM 2

An investor has $4000 to invest. Suppose she invests $2000 at 8% interest. At what rate must she invest the other $2000 so that the two investments together yield more than $270 of yearly interest?

Solution

Let r represent the unknown rate of interest. We can use the following guideline to set up an inequality.

Interest from 8% investment	+	Interest from r percent investment	>	$270
↓		↓		↓
(8%)($2000)	+	r($2000)	>	$270

Solving this inequality yields

$$160 + 2000r > 270$$

$$2000r > 110$$

$$r > \frac{110}{2000}$$

$$r > 0.055 \qquad \text{Change to a decimal.}$$

She must invest the other $2000 at a rate greater than 5.5%. ■

PROBLEM 3

If the temperature for a 24-hour period ranged between 41°F and 59°F, inclusive (that is, $41 \leq F \leq 59$), what was the range in Celsius degrees?

Solution

Use the formula $F = \frac{9}{5}C + 32$, to solve the following compound inequality.

$$41 \leq \frac{9}{5}C + 32 \leq 59$$

Solving this yields

$$9 \leq \frac{9}{5}C \leq 27 \qquad \text{Add} -32.$$

$$\frac{5}{9}(9) \leq \frac{5}{9}\left(\frac{9}{5}C\right) \leq \frac{5}{9}(27) \qquad \text{Multiply by } \frac{5}{9}.$$

$$5 \leq C \leq 15$$

The range was between 5°C and 15°C, inclusive. ■

CONCEPT QUIZ

For Problems 1–5, match the inequality statements with the solution set expressed in interval notation.

1. $-2x > 6$ **A.** $(3, \infty)$

2. $x + 3 > 6$ **B.** $[-1, 3]$

3. $x + 1 \geq 0$ and $x \geq 3$ **C.** $[3, \infty)$

4. $x + 1 \geq 0$ or $x \geq 3$ **D.** $(-\infty, -3)$

5. $-2 \leq 2x \leq 6$ **E.** $[-1, \infty)$

PROBLEM SET 8.2

For Problems 1–24, solve each inequality and express the solution set using interval notation.

1. $6x - 2 > 4x - 14$ **2.** $9x + 5 < 6x - 10$

3. $2x - 7 < 6x + 13$ **4.** $2x - 3 > 7x + 22$

5. $4(x - 3) \leq -2(x + 1)$

6. $3(x - 1) \geq -(x + 4)$

7. $5(x - 4) - 6(x + 2) < 4$

8. $3(x + 2) - 4(x - 1) < 6$

9. $-3(3x + 2) - 2(4x + 1) \geq 0$

10. $-4(2x - 1) - 3(x + 2) \geq 0$

11. $-(x - 3) + 2(x - 1) < 3(x + 4)$

12. $3(x - 1) - (x - 2) > -2(x + 4)$

13. $7(x + 1) - 8(x - 2) < 0$

14. $5(x - 6) - 6(x + 2) < 0$

15. $\dfrac{x + 3}{8} - \dfrac{x + 5}{5} \geq \dfrac{3}{10}$

16. $\dfrac{x - 4}{6} - \dfrac{x - 2}{9} \leq \dfrac{5}{18}$

17. $\dfrac{4x - 3}{6} - \dfrac{2x - 1}{12} < -2$

18. $\dfrac{3x + 2}{9} - \dfrac{2x + 1}{3} > -1$

19. $0.06x + 0.08(250 - x) \geq 19$

20. $0.08x + 0.09(2x) \geq 130$

21. $0.09x + 0.1(x + 200) > 77$

22. $0.07x + 0.08(x + 100) > 38$

23. $x \geq 3.4 + 0.15x$

24. $x \geq 2.1 + 0.3x$

For Problems 25–30, solve each compound inequality and graph the solution sets. Express the solution sets in interval notation.

25. $2x - 1 \geq 5$ and $x > 0$

26. $3x + 2 > 17$ and $x \geq 0$

27. $5x - 2 < 0$ and $3x - 1 > 0$

28. $x + 1 > 0$ and $3x - 4 < 0$

29. $3x + 2 < -1$ or $3x + 2 > 1$

30. $5x - 2 < -2$ or $5x - 2 > 2$

For Problems 31–36, solve each compound inequality using the compact form. Express the solution sets in interval notation.

31. $-6 < 4x - 5 < 6$ **32.** $-2 < 3x + 4 < 2$

33. $-4 \leq \dfrac{x - 1}{3} \leq 4$ **34.** $-1 \leq \dfrac{x + 2}{4} \leq 1$

35. $-3 < 2 - x < 3$ **36.** $-4 < 3 - x < 4$

For Problems 37–44, solve each problem by setting up and solving an appropriate inequality.

37. Mona invests $1000 at 8% yearly interest. How much does she have to invest at 9% so that the total yearly interest from the two investments exceeds $98?

38. Marsha bowled 142 and 170 in her first two games. What must she bowl in the third game to have an average of at least 160 for the three games?

39. Candace had scores of 95, 82, 93, and 84 on her first four exams of the semester. What score must she obtain on the fifth exam to have an average of 90 or better for the five exams?

40. Suppose that Derwin shot rounds of 82, 84, 78, and 79 on the first four days of a golf tournament. What must he shoot on the fifth day of the tournament to average 80 or less for the five days?

41. The temperatures for a 24-hour period ranged between $-4°F$ and $23°F$, inclusive. What was the range in Celsius degrees? (Use $F = \dfrac{9}{5}C + 32$.)

42. Oven temperatures for baking various foods usually range between $325°F$ and $425°F$, inclusive. Express this range in Celsius degrees. (Round answers to the nearest degree.)

43. A person's intelligence quotient (I) is found by dividing mental age (M), as indicated by standard tests, by chronological age (C) and then multiplying this ratio by 100. The formula $I = \dfrac{100M}{C}$ can be used. If the I range of a group of 11-year-olds is given by $80 \le I \le 140$, find the range of the mental age of this group.

44. Repeat Problem 43 for an I range of 70 to 125, inclusive, for a group of 9-year-olds.

■ ■ ■ **Thoughts into words**

45. Do the *less than* and *greater than* relations possess a symmetric property similar to the symmetric property of equality? Defend your answer.

46. Explain the difference between a conjunction and a disjunction. Give an example of each (outside the field of mathematics).

47. How do you know by inspection that the solution set of the inequality $x + 3 > x + 2$ is the entire set of real numbers?

48. Find the solution set for each of the following compound statements, and in each case explain your reasoning.

a. $x < 3$ and $5 > 2$

b. $x < 3$ or $5 > 2$

c. $x < 3$ and $6 < 4$

d. $x < 3$ or $6 < 4$

Answers to Concept Quiz

1. D   **2.** A   **3.** C   **4.** E   **5.** B

8.3   Equations and Inequalities Involving Absolute Value

Objectives

■   Know the definition and properties of absolute value.

■   Solve absolute value equations.

■   Solve absolute value inequalities.

Absolute Value

In Chapter 1 we used the concept of absolute value to describe precisely how to operate with positive and negative numbers. At that time we gave a geometric description of absolute value as the distance between a number and zero on the number line.

For example, using vertical bars to denote absolute value, we can state that $|-3| = 3$ because the distance between -3 and 0 on the number line is 3 units. Likewise, $|2| = 2$ because the distance between 2 and 0 on the number line is 2 units. Using the distance interpretation, we can also state that $|0| = 0$ (Figure 8.8).

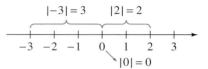

Figure 8.8

More formally, we define the concept of absolute value as follows:

DEFINITION 8.3

For all real numbers a,

  1. If $a \geq 0$, then $|a| = a$.
  2. If $a < 0$, then $|a| = -a$.

Applying Definition 8.3, we obtain the following results:

$|6| = 6$	By applying part 1 of Definition 8.3
$|0| = 0$	By applying part 1 of Definition 8.3
$|-7| = -(-7) = 7$	By applying part 2 of Definition 8.3

Note the following ideas about absolute value:

  1. The absolute value of a positive number is the number itself.
  2. The absolute value of a negative number is its opposite.
  3. The absolute value of any number except zero is always positive.
  4. The absolute value of zero is zero.
  5. A number and its opposite have the same absolute value.

We summarize these ideas in the following properties.

Properties of Absolute Value

The variables a and b represent any real number.

  1. $|a| \geq 0$
  2. $|a| = |-a|$
  3. $|a - b| = |b - a|$ $a - b$ and $b - a$ are opposites of each other.

Equations Involving Absolute Value

The interpretation of absolute value as distance on a number line provides a straightforward approach to solving a variety of equations and inequalities involving absolute value. First, let's consider some equations.

EXAMPLE 1   Solve $|x| = 2$.

Solution

Think in terms of *distance between the number and zero,* and you will see that x must be 2 or -2. That is, the equation $|x| = 2$ is equivalent to

$$x = -2 \quad \text{or} \quad x = 2$$

The solution set is $\{-2, 2\}$.

EXAMPLE 2   Solve $|x + 2| = 5$.

Solution

The number, $x + 2$, must be -5 or 5. Thus $|x + 2| = 5$ is equivalent to

$$x + 2 = -5 \quad \text{or} \quad x + 2 = 5$$

Solving each equation of the disjunction yields

$$x + 2 = -5 \quad \text{or} \quad x + 2 = 5$$
$$x = -7 \quad \text{or} \quad x = 3$$

The solution set is $\{-7, 3\}$.

 Check

$$
\begin{array}{ll}
|x + 2| = 5 & |x + 2| = 5 \\
|-7 + 2| \overset{?}{=} 5 & |3 + 2| \overset{?}{=} 5 \\
|-5| \overset{?}{=} 5 & |5| \overset{?}{=} 5 \\
5 = 5 & 5 = 5
\end{array}
$$

The following general property should seem reasonable from the distance interpretation of absolute value.

PROPERTY 8.3

$|ax + b| = k$ is equivalent to $ax + b = -k$ or $ax + b = k$, where k is a positive number.

Example 3 demonstrates our format for solving equations of the form $|ax + b| = k$.

EXAMPLE 3 Solve $|5x + 3| = 7$.

Solution

$$|5x + 3| = 7$$

$$5x + 3 = -7 \quad \text{or} \quad 5x + 3 = 7$$

$$5x = -10 \quad \text{or} \quad 5x = 4$$

$$x = -2 \quad \text{or} \quad x = \frac{4}{5}$$

The solution set is $\left\{-2, \dfrac{4}{5}\right\}$. Check these solutions! ■

Inequalities Involving Absolute Value

The *distance interpretation* for absolute value also provides a good basis for solving some inequalities that involve absolute value. Consider the following examples.

EXAMPLE 4 Solve $|x| < 2$ and graph the solution set.

Solution

The number, x, must be *less than 2 units away from zero*. Thus $|x| < 2$ is equivalent to

$$x > -2 \quad \text{and} \quad x < 2$$

The solution set is $(-2, 2)$, and its graph is shown in Figure 8.9.

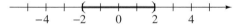

Figure 8.9 ■

EXAMPLE 5 Solve $|x + 3| < 1$ and graph the solutions.

Solution

Let's continue to think in terms of *distance* on a number line. The number, $x + 3$, must be *less than 1 unit away from zero*. Thus $|x + 3| < 1$ is equivalent to

$$x + 3 > -1 \quad \text{and} \quad x + 3 < 1$$

Solving this conjunction yields

$$x + 3 > -1 \quad \text{and} \quad x + 3 < 1$$
$$x > -4 \quad \text{and} \quad x < -2$$

The solution set is $(-4, -2)$ and its graph is shown in Figure 8.10.

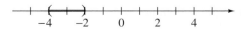

Figure 8.10 ■

Take another look at Examples 4 and 5. The following general property should seem reasonable.

> ### PROPERTY 8.4
>
> $|ax + b| < k$ is equivalent to $ax + b > -k$ and $ax + b < k$, where k is a positive number.

Remember that we can write a conjunction such as $ax + b > -k$ and $ax + b < k$ in the compact form $-k < ax + b < k$. The compact form provides a very convenient format for solving inequalities such as $|3x - 1| < 8$, as Example 6 illustrates.

EXAMPLE 6   Solve $|3x - 1| < 8$ and graph the solutions.

Solution

$$|3x - 1| < 8$$
$$-8 < 3x - 1 < 8$$
$$-7 < 3x < 9 \qquad \text{Add 1 to left side, middle, and right side.}$$
$$\frac{1}{3}(-7) < \frac{1}{3}(3x) < \frac{1}{3}(9) \qquad \text{Multiply through by } \frac{1}{3}.$$
$$-\frac{7}{3} < x < 3$$

The solution set is $\left(-\frac{7}{3}, 3\right)$, and its graph is shown in Figure 8.11.

Figure 8.11

The distance interpretation also clarifies a property that pertains to *greater than* situations involving absolute value. Consider the following examples.

EXAMPLE 7   Solve $|x| > 1$ and graph the solutions.

Solution

The number, x, must be *more than 1 unit away from zero*. Thus $|x| > 1$ is equivalent to

$$x < -1 \qquad \text{or} \qquad x > 1$$

The solution set is $(-\infty, -1) \cup (1, \infty)$, and its graph is shown in Figure 8.12.

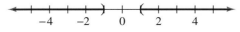

Figure 8.12

EXAMPLE 8   Solve $|x - 1| > 3$ and graph the solutions.

Solution

The number, $x - 1$, must be *more than 3 units away from zero*. Thus $|x - 1| > 3$ is equivalent to

$$x - 1 < -3 \qquad \text{or} \qquad x - 1 > 3$$

Solving this disjunction yields

$$x - 1 < -3 \qquad \text{or} \qquad x - 1 > 3$$
$$x < -2 \qquad \text{or} \qquad x > 4$$

The solution set is $(-\infty, -2) \cup (4, \infty)$, and its graph is shown in Figure 8.13.

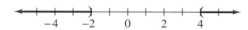

Figure 8.13

Examples 7 and 8 illustrate the following general property.

PROPERTY 8.5

$|ax + b| > k$ is equivalent to $ax + b < -k$ and $ax + b > k$, where k is a positive number.

Therefore, solving inequalities of the form $|ax + b| > k$ can take on the format shown in Example 9.

EXAMPLE 9   Solve $|3x - 1| > 2$ and graph the solutions.

Solution

$$|3x - 1| > 2$$
$$3x - 1 < -2 \qquad \text{or} \qquad 3x - 1 > 2$$
$$3x < -1 \qquad \text{or} \qquad 3x > 3$$
$$x < -\frac{1}{3} \qquad \text{or} \qquad x > 1$$

The solution set is $\left(-\infty, -\frac{1}{3}\right) \cup (1, \infty)$, and its graph is shown in Figure 8.14.

Figure 8.14

Properties 8.3, 8.4, and 8.5 provide the basis for solving a variety of equations and inequalities that involve absolute value. However, if at any time you become doubtful about what property applies, don't forget the distance interpretation. Furthermore, note that in each of the properties, k is a positive number. If k is a nonpositive number, we can determine the solution sets by inspection, as indicated by the following examples.

The solution set of $|x + 3| = 0$ is $\{-3\}$ because the number $x + 3$ has to be 0.

$|2x - 5| = -3$ has *no solutions* because the absolute value (distance) cannot be negative. (The solution set is $\varnothing$, the null set.)

$|x - 7| < -4$ has *no solutions* because we cannot obtain an absolute value less than -4. (The solution set is $\varnothing$.)

$|2x - 1| > -1$ is *satisfied by all real numbers* because the absolute value of $(2x - 1)$, regardless of what number is substituted for x, will always be greater than -1. [The solution set is the set of all real numbers, which we can express in interval notation as $(-\infty, \infty)$].

CONCEPT QUIZ

For Problems 1–8, answer true or false.

1. The absolute value of a negative number is the opposite of the number.

2. The absolute value of a number is always positive or zero.

3. The absolute value of a number is equal to the absolute value of its opposite.

4. The compound statement $x < 1$ or $x > 3$ can be written in compact form $3 < x < 1$.

5. The solution set for the equation $|x + 5| = 0$ is the null set, $\varnothing$.

6. The solution set for $|x - 2| \geq -6$ is all real numbers.

7. The solution set for $|x + 1| < -3$ is all real numbers.

8. The solution set for $|x - 4| \leq 0$ is $\{4\}$.

PROBLEM SET 8.3

For Problems 1–14, solve each inequality and graph the solutions.

1. $|x| < 5$

2. $|x| < 1$

3. $|x| \leq 2$

4. $|x| \leq 4$

5. $|x| > 2$

6. $|x| > 3$

7. $|x - 1| < 2$

8. $|x - 2| < 4$

9. $|x + 2| \leq 4$

10. $|x + 1| \leq 1$

11. $|x + 2| > 1$

12. $|x + 1| > 3$

13. $|x - 3| \geq 2$

14. $|x - 2| \geq 1$

For Problems 15–50, solve each equation or inequality.

15. $|x - 1| = 8$

16. $|x + 2| = 9$

17. $|x - 2| > 6$

18. $|x - 3| > 9$

19. $|x + 3| < 5$

20. $|x + 1| < 8$

21. $|2x - 4| = 6$

22. $|3x - 4| = 14$

23. $|2x - 1| \leq 9$

24. $|3x + 1| \leq 13$

25. $|4x + 2| \geq 12$

26. $|5x - 2| \geq 10$

27. $|3x + 4| = 11$

28. $|5x - 7| = 14$

29. $|4 - 2x| = 6$

30. $|3 - 4x| = 8$

31. $|2 - x| > 4$

32. $|4 - x| > 3$

33. $|1 - 2x| < 2$

34. $|2 - 3x| < 5$

35. $|5x + 9| \leq 16$

36. $|7x - 6| \geq 22$

37. $\left| x - \dfrac{3}{4} \right| = \dfrac{2}{3}$

38. $\left| x + \dfrac{1}{2} \right| = \dfrac{3}{5}$

39. $|-2x + 7| \leq 13$

40. $|-3x - 4| \leq 15$

41. $\left| \dfrac{x - 3}{4} \right| < 2$

42. $\left| \dfrac{x + 2}{3} \right| < 1$

43. $\left| \dfrac{2x + 1}{2} \right| > 1$

44. $\left| \dfrac{3x - 1}{4} \right| > 3$

45. $|2x - 3| + 2 = 5$

46. $|3x - 1| - 1 = 9$

47. $|x + 7| - 3 \geq 4$

48. $|x - 2| + 4 \geq 10$

49. $|2x - 1| + 1 \leq 6$

50. $|4x + 3| - 2 \leq 5$

For Problems 51–60, solve each equation and inequality *by inspection*.

51. $|2x + 1| = -4$

52. $|5x - 1| = -2$

53. $|3x - 1| > -2$

54. $|4x + 3| < -4$

55. $|5x - 2| = 0$

56. $|3x - 1| = 0$

57. $|4x - 6| < -1$

58. $|x + 9| > -6$

59. $|x + 4| < 0$

60. $|x + 6| > 0$

■■■ **Thoughts into words**

61. Explain how you would solve the inequality $|2x + 5| > -3$.

62. Why is 2 the only solution for $|x - 2| \leq 0$?

63. Explain how you would solve the equation $|2x - 3| = 0$.

■■■ **Further investigations**

For Problems 64–69, solve each equation.

64. $|3x + 1| = |2x + 3|$
    [*Hint:* $3x + 1 = 2x + 3$ or $3x + 1 = -(2x + 3)$]

65. $|-2x - 3| = |x + 1|$

66. $|2x - 1| = |x - 3|$

67. $|x - 2| = |x + 6|$

68. $|x + 1| = |x - 4|$

69. $|x + 1| = |x - 1|$

70. Use the definition of absolute value to help prove Property 8.3.

71. Use the definition of absolute value to help prove Property 8.4.

72. Use the definition of absolute value to help prove Property 8.5.

Answers to Concept Quiz

1. True   **2.** True   **3.** True   **4.** False   **5.** False   **6.** True   **7.** False   **8.** True

8.4   Polynomials: A Brief Review and Binomial Expansions

Objectives

■   Review the addition and subtraction of polynomials.

■   Review the properties of exponents.

■   Review the multiplication of polynomials.

■   Review the division of monomials.

■   Write binomial expansions.

Polynomials

Recall that algebraic expressions such as $5x$, $-6y^2$, $2x^{-1}y^{-2}$, $14a^2b$, $5x^{-4}$, and $-17ab^2c^3$ are called **terms.** Terms that contain variables with only nonnegative integers as exponents are called **monomials.** Of the previously listed terms, $5x$, $-6y^2$, $14a^2b$, and $-17ab^2c^3$ are monomials. The **degree** of a monomial is the sum of the exponents of the literal factors. For example, $7xy$ is of degree 2, whereas $14a^2b$ is of degree 3, and $-17ab^2c^3$ is of degree 6. If the monomial contains only one variable, then the exponent of that variable is the degree of the monomial. For example, $5x^3$ is of degree 3, and $-8y^4$ is of degree 4. Any nonzero constant term, such as 8, is of degree zero.

A **polynomial** is a monomial or a finite sum of monomials. Thus all of the following are polynomials.

$$4x^2 \qquad\qquad 3x^2 - 2x - 4 \qquad 7x^4 - 6x^3 + 5x^2 + 2x - 1$$

$$3x^2y - 2y \qquad \frac{1}{5}a^2 - \frac{2}{3}b^2 \qquad 14$$

In addition to calling a polynomial with one term a monomial, we also classify polynomials with two terms as **binomials** and those with three terms as **trinomials.** The **degree of a polynomial** is the degree of the term with the highest degree in the polynomial. The following examples illustrate some of this terminology.

The polynomial $4x^3y^4$ is a monomial in two variables of degree 7.

The polynomial $4x^2y - 2xy$ is a binomial in two variables of degree 3.

The polynomial $9x^2 - 7x - 1$ is a trinomial in one variable of degree 2.

Addition and Subtraction of Polynomials

Both adding polynomials and subtracting them rely on basically the same ideas. The commutative, associative, and distributive properties provide the basis for rearranging, regrouping, and combining similar terms. Consider the following addition problems.

$$(4x^2 + 5x + 1) + (7x^2 - 9x + 4) = (4x^2 + 7x^2) + (5x - 9x) + (1 + 4)$$
$$= 11x^2 - 4x + 5$$
$$(5x - 3) + (3x + 2) + (8x + 6) = (5x + 3x + 8x) + (-3 + 2 + 6)$$
$$= 16x + 5$$

The definition of subtraction as *adding the opposite* $[a - b = a + (-b)]$ extends to polynomials in general. The opposite of a polynomial can be formed by taking the opposite of each term. For example, the opposite of $3x^2 - 7x + 1$ is $-3x^2 + 7x - 1$. Symbolically, this is expressed as

$$-(3x^2 - 7x + 1) = -3x^2 + 7x - 1$$

You can also think in terms of the property $-x = -1(x)$ and the distributive property. Therefore

$$-(3x^2 - 7x + 1) = -1(3x^2 - 7x + 1) = -3x^2 + 7x - 1$$

Now consider the following subtraction problems.

$$
\begin{aligned}
(7x^2 - 2x - 4) - (3x^2 + 7x - 1) &= (7x^2 - 2x - 4) + (-3x^2 - 7x + 1) \\
&= (7x^2 - 3x^2) + (-2x - 7x) + (-4 + 1) \\
&= 4x^2 - 9x - 3
\end{aligned}
$$

$$
\begin{aligned}
(4y^2 + 7) - (-3y^2 + y - 2) &= (4y^2 + 7) + (3y^2 - y + 2) \\
&= (4y^2 + 3y^2) + (-y) + (7 + 2) \\
&= 7y^2 - y + 9
\end{aligned}
$$

As we will see in Section 8.5, sometimes a vertical format is used, especially for subtraction of polynomials. Suppose, for example, that we want to subtract $4x^2 - 7xy + 5y^2$ from $3x^2 - 2xy + y^2$.

$$
\begin{array}{l}
3x^2 - 2xy + y^2 \\
\underline{4x^2 - 7xy + 5y^2}
\end{array}
$$ Note which polynomial goes on the bottom and how the similar terms are aligned.

Now we can *mentally form the opposite of the bottom polynomial* and add.

$$
\begin{array}{l}
3x^2 - 2xy + y^2 \\
\underline{4x^2 - 7xy + 5y^2} \\
-x^2 + 5xy - 4y^2
\end{array}
$$ The opposite of $4x^2 - 7xy + 5y^2$ is $-4x^2 + 7xy - 5y^2$.

Products and Quotients of Monomials

Some basic properties of exponents play an important role in the multiplying and dividing of polynomials. These properties were introduced in Chapter 6, so at this time let's restate them and include (at the right) a "name tag." The name tags can be used for reference purposes, and they help reinforce the meaning of each specific part of the property.

PROPERTY 8.6

If m and n are integers and a and b are real numbers, with $b \neq 0$ whenever it appears in a denominator, then

1. $b^n \cdot b^m = b^{n+m}$ Product of two like bases with powers
2. $(b^n)^m = b^{mn}$ Power of a power

$$3.\ (ab)^n = a^n b^n \qquad\qquad \text{Power of a product}$$

$$4.\ \left(\dfrac{a}{b}\right)^n = \dfrac{a^n}{b^n} \qquad\qquad \text{Power of a quotient}$$

$$5.\ \dfrac{b^n}{b^m} = b^{n-m} \qquad\qquad \text{Quotient of two like bases with powers}$$

Part 1 of Property 8.6, along with the commutative and associative properties of multiplication, form the basis for multiplying monomials. In the following examples, the steps enclosed in dashed boxes can be performed mentally whenever you feel comfortable with the process.

$$(-5a^3 b^4)(7a^2 b^5) = -5 \cdot 7 \cdot a^3 \cdot a^2 \cdot b^4 \cdot b^5$$
$$= -35a^{3+2}b^{4+5}$$
$$= -35a^5 b^9$$

$$(3x^2 y)(4x^3 y^2) = 3 \cdot 4 \cdot x^2 \cdot x^3 \cdot y \cdot y^2$$
$$= 12x^{2+3}y^{1+2}$$
$$= 12x^5 y^3$$

$$(-ab^2)(-5a^2 b) = (-1)(-5)(a)(a^2)(b^2)(b)$$
$$= 5a^{1+2}b^{2+1}$$
$$= 5a^3 b^3$$

$$(2x^2 y^2)(3x^2 y)(4y^3) = 2 \cdot 3 \cdot 4 \cdot x^2 \cdot x^2 \cdot y^2 \cdot y \cdot y^3$$
$$= 24x^{2+2}y^{2+1+3}$$
$$= 24x^4 y^6$$

The following examples show how part 2 of Property 8.6 is used to find "a power of a power."

$$(x^4)^5 = x^{5(4)} = x^{20} \qquad (y^6)^3 = y^{3(6)} = y^{18}$$
$$(2^3)^7 = 2^{7(3)} = 2^{21}$$

Parts 2 and 3 of Property 8.6 form the basis for raising a monomial to a power, as in the next examples.

$$(x^2 y^3)^4 = (x^2)^4 (y^3)^4 \qquad (3a^5)^3 = (3)^3 (a^5)^3$$
$$= x^8 y^{12} \qquad\qquad = 27a^{15}$$
$$(-2xy^4)^5 = (-2)^5 (x)^5 (y^4)^5$$
$$= -32x^5 y^{20}$$

Dividing Monomials

Part 5 of Property 8.6, along with our knowledge of dividing integers, provides the basis for dividing monomials. The following examples demonstrate the process.

$$\frac{24x^5}{3x^2} = 8x^{5-2} = 8x^3 \qquad\qquad \frac{-36a^{13}}{-12a^5} = 3a^{13-5} = 3a^8$$

$$\frac{-56x^9}{7x^4} = -8x^{9-4} = -8x^5 \qquad \frac{72b^5}{8b^5} = 9 \quad \left(\frac{b^5}{b^5} = 1\right)$$

$$\frac{48y^7}{-12y} = -4y^{7-1} = -4y^6 \qquad \frac{12x^4y^7}{2x^2y^4} = 6x^{4-2}y^{7-4} = 6x^2y^3$$

Multiplying Polynomials

The distributive property is usually stated as $a(b + c) = ab + ac$, but it can be extended as follows.

$$a(b + c + d) = ab + ac + ad$$
$$a(b + c + d + e) = ab + ac + ad + ae \qquad \text{etc.}$$

The commutative and associative properties, the properties of exponents, and the distributive property work together to form a basis for finding the product of a monomial and a polynomial. The following example illustrates this idea.

$$3x^2(2x^2 + 5x + 3) = 3x^2(2x^2) + 3x^2(5x) + 3x^2(3)$$
$$= 6x^4 + 15x^3 + 9x^2$$

Extending the method of finding the product of a monomial and a polynomial to finding the product of two polynomials is again based on the distributive property.

$$(x + 2)(y + 5) = x(y + 5) + 2(y + 5)$$
$$= x(y) + x(5) + 2(y) + 2(5)$$
$$= xy + 5x + 2y + 10$$

Note that we are multiplying each term of the first polynomial times each term of the second polynomial.

$$(x - 3)(y + z + 3) = x(y + z + 3) - 3(y + z + 3)$$
$$= xy + xz + 3x - 3y - 3z - 9$$

Frequently, multiplying polynomials produces similar terms that can be combined, which simplifies the resulting polynomial.

$$(x + 5)(x + 7) = x(x + 7) + 5(x + 7)$$
$$= x^2 + 7x + 5x + 35$$
$$= x^2 + 12x + 35$$

$$(x - 2)(x^2 - 3x + 4) = x(x^2 - 3x + 4) - 2(x^2 - 3x + 4)$$
$$= x^3 - 3x^2 + 4x - 2x^2 + 6x - 8$$
$$= x^3 - 5x^2 + 10x - 8$$

It helps to be able to find the product of two binomials without showing all of the intermediate steps. This is quite easy to do with the three-step shortcut pattern demonstrated (Figure 8.15) in the following example.

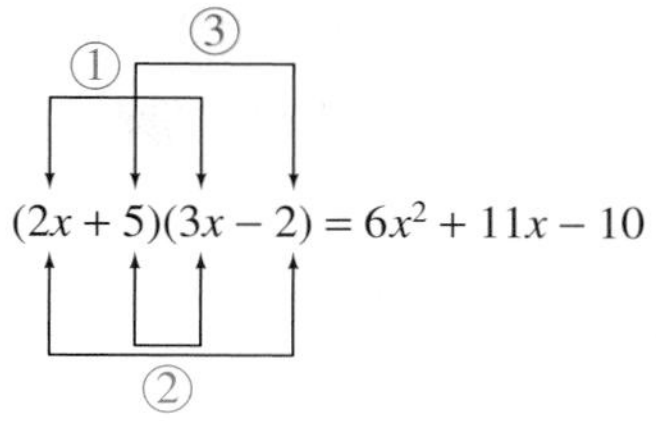

(2x + 5)(3x − 2) = 6x² + 11x − 10	**STEP 1** Multiply $(2x)(3x)$. **STEP 2** Multiply $(5)(3x)$ and $(2x)(-2)$ and combine. **STEP 3** Multiply $(5)(-2)$.

Figure 8.15

Now see whether you can use the pattern to find the following products.

$$(x + 2)(x + 6) = ?$$
$$(x - 3)(x + 5) = ?$$
$$(2x + 5)(3x + 7) = ?$$
$$(3x - 1)(4x - 3) = ?$$

Your answers should be $x^2 + 8x + 12$, $x^2 + 2x - 15$, $6x^2 + 29x + 35$, and $12x^2 - 13x + 3$. Keep in mind that this shortcut pattern applies only to finding the product of two binomials.

REMARK: Shortcuts can be very helpful for certain manipulations in mathematics. But a word of caution: Do not lose the understanding of what you are doing. Make sure that you are able to do the manipulation without the shortcut.

Exponents can also be used to indicate repeated multiplication of polynomials. For example, $(3x - 4y)^2$ means $(3x - 4y)(3x - 4y)$, and $(x + 4)^3$ means $(x + 4)(x + 4)(x + 4)$. Therefore, raising a polynomial to a power is merely another multiplication problem.

$$(3x - 4y)^2 = (3x - 4y)(3x - 4y)$$
$$= 9x^2 - 24xy + 16y^2$$

[*Hint:* When squaring a binomial, be careful not to forget the middle term. That is, $(x + 5)^2 \neq x^2 + 25$; instead, $(x + 5)^2 = x^2 + 10x + 25$.]

$$(x + 4)^3 = (x + 4)(x + 4)(x + 4)$$
$$= (x + 4)(x^2 + 8x + 16)$$
$$= x(x^2 + 8x + 16) + 4(x^2 + 8x + 16)$$
$$= x^3 + 8x^2 + 16x + 4x^2 + 32x + 64$$
$$= x^3 + 12x^2 + 48x + 64$$

Special Patterns

When multiplying binomials, some special patterns occur that you should learn to recognize. These patterns can be used to find products, and some of them will be helpful later when you are factoring polynomials.

$$(a + b)^2 = a^2 + 2ab + b^2$$
$$(a - b)^2 = a^2 - 2ab + b^2$$

$$(a + b)(a - b) = a^2 - b^2$$
$$(a + b)^3 = a^3 + 3a^2b + 3ab^2 + b^3$$
$$(a - b)^3 = a^3 - 3a^2b + 3ab^2 - b^3$$

The three following examples illustrate the first three patterns, respectively.

$$(2x + 3)^2 = (2x)^2 + 2(2x)(3) + (3)^2$$
$$= 4x^2 + 12x + 9$$

$$(5x - 2)^2 = (5x)^2 - 2(5x)(2) + (2)^2$$
$$= 25x^2 - 20x + 4$$

$$(3x + 2y)(3x - 2y) = (3x)^2 - (2y)^2 = 9x^2 - 4y^2$$

In the first two examples, the resulting trinomial is called a **perfect-square trinomial;** it is the result of squaring a binomial. In the third example, the resulting binomial is called the **difference of two squares.** Later we will use both of these patterns when factoring polynomials.

The cubing-of-a-binomial patterns are helpful primarily when you are multiplying. These patterns can shorten the work of cubing a binomial, as the next two examples illustrate.

$$(3x + 2)^3 = (3x)^3 + 3(3x)^2(2) + 3(3x)(2)^2 + (2)^3$$
$$= 27x^3 + 54x^2 + 36x + 8$$

$$(5x - 2y)^3 = (5x)^3 - 3(5x)^2(2y) + 3(5x)(2y)^2 - (2y)^3$$
$$= 125x^3 - 150x^2y + 60xy^2 - 8y^3$$

Keep in mind that these multiplying patterns are useful shortcuts, but if you forget them, you can simply revert to applying the distributive property.

Binomial Expansion Pattern

It is possible to write the expansion of $(a + b)^n$, where n is *any* positive integer, without showing all of the intermediate steps of multiplying and combining similar terms. To do this, let's observe some patterns in the following examples; each one can be verified by direct multiplication.

$$(a + b)^1 = a + b$$
$$(a + b)^2 = a^2 + 2ab + b^2$$
$$(a + b)^3 = a^3 + 3a^2b + 3ab^2 + b^3$$
$$(a + b)^4 = a^4 + 4a^3b + 6a^2b^2 + 4ab^3 + b^4$$
$$(a + b)^5 = a^5 + 5a^4b + 10a^3b^2 + 10a^2b^3 + 5ab^4 + b^5$$

First, note the patterns of the exponents for a and b on a term-by-term basis. The exponents of a begin with the exponent of the binomial and decrease by 1, term by term, until the last term, which has $a^0 = 1$. The exponents of b begin with zero ($b^0 = 1$) and increase by 1, term-by-term, until the last term, which contains b to the power of the original binomial. In other words, the variables in the expansion of $(a + b)^n$ have the pattern

$$a^n, \quad a^{n-1}b, \quad a^{n-2}b^2, \quad \ldots, \quad ab^{n-1}, \quad b^n$$

where for each term, the *sum* of the exponents of a and b is n.

Next, let's arrange the *coefficients* in a triangular formation; this yields an easy-to-remember pattern.

```
            1         1
       1         2         1
   1       3         3        1
 1     4        6        4       1
1    5       10       10       5       1
```

Row number n in the formation contains the coefficients of the expansion of $(a + b)^n$. For example, the fifth row contains 1   5   10   10   5   1, and these numbers are the coefficients of the terms in the expansion of $(a + b)^5$. Furthermore, each can be formed from the previous row as follows:

1. Start and end each row with 1.
2. All other entries result from adding the two numbers in the row immediately above, one number to the left and one number to the right.

Thus from row 5, we can form row 6.

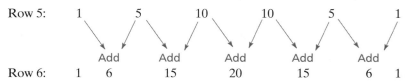

Row 5:   1   5   10   10   5   1

                Add   Add   Add   Add   Add
Row 6:   1   6   15   20   15   6   1

Now we can use these seven coefficients and our discussion about the exponents to write out the expansion for $(a + b)^6$.

$$(a + b)^6 = a^6 + 6a^5b + 15a^4b^2 + 20a^3b^3 + 15a^2b^4 + 6ab^5 + b^6$$

REMARK: The triangular formation of numbers that we have been discussing is often referred to as **Pascal's triangle.** This is in honor of Blaise Pascal, a 17th-century mathematician, to whom the discovery of this pattern is attributed.

Let's consider two more examples using Pascal's triangle and the exponent relationships.

E X A M P L E   1

Expand $(a - b)^4$.

Solution

We can treat $a - b$ as $a + (-b)$ and use the fourth row of Pascal's triangle to obtain the coefficients.

$$[a + (-b)]^4 = a^4 + 4a^3(-b) + 6a^2(-b)^2 + 4a(-b)^3 + (-b)^4$$
$$= a^4 - 4a^3b + 6a^2b^2 - 4ab^3 + b^4$$

EXAMPLE 2   Expand $(2x + 3y)^5$.

Solution

Let $2x = a$ and $3y = b$. The coefficients come from the fifth row of Pascal's triangle.

$$(2x + 3y)^5 = (2x)^5 + 5(2x)^4(3y) + 10(2x)^3(3y)^2 + 10(2x)^2(3y)^3 + 5(2x)(3y)^4 + (3y)^5$$
$$= 32x^5 + 240x^4y + 720x^3y^2 + 1080x^2y^3 + 810xy^4 + 243y^5$$ ■

CONCEPT QUIZ

For Problems 1–8, answer true or false.

1. The degree of the polynomial $3x^4 - 2x^3 + 7x^2 + 5x - 1$ is 4.

2. The polynomial $4x^2y^3$ is a binomial in two variables with degree 5.

3. $(-5xy^3)^2 = (-5)^2xy^6$

4. $\dfrac{-15x^8}{3x^2} = -5x^4$.

5. The pattern for squaring a binomial is $(a + b)^2 = a^2 + 2ab + b^2$.

6. $4x^2 + 20x + 25$ is a perfect square binomial.

7. The last term of the binomial expansion for $(2x + 3y)^4$ is $3y^4$.

8. Pascal's triangle is used to find the exponents in a binomial expansion.

PROBLEM SET 8.4

For Problems 1–40, find each indicated product.

1. $\left(-\dfrac{1}{2}xy\right)\left(\dfrac{1}{3}x^2y^3\right)$

2. $\left(\dfrac{3}{4}x^4y^5\right)(-x^2y)$

3. $(3x)(-2x^2)(-5x^3)$

4. $(-2x)(-6x^3)(x^2)$

5. $(-6x^2)(3x^3)(x^4)$

6. $(-7x^2)(3x)(4x^3)$

7. $(x^2y)(-3xy^2)(x^3y^3)$

8. $(xy^2)(-5xy)(x^2y^4)$

9. $(-3ab^3)^4$

10. $(-2a^2b^4)^4$

11. $-(2ab)^4$

12. $-(3ab)^4$

13. $(4x + 5)(x + 7)$

14. $(6x + 5)(x + 3)$

15. $(3y - 1)(3y + 1)$

16. $(5y - 2)(5y + 2)$

17. $(7x - 2)(2x + 1)$

18. $(6x - 1)(3x + 2)$

19. $(1 + t)(5 - 2t)$

20. $(3 - t)(2 + 4t)$

21. $(3t + 7)^2$

22. $(4t + 6)^2$

23. $(2 - 5x)(2 + 5x)$

24. $(6 - 3x)(6 + 3x)$

25. $(x + 1)(x - 2)(x - 3)$

26. $(x - 1)(x + 4)(x - 6)$

27. $(x - 3)(x + 3)(x - 1)$

28. $(x - 5)(x + 5)(x - 8)$

29. $(x - 4)(x^2 + 5x - 4)$

30. $(x + 6)(2x^2 - x - 7)$

31. $(2x - 3)(x^2 + 6x + 10)$

32. $(3x + 4)(2x^2 - 2x - 6)$

33. $(4x - 1)(3x^2 - x + 6)$

34. $(5x - 2)(6x^2 + 2x - 1)$

35. $(x^2 + 2x + 1)(x^2 + 3x + 4)$

36. $(x^2 - x + 6)(x^2 - 5x - 8)$

37. $(4x - 1)^3$

38. $(3x - 2)^3$

39. $(5x + 2)^3$

40. $(4x - 5)^3$

For Problems 41–50, find the indicated products. Assume all variables that appear as exponents represent positive integers.

41. $(x^n - 4)(x^n + 4)$

42. $(x^{3a} - 1)(x^{3a} + 1)$

43. $(x^a + 6)(x^a - 2)$

44. $(x^a + 4)(x^a - 9)$

45. $(2x^n + 5)(3x^n - 7)$

46. $(3x^n + 5)(4x^n - 9)$

47. $(x^{2a} - 7)(x^{2a} - 3)$

48. $(x^{2a} + 6)(x^{2a} - 4)$

49. $(2x^n + 5)^2$

50. $(3x^n - 7)^2$

For Problems 51–60, find each quotient.

51. $\dfrac{9x^4 y^5}{3xy^2}$

52. $\dfrac{12x^2 y^7}{6x^2 y^3}$

53. $\dfrac{25x^5 y^6}{-5x^2 y^4}$

54. $\dfrac{56x^6 y^4}{-7x^2 y^3}$

55. $\dfrac{-54ab^2 c^3}{-6abc}$

56. $\dfrac{-48a^3 bc^5}{-6a^2 c^4}$

57. $\dfrac{-18x^2 y^2 z^6}{xyz^2}$

58. $\dfrac{-32x^4 y^5 z^8}{x^2 yz^3}$

59. $\dfrac{a^3 b^4 c^7}{-abc^5}$

60. $\dfrac{-a^4 b^5 c}{a^2 b^4 c}$

For Problems 61–72, use Pascal's triangle to help expand each of the following.

61. $(a + b)^7$

62. $(a + b)^8$

63. $(x - y)^5$

64. $(x - y)^6$

65. $(x + 2y)^4$

66. $(2x + y)^5$

67. $(2a - b)^6$

68. $(3a - b)^4$

69. $(x^2 + y)^7$

70. $(x + 2y^2)^7$

71. $(2a - 3b)^5$

72. $(4a - 3b)^3$

■ ■ ■ **Thoughts into words**

73. Determine the number of terms in the product of $(x + y)$ and $(a + b + c + d)$ without doing the multiplication. Explain how you arrived at your answer.

74. How would you convince someone that $x^6 \div x^2$ is x^4 and not x^3?

Answers to Concept Quiz

1. True   **2.** False   **3.** False   **4.** False   **5.** True   **6.** True   **7.** False   **8.** False

8.5   Dividing Polynomials: Synthetic Division

Objectives

■   Review the division of a polynomial.

■   Perform synthetic division.

In the previous section, we reviewed the process of dividing monomials by monomials. In Section 2.2, we used $\dfrac{a}{b} + \dfrac{c}{b} = \dfrac{a + c}{b}$ and $\dfrac{a}{b} - \dfrac{c}{b} = \dfrac{a - c}{b}$ as the basis for adding and subtracting rational numbers and rational expressions. These same equalities, viewed as $\dfrac{a + c}{b} = \dfrac{a}{b} + \dfrac{c}{b}$ and $\dfrac{a - c}{b} = \dfrac{a}{b} - \dfrac{c}{b}$, along with our knowledge of dividing monomials, provide the basis for dividing polynomials by monomials. Consider the following examples:

$$\frac{18x^3 + 24x^2}{6x} = \frac{18x^3}{6x} + \frac{24x^2}{6x} = 3x^2 + 4x$$

$$\frac{35x^2y^3 - 55x^3y^4}{5xy^2} = \frac{35x^2y^3}{5xy^2} - \frac{55x^3y^4}{5xy^2} = 7xy - 11x^2y^2$$

To divide a polynomial by a monomial, we divide each term of the polynomial by the monomial. As with many skills, once you feel comfortable with the process, you may then want to perform some of the steps mentally. Your work could take on the following format:

$$\frac{40x^4y^5 + 72x^5y^7}{8x^2y} = 5x^2y^4 + 9x^3y^6 \qquad \frac{36a^3b^4 - 45a^4b^6}{-9a^2b^3} = -4ab + 5a^2b^3$$

In Section 6.5, we used some examples to introduce the process of dividing a polynomial by a binomial. The first example of that section gave a detailed step-by-step procedure for this division process. The following example uses some "think steps" to help you review that procedure.

EXAMPLE 1   Divide $5x^2 + 6x - 8$ by $x + 2$.

Solution

$$\begin{array}{r}
5x - 4 \\
x + 2 \overline{)\,5x^2 + 6x - 8} \\
\underline{5x^2 + 10x} \\
-4x - 8 \\
\underline{-4x - 8} \\
0
\end{array}$$

Think Steps

1. $\dfrac{5x^2}{x} = 5x$.

2. $5x(x + 2) = 5x^2 + 10x$.

3. $(5x^2 + 6x - 8) - (5x^2 + 10x) = -4x - 8$.

4. $\dfrac{-4x}{x} = -4$.

5. $-4(x + 2) = -4x - 8$.   ∎

Recall that to check a division problem, we can multiply the divisor times the quotient and add the remainder. In other words,

Dividend = (Divisor)(Quotient) + (Remainder)

Sometimes the remainder is expressed as a fractional part of the divisor. The relationship then becomes

$$\frac{\text{Dividend}}{\text{Divisor}} = \text{Quotient} + \frac{\text{Remainder}}{\text{Divisor}}$$

E X A M P L E   2

Divide $2x^2 - 3x + 1$ by $x - 5$.

Solution

$$
\begin{array}{r}
2x \;+\; 7 \\
x - 5 \overline{)\, 2x^2 \;-\; 3x \;+\; 1\,} \\
\underline{2x^2 \;-\; 10x} \\
7x \;+\; 1 \\
\underline{7x \;-\; 35} \\
36 \quad \longleftarrow \text{Remainder}
\end{array}
$$

Thus

$$
\frac{2x^2 - 3x + 1}{x - 5} = 2x + 7 + \frac{36}{x - 5}, \qquad x \neq 5
$$

E X A M P L E   3

Divide $t^3 - 8$ by $t - 2$.

Solution

$$
\begin{array}{r}
t^2 + 2t \;\;\; + 4 \\
t - 2 \overline{)\, t^3 + 0t^2 + 0t - 8\,} \\
\underline{t^3 - 2t^2} \\
2t^2 + 0t - 8 \\
\underline{2t^2 - 4t} \\
4t - 8 \\
\underline{4t - 8} \\
0
\end{array}
$$

$\longleftarrow$ Note the insertion of a "t-squared" term and a "t term" with zero coefficients.

Thus we can say that the quotient is $t^2 + 2t + 4$ and the remainder is 0. ■

E X A M P L E   4

Divide $y^3 + 3y^2 - 2y - 1$ by $y^2 + 2y$.

Solution

$$
\begin{array}{r}
y \;+\; 1 \\
y^2 + 2y \overline{)\, y^3 + 3y^2 - 2y - 1\,} \\
\underline{y^3 + 2y^2} \\
y^2 - 2y - 1 \\
\underline{y^2 + 2y} \\
-4y - 1 \quad \longleftarrow \text{Remainder of } -4y - 1
\end{array}
$$

The division process is complete when the degree of the remainder is less than the degree of the divisor. Thus the quotient is $y + 1$ and the remainder is $-4y - 1$. ■

Synthetic Division

If the divisor is of the form $x - c$, where c is a constant, then the typical long-division algorithm can be simplified to a process called **synthetic division.** First, let's consider another division problem and use the regular-division algorithm. Then, in a step-by-step fashion, we will demonstrate some shortcuts that will lead us into the synthetic-division procedure. Consider the division problem $(2x^4 + x^3 - 17x^2 + 13x + 2) \div (x - 2)$.

$$
\begin{array}{r}
2x^3 + 5x^2 - \ 7x \ - 1 \\
x - 2)\overline{2x^4 + \ \ x^3 - 17x^2 + 13x + 2} \\
\underline{2x^4 - 4x^3} \\
5x^3 - 17x^2 \\
\underline{5x^3 - 10x^2} \\
-7x^2 + 13x \\
\underline{-7x^2 + 14x} \\
-x + 2 \\
\underline{-x + 2}
\end{array}
$$

Because the dividend is written in descending powers of x, the quotient is produced in descending powers of x. In other words, the numerical coefficients are the *key issues,* so let's rewrite the problem in terms of its coefficients.

$$
\begin{array}{r}
2 \quad 5 \quad -7 \quad -1 \\
1 - 2)\overline{2 \quad \ 1 \quad -17 \quad \ 13 \quad \ 2} \\
②\quad -4 \\
5 \quad ⟨-17⟩ \\
⑤\quad -10 \\
-7 \quad ⟨13⟩ \\
⟨-7⟩ \quad 14 \\
-1 \quad ② \\
⟨-1⟩ \quad 2
\end{array}
$$

Now observe that the circled numbers are simply repetitions of the numbers directly above them in the format. Thus the circled numbers can be omitted, and the format will be as follows (disregard the arrows for the moment).

$$
\begin{array}{r}
2 \quad 5 \quad -7 \quad -1 \\
1 - 2)\overline{2 \quad \ 1 \quad -17 \quad 13 \quad 2} \\
-4 \\
5 \\
-10 \\
-7 \\
14 \\
-1 \\
2
\end{array}
$$

Next, by moving some numbers up (indicated by the arrows) and by not writing the 1 that is the coefficient of x in the divisor, we obtain the following more compact form.

$$
\begin{array}{r}
2 \quad\; 5 \quad -7 \quad -1 \\
\hline
-2)\,2 \quad\; 1 \quad -17 \quad 13 \quad 2 \\
-4 \quad -10 \quad 14 \quad 2 \\
\hline
5 \quad -7 \quad -1 \quad 0
\end{array}
$$

(1)
(2)
(3)
(4)

Note that line 4 reveals all of the coefficients of the quotient (line 1) except for the first coefficient, 2. Thus we can omit line 1, begin line 4 with the first coefficient, and then use the following form.

$$
\begin{array}{r}
-2)\,2 \quad\; 1 \quad -17 \quad 13 \quad 2 \\
-4 \quad -10 \quad 14 \quad 2 \\
\hline
2 \quad\; 5 \quad -7 \quad -1 \quad 0
\end{array}
$$

(5)
(6)
(7)

Line 7 contains the coefficients of the quotient, where the zero indicates the remainder. Finally, by changing the constant in the divisor to 2 (instead of -2), which changes the signs of the numbers in line 6, we can *add* the corresponding entries in lines 5 and 6 rather than subtracting. Thus the final synthetic-division form for this problem is

$$
\begin{array}{r}
2)\,2 \quad\; 1 \quad -17 \quad\; 13 \quad\;\; 2 \\
4 \quad\;\; 10 \quad -14 \quad -2 \\
\hline
2 \quad\; 5 \quad\; -7 \quad\; -1 \quad\;\; 0
\end{array}
\qquad \textbf{(6)}
$$

The first four entries of the bottom row are the coefficients and the constant term of the quotient $(2x^3 + 5x^2 - 7x - 1)$, and the last entry is the remainder (0).

Now we will consider another problem and indicate a step-by-step procedure for setting up and carrying out the synthetic-division process. Suppose that we want to do the division problem

$$
x + 4)\overline{2x^3 + 5x^2 - 13x - 2}
$$

STEP 1   Write the coefficients of the dividend as follows:

$$
)\overline{2 \quad\; 5 \quad -13 \quad -2}
$$

STEP 2   In the divisor, use -4 instead of 4 so that later we can add rather than subtract.

$$
-4)\overline{2 \quad\; 5 \quad -13 \quad -2}
$$

STEP 3   Bring down the first coefficient of the dividend.

$$
-4)\overline{2 \quad\; 5 \quad -13 \quad -2}
$$
$$
\overline{}
$$
$$
2
$$

STEP 4   Multiply that first coefficient times the divisor, which yields $2(-4)$
$= -8$. Add this result to the second coefficient of the dividend.

$$
\begin{array}{r|rrrr}
-4) & 2 & 5 & -13 & -2 \\
& & -8 & & \\
\hline
& 2 & -3 & &
\end{array}
$$

STEP 5   Multiply $(-3)(-4)$, which yields 12. Add this result to the third coefficient of the dividend.

$$
\begin{array}{r|rrrr}
-4) & 2 & 5 & -13 & -2 \\
& & -8 & 12 & \\
\hline
& 2 & -3 & -1 &
\end{array}
$$

STEP 6   Multiply $(-1)(-4)$, which yields 4. Add this result to the last term
of the dividend.

$$
\begin{array}{r|rrrr}
-4) & 2 & 5 & -13 & -2 \\
& & -8 & 12 & 4 \\
\hline
& 2 & -3 & -1 & 2
\end{array}
$$

The last row indicates a quotient of $2x^2 - 3x - 1$ and a remainder
of 2.

Now let's consider some examples in which we show only the final compact
form of synthetic division.

E X A M P L E   5   Find the quotient and the remainder for $(x^3 + 8x^2 + 13x - 6) \div (x + 3)$.

Solution

$$
\begin{array}{r|rrrr}
-3) & 1 & 8 & 13 & -6 \\
& & -3 & -15 & 6 \\
\hline
& 1 & 5 & -2 & 0
\end{array}
$$

Thus the quotient is $x^2 + 5x - 2$, and the remainder is zero.   ■

E X A M P L E   6   Find the quotient and the remainder for $(3x^4 + 5x^3 - 29x^2 - 45x + 14) \div (x - 3)$.

Solution

$$
\begin{array}{r|rrrrr}
3) & 3 & 5 & -29 & -45 & 14 \\
& & 9 & 42 & 39 & -18 \\
\hline
& 3 & 14 & 13 & -6 & -4
\end{array}
$$

Thus the quotient is $3x^3 + 14x^2 + 13x - 6$, and the remainder is -4.   ■

EXAMPLE 7   Find the quotient and the remainder for $(4x^4 - 2x^3 + 6x - 1) \div (x - 1)$.

Solution

$$
\begin{array}{r|rrrrr}
1) & 4 & -2 & 0 & 6 & -1 \\
  &   & 4 & 2 & 2 & 8 \\
\hline
  & 4 & 2 & 2 & 8 & 7
\end{array}
$$

Note that a zero has been inserted as the coefficient of the missing x^2 term.

Thus the quotient is $4x^3 + 2x^2 + 2x + 8$, and the remainder is 7.   ■

EXAMPLE 8   Find the quotient and the remainder for $(x^4 + 16) \div (x + 2)$.

Solution

$$
\begin{array}{r|rrrrr}
-2) & 1 & 0 & 0 & 0 & 16 \\
    &   & -2 & 4 & -8 & 16 \\
\hline
    & 1 & -2 & 4 & -8 & 32
\end{array}
$$

Note that zeros have been inserted as coefficients of the missing terms in the dividend.

Thus the quotient is $x^3 - 2x^2 + 4x - 8$, and the remainder is 32.   ■

CONCEPT QUIZ

For Problems 1–5, answer true or false.

1. To divide a polynomial by a monomial, we divide each term of the polynomial by the monomial.

2. To check a division problem, we can multiply the divisor by the quotient and add the remainder. The result should equal the dividend.

3. Synthetic division is used to simplify the process of dividing a polynomial by a monomial.

4. Synthetic division is used when the divisor is of the form $x - c$, where c is a constant.

5. The synthetic division process is used only when the remainder is zero.

PROBLEM SET 8.5

For Problems 1–10, perform the indicated divisions of polynomials by monomials.

1. $\dfrac{9x^4 + 18x^3}{3x}$

2. $\dfrac{12x^3 - 24x^2}{6x^2}$

3. $\dfrac{-24x^6 + 36x^8}{4x^2}$

4. $\dfrac{-35x^5 - 42x^3}{-7x^2}$

5. $\dfrac{15a^3 - 25a^2 - 40a}{5a}$

6. $\dfrac{-16a^4 + 32a^3 - 56a^2}{-8a}$

7. $\dfrac{13x^3 - 17x^2 + 28x}{-x}$

8. $\dfrac{14xy - 16x^2y^2 - 20x^3y^4}{-xy}$

9. $\dfrac{-18x^2y^2 + 24x^3y^2 - 48x^2y^3}{6xy}$

10. $\dfrac{-27a^3b^4 - 36a^2b^3 + 72a^2b^5}{9a^2b^2}$

For Problems 11–22, find the quotient and remainder for each division problem.

11. $(12x^2 + 7x - 10) \div (3x - 2)$

12. $(20x^2 - 39x + 18) \div (5x - 6)$

13. $(3t^3 + 7t^2 - 10t - 4) \div (3t + 1)$

14. $(4t^3 - 17t^2 + 7t + 10) \div (4t - 5)$

15. $(6x^2 + 19x + 11) \div (3x + 2)$

16. $(20x^2 + 3x - 1) \div (5x + 2)$

17. $(3x^3 + 2x^2 - 5x - 1) \div (x^2 + 2x)$

18. $(4x^3 - 5x^2 + 2x - 6) \div (x^2 - 3x)$

19. $(5y^3 - 6y^2 - 7y - 2) \div (y^2 - y)$

20. $(8y^3 - y^2 - y + 5) \div (y^2 + y)$

21. $(4a^3 - 2a^2 + 7a - 1) \div (a^2 - 2a + 3)$

22. $(5a^3 + 7a^2 - 2a - 9) \div (a^2 + 3a - 4)$

For Problems 23–46, use *synthetic division* to determine the quotient and remainder for each division problem.

23. $(3x^2 + x - 4) \div (x - 1)$

24. $(2x^2 - 5x - 3) \div (x - 3)$

25. $(x^2 + 2x - 10) \div (x - 4)$

26. $(x^2 - 10x + 15) \div (x - 8)$

27. $(4x^2 + 5x - 4) \div (x + 2)$

28. $(5x^2 + 18x - 8) \div (x + 4)$

29. $(x^3 - 2x^2 - x + 2) \div (x - 2)$

30. $(x^3 - 5x^2 + 2x + 8) \div (x + 1)$

31. $(3x^4 - x^3 + 2x^2 - 7x - 1) \div (x + 1)$

32. $(2x^3 - 5x^2 - 4x + 6) \div (x - 2)$

33. $(x^3 - 7x - 6) \div (x + 2)$

34. $(x^3 + 6x^2 - 5x - 1) \div (x - 1)$

35. $(x^4 + 4x^3 - 7x - 1) \div (x - 3)$

36. $(2x^4 + 3x^2 + 3) \div (x + 2)$

37. $(x^3 + 6x^2 + 11x + 6) \div (x + 3)$

38. $(x^3 - 4x^2 - 11x + 30) \div (x - 5)$

39. $(x^5 - 1) \div (x - 1)$

40. $(x^5 - 1) \div (x + 1)$

41. $(x^5 + 1) \div (x - 1)$

42. $(x^5 + 1) \div (x + 1)$

43. $(2x^3 + 3x^2 - 2x + 3) \div \left(x + \dfrac{1}{2} \right)$

44. $(9x^3 - 6x^2 + 3x - 4) \div \left(x - \dfrac{1}{3} \right)$

45. $(4x^4 - 5x^2 + 1) \div \left(x - \dfrac{1}{2} \right)$

46. $(3x^4 - 2x^3 + 5x^2 - x - 1) \div \left(x + \dfrac{1}{3} \right)$

■ ■ ■ **Thoughts into words**

47. How do you know by inspection that a quotient of $3x^2 + 5x + 1$ and a remainder of 0 cannot be the correct answer for the division problem $(3x^3 - 7x^2 - 22x + 8) \div (x - 4)$?

48. Why is synthetic division restricted to situations where the divisor is of the format $x - c$?

Answers to Concept Quiz

1. True   **2.** True   **3.** False   **4.** True   **5.** False

8.6 Factoring: A Brief Review and a Step Further

Objectives

■ Review factoring techniques.

■ Factor the sum or difference of two cubes.

■ Factor trinomials by a systematic technique.

■ Review solving equations and word problems that involve factoring.

Chapter 7 was organized as follows: First a factoring technique was introduced. Next some equations were solved using that factoring technique. Then this type of equation was used to solve some word problems. In this section we will briefly review and expand upon that material by first considering all of the factoring techniques, then solving a few equations, and finally solving some word problems.

Factoring: Use of the Distributive Property

In general, factoring is the reverse of multiplication. Previously, we have used the distributive property to find the product of a monomial and a polynomial, as in the next examples.

$$3(x + 2) = 3(x) + 3(2) = 3x + 6$$
$$5(2x - 1) = 5(2x) - 5(1) = 10x - 5$$
$$x(x^2 + 6x - 4) = x(x^2) + x(6x) - x(4) = x^3 + 6x^2 - 4x$$

We shall also use the distributive property (in the form $ab + ac = a(b + c)$) to reverse the process — that is, to factor a given polynomial. Consider the following examples. (The steps in the dashed boxes can be done mentally.)

$$3x + 6 = \boxed{3(x) + 3(2)} = 3(x + 2)$$
$$10x - 5 = \boxed{5(2x) - 5(1)} = 5(2x - 1)$$
$$x^3 + 6x^2 - 4x = \boxed{x(x^2) + x(6x) - x(4)} = x(x^2 + 6x - 4)$$

Note that in each example a given polynomial has been factored into the product of a monomial and a polynomial. Obviously, polynomials could be factored in a variety of ways. Consider some factorizations of $3x^2 + 12x$.

$$3x^2 + 12x = 3x(x + 4) \qquad \text{or} \qquad 3x^2 + 12x = 3(x^2 + 4x) \qquad \text{or}$$

$$3x^2 + 12x = x(3x + 12) \qquad \text{or} \qquad 3x^2 + 12x = \frac{1}{2}(6x^2 + 24x)$$

We are, however, primarily interested in the first of the previous factorization forms, which we refer to as the **completely factored form.** A polynomial with integral coefficients is in completely factored form if:

> **1.** It is expressed as a product of polynomials with *integral coefficients,* and
> **2.** No polynomial, other than a monomial, within the factored form can be further factored into polynomials with integral coefficients.

Do you see why only the first of the above factored forms of $3x^2 + 12x$ is said to be in completely factored form? In each of the other three forms, the polynomial inside the parentheses can be factored further. Moreover, in the last form, $\frac{1}{2}(6x^2 + 24x)$, the condition of using only integral coefficients is violated.

This factoring process, $ab + ac = a(b + c)$, is referred to as **factoring out the highest common factor.** The key idea is to recognize the largest monomial factor that is common to all terms. For example, we observe that each term of $2x^3 + 4x^2 + 6x$ has a factor of $2x$. Thus we write

$$2x^3 + 4x^2 + 6x = 2x(\underline{\hspace{2cm}})$$

and insert within the parentheses the result of dividing $2x^3 + 4x^2 + 6x$ by $2x$.

$$2x^3 + 4x^2 + 6x = 2x(x^2 + 2x + 3)$$

The following examples further demonstrate this process of factoring out the highest common monomial factor.

$$12x^3 + 16x^2 = 4x^2(3x + 4) \qquad 6x^2y^3 + 27xy^4 = 3xy^3(2x + 9y)$$
$$8ab - 18b = 2b(4a - 9) \qquad 8y^3 + 4y^2 = 4y^2(2y + 1)$$
$$30x^3 + 42x^4 - 24x^5 = 6x^3(5 + 7x - 4x^2)$$

Note that in each example, the common monomial factor itself is not in a completely factored form. For example, $4x^2(3x + 4)$ is not written as $2 \cdot 2x \cdot x \cdot (3x + 4)$.

Sometimes there may be a common binomial factor rather than a common monomial factor. For example, each of the two terms of the expression $x(y + 2) + z(y + 2)$ has a binomial factor of $(y + 2)$. Thus we can factor $(y + 2)$ from each term, and our result is

$$x(y + 2) + z(y + 2) = (y + 2)(x + z)$$

It may be that the original polynomial exhibits no apparent common monomial or binomial factor, which is the case with $ab + 3a + bc + 3c$. However, by factoring a from the first two terms and c from the last two terms, we get

$$ab + 3a + bc + 3c = a(b + 3) + c(b + 3)$$

Now a common binomial factor of $(b + 3)$ is obvious, and we can proceed as before.

$$a(b + 3) + c(b + 3) = (b + 3)(a + c)$$

We refer to this factoring process as *factoring by grouping.* Let's consider a few more examples of this type.

$$ab^2 - 4b^2 + 3a - 12 = b^2(a - 4) + 3(a - 4) \qquad \text{Factor } b^2 \text{ from the first two terms and 3 from the last two terms.}$$

$$= (a - 4)(b^2 + 3) \qquad \text{Factor the common binomial from both terms.}$$

$$x^2 - x + 5x - 5 = x(x - 1) + 5(x - 1)$$ Factor x from the first two terms and 5 from the last two terms.

$$= (x - 1)(x + 5)$$ Factor the common binomial from both terms.

$$x^2 + 2x - 3x - 6 = x(x + 2) - 3(x + 2)$$ Factor x from the first two terms and -3 from the last two terms.

$$= (x + 2)(x - 3)$$ Factor the common binomial from both terms.

It may be necessary to rearrange some terms before applying the distributive property. Terms that contain common factors need to be grouped together, and this may be done in more than one way. The next example illustrates this idea.

$$4a^2 - bc^2 - a^2b + 4c^2 = 4a^2 - a^2b + 4c^2 - bc^2$$
$$= a^2(4 - b) + c^2(4 - b)$$
$$= (4 - b)(a^2 + c^2) \quad \text{or}$$
$$4a^2 - bc^2 - a^2b + 4c^2 = 4a^2 + 4c^2 - bc^2 - a^2b$$
$$= 4(a^2 + c^2) - b(c^2 + a^2)$$
$$= 4(a^2 + c^2) - b(a^2 + c^2)$$
$$= (a^2 + c^2)(4 - b)$$

Factoring: Difference of Two Squares

In Section 6.3, we examined some special multiplication patterns. One of these patterns was

$$(a + b)(a - b) = a^2 - b^2$$

This same pattern, viewed as a factoring pattern, is referred to as the difference of two squares.

Difference of Two Squares

$$a^2 - b^2 = (a + b)(a - b)$$

Applying the pattern is fairly simple, as these next examples demonstrate. Again, the steps in dashed boxes are usually performed mentally.

$$x^2 - 16 = (x)^2 - (4)^2 = (x + 4)(x - 4)$$
$$4x^2 - 25 = (2x)^2 - (5)^2 = (2x + 5)(2x - 5)$$
$$16x^2 - 9y^2 = (4x)^2 - (3y)^2 = (4x + 3y)(4x - 3y)$$
$$1 - a^2 = (1)^2 - (a)^2 = (1 + a)(1 - a)$$

Multiplication is commutative, so the order in which we write the factors is not important. For example, $(x + 4)(x - 4)$ can also be written as $(x - 4)(x + 4)$.

You must be careful not to assume an analogous factoring pattern for the *sum* of two squares; *it does not exist.* For example, $x^2 + 4 \neq (x + 2)(x + 2)$ because $(x + 2)(x + 2) = x^2 + 4x + 4$. We say that a polynomial such as $x^2 + 4$ is a **prime polynomial** or that it is *not factorable using integers.*

Sometimes the difference-of-two-squares pattern can be applied more than once, as the next examples illustrate.

$$x^4 - y^4 = (x^2 + y^2)(x^2 - y^2) = (x^2 + y^2)(x + y)(x - y)$$
$$16x^4 - 81y^4 = (4x^2 + 9y^2)(4x^2 - 9y^2) = (4x^2 + 9y^2)(2x + 3y)(2x - 3y)$$

It may also be that the squares are other than simple monomial squares, as in the next three examples.

$$(x + 3)^2 - y^2 = [(x + 3) + y][(x + 3) - y] = (x + 3 + y)(x + 3 - y)$$
$$4x^2 - (2y + 1)^2 = [2x + (2y + 1)][(2x - (2y + 1)]$$
$$= (2x + 2y + 1)(2x - 2y - 1)$$
$$(x - 1)^2 - (x + 4)^2 = [(x - 1) + (x + 4)][(x - 1) - (x + 4)]$$
$$= (x - 1 + x + 4)(x - 1 - x - 4)$$
$$= (2x + 3)(-5)$$

It is possible to apply both the technique of *factoring out a common monomial factor* and the pattern of the *difference of two squares* to the same problem. *In general, it is best to look first for a common monomial factor.* Consider the following examples.

$$2x^2 - 50 = 2(x^2 - 25)$$
$$= 2(x + 5)(x - 5)$$
$$48y^3 - 27y = 3y(16y^2 - 9)$$
$$= 3y(4y + 3)(4y - 3)$$
$$9x^2 - 36 = 9(x^2 - 4)$$
$$= 9(x + 2)(x - 2)$$

WORD OF CAUTION   The polynomial $9x^2 - 36$ can be factored as follows:

$$9x^2 - 36 = (3x + 6)(3x - 6)$$
$$= 3(x + 2)(3)(x - 2)$$
$$= 9(x + 2)(x - 2)$$

However, when one is taking this approach, there seems to be a tendency to stop at the step $(3x + 6)(3x - 6)$. Therefore, remember the suggestion to *look first for a common monomial factor.*

Factoring: Sum or Difference of Two Cubes

As we pointed out before, there exists no sum-of-squares pattern analogous to the difference-of-squares factoring pattern. That is, a polynomial such as $x^2 + 9$ is not factorable using integers. However, patterns do exist for both *the sum and the difference of two cubes*. These patterns are as follows:

Sum and Difference of Two Cubes
$a^3 + b^3 = (a + b)(a^2 - ab + b^2)$
$a^3 - b^3 = (a - b)(a^2 + ab + b^2)$

Note how we apply these patterns in the next four examples.

$$x^3 + 27 = (x)^3 + (3)^3 = (x + 3)(x^2 - 3x + 9)$$
$$8a^3 + 125b^3 = (2a)^3 + (5b)^3 = (2a + 5b)(4a^2 - 10ab + 25b^2)$$
$$x^3 - 1 = (x)^3 - (1)^3 = (x - 1)(x^2 + x + 1)$$
$$27y^3 - 64x^3 = (3y)^3 - (4x)^3 = (3y - 4x)(9y^2 + 12xy + 16x^2)$$

Factoring: Trinomials of the Form $x^2 + bx + c$

To factor trinomials of the form $x^2 + bx + c$ (that is, trinomials for which the coefficient of the squared term is 1), we can use the result of the following multiplication problem.

$$(x + a)(x + b) = x(x + b) + a(x + b)$$
$$= x(x) + x(b) + a(x) + a(b)$$
$$= x^2 + (b + a)x + ab$$
$$= x^2 + (a + b)x + ab$$

Note that the coefficient of x is the *sum of a and b*, and the last term is the *product of a and b.* Let's consider some examples to review the use of these ideas.

E X A M P L E   1

Factor $x^2 + 8x + 12$.

Solution

We need to complete the following with two integers whose product is 12 and whose sum is 8.

$$x^2 + 8x + 12 = (x + \underline{})(x + \underline{})$$

The possible pairs of factors of 12 are 1(12), 2(6), and 3(4). Because $6 + 2 = 8$, we can complete the factoring as follows:

$$x^2 + 8x + 12 = (x + 6)(x + 2)$$

To *check* our answer, we find the product of $(x + 6)$ and $(x + 2)$.   ■

EXAMPLE 2   Factor $x^2 - 10x + 24$.

Solution

We need two integers whose product is 24 and whose sum is -10. Let's use a small table to organize our thinking.

Product	Sum
$(-1)(-24) = 24$	$-1 + (-24) = -25$
$(-2)(-12) = 24$	$-2 + (-12) = -14$
$(-3)(-8) = 24$	$-3 + (-8) = -11$
$(-4)(-6) = 24$	$-4 + (-6) = -10$

The bottom line contains the numbers that we need. Thus

$$x^2 - 10x + 24 = (x - 4)(x - 6)$$

EXAMPLE 3   Factor $x^2 + 7x - 30$.

Solution

We need two integers whose product is -30 and whose sum is 7.

Product	Sum
$(-1)(30) = -30$	$-1 + 30 = 29$
$1(-30) = -30$	$1 + (-30) = -29$
$2(-15) = -30$	$2 + (-15) = -13$
$-2(15) = -30$	$-2 + 15 = 13$
$-3(10) = -30$	$-3 + 10 = 7$

No need to search any further.

The numbers that we need are -3 and 10, and we can complete the factoring.

$$x^2 + 7x - 30 = (x + 10)(x - 3)$$

EXAMPLE 4   Factor $x^2 + 7x + 16$.

Solution

We need two integers whose product is 16 and whose sum is 7.

Product	Sum
$1(16) = 16$	$1 + 16 = 17$
$2(8) = 16$	$2 + 8 = 10$
$4(4) = 16$	$4 + 4 = 8$

We have exhausted all possible pairs of factors of 16, and no two factors have a sum of 7, so we conclude that $x^2 + 7x + 16$ *is not factorable using integers.*

The tables in Examples 2, 3, and 4 were used to illustrate one way of organizing your thoughts for such problems. Normally you would probably factor such problems mentally without taking the time to formulate a table. Note, however, that in Example 4 the table helped us to be absolutely sure that we tried all the possibilities. Whether or not you use the table, keep in mind that the key ideas are the product and sum relationships.

EXAMPLE 5   Factor $t^2 + 2t - 168$.

Solution

We need two integers whose product is -168 and whose sum is 2. Because the absolute value of the constant term is rather large, it might help to look at it in prime factored form.

$$168 = 2 \cdot 2 \cdot 2 \cdot 3 \cdot 7$$

Now we can mentally form two numbers by using all of these factors in different combinations. Using two 2s and a 3 in one number and the other 2 and the 7 in the second number produces $2 \cdot 2 \cdot 3 = 12$ and $2 \cdot 7 = 14$. The coefficient of the middle term of the trinomial is 2, so we know that we must use 14 and -12. Thus we obtain

$$t^2 + 2t - 168 = (t + 14)(t - 12)$$ ∎

Factoring: Trinomials of the Form $ax^2 + bx + c$

Now let's consider factoring trinomials where the coefficient of the squared term is not 1. In Section 7.4, we used an informal trial-and-error process for such trinomials. This technique is based on our knowledge of multiplication of binomials and works quite well for certain trinomials. Let's review the process with an example.

EXAMPLE 6   Factor $5x^2 - 18x - 8$.

Solution

The first term, $5x^2$, can be written as $x \cdot 5x$. The last term, -8, can be written as $(-2)(4)$, $(2)(-4)$, $(-1)(8)$, or $(1)(-8)$. Therefore, we have the following possibilities to try.

$$(x - 2)(5x + 4) \qquad (x + 4)(5x - 2)$$
$$(x + 2)(5x - 4) \qquad (x - 4)(5x + 2)$$
$$(x - 1)(5x + 8) \qquad (x + 8)(5x - 1)$$
$$(x + 1)(5x - 8) \qquad (x - 8)(5x + 1)$$

By checking the middle terms, we find that $(x - 4)(5x + 2)$ yields the desired middle term of $-18x$. Thus

$$5x^2 - 18x - 8 = (x - 4)(5x + 2)$$ ∎

Certainly, as the number of possibilities increases, this trial-and-error technique for factoring becomes more tedious. The key idea is to organize your work so that all possibilities are considered. We have suggested one possible format in the previous examples. However, as you practice such problems, you may devise a format that works better for you. Whatever works best for you is the right approach.

There is another, more systematic technique that you may wish to use with some trinomials. It is an extension of the technique we used earlier with trinomials where the coefficient of the squared term was 1. To see the basis of this technique, consider the following general product:

$$(px + r)(qx + s) = px(qx) + px(s) + r(qx) + r(s)$$
$$= (pq)x^2 + ps(x) + rq(x) + rs$$
$$= (pq)x^2 + (ps + rq)x + rs$$

Note that the product of the coefficient of x^2 and the constant term is $pqrs$. Likewise, the product of the two coefficients of x (ps and rq) is also $pqrs$. Therefore, the coefficient of x must be a sum of the form $ps + rq$, such that the product of the coefficient of x^2 and the constant term is $pqrs$. Now let's see how this works in some specific examples.

EXAMPLE 7

Factor $6x^2 + 17x + 5$.

Solution

$$6x^2 + 17x + 5 \qquad \text{Sum of 17}$$

Product of $6 \cdot 5 = 30$

We need two integers whose sum is 17 and whose product is 30. The integers 2 and 15 satisfy these conditions. Therefore the middle term, $17x$, of the given trinomial can be expressed as $2x + 15x$, and we can proceed as follows:

$$6x^2 + 17x + 5 = 6x^2 + 2x + 15x + 5$$
$$= 2x(3x + 1) + 5(3x + 1) \qquad \text{Factor by grouping.}$$
$$= (3x + 1)(2x + 5)$$

EXAMPLE 8

Factor $5x^2 - 18x - 8$.

Solution

$$5x^2 - 18x - 8 \qquad \text{Sum of } -18$$

Product of $5(-8) = -40$

We need two integers whose sum is -18 and whose product is -40. The integers -20 and 2 satisfy these conditions. Therefore the middle term, $-18x$, of the trinomial can be written as $-20x + 2x$, and we can factor as follows:

$$5x^2 - 18x - 8 = 5x^2 - 20x + 2x - 8$$
$$= 5x(x - 4) + 2(x - 4) \qquad \text{Factor by grouping.}$$
$$= (x - 4)(5x + 2)$$

EXAMPLE 9

Factor $24x^2 + 2x - 15$.

Solution

$$24x^2 + 2x - 15 \qquad \text{Sum of 2}$$

Product of $24(-15) = -360$

We need two integers whose sum is 2 and whose product is -360. To help find these integers, let's factor 360 into primes.

$$360 = 2 \cdot 2 \cdot 2 \cdot 3 \cdot 3 \cdot 5$$

Now by grouping these factors in various ways, we find that $2 \cdot 2 \cdot 5 = 20$ and $2 \cdot 3 \cdot 3 = 18$, so we can use the integers 20 and -18 to produce a sum of 2 and a product of -360. Therefore, the middle term, $2x$, of the trinomial can be expressed as $20x - 18x$, and we can proceed as follows:

$$24x^2 + 2x - 15 = 24x^2 + 20x - 18x - 15$$
$$= 4x(6x + 5) - 3(6x + 5)$$
$$= (6x + 5)(4x - 3)$$

Factoring: Perfect-Square Trinomials

In Section 6.3 we used the following two patterns to square binomials.

$$(a + b)^2 = a^2 + 2ab + b^2 \qquad \text{and} \qquad (a - b)^2 = a^2 - 2ab + b^2$$

These patterns can also be used for factoring purposes.

$$a^2 + 2ab + b^2 = (a + b)^2 \qquad \text{and} \qquad a^2 - 2ab + b^2 = (a - b)^2$$

The trinomials on the left sides are called **perfect-square trinomials;** they are the result of squaring a binomial. We can always factor perfect-square trinomials using the usual techniques for factoring trinomials. However, they are easily recognized by the nature of their terms. For example, $4x^2 + 12x + 9$ is a perfect-square trinomial because

1. The first term is a perfect square. $(2x)^2$
2. The last term is a perfect square. $(3)^2$

3. The middle term is twice the product of the quantities being squared in the first and last terms.

$2(2x)(3)$

Likewise, $9x^2 - 30x + 25$ is a perfect-square trinomial because

1. The first term is a perfect square. $(3x)^2$

2. The last term is a perfect square. $(5)^2$

3. The middle term is the negative of twice the product of the quantities being squared in the first and last terms. $-2(3x)(5)$

Once we know that we have a perfect-square trinomial, then the factors follow immediately from the two basic patterns. Thus

$$4x^2 + 12x + 9 = (2x + 3)^2 \qquad 9x^2 - 30x + 25 = (3x - 5)^2$$

Here are some additional examples of perfect-square trinomials and their factored forms.

$$x^2 + 14x + 49 = (x)^2 + 2(x)(7) + (7) = (x + 7)^2$$
$$n^2 - 16n + 64 = (n)^2 - 2(n)(8) + (8)^2 = (n - 8)^2$$
$$36a^2 + 60ab + 25b^2 = (6a)^2 + 2(6a)(5b) + (5b)^2 = (6a + 5b)^2$$
$$16x^2 - 8xy + y^2 = (4x)^2 - 2(4x)(y) + (y)^2 = (4x - y)^2$$

Perhaps you will want to do this step mentally after you feel comfortable with the process.

Solving Equations

One reason why factoring is an important algebraic skill is that it extends our techniques for solving equations. Each factoring technique provides us with more power to solve equations. Let's review this process with two examples.

EXAMPLE 10   Solve $x^3 - 49x = 0$.

Solution

$$x^3 - 49x = 0$$
$$x(x^2 - 49) = 0$$
$$x(x + 7)(x - 7) = 0$$
$$x = 0 \quad \text{or} \quad x + 7 = 0 \quad \text{or} \quad x - 7 = 0$$
$$x = 0 \quad \text{or} \quad x = -7 \quad \text{or} \quad x = 7$$

The solution set is $\{-7, 0, 7\}$.

EXAMPLE 11

Solve $9a(a + 1) = 4$.

Solution

$$9a(a + 1) = 4$$
$$9a^2 + 9a = 4$$
$$9a^2 + 9a - 4 = 0$$
$$(3a + 4)(3a - 1) = 0$$

$$3a + 4 = 0 \qquad \text{or} \qquad 3a - 1 = 0$$
$$3a = -4 \qquad \text{or} \qquad 3a = 1$$
$$a = -\frac{4}{3} \qquad \text{or} \qquad a = \frac{1}{3}$$

The solution set is $\left\{ -\dfrac{4}{3}, \dfrac{1}{3} \right\}$. ◼

Problem Solving

Finally, one of the end results of being able to factor and solve equations is that we can use these skills to help solve problems. Let's conclude this section with two problem-solving situations.

PROBLEM 1

A room contains 78 chairs. The number of chairs per row is 1 more than twice the number of rows. Find the number of rows and the number of chairs per row.

Solution

Let r represent the number of rows. Then $2r + 1$ represents the number of chairs per row.

$$r(2r + 1) = 78 \qquad \text{The number of rows times the numbers of chairs}$$
$$2r^2 + r = 78 \qquad \text{per row yields the total number of chairs.}$$
$$2r^2 + r - 78 = 0$$
$$(2r + 13)(r - 6) = 0$$

$$2r + 13 = 0 \qquad \text{or} \qquad r - 6 = 0$$
$$2r = -13 \qquad \text{or} \qquad r = 6$$
$$r = -\frac{13}{2} \qquad \text{or} \qquad r = 6$$

The solution $-\dfrac{13}{2}$ must be disregarded, so there are 6 rows and $2r + 1$ or $2(6) + 1 = 13$ chairs per row. ◼

PROBLEM 2   Suppose that the volume of a right circular cylinder is numerically equal to the total surface area of the cylinder. If the height of the cylinder is equal to the length of a radius of the base, find the height.

Solution

Because $r = h$, the formula for volume $V = \pi r^2 h$ becomes $V = \pi r^3$ and the formula for the total surface area $S = 2\pi r^2 + 2\pi rh$ becomes $S = 2\pi r^2 + 2\pi r^2$, or $S = 4\pi r^2$. Therefore, we can set up and solve the following equation.

$$\pi r^3 = 4\pi r^2$$
$$\pi r^3 - 4\pi r^2 = 0$$
$$\pi r^2(r - 4) = 0$$
$$\pi r^2 = 0 \quad \text{or} \quad r - 4 = 0$$
$$r = 0 \quad \text{or} \quad r = 4$$

0 is not a reasonable answer, so the height must be 4 units.   ■

CONCEPT QUIZ

For Problems 1–5, answer true or false.

1. The factored form $4xy^2(2x + 5y)$ is factored completely.

2. The polynomial $ax + ay - bx - by$ can be factored by grouping but its equivalent $ax - bx + ay - by$ cannot be factored by grouping.

3. If a polynomial is not factorable using integers it is referred to as a prime polynomial.

4. The polynomial $x^3 + 64$ can be factored into $(x + 3)(x + 3)(x + 3)$.

5. The polynomial $x^2 + 10x + 24$ is not factorable using integers.

PROBLEM SET 8.6

For Problems 1–50, factor completely each of the polynomials. Indicate any that are not factorable using integers.

1. $6xy - 8xy^2$

2. $4a^2b^2 + 12ab^3$

3. $x(z + 3) + y(z + 3)$

4. $5(x + y) + a(x + y)$

5. $3x + 3y + ax + ay$

6. $ac + bc + a + b$

7. $ax - ay - bx + by$

8. $2a^2 - 3bc - 2ab + 3ac$

9. $9x^2 - 25$

10. $4x^2 + 9$

11. $1 - 81n^2$

12. $9x^2y^2 - 64$

13. $(x + 4)^2 - y^2$

14. $x^2 - (y - 1)^2$

15. $9s^2 - (2t - 1)^2$

16. $4a^2 - (3b + 1)^2$

17. $x^2 - 5x - 14$

18. $a^2 + 5a - 24$

19. $15 - 2x - x^2$

20. $40 - 6x - x^2$

21. $x^2 + 7x - 36$

22. $x^2 - 4xy - 5y^2$

23. $3x^2 - 11x + 10$

24. $2x^2 - 7x - 30$

25. $10x^2 - 33x - 7$

26. $8y^2 + 22y - 21$

27. $x^3 - 8$

28. $x^3 + 64$

29. $64x^3 + 27y^3$

30. $27x^3 - 8y^3$

31. $4x^2 + 16$

32. $n^3 - 49n$

33. $x^3 - 9x$

34. $12n^2 + 59n + 72$

35. $9a^2 - 42a + 49$

36. $1 - 16x^4$

37. $2n^3 + 6n^2 + 10n$

38. $x^2 - (y - 7)^2$

39. $10x^2 + 39x - 27$

40. $3x^2 + x - 5$

41. $36a^2 - 12a + 1$

42. $18n^3 + 39n^2 - 15n$

43. $8x^2 + 2xy - y^2$

44. $12x^2 + 7xy - 10y^2$

45. $2n^2 - n - 5$

46. $25t^2 - 100$

47. $2n^3 + 14n^2 - 20n$

48. $25n^2 + 64$

49. $4x^3 + 32$

50. $2x^3 - 54$

For Problems 51–80, solve each equation.

51. $x^2 + 4x + 3 = 0$

52. $x^2 + 7x + 10 = 0$

53. $x^2 + 18x + 72 = 0$

54. $n^2 + 20n + 91 = 0$

55. $n^2 - 13n + 36 = 0$

56. $n^2 - 10n + 16 = 0$

57. $x^2 + 4x - 12 = 0$

58. $x^2 + 7x - 30 = 0$

59. $w^2 - 4w = 5$

60. $s^2 - 4s = 21$

61. $n^2 + 25n + 156 = 0$

62. $n(n - 24) = -128$

63. $3t^2 + 14t - 5 = 0$

64. $4t^2 - 19t - 30 = 0$

65. $6x^2 + 25x + 14 = 0$

66. $25x^2 + 30x + 8 = 0$

67. $3t(t - 4) = 0$

68. $4x^2 + 12x + 9 = 0$

69. $-6n^2 + 13n - 2 = 0$

70. $(x + 1)^2 - 4 = 0$

71. $2n^3 = 72n$

72. $a(a - 1) = 2$

73. $(x - 5)(x + 3) = 9$

74. $3w^3 - 24w^2 + 36w = 0$

75. $9x^2 - 6x + 1 = 0$

76. $16t^2 - 72t + 81 = 0$

77. $n^2 + 7n - 44 = 0$

78. $2x^3 = 50x$

79. $3x^2 = 75$

80. $x^2 + x - 2 = 0$

For Problems 81–90, set up an equation and solve each problem.

81. Suppose that the volume of a sphere is numerically equal to twice the surface area of the sphere. Find the length of a radius of the sphere.

82. Suppose that a radius of a sphere is equal in length to a radius of a circle. If the volume of the sphere is numerically equal to four times the area of the circle, find the length of a radius for both the sphere and the circle.

83. Find two integers whose product is 104 such that one of the integers is 3 less than twice the other integer.

84. The perimeter of a rectangle is 32 inches, and the area is 60 square inches. Find the length and width of the rectangle.

85. The lengths of the three sides of a right triangle are represented by consecutive even whole numbers. Find the lengths of the three sides.

86. The area of a triangular sheet of paper is 28 square inches. One side of the triangle is 2 inches more than three times the length of the altitude to that side. Find the length of that side and the altitude to that side.

87. The total surface area of a right circular cylinder is 54π square inches. If the altitude of the cylinder is twice the length of a radius, find the altitude of the cylinder.

88. The Ortegas have an apple orchard that contains 90 trees. The number of trees in each row is 3 more than twice the number of rows. Find the number of rows and the number of trees per row.

89. The combined area of a square and a rectangle is 64 square centimeters. The width of the rectangle is 2 centimeters more than the length of a side of the square, and the length of the rectangle is 2 centimeters more than its width. Find the dimensions of the square and the rectangle.

90. The cube of a number equals nine times the same number. Find the number.

■ ■ ■ **Thoughts into words**

91. Suppose that your friend factors $36x^2y + 48xy^2$ as follows:

$$36x^2y + 48xy^2 = (4xy)(9x + 12y)$$
$$= (4xy)(3)(3x + 4y)$$
$$= 12xy(3x + 4y)$$

Is this a correct approach? Would you have any suggestion to offer your friend?

92. Your classmate solves the equation $3ax + bx = 0$ for x as follows:

$$3ax + bx = 0$$
$$3ax = -bx$$
$$x = \frac{-bx}{3a}$$

How should he know that the solution is incorrect? How would you help him obtain the correct solution?

93. Consider the following solution:

$$6x^2 - 24 = 0$$
$$6(x^2 - 4) = 0$$
$$6(x + 2)(x - 2) = 0$$
$$6 = 0 \quad \text{or} \quad x + 2 = 0 \quad \text{or} \quad x - 2 = 0$$
$$6 = 0 \quad \text{or} \quad x = -2 \quad \text{or} \quad x = 2$$

The solution set is $\{-2, 2\}$.

Is this a correct solution? Would you have any suggestion to offer the person who used this approach?

94. Explain how you would solve the equation $(x + 6)(x - 4) = 0$ and also how you would solve $(x + 6)(x - 4) = -16$.

95. Explain how you would solve the equation $3(x - 1)(x + 2) = 0$ and also how you would solve the equation $x(x - 1)(x + 2) = 0$.

96. Consider the following two solutions for the equation $(x + 3)(x - 4) = (x + 3)(2x - 1)$.

Solution A

$$(x + 3)(x - 4) = (x + 3)(2x - 1)$$
$$(x + 3)(x - 4) - (x + 3)(2x - 1) = 0$$
$$(x + 3)[x - 4 - (2x - 1)] = 0$$
$$(x + 3)(x - 4 - 2x + 1) = 0$$
$$(x + 3)(-x - 3) = 0$$
$$x + 3 = 0 \quad \text{or} \quad -x - 3 = 0$$
$$x = -3 \quad \text{or} \quad -x = 3$$
$$x = -3 \quad \text{or} \quad x = -3$$

The solution set is $\{-3\}$.

Solution B

$$(x + 3)(x - 4) = (x + 3)(2x - 1)$$
$$x^2 - x - 12 = 2x^2 + 5x - 3$$
$$0 = x^2 + 6x + 9$$
$$0 = (x + 3)^2$$
$$x + 3 = 0$$
$$x = -3$$

The solution set is $\{-3\}$.

Are both approaches correct? Which approach would you use, and why?

SUMMARY

(8.1) Solving an algebraic equation refers to the process of finding the number (or numbers) that make(s) the algebraic equation a true numerical statement. We call such numbers the **solutions** or **roots** of the equation that **satisfy** the equation. We call the set of all solutions of an equation the **solution set.** The general procedure for solving an equation is to continue replacing the given equation with equivalent, *but simpler,* equations until we arrive at one that can be solved by inspection. Two properties of equality play an important role in the process of solving equations.

Addition Property of Equality
$a = b$ if and only if $a + c = b + c$.
Multiplication Property of Equality
For $c \neq 0$, $a = b$ if and only if $ac = bc$.

To solve an equation involving fractions, first **clear the equation of all fractions.** It is usually easiest to begin by multiplying both sides of the equation by the least common multiple of all of the denominators in the equation (by the *least common denominator,* or LCD).

To solve equations that contain decimals, you can **clear the equation of all decimals** by multiplying both sides by an appropriate power of 10.

If an equation is a proportion, then it can be solved by equating the cross products.

An equation that is satisfied by all numbers for which both sides of the equation are defined is called an **algebraic identity.**

Keep the following suggestions in mind as you solve word problems.

1. Read the problem carefully.
2. Sketch any figure, diagram, or chart that might be helpful.
3. Choose a meaningful variable.
4. Look for a guideline.
5. Form an equation or inequality.
6. Solve the equation or inequality.
7. Check your answers.

(8.2) Solving an algebraic inequality refers to the process of finding the numbers that make the algebraic inequality a true numerical statement. We call such numbers the **solutions,** and we call the set of all solutions the **solution set.**

The general procedure for solving an inequality is to continue replacing the given inequality with equivalent, *but simpler,* inequalities until we arrive at one that we can solve by inspection. The following properties form the basis for solving algebraic inequalities.

1. $a > b$ if and only if $a + c > b + c$.
   (Addition property)
2. **a.** For $c > 0$, $a > b$ if and only if $ac > bc$.
   (Multiplication properties)
   b. For $c < 0$, $a > b$ if and only if $ac < bc$.

To solve compound sentences that involve inequalities, we proceed as follows:

1. Solve separately each inequality in the compound sentence.
2. If it is a **conjunction,** the solution set is the **intersection** of the solution sets of each inequality.
3. If it is a **disjunction,** the solution set is the **union** of the solution sets of each inequality.

We define the intersection and union of two sets as follows:

Intersection $A \cap B = \{x | x \in A \quad and \quad x \in B\}$
Union $A \cup B = \{x | x \in A \quad or \quad x \in B\}$

The following chart summarizes the use of interval notation.

Set	Graph	Interval notation
$\{x \mid x > a\}$		(a, ∞)
$\{x \mid x \geq a\}$		$[a, \infty)$
$\{x \mid x < b\}$		$(-\infty, b)$
$\{x \mid x \leq b\}$		$(-\infty, b]$
$\{x \mid a < x < b\}$		(a, b)
$\{x \mid a \leq x < b\}$		$[a, b)$
$\{x \mid a < x \leq b\}$		$(a, b]$
$\{x \mid a \leq x \leq b\}$		$[a, b]$
$\{x \mid x \text{ is a real number}\}$		$(-\infty, \infty)$

Figure 8.16

(8.3) We can interpret the **absolute value** of a number on the number line as the distance between that number and zero. The following properties form the basis for solving equations and inequalities involving absolute value.

1. $|ax + b| = k$ is equivalent to
   $ax + b = -k$ or $ax + b = k$
2. $|ax + b| < k$ is equivalent to
   $-k < ax + b < k$ $\Big\}$ $k > 0$
3. $|ax + b| > k$ is equivalent to
   $ax + b < -k$ or $ax + b > k$

(8.4) A **term** is an indicated product and may contain any number of factors. The variables involved in a term are called **literal factors,** and the numerical factor is called the **numerical coefficient.** Terms that contain variables with only nonnegative integers as exponents are called **monomials.** The **degree** of a monomial is the sum of the exponents of the literal factors.

A **polynomial** is a monomial or a finite sum (or difference) of monomials. We classify polynomials as follows:

Polynomial with one term $\longrightarrow$ Monomial
Polynomial with two terms $\longrightarrow$ Binomial
Polynomial with three terms $\longrightarrow$ Trinomial

Similar terms, or **like terms,** have the same literal factors. The commutative, associative, and distributive properties provide the basis for rearranging, regrouping, and combining similar terms.

The following properties provide the basis for multiplying and dividing monomials. If m and n are integers, and a and b are real numbers, with $b \neq 0$ whenever it appears in a denominator, then

1. $b^n \cdot b^m = b^{n+m}$ Product of two like bases with powers
2. $(b^n)^m = b^{mn}$ Power of a power
3. $(ab)^n = a^n b^n$ Power of a product

4. $\left(\dfrac{a}{b}\right)^n = \dfrac{a^n}{b^n}$ Power of a quotient

5. $\dfrac{b^n}{b^m} = b^{n-m}$ Quotient of two powers

The commutative and associative properties, the properties of exponents, and the distributive property work together to form a basis for multiplying polynomials. The following can be used as multiplication patterns:

$$(a + b)^2 = a^2 + 2ab + b^2$$
$$(a - b)^2 = a^2 - 2ab + b^2$$
$$(a + b)(a - b) = a^2 - b^2$$
$$(a + b)^3 = a^3 + 3a^2b + 3ab^2 + b^3$$
$$(a - b)^3 = a^3 - 3a^2b + 3ab^2 - b^3$$

The expansion of a binomial such as $(a + b)^7$ can be accomplished by getting the coefficients from the seventh row of Pascal's triangle and getting the exponents for a and b from the following pattern:

$$a^n, a^{n-1}b, a^{n-2}b^2, \ldots, ab^{n-1}, b^n$$

(8.5) If the divisor is of the form $x - c$, where c is a constant, then the typical long-division format for dividing polynomials can be simplified to a process called **synthetic division.** Review this process by studying the examples of this section.

(8.6) The distributive property in the form $ab + ac = a(b + c)$ is the basis for factoring out the **highest common monomial factor.**

An expression such as $ax + bx + ay + by$ can be factored as follows:

$$ax + bx + ay + by = x(a + b) + y(a + b)$$
$$= (a + b)(x + y)$$

This is called **factoring by grouping.**

The factoring pattern

$$a^2 - b^2 = (a + b)(a - b)$$

is called the **difference of two squares.**

The factoring patterns

$$a^3 + b^3 = (a + b)(a^2 - ab + b^2) \quad \text{and}$$
$$a^3 - b^3 = (a - b)(a^2 + ab + b^2)$$

are called the **sum of two cubes** and the **difference of two cubes.**

Expressing a trinomial (for which the coefficient of the squared term is 1) as a product of two binomials is based on the following relationship:

$$(x + a)(x + b) = x^2 + (a + b)x + ab$$

The coefficient of the middle term is the *sum of a and b,* and the last term is the *product of a and b.*

If the coefficient of the squared term of a trinomial does not equal 1, then the following relationship holds.

$$(pq)x^2 + (ps + rq)x + rs = (pq)x^2 + psx + rqx + rs$$
$$= (px + r)(qx + s)$$

The two coefficients of x, ps and rq, must have a sum of $(ps) + (rq)$ and a product of $pqrs$. Thus to factor something like $6x^2 + 7x - 3$, we need to find two integers whose product is $6(-3) = -18$ and whose sum is 7. The integers are 9 and -2, and we can factor as follows:

$$6x^2 + 7x - 3 = 6x^2 + 9x - 2x - 3$$
$$= 3x(2x + 3) - 1(2x + 3)$$
$$= (2x + 3)(3x - 1)$$

A **perfect-square trinomial** is the result of squaring a binomial. There are two basic perfect-square-trinomial factoring patterns:

$$a^2 + 2ab + b^2 = (a + b)^2$$
$$a^2 - 2ab + b^2 = (a - b)^2$$

The factoring techniques we discussed in this chapter, along with the property $ab = 0$ if and only if $a = 0$ or $b = 0$ provide the basis for expanding our repertoire of equation-solving processes.

The ability to solve more types of equations increases our capabilities for problem solving.

CHAPTER 8 REVIEW PROBLEM SET

For Problems 1–24, solve each of the equations.

1. $5(x - 6) = 3(x + 2)$

2. $2(2x + 1) - (x - 4) = 4(x + 5)$

3. $-(2n - 1) + 3(n + 2) = 7$

4. $2(3n - 4) + 3(2n - 3) = -2(n + 5)$

5. $\dfrac{3t - 2}{4} = \dfrac{2t + 1}{3}$ **6.** $\dfrac{x + 6}{5} + \dfrac{x - 1}{4} = 2$

7. $1 - \dfrac{2x - 1}{6} = \dfrac{3x}{8}$

8. $\dfrac{2x + 1}{3} + \dfrac{3x - 1}{5} = \dfrac{1}{10}$

9. $\dfrac{3n - 1}{2} - \dfrac{2n + 3}{7} = 1$

10. $|3x - 1| = 11$

11. $0.06x + 0.08(x + 100) = 15$

12. $0.4(t - 6) = 0.3(2t + 5)$

13. $0.1(n + 300) = 0.09n + 32$

14. $0.2(x - 0.5) - 0.3(x + 1) = 0.4$

15. $4x^2 - 36 = 0$ **16.** $x^2 + 5x - 6 = 0$

17. $49n^2 - 28n + 4 = 0$ **18.** $(3x - 1)(5x + 2) = 0$

19. $(3x - 4)^2 - 25 = 0$ **20.** $6a^3 = 54a$

21. $7n(7n + 2) = 8$ **22.** $30w^2 - w - 20 = 0$

23. $3t^3 - 27t^2 + 24t = 0$ **24.** $-4n^2 - 39n + 10 = 0$

For Problems 25–29, solve each equation for x.

25. $ax - b = b + 2$ **26.** $ax = bx + c$

27. $m(x + a) = p(x + b)$

28. $5x - 7y = 11$ **29.** $\dfrac{x - a}{b} = \dfrac{y + 1}{c}$

For Problems 30–39, solve each inequality and express the solutions using interval notation.

30. $5x - 2 \geq 4x - 7$ **31.** $3 - 2x < -5$

32. $2(3x - 1) - 3(x - 3) > 0$

33. $3(x + 4) \leq 5(x - 1)$

34. $\dfrac{5}{6}n - \dfrac{1}{3}n < \dfrac{1}{6}$

35. $\dfrac{n - 4}{5} + \dfrac{n - 3}{6} > \dfrac{7}{15}$

36. $s \geq 4.5 + 0.25s$

37. $0.07x + 0.09(500 - x) \geq 43$

38. $|2x - 1| < 11$ **39.** $|3x + 1| > 10$

For Problems 40–43, graph the solutions of each compound inequality.

40. $x > -1$ and $x < 1$ **41.** $x > 2$ or $x \leq -3$

42. $x > 2$ and $x > 3$ **43.** $x < 2$ or $x > -1$

For Problems 44–65, perform the indicated operations and simplify each of the following.

44. $(3x - 2) + (4x - 6) + (-2x + 5)$

45. $(8x^2 + 9x - 3) - (5x^2 - 3x - 1)$

46. $(6x^2 - 2x - 1) + (4x^2 + 2x + 5) - (-2x^2 + x - 1)$

47. $(-5x^2y^3)(4x^3y^4)$ **48.** $(-2a^2)(3ab^2)(a^2b^3)$

49. $5a^2(3a^2 - 2a - 1)$ **50.** $(4x - 3y)(6x + 5y)$

51. $(x + 4)(3x^2 - 5x - 1)$ **52.** $(4x^2y^3)^4$

53. $(3x - 2y)^2$ **54.** $(-2x^2y^3z)^3$

55. $\dfrac{-39x^3y^4}{3xy^3}$

56. $[3x - (2x - 3y + 1)] - [2y - (x - 1)]$

57. $(x^2 - 2x - 5)(x^2 + 3x - 7)$

58. $(7 - 3x)(3 + 5x)$ **59.** $-(3ab)(2a^2b^3)^2$

60. $\left(\dfrac{1}{2}ab\right)(8a^3b^2)(-2a^3)$ **61.** $(7x - 9)(x + 4)$

62. $(3x + 2)(2x^2 - 5x + 1)$ **63.** $(3x^{n+1})(2x^{3n-1})$

64. $(2x + 5y)^2$ **65.** $(x - 2)^3$

For Problems 66–69, use synthetic division to find the quotient and remainder for each of the following.

66. $(3x^3 - 10x^2 + 2x + 41) \div (x - 2)$

67. $(5x^3 + 8x^2 + x + 6) \div (x + 1)$

68. $(x^4 - x^3 - 19x^2 - 22x - 3) \div (x + 3)$

69. $(2x^4 - 5x^3 - 16x^2 + 14x + 8) \div (x - 4)$

For Problems 70–91, factor each polynomial completely. Indicate any that are not factorable using integers.

70. $x^2 + 3x - 28$ **71.** $2t^2 - 18$

72. $4n^2 + 9$ **73.** $12n^2 - 7n + 1$

74. $x^6 - x^2$ **75.** $x^3 - 6x^2 - 72x$

76. $6a^3b + 4a^2b^2 - 2a^2bc$ **77.** $x^2 - (y - 1)^2$

78. $8x^2 + 12$ **79.** $12x^2 + x - 35$

80. $16n^2 - 40n + 25$ **81.** $4n^2 - 8n$

82. $3w^3 + 18w^2 - 24w$ **83.** $20x^2 + 3xy - 2y^2$

84. $16a^2 - 64a$ **85.** $3x^3 - 15x^2 - 18x$

86. $n^2 - 8n - 128$ **87.** $t^4 - 22t^2 - 75$

88. $35x^2 - 11x - 6$ **89.** $15 - 14x + 3x^2$

90. $64n^3 - 27$ **91.** $16x^3 + 250$

Solve each of Problems 92–107 by setting up and solving an appropriate equation, inequality, or system of equations.

92. The width of a rectangle is 2 meters more than one-third of the length. The perimeter of the rectangle is 44 meters. Find the length and width of the rectangle.

93. A total of $5000 was invested, part of it at 7% interest and the remainder at 8%. If the total yearly interest from both investments amounted to $380, how much was invested at each rate?

94. Susan's average score for her first three psychology exams is 84. What must she get on the fourth exam so that her average for the four exams is 85 or better?

95. Find three consecutive integers such that the sum of one-half of the smallest and one-third of the largest is one less than the other integer.

96. Pat is paid time-and-a-half for each hour he works over 36 hours in a week. Last week he worked 42 hours for a total of $472.50. What is his normal hourly rate?

97. Marcela has a collection of nickels, dimes, and quarters worth $24.75. The number of dimes is 10 more than twice the number of nickels, and the number of quarters is 25 more than the number of dimes. How many coins of each kind does she have?

98. If the complement of an angle is one-tenth of the supplement of the angle, find the measure of the angle.

99. A retailer has some sweaters that cost her $38 each. She wants to sell them at a profit of 20% of her cost. What price should she charge for the sweaters?

100. Nora scored 16, 22, 18, and 14 points for each of the first four basketball games. How many points does she need to score in the fifth game so that her average for the first five games is at least 20 points per game?

101. Gladys leaves a town driving at a rate of 40 miles per hour. Two hours later, Reena leaves from the same place traveling the same route. She catches Gladys in 5 hours and 20 minutes. How fast was Reena traveling?

102. In $1\frac{1}{4}$ hours more time, Rita, riding her bicycle at 12 miles per hour, rode 2 miles farther than Sonya, who was riding her bicycle at 16 miles per hour. How long did each girl ride?

103. How many cups of orange juice must be added to 50 cups of a punch that is 10% orange juice to obtain a punch that is 20% orange juice?

104. Two cars leave an intersection at the same time, one traveling north and the other traveling east. Some time later, they are 20 miles apart and the car going east has traveled 4 miles farther than the other car. How far has each car traveled?

105. The perimeter of a rectangle is 32 meters and its area is 48 square meters. Find the length and width of the rectangle.

106. A room contains 144 chairs. The number of chairs per row is two less than twice the number of rows. Find the number of rows and the number of chairs per row.

107. The area of a triangle is 39 square feet. The length of one side is 1 foot more than twice the altitude to that side. Find the length of that side and the altitude to the side.

TEST

 applies to all problems in this Chapter Test.

For Problems 1–4, perform the indicated operations and simplify each expression.

1. $(-3x - 1) + (9x - 2) - (4x + 8)$

2. $(5x - 7)(4x + 9)$

3. $(x + 6)(2x^2 - x - 5)$

4. $(x - 4y)^3$

5. Find the quotient and remainder for the division problem $(6x^3 - 19x^2 + 3x + 20) \div (3x - 5)$.

6. Find the quotient and remainder for the division problem $(3x^4 + 8x^3 - 5x^2 - 12x - 15) \div (x + 3)$.

7. Factor $x^2 - xy + 4x - 4y$ completely.

8. Factor $12x^2 - 3$ completely.

For Problems 9–18, solve each equation.

9. $3(2x - 1) - 2(x + 5) = -(x - 3)$

10. $\dfrac{3t - 2}{4} = \dfrac{5t + 1}{5}$

11. $|4x - 3| = 9$

12. $\dfrac{1 - 3x}{4} + \dfrac{2x + 3}{3} = 1$

13. $0.05x + 0.06(1500 - x) = 83.5$

14. $4n^2 = n$

15. $4x^2 - 12x + 9 = 0$

16. $3x^3 + 21x^2 - 54x = 0$

17. $12 + 13x - 35x^2 = 0$

18. $n(3n - 5) = 2$

For Problems 19–21, solve each inequality and use interval notation to express the solutions.

19. $|6x - 4| < 10$

20. $\dfrac{x - 2}{6} - \dfrac{x + 3}{9} > -\dfrac{1}{2}$

21. $2(x - 1) - 3(3x + 1) \geq -6(x - 5)$

For Problems 22–25, use an equation, an inequality, or a system of equations to help solve each problem.

22. How many cups of grapefruit juice must be added to 30 cups of a punch that is 8% grapefruit juice to obtain a punch that is 10% grapefruit juice?

23. Rex has scores of 85, 92, 87, 88, and 91 on the first five exams. What score must he make on the sixth exam to have an average of 90 or better for all six exams?

24. If the complement of an angle is $\dfrac{2}{11}$ of the supplement of the angle, find the measure of the angle.

25. The combined area of a square and a rectangle is 57 square feet. The width of the rectangle is 3 feet more than the length of a side of the square, and the length of the rectangle is 5 feet more than the length of a side of the square. Find the length of the rectangle.

Computers often work together to compile large processing jobs. Rational numbers are used to express the rate of the processing speed of a computer.

AP /Wide World Photos

Rational Expressions

*I*t takes Pat 12 hours to complete a task. After he had been working on this task for 3 hours, he was joined by his brother, Liam, and together they finished the job in 5 hours. How long would it take Liam to do the job by himself? We can use the **fractional equation** $\frac{5}{12} + \frac{5}{h} = \frac{3}{4}$ to determine that Liam could do the entire job by himself in 15 hours.

Rational expressions are to algebra what rational numbers are to arithmetic. Most of the work we will do with rational expressions in this chapter parallels the work you have previously done with rational numbers in arithmetic. The same basic properties that we use to explain reducing, adding, subtracting, multiplying, and dividing arithmetic fractions will serve as a basis for our work with rational expressions. The factoring techniques we introduced in Chapter 7 and reviewed in Chapter 8 will also play an important role in our discussions. As always, learning some new skills will provide the basis for solving more equations, which in turn will give us more problem-solving power.

403

In this chapter we will begin using graphical displays to enhance our work with algebraic concepts. After doing an algebraic computational problem, we will occasionally show a graph of the situation to support our algebraic work. For example, suppose we simplify the expression $\dfrac{4x^2 + 12x + 9}{2x + 3}$ and obtain a result of $2x + 3$. By graphing, $Y_1 = \dfrac{4x^2 + 12x + 9}{2x + 3}$ and $Y_2 = 2x + 3$ on the same set of axes, and getting the same graph for both equations (Figure 9.1), we should feel pretty good about our algebraic computation. Starting with Problem Set 9.1, many of the problem sets contain a section of problems called Graphing Calculator Activities.

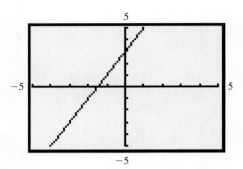

Figure 9.1

InfoTrac Project

Do a subject guide search on pi. Find a periodical article called "Record-Breaking Pi." Write a short summary of the article. To do the task, the work was divided among several computers. Suppose three computers worked on the project together for 85 hours. If the first computer would need 255 hours to complete the job alone and the second computer would need 340 hours to do the job alone, how long would it take the third computer to do the job alone? (*Hint:* After you write your equation, reduce your fractions, if possible.)

9.1 Simplifying Rational Expressions

Objectives

■ Express rational numbers in reduced form.

■ Reduce rational monomial expressions.

■ Simplify rational expressions using factoring techniques.

We reviewed the basic operations with rational numbers in an informal setting in Chapter 2. In this review, we relied primarily on your knowledge of arithmetic. At this time, we want to become a little more formal with our review so that we can use the work with rational numbers as a basis for operating with rational expressions. We will define a rational expression shortly.

You will recall that any number that can be written in the form $\dfrac{a}{b}$, where a and b are integers and $b \neq 0$, is called a rational number. The following are examples of rational numbers.

$$\frac{1}{2} \qquad \frac{3}{4} \qquad \frac{15}{7} \qquad \frac{-5}{6} \qquad \frac{7}{-8} \qquad \frac{-12}{-17}$$

Numbers such as $6, -4, 0, 4\dfrac{1}{2}, 0.7$, and 0.21 are also rational because we can express them as the indicated quotient of two integers. For example,

$$6 = \frac{6}{1} = \frac{12}{2} = \frac{18}{3} \text{ and so on} \qquad 4\frac{1}{2} = \frac{9}{2}$$

$$-4 = \frac{4}{-1} = \frac{-4}{1} = \frac{8}{-2} \text{ and so on} \qquad 0.7 = \frac{7}{10}$$

$$0 = \frac{0}{1} = \frac{0}{2} = \frac{0}{3} \text{ and so on} \qquad 0.21 = \frac{21}{100}$$

Our work with division of integers helps with the next examples.

$$\frac{8}{-2} = \frac{-8}{2} = -\frac{8}{2} = -4 \qquad \frac{12}{3} = \frac{-12}{-3} = 4$$

Observe the following general properties.

PROPERTY 9.1

1. $\dfrac{-a}{b} = \dfrac{a}{-b} = -\dfrac{a}{b}$, where $b \neq 0$

2. $\dfrac{-a}{-b} = \dfrac{a}{b}$, where $b \neq 0$

Therefore, a rational number such as $\dfrac{-2}{5}$ can also be written as $\dfrac{2}{-5}$ or $-\dfrac{2}{5}$.

We use the following property, often referred to as the **fundamental principle of fractions,** to reduce fractions to lowest terms or express fractions in simplest or reduced form.

PROPERTY 9.2

If b and k are nonzero integers and a is any integer, then

$$\frac{a \cdot k}{b \cdot k} = \frac{a}{b}$$

Let's apply Properties 9.1 and 9.2 to the following examples.

EXAMPLE 1   Reduce $\dfrac{18}{24}$ to lowest terms.

Solution

$$\frac{18}{24} = \frac{3 \cdot 6}{4 \cdot 6} = \frac{3}{4}$$

■

EXAMPLE 2   Change $\dfrac{40}{48}$ to simplest form.

Solution

$$\frac{\overset{5}{\cancel{40}}}{\underset{6}{\cancel{48}}} = \frac{5}{6} \qquad \text{A common factor of 8 was divided out of both} \\ \text{numerator and denominator.}$$

■

EXAMPLE 3   Express $\dfrac{-36}{63}$ in reduced form.

Solution

$$\frac{-36}{63} = -\frac{36}{63} = -\frac{4 \cdot 9}{7 \cdot 9} = -\frac{4}{7}$$

■

EXAMPLE 4   Reduce $\dfrac{72}{-90}$ to simplest form.

Solution

$$\frac{72}{-90} = -\frac{72}{90} = -\frac{2 \cdot 2 \cdot 2 \cdot \cancel{3} \cdot \cancel{3}}{2 \cdot \cancel{3} \cdot \cancel{3} \cdot 5} = -\frac{4}{5}$$

■

Note the different terminology used in Examples 1–4. Regardless of the terminology, keep in mind that the number is not being changed, but the form of the numeral representing the number is being changed. In Example 1, $\dfrac{18}{24}$ and $\dfrac{3}{4}$ are equivalent fractions; they name the same number. Also note the use of prime factors in Example 4.

Rational Expressions

A **rational expression** is the indicated quotient of two polynomials. The following are examples of rational expressions.

$$\frac{3x^2}{5} \qquad \frac{x-2}{x+3} \qquad \frac{x^2 + 5x - 1}{x^2 - 9} \qquad \frac{xy^2 + x^2y}{xy} \qquad \frac{a^3 - 3a^2 - 5a - 1}{a^4 + a^3 + 6}$$

Because we must avoid division by zero, no values that create a denominator of zero can be assigned to variables. Thus the rational expression $\dfrac{x-2}{x+3}$ is meaningful for all values of x except for $x = -3$. Rather than making restrictions for each individual expression, we will merely assume that all denominators represent nonzero real numbers.

Property 9.2 $\left(\dfrac{a \cdot k}{b \cdot k} = \dfrac{a}{b}\right)$ serves as the basis for simplifying rational expressions, as the next examples illustrate.

EXAMPLE 5   Simplify $\dfrac{15xy}{25y}$.

Solution

$$\frac{15xy}{25y} = \frac{3 \cdot \cancel{5} \cdot x \cdot \cancel{y}}{\cancel{5} \cdot 5 \cdot \cancel{y}} = \frac{3x}{5}$$

EXAMPLE 6   Simplify $\dfrac{-9}{18x^2y}$.

Solution

$$\frac{-9}{18x^2y} = -\frac{\overset{1}{\cancel{9}}}{\underset{2}{\cancel{18}}x^2y} = -\frac{1}{2x^2y} \qquad \text{A common factor of 9 was divided out of numerator and denominator.}$$

EXAMPLE 7   Simplify $\dfrac{-28a^2b^2}{-63a^2b^3}$.

Solution

$$\frac{-28a^2b^2}{-63a^2b^3} = \frac{4 \cdot \cancel{7} \cdot \cancel{a^2} \cdot \cancel{b^2}}{9 \cdot \cancel{7} \cdot \cancel{a^2} \cdot \underset{b}{\cancel{b^3}}} = \frac{4}{9b}$$

The factoring techniques from Chapter 7 can be used to factor numerators and/or denominators so that we can apply the property $\dfrac{a \cdot k}{b \cdot k} = \dfrac{a}{b}$. Examples 8–12 should clarify this process.

EXAMPLE 8   Simplify $\dfrac{x^2 + 4x}{x^2 - 16}$.

Solution

$$\frac{x^2 + 4x}{x^2 - 16} = \frac{x\cancel{(x+4)}}{(x-4)\cancel{(x+4)}} = \frac{x}{x-4}$$

EXAMPLE 9 Simplify $\dfrac{4x^2 + 12x + 9}{2x + 3}$.

Solution

$$\frac{4x^2 + 12x + 9}{2x + 3} = \frac{(2x+3)(2x+3)}{1(2x+3)} = \frac{2x + 3}{1} = 2x + 3 \qquad \blacksquare$$

Recall that we used the rational expression in Example 9 in our introductory remarks for this chapter. We showed that the graphs of $Y_1 = \dfrac{4x^2 + 12x + 9}{2x + 3}$ and $Y_2 = 2x + 3$ appear to be identical. Thus we have given some graphical support for our algebraic simplification process. One point should be made at this time. Because $-\dfrac{3}{2}$ makes the denominator zero, the original rational expression is defined for all real numbers except $-\dfrac{3}{2}$. Thus the graph of $Y_1 = \dfrac{4x^2 + 12x + 9}{2x + 3}$ actually has a hole at $x = -\dfrac{3}{2}$, which may not be visible on the graph done by a graphing calculator. For all other values, the two graphs are identical.

EXAMPLE 10 Simplify $\dfrac{5n^2 + 6n - 8}{10n^2 - 3n - 4}$.

Solution

$$\frac{5n^2 + 6n - 8}{10n^2 - 3n - 4} = \frac{(5n-4)(n + 2)}{(5n-4)(2n + 1)} = \frac{n + 2}{2n + 1} \qquad \blacksquare$$

EXAMPLE 11 Simplify $\dfrac{6x^3y - 6xy}{x^3 + 5x^2 + 4x}$.

Solution

$$\frac{6x^3y - 6xy}{x^3 + 5x^2 + 4x} = \frac{6xy(x^2 - 1)}{x(x^2 + 5x + 4)} = \frac{6xy(x+1)(x - 1)}{x(x+1)(x + 4)} = \frac{6y(x - 1)}{x + 4} \qquad \blacksquare$$

Note that in Example 11 we left the numerator of the final fraction in factored form. This is often done if expressions other than monomials are involved. Either $\dfrac{6y(x - 1)}{x + 4}$ or $\dfrac{6xy - 6y}{x + 4}$ is an acceptable answer.

Remember that the quotient of any nonzero real number and its opposite is -1. For example, $\dfrac{6}{-6} = -1$ and $\dfrac{-8}{8} = -1$. Likewise, the indicated quotient of any polynomial and its opposite is equal to -1. For example,

$$\frac{a}{-a} = -1 \quad \text{because } a \text{ and } -a \text{ are opposites}$$

$$\frac{a - b}{b - a} = -1 \quad \text{because } a - b \text{ and } b - a \text{ are opposites}$$

$$\frac{x^2 - 4}{4 - x^2} = -1 \quad \text{because } x^2 - 4 \text{ and } 4 - x^2 \text{ are opposites}$$

Example 12 shows how we use this idea when simplifying rational expressions.

EXAMPLE 12   Simplify $\dfrac{6a^2 - 7a + 2}{10a - 15a^2}$.

Solution

$$\frac{6a^2 - 7a + 2}{10a - 15a^2} = \frac{(2a - 1)(3a - 2)}{5a \,(2 - 3a)} \qquad \frac{3a - 2}{2 - 3a} = -1$$

$$= (-1)\left(\frac{2a - 1}{5a}\right)$$

$$= -\frac{2a - 1}{5a} \quad \text{or} \quad \frac{1 - 2a}{5a}$$

CONCEPT QUIZ

For Problems 1–5, answer true or false.

1. When a rational number is being reduced, the form of the numeral is being changed but not the number it represents.

2. The rational expression $\dfrac{x + 2}{x - 3}$ is meaningful for all values of x except for $x = -2$ and $x = 3$.

3. The binomials $x - y$ and $y - x$ are opposites.

4. The binomials $x + 3$ and $x - 3$ are opposites.

5. The rational expression $\dfrac{2 - x}{x + 2}$ reduces to -1.

PROBLEM SET 9.1

For Problems 1–8, express each rational number in reduced form.

1. $\dfrac{27}{36}$ **2.** $\dfrac{14}{21}$ **3.** $\dfrac{45}{54}$

4. $\dfrac{-14}{42}$ **5.** $\dfrac{24}{-60}$ **6.** $\dfrac{45}{-75}$

7. $\dfrac{-16}{-56}$ **8.** $\dfrac{-30}{-42}$

For Problems 9–50, simplify each rational expression.

9. $\dfrac{12xy}{42y}$ **10.** $\dfrac{21xy}{35x}$ **11.** $\dfrac{18a^2}{45ab}$

12. $\dfrac{48ab}{84b^2}$ **13.** $\dfrac{-14y^3}{56xy^2}$ **14.** $\dfrac{-14x^2y^3}{63xy^2}$

15. $\dfrac{54c^2d}{-78cd^2}$ **16.** $\dfrac{60x^3z}{-64xyz^2}$ **17.** $\dfrac{-40x^3y}{-24xy^4}$

18. $\dfrac{-30x^2y^2z^2}{-35xz^3}$ **19.** $\dfrac{x^2 - 4}{x^2 + 2x}$ **20.** $\dfrac{xy + y^2}{x^2 - y^2}$

21. $\dfrac{18x + 12}{12x - 6}$ **22.** $\dfrac{20x + 50}{15x - 30}$

23. $\dfrac{a^2 + 7a + 10}{a^2 - 7a - 18}$ **24.** $\dfrac{a^2 + 4a - 32}{3a^2 + 26a + 16}$

25. $\dfrac{2n^2 + n - 21}{10n^2 + 33n - 7}$ **26.** $\dfrac{4n^2 - 15n - 4}{7n^2 - 30n + 8}$

27. $\dfrac{5x^2 + 7}{10x}$ **28.** $\dfrac{12x^2 + 11x - 15}{20x^2 - 23x + 6}$

29. $\dfrac{6x^2 + x - 15}{8x^2 - 10x - 3}$ **30.** $\dfrac{4x^2 + 8x}{x^3 + 8}$

31. $\dfrac{3x^2 - 12x}{x^3 - 64}$ **32.** $\dfrac{x^2 - 14x + 49}{6x^2 - 37x - 35}$

33. $\dfrac{3x^2 + 17x - 6}{9x^2 - 6x + 1}$ **34.** $\dfrac{9y^2 - 1}{3y^2 + 11y - 4}$

35. $\dfrac{2x^3 + 3x^2 - 14x}{x^2y + 7xy - 18y}$ **36.** $\dfrac{3x^3 + 12x}{9x^2 + 18x}$

37. $\dfrac{5y^2 + 22y + 8}{25y^2 - 4}$ **38.** $\dfrac{16x^3y + 24x^2y^2 - 16xy^3}{24x^2y + 12xy^2 - 12y^3}$

39. $\dfrac{15x^3 - 15x^2}{5x^3 + 5x}$ **40.** $\dfrac{5n^2 + 18n - 8}{3n^2 + 13n + 4}$

41. $\dfrac{4x^2y + 8xy^2 - 12y^3}{18x^3y - 12x^2y^2 - 6xy^3}$ **42.** $\dfrac{3 + x - 2x^2}{2 + x - x^2}$

43. $\dfrac{3n^2 + 14n - 24}{7n^2 + 44n + 12}$ **44.** $\dfrac{x^4 - 2x^2 - 15}{2x^4 + 9x^2 + 9}$

45. $\dfrac{8 + 18x - 5x^2}{10 + 31x + 15x^2}$ **46.** $\dfrac{6x^4 - 11x^2 + 4}{2x^4 + 17x^2 - 9}$

47. $\dfrac{27x^4 - x}{6x^3 + 10x^2 - 4x}$ **48.** $\dfrac{64x^4 + 27x}{12x^3 - 27x^2 - 27x}$

49. $\dfrac{-40x^3 + 24x^2 + 16x}{20x^3 + 28x^2 + 8x}$ **50.** $\dfrac{-6x^3 - 21x^2 + 12x}{-18x^3 - 42x^2 + 120x}$

For Problems 51–58, simplify each rational expression. You will need to use factoring by grouping.

51. $\dfrac{xy + ay + bx + ab}{xy + ay + cx + ac}$ **52.** $\dfrac{xy + 2y + 3x + 6}{xy + 2y + 4x + 8}$

53. $\dfrac{ax - 3x + 2ay - 6y}{2ax - 6x + ay - 3y}$ **54.** $\dfrac{x^2 - 2x + ax - 2a}{x^2 - 2x + 3ax - 6a}$

55. $\dfrac{5x^2 + 5x + 3x + 3}{5x^2 + 3x - 30x - 18}$ **56.** $\dfrac{x^2 + 3x + 4x + 12}{2x^2 + 6x - x - 3}$

57. $\dfrac{2st - 30 - 12s + 5t}{3st - 6 - 18s + t}$ **58.** $\dfrac{nr - 6 - 3n + 2r}{nr + 10 + 2r + 5n}$

For Problems 59–68, simplify each rational expression. You may want to refer to Example 12 of this section.

59. $\dfrac{5x - 7}{7 - 5x}$ **60.** $\dfrac{4a - 9}{9 - 4a}$

61. $\dfrac{n^2 - 49}{7 - n}$ **62.** $\dfrac{9 - y}{y^2 - 81}$

63. $\dfrac{2y - 2xy}{x^2y - y}$ **64.** $\dfrac{3x - x^2}{x^2 - 9}$

65. $\dfrac{2x^3 - 8x}{4x - x^3}$ **66.** $\dfrac{x^2 - (y - 1)^2}{(y - 1)^2 - x^2}$

67. $\dfrac{n^2 - 5n - 24}{40 + 3n - n^2}$ **68.** $\dfrac{x^2 + 2x - 24}{20 - x - x^2}$

■ ■ ■ **Thoughts into words**

69. Compare the concept of a rational number in arithmetic to the concept of a rational expression in algebra.

70. What role does factoring play in the simplifying of rational expressions?

71. Why is the rational expression $\dfrac{x + 3}{x^2 - 4}$ undefined for $x = 2$ and $x = -2$ but defined for $x = -3$?

72. How would you convince someone that $\dfrac{x - 4}{4 - x} = -1$ for all real numbers except 4?

Graphing calculator activities

This is the first of many appearances of a group of problems called Graphing Calculator Activities. These problems are specifically designed for those of you who have access to a graphing calculator or a computer with an appropriate software package. Within the framework of these problems, you will be given the opportunity to reinforce concepts we discussed in the text; lay groundwork for concepts we will introduce later in the text; predict shapes and locations of graphs on the basis of your previous graphing experiences; solve problems that are unreasonable (or perhaps impossible) to solve without a graphing utility; and in general become familiar with the capabilities and limitations of your graphing utility.

This first set of activities is designed to help you get started with your graphing utility by setting different boundaries for the viewing rectangle; you will notice the effect on the graphs produced. These boundaries are usually set by using a menu displayed by a key marked either WINDOW or RANGE. You may need to consult the user's manual for specific key-punching instructions.

73. Graph the equation $y = \dfrac{1}{x}$ using the following boundaries.

   a. $-15 \le x \le 15$ and $-10 \le y \le 10$

   b. $-10 \le x \le 10$ and $-10 \le y \le 10$

   c. $-5 \le x \le 5$ and $-5 \le y \le 5$

74. Graph the equation $y = \dfrac{-2}{x^2}$ using the following boundaries.

   a. $-15 \le x \le 15$ and $-10 \le y \le 10$

   b. $-5 \le x \le 5$ and $-10 \le y \le 10$

   c. $-5 \le x \le 5$ and $-10 \le y \le 1$

75. Graph the two equations $y = \pm\sqrt{x}$ on the same set of axes using the following boundaries. Let $Y_1 = \sqrt{x}$ and $Y_2 = -\sqrt{x}$.

   a. $-15 \le x \le 15$ and $-10 \le y \le 10$

   b. $-1 \le x \le 15$ and $-10 \le y \le 10$

   c. $-1 \le x \le 15$ and $-5 \le y \le 5$

76. Graph $y = \dfrac{1}{x}$, $y = \dfrac{5}{x}$, $y = \dfrac{10}{x}$, and $y = \dfrac{20}{x}$ on the same set of axes. (Choose your own boundaries.) What effect does increasing the constant seem to have on the graph?

77. Graph $y = \dfrac{10}{x}$ and $y = \dfrac{-10}{x}$ on the same set of axes. What relationship exists between the two graphs?

78. Graph $y = \dfrac{10}{x^2}$ and $y = \dfrac{-10}{x^2}$ on the same set of axes. What relationship exists between the two graphs?

79. Use a graphing calculator to give visual support for your answers for Problems 21–30.

80. Use a graphing calculator to give visual support for your answers for Problems 59–62.

9.2 Multiplying and Dividing Rational Expressions

Objectives

■   Multiply rational expressions.

■   Divide rational expressions.

We define multiplication of rational numbers in common fraction form as follows:

DEFINITION 9.1

If a, b, c, and d are integers, and b and d are not equal to zero, then

$$\frac{a}{b} \cdot \frac{c}{d} = \frac{a \cdot c}{b \cdot d} = \frac{ac}{bd}$$

To multiply rational numbers in common-fraction form, we merely **multiply numerators and multiply denominators,** as the next examples demonstrate. (The steps in the dashed boxes are usually done mentally.)

$$\frac{2}{3} \cdot \frac{4}{5} = \frac{2 \cdot 4}{3 \cdot 5} = \frac{8}{15}$$

$$\frac{-3}{4} \cdot \frac{5}{7} = \frac{-3 \cdot 5}{4 \cdot 7} = \frac{-15}{28} = -\frac{15}{28}$$

$$-\frac{5}{6} \cdot \frac{13}{3} = \frac{-5}{6} \cdot \frac{13}{3} = \frac{-5 \cdot 13}{6 \cdot 3} = \frac{-65}{18} = -\frac{65}{18}$$

We also agree, when multiplying rational numbers, to express the final product in reduced form. The following examples show some different formats used to *multiply and simplify* rational numbers.

$$\frac{3}{4} \cdot \frac{4}{7} = \frac{3 \cdot 4}{4 \cdot 7} = \frac{3}{7}$$

$$\frac{8}{9} \cdot \frac{27}{32} = \frac{\overset{1}{8}}{\underset{1}{9}} \cdot \frac{\overset{3}{27}}{\underset{4}{32}} = \frac{3}{4} \qquad \begin{array}{l}\text{A common factor of 9 was divided out of}\\ \text{9 and 27, and a common factor of 8 was}\\ \text{divided out of 8 and 32.}\end{array}$$

$$\left(-\frac{28}{25}\right)\left(-\frac{65}{78}\right) = \frac{2 \cdot 2 \cdot 7 \cdot 5 \cdot 13}{5 \cdot 5 \cdot 2 \cdot 3 \cdot 13} = \frac{14}{15}. \qquad \begin{array}{l}\text{We should recognize that a}\\ \textit{negative times a negative is}\\ \textit{positive.} \text{ Also, note the use}\\ \text{of prime factors to help us}\\ \text{recognize common factors.}\end{array}$$

Multiplication of rational expressions follows the same basic pattern as multiplication of rational numbers in common-fraction form. That is to say, *we multiply numerators and multiply denominators and express the final product in simplified or reduced form.* Let's consider some examples.

$$\frac{3x}{4y} \cdot \frac{8y^2}{9x} = \frac{3 \cdot \overset{2}{8} \cdot x \cdot \overset{y}{y^2}}{\underset{3}{4} \cdot 9 \cdot x \cdot y} = \frac{2y}{3} \qquad \begin{array}{l}\text{Note that we use the commutative property}\\ \text{of multiplication to rearrange the factors in}\\ \text{a form that allows us to identify common}\\ \text{factors of the numerator and denominator.}\end{array}$$

$$\frac{-4a}{6a^2b^2} \cdot \frac{9ab}{12a^2} = -\frac{4 \cdot \overset{3}{9} \cdot a^2 \cdot b}{\underset{2}{6} \cdot \underset{3}{12} \cdot \underset{a^2}{a^4} \cdot \underset{b}{b^2}} = -\frac{1}{2a^2b}$$

$$\frac{12x^2y}{-18xy} \cdot \frac{-24xy^2}{56y^3} = \frac{\overset{2}{\cancel{12}} \cdot \overset{8}{\cancel{24}} \cdot \overset{x^2}{\cancel{x^3}} \cdot \cancel{y^3}}{\underset{3}{\cancel{18}} \cdot \underset{7}{\cancel{56}} \cdot \cancel{x} \cdot \underset{y}{\cancel{y^4}}} = \frac{2x^2}{7y}$$

You should recognize that the first fraction is equivalent to $\frac{12x^2y}{18xy}$ and the second to $-\frac{24xy^2}{56y^3}$; thus the product is positive.

If the rational expressions contain polynomials (other than monomials) that are factorable, then our work may take on the following format.

EXAMPLE 1   Multiply and simplify $\dfrac{y}{x^2 - 4} \cdot \dfrac{x + 2}{y^2}$.

Solution

$$\frac{y}{x^2 - 4} \cdot \frac{x + 2}{y^2} = \frac{\cancel{y}(\cancel{x+2})}{\underset{y}{\cancel{y^2}}(\cancel{x+2})(x - 2)} = \frac{1}{y(x - 2)} \qquad \blacksquare$$

In Example 1, note that we combined the steps of multiplying numerators and denominators and factoring the polynomials. Also note that we left the final answer in factored form. Either $\dfrac{1}{y(x - 2)}$ or $\dfrac{1}{xy - 2y}$ would be an acceptable answer.

EXAMPLE 2   Multiply and simplify $\dfrac{x^2 - x}{x + 5} \cdot \dfrac{x^2 + 5x + 4}{x^4 - x^2}$.

Solution

$$\frac{x^2 - x}{x + 5} \cdot \frac{x^2 + 5x + 4}{x^4 - x^2} = \frac{\cancel{x}(\cancel{x-1})(\cancel{x+1})(x + 4)}{(x + 5)(\cancel{x^2})(\cancel{x-1})(\cancel{x+1})} = \frac{x + 4}{x(x + 5)} \qquad \blacksquare$$

Let's pause for a moment and consider again the idea of "checking algebraic manipulations." In Example 1, we are claiming that $\dfrac{y}{x^2 - 4} \cdot \dfrac{x + 2}{y^2} = \dfrac{1}{y(x - 2)}$ for all real numbers except -2 and 2 for x and 0 for y. In Figure 9.2, we show a calculator check for $y = 2$ and $x = 5$. Remember that this is only a partial check.

```
2→Y
              2
5→X
              5
(Y/(X²-4))*((X+2
)/Y²)
      .1666666667
1/(Y(X-2))
      .1666666667
```

Figure 9.2

In Example 2, we are claiming that

$$\frac{x^2 - x}{x + 5} \cdot \frac{x^2 + 5x + 4}{x^4 - x^2} = \frac{x + 4}{x(x + 5)}$$

for all real numbers except -5, -1, 0, and 1. Figure 9.3 is the result of graphing $Y_1 = \dfrac{x^2 - x}{x + 5} \cdot \dfrac{x^2 + 5x + 4}{x^4 - x^2}$ and $Y_2 = \dfrac{x + 4}{x(x + 5)}$ on the same set of axes. The graphs appear to be identical, which certainly gives visual support for our original claim.

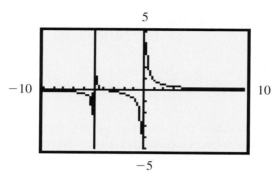

Figure 9.3

E X A M P L E 3

Multiply and simplify $\dfrac{6n^2 + 7n - 5}{n^2 + 2n - 24} \cdot \dfrac{4n^2 + 21n - 18}{12n^2 + 11n - 15}$.

Solution

$$\frac{6n^2 + 7n - 5}{n^2 + 2n - 24} \cdot \frac{4n^2 + 21n - 18}{12n^2 + 11n - 15}$$

$$= \frac{(3n + 5)(2n - 1)(4n - 3)(n + 6)}{(n + 6)(n - 4)(3n + 5)(4n - 3)} = \frac{2n - 1}{n - 4} \qquad ∎$$

Dividing Rational Expressions

We define division of rational numbers in common fraction form as follows:

> ### DEFINITION 9.2
>
> If a, b, c, and d are integers, and b, c, and d are not equal to zero, then
>
> $$\frac{a}{b} \div \frac{c}{d} = \frac{a}{b} \cdot \frac{d}{c} = \frac{ad}{bc}$$

Definition 9.2 states that to divide two rational numbers in fraction form, we **invert the divisor and multiply.** We call the numbers $\dfrac{c}{d}$ and $\dfrac{d}{c}$ **reciprocals** or **multiplicative**

inverses of each other because their product is 1. Thus we can describe division by saying **to divide by a fraction, multiply by its reciprocal.** The following examples demonstrate the use of Definition 9.2.

$$\frac{7}{8} \div \frac{5}{6} = \frac{7}{\underset{4}{8}} \cdot \frac{\overset{3}{6}}{5} = \frac{21}{20}, \qquad \frac{-5}{9} \div \frac{15}{18} = -\frac{5}{9} \cdot \frac{\overset{2}{18}}{\underset{3}{15}} = -\frac{2}{3}$$

$$\frac{14}{-19} \div \frac{21}{-38} = \left(-\frac{14}{19}\right) \div \left(-\frac{21}{38}\right) = \left(-\frac{\overset{2}{14}}{19}\right)\left(-\frac{\overset{2}{38}}{\underset{3}{21}}\right) = \frac{4}{3}$$

We define division of algebraic rational expressions in the same way that we define division of rational numbers. That is, the quotient of two rational expressions is the product we obtain when we multiply the first expression by the reciprocal of the second. Consider the following examples.

EXAMPLE 4 Divide and simplify $\dfrac{16x^2y}{24xy^3} \div \dfrac{9xy}{8x^2y^2}$.

Solution

$$\frac{16x^2y}{24xy^3} \div \frac{9xy}{8x^2y^2} = \frac{16x^2y}{24xy^3} \cdot \frac{8x^2y^2}{9xy} = \frac{16 \cdot 8 \cdot \overset{x^2}{x^4} \cdot y^3}{\underset{3}{24} \cdot 9 \cdot x^2 \cdot \underset{y}{y^4}} = \frac{16x^2}{27y}$$

EXAMPLE 5 Divide and simplify $\dfrac{3a^2 + 12}{3a^2 - 15a} \div \dfrac{a^4 - 16}{a^2 - 3a - 10}$.

Solution

$$\frac{3a^2 + 12}{3a^2 - 15a} \div \frac{a^4 - 16}{a^2 - 3a - 10} = \frac{3a^2 + 12}{3a^2 - 15a} \cdot \frac{a^2 - 3a - 10}{a^4 - 16}$$

$$= \frac{3(a^2 + 4)(a - 5)(a + 2)}{3a(a - 5)(a^2 + 4)(a + 2)(a - 2)}$$

$$= \frac{1}{a(a - 2)}$$

EXAMPLE 6 Divide and simplify $\dfrac{28t^3 - 51t^2 - 27t}{49t^2 + 42t + 9} \div (4t - 9)$.

Solution

$$\frac{28t^3 - 51t^2 - 27t}{49t^2 + 42t + 9} \div \frac{4t - 9}{1} = \frac{28t^3 - 51t^2 - 27t}{49t^2 + 42t + 9} \cdot \frac{1}{4t - 9}$$

$$= \frac{t(7t + 3)(4t - 9)}{(7t + 3)(7t + 3)(4t - 9)}$$

$$= \frac{t}{7t + 3}$$

In a problem such as Example 6, it may be helpful to write the divisor with a denominator of 1. Thus we write $4t - 9$ as $\dfrac{4t - 9}{1}$; its reciprocal is obviously $\dfrac{1}{4t - 9}$.

Let's consider one final example that involves both multiplication and division.

EXAMPLE 7   Perform the indicated operations and simplify.

$$\frac{x^2 + 5x}{3x^2 - 4x - 20} \cdot \frac{x^2y + y}{2x^2 + 11x + 5} \div \frac{xy^2}{6x^2 - 17x - 10}$$

Solution

$$\frac{x^2 + 5x}{3x^2 - 4x - 20} \cdot \frac{x^2y + y}{2x^2 + 11x + 5} \div \frac{xy^2}{6x^2 - 17x - 10}$$

$$= \frac{x^2 + 5x}{3x^2 - 4x - 20} \cdot \frac{x^2y + y}{2x^2 + 11x + 5} \cdot \frac{6x^2 - 17x - 10}{xy^2}$$

$$= \frac{\cancel{x}(x+5)(\cancel{y})(x^2 + 1)(2x+1)(3x-10)}{(3x-10)(x + 2)(2x+1)(x+5)(\cancel{x})(\cancel{y^2}_{y})} = \frac{x^2 + 1}{y(x + 2)}$$ ∎

CONCEPT QUIZ

For Problems 1–5, answer true or false.

1. To multiply two rational numbers in fraction form, we need to change to equivalent fractions with a common denominator.

2. When multiplying rational expressions that contain polynomials, the polynomials are factored so that common factors can be divided out.

3. In the division problem $\dfrac{2x^2y}{3z} \div \dfrac{4x^3}{5y^2}$, the fraction $\dfrac{4x^3}{5y^2}$ is the divisor.

4. The numbers $-\dfrac{2}{3}$ and $\dfrac{3}{2}$ are multiplicative inverses.

5. To divide two numbers in fraction form, we invert the divisor and multiply.

PROBLEM SET 9.2

For Problems 1–12, perform the indicated operations involving rational numbers. Express final answers in reduced form.

1. $\dfrac{7}{12} \cdot \dfrac{6}{35}$

2. $\dfrac{5}{8} \cdot \dfrac{12}{20}$

3. $\dfrac{-4}{9} \cdot \dfrac{18}{30}$

4. $\dfrac{-6}{9} \cdot \dfrac{36}{48}$

5. $\dfrac{3}{-8} \cdot \dfrac{-6}{12}$

6. $\dfrac{-12}{16} \cdot \dfrac{18}{-32}$

7. $\left(-\dfrac{5}{7}\right) \div \dfrac{6}{7}$

8. $\left(-\dfrac{5}{9}\right) \div \dfrac{10}{3}$

9. $\dfrac{-9}{5} \div \dfrac{27}{10}$

10. $\dfrac{4}{7} \div \dfrac{16}{-21}$

11. $\dfrac{4}{9} \cdot \dfrac{6}{11} \div \dfrac{4}{15}$

12. $\dfrac{2}{3} \cdot \dfrac{6}{7} \div \dfrac{8}{3}$

For Problems 13–50, perform the indicated operations involving rational expressions. Express final answers in simplest form.

13. $\dfrac{6xy}{9y^4} \cdot \dfrac{30x^3y}{-48x}$

14. $\dfrac{-14xy^4}{18y^2} \cdot \dfrac{24x^2y^3}{35y^2}$

15. $\dfrac{5a^2b^2}{11ab} \cdot \dfrac{22a^3}{15ab^2}$

16. $\dfrac{10a^2}{5b^2} \cdot \dfrac{15b^3}{2a^4}$

17. $\dfrac{5xy}{8y^2} \cdot \dfrac{18x^2y}{15}$

18. $\dfrac{4x^2}{5y^2} \cdot \dfrac{15xy}{24x^2y^2}$

19. $\dfrac{5x^4}{12x^2y^3} \div \dfrac{9}{5xy}$

20. $\dfrac{7x^2y}{9xy^3} \div \dfrac{3x^4}{2x^2y^2}$

21. $\dfrac{9a^2c}{12bc^2} \div \dfrac{21ab}{14c^3}$

22. $\dfrac{3ab^3}{4c} \div \dfrac{21ac}{12bc^3}$

23. $\dfrac{9x^2y^3}{14x} \cdot \dfrac{21y}{15xy^2} \cdot \dfrac{10x}{12y^3}$

24. $\dfrac{5xy}{7a} \cdot \dfrac{14a^2}{15x} \cdot \dfrac{3a}{8y}$

25. $\dfrac{3x+6}{5y} \cdot \dfrac{x^2+4}{x^2+10x+16}$

26. $\dfrac{5xy}{x+6} \cdot \dfrac{x^2-36}{x^2-6x}$

27. $\dfrac{5a^2+20a}{a^3-2a^2} \cdot \dfrac{a^2-a-12}{a^2-16}$

28. $\dfrac{2a^2+6}{a^2-a} \cdot \dfrac{a^3-a^2}{8a-4}$

29. $\dfrac{3n^2+15n-18}{3n^2+10n-48} \cdot \dfrac{12n^2-17n-40}{8n^2+2n-10}$

30. $\dfrac{10n^2+21n-10}{5n^2+33n-14} \cdot \dfrac{2n^2+6n-56}{2n^2-3n-20}$

31. $\dfrac{9y^2}{x^2+12x+36} \div \dfrac{12y}{x^2+6x}$

32. $\dfrac{7xy}{x^2-4x+4} \div \dfrac{14y}{x^2-4}$

33. $\dfrac{x^2-4xy+4y^2}{7xy^2} \div \dfrac{4x^2-3xy-10y^2}{20x^2y+25xy^2}$

34. $\dfrac{x^2+5xy-6y^2}{xy^2-y^3} \cdot \dfrac{2x^2+15xy+18y^2}{xy+4y^2}$

35. $\dfrac{5-14n-3n^2}{1-2n-3n^2} \cdot \dfrac{9+7n-2n^2}{27-15n+2n^2}$

36. $\dfrac{6-n-2n^2}{12-11n+2n^2} \cdot \dfrac{24-26n+5n^2}{2+3n+n^2}$

37. $\dfrac{3x^4+2x^2-1}{3x^4+14x^2-5} \cdot \dfrac{x^4-2x^2-35}{x^4-17x^2+70}$

38. $\dfrac{2x^4+x^2-3}{2x^4+5x^2+2} \cdot \dfrac{3x^4+10x^2+8}{3x^4+x^2-4}$

39. $\dfrac{6x^2-35x+25}{4x^2-11x-45} \div \dfrac{18x^2+9x-20}{24x^2+74x+45}$

40. $\dfrac{21t^2+22t-8}{5t^2-43t-18} \div \dfrac{12t^2+7t-12}{20t^2-7t-6}$

41. $\dfrac{10t^3+25t}{20t+10} \cdot \dfrac{2t^2-t-1}{t^5-t}$

42. $\dfrac{t^4-81}{t^2-6t+9} \cdot \dfrac{6t^2-11t-21}{5t^2+8t-21}$

43. $\dfrac{4t^2+t-5}{t^3-t^2} \cdot \dfrac{t^4+6t^3}{16t^2+40t+25}$

44. $\dfrac{9n^2-12n+4}{n^2-4n-32} \cdot \dfrac{n^2+4n}{3n^3-2n^2}$

45. $\dfrac{nr+3n+2r+6}{nr+3n-3r-9} \cdot \dfrac{n^2-9}{n^3-4n}$

46. $\dfrac{xy+xc+ay+ac}{xy-2xc+ay-2ac} \cdot \dfrac{2x^3-8x}{12x^3+20x^2-8x}$

47. $\dfrac{x^2-x}{4y} \cdot \dfrac{10xy^2}{2x-2} \div \dfrac{3x^2+3x}{15x^2y^2}$

48. $\dfrac{4xy^2}{7x} \cdot \dfrac{14x^3y}{12y} \div \dfrac{7y}{9x^3}$

49. $\dfrac{a^2-4ab+4b^2}{6a^2-4ab} \cdot \dfrac{3a^2+5ab-2b^2}{6a^2+ab-b^2} \div \dfrac{a^2-4b^2}{8a+4b}$

50. $\dfrac{2x^2+3x}{2x^3-10x^2} \cdot \dfrac{x^2-8x+15}{3x^3-27x} \div \dfrac{14x+21}{x^2-6x-27}$

■ ■ ▨ Thoughts into words

51. Explain in your own words how to divide two rational expressions.

52. Suppose that your friend missed class the day the material in this section was discussed. How could you use her background in arithmetic to explain how to multiply and divide rational expressions?

53. Give a step-by-step description of how to do the following multiplication problem.

$$\frac{x^2 + 5x + 6}{x^2 - 2x - 8} \cdot \frac{x^2 - 16}{16 - x^2}$$

 Graphing calculator activities

54. Use your graphing calculator to check Examples 3–7.

55. Use your graphing calculator to check your answers for Problems 27–34.

56. Use your graphing calculator to graph $y = \dfrac{1}{x}$ again. Then predict the graphs of $y = \dfrac{4}{x}$, $y = \dfrac{2}{x}$, and $y = \dfrac{8}{x}$. Finally, using a graphing calculator, graph all four equations on the same set of axes to check your predictions.

57. Draw rough sketches of the graphs of $y = \dfrac{1}{x^2}$, $y = \dfrac{4}{x^2}$, and $y = \dfrac{6}{x^2}$. Then check your sketches by using your graphing calculator to graph all three equations on the same set of axes.

58. Use your graphing calculator to graph $y = \dfrac{1}{x^2 + 1}$. Now predict the graphs of $y = \dfrac{3}{x^2 + 1}$, $y = \dfrac{5}{x^2 + 1}$, and $y = \dfrac{8}{x^2 + 1}$. Finally, check your predictions by using your graphing calculator to graph all four equations on the same set of axes.

Answers to Concept Quiz

1. False   **2.** True   **3.** True   **4.** False   **5.** True

9.3   Adding and Subtracting Rational Expressions

Objectives

- Combine rational expressions with common denominators.
- Find the lowest common denominators.
- Add and subtract rational expressions with different denominators.

We can define addition and subtraction of rational numbers as follows:

DEFINITION 9.3

If a, b, and c are integers and b is not zero, then

$$\frac{a}{b} + \frac{c}{b} = \frac{a+c}{b} \qquad \text{Addition}$$

$$\frac{a}{b} - \frac{c}{b} = \frac{a-c}{b} \qquad \text{Subtraction}$$

We can add or subtract rational numbers with a common denominator by adding or subtracting the numerators and placing the result over the common denominator. The following examples illustrate Definition 9.3:

$$\frac{2}{9} + \frac{3}{9} = \frac{2+3}{9} = \frac{5}{9}$$

$$\frac{7}{8} - \frac{3}{8} = \frac{7-3}{8} = \frac{4}{8} = \frac{1}{2} \qquad \text{Don't forget to reduce!}$$

$$\frac{4}{6} + \frac{-5}{6} = \frac{4+(-5)}{6} = \frac{-1}{6} = -\frac{1}{6}$$

$$\frac{7}{10} + \frac{4}{-10} = \frac{7}{10} + \frac{-4}{10} = \frac{7+(-4)}{10} = \frac{3}{10}$$

We use this same *common denominator* approach when adding or subtracting rational expressions, as in these next examples:

$$\frac{3}{x} + \frac{9}{x} = \frac{3+9}{x} = \frac{12}{x}$$

$$\frac{8}{x-2} - \frac{3}{x-2} = \frac{8-3}{x-2} = \frac{5}{x-2}$$

$$\frac{9}{4y} + \frac{5}{4y} = \frac{9+5}{4y} = \frac{14}{4y} = \frac{7}{2y} \qquad \text{Don't forget to simplify the final answer!}$$

$$\frac{n^2}{n-1} - \frac{1}{n-1} = \frac{n^2-1}{n-1} = \frac{(n+1)(n-1)}{n-1} = n+1$$

$$\frac{6a^2}{2a+1} + \frac{13a+5}{2a+1} = \frac{6a^2+13a+5}{2a+1} = \frac{(2a+1)(3a+5)}{2a+1} = 3a+5$$

In each of the previous examples that involve rational expressions, we should technically restrict the variables to exclude division by zero. For example, $\frac{3}{x} + \frac{9}{x} = \frac{12}{x}$ is true for all real number values for x, *except $x = 0$*. Likewise, $\frac{8}{x-2} - \frac{3}{x-2} = \frac{5}{x-2}$ as long as x does not equal 2. Rather than taking the time and space to write down restrictions for each problem, we will merely assume that such restrictions exist.

If rational numbers that do not have a common denominator are to be added or subtracted, then we apply the fundamental principle of fractions $\left(\dfrac{a}{b} = \dfrac{ak}{bk} \right)$ **to obtain equivalent fractions with a common denominator.** Equivalent fractions are fractions such as $\dfrac{1}{2}$ and $\dfrac{2}{4}$ that name the same number. Consider the following example.

$$\frac{1}{2} + \frac{1}{3} = \frac{3}{6} + \frac{2}{6} = \frac{3 + 2}{6} = \frac{5}{6}$$

$$\left(\begin{array}{l} \frac{1}{2} \text{ and } \frac{3}{6} \\ \text{are equivalent} \\ \text{fractions.} \end{array} \right) \left(\begin{array}{l} \frac{1}{3} \text{ and } \frac{2}{6} \\ \text{are equivalent} \\ \text{fractions.} \end{array} \right)$$

Note that we chose 6 as our common denominator and 6 is the **least common multiple** of the original denominators 2 and 3. (The least common multiple of a set of whole numbers is the smallest nonzero whole number divisible by each of the numbers.) In general, we use the least common multiple of the denominators of the fractions to be added or subtracted as a **least common denominator** (LCD).

A least common denominator may be found by inspection or by using the prime-factored forms of the numbers. Let's consider some examples and use each of these techniques.

EXAMPLE 1 Subtract $\dfrac{5}{6} - \dfrac{3}{8}$.

Solution

By inspection, we can see that the LCD is 24. Thus both fractions can be changed to equivalent fractions, each with a denominator of 24.

$$\frac{5}{6} - \frac{3}{8} = \left(\frac{5}{6}\right)\left(\frac{4}{4}\right) - \left(\frac{3}{8}\right)\left(\frac{3}{3}\right) = \frac{20}{24} - \frac{9}{24} = \frac{11}{24}$$

Form of 1   Form of 1   ∎

In Example 1, note that the fundamental principle of fractions, $\dfrac{a}{b} = \dfrac{a \cdot k}{b \cdot k}$, can be written as $\dfrac{a}{b} = \left(\dfrac{a}{b}\right)\left(\dfrac{k}{k}\right)$. This latter form emphasizes the fact that 1 is the multiplication identity element.

EXAMPLE 2 Perform the indicated operations $\dfrac{3}{5} + \dfrac{1}{6} - \dfrac{13}{15}$.

Solution

Again by inspection, we can determine that the LCD is 30. Thus we can proceed as follows:

$$\frac{3}{5} + \frac{1}{6} - \frac{13}{15} = \left(\frac{3}{5}\right)\left(\frac{6}{6}\right) + \left(\frac{1}{6}\right)\left(\frac{5}{5}\right) - \left(\frac{13}{15}\right)\left(\frac{2}{2}\right)$$

$$= \frac{18}{30} + \frac{5}{30} - \frac{26}{30} = \frac{18 + 5 - 26}{30}$$

$$= \frac{-3}{30} = -\frac{1}{10} \qquad \text{Don't forget to reduce!} \qquad \blacksquare$$

EXAMPLE 3 Add $\dfrac{7}{18} + \dfrac{11}{24}$.

Solution

Let's use the prime-factored forms of the denominators to help find the LCD.

$$18 = 2 \cdot 3 \cdot 3 \qquad 24 = 2 \cdot 2 \cdot 2 \cdot 3$$

The LCD must contain three factors of 2 because 24 contains three 2s. The LCD must also contain two factors of 3 because 18 has two 3s. Thus the LCD $= 2 \cdot 2 \cdot 2 \cdot 3 \cdot 3 = 72$. Now we can proceed as usual.

$$\frac{7}{18} + \frac{11}{24} = \left(\frac{7}{18}\right)\left(\frac{4}{4}\right) + \left(\frac{11}{24}\right)\left(\frac{3}{3}\right) = \frac{28}{72} + \frac{33}{72} = \frac{61}{72} \qquad \blacksquare$$

To add and subtract rational expressions with different denominators, follow the same basic routine as you follow when you add or subtract rational numbers with different denominators. Study the following examples carefully and note the similarity to our previous work with rational numbers.

EXAMPLE 4 Add $\dfrac{x + 2}{4} + \dfrac{3x + 1}{3}$.

Solution

By inspection, we see that the LCD is 12.

$$\frac{x + 2}{4} + \frac{3x + 1}{3} = \left(\frac{x + 2}{4}\right)\left(\frac{3}{3}\right) + \left(\frac{3x + 1}{3}\right)\left(\frac{4}{4}\right)$$

$$= \frac{3(x + 2)}{12} + \frac{4(3x + 1)}{12}$$

$$= \frac{3(x + 2) + 4(3x + 1)}{12}$$

$$= \frac{3x + 6 + 12x + 4}{12}$$

$$= \frac{15x + 10}{12}$$

Note the final result in Example 4. The numerator, $15x + 10$, could be factored as $5(3x + 2)$. However, because this produces no common factors with the denominator, the fraction cannot be simplified. Thus the final answer can be left as $\frac{15x + 10}{12}$. It would also be acceptable to express it as $\frac{5(3x + 2)}{12}$.

EXAMPLE 5 Subtract $\dfrac{a - 2}{2} - \dfrac{a - 6}{6}$.

Solution

By inspection, we see that the LCD is 6.

$$\frac{a - 2}{2} - \frac{a - 6}{6} = \left(\frac{a - 2}{2}\right)\left(\frac{3}{3}\right) - \frac{a - 6}{6}$$

$$= \frac{3(a - 2) - (a - 6)}{6} \qquad \text{Be careful with this sign as you move to the next step!}$$

$$= \frac{3a - 6 - a + 6}{6}$$

$$= \frac{2a}{6} = \frac{a}{3} \qquad \text{Don't forget to simplify.}$$

EXAMPLE 6 Perform the indicated operations: $\dfrac{x + 3}{10} + \dfrac{2x + 1}{15} - \dfrac{x - 2}{18}$

Solution

If you cannot determine the LCD by inspection, then use the prime-factored forms of the denominators.

$$10 = 2 \cdot 5 \qquad 15 = 3 \cdot 5 \qquad 18 = 2 \cdot 3 \cdot 3$$

The LCD must contain one factor of 2, two factors of 3, and one factor of 5. Thus the LCD is $2 \cdot 3 \cdot 3 \cdot 5 = 90$.

$$\frac{x + 3}{10} + \frac{2x + 1}{15} - \frac{x - 2}{18} = \left(\frac{x + 3}{10}\right)\left(\frac{9}{9}\right) + \left(\frac{2x + 1}{15}\right)\left(\frac{6}{6}\right) - \left(\frac{x - 2}{18}\right)\left(\frac{5}{5}\right)$$

$$= \frac{9(x + 3)}{90} + \frac{6(2x + 1)}{90} - \frac{5(x - 2)}{90}$$

$$= \frac{9(x + 3) + 6(2x + 1) - 5(x - 2)}{90}$$

$$= \frac{9x + 27 + 12x + 6 - 5x + 10}{90}$$

$$= \frac{16x + 43}{90}$$ ■

Don't forget that at any time, we can give visual support by graphing for a problem containing one variable, such as Example 6. In Figure 9.4, we graphed $Y_1 = \frac{x + 3}{10} + \frac{2x + 1}{15} - \frac{x - 2}{18}$ and $Y_2 = \frac{16x + 43}{90}$ on the same set of axes. Because the graphs appear to be identical, we have visual support for the answer in Example 6.

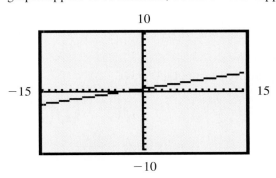

Figure 9.4

A denominator that contains variables does not create any serious difficulties; our approach remains basically the same.

EXAMPLE 7   Add $\frac{3}{2x} + \frac{5}{3y}$.

Solution

Using an LCD of $6xy$, we can proceed as follows:

$$\frac{3}{2x} + \frac{5}{3y} = \left(\frac{3}{2x}\right)\left(\frac{3y}{3y}\right) + \left(\frac{5}{3y}\right)\left(\frac{2x}{2x}\right)$$

$$= \frac{9y}{6xy} + \frac{10x}{6xy}$$

$$= \frac{9y + 10x}{6xy}$$ ■

EXAMPLE 8   Subtract $\frac{7}{12ab} - \frac{11}{15a^2}$.

Solution

We can prime-factor the numerical coefficients of the denominators to help find the LCD.

$$\left.\begin{array}{l} 12ab = 2 \cdot 2 \cdot 3 \cdot a \cdot b \\ 15a^2 = 3 \cdot 5 \cdot a^2 \end{array}\right\} \longrightarrow \text{LCD} = 2 \cdot 2 \cdot 3 \cdot 5 \cdot a^2 \cdot b = 60a^2b$$

$$\frac{7}{12ab} - \frac{11}{15a^2} = \left(\frac{7}{12ab}\right)\left(\frac{5a}{5a}\right) - \left(\frac{11}{15a^2}\right)\left(\frac{4b}{4b}\right)$$

$$= \frac{35a}{60a^2b} - \frac{44b}{60a^2b}$$

$$= \frac{35a - 44b}{60a^2b}$$

EXAMPLE 9   Add $\dfrac{x}{x-3} + \dfrac{4}{x}$.

Solution

By inspection, the LCD is $x(x - 3)$.

$$\frac{x}{x-3} + \frac{4}{x} = \left(\frac{x}{x-3}\right)\left(\frac{x}{x}\right) + \left(\frac{4}{x}\right)\left(\frac{x-3}{x-3}\right)$$

$$= \frac{x^2}{x(x-3)} + \frac{4(x-3)}{x(x-3)}$$

$$= \frac{x^2 + 4(x-3)}{x(x-3)}$$

$$= \frac{x^2 + 4x - 12}{x(x-3)} \qquad \text{or} \qquad \frac{(x+6)(x-2)}{x(x-3)}$$

EXAMPLE 10   Subtract $\dfrac{2x}{x+1} - 3$.

Solution

Using an LCD of $x + 1$ and writing 3 as $\dfrac{3}{1}$, we can proceed as follows:

$$\frac{2x}{x+1} - 3 = \frac{2x}{x+1} - \frac{3}{1}\left(\frac{x+1}{x+1}\right)$$

$$= \frac{2x}{x+1} - \frac{3(x+1)}{x+1}$$

$$= \frac{2x - 3(x+1)}{x+1}$$

$$= \frac{2x - 3x - 3}{x+1}$$

$$= \frac{-x - 3}{x+1}$$

For Problems 1–5, answer true or false.

1. The addition problem $\dfrac{2x}{x+4} + \dfrac{1}{x+4}$ is equal to $\dfrac{2x+1}{x+4}$ for all values of x except $x = -\dfrac{1}{2}$ and $x = -4$.

2. Any common denominator can be used to add rational expressions, but typically we use the least common denominator.

3. The fractions $\dfrac{2x^2}{3y}$ and $\dfrac{10x^2z}{15yz}$ are equivalent fractions.

4. The least common multiple of the denominators is always the lowest common denominator.

5. To simplify the expression $\dfrac{5}{2x-1} + \dfrac{3}{1-2x}$ we could use $2x - 1$ for the common denominator.

PROBLEM SET 9.3

For Problems 1–12, perform the indicated operations involving rational numbers. Be sure to express your answers in reduced form.

1. $\dfrac{1}{4} + \dfrac{5}{6}$ **2.** $\dfrac{3}{5} + \dfrac{1}{6}$ **3.** $\dfrac{7}{8} - \dfrac{3}{5}$

4. $\dfrac{7}{9} - \dfrac{1}{6}$ **5.** $\dfrac{6}{5} + \dfrac{1}{-4}$ **6.** $\dfrac{7}{8} + \dfrac{5}{-12}$

7. $\dfrac{8}{15} + \dfrac{3}{25}$ **8.** $\dfrac{5}{9} - \dfrac{11}{12}$ **9.** $\dfrac{1}{5} + \dfrac{5}{6} - \dfrac{7}{15}$

10. $\dfrac{2}{3} - \dfrac{7}{8} + \dfrac{1}{4}$ **11.** $\dfrac{1}{3} - \dfrac{1}{4} - \dfrac{3}{14}$ **12.** $\dfrac{5}{6} - \dfrac{7}{9} - \dfrac{3}{10}$

For Problems 13–66, add or subtract the rational expressions as indicated. Be sure to express your answers in simplest form.

13. $\dfrac{2x}{x-1} + \dfrac{4}{x-1}$ **14.** $\dfrac{3x}{2x+1} - \dfrac{5}{2x+1}$

15. $\dfrac{4a}{a+2} + \dfrac{8}{a+2}$ **16.** $\dfrac{6a}{a-3} - \dfrac{18}{a-3}$

17. $\dfrac{3(y-2)}{7y} + \dfrac{4(y-1)}{7y}$ **18.** $\dfrac{2x-1}{4x^2} + \dfrac{3(x-2)}{4x^2}$

19. $\dfrac{x-1}{2} + \dfrac{x+3}{3}$ **20.** $\dfrac{x-2}{4} + \dfrac{x+6}{5}$

21. $\dfrac{2a-1}{4} + \dfrac{3a+2}{6}$ **22.** $\dfrac{a-4}{6} + \dfrac{4a-1}{8}$

23. $\dfrac{n+2}{6} - \dfrac{n-4}{9}$ **24.** $\dfrac{2n+1}{9} - \dfrac{n+3}{12}$

25. $\dfrac{3x-1}{3} - \dfrac{5x+2}{5}$ **26.** $\dfrac{4x-3}{6} - \dfrac{8x-2}{12}$

27. $\dfrac{x-2}{5} - \dfrac{x+3}{6} + \dfrac{x+1}{15}$

28. $\dfrac{x+1}{4} + \dfrac{x-3}{6} - \dfrac{x-2}{8}$

29. $\dfrac{3}{8x} + \dfrac{7}{10x}$ **30.** $\dfrac{5}{6x} - \dfrac{3}{10x}$

31. $\dfrac{5}{7x} - \dfrac{11}{4y}$ **32.** $\dfrac{5}{12x} - \dfrac{9}{8y}$

33. $\dfrac{4}{3x} + \dfrac{5}{4y} - 1$ **34.** $\dfrac{7}{3x} - \dfrac{8}{7y} - 2$

35. $\dfrac{7}{10x^2} + \dfrac{11}{15x}$ **36.** $\dfrac{7}{12a^2} - \dfrac{5}{16a}$

37. $\dfrac{10}{7n} - \dfrac{12}{4n^2}$

38. $\dfrac{6}{8n^2} - \dfrac{3}{5n}$

39. $\dfrac{3}{n^2} - \dfrac{2}{5n} + \dfrac{4}{3}$

40. $\dfrac{1}{n^2} + \dfrac{3}{4n} - \dfrac{5}{6}$

41. $\dfrac{3}{x} - \dfrac{5}{3x^2} - \dfrac{7}{6x}$

42. $\dfrac{7}{3x^2} - \dfrac{9}{4x} - \dfrac{5}{2x}$

43. $\dfrac{6}{5t^2} - \dfrac{4}{7t^3} + \dfrac{9}{5t^3}$

44. $\dfrac{5}{7t} + \dfrac{3}{4t^2} + \dfrac{1}{14t}$

45. $\dfrac{5b}{24a^2} - \dfrac{11a}{32b}$

46. $\dfrac{9}{14x^2y} - \dfrac{4x}{7y^2}$

47. $\dfrac{7}{9xy^3} - \dfrac{4}{3x} + \dfrac{5}{2y^2}$

48. $\dfrac{7}{16a^2b} + \dfrac{3a}{20b^2}$

49. $\dfrac{2x}{x-1} + \dfrac{3}{x}$

50. $\dfrac{3x}{x-4} - \dfrac{2}{x}$

51. $\dfrac{a-2}{a} - \dfrac{3}{a+4}$

52. $\dfrac{a+1}{a} - \dfrac{2}{a+1}$

53. $\dfrac{-3}{4n+5} - \dfrac{8}{3n+5}$

54. $\dfrac{-2}{n-6} - \dfrac{6}{2n+3}$

55. $\dfrac{-1}{x+4} + \dfrac{4}{7x-1}$

56. $\dfrac{-3}{4x+3} + \dfrac{5}{2x-5}$

57. $\dfrac{7}{3x-5} - \dfrac{5}{2x+7}$

58. $\dfrac{5}{x-1} - \dfrac{3}{2x-3}$

59. $\dfrac{5}{3x-2} + \dfrac{6}{4x+5}$

60. $\dfrac{3}{2x+1} + \dfrac{2}{3x+4}$

61. $\dfrac{3x}{2x+5} + 1$

62. $2 + \dfrac{4x}{3x-1}$

63. $\dfrac{4x}{x-5} - 3$

64. $\dfrac{7x}{x+4} - 2$

65. $-1 - \dfrac{3}{2x+1}$

66. $-2 - \dfrac{5}{4x-3}$

67. Recall that the indicated quotient of a polynomial and its opposite is -1. For example, $\dfrac{x-2}{2-x}$ simplifies to -1. Keep this idea in mind as you add or subtract the following rational expressions.

  a. $\dfrac{1}{x-1} - \dfrac{x}{x-1}$ b. $\dfrac{3}{2x-3} - \dfrac{2x}{2x-3}$

  c. $\dfrac{4}{x-4} - \dfrac{x}{x-4} + 1$ d. $-1 + \dfrac{2}{x-2} - \dfrac{x}{x-2}$

68. Consider the addition problem $\dfrac{8}{x-2} + \dfrac{5}{2-x}$. Note that the denominators are opposites of each other. If the property $\dfrac{a}{-b} = -\dfrac{a}{b}$ is applied to the second fraction, we have $\dfrac{5}{2-x} = -\dfrac{5}{x-2}$. Thus we proceed as follows:

$$\dfrac{8}{x-2} + \dfrac{5}{2-x} = \dfrac{8}{x-2} - \dfrac{5}{x-2} = \dfrac{8-5}{x-2} = \dfrac{3}{x-2}$$

Use this approach to do the following problems.

  a. $\dfrac{7}{x-1} + \dfrac{2}{1-x}$ b. $\dfrac{5}{2x-1} + \dfrac{8}{1-2x}$

  c. $\dfrac{4}{a-3} - \dfrac{1}{3-a}$ d. $\dfrac{10}{a-9} - \dfrac{5}{9-a}$

  e. $\dfrac{x^2}{x-1} - \dfrac{2x-3}{1-x}$ f. $\dfrac{x^2}{x-4} - \dfrac{3x-28}{4-x}$

■ ■ ■ **Thoughts into words**

69. What is the difference between the concept of **least common multiple** and the concept of **least common denominator**?

70. A classmate tells you that she finds the least common multiple of two counting numbers by listing the multiples of each number and then choosing the smallest number that appears in both lists. Is this a correct procedure? What is the weakness of this procedure?

71. For which real numbers does $\dfrac{x}{x-3} + \dfrac{4}{x}$ equal $\dfrac{(x+6)(x-2)}{x(x-3)}$? Explain your answer.

72. Suppose that your friend does an addition problem as follows:

$$\dfrac{5}{8} + \dfrac{7}{12} = \dfrac{5(12) + 8(7)}{8(12)} = \dfrac{60+56}{96} = \dfrac{116}{96} = \dfrac{29}{24}$$

Is this answer correct? If not, what advice would you offer your friend?

Graphing calculator activities

73. Use your graphing calculator to check your answers for Problems 19–28.

74. There is another way to use the graphing calculator to check some real numbers in two one-variable algebraic expressions that we claim are equal. In Example 4 we claim that

$$\frac{x + 2}{4} + \frac{3x + 1}{3} = \frac{15x + 10}{12}$$

for all real numbers. Let's check this claim for

$x = 3$, 7, and -5. Enter $Y_1 = \dfrac{x + 2}{4} + \dfrac{3x + 1}{3}$ and $Y_2 = \dfrac{15x + 10}{12}$ as though we were going to graph them (Figure 9.5).

Figure 9.5

Return to the home screen and store $x = 3$ and evaluate Y_1 and Y_2; then store $x = 7$ and evaluate Y_1 and Y_2; finally, store $x = -5$ and evaluate Y_1 and Y_2. Part of this procedure is shown in Figure 9.6.

Figure 9.6

Use this technique to check at least three values of x for Problems 57–66.

75. Once again, let's start with the graph of $y = \dfrac{1}{x}$. Now draw rough sketches of $y = \dfrac{1}{x - 2}$, $y = \dfrac{1}{x - 4}$, and $y = \dfrac{1}{x + 2}$. Finally, use your graphing calculator to graph all four equations on the same set of axes.

76. Use your graphing calculator to graph $y = \dfrac{1}{x^3}$. Then predict the graphs of $y = \dfrac{1}{(x - 1)^3}$, $y = \dfrac{1}{(x - 4)^3}$, and $y = \dfrac{1}{(x + 2)^3}$. Finally, use your graphing calculator to graph all four equations on the same set of axes.

77. Use your graphing calculator to obtain the graph of $y = \dfrac{1}{x^2 + 1}$. Then predict the graphs of $y = \dfrac{1}{(x - 2)^2 + 1}$, $y = \dfrac{1}{(x - 4)^2 + 1}$, and $y = \dfrac{1}{(x + 3)^2 + 1}$. Finally, use your graphing calculator to check your predictions.

Answers to Concept Quiz

1. False   **2.** True   **3.** True   **4.** True   **5.** True

9.4 More on Rational Expressions and Complex Fractions

Objectives

■   Add or subtract rational expressions where some denominators can be factored.

■   Simplify complex fractions.

In this section, we expand our work with adding and subtracting rational expressions, and we discuss the process of simplifying complex fractions. Before we begin, however, this seems like an appropriate time to offer a bit of advice regarding your study of algebra. Success in algebra depends on having a good understanding of the concepts, as well as being able to perform the various computations. As for the computational work, you should adopt a carefully organized format that shows as many steps as you need in order to minimize the chances of making careless errors. Don't be eager to find shortcuts for certain computations before you have a thorough understanding of the steps involved in the process. This advice is especially appropriate at the beginning of this section.

Study Examples 1–4 very carefully. Note that the same basic procedure is followed in solving each problem:

STEP 1   Factor the denominators.

STEP 2   Find the LCD.

STEP 3   Change each fraction to an equivalent fraction that has the LCD as its denominator.

STEP 4   Combine the numerators and place over the LCD.

STEP 5   Simplify by performing the addition or subtraction.

STEP 6   Look for ways to reduce the resulting fraction.

EXAMPLE 1   Add $\dfrac{8}{x^2 - 4x} + \dfrac{2}{x}$.

Solution

$$\frac{8}{x^2 - 4x} + \frac{2}{x} = \frac{8}{x(x - 4)} + \frac{2}{x}$$ Factor the denominators.

The LCD is $x(x - 4)$.   Find the LCD.

$$= \frac{8}{x(x - 4)} + \left(\frac{2}{x}\right)\left(\frac{x - 4}{x - 4}\right)$$ Change each fraction to an equivalent fraction that has the LCD as its denominator.

$$= \frac{8 + 2(x - 4)}{x(x - 4)}$$ Combine numerators and place over the LCD.

$$= \frac{8 + 2x - 8}{x(x - 4)}$$ Simplify performing the addition or subtraction.

$$= \frac{2x}{x(x - 4)}$$

$$= \frac{2}{x - 4}$$ Reduce.   ■

E X A M P L E   2   Subtract $\dfrac{a}{a^2 - 4} - \dfrac{3}{a + 2}$.

Solution

$$\frac{a}{a^2 - 4} - \frac{3}{a + 2} = \frac{a}{(a + 2)(a - 2)} - \frac{3}{a + 2}$$ Factor the denominators.

The LCD is $(a + 2)(a - 2)$.   Find the LCD.

$$= \frac{a}{(a + 2)(a - 2)} - \left(\frac{3}{a + 2}\right)\left(\frac{a - 2}{a - 2}\right)$$ Change each fraction to an equivalent fraction that has the LCD as its denominator.

$$= \frac{a - 3(a - 2)}{(a + 2)(a - 2)}$$ Combine numerators and place over the LCD.

$$= \frac{a - 3a + 6}{(a + 2)(a - 2)}$$ Simplify performing the addition or subtraction.

$$= \frac{-2a + 6}{(a + 2)(a - 2)} \text{ or } \frac{-2(a - 3)}{(a + 2)(a - 2)}$$ ■

E X A M P L E   3   Add $\dfrac{3n}{n^2 + 6n + 5} + \dfrac{4}{n^2 - 7n - 8}$.

Solution

$$\frac{3n}{n^2 + 6n + 5} + \frac{4}{n^2 - 7n - 8}$$

$$= \frac{3n}{(n + 5)(n + 1)} + \frac{4}{(n - 8)(n + 1)}$$ Factor the denominators.

The LCD is $(n + 5)(n + 1)(n - 8)$.   Find the LCD.

$$= \left(\frac{3n}{(n + 5)(n + 1)}\right)\left(\frac{n - 8}{n - 8}\right)$$

$$+ \left(\frac{4}{(n - 8)(n + 1)}\right)\left(\frac{n + 5}{n + 5}\right)$$ Change each fraction to an equivalent fraction that has the LCD as its denominator.

$$= \frac{3n(n-8) + 4(n+5)}{(n+5)(n+1)(n-8)}$$ Combine numerators and place over the LCD.

$$= \frac{3n^2 - 24n + 4n + 20}{(n+5)(n+1)(n-8)}$$ Simplify performing the addition or subtraction.

$$= \frac{3n^2 - 20n + 20}{(n+5)(n+1)(n-8)}$$ ■

E X A M P L E   4   Perform the indicated operations.

$$\frac{2x^2}{x^4 - 1} + \frac{x}{x^2 - 1} - \frac{1}{x - 1}$$

Solution

$$\frac{2x^2}{x^4 - 1} + \frac{x}{x^2 - 1} - \frac{1}{x - 1}$$

$$= \frac{2x^2}{(x^2 + 1)(x + 1)(x - 1)} + \frac{x}{(x + 1)(x - 1)} - \frac{1}{x - 1}$$ Factor the denominators.

The LCD is $(x^2 + 1)(x + 1)(x - 1)$.   Find the LCD.

$$= \frac{2x^2}{(x^2 + 1)(x + 1)(x - 1)}$$

$$+ \left(\frac{x}{(x + 1)(x - 1)} \right) \left(\frac{x^2 + 1}{x^2 + 1} \right)$$

$$- \left(\frac{1}{x - 1} \right) \frac{(x^2 + 1)(x + 1)}{(x^2 + 1)(x + 1)}$$ Change each fraction to an equivalent fraction that has the LCD as its denominator.

$$= \frac{2x^2 + x(x^2 + 1) - (x^2 + 1)(x + 1)}{(x^2 + 1)(x + 1)(x - 1)}$$ Combine numerators and place over the LCD.

$$= \frac{2x^2 + x^3 + x - x^3 - x^2 - x - 1}{(x^2 + 1)(x + 1)(x - 1)}$$ Simplify performing the addition or subtraction.

$$= \frac{x^2 - 1}{(x^2 + 1)(x + 1)(x - 1)}$$

$$= \frac{(x + 1)(x - 1)}{(x^2 + 1)(x + 1)(x - 1)}$$

$$= \frac{1}{x^2 + 1}$$ Reduce.   ■

Example 4 contains a significant amount of algebraic computation. Let's give our confidence a boost by getting some visual support for our answer. Figure 9.7 shows the graphs of $Y_1 = \dfrac{2x^2}{x^4 - 1} + \dfrac{x}{x^2 - 1} - \dfrac{1}{x - 1}$ and $Y_2 = \dfrac{1}{x^2 + 1}$. The graphs appear to be identical, so we feel pretty good about our computational work.

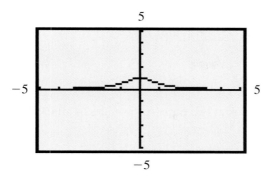

Figure 9.7

Complex Fractions

Complex fractions are fractional forms that contain rational numbers or rational expressions in the numerators and/or denominators. The following are examples of complex fractions.

$$\frac{\dfrac{4}{x}}{\dfrac{2}{xy}} \qquad \frac{\dfrac{1}{2}+\dfrac{3}{4}}{\dfrac{5}{6}-\dfrac{3}{8}} \qquad \frac{\dfrac{3}{x}+\dfrac{2}{y}}{\dfrac{5}{x}-\dfrac{6}{y^2}} \qquad \frac{\dfrac{1}{x}+\dfrac{1}{y}}{2} \qquad \frac{-3}{\dfrac{2}{x}-\dfrac{3}{y}}$$

It is often necessary to **simplify** a complex fraction. We will take each of these five examples and examine some techniques for simplifying complex fractions.

EXAMPLE 5   Simplify $\dfrac{\dfrac{4}{x}}{\dfrac{2}{xy}}$.

Solution

This type of problem is a simple division problem.

$$\frac{\dfrac{4}{x}}{\dfrac{2}{xy}} = \frac{4}{x} \div \frac{2}{xy}$$

$$= \frac{\overset{2}{\cancel{4}}}{x} \cdot \frac{xy}{2} = 2y$$

E X A M P L E   6   Simplify $\dfrac{\dfrac{1}{2} + \dfrac{3}{4}}{\dfrac{5}{6} - \dfrac{3}{8}}$.

Solution A

Let's look at two possible "avenues of attack" for such a problem.

$$\frac{\dfrac{1}{2} + \dfrac{3}{4}}{\dfrac{5}{6} - \dfrac{3}{8}} = \frac{\dfrac{2}{4} + \dfrac{3}{4}}{\dfrac{20}{24} - \dfrac{9}{24}}$$

$$= \frac{\dfrac{5}{4}}{\dfrac{11}{24}} = \frac{5}{\overset{}{4}} \cdot \frac{\overset{6}{24}}{11}$$

$$= \frac{30}{11}$$

Solution B

The LCD of all four denominators (2, 4, 6, and 8) is 24. Multiply the entire complex fraction by a form of 1, specifically $\dfrac{24}{24}$.

$$\frac{\dfrac{1}{2} + \dfrac{3}{4}}{\dfrac{5}{6} - \dfrac{3}{8}} = \left(\frac{24}{24}\right)\left(\frac{\dfrac{1}{2} + \dfrac{3}{4}}{\dfrac{5}{6} - \dfrac{3}{8}}\right)$$

$$= \frac{24\left(\dfrac{1}{2} + \dfrac{3}{4}\right)}{24\left(\dfrac{5}{6} - \dfrac{3}{8}\right)}$$

$$= \frac{24\left(\dfrac{1}{2}\right) + 24\left(\dfrac{3}{4}\right)}{24\left(\dfrac{5}{6}\right) - 24\left(\dfrac{3}{8}\right)}$$

$$= \frac{12 + 18}{20 - 9} = \frac{30}{11}$$ ∎

E X A M P L E   7   Simplify $\dfrac{\dfrac{3}{x} + \dfrac{2}{y}}{\dfrac{5}{x} - \dfrac{6}{y^2}}$.

Solution A

$$\frac{\dfrac{3}{x} + \dfrac{2}{y}}{\dfrac{5}{x} - \dfrac{6}{y^2}} = \frac{\left(\dfrac{3}{x}\right)\left(\dfrac{y}{y}\right) + \left(\dfrac{2}{y}\right)\left(\dfrac{x}{x}\right)}{\left(\dfrac{5}{x}\right)\left(\dfrac{y^2}{y^2}\right) - \left(\dfrac{6}{y^2}\right)\left(\dfrac{x}{x}\right)}$$

$$= \frac{\dfrac{3y}{xy} + \dfrac{2x}{xy}}{\dfrac{5y^2}{xy^2} - \dfrac{6x}{xy^2}}$$

$$= \frac{\dfrac{3y + 2x}{xy}}{\dfrac{5y^2 - 6x}{xy^2}}$$

$$= \frac{3y + 2x}{xy} \div \frac{5y^2 - 6x}{xy^2}$$

$$= \frac{3y + 2x}{\cancel{xy}} \cdot \frac{\overset{y}{\cancel{xy^2}}}{5y^2 - 6x}$$

$$= \frac{y(3y + 2x)}{5y^2 - 6x} \qquad \blacksquare$$

Solution B

The LCD of all four denominators (x, y, x, and y^2) is xy^2. Multiply the entire complex fraction by a form of 1, specifically $\dfrac{xy^2}{xy^2}$.

$$\frac{\dfrac{3}{x} + \dfrac{2}{y}}{\dfrac{5}{x} - \dfrac{6}{y^2}} = \left(\frac{xy^2}{xy^2}\right)\left(\frac{\dfrac{3}{x} + \dfrac{2}{y}}{\dfrac{5}{x} - \dfrac{6}{y^2}}\right)$$

$$= \frac{xy^2\left(\dfrac{3}{x} + \dfrac{2}{y}\right)}{xy^2\left(\dfrac{5}{x} - \dfrac{6}{y^2}\right)}$$

$$= \frac{xy^2\left(\dfrac{3}{x}\right) + xy^2\left(\dfrac{2}{y}\right)}{xy^2\left(\dfrac{5}{x}\right) - xy^2\left(\dfrac{6}{y^2}\right)}$$

$$= \frac{3y^2 + 2xy}{5y^2 - 6x} \quad \text{or} \quad \frac{y(3y + 2x)}{5y^2 - 6x} \qquad \blacksquare$$

Certainly either approach (Solution A or Solution B) will work with problems such as Examples 6 and 7. Examine Solution B in both examples carefully. This approach works effectively with complex fractions where the LCD of all the denominators is easy to find. (Don't be discouraged by the length of Solution B for Example 6; we were especially careful to show every step.)

EXAMPLE 8   Simplify $\dfrac{\dfrac{1}{x} + \dfrac{1}{y}}{2}$.

Solution

The number 2 can be written as $\dfrac{2}{1}$; thus the LCD of all three denominators (x, y, and 1) is xy. Therefore, let's multiply the entire complex fraction by a form of 1, specifically $\dfrac{xy}{xy}$.

$$\left(\dfrac{\dfrac{1}{x} + \dfrac{1}{y}}{\dfrac{2}{1}}\right)\left(\dfrac{xy}{xy}\right) = \dfrac{xy\left(\dfrac{1}{x}\right) + xy\left(\dfrac{1}{y}\right)}{2xy}$$

$$= \dfrac{y + x}{2xy} \qquad\blacksquare$$

EXAMPLE 9   Simplify $\dfrac{-3}{\dfrac{2}{x} - \dfrac{3}{y}}$.

Solution

$$\left(\dfrac{\dfrac{-3}{1}}{\dfrac{2}{x} - \dfrac{3}{y}}\right)\left(\dfrac{xy}{xy}\right) = \dfrac{-3(xy)}{xy\left(\dfrac{2}{x}\right) - xy\left(\dfrac{3}{y}\right)}$$

$$= \dfrac{-3xy}{2y - 3x} \qquad\blacksquare$$

Let's conclude this section with an example that has a complex fraction as part of an algebraic expression.

EXAMPLE 10   Simplify $1 - \dfrac{n}{1 - \dfrac{1}{n}}$.

Solution

First simplify the complex fraction $\dfrac{n}{1 - \dfrac{1}{n}}$ by multiplying by $\dfrac{n}{n}$.

$$\left(\dfrac{n}{1 - \dfrac{1}{n}}\right)\left(\dfrac{n}{n}\right) = \dfrac{n^2}{n - 1}$$

Now we can perform the subtraction.

$$1 - \dfrac{n^2}{n - 1} = \left(\dfrac{n - 1}{n - 1}\right)\left(\dfrac{1}{1}\right) - \dfrac{n^2}{n - 1}$$

$$= \dfrac{n - 1}{n - 1} - \dfrac{n^2}{n - 1}$$

$$= \dfrac{n - 1 - n^2}{n - 1} \quad \text{or} \quad \dfrac{-n^2 + n - 1}{n - 1}$$

CONCEPT QUIZ

For Problems 1–4, answer true or false.

1. A complex fraction can be described as a fraction within a fraction.

2. Division can simplify the complex fraction $\dfrac{\dfrac{2y}{x}}{\dfrac{6}{x^2}}$.

3. The complex fraction $\dfrac{\dfrac{3}{x - 2} + \dfrac{2}{x + 2}}{\dfrac{7x}{(x + 2)(x - 2)}}$ simplifies to $\dfrac{5x + 2}{7x}$ for all values of x except $x = 0$.

4. One method for simplifying a complex fraction is to multiply the entire fraction by a form of 1.

5. Arrange in order the following steps for adding rational expressions.
   A. Combine numerators and place over the LCD.
   B. Find the LCD.
   C. Reduce.
   D. Factor the denominators.
   E. Simplify by performing addition or subtraction.
   F. Change each fraction to an equivalent fraction that has the LCD as its denominator.

PROBLEM SET 9.4

For Problems 1–36, perform the indicated operations and express your answers in simplest form.

1. $\dfrac{2x}{x^2 + 4x} + \dfrac{5}{x}$

2. $\dfrac{3x}{x^2 - 6x} + \dfrac{4}{x}$

3. $\dfrac{4}{x^2 + 7x} - \dfrac{1}{x}$

4. $\dfrac{-10}{x^2 - 9x} - \dfrac{2}{x}$

5. $\dfrac{x}{x^2 - 1} + \dfrac{5}{x + 1}$

6. $\dfrac{2x}{x^2 - 16} + \dfrac{7}{x - 4}$

7. $\dfrac{6a + 4}{a^2 - 1} - \dfrac{5}{a - 1}$

8. $\dfrac{4a - 4}{a^2 - 4} - \dfrac{3}{a + 2}$

9. $\dfrac{2n}{n^2 - 25} - \dfrac{3}{4n + 20}$

10. $\dfrac{3n}{n^2 - 36} - \dfrac{2}{5n + 30}$

11. $\dfrac{5}{x} - \dfrac{5x - 30}{x^2 + 6x} + \dfrac{x}{x + 6}$

12. $\dfrac{3}{x + 1} + \dfrac{x + 5}{x^2 - 1} - \dfrac{3}{x - 1}$

13. $\dfrac{3}{x^2 + 9x + 14} + \dfrac{5}{2x^2 + 15x + 7}$

14. $\dfrac{6}{x^2 + 11x + 24} + \dfrac{4}{3x^2 + 13x + 12}$

15. $\dfrac{1}{a^2 - 3a - 10} - \dfrac{4}{a^2 + 4a - 45}$

16. $\dfrac{6}{a^2 - 3a - 54} - \dfrac{10}{a^2 + 5a - 6}$

17. $\dfrac{3a}{20a^2 - 11a - 3} + \dfrac{1}{12a^2 + 7a - 12}$

18. $\dfrac{2a}{6a^2 + 11a - 10} + \dfrac{a}{2a^2 - 3a - 20}$

19. $\dfrac{5}{x^2 + 3} - \dfrac{2}{x^2 + 4x - 21}$

20. $\dfrac{7}{x^2 + 1} - \dfrac{3}{x^2 + 7x - 60}$

21. $\dfrac{2}{y^2 + 6y - 16} - \dfrac{4}{y + 8} - \dfrac{3}{y - 2}$

22. $\dfrac{7}{y - 6} - \dfrac{10}{y + 12} + \dfrac{4}{y^2 + 6y - 72}$

23. $x - \dfrac{x^2}{x - 2} + \dfrac{3}{x^2 - 4}$

24. $x + \dfrac{5}{x^2 - 25} - \dfrac{x^2}{x + 5}$

25. $\dfrac{x + 3}{x + 10} + \dfrac{4x - 3}{x^2 + 8x - 20} + \dfrac{x - 1}{x - 2}$

26. $\dfrac{2x - 1}{x + 3} + \dfrac{x + 4}{x - 6} + \dfrac{3x - 1}{x^2 - 3x - 18}$

27. $\dfrac{n}{n - 6} + \dfrac{n + 3}{n + 8} + \dfrac{12n + 26}{n^2 + 2n - 48}$

28. $\dfrac{n - 1}{n + 4} + \dfrac{n}{n + 6} + \dfrac{2n + 18}{n^2 + 10n + 24}$

29. $\dfrac{4x - 3}{2x^2 + x - 1} - \dfrac{2x + 7}{3x^2 + x - 2} - \dfrac{3}{3x - 2}$

30. $\dfrac{2x + 5}{x^2 + 3x - 18} - \dfrac{3x - 1}{x^2 + 4x - 12} + \dfrac{5}{x - 2}$

31. $\dfrac{n}{n^2 + 1} + \dfrac{n^2 + 3n}{n^4 - 1} - \dfrac{1}{n - 1}$

32. $\dfrac{2n^2}{n^4 - 16} - \dfrac{n}{n^2 - 4} + \dfrac{1}{n + 2}$

33. $\dfrac{15x^2 - 10}{5x^2 - 7x + 2} - \dfrac{3x + 4}{x - 1} - \dfrac{2}{5x - 2}$

34. $\dfrac{32x + 9}{12x^2 + x - 6} - \dfrac{3}{4x + 3} - \dfrac{x + 5}{3x - 2}$

35. $\dfrac{t + 3}{3t - 1} + \dfrac{8t^2 + 8t + 2}{3t^2 - 7t + 2} - \dfrac{2t + 3}{t - 2}$

36. $\dfrac{t - 3}{2t + 1} + \dfrac{2t^2 + 19t - 46}{2t^2 - 9t - 5} - \dfrac{t + 4}{t - 5}$

For Problems 37–60, simplify each complex fraction.

37. $\dfrac{\dfrac{1}{2} - \dfrac{1}{4}}{\dfrac{5}{8} + \dfrac{3}{4}}$

38. $\dfrac{\dfrac{3}{8} + \dfrac{3}{4}}{\dfrac{5}{8} - \dfrac{7}{12}}$

39. $\dfrac{\dfrac{3}{28} - \dfrac{5}{14}}{\dfrac{5}{7} + \dfrac{1}{4}}$

40. $\dfrac{\dfrac{5}{9} + \dfrac{7}{36}}{\dfrac{3}{18} - \dfrac{5}{12}}$

41. $\dfrac{\dfrac{5}{6y}}{\dfrac{10}{3xy}}$

42. $\dfrac{\dfrac{9}{8xy^2}}{\dfrac{5}{4x^2}}$

43. $\dfrac{\dfrac{3}{x} - \dfrac{2}{y}}{\dfrac{4}{y} - \dfrac{7}{xy}}$

44. $\dfrac{\dfrac{9}{x} + \dfrac{7}{x^2}}{\dfrac{5}{y} + \dfrac{3}{y^2}}$

45. $\dfrac{\dfrac{6}{a} - \dfrac{5}{b^2}}{\dfrac{12}{a^2} + \dfrac{2}{b}}$

46. $\dfrac{\dfrac{4}{ab} - \dfrac{3}{b^2}}{\dfrac{1}{a} + \dfrac{3}{b}}$

47. $\dfrac{\dfrac{2}{x} - 3}{\dfrac{3}{y} + 4}$

48. $\dfrac{1 + \dfrac{3}{x}}{1 - \dfrac{6}{x}}$

54. $\dfrac{\dfrac{-2}{x} - \dfrac{4}{x + 2}}{\dfrac{3}{x^2 + 2x} + \dfrac{3}{x}}$

55. $\dfrac{\dfrac{2}{x - 3} - \dfrac{3}{x + 3}}{\dfrac{5}{x^2 - 9} - \dfrac{2}{x - 3}}$

49. $\dfrac{3 + \dfrac{2}{n + 4}}{5 - \dfrac{1}{n + 4}}$

50. $\dfrac{4 + \dfrac{6}{n - 1}}{7 - \dfrac{4}{n - 1}}$

51. $\dfrac{5 - \dfrac{2}{n - 3}}{4 - \dfrac{1}{n - 3}}$

56. $\dfrac{\dfrac{2}{x - y} + \dfrac{3}{x + y}}{\dfrac{5}{x + y} - \dfrac{1}{x^2 - y^2}}$

57. $\dfrac{3a}{2 - \dfrac{1}{a}} - 1$

52. $\dfrac{\dfrac{3}{n - 5} - 2}{1 - \dfrac{4}{n - 5}}$

53. $\dfrac{\dfrac{-1}{y - 2} + \dfrac{5}{x}}{\dfrac{3}{x} - \dfrac{4}{xy - 2x}}$

58. $\dfrac{a}{\dfrac{1}{a} + 4} + 1$

59. $2 - \dfrac{x}{3 - \dfrac{2}{x}}$

60. $1 + \dfrac{x}{1 + \dfrac{1}{x}}$

■■■ Thoughts into words

61. Which of the two techniques presented in the text would you use to simplify $\dfrac{\dfrac{1}{4} + \dfrac{1}{3}}{\dfrac{3}{4} - \dfrac{1}{6}}$? Which technique would you use to simplify $\dfrac{\dfrac{3}{8} - \dfrac{5}{7}}{\dfrac{7}{9} + \dfrac{6}{25}}$? Explain your choice for each problem.

62. Give a step-by-step description of how to do the following addition problem.

$$\frac{3x + 4}{8} + \frac{5x - 2}{12}$$

 ### Graphing calculator activities

63. Before doing this problem, refer to Problem 74 of Problem Set 9.3. Now check your answers for Problems 13–22, using both a graphical approach and the approach described in Problem 74 of Problem Set 9.3.

64. Again, let's start with the graph of $y = \dfrac{1}{x}$. Draw rough sketches of the graphs of $y = -\dfrac{1}{x}$, $y = \dfrac{3}{x}$, and $y = -\dfrac{5}{x}$. Then use your graphing calculator to graph all four equations on the same set of axes.

65. Let's start with the graph of $y = \dfrac{1}{x^2}$. Draw rough sketches of the graphs of $y = -\dfrac{1}{x^2}$, $y = -\dfrac{4}{x^2}$, and $y = -\dfrac{6}{x^2}$. Then use your graphing calculator to graph all four equations on the same set of axes.

66. Use your graphing calculator to graph $y = \dfrac{1}{(x - 1)^2}$. Then draw rough sketches of $y = -\dfrac{1}{(x - 1)^2}$, $y = \dfrac{3}{(x - 1)^2}$, and $y = -\dfrac{3}{(x - 1)^2}$. Finally, use your graphing calculator to graph all four equations on the same set of axes.

9.5   Equations Containing Rational Expressions

Objectives

▪   Solve rational equations.

▪   Solve rational equations that are in the form of a proportion.

▪   Solve proportion word problems.

▪   Solve word problems that involve relationships from the division process.

Equations that contain rational expressions are referred to as **rational equations.** In Chapter 3 we considered rational equations that involve only constants in the denominators. Let's briefly review our approach to solving such equations, because we will be using that same basic technique to solve any type of rational equation.

E X A M P L E   1   Solve $\dfrac{x-2}{3} + \dfrac{x+1}{4} = \dfrac{1}{6}$.

Solution

$$\frac{x-2}{3} + \frac{x+1}{4} = \frac{1}{6}$$

$$12\left(\frac{x-2}{3} + \frac{x+1}{4}\right) = 12\left(\frac{1}{6}\right) \qquad \text{Multiply both sides by 12, which is the LCD of all of the denominators.}$$

$$4(x-2) + 3(x+1) = 2$$

$$4x - 8 + 3x + 3 = 2$$

$$7x - 5 = 2$$

$$7x = 7$$

$$x = 1$$

The solution set is {1}. Check it!   ▪

Let's pause for a moment and consider a graphical analysis of the equation in Example 1. The given equation is equivalent to $\dfrac{x-2}{3} + \dfrac{x+1}{4} - \dfrac{1}{6} = 0$. Now suppose we graph the equation $y = \dfrac{x-2}{3} + \dfrac{x+1}{4} - \dfrac{1}{6}$ as in Figure 9.8. Remember that the x intercepts of a graph are found by letting $y = 0$ and solving the resulting equation for x. In other words, the x intercept of $y = \dfrac{x-2}{3} + \dfrac{x+1}{4} - \dfrac{1}{6}$ is the solution of the equation $\dfrac{x-2}{3} + \dfrac{x+1}{4} - \dfrac{1}{6} = 0$. Using the $\boxed{\text{TRACE}}$ function of the

graphing calculator, we can establish that $y = 0$ at $x = 1$. Thus our solution set of {1} in Example 1 has been verified.

    If an equation contains a variable (or variables) in one or more denominators, then we proceed in essentially the same way as in Example 1 **except that we must avoid any value of the variable that makes a denominator zero.** Consider the following examples.

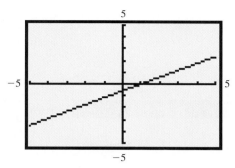

Figure 9.8

E X A M P L E   2

Solve $\dfrac{5}{n} + \dfrac{1}{2} = \dfrac{9}{n}$.

Solution

First, we need to realize that **n cannot equal zero.** (Let's indicate this restriction so that it is not forgotten!) Then we can proceed.

$$\dfrac{5}{n} + \dfrac{1}{2} = \dfrac{9}{n}, \qquad n \neq 0$$

$$2n\left(\dfrac{5}{n} + \dfrac{1}{2}\right) = 2n\left(\dfrac{9}{n}\right) \quad \text{Multiply both sides by the LCD, which is } 2n.$$

$$10 + n = 18$$

$$n = 8$$

The solution set is {8}. Check it!   ■

E X A M P L E   3

Solve $\dfrac{35-x}{x} = 7 + \dfrac{3}{x}$.

Solution

$$\dfrac{35-x}{x} = 7 + \dfrac{3}{x}, \qquad x \neq 0$$

$$x\left(\dfrac{35-x}{x}\right) = x\left(7 + \dfrac{3}{x}\right) \quad \text{Multiply both sides by } x.$$

$$35 - x = 7x + 3$$

$$32 = 8x$$

$$4 = x$$

The solution set is {4}.   ■

In Chapter 4 we introduced the concept of a proportion and used it to solve some consumer-type problems. Then in Chapter 8 we reviewed proportions and used the property "$\dfrac{a}{b} = \dfrac{c}{d}$ if and only if $ad = bc$" to solve some equations. Here again, that same property can be used to solve some rational equations that are in the form of a proportion.

E X A M P L E   4

Solve $\dfrac{3}{a - 2} = \dfrac{4}{a + 1}$.

Solution

$$\dfrac{3}{a - 2} = \dfrac{4}{a + 1}, \qquad a \neq 2 \text{ and } a \neq -1$$

$$3(a + 1) = 4(a - 2) \qquad \text{Cross products are equal.}$$

$$3a + 3 = 4a - 8$$

$$11 = a$$

The solution set is {11}. ■

The equation in Example 4 could also be solved by multiplying both sides by $(a - 2)(a + 1)$. Also keep in mind that listing the restrictions at the beginning of a problem does not replace *checking* the potential solutions. In Example 4, the solution 11 needs to be checked in the original equation.

E X A M P L E   5

Solve $\dfrac{a}{a - 2} + \dfrac{2}{3} = \dfrac{2}{a - 2}$.

Solution

$$\dfrac{a}{a - 2} + \dfrac{2}{3} = \dfrac{2}{a - 2}, \qquad a \neq 2$$

$$3(a - 2)\left(\dfrac{a}{a - 2} + \dfrac{2}{3}\right) = 3(a - 2)\left(\dfrac{2}{a - 2}\right) \qquad \begin{matrix}\text{Multiply both sides} \\ \text{by } 3(a - 2).\end{matrix}$$

$$3a + 2(a - 2) = 6$$

$$3a + 2a - 4 = 6$$

$$5a = 10$$

$$a = 2$$

Because our initial restriction was $a \neq 2$, we conclude that this equation *has no solution.* Thus the solution set is $\varnothing$. ■

REMARK: Example 5 demonstrates the importance of recognizing the restrictions that must be made to exclude division by zero.

The solution set in Example 5 is a bit unusual, so let's get some visual support. Figure 9.9 shows the graph of $Y = \dfrac{x}{x-2} + \dfrac{2}{3} - \dfrac{2}{x-2}$. It appears that the graph is a straight line parallel to the x axis; in other words, there is no x intercept. Therefore, our solution set (the null set) seems very reasonable.

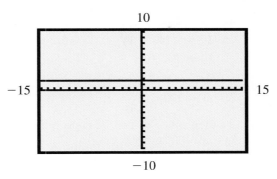

Figure 9.9

Back to Problem Solving

The ability to solve rational equations broadens our base for solving word problems. We are now ready to tackle some word problems that translate into rational equations.

P R O B L E M   1

A sum of $750 is to be divided between two people in the ratio of 2 to 3. How much does each person receive?

Solution

Let d represent the amount of money that one person receives. Then $750 - d$ represents the amount for the other person.

$$\frac{d}{750 - d} = \frac{2}{3}, \qquad d \neq 750$$

$$3d = 2(750 - d) \qquad \text{Cross-multiplication property}$$

$$3d = 1500 - 2d$$

$$5d = 1500$$

$$d = 300$$

If $d = 300$, then $750 - d$ equals 450. Therefore, one person receives $300 and the other person receives $450. ■

P R O B L E M   2

One angle of a triangle has a measure of $40°$, and the measures of the other two angles are in the ratio of 5 to 2. Find the measures of the other two angles.

Solution

The sum of the measures of the other two angles is $180° - 40° = 140°$. Let y represent the measure of one angle. Then $140 - y$ represents the measure of the other angle.

$$\frac{y}{140 - y} = \frac{5}{2}, \qquad y \neq 140$$

$$2y = 5(140 - y) \qquad \text{Cross products are equal.}$$

$$2y = 700 - 5y$$

$$7y = 700$$

$$y = 100$$

If $y = 100$, then $140 - y = 40$. Therefore the measures of the other two angles of the triangle are $100°$ and $40°$.   ■

P R O B L E M   3

On a certain map $1\frac{1}{2}$ inches represents 25 miles. If two cities are $5\frac{1}{4}$ inches apart on the map, find the number of miles between the cities (see Figure 9.10).

Solution

Let m represent the number of miles between the two cities. Set up the following proportion and solve the problem.

$$\frac{1\frac{1}{2}}{25} = \frac{5\frac{1}{4}}{m}, \qquad m \neq 0$$

$$\frac{\frac{3}{2}}{25} = \frac{\frac{21}{4}}{m}$$

$$\frac{3}{2}m = 25\left(\frac{21}{4}\right) \qquad \text{Cross-multiplication property}$$

$$\frac{2}{3}\left(\frac{3}{2}m\right) = \frac{2}{3}(25)\left(\frac{\overset{7}{\cancel{21}}}{\underset{2}{\cancel{4}}}\right) \qquad \text{Multiply both sides by } \frac{2}{3}.$$

$$m = \frac{175}{2}$$

$$= 87\frac{1}{2}$$

The distance between the two cities is $87\frac{1}{2}$ miles.   ■

Newton

Kenmore

East Islip

$5\frac{1}{4}$ inches

Islip

Windham

Descartes

Figure 9.10

PROBLEM 4

The sum of two numbers is 52. If the larger is divided by the smaller, the quotient is 9 and the remainder is 2. Find the numbers.

Solution

Let n represent the smaller number. Then $52 - n$ represents the larger number. Let's use the relationship we discussed previously as a guideline and proceed as follows:

$$\frac{\text{Dividend}}{\text{Divisor}} = \text{Quotient} + \frac{\text{Remainder}}{\text{Divisor}}$$

$$\frac{52 - n}{n} = 9 + \frac{2}{n}, \qquad n \neq 0$$

$$n\left(\frac{52 - n}{n}\right) = n\left(9 + \frac{2}{n}\right)$$

$$52 - n = 9n + 2$$

$$50 = 10n$$

$$5 = n$$

If $n = 5$, then $52 - n$ equals 47. The numbers are 5 and 47.

CONCEPT QUIZ

For Problems 1–3, answer true or false.

1. In solving rational equations, any value of the variable that makes a denominator zero cannot be a solution of the equation.

2. One method to solve rational equations is to multiply both sides of the equation by the lowest common denominator of the fractions in the equation.

3. In solving a rational equation that is a proportion, cross products can be set equal to each other.

4. Identify the following equations as a proportion or not a proportion.

   A. $\dfrac{2x}{x + 1} + x = \dfrac{7}{x + 1}$ **B.** $\dfrac{x - 8}{2x + 5} = \dfrac{7}{9}$ **C.** $5 + \dfrac{2x}{x + 6} = \dfrac{x - 3}{x + 4}$

5. Select all the equations that could represent the following problem. John bought 3 bottles of energy drink for $5.07. If the price remains the same, what will 8 bottles of energy drink cost?

   A. $\dfrac{3}{5.07} = \dfrac{x}{8}$ **B.** $\dfrac{5.07}{8} = \dfrac{x}{3}$ **C.** $\dfrac{3}{8} = \dfrac{5.07}{x}$ **D.** $\dfrac{5.07}{3} = \dfrac{x}{8}$

PROBLEM SET 9.5

For Problems 1–42, solve each equation.

1. $\dfrac{x+1}{4} + \dfrac{x-2}{6} = \dfrac{3}{4}$

2. $\dfrac{x+2}{5} + \dfrac{x-1}{6} = \dfrac{3}{5}$

3. $\dfrac{x+3}{2} - \dfrac{x-4}{7} = 1$

4. $\dfrac{x+4}{3} - \dfrac{x-5}{9} = 1$

5. $\dfrac{5}{n} + \dfrac{1}{3} = \dfrac{7}{n}$

6. $\dfrac{3}{n} + \dfrac{1}{6} = \dfrac{11}{3n}$

7. $\dfrac{7}{2x} + \dfrac{3}{5} = \dfrac{2}{3x}$

8. $\dfrac{9}{4x} + \dfrac{1}{3} = \dfrac{5}{2x}$

9. $\dfrac{3}{4x} + \dfrac{5}{6} = \dfrac{4}{3x}$

10. $\dfrac{5}{7x} - \dfrac{5}{6} = \dfrac{1}{6x}$

11. $\dfrac{47-n}{n} = 8 + \dfrac{2}{n}$

12. $\dfrac{45-n}{n} = 6 + \dfrac{3}{n}$

13. $\dfrac{n}{65-n} = 8 + \dfrac{2}{65-n}$

14. $\dfrac{n}{70-n} = 7 + \dfrac{6}{70-n}$

15. $n + \dfrac{1}{n} = \dfrac{17}{4}$

16. $n + \dfrac{1}{n} = \dfrac{37}{6}$

17. $n - \dfrac{2}{n} = \dfrac{23}{5}$

18. $n - \dfrac{3}{n} = \dfrac{26}{3}$

19. $\dfrac{5}{7x-3} = \dfrac{3}{4x-5}$

20. $\dfrac{3}{2x-1} = \dfrac{5}{3x+2}$

21. $\dfrac{-2}{x-5} = \dfrac{1}{x+9}$

22. $\dfrac{5}{2a-1} = \dfrac{-6}{3a+2}$

23. $\dfrac{x}{x+1} - 2 = \dfrac{3}{x-3}$

24. $\dfrac{x}{x-2} + 1 = \dfrac{8}{x-1}$

25. $\dfrac{a}{a+5} - 2 = \dfrac{3a}{a+5}$

26. $\dfrac{a}{a-3} - \dfrac{3}{2} = \dfrac{3}{a-3}$

27. $\dfrac{5}{x+6} = \dfrac{6}{x-3}$

28. $\dfrac{3}{x-1} = \dfrac{4}{x+2}$

29. $\dfrac{3x-7}{10} = \dfrac{2}{x}$

30. $\dfrac{x}{-4} = \dfrac{3}{12x-25}$

31. $\dfrac{x}{x-6} - 3 = \dfrac{6}{x-6}$

32. $\dfrac{x}{x+1} + 3 = \dfrac{4}{x+1}$

33. $\dfrac{3s}{s+2} + 1 = \dfrac{35}{2(3s+1)}$

34. $\dfrac{s}{2s-1} - 3 = \dfrac{-32}{3(s+5)}$

35. $2 - \dfrac{3x}{x-4} = \dfrac{14}{x+7}$

36. $-1 + \dfrac{2x}{x+3} = \dfrac{-4}{x+4}$

37. $\dfrac{n+6}{27} = \dfrac{1}{n}$

38. $\dfrac{n}{5} = \dfrac{10}{n-5}$

39. $\dfrac{3n}{n-1} - \dfrac{1}{3} = \dfrac{-40}{3n-18}$

40. $\dfrac{n}{n+1} + \dfrac{1}{2} = \dfrac{-2}{n+2}$

41. $\dfrac{-3}{4x+5} = \dfrac{2}{5x-7}$

42. $\dfrac{7}{x+4} = \dfrac{3}{x-8}$

For Problems 43–58, set up an algebraic equation and solve each problem.

43. A sum of $1750 is to be divided between two people in the ratio of 3 to 4. How much does each person receive?

44. A home decor center was given a drawing on which a rectangular room measured $3\dfrac{1}{2}$ inches by $5\dfrac{3}{4}$ inches. The scale of the drawing was 1 inch represented 5 feet. Find the dimensions of the room.

45. One angle of a triangle has a measure of 60° and the measures of the other two angles are in a ratio of 2 to 3. Find the measures of the other two angles.

46. The measure of angle A of a triangle is 20° more than the measure of angle B. The measures of the angles are in a ratio of 3 to 4. Find the measure of each angle.

47. The ratio of the complement of an angle to its supplement is 1 to 4. Find the measure of the angle.

48. The sum of two numbers is 80. If the larger is divided by the smaller, the quotient is 7 and the remainder is 8. Find the numbers.

49. If a home valued at $150,000 is assessed $1900 in real estate taxes, then how much, at the same rate, are the taxes on a home valued at $180,000?

50. The ratio of male students to female students at a certain university is 5 to 7. If there is a total of 16,200 students, find the number of male students and the number of female students.

51. Suppose that, together, Laura and Tammy sold $210.00 worth of candy for the annual school fair. If the ratio of Tammy's sales to Laura's sales was 4 to 3, how much did each sell?

52. The total value of a house and a lot is $168,000. If the ratio of the value of the house to the value of the lot is 7 to 1, find the value of the house.

53. The sum of two numbers is 90. If the larger is divided by the smaller, the quotient is 10 and the remainder is 2. Find the numbers.

54. What number must be added to the numerator and denominator of $\dfrac{2}{5}$ to produce a rational number that is equivalent to $\dfrac{7}{8}$?

55. A 20-foot board is to be cut into two pieces whose lengths are in the ratio of 7 to 3. Find the lengths of the two pieces.

56. An inheritance of $300,000 is to be divided between a son and the local heart fund in the ratio of 3 to 1. How much money will the son receive?

57. Suppose that in a certain precinct, 21,150 people voted in the last presidential election. If the ratio of female voters to male voters was 3 to 2, how many females and how many males voted?

58. The perimeter of a rectangle is 114 centimeters. If the ratio of its width to its length is 7 to 12, find the dimensions of the rectangle.

■ ■ ■ Thoughts into words

59. How could you do Problem 55 without using algebra?

60. Now do Problem 57 using the same approach that you used in Problem 59. What difficulties do you encounter?

61. How can you tell by inspection that the equation $\dfrac{x}{x + 2} = \dfrac{-2}{x + 2}$ has no solution?

62. How would you help someone solve the equation $\dfrac{3}{x} - \dfrac{4}{x} = \dfrac{-1}{x}$?

 ### Graphing calculator activities

63. Use your graphing calculator and supply a partial check for Problems 25–34.

64. Use your graphing calculator to help solve each of the following equations. Be sure to check your answers.

a. $\dfrac{1}{6} + \dfrac{1}{x} = \dfrac{5}{18}$

b. $\dfrac{1}{50} + \dfrac{1}{40} = \dfrac{1}{x}$

c. $\dfrac{2050}{x + 358} = \dfrac{260}{x}$

d. $\dfrac{280}{x} = \dfrac{300}{x + 2} + 20$

e. $\dfrac{x}{2x - 8} + \dfrac{16}{x^2 - 16} = \dfrac{1}{2}$

f. $\dfrac{3}{x - 5} - \dfrac{2}{2x + 1} = \dfrac{x + 3}{2x^2 - 9x - 5}$

g. $2 + \dfrac{4}{x - 2} = \dfrac{8}{x^2 - 2x}$

Answers to Concept Quiz

1. True   **2.** True   **3.** True   **4. A.** Not a proportion   **B.** Proportion   **C.** Not a proportion   **5.** C, D

9.6 More on Rational Equations and Applications

Objectives

■ Solve rational equations with denominators that are factorable.

■ Solve formulas that are in the form of rational equations.

■ Solve word problems that involve uniform-motion rate-time relationships.

■ Solve word problems that involve rate-time relationships.

Let's begin this section by considering a few more rational equations. We will continue to solve them using the same basic techniques as in the previous section. That is, we will multiply both sides of the equation by the least common denominator of all of the denominators in the equation, imposing the restrictions necessary to avoid division by zero. Some of the denominators in these problems will require factoring before we can determine a least common denominator.

EXAMPLE 1 Solve $\dfrac{x}{2x-8} + \dfrac{16}{x^2-16} = \dfrac{1}{2}$.

Solution

$$\frac{x}{2x-8} + \frac{16}{x^2-16} = \frac{1}{2}$$

$$\frac{x}{2(x-4)} + \frac{16}{(x+4)(x-4)} = \frac{1}{2}, \qquad x \neq 4 \text{ and } x \neq -4$$

$$2(x-4)(x+4)\left(\frac{x}{2(x-4)} + \frac{16}{(x+4)(x-4)}\right) = 2(x+4)(x-4)\left(\frac{1}{2}\right) \qquad \begin{array}{l}\text{Multiply both}\\\text{sides by the LCD,}\\2(x-4)(x+4).\end{array}$$

$$x(x+4) + 2(16) = (x+4)(x-4)$$

$$x^2 + 4x + 32 = x^2 - 16$$

$$4x = -48$$

$$x = -12$$

The solution set is $\{-12\}$. Perhaps you should check it! ■

In Example 1, note that the restrictions were not indicated until the denominators were expressed in factored form. It is usually easier to determine the necessary restrictions at this step.

EXAMPLE 2 Solve $\dfrac{3}{n-5} - \dfrac{2}{2n+1} = \dfrac{n+3}{2n^2-9n-5}$.

Solution

$$\frac{3}{n-5} - \frac{2}{2n+1} = \frac{n+3}{2n^2-9n-5}$$

$$\frac{3}{n-5} - \frac{2}{2n+1} = \frac{n+3}{(2n+1)(n-5)}, \qquad n \neq -\frac{1}{2} \text{ and } n \neq 5$$

$$(2n+1)(n-5)\left(\frac{3}{n-5} - \frac{2}{2n+1}\right) = (2n+1)(n-5)\left(\frac{n+3}{(2n+1)(n-5)}\right) \quad \begin{array}{l}\text{Multiply both} \\ \text{sides by the LCD,} \\ (2n+1)(n-5).\end{array}$$

$$3(2n+1) - 2(n-5) = n+3$$

$$6n + 3 - 2n + 10 = n+3$$

$$4n + 13 = n+3$$

$$3n = -10$$

$$n = -\frac{10}{3}$$

The solution set is $\left\{-\dfrac{10}{3}\right\}$. ■

EXAMPLE 3   Solve $2 + \dfrac{4}{x-2} = \dfrac{8}{x^2 - 2x}$.

Solution

$$2 + \frac{4}{x-2} = \frac{8}{x^2 - 2x}$$

$$2 + \frac{4}{x-2} = \frac{8}{x(x-2)}, \qquad x \neq 0 \text{ and } x \neq 2$$

$$x(x-2)\left(2 + \frac{4}{x-2}\right) = x(x-2)\left(\frac{8}{x(x-2)}\right) \quad \begin{array}{l}\text{Multiply both sides} \\ \text{by the LCD, } x(x-2).\end{array}$$

$$2x(x-2) + 4x = 8$$

$$2x^2 - 4x + 4x = 8$$

$$2x^2 = 8$$

$$x^2 = 4$$

$$x^2 - 4 = 0$$

$$(x+2)(x-2) = 0$$

$$x + 2 = 0 \qquad \text{or} \qquad x - 2 = 0$$

$$x = -2 \qquad \text{or} \qquad x = 2$$

Because our initial restriction indicated that $x \neq 2$, the *only solution* is -2. Thus the solution set is $\{-2\}$. ■

In Section 4.3, we discussed using the properties of equality to change the form of various formulas. For example, we considered the simple interest formula $A = P + Prt$ and changed its form by solving for P as follows:

$$A = P + Prt$$

$$A = P(1 + rt)$$

$$\frac{A}{1 + rt} = P \qquad \text{Multiply both sides by } \frac{1}{1 + rt}.$$

If the formula is in the form of a rational equation, then the techniques of these last two sections are applicable. Consider the following example.

EXAMPLE 4

If the original cost of some business property is C dollars and it is depreciated linearly over N years, then its value, V, at the end of T years is given by

$$V = C\left(1 - \frac{T}{N}\right)$$

Solve this formula for N in terms of V, C, and T.

Solution

$$V = C\left(1 - \frac{T}{N}\right)$$

$$V = C - \frac{CT}{N}$$

$$N(V) = N\left(C - \frac{CT}{N}\right) \qquad \text{Multiply both sides by } N.$$

$$NV = NC - CT$$

$$NV - NC = -CT$$

$$N(V - C) = -CT$$

$$N = \frac{-CT}{V - C}$$

$$N = -\frac{CT}{V - C}$$

Problem Solving

In Chapter 4 we solved some uniform-motion problems. The formula $d = rt$ was used in the analysis of these problems, and we used guidelines that involved distance relationships. Now let's consider some uniform-motion problems wherein guidelines that involve either times or rates are appropriate. These problems will generate rational equations to solve.

PROBLEM 1

An airplane travels 2050 miles in the same time that a car travels 260 miles. If the rate of the plane is 358 miles per hour greater than the rate of the car, find the rate of each.

Solution

Let r represent the rate of the car. Then $r + 358$ represents the rate of the plane. The fact that the times are equal can be a guideline. Remember from the basic formula, $d = rt$, that $t = \frac{d}{r}$.

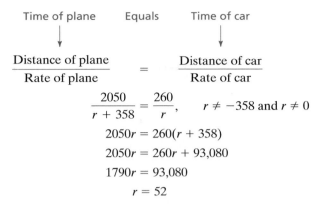

$$\frac{2050}{r + 358} = \frac{260}{r}, \qquad r \neq -358 \text{ and } r \neq 0$$

$$2050r = 260(r + 358)$$

$$2050r = 260r + 93{,}080$$

$$1790r = 93{,}080$$

$$r = 52$$

If $r = 52$, then $r + 358$ equals 410. Thus the rate of the car is 52 miles per hour, and the rate of the plane is 410 miles per hour.   ◼

PROBLEM 2

It takes a freight train 2 hours longer to travel 300 miles than it takes an express train to travel 280 miles. The rate of the express train is 20 miles per hour greater than the rate of the freight train. Find the times and rates of both trains.

Solution

Let t represent the time of the express train. Then $t + 2$ represents the time of the freight train. Let's record the information of this problem in a table.

	Distance	Time	Rate $= \dfrac{\text{distance}}{\text{time}}$
Express train	280	t	$\dfrac{280}{t}$
Freight train	300	$t + 2$	$\dfrac{300}{t + 2}$

The fact that the rate of the express train is 20 miles per hour greater than the rate of the freight train can be a guideline.

Rate of express   Equals   Rate of freight train plus 20

$$\frac{280}{t} = \frac{300}{t + 2} + 20, \qquad t \neq 0 \text{ and } t \neq -2$$

$$t(t + 2)\left(\frac{280}{t}\right) = t(t + 2)\left(\frac{300}{t + 2} + 20\right)$$

$$280(t + 2) = 300t + 20t(t + 2)$$

$$280t + 560 = 300t + 20t^2 + 40t$$

$$280t + 560 = 340t + 20t^2$$

$$0 = 20t^2 + 60t - 560$$
$$0 = t^2 + 3t - 28$$
$$0 = (t + 7)(t - 4)$$
$$t + 7 = 0 \qquad \text{or} \qquad t - 4 = 0$$
$$t = -7 \qquad \text{or} \qquad t = 4$$

The negative solution must be discarded, so the time of the express train (t) is 4 hours, and the time of the freight train ($t + 2$) is 6 hours. The rate of the express train $\left(\dfrac{280}{t}\right)$ is $\dfrac{280}{4} = 70$ miles per hour, and the rate of the freight train $\left(\dfrac{300}{t + 2}\right)$ is $\dfrac{300}{6} = 50$ miles per hour.   ■

REMARK: Note that to solve Problem 1 we went directly to a guideline without the use of a table, but for Problem 2 we used a table. Again, remember that this is a personal preference; we are merely introducing you to a variety of techniques.

Uniform motion problems are a special case of a larger group of problems we refer to as **rate-time problems.** For example, if a certain machine can produce 150 items in 10 minutes, then we say that the machine is producing at a rate of $\dfrac{150}{10} = 15$ items per minute. Likewise, if a person can do a certain job in 3 hours, then, assuming a constant rate of work, we say that the person is working at a rate of $\dfrac{1}{3}$ of the job per hour. In general, if Q is the quantity of something done in t units of time, then the rate, r, is given by $r = \dfrac{Q}{t}$. We state the rate in terms of *so much quantity per unit of time.* (In uniform motion problems the "quantity" is distance.) Let's consider some examples of rate-time problems.

P R O B L E M   3

If Jim can mow a lawn in 50 minutes and his son, Todd, can mow the same lawn in 40 minutes, how long will it take them to mow the lawn if they work together?

Solution

Jim's rate is $\dfrac{1}{50}$ of the lawn per minute and Todd's rate is $\dfrac{1}{40}$ of the lawn per minute.

If we let m represent the number of minutes that they work together, then $\dfrac{1}{m}$ represents their rate when working together. Therefore, because the sum of the individual rates must equal the rate working together, we can set up and solve the following equation.

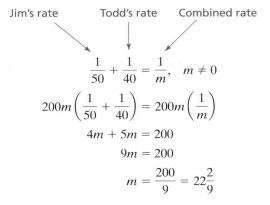

$$\frac{1}{50} + \frac{1}{40} = \frac{1}{m}, \quad m \neq 0$$

$$200m\left(\frac{1}{50} + \frac{1}{40}\right) = 200m\left(\frac{1}{m}\right)$$

$$4m + 5m = 200$$

$$9m = 200$$

$$m = \frac{200}{9} = 22\frac{2}{9}$$

It should take them $22\frac{2}{9}$ minutes.   ■

PROBLEM 4   Working together, Lucia and Kate can type a term paper in $3\frac{3}{5}$ hours. Lucia can type the paper by herself in 6 hours. How long would it take Kate to type the paper by herself?

Solution

Their rate working together is $\dfrac{1}{3\frac{3}{5}} = \dfrac{1}{\frac{18}{5}} = \dfrac{5}{18}$ of the job per hour, and Lucia's rate is $\dfrac{1}{6}$ of the job per hour. If we let h represent the number of hours that it would take Kate by herself, then her rate is $\dfrac{1}{h}$ of the job per hour. Thus we have

Lucia's rate   Kate's rate   Combined rate

$$\frac{1}{6} \quad + \quad \frac{1}{h} \quad = \quad \frac{5}{18}, \quad h \neq 0$$

Solving this equation yields

$$18h\left(\frac{1}{6} + \frac{1}{h}\right) = 18h\left(\frac{5}{18}\right)$$

$$3h + 18 = 5h$$

$$18 = 2h$$

$$9 = h$$

It would take Kate 9 hours to type the paper by herself.   ■

Our final example of this section illustrates another approach that some people find meaningful for rate-time problems. For this approach, think in terms of fractional parts of the job. For example, if a person can do a certain job in 5 hours, then at the end of 2 hours, he or she has done $\frac{2}{5}$ of the job. (Again, assume a constant rate of work.) At the end of 4 hours, he or she has finished $\frac{4}{5}$ of the job; and, in general, at the end of h hours, he or she has done $\frac{h}{5}$ of the job. Let's see how this works in a problem.

PROBLEM 5

It takes Pat 12 hours to complete a task. After he had been working for 3 hours, he was joined by his brother, Mike, and together they finished the task in 5 hours. How long would it take Mike to do the job by himself?

Solution

Let h represent the number of hours that it would take Mike by himself. The fractional part of the job that Pat does equals his working rate times his time. Because it takes Pat 12 hours to do the entire job, his working rate is $\frac{1}{12}$. He works for 8 hours (3 hours before Mike and then 5 hours with Mike). Therefore, Pat's part of the job is $\frac{1}{12}(8) = \frac{8}{12}$. The fractional part of the job that Mike does equals his working rate times his time. Because h represents Mike's time to do the entire job, his working rate is $\frac{1}{h}$. He works for 5 hours. Therefore, Mike's part of the job is $\frac{1}{h}(5) = \frac{5}{h}$. Adding the two fractional parts together results in 1 entire job being done. Let's also show this information in chart form and set up our guideline. Then we can set up and solve the equation.

	Time to do entire job	Working rate	Time working	Fractional part of the job done
Pat	12	$\frac{1}{12}$	8	$\frac{8}{12}$
Mike	h	$\frac{1}{h}$	5	$\frac{5}{h}$

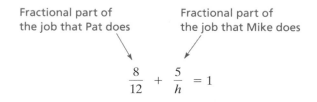

Fractional part of the job that Pat does

Fractional part of the job that Mike does

$$\frac{8}{12} + \frac{5}{h} = 1$$

$$12h\left(\frac{8}{12} + \frac{5}{h}\right) = 12h(1)$$

$$12h\left(\frac{8}{12}\right) + 12h\left(\frac{5}{h}\right) = 12h$$

$$8h + 60 = 12h$$

$$60 = 4h$$

$$15 = h$$

It would take Mike 15 hours to do the entire job by himself.

CONCEPT QUIZ

For Problems 1–5, answer true or false.

1. Assuming uniform motion, the rate at which a car travels is equal to the time traveled divided by the distance traveled.

2. If a worker can lay 640 square feet of tile in 8 hours, we can say his rate of work is 80 square feet per hour.

3. If a person can complete 2 jobs in 5 hours, then the person is working at the rate of $\frac{5}{2}$ of the job per hour.

4. In a time-rate problem involving two workers, the sum of their individual rates must equal the rate working together.

5. If a person works at the rate of $\frac{2}{15}$ of the job per hour, then at the end of 3 hours the job would be $\frac{6}{15}$ completed.

PROBLEM SET 9.6

For Problems 1–30, solve each equation.

1. $\dfrac{x}{4x - 4} + \dfrac{5}{x^2 - 1} = \dfrac{1}{4}$

2. $\dfrac{x}{3x - 6} + \dfrac{4}{x^2 - 4} = \dfrac{1}{3}$

3. $3 + \dfrac{6}{t - 3} = \dfrac{6}{t^2 - 3t}$

4. $2 + \dfrac{4}{t - 1} = \dfrac{4}{t^2 - t}$

5. $\dfrac{3}{n - 5} + \dfrac{4}{n + 7} = \dfrac{2n + 11}{n^2 + 2n - 35}$

6. $\dfrac{2}{n + 3} + \dfrac{3}{n - 4} = \dfrac{2n - 1}{n^2 - n - 12}$

7. $\dfrac{5x}{2x + 6} - \dfrac{4}{x^2 - 9} = \dfrac{5}{2}$

8. $\dfrac{3x}{5x + 5} - \dfrac{2}{x^2 - 1} = \dfrac{3}{5}$

9. $1 + \dfrac{1}{n - 1} = \dfrac{1}{n^2 - n}$

10. $3 + \dfrac{9}{n - 3} = \dfrac{27}{n^2 - 3n}$

11. $\dfrac{2}{n - 2} - \dfrac{n}{n + 5} = \dfrac{10n + 15}{n^2 + 3n - 10}$

12. $\dfrac{n}{n + 3} + \dfrac{1}{n - 4} = \dfrac{11 - n}{n^2 - n - 12}$

13. $\dfrac{2}{2x - 3} - \dfrac{2}{10x^2 - 13x - 3} = \dfrac{x}{5x + 1}$

14. $\dfrac{1}{3x + 4} + \dfrac{6}{6x^2 + 5x - 4} = \dfrac{x}{2x - 1}$

15. $\dfrac{2x}{x + 3} - \dfrac{3}{x - 6} = \dfrac{29}{x^2 - 3x - 18}$

16. $\dfrac{x}{x - 4} - \dfrac{2}{x + 8} = \dfrac{63}{x^2 + 4x - 32}$

17. $\dfrac{a}{a - 5} + \dfrac{2}{a - 6} = \dfrac{2}{a^2 - 11a + 30}$

18. $\dfrac{a}{a + 2} + \dfrac{3}{a + 4} = \dfrac{14}{a^2 + 6a + 8}$

19. $\dfrac{-1}{2x - 5} + \dfrac{2x - 4}{4x^2 - 25} = \dfrac{5}{6x + 15}$

20. $\dfrac{-2}{3x + 2} + \dfrac{x - 1}{9x^2 - 4} = \dfrac{3}{12x - 8}$

21. $\dfrac{7y + 2}{12y^2 + 11y - 15} - \dfrac{1}{3y + 5} = \dfrac{2}{4y - 3}$

22. $\dfrac{5y - 4}{6y^2 + y - 12} - \dfrac{2}{2y + 3} = \dfrac{5}{3y - 4}$

23. $\dfrac{2n}{6n^2 + 7n - 3} - \dfrac{n - 3}{3n^2 + 11n - 4} = \dfrac{5}{2n^2 + 11n + 12}$

24. $\dfrac{x + 1}{2x^2 + 7x - 4} - \dfrac{x}{2x^2 - 7x + 3} = \dfrac{1}{x^2 + x - 12}$

25. $\dfrac{1}{2x^2 - x - 1} + \dfrac{3}{2x^2 + x} = \dfrac{2}{x^2 - 1}$

26. $\dfrac{2}{n^2 + 4n} + \dfrac{3}{n^2 - 3n - 28} = \dfrac{5}{n^2 - 6n - 7}$

27. $\dfrac{x + 1}{x^3 - 9x} - \dfrac{1}{2x^2 + x - 21} = \dfrac{1}{2x^2 + 13x + 21}$

28. $\dfrac{x}{2x^2 + 5x} - \dfrac{x}{2x^2 + 7x + 5} = \dfrac{2}{x^2 + x}$

29. $\dfrac{4t}{4t^2 - t - 3} + \dfrac{2 - 3t}{3t^2 - t - 2} = \dfrac{1}{12t^2 + 17t + 6}$

30. $\dfrac{2t}{2t^2 + 9t + 10} + \dfrac{1 - 3t}{3t^2 + 4t - 4} = \dfrac{4}{6t^2 + 11t - 10}$

For Problems 31–44, solve each equation for the indicated variable.

31. $y = \dfrac{5}{6}x + \dfrac{2}{9}$ for x **32.** $y = \dfrac{3}{4}x - \dfrac{2}{3}$ for x

33. $\dfrac{-2}{x - 4} = \dfrac{5}{y - 1}$ for y **34.** $\dfrac{7}{y - 3} = \dfrac{3}{x + 1}$ for y

35. $I = \dfrac{100M}{C}$ for M **36.** $V = C\left(1 - \dfrac{T}{N}\right)$ for T

37. $\dfrac{R}{S} = \dfrac{T}{S + T}$ for R **38.** $\dfrac{1}{R} = \dfrac{1}{S} + \dfrac{1}{T}$ for R

39. $\dfrac{y - 1}{x - 3} = \dfrac{b - 1}{a - 3}$ for y **40.** $y = -\dfrac{a}{b}x + \dfrac{c}{d}$ for x

41. $\dfrac{x}{a} + \dfrac{y}{b} = 1$ for y **42.** $\dfrac{y - b}{x} = m$ for y

43. $\dfrac{y - 1}{x + 6} = \dfrac{-2}{3}$ for y **44.** $\dfrac{y + 5}{x - 2} = \dfrac{3}{7}$ for y

Set up an equation and solve each of the following problems.

45. Kent drives his Mazda 270 miles in the same time that Dave drives his Nissan 250 miles. If Kent averages 4 miles per hour faster than Dave, find their rates.

46. Suppose that Wendy rides her bicycle 30 miles in the same time that it takes Kim to ride her bicycle 20 miles. If Wendy rides 5 miles per hour faster than Kim, find the rate of each.

47. An inlet pipe can fill a tank (see Figure 9.11) in 10 minutes. A drain can empty the tank in 12 minutes. If the tank is empty and both the pipe and the drain are open, how long will it take before the tank overflows?

Figure 9.11

48. Barry can do a certain job in 3 hours, whereas it takes Sanchez 5 hours to do the same job. How long would it take them to do the job working together?

49. Connie can type 600 words in 5 minutes less than it takes Katie to type 600 words. If Connie types at a rate of 20 words per minute faster than Katie types, find the typing rate of each woman.

50. Ryan can mow a lawn in 1 hour, and his son, Malik, can mow the same lawn in 50 minutes. One day Malik started mowing the lawn by himself and worked for 30 minutes. Then Ryan joined him and they finished the lawn. How long did it take them to finish mowing the lawn after Ryan started to help?

51. Plane A can travel 1400 miles in 1 hour less time than it takes plane B to travel 2000 miles. The rate of plane B is 50 miles per hour greater than the rate of plane A. Find the times and rates of both planes.

52. To travel 60 miles, it takes Sue, riding a moped, 2 hours less time than it takes Doreen to travel 50 miles riding a bicycle. Sue travels 10 miles per hour faster than Doreen. Find the times and rates of both girls.

53. It takes Amy twice as long to clean the office as it does Nancy. How long would it take each girl to clean the office by herself if they can clean the office together in 40 minutes?

54. If two inlet pipes are both open, they can fill a pool in 1 hour and 12 minutes. One of the pipes can fill the pool by itself in 2 hours. How long would it take the other pipe to fill the pool by itself?

55. Rod agreed to mow a vacant lot for $12. It took him an hour longer than what he had anticipated, so he earned $1 per hour less than he originally calculated. How long had he anticipated that it would take him to mow the lot?

56. Last week Al bought some golf balls for $20. The next day they were on sale for $.50 per ball less, and he bought $22.50 worth of balls. If he purchased 5 more balls on the second day than he did on the first day, how many did he buy each day and at what price per ball?

57. Debbie rode her bicycle out into the country for a distance of 24 miles. On the way back, she took a much shorter route of 12 miles and made the return trip in one-half hour less time. If her rate out into the country was 4 miles per hour faster than her rate on the return trip, find both rates.

58. Felipe jogs for 10 miles and then walks another 10 miles. He jogs $2\frac{1}{2}$ miles per hour faster than he walks, and the entire distance of 20 miles takes 6 hours. Find the rate at which he walks and the rate at which he jogs.

■ ■ ▨ Thoughts into words

59. Why is it important to consider more than one way to do a problem?

60. Write a paragraph or two summarizing the new ideas about problem solving you have acquired thus far in this course.

 Graphing calculator activities

In Section 4.5 we solved mixture-of-solution problems. You can use the graphing calculator in a more general approach to problems of this type: How much pure alcohol should be added to 6 liters of a 40% alcohol solution to raise it to a 60% alcohol solution?

We let x represent the amount of pure alcohol to be added to the solution. For this more general approach we want to write a rational expression that represents the concentration of pure alcohol in the final solution. The amount of pure alcohol we are starting with is 40% of the 6 liters, which equals $0.40(6) = 2.4$ liters. Because we are adding x liters of pure alcohol to the solution, the expression $2.4 + x$ represents the amount of pure alcohol in the final solution. The final amount of solution is $6 + x$. The rational expression $\frac{2.4 + x}{6 + x}$ represents the concentration of pure alcohol in the final solution.

Let's graph the equation $y = \frac{2.4 + x}{6 + x}$ as shown in Figure 9.12.

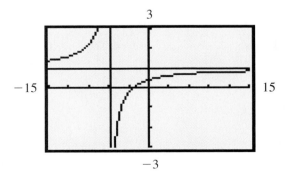

Figure 9.12

The y axis is the concentration of alcohol, so that will be a number between 0.40 and 1.0. The x axis is the amount of alcohol to be added, so x will be a nonnegative number. Therefore let's change the viewing window so that $0 \le x \le 15$ and $0 \le y \le 2$ to obtain Figure 9.13. Now we can use the graph to answer a variety of questions about this problem.

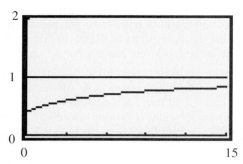

Figure 9.13

1. How much pure alcohol needs to be added to raise the 40% solution to a 60% alcohol solution? (*Answer:* Using the trace feature of the graphing utility, we find that $y = 0.6$ when $x = 3$. Therefore, 3 liters of pure alcohol need to be added.)

2. How much pure alcohol needs to be added to raise the 40% solution to a 70% alcohol solution? (*Answer:* Using the trace feature, we find that $y = 0.7$ when $x = 6$. Therefore, 6 liters of pure alcohol need to be added.)

3. What percent of alcohol do we have if we add 9 liters of pure alcohol to the 6 liters of a 40% solution? (*Answer:* Using the trace feature, we find that $y = 0.76$ when $x = 9$. Therefore, adding 9 liters of pure alcohol will give us a 76% alcohol solution.)

Now use this approach with your graphing utility to solve the following problems.

61. Suppose that x ounces of pure acid have been added to 14 ounces of a 15% acid solution.

   a. Set up the rational expression that represents the concentration of pure acid in the final solution.

   b. Graph the rational equation that displays the level of concentration.

   c. How many ounces of pure acid need to be added to the 14 ounces of a 15% acid solution to raise it to a 40.5% acid solution? Check your answer.

   d. How many ounces of pure acid need to be added to the 14 ounces of 15% acid solution to raise it to a 50% acid solution? Check your answer.

   e. What percent of acid do we obtain if we add 12 ounces of pure acid to the 14 ounces of 15% acid solution? Check your answer.

62. Solve the following problem both algebraically and graphically: One solution contains 50% alcohol and another solution contains 80% alcohol. How many liters of each solution should be mixed to produce 10.5 liters of a 70% alcohol solution? Check your answer.

Answers to Concept Quiz

1. False   **2.** True   **3.** False   **4.** True   **5.** True

CHAPTER 9

SUMMARY

(9.1) Any number that can be written in the form $\frac{a}{b}$, where a and b are integers and $b \neq 0$, is called a **rational number.**

A **rational expression** is defined as the indicated quotient of two polynomials. The following properties pertain to rational numbers and rational expressions.

1. $\dfrac{-a}{b} = \dfrac{a}{-b} = -\dfrac{a}{b}$

2. $\dfrac{-a}{-b} = \dfrac{a}{b}$

3. $\dfrac{a \cdot k}{b \cdot k} = \dfrac{a}{b}$ Fundamental principle of fractions

(9.2) Multiplication and division of rational expressions are based on the following definitions:

1. $\dfrac{a}{b} \cdot \dfrac{c}{d} = \dfrac{ac}{bd}$ Multiplication

2. $\dfrac{a}{b} \div \dfrac{c}{d} = \dfrac{a}{b} \cdot \dfrac{d}{c} = \dfrac{ad}{bc}$ Division

(9.3) Addition and subtraction of rational expressions are based on the following definitions:

1. $\dfrac{a}{b} + \dfrac{c}{b} = \dfrac{a + c}{b}$ Addition

2. $\dfrac{a}{b} - \dfrac{c}{b} = \dfrac{a - c}{b}$ Subtraction

(9.4) The following basic procedure is used to add or subtract rational expressions.

1. Find the LCD of all denominators.

2. Change each fraction to an equivalent fraction that has the LCD as its denominator.

3. Add or subtract numerators and place this result over the LCD.

4. Look for possibilities to simplify the resulting fraction.

Fractional forms that contain rational numbers or rational expressions in the numerators and/or denominators are called **complex fractions.** The fundamental principle of fractions serves as a basis for simplifying complex fractions.

(9.5) To solve a rational equation, it is often easiest to begin by multiplying both sides of the equation by the LCD of all of the denominators in the equation. If an equation contains a variable in one or more denominators, then we must be careful to avoid any value of the variable that makes the denominator zero.

A ratio is the comparison of two numbers by division. A statement of equality between two ratios is a proportion.

We can treat some rational equations as proportions, and we can solve them by applying the following property.

$$\frac{a}{b} = \frac{c}{d} \quad \text{if and only if } ad = bc$$

(9.6) The techniques that we use to solve rational equations can also be used to change the form of formulas containing rational expressions so that we can use those formulas to solve problems.

CHAPTER 9 REVIEW PROBLEM SET

For Problems 1–6, simplify each of the rational expressions.

1. $\dfrac{26x^2y^3}{39x^4y^2}$

2. $\dfrac{a^2 - 9}{a^2 + 3a}$

3. $\dfrac{n^2 - 3n - 10}{n^2 + n - 2}$

4. $\dfrac{x^4 - 1}{x^3 - x}$

5. $\dfrac{8x^3 - 2x^2 - 3x}{12x^2 - 9x}$

6. $\dfrac{x^4 - 7x^2 - 30}{2x^4 + 7x^2 + 3}$

For Problems 7–10, simplify each complex fraction.

7. $\dfrac{\dfrac{5}{8} - \dfrac{1}{2}}{\dfrac{1}{6} + \dfrac{3}{4}}$

8. $\dfrac{\dfrac{3}{2x} + \dfrac{5}{3y}}{\dfrac{4}{x} - \dfrac{3}{4y}}$

9. $\dfrac{\dfrac{3}{x-2} - \dfrac{4}{x^2-4}}{\dfrac{2}{x+2} + \dfrac{1}{x-2}}$

10. $1 - \dfrac{1}{2 - \dfrac{1}{x}}$

For Problems 11–22, perform the indicated operations and express your answers in simplest form.

11. $\dfrac{6xy^2}{7y^3} \div \dfrac{15x^2y}{5x^2}$

12. $\dfrac{9ab}{3a+6} \cdot \dfrac{a^2-4a-12}{a^2-6a}$

13. $\dfrac{n^2+10n+25}{n^2-n} \cdot \dfrac{5n^3-3n^2}{5n^2+22n-15}$

14. $\dfrac{x^2-2xy-3y^2}{x^2+9y^2} \div \dfrac{2x^2+xy-y^2}{2x^2-xy}$

15. $\dfrac{2x+1}{5} + \dfrac{3x-2}{4}$

16. $\dfrac{3}{2n} + \dfrac{5}{3n} - \dfrac{1}{9}$

17. $\dfrac{3x}{x+7} - \dfrac{2}{x}$

18. $\dfrac{10}{x^2-5x} + \dfrac{2}{x}$

19. $\dfrac{3}{n^2-5n-36} + \dfrac{2}{n^2+3n-4}$

20. $\dfrac{3}{2y+3} + \dfrac{5y-2}{2y^2-9y-18} - \dfrac{1}{y-6}$

21. $\dfrac{3}{x+1} + \dfrac{x+5}{x^2-1} - \dfrac{3}{x-1}$

22. $\dfrac{3n}{n^2+6n+5} + \dfrac{4}{n^2-7n-8}$

For Problems 23–32, solve each equation.

23. $\dfrac{4x+5}{3} + \dfrac{2x-1}{5} = 2$

24. $\dfrac{3}{4x} + \dfrac{4}{5} = \dfrac{9}{10x}$

25. $\dfrac{a}{a-2} - \dfrac{3}{2} = \dfrac{2}{a-2}$

26. $\dfrac{4}{5y-3} = \dfrac{2}{3y+7}$

27. $n + \dfrac{1}{n} = \dfrac{53}{14}$

28. $\dfrac{1}{2x-7} + \dfrac{x-5}{4x^2-49} = \dfrac{4}{6x-21}$

29. $\dfrac{x}{2x+1} - 1 = \dfrac{-4}{7(x-2)}$

30. $\dfrac{2x}{-5} = \dfrac{3}{4x-13}$

31. $\dfrac{2n}{2n^2+11n-21} - \dfrac{n}{n^2+5n-14} = \dfrac{3}{n^2+5n-14}$

32. $\dfrac{2}{t^2-t-6} + \dfrac{t+1}{t^2+t-12} = \dfrac{t}{t^2+6t+8}$

33. Solve $\dfrac{y-6}{x+1} = \dfrac{3}{4}$ for y.

34. Solve $\dfrac{x}{a} - \dfrac{y}{b} = 1$ for y.

For Problems 35–40, set up an equation and solve the problem.

35. A sum of $1400 is to be divided between two people in the ratio of $\dfrac{3}{5}$. How much does each person receive?

36. Working together, Dan and Julio can mow a lawn in 12 minutes. Julio can mow the lawn by himself in 10 minutes less time than it takes Dan by himself. How long does it take each of them to mow the lawn alone?

37. Suppose that car A can travel 250 miles in 3 hours less time than it takes car B to travel 440 miles. The rate of car B is 5 miles per hour faster than that of car A. Find the rates of both cars.

38. Mark can overhaul an engine in 20 hours, and Phil can do the same job by himself in 30 hours. If they both work together for a time and then Mark finishes the job by himself in 5 hours, how long did they work together?

39. Kelly contracted to paint a house for $640. It took him 20 hours longer than he had anticipated, so he earned $1.60 per hour less than he had calculated. How long had he anticipated that it would take him to paint the house?

40. Nasser rode his bicycle 66 miles in $4\dfrac{1}{2}$ hours. For the first 40 miles he averaged a certain rate, and then for the last 26 miles he reduced his rate by 3 miles per hour. Find his rate for the last 26 miles.

CHAPTER 9

TEST

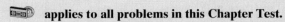 applies to all problems in this Chapter Test.

For Problems 1–4, simplify each rational expression.

1. $\dfrac{39x^2y^3}{72x^3y}$

2. $\dfrac{3x^2 + 17x - 6}{x^3 - 36x}$

3. $\dfrac{6n^2 - 5n - 6}{3n^2 + 14n + 8}$

4. $\dfrac{2x - 2x^2}{x^2 - 1}$

For Problems 5–14, perform the indicated operations and express your answers in simplest form.

5. $\dfrac{5x^2y}{8x} \cdot \dfrac{12y^2}{20xy}$

6. $\dfrac{5a + 5b}{20a + 10b} \cdot \dfrac{a^2 - ab}{2a^2 + 2ab}$

7. $\dfrac{3x^2 + 10x - 8}{5x^2 + 19x - 4} \div \dfrac{3x^2 - 23x + 14}{x^2 - 3x - 28}$

8. $\dfrac{3x - 1}{4} + \dfrac{2x + 5}{6}$

9. $\dfrac{5x - 6}{3} - \dfrac{x - 12}{6}$

10. $\dfrac{3}{5n} + \dfrac{2}{3} - \dfrac{7}{3n}$

11. $\dfrac{3x}{x - 6} + \dfrac{2}{x}$

12. $\dfrac{9}{x^2 - x} - \dfrac{2}{x}$

13. $\dfrac{3}{2n^2 + n - 10} + \dfrac{5}{n^2 + 5n - 14}$

14. $\dfrac{5}{2x^2 - 6x} + \dfrac{4}{3x^2 + 6x}$

15. Simplify the complex fraction $\dfrac{\dfrac{3}{2x} - \dfrac{1}{6}}{\dfrac{2}{3x} + \dfrac{3}{4}}$.

16. Solve $\dfrac{x + 2}{y - 4} = \dfrac{3}{4}$ for y.

For Problems 17–22, solve each equation.

17. $\dfrac{x - 1}{2} - \dfrac{x + 2}{5} = -\dfrac{3}{5}$

18. $\dfrac{5}{4x} + \dfrac{3}{2} = \dfrac{7}{5x}$

19. $\dfrac{-3}{4n - 1} = \dfrac{-2}{3n + 11}$

20. $n - \dfrac{5}{n} = 4$

21. $\dfrac{6}{x - 4} - \dfrac{4}{x + 3} = \dfrac{8}{x - 4}$

22. $\dfrac{1}{3x - 1} + \dfrac{x - 2}{9x^2 - 1} = \dfrac{7}{6x - 2}$

For Problems 23–25, set up an equation and solve the problem.

23. The denominator of a rational number is 9 less than three times the numerator. The number in simplest form is $\dfrac{3}{8}$. Find the number.

24. It takes Jodi three times as long to wash the car as it does Jannie. Together they can wash the car in 15 minutes. How long would it take Jodi by herself?

25. René can ride her bike 60 miles in 1 hour less time than it takes Sue to ride 60 miles. René's rate is 3 miles per hour faster than Sue's rate. Find René's rate.

CUMULATIVE PRACTICE TEST *Chapters 1-9*

For Problems 1–3, evaluate each algebraic expression for the given values of the variables.

1. $3(x - 4) - 4(2x + 1) - 6(4 - 3x)$ for $x = -19$

2. $\dfrac{3x^2 + 2x - 1}{3x^2 - 4x + 1}$ for $x = 15$

3. $\dfrac{9x^2 y^3}{3x^{-2} y^{-1}}$ for $x = -1$ and $y = -2$

4. Find the quotient and remainder for the division problem $(2x^4 - x^3 - 22x^2 + 15x + 21) \div (x - 3)$.

5. Solve the system $\begin{pmatrix} 3x - 4y = -25 \\ 2x + 5y = 14 \end{pmatrix}$.

6. Solve the system $\begin{pmatrix} 7x - 4y = 59 \\ 3x + y = 9 \end{pmatrix}$.

7. Graph the equation $y = -x^2 + 1$.

8. Graph the equation $y = -x^3 + 1$.

9. Graph the inequality $-2x - 4y \geq -4$.

10. Evaluate $\left(\dfrac{1}{2} - \dfrac{1}{3} \right)^{-2}$.

11. Express $\dfrac{9}{4}$ as a percent.

12. Express 0.00013 in scientific notation.

For Problems 13–17, solve each equation.

13. $-3(2x - 1) - (x + 4) = -5(x - 3)$

14. $\dfrac{5 - x}{2 - x} - \dfrac{3 - 2x}{2x} = 1$

15. $15x^3 + x^2 - 2x = 0$

16. $|5x - 1| = 7$

17. $(3x - 1)^2 = 16$

For Problems 18–20, solve each inequality and express the solutions using interval notation.

18. $-16 \leq 7x - 2 \leq 5$

19. $\dfrac{x - 1}{3} - \dfrac{2x + 1}{4} > \dfrac{1}{6}$

20. $|2x - 1| > 1$

For Problems 21–25, use an equation, an inequality, or a system of equations to help solve each problem.

21. A retailer has some shirts that cost him $14 each. He wants to sell them to make a profit of 30% of the selling price. What price should he charge for the shirts?

22. How many gallons of a solution of glycerine and water containing 55% glycerine should be added to 15 gallons of a 20% solution to give a 40% solution?

23. Russ started to mow the lawn, a task that usually takes him 40 minutes. After he had been working for 15 minutes, his friend Jay came along with his mower and began to help Russ. Working together, they finished the lawn in 10 minutes. How long would it have taken Jay to mow the lawn by himself?

24. One leg of a right triangle is 5 centimeters longer than the other leg. The hypotenuse is 25 centimeters long. Find the length of each leg.

25. Regina had scores of 93, 88, 89, and 95 on her first four math exams. What score must she get on the fifth exam to have an average of 92 or better for the five exams?

CHAPTER 10

REDUCE SPEED

SNOW AHEAD

SPEED LIMIT 45

© Omni Photo Communications Inc./Index Stock

After an auto accident, law enforcement agencies can use the formula $S = \sqrt{30Df}$ to determine the speed of a vehicle by measuring the length of the skid marks and knowing the coefficient of the friction of the pavement.

Exponents and Radicals

Suppose that a car is traveling at 65 miles per hour on a highway during a rainstorm. Suddenly, something darts across the highway, and the driver hits the brake pedal. How far will the car skid on the wet pavement? We can use the formula $S = \sqrt{30Df}$, where S represents the speed of the car, D the length of skid marks, and f a coefficient of friction, to determine that the car will skid approximately 400 feet.

In Section 2.3 we used $\sqrt{2}$, $\sqrt{3}$, and π as examples of irrational numbers. Irrational numbers in decimal form are nonrepeating decimals. For example, $\sqrt{2} = 1.414213562373\ldots$, where the three dots at the end of the number indicates that the expansion continues indefinitely. In Chapter 2, we stated that we would return to the irrationals in Chapter 10. The time has come for us to expand our skills relative to the set of irrational numbers.

It is not uncommon in mathematics to find two separately developed concepts that are closely related to each other. In this chapter, we will first develop the concepts of exponent and root individually and then show how they merge to become even more functional as a unified idea.

461

InfoTrac Project

Do a subject guide search on Pythagoras. Although he is best remembered for the Pythagorean theorem, find another important discovery he made. Before Pythagoras, who were the earliest known users of the equation $a^2 + b^2 = c^2$? A ladder is leaning up against a house. Solve the following radical equation for x to find the distance from the top of the ladder (in feet) to the bottom of the house:

$$\sqrt{x^2 + 10x - 23} = x + 3$$

If the distance from the bottom of the house to the bottom of the ladder is 6 feet, use the Pythagorean theorem to find the length of the ladder. Draw a picture to help you.

10.1 Integral Exponents and Scientific Notation Revisited

Objectives

- Simplify numerical expressions with integer exponents.
- Simplify monomials with integer exponents.
- Write numbers in scientific notation.
- Use scientific notation to multiply and divide numbers.

In Section 6.6 we used the following definitions to extend our work with exponents from the positive integers to all integers.

RESTATEMENT OF DEFINITION 6.2

If b is a nonzero real number, then

$$b^0 = 1$$

RESTATEMENT OF DEFINITION 6.3

If n is a positive integer and b is a nonzero real number, then

$$b^{-n} = \frac{1}{b^n}$$

Using either Definition 6.2 or Definition 6.3, the following statements can be made.

$$(-48)^0 = 1, \qquad \left(\frac{2}{3}\right)^0 = 1, \qquad x^0 = 1 \quad \text{for } x \neq 0,$$

$$3^{-2} = \frac{1}{3^2} = \frac{1}{9}, \qquad 10^{-3} = \frac{1}{10^3} = \frac{1}{1000} = 0.001,$$

$$\left(\frac{3}{4}\right)^{-2} = \frac{1}{\left(\frac{3}{4}\right)^2} = \frac{1}{\frac{9}{16}} = \frac{16}{9}, \qquad x^{-5} = \frac{1}{x^5} \quad \text{for } x \neq 0$$

The following properties of exponents were stated both in Chapter 6 and in Chapter 8, but we will restate them here for your convenience.

RESTATEMENT OF PROPERTY 8.6

If m and n are integers, and a and b are real numbers (and $b \neq 0$ whenever it appears in a denominator), then

1. $b^n \cdot b^m = b^{n+m}$ Product of two like bases with powers

2. $(b^n)^m = b^{mn}$ Power of a power

3. $(ab)^n = a^n b^n$ Power of a product

4. $\left(\dfrac{a}{b}\right)^n = \dfrac{a^n}{b^n}$ Power of a quotient

5. $\dfrac{b^n}{b^m} = b^{n-m}$ Quotient of two like bases with powers

Having the use of all integers as exponents allows us to work with a large variety of numerical and algebraic expressions. Let's consider some examples that illustrate the use of the various parts of Property 8.6.

EXAMPLE 1

Simplify each of the following; express final results without using zero or negative integers as exponents.

    (a) $x^2 \cdot x^{-5}$ **(b)** $(x^{-2})^4$ **(c)** $(x^2 y^{-3})^{-4}$

    (d) $\left(\dfrac{a^3}{b^{-5}}\right)^{-2}$ **(e)** $\dfrac{x^{-4}}{x^{-2}}$

Solution

    (a) $x^2 \cdot x^{-5} = x^{2+(-5)}$ Product of two like bases with powers

                 $= x^{-3}$

                 $= \dfrac{1}{x^3}$

    (b) $(x^{-2})^4 = x^{4(-2)}$ Power of a power

             $= x^{-8}$

             $= \dfrac{1}{x^8}$

    (c) $(x^2 y^{-3})^{-4} = (x^2)^{-4}(y^{-3})^{-4}$ Power of a product

                 $= x^{-4(2)} y^{-4(-3)}$

                 $= x^{-8} y^{12}$

                 $= \dfrac{y^{12}}{x^8}$

(d) $\left(\dfrac{a^3}{b^{-5}}\right)^{-2} = \dfrac{(a^3)^{-2}}{(b^{-5})^{-2}}$ Power of a quotient

$\qquad\qquad = \dfrac{a^{-6}}{b^{10}}$

$\qquad\qquad = \dfrac{1}{a^6 b^{10}}$

(e) $\dfrac{x^{-4}}{x^{-2}} = x^{-4-(-2)}$ Quotient of two like bases with powers

$\qquad\quad = x^{-2}$

$\qquad\quad = \dfrac{1}{x^2}$

E X A M P L E   2

Find the indicated products and quotients; express your results using positive integral exponents only.

(a) $(3x^2 y^{-4})(4x^{-3}y)$ **(b)** $\dfrac{12a^3 b^2}{-3a^{-1}b^5}$ **(c)** $\left(\dfrac{15x^{-1}y^2}{5xy^{-4}}\right)^{-1}$

Solution

(a) $(3x^2 y^{-4})(4x^{-3}y) = 12x^{2+(-3)}y^{-4+1}$

$\qquad\qquad\qquad\qquad = 12x^{-1}y^{-3}$

$\qquad\qquad\qquad\qquad = \dfrac{12}{xy^3}$

(b) $\dfrac{12a^3 b^2}{-3a^{-1}b^5} = -4a^{3-(-1)}b^{2-5}$

$\qquad\qquad\quad = -4a^4 b^{-3}$

$\qquad\qquad\quad = -\dfrac{4a^4}{b^3}$

(c) $\left(\dfrac{15x^{-1}y^2}{5xy^{-4}}\right)^{-1} = (3x^{-1-1}y^{2-(-4)})^{-1}$ Note that we are first simplifying inside the parentheses.

$\qquad\qquad\qquad = (3x^{-2}y^6)^{-1}$

$\qquad\qquad\qquad = 3^{-1}x^2 y^{-6}$

$\qquad\qquad\qquad = \dfrac{x^2}{3y^6}$

E X A M P L E   3

Simplify $2^{-3} + 3^{-1}$.

Solution

$$2^{-3} + 3^{-1} = \dfrac{1}{2^3} + \dfrac{1}{3^1}$$

$$= \frac{1}{8} + \frac{1}{3}$$

$$= \frac{3}{24} + \frac{8}{24}$$

$$= \frac{11}{24}$$

■

EXAMPLE 4   Simplify $(4^{-1} - 3^{-2})^{-1}$.

Solution

$$(4^{-1} - 3^{-2})^{-1} = \left(\frac{1}{4^1} - \frac{1}{3^2} \right)^{-1}$$ Apply $b^{-n} = \frac{1}{b^n}$ to 4^{-1} and to 3^{-2}.

$$= \left(\frac{1}{4} - \frac{1}{9} \right)^{-1}$$

$$= \left(\frac{9}{36} - \frac{4}{36} \right)^{-1}$$ Change to equivalent fraction with LCD = 36.

$$= \left(\frac{5}{36} \right)^{-1}$$

$$= \frac{1}{\left(\frac{5}{36} \right)^1}$$ Apply $b^{-n} = \frac{1}{b^n}$.

$$= \frac{1}{\frac{5}{36}} = \frac{36}{5}$$

■

EXAMPLE 5   Express $a^{-1} + b^{-2}$ as a single fraction involving positive exponents only.

Solution

$$a^{-1} + b^{-2} = \frac{1}{a^1} + \frac{1}{b^2}$$

$$= \left(\frac{1}{a} \right)\left(\frac{b^2}{b^2} \right) + \left(\frac{1}{b^2} \right)\left(\frac{a}{a} \right)$$ Use ab^2 as the LCD.

$$= \frac{b^2}{ab^2} + \frac{a}{ab^2}$$

$$= \frac{b^2 + a}{ab^2}$$

■

Scientific Notation

In symbols, a number in scientific notation has the form $(N)(10)^k$, where $1 \le N < 10$ and k is an integer. For example, 617 can be written as $(6.17)(10)^2$, and 0.0014 can be written as $(1.4)(10)^{-3}$. To switch from ordinary decimal notation to scientific notation, you can use the following procedure.

> Write the given number as the product of a number greater than or equal to 1 and less than 10, and a power of 10. The exponent of 10 is determined by counting the number of places that the decimal point was moved when going from the original number to the number greater than or equal to 1 and less than 10. This exponent is (a) negative if the original number is less than 1, (b) positive if the original number is greater than 10, and (c) 0 if the original number itself is between 1 and 10.

Thus we can write

$$0.00467 = (4.67)(10)^{-3}$$
$$87{,}000 = (8.7)(10)^4$$
$$3.1416 = (3.1416)(10)^0$$

To switch from scientific notation to ordinary decimal notation, you can use the following procedure.

> Move the decimal point the number of places indicated by the exponent of 10. The decimal point is moved to the right if the exponent is positive and to the left if it is negative.

Thus we can write

$$(4.78)(10)^4 = 47{,}800$$
$$(8.4)(10)^{-3} = 0.0084$$

Scientific notation can frequently be used to simplify numerical calculations. We merely change the numbers to scientific notation and use the appropriate properties of exponents. Consider the following examples.

EXAMPLE 6   Perform the indicated operations.

(a) $(0.00024)(20{,}000)$

(b) $\dfrac{7{,}800{,}000}{0.0039}$

(c) $\dfrac{(0.00069)(0.0034)}{(0.0000017)(0.023)}$

Solution

(a) $(0.00024)(20,000) = (2.4)(10)^{-4}(2)(10)^4$

$$= (2.4)(2)(10)^{-4}(10)^4$$

$$= (4.8)(10)^0$$

$$= (4.8)(1)$$

$$= 4.8$$

(b) $\dfrac{7,800,000}{0.0039} = \dfrac{(7.8)(10)^6}{(3.9)(10)^{-3}}$

$$= (2)(10)^9$$

$$= 2,000,000,000$$

(c) $\dfrac{(0.00069)(0.0034)}{(0.0000017)(0.023)} = \dfrac{(6.9)(10)^{-4}(3.4)(10)^{-3}}{(1.7)(10)^{-6}(2.3)(10)^{-2}}$

$$= \dfrac{\overset{3}{(6.9)}\overset{2}{(3.4)}(10)^{-7}}{(1.7)(2.3)(10)^{-8}}$$

$$= (6)(10)^1$$

$$= 60 \qquad \blacksquare$$

Many calculators are equipped to display numbers in scientific notation. The display panel shows the number between 1 and 10 and the appropriate exponent of 10. For example, evaluating $(3,800,000)^2$ yields

$$\boxed{\text{1.444E13}}$$

Thus $(3,800,000)^2 = (1.444)(10)^{13} = 14,440,000,000,000$.
Similarly, the answer for $(0.000168)^2$ is displayed as

$$\boxed{\text{2.8224E-8}}$$

Thus $(0.000168)^2 = (2.8224)(10)^{-8} = 0.000000028224$.
Calculators vary as to the number of digits displayed in the number between 1 and 10 when scientific notation is used. For example, we used two different calculators to estimate $(6729)^6$ and obtained the following results.

$$\boxed{\text{9.2833E22}}$$

$$\boxed{\text{9.283316768E22}}$$

Obviously, you need to know the capabilities of your calculator when working with problems in scientific notation. Many calculators also allow the entry of a number in scientific notation. Such calculators are equipped with an enter-the-exponent key (often labeled as $\boxed{EE}$ or $\boxed{EEX}$). Thus a number such as $(3.14)(10)^8$ might be entered as follows:

Enter	Press	Display
3.14	$\boxed{EE}$	3.14E
8		3.14E8

A $\boxed{MODE}$ key is often used on calculators to let you choose normal decimal notation, scientific notation, or engineering notation. (The abbreviations Norm, Sci, and Eng are commonly used.) If the calculator is in scientific mode, then a number can be entered and changed to scientific form by pressing the $\boxed{ENTER}$ key. For example, when we enter 589 and press the $\boxed{ENTER}$ key, the display will show 5.89E2. Likewise, when the calculator is in scientific mode, the answers to computational problems are given in scientific form. For example, the answer for $(76)(533)$ is given as 4.0508E4.

It should be evident from this brief discussion that even when you are using a calculator, you need to have a thorough understanding of scientific notation.

CONCEPT QUIZ

For Problems 1–5, answer true or false.

1. $\left(\dfrac{2}{5}\right)^{-2} = \left(\dfrac{5}{2}\right)^{2}$

2. $(3)^0(3)^2 = 9^2$

3. $(2)^{-4}(2)^4 = 2$

4. $0.000037 = (3.7)(10)^{-4}$.

5. In scientific notation, a number has the form $(N)(10)^k$, where N is between 0 and 10 and k is an integer.

PROBLEM SET 10.1

For Problems 1–42, simplify each numerical expression.

1. 3^{-3}

2. 2^{-4}

3. -10^{-2}

4. 10^{-3}

5. $\dfrac{1}{3^{-4}}$

6. $\dfrac{1}{2^{-6}}$

7. $-\left(\dfrac{1}{3}\right)^{-3}$

8. $\left(\dfrac{1}{2}\right)^{-3}$

9. $\left(-\dfrac{1}{2}\right)^{-3}$

10. $\left(\dfrac{2}{7}\right)^{-2}$

11. $\left(-\dfrac{3}{4}\right)^{0}$

12. $\dfrac{1}{\left(\dfrac{4}{5}\right)^{-2}}$

13. $\dfrac{1}{\left(\dfrac{3}{7}\right)^{-2}}$

14. $-\left(\dfrac{5}{6}\right)^{0}$

15. $2^7 \cdot 2^{-3}$

16. $3^{-4} \cdot 3^6$

17. $10^{-5} \cdot 10^2$

18. $10^4 \cdot 10^{-6}$

19. $10^{-1} \cdot 10^{-2}$

20. $10^{-2} \cdot 10^{-2}$

21. $(3^{-1})^{-3}$

22. $(2^{-2})^{-4}$

23. $(5^3)^{-1}$

24. $(3^{-1})^3$

25. $(2^3 \cdot 3^{-2})^{-1}$

26. $(2^{-2} \cdot 3^{-1})^{-3}$

27. $(4^2 \cdot 5^{-1})^2$

28. $(2^{-3} \cdot 4^{-1})^{-1}$

29. $\left(\dfrac{2^{-1}}{5^{-2}}\right)^{-1}$

30. $\left(\dfrac{2^{-4}}{3^{-2}}\right)^{-2}$

31. $\left(\dfrac{2^{-1}}{3^{-2}}\right)^2$

32. $\left(\dfrac{3^2}{5^{-1}}\right)^{-1}$

33. $\dfrac{3^3}{3^{-1}}$

34. $\dfrac{2^{-2}}{2^3}$

35. $\dfrac{10^{-2}}{10^2}$

36. $\dfrac{10^{-2}}{10^{-5}}$

37. $2^{-2} + 3^{-2}$

38. $2^{-4} + 5^{-1}$

39. $\left(\dfrac{1}{3}\right)^{-1} - \left(\dfrac{2}{5}\right)^{-1}$

40. $\left(\dfrac{3}{2}\right)^{-1} - \left(\dfrac{1}{4}\right)^{-1}$

41. $(2^{-3} + 3^{-2})^{-1}$

42. $(5^{-1} - 2^{-3})^{-1}$

For Problems 43–62, simplify each expression. Express final results without using zero or negative integers as exponents.

43. $x^2 \cdot x^{-8}$

44. $x^{-3} \cdot x^{-4}$

45. $a^3 \cdot a^{-5} \cdot a^{-1}$

46. $b^{-2} \cdot b^3 \cdot b^{-6}$

47. $(a^{-4})^2$

48. $(b^4)^{-3}$

49. $(x^2 y^{-6})^{-1}$

50. $(x^5 y^{-1})^{-3}$

51. $(ab^3 c^{-2})^{-4}$

52. $(a^3 b^{-3} c^{-2})^{-5}$

53. $(2x^3 y^{-4})^{-3}$

54. $(4x^5 y^{-2})^{-2}$

55. $\left(\dfrac{x^{-1}}{y^{-4}}\right)^{-3}$

56. $\left(\dfrac{y^3}{x^{-4}}\right)^{-2}$

57. $\left(\dfrac{3a^{-2}}{2b^{-1}}\right)^{-2}$

58. $\left(\dfrac{2xy^2}{5a^{-1}b^{-2}}\right)^{-1}$

59. $\dfrac{x^{-6}}{x^{-4}}$

60. $\dfrac{a^{-2}}{a^2}$

61. $\dfrac{a^3 b^{-2}}{a^{-2}b^{-4}}$

62. $\dfrac{x^{-3}y^{-4}}{x^2 y^{-1}}$

For Problems 63–74, find the indicated products and quotients. Express final results using positive integral exponents only.

63. $(2xy^{-1})(3x^{-2}y^4)$

64. $(-4x^{-1}y^2)(6x^3 y^{-4})$

65. $(-7a^2 b^{-5})(-a^{-2}b^7)$

66. $(-9a^{-3}b^{-6})(-12a^{-1}b^4)$

67. $\dfrac{28x^{-2}y^{-3}}{4x^{-3}y^{-1}}$

68. $\dfrac{63x^2 y^{-4}}{7xy^{-4}}$

69. $\dfrac{-72a^2 b^{-4}}{6a^3 b^{-7}}$

70. $\dfrac{108a^{-5}b^{-4}}{9a^{-2}b}$

71. $\left(\dfrac{35x^{-1}y^{-2}}{7x^4 y^3}\right)^{-1}$

72. $\left(\dfrac{-48ab^2}{-6a^3 b^5}\right)^{-2}$

73. $\left(\dfrac{-36a^{-1}b^{-6}}{4a^{-1}b^4}\right)^{-2}$

74. $\left(\dfrac{8xy^3}{-4x^4 y}\right)^{-3}$

For Problems 75–84, express each of the following as a single fraction involving positive exponents only.

75. $x^{-2} + x^{-3}$

76. $x^{-1} + x^{-5}$

77. $x^{-3} - y^{-1}$

78. $2x^{-1} - 3y^{-2}$

79. $3a^{-2} + 4b^{-1}$

80. $a^{-1} + a^{-1}b^{-3}$

81. $x^{-1}y^{-2} - xy^{-1}$

82. $x^2 y^{-2} - x^{-1}y^{-3}$

83. $2x^{-1} - 3x^{-2}$

84. $5x^{-2}y + 6x^{-1}y^{-2}$

For Problems 85–92, write each of the following in scientific notation. For example

$$27{,}800 = (2.78)(10)^4$$

85. 40,000,000

86. 500,000,000

87. 376.4

88. 9126.21

89. 0.347

90. 0.2165

91. 0.0214

92. 0.0037

For Problems 93–100, write each of the following in ordinary decimal notation. For example,

$$(3.18)(10)^2 = 318$$

93. $(3.14)(10)^{10}$

94. $(2.04)(10)^{12}$

95. $(4.3)(10)^{-1}$

96. $(5.2)(10)^{-2}$

97. $(9.14)(10)^{-4}$

98. $(8.76)(10)^{-5}$

99. $(5.123)(10)^{-8}$

100. $(6)(10)^{-9}$

For Problems 101–104, use scientific notation and the properties of exponents to help you perform the following operations.

101. $\dfrac{(60,000)(0.006)}{(0.0009)(400)}$

102. $\dfrac{(0.00063)(960,000)}{(3,200)(0.0000021)}$

103. $\dfrac{(0.0045)(60,000)}{(1800)(0.00015)}$

104. $\dfrac{(0.00016)(300)(0.028)}{0.064}$

▪▪▪ Thoughts into words

105. Is the following simplification process correct?

$$(3^{-2})^{-1} = \left(\frac{1}{3^2}\right)^{-1} = \left(\frac{1}{9}\right)^{-1} = \frac{1}{\left(\frac{1}{9}\right)^1} = 9$$

Could you suggest a better way to do the problem?

106. Explain how to simplify $(2^{-1} \cdot 3^{-2})^{-1}$ and also how to simplify $(2^{-1} + 3^{-2})^{-1}$.

107. Why do we need scientific notation even when using calculators and computers?

▦ Calculator activities

108. Use your calculator to check your answers for Problems 37–42 and 85–104.

109. Use a calculator to evaluate each of the following. Then evaluate each one without a calculator.

a. $(2^{-3} + 3^{-3})^{-2}$

b. $(4^{-3} - 2^{-1})^{-2}$

c. $(5^{-3} - 3^{-5})^{-1}$

d. $(6^{-2} + 7^{-4})^{-2}$

e. $(7^{-3} - 2^{-4})^{-2}$

f. $(3^{-4} + 2^{-3})^{-3}$

110. Use your calculator to evaluate each of the following. Express final answers in ordinary notation.

a. $(27,000)^2$ **b.** $(450,000)^2$

c. $(14,800)^2$ **d.** $(1700)^3$

e. $(900)^4$ **f.** $(60)^5$

g. $(0.0213)^2$ **h.** $(0.000213)^2$

i. $(0.000198)^2$ **j.** $(0.000009)^3$

111. Use your calculator to estimate each of the following. Express final answers in scientific notation with the number between 1 and 10 rounded to the nearest one-thousandth.

a. $(4576)^4$ **b.** $(719)^{10}$

c. $(28)^{12}$ **d.** $(8619)^6$

e. $(314)^5$ **f.** $(145,723)^2$

112. Use your calculator to estimate each of the following. Express final answers in ordinary notation rounded to the nearest one-thousandth.

a. $(1.09)^5$ **b.** $(1.08)^{10}$

c. $(1.14)^7$ **d.** $(1.12)^{20}$

e. $(0.785)^4$ **f.** $(0.492)^5$

Answers to Concept Quiz

1. True **2.** False **3.** False **4.** False **5.** False

10.2   Roots and Radicals

Objectives

■   Evaluate roots of numerical expressions.

■   Simplify radicals of numerical expressions.

■   Rationalize the denominator of radical expressions.

■   Use formulas involving radicals.

To **square a number** means to raise it to the second power — that is, to use the number as a factor twice.

$$4^2 = 4 \cdot 4 = 16 \qquad \text{Read "four squared equals sixteen"}$$

$$10^2 = 10 \cdot 10 = 100$$

$$\left(\frac{1}{2}\right)^2 = \frac{1}{2} \cdot \frac{1}{2} = \frac{1}{4}$$

$$(-3)^2 = (-3)(-3) = 9$$

A **square root of a number** is one of its two equal factors. Thus 4 is a square root of 16 because $4 \cdot 4 = 16$. Likewise, -4 is also a square root of 16 because $(-4)(-4) = 16$. In general, a is a square root of b if $a^2 = b$. The following generalizations are a direct consequence of the previous statement.

1. Every positive real number has two square roots; one is positive and the other is negative. They are opposites of each other.
2. Negative real numbers have no real number square roots because any nonzero real number is positive when squared.
3. The square root of 0 is 0.

The symbol $\sqrt{}$, called a **radical sign,** is used to designate the nonnegative square root. The number under the radical sign is called the **radicand.** The entire expression, such as $\sqrt{16}$, is called a **radical.**

$$\sqrt{16} = 4 \qquad \sqrt{16} \text{ indicates the nonnegative or \textbf{principal square root} of 16.}$$

$$-\sqrt{16} = -4 \qquad -\sqrt{16} \text{ indicates the negative square root of 16.}$$

$$\sqrt{0} = 0 \qquad \text{Zero has only one square root. Technically, we could write } -\sqrt{0} = -0 = 0.$$

$\sqrt{-4}$ is not a real number.

$-\sqrt{-4}$ is not a real number.

In general, the following definition is useful.

DEFINITION 10.1

If $a \geq 0$ and $b \geq 0$, then $\sqrt{b} = a$ if and only if $a^2 = b$; a is called the **principal square root of b.**

To **cube a number** means to raise it to the third power — that is, to use the number as a factor three times.

$$2^3 = 2 \cdot 2 \cdot 2 = 8 \qquad \text{Read "two cubed equals eight"}$$
$$4^3 = 4 \cdot 4 \cdot 4 = 64$$
$$\left(\frac{2}{3}\right)^3 = \frac{2}{3} \cdot \frac{2}{3} \cdot \frac{2}{3} = \frac{8}{27}$$
$$(-2)^3 = (-2)(-2)(-2) = -8$$

A **cube root of a number** is one of its three equal factors. Thus 2 is a cube root of 8 because $2 \cdot 2 \cdot 2 = 8$. (In fact, 2 is the only real number that is a cube root of 8.) Furthermore, -2 is a cube root of -8 because $(-2)(-2)(-2) = -8$. (In fact, -2 is the only real number that is a cube root of -8.)

In general, a is a cube root of b if $a^3 = b$. The following generalizations are a direct consequence of the previous statement.

1. Every positive real number has one positive real number cube root.
2. Every negative real number has one negative real number cube root.
3. The cube root of 0 is 0.

REMARK: Technically, every nonzero real number has three cube roots, but only one of them is a real number. The other two roots are classified as complex numbers. We are restricting our work at this time to the set of real numbers.

The symbol $\sqrt[3]{}$ designates the cube root of a number. Thus we can write

$$\sqrt[3]{8} = 2 \qquad\qquad \sqrt[3]{\frac{1}{27}} = \frac{1}{3}$$

$$\sqrt[3]{-8} = -2 \qquad\qquad \sqrt[3]{-\frac{1}{27}} = -\frac{1}{3}$$

In general, the following definition is useful:

DEFINITION 10.2

$\sqrt[3]{b} = a$ if and only if $a^3 = b$.

In Definition 10.2, if $b \geq 0$ then $a \geq 0$, whereas if $b < 0$ then $a < 0$. The number a is called **the principal cube root of b** or simply **the cube root of b.**

The concept of root can be extended to fourth roots, fifth roots, sixth roots, and, in general, nth roots.

DEFINITION 10.3

$\sqrt[n]{b} = a$ if and only if $a^n = b$.

We can make the following generalizations.

If n is an even positive integer, then the following statements are true.

1. Every positive real number has exactly two real nth roots — one positive and one negative. For example, the real fourth roots of 16 are 2 and -2.
2. Negative real numbers do not have real nth roots. For example, there are no real fourth roots of -16.

If n is an odd positive integer greater than 1, then the following statements are true.

1. Every real number has exactly one real nth root.
2. The real nth root of a positive number is positive. For example, the fifth root of 32 is 2.
3. The real nth root of a negative number is negative. For example, the fifth root of -32 is -2.

The symbol $\sqrt[n]{}$ designates the principal nth root. To complete our terminology, the n in the radical $\sqrt[n]{b}$ is called the index of the radical. If $n = 2$, we commonly write $\sqrt{b}$ instead of $\sqrt[2]{b}$. In the future as we use symbols, such as $\sqrt[n]{b}$, $\sqrt[m]{y}$, and $\sqrt[r]{x}$, we will assume the previous agreements relative to the existence of real roots (without listing the various restrictions) unless a special restriction is necessary.

The following chart can help summarize this information with respect to $\sqrt[n]{b}$, where n is a positive integer greater than 1.

	If *b* is		
	Positive	**Zero**	**Negative**
***n* is even**	$\sqrt[n]{b}$ is a positive real number	$\sqrt[n]{b} = 0$	$\sqrt[n]{b}$ is not a real number
***n* is odd**	$\sqrt[n]{b}$ is a positive real number	$\sqrt[n]{b} = 0$	$\sqrt[n]{b}$ is a negative real number

Consider the following examples.

$\sqrt[4]{81} = 3$ because $3^4 = 81$
$\sqrt[5]{32} = 2$ because $2^5 = 32$
$\sqrt[5]{-32} = -2$ because $(-2)^5 = -32$

The following property is a direct consequence of Definition 10.3.

PROPERTY 10.1

1. $(\sqrt[n]{b})^n = b$ n is any positive integer greater than 1.

2. $\sqrt[n]{b^n} = b$ n is any positive integer greater than 1 if $b \geq 0$; n is an odd positive integer greater than 1 if $b < 0$.

Because the radical expressions in parts (1) and (2) of Property 10.1 are both equal to b, by the transitive property they are equal to each other. Hence $\sqrt[n]{b^n} = (\sqrt[n]{b})^n$. The arithmetic is usually easier to simplify when we use the form $(\sqrt[n]{b})^n$. The following examples demonstrate the use of Property 10.1.

$$\sqrt{144^2} = (\sqrt{144})^2 = 12^2 = 144$$
$$\sqrt[3]{64^3} = (\sqrt[3]{64})^3 = 4^3 = 64$$
$$\sqrt[4]{16^4} = (\sqrt[4]{16})^4 = 2^4 = 16$$

Let's use some examples to lead into the next very useful property of radicals.

$$\sqrt{4 \cdot 9} = \sqrt{36} = 6 \quad \text{and} \quad \sqrt{4} \cdot \sqrt{9} = 2 \cdot 3 = 6$$
$$\sqrt{16 \cdot 25} = \sqrt{400} = 20 \quad \text{and} \quad \sqrt{16} \cdot \sqrt{25} = 4 \cdot 5 = 20$$
$$\sqrt[3]{8 \cdot 27} = \sqrt[3]{216} = 6 \quad \text{and} \quad \sqrt[3]{8} \cdot \sqrt[3]{27} = 2 \cdot 3 = 6$$
$$\sqrt[3]{(-8)(27)} = \sqrt[3]{-216} = -6 \quad \text{and} \quad \sqrt[3]{-8} \cdot \sqrt[3]{27} = (-2)(3) = -6$$

In general, we can state the following property:

PROPERTY 10.2

$\sqrt[n]{bc} = \sqrt[n]{b}\sqrt[n]{c}$ $\sqrt[n]{b}$ and $\sqrt[n]{c}$ are real numbers

---

Property 10.2 states that **the nth root of a product is equal to the product of the nth roots.**

Simplest Radical Form

The definition of nth root, along with Property 10.2, provide the basis for changing radicals to simplest radical form. The concept of **simplest radical form** takes on additional meaning as we encounter more complicated expressions, but for now it simply means that the radicand is not to contain any perfect powers of the index. Let's consider some examples to clarify this idea.

E X A M P L E   1

Express each of the following in simplest radical form.

     (a) $\sqrt{8}$ **(b)** $\sqrt{45}$ **(c)** $\sqrt[3]{24}$ **(d)** $\sqrt[3]{54}$

Solution

     (a) $\sqrt{8} = \sqrt{4 \cdot 2} = \sqrt{4}\sqrt{2} = 2\sqrt{2}$

                ↑

           4 is a
           perfect
           square.

(b) $\sqrt{45} = \sqrt{9 \cdot 5} = \sqrt{9}\sqrt{5} = 3\sqrt{5}$

9 is a
perfect
square.

(c) $\sqrt[3]{24} = \sqrt[3]{8 \cdot 3} = \sqrt[3]{8}\sqrt[3]{3} = 2\sqrt[3]{3}$

8 is a
perfect
cube.

(d) $\sqrt[3]{54} = \sqrt[3]{27 \cdot 2} = \sqrt[3]{27}\sqrt[3]{2} = 3\sqrt[3]{2}$

27 is a
perfect
cube.

■

The first step in each example is to express the radicand of the given radical as the product of two factors, one of which must be a perfect nth power other than 1. Also, observe the radicands of the final radicals. In each case, the radicand *cannot* be the product of two factors; one must be a perfect nth power other than 1. We say that the final radicals $2\sqrt{2}, 3\sqrt{5}, 2\sqrt[3]{3}$, and $3\sqrt[3]{2}$ are in **simplest radical form.**

You may vary the steps somewhat in changing to simplest radical form, but the final result should be the same. Consider some different approaches to changing $\sqrt{72}$ to simplest form.

$$\sqrt{72} = \sqrt{9}\sqrt{8} = 3\sqrt{8} = 3\sqrt{4}\sqrt{2} = 3 \cdot 2\sqrt{2} = 6\sqrt{2} \quad \text{or}$$
$$\sqrt{72} = \sqrt{4}\sqrt{18} = 2\sqrt{18} = 2\sqrt{9}\sqrt{2} = 2 \cdot 3\sqrt{2} = 6\sqrt{2} \quad \text{or}$$
$$\sqrt{72} = \sqrt{36}\sqrt{2} = 6\sqrt{2}$$

Another variation of the technique for changing radicals to simplest form is to prime-factor the radicand and then to look for perfect nth powers in exponential form. The following example illustrates the use of this technique.

EXAMPLE 2   Express each of the following in simplest radical form.

   (a) $\sqrt{50}$ **(b)** $3\sqrt{80}$ **(c)** $\sqrt[3]{108}$

Solution

   (a) $\sqrt{50} = \sqrt{2 \cdot 5 \cdot 5} = \sqrt{5^2}\sqrt{2} = 5\sqrt{2}$
   (b) $3\sqrt{80} = 3\sqrt{2 \cdot 2 \cdot 2 \cdot 2 \cdot 5} = 3\sqrt{2^4}\sqrt{5} = 3 \cdot 2^2\sqrt{5} = 12\sqrt{5}$
   (c) $\sqrt[3]{108} = \sqrt[3]{2 \cdot 2 \cdot 3 \cdot 3 \cdot 3} = \sqrt[3]{3^3}\sqrt[3]{4} = 3\sqrt[3]{4}$

■

Another property of nth roots is demonstrated by the following examples.

$$\sqrt{\frac{36}{9}} = \sqrt{4} = 2 \qquad \text{and} \qquad \frac{\sqrt{36}}{\sqrt{9}} = \frac{6}{3} = 2$$

$$\sqrt[3]{\frac{64}{8}} = \sqrt[3]{8} = 2 \qquad \text{and} \qquad \frac{\sqrt[3]{64}}{\sqrt[3]{8}} = \frac{4}{2} = 2$$

$$\sqrt[3]{\frac{-8}{64}} = \sqrt[3]{-\frac{1}{8}} = -\frac{1}{2} \qquad \text{and} \qquad \frac{\sqrt[3]{-8}}{\sqrt[3]{64}} = \frac{-2}{4} = -\frac{1}{2}$$

In general, we can state the following property.

PROPERTY 10.3

$$\sqrt[n]{\frac{b}{c}} = \frac{\sqrt[n]{b}}{\sqrt[n]{c}} \qquad \sqrt[n]{b} \text{ and } \sqrt[n]{c} \text{ are real numbers and } c \neq 0.$$

Property 10.3 states that **the nth root of a quotient is equal to the quotient of the nth roots.**

To evaluate radicals such as $\sqrt{\dfrac{4}{25}}$ and $\sqrt[3]{\dfrac{27}{8}}$, for which the numerator and denominator of the fractional radicand are perfect nth powers, you may use Property 10.3 or merely rely on the definition of nth root.

$$\sqrt{\frac{4}{25}} = \frac{\sqrt{4}}{\sqrt{25}} = \frac{2}{5} \qquad \text{or} \qquad \sqrt{\frac{4}{25}} = \frac{2}{5} \quad \text{because} \quad \frac{2}{5} \cdot \frac{2}{5} = \frac{4}{25}$$

$$\uparrow \qquad\qquad\qquad\qquad\qquad \uparrow$$
$$\text{Property 6.5} \qquad\qquad\qquad \text{Definition of } n\text{th root}$$
$$\downarrow \qquad\qquad\qquad\qquad\qquad \downarrow$$

$$\sqrt[3]{\frac{27}{8}} = \frac{\sqrt[3]{27}}{\sqrt[3]{8}} = \frac{3}{2} \qquad \text{or} \qquad \sqrt[3]{\frac{27}{8}} = \frac{3}{2} \quad \text{because} \quad \frac{3}{2} \cdot \frac{3}{2} \cdot \frac{3}{2} = \frac{27}{8}$$

Radicals such as $\sqrt{\dfrac{28}{9}}$ and $\sqrt[3]{\dfrac{24}{27}}$, in which only the denominators of the radicand are perfect nth powers, can be simplified as follows:

$$\sqrt{\frac{28}{9}} = \frac{\sqrt{28}}{\sqrt{9}} = \frac{\sqrt{28}}{3} = \frac{\sqrt{4}\sqrt{7}}{3} = \frac{2\sqrt{7}}{3}$$

$$\sqrt[3]{\frac{24}{27}} = \frac{\sqrt[3]{24}}{\sqrt[3]{27}} = \frac{\sqrt[3]{24}}{3} = \frac{\sqrt[3]{8}\sqrt[3]{3}}{3} = \frac{2\sqrt[3]{3}}{3}$$

Before we consider more examples, let's summarize some ideas that pertain to the simplifying of radicals. A radical is said to be in **simplest radical form** if the following conditions are satisfied.

1. No fraction appears with a radical sign. $\sqrt{\dfrac{3}{4}}$ violates this condition.

2. No radical appears in the denominator. $\dfrac{\sqrt{2}}{\sqrt{3}}$ violates this condition.

3. No radicand, when expressed in prime-factored form, contains a factor raised to a power equal to or greater than the index.
$$\sqrt{2^3 \cdot 5} \text{ violates this condition.}$$

Now let's consider an example in which neither the numerator nor the denominator of the radicand is a perfect nth power.

EXAMPLE 3 Simplify $\sqrt{\dfrac{2}{3}}$.

Solution

$$\sqrt{\frac{2}{3}} = \frac{\sqrt{2}}{\sqrt{3}} = \frac{\sqrt{2}}{\sqrt{3}} \cdot \frac{\sqrt{3}}{\sqrt{3}} = \frac{\sqrt{6}}{3}$$

Form of 1

We refer to the process we used to simplify the radical in Example 3 as **rationalizing the denominator.** Note that the denominator becomes a rational number. The process of rationalizing the denominator can often be accomplished in more than one way, as we will see in the next example.

EXAMPLE 4 Simplify $\dfrac{\sqrt{5}}{\sqrt{8}}$.

Solution A

$$\frac{\sqrt{5}}{\sqrt{8}} = \frac{\sqrt{5}}{\sqrt{8}} \cdot \frac{\sqrt{8}}{\sqrt{8}} = \frac{\sqrt{40}}{8} = \frac{\sqrt{4}\sqrt{10}}{8} = \frac{2\sqrt{10}}{8} = \frac{\sqrt{10}}{4}$$

Solution B

$$\frac{\sqrt{5}}{\sqrt{8}} = \frac{\sqrt{5}}{\sqrt{8}} \cdot \frac{\sqrt{2}}{\sqrt{2}} = \frac{\sqrt{10}}{\sqrt{16}} = \frac{\sqrt{10}}{4}$$

Solution C

$$\frac{\sqrt{5}}{\sqrt{8}} = \frac{\sqrt{5}}{\sqrt{4}\sqrt{2}} = \frac{\sqrt{5}}{2\sqrt{2}} = \frac{\sqrt{5}}{2\sqrt{2}} \cdot \frac{\sqrt{2}}{\sqrt{2}} = \frac{\sqrt{10}}{2\sqrt{4}} = \frac{\sqrt{10}}{2(2)} = \frac{\sqrt{10}}{4}$$

The three approaches to Example 4 again illustrate the need to think first and then push the pencil. You may find one approach easier than another. To conclude this section, study the following examples and check the final radicals against the three conditions previously listed for **simplest radical form.**

EXAMPLE 5

Simplify each of the following.

(a) $\dfrac{3\sqrt{2}}{5\sqrt{3}}$ (b) $\dfrac{3\sqrt{7}}{2\sqrt{18}}$ (c) $\sqrt[3]{\dfrac{5}{9}}$ (d) $\dfrac{\sqrt[3]{5}}{\sqrt[3]{16}}$

Solution

(a) $\dfrac{3\sqrt{2}}{5\sqrt{3}} = \dfrac{3\sqrt{2}}{5\sqrt{3}} \cdot \dfrac{\sqrt{3}}{\sqrt{3}} = \dfrac{3\sqrt{6}}{5\sqrt{9}} = \dfrac{3\sqrt{6}}{15} = \dfrac{\sqrt{6}}{5}$

$\uparrow$
Form of 1

(b) $\dfrac{3\sqrt{7}}{2\sqrt{18}} = \dfrac{3\sqrt{7}}{2\sqrt{18}} \cdot \dfrac{\sqrt{2}}{\sqrt{2}} = \dfrac{3\sqrt{14}}{2\sqrt{36}} = \dfrac{3\sqrt{14}}{12} = \dfrac{\sqrt{14}}{4}$

$\uparrow$
Form of 1

(c) $\sqrt[3]{\dfrac{5}{9}} = \dfrac{\sqrt[3]{5}}{\sqrt[3]{9}} = \dfrac{\sqrt[3]{5}}{\sqrt[3]{9}} \cdot \dfrac{\sqrt[3]{3}}{\sqrt[3]{3}} = \dfrac{\sqrt[3]{15}}{\sqrt[3]{27}} = \dfrac{\sqrt[3]{15}}{3}$

$\uparrow$
Form of 1

(d) $\dfrac{\sqrt[3]{5}}{\sqrt[3]{16}} = \dfrac{\sqrt[3]{5}}{\sqrt[3]{16}} \cdot \dfrac{\sqrt[3]{4}}{\sqrt[3]{4}} = \dfrac{\sqrt[3]{20}}{\sqrt[3]{64}} = \dfrac{\sqrt[3]{20}}{4}$

$\uparrow$
Form of 1

Applications of Radicals

Many real-world applications involve radical expressions. For example, police often use the formula $S = \sqrt{30Df}$ to estimate the speed of a car on the basis of the length of skid marks. In this formula, S represents the speed of the car in miles per hour, D represents the length of skid marks measured in feet, and f represents a coefficient of friction. For a particular situation, the coefficient of friction is a constant that depends on the type and condition of the road surface.

E X A M P L E   6

Using 0.35 as a coefficient of friction, determine how fast a car was traveling if it skidded 325 feet.

Solution
Substitute 0.35 for f and 325 for D in the formula.

$$S = \sqrt{30Df} = \sqrt{30(325)(0.35)} = 58, \quad \text{to the nearest whole number}$$

The car was traveling at approximately 58 miles per hour.   ■

The **period** of a pendulum is the time it takes to swing from one side to the other side and back. The formula

$$T = 2\pi\sqrt{\frac{L}{32}}$$

where T represents the time in seconds and L the length in feet, can be used to determine the period of a pendulum (see Figure 10.1).

E X A M P L E   7

Find, to the nearest tenth of a second, the period of a pendulum of length 3.5 feet.

Solution
Let's use 3.14 as an approximation for π and substitute 3.5 for L in the formula.

$$T = 2\pi\sqrt{\frac{L}{32}} = 2(3.14)\sqrt{\frac{3.5}{32}} = 2.1, \quad \text{to the nearest tenth}$$

The period is approximately 2.1 seconds.   ■

Radical expressions are also used in some geometric applications. For example, if a, b, and c represent the lengths of the three sides of a triangle, the formula $K = \sqrt{s(s-a)(s-b)(s-c)}$, known as Heron's formula, can be used to determine the area (K) of the triangle. The letter s represents the semiperimeter of the triangle; that is, $s = \dfrac{a+b+c}{2}$.

Figure 10.1

E X A M P L E   8

Find the area of a triangular piece of sheet metal that has sides of lengths 17 inches, 19 inches, and 26 inches.

Solution
First, let's find the value of s.

$$s = \frac{17 + 19 + 26}{2} = 31$$

Now we can use Heron's formula.

$$K = \sqrt{s(s - a)(s - b)(s - c)} = \sqrt{31(31 - 17)(31 - 19)(31 - 26)}$$
$$= \sqrt{31(14)(12)(5)}$$
$$= \sqrt{20{,}640}$$
$$= 161.4, \quad \text{to the nearest tenth}$$

Thus the area of the piece of sheet metal is approximately 161.4 square inches. ■

REMARK: Note that in Examples 6–8, we did not simplify the radicals. When you are using a calculator to approximate the square roots, there is no need to simplify first.

CONCEPT QUIZ

For Problems 1–8, answer true or false.

1. The cube root of a number is one of its three equal factors.

2. Every positive real number has one positive real number square root.

3. The principal square root of a number is the positive square root of the number.

4. The symbol $\sqrt{}$ is called a radical.

5. The square root of 0 is not a real number.

6. The number under the radical sign is called the radicand.

7. Every positive real number has two square roots.

8. The n in the radical $\sqrt[n]{a}$ is called the index of the radical.

PROBLEM SET 10.2

For Problems 1–20, evaluate each of the following. For example, $\sqrt{25} = 5$.

1. $\sqrt{64}$

2. $\sqrt{49}$

3. $-\sqrt{100}$

4. $-\sqrt{81}$

5. $\sqrt[3]{27}$

6. $\sqrt[3]{216}$

7. $\sqrt[3]{-64}$

8. $\sqrt[3]{-125}$

9. $\sqrt[4]{81}$

10. $-\sqrt[4]{16}$

11. $\sqrt{\dfrac{16}{25}}$

12. $\sqrt{\dfrac{25}{64}}$

13. $-\sqrt{\dfrac{36}{49}}$

14. $\sqrt{\dfrac{16}{64}}$

15. $\sqrt{\dfrac{9}{36}}$

16. $\sqrt{\dfrac{144}{36}}$

17. $\sqrt[3]{\dfrac{27}{64}}$

18. $\sqrt[3]{-\dfrac{8}{27}}$

19. $\sqrt[3]{8^3}$

20. $\sqrt[4]{16^4}$

For Problems 21–74, change each radical to simplest radical form.

21. $\sqrt{27}$

22. $\sqrt{48}$

23. $\sqrt{32}$

24. $\sqrt{98}$

25. $\sqrt{80}$

26. $\sqrt{125}$

27. $\sqrt{160}$

28. $\sqrt{112}$

29. $4\sqrt{18}$

30. $5\sqrt{32}$

31. $-6\sqrt{20}$

32. $-4\sqrt{54}$

33. $\dfrac{2}{5}\sqrt{75}$

34. $\dfrac{1}{3}\sqrt{90}$

35. $\dfrac{3}{2}\sqrt{24}$

36. $\dfrac{3}{4}\sqrt{45}$

37. $-\dfrac{5}{6}\sqrt{28}$

38. $-\dfrac{2}{3}\sqrt{96}$

39. $\sqrt{\dfrac{19}{4}}$

40. $\sqrt{\dfrac{22}{9}}$

41. $\sqrt{\dfrac{27}{16}}$

42. $\sqrt{\dfrac{8}{25}}$

43. $\sqrt{\dfrac{75}{81}}$

44. $\sqrt{\dfrac{24}{49}}$

45. $\sqrt{\dfrac{2}{7}}$

46. $\sqrt{\dfrac{3}{8}}$

47. $\sqrt{\dfrac{2}{3}}$

48. $\sqrt{\dfrac{7}{12}}$

49. $\dfrac{\sqrt{5}}{\sqrt{12}}$

50. $\dfrac{\sqrt{3}}{\sqrt{7}}$

51. $\dfrac{\sqrt{11}}{\sqrt{24}}$

52. $\dfrac{\sqrt{5}}{\sqrt{48}}$

53. $\dfrac{\sqrt{18}}{\sqrt{27}}$

54. $\dfrac{\sqrt{10}}{\sqrt{20}}$

55. $\dfrac{\sqrt{35}}{\sqrt{7}}$

56. $\dfrac{\sqrt{42}}{\sqrt{6}}$

57. $\dfrac{2\sqrt{3}}{\sqrt{7}}$

58. $\dfrac{3\sqrt{2}}{\sqrt{6}}$

59. $-\dfrac{4\sqrt{12}}{\sqrt{5}}$

60. $\dfrac{-6\sqrt{5}}{\sqrt{18}}$

61. $\dfrac{3\sqrt{2}}{4\sqrt{3}}$

62. $\dfrac{6\sqrt{5}}{5\sqrt{12}}$

63. $\dfrac{-8\sqrt{18}}{10\sqrt{50}}$

64. $\dfrac{4\sqrt{45}}{-6\sqrt{20}}$

65. $\sqrt[3]{16}$

66. $\sqrt[3]{40}$

67. $2\sqrt[3]{81}$

68. $-3\sqrt[3]{54}$

69. $\dfrac{2}{\sqrt[3]{9}}$

70. $\dfrac{3}{\sqrt[3]{3}}$

71. $\dfrac{\sqrt[3]{27}}{\sqrt[3]{4}}$

72. $\dfrac{\sqrt[3]{8}}{\sqrt[3]{16}}$

73. $\dfrac{\sqrt[3]{6}}{\sqrt[3]{4}}$

74. $\dfrac{\sqrt[3]{4}}{\sqrt[3]{2}}$

75. Use a coefficient of friction of 0.4 in the formula from Example 6 and find the speeds of cars that left skid marks of lengths 150 feet, 200 feet, and 350 feet. Express your answers to the nearest mile per hour.

76. Use the formula from Example 7 and find the periods of pendulums of lengths 2 feet, 3 feet, and 4.5 feet. Express your answers to the nearest tenth of a second.

77. Find, to the nearest square centimeter, the area of a triangle that measures 14 centimeters by 16 centimeters by 18 centimeters.

78. Find, to the nearest square yard, the area of a triangular plot of ground that measures 45 yards by 60 yards by 75 yards.

79. Find the area of an equilateral triangle, each of whose sides is 18 inches long. Express the area to the nearest square inch.

80. Find, to the nearest square inch, the area of the quadrilateral shown in Figure 10.2.

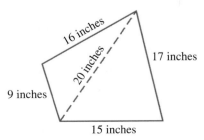

Figure 10.2

■ ■ ■ **Thoughts into words**

81. Why is $\sqrt{-9}$ not a real number?

82. Why do we say that 25 has two square roots (5 and -5), but we write $\sqrt{25} = 5$?

83. How is the multiplication property of 1 used when simplifying radicals?

84. How could you find a whole number approximation for $\sqrt{2750}$ if you did not have a calculator or table available?

 Calculator activities

85. Use your calculator to find a rational approximation, to the nearest thousandth, for each radical.

   a. $\sqrt{2}$ **b.** $\sqrt{75}$ **c.** $\sqrt{156}$

   d. $\sqrt{691}$ **e.** $\sqrt{3249}$ **f.** $\sqrt{45{,}123}$

   g. $\sqrt{0.14}$ **h.** $\sqrt{0.023}$ **i.** $\sqrt{0.8649}$

86. Sometimes a fairly good estimate can be made of a radical expression by using whole number approximations. For example, $5\sqrt{35} + 7\sqrt{50}$ is approximately $5(6) + 7(7) = 79$. Using a calculator, we find that $5\sqrt{35} + 7\sqrt{50} = 79.1$ to the nearest tenth. In this case our whole number estimate is very good. For a–f, first make a whole number estimate, and then use your calculator to see how well you estimated.

   a. $3\sqrt{10} - 4\sqrt{24} + 6\sqrt{65}$

   b. $9\sqrt{27} + 5\sqrt{37} - 3\sqrt{80}$

   c. $12\sqrt{5} + 13\sqrt{18} + 9\sqrt{47}$

   d. $3\sqrt{98} - 4\sqrt{83} - 7\sqrt{120}$

   e. $4\sqrt{170} + 2\sqrt{198} + 5\sqrt{227}$

   f. $-3\sqrt{256} - 6\sqrt{287} + 11\sqrt{321}$

Answers to Concept Quiz

  1. True   **2.** True   **3.** True   **4.** False   **5.** False   **6.** True   **7.** True   **8.** True

---

10.3   Simplifying and Combining Radicals

Objectives

■   Simplify radicals that contain variables.

■   Use addition and subtraction to combine radicals.

Recall our use of the distributive property as the basis for combining similar terms. For example,

$$3x + 2x = (3 + 2)x = 5x$$

$$8y - 5y = (8 - 5)y = 3y$$

$$\frac{2}{3}a^2 + \frac{3}{4}a^2 = \left(\frac{2}{3} + \frac{3}{4}\right)a^2 = \left(\frac{8}{12} + \frac{9}{12}\right)a^2 = \frac{17}{12}a^2$$

In a like manner, expressions that contain radicals can often be simplified by using the distributive property, as follows:

$$3\sqrt{2} + 5\sqrt{2} = (3 + 5)\sqrt{2} = 8\sqrt{2}$$
$$7\sqrt[3]{5} - 3\sqrt[3]{5} = (7 - 3)\sqrt[3]{5} = 4\sqrt[3]{5}$$
$$4\sqrt{7} + 5\sqrt{7} + 6\sqrt{11} - 2\sqrt{11} = (4 + 5)\sqrt{7} + (6 - 2)\sqrt{11} = 9\sqrt{7} + 4\sqrt{11}$$

Note that *in order to be added or subtracted, radicals must have the same index and the same radicand.* Thus we cannot simplify an expression such as $5\sqrt{2} + 7\sqrt{11}$.

Simplifying by combining radicals sometimes requires that you first express the given radicals in simplest form and then apply the distributive property. The following examples illustrate this idea.

EXAMPLE 1   Simplify $3\sqrt{8} + 2\sqrt{18} - 4\sqrt{2}$.

Solution

$$3\sqrt{8} + 2\sqrt{18} - 4\sqrt{2} = 3\sqrt{4}\sqrt{2} + 2\sqrt{9}\sqrt{2} - 4\sqrt{2}$$
$$= 6\sqrt{2} + 6\sqrt{2} - 4\sqrt{2}$$
$$= (6 + 6 - 4)\sqrt{2} = 8\sqrt{2} \quad\blacksquare$$

EXAMPLE 2   Simplify $\frac{1}{4}\sqrt{45} + \frac{1}{3}\sqrt{20}$.

Solution

$$\frac{1}{4}\sqrt{45} + \frac{1}{3}\sqrt{20} = \frac{1}{4}\sqrt{9}\sqrt{5} + \frac{1}{3}\sqrt{4}\sqrt{5}$$
$$= \frac{1}{4} \cdot 3 \cdot \sqrt{5} + \frac{1}{3} \cdot 2 \cdot \sqrt{5}$$
$$= \frac{3}{4}\sqrt{5} + \frac{2}{3}\sqrt{5} = \left(\frac{3}{4} + \frac{2}{3}\right)\sqrt{5}$$
$$= \left(\frac{9}{12} + \frac{8}{12}\right)\sqrt{5} = \frac{17}{12}\sqrt{5} \quad\blacksquare$$

EXAMPLE 3   Simplify $5\sqrt[3]{2} - 2\sqrt[3]{16} - 6\sqrt[3]{54}$.

Solution

$$5\sqrt[3]{2} - 2\sqrt[3]{16} - 6\sqrt[3]{54} = 5\sqrt[3]{2} - 2\sqrt[3]{8}\sqrt[3]{2} - 6\sqrt[3]{27}\sqrt[3]{2}$$
$$= 5\sqrt[3]{2} - 2 \cdot 2 \cdot \sqrt[3]{2} - 6 \cdot 3 \cdot \sqrt[3]{2}$$
$$= 5\sqrt[3]{2} - 4\sqrt[3]{2} - 18\sqrt[3]{2}$$
$$= (5 - 4 - 18)\sqrt[3]{2}$$
$$= -17\sqrt[3]{2} \quad\blacksquare$$

Radicals That Contain Variables

Before we discuss the process of simplifying *radicals that contain variables,* there is one technicality that we should call to your attention. Let's look at some examples to clarify the point. Consider the radical $\sqrt{x^2}$.

$$\text{Let } x = 3; \quad \text{then } \sqrt{x^2} = \sqrt{3^2} = \sqrt{9} = 3.$$
$$\text{Let } x = -3; \quad \text{then } \sqrt{x^2} = \sqrt{(-3)^2} = \sqrt{9} = 3.$$

Thus if $x \geq 0$, then $\sqrt{x^2} = x$, *but if* $x < 0$, then $\sqrt{x^2} = -x$. Using the concept of absolute value, we can state that *for all real numbers,* $\sqrt{x^2} = |x|$.

Now consider the radical $\sqrt{x^3}$. Because x^3 is negative when x is negative, we need to restrict x to the nonnegative reals when working with $\sqrt{x^3}$. Thus we can write: If $x \geq 0$, then $\sqrt{x^3} = \sqrt{x^2}\sqrt{x} = x\sqrt{x}$, and no absolute value sign is necessary. Finally, let's consider the radical $\sqrt[3]{x^3}$.

$$\text{Let } x = 2; \quad \text{then } \sqrt[3]{x^3} = \sqrt[3]{2^3} = \sqrt[3]{8} = 2.$$
$$\text{Let } x = -2; \quad \text{then } \sqrt[3]{x^3} = \sqrt[3]{(-2)^3} = \sqrt[3]{-8} = -2.$$

Thus it is correct to write $\sqrt[3]{x^3} = x$ for all real numbers, and again no absolute value sign is necessary.

The previous discussion indicates that technically, every radical expression involving variables in the radicand needs to be analyzed individually to determine whether it is necessary to impose restrictions on the variables. However, to avoid considering such restrictions on a problem-to-problem basis, we shall merely *assume that all variables represent positive real numbers.*

Let's consider the process of simplifying radicals that contain variables in the radicand. Study the following examples and note that this process is the same basic approach we used in Section 10.2.

EXAMPLE 4   Simplify each of the following.

(a) $\sqrt{8x^3}$ (b) $\sqrt{45x^3y^7}$ (c) $\sqrt{180a^4b^3}$ (d) $\sqrt[3]{40x^4y^8}$

Solution

(a) $\sqrt{8x^3} = \sqrt{4x^2}\sqrt{2x} = 2x\sqrt{2x}$

$4x^2$ is a
perfect square.

(b) $\sqrt{45x^3y^7} = \sqrt{9x^2y^6}\sqrt{5xy} = 3xy^3\sqrt{5xy}$

$9x^2y^6$ is a
perfect square.

(c) If the numerical coefficient of the radicand is quite large, you may want to look at it in the prime-factored form.

$$\sqrt{180a^4b^3} = \sqrt{2 \cdot 2 \cdot 3 \cdot 3 \cdot 5 \cdot a^4 \cdot b^3}$$
$$= \sqrt{36 \cdot 5 \cdot a^4 \cdot b^3}$$
$$= \sqrt{36a^4b^2}\sqrt{5b}$$
$$= 6a^2b\sqrt{5b}$$

(d) $\sqrt[3]{40x^4y^8} = \sqrt[3]{8x^3y^6}\sqrt[3]{5xy^2} = 2xy^2\sqrt[3]{5xy^2}$

$8x^3y^6$ is a perfect cube.

Before we consider more examples, let's restate (so as to include radicands containing variables) the conditions necessary for a radical to be in *simplest radical form*.

1. No radicand, when expressed in prime-factored form, contains a polynomial factor raised to a power equal to or greater than the index of the radical. $\sqrt{x^3}$ violates this condition.

2. No fraction appears within a radical sign. $\sqrt{\dfrac{2x}{3y}}$ violates this condition.

3. No radical appears in the denominator. $\dfrac{3}{\sqrt[3]{4x}}$ violates this condition.

E X A M P L E   5

Express each of the following in simplest radical form.

(a) $\sqrt{\dfrac{2x}{3y}}$ **(b)** $\dfrac{\sqrt{5}}{\sqrt{12a^3}}$ **(c)** $\dfrac{\sqrt{8x^2}}{\sqrt{27y^5}}$

(d) $\dfrac{3}{\sqrt[3]{4x}}$ **(e)** $\dfrac{\sqrt[3]{16x^2}}{\sqrt[3]{9y^5}}$

Solution

(a) $\sqrt{\dfrac{2x}{3y}} = \dfrac{\sqrt{2x}}{\sqrt{3y}} = \dfrac{\sqrt{2x}}{\sqrt{3y}} \cdot \dfrac{\sqrt{3y}}{\sqrt{3y}} = \dfrac{\sqrt{6xy}}{3y}$

Form of 1

(b) $\dfrac{\sqrt{5}}{\sqrt{12a^3}} = \dfrac{\sqrt{5}}{\sqrt{12a^3}} \cdot \dfrac{\sqrt{3a}}{\sqrt{3a}} = \dfrac{\sqrt{15a}}{\sqrt{36a^4}} = \dfrac{\sqrt{15a}}{6a^2}$

Form of 1

(c) $\dfrac{\sqrt{8x^2}}{\sqrt{27y^5}} = \dfrac{\sqrt{4x^2}\sqrt{2}}{\sqrt{9y^4}\sqrt{3y}} = \dfrac{2x\sqrt{2}}{3y^2\sqrt{3y}} = \dfrac{2x\sqrt{2}}{3y^2\sqrt{3y}} \cdot \dfrac{\sqrt{3y}}{\sqrt{3y}}$

$\qquad = \dfrac{2x\sqrt{6y}}{(3y^2)(3y)} = \dfrac{2x\sqrt{6y}}{9y^3}$

(d) $\dfrac{3}{\sqrt[3]{4x}} = \dfrac{3}{\sqrt[3]{4x}} \cdot \dfrac{\sqrt[3]{2x^2}}{\sqrt[3]{2x^2}} = \dfrac{3\sqrt[3]{2x^2}}{\sqrt[3]{8x^3}} = \dfrac{3\sqrt[3]{2x^2}}{2x}$

(e) $\dfrac{\sqrt[3]{16x^2}}{\sqrt[3]{9y^5}} = \dfrac{\sqrt[3]{16x^2}}{\sqrt[3]{9y^5}} \cdot \dfrac{\sqrt[3]{3y}}{\sqrt[3]{3y}} = \dfrac{\sqrt[3]{48x^2y}}{\sqrt[3]{27y^6}} = \dfrac{\sqrt[3]{8}\sqrt[3]{6x^2y}}{3y^2} = \dfrac{2\sqrt[3]{6x^2y}}{3y^2}$ ∎

Note that in part **(c)** we did some simplifying first before rationalizing the denominator, whereas in part **(b)** we proceeded immediately to rationalize the denominator. This is an individual choice, and you should probably do it both ways a few times to decide which you prefer.

CONCEPT QUIZ

For Problems 1–8, answer true or false.

1. In order to be combined when adding, radicals must have the same index and the same radicand.

2. If $x \geq 0$, then $\sqrt{x^2} = x$.

3. For all real numbers, $\sqrt{x^2} = x$.

4. For all real numbers, $\sqrt[3]{x^3} = x$.

5. A radical is not in simplest radical form if it has a fraction within the radical sign.

6. If a radical contains a factor raised to a power that is equal to the index of the radical, then the radical is not in simplest radical form.

7. The radical $\dfrac{1}{\sqrt{x}}$ is in simplest radical form.

8. $3\sqrt{2} + 4\sqrt{3} = 7\sqrt{5}$

PROBLEM SET 10.3

For Problems 1–20, use the distributive property to help simplify each of the following. For example,

$3\sqrt{8} - \sqrt{32} = 3\sqrt{4}\sqrt{2} - \sqrt{16}\sqrt{2}$

$\qquad = 3(2)\sqrt{2} - 4\sqrt{2}$

$\qquad = 6\sqrt{2} - 4\sqrt{2}$

$\qquad = (6 - 4)\sqrt{2} = 2\sqrt{2}$

1. $5\sqrt{18} - 2\sqrt{2}$

2. $7\sqrt{12} + 4\sqrt{3}$

3. $7\sqrt{12} + 10\sqrt{48}$

4. $6\sqrt{8} - 5\sqrt{18}$

5. $-2\sqrt{50} - 5\sqrt{32}$

6. $-2\sqrt{20} - 7\sqrt{45}$

7. $3\sqrt{20} - \sqrt{5} - 2\sqrt{45}$

8. $6\sqrt{12} + \sqrt{3} - 2\sqrt{48}$

9. $-9\sqrt{24} + 3\sqrt{54} - 12\sqrt{6}$

10. $13\sqrt{28} - 2\sqrt{63} - 7\sqrt{7}$

11. $\frac{3}{4}\sqrt{7} - \frac{2}{3}\sqrt{28}$ **12.** $\frac{3}{5}\sqrt{5} - \frac{1}{4}\sqrt{80}$

13. $\frac{3}{5}\sqrt{40} + \frac{5}{6}\sqrt{90}$ **14.** $\frac{3}{8}\sqrt{96} - \frac{2}{3}\sqrt{54}$

15. $\frac{3\sqrt{18}}{5} - \frac{5\sqrt{72}}{6} + \frac{3\sqrt{98}}{4}$

16. $\frac{-2\sqrt{20}}{3} + \frac{3\sqrt{45}}{4} - \frac{5\sqrt{80}}{6}$

17. $5\sqrt[3]{3} + 2\sqrt[3]{24} - 6\sqrt[3]{81}$

18. $-3\sqrt[3]{2} - 2\sqrt[3]{16} + \sqrt[3]{54}$

19. $-\sqrt[3]{16} + 7\sqrt[3]{54} - 9\sqrt[3]{2}$

20. $4\sqrt[3]{24} - 6\sqrt[3]{3} + 13\sqrt[3]{81}$

For Problems 21–64, express each of the following in simplest radical form. All variables represent positive real numbers.

21. $\sqrt{32x}$ **22.** $\sqrt{50y}$

23. $\sqrt{75x^2}$ **24.** $\sqrt{108y^2}$

25. $\sqrt{20x^2y}$ **26.** $\sqrt{80xy^2}$

27. $\sqrt{64x^3y^7}$ **28.** $\sqrt{36x^5y^6}$

29. $\sqrt{54a^4b^3}$ **30.** $\sqrt{96a^7b^8}$

31. $\sqrt{63x^6y^8}$ **32.** $\sqrt{28x^4y^{12}}$

33. $2\sqrt{40a^3}$ **34.** $4\sqrt{90a^5}$

35. $\frac{2}{3}\sqrt{96xy^3}$ **36.** $\frac{4}{5}\sqrt{125x^4y}$

37. $\sqrt{\frac{2x}{5y}}$ **38.** $\sqrt{\frac{3x}{2y}}$ **39.** $\sqrt{\frac{5}{12x^4}}$

40. $\sqrt{\frac{7}{8x^2}}$ **41.** $\frac{5}{\sqrt{18y}}$ **42.** $\frac{3}{\sqrt{12x}}$

43. $\frac{\sqrt{7x}}{\sqrt{8y^5}}$ **44.** $\frac{\sqrt{5y}}{\sqrt{18x^3}}$ **45.** $\frac{\sqrt{18y^3}}{\sqrt{16x}}$

46. $\frac{\sqrt{2x^3}}{\sqrt{9y}}$ **47.** $\frac{\sqrt{24a^2b^3}}{\sqrt{7ab^6}}$ **48.** $\frac{\sqrt{12a^2b}}{\sqrt{5a^3b^3}}$

49. $\sqrt[3]{24y}$ **50.** $\sqrt[3]{16x^2}$ **51.** $\sqrt[3]{16x^4}$

52. $\sqrt[3]{54x^3}$ **53.** $\sqrt[3]{56x^6y^8}$ **54.** $\sqrt[3]{81x^5y^6}$

55. $\sqrt[3]{\frac{7}{9x^2}}$ **56.** $\sqrt[3]{\frac{5}{2x}}$ **57.** $\frac{\sqrt[3]{3y}}{\sqrt[3]{16x^4}}$

58. $\frac{\sqrt[3]{2y}}{\sqrt[3]{3x}}$ **59.** $\frac{\sqrt[3]{12xy}}{\sqrt[3]{3x^2y^5}}$ **60.** $\frac{5}{\sqrt[3]{9xy^2}}$

61. $\sqrt{8x + 12y}$ [*Hint:* $\sqrt{8x + 12y} = \sqrt{4(2x + 3y)}$]

62. $\sqrt{4x + 4y}$ **63.** $\sqrt{16x + 48y}$

64. $\sqrt{27x + 18y}$

For Problems 65–74, use the distributive property to help simplify each of the following. All variables represent positive real numbers.

65. $-3\sqrt{4x} + 5\sqrt{9x} + 6\sqrt{16x}$

66. $-2\sqrt{25x} - 4\sqrt{36x} + 7\sqrt{64x}$

67. $2\sqrt{18x} - 3\sqrt{8x} - 6\sqrt{50x}$

68. $4\sqrt{20x} + 5\sqrt{45x} - 10\sqrt{80x}$

69. $5\sqrt{27n} - \sqrt{12n} - 6\sqrt{3n}$

70. $4\sqrt{8n} + 3\sqrt{18n} - 2\sqrt{72n}$

71. $7\sqrt{4ab} - \sqrt{16ab} - 10\sqrt{25ab}$

72. $4\sqrt{ab} - 9\sqrt{36ab} + 6\sqrt{49ab}$

73. $-3\sqrt{2x^3} + 4\sqrt{8x^3} - 3\sqrt{32x^3}$

74. $2\sqrt{40x^5} - 3\sqrt{90x^5} + 5\sqrt{160x^5}$

■ ■ ■ **Thoughts into words**

75. Is the expression $3\sqrt{2} + \sqrt{50}$ in simplest radical form? Defend your answer.

76. Your friend simplified $\frac{\sqrt{6}}{\sqrt{8}}$ as follows:

$$\frac{\sqrt{6}}{\sqrt{8}} \cdot \frac{\sqrt{8}}{\sqrt{8}} = \frac{\sqrt{48}}{8} = \frac{\sqrt{16}\sqrt{3}}{8} = \frac{4\sqrt{3}}{8} = \frac{\sqrt{3}}{2}$$

Is this a correct procedure? Can you show her a better way to do this problem?

77. Does $\sqrt{x + y}$ equal $\sqrt{x} + \sqrt{y}$? Defend your answer.

■ ■ ■ **Further investigations**

78. Do the following problems, where the variable could be any real number as long as the radical represents a real number. Use absolute-value signs in the answers as necessary.

(a) $\sqrt{125x^2}$

(b) $\sqrt{16x^4}$

(c) $\sqrt{8b^3}$

(d) $\sqrt{3y^5}$

(e) $\sqrt{288x^6}$

(f) $\sqrt{28m^8}$

(g) $\sqrt{128c^{10}}$

(h) $\sqrt{18d^7}$

(i) $\sqrt{49x^2}$

(j) $\sqrt{80n^{20}}$

(k) $\sqrt{81h^3}$

Answers to Concept Quiz

1. True   **2.** True   **3.** False   **4.** True   **5.** True   **6.** True   **7.** False   **8.** False

10.4   Products and Quotients of Radicals

Objectives

■   Multiply radical expressions.

■   Rationalize binomial denominators.

As we have seen, Property 10.2 ($\sqrt[n]{bc} = \sqrt[n]{b}\sqrt[n]{c}$) is used to express one radical as the product of two radicals and also to express the product of two radicals as one radical. In fact, we have used the property for both purposes within the framework of simplifying radicals. For example,

$$\frac{\sqrt{3}}{\sqrt{32}} = \frac{\sqrt{3}}{\sqrt{16}\sqrt{2}} = \frac{\sqrt{3}}{4\sqrt{2}} = \frac{\sqrt{3}}{4\sqrt{2}} \cdot \frac{\sqrt{2}}{\sqrt{2}} = \frac{\sqrt{6}}{8}$$

$$\sqrt[n]{bc} = \sqrt[n]{b}\sqrt[n]{c} \qquad \sqrt[n]{b}\sqrt[n]{c} = \sqrt[n]{bc}$$

The following examples demonstrate the use of Property 10.2 to multiply radicals and to express the product in simplest form.

EXAMPLE 1   Multiply and simplify where possible.

(a) $(2\sqrt{3})(3\sqrt{5})$

(b) $(3\sqrt{8})(5\sqrt{2})$

(c) $(7\sqrt{6})(3\sqrt{8})$

(d) $(2\sqrt[3]{6})(5\sqrt[3]{4})$

Solution

(a) $(2\sqrt{3})(3\sqrt{5}) = 2 \cdot 3 \cdot \sqrt{3} \cdot \sqrt{5} = 6\sqrt{15}$

(b) $(3\sqrt{8})(5\sqrt{2}) = 3 \cdot 5 \cdot \sqrt{8} \cdot \sqrt{2} = 15\sqrt{16} = 15 \cdot 4 = 60$

(c) $(7\sqrt{6})(3\sqrt{8}) = 7 \cdot 3 \cdot \sqrt{6} \cdot \sqrt{8} = 21\sqrt{48} = 21\sqrt{16}\sqrt{3}$
$$= 21 \cdot 4 \cdot \sqrt{3} = 84\sqrt{3}$$

(d) $(2\sqrt[3]{6})(5\sqrt[3]{4}) = 2 \cdot 5 \cdot \sqrt[3]{6} \cdot \sqrt[3]{4} = 10\sqrt[3]{24}$
$$= 10\sqrt[3]{8}\sqrt[3]{3}$$
$$= 10 \cdot 2 \cdot \sqrt[3]{3}$$
$$= 20\sqrt[3]{3} \qquad \blacksquare$$

Recall the use of the distributive property in finding the product of a monomial and a polynomial. For example, $3x^2(2x + 7) = 3x^2(2x) + 3x^2(7) = 6x^3 + 21x^2$. In a similar manner, the distributive property and Property 10.2 provide the basis for finding certain special products that involve radicals. The following examples illustrate this idea.

E X A M P L E   2

Multiply and simplify where possible.

(a) $\sqrt{3}(\sqrt{6} + \sqrt{12})$ **(b)** $2\sqrt{2}(4\sqrt{3} - 5\sqrt{6})$

(c) $\sqrt{6x}(\sqrt{8x} + \sqrt{12xy})$ **(d)** $\sqrt[3]{2}(5\sqrt[3]{4} - 3\sqrt[3]{16})$

Solution

(a) $\sqrt{3}(\sqrt{6} + \sqrt{12}) = \sqrt{3}\sqrt{6} + \sqrt{3}\sqrt{12}$
$$= \sqrt{18} + \sqrt{36}$$
$$= \sqrt{9}\sqrt{2} + 6$$
$$= 3\sqrt{2} + 6$$

(b) $2\sqrt{2}(4\sqrt{3} - 5\sqrt{6}) = (2\sqrt{2})(4\sqrt{3}) - (2\sqrt{2})(5\sqrt{6})$
$$= 8\sqrt{6} - 10\sqrt{12}$$
$$= 8\sqrt{6} - 10\sqrt{4}\sqrt{3}$$
$$= 8\sqrt{6} - 20\sqrt{3}$$

(c) $\sqrt{6x}(\sqrt{8x} + \sqrt{12xy}) = (\sqrt{6x})(\sqrt{8x}) + (\sqrt{6x})(\sqrt{12xy})$
$$= \sqrt{48x^2} + \sqrt{72x^2y}$$
$$= \sqrt{16x^2}\sqrt{3} + \sqrt{36x^2}\sqrt{2y}$$
$$= 4x\sqrt{3} + 6x\sqrt{2y}$$

(d) $\sqrt[3]{2}(5\sqrt[3]{4} - 3\sqrt[3]{16}) = (\sqrt[3]{2})(5\sqrt[3]{4}) - (\sqrt[3]{2})(3\sqrt[3]{16})$
$$= 5\sqrt[3]{8} - 3\sqrt[3]{32}$$
$$= 5 \cdot 2 - 3\sqrt[3]{8}\sqrt[3]{4}$$
$$= 10 - 6\sqrt[3]{4} \qquad \blacksquare$$

The distributive property also plays a central role in determining the product of two binomials. For example, $(x + 2)(x + 3) = x(x + 3) + 2(x + 3) = x^2 + 3x + 2x + 6 = x^2 + 5x + 6$. Finding the product of two binomial expressions that involve radicals can be handled in a similar fashion, as in the next examples.

EXAMPLE 3   Find the following products and simplify.

(a) $(\sqrt{3} + \sqrt{5})(\sqrt{2} + \sqrt{6})$ **(b)** $(2\sqrt{2} - \sqrt{7})(3\sqrt{2} + 5\sqrt{7})$

(c) $(\sqrt{8} + \sqrt{6})(\sqrt{8} - \sqrt{6})$ **(d)** $(\sqrt{x} + \sqrt{y})(\sqrt{x} - \sqrt{y})$

Solution

(a) $(\sqrt{3} + \sqrt{5})(\sqrt{2} + \sqrt{6}) = \sqrt{3}(\sqrt{2} + \sqrt{6}) + \sqrt{5}(\sqrt{2} + \sqrt{6})$
$= \sqrt{3}\sqrt{2} + \sqrt{3}\sqrt{6} + \sqrt{5}\sqrt{2} + \sqrt{5}\sqrt{6}$
$= \sqrt{6} + \sqrt{18} + \sqrt{10} + \sqrt{30}$
$= \sqrt{6} + 3\sqrt{2} + \sqrt{10} + \sqrt{30}$

(b) $(2\sqrt{2} - \sqrt{7})(3\sqrt{2} + 5\sqrt{7}) = 2\sqrt{2}(3\sqrt{2} + 5\sqrt{7})$
$- \sqrt{7}(3\sqrt{2} + 5\sqrt{7})$
$= (2\sqrt{2})(3\sqrt{2}) + (2\sqrt{2})(5\sqrt{7})$
$- (\sqrt{7})(3\sqrt{2}) - (\sqrt{7})(5\sqrt{7})$
$= 12 + 10\sqrt{14} - 3\sqrt{14} - 35$
$= -23 + 7\sqrt{14}$

(c) $(\sqrt{8} + \sqrt{6})(\sqrt{8} - \sqrt{6}) = \sqrt{8}(\sqrt{8} - \sqrt{6}) + \sqrt{6}(\sqrt{8} - \sqrt{6})$
$= \sqrt{8}\sqrt{8} - \sqrt{8}\sqrt{6} + \sqrt{6}\sqrt{8} - \sqrt{6}\sqrt{6}$
$= 8 - \sqrt{48} + \sqrt{48} - 6$
$= 2$

(d) $(\sqrt{x} + \sqrt{y})(\sqrt{x} - \sqrt{y}) = \sqrt{x}(\sqrt{x} - \sqrt{y}) + \sqrt{y}(\sqrt{x} - \sqrt{y})$
$= \sqrt{x}\sqrt{x} - \sqrt{x}\sqrt{y} + \sqrt{y}\sqrt{x} - \sqrt{y}\sqrt{y}$
$= x - \sqrt{xy} + \sqrt{xy} - y$
$= x - y$ ∎

Notice parts **(c)** and **(d)** of Example 3; they fit the special product pattern $(a + b)(a - b) = a^2 - b^2$. Furthermore, in each case the final product is in rational form. (The factors $a + b$ and $a - b$ are called conjugates.) This suggests a way of rationalizing the denominator in an expression that contains a binomial denominator with radicals. We will multiply by the conjugate of the binomial denominator. Consider the following example.

EXAMPLE 4

Simplify $\dfrac{4}{\sqrt{5} + \sqrt{2}}$ by rationalizing the denominator.

Solution

$$\frac{4}{\sqrt{5} + \sqrt{2}} = \frac{4}{\sqrt{5} + \sqrt{2}} \cdot \left(\frac{\sqrt{5} - \sqrt{2}}{\sqrt{5} - \sqrt{2}}\right) \qquad \text{A form of 1.}$$

$$= \frac{4(\sqrt{5} - \sqrt{2})}{(\sqrt{5} + \sqrt{2})(\sqrt{5} - \sqrt{2})} = \frac{4(\sqrt{5} - \sqrt{2})}{5 - 2}$$

$$= \frac{4(\sqrt{5} - \sqrt{2})}{3} \qquad \text{or} \qquad \frac{4\sqrt{5} - 4\sqrt{2}}{3}$$

Either answer
is acceptable.

The next examples further illustrate the process of rationalizing and simplifying expressions that contain binomial denominators.

EXAMPLE 5

For each of the following, rationalize the denominator and simplify.

(a) $\dfrac{\sqrt{3}}{\sqrt{6} - 9}$

(b) $\dfrac{7}{3\sqrt{5} + 2\sqrt{3}}$

(c) $\dfrac{\sqrt{x} + 2}{\sqrt{x} - 3}$

(d) $\dfrac{2\sqrt{x} - 3\sqrt{y}}{\sqrt{x} + \sqrt{y}}$

Solution

(a) $\dfrac{\sqrt{3}}{\sqrt{6} - 9} = \dfrac{\sqrt{3}}{\sqrt{6} - 9} \cdot \dfrac{\sqrt{6} + 9}{\sqrt{6} + 9}$

$$= \frac{\sqrt{3}(\sqrt{6} + 9)}{(\sqrt{6} - 9)(\sqrt{6} + 9)}$$

$$= \frac{\sqrt{18} + 9\sqrt{3}}{6 - 81}$$

$$= \frac{3\sqrt{2} + 9\sqrt{3}}{-75}$$

$$= \frac{3(\sqrt{2} + 3\sqrt{3})}{(-3)(25)}$$

$$= -\frac{\sqrt{2} + 3\sqrt{3}}{25} \qquad \text{or} \qquad \frac{-\sqrt{2} - 3\sqrt{3}}{25}$$

(b)
$$\frac{7}{3\sqrt{5}+2\sqrt{3}} = \frac{7}{3\sqrt{5}+2\sqrt{3}} \cdot \frac{3\sqrt{5}-2\sqrt{3}}{3\sqrt{5}-2\sqrt{3}}$$
$$= \frac{7(3\sqrt{5}-2\sqrt{3})}{(3\sqrt{5}+2\sqrt{3})(3\sqrt{5}-2\sqrt{3})}$$
$$= \frac{7(3\sqrt{5}-2\sqrt{3})}{45-12}$$
$$= \frac{7(3\sqrt{5}-2\sqrt{3})}{33} \quad \text{or} \quad \frac{21\sqrt{5}-14\sqrt{3}}{33}$$

(c)
$$\frac{\sqrt{x}+2}{\sqrt{x}-3} = \frac{\sqrt{x}+2}{\sqrt{x}-3} \cdot \frac{\sqrt{x}+3}{\sqrt{x}+3} = \frac{(\sqrt{x}+2)(\sqrt{x}+3)}{(\sqrt{x}-3)(\sqrt{x}+3)}$$
$$= \frac{x+3\sqrt{x}+2\sqrt{x}+6}{x-9}$$
$$= \frac{x+5\sqrt{x}+6}{x-9}$$

(d)
$$\frac{2\sqrt{x}-3\sqrt{y}}{\sqrt{x}+\sqrt{y}} = \frac{2\sqrt{x}-3\sqrt{y}}{\sqrt{x}+\sqrt{y}} \cdot \frac{\sqrt{x}-\sqrt{y}}{\sqrt{x}-\sqrt{y}}$$
$$= \frac{(2\sqrt{x}-3\sqrt{y})(\sqrt{x}-\sqrt{y})}{(\sqrt{x}+\sqrt{y})(\sqrt{x}-\sqrt{y})}$$
$$= \frac{2x-2\sqrt{xy}-3\sqrt{xy}+3y}{x-y}$$
$$= \frac{2x-5\sqrt{xy}+3y}{x-y}$$

CONCEPT QUIZ

For Problems 1–5, answer true or false.

1. The property $\sqrt[n]{x}\sqrt[n]{y} = \sqrt[n]{xy}$ can be used to express the product of two radicals as one radical.

2. The product of two radicals always results in an expression that has a radical even after simplifying.

3. The conjugate of $5 + \sqrt{3}$ is $-5 - \sqrt{3}$.

4. The product of $2 - \sqrt{7}$ and $2 + \sqrt{7}$ is a rational number.

5. To rationalize the denominator for the expression $\dfrac{2\sqrt{5}}{4-\sqrt{5}}$, we would multiply by $\dfrac{\sqrt{5}}{\sqrt{5}}$.

PROBLEM SET 10.4

For Problems 1–14, multiply and simplify where possible.

1. $\sqrt{6}\sqrt{12}$ **2.** $\sqrt{8}\sqrt{6}$

3. $(3\sqrt{3})(2\sqrt{6})$ **4.** $(5\sqrt{2})(3\sqrt{12})$

5. $(4\sqrt{2})(-6\sqrt{5})$ **6.** $(-7\sqrt{3})(2\sqrt{5})$

7. $(-3\sqrt{3})(-4\sqrt{8})$ **8.** $(-5\sqrt{8})(-6\sqrt{7})$

9. $(5\sqrt{6})(4\sqrt{6})$ **10.** $(3\sqrt{7})(2\sqrt{7})$

11. $(2\sqrt[3]{4})(6\sqrt[3]{2})$ **12.** $(4\sqrt[3]{3})(5\sqrt[3]{9})$

13. $(4\sqrt[3]{6})(7\sqrt[3]{4})$ **14.** $(9\sqrt[3]{6})(2\sqrt[3]{9})$

For Problems 15–52, find the following products and express answers in simplest radical form. All variables represent nonnegative real numbers.

15. $\sqrt{2}(\sqrt{3} + \sqrt{5})$ **16.** $\sqrt{3}(\sqrt{7} + \sqrt{10})$

17. $3\sqrt{5}(2\sqrt{2} - \sqrt{7})$ **18.** $5\sqrt{6}(2\sqrt{5} - 3\sqrt{11})$

19. $2\sqrt{6}(3\sqrt{8} - 5\sqrt{12})$ **20.** $4\sqrt{2}(3\sqrt{12} + 7\sqrt{6})$

21. $-4\sqrt{5}(2\sqrt{5} + 4\sqrt{12})$ **22.** $-5\sqrt{3}(3\sqrt{12} - 9\sqrt{8})$

23. $3\sqrt{x}(5\sqrt{2} + \sqrt{y})$ **24.** $\sqrt{2x}(3\sqrt{y} - 7\sqrt{5})$

25. $\sqrt{xy}(5\sqrt{xy} - 6\sqrt{x})$ **26.** $4\sqrt{x}(2\sqrt{xy} + 2\sqrt{x})$

27. $\sqrt{5y}(\sqrt{8x} + \sqrt{12y^2})$ **28.** $\sqrt{2x}(\sqrt{12xy} - \sqrt{8y})$

29. $5\sqrt{3}(2\sqrt{8} - 3\sqrt{18})$ **30.** $2\sqrt{2}(3\sqrt{12} - \sqrt{27})$

31. $(\sqrt{3} + 4)(\sqrt{3} - 7)$ **32.** $(\sqrt{2} + 6)(\sqrt{2} - 2)$

33. $(\sqrt{5} - 6)(\sqrt{5} - 3)$ **34.** $(\sqrt{7} - 2)(\sqrt{7} - 8)$

35. $(3\sqrt{5} - 2\sqrt{3})(2\sqrt{7} + \sqrt{2})$

36. $(\sqrt{2} + \sqrt{3})(\sqrt{5} - \sqrt{7})$

37. $(2\sqrt{6} + 3\sqrt{5})(\sqrt{8} - 3\sqrt{12})$

38. $(5\sqrt{2} - 4\sqrt{6})(2\sqrt{8} + \sqrt{6})$

39. $(2\sqrt{6} + 5\sqrt{5})(3\sqrt{6} - \sqrt{5})$

40. $(7\sqrt{3} - \sqrt{7})(2\sqrt{3} + 4\sqrt{7})$

41. $(3\sqrt{2} - 5\sqrt{3})(6\sqrt{2} - 7\sqrt{3})$

42. $(\sqrt{8} - 3\sqrt{10})(2\sqrt{8} - 6\sqrt{10})$

43. $(\sqrt{6} + 4)(\sqrt{6} - 4)$

44. $(\sqrt{7} - 2)(\sqrt{7} + 2)$

45. $(\sqrt{2} + \sqrt{10})(\sqrt{2} - \sqrt{10})$

46. $(2\sqrt{3} + \sqrt{11})(2\sqrt{3} - \sqrt{11})$

47. $(\sqrt{2x} + \sqrt{3y})(\sqrt{2x} - \sqrt{3y})$

48. $(2\sqrt{x} - 5\sqrt{y})(2\sqrt{x} + 5\sqrt{y})$

49. $2\sqrt[3]{3}(5\sqrt[3]{4} + \sqrt[3]{6})$

50. $2\sqrt[3]{2}(3\sqrt[3]{6} - 4\sqrt[3]{5})$

51. $3\sqrt[3]{4}(2\sqrt[3]{2} - 6\sqrt[3]{4})$

52. $3\sqrt[3]{3}(4\sqrt[3]{9} + 5\sqrt[3]{7})$

For Problems 53–76, rationalize the denominator and simplify. All variables represent positive real numbers.

53. $\dfrac{2}{\sqrt{7} + 1}$ **54.** $\dfrac{6}{\sqrt{5} + 2}$

55. $\dfrac{3}{\sqrt{2} - 5}$ **56.** $\dfrac{-4}{\sqrt{6} - 3}$

57. $\dfrac{1}{\sqrt{2} + \sqrt{7}}$ **58.** $\dfrac{3}{\sqrt{3} + \sqrt{10}}$

59. $\dfrac{\sqrt{2}}{\sqrt{10} - \sqrt{3}}$ **60.** $\dfrac{\sqrt{3}}{\sqrt{7} - \sqrt{2}}$

61. $\dfrac{\sqrt{3}}{2\sqrt{5} + 4}$ **62.** $\dfrac{\sqrt{7}}{3\sqrt{2} - 5}$

63. $\dfrac{6}{3\sqrt{7} - 2\sqrt{6}}$ **64.** $\dfrac{5}{2\sqrt{5} + 3\sqrt{7}}$

65. $\dfrac{\sqrt{6}}{3\sqrt{2} + 2\sqrt{3}}$ **66.** $\dfrac{3\sqrt{6}}{5\sqrt{3} - 4\sqrt{2}}$

67. $\dfrac{2}{\sqrt{x} + 4}$ **68.** $\dfrac{3}{\sqrt{x} + 7}$

69. $\dfrac{\sqrt{x}}{\sqrt{x} - 5}$ **70.** $\dfrac{\sqrt{x}}{\sqrt{x} - 1}$

71. $\dfrac{\sqrt{x} - 2}{\sqrt{x} + 6}$ **72.** $\dfrac{\sqrt{x} + 1}{\sqrt{x} - 10}$

73. $\dfrac{\sqrt{x}}{\sqrt{x} + 2\sqrt{y}}$ **74.** $\dfrac{\sqrt{y}}{2\sqrt{x} - \sqrt{y}}$

75. $\dfrac{3\sqrt{y}}{2\sqrt{x} - 3\sqrt{y}}$ **76.** $\dfrac{2\sqrt{x}}{3\sqrt{x} + 5\sqrt{y}}$

■ ■ ▪ **Thoughts into words**

77. How would you help someone rationalize the denominator and simplify $\dfrac{4}{\sqrt{8} + \sqrt{12}}$?

78. Discuss how the distributive property has been used thus far in this chapter.

79. How would you simplify the expression $\dfrac{\sqrt{8} + \sqrt{12}}{\sqrt{2}}$?

 Calculator activities

80. Use your calculator to evaluate each expression in Problems 53–66. Then evaluate the results you obtained when you did the problems.

Answers to Concept Quiz

1. True   **2.** False   **3.** False   **4.** True   **5.** False

---

10.5 Radical Equations

Objectives

■   Find the solution sets for radical equations.

■   Check the solutions for radical equations.

■   Solve formulas that are radical equations.

We often refer to equations that contain radicals with variables in a radicand as **radical equations.** In this section we discuss techniques for solving such equations that contain one or more radicals. To solve radical equations, we need the following property of equality:

> **PROPERTY 10.4**
>
> Let a and b be real numbers and n be a positive integer.
>
>   If $a = b$,   then $a^n = b^n$.

Property 10.4 states that we can *raise both sides of an equation to a positive integral power.* However, raising both sides of an equation to a positive integral power sometimes produces results that do not satisfy the original equation. Let's consider two examples to illustrate this point.

EXAMPLE 1   Solve $\sqrt{2x - 5} = 7$.

Solution

$$\sqrt{2x - 5} = 7$$
$$(\sqrt{2x - 5})^2 = 7^2 \qquad \text{Square both sides.}$$
$$2x - 5 = 49$$
$$2x = 54$$
$$x = 27$$

 Check

$$\sqrt{2x - 5} = 7$$
$$\sqrt{2(27) - 5} \stackrel{?}{=} 7$$
$$\sqrt{49} \stackrel{?}{=} 7$$
$$7 = 7$$

The solution set for $\sqrt{2x - 5} = 7$ is $\{27\}$.

EXAMPLE 2   Solve $\sqrt{3a + 4} = -4$.

Solution

$$\sqrt{3a + 4} = -4$$
$$(\sqrt{3a + 4})^2 = (-4)^2 \qquad \text{Square both sides.}$$
$$3a + 4 = 16$$
$$3a = 12$$
$$a = 4$$

 Check

$$\sqrt{3a + 4} = -4$$
$$\sqrt{3(4) + 4} \stackrel{?}{=} -4$$
$$\sqrt{16} \stackrel{?}{=} -4$$
$$4 \neq -4$$

Because 4 does not check, the original equation *has no real number solution.* Thus the solution set is $\varnothing$.

In general, raising both sides of an equation to a positive integral power produces an equation that has all of the solutions of the original equation, but it may also have some extra solutions that do not satisfy the original equation. Such extra solutions are called **extraneous solutions.** Therefore, when using Property 10.4, *you must check each potential solution in the original equation.*

Let's consider some examples to illustrate different situations that arise when we are solving radical equations.

EXAMPLE 3   Solve $\sqrt{2t - 4} = t - 2$.

Solution

$$\sqrt{2t - 4} = t - 2$$
$$(\sqrt{2t - 4})^2 = (t - 2)^2 \qquad \text{Square both sides.}$$
$$2t - 4 = t^2 - 4t + 4$$
$$0 = t^2 - 6t + 8$$
$$0 = (t - 2)(t - 4) \qquad \text{Factor the right side.}$$

$t - 2 = 0$ or $t - 4 = 0$ Apply: $ab = 0$ if and only if
$t = 2$ or $t = 4$ $a = 0$ or $b = 0$.

 Check

$$\sqrt{2t - 4} = t - 2 \qquad\qquad \sqrt{2t - 4} = t - 2$$
$$\sqrt{2(2) - 4} \overset{?}{=} 2 - 2, \quad \text{when } t = 2 \quad \text{or} \quad \sqrt{2(4) - 4} \overset{?}{=} 4 - 2, \quad \text{when } t = 4$$
$$\sqrt{0} \overset{?}{=} 0 \qquad\qquad\qquad \sqrt{4} \overset{?}{=} 2$$
$$0 = 0 \qquad\qquad\qquad\qquad 2 = 2$$

The solution set is $\{2, 4\}$.   ∎

EXAMPLE 4   Solve $\sqrt{y} + 6 = y$.

Solution

$$\sqrt{y} + 6 = y$$
$$\sqrt{y} = y - 6 \qquad \text{Isolate the radical.}$$
$$(\sqrt{y})^2 = (y - 6)^2 \qquad \text{Square both sides.}$$
$$y = y^2 - 12y + 36$$
$$0 = y^2 - 13y + 36$$
$$0 = (y - 4)(y - 9) \qquad \text{Factor the right side.}$$

$y - 4 = 0$ or $y - 9 = 0$ Apply: $ab = 0$ if and
$y = 4$ or $y = 9$ only if $a = 0$ or $b = 0$.

 Check

$$\sqrt{y} + 6 = y \qquad\qquad \sqrt{y} + 6 = y$$
$$\sqrt{4} + 6 \overset{?}{=} 4, \quad \text{when } y = 4 \quad \text{or} \quad \sqrt{9} + 6 \overset{?}{=} 9, \quad \text{when } y = 9$$
$$2 + 6 \overset{?}{=} 4 \qquad\qquad\qquad 3 + 6 \overset{?}{=} 9$$
$$8 \neq 4 \qquad\qquad\qquad\qquad 9 = 9$$

The only solution is 9; the solution set is $\{9\}$.   ∎

In Example 4, note that we changed the form of the original equation $\sqrt{y} + 6 = y$ to $\sqrt{y} = y - 6$ before we squared both sides. Squaring both sides of $\sqrt{y} + 6 = y$ produces $y + 12\sqrt{y} + 36 = y^2$, which is a much more complex equation that still contains a radical. Here again, it pays to think ahead a few steps before carrying out the details. Now let's consider an example involving a cube root.

EXAMPLE 5   Solve $\sqrt[3]{n^2 - 1} = 2$.

Solution

$$\sqrt[3]{n^2 - 1} = 2$$
$$(\sqrt[3]{n^2 - 1})^3 = 2^3 \qquad \text{Cube both sides.}$$
$$n^2 - 1 = 8$$
$$n^2 - 9 = 0$$
$$(n + 3)(n - 3) = 0$$
$$n + 3 = 0 \qquad \text{or} \qquad n - 3 = 0$$
$$n = -3 \qquad \text{or} \qquad n = 3$$

 Check

$$\sqrt[3]{n^2 - 1} = 2 \qquad\qquad\qquad \sqrt[3]{n^2 - 1} = 2$$
$$\sqrt[3]{(-3)^2 - 1} \stackrel{?}{=} 2, \quad \text{when } n = -3 \qquad \text{or} \qquad \sqrt[3]{3^2 - 1} \stackrel{?}{=} 2, \quad \text{when } n = 3$$
$$\sqrt[3]{8} \stackrel{?}{=} 2 \qquad\qquad\qquad\qquad \sqrt[3]{8} \stackrel{?}{=} 2$$
$$2 = 2 \qquad\qquad\qquad\qquad\qquad 2 = 2$$

The solution set is $\{-3, 3\}$. ■

It may be necessary to square both sides of an equation, simplify the resulting equation, and then square both sides again. The next example illustrates this type of problem.

EXAMPLE 6   Solve $\sqrt{x + 2} = 7 - \sqrt{x + 9}$.

 Solution

$$\sqrt{x + 2} = 7 - \sqrt{x + 9}$$
$$(\sqrt{x + 2})^2 = (7 - \sqrt{x + 9})^2 \qquad\qquad \text{Square both sides.}$$
$$x + 2 = 49 - 14\sqrt{x + 9} + x + 9$$
$$x + 2 = x + 58 - 14\sqrt{x + 9}$$
$$-56 = -14\sqrt{x + 9}$$
$$4 = \sqrt{x + 9}$$
$$(4)^2 = (\sqrt{x + 9})^2 \qquad\qquad \text{Square both sides.}$$
$$16 = x + 9$$
$$7 = x$$

Check

$$\sqrt{x + 2} = 7 - \sqrt{x + 9}$$
$$\sqrt{7 + 2} \overset{?}{=} 7 - \sqrt{7 + 9}$$
$$\sqrt{9} \overset{?}{=} 7 - \sqrt{16}$$
$$3 \overset{?}{=} 7 - 4$$
$$3 = 3$$

The solution set is {7}. ■

Another Look at Applications

In Section 10.2 we used the formula $S = \sqrt{30Df}$ to approximate, on the basis of the length of its skid marks, how fast a car was traveling when the brakes were applied. (Remember that S represents the speed of the car in miles per hour, D represents the length of the skid marks measured in feet, and f represents a coefficient of friction.) This same formula can be used to estimate the length of skid marks that are produced by cars traveling at different rates on various types of road surfaces. To use the formula for this purpose, let's change the form of the equation by solving for D.

$$\sqrt{30Df} = S$$
$$30Df = S^2 \quad \text{The result of squaring both sides of the original equation}$$
$$D = \frac{S^2}{30f} \quad \begin{array}{l}\text{D, S, and f are positive numbers, so this final equation} \\ \text{and the original one are equivalent.}\end{array}$$

EXAMPLE 7

Suppose that for a particular road surface, the coefficient of friction is 0.35. How far will a car skid when the brakes are applied at 60 miles per hour?

 Solution

We can substitute 0.35 for f and 60 for S in the formula $D = \frac{S^2}{30f}$.

$$D = \frac{60^2}{30(0.35)} = 343, \quad \text{to the nearest whole number}$$

The car will skid approximately 343 feet. ■

CONCEPT QUIZ

For Problems 1–5, answer true or false.

1. To solve a radical equation, we can raise each side of the equation to a positive integer power.

2. Solving the equation that results from squaring each side of an original equation may not give all the solutions of the original equation.

3. The equation $\sqrt[3]{x - 1} = -2$ has a solution.

4. Potential solutions that do not satisfy the original equation are called extraneous solutions.

5. The equation $\sqrt{x + 1} = -2$ has no solutions.

PROBLEM SET 10.5

For Problems 1–52, solve each equation. Don't forget to check each of your potential solutions.

1. $\sqrt{5x} = 10$

2. $\sqrt{3x} = 9$

3. $\sqrt{2x} + 4 = 0$

4. $\sqrt{4x} + 5 = 0$

5. $2\sqrt{n} = 5$

6. $5\sqrt{n} = 3$

7. $3\sqrt{n} - 2 = 0$

8. $2\sqrt{n} - 7 = 0$

9. $\sqrt{3y + 1} = 4$

10. $\sqrt{2y - 3} = 5$

11. $\sqrt{4y - 3} - 6 = 0$

12. $\sqrt{3y + 5} - 2 = 0$

13. $\sqrt{2x - 5} = -1$

14. $\sqrt{4x - 3} = -4$

15. $\sqrt{5x + 2} = \sqrt{6x + 1}$

16. $\sqrt{4x + 2} = \sqrt{3x + 4}$

17. $\sqrt{3x + 1} = \sqrt{7x - 5}$

18. $\sqrt{6x + 5} = \sqrt{2x + 10}$

19. $\sqrt{3x - 2} - \sqrt{x + 4} = 0$

20. $\sqrt{7x - 6} - \sqrt{5x + 2} = 0$

21. $5\sqrt{t - 1} = 6$

22. $4\sqrt{t + 3} = 6$

23. $\sqrt{x^2 + 7} = 4$

24. $\sqrt{x^2 + 3} - 2 = 0$

25. $\sqrt{x^2 + 13x + 37} = 1$

26. $\sqrt{x^2 + 5x - 20} = 2$

27. $\sqrt{x^2 - x + 1} = x + 1$

28. $\sqrt{n^2 - 2n - 4} = n$

29. $\sqrt{x^2 + 3x + 7} = x + 2$

30. $\sqrt{x^2 + 2x + 1} = x + 3$

31. $\sqrt{-4x + 17} = x - 3$

32. $\sqrt{2x - 1} = x - 2$

33. $\sqrt{n + 4} = n + 4$

34. $\sqrt{n + 6} = n + 6$

35. $\sqrt{3y} = y - 6$

36. $2\sqrt{n} = n - 3$

37. $4\sqrt{x + 5} = x$

38. $\sqrt{-x - 6} = x$

39. $\sqrt[3]{x - 2} = 3$

40. $\sqrt[3]{x + 1} = 4$

41. $\sqrt[3]{2x + 3} = -3$

42. $\sqrt[3]{3x - 1} = -4$

43. $\sqrt[3]{2x + 5} = \sqrt[3]{4 - x}$

44. $\sqrt[3]{3x - 1} = \sqrt[3]{2 - 5x}$

45. $\sqrt{x + 19} - \sqrt{x + 28} = -1$

46. $\sqrt{x + 4} = \sqrt{x - 1} + 1$

47. $\sqrt{3x + 1} + \sqrt{2x + 4} = 3$

48. $\sqrt{2x - 1} - \sqrt{x + 3} = 1$

49. $\sqrt{n - 4} + \sqrt{n + 4} = 2\sqrt{n - 1}$

50. $\sqrt{n - 3} + \sqrt{n + 5} = 2\sqrt{n}$

51. $\sqrt{t + 3} - \sqrt{t - 2} = \sqrt{7 - t}$

52. $\sqrt{t + 7} - 2\sqrt{t - 8} = \sqrt{t - 5}$

53. Use the formula given in Example 7 with a coefficient of friction of 0.95. How far will a car skid at 40 miles per hour? At 55 miles per hour? At 65 miles per hour? Express the answers to the nearest foot.

54. Solve the formula $T = 2\pi\sqrt{\dfrac{L}{32}}$ for L. (Remember that in this formula, which was used in Section 10.2, T represents the period of a pendulum expressed in seconds, and L represents the length of the pendulum in feet.)

55. In Problem 54, you should have obtained the equation $L = \dfrac{8T^2}{\pi^2}$. What is the length of a pendulum that has a period of 2 seconds? Of 2.5 seconds? Of 3 seconds? Express your answers to the nearest tenth of a foot.

■ ▨ ▨ Thoughts into words

56. Explain the concept of extraneous solutions.

57. Explain why possible solutions for radical equations *must* be checked.

58. Your friend makes an effort to solve the equation $3 + 2\sqrt{x} = x$ as follows:

$$(3 + 2\sqrt{x})^2 = x^2$$

$$9 + 12\sqrt{x} + 4x = x^2$$

At this step he stops and doesn't know how to proceed. What help would you give him?

Graphing calculator activities

59. Plot a few points and sketch the graph of $y = \sqrt{x}$. Then use your graphing calculator to graph $y = \sqrt{x}$.

60. On the basis of the graph of $y = \sqrt{x}$ in Problem 59, sketch the graph of each of the following equations. Then use your graphing calculator to check your sketches.

a. $y = \sqrt{x} + 3$ **b.** $y = \sqrt{x} - 2$

c. $y = 2\sqrt{x}$

d. $\dfrac{1}{2}\sqrt{x}$

e. $y = \sqrt{x} - 4$

f. $y = \sqrt{x} + 3$

g. $y = -\sqrt{x}$

h. $y = -\sqrt{x} - 2$

i. $y = 2\sqrt{x - 1} + 4$ **j.** $y = -2\sqrt{x + 3} - 2$

Answers to Concept Quiz

1. True   **2.** False   **3.** True   **4.** True   **5.** True

10.6   Merging Exponents and Roots

Objectives

- Convert rational exponent expressions into radical form.
- Write radical expressions using positive rational exponents.
- Simplify expressions with rational exponents.
- Multiply and divide radical expressions with different indexes.

Recall that the basic properties of positive integral exponents led to a definition for the use of negative integers as exponents. In this section, the properties of integral exponents are used to form definitions for the use of rational numbers as exponents. These definitions will tie together the concepts of *exponent* and *root*.

Let's consider the following comparisons.

From our study of radicals, we know that

If $(b^n)^m = b^{mn}$ is to hold when n equals a rational number of the form $\dfrac{1}{p}$, where p is a positive integer greater than 1, then

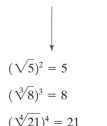

$$(\sqrt{5})^2 = 5$$
$$(\sqrt[3]{8})^3 = 8$$
$$(\sqrt[4]{21})^4 = 21$$

$$\left(5^{\frac{1}{2}}\right)^2 = 5^{2\left(\frac{1}{2}\right)} = 5^1 = 5$$
$$\left(8^{\frac{1}{3}}\right)^3 = 8^{3\left(\frac{1}{3}\right)} = 8^1 = 8$$
$$\left(21^{\frac{1}{4}}\right)^4 = 21^{4\left(\frac{1}{4}\right)} = 21^1 = 21$$

It would seem reasonable to make the following definition.

DEFINITION 10.4

If b is a real number, n is a positive integer greater than 1, and $\sqrt[n]{b}$ exists, then

$$b^{\frac{1}{n}} = \sqrt[n]{b}$$

Definition 10.4 states that $b^{\frac{1}{n}}$ *means the nth root of b*. We shall assume that b and n are chosen in such a way that $\sqrt[n]{b}$ exists. For example, $(-25)^{\frac{1}{2}}$ is not meaningful at this time because $\sqrt{-25}$ is not a real number. Consider the following examples, which demonstrate the use of Definition 10.4.

$$25^{\frac{1}{2}} = \sqrt{25} = 5 \qquad\qquad 16^{\frac{1}{4}} = \sqrt[4]{16} = 2$$

$$8^{\frac{1}{3}} = \sqrt[3]{8} = 2 \qquad\qquad \left(\frac{36}{49}\right)^{\frac{1}{2}} = \sqrt{\frac{36}{49}} = \frac{6}{7}$$

$$(-27)^{\frac{1}{3}} = \sqrt[3]{-27} = -3$$

The following definition provides the basis for the use of *all* rational numbers as exponents.

DEFINITION 10.5

If $\dfrac{m}{n}$ is a rational number, where n is a positive integer greater than 1, and b is a real number such that $\sqrt[n]{b}$ exists, then

$$b^{\frac{m}{n}} = \sqrt[n]{b^m} = (\sqrt[n]{b})^m$$

In Definition 10.5, note that the denominator of the exponent is the index of the radical and that the numerator of the exponent is either the exponent of the radicand or the exponent of the root.

Whether we use the form $\sqrt[n]{b^m}$ or the form $(\sqrt[n]{b})^m$ for computational purposes depends somewhat on the magnitude of the problem. Let's use both forms on two problems to illustrate this point.

$$8^{\frac{2}{3}} = \sqrt[3]{8^2} \qquad \text{or} \qquad 8^{\frac{2}{3}} = (\sqrt[3]{8})^2$$
$$= \sqrt[3]{64} \qquad\qquad\qquad = 2^2$$
$$= 4 \qquad\qquad\qquad\quad = 4$$

$$27^{\frac{2}{3}} = \sqrt[3]{27^2} \qquad \text{or} \qquad 27^{\frac{2}{3}} = (\sqrt[3]{27})^2$$
$$= \sqrt[3]{729} \qquad\qquad\qquad = 3^2$$
$$= 9 \qquad\qquad\qquad\quad = 9$$

To compute $8^{\frac{2}{3}}$, either form seems to work about as well as the other one. However, to compute $27^{\frac{2}{3}}$, it should be obvious that $(\sqrt[3]{27})^2$ is much easier to handle than $\sqrt[3]{27^2}$.

EXAMPLE 1   Simplify each of the following numerical expressions.

$$\textbf{(a)} \ 25^{\frac{3}{2}} \qquad\qquad \textbf{(b)} \ 16^{\frac{3}{4}} \qquad\qquad \textbf{(c)} \ (32)^{-\frac{2}{5}}$$

$$\textbf{(d)} \ (-64)^{\frac{2}{3}} \qquad\qquad \textbf{(e)} \ -8^{\frac{1}{3}}$$

Solution

$$\textbf{(a)} \ 25^{\frac{3}{2}} = (\sqrt{25})^3 = 5^3 = 125$$

$$\textbf{(b)} \ 16^{\frac{3}{4}} = (\sqrt[4]{16})^3 = 2^3 = 8$$

$$\textbf{(c)} \ (32)^{-\frac{2}{5}} = \frac{1}{(32)^{\frac{2}{5}}} = \frac{1}{(\sqrt[5]{32})^2} = \frac{1}{2^2} = \frac{1}{4}$$

$$\textbf{(d)} \ (-64)^{\frac{2}{3}} = (\sqrt[3]{-64})^2 = (-4)^2 = 16$$

$$\textbf{(e)} \ -8^{\frac{1}{3}} = -\sqrt[3]{8} = -2 \qquad\qquad \blacksquare$$

The basic laws of exponents that we stated in Property 8.6 are true for all rational exponents. Therefore, from now on we will use Property 8.6 for rational as well as integral exponents.

Some problems can be handled better in exponential form and others in radical form. Thus we must be able to switch forms with a certain amount of ease. Let's consider some examples where we switch from one form to the other.

EXAMPLE 2   Write each of the following expressions in radical form.

$$\textbf{(a)} \ x^{\frac{3}{4}} \qquad \textbf{(b)} \ 3y^{\frac{2}{5}} \qquad \textbf{(c)} \ x^{\frac{1}{4}}y^{\frac{3}{4}} \qquad \textbf{(d)} \ (x+y)^{\frac{2}{3}}$$

Solution

$$\textbf{(a)} \ x^{\frac{3}{4}} = \sqrt[4]{x^3} \qquad\qquad\qquad \textbf{(b)} \ 3y^{\frac{2}{5}} = 3\sqrt[5]{y^2}$$

$$\textbf{(c)} \ x^{\frac{1}{4}}y^{\frac{3}{4}} = (xy^3)^{\frac{1}{4}} = \sqrt[4]{xy^3} \qquad \textbf{(d)} \ (x+y)^{\frac{2}{3}} = \sqrt[3]{(x+y)^2} \qquad \blacksquare$$

EXAMPLE 3   Write each of the following using positive rational exponents.

$$\textbf{(a)} \ \sqrt{xy} \qquad \textbf{(b)} \ \sqrt[4]{a^3b} \qquad \textbf{(c)} \ 4\sqrt[3]{x^2} \qquad \textbf{(d)} \ \sqrt[5]{(x+y)^4}$$

Solution

$$\textbf{(a)} \ \sqrt{xy} = (xy)^{\frac{1}{2}} = x^{\frac{1}{2}}y^{\frac{1}{2}} \qquad\qquad \textbf{(b)} \ \sqrt[4]{a^3b} = (a^3b)^{\frac{1}{4}} = a^{\frac{3}{4}}b^{\frac{1}{4}}$$

$$\textbf{(c)} \ 4\sqrt[3]{x^2} = 4x^{\frac{2}{3}} \qquad\qquad\qquad \textbf{(d)} \ \sqrt[5]{(x+y)^4} = (x+y)^{\frac{4}{5}} \qquad \blacksquare$$

The basic properties of exponents provide the basis for simplifying algebraic expressions that contain rational exponents, as these next examples illustrate.

EXAMPLE 4   Simplify each of the following. Express final results using positive exponents only.

(a) $\left(3x^{\frac{1}{2}}\right)\left(4x^{\frac{2}{3}}\right)$ (b) $\left(5a^{\frac{1}{3}}b^{\frac{1}{2}}\right)^2$ (c) $\dfrac{12y^{\frac{1}{3}}}{6y^{\frac{1}{2}}}$ (d) $\left(\dfrac{3x^{\frac{2}{5}}}{2y^{\frac{2}{3}}}\right)^4$

Solution

(a) $\left(3x^{\frac{1}{2}}\right)\left(4x^{\frac{2}{3}}\right) = 3\cdot 4\cdot x^{\frac{1}{2}}\cdot x^{\frac{2}{3}}$

$\qquad = 12x^{\frac{1}{2}+\frac{2}{3}}$ $b^n\cdot b^m = b^{n+m}$

$\qquad = 12x^{\frac{3}{6}+\frac{4}{6}}$ Use 6 as the LCD.

$\qquad = 12x^{\frac{7}{6}}$

(b) $\left(5a^{\frac{1}{3}}b^{\frac{1}{2}}\right)^2 = 5^2\cdot\left(a^{\frac{1}{3}}\right)^2\cdot\left(b^{\frac{1}{2}}\right)^2$ $(ab)^n = a^n b^n$

$\qquad = 25a^{\frac{2}{3}}b$ $(b^n)^m = b^{mn}$

(c) $\dfrac{12y^{\frac{1}{3}}}{6y^{\frac{1}{2}}} = 2y^{\frac{1}{3}-\frac{1}{2}}$ $\dfrac{b^n}{b^m} = b^{n-m}$

$\qquad = 2y^{\frac{2}{6}-\frac{3}{6}}$

$\qquad = 2y^{-\frac{1}{6}}$

$\qquad = \dfrac{2}{y^{\frac{1}{6}}}$

(d) $\left(\dfrac{3x^{\frac{2}{5}}}{2y^{\frac{2}{3}}}\right)^4 = \dfrac{\left(3x^{\frac{2}{5}}\right)^4}{\left(2y^{\frac{2}{3}}\right)^4}$ $\left(\dfrac{a}{b}\right)^n = \dfrac{a^n}{b^n}$

$\qquad = \dfrac{3^4\cdot\left(x^{\frac{2}{5}}\right)^4}{2^4\cdot\left(y^{\frac{2}{3}}\right)^4}$ $(ab)^n = a^n b^n$

$\qquad = \dfrac{81x^{\frac{8}{5}}}{16y^{\frac{8}{3}}}$ $(b^n)^m = b^{mn}$ ∎

The link between exponents and roots also provides a basis for multiplying and dividing some radicals even if they have different indexes. The general procedure is as follows:

1. Change from radical form to exponential form.
2. Apply the properties of exponents.
3. Then change back to radical form.

The three parts of Example 5 illustrate this process.

EXAMPLE 5 Perform the indicated operations and express the answer in simplest radical form.

$$\text{(a) } \sqrt{2}\sqrt[3]{2} \qquad \text{(b) } \frac{\sqrt{5}}{\sqrt[3]{5}} \qquad \text{(c) } \frac{\sqrt{4}}{\sqrt[3]{2}}$$

Solution

(a) $\sqrt{2}\sqrt[3]{2} = 2^{\frac{1}{2}} \cdot 2^{\frac{1}{3}}$

$\qquad = 2^{\frac{1}{2}+\frac{1}{3}}$

$\qquad = 2^{\frac{3}{6}+\frac{2}{6}}$

$\qquad = 2^{\frac{5}{6}}$

$\qquad = \sqrt[6]{2^5} = \sqrt[6]{32}$

(b) $\dfrac{\sqrt{5}}{\sqrt[3]{5}} = \dfrac{5^{\frac{1}{2}}}{5^{\frac{1}{3}}}$

$\qquad = 5^{\frac{1}{2}-\frac{1}{3}}$

$\qquad = 5^{\frac{3}{6}-\frac{2}{6}}$

$\qquad = 5^{\frac{1}{6}} = \sqrt[6]{5}$

(c) $\dfrac{\sqrt{4}}{\sqrt[3]{2}} = \dfrac{4^{\frac{1}{2}}}{2^{\frac{1}{3}}}$

$\qquad = \dfrac{(2^2)^{\frac{1}{2}}}{2^{\frac{1}{3}}}$

$\qquad = \dfrac{2^1}{2^{\frac{1}{3}}}$

$\qquad = 2^{1-\frac{1}{3}}$

$\qquad = 2^{\frac{2}{3}} = \sqrt[3]{2^2} = \sqrt[3]{4}$

CONCEPT QUIZ

For Problems 1–5, answer true or false.

1. Assuming the nth root of x exists, $\sqrt[n]{x}$ can be written as $x^{\frac{1}{n}}$.

2. An exponent of $\dfrac{1}{3}$ means that we need to find the cube root of the number.

3. To evaluate $16^{\frac{2}{3}}$ we would find the square root of 16 and then cube the result.

4. When an expression with a rational exponent is written as a radical expression, the denominator of the rational exponent is the index of the radical.

5. The expression $\sqrt[n]{x^m}$ is equivalent to $(\sqrt[n]{x})^m$.

PROBLEM SET 10.6

For Problems 1–30, evaluate each numerical expression.

1. $81^{\frac{1}{2}}$

2. $64^{\frac{1}{2}}$

3. $27^{\frac{1}{3}}$

4. $(-32)^{\frac{1}{5}}$

5. $(-8)^{\frac{1}{3}}$

6. $\left(-\dfrac{27}{8}\right)^{\frac{1}{3}}$

7. $-25^{\frac{1}{2}}$

8. $-64^{\frac{1}{3}}$

9. $36^{-\frac{1}{2}}$

10. $81^{-\frac{1}{2}}$

11. $\left(\dfrac{1}{27}\right)^{-\frac{1}{3}}$

12. $\left(-\dfrac{8}{27}\right)^{-\frac{1}{3}}$

13. $4^{\frac{3}{2}}$

14. $64^{\frac{2}{3}}$

15. $27^{\frac{4}{3}}$

16. $4^{\frac{7}{2}}$

17. $(-1)^{\frac{7}{3}}$

18. $(-8)^{\frac{4}{3}}$

19. $-4^{\frac{5}{2}}$

20. $-16^{\frac{3}{2}}$

21. $\left(\dfrac{27}{8}\right)^{\frac{4}{3}}$

22. $\left(\dfrac{8}{125}\right)^{\frac{2}{3}}$

23. $\left(\dfrac{1}{8}\right)^{-\frac{2}{3}}$

24. $\left(-\dfrac{1}{27}\right)^{-\frac{2}{3}}$

25. $64^{-\frac{7}{6}}$

26. $32^{-\frac{4}{5}}$

27. $-25^{\frac{3}{2}}$

28. $-16^{\frac{3}{4}}$

29. $125^{\frac{4}{3}}$

30. $81^{\frac{5}{4}}$

For Problems 31–44, write each of the following in radical form. For example,

$$3x^{\frac{2}{3}} = 3\sqrt[3]{x^2}$$

31. $x^{\frac{4}{3}}$

32. $x^{\frac{2}{5}}$

33. $3x^{\frac{1}{2}}$

34. $5x^{\frac{1}{4}}$

35. $(2y)^{\frac{1}{3}}$

36. $(3xy)^{\frac{1}{2}}$

37. $(2x - 3y)^{\frac{1}{2}}$

38. $(5x + y)^{\frac{1}{3}}$

39. $(2a - 3b)^{\frac{2}{3}}$

40. $(5a + 7b)^{\frac{3}{5}}$

41. $x^{\frac{2}{3}}y^{\frac{1}{3}}$

42. $x^{\frac{3}{7}}y^{\frac{5}{7}}$

43. $-3x^{\frac{1}{5}}y^{\frac{2}{5}}$

44. $-4x^{\frac{3}{4}}y^{\frac{1}{4}}$

For Problems 45–58, write each of the following using positive rational exponents. For example,

$$\sqrt{ab} = (ab)^{\frac{1}{2}} = a^{\frac{1}{2}}b^{\frac{1}{2}}$$

45. $\sqrt{5y}$

46. $\sqrt{2xy}$

47. $3\sqrt{y}$

48. $5\sqrt{ab}$

49. $\sqrt[3]{xy^2}$

50. $\sqrt[5]{x^2y^4}$

51. $\sqrt[4]{a^2b^3}$

52. $\sqrt[6]{ab^5}$

53. $\sqrt[5]{(2x - y)^3}$

54. $\sqrt[7]{(3x - y)^4}$

55. $5x\sqrt{y}$

56. $4y\sqrt[3]{x}$

57. $-\sqrt[3]{x + y}$

58. $-\sqrt[5]{(x - y)^2}$

For Problems 59–80, simplify each of the following. Express final results using positive exponents only. For example,

$$\left(2x^{\frac{1}{2}}\right)\left(3x^{\frac{1}{3}}\right) = 6x^{\frac{5}{6}}$$

59. $\left(2x^{\frac{2}{5}}\right)\left(6x^{\frac{1}{4}}\right)$

60. $\left(3x^{\frac{1}{4}}\right)\left(5x^{\frac{1}{3}}\right)$

61. $\left(y^{\frac{2}{3}}\right)\left(y^{-\frac{1}{4}}\right)$

62. $\left(y^{\frac{3}{4}}\right)\left(y^{-\frac{1}{2}}\right)$

63. $\left(x^{\frac{2}{5}}\right)\left(4x^{-\frac{1}{2}}\right)$

64. $\left(2x^{\frac{1}{3}}\right)\left(x^{-\frac{1}{2}}\right)$

65. $\left(4x^{\frac{1}{2}}y\right)^2$

66. $\left(3x^{\frac{1}{4}}y^{\frac{1}{5}}\right)^3$

67. $(8x^6y^3)^{\frac{1}{3}}$

68. $(9x^2y^4)^{\frac{1}{2}}$

69. $\dfrac{24x^{\frac{3}{5}}}{6x^{\frac{1}{3}}}$

70. $\dfrac{18x^{\frac{1}{2}}}{9x^{\frac{1}{3}}}$

71. $\dfrac{48b^{\frac{1}{3}}}{12b^{\frac{3}{4}}}$

72. $\dfrac{56a^{\frac{1}{6}}}{8a^{\frac{1}{4}}}$

73. $\left(\dfrac{6x^{\frac{2}{5}}}{7y^{\frac{2}{3}}}\right)^2$

74. $\left(\dfrac{2x^{\frac{1}{3}}}{3y^{\frac{1}{4}}}\right)^4$

75. $\left(\dfrac{x^2}{y^3}\right)^{-\frac{1}{2}}$

76. $\left(\dfrac{a^3}{b^{-2}}\right)^{-\frac{1}{3}}$

77. $\left(\dfrac{18x^{\frac{1}{3}}}{9x^{\frac{1}{4}}}\right)^2$

78. $\left(\dfrac{72x^{\frac{3}{4}}}{6x^{\frac{1}{2}}}\right)^2$

79. $\left(\dfrac{60a^{\frac{1}{5}}}{15a^{\frac{3}{4}}}\right)^2$

80. $\left(\dfrac{64a^{\frac{1}{3}}}{16a^{\frac{5}{9}}}\right)^3$

For Problems 81–90, perform the indicated operations and express answers in simplest radical form. (See Example 5.)

81. $\sqrt[3]{3}\sqrt{3}$

82. $\sqrt{2}\sqrt[4]{2}$

83. $\sqrt[4]{6}\sqrt{6}$

84. $\sqrt[3]{5}\sqrt{5}$

85. $\dfrac{\sqrt[3]{3}}{\sqrt[4]{3}}$

86. $\dfrac{\sqrt{2}}{\sqrt[3]{2}}$

87. $\dfrac{\sqrt[3]{8}}{\sqrt[4]{4}}$

88. $\dfrac{\sqrt{9}}{\sqrt[3]{3}}$

89. $\dfrac{\sqrt[4]{27}}{\sqrt{3}}$

90. $\dfrac{\sqrt[3]{16}}{\sqrt[6]{4}}$

■ ■ ■ Thoughts into words

91. Your friend keeps getting an error message when evaluating $-4^{\frac{5}{2}}$ on his calculator. What error is he probably making?

92. Explain how you would evaluate $27^{\frac{2}{3}}$ without a calculator.

 Calculator activities

93. Use your calculator to evaluate each of the following.

a. $\sqrt[3]{1728}$

b. $\sqrt[3]{5832}$

c. $\sqrt[4]{2401}$

d. $\sqrt[4]{65,536}$

e. $\sqrt[5]{161,051}$

f. $\sqrt[5]{6,436,343}$

94. Definition 10.5 states that

$$b^{\frac{m}{n}} = \sqrt[n]{b^m} = (\sqrt[n]{b})^m$$

Use your calculator to verify each of the following.

a. $\sqrt[3]{27^2} = (\sqrt[3]{27})^2$

b. $\sqrt[3]{8^5} = (\sqrt[3]{8})^5$

c. $\sqrt[4]{16^3} = (\sqrt[4]{16})^3$

d. $\sqrt[3]{16^2} = (\sqrt[3]{16})^2$

e. $\sqrt[5]{9^4} = (\sqrt[5]{9})^4$

f. $\sqrt[3]{12^4} = (\sqrt[3]{12})^4$

95. Use your calculator to evaluate each of the following.

a. $16^{\frac{5}{2}}$

b. $25^{\frac{7}{2}}$

c. $16^{\frac{9}{4}}$

d. $27^{\frac{5}{3}}$

e. $343^{\frac{2}{3}}$

f. $512^{\frac{4}{3}}$

96. Use your calculator to estimate each of the following to the nearest thousandth.

a. $7^{\frac{4}{3}}$

b. $10^{\frac{4}{5}}$

c. $12^{\frac{3}{5}}$

d. $19^{\frac{2}{5}}$

e. $7^{\frac{3}{4}}$

f. $10^{\frac{5}{4}}$

97. a. Because $\dfrac{4}{5} = 0.8$, we can evaluate $10^{\frac{4}{5}}$ by evaluating $10^{0.8}$, which involves a shorter sequence of "calculator steps." Evaluate parts b, c, d, e, and f of Problem 96 and take advantage of decimal exponents.

b. What problem is created when we try to evaluate $7^{\frac{4}{3}}$ by changing the exponent to decimal form?

SUMMARY

(10.1) The following properties form the basis for manipulating with exponents.

1. $b^n \cdot b^m = b^{n+m}$ Product of two like bases with powers
2. $(b^n)^m = b^{mn}$ Power of a power
3. $(ab)^n = a^n b^n$ Power of a product
4. $\left(\dfrac{a}{b}\right)^n = \dfrac{a^n}{b^n}$ Power of a quotient
5. $\dfrac{b^n}{b^m} = b^{n-m}$ Quotient of two like bases with powers

The **scientific form** of a number is expressed as

$$(N)(10)^k$$

where N is a number between 1 and 10 (including 1) written in decimal form, and k is an integer. Scientific notation is often convenient to use with very small and very large numbers. For example, 0.000046 can be expressed as $(4.6)(10^{-5})$, and 92,000,000 can be written as $(9.2)(10)^7$.

Scientific notation can often be used to simplify numerical calculations. For example,

$$(0.000016)(30,000) = (1.6)(10)^{-5}(3)(10)^4$$
$$= (4.8)(10)^{-1} = 0.48$$

(10.2) and (10.3) The **principal nth root of b** is designated by $\sqrt[n]{b}$, where n is the **index,** and b is the **radicand.**

A radical expression is in **simplest radical form** if

1. A radicand contains no polynomial factor raised to a power equal to or greater than the index of the radical.
2. No fraction appears within a radical sign.
3. No radical appears in the denominator.

The following properties are used to express radicals in simplest form.

$$\sqrt[n]{bc} = \sqrt[n]{b}\sqrt[n]{c} \qquad \sqrt[n]{\dfrac{b}{c}} = \dfrac{\sqrt[n]{b}}{\sqrt[n]{c}}$$

Simplifying by combining radicals sometimes requires that we first express the given radicals in simplest form and then apply the distributive property.

(10.4) The distributive property and the property $\sqrt[n]{b}\sqrt[n]{c} = \sqrt[n]{bc}$ are used to find products of expressions that involve radicals.

The special product pattern $(a + b)(a - b) = a^2 - b^2$ suggests a procedure for **rationalizing the denominator** of an expression that contains a binomial denominator with radicals.

(10.5) Equations that contain radicals with variables in a radicand are called **radical equations.** The property "if $a = b$, then $a^n = b^n$" forms the basis for solving radical equations. Raising both sides of an equation to a positive integral power may produce **extraneous solutions**—that is, solutions that do not satisfy the original equation. Therefore, *you must check* each potential solution.

(10.6) If b is a real number, n is a positive integer greater than 1, and $\sqrt[n]{b}$ exists, then

$$b^{\frac{1}{n}} = \sqrt[n]{b}$$

Thus $b^{\frac{1}{n}}$ means **the nth root of b.**

If $\dfrac{m}{n}$ is a rational number, n is a positive integer greater than 1, and b is a real number such that $\sqrt[n]{b}$ exists, then

$$b^{\frac{m}{n}} = \sqrt[n]{b^m} = (\sqrt[n]{b})^m$$

Both $\sqrt[n]{b^m}$ and $(\sqrt[n]{b})^m$ can be used for computational purposes.

We need to be able to switch back and forth between **exponential form** and **radical form.** The link between exponents and roots provides a basis for multiplying and dividing some radicals, even if they have different indexes.

CHAPTER 10 REVIEW PROBLEM SET

For Problems 1–12, evaluate each of the following numerical expressions.

1. 4^{-3}

2. $\left(\dfrac{2}{3}\right)^{-2}$

3. $(3^2 \cdot 3^{-3})^{-1}$

4. $\sqrt[3]{-8}$

5. $\sqrt[4]{\dfrac{16}{81}}$

6. $4^{\frac{5}{2}}$

7. $(-1)^{\frac{2}{3}}$

8. $\left(\dfrac{8}{27}\right)^{\frac{2}{3}}$

9. $-16^{\frac{3}{2}}$

10. $\dfrac{2^3}{2^{-2}}$

11. $(4^{-2} \cdot 4^2)^{-1}$

12. $\left(\dfrac{3^{-1}}{3^2}\right)^{-1}$

For Problems 13–24, express each of the following radicals in simplest radical form.

13. $\sqrt{54}$

14. $\sqrt{48x^3y}$

15. $\dfrac{4\sqrt{3}}{\sqrt{6}}$

16. $\sqrt{\dfrac{5}{12x^3}}$

17. $\sqrt[3]{56}$

18. $\dfrac{\sqrt[3]{2}}{\sqrt[3]{9}}$

19. $\sqrt{\dfrac{9}{5}}$

20. $\sqrt{\dfrac{3x^3}{7}}$

21. $\sqrt[3]{108x^4y^8}$

22. $\dfrac{3}{4}\sqrt{150}$

23. $\dfrac{2}{3}\sqrt{45xy^3}$

24. $\dfrac{\sqrt{8x^2}}{\sqrt{2x}}$

For Problems 25–32, multiply and simplify.

25. $(3\sqrt{8})(4\sqrt{5})$

26. $(5\sqrt[3]{2})(6\sqrt[3]{4})$

27. $3\sqrt{2}(4\sqrt{6} - 2\sqrt{7})$

28. $(\sqrt{x} + 3)(\sqrt{x} - 5)$

29. $(2\sqrt{5} - \sqrt{3})(2\sqrt{5} + \sqrt{3})$

30. $(3\sqrt{2} + \sqrt{6})(5\sqrt{2} - 3\sqrt{6})$

31. $(2\sqrt{a} + \sqrt{b})(3\sqrt{a} - 4\sqrt{b})$

32. $(4\sqrt{8} - \sqrt{2})(\sqrt{8} + 3\sqrt{2})$

For Problems 33–36, rationalize the denominator and simplify.

33. $\dfrac{4}{\sqrt{7} - 1}$

34. $\dfrac{\sqrt{3}}{\sqrt{8} + \sqrt{5}}$

35. $\dfrac{3}{2\sqrt{3} + 3\sqrt{5}}$

36. $\dfrac{3\sqrt{2}}{2\sqrt{6} - \sqrt{10}}$

For Problems 37–42, simplify each of the following and express the final results using positive exponents.

37. $(x^{-3}y^4)^{-2}$

38. $\left(\dfrac{2a^{-1}}{3b^4}\right)^{-3}$

39. $(4x^{\frac{1}{2}})(5x^{\frac{1}{5}})$

40. $\dfrac{42a^{\frac{3}{4}}}{6a^{\frac{1}{3}}}$

41. $\left(\dfrac{x^3}{y^4}\right)^{-\frac{1}{3}}$

42. $\left(\dfrac{6x^{-2}}{2x^4}\right)^{-2}$

For Problems 43–46, use the distributive property to help simplify each of the following.

43. $3\sqrt{45} - 2\sqrt{20} - \sqrt{80}$

44. $4\sqrt[3]{24} + 3\sqrt[3]{3} - 2\sqrt[3]{81}$

45. $3\sqrt{24} - \dfrac{2\sqrt{54}}{5} + \dfrac{\sqrt{96}}{4}$

46. $-2\sqrt{12x} + 3\sqrt{27x} - 5\sqrt{48x}$

For Problems 47 and 48, express each as a single fraction involving positive exponents only.

47. $x^{-2} + y^{-1}$

48. $a^{-2} - 2a^{-1}b^{-1}$

For Problems 49–56, solve each equation.

49. $\sqrt{7x - 3} = 4$

50. $\sqrt{2y + 1} = \sqrt{5y - 11}$

51. $\sqrt{2x} = x - 4$

52. $\sqrt{n^2 - 4n - 4} = n$

53. $\sqrt[3]{2x - 1} = 3$

54. $\sqrt{t^2 + 9t - 1} = 3$

55. $\sqrt{x^2 + 3x - 6} = x$

56. $\sqrt{x + 1} - \sqrt{2x} = -1$

For Problems 57–64, use scientific notation and the properties of exponents to help perform the following calculations.

57. $(0.00002)(0.0003)$

58. $(120,000)(300,000)$

59. $(0.000015)(400,000)$

60. $\dfrac{0.000045}{0.0003}$

61. $\dfrac{(0.00042)(0.0004)}{0.006}$

62. $\sqrt{0.000004}$

63. $\sqrt[3]{0.000000008}$

64. $(4,000,000)^{\frac{3}{2}}$

TEST

  applies to all problems in this Chapter Test.

For Problems 1–4, simplify each of the numerical expressions.

1. $(4)^{-\frac{5}{2}}$

2. $-16^{\frac{5}{4}}$

3. $\left(\dfrac{2}{3}\right)^{-4}$

4. $\left(\dfrac{2^{-1}}{2^{-2}}\right)^{-2}$

For Problems 5–9, express each radical expression in simplest radical form.

5. $\sqrt{63}$

6. $\sqrt[3]{108}$

7. $\sqrt{52x^4y^3}$

8. $\dfrac{5\sqrt{18}}{3\sqrt{12}}$

9. $\sqrt{\dfrac{7}{24x^3}}$

10. Multiply and simplify: $(4\sqrt{6})(3\sqrt{12})$

11. Multiply and simplify: $(3\sqrt{2} + \sqrt{3})(\sqrt{2} - 2\sqrt{3})$

12. Simplify by combining similar radicals:
$2\sqrt{50} - 4\sqrt{18} - 9\sqrt{32}$

13. Rationalize the denominator and simplify:
$$\dfrac{3\sqrt{2}}{4\sqrt{3} - \sqrt{8}}$$

14. Simplify and express the answer using positive exponents: $\left(\dfrac{2x^{-1}}{3y}\right)^{-2}$

15. Simplify and express the answer using positive exponents: $\dfrac{-84a^{\frac{1}{2}}}{7a^{\frac{4}{5}}}$

16. Express $x^{-1} + y^{-3}$ as a single fraction involving positive exponents.

17. Multiply and express the answer using positive exponents: $\left(3x^{-\frac{1}{2}}\right)\left(-4x^{\frac{3}{4}}\right)$

18. Multiply and simplify:
$(3\sqrt{5} - 2\sqrt{3})(3\sqrt{5} + 2\sqrt{3})$

For Problems 19 and 20, use scientific notation and the properties of exponents to help with the calculations.

19. $\dfrac{(0.00004)(300)}{0.00002}$

20. $\sqrt{0.000009}$

For Problems 21–25, solve each equation.

21. $\sqrt{3x + 1} = 3$

22. $\sqrt[3]{3x + 2} = 2$

23. $\sqrt{x} = x - 2$

24. $\sqrt{5x - 2} = \sqrt{3x + 8}$

25. $\sqrt{x^2 - 10x + 28} = 2$

CUMULATIVE PRACTICE TEST *Chapters 1-10*

1. Evaluate each of the following numerical expressions.

   a. $\left(\dfrac{3}{4} - \dfrac{1}{3}\right)^{-1}$

   b. $(2^{-1} + 3^{-2})^{-2}$

   c. $\sqrt[3]{-\dfrac{1}{8}}$

   d. $(-27)^{\frac{4}{3}}$

For Problems 2–5, perform the indicated operations and express results using positive exponents only.

2. $\dfrac{-64x^{-1}y^3}{16x^3y^{-2}}$

3. $(-4x^{-2}y^{-3})(3x^2y^{-1})$

4. $(3x - 1)(2x^2 + 6x - 4)$

5. $(4x^3 + 17x^2 - 44x - 12) \div (x + 6)$

6. Solve the system $\left(\begin{array}{l} 5x - 3y = -31 \\ 4x + 7y = 41 \end{array}\right)$.

7. Express each of the following in simplest radical form.

   a. $3\sqrt{56}$

   b. $2\sqrt[3]{56}$

   c. $\dfrac{3\sqrt{2}}{4\sqrt{6}}$

   d. $\sqrt{\dfrac{6}{8}}$

8. Using 1.414 as an approximation for $\sqrt{2}$, evaluate the expression $3\sqrt{2} - 2\sqrt{8} + 9\sqrt{18} - \sqrt{32}$ to the nearest hundredth.

9. Twenty-five percent of what number is 18?

10. Evaluate $-4(2a - b) + 6(b - 2a) - (3a + 4b)$ for $a = -15$ and $b = 14$.

11. Evaluate $\dfrac{56x^{-1}y^{-3}}{8x^{-2}y^{-4}}$ for $x = -\dfrac{1}{2}$ and $y = \dfrac{2}{3}$.

12. Prime-factor each of the following composite numbers.

   a. 52

   b. 80

   c. 91

   d. 78

13. Simplify the complex fraction $\dfrac{\dfrac{5}{3x} - \dfrac{2}{y}}{\dfrac{3}{x} + \dfrac{5}{3y}}$.

For Problems 14–16, perform the indicated operations and express your answers in simplest form.

14. $\dfrac{3x - 7}{5} - \dfrac{2x + 1}{4}$

15. $\dfrac{5x^2y}{7xy} \div \dfrac{15y}{14x}$

16. $\left(\dfrac{x^2 + 5x}{3x^2 + 2x - 8}\right)\left(\dfrac{2x^2 - 8}{x^3 + 3x^2 - 10x}\right)$

For Problems 17–21, solve each equation.

17. $x(x - 4) - 3(x - 4) = 0$

18. $(x + 2)(x - 5) = -6$

19. $(3x - 5)(2x + 7) = 0$

20. $|2x - 5| = -4$

21. $\sqrt{5 + 2x} = 1 + \sqrt{2x}$

For Problems 22–25, use an equation or a system of equations to help solve each problem.

22. The area of a triangle is 51 square inches. The length of one side of the triangle is 1 inch less than three times the length of the altitude to that side. Find the length of that side and the length of the altitude to that side.

23. Brad is 6 years older than Pedro. Five years ago Pedro's age was three-fourths of Brad's age at that time. Find the present ages of Brad and Pedro.

24. Karla sold an autographed sports card for $97.50. This selling price represented a 30% profit for her, on the basis of what she originally paid for the card. Find Karla's original cost for the autographed sports card.

25. A rectangular piece of cardboard is 4 inches longer than it is wide. From each of its corners, a square piece 2 inches on a side is cut out. The flaps are then turned up to form an open box, which has a volume of 42 cubic inches. Find the length and width of the original piece of cardboard.

Craftspeople in the construction industry use quadratic equations when they apply the Pythagorean theorem in right triangle situations that arise when they measure and design their projects.

© Jeff Greenberg/PhotoEdit

Quadratic Equations and Inequalities

A page in a magazine contains 70 square inches of type. The height of the page is twice the width. If the margin around the type is 2 inches uniformly, what are the dimensions of a page? We can use the quadratic equation $(x - 4)(2x - 4) = 70$ to determine that the page measures 9 inches by 18 inches.

Solving equations is one of the central themes of this text. Let's pause for a moment and reflect on the different types of equations that we have solved thus far in this text, as shown in the chart on the next page.

Type of equation	Examples
First-degree equations in one variable	$3x + 2x = x - 4$; $5(x + 4) = 12$; $\dfrac{x + 2}{3} + \dfrac{x - 1}{4} = 2$
Second-degree equations in one variable *that are factorable*	$x^2 + 5x = 0$; $x^2 + 5x + 6 = 0$; $x^2 - 9 = 0$; $x^2 - 10x + 25 = 0$
Rational equations	$\dfrac{2}{x} + \dfrac{3}{x} = 4$; $\dfrac{5}{a - 1} = \dfrac{6}{a - 2}$; $\dfrac{2}{x^2 - 9} + \dfrac{3}{x + 3} = \dfrac{4}{x - 3}$
Radical equations	$\sqrt{x} = 2$; $\sqrt{3x - 2} = 5$; $\sqrt{5y + 1} = \sqrt{3y + 4}$

As this chart shows, we have solved second-degree equations in one variable, but only those for which the polynomial is factorable. In this chapter we will expand our work to include more general types of second-degree equations, as well as inequalities in one variable.

InfoTrac Project

Do a subject guide search on area measurement and find a periodical article on measuring land. Write a brief summary of the article. How did we get the measurement called *mile*? A piece of land is in the shape of a right triangle. One leg of the right triangle is 3 kilometers longer than the other. If the hypotenuse is 6 kilometers longer than the shorter leg, how long is each side of the piece of land?

11.1 Complex Numbers

Objectives

- Define complex numbers.
- Know the terminology associated with complex numbers.
- Add and subtract complex numbers.
- Multiply and divide complex numbers.

Because the square of any real number is nonnegative, a simple equation such as $x^2 = -4$ has no solutions in the set of real numbers. To handle this situation, we can expand the set of real numbers into a larger set called the **complex numbers.** In this section we will consider how to manipulate complex numbers.

To provide a solution for the equation $x^2 + 1 = 0$, we use the number i, such that

$$i^2 = -1$$

The number i is not a real number and is often called the **imaginary unit,** but the number i^2 is the real number -1. The imaginary unit i is used to define a complex number as follows:

DEFINITION 11.1

A **complex number** is any number that can be expressed in the form

$$a + bi$$

where a and b are real numbers, and $i^2 = -1$.

The form $a + bi$ is called the **standard form** of a complex number. The real number a is called the **real part** of the complex number, and b is called the **imaginary part.** (Note that b is a real number even though it is called the imaginary part.) The following list exemplifies this terminology.

1. The number $7 + 5i$ is a complex number that has a real part of 7 and an imaginary part of 5.

2. The number $\frac{2}{3} + i\sqrt{2}$ is a complex number that has a real part of $\frac{2}{3}$ and an imaginary part of $\sqrt{2}$. (It is easy to mistake $\sqrt{2}i$ for $\sqrt{2i}$. Thus it is customary to write $i\sqrt{2}$ instead of $\sqrt{2}i$ to avoid any difficulties with the radical sign.)

3. The number $-4 - 3i$ can be written in the standard form $-4 + (-3i)$ and therefore is a complex number that has a real part of -4 and an imaginary part of -3. (The form $-4 - 3i$ is often used, but we know that it means $-4 + (-3i)$.)

4. The number $-9i$ can be written as $0 + (-9i)$; thus it is a complex number that has a real part of 0 and an imaginary part of -9. (Complex numbers, such as $-9i$, for which $a = 0$ and $b \neq 0$ are called **pure imaginary numbers.**)

5. The real number 4 can be written as $4 + 0i$ and is thus a complex number that has a real part of 4 and an imaginary part of 0.

Look at item 5 in this list. We see that the set of real numbers is a subset of the set of complex numbers. The following diagram indicates the organizational format of the complex numbers.

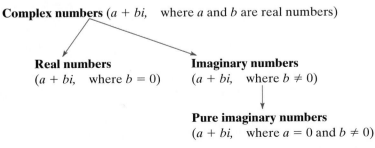

Complex numbers ($a + bi$, where a and b are real numbers)

Real numbers
($a + bi$, where $b = 0$)

Imaginary numbers
($a + bi$, where $b \neq 0$)

Pure imaginary numbers
($a + bi$, where $a = 0$ and $b \neq 0$)

Two complex numbers $a + bi$ and $c + di$ are said to be **equal** if and only if $a = c$ and $b = d$.

Adding and Subtracting Complex Numbers

To **add complex numbers,** we simply add their real parts and add their imaginary parts. Thus

$$(a + bi) + (c + di) = (a + c) + (b + d)i$$

The following examples show addition of two complex numbers.

1. $(4 + 3i) + (5 + 9i) = (4 + 5) + (3 + 9)i = 9 + 12i$

2. $(-6 + 4i) + (8 - 7i) = (-6 + 8) + (4 - 7)i$
$$= 2 - 3i$$

3. $\left(\dfrac{1}{2} + \dfrac{3}{4}i\right) + \left(\dfrac{2}{3} + \dfrac{1}{5}i\right) = \left(\dfrac{1}{2} + \dfrac{2}{3}\right) + \left(\dfrac{3}{4} + \dfrac{1}{5}\right)i$

$$= \left(\dfrac{3}{6} + \dfrac{4}{6}\right) + \left(\dfrac{15}{20} + \dfrac{4}{20}\right)i$$

$$= \dfrac{7}{6} + \dfrac{19}{20}i$$

The set of complex numbers is closed with respect to addition; that is, the sum of two complex numbers is a complex number. Furthermore, the commutative and associative properties of addition hold for all complex numbers. The addition identity element is $0 + 0i$ (or simply the real number 0). The additive inverse of $a + bi$ is $-a - bi$, because

$$(a + bi) + (-a - bi) = 0$$

To **subtract complex numbers,** $c + di$ from $a + bi$, add the additive inverse of $c + di$. Thus

$$(a + bi) - (c + di) = (a + bi) + (-c - di)$$
$$= (a - c) + (b - d)i$$

In other words, we subtract the real parts and subtract the imaginary parts, as in the next examples.

1. $(9 + 8i) - (5 + 3i) = (9 - 5) + (8 - 3)i$
$$= 4 + 5i$$

2. $(3 - 2i) - (4 - 10i) = (3 - 4) + (-2 - (-10))i$
$$= -1 + 8i$$

Products and Quotients of Complex Numbers

Because $i^2 = -1$, i is a square root of -1, so we let $i = \sqrt{-1}$. It should also be evident that $-i$ is a square root of -1, because

$$(-i)^2 = (-i)(-i) = i^2 = -1$$

Thus in the set of complex numbers, -1 has two square roots, i and $-i$. We express these symbolically as

$$\sqrt{-1} = i \quad \text{and} \quad -\sqrt{-1} = -i$$

Let us extend our definition so that in the set of complex numbers every negative real number has two square roots. We simply define $\sqrt{-b}$, where b is a positive real number, to be the number whose square is $-b$. Thus

$$(\sqrt{-b})^2 = -b, \quad \text{for } b > 0$$

Furthermore because $(i\sqrt{b})(i\sqrt{b}) = i^2(b) = -1(b) = -b$ we see that

$$\sqrt{-b} = i\sqrt{b}$$

In other words, a square root of any negative real number can be represented as the product of a real number and the imaginary unit i. Consider the following examples.

$$\sqrt{-4} = i\sqrt{4} = 2i$$
$$\sqrt{-17} = i\sqrt{17}$$
$$\sqrt{-24} = i\sqrt{24} = i\sqrt{4}\sqrt{6} = 2i\sqrt{6} \qquad \text{Note that we simplified the radical } \sqrt{24} \text{ to } 2\sqrt{6}.$$

We should also observe that $-\sqrt{-b}$, where $b > 0$, is a square root of $-b$ because

$$(-\sqrt{-b})^2 = (-i\sqrt{b})^2 = i^2(b) = -1(b) = -b$$

Thus, in the set of complex numbers, $-b$ (where $b > 0$) has two square roots, $i\sqrt{b}$ and $-i\sqrt{b}$. We express these in symbols as

$$\sqrt{-b} = i\sqrt{b} \quad \text{and} \quad -\sqrt{-b} = -i\sqrt{b}$$

We must be very careful with the use of the symbol $\sqrt{-b}$, where $b > 0$. Some relationships that involve the square root symbol and are true in the set of real numbers do not hold if the square root symbol does not represent a real number. For example, $\sqrt{a}\sqrt{b} = \sqrt{ab}$ *does not hold* if a and b are both negative numbers.

Correct $\quad \sqrt{-4}\sqrt{-9} = (2i)(3i) = 6i^2 = 6(-1) = -6$

Incorrect $\quad \sqrt{-4}\sqrt{-9} = \sqrt{(-4)(-9)} = \sqrt{36} = 6$

To avoid difficulty with this idea, you should rewrite all expressions of the form $\sqrt{-b}$, where $b > 0$, in the form $i\sqrt{b}$ before doing *any computations*. The following examples further demonstrate this point.

1. $\sqrt{-5}\sqrt{-7} = (i\sqrt{5})(i\sqrt{7}) = i^2\sqrt{35} = (-1)\sqrt{35} = -\sqrt{35}$

2. $\sqrt{-2}\sqrt{-8} = (i\sqrt{2})(i\sqrt{8}) = i^2\sqrt{16} = (-1)(4) = -4$

3. $\sqrt{-6}\sqrt{-8} = (i\sqrt{6})(i\sqrt{8}) = i^2\sqrt{48} = (-1)\sqrt{16}\sqrt{3} = -4\sqrt{3}$

4. $\dfrac{\sqrt{-75}}{\sqrt{-3}} = \dfrac{i\sqrt{75}}{i\sqrt{3}} = \dfrac{\sqrt{75}}{\sqrt{3}} = \sqrt{\dfrac{75}{3}} = \sqrt{25} = 5$

5. $\dfrac{\sqrt{-48}}{\sqrt{12}} = \dfrac{i\sqrt{48}}{\sqrt{12}} = i\sqrt{\dfrac{48}{12}} = i\sqrt{4} = 2i$

Complex numbers have a *binomial form,* so we find the *product* of two complex numbers in the same way that we find the product of two binomials. Then, by replacing i^2 with -1, we are able to simplify and express the final result in standard form. Consider the following examples.

6. $(2 + 3i)(4 + 5i) = 2(4 + 5i) + 3i(4 + 5i)$

$\qquad\qquad\qquad = 8 + 10i + 12i + 15i^2$

$\qquad\qquad\qquad = 8 + 22i + 15i^2$

$\qquad\qquad\qquad = 8 + 22i + 15(-1) = -7 + 22i$

7. $(-3 + 6i)(2 - 4i) = -3(2 - 4i) + 6i(2 - 4i)$

$\qquad\qquad\qquad = -6 + 12i + 12i - 24i^2$

$\qquad\qquad\qquad = -6 + 24i - 24(-1)$

$\qquad\qquad\qquad = -6 + 24i + 24 = 18 + 24i$

8. $(1 - 7i)^2 = (1 - 7i)(1 - 7i)$

$\qquad\qquad\quad = 1(1 - 7i) - 7i(1 - 7i)$

$\qquad\qquad\quad = 1 - 7i - 7i + 49i^2$

$\qquad\qquad\quad = 1 - 14i + 49(-1)$

$\qquad\qquad\quad = 1 - 14i - 49$

$\qquad\qquad\quad = -48 - 14i$

9. $(2 + 3i)(2 - 3i) = 2(2 - 3i) + 3i(2 - 3i)$

$\qquad\qquad\qquad = 4 - 6i + 6i - 9i^2$

$\qquad\qquad\qquad = 4 - 9(-1)$

$\qquad\qquad\qquad = 4 + 9$

$\qquad\qquad\qquad = 13$

Example 9 illustrates an important situation: The complex numbers $2 + 3i$ and $2 - 3i$ are conjugates of each other. In general, two complex numbers $a + bi$ and $a - bi$ are called **conjugates** of each other. *The product of a complex number and its conjugate is always a real number,* which can be shown as follows:

$(a + bi)(a - bi) = a(a - bi) + bi(a - bi)$

$\qquad\qquad\qquad = a^2 - abi + abi - b^2i^2$

$\qquad\qquad\qquad = a^2 - b^2(-1)$

$\qquad\qquad\qquad = a^2 + b^2$

We use conjugates to *simplify expressions* such as $\dfrac{3i}{5+2i}$ that *indicate the quotient* of two complex numbers. To eliminate i in the denominator and change the indicated quotient to the standard form of a complex number, we can multiply both the numerator and the denominator by the conjugate of the denominator as follows:

$$\frac{3i}{5+2i} = \frac{3i(5-2i)}{(5+2i)(5-2i)} \qquad \text{$5-2i$ is the conjugate of $5+2i$.}$$

$$= \frac{15i - 6i^2}{25 - 4i^2}$$

$$= \frac{15i - 6(-1)}{25 - 4(-1)}$$

$$= \frac{15i + 6}{29}$$

$$= \frac{6}{29} + \frac{15}{29}i$$

The following examples further clarify the process of *dividing* complex numbers.

10. $\dfrac{2-3i}{4-7i} = \dfrac{(2-3i)(4+7i)}{(4-7i)(4+7i)} \qquad \text{$4+7i$ is the conjugate of $4-7i$.}$

$$= \frac{8 + 14i - 12i - 21i^2}{16 - 49i^2}$$

$$= \frac{8 + 2i - 21(-1)}{16 - 49(-1)}$$

$$= \frac{8 + 2i + 21}{16 + 49}$$

$$= \frac{29 + 2i}{65}$$

$$= \frac{29}{65} + \frac{2}{65}i$$

11. $\dfrac{4-5i}{2i} = \dfrac{(4-5i)(-2i)}{(2i)(-2i)} \qquad \text{$-2i$ is the conjugate of $2i$.}$

$$= \frac{-8i + 10i^2}{-4i^2}$$

$$= \frac{-8i + 10(-1)}{-4(-1)}$$

$$= \frac{-8i - 10}{4}$$

$$= -\frac{5}{2} - 2i$$

In Example 11, where the denominator is a pure imaginary number, we can change to standard form by choosing a multiplier other than the conjugate. Consider the following alternative approach for Example 11.

$$\frac{4 - 5i}{2i} = \frac{(4 - 5i)(i)}{(2i)(i)}$$

$$= \frac{4i - 5i^2}{2i^2}$$

$$= \frac{4i - 5(-1)}{2(-1)}$$

$$= \frac{4i + 5}{-2}$$

$$= -\frac{5}{2} - 2i$$

CONCEPT QUIZ

For Problems 1–10, answer true or false.

1. The number i is a real number and is called the imaginary unit.

2. The number $4 + 2i$ is a complex number that has a real part of 4.

3. The number $-3 - 5i$ is a complex number that has an imaginary part of 5.

4. Complex numbers that have a real part of 0 are called pure imaginary numbers.

5. The set of real numbers is a subset of the set of complex numbers.

6. Any real number x can be written as the complex number $x + 0i$.

7. By definition, i^2 is equal to -1.

8. The complex numbers $-2 + 5i$ and $2 - 5i$ are conjugates.

9. The product of two complex numbers is never a real number.

10. In the set of complex numbers, -16 has two square roots.

PROBLEM SET 11.1

For Problems 1–8, label each statement true or false.

1. Every complex number is a real number.

2. Every real number is a complex number.

3. The real part of the complex number $6i$ is 0.

4. Every complex number is a pure imaginary number.

5. The sum of two complex numbers is always a complex number.

6. The imaginary part of the complex number 7 is 0.

7. The sum of two complex numbers is sometimes a real number.

8. The sum of two pure imaginary numbers is always a pure imaginary number.

For Problems 9–26, add or subtract as indicated.

9. $(6 + 3i) + (4 + 5i)$ **10.** $(5 + 2i) + (7 + 10i)$

11. $(-8 + 4i) + (2 + 6i)$ **12.** $(5 - 8i) + (-7 + 2i)$

13. $(3 + 2i) - (5 + 7i)$ **14.** $(1 + 3i) - (4 + 9i)$

15. $(-7 + 3i) - (5 - 2i)$ **16.** $(-8 + 4i) - (9 - 4i)$

17. $(-3 - 10i) + (2 - 13i)$ **18.** $(-4 - 12i) + (-3 + 16i)$

19. $(4 - 8i) - (8 - 3i)$ **20.** $(12 - 9i) - (14 - 6i)$

21. $(-1 - i) - (-2 - 4i)$ **22.** $(-2 - 3i) - (-4 - 14i)$

23. $\left(\dfrac{3}{2} + \dfrac{1}{3}i\right) + \left(\dfrac{1}{6} - \dfrac{3}{4}i\right)$ **24.** $\left(\dfrac{2}{3} - \dfrac{1}{5}i\right) + \left(\dfrac{3}{5} - \dfrac{3}{4}i\right)$

25. $\left(-\dfrac{5}{9} + \dfrac{3}{5}i\right) - \left(\dfrac{4}{3} - \dfrac{1}{6}i\right)$ **26.** $\left(\dfrac{3}{8} - \dfrac{5}{2}i\right) - \left(\dfrac{5}{6} + \dfrac{1}{7}i\right)$

For Problems 27–42, write each of the following in terms of i and simplify. For example,

$$\sqrt{-20} = i\sqrt{20} = i\sqrt{4}\sqrt{5} = 2i\sqrt{5}$$

27. $\sqrt{-81}$ **28.** $\sqrt{-49}$

29. $\sqrt{-14}$ **30.** $\sqrt{-33}$

31. $\sqrt{-\dfrac{16}{25}}$ **32.** $\sqrt{-\dfrac{64}{36}}$

33. $\sqrt{-18}$ **34.** $\sqrt{-84}$

35. $\sqrt{-75}$ **36.** $\sqrt{-63}$

37. $3\sqrt{-28}$ **38.** $5\sqrt{-72}$

39. $-2\sqrt{-80}$ **40.** $-6\sqrt{-27}$

41. $12\sqrt{-90}$ **42.** $9\sqrt{-40}$

For Problems 43–60, write each of the following in terms of i, perform the indicated operations, and simplify. For example,

$$\sqrt{-3}\sqrt{-8} = (i\sqrt{3})(i\sqrt{8})$$
$$= i^2\sqrt{24}$$
$$= (-1)\sqrt{4}\sqrt{6}$$
$$= -2\sqrt{6}$$

43. $\sqrt{-4}\sqrt{-16}$ **44.** $\sqrt{-81}\sqrt{-25}$

45. $\sqrt{-3}\sqrt{-5}$ **46.** $\sqrt{-7}\sqrt{-10}$

47. $\sqrt{-9}\sqrt{-6}$ **48.** $\sqrt{-8}\sqrt{-16}$

49. $\sqrt{-15}\sqrt{-5}$ **50.** $\sqrt{-2}\sqrt{-20}$

51. $\sqrt{-2}\sqrt{-27}$ **52.** $\sqrt{-3}\sqrt{-15}$

53. $\sqrt{6}\sqrt{-8}$ **54.** $\sqrt{-75}\sqrt{3}$

55. $\dfrac{\sqrt{-25}}{\sqrt{-4}}$ **56.** $\dfrac{\sqrt{-81}}{\sqrt{-9}}$

57. $\dfrac{\sqrt{-56}}{\sqrt{-7}}$ **58.** $\dfrac{\sqrt{-72}}{\sqrt{-6}}$

59. $\dfrac{\sqrt{-24}}{\sqrt{6}}$ **60.** $\dfrac{\sqrt{-96}}{\sqrt{2}}$

For Problems 61–84, find each of the products and express the answer in the standard form of a complex number.

61. $(5i)(4i)$ **62.** $(-6i)(9i)$

63. $(7i)(-6i)$ **64.** $(-5i)(-12i)$

65. $(3i)(2 - 5i)$ **66.** $(7i)(-9 + 3i)$

67. $(-6i)(-2 - 7i)$ **68.** $(-9i)(-4 - 5i)$

69. $(3 + 2i)(5 + 4i)$ **70.** $(4 + 3i)(6 + i)$

71. $(6 - 2i)(7 - i)$ **72.** $(8 - 4i)(7 - 2i)$

73. $(-3 - 2i)(5 + 6i)$ **74.** $(-5 - 3i)(2 - 4i)$

75. $(9 + 6i)(-1 - i)$ **76.** $(10 + 2i)(-2 - i)$

77. $(4 + 5i)^2$ **78.** $(5 - 3i)^2$

79. $(-2 - 4i)^2$ **80.** $(-3 - 6i)^2$

81. $(6 + 7i)(6 - 7i)$ **82.** $(5 - 7i)(5 + 7i)$

83. $(-1 + 2i)(-1 - 2i)$ **84.** $(-2 - 4i)(-2 + 4i)$

For Problems 85–100, find each of the following quotients and express the answer in the standard form of a complex number.

85. $\dfrac{3i}{2 + 4i}$ **86.** $\dfrac{4i}{5 + 2i}$

87. $\dfrac{-2i}{3 - 5i}$ **88.** $\dfrac{-5i}{2 - 4i}$

89. $\dfrac{-2 + 6i}{3i}$ **90.** $\dfrac{-4 - 7i}{6i}$

91. $\dfrac{2}{7i}$ **92.** $\dfrac{3}{10i}$

93. $\dfrac{2 + 6i}{1 + 7i}$ **94.** $\dfrac{5 + i}{2 + 9i}$ **97.** $\dfrac{-2 + 7i}{-1 + i}$ **98.** $\dfrac{-3 + 8i}{-2 + i}$

95. $\dfrac{3 + 6i}{4 - 5i}$ **96.** $\dfrac{7 - 3i}{4 - 3i}$ **99.** $\dfrac{-1 - 3i}{-2 - 10i}$ **100.** $\dfrac{-3 - 4i}{-4 - 11i}$

■ ■ ■ **Thoughts into words**

101. Why is the set of real numbers a subset of the set of complex numbers?

102. Can the sum of two nonreal complex numbers be a real number? Defend your answer.

103. Can the product of two nonreal complex numbers be a real number? Defend your answer.

Answers to Concept Quiz

1. False   **2.** True   **3.** False   **4.** True   **5.** True   **6.** True   **7.** True   **8.** False   **9.** False   **10.** True

11.2 Quadratic Equations

Objectives

■ Solve quadratic equations of the form $x^2 = a$.

■ Solve word problems involving the Pythagorean theorem and 30°–60° right triangles.

A second-degree equation in one variable contains the variable with an exponent of 2, but no higher power. Such equations are also called **quadratic equations.** The following are examples of quadratic equations.

$$x^2 = 36 \qquad\qquad y^2 + 4y = 0 \qquad\qquad x^2 + 5x - 2 = 0$$
$$3n^2 + 2n - 1 = 0 \qquad 5x^2 + x + 2 = 3x^2 - 2x - 1$$

A quadratic equation in the variable x can also be defined as any equation that can be written in the form

$$ax^2 + bx + c = 0$$

where a, b, and c are real numbers and $a \neq 0$. The form $ax^2 + bx + c = 0$ is called the **standard form** of a quadratic equation.

In previous chapters you solved quadratic equations (the term *quadratic* was not used at that time) by factoring and applying the property "$ab = 0$ if and only if $a = 0$ or $b = 0$." Let's review a few such examples.

E X A M P L E   1

Solve $3n^2 + 14n - 5 = 0$.

Solution

$$3n^2 + 14n - 5 = 0$$
$$(3n - 1)(n + 5) = 0 \qquad \text{Factor the left side.}$$
$$3n - 1 = 0 \qquad \text{or} \qquad n + 5 = 0 \qquad \text{Apply "}ab = 0 \text{ if and}$$
$$\text{only if } a = 0 \text{ or } b = 0\text{."}$$
$$3n = 1 \qquad \text{or} \qquad n = -5$$
$$n = \frac{1}{3} \qquad \text{or} \qquad n = -5$$

The solution set is $\left\{ -5, \frac{1}{3} \right\}$. ■

E X A M P L E   2

Solve $x^2 + 3kx - 10k^2 = 0$ for x.

Solution

$$x^2 + 3kx - 10k^2 = 0$$
$$(x + 5k)(x - 2k) = 0 \qquad \text{Factor the left side.}$$
$$x + 5k = 0 \qquad \text{or} \qquad x - 2k = 0 \qquad \text{Apply "}ab = 0 \text{ if and}$$
$$\text{only if } a = 0 \text{ or } b = 0\text{."}$$
$$x = -5k \qquad \text{or} \qquad x = 2k$$

The solution set is $\{-5k, 2k\}$. ■

E X A M P L E   3

Solve $2\sqrt{x} = x - 8$.

Solution

$$2\sqrt{x} = x - 8$$
$$(2\sqrt{x})^2 = (x - 8)^2 \qquad \text{Square both sides.}$$
$$4x = x^2 - 16x + 64$$
$$0 = x^2 - 20x + 64$$
$$0 = (x - 16)(x - 4) \qquad \text{Factor the right side.}$$
$$x - 16 = 0 \qquad \text{or} \qquad x - 4 = 0 \qquad \text{Apply "}ab = 0 \text{ if and}$$
$$\text{only if } a = 0 \text{ or } b = 0\text{."}$$
$$x = 16 \qquad \text{or} \qquad x = 4$$

 Check

$$2\sqrt{x} = x - 8 \qquad\qquad\qquad 2\sqrt{x} = x - 8$$
$$2\sqrt{16} \overset{?}{=} 16 - 8 \qquad \text{or} \qquad 2\sqrt{4} \overset{?}{=} 4 - 8$$
$$2(4) \overset{?}{=} 8 \qquad\qquad\qquad 2(2) \overset{?}{=} -4$$
$$8 = 8 \qquad\qquad\qquad\qquad 4 \neq -4$$

The solution set is $\{16\}$. ■

We should make two comments about Example 3. First, remember that applying the property "if $a = b$, then $a^n = b^n$" might produce extraneous solutions. Therefore, we *must* check all potential solutions. Second, the equation $2\sqrt{x} = x - 8$ is said to be of **quadratic form** because it can be written as $2x^{\frac{1}{2}} = \left(x^{\frac{1}{2}}\right)^2 - 8$. More will be said about the phrase "quadratic form" later.

Let's consider quadratic equations of the form $x^2 = a$, where x is the variable and a is any real number. We can solve $x^2 = a$ as follows:

$$x^2 = a$$
$$x^2 - a = 0$$
$$x^2 - (\sqrt{a})^2 = 0 \qquad\qquad a = (\sqrt{a})^2$$
$$(x - \sqrt{a})(x + \sqrt{a}) = 0 \qquad\qquad \text{Factor the left side.}$$
$$x - \sqrt{a} = 0 \qquad \text{or} \qquad x + \sqrt{a} = 0 \qquad \text{Apply "} ab = 0 \text{ if and}$$
$$\qquad\qquad\qquad\qquad\qquad\qquad\qquad\qquad \text{only if } a = 0 \text{ or } b = 0."$$
$$x = \sqrt{a} \qquad \text{or} \qquad x = -\sqrt{a}.$$

The solutions are $\sqrt{a}$ and $-\sqrt{a}$. We can state this result as a general property and use it to solve certain types of quadratic equations.

PROPERTY 11.1

For any real number a,

$$x^2 = a \quad \text{if and only if } x = \sqrt{a} \text{ or } x = -\sqrt{a}$$

(The statement $x = \sqrt{a}$ or $x = -\sqrt{a}$ can be written as $x = \pm\sqrt{a}$.)

Property 11.1, along with our knowledge of square roots, make it very easy to solve quadratic equations of the form $x^2 = a$.

EXAMPLE 4   Solve $x^2 = 45$.

Solution

$$x^2 = 45$$
$$x = \pm\sqrt{45}$$
$$x = \pm 3\sqrt{5} \qquad \sqrt{45} = \sqrt{9}\sqrt{5} = 3\sqrt{5}$$

The solution set is $\{\pm 3\sqrt{5}\}$.

EXAMPLE 5   Solve $x^2 = -9$.

Solution

$$x^2 = -9$$
$$x = \pm\sqrt{-9}$$
$$x = \pm 3i \qquad \sqrt{-9} = i\sqrt{9} = 3i$$

Thus the solution set is $\{\pm 3i\}$.

EXAMPLE 6

Solve $7n^2 = 12$.

Solution

$$7n^2 = 12$$

$$n^2 = \frac{12}{7}$$

$$n = \pm\sqrt{\frac{12}{7}}$$

$$n = \pm\frac{2\sqrt{21}}{7} \qquad \sqrt{\frac{12}{7}} = \frac{\sqrt{12}}{\sqrt{7}} \cdot \frac{\sqrt{7}}{\sqrt{7}} = \frac{\sqrt{84}}{7} = \frac{\sqrt{4}\sqrt{21}}{7} = \frac{2\sqrt{21}}{7}$$

The solution set is $\left\{\pm\dfrac{2\sqrt{21}}{7}\right\}$.

■

EXAMPLE 7

Solve $(3n + 1)^2 = 25$.

Solution

$$(3n + 1)^2 = 25$$
$$(3n + 1) = \pm\sqrt{25}$$
$$3n + 1 = \pm5$$
$$3n + 1 = 5 \qquad \text{or} \qquad 3n + 1 = -5$$
$$3n = 4 \qquad \text{or} \qquad 3n = -6$$
$$n = \frac{4}{3} \qquad \text{or} \qquad n = -2$$

The solution set is $\left\{-2, \dfrac{4}{3}\right\}$.

■

EXAMPLE 8

Solve $(x - 3)^2 = -10$.

Solution

$$(x - 3)^2 = -10$$
$$x - 3 = \pm\sqrt{-10}$$
$$x - 3 = \pm i\sqrt{10}$$
$$x = 3 \pm i\sqrt{10}$$

Thus the solution set is $\{3 \pm i\sqrt{10}\}$.

■

REMARK: Take another look at the equations in Examples 5 and 8. We should immediately realize that the solution sets will consist only of nonreal complex numbers, because any nonzero real number squared is positive.

Sometimes it may be necessary to change the form before we can apply Property 11.1. Let's consider one example to illustrate this idea.

EXAMPLE 9

Solve $3(2x - 3)^2 + 8 = 44$.

Solution

$$3(2x - 3)^2 + 8 = 44$$
$$3(2x - 3)^2 = 36 \qquad \text{Add } -8 \text{ to both sides.}$$
$$(2x - 3)^2 = 12 \qquad \text{Divide both sides by 3.}$$
$$2x - 3 = \pm\sqrt{12}$$
$$2x - 3 = \pm 2\sqrt{3}$$
$$2x = 3 \pm 2\sqrt{3}$$
$$x = \frac{3 \pm 2\sqrt{3}}{2}$$

The solution set is $\left\{ \dfrac{3 \pm 2\sqrt{3}}{2} \right\}$. ■

Back to the Pythagorean Theorem

Our work with radicals, Property 11.1, and the Pythagorean theorem form a basis for solving a variety of problems that pertain to right triangles.

EXAMPLE 10

A 50-foot rope hangs from the top of a flagpole. When pulled taut to its full length, the rope reaches a point on the ground 18 feet from the base of the pole. Find the height of the pole to the nearest tenth of a foot.

Solution

Let's make a sketch (Figure 11.1) and record the given information. Use the Pythagorean theorem to solve for p as follows:

$$p^2 + 18^2 = 50^2$$
$$p^2 + 324 = 2500$$
$$p^2 = 2176$$
$$p = \sqrt{2176} = 46.6, \quad \text{to the nearest tenth}$$

The height of the flagpole is approximately 46.6 feet. ■

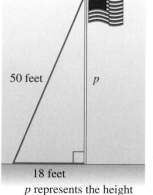

50 feet

p

18 feet

p represents the height of the flagpole.

Figure 11.1

There are two special kinds of right triangles that we use extensively in later mathematics courses. The first is the **isosceles right triangle,** which is a right triangle that has both legs of the same length. Let's consider a problem that involves an isosceles right triangle.

E X A M P L E   1 1

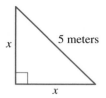

Figure 11.2

Find the length of each leg of an isosceles right triangle that has a hypotenuse of length 5 meters.

Solution

Let's sketch an isosceles right triangle and let x represent the length of each leg (Figure 11.2). Then we can apply the Pythagorean theorem.

$$x^2 + x^2 = 5^2$$
$$2x^2 = 25$$
$$x^2 = \frac{25}{2}$$
$$x = \pm\sqrt{\frac{25}{2}} = \pm\frac{5}{\sqrt{2}} = \pm\frac{5\sqrt{2}}{2}$$

Each leg is $\dfrac{5\sqrt{2}}{2}$ meters long.   ■

REMARK: In Example 10 we made no attempt to express $\sqrt{2176}$ in simplest radical form, because the answer was to be given as a rational approximation to the nearest tenth. However, in Example 11 we left the final answer in radical form and therefore expressed it in simplest radical form.

The second special kind of right triangle that we use frequently is one that contains acute angles of 30° and 60°. In such a right triangle, which we refer to as a **30°– 60° right triangle,** the side opposite the 30° angle is equal in length to one-half of the length of the hypotenuse. This relationship, along with the Pythagorean theorem, provides us with another problem-solving technique.

E X A M P L E   1 2

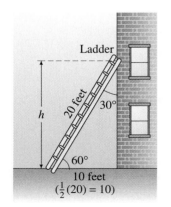

Figure 11.3

Suppose that a 20-foot ladder is leaning against a building and makes an angle of 60° with the ground. How far up the building does the top of the ladder reach? Express your answer to the nearest tenth of a foot.

Solution

Figure 11.3 depicts this situation. The side opposite the 30° angle equals one-half of the hypotenuse, so it is of length $\frac{1}{2}(20) = 10$ feet. Now we can apply the Pythagorean theorem.

$$h^2 + 10^2 = 20^2$$
$$h^2 + 100 = 400$$
$$h^2 = 300$$
$$h = \sqrt{300} = 17.3, \quad \text{to the nearest tenth}$$

The top of the ladder touches the building at a point approximately 17.3 feet from the ground.   ■

CONCEPT QUIZ

For Problems 1–5, answer true or false.

1. The quadratic equation $-3x^2 + 5x - 8 = 0$ is in standard form.

2. The solution set of the equation $(x + 1)^2 = -25$ will consist only of nonreal complex numbers.

3. An isosceles right triangle is a right triangle that has a hypotenuse of the same length as one of the legs.

4. In a 30°–60° right triangle, the hypotenuse is equal in length to twice the length of the side opposite the 30° angle.

5. The equation $2x^2 + x^3 - x + 4 = 0$ is a quadratic equation.

PROBLEM SET 11.2

For Problems 1–20, solve each of the quadratic equations by factoring and applying the property "$ab = 0$ if and only if $a = 0$ or $b = 0$." If necessary, return to Chapter 7 and review the factoring techniques presented there.

1. $x^2 - 9x = 0$

2. $x^2 + 5x = 0$

3. $x^2 = -3x$

4. $x^2 = 15x$

5. $3y^2 + 12y = 0$

6. $6y^2 - 24y = 0$

7. $5n^2 - 9n = 0$

8. $4n^2 + 13n = 0$

9. $x^2 + x - 30 = 0$

10. $x^2 - 8x - 48 = 0$

11. $x^2 - 19x + 84 = 0$

12. $x^2 - 21x + 104 = 0$

13. $2x^2 + 19x + 24 = 0$

14. $4x^2 + 29x + 30 = 0$

15. $15x^2 + 29x - 14 = 0$

16. $24x^2 + x - 10 = 0$

17. $25x^2 - 30x + 9 = 0$

18. $16x^2 - 8x + 1 = 0$

19. $6x^2 - 5x - 21 = 0$

20. $12x^2 - 4x - 5 = 0$

For Problems 21–26, solve each radical equation. Don't forget that you *must* check potential solutions.

21. $3\sqrt{x} = x + 2$

22. $3\sqrt{2x} = x + 4$

23. $\sqrt{2x} = x - 4$

24. $\sqrt{x} = x - 2$

25. $\sqrt{3x + 6} = x$

26. $\sqrt{5x + 10} = x$

For Problems 27–34, solve each equation for x by factoring and applying the property "$ab = 0$ if and only if $a = 0$ or $b = 0$."

27. $x^2 - 5kx = 0$

28. $x^2 + 7kx = 0$

29. $x^2 = 16k^2x$

30. $x^2 = 25k^2x$

31. $x^2 - 12kx + 35k^2 = 0$

32. $x^2 - 3kx - 18k^2 = 0$

33. $2x^2 + 5kx - 3k^2 = 0$

34. $3x^2 - 20kx - 7k^2 = 0$

For Problems 35–70, use Property 11.1 to help solve each quadratic equation.

35. $x^2 = 1$

36. $x^2 = 81$

37. $x^2 = -36$

38. $x^2 = -49$

39. $x^2 = 14$

40. $x^2 = 22$

41. $n^2 - 28 = 0$

42. $n^2 - 54 = 0$

43. $3t^2 = 54$

44. $4t^2 = 108$

45. $2t^2 = 7$

46. $3t^2 = 8$

47. $15y^2 = 20$

48. $14y^2 = 80$

49. $10x^2 + 48 = 0$

50. $12x^2 + 50 = 0$

51. $24x^2 = 36$

52. $12x^2 = 49$

53. $(x - 2)^2 = 9$

54. $(x + 1)^2 = 16$

55. $(x + 3)^2 = 25$

56. $(x - 2)^2 = 49$

57. $(x + 6)^2 = -4$

58. $(3x + 1)^2 = 9$

59. $(2x - 3)^2 = 1$

60. $(2x + 5)^2 = -4$

61. $(n - 4)^2 = 5$

62. $(n - 7)^2 = 6$

63. $(t + 5)^2 = 12$

64. $(t - 1)^2 = 18$

65. $(3y - 2)^2 = -27$

66. $(4y + 5)^2 = 80$

67. $3(x + 7)^2 + 4 = 79$

68. $2(x + 6)^2 - 9 = 63$

69. $2(5x - 2)^2 + 5 = 25$

70. $3(4x - 1)^2 + 1 = -17$

For Problems 71–76, a and b represent the lengths of the legs of a right triangle, and c represents the length of the hypotenuse. Express answers in simplest radical form.

71. Find c if $a = 4$ centimeters and $b = 6$ centimeters.

72. Find c if $a = 3$ meters and $b = 7$ meters.

73. Find a if $c = 12$ inches and $b = 8$ inches.

74. Find a if $c = 8$ feet and $b = 6$ feet.

75. Find b if $c = 17$ yards and $a = 15$ yards.

76. Find b if $c = 14$ meters and $a = 12$ meters.

For Problems 77–80, use the isosceles right triangle in Figure 11.4. Express your answers in simplest radical form.

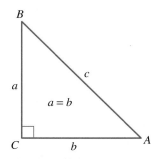

Figure 11.4

77. If $b = 6$ inches, find c.

78. If $a = 7$ centimeters, find c.

79. If $c = 8$ meters, find a and b.

80. If $c = 9$ feet, find a and b.

For Problems 81–86, use the triangle in Figure 11.5. Express your answers in simplest radical form.

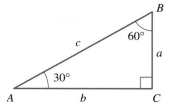

Figure 11.5

81. If $a = 3$ inches, find b and c.

82. If $a = 6$ feet, find b and c.

83. If $c = 14$ centimeters, find a and b.

84. If $c = 9$ centimeters, find a and b.

85. If $b = 10$ feet, find a and c.

86. If $b = 8$ meters, find a and c.

87. A 24-foot ladder resting against a house reaches a windowsill 16 feet above the ground. How far is the foot of the ladder from the foundation of the house? Express your answer to the nearest tenth of a foot.

88. A 62-foot guy-wire makes an angle of 60° with the ground and is attached to a telephone pole (see Figure 11.6). Find the distance from the base of the pole to the point on the pole where the wire is attached. Express your answer to the nearest tenth of a foot.

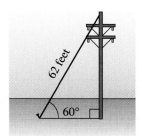

Figure 11.6

89. A rectangular plot measures 16 meters by 34 meters. Find, to the nearest meter, the distance from one corner of the plot to the corner diagonally opposite.

90. Consecutive bases of a square-shaped baseball diamond are 90 feet apart (see Figure 11.7). Find, to the nearest tenth of a foot, the distance from first base diagonally across the diamond to third base.

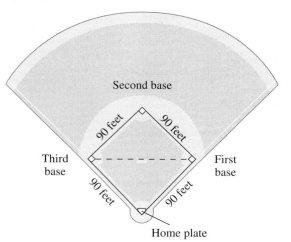

Second base

90 feet   90 feet

Third
base

First
base

90 feet   90 feet

Home plate

Figure 11.7

91. A diagonal of a square parking lot is 75 meters. Find, to the nearest meter, the length of a side of the lot.

■ ■ ▫ **Thoughts into words**

92. Explain why the equation $(x + 2)^2 + 5 = 1$ has no real number solutions.

93. Suppose that your friend solved the equation $(x + 3)^2 = 25$ as follows:

$$(x + 3)^2 = 25$$
$$x^2 + 6x + 9 = 25$$
$$x^2 + 6x - 16 = 0$$
$$(x + 8)(x - 2) = 0$$

$$x + 8 = 0 \qquad \text{or} \qquad x - 2 = 0$$
$$x = -8 \qquad \text{or} \qquad x = 2$$

Is this a correct approach to the problem? Would you offer any suggestion about an easier approach to the problem?

■ ■ ▫ **Further investigations**

94. Suppose that we are given a cube with edges 12 centimeters in length. Find the length of a diagonal from a lower corner to the diagonally opposite upper corner. Express your answer to the nearest tenth of a centimeter.

95. Suppose that we are given a rectangular box with a length of 8 centimeters, a width of 6 centimeters, and a height of 4 centimeters. Find the length of a diagonal from a lower corner to the upper corner diagonally opposite. Express your answer to the nearest tenth of a centimeter.

96. The converse of the Pythagorean theorem is also true. It states that "if the measures a, b, and c of the sides of a triangle are such that $a^2 + b^2 = c^2$, then the triangle is a right triangle with a and b the measures of the legs and c the measure of the hypotenuse." Use the converse of the Pythagorean theorem to determine which of the triangles with sides of the following measures are right triangles.

a. 9, 40, 41

b. 20, 48, 52

c. 19, 21, 26

d. 32, 37, 49

e. 65, 156, 169

f. 21, 72, 75

97. Find the length of the hypotenuse (h) of an isosceles right triangle if each leg is s units long. Then use this relationship and redo Problems 77–80.

98. Suppose that the side opposite the 30° angle in a 30°–60° right triangle is s units long. Express the length of the hypotenuse and the length of the other leg in terms of s. Then use these relationships and redo Problems 81–86.

Answers to Concept Quiz

1. True   **2.** True   **3.** False   **4.** True   **5.** False

11.3   Completing the Square

Objective

■   Solve quadratic equations by completing the square.

Thus far we have solved quadratic equations by factoring and applying the property $ab = 0$ *if and only if* $a = 0$ *or* $b = 0$, or by applying the property $x^2 = a$ *if and only if* $x = \pm\sqrt{a}$. In this section we examine another method, called **completing the square,** which will give us the power to solve *any* quadratic equation.

A factoring technique we studied in Chapter 7 relied on recognizing **perfect-square trinomials.** In each of the following, the perfect-square trinomial on the right side is the result of squaring the binomial on the left side.

$$(x + 4)^2 = x^2 + 8x + 16 \qquad (x - 6)^2 = x^2 - 12x + 36$$
$$(x + 7)^2 = x^2 + 14x + 49 \qquad (x - 9)^2 = x^2 - 18x + 81$$
$$(x + a)^2 = x^2 + 2ax + a^2$$

Note that in each of the square trinomials, *the constant term is equal to the square of one-half of the coefficient of the x term.* This relationship enables us to form a perfect-square trinomial by adding a proper constant term. For example, suppose that we want to form a perfect-square trinomial from $x^2 + 10x$. Because $\frac{1}{2}(10) = 5$ and $5^2 = 25$, the perfect-square trinomial $x^2 + 10x + 25$ can be formed. Let's use the previous ideas to help solve some quadratic equations.

EXAMPLE 1

Solve $x^2 + 10x - 2 = 0$.

Solution

$$x^2 + 10x - 2 = 0$$

$$x^2 + 10x = 2 \qquad \text{Isolate the } x^2 \text{ and } x \text{ terms.}$$

$$\frac{1}{2}(10) = 5 \text{ and } 5^2 = 25 \qquad \text{Take } \frac{1}{2} \text{ of the coefficient of the } x \text{ term and then square the result.}$$

$$x^2 + 10x + 25 = 2 + 25 \qquad \text{Add 25 to } \textit{both} \text{ sides of the equation.}$$

$$(x + 5)^2 = 27 \qquad \text{Factor the perfect-square trinomial.}$$

$$x + 5 = \pm\sqrt{27} \qquad \text{Now solve by applying Property 11.1.}$$

$$x + 5 = \pm 3\sqrt{3}$$

$$x = -5 \pm 3\sqrt{3}$$

The solution set is $\{-5 \pm 3\sqrt{3}\}$. ■

Note from Example 1 that the method of completing the square to solve a quadratic equation is exactly what the name implies. A perfect-square trinomial is formed, and then the equation can be changed to the necessary form for applying the property, $x^2 = a$ *if and only if* $x = \pm\sqrt{a}$. Let's consider another example.

EXAMPLE 2

Solve $x^2 + 4x + 7 = 0$.

Solution

$$x^2 + 4x + 7 = 0$$

$$x^2 + 4x = -7 \qquad \text{Isolate the } x^2 \text{ and } x \text{ terms.}$$

$$x^2 + 4x + 4 = -7 + 4 \qquad \frac{1}{2}(4) = 2 \text{ and } 2^2 = 4$$

$$(x + 2)^2 = -3 \qquad \text{Factor the perfect-square trinomial.}$$

$$x + 2 = \pm\sqrt{-3} \qquad \text{Now solve by applying Property 11.1.}$$

$$x + 2 = \pm i\sqrt{3}$$

$$x = -2 \pm i\sqrt{3}$$

The solution set is $\{-2 \pm i\sqrt{3}\}$. ■

Let's pause for a moment and give a little visual support for our answer in Example 2. Figure 11.8 shows the graph of $y = x^2 + 4x + 7$. Because it does not intersect the x axis, the equation $x^2 + 4x + 7 = 0$ has no real number solutions. This supports our solution set of two nonreal complex numbers. To be absolutely sure that we have the correct complex numbers, we should substitute them back into the original equation.

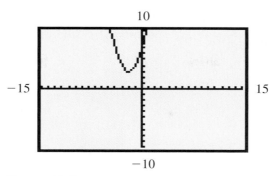

10

−15   15

−10

Figure 11.8

EXAMPLE 3

Solve $x^2 - 3x + 1 = 0$.

Solution

$$x^2 - 3x + 1 = 0$$

$$x^2 - 3x = -1 \qquad \text{Isolate the } x^2 \text{ and } x \text{ terms.}$$

$$x^2 - 3x + \frac{9}{4} = -1 + \frac{9}{4} \qquad \frac{1}{2}(3) = \frac{3}{2} \text{ and } \left(\frac{3}{2}\right)^2 = \frac{9}{4}$$

$$\left(x - \frac{3}{2}\right)^2 = \frac{5}{4}$$

$$x - \frac{3}{2} = \pm\sqrt{\frac{5}{4}}$$

$$x - \frac{3}{2} = \pm\frac{\sqrt{5}}{2}$$

$$x = \frac{3}{2} \pm \frac{\sqrt{5}}{2}$$

$$x = \frac{3 \pm \sqrt{5}}{2}$$

The solution set is $\left\{\dfrac{3 \pm \sqrt{5}}{2}\right\}$.

In Example 3, note that because the coefficient of the x term is odd, we are forced into the realm of fractions. The use of common fractions rather than decimals enables us to apply our previous work with radicals.

The relationship for a perfect-square trinomial that states that "the constant term is equal to the square of one-half of the coefficient of the x term" holds only if the coefficient of x^2 is 1. Thus we must make an adjustment when solving quadratic equations that have a coefficient of x^2 other than 1. The next example shows how to make this adjustment.

E X A M P L E 4 Solve $2x^2 + 12x - 5 = 0$.

Solution

$$2x^2 + 12x - 5 = 0$$
$$2x^2 + 12x = 5$$
$$x^2 + 6x = \frac{5}{2} \qquad \text{Multiply both sides by } \frac{1}{2}.$$
$$x^2 + 6x + 9 = \frac{5}{2} + 9 \qquad \frac{1}{2}(6) = 3 \text{ and } 3^2 = 9$$
$$x^2 + 6x + 9 = \frac{23}{2}$$
$$(x + 3)^2 = \frac{23}{2}$$
$$x + 3 = \pm\sqrt{\frac{23}{2}}$$
$$x + 3 = \pm\frac{\sqrt{46}}{2} \qquad \sqrt{\frac{23}{2}} = \frac{\sqrt{23}}{\sqrt{2}} \cdot \frac{\sqrt{2}}{\sqrt{2}} = \frac{\sqrt{46}}{2}$$
$$x = -3 \pm \frac{\sqrt{46}}{2}$$
$$x = \frac{-6}{2} \pm \frac{\sqrt{46}}{2} \qquad \text{Common denominator of 2}$$
$$x = \frac{-6 \pm \sqrt{46}}{2}$$

The solution set is $\left\{\dfrac{-6 \pm \sqrt{46}}{2}\right\}$. ∎

As we mentioned earlier, we can use the method of completing the square to solve *any* quadratic equation. To illustrate, let's use it to solve an equation that could also be solved by factoring.

E X A M P L E 5 Solve $x^2 - 2x - 8 = 0$ by completing the square.

Solution

$$x^2 - 2x - 8 = 0$$
$$x^2 - 2x = 8$$
$$x^2 - 2x + 1 = 8 + 1 \qquad \frac{1}{2}(-2) = -1 \text{ and } (-1)^2 = 1$$
$$(x - 1)^2 = 9$$
$$x - 1 = \pm 3$$
$$x - 1 = 3 \quad \text{ or } \quad x - 1 = -3$$
$$x = 4 \quad \text{ or } \quad x = -2$$

The solution set is $\{-2, 4\}$. ∎

We make no claim that using the method of completing the square with an equation such as the one in Example 5 is easier than the factoring technique. However, you should recognize that the method of completing the square will work with any quadratic equation.

CONCEPT QUIZ

For Problems 1–5, answer true or false.

1. In a perfect-square trinomial, the constant term is equal to one-half the coefficient of the x term.

2. The method of completing the square will solve any quadratic equation.

3. Every quadratic equation solved by completing the square will have real number solutions.

4. The completing-the-square method cannot be used if factoring could solve the quadratic equation.

5. To use the completing-the-square method for solving the equation $3x^2 + 2x = 5$ we would first divide both sides of the equation by 3.

PROBLEM SET 11.3

For Problems 1–14, solve each quadratic equation by using (a) the factoring method and (b) the method of completing the square.

1. $x^2 - 4x - 60 = 0$ **2.** $x^2 + 6x - 16 = 0$

3. $x^2 - 14x = -40$ **4.** $x^2 - 18x = -72$

5. $x^2 - 5x - 50 = 0$ **6.** $x^2 + 3x - 18 = 0$

7. $x(x + 7) = 8$ **8.** $x(x - 1) = 30$

9. $2n^2 - n - 15 = 0$ **10.** $3n^2 + n - 14 = 0$

11. $3n^2 + 7n - 6 = 0$ **12.** $2n^2 + 7n - 4 = 0$

13. $n(n + 6) = 160$ **14.** $n(n - 6) = 216$

For Problems 15–38, use the method of completing the square to solve each quadratic equation.

15. $x^2 + 4x - 2 = 0$ **16.** $x^2 + 2x - 1 = 0$

17. $x^2 + 6x - 3 = 0$ **18.** $x^2 + 8x - 4 = 0$

19. $y^2 - 10y = 1$ **20.** $y^2 - 6y = -10$

21. $n^2 - 8n + 17 = 0$ **22.** $n^2 - 4n + 2 = 0$

23. $n(n + 12) = -9$ **24.** $n(n + 14) = -4$

25. $n^2 + 2n + 6 = 0$ **26.** $n^2 + n - 1 = 0$

27. $x^2 + 3x - 2 = 0$ **28.** $x^2 + 5x - 3 = 0$

29. $x^2 + 5x + 1 = 0$ **30.** $x^2 + 7x + 2 = 0$

31. $y^2 - 7y + 3 = 0$ **32.** $y^2 - 9y + 30 = 0$

33. $2x^2 + 4x - 3 = 0$ **34.** $2t^2 - 4t + 1 = 0$

35. $3n^2 - 6n + 5 = 0$ **36.** $3x^2 + 12x - 2 = 0$

37. $3x^2 + 5x - 1 = 0$ **38.** $2x^2 + 7x - 3 = 0$

For Problems 39–60, solve each quadratic equation using the method that seems most appropriate.

39. $x^2 + 8x - 48 = 0$ **40.** $x^2 + 5x - 14 = 0$

41. $2n^2 - 8n = -3$ **42.** $3x^2 + 6x = 1$

43. $(3x - 1)(2x + 9) = 0$ **44.** $(5x + 2)(x - 4) = 0$

45. $(x + 2)(x - 7) = 10$ **46.** $(x - 3)(x + 5) = -7$

47. $(x - 3)^2 = 12$ **48.** $x^2 = 16x$

49. $3n^2 - 6n + 4 = 0$ **50.** $2n^2 - 2n - 1 = 0$

51. $n(n + 8) = 240$ **52.** $t(t - 26) = -160$

53. $3x^2 + 29x = -66$

54. $6x^2 - 13x = 28$

55. $6n^2 + 23n + 21 = 0$

56. $6n^2 + n - 2 = 0$

57. $x^2 + 12x = 4$

58. $x^2 + 6x = -11$

59. $12n^2 - 7n + 1 = 0$

60. $5(x + 2)^2 + 1 = 16$

61. Use the method of completing the square to solve $ax^2 + bx + c = 0$ for x, where a, b, and c are real numbers and $a \neq 0$.

■ ■ ▪ **Thoughts into words**

62. Explain the process of *completing the square* to solve a quadratic equation.

63. Give a step-by-step description of how to solve $3x^2 + 9x - 4 = 0$ by completing the square.

■ ■ ▪ **Further investigations**

Solve Problems 64–67 for the indicated variable. Assume that all letters represent positive numbers.

64. $\dfrac{x^2}{a^2} - \dfrac{y^2}{b^2} = 1$ for y

65. $\dfrac{x^2}{a^2} + \dfrac{y^2}{b^2} = 1$ for x

66. $s = \dfrac{1}{2}gt^2$ for t

67. $A = \pi r^2$ for r

Solve each of the following equations for x.

68. $x^2 + 8ax + 15a^2 = 0$

69. $x^2 - 5ax + 6a^2 = 0$

70. $10x^2 - 31ax - 14a^2 = 0$

71. $6x^2 + ax - 2a^2 = 0$

72. $4x^2 + 4bx + b^2 = 0$

73. $9x^2 - 12bx + 4b^2 = 0$

▦ **Graphing calculator activities**

74. Use your graphing calculator to graph the appropriate equation to give visual support for our answers for Examples 1, 3, 4, and 5 of this section.

75. Use your graphing calculator to graph the appropriate equation to give visual support for your answers for Problems 51–60.

76. Use your graphing calculator to predict whether each of the following quadratic equations has two nonreal complex solutions, one real solution, or two real solutions. (Keep these results so that you can use them in the next problem set.)

a. $x^2 + 4x - 21 = 0$ **b.** $x^2 - 3x - 54 = 0$

c. $9x^2 - 6x + 1 = 0$ **d.** $4x^2 + 20x + 25 = 0$

e. $x^2 - 7x + 13 = 0$ **f.** $2x^2 - x + 5 = 0$

g. $15x^2 + 17x - 4 = 0$ **h.** $8x^2 + 18x - 5 = 0$

i. $3x^2 + 4x = 2$ **j.** $2x^2 - 6x = -1$

Answers to Concept Quiz

1. False   **2.** True   **3.** False   **4.** False   **5.** True

11.4 Quadratic Formula

Objectives

- ■ Solve quadratic equations by using the quadratic formula.
- ■ Use the discriminant to indicate the type of solutions for the equation.
- ■ Check solutions by using the sum and product of solutions.

As we saw in the last section, the method of completing the square can be used to solve any quadratic equation. Thus if we apply the method of completing the square to the equation $ax^2 + bx + c = 0$, where a, b, and c are real numbers and $a \neq 0$, we can produce a formula for solving quadratic equations. This formula can then be used to solve any quadratic equation. Let's solve $ax^2 + bx + c = 0$ by completing the square.

$$ax^2 + bx + c = 0$$

$$ax^2 + bx = -c \qquad \text{Isolate the } x^2 \text{ and } x \text{ terms.}$$

$$x^2 + \frac{b}{a}x = -\frac{c}{a} \qquad \text{Multiply both sides by } \frac{1}{a}.$$

$$x^2 + \frac{b}{a}x + \frac{b^2}{4a^2} = -\frac{c}{a} + \frac{b^2}{4a^2} \qquad \frac{1}{2}\left(\frac{b}{a}\right) = \frac{b}{2a} \text{ and } \left(\frac{b}{2a}\right)^2 = \frac{b^2}{4a^2}$$

Complete the square by adding $\frac{b^2}{4a^2}$ to both sides.

$$x^2 + \frac{b}{a}x + \frac{b^2}{4a^2} = -\frac{4ac}{4a^2} + \frac{b^2}{4a^2} \qquad \text{Common denominator of } 4a^2 \text{ on right side}$$

$$x^2 + \frac{b}{a}x + \frac{b^2}{4a^2} = \frac{b^2}{4a^2} - \frac{4ac}{4a^2} \qquad \text{Commutative property}$$

$$\left(x + \frac{b}{2a}\right)^2 = \frac{b^2 - 4ac}{4a^2} \qquad \begin{array}{l}\text{The right side is combined into a single}\\\text{fraction.}\end{array}$$

$$x + \frac{b}{2a} = \pm\sqrt{\frac{b^2 - 4ac}{4a^2}}$$

$$x + \frac{b}{2a} = \pm\frac{\sqrt{b^2 - 4ac}}{\sqrt{4a^2}} \qquad \begin{array}{l}\sqrt{4a^2} = |2a| \text{ but } 2a \text{ can be used}\\\text{because of the use of } \pm.\end{array}$$

$$x + \frac{b}{2a} = \pm\frac{\sqrt{b^2 - 4ac}}{2a}$$

$$x + \frac{b}{2a} = \frac{\sqrt{b^2 - 4ac}}{2a} \qquad \text{or} \qquad x + \frac{b}{2a} = -\frac{\sqrt{b^2 - 4ac}}{2a}$$

$$x = -\frac{b}{2a} + \frac{\sqrt{b^2 - 4ac}}{2a} \qquad \text{or} \qquad x = -\frac{b}{2a} - \frac{\sqrt{b^2 - 4ac}}{2a}$$

$$x = \frac{-b + \sqrt{b^2 - 4ac}}{2a} \qquad \text{or} \qquad x = \frac{-b - \sqrt{b^2 - 4ac}}{2a}$$

The quadratic formula is usually stated as follows:

Quadratic Formula

$$x = \frac{-b \pm \sqrt{b^2 - 4ac}}{2a}, \qquad a \neq 0$$

We can use the quadratic formula to solve *any* quadratic equation by expressing the equation in the standard form $ax^2 + bx + c = 0$ and substituting the values for a, b, and c into the formula. Let's consider some examples, and let's use a graphical approach first to predict approximate solutions whenever possible.

E X A M P L E   1   Solve $x^2 + 5x + 2 = 0$.

Solution

Figure 11.9 shows the graph of the equation.

$$y = x^2 + 5x + 2$$

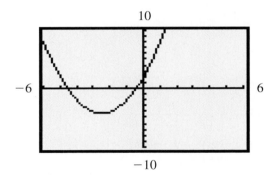

One *x* intercept appears to be between -1 and 0 and the other between -5 and -4.

Figure 11.9

The given equation is in standard form with $a = 1$, $b = 5$, and $c = 2$. Let's substitute these values into the formula and simplify.

$$x = \frac{-b \pm \sqrt{b^2 - 4ac}}{2a}$$

$$x = \frac{-5 \pm \sqrt{5^2 - 4(1)(2)}}{2(1)}$$

$$= \frac{-5 \pm \sqrt{25 - 8}}{2}$$

$$= \frac{-5 \pm \sqrt{17}}{2}$$

The solution set is $\left\{ \dfrac{-5 \pm \sqrt{17}}{2} \right\}$.

We can check the solutions, $\dfrac{-5 + \sqrt{17}}{2}$ and $\dfrac{-5 - \sqrt{17}}{2}$, by comparing them with our estimates from the graphical approach. The decimal approximations of the solutions are -0.44 and -4.56, rounded to the hundredths place. These results agree with our graphical-approach prediction of a solution between -1 and 0 and another solution between -5 and -4. ■

EXAMPLE 2

Solve $x^2 - 2x - 4 = 0$.

Solution

The graph of the equation $y = x^2 - 2x - 4$ is shown in Figure 11.10.

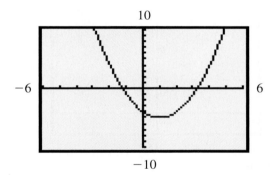

One x intercept appears to be between -2 and -1 and the other between 3 and 4.

Figure 11.10

We need to think of $x^2 - 2x - 4 = 0$ as $x^2 + (-2)x + (-4) = 0$ to determine the values $a = 1$, $b = -2$, and $c = -4$. Let's substitute these values into the quadratic formula and simplify.

$$x = \frac{-b \pm \sqrt{b^2 - 4ac}}{2a}$$

$$x = \frac{-(-2) \pm \sqrt{(-2)^2 - 4(1)(-4)}}{2(1)}$$

$$= \frac{2 \pm \sqrt{4 + 16}}{2}$$

$$= \frac{2 \pm \sqrt{20}}{2}$$

$$= \frac{2 \pm 2\sqrt{5}}{2}$$

$$= \frac{\cancel{2}(1 \pm \sqrt{5})}{\cancel{2}}$$

The solution set is $\{1 \pm \sqrt{5}\}$.

The decimal approximations for the solutions $1 - \sqrt{5}$ and $1 + \sqrt{5}$ are -1.24 and 3.24, rounded to the hundredths place. These results agree with our graphical-approach prediction of a solution between -2 and -1 and another solution between 3 and 4. ■

E X A M P L E   3   Solve $x^2 - 2x + 19 = 0$.

Solution

Figure 11.11 shows the graph of the equation $y = x^2 - 2x + 19$.

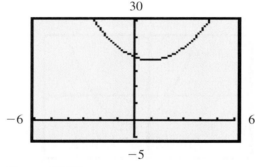

There are no x intercepts. (The solutions are nonreal complex numbers.)

Figure 11.11

$$x^2 - 2x + 19 = 0$$

$$x = \frac{-(-2) \pm \sqrt{(-2)^2 - 4(1)(19)}}{2(1)}$$

$$= \frac{2 \pm \sqrt{4 - 76}}{2}$$

$$= \frac{2 \pm \sqrt{-72}}{2}$$

$$= \frac{2 \pm 6i\sqrt{2}}{2} \qquad \sqrt{-72} = i\sqrt{72} = i\sqrt{36}\sqrt{2} = 6i\sqrt{2}$$

$$= \frac{2(1 \pm 3i\sqrt{2})}{2}$$

$$= 1 \pm 3i\sqrt{2}$$

The solution set is $\{1 \pm 3i\sqrt{2}\}$. ■

E X A M P L E   4   Solve $2x^2 + 4x - 3 = 0$.

Solution

Let's begin by estimating the solutions using a graphical approach. The graph of the equation $y = 2x^2 + 4x - 3$ is shown in Figure 11.12.

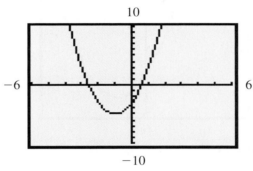

One x intercept is between -3 and -2, and the other is between 0 and 1.

Figure 11.12

$$x = \frac{-b \pm \sqrt{b^2 - 4ac}}{2a}$$

$$x = \frac{-4 \pm \sqrt{4^2 - 4(2)(-3)}}{2(2)}$$

$$x = \frac{-4 \pm \sqrt{16 + 24}}{4}$$

$$x = \frac{-4 \pm \sqrt{40}}{4}$$

$$x = \frac{-4 \pm 2\sqrt{10}}{4}$$

$$x = \frac{-2 \pm \sqrt{10}}{2}$$

The solution set is $\left\{ \dfrac{-2 \pm \sqrt{10}}{2} \right\}$.

The decimal approximations for the solutions, $\dfrac{-2 - \sqrt{10}}{2}$ and $\dfrac{-2 + \sqrt{10}}{2}$, are -2.58 and 0.58 rounded to the hundredths place. These results agree with our graphical-approach prediction of a solution between -3 and -2 and another solution between 0 and 1. ∎

E X A M P L E   5   Solve $n(3n - 10) = 25$.

Solution

Figure 11.13 shows the graph of the equation $y = x(3x - 10) - 25$.

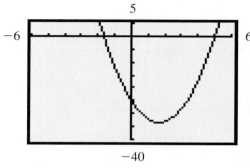

One x intercept is between -2 and -1, and the other appears to be 5.

Figure 11.13

First, we need to change the equation to the standard form $an^2 + bn + c = 0$.

$$n(3n - 10) = 25$$
$$3n^2 - 10n = 25$$
$$3n^2 - 10n - 25 = 0$$

Now we can substitute $a = 3$, $b = -10$, and $c = -25$ into the quadratic formula.

$$n = \frac{-b \pm \sqrt{b^2 - 4ac}}{2a}$$

$$= \frac{-(-10) \pm \sqrt{(-10)^2 - 4(3)(-25)}}{2(3)}$$

$$= \frac{10 \pm \sqrt{100 + 300}}{2(3)}$$

$$= \frac{10 \pm \sqrt{400}}{6}$$

$$= \frac{10 \pm 20}{6}$$

$$n = \frac{10 + 20}{6} \quad \text{or} \quad n = \frac{10 - 20}{6}$$

$$n = 5 \quad \text{or} \quad n = -\frac{5}{3}$$

The solution set is $\left\{ -\frac{5}{3}, 5 \right\}$.

The solutions are $-1\dfrac{2}{3}$ and 5. These results agree with our graphical approach of a solution between -2 and -1 and another solution of 5. ■

In Example 5, note that we used the variable n. The quadratic formula is usually stated in terms of x, but it certainly can be applied to quadratic equations in other variables. Also note in Example 5 that the polynomial $3n^2 - 10n - 25$ can be factored as $(3n + 5)(n - 5)$. Therefore, we could also solve the equation $3n^2 - 10n - 25 = 0$ by using the factoring approach. In the next section, we will give you some guidance on which approach to use for a particular equation.

Nature of Roots

The quadratic formula makes it easy to determine the nature of the roots of a quadratic equation without completely solving the equation. The number

$$b^2 - 4ac$$

which appears under the radical sign in the quadratic formula, is called the **discriminant** of the quadratic equation. The discriminant is the indicator of the kind of roots the equation has. For example, suppose that we start to solve the equation $x^2 - 4x + 7 = 0$ as follows:

$$x = \frac{-b \pm \sqrt{b^2 - 4ac}}{2a}$$

$$x = \frac{-(-4) \pm \sqrt{(-4)^2 - 4(1)(7)}}{2(1)}$$

$$= \frac{4 \pm \sqrt{16 - 28}}{2}$$

$$= \frac{4 \pm \sqrt{-12}}{2}$$

At this stage you should be able to look ahead and realize that you will obtain two complex solutions for the equation. (By the way, observe that these solutions are complex conjugates.) In other words, the discriminant, -12, indicates what type of roots you will obtain.

We make the following general statements relative to the roots of a quadratic equation of the form $ax^2 + bx + c = 0$.

1. If $b^2 - 4ac < 0$, then the equation has two nonreal complex solutions.
2. If $b^2 - 4ac = 0$, then the equation has one real solution.
3. If $b^2 - 4ac > 0$, then the equation has two real solutions.

The following examples illustrate each of these situations. (You may want to solve the equations completely to verify the conclusions.)

Equation	Discriminant	Nature of roots
$x^2 - 3x + 7 = 0$	$b^2 - 4ac = (-3)^2 - 4(1)(7)$ $= 9 - 28$ $= -19$	Two nonreal complex solutions
$9x^2 - 12x + 4 = 0$	$b^2 - 4ac = (-12)^2 - 4(9)(4)$ $= 144 - 144$ $= 0$	One real solution
$2x^2 + 5x - 3 = 0$	$b^2 - 4ac = (5)^2 - 4(2)(-3)$ $= 25 + 24$ $= 49$	Two real solutions

There is another very useful relationship that involves the roots of a quadratic equation and the numbers a, b, and c of the general form $ax^2 + bx + c = 0$. Suppose that we let x_1 and x_2 be the two roots generated by the quadratic formula. Thus we have

$$x_1 = \frac{-b + \sqrt{b^2 - 4ac}}{2a} \quad \text{and} \quad x_2 = \frac{-b - \sqrt{b^2 - 4ac}}{2a}$$

REMARK: A clarification is called for at this time. Previously, we made the statement that if $b^2 - 4ac = 0$, then the equation has one real solution. Technically, such an equation has two solutions, but they are equal. For example, each factor of $(x - 2)(x - 2) = 0$ produces a solution, but both solutions are the number 2. We sometimes refer to this as one real solution with a *multiplicity of two*. Using the idea of multiplicity of roots, we can say that every quadratic equation has two roots.

Now let's consider the sum and product of the two roots.

$$\textbf{Sum} \quad x_1 + x_2 = \frac{-b + \sqrt{b^2 - 4ac}}{2a} + \frac{-b - \sqrt{b^2 - 4ac}}{2a} = \frac{-2b}{2a} = \boxed{-\frac{b}{a}}$$

$$\textbf{Product} \quad (x_1)(x_2) = \left(\frac{-b + \sqrt{b^2 - 4ac}}{2a}\right)\left(\frac{-b - \sqrt{b^2 - 4ac}}{2a}\right)$$

$$= \frac{b^2 - (b^2 - 4ac)}{4a^2}$$

$$= \frac{b^2 - b^2 + 4ac}{4a^2}$$

$$= \frac{4ac}{4a^2} = \boxed{\frac{c}{a}}$$

These relationships provide another way of checking potential solutions when solving quadratic equations. For instance, back in Example 3 we solved the equation $x^2 - 2x + 19 = 0$ and obtained solutions of $1 + 3i\sqrt{2}$ and $1 - 3i\sqrt{2}$. Let's check these solutions by using the sum and product relationships.

Check for Example 3

Sum of roots $(1 + 3i\sqrt{2}) + (1 - 3i\sqrt{2}) = 2$ and $-\dfrac{b}{a} = -\dfrac{-2}{1} = 2$

Product of roots $(1 + 3i\sqrt{2})(1 - 3i\sqrt{2}) = 1 - 18i^2 = 1 + 18 = 19$

and $\dfrac{c}{a} = \dfrac{19}{1} = 19$

Likewise, a check for Example 4 is as follows:

Check for Example 4

Sum of roots $\left(\dfrac{-2 + \sqrt{10}}{2}\right) + \left(\dfrac{-2 - \sqrt{10}}{2}\right) = -\dfrac{4}{2} = -2$

and $-\dfrac{b}{a} = -\dfrac{4}{2} = -2$

Product of roots $\left(\dfrac{-2 + \sqrt{10}}{2}\right)\left(\dfrac{-2 - \sqrt{10}}{2}\right) = -\dfrac{6}{4} = -\dfrac{3}{2}$

and $\dfrac{c}{a} = \dfrac{-3}{2} = -\dfrac{3}{2}$

Note that for both Examples 3 and 4, it was much easier to check by using the sum and product relationships than it would have been to check by substituting back into the original equation. Don't forget that the values for a, b, and c come from a quadratic equation of the form $ax^2 + bx + c = 0$. Therefore, if we are going to check the potential solutions to Example 5 by using the sum and product relationships, we must be certain that we made no errors when changing the given equation $n(3n - 10) = 25$ to the form $3n^2 - 10n - 25 = 0$.

CONCEPT QUIZ

For Problems 1–6, answer true or false.

1. The quadratic formula can be used to solve any quadratic equation.

2. The number $\sqrt{b^2 - 4ac}$ is called the discriminant of the quadratic equation.

3. Every quadratic equation will have two solutions.

4. The quadratic formula cannot be used if the quadratic equation can be solved by factoring.

5. To use the quadratic formula for solving the equation $3x^2 + 2x - 5 = 0$ you must first divide both sides of the equation by 3.

6. The sum of the roots of a quadratic equation is equal to $\dfrac{c}{a}$.

PROBLEM SET 11.4

For each quadratic equation in Problems 1–10, first use the discriminant to determine whether the equation has two nonreal complex solutions, one real solution with a multiplicity of two, or two real solutions. Then solve the equation.

1. $x^2 + 4x - 21 = 0$

2. $x^2 - 3x - 54 = 0$

3. $9x^2 - 6x + 1 = 0$

4. $4x^2 + 20x + 25 = 0$

5. $x^2 - 7x + 13 = 0$

6. $2x^2 - x + 5 = 0$

7. $15x^2 + 17x - 4 = 0$

8. $8x^2 + 18x - 5 = 0$

9. $3x^2 + 4x = 2$

10. $2x^2 - 6x = -1$

For Problems 11–50, use the quadratic formula to solve each of the quadratic equations. Check your solutions by using the *sum and product relationships*.

11. $x^2 + 2x - 1 = 0$

12. $x^2 + 4x - 1 = 0$

13. $n^2 + 5n - 3 = 0$

14. $n^2 + 3n - 2 = 0$

15. $a^2 - 8a = 4$

16. $a^2 - 6a = 2$

17. $n^2 + 5n + 8 = 0$

18. $2n^2 - 3n + 5 = 0$

19. $x^2 - 18x + 80 = 0$

20. $x^2 + 19x + 70 = 0$

21. $-y^2 = -9y + 5$

22. $-y^2 + 7y = 4$

23. $2x^2 + x - 4 = 0$

24. $2x^2 + 5x - 2 = 0$

25. $4x^2 + 2x + 1 = 0$

26. $3x^2 - 2x + 5 = 0$

27. $3a^2 - 8a + 2 = 0$

28. $2a^2 - 6a + 1 = 0$

29. $-2n^2 + 3n + 5 = 0$

30. $-3n^2 - 11n + 4 = 0$

31. $3x^2 + 19x + 20 = 0$

32. $2x^2 - 17x + 30 = 0$

33. $36n^2 - 60n + 25 = 0$

34. $9n^2 + 42n + 49 = 0$

35. $4x^2 - 2x = 3$

36. $6x^2 - 4x = 3$

37. $5x^2 - 13x = 0$

38. $7x^2 + 12x = 0$

39. $3x^2 = 5$

40. $4x^2 = 3$

41. $6t^2 + t - 3 = 0$

42. $2t^2 + 6t - 3 = 0$

43. $n^2 + 32n + 252 = 0$

44. $n^2 - 4n - 192 = 0$

45. $12x^2 - 73x + 110 = 0$

46. $6x^2 + 11x - 255 = 0$

47. $-2x^2 + 4x - 3 = 0$

48. $-2x^2 + 6x - 5 = 0$

49. $-6x^2 + 2x + 1 = 0$

50. $-2x^2 + 4x + 1 = 0$

■ ■ ■ Thoughts into words

51. Your friend states that the equation $-2x^2 + 4x - 1 = 0$ must be changed to $2x^2 - 4x + 1 = 0$ (by multiplying both sides by -1) before the quadratic formula can be applied. Is she right about this? If not, how would you convince her she is wrong?

52. Another of your friends claims that the quadratic formula can be used to solve the equation $x^2 - 9 = 0$. How would you react to this claim?

53. Why must we change the equation $3x^2 - 2x = 4$ to $3x^2 - 2x - 4 = 0$ before applying the quadratic formula?

■ ■ ■ Further investigations

The solution set for $x^2 - 4x - 37 = 0$ is $\{2 \pm \sqrt{41}\}$. With a calculator, we found a rational approximation, to the nearest thousandth, for each of these solutions.

$$2 - \sqrt{41} = -4.403 \quad \text{and} \quad 2 + \sqrt{41} = 8.403$$

Thus the solution set is $\{-4.403, 8.403\}$, with the answers rounded to the nearest thousandth.

Solve each of the equations in Problems 54–63, expressing solutions to the nearest thousandth.

54. $x^2 - 6x - 10 = 0$

55. $x^2 - 16x - 24 = 0$

56. $x^2 + 6x - 44 = 0$

57. $x^2 + 10x - 46 = 0$

58. $x^2 + 8x + 2 = 0$

59. $x^2 + 9x + 3 = 0$

60. $4x^2 - 6x + 1 = 0$ **61.** $5x^2 - 9x + 1 = 0$

62. $2x^2 - 11x - 5 = 0$ **63.** $3x^2 - 12x - 10 = 0$

For Problems 64–66, use the discriminant to help solve each problem.

64. Determine k so that the solutions of $x^2 - 2x + k = 0$ are complex but nonreal.

65. Determine k so that $4x^2 - kx + 1 = 0$ has two equal real solutions.

66. Determine k so that $3x^2 - kx - 2 = 0$ has real solutions.

 Graphing calculator activities

67. Use your graphing calculator to verify your answers for Problems 1–10.

68. Use your graphing calculator to give visual support for your answers for Problems 17, 18, 25, 26, 47, and 48.

69. Use your graphing calculator to verify your solutions, to the nearest tenth, for Problems 13, 14, 15, 16, 21, 22, 23, 24, 35, 36, 41, 42, 49, and 50.

Answers to Concept Quiz

1. True   **2.** False   **3.** True   **4.** False   **5.** False   **6.** False

---

11.5   More Quadratic Equations and Applications

Objectives

- Choose the most appropriate method for solving a quadratic equation.
- Solve equations that are quadratic in form.
- Use quadratic equations to solve word problems.
- Solve word problems involving interest compounded annually.

Which method should be used to solve a particular quadratic equation? There is no hard and fast answer to that question; it depends on the *type* of equation and on your personal preference. In the following examples we will state reasons for choosing a specific technique. However, keep in mind that usually this is a decision *you* must make as the need arises. That's why you need to be familiar with the strengths and weaknesses of each method.

EXAMPLE 1   Solve $2x^2 - 3x - 1 = 0$.

Solution

Because of the leading coefficient of 2 and the constant term of -1, there are very few factoring possibilities to consider. Therefore, with such problems, first try the factoring approach. Unfortunately, this particular polynomial is not factorable using integers. Thus let's use the quadratic formula to solve the equation.

$$x = \frac{-b \pm \sqrt{b^2 - 4ac}}{2a}$$

$$x = \frac{-(-3) \pm \sqrt{(-3)^2 - 4(2)(-1)}}{2(2)}$$

$$= \frac{3 \pm \sqrt{9 + 8}}{4}$$

$$= \frac{3 \pm \sqrt{17}}{4}$$

Check

We can use the *sum-of-roots* and the *product-of-roots* relationships for our checking purposes.

Sum of roots $\dfrac{3 + \sqrt{17}}{4} + \dfrac{3 - \sqrt{17}}{4} = \dfrac{6}{4} = \dfrac{3}{2}$ and $-\dfrac{b}{a} = -\dfrac{-3}{2} = \dfrac{3}{2}$

Product of roots $\left(\dfrac{3 + \sqrt{17}}{4}\right)\left(\dfrac{3 - \sqrt{17}}{4}\right) = \dfrac{9 - 17}{16} = -\dfrac{8}{16} = -\dfrac{1}{2}$ and

$$\frac{c}{a} = \frac{-1}{2} = -\frac{1}{2}$$

The solution set is $\left\{\dfrac{3 \pm \sqrt{17}}{4}\right\}$.   ■

EXAMPLE 2   Solve $\dfrac{3}{n} + \dfrac{10}{n + 6} = 1$.

Solution

$$\frac{3}{n} + \frac{10}{n + 6} = 1, \qquad n \neq 0 \text{ and } n \neq -6$$

$$n(n + 6)\left(\frac{3}{n} + \frac{10}{n + 6}\right) = 1(n)(n + 6) \qquad \text{Multiply both sides by } n(n + 6),\text{ which is the LCD.}$$

$$3(n + 6) + 10n = n(n + 6)$$

$$3n + 18 + 10n = n^2 + 6n$$

$$13n + 18 = n^2 + 6n$$

$$0 = n^2 - 7n - 18$$

This equation is an easy one to consider for possible factoring, and it factors as follows:

$$0 = (n - 9)(n + 2)$$

$$n - 9 = 0 \qquad \text{or} \qquad n + 2 = 0$$

$$n = 9 \qquad \text{or} \qquad n = -2$$

Check

Substituting 9 and -2 back into the original equation, we obtain

$$\frac{3}{n} + \frac{10}{n+6} = 1 \qquad\qquad\qquad \frac{3}{n} + \frac{10}{n+6} = 1$$

$$\frac{3}{9} + \frac{10}{9+6} \overset{?}{=} 1 \qquad\qquad\qquad \frac{3}{-2} + \frac{10}{-2+6} \overset{?}{=} 1$$

$$\frac{1}{3} + \frac{10}{15} \overset{?}{=} 1 \qquad\qquad\qquad -\frac{3}{2} + \frac{10}{4} \overset{?}{=} 1$$

$$\frac{1}{3} + \frac{2}{3} \overset{?}{=} 1 \qquad\qquad\qquad -\frac{3}{2} + \frac{5}{2} \overset{?}{=} 1$$

$$1 = 1 \qquad\qquad\qquad\qquad \frac{2}{2} = 1$$

The solution set is $\{-2, 9\}$. ■

We should make two comments about Example 2. First, note the indication of the initial restrictions $n \neq 0$ and $n \neq -6$. Remember that we need to do this when solving fractional equations. Second, the *sum-of-roots* and *product-of-roots* relationships were not used for checking purposes in this problem. Those relationships would check the validity of our work only from the step $0 = n^2 - 7n - 18$ to the finish. In other words, an error made in changing the original equation to quadratic form would not be detected by checking the sum and product of potential roots. With such a problem, the only *absolute check* is to substitute the potential solutions back into the *original equation*.

E X A M P L E   3   Solve $x^2 + 22x + 112 = 0$.

 Solution

The size of the constant term makes the factoring approach a little cumbersome for this problem. Furthermore, because the leading coefficient is 1 and the coefficient of the x term is even, the method of completing the square will work effectively.

$$x^2 + 22x + 112 = 0$$
$$x^2 + 22x = -112$$
$$x^2 + 22x + 121 = -112 + 121$$
$$(x + 11)^2 = 9$$
$$x + 11 = \pm\sqrt{9}$$
$$x + 11 = \pm 3$$
$$x + 11 = 3 \qquad \text{or} \qquad x + 11 = -3$$
$$x = -8 \qquad \text{or} \qquad x = -14$$

Check

Sum of roots $-8 + (-14) = -22$ and $-\dfrac{b}{a} = -22$

Product of roots $(-8)(-14) = 112$ and $\dfrac{c}{a} = 112$

The solution set is $\{-14, -8\}$.   ▪

EXAMPLE 4

Solve $x^4 - 4x^2 - 96 = 0$.

Solution

An equation such as $x^4 - 4x^2 - 96 = 0$ is not a quadratic equation, but we can solve it using the techniques that we use on quadratic equations. That is, we can factor the polynomial and apply the property "$ab = 0$ if *and only if* $a = 0$ or $b = 0$" as follows:

$$x^4 - 4x^2 - 96 = 0$$
$$(x^2 - 12)(x^2 + 8) = 0$$
$$x^2 - 12 = 0 \qquad \text{or} \qquad x^2 + 8 = 0$$
$$x^2 = 12 \qquad \text{or} \qquad x^2 = -8$$
$$x = \pm\sqrt{12} \qquad \text{or} \qquad x = \pm\sqrt{-8}$$
$$x = \pm 2\sqrt{3} \qquad \text{or} \qquad x = \pm 2i\sqrt{2}$$

The solution set is $\{\pm 2\sqrt{3}, \pm 2i\sqrt{2}\}$. (We will leave the check for this problem for you to do!)   ▪

REMARK: Another approach to Example 4 would be to substitute y for x^2 and y^2 for x^4. The equation $x^4 - 4x^2 - 96 = 0$ becomes the quadratic equation $y^2 - 4y - 96 = 0$. Thus we say that $x^4 - 4x^2 - 96 = 0$ is of *quadratic form*. Then we could solve the quadratic equation $y^2 - 4y - 96 = 0$ and use the equation $y = x^2$ to determine the solutions for x.

Applications

Before we conclude this section with some word problems that can be solved using quadratic equations, let's restate the suggestions we made, in an earlier chapter, for solving word problems.

Suggestions for Solving Word Problems

1. Read the problem carefully and make certain that you understand the meanings of all the words. Be especially alert for any technical terms used in the statement of the problem.

2. Read the problem a second time (perhaps even a third time) to get an overview of the situation being described and to determine the known facts, as well as what is to be found.

3. Sketch any figure, diagram, or chart that might be helpful in analyzing the problem.
4. Choose a meaningful variable to represent an unknown quantity in the problem (perhaps *l*, if the length of a rectangle is an unknown quantity) and represent any other unknowns in terms of that variable.
5. Look for a *guideline* that you can use to set up an equation. A guideline might be a formula such as $A = lw$ or a relationship such as *the fractional part of a job done by Bill plus the fractional part of the job done by Mary equals the total job.*
6. Form an equation that contains the variable and translates the conditions of the guideline from English into algebra.
7. Solve the equation and use the solutions to determine all facts requested in the problem.
8. **Check all answers back into the original statement of the problem.**

Keep these suggestions in mind as we consider some word problems.

PROBLEM 1

A page for a magazine contains 70 square inches of type. The height of a page is twice the width. If the margin around the type is to be 2 inches uniformly, what are the dimensions of a page?

Solution
Let x represent the width of a page. Then $2x$ represents the height of a page. Now let's draw and label a model of a page (Figure 11.14).

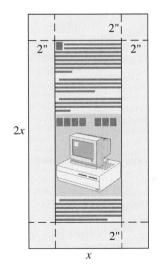

Figure 11.14

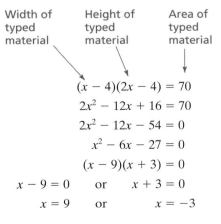

Width of typed material   Height of typed material   Area of typed material

$$(x - 4)(2x - 4) = 70$$
$$2x^2 - 12x + 16 = 70$$
$$2x^2 - 12x - 54 = 0$$
$$x^2 - 6x - 27 = 0$$
$$(x - 9)(x + 3) = 0$$
$$x - 9 = 0 \quad \text{or} \quad x + 3 = 0$$
$$x = 9 \quad \text{or} \quad x = -3$$

Disregard the negative solution; the page must be 9 inches wide, and its height is $2(9) = 18$ inches.

Let's use our knowledge of quadratic equations to analyze some applications in the business world. For example, if P dollars is invested at r rate of interest compounded annually for t years, then the amount of money, A, accumulated at the end of t years is given by the formula

$$A = P(1 + r)^t$$

This compound interest formula serves as a guideline for the next problem.

P R O B L E M   2   Suppose that $100 is invested at a certain rate of interest compounded annually for 2 years. If the accumulated value at the end of 2 years is $121, find the rate of interest.

Solution

Let r represent the rate of interest. Substitute the known values into the compound interest formula.

$$A = P(1 + r)^t$$
$$121 = 100(1 + r)^2$$

Solving this equation yields

$$\frac{121}{100} = (1 + r)^2$$

$$\pm\sqrt{\frac{121}{100}} = (1 + r)$$

$$\pm\frac{11}{10} = 1 + r$$

$$1 + r = \frac{11}{10} \qquad \text{or} \qquad 1 + r = -\frac{11}{10}$$

$$r = -1 + \frac{11}{10} \qquad \text{or} \qquad r = -1 - \frac{11}{10}$$

$$r = \frac{1}{10} \qquad \text{or} \qquad r = -\frac{21}{10}$$

We must disregard the negative solution, so $r = \frac{1}{10}$ is the only solution. Change $\frac{1}{10}$ to a percent, and the rate of interest is 10%.   ■

P R O B L E M   3   On a 130-mile trip from Orlando to Sarasota, Roberto encountered a heavy thunderstorm for the last 40 miles of the trip. During the thunderstorm he averaged 20 miles per hour slower than before the storm. The entire trip took $2\frac{1}{2}$ hours. How fast did he travel before the storm?

Solution

Let x represent Roberto's rate before the thunderstorm. Then $x - 20$ represents his speed during the thunderstorm. Because $t = \frac{d}{r}$, then $\frac{90}{x}$ represents the time traveling before the storm and $\dfrac{40}{x - 20}$ represents the time traveling during the storm. The following guideline sums up the situation.

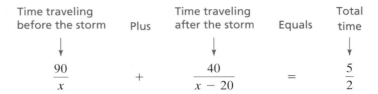

Time traveling before the storm	Plus	Time traveling after the storm	Equals	Total time
↓		↓		↓
$\dfrac{90}{x}$	$+$	$\dfrac{40}{x - 20}$	$=$	$\dfrac{5}{2}$

Solving this equation, we obtain

$$2x(x - 20)\left(\frac{90}{x} + \frac{40}{x - 20}\right) = 2x(x - 20)\left(\frac{5}{2}\right)$$

$$2x(x - 20)\left(\frac{90}{x}\right) + 2x(x - 20)\left(\frac{40}{x - 20}\right) = 2x(x - 20)\left(\frac{5}{2}\right)$$

$$180(x - 20) + 2x(40) = 5x(x - 20)$$

$$180x - 3600 + 80x = 5x^2 - 100x$$

$$0 = 5x^2 - 360x + 3600$$

$$0 = 5(x^2 - 72x + 720)$$

$$0 = 5(x - 60)(x - 12)$$

$$x - 60 = 0 \quad \text{or} \quad x - 12 = 0$$

$$x = 60 \quad \text{or} \quad x = 12$$

We discard the solution of 12 because it would be impossible to drive 20 miles per hour slower than 12 miles per hour; thus Roberto's rate before the thunderstorm was 60 miles per hour.   ■

PROBLEM 4

A businesswoman bought a parcel of land on speculation for $120,000. She subdivided the land into lots, and when she had sold all but 18 lots at a profit of $6000 per lot, she had regained the entire cost of the land. How many lots were sold and at what price per lot?

Solution

Let x represent the number of lots sold. Then $x + 18$ represents the total number of lots. Therefore, $\dfrac{120{,}000}{x}$ represents the selling price per lot, and $\dfrac{120{,}000}{x + 18}$ represents the cost per lot. The following equation sums up the situation.

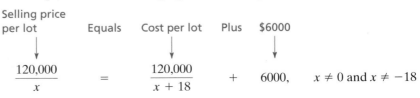

Selling price per lot	Equals	Cost per lot	Plus	$6000	
↓		↓		↓	
$\dfrac{120{,}000}{x}$	$=$	$\dfrac{120{,}000}{x + 18}$	$+$	$6000,$	$x \neq 0$ and $x \neq -18$

Solving this equation, we obtain

$$x(x + 18)\left(\frac{120{,}000}{x}\right) = \left(\frac{120{,}000}{x + 18} + 6000\right)(x)(x + 18)$$

$$120{,}000(x + 18) = 120{,}000x + 6000x(x + 18)$$

$$120{,}000x + 2{,}160{,}000 = 120{,}000x + 6000x^2 + 108{,}000x$$

$$0 = 6000x^2 + 108{,}000x - 2{,}160{,}000$$

$$0 = x^2 + 18x - 360$$

The method of completing the square works very well with this equation.

$$x^2 + 18x = 360$$

$$x^2 + 18x + 81 = 441$$

$$(x + 9)^2 = 441$$

$$x + 9 = \pm\sqrt{441}$$

$$x + 9 = \pm 21$$

$$x + 9 = 21 \quad \text{or} \quad x + 9 = -21$$

$$x = 12 \quad \text{or} \quad x = -30$$

We discard the negative solution; thus 12 lots were sold at $\dfrac{120{,}000}{x} = \dfrac{120{,}000}{12} =$ $10,000 per lot. ∎

Barry bought a number of shares of stock for $600. A week later the value of the stock had increased $3 per share, and he sold all but 10 shares and regained his original investment of $600. How many shares did he sell and at what price per share?

Solution

Let s represent the number of shares Barry sold. Then $s + 10$ represents the number of shares purchased. Therefore, $\dfrac{600}{s}$ represents the selling price per share, and $\dfrac{600}{s + 10}$ represents the cost per share.

$$\underbrace{\frac{600}{s}}_{\substack{\text{Selling price} \\ \text{per share}}} = \underbrace{\frac{600}{s + 10}}_{\text{Cost per share}} + 3, \qquad s \neq 0 \text{ and } s \neq -10$$

Solving this equation yields

$$s(s + 10)\left(\frac{600}{s}\right) = \left(\frac{600}{s + 10} + 3\right)(s)(s + 10)$$

$$600(s + 10) = 600s + 3s(s + 10)$$
$$600s + 6000 = 600s + 3s^2 + 30s$$
$$0 = 3s^2 + 30s - 6000$$
$$0 = s^2 + 10s - 2000$$

Use the quadratic formula to obtain

$$s = \frac{-10 \pm \sqrt{10^2 - 4(1)(-2000)}}{2(1)}$$

$$= \frac{-10 \pm \sqrt{100 + 8000}}{2}$$

$$= \frac{-10 \pm \sqrt{8100}}{2}$$

$$= \frac{-10 \pm 90}{2}$$

$$s = \frac{-10 + 90}{2} \qquad \text{or} \qquad s = \frac{-10 - 90}{2}$$

$$s = 40 \qquad \qquad \text{or} \qquad s = -50$$

We discard the negative solution, and we know that 40 shares were sold at $\dfrac{600}{s} = \dfrac{600}{40} = \$15$ per share.   ∎

This next problem set contains a wide variety of word problems. Not only are there some business applications similar to those we discussed in this section, but there are also more problems of the types we discussed back in Chapters 4 and 5. Try to give them your best shot without referring to the examples in earlier chapters.

CONCEPT QUIZ

For Problems 1–7, choose the method that you think is most appropriate for solving the given equation.

1. $2x^2 + 6x - 3 = 0$ **A.** Factoring

2. $(x + 1)^2 = 36$ **B.** Square root property (Property 11.1)

3. $x^2 - 3x + 2 = 0$ **C.** Completing the square

4. $x^2 + 6x = 19$ **D.** Quadratic formula

5. $4x^2 + 2x - 5 = 0$

6. $4x^2 = 3$

7. $x^2 - 4x - 12 = 0$

PROBLEM SET 11.5

For Problems 1–20, solve each quadratic equation using the method that seems most appropriate to you.

1. $x^2 - 4x - 6 = 0$

2. $x^2 - 8x - 4 = 0$

3. $3x^2 + 23x - 36 = 0$

4. $n^2 + 22n + 105 = 0$

5. $x^2 - 18x = 9$

6. $x^2 + 20x = 25$

7. $2x^2 - 3x + 4 = 0$

8. $3y^2 - 2y + 1 = 0$

9. $135 + 24n + n^2 = 0$

10. $28 - x - 2x^2 = 0$

11. $(x - 2)(x + 9) = -10$

12. $(x + 3)(2x + 1) = -3$

13. $2x^2 - 4x + 7 = 0$

14. $3x^2 - 2x + 8 = 0$

15. $x^2 - 18x + 15 = 0$

16. $x^2 - 16x + 14 = 0$

17. $20y^2 + 17y - 10 = 0$

18. $12x^2 + 23x - 9 = 0$

19. $4t^2 + 4t - 1 = 0$

20. $5t^2 + 5t - 1 = 0$

For Problems 21–40, solve each equation.

21. $n + \dfrac{3}{n} = \dfrac{19}{4}$

22. $n - \dfrac{2}{n} = -\dfrac{7}{3}$

23. $\dfrac{3}{x} + \dfrac{7}{x - 1} = 1$

24. $\dfrac{2}{x} + \dfrac{5}{x + 2} = 1$

25. $\dfrac{12}{x - 3} + \dfrac{8}{x} = 14$

26. $\dfrac{16}{x + 5} - \dfrac{12}{x} = -2$

27. $\dfrac{3}{x - 1} - \dfrac{2}{x} = \dfrac{5}{2}$

28. $\dfrac{4}{x + 1} + \dfrac{2}{x} = \dfrac{5}{3}$

29. $\dfrac{6}{x} + \dfrac{40}{x + 5} = 7$

30. $\dfrac{12}{t} + \dfrac{18}{t + 8} = \dfrac{9}{2}$

31. $\dfrac{5}{n - 3} - \dfrac{3}{n + 3} = 1$

32. $\dfrac{3}{t + 2} + \dfrac{4}{t - 2} = 2$

33. $x^4 - 18x^2 + 72 = 0$

34. $x^4 - 21x^2 + 54 = 0$

35. $3x^4 - 35x^2 + 72 = 0$

36. $5x^4 - 32x^2 + 48 = 0$

37. $3x^4 + 17x^2 + 20 = 0$

38. $4x^4 + 11x^2 - 45 = 0$

39. $6x^4 - 29x^2 + 28 = 0$

40. $6x^4 - 31x^2 + 18 = 0$

For Problems 41–72, set up an equation and solve each problem.

41. Find two consecutive whole numbers such that the sum of their squares is 145.

42. Find two consecutive odd whole numbers such that the sum of their squares is 74.

43. Two positive integers differ by 3, and their product is 108. Find the numbers.

44. Suppose that the sum of two numbers is 20, and the sum of their squares is 232. Find the numbers.

45. Find two numbers such that their sum is 10 and their product is 22.

46. Find two numbers such that their sum is 6 and their product is 7.

47. Suppose that the sum of two whole numbers is 9 and the sum of their reciprocals is $\dfrac{1}{2}$. Find the numbers.

48. The difference between two whole numbers is 8 and the difference between their reciprocals is $\dfrac{1}{6}$. Find the two numbers.

49. The sum of the lengths of the two legs of a right triangle is 21 inches. If the length of the hypotenuse is 15 inches, find the length of each leg.

50. The length of a rectangular floor is 1 meter less than twice its width. If a diagonal of the rectangle is 17 meters, find the length and width of the floor.

51. A rectangular plot of ground measuring 12 meters by 20 meters is surrounded by a sidewalk of a uniform width (see Figure 11.15). The area of the sidewalk is 68 square meters. Find the width of the sidewalk.

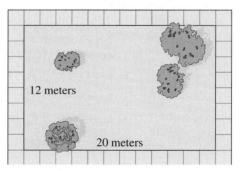

12 meters

20 meters

Figure 11.15

52. A 5-inch-by-7-inch picture is surrounded by a frame of uniform width. The area of the picture and frame together is 80 square inches. Find the width of the frame.

53. The perimeter of a rectangle is 44 inches, and its area is 112 square inches. Find the length and width of the rectangle.

54. A rectangular piece of cardboard is 2 units longer than it is wide. From each of its corners a square piece 2 units on a side is cut out. The flaps are then turned up to form an open box that has a volume of 70 cubic units. Find the length and width of the original piece of cardboard.

55. Charlotte traveled 250 miles in 1 hour more time than it took Lorraine to travel 180 miles. Charlotte drove 5 miles per hour faster than Lorraine. How fast did each one travel?

56. Larry drove 156 miles in 1 hour more than it took Terrell to drive 108 miles. Terrell drove at an average rate of 2 miles per hour faster than Larry. How fast did each one travel?

57. On a 570-mile trip, Andy averaged 5 miles per hour faster for the last 240 miles than he did for the first 330 miles. The entire trip took 10 hours. How fast did he travel for the first 330 miles?

58. On a 135-mile bicycle excursion, Maria averaged 5 miles per hour faster for the first 60 miles than she did for the last 75 miles. The entire trip took 8 hours. Find her rate for the first 60 miles.

59. It takes Terry 2 hours longer to do a certain job than it takes Tom. They worked together for 3 hours; then Tom left and Terry finished the job in 1 hour. How long would it take each of them to do the job alone?

60. Suppose that Arlene can mow the entire lawn in 40 minutes less time with the power mower than she can with the push mower. One day the power mower broke down after she had been mowing for 30 minutes. She finished the lawn with the push mower in 20 minutes. How long does it take Arlene to mow the entire lawn with the power mower?

61. A man did a job for $360. It took him 6 hours longer than he expected, and therefore he earned $2 per hour less than he anticipated. How long did he expect that it would take to do the job?

62. A group of students agreed that each would chip in the same amount to pay for a party that would cost $100. Then they found 5 more students interested in the party and in sharing the expenses. This decreased the amount each had to pay by $1. How many students were involved in the party and how much did each student have to pay?

63. A group of customers agreed that each would contribute the same amount to buy their favorite waitress a $100 birthday gift. At the last minute, 2 of the people decided not to chip in. This increased the amount that the remaining people had to pay by $2.50 per person. How many people actually contributed to the gift?

64. A retailer bought a number of special mugs for $48. Two of the mugs were broken in the store, but by selling each of the other mugs for $3 above the original cost per mug, she made a total profit of $22. How many mugs did she buy and at what price per mug did she sell them?

65. Tony bought a number of shares of stock for $720. A month later the value of the stock increased by $8 per share, and he sold all but 20 shares and regained his original investment plus a profit of $80. How many shares did he sell and at what price per share?

66. The formula $D = \dfrac{n(n-3)}{2}$ yields the number of diagonals, D, in a polygon of n sides. Find the number of sides of a polygon that has 54 diagonals.

67. The formula $S = \dfrac{n(n+1)}{2}$ yields the sum, S, of the first n natural numbers $1, 2, 3, 4, \ldots$. How many consecutive natural numbers, starting with 1, will give a sum of 1275?

68. At a point 16 yards from the base of a tower, the distance to the top of the tower is 4 yards more than the height of the tower (see Figure 11.16). Find the height of the tower.

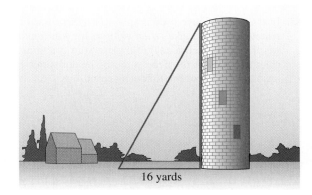

16 yards

Figure 11.16

69. Suppose that $5000 is invested at a certain rate of interest compounded annually for 2 years. If the accumulated value at the end of 2 years is $5724.50, find the rate of interest.

70. Suppose that $10,000 is invested at a certain rate of interest compounded annually for 2 years. If the accumulated value at the end of 2 years is $12,544, find the rate of interest.

71. What rate of interest compounded annually is needed to make an investment of $8000 accumulate to $8988.80 at the end of two years?

72. What rate of interest compounded annually is needed to make an investment of $6500 accumulate to $7166.25 at the end of two years?

■ ■ ▨ **Thoughts into words**

73. How would you solve the equation $x^2 - 4x = 252$? Explain your choice of the method that you would use.

74. Explain how you would solve $(x - 2)(x - 7) = 0$ and also how you would solve $(x - 2)(x - 7) = 4$.

75. One of our problem-solving suggestions is to *look for a guideline that can be used to help determine an equation.* What does this suggestion mean to you?

76. Can a quadratic equation with integral coefficients have exactly one nonreal complex solution? Explain your answer.

■ ■ ▨ **Further investigations**

For Problems 77–83, solve each equation.

77. $x - 9\sqrt{x} + 18 = 0$ [*Hint:* Let $y = \sqrt{x}$.]

78. $x - 4\sqrt{x} + 3 = 0$

79. $x + \sqrt{x} - 2 = 0$

80. $x^{\frac{2}{3}} + x^{\frac{1}{3}} - 6 = 0$ [*Hint:* Let $y = x^{\frac{1}{3}}$.]

81. $6x^{\frac{2}{3}} - 5x^{\frac{1}{3}} - 6 = 0$

82. $x^{-2} + 4x^{-1} - 12 = 0$

83. $12x^{-2} - 17x^{-1} - 5 = 0$

▦ **Graphing calculator activities**

84. Use your graphing calculator to give visual support for your answers for Problems 33–40.

85. In the text we have used graphs to *predict* solutions of equations and to give *visual support* for solutions that we obtained algebraically. Now use your graphing calculator to *find*, to the nearest tenth, the real number solutions for each of the following equations.

a. $x^2 - 8x + 14 = 0$ **b.** $x^2 + 2x - 2 = 0$

c. $x^2 + 4x - 16 = 0$ **d.** $x^2 + 2x - 111 = 0$

e. $x^4 + x^2 - 12 = 0$ **f.** $x^4 - 19x^2 - 20 = 0$

g. $2x^2 - x - 2 = 0$ **h.** $3x^2 - 2x - 2 = 0$

Answers to Concept Quiz

Answers for these questions may vary.
1. D   **2.** B   **3.** A   **4.** C   **5.** D   **6.** B   **7.** A

11.6   Quadratic and Other Nonlinear Inequalities

Objectives

■   Determine the solution set for quadratic inequalities.

■   Solve rational inequalities.

We refer to the equation $ax^2 + bx + c = 0$ as the standard form of a quadratic equation in one variable. Similarly, the following forms express **quadratic inequalities** in one variable.

$$ax^2 + bx + c > 0 \qquad\qquad ax^2 + bx + c < 0$$
$$ax^2 + bx + c \geq 0 \qquad\qquad ax^2 + bx + c \leq 0$$

We can use the number line very effectively to help solve quadratic inequalities where the quadratic polynomial is factorable. Let's consider some examples to illustrate the procedure.

E X A M P L E   1

Solve and graph the solutions for $x^2 + 2x - 8 > 0$.

Solution

First, let's factor the polynomial.

$$x^2 + 2x - 8 > 0$$
$$(x + 4)(x - 2) > 0$$

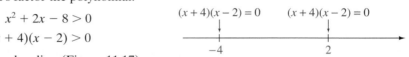

Figure 11.17

On a number line (Figure 11.17), we indicate that at $x = 2$ and $x = -4$, the product $(x + 4)(x - 2)$ equals zero. The numbers -4 and 2 divide the number line into three intervals: (1) the numbers less than -4, (2) the numbers between -4 and 2, and (3) the numbers greater than 2. We can choose a **test number** from each of these intervals and see how it affects the signs of the factors $x + 4$ and $x - 2$ and, consequently, the sign of the product of these factors. For example, if $x < -4$ (try $x = -5$), then $x + 4$ is negative and $x - 2$ is negative, so their product is positive. If $-4 < x < 2$ (try $x = 0$), then $x + 4$ is positive and $x - 2$ is negative, so their product is negative. If $x > 2$ (try $x = 3$), then $x + 4$ is positive and $x - 2$ is positive, so their product is positive. This information can be conveniently arranged using a number line as shown in Figure 11.18. Note the open circles at -4 and 2 to indicate that they are not included in the solution set.

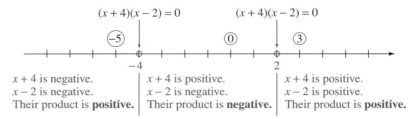

$x + 4$ is negative. $x + 4$ is positive. $x + 4$ is positive.
$x - 2$ is negative. $x - 2$ is negative. $x - 2$ is positive.
Their product is **positive.**   Their product is **negative.**   Their product is **positive.**

Figure 11.18

Thus the given inequality, $x^2 + 2x - 8 > 0$, is satisfied by numbers less than -4 along with numbers greater than 2. Using interval notation, the solution set is $(-\infty, -4) \cup (2, \infty)$. These solutions can be shown on a number line (Figure 11.19).

Figure 11.19

How can we give graphical support for a problem such as Example 1? Suppose that we graph $y = x^2 + 2x - 8$, as in Figure 11.20. Because $y = x^2 + 2x - 8$ and we want $x^2 + 2x - 8$ to be greater than zero, let's consider the part of the graph above the x axis — in other words, where y is positive. The x intercepts appear to be -4 and 2. Therefore, y is positive when $x < -4$ and also when $x > 2$. Thus the solution set given in Example 1 appears to be correct.

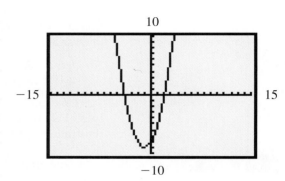

Figure 11.20

We refer to numbers such as -4 and 2 in the preceding example (where the given polynomial or algebraic expression equals zero or is undefined) as **critical numbers.** Let's consider some additional examples that make use of critical numbers and test numbers.

E X A M P L E   2

Solve and graph the solutions for $x^2 + 2x - 3 \leq 0$.

Solution

For this problem, let's use a graphical approach to *predict* the solution set. Let's graph $y = x^2 + 2x - 3$, as shown in Figure 11.21. The x intercepts appear to be -3 and 1.

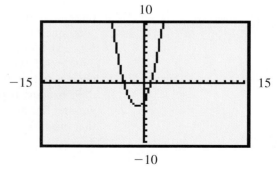

Figure 11.21

Therefore, the graph is on or below the x axis (y is nonpositive) when $-3 \leq x \leq 1$. Now let's solve the given inequality using a number-line analysis.

First, factor the polynomial.

$$x^2 + 2x - 3 \leq 0$$
$$(x + 3)(x - 1) \leq 0$$

Second, locate the values for which $(x + 3)(x - 1)$ equals zero. We put dots at -3 and 1 to remind ourselves that these two numbers are to be included in the solution set because the given statement includes equality. Now let's choose a test number from each of the three intervals and record the sign behavior of the factors $(x + 3)$ and $(x - 1)$ (Figure 11.22).

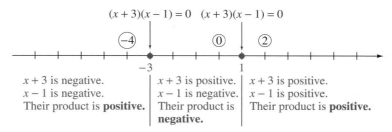

Figure 11.22

Therefore, the solution set is $[-3, 1]$ and it can be graphed as in Figure 11.23.

Figure 11.23

Examples 1 and 2 have indicated a systematic approach for solving quadratic inequalities where the polynomial is factorable. This same type of number-line analysis can also be used to solve indicated quotients such as $\dfrac{x + 1}{x - 5} > 0$.

EXAMPLE 3   Solve and graph the solutions for $\dfrac{x + 1}{x - 5} > 0$.

Solution

First, indicate that at $x = -1$ the given quotient equals zero, and at $x = 5$ the quotient is undefined. Second, choose test numbers from each of the three intervals, and record the sign behavior of $(x + 1)$ and $(x - 5)$ as in Figure 11.24.

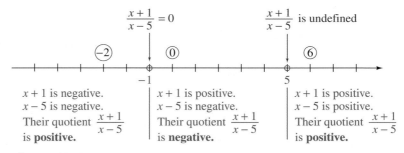

Figure 11.24

Therefore, the solution set is $(-\infty, -1) \cup (5, \infty)$, and its graph is shown in Figure 11.25.

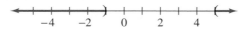

Figure 11.25

EXAMPLE 4

Solve $\dfrac{x + 2}{x + 4} \le 0$.

Solution

The indicated quotient equals zero at $x = -2$ and is undefined at $x = -4$. (Note that -2 is to be included in the solution set, but -4 is not to be included.) Now let's choose some test numbers and record the sign behavior of $(x + 2)$ and $(x + 4)$ as in Figure 11.26.

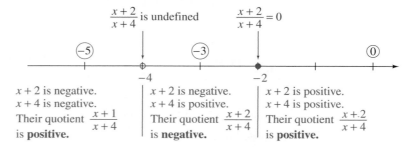

Figure 11.26

Therefore, the solution set is $(-4, -2]$.

The final example illustrates that sometimes we need to change the form of the given inequality before we use the number-line analysis.

EXAMPLE 5

Solve $\dfrac{x}{x + 2} \ge 3$.

Solution

First, let's change the form of the given inequality as follows.

$$\frac{x}{x + 2} \ge 3$$

$$\frac{x}{x + 2} - 3 \ge 0 \qquad \text{Add } -3 \text{ to both sides.}$$

$$\frac{x - 3(x + 2)}{x + 2} \ge 0 \qquad \text{Express the left side over a common denominator.}$$

$$\frac{x - 3x - 6}{x + 2} \geq 0$$

$$\frac{-2x - 6}{x + 2} \geq 0$$

Now we can proceed as we did with the previous examples. If $x = -3$, then $\frac{-2x - 6}{x + 2}$ equals zero; and if $x = -2$, then $\frac{-2x - 6}{x + 2}$ is undefined. Then, choosing test numbers, we can record the sign behavior of $(-2x - 6)$ and $(x + 2)$ as in Figure 11.27.

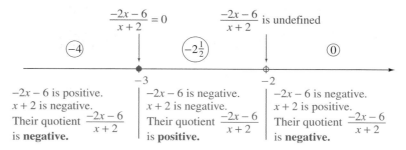

Figure 11.27

Therefore, the solution set is $[-3, -2)$. Perhaps you should check a few numbers from this solution set back into the original inequality! ∎

CONCEPT QUIZ

For Problems 1–5, answer true or false.

1. When solving the inequality $(x + 3)(x - 2) > 0$, we are finding values of x that make the product of $(x + 3)$ and $(x - 2)$ a positive number.

2. The solution set of the inequality $x^2 + 4 > 0$ is all real numbers.

3. The solution set of the inequality $x^2 \leq 0$ is the null set.

4. The critical numbers for the inequality $(x + 4)(x - 1) \leq 0$ are -4 and -1.

5. The number 2 is included in the solution set of the inequality $\frac{x + 4}{x - 2} \geq 0$.

PROBLEM SET 11.6

For Problems 1–20, solve each inequality and graph its solution set on a number line.

1. $(x + 2)(x - 1) > 0$

2. $(x - 2)(x + 3) > 0$

3. $(x + 1)(x + 4) < 0$

4. $(x - 3)(x - 1) < 0$

5. $(2x - 1)(3x + 7) \geq 0$

6. $(3x + 2)(2x - 3) \geq 0$

7. $(x + 2)(4x - 3) \leq 0$

8. $(x - 1)(2x - 7) \leq 0$

9. $(x + 1)(x - 1)(x - 3) > 0$

10. $(x + 2)(x + 1)(x - 2) > 0$

11. $x(x + 2)(x - 4) \leq 0$

12. $x(x + 3)(x - 3) \leq 0$

13. $\dfrac{x + 1}{x - 2} > 0$

14. $\dfrac{x - 1}{x + 2} > 0$

15. $\dfrac{x - 3}{x + 2} < 0$

16. $\dfrac{x + 2}{x - 4} < 0$

17. $\dfrac{2x - 1}{x} \geq 0$

18. $\dfrac{x}{3x + 7} \geq 0$

19. $\dfrac{-x + 2}{x - 1} \leq 0$

20. $\dfrac{3 - x}{x + 4} \leq 0$

For Problems 21–46, solve each inequality.

21. $x^2 + 2x - 35 < 0$

22. $x^2 + 3x - 54 < 0$

23. $x^2 - 11x + 28 > 0$

24. $x^2 + 11x + 18 > 0$

25. $3x^2 + 13x - 10 \leq 0$

26. $4x^2 - x - 14 \leq 0$

27. $8x^2 + 22x + 5 \geq 0$

28. $12x^2 - 20x + 3 \geq 0$

29. $x(5x - 36) > 32$

30. $x(7x + 40) < 12$

31. $x^2 - 14x + 49 \geq 0$

32. $(x + 9)^2 \geq 0$

33. $4x^2 + 20x + 25 \leq 0$

34. $9x^2 - 6x + 1 \leq 0$

35. $(x + 1)(x - 3)^2 > 0$

36. $(x - 4)^2(x - 1) \leq 0$

37. $\dfrac{2x}{x + 3} > 4$

38. $\dfrac{x}{x - 1} > 2$

39. $\dfrac{x - 1}{x - 5} \leq 2$

40. $\dfrac{x + 2}{x + 4} \leq 3$

41. $\dfrac{x + 2}{x - 3} > -2$

42. $\dfrac{x - 1}{x - 2} < -1$

43. $\dfrac{3x + 2}{x + 4} \leq 2$

44. $\dfrac{2x - 1}{x + 2} \geq -1$

45. $\dfrac{x + 1}{x - 2} < 1$

46. $\dfrac{x + 3}{x - 4} \geq 1$

■ ■ ■ **Thoughts into words**

47. Explain how to solve the inequality $(x + 1)(x - 2)$ $(x - 3) > 0$.

48. Explain how to solve the inequality $(x - 2)^2 > 0$ by inspection.

49. Your friend looks at the inequality $1 + \dfrac{1}{x} > 2$ and without any computation states that the solution set is all real numbers between 0 and 1. How can she do that?

50. Why is the solution set for $(x - 2)^2 \geq 0$ the set of all real numbers?

51. Why is the solution set for $(x - 2)^2 \leq 0$ the set $\{2\}$?

■ ■ ■ **Further investigations**

52. The product $(x - 2)(x + 3)$ is positive if both factors are negative *or* if both factors are positive. Therefore, we can solve $(x - 2)(x + 3) > 0$ as follows:

$(x - 2 < 0 \text{ and } x + 3 < 0) \text{ or } (x - 2 > 0 \text{ and } x + 3 > 0)$

$(x < 2 \text{ and } x < -3) \text{ or } (x > 2 \text{ and } x > -3)$

$x < -3 \text{ or } x > 2$

The solution set is $(-\infty, -3) \cup (2, \infty)$. Use this type of analysis to solve each of the following.

a. $(x - 2)(x + 7) > 0$ **b.** $(x - 3)(x + 9) \geq 0$

c. $(x + 1)(x - 6) \leq 0$ **d.** $(x + 4)(x - 8) < 0$

e. $\dfrac{x + 4}{x - 7} > 0$ **f.** $\dfrac{x - 5}{x + 8} \leq 0$

Graphing calculator activities

53. Use your graphing calculator to give visual support for our answers for Examples 3, 4, and 5.

54. Use your graphing calculator to give visual support for your answers for Problems 30–39.

55. Use your graphing calculator to help determine the solution sets for the following inequalities. Express the solution sets in interval notation.

    a. $x^2 + x - 2 \geq 0$ **b.** $(3 - x)(x + 4) < 0$

    c. $x^3 - 3x^2 - x + 3 > 0$ **d.** $-x^3 + 7x - 6 \leq 0$

    e. $x^3 - 21x - 20 < 0$

Answers to Concept Quiz

1. True   **2.** True   **3.** False   **4.** False   **5.** False

CHAPTER 11
SUMMARY

(11.1) A number of the form $a + bi$, where a and b are real numbers and i is the imaginary unit defined by $i = \sqrt{-1}$, is a **complex number.**

Two complex numbers $a + bi$ and $c + di$ are said to be *equal* if and only if $a = c$ and $b = d$.

We describe addition and subtraction of complex numbers as follows:

$$(a + bi) + (c + di) = (a + c) + (b + d)i$$

$$(a + bi) - (c + di) = (a - c) + (b - d)i$$

We can represent a square root of any negative real number as the product of a real number and the imaginary unit i. That is,

$$\sqrt{-b} = i\sqrt{b}, \quad \text{where } b \text{ is a positive real number}$$

The product of two complex numbers conforms with the product of two binomials. The **conjugate** of $a + bi$ is $a - bi$. The product of a complex number and its conjugate is a real number. Therefore, conjugates are used to simplify expressions such as $\dfrac{4 + 3i}{5 - 2i}$, which indicate the quotient of two complex numbers.

(11.2) The **standard form for a quadratic equation** in one variable is

$$ax^2 + bx + c = 0$$

where a, b, and c are real numbers and $a \neq 0$.

Some quadratic equations can be solved by *factoring* and applying the property "$ab = 0$ if and only if $a = 0$ or $b = 0$."

Don't forget that applying the property "if $a = b$, then $a^n = b^n$" might produce extraneous solutions. Therefore, we *must check* all potential solutions.

We can solve some quadratic equations by applying the property "$x^2 = a$ if and only if $x = \pm\sqrt{a}$."

(11.3) To solve a quadratic equation of the form $x^2 + bx = k$ by **completing the square,** we (1) add $\left(\dfrac{b}{2}\right)^2$ to both sides, (2) factor the left side, and (3) apply the property "$x^2 = a$ if and only if $x = \pm\sqrt{a}$."

(11.4) We can solve any quadratic equation of the form $ax^2 + bx + c = 0$ by the **quadratic formula,** which we usually state as

$$x = \frac{-b \pm \sqrt{b^2 - 4ac}}{2a}$$

The **discriminant,** $b^2 - 4ac$, can be used to determine the nature of the roots of a quadratic equation as follows:

1. If $b^2 - 4ac < 0$, then the equation has two nonreal complex solutions.
2. If $b^2 - 4ac = 0$, then the equation has two equal real solutions.
3. If $b^2 - 4ac > 0$, then the equation has two unequal real solutions.

If x_1 and x_2 are roots of a quadratic equation, then the following relationships exist.

$$x_1 + x_2 = -\frac{b}{a} \quad \text{and} \quad (x_1)(x_2) = \frac{c}{a}$$

These **sum-of-roots** and **product-of-roots relationships** can be used to check potential solutions of quadratic equations.

(11.5) To review the strengths and weaknesses of the three basic methods for solving a quadratic equation (factoring, completing the square, and the quadratic formula), go back over the examples in this section.

Keep the following suggestions in mind as you solve word problems.

1. Read the problem carefully.
2. Sketch any figure, diagram, or chart that might help you organize and analyze the problem.
3. Choose a meaningful variable.

4. Look for a guideline that can be used to set up an equation.

5. Form an equation that translates the guideline from English into algebra.

6. Solve the equation and use the solutions to determine all facts requested in the problem.

7. Check all answers back into the original statement of the problem.

(11.6) The number line, along with **critical numbers** and **test numbers,** provides a good basis for solving **quadratic inequalities** where the polynomial is factorable. We can use this same basic approach to solve inequalities, such as $\dfrac{3x + 1}{x - 4} > 0$, that indicate quotients.

CHAPTER 11 REVIEW PROBLEM SET

For Problems 1–8, perform the indicated operations and express the answers in the standard form of a complex number.

1. $(-7 + 3i) + (9 - 5i)$ **2.** $(4 - 10i) - (7 - 9i)$

3. $5i(3 - 6i)$ **4.** $(5 - 7i)(6 + 8i)$

5. $(-2 - 3i)(4 - 8i)$ **6.** $(4 - 3i)(4 + 3i)$

7. $\dfrac{4 + 3i}{6 - 2i}$ **8.** $\dfrac{-1 - i}{-2 + 5i}$

For Problems 9–12, find the discriminant of each equation and determine whether the equation has two nonreal complex solutions, one real solution with a multiplicity of two, or two real solutions. Do not solve the equations.

9. $4x^2 - 20x + 25 = 0$ **10.** $5x^2 - 7x + 31 = 0$

11. $7x^2 - 2x - 14 = 0$ **12.** $5x^2 - 2x = 4$

For Problems 13–31, solve each equation.

13. $x^2 - 17x = 0$ **14.** $(x - 2)^2 = 36$

15. $(2x - 1)^2 = -64$ **16.** $x^2 - 4x - 21 = 0$

17. $x^2 + 2x - 9 = 0$ **18.** $x^2 - 6x = -34$

19. $4\sqrt{x} = x - 5$ **20.** $3n^2 + 10n - 8 = 0$

21. $n^2 - 10n = 200$ **22.** $3a^2 + a - 5 = 0$

23. $x^2 - x + 3 = 0$ **24.** $2x^2 - 5x + 6 = 0$

25. $2a^2 + 4a - 5 = 0$ **26.** $t(t + 5) = 36$

27. $x^2 + 4x + 9 = 0$ **28.** $(x - 4)(x - 2) = 80$

29. $\dfrac{3}{x} + \dfrac{2}{x + 3} = 1$ **30.** $2x^4 - 23x^2 + 56 = 0$

31. $\dfrac{3}{n - 2} = \dfrac{n + 5}{4}$

For Problems 32–35, solve each inequality, and indicate the solution set using interval notation.

32. $x^2 + 3x - 10 > 0$ **33.** $2x^2 + x - 21 \le 0$

34. $\dfrac{x - 4}{x + 6} \ge 0$ **35.** $\dfrac{2x - 1}{x + 1} > 4$

For Problems 36–43, set up an equation, and solve each problem.

36. Find two numbers whose sum is 6 and whose product is 2.

37. Sherry bought a number of shares of stock for $250. Six months later the value of the stock had increased by $5 per share, and she sold all but 5 shares and regained her original investment plus a profit of $50. How many shares did she sell and at what price per share?

38. Andre traveled 270 miles in 1 hour more time than it took Sandy to travel 260 miles. Sandy drove 7 miles per hour faster than Andre. How fast did each one travel?

39. The area of a square is numerically equal to twice its perimeter. Find the length of a side of the square.

40. Find two consecutive even whole numbers such that the sum of their squares is 164.

41. The perimeter of a rectangle is 38 inches, and its area is 84 square inches. Find the length and width of the rectangle.

42. It takes Billy 2 hours longer to do a certain job than it takes Reena. They worked together for 2 hours; then Reena left and Billy finished the job in 1 hour. How long would it take each of them to do the job alone?

43. A company has a rectangular parking lot 40 meters wide and 60 meters long. The company plans to increase the area of the lot by 1100 square meters by adding a strip of equal width to one side and one end. Find the width of the strip to be added.

CHAPTER 11

TEST

🖥 **applies to all problems in this Chapter Test.**

1. Find the product $(3 - 4i)(5 + 6i)$ and express the result in the standard form of a complex number.

2. Find the quotient $\dfrac{2 - 3i}{3 + 4i}$ and express the result in the standard form of a complex number.

For Problems 3–15, solve each equation.

3. $x^2 = 7x$

4. $(x - 3)^2 = 16$

5. $x^2 + 3x - 18 = 0$

6. $x^2 - 2x - 1 = 0$

7. $5x^2 - 2x + 1 = 0$

8. $x^2 + 30x = -224$

9. $(3x - 1)^2 + 36 = 0$

10. $(5x - 6)(4x + 7) = 0$

11. $(2x + 1)(3x - 2) = 55$

12. $n(3n - 2) = 40$

13. $x^4 + 12x^2 - 64 = 0$

14. $\dfrac{3}{x} + \dfrac{2}{x + 1} = 4$

15. $3x^2 - 2x - 3 = 0$

16. Does the equation $4x^2 + 20x + 25 = 0$ have two nonreal complex solutions, two equal real solutions, or two unequal real solutions?

17. Does the equation $4x^2 - 3x = -5$ have two nonreal complex solutions, two equal real solutions, or two unequal real solutions?

For Problems 18–20, solve each inequality and express the solution set using interval notation.

18. $x^2 - 3x - 54 \le 0$

19. $\dfrac{3x - 1}{x + 2} > 0$

20. $\dfrac{x - 2}{x + 6} \ge 3$

For Problems 21–25, set up an equation and solve each problem.

21. A 24-foot ladder leans against a building and makes an angle of 60° with the ground. How far up on the building does the top of the ladder reach? Express your answer to the nearest tenth of a foot.

22. A rectangular plot of ground measures 16 meters by 24 meters. Find, to the nearest meter, the distance from one corner of the plot to the diagonally opposite corner.

23. Dana bought a number of shares of stock for a total of $3000. Three months later the stock had increased in value by $5 per share, and she sold all but 50 shares and regained her original investment of $3000. How many shares did she sell?

24. The perimeter of a rectangle is 41 inches and its area is 91 square inches. Find the length of its shortest side.

25. The sum of two numbers is 6 and their product is 4. Find the larger of the two numbers.

CUMULATIVE REVIEW PROBLEM SET *Chapters 1-11*

For Problems 1–5, evaluate each algebraic expression for the given values of the variables.

1. $\dfrac{4a^2b^3}{12a^3b}$ for $a = 5$ and $b = -8$

2. $\dfrac{\dfrac{1}{x} + \dfrac{1}{y}}{\dfrac{1}{x} - \dfrac{1}{y}}$ for $x = 4$ and $y = 7$

3. $\dfrac{3}{n} + \dfrac{5}{2n} - \dfrac{4}{3n}$ for $n = 25$

4. $\dfrac{4}{x - 1} - \dfrac{2}{x + 2}$ for $x = \dfrac{1}{2}$

5. $2\sqrt{2x + y} - 5\sqrt{3x - y}$ for $x = 5$ and $y = 6$

For Problems 6–17, perform the indicated operations and express the answers in simplified form.

6. $(3a^2b)(-2ab)(4ab^3)$

7. $(x + 3)(2x^2 - x - 4)$

8. $\dfrac{6xy^2}{14y} \cdot \dfrac{7x^2y}{8x}$

9. $\dfrac{a^2 + 6a - 40}{a^2 - 4a} \div \dfrac{2a^2 + 19a - 10}{a^3 + a^2}$

10. $\dfrac{3x + 4}{6} - \dfrac{5x - 1}{9}$

11. $\dfrac{4}{x^2 + 3x} + \dfrac{5}{x}$

12. $\dfrac{3n^2 + n}{n^2 + 10n + 16} \cdot \dfrac{2n^2 - 8}{3n^3 - 5n^2 - 2n}$

13. $\dfrac{3}{5x^2 + 3x - 2} - \dfrac{2}{5x^2 - 22x + 8}$

14. $\dfrac{y^3 - 7y^2 + 16y - 12}{y - 2}$

15. $(4x^3 - 17x^2 + 7x + 10) \div (4x - 5)$

16. $(3\sqrt{2} + 2\sqrt{5})(5\sqrt{2} - \sqrt{5})$

17. $(\sqrt{x} - 3\sqrt{y})(2\sqrt{x} + 4\sqrt{y})$

For Problems 18–25, evaluate each of the numerical expressions.

18. $-\sqrt{\dfrac{9}{64}}$

19. $\sqrt[3]{-\dfrac{8}{27}}$

20. $\sqrt[3]{0.008}$

21. $32^{-\frac{1}{5}}$

22. $3^0 + 3^{-1} + 3^{-2}$

23. $-9^{\frac{3}{2}}$

24. $\left(\dfrac{3}{4}\right)^{-2}$

25. $\dfrac{1}{\left(\dfrac{2}{3}\right)^{-3}}$

For Problems 26–31, factor each of the algebraic expressions completely.

26. $3x^4 + 81x$

27. $6x^2 + 19x - 20$

28. $12 + 13x - 14x^2$

29. $9x^4 + 68x^2 - 32$

30. $2ax - ay - 2bx + by$

31. $27x^3 - 8y^3$

For Problems 32–55, solve each of the equations.

32. $3(x - 2) - 2(3x + 5) = 4(x - 1)$

33. $0.06n + 0.08(n + 50) = 25$

34. $4\sqrt{x} + 5 = x$

35. $\sqrt[3]{n^2 - 1} = -1$

36. $6x^2 - 24 = 0$

37. $a^2 + 14a + 49 = 0$

38. $3n^2 + 14n - 24 = 0$

39. $\dfrac{2}{5x - 2} = \dfrac{4}{6x + 1}$

40. $\sqrt{2x - 1} - \sqrt{x + 2} = 0$

41. $5x - 4 = \sqrt{5x - 4}$

42. $|3x - 1| = 11$

43. $(3x - 2)(4x - 1) = 0$

44. $(2x + 1)(x - 2) = 7$

45. $\dfrac{5}{6x} - \dfrac{2}{3} = \dfrac{7}{10x}$

46. $\dfrac{3}{y + 4} + \dfrac{2y - 1}{y^2 - 16} = \dfrac{-2}{y - 4}$

47. $6x^4 - 23x^2 - 4 = 0$

48. $3n^3 + 3n = 0$

49. $n^2 - 13n - 114 = 0$

50. $12x^2 + x - 6 = 0$

51. $x^2 - 2x + 26 = 0$

52. $(x + 2)(x - 6) = -15$

53. $(3x - 1)(x + 4) = 0$

54. $x^2 + 4x + 20 = 0$

55. $2x^2 - x - 4 = 0$

For Problems 56–65, solve each inequality and express the solution set using interval notation.

56. $6 - 2x \geq 10$

57. $4(2x - 1) < 3(x + 5)$

58. $\dfrac{n + 1}{4} + \dfrac{n - 2}{12} > \dfrac{1}{6}$

59. $|2x - 1| < 5$

60. $|3x + 2| > 11$

61. $\dfrac{1}{2}(3x - 1) - \dfrac{2}{3}(x + 4) \leq \dfrac{3}{4}(x - 1)$

62. $x^2 - 2x - 8 \leq 0$

63. $3x^2 + 14x - 5 > 0$

64. $\dfrac{x + 2}{x - 7} \geq 0$

65. $\dfrac{2x - 1}{x + 3} < 1$

For Problems 66–68, solve each system.

66. $\begin{pmatrix} 4x - 9y = 14 \\ x + 5y = -11 \end{pmatrix}$

67. $\begin{pmatrix} \dfrac{1}{2}x - \dfrac{1}{3}y = 7 \\[2mm] \dfrac{3}{4}x + \dfrac{2}{3}y = 0 \end{pmatrix}$

68. $\begin{pmatrix} 7x - 2y = 16 \\ 3x + 5y = 42 \end{pmatrix}$

69. Find the product $(-6x^{-3}y^{-4})(5xy^6)$ and express the result using positive exponents only.

70. Express each of the following in simplest radical form.

    a. $4\sqrt{54}$ **b.** $2\sqrt[3]{54}$

    c. $\dfrac{3\sqrt{6}}{4\sqrt{10}}$ **d.** $\sqrt{24x^3y^5}$

71. Express each of the following in scientific notation.

    a. 7652   **b.** 0.000026

    c. 1.414   **d.** 1000

For Problems 72–85, solve each problem by setting up and solving an appropriate equation, inequality, or system of equations.

72. How many liters of a 60% acid solution must be added to 14 liters of a 10% acid solution to produce a 25% acid solution?

73. A sum of $2250 is to be divided between two people in the ratio of 2 to 3. How much does each person receive?

74. The length of a picture without its border is 7 inches less than twice its width. If the border is 1 inch wide and its area is 62 square inches, what are the dimensions of the picture alone?

75. Working together, Lolita and Doug can paint a shed in 3 hours and 20 minutes. If Doug can paint the shed by himself in 10 hours, how long would it take Lolita to paint the shed by herself?

76. Angie bought some golf balls for $14. If each ball had cost $.25 less, she could have purchased one more ball for the same amount of money. How many golf balls did Angie buy?

77. A jogger who can run an 8-minute mile starts a half-mile ahead of a jogger who can run a 6-minute mile. How long will it take the faster jogger to catch the slower jogger?

78. Suppose that $1000 is invested at a certain rate of interest compounded annually for 2 years. If the accumulated value at the end of 2 years is $1149.90, find the rate of interest.

79. A theater contains 120 chairs arranged in rows. The number of chairs per row is one less than twice the number of rows. Find the number of chairs per row.

80. Bjorn bought a number of shares of stock for $2800. A month later the value of the stock had increased $6 per share, and he sold all but 60 shares and regained his original investment of $2800. How many shares did he sell?

81. One angle of a triangle has a measure of 40°, and the measures of the other two angles are in a ratio of 3 to 4. Find the measures of the other two angles.

82. The supplement of an angle is 10° more than five times its complement. Find the measure of the angle.

83. Megan has $5000 to invest. If she invests $3000 at 7% interest, at what rate must she invest the other $2000 so that the two investments together yield more than $380 in yearly interest?

84. Wilma went to the local market and bought 3 lemons and 5 oranges for $2.40. The same day and for the same prices, Fran bought 4 lemons and 7 oranges for $3.31. What was the price per lemon and the price per orange?

85. Larry has an oil painting that he bought for $700 and now wants to sell. How much more profit could he gain by fixing a 30% profit based on the selling price rather than a 30% profit based on the cost?

René Descartes, a philosopher and mathematician, developed a system for locating a point on a plane. This system is our current rectangular coordinate grid used for graphing and is named the Cartesian coordinate system.

Coordinate Geometry: Lines, Parabolas, Circles, Ellipses, and Hyperbolas

The graph in Figure 12.1 shows the number of gallons of lemonade sold each day at Marci's Little Lemonade Stand depending on how high the temperature was each day. The owner of the stand would like to determine an algebraic equation that shows the relationship between the daily temperature and the number of gallons of lemonade sold each day.

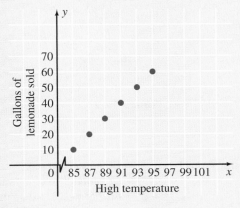

Figure 12.1

571

The owner of Marci's Little Lemonade Stand decides that the relationship is linear and knows that two pairs of values can determine the equation for the relationship. For example, 20 gallons were sold when the temperature was 87°, and 60 gallons were sold when the temperature was 95°. Therefore it can be determined that the equation $y = 5x - 415$ describes the relationship in which x is the daily high temperature, and y is the number of gallons sold. So from the geometric information in the graph, Figure 12.1, we are able to determine an algebraic equation for the relationship.

It was René Descartes who connected algebraic and geometric ideas to found the branch of mathematics called analytic geometry — today more commonly called coordinate geometry. Basically, there are two kinds of problems in coordinate geometry: Given an algebraic equation, find its geometric graph; and given a set of conditions pertaining to a geometric graph, find its algebraic equation. We discuss problems of both types in this chapter, but our strong emphasis will be on developing graphing techniques.

With the graphing techniques developed in this chapter we will graph parabolas, circles, ellipses, and hyperbolas; we often refer to these curves as **conic sections.** Conic sections can be formed when a plane intersects a conical surface as shown in Figure 12.2. A flashlight produces a "cone of light" that can be cut by the plane of a wall to illustrate the conic sections. Try shining a flashlight against a wall at different angles to produce a circle, an ellipse, a parabola, and one branch of a hyperbola. (You may find it difficult to distinguish between a parabola and a branch of a hyperbola.)

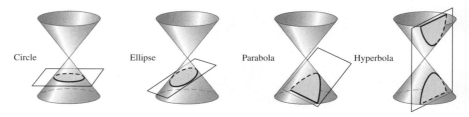

Circle   Ellipse   Parabola   Hyperbola

Figure 12.2

InfoTrac Project   Do a keyword search on circumference. Look for articles discussing how the circumference of a patient's leg, waist, or neck is being used when studying his/her health problems. Write a short paragraph outlining what you found. Using 12 other people, choose one of those body parts, (neck, leg, or waist) and measure the height of the person and the circumference of the body part you have chosen. Make a table of values, using height as your independent variable. Make a conjecture about whether your graph will be increasing or decreasing. Graph your values. What conclusions might you draw from your graph?

Now, using weight and the circumference measurement, make a table of values and create a graph. What are your conclusions this time? Choose two data points from your data and develop the equation of the line passing through these two points. Connect the points on the graph. Does the line seem to fit the data? Choosing the weight measurement from another data point, determine what the circumference would have been if your equation had fit the data. What is the difference between the actual measurement and the predicted value for your equation?

12.1   Distance, Slope, and Graphing Techniques

Objectives

■   Find the distance between two points in the coordinate plane.

■   Determine the slope of a line.

■   Find x and y intercepts for graphs of equations.

■   Determine the type of symmetries the graph of an equation will exhibit.

■   Graph equations using intercepts, symmetries, and plotting points.

We introduced the rectangular coordinate system in Chapter 5. Most of our work at that time concentrated on graphing linear equations (straight line graphs) including a graphical approach to solving systems of two linear equations. Then in Chapter 9 we began to use graphs occasionally to give some visual support for an algebraic computation. These graphs were calculator-generated, and no attempt was made to introduce any graphing techniques for sketching graphs. Now in this chapter, we will begin to develop some specific techniques to aid in the free-hand sketching of graphs. Let's begin by briefly reviewing some basic ideas pertaining to the rectangular coordinate system.

Consider two number lines, one vertical and one horizontal, perpendicular to each other at the point we associate with zero on both lines (Figure 12.3). We refer to these number lines as the **horizontal and vertical axes** or, together, as the **coordinate axes.** They partition the plane into four regions called **quadrants.** The quadrants are numbered counterclockwise from I through IV, as indicated in Figure 12.3. The point of intersection of the two axes is called the **origin.**

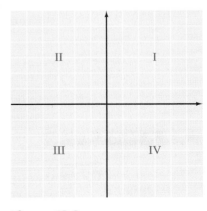

Figure 12.3

In general we refer to the real numbers a and b in an ordered pair (a, b), associated with a point, as the **coordinates of the point.** The first number, a, called the **abscissa,** is the directed distance of the point from the vertical axis measured parallel to the horizontal axis. The second number, b, called the **ordinate,** is the directed distance of the point from the horizontal axis measured parallel to the vertical axis (Figure 12.4). This system of associating points in a plane with pairs of real numbers is called the **rectangular coordinate system** or the **Cartesian coordinate system.**

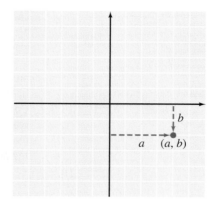

Figure 12.4

Distance Between Two Points

As we work with the rectangular coordinate system, it is sometimes necessary to express the length of certain line segments. In other words, we need to be able to find the distance *between two points*. Let's first consider two specific examples and then develop the general distance formula.

EXAMPLE 1 Find the distance between the points $A(2, 2)$ and $B(5, 2)$ and also the distance between the points $C(-2, 5)$ and $D(-2, -4)$.

Solution

Let's plot the points and draw $\overline{AB}$ as in Figure 12.5. Because $\overline{AB}$ is parallel to the x axis, its length can be expressed as $|5 - 2|$ or $|2 - 5|$. (The absolute-value symbol is used to ensure a nonnegative value.) Thus the length of $\overline{AB}$ is 3 units. Likewise, the length of $\overline{CD}$ is $|5 - (-4)| = |-4 - 5| = 9$ units.

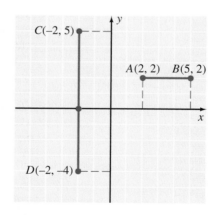

Figure 12.5

EXAMPLE 2 Find the distance between the points $A(2, 3)$ and $B(5, 7)$.

Solution

Let's plot the points and form a right triangle as indicated in Figure 12.6. Note that the coordinates of point C are $(5, 3)$. Because $\overline{AC}$ is parallel to the horizontal axis, its length is easily determined to be 3 units. Likewise, $\overline{CB}$ is parallel to the vertical axis, and its length is 4 units. Let d represent the length of $\overline{AB}$ and apply the Pythagorean theorem to obtain

$$d^2 = 3^2 + 4^2$$
$$d^2 = 9 + 16$$
$$d^2 = 25$$
$$d = \pm\sqrt{25} = \pm 5$$

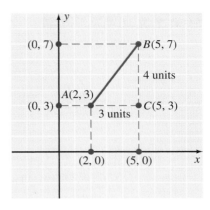

"Distance between" is a nonnegative value, so the length of $\overline{AB}$ is 5 units.

Figure 12.6

We can use the approach we used in Example 2 to develop a general distance formula for finding the distance between any two points in a coordinate plane. The development proceeds as follows:

1. Let $P_1(x_1, y_1)$ and $P_2(x_2, y_2)$ represent any two points in a coordinate plane.
2. Form a right triangle as indicated in Figure 12.7. The coordinates of the vertex of the right angle, point R, are (x_2, y_1).

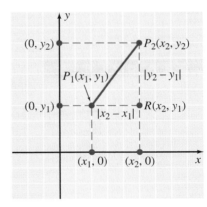

The length of $\overline{P_1 R}$ is $|x_2 - x_1|$, and the length of $\overline{RP_2}$ is $|y_2 - y_1|$. (The absolute-value symbol is used to ensure a nonnegative value.) Let d represent the length of P_1P_2 and apply the Pythagorean theorem to obtain

Figure 12.7

$$d^2 = |x_2 - x_1|^2 + |y_2 - y_1|^2$$

Because $|a|^2 = a^2$, the **distance formula** can be stated as

$$d = \sqrt{(x_2 - x_1)^2 + (y_2 - y_1)^2}$$

It makes no difference which point you call P_1 and which you call P_2 when using the distance formula. If you forget the formula, don't panic. Just form a right triangle and apply the Pythagorean theorem as we did in Example 2. Let's consider an example that demonstrates the use of the distance formula.

EXAMPLE 3   Find the distance between $(-1, 4)$ and $(1, 2)$.

Solution
Let $(-1, 4)$ be P_1 and $(1, 2)$ be P_2. Using the distance formula, we obtain

$$
\begin{aligned}
d &= \sqrt{(1 - (-1))^2 + (2 - 4)^2} \\
&= \sqrt{2^2 + (-2)^2} \\
&= \sqrt{4 + 4} \\
&= \sqrt{8} = 2\sqrt{2} \qquad \text{Express the answer in simplest radical form.}
\end{aligned}
$$

The distance between the two points is $2\sqrt{2}$ units.   ■

REMARK: In the solution of Example 3 we expressed the answer in lowest radical form. If you need a brief review regarding lowest radical form, please turn to Section 10.3.

In Example 3, we did not sketch a figure because of the simplicity of the problem. However, sometimes it is helpful to use a figure to organize the given information and aid in the analysis of the problem, as we see in the next example.

EXAMPLE 4   Verify that the points $(-3, 6)$, $(3, 4)$, and $(1, -2)$ are vertices of an isosceles triangle. (An isosceles triangle has two sides of the same length.)

Solution
Let's plot the points and draw the triangle (Figure 12.8). Use the distance formula to find the lengths d_1, d_2, and d_3, as follows:

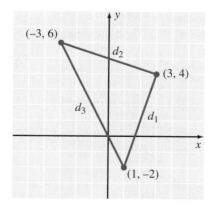

$$
\begin{aligned}
d_1 &= \sqrt{(3 - 1)^2 + (4 - (-2))^2} \\
&= \sqrt{2^2 + 6^2} = \sqrt{40} = 2\sqrt{10} \\
d_2 &= \sqrt{(-3 - 3)^2 + (6 - 4)^2} \\
&= \sqrt{(-6)^2 + 2^2} = \sqrt{40} \\
&= 2\sqrt{10} \\
d_3 &= \sqrt{(-3 - 1)^2 + (6 - (-2))^2} \\
&= \sqrt{(-4)^2 + 8^2} = \sqrt{80} = 4\sqrt{5}
\end{aligned}
$$

Figure 12.8

Because $d_1 = d_2$, we know that it is an isosceles triangle.   ■

Slope of a Line

In Chapter 5, we introduced the concept of slope of a line; in this section we will review the basic concepts of slope. In coordinate geometry, the concept of **slope** is used to describe the "steepness" of lines. The slope of a line is the ratio of the vertical change to the horizontal change as we move from one point on a line to another point.

A precise definition for slope can be given by considering the coordinates of the points P_1, P_2, and R as indicated in Figure 12.9. The horizontal change as we move from P_1 to P_2 is $x_2 - x_1$ and the vertical change is $y_2 - y_1$. The following definition for slope, Definition 12.1, was originally introduced in Chapter 5.

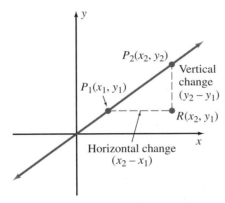

Figure 12.9

DEFINITION 12.1

If points P_1 and P_2 with coordinates (x_1, y_1) and (x_2, y_2), respectively, are any two different points on a line, then the slope of the line (denoted by m) is

$$m = \frac{y_2 - y_1}{x_2 - x_1}, \qquad x_1 \neq x_2$$

Because $\dfrac{y_2 - y_1}{x_2 - x_1} = \dfrac{y_1 - y_2}{x_1 - x_2}$, how we designate P_1 and P_2 is not important. Let's use Definition 12.1 to find the slopes of some lines.

EXAMPLE 5

Find the slope of the line determined by each of the following pairs of points, and graph the lines.

(a) $(-1, 1)$ and $(3, 2)$ **(b)** $(4, -2)$ and $(-1, 5)$

(c) $(2, -3)$ and $(-3, -3)$

Solution

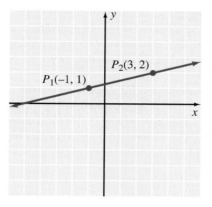

Figure 12.10

(a) Let $(-1, 1)$ be P_1 and $(3, 2)$ be P_2 (Figure 12.10).

$$m = \frac{y_2 - y_1}{x_2 - x_1} = \frac{2 - 1}{3 - (-1)} = \frac{1}{4}$$

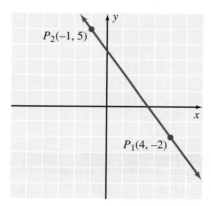

Figure 12.11

(b) Let $(4, -2)$ be P_1 and $(-1, 5)$ be P_2 (Figure 12.11).

$$m = \frac{y_2 - y_1}{x_2 - x_1} = \frac{5 - (-2)}{-1 - 4} = \frac{7}{-5} = -\frac{7}{5}$$

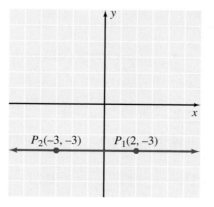

Figure 12.12

(c) Let $(2, -3)$ be P_1 and $(-3, -3)$ be P_2 (Figure 12.12).

$$m = \frac{y_2 - y_1}{x_2 - x_1}$$

$$= \frac{-3 - (-3)}{-3 - 2}$$

$$= \frac{0}{-5} = 0$$

The three parts of Example 5 represent the three basic possibilities for slope; that is, the slope of a line can be positive, negative, or zero. A line that has a positive slope rises as we move from left to right, as in Figure 12.10. A line that has a negative slope falls as we move from left to right, as in Figure 12.11. A horizontal line, as in Figure 12.12, has a slope of zero. Finally, we need to realize that *the concept of slope is undefined for vertical lines.* This is due to the fact that for any vertical line, the horizontal change as we move from one point on the line to another is zero. Thus the ratio $\dfrac{y_2 - y_1}{x_2 - x_1}$ will have a denominator of zero and be undefined. Accordingly, the restriction $x_2 \neq x_1$ is imposed in Definition 12.1.

Graphing Techniques

As stated in the introductory remarks for this chapter, there are two kinds of problems in coordinate geometry:

1. Given an algebraic equation, determine its geometric graph.
2. Given a set of conditions pertaining to a geometric figure, determine its algebraic equation.

We will work with both kinds of problems in this chapter, with an emphasis in this section on curve-sketching techniques.

One very important graphing technique is to be able to recognize that a certain kind of algebraic equation produces a certain kind of geometric graph. For example, from our work in Chapter 5, we know that any equation of the form $Ax + By = C$, where A, B, and C are real numbers (A and B not both zero) and x and y are variables, is a **linear equation** in two variables, and its graph is a straight line. Because two points determine a straight line, graphing linear equations is a simple process. We find two solutions, plot the corresponding points, and connect the points with a straight line. Let's consider an example.

EXAMPLE   6   Graph $3y = 6 - 2x$.

Solution

First, we need to realize that $3y = 6 - 2x$ is equivalent to $2x + 3y = 6$ and therefore fits the form of a linear equation. Now we can determine two points. Let $x = 0$ in the original equation.

$$3y = 6 - 2(0)$$
$$3y = 6$$
$$y = 2$$

Thus the point $(0, 2)$ is on the line. Then let $y = 0$.

$$3(0) = 6 - 2x$$
$$2x = 6$$
$$x = 3$$

Thus the point (3, 0) is on the line. Plotting the two points (0, 2) and (3, 0) and connecting them with a straight line produces Figure 12.13.

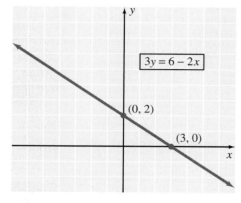

$3y = 6 - 2x$

(0, 2)

(3, 0)

Figure 12.13

The points (3, 0) and (0, 2) in Figure 12.13 are special points. They are the points of the graph that are on the coordinate axes. That is, they yield the x intercept and the y intercept of the graph. Let's define in general the *intercepts* of a graph.

> The x coordinates of the points that a graph has in common with the x axis are called the x **intercepts** of the graph. (To compute the x intercepts, let $y = 0$ and solve for x.)
>
> The y coordinates of the points that a graph has in common with the y axis are called the y **intercepts** of the graph. (To compute the y intercepts, let $x = 0$ and solve for y.)

Each of the following examples, along with the follow-up discussion, will introduce another aspect of curve sketching. Then toward the end of the section, we will summarize these techniques for you.

E X A M P L E   7   Graph $y = (x + 2)(x - 2)$.

Solution

Let's begin by finding the intercepts. If $x = 0$, then

$$y = (0 + 2)(0 - 2)$$
$$y = -4$$

The point $(0, -4)$ is on the graph. If $y = 0$, then

$$(x + 2)(x - 2) = 0$$
$$x + 2 = 0 \quad \text{or} \quad x - 2 = 0$$
$$x = -2 \quad \text{or} \quad x = 2$$

The points $(-2, 0)$ and $(2, 0)$ are on the graph. The given equation is in a convenient form for setting up a table of values.

x	y	
0	−4	Intercepts
−2	0	
2	0	
1	−3	
−1	−3	
3	5	Other points
−3	5	

Plotting these points and connecting them with a smooth curve produces Figure 12.14.

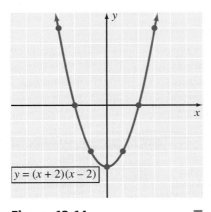

$$y = (x + 2)(x - 2)$$

Figure 12.14

The curve in Figure 12.14 is called a parabola; we will study parabolas in more detail in a later section. However, at this time we want to emphasize that the parabola in Figure 12.14 is said to be *symmetric with respect to the y axis*. In other words, the y axis is a line of symmetry. Each half of the curve is a mirror image of the other half through the y axis. Note, in the table of values, that for each ordered pair (x, y), the ordered pair $(-x, y)$ is also a solution. A general test for y axis symmetry can be stated as follows:

y Axis Symmetry

The graph of an equation is symmetric with respect to the y axis if replacing x with $-x$ results in an equivalent equation.

The equation $y = x^2 - 4$ exhibits y axis symmetry because replacing x with $-x$ produces $y = (-x)^2 - 4 = x^2 - 4$. Likewise, the equations $y = -x^2 + 2$, $y = 2x^2 + 5$, and $y = x^4 - x^2$ exhibit y axis symmetry.

EXAMPLE 8

Graph $x = y^2$.

Solution

First, we see that $(0, 0)$ is on the graph and determines both intercepts. Second, the given equation is in a convenient form for setting up a table of values.

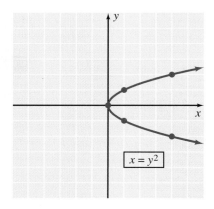

x	y	
0	0	Intercepts
1	1	
1	−1	Other points
4	2	
4	−2	

Plotting these points and connecting them with a smooth curve produces Figure 12.15.

Figure 12.15

The parabola in Figure 12.15 is said to be *symmetric with respect to the x axis*. Each half of the curve is a mirror image of the other half through the *x* axis. Also note, in the table of values, that for each ordered pair (x, y), the ordered pair $(x, -y)$ is a solution. A general test for *x* axis symmetry can be stated as follows:

x Axis Symmetry

The graph of an equation is symmetric with respect to the *x* axis if replacing *y* with $-y$ results in an equivalent equation.

The equation $x = y^2$ exhibits *x* axis symmetry because replacing *y* with $-y$ produces $x = (-y)^2 = y^2$. Likewise, the equations $x + y^2 = 5$, $x = y^2 - 5$, and $x = 2y^4 - 3y^2$ exhibit *x* axis symmetry.

EXAMPLE 9   Graph $y = \dfrac{1}{x}$.

Solution

First, let's find the intercepts. Let $x = 0$; then $y = \dfrac{1}{x}$ becomes $y = \dfrac{1}{0}$, and $\dfrac{1}{0}$ is undefined. Thus there is no *y* intercept. Let $y = 0$; then $y = \dfrac{1}{x}$ becomes $0 = \dfrac{1}{x}$, and there are no values of *x* that will satisfy this equation. In other words, this graph has no points on either the *x* axis or the *y* axis. Second, let's set up a table of values and keep in mind that neither *x* nor *y* can equal zero.

x	y
$\frac{1}{2}$	2
1	1
2	$\frac{1}{2}$
3	$\frac{1}{3}$
$-\frac{1}{2}$	-2
-1	-1
-2	$-\frac{1}{2}$
-3	$-\frac{1}{3}$

In Figure 12.16(a) we plotted the points associated with the solutions from the table. Because the graph does not intersect either axis, it must consist of two branches. Thus connecting the points in the first quadrant with a smooth curve and then connecting the points in the third quadrant with a smooth curve, we obtain the graph in Figure 12.16(b).

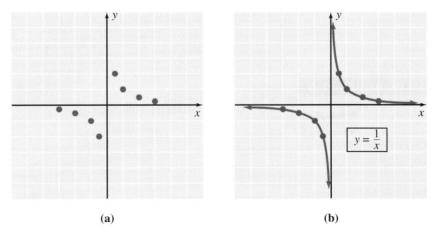

(a)   (b)

Figure 12.16

The curve in Figure 12.16(b) is said to be *symmetric with respect to the origin.* Each half of the curve is a mirror image of the other half through the origin. Note, in the table of values, that for each ordered pair (x, y), the ordered pair $(-x, -y)$ is also a solution. A general test for origin symmetry can be stated as follows:

Origin Symmetry

The graph of an equation is symmetric with respect to the origin if replacing x with $-x$ and y with $-y$ results in an equivalent equation.

The equation $y = \dfrac{1}{x}$ exhibits origin symmetry because replacing x with $-x$ and y with $-y$ produces $-y = \dfrac{1}{-x}$, which is equivalent to $y = \dfrac{1}{x}$. $\left(\text{We can mul-}\right.$ tiply both sides of $-y = \dfrac{1}{-x}$ by -1 and obtain $y = \dfrac{1}{x}.\left.\right)$ Likewise, the equations $xy = 4$, $y = x^3$, and $y = x^5$ exhibit origin symmetry.

Let's pause for a moment and pull together the graphing techniques that we have introduced thus far. Following is a list of graphing suggestions. The order of the suggestions indicates the order in which we usually attack a new graphing problem.

1. Determine what type of symmetry the equation exhibits.
2. Find the intercepts.
3. Solve the equation for y in terms of x or for x in terms of y if it is not already in such a form.
4. Set up a table of ordered pairs that satisfy the equation. The type of symmetry will affect your choice of values in the table. (We will illustrate this in a moment.)
5. Plot the points associated with the ordered pairs from the table, and connect them with a smooth curve. Then, if appropriate, reflect this part of the curve according to the symmetry shown by the equation.

EXAMPLE 10   Graph $x^2 y = -2$.

Solution

Because replacing x with $-x$ produces $(-x)^2 y = -2$ or, equivalently, $x^2 y = -2$, the equation exhibits y axis symmetry. There are no intercepts because neither x nor y can equal 0. Solving the equation for y produces $y = \dfrac{-2}{x^2}$. The equation exhibits y axis symmetry, so let's use only positive values for x, and then reflect the curve across the y axis.

x	y
1	-2
2	$-\dfrac{1}{2}$
3	$-\dfrac{2}{9}$
4	$-\dfrac{1}{8}$
$\dfrac{1}{2}$	-8

Let's plot the points determined by the table, connect them with a smooth curve, and reflect this portion of the curve across the y axis. Figure 12.17 is the result of this process.

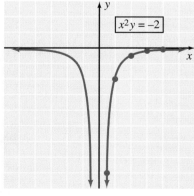

Figure 12.17

EXAMPLE 11

Graph $x = y^3$.

Solution

Because replacing x with $-x$ and y with $-y$ produces $-x = (-y)^3 = -y^3$, which is equivalent to $x = y^3$, the given equation exhibits origin symmetry. If $x = 0$, then $y = 0$, so the origin is a point of the graph. The given equation is in an easy form for deriving a table of values.

x	y
0	0
8	2
$\dfrac{1}{8}$	$\dfrac{1}{2}$
$\dfrac{27}{64}$	$\dfrac{3}{4}$

Let's plot the points determined by the table, connect them with a smooth curve, and reflect this portion of the curve through the origin to produce Figure 12.18.

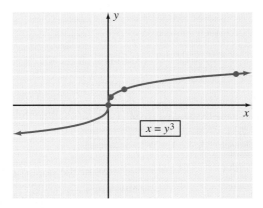

Figure 12.18

CONCEPT QUIZ

For Problems 1–8, answer true or false.

1. The point $(-2, 4)$ is located in the second quadrant.

2. The distance between P_2 and P_1 is the opposite of the distance between P_1 and P_2.

3. The slope of a line is the ratio of the vertical change to the horizontal change when moving from one point to another point.

4. The slope is always a positive number.

5. A slope of 0 means there is no change in the vertical direction moving from one point to another point.

6. When replacing y with $-y$ in an equation results in an equivalent equation, then the graph of the equation is symmetric with respect to the x axis.

7. If the graph of an equation is symmetric with respect to the x axis, then it cannot be symmetric with respect to the y axis.

8. If for each ordered pair (x, y) that is a solution of the equation, the ordered pair $(-x, -y)$ is also a solution, then the graph of the equation is symmetric with respect to the origin.

PROBLEM SET 12.1

For Problems 1–12, find the distance between each of the pairs of points. Express answers in simplest radical form.

1. $(-2, -1), (7, 11)$

2. $(2, 1), (10, 7)$

3. $(1, -1), (3, -4)$

4. $(-1, 3), (2, -2)$

5. $(6, -4), (9, -7)$

6. $(-5, 2), (-1, 6)$

7. $(-3, 3), (0, -3)$

8. $(-2, -4), (4, 0)$

9. $(1, -6), (-5, -6)$

10. $(-2, 3), (-2, -7)$

11. $(1, 7), (4, -2)$

12. $(6, 4), (-4, -8)$

13. Verify that the points $(-3, 1), (5, 7)$, and $(8, 3)$ are vertices of a right triangle. [*Hint:* If $a^2 + b^2 = c^2$, then it is a right triangle with the right angle opposite side c.]

14. Verify that the points $(0, 3), (2, -3)$, and $(-4, -5)$ are vertices of an isosceles triangle.

15. Verify that the points $(7, 12)$ and $(11, 18)$ divide the line segment joining $(3, 6)$ and $(15, 24)$ into three segments of equal length.

16. Verify that $(3, 1)$ is the midpoint of the line segment joining $(-2, 6)$ and $(8, -4)$.

For Problems 17–28, graph the line determined by the two points and find the slope of the line.

17. $(1, 2), (4, 6)$

18. $(3, 1), (-2, -2)$

19. $(3, 4), (3, 8)$

20. $(-2, 5), (3, -1)$

21. $(2, 6), (6, -2)$

22. $(-2, -1), (2, -5)$

23. $(-6, 1), (-1, 4)$

24. $(-3, 3), (2, 3)$

25. $(-2, -4), (2, -4)$

26. $(1, -5), (1, -1)$

27. $(0, -2), (4, 0)$

28. $(-4, 0), (0, -6)$

29. Find x if the line through $(-2, 4)$ and $(x, 6)$ has a slope of $\dfrac{2}{9}$.

30. Find y if the line through $(1, y)$ and $(4, 2)$ has a slope of $\dfrac{5}{3}$.

31. Find x if the line through $(x, 4)$ and $(2, -5)$ has a slope of $-\dfrac{9}{4}$.

32. Find y if the line through $(5, 2)$ and $(-3, y)$ has a slope of $-\dfrac{7}{8}$.

For each of the points in Problems 33–37, determine the points that are symmetric with respect to (a) the x axis, (b) the y axis, and (c) the origin.

33. $(-3, 1)$

34. $(-2, -4)$

35. $(7, -2)$

36. $(0, -4)$

37. $(5, 0)$

For Problems 38–51, determine the type(s) of symmetry (symmetry with respect to the x axis, y axis, and/or origin) exhibited by the graph of each of the following equations. *Do not* sketch the graph.

38. $x^2 + 2y = 4$

39. $-3x + 2y^2 = -4$

40. $x = -y^2 + 5$

41. $y = 4x^2 + 13$

42. $xy = -6$

43. $2x^2y^2 = 5$

44. $2x^2 + 3y^2 = 9$

45. $x^2 - 2x - y^2 = 4$

46. $y = x^2 - 6x - 4$

47. $y = 2x^2 - 7x - 3$

48. $y = x$ **49.** $y = 2x$

50. $y = x^4 + 4$ **51.** $y = x^4 - x^2 + 2$

For Problems 52–79, graph each of the equations.

52. $y = x + 1$ **53.** $y = x - 4$

54. $y = 3x - 6$ **55.** $y = 2x + 4$

56. $y = -2x + 1$ **57.** $y = -3x - 1$

58. $y = x^2 - 1$ **59.** $y = x^2 + 2$

60. $y = -x^3$ **61.** $y = x^3$

62. $y = \dfrac{2}{x^2}$ **63.** $y = \dfrac{-1}{x^2}$

64. $2x + y = 6$ **65.** $2x - y = 4$

66. $y = 2x^2$ **67.** $y = -3x^2$

68. $xy = -3$ **69.** $xy = 2$

70. $x^2y = 4$ **71.** $xy^2 = -4$

72. $y^3 = x^2$ **73.** $y^2 = x^3$

74. $y = \dfrac{-2}{x^2 + 1}$ **75.** $y = \dfrac{4}{x^2 + 1}$

76. $x = -y^3$ **77.** $y = x^4$

78. $y = -x^4$ **79.** $x = -y^3 + 2$

■ ■ ■ **Thoughts into words**

80. If one line has a slope of $\dfrac{2}{5}$ and another line has a slope

of $\dfrac{3}{7}$, which line is steeper? Explain your answer.

81. Suppose that a line has a slope of $\dfrac{2}{3}$ and contains the

point $(4, 7)$. Are the points $(7, 9)$ and $(1, 3)$ also on the line? Explain your answer.

82. What is the graph of $x = 0$? What is the graph of $y = 0$? Explain your answers.

83. Is a graph symmetric with respect to the origin if it is symmetric with respect to both axes? Defend your answer.

84. Is a graph symmetric with respect to both axes if it is symmetric with respect to the origin? Defend your answer.

■ ■ ■ **Further investigations**

85. Sometimes it is necessary to find the coordinate of a point that is located somewhere between two given points on a number line. For example, suppose that we want to find the coordinate (x) of the point located two-thirds of the distance *from 2 to* 8. Because the total distance from 2 to 8 is $8 - 2 = 6$ units, we can start at 2 and

move $\dfrac{2}{3}(6) = 4$ units toward 8. Thus $x = 2 + \dfrac{2}{3}(6) =$

$2 + 4 = 6$.

For each of the following, find the coordinate of the indicated point on a number line.

    a. Two-thirds of the distance from 1 to 10

    b. Three-fourths of the distance from -2 to 14

    c. One-third of the distance from -3 to 7

    d. Two-fifths of the distance from -5 to 6

    e. Three-fifths of the distance from -1 to -11

    f. Five-sixths of the distance from 3 to -7

86. Now suppose that we want to find the coordinates of point P, which is located two-thirds of the distance from $A(1, 2)$ to $B(7, 5)$ in a coordinate plane. We have plotted the given points A and B in Figure 12.19 to help

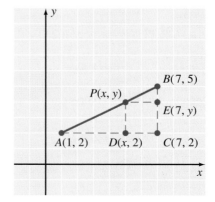

Figure 12.19

with the analysis of this problem. Point D is two-thirds of the distance from A to C, because parallel lines cut off proportional segments on every transversal that intersects the lines. Thus $\overline{AC}$ can be treated as a segment of a number line, as shown in Figure 12.20.

1 ●————————— x —————— 7 ●
A D C

Figure 12.20

Therefore,

$$x = 1 + \frac{2}{3}(7-1) = 1 + \frac{2}{3}(6) = 5$$

Similarly, $\overline{CB}$ can be treated as a segment of a number line, as shown in Figure 12.21. Therefore,

B ● 5
E ● y $$y = 2 + \frac{2}{3}(5-2) = 2 + \frac{2}{3}(3) = 4$$

                  The coordinates of point P are $(5, 4)$.
C ● 2

Figure 12.21

For each of the following, find the coordinates of the indicated point in the xy plane.

a. One-third of the distance from $(2, 3)$ to $(5, 9)$

b. Two-thirds of the distance from $(1, 4)$ to $(7, 13)$

c. Two-fifths of the distance from $(-2, 1)$ to $(8, 11)$

d. Three-fifths of the distance from $(2, -3)$ to $(-3, 8)$

e. Five-eighths of the distance from $(-1, -2)$ to $(4, -10)$

f. Seven-eighths of the distance from $(-2, 3)$ to $(-1, -9)$

Graphing calculator activities

For Problems 87–93, answer the "How does . . . ." question and then use your calculator to graph both equations on the same set of axes to check your answer.

87. How does the graph of $y = -\frac{5}{x}$ compare to the graph of $y = \frac{5}{x}$?

88. How does the graph of $-y = -\frac{5}{x}$ compare to the graph of $y = \frac{5}{x}$?

89. How does the graph of $y = -\frac{2}{x^2}$ compare to the graph of $y = \frac{2}{x^2}$?

90. How does the graph of $y = -2x$ compare to the graph of $y = 2x$?

91. How does the graph of $y = \frac{1}{x} + 2$ compare to the graph of $y = \frac{1}{x}$?

92. How does the graph of $y = \frac{1}{x^2} - 3$ compare to the graph of $y = \frac{1}{x^2}$?

93. How does the graph of $y = \frac{1}{x-2}$ compare to the graph of $y = \frac{1}{x}$?

Answers to Concept Quiz

1. True   **2.** False   **3.** True   **4.** False   **5.** True   **6.** True   **7.** False   **8.** True

12.2 Writing Equations of Lines

Objectives

■ Become familiar with the point-slope form and the slope-intercept form of the equation of a straight line.

■ Know the relationships for slopes of parallel and perpendicular lines.

■ Find the equation of a line given
 1. a slope and a point.
 2. two points on the line.
 3. a point on the line and that the line is parallel or perpendicular to another line.

To review, there are basically two types of problems to solve in coordinate geometry:

 1. Given an algebraic equation, find its geometric graph.
 2. Given a set of conditions pertaining to a geometric figure, find its algebraic equation.

Problems of type 1 have been our primary concern thus far in this chapter. Now let's analyze some problems of type 2 that deal specifically with straight lines. Given certain facts about a line, we need to be able to determine its algebraic equation. Let's consider some examples.

E X A M P L E   1

Find the equation of the line that has a slope of $\dfrac{2}{3}$ and contains the point $(1, 2)$.

Solution

First draw the line and record the given information. Then choose a point (x, y) that represents any point on the line other than the given point $(1, 2)$. (See Figure 12.22.)

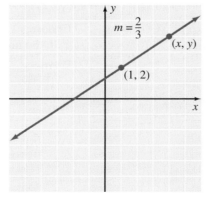

The slope determined by $(1, 2)$ and (x, y) is $\dfrac{2}{3}$. Thus

$$\frac{y - 2}{x - 1} = \frac{2}{3}$$

$$2(x - 1) = 3(y - 2)$$

$$2x - 2 = 3y - 6$$

$$2x - 3y = -4$$

Figure 12.22

EXAMPLE 2 Find the equation of the line that contains $(3, 2)$ and $(-2, 5)$.

Solution

First, let's draw the line determined by the given points (Figure 12.23).

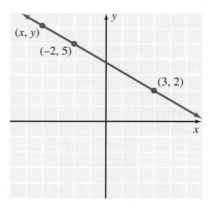

Because we know two points, we can find the slope.

$$m = \frac{y_2 - y_1}{x_2 - x_1} = \frac{3}{-5} = -\frac{3}{5}$$

Figure 12.23

Now we can use the same approach as in Example 1. Form an equation using a variable point (x, y), one of the two given points, and the slope, $-\frac{3}{5}$.

$$\frac{y - 5}{x + 2} = \frac{3}{-5} \qquad \left(-\frac{3}{5} = \frac{3}{-5}\right)$$
$$3(x + 2) = -5(y - 5)$$
$$3x + 6 = -5y + 25$$
$$3x + 5y = 19$$

EXAMPLE 3 Find the equation of the line that has a slope of $\frac{1}{4}$ and a y intercept of 2.

Solution

A y intercept of 2 means that the point $(0, 2)$ is on the line (Figure 12.24).

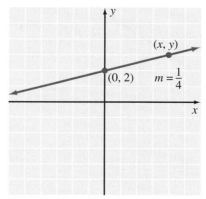

Choose a variable point (x, y) and proceed as in the previous examples.

$$\frac{y - 2}{x - 0} = \frac{1}{4}$$
$$1(x - 0) = 4(y - 2)$$
$$x = 4y - 8$$
$$x - 4y = -8$$

Figure 12.24

Perhaps it would be helpful to pause a moment and look back over Examples 1, 2, and 3. Note that we used the same basic approach in all three situations. We chose a variable point (x, y) and used it to determine the equation that satisfies the conditions given in the problem. The approach we took in the previous examples can be generalized to produce some special forms of equations of straight lines.

Point-Slope Form

E X A M P L E   4

Find the equation of the line that has a slope of m and contains the point (x_1, y_1).

Solution

Choose (x, y) to represent any other point on the line (Figure 12.25), and the slope of the line is therefore given by

$$m = \frac{y - y_1}{x - x_1}, \qquad x \neq x_1$$

from which we can obtain the equivalent equation,

$$y - y_1 = m(x - x_1).$$

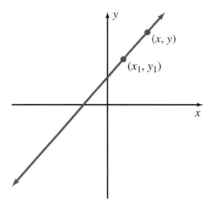

Figure 12.25

We refer to the equation

$$y - y_1 = m(x - x_1)$$

as the **point-slope form** of the equation of a straight line. Instead of the approach we used in Example 1, we could use the point-slope form to write the equation of a line with a given slope that contains a given point. For example, we can determine the equation of the line that has a slope of $\frac{3}{5}$ and contains the point $(2, 4)$ as follows:

$$y - y_1 = m(x - x_1)$$

Substitute $(2, 4)$ for (x_1, y_1) and $\dfrac{3}{5}$ for m.

$$y - 4 = \frac{3}{5}(x - 2)$$
$$5(y - 4) = 3(x - 2)$$
$$5y - 20 = 3x - 6$$
$$-14 = 3x - 5y$$

Slope-Intercept Form

E X A M P L E 5

Find the equation of the line that has a slope of m and a y intercept of b.

Solution

A y intercept of b means that the line contains the point $(0, b)$, as in Figure 12.26. Therefore, we can use the point-slope form as follows:

$$y - y_1 = m(x - x_1)$$
$$y - b = m(x - 0)$$
$$y - b = mx$$
$$y = mx + b$$

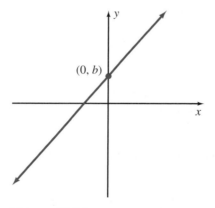

Figure 12.26

We refer to the equation

$$y = mx + b$$

as the **slope-intercept form** of the equation of a straight line. We use it for three primary purposes, as the next three examples illustrate.

EXAMPLE 6   Find the equation of the line that has a slope of $\dfrac{1}{4}$ and a y intercept of 2.

Solution

This is a restatement of Example 3, but this time we will use the slope-intercept form ($y = mx + b$) of a line to write its equation. Because $m = \dfrac{1}{4}$ and $b = 2$, we can substitute these values into $y = mx + b$.

$$y = mx + b$$

$$y = \frac{1}{4}x + 2$$

$$4y = x + 8 \qquad \text{Multiply both sides by 4.}$$
$$x - 4y = -8 \qquad \text{Same result as in Example 3.} \qquad \blacksquare$$

EXAMPLE 7   Find the slope of the line when the equation is $3x + 2y = 6$.

Solution

We can solve the equation for y in terms of x, and then compare it to the slope-intercept form to determine its slope. Thus

$$3x + 2y = 6$$
$$2y = -3x + 6$$
$$y = -\frac{3}{2}x + 3$$
$$y = -\frac{3}{2}x + 3 \qquad y = mx + b$$

The slope of the line is $-\dfrac{3}{2}$. Furthermore, the y intercept is 3. $\qquad \blacksquare$

EXAMPLE 8   Graph the line determined by the equation $y = \dfrac{2}{3}x - 1$.

Solution

Comparing the given equation to the general slope-intercept form, we see that the slope of the line is $\dfrac{2}{3}$ and the y intercept is -1. Because the y intercept is -1, we can plot the point $(0, -1)$. Then because the slope is $\dfrac{2}{3}$, let's move 3 units to the right and 2 units up from $(0, -1)$ to locate the point $(3, 1)$. The two points $(0, -1)$ and $(3, 1)$ determine the line in Figure 12.27.

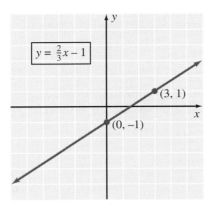

Figure 12.27

In general, if the equation of a nonvertical line is written in slope-intercept form ($y = mx + b$), the coefficient of x is the slope of the line, and the constant term is the y intercept. (Remember that the concept of slope is not defined for a vertical line.)

Parallel and Perpendicular Lines

We can use two important relationships between lines and their slopes to solve certain kinds of problems. It can be shown that nonvertical parallel lines have the same slope and that two nonvertical lines are perpendicular if the product of their slopes is -1. (Details for verifying these facts are left to another course.) In other words, if two lines have slopes m_1 and m_2, respectively, then

1. The two lines are parallel if and only if $m_1 = m_2$.
2. The two lines are perpendicular if and only if $(m_1)(m_2) = -1$.

The following examples demonstrate the use of these properties.

EXAMPLE 9

(a) Verify that the graphs of $2x + 3y = 7$ and $4x + 6y = 11$ are parallel lines.
(b) Verify that the graphs of $8x - 12y = 3$ and $3x + 2y = 2$ are perpendicular lines.

Solution

(a) Let's change each equation to slope-intercept form.

$$2x + 3y = 7 \quad \rightarrow \quad 3y = -2x + 7$$
$$y = -\frac{2}{3}x + \frac{7}{3}$$

$$4x + 6y = 11 \quad \rightarrow \quad 6y = -4x + 11$$
$$y = -\frac{4}{6}x + \frac{11}{6}$$
$$y = -\frac{2}{3}x + \frac{11}{6}$$

Both lines have a slope of $-\dfrac{2}{3}$, but they have different y intercepts. There-fore, the two lines are parallel.

(b) Solving each equation for y in terms of x, we obtain

$$8x - 12y = 3 \qquad \longrightarrow \qquad -12y = -8x + 3$$

$$y = \frac{8}{12}x - \frac{3}{12}$$

$$y = \frac{2}{3}x - \frac{1}{4}$$

$$3x + 2y = 2 \qquad \longrightarrow \qquad 2y = -3x + 2$$

$$y = -\frac{3}{2}x + 1$$

Because $\left(\dfrac{2}{3}\right)\left(-\dfrac{3}{2}\right) = -1$ (the product of the two slopes is -1), the lines are perpendicular.   ■

REMARK: The statement "the product of two slopes is -1" is the same as saying that the two slopes are negative reciprocals of each other; that is, $m_1 = -\dfrac{1}{m_2}$.

EXAMPLE   10   Find the equation of the line that contains the point $(1, 4)$ and is parallel to the line determined by $x + 2y = 5$.

Solution

First, let's draw a figure to help in our analysis of the problem (Figure 12.28). Be-cause the line through $(1, 4)$ is to be parallel to the line determined by $x + 2y = 5$, it must have the same slope. Let's find the slope by changing $x + 2y = 5$ to the slope-intercept form.

$$x + 2y = 5$$

$$2y = -x + 5$$

$$y = -\frac{1}{2}x + \frac{5}{2}$$

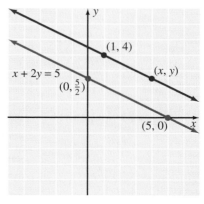

Figure 12.28

The slope of both lines is $-\dfrac{1}{2}$. Now we can choose a variable point (x, y) on the line through $(1, 4)$ and proceed as we did in earlier examples.

$$\frac{y - 4}{x - 1} = \frac{1}{-2}$$
$$1(x - 1) = -2(y - 4)$$
$$x - 1 = -2y + 8$$
$$x + 2y = 9$$

 EXAMPLE 11 Find the equation of the line that contains the point $(-1, -2)$ and is perpendicular to the line determined by $2x - y = 6$.

Solution

First, let's draw a figure to help in our analysis of the problem (Figure 12.29). Because the line through $(-1, -2)$ is to be perpendicular to the line determined by $2x - y = 6$, its slope must be the negative reciprocal of the slope of $2x - y = 6$. Let's find the slope of $2x - y = 6$ by changing it to the slope-intercept form.

$$2x - y = 6$$
$$-y = -2x + 6$$
$$y = 2x - 6 \qquad \text{The slope is 2.}$$

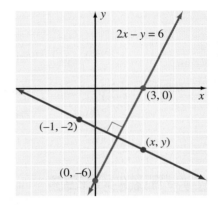

Figure 12.29

The slope of the desired line is $-\dfrac{1}{2}$ (the negative reciprocal of 2), and we can proceed as before by using a variable point (x, y).

$$\frac{y + 2}{x + 1} = \frac{1}{-2}$$
$$1(x + 1) = -2(y + 2)$$
$$x + 1 = -2y - 4$$
$$x + 2y = -5$$

We use two forms of equations of straight lines extensively. They are the **standard form** and the **slope-intercept form,** and we describe them as follows.

Standard Form $Ax + By = C$, where B and C are integers and A is a nonnegative integer (A and B not both zero).

Slope-Intercept Form $y = mx + b$, where m is a real number representing the slope and b is a real number representing the y intercept.

CONCEPT QUIZ

For Problems 1–5, answer true or false.

1. If two lines have the same slope, then the lines are parallel.

2. If the slopes of two lines are reciprocals, then the lines are perpendicular.

3. In the standard form of the equation of a line $Ax + By = C$, A can be a rational number in fractional form.

4. In the slope-intercept form of an equation of a line $y = mx + b$, m is the slope.

5. In the standard form of the equation of a line $Ax + By = C$, A is the slope.

PROBLEM SET 12.2

For Problems 1–8, write the equation of the line that has the indicated slope and contains the indicated point. Express final equations in standard form.

1. $m = \dfrac{1}{2}$, $(3, 5)$

2. $m = \dfrac{1}{3}$, $(2, 3)$

3. $m = 3$, $(-2, 4)$

4. $m = -2$, $(-1, 6)$

5. $m = -\dfrac{3}{4}$, $(-1, -3)$

6. $m = -\dfrac{3}{5}$, $(-2, -4)$

7. $m = \dfrac{5}{4}$, $(4, -2)$

8. $m = \dfrac{3}{2}$, $(8, -2)$

For Problems 9–18, write the equation of the line that contains the indicated pair of points. Express final equations in standard form.

9. $(2, 1), (6, 5)$

10. $(-1, 2), (2, 5)$

11. $(-2, -3), (2, 7)$

12. $(-3, -4), (1, 2)$

13. $(-3, 2), (4, 1)$

14. $(-2, 5), (3, -3)$

15. $(-1, -4), (3, -6)$

16. $(3, 8), (7, 2)$

17. $(0, 0), (5, 7)$

18. $(0, 0), (-5, 9)$

For Problems 19–26, write the equation of the line that has the indicated slope (m) and y intercept (b). Express final equations in slope-intercept form.

19. $m = \dfrac{3}{7}$, $b = 4$

20. $m = \dfrac{2}{9}$, $b = 6$

21. $m = 2$, $b = -3$

22. $m = -3$, $b = -1$

23. $m = -\dfrac{2}{5}$, $b = 1$

24. $m = -\dfrac{3}{7}$, $b = 4$

25. $m = 0$, $b = -4$

26. $m = \dfrac{1}{5}$, $b = 0$

For Problems 27–42, write the equation of the line that satisfies the given conditions. Express final equations in standard form.

27. x intercept of 2 and y intercept of -4

28. x intercept of -1 and y intercept of -3

29. x intercept of -3 and slope of $-\dfrac{5}{8}$

30. x intercept of 5 and slope of $-\dfrac{3}{10}$

31. Contains the point $(2, -4)$ and is parallel to the y axis

32. Contains the point $(-3, -7)$ and is parallel to the x axis

33. Contains the point $(5, 6)$ and is perpendicular to the y axis

34. Contains the point $(-4, 7)$ and is perpendicular to the x axis

35. Contains the point $(1, 3)$ and is parallel to the line $x + 5y = 9$

36. Contains the point $(-1, 4)$ and is parallel to the line $x - 2y = 6$

37. Contains the origin and is parallel to the line $4x - 7y = 3$

38. Contains the origin and is parallel to the line $-2x - 9y = 4$

39. Contains the point $(-1, 3)$ and is perpendicular to the line $2x - y = 4$

40. Contains the point $(-2, -3)$ and is perpendicular to the line $x + 4y = 6$

41. Contains the origin and is perpendicular to the line $-2x + 3y = 8$

42. Contains the origin and is perpendicular to the line $y = -5x$

For Problems 43–48, change the equation to slope-intercept form and determine the slope and y intercept of the line.

43. $3x + y = 7$

44. $5x - y = 9$

45. $3x + 2y = 9$

46. $x - 4y = 3$

47. $x = 5y + 12$

48. $-4x - 7y = 14$

For Problems 49–56, use the slope-intercept form to graph the following lines.

49. $y = \dfrac{2}{3}x - 4$

50. $y = \dfrac{1}{4}x + 2$

51. $y = 2x + 1$

52. $y = 3x - 1$

53. $y = -\dfrac{3}{2}x + 4$

54. $y = -\dfrac{5}{3}x + 3$

55. $y = -x + 2$

56. $y = -2x + 4$

For Problems 57–60, the situations can be described by the use of linear equations in two variables. If two pairs of values are known, then we can determine the equation by using the approach we used in Example 2 of this section. For each of the following, assume that the relationship can be expressed as a linear equation in two variables, and use the given information to determine the equation. Express the equation in slope-intercept form.

57. A company uses 7 pounds of fertilizer for a lawn that measures 5000 square feet and 12 pounds for a lawn that measures 10,000 square feet. Let y represent the pounds of fertilizer and x the square footage of the lawn.

58. A new diet fad claims that a person weighing 140 pounds should consume 1490 daily calories and that a 200-pound person should consume 1700 calories. Let y represent the calories and x the weight of the person in pounds.

59. Two banks on opposite corners of a town square had signs that displayed the current temperature. One bank displayed the temperature in degrees Celsius and the other in degrees Fahrenheit. A temperature of $10°C$ was displayed at the same time as a temperature of $50°F$. On another day, a temperature of $-5°C$ was displayed at the same time as a temperature of $23°F$. Let y represent the temperature in degrees Fahrenheit and x the temperature in degrees Celsius.

60. An accountant has a schedule of depreciation for some business equipment. The schedule shows that after 12 months the equipment is worth $7600 and that after 20 months it is worth $6000. Let y represent the worth and x represent the time in months.

■■■ **Thoughts into words**

61. What does it mean to say that two points *determine* a line?

62. How would you help a friend determine the equation of the line that is perpendicular to $x - 5y = 7$ and contains the point $(5, 4)$?

63. Explain how you would find the slope of the line $y = 4$.

■■■ **Further investigations**

64. The equation of a line that contains the two points (x_1, y_1) and (x_2, y_2) is $\dfrac{y - y_1}{x - x_1} = \dfrac{y_2 - y_1}{x_2 - x_1}$. We often refer to this as the **two-point form** of the equation of a straight line. Use the two-point form and write the equation of the line that contains each of the indicated pairs of points. Express final equations in standard form.

a. $(1, 1)$ and $(5, 2)$

b. $(2, 4)$ and $(-2, -1)$

c. $(-3, 5)$ and $(3, 1)$

d. $(-5, 1)$ and $(2, -7)$

65. Let $Ax + By = C$ and $A'x + B'y = C'$ represent two lines. Change both of these equations to slope-intercept form, and then verify each of the following properties.

a. If $\dfrac{A}{A'} = \dfrac{B}{B'} \neq \dfrac{C}{C'}$, then the lines are parallel.

b. If $AA' = -BB'$, then the lines are perpendicular.

66. The properties in Problem 65 provide us with another way to write the equation of a line that is parallel or perpendicular to a given line and contains a given point not on the line. For example, suppose we want the equation of the line that is perpendicular to $3x + 4y = 6$ and contains the point $(1, 2)$. The form $4x - 3y = k$, where k is

a constant, represents a family of lines perpendicular to $3x + 4y = 6$ because we have satisfied the condition $AA' = -BB'$. Therefore, to find the specific line of the family that contains $(1, 2)$, we substitute 1 for x and 2 for y to determine k.

$$4x - 3y = k$$
$$4(1) - 3(2) = k$$
$$-2 = k$$

Thus the equation of the desired line is $4x - 3y = -2$.

Use the properties from Problem 65 to help write the equation of each of the following lines.

a. Contains $(1, 8)$ and is parallel to $2x + 3y = 6$

b. Contains $(-1, 4)$ and is parallel to $x - 2y = 4$

c. Contains $(2, -7)$ and is perpendicular to $3x - 5y = 10$

d. Contains $(-1, -4)$ and is perpendicular to $2x + 5y = 12$

 Graphing calculator activities

67. Predict whether each of the following pairs of equations represents parallel lines, perpendicular lines, or lines that intersect but are not perpendicular. Then graph each pair of lines to check your predictions. Use a zoom square option for your viewing window. (The properties presented in Problem 65 should be very helpful.)

a. $5.2x + 3.3y = 9.4$ and $5.2x + 3.3y = 12.6$

b. $1.3x - 4.7y = 3.4$ and $1.3x - 4.7y = 11.6$

c. $2.7x + 3.9y = 1.4$ and $2.7x - 3.9y = 8.2$

d. $5x - 7y = 17$ and $7x + 5y = 19$

e. $9x + 2y = 14$ and $2x + 9y = 17$

f. $2.1x + 3.4y = 11.7$ and $3.4x - 2.1y = 17.3$

g. $7.1x - 2.3y = 6.2$ and $2.3x + 7.1y = 9.9$

h. $-3x + 9y = 12$ and $9x - 3y = 14$

i. $2.6x - 5.3y = 3.4$ and $5.2x - 10.6y = 19.2$

j. $4.8x - 5.6y = 3.4$ and $6.1x + 7.6y = 12.3$

Answers to Concept Quiz

1. True **2.** False **3.** False **4.** True **5.** False

12.3 Graphing Parabolas

Objectives

- Graph parabolas by plotting points and using symmetry.

- Know the effects that a, h, and k have on the graph of a parabola for the equation $y = a(x - h)^2 + k$.

- Graph parabolas by comparing their equations to the basic equation $y = x^2$.

In Chapter 11, we used a graphing calculator to graph some equations of the form $y = ax^2 + bx + c$. We did this to predict approximate solutions of quadratic equations and also to give visual support for solutions that we obtained algebraically. In each case, the graph was a curve called a **parabola.** In general, the graph of any equation of the form $y = ax^2 + bx + c$, where a, b, and c are real numbers and $a \neq 0$, is a parabola.

At this time, we want to develop a very easy and systematic way of graphing parabolas without the use of a graphing calculator. As we work with parabolas, we will use the vocabulary indicated in Figure 12.30.

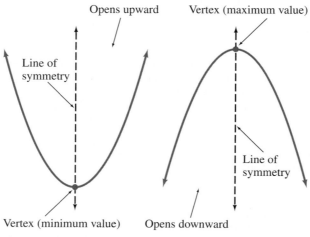

Figure 12.30

Let's begin by using the concepts of intercepts and symmetry to help sketch the graph of the equation $y = x^2$.

EXAMPLE 1 Graph $y = x^2$.

Solution

If we replace x with $-x$, the given equation becomes $y = (-x)^2 = x^2$; therefore, we have y-axis symmetry. The origin, $(0, 0)$, is a point of the graph. Now we can set up a table of values that uses nonnegative values for x. Plot the points determined by the table, connect them with a smooth curve, and reflect that portion of the curve across the y axis to produce Figure 12.31.

x	y
0	0
1	1
2	4
3	9
$\frac{1}{2}$	$\frac{1}{4}$

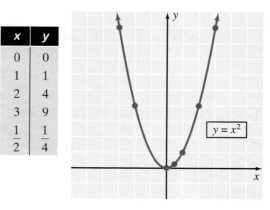

Figure 12.31

To graph parabolas, we want to be able to:

1. Find the vertex.
2. Determine whether the parabola opens upward or downward.
3. Locate two points on opposite sides of the line of symmetry.
4. Compare the parabola to the basic parabola $y = x^2$.

To graph parabolas produced by the various types of equations such as $y = x^2 + k$, $y = ax^2$, $y = (x - h)^2$, and $y = a(x - h)^2 + k$, we can compare these equations to that of the basic parabola, $y = x^2$. First, let's consider some equations of the form $y = x^2 + k$, where k is a constant.

EXAMPLE 2   Graph $y = x^2 + 1$.

Solution

Set up a table of values to compare y values for $y = x^2 + 1$ to the corresponding y values for $y = x^2$.

x	$y = x^2$	$y = x^2 + 1$
0	0	1
1	1	2
2	4	5
−1	1	2
−2	4	5

It should be evident that y values for $y = x^2$ + 1 are *one greater than* corresponding y values for $y = x^2$. For example, for the equation $y = x^2$, if $x = 2$ then $y = 4$; but for the equation $y = x^2 + 1$, if $x = 2$ then $y = 5$. Thus the graph of $y = x^2 + 1$ is the same as the graph of $y = x^2$, but *moved up 1 unit* (Figure 12.32).

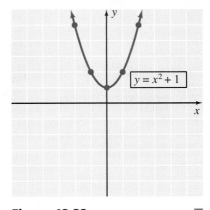

$y = x^2 + 1$

Figure 12.32

EXAMPLE 3   Graph $y = x^2 - 2$.

Solution

The y values for $y = x^2 - 2$ are *two less than* the corresponding y values for $y = x^2$, as indicated in the following table.

x	$y = x^2$	$y = x^2 - 2$
0	0	−2
1	1	−1
2	4	2
−1	1	−1
−2	4	2

Thus the graph of $y = x^2 - 2$ is the same as the graph of $y = x^2$ but *moved down 2 units* (Figure 12.33).

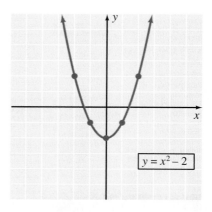

$y = x^2 - 2$

Figure 12.33

In general, the graph of a quadratic equation of the form $y = x^2 + k$ is the same as the graph of $y = x^2$ but moved up or down $|k|$ units, depending on whether k is positive or negative.

Now, let's consider some quadratic equations of the form $y = ax^2$, where a is a nonzero constant.

E X A M P L E   4

Graph $y = 2x^2$.

Solution

Again, let's use a table to make some comparisons of y values.

x	y = x²	y = 2x²
0	0	0
1	1	2
2	4	8
−1	1	2
−2	4	8

Obviously, the y values for $y = 2x^2$ are *twice* the corresponding y values for $y = x^2$. Thus the parabola associated with $y = 2x^2$ has the same vertex (the origin) as the graph of $y = x^2$, but it is *narrower* (Figure 12.34).

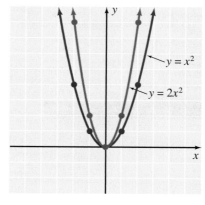

$y = x^2$

$y = 2x^2$

Figure 12.34

EXAMPLE 5

Graph $y = \frac{1}{2}x^2$.

Solution

The following table indicates some comparisons of y values.

x	$y = x^2$	$y = \frac{1}{2}x^2$
0	0	0
1	1	$\frac{1}{2}$
2	4	2
−1	1	$\frac{1}{2}$
−2	4	2

The y values for $y = \frac{1}{2}x^2$ are *one-half* of the corresponding y values for $y = x^2$. Therefore, the graph of $y = \frac{1}{2}x^2$ is *wider* than that of $y = x^2$, which we can call the basic parabola (Figure 12.35).

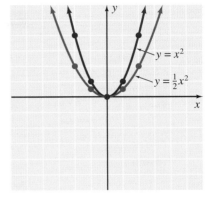

Figure 12.35

EXAMPLE 6

Graph $y = -x^2$.

Solution

x	$y = x^2$	$y = -x^2$
0	0	0
1	1	−1
2	4	−4
−1	1	−1
−2	4	−4

The y values for $y = -x^2$ are the *opposites* of the corresponding y values for $y = x^2$. Thus the graph of $y = -x^2$ is a *reflection across the x axis* of the basic parabola (Figure 12.36).

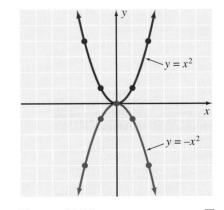

Figure 12.36

In general, the graph of a quadratic equation of the form $y = ax^2$ has its vertex at the origin and opens upward if a is positive and downward if a is negative. The parabola is narrower than the basic parabola if $|a| > 1$ and wider if $|a| < 1$.

Let's continue our investigation of quadratic equations by considering those of the form $y = (x - h)^2$, where h is a nonzero constant.

EXAMPLE 7   Graph $y = (x - 2)^2$.

Solution

A fairly extensive table of values reveals a pattern.

x	$y = x^2$	$y = (x - 2)^2$
-2	4	16
-1	1	9
0	0	4
1	1	1
2	4	0
3	9	1
4	16	4
5	25	9

Note that $y = (x - 2)^2$ and $y = x^2$ take on the same y values, *but* for different values of x. More specifically, if $y = x^2$ achieves a certain y value at x equals a constant, then $y = (x - 2)^2$ achieves the same y value at x equals the *constant plus two*. In other words, the graph of $y = (x - 2)^2$ is the same as the graph of $y = x^2$ but *moved 2 units to the right* (Figure 12.37).

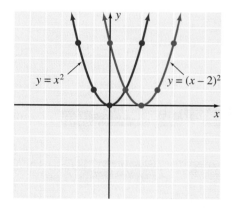

Figure 12.37

EXAMPLE 8   Graph $y = (x + 3)^2$.

Solution

If $y = x^2$ achieves a certain y value at x equals a constant, then $y = (x + 3)^2$ achieves that same y value at x equals that *constant minus three*. Therefore, the graph of $y = (x + 3)^2$ is the same as the graph of $y = x^2$ but *moved 3 units to the left* (Figure 12.38).

x	y = x²	y = (x + 3)²
−3	9	0
−2	4	1
−1	1	4
0	0	9
1	1	16
2	4	25
3	9	36

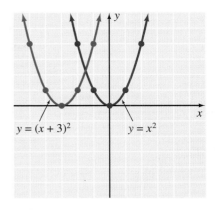

Figure 12.38

In general, the graph of a quadratic equation of the form $y = (x - h)^2$ is the same as the graph of $y = x^2$ but moved to the right h units if h is positive or moved to the left $|h|$ units if h is negative.

$y = (x - 4)^2$ →   Moved to the *right* 4 units

$y = (x + 2)^2 = (x - (-2))^2$ →   Moved to the *left* 2 units

The following diagram summarizes our work with graphing quadratic equations.

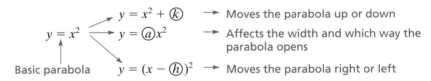

Equations of the form $y = x^2 + k$ and $y = ax^2$ are symmetric about the y axis. The next two examples of this section show how we can combine these ideas to graph a quadratic equation of the form $y = a(x - h)^2 + k$.

E X A M P L E   9

Graph $y = 2(x - 3)^2 + 1$.

Solution

$$y = 2(x - 3)^2 + 1$$

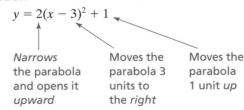

Narrows the parabola and opens it *upward*

Moves the parabola 3 units to the *right*

Moves the parabola 1 unit *up*

The parabola is drawn in Figure 12.39. In addition to the vertex, two points are located to determine the parabola.

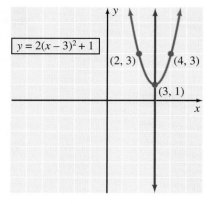

$y = 2(x - 3)^2 + 1$

(2, 3) (4, 3)

(3, 1)

Figure 12.39

EXAMPLE 10   Graph $y = -\dfrac{1}{2}(x + 1)^2 - 2$.

Solution

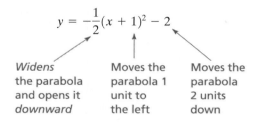

$$y = -\frac{1}{2}(x + 1)^2 - 2$$

| *Widens* the parabola and opens it *downward* | Moves the parabola 1 unit to the left | Moves the parabola 2 units down |

The parabola is drawn in Figure 12.40.

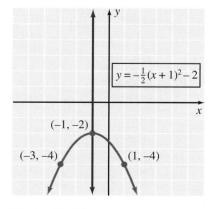

$y = -\frac{1}{2}(x + 1)^2 - 2$

(−1, −2)

(−3, −4)   (1, −4)

Figure 12.40

Finally, we can use a graphing utility to demonstrate some of the ideas of this section. Let's graph $y = x^2$, $y = -3(x - 7)^2 - 1$, $y = 2(x + 9)^2 + 5$, and $y = -0.2(x + 8)^2 - 3.5$ on the same set of axes, as shown in Figure 12.41. Certainly, Figure 12.41 is consistent with the ideas we presented in this section.

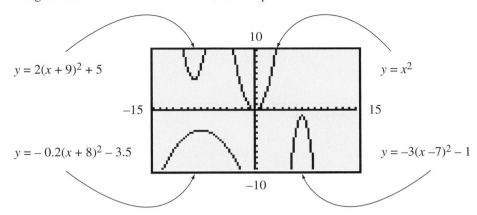

$y = 2(x + 9)^2 + 5$

$y = x^2$

$y = -0.2(x + 8)^2 - 3.5$

$y = -3(x - 7)^2 - 1$

Figure 12.41

CONCEPT QUIZ

For Problems 1–6, answer true or false.

1. The graph of $y = (x - 3)^2$ is the same as the graph of $y = x^2$ but moved 3 units to the right.

2. The graph of $y = x^2 - 4$ is the same as the graph of $y = x^2$ but moved 4 units to the right.

3. The graph of $y = x^2 + 1$ is the same as the graph of $y = x^2$ but moved 1 unit up.

4. The graph of $y = -x^2$ is the same as the graph of $y = x^2$ but is reflected across the y axis.

5. The vertex of the parabola given by the equation $y = (x + 2)^2 - 5$ is located at $(-2, -5)$.

6. The graph of $y = \frac{1}{3}x^2$ is narrower than the graph of $y = x^2$.

PROBLEM SET 12.3

For Problems 1–30, graph each parabola.

1. $y = x^2 + 2$

2. $y = x^2 + 3$

3. $y = x^2 - 1$

4. $y = x^2 - 5$

5. $y = 4x^2$

6. $y = 3x^2$

7. $y = -3x^2$

8. $y = -4x^2$

9. $y = \frac{1}{3}x^2$

10. $y = \frac{1}{4}x^2$

11. $y = -\dfrac{1}{2}x^2$

12. $y = -\dfrac{2}{3}x^2$

13. $y = (x - 1)^2$

14. $y = (x - 3)^2$

15. $y = (x + 4)^2$

16. $y = (x + 2)^2$

17. $y = 3x^2 + 2$

18. $y = 2x^2 + 3$

19. $y = -2x^2 - 2$

20. $y = \dfrac{1}{2}x^2 - 2$

21. $y = (x - 1)^2 - 2$

22. $y = (x - 2)^2 + 3$

23. $y = (x + 2)^2 + 1$

24. $y = (x + 1)^2 - 4$

25. $y = 3(x - 2)^2 - 4$

26. $y = 2(x + 3)^2 - 1$

27. $y = -(x + 4)^2 + 1$

28. $y = -(x - 1)^2 + 1$

29. $y = -\dfrac{1}{2}(x + 1)^2 - 2$

30. $y = -3(x - 4)^2 - 2$

■ ■ ■ Thoughts into words

31. Write a few paragraphs that summarize the ideas we presented in this section for someone who was absent from class that day.

32. How would you convince someone that $y = (x + 3)^2$ is the basic parabola moved 3 units to the *left* but that $y = (x - 3)^2$ is the basic parabola moved 3 units to the *right*?

33. How does the graph of $-y = x^2$ compare to the graph of $y = x^2$? Explain your answer.

34. How does the graph of $y = 4x^2$ compare to the graph of $y = 2x^2$? Explain your answer.

 ## Graphing calculator activities

35. Use a graphing calculator to check your graphs for Problems 21–30.

36. a. Graph $y = x^2$, $y = 2x^2$, $y = 3x^2$, and $y = 4x^2$ on the same set of axes.

b. Graph $y = x^2$, $y = \dfrac{3}{4}x^2$, $y = \dfrac{1}{2}x^2$, and $y = \dfrac{1}{5}x^2$ on the same set of axes.

c. Graph $y = x^2$, $y = -x^2$, $y = -3x^2$, and $y = -\dfrac{1}{4}x^2$ on the same set of axes.

37. a. Graph $y = x^2$, $y = (x - 2)^2$, $y = (x - 3)^2$, and $y = (x - 5)^2$ on the same set of axes.

b. Graph $y = x^2$, $y = (x + 1)^2$, $y = (x + 3)^2$, and $y = (x + 6)^2$ on the same set of axes.

38. a. Graph $y = x^2$, $y = (x - 2)^2 + 3$, $y = (x + 4)^2 - 2$, and $y = (x - 6)^2 - 4$ on the same set of axes.

b. Graph $y = x^2$, $y = 2(x + 1)^2 + 4$, $y = 3(x - 1)^2 - 3$, and $y = \dfrac{1}{2}(x - 5)^2 + 2$ on the same set of axes.

c. Graph $y = x^2$, $y = -(x - 4)^2 - 3$, $y = -2(x + 3)^2 - 1$, and $y = -\dfrac{1}{2}(x - 2)^2 + 6$ on the same set of axes.

39. a. Graph $y = x^2 - 12x + 41$ and $y = x^2 + 12x + 41$ on the same set of axes. What relationship seems to exist between the two graphs?

b. Graph $y = x^2 - 8x + 22$ and $y = -x^2 + 8x - 22$ on the same set of axes. What relationship seems to exist between the two graphs?

c. Graph $y = x^2 + 10x + 29$ and $y = -x^2 + 10x - 29$ on the same set of axes. What relationship seems to exist between the two graphs?

d. Summarize your findings for parts **a** through **c**.

Answers to Concept Quiz

1. True   **2.** False   **3.** True   **4.** False   **5.** True   **6.** False

12.4   **More Parabolas and Some Circles**

Objectives

■   Complete the square on equations of parabolas to obtain the form
   $y = a(x - h)^2 + k$.

■   Know the standard form of the equation of a circle, $(x - h)^2 + (y - k)^2 = r^2$.

■   Find the center and length of a radius given the equation of a circle.

■   Graph circles.

■   Write the equation of a circle given specific conditions.

We are now ready to graph quadratic equations of the form $y = ax^2 + bx + c$, where
a, b, and c are real numbers and $a \neq 0$. The general approach is one of changing
equations of the form $y = ax^2 + bx + c$ to the form $y = a(x - h)^2 + k$. Then we can
proceed to graph them as we did in the previous section. The process of *completing the square* is used to make the necessary change in the form of the equations.
Let's consider some examples to illustrate the details.

EXAMPLE 1

Graph $y = x^2 + 6x + 8$.

Solution

$$y = x^2 + 6x + 8$$

$$y = (x^2 + 6x + \underline{}) - (\underline{}) + 8 \qquad \text{Complete the square.}$$

$$y = (x^2 + 6x + 9) - (9) + 8 \qquad \frac{1}{2}(6) = 3 \text{ and } 3^2 = 9. \text{ Add 9 and also}$$

$$y = (x + 3)^2 - 1 \qquad\qquad \text{subtract 9 to compensate for the 9 that}$$
$$\text{was added.}$$

The graph of $y = (x + 3)^2 - 1$ is the basic parabola moved 3 units to the left and
1 unit down (Figure 12.42).

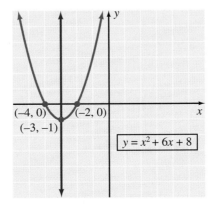

Figure 12.42

E X A M P L E   2   Graph $y = x^2 - 3x - 1$.

Solution

$$y = x^2 - 3x - 1$$

$$y = (x^2 - 3x + \underline{}) - (\underline{}) - 1 \qquad \text{Complete the square.}$$

$$y = \left(x^2 - 3x + \frac{9}{4} \right) - \frac{9}{4} - 1 \qquad \frac{1}{2}(-3) = -\frac{3}{2} \text{ and } \left(-\frac{3}{2} \right)^2 = \frac{9}{4}. \text{ Add and}$$

$$\text{subtract } \frac{9}{4}.$$

$$y = \left(x - \frac{3}{2} \right)^2 - \frac{13}{4}$$

The graph of $y = \left(x - \dfrac{3}{2} \right)^2 - \dfrac{13}{4}$ is the

basic parabola moved $1\dfrac{1}{2}$ units to the right

and $3\dfrac{1}{4}$ units down (Figure 12.43).

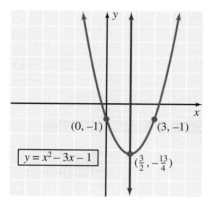

(0, −1)   (3, −1)

$y = x^2 - 3x - 1$ $(\tfrac{3}{2}, -\tfrac{13}{4})$

Figure 12.43   ■

If the coefficient of x^2 is not 1, then a slight adjustment has to be made before we apply the process of completing the square. The next two examples illustrate this situation.

E X A M P L E   3   Graph $y = 2x^2 + 8x + 9$.

Solution

$$y = 2x^2 + 8x + 9$$

$$y = 2(x^2 + 4x) + 9 \qquad \text{Factor a 2 from the } x\text{-variable terms.}$$

$$y = 2(x^2 + 4x + \underline{}) - (2)(\underline{}) + 9 \qquad \text{Complete the square. Note that the number being subtracted will be multiplied by a factor of 2.}$$

$$y = 2(x^2 + 4x + 4) - 2(4) + 9 \qquad \frac{1}{2}(4) = 2 \text{ and } 2^2 = 4$$

$$y = 2(x^2 + 4x + 4) - 8 + 9$$

$$y = 2(x + 2)^2 + 1$$

See Figure 12.44 for the graph of
$y = 2(x + 2)^2 + 1$.

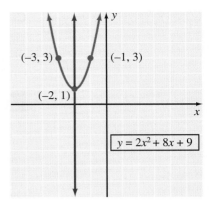

Figure 12.44

■

EXAMPLE 4

Graph $y = -3x^2 + 6x - 5$.

Solution

$$y = -3x^2 + 6x - 5$$

$$y = -3(x^2 - 2x) - 5 \qquad \text{Factor } -3 \text{ from the } x\text{-variable terms.}$$

$$y = -3(x^2 - 2x + \underline{}) - (-3)(\underline{}) - 5 \qquad \text{Complete the square. Note that the number being subtracted will be multiplied by a factor of } -3.$$

$$y = -3(x^2 - 2x + 1) - (-3)(1) - 5 \qquad \tfrac{1}{2}(-2) = -1 \text{ and } (-1)^2 = 1$$

$$y = -3(x^2 - 2x + 1) + 3 - 5$$

$$y = -3(x - 1)^2 - 2$$

The graph of $y = -3(x - 1)^2 - 2$
is shown in Figure 12.45.

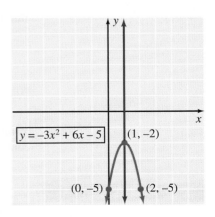

Figure 12.45

■

Circles

The distance formula, $d = \sqrt{(x_2 - x_1)^2 + (y_2 - y_1)^2}$ (developed in Section 12.1), when it applies to the definition of a circle, produces what is known as the **standard equation of a circle.** We start with a precise definition of a circle.

DEFINITION 12.2

A **circle** is the set of all points in a plane equidistant from a given fixed point called the **center.** A line segment determined by the center and any point on the circle is called a **radius.**

Let's consider a circle that has a radius of length r and a center at (h, k) on a coordinate system (Figure 12.46).

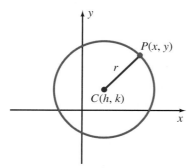

Figure 12.46

For any point P on the circle with coordinates (x, y), the length of a radius (denoted by r) can be expressed as

$$r = \sqrt{(x - h)^2 + (y - k)^2}$$

Thus, squaring both sides of the equation, we obtain the **standard form of the equation of a circle:**

$$(x - h)^2 + (y - k)^2 = r^2$$

We can use the standard form of the equation of a circle to solve two basic kinds of circle problems: (1) Given the coordinates of the center and the length of a radius of a circle, find its equation; and (2) given the equation of a circle, find its center and the length of a radius. Let's look at some examples of such problems.

EXAMPLE 5   Write the equation of a circle that has its center at $(3, -5)$ and a radius of length 6 units.

Solution

Substitute 3 for h, -5 for k, and 6 for r into the standard form $(x - h)^2 + (y - k)^2 = r^2$. It becomes $(x - 3)^2 + (y + 5)^2 = 6^2$, which we can simplify as follows:

$$(x - 3)^2 + (y + 5)^2 = 6^2$$
$$x^2 - 6x + 9 + y^2 + 10y + 25 = 36$$
$$x^2 + y^2 - 6x + 10y - 2 = 0$$

Note in Example 5 that we simplified the equation to the form $x^2 + y^2 + Dx + Ey + F = 0$, where D, E, and F are integers. This is another form that we commonly use when working with circles.

EXAMPLE 6

Graph $x^2 + y^2 + 4x - 6y + 9 = 0$.

Solution

This equation is of the form $x^2 + y^2 + Dx + Ey + F = 0$, so its graph is a circle. We can change the given equation into the form $(x - h)^2 + (y - k)^2 = r^2$ by completing the square on x and on y as follows:

$$x^2 + y^2 + 4x - 6y + 9 = 0$$
$$(x^2 + 4x + \underline{\quad}) + (y^2 - 6y + \underline{\quad}) = -9$$
$$(x^2 + 4x + 4) + (y^2 - 6y + 9) = -9 + 4 + 9$$

| Added 4 to complete the square on x | Added 9 to complete the square on y | Added 4 and 9 to compensate for the 4 and 9 added on the left side |

$$(x + 2)^2 + (y - 3)^2 = 4$$
$$(x - (-2))^2 + (y - 3)^2 = 2^2$$

$$\underset{h}{\uparrow} \qquad \underset{k}{\uparrow} \quad \underset{r}{\uparrow}$$

The center of the circle is at $(-2, 3)$ and the length of a radius is 2 (Figure 12.47).

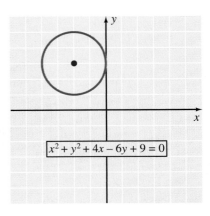

$$x^2 + y^2 + 4x - 6y + 9 = 0$$

Figure 12.47

As demonstrated by Examples 5 and 6, both forms, $(x - h)^2 + (y - k)^2 = r^2$ and $x^2 + y^2 + Dx + Ey + F = 0$, play an important role when we are solving problems that deal with circles.

Finally, we need to recognize that the standard form of a circle that has its center at the origin is $x^2 + y^2 = r^2$. This is merely the result of letting $h = 0$ and $k = 0$ in the general standard form.

$$(x - h)^2 + (y - k)^2 = r^2$$
$$(x - 0)^2 + (y - 0)^2 = r^2$$
$$x^2 + y^2 = r^2$$

Thus by inspection we can recognize that $x^2 + y^2 = 9$ is a circle with its center at the origin; the length of a radius is 3 units. Likewise, the equation of a circle that has its center at the origin and a radius of length 6 units is $x^2 + y^2 = 36$.

When using a graphing utility to graph a circle, we need to solve the equation for y in terms of x. This will produce two equations that can be graphed on the same set of axes. Furthermore, as with any graph, it may be necessary to change the boundaries on x or y (or both) to obtain a complete graph. Let's consider an example.

EXAMPLE 7

Use a graphing utility to graph $x^2 - 40x + y^2 + 351 = 0$.

Solution

First, we need to solve for y in terms of x.

$$x^2 - 40x + y^2 + 351 = 0$$
$$y^2 = -x^2 + 40x - 351$$
$$y = \pm\sqrt{-x^2 + 40x - 351}$$

Now we can make the following assignments.

$$Y_1 = \sqrt{-x^2 + 40x - 351}$$
$$Y_2 = -Y_1$$

(Note that we assigned Y_2 in terms of Y_1. By doing this we avoid repetitive key strokes and may thus reduce the chance for errors. You may need to consult your user's manual for instructions on how to keystroke $-Y_1$.) Figure 12.48 shows the graph. Because we know from the original equation that this graph should be a circle, we need to

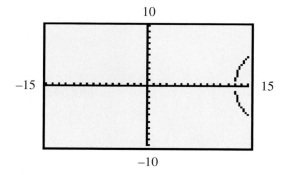

Figure 12.48

make some adjustments on the boundaries in order to get a complete graph. This can be done by completing the square on the original equation to change its form to $(x - 20)^2 + y^2 = 49$ or simply by a trial-and-error process. By changing the boundaries on x such that $-15 \leq x \leq 30$, we obtain Figure 12.49.

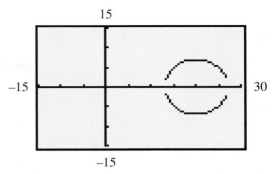

Figure 12.49

CONCEPT QUIZ

For Problems 1–6, answer true or false.

1. Equations of the form $y = ax^2 + bx + c$ can be changed to the form $y = a(x - h)^2 + k$ by completing the square.

2. A circle is the set of points in a plane that are equidistant from a given fixed point.

3. A line segment determined by the center and any point on the circle is called the diameter.

4. The circle $(x + 2)^2 + (y - 5)^2 = 20$ has its center at $(2, -5)$.

5. The circle $(x - 4)^2 + (y + 3)^2 = 10$ has a radius of length 10.

6. The circle $x^2 + y^2 = 16$ has its center at the origin.

PROBLEM SET 12.4

For Problems 1–22, graph each parabola.

1. $y = x^2 - 6x + 13$

2. $y = x^2 - 4x + 7$

3. $y = x^2 + 2x + 6$

4. $y = x^2 + 8x + 14$

5. $y = x^2 - 5x + 3$

6. $y = x^2 + 3x + 1$

7. $y = x^2 + 7x + 14$

8. $y = x^2 - x - 1$

9. $y = 3x^2 - 6x + 5$

10. $y = 2x^2 + 4x + 7$

11. $y = 4x^2 - 24x + 32$

12. $y = 3x^2 + 24x + 49$

13. $y = -2x^2 - 4x - 5$

14. $y = -2x^2 + 8x - 5$

15. $y = -x^2 + 8x - 21$

16. $y = -x^2 - 6x - 7$

17. $y = 2x^2 - x + 2$

18. $y = 2x^2 + 3x + 1$

19. $y = 3x^2 + 2x + 1$

20. $y = 3x^2 - x - 1$

21. $y = -3x^2 - 7x - 2$

22. $y = -2x^2 + x - 2$

For Problems 23–34, find the center and the length of a radius of each circle.

23. $x^2 + y^2 - 2x - 6y - 6 = 0$

24. $x^2 + y^2 + 4x - 12y + 39 = 0$

25. $x^2 + y^2 + 6x + 10y + 18 = 0$

26. $x^2 + y^2 - 10x + 2y + 1 = 0$

27. $x^2 + y^2 = 10$

28. $x^2 + y^2 + 4x + 14y + 50 = 0$

29. $x^2 + y^2 - 16x + 6y + 71 = 0$

30. $x^2 + y^2 = 12$

31. $x^2 + y^2 + 6x - 8y = 0$

32. $x^2 + y^2 - 16x + 30y = 0$

33. $4x^2 + 4y^2 + 4x - 32y + 33 = 0$

34. $9x^2 + 9y^2 - 6x - 12y - 40 = 0$

For Problems 35–44, graph each circle.

35. $x^2 + y^2 = 25$

36. $x^2 + y^2 = 36$

37. $(x - 1)^2 + (y + 2)^2 = 9$

38. $(x + 3)^2 + (y - 2)^2 = 1$

39. $x^2 + y^2 + 6x - 2y + 6 = 0$

40. $x^2 + y^2 - 4x - 6y - 12 = 0$

41. $x^2 + y^2 + 4y - 5 = 0$

42. $x^2 + y^2 - 4x + 3 = 0$

43. $x^2 + y^2 + 4x + 4y - 8 = 0$

44. $x^2 + y^2 - 6x + 6y + 2 = 0$

For Problems 45–54, write the equation of each circle. Express the final equation in the form $x^2 + y^2 + Dx + Ey + F = 0$.

45. Center at $(3, 5)$ and $r = 5$

46. Center at $(2, 6)$ and $r = 7$

47. Center at $(-4, 1)$ and $r = 8$

48. Center at $(-3, 7)$ and $r = 6$

49. Center at $(-2, -6)$ and $r = 3\sqrt{2}$

50. Center at $(-4, -5)$ and $r = 2\sqrt{3}$

51. Center at $(0, 0)$ and $r = 2\sqrt{5}$

52. Center at $(0, 0)$ and $r = \sqrt{7}$

53. Center at $(5, -8)$ and $r = 4\sqrt{6}$

54. Center at $(4, -10)$ and $r = 8\sqrt{2}$

55. Find the equation of the circle that passes through the origin and has its center at $(0, 4)$.

56. Find the equation of the circle that passes through the origin and has its center at $(-6, 0)$.

57. Find the equation of the circle that passes through the origin and has its center at $(-4, 3)$.

58. Find the equation of the circle that passes through the origin and has its center at $(8, -15)$.

■ ■ ■ **Thoughts into words**

59. What is the graph of $x^2 + y^2 = -4$? Explain your answer.

60. On which axis is the center of the circle $x^2 + y^2 - 8y + 7 = 0$? Defend your answer.

61. Give a step-by-step description of how you would help someone graph the parabola $y = 2x^2 - 12x + 9$.

■ ■ ■ **Further investigations**

62. The points (x, y) and (y, x) are mirror images of each other across the line $y = x$. Therefore, by interchanging x and y in the equation $y = ax^2 + bx + c$, we obtain the equation of its mirror image across the line $y = x$; namely, $x = ay^2 + by + c$. Thus to graph $x = y^2 + 2$, we can first graph $y = x^2 + 2$ and then reflect it across the line $y = x$, as indicated in Figure 12.50.

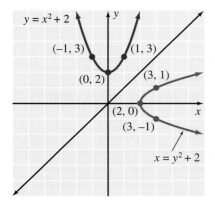

Figure 12.50

Graph each of the following parabolas.

a. $x = y^2$ **b.** $x = -y^2$

c. $x = y^2 - 1$ **d.** $x = -y^2 + 3$

e. $x = -2y^2$ **f.** $x = 3y^2$

g. $x = y^2 + 4y + 7$ **h.** $x = y^2 - 2y - 3$

63. By expanding $(x - h)^2 + (y - k)^2 = r^2$, we obtain $x^2 - 2hx + h^2 + y^2 - 2ky + k^2 - r^2 = 0$. When we compare this result to the form $x^2 + y^2 + Dx + Ey + F = 0$, we see that $D = -2h$, $E = -2k$, and $F = h^2 + k^2 - r^2$. Therefore, the center and length of a radius of a circle can be found by using $h = \dfrac{D}{-2}$, $k = \dfrac{E}{-2}$, and $r = \sqrt{h^2 + k^2 - F}$. Use these relationships to find the center and the length of a radius of each of the following circles.

a. $x^2 + y^2 - 2x - 8y + 8 = 0$

b. $x^2 + y^2 + 4x - 14y + 49 = 0$

c. $x^2 + y^2 + 12x + 8y - 12 = 0$

d. $x^2 + y^2 - 16x + 20y + 115 = 0$

e. $x^2 + y^2 - 12y - 45 = 0$

f. $x^2 + y^2 + 14x = 0$

⌗ **Graphing calculator activities**

64. Use a graphing calculator to check your graphs for Problems 1–22.

65. Use a graphing calculator to graph the circles in Problems 23–26. Be sure that your graphs are consistent with the center and the length of a radius that you found when you did the problems.

66. Graph each of the following parabolas and circles. Be sure to set your boundaries so that you get a complete graph.

a. $x^2 + 24x + y^2 + 135 = 0$

b. $y = x^2 - 4x + 18$

c. $x^2 + y^2 - 18y + 56 = 0$

d. $x^2 + y^2 + 24x + 28y + 336 = 0$

e. $y = -3x^2 - 24x - 58$

f. $y = x^2 - 10x + 3$

Answers to Concept Quiz

1. True   **2.** True   **3.** False   **4.** False   **5.** False   **6.** True

12.5 Graphing Ellipses

Objectives

■ Graph an ellipse with its center at the origin.

■ Find the center and endpoints of the axes for an ellipse with its center not at the origin, and graph the ellipse.

In the previous section, we found that the graph of the equation $x^2 + y^2 = 36$ is a circle of radius 6 units with its center at the origin. More generally, it is true that any equation of the form $Ax^2 + By^2 = C$, where $A = B$ and A, B, and C are nonzero constants that have the same sign, is a circle with its center at the origin. For example, $3x^2 + 3y^2 = 12$ is equivalent to $x^2 + y^2 = 4$ (divide both sides of the given equation by 3), and thus it is a circle of radius 2 units with its center at the origin.

The general equation $Ax^2 + By^2 = C$ can be used to describe other geometric figures by changing the restrictions on A and B. For example, if A, B, and C are of the same sign, but $A \neq B$, then the graph of the equation $Ax^2 + By^2 = C$ is an **ellipse.** Let's consider two examples.

EXAMPLE 1

Graph $4x^2 + 25y^2 = 100$.

Solution

First find the x and y intercepts. Let $x = 0$; then

$$4(0)^2 + 25y^2 = 100$$
$$25y^2 = 100$$
$$y^2 = 4$$
$$y = \pm 2$$

Thus the points $(0, 2)$ and $(0, -2)$ are on the graph. Let $y = 0$; then

$$4x^2 + 25(0)^2 = 100$$
$$4x^2 = 100$$
$$x^2 = 25$$
$$x = \pm 5$$

Thus the points $(5, 0)$ and $(-5, 0)$ are also on the graph. We know that this figure is an ellipse, so we plot the four points and we get a pretty good sketch of the figure (Figure 12.51).

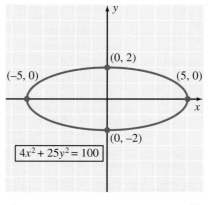

Figure 12.51

In Figure 12.51, the line segment with endpoints at $(-5, 0)$ and $(5, 0)$ is called the **major axis** of the ellipse. The shorter line segment with endpoints at $(0, -2)$ and $(0, 2)$ is called the **minor axis.** Establishing the endpoints of the major and minor axes provides a basis for sketching an ellipse. The point of intersection of the major and minor axes is called the **center** of the ellipse.

EXAMPLE 2

Graph $9x^2 + 4y^2 = 36$.

Solution

Again, we first find the x and y intercepts. Let $x = 0$; then

$$9(0)^2 + 4y^2 = 36$$
$$4y^2 = 36$$
$$y^2 = 9$$
$$y = \pm 3$$

Thus the points $(0, 3)$ and $(0, -3)$ are on the graph. Let $y = 0$; then

$$9x^2 + 4(0)^2 = 36$$
$$9x^2 = 36$$
$$x^2 = 4$$
$$x = \pm 2$$

Thus the points $(2, 0)$ and $(-2, 0)$ are also on the graph. The ellipse is sketched in Figure 12.52.

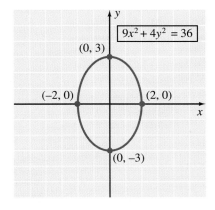

Figure 12.52

In Figure 12.52, the major axis has endpoints at $(0, -3)$ and $(0, 3)$, and the minor axis has endpoints at $(-2, 0)$ and $(2, 0)$. The ellipses in Figures 12.51 and 12.52 are symmetric about the x axis and about the y axis. In other words, both the x axis and the y axis serve as **axes of symmetry.**

Now let's consider some ellipses whose centers are not at the origin but whose major and minor axes are parallel to the x axis and the y axis. We can graph such ellipses in much the same way as we handled circles in Section 12.4. Let's consider two examples to illustrate the procedure.

EXAMPLE 3   Graph $4x^2 + 24x + 9y^2 - 36y + 36 = 0$.

Solution

Let's complete the square on x and y as follows:

$$4x^2 + 24x + 9y^2 - 36y + 36 = 0$$
$$4(x^2 + 6x + \underline{}) + 9(y^2 - 4y + \underline{}) = -36$$
$$4(x^2 + 6x + 9) + 9(y^2 - 4y + 4) = -36 + 36 + 36$$
$$4(x + 3)^2 + 9(y - 2)^2 = 36$$
$$4(x - (-3))^2 + 9(y - 2)^2 = 36$$

Because 4, 9, and 36 are of the same sign and $4 \neq 9$, the graph is an ellipse. The center of the ellipse is at $(-3, 2)$. We can find the endpoints of the major and minor axes as follows: Use the equation $4(x + 3)^2 + 9(y - 2)^2 = 36$ and let $y = 2$.

$$4(x + 3)^2 + 9(2 - 2)^2 = 36$$
$$4(x + 3)^2 = 36$$
$$(x + 3)^2 = 9$$
$$x + 3 = \pm 3$$

$$x + 3 = 3 \quad \text{or} \quad x + 3 = -3$$
$$x = 0 \quad \text{or} \quad x = -6$$

The endpoints of the major axis are at $(0, 2)$ and $(-6, 2)$. Now let $x = -3$.

$$4(-3 + 3)^2 + 9(y - 2)^2 = 36$$
$$9(y - 2)^2 = 36$$
$$(y - 2)^2 = 4$$
$$y - 2 = \pm 2$$

$$y - 2 = 2 \quad \text{or} \quad y - 2 = -2$$
$$y = 4 \quad \text{or} \quad y = 0$$

The endpoints of the minor axis are at $(-3, 4)$ and $(-3, 0)$. The ellipse is shown in Figure 12.53.

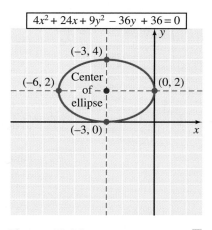

Figure 12.53

E X A M P L E   4

Graph $4x^2 - 16x + y^2 + 6y + 9 = 0$.

Solution

First, complete the square on x and on y.

$$4x^2 - 16x + y^2 + 6y + 9 = 0$$
$$4(x^2 - 4x + \underline{}) + (y^2 + 6y + \underline{}) = -9$$
$$4(x^2 - 4x + 4) + (y^2 + 6y + 9) = -9 + 16 + 9$$
$$4(x - 2)^2 + (y + 3)^2 = 16$$

The center of the ellipse is at $(2, -3)$. Now let $x = 2$.

$$4(2 - 2)^2 + (y + 3)^2 = 16$$
$$(y + 3)^2 = 16$$
$$y + 3 = \pm 4$$

$$y + 3 = -4 \quad \text{or} \quad y + 3 = 4$$
$$y = -7 \quad \text{or} \quad y = 1$$

The endpoints of the major axis are at $(2, -7)$ and $(2, 1)$. Now let $y = -3$.

$$4(x - 2)^2 + (-3 + 3)^2 = 16$$
$$4(x - 2)^2 = 16$$
$$(x - 2)^2 = 4$$
$$x - 2 = \pm 2$$

$$x - 2 = -2 \quad \text{or} \quad x - 2 = 2$$
$$x = 0 \quad \text{or} \quad x = 4$$

The endpoints of the minor axis are at $(0, -3)$ and $(4, -3)$. The ellipse is shown in Figure 12.54.

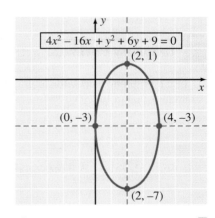

$4x^2 - 16x + y^2 + 6y + 9 = 0$

$(2, 1)$

$(0, -3)$ $(4, -3)$

$(2, -7)$

Figure 12.54

C O N C E P T   Q U I Z

For Problems 1–5, answer true or false.

1. The length of the major axis of an ellipse is always greater than the length of the minor axis.

2. The major axis of an ellipse is always parallel to the x axis.

3. The axes of symmetry for an ellipse pass through the center of the ellipse.

4. The ellipse $9(x - 1)^2 + 4(y + 5)^2 = 36$ has its center at $(1, 5)$.

5. The x and y intercepts of the graph of an ellipse centered at the origin and symmetric to both axes are the endpoints of its axes.

PROBLEM SET 12.5

For Problems 1–16, graph each ellipse.

1. $x^2 + 4y^2 = 36$

2. $x^2 + 4y^2 = 16$

3. $9x^2 + y^2 = 36$

4. $16x^2 + 9y^2 = 144$

5. $4x^2 + 3y^2 = 12$

6. $5x^2 + 4y^2 = 20$

7. $16x^2 + y^2 = 16$

8. $9x^2 + 2y^2 = 18$

9. $25x^2 + 2y^2 = 50$

10. $12x^2 + y^2 = 36$

11. $4x^2 + 8x + 16y^2 - 64y + 4 = 0$

12. $9x^2 - 36x + 4y^2 - 24y + 36 = 0$

13. $x^2 + 8x + 9y^2 + 36y + 16 = 0$

14. $4x^2 - 24x + y^2 + 4y + 24 = 0$

15. $4x^2 + 9y^2 - 54y + 45 = 0$

16. $x^2 + 2x + 4y^2 - 15 = 0$

■ ■ ■ Thoughts into words

17. Is the graph of $x^2 + y^2 = 4$ the same as the graph of $y^2 + x^2 = 4$? Explain your answer.

18. Is the graph of $x^2 + y^2 = 0$ a circle? If so, what is the length of a radius?

19. Is the graph of $4x^2 + 9y^2 = 36$ the same as the graph of $9x^2 + 4y^2 = 36$? Explain your answer.

20. What is the graph of $x^2 + 2y^2 = -16$? Explain your answer.

 Graphing calculator activities

21. Use a graphing calculator to graph the ellipses in Examples 1–4 of this section.

22. Use a graphing calculator to check your graphs for Problems 11–16.

Answers to Concept Quiz

1. True   **2.** False   **3.** True   **4.** False   **5.** True

---

12.6   Graphing Hyperbolas

Objectives

■   Graph a hyperbola with its center at the origin.

■   Find the center and asymptotes for a hyperbola with its center not at the origin, and graph the hyperbola.

The graph of an equation of the form $Ax^2 + By^2 = C$, where A, B, and C are nonzero real numbers, and A and B are of unlike signs, is a **hyperbola.** Let's use some examples to illustrate a procedure for graphing hyperbolas.

E X A M P L E   1

Graph $x^2 - y^2 = 9$.

Solution

If we let $y = 0$, we obtain

$$x^2 - 0^2 = 0$$
$$x^2 = 9$$
$$x = \pm 3$$

Thus the points $(3, 0)$ and $(-3, 0)$ are on the graph. If we let $x = 0$, we obtain

$$0^2 - y^2 = 9$$
$$-y^2 = 9$$
$$y^2 = -9$$

Because $y^2 = -9$ has no real number solutions, there are no points of the y axis on this graph. That is, the graph does not intersect the y axis. Now let's solve the given equation for y so that we have a more convenient form for finding other solutions.

$$x^2 - y^2 = 9$$
$$-y^2 = 9 - x^2$$
$$y^2 = x^2 - 9$$
$$y = \pm\sqrt{x^2 - 9}$$

The radicand, $x^2 - 9$, must be nonnegative, so the values we choose for x must be greater than or equal to 3, or less than or equal to -3. With this in mind, we can form the following table of values.

x	y	
3	0	Intercepts
−3	0	
4	$\pm\sqrt{7}$	
−4	$\pm\sqrt{7}$	
5	± 4	Other points
−5	± 4	

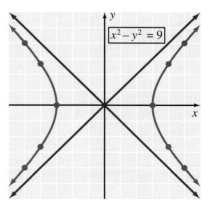

Figure 12.55

We plot these points and draw the hyperbola as in Figure 12.55. (This graph is symmetric about both axes, and the center of the hyperbola is at the origin.)

Note the blue lines in Figure 12.55; they are called **asymptotes.** Each branch of the hyperbola approaches one of these lines but does not intersect it. Therefore, the ability to sketch the asymptotes of a hyperbola is very helpful when we are graphing the hyperbola. Fortunately, the equations of the asymptotes are easy to determine. They can be found by replacing the constant term in the given equation of the hyperbola with 0 and solving for y. (The reason why this works will become evident in a later course.) Thus for the hyperbola in Example 1, we obtain

$$x^2 - y^2 = 0$$
$$y^2 = x^2$$
$$y = \pm x$$

Thus the two lines $y = x$ and $y = -x$ are the asymptotes indicated by the blue lines in Figure 12.55.

E X A M P L E   2   Graph $y^2 - 5x^2 = 4$.

Solution

If we let $x = 0$, we obtain

$$y^2 - 5(0)^2 = 4$$
$$y^2 = 4$$
$$y = \pm 2$$

The points $(0, 2)$ and $(0, -2)$ are on the graph. If we let $y = 0$, we obtain

$$0^2 - 5x^2 = 4$$
$$-5x^2 = 4$$
$$x^2 = -\frac{4}{5}$$

Because $x^2 = -\dfrac{4}{5}$ has no real number solutions, we know that this hyperbola does not intersect the x axis. Solving the given equation for y yields

$$y^2 - 5x^2 = 4$$
$$y^2 = 5x^2 + 4$$
$$y = \sqrt{5x^2 + 4}$$

The following table shows some additional solutions for the equation.

x	y	
0	2	Intercepts
0	-2	
1	± 3	Other points
-1	± 3	
2	$\pm\sqrt{24}$	
-2	$\pm\sqrt{24}$	

The equations of the asymptotes are determined as follows:

$$y^2 - 5x^2 = 0$$
$$y^2 = 5x^2$$
$$y = \pm\sqrt{5}x$$

Sketch the asymptotes and plot the points listed in the table of values to determine the hyperbola in Figure 12.56. (Note that this hyperbola is also symmetric about both axes.)

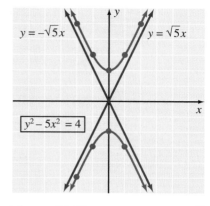

Figure 12.56

EXAMPLE 3

Graph $4x^2 - 9y^2 = 36$.

Solution

If we let $x = 0$, we obtain

$$4(0)^2 - 9y^2 = 36$$
$$-9y^2 = 36$$
$$y^2 = -4$$

Because $y^2 = -4$ has no real number solutions, we know that this hyperbola does not intersect the y axis. If we let $y = 0$, we obtain

$$4x^2 - 9(0)^2 = 36$$
$$4x^2 = 36$$
$$x^2 = 9$$
$$x = \pm 3$$

Thus the points $(3, 0)$ and $(-3, 0)$ are on the graph. Now let's solve the equation for y in terms of x and set up a table of values.

$$4x^2 - 9y^2 = 36$$
$$-9y^2 = 36 - 4x^2$$
$$9y^2 = 4x^2 - 36$$
$$y^2 = \frac{4x^2 - 36}{9}$$
$$y = \pm\frac{\sqrt{4x^2 - 36}}{3}$$

x	y	
3	0	Intercepts
−3	0	
4	$\pm\dfrac{2\sqrt{7}}{3}$	
−4	$\pm\dfrac{2\sqrt{7}}{3}$	Other points
5	$\pm\dfrac{8}{3}$	
−5	$\pm\dfrac{8}{3}$	

The equations of the asymptotes are found as follows:

$$4x^2 - 9y^2 = 0$$
$$-9y^2 = -4x^2$$
$$9y^2 = 4x^2$$
$$y^2 = \frac{4x^2}{9}$$
$$y = \pm\frac{2}{3}x$$

Sketch the asymptotes and plot the points listed in the table to determine the hyperbola as shown in Figure 12.57.

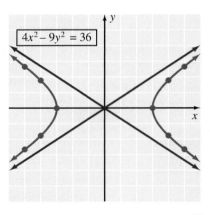

$$4x^2 - 9y^2 = 36$$

Figure 12.57

Now let's consider hyperbolas that are not symmetric with respect to the origin but are symmetric with respect to lines parallel to one of the axes — that is, vertical and horizontal lines. Again, let's use examples to illustrate a procedure for graphing such hyperbolas.

E X A M P L E   4

Graph $4x^2 - 8x - y^2 - 4y - 16 = 0$.

Solution

Completing the square on x and y, we obtain

$$4x^2 - 8x - y^2 - 4y - 16 = 0$$
$$4(x^2 - 2x + \underline{}) - (y^2 + 4y + \underline{}) = 16$$
$$4(x^2 - 2x + 1) - (y^2 + 4y + 4) = 16 + 4 - 4$$
$$4(x - 1)^2 - (y + 2)^2 = 16$$
$$4(x - 1)^2 - 1(y - (-2))^2 = 16$$

Because 4 and -1 are of opposite signs, the graph is a hyperbola. The center of the hyperbola is at $(1, -2)$.

Now, using the equation $4(x - 1)^2 - (y + 2)^2 = 16$, we can proceed as follows:
Let $y = -2$; then

$$4(x - 1)^2 - (-2 + 2)^2 = 16$$
$$4(x - 1)^2 = 16$$
$$(x - 1)^2 = 4$$
$$x - 1 = \pm 2$$

$$x - 1 = 2 \qquad \text{or} \qquad x - 1 = -2$$
$$x = 3 \qquad \text{or} \qquad x = -1$$

Thus the hyperbola intersects the horizontal line $y = -2$ at $(3, -2)$ and at $(-1, -2)$.
Let $x = 1$; then

$$4(1 - 1)^2 - (y + 2)^2 = 16$$
$$-(y + 2)^2 = 16$$
$$(y + 2)^2 = -16$$

Because $(y + 2)^2 = -16$ has no real number solutions, we know that the hyperbola does not intersect the vertical line $x = 1$. Replacing the constant term of $4(x - 1)^2 - (y + 2)^2 = 16$ with 0 and solving for y, we obtain the equations of the asymptotes as follows:

$$4(x - 1)^2 - (y + 2)^2 = 0$$

The left side can be factored using the pattern of the difference of squares.

$$(2(x - 1) + (y + 2))(2(x - 1) - (y + 2)) = 0$$
$$(2x - 2 + y + 2)(2x - 2 - y - 2) = 0$$
$$(2x + y)(2x - y - 4) = 0$$
$$2x + y = 0 \qquad \text{or} \qquad 2x - y - 4 = 0$$
$$y = -2x \qquad \text{or} \qquad 2x - 4 = y$$

Thus the equations of the asymptotes are $y = -2x$ and $y = 2x - 4$. Sketching the asymptotes and plotting the two points $(3, -2)$ and $(-1, -2)$, we can draw the hyperbola as shown in Figure 12.58.

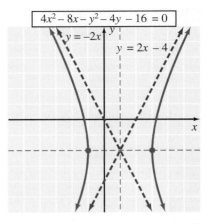

Figure 12.58

EXAMPLE 5

Graph $y^2 - 4y - 4x^2 - 24x - 36 = 0$.

Solution

First, complete the square on x and on y.

$$y^2 - 4y - 4x^2 - 24x - 36 = 0$$
$$(y^2 - 4y + \_\_) - 4(x^2 + 6x + \_\_) = 36$$
$$(y^2 - 4y + 4) - 4(x^2 + 6x + 9) = 36 + 4 - 36$$
$$(y - 2)^2 - 4(x + 3)^2 = 4$$

The center of the hyperbola is at $(-3, 2)$. Now let $y = 2$.

$$(2 - 2)^2 - 4(x + 3)^2 = 4$$
$$-4(x + 3)^2 = 4$$
$$(x + 3)^2 = -1$$

Because $(x + 3)^2 = -1$ has no real number solutions, the graph does not intersect the line $y = 2$. Now let $x = -3$.

$$(y - 2)^2 - 4(-3 + 3)^2 = 4$$
$$(y - 2)^2 = 4$$
$$y - 2 = \pm 2$$
$$y - 2 = -2 \quad \text{or} \quad y - 2 = 2$$
$$y = 0 \quad \text{or} \quad y = 4$$

Therefore, the hyperbola intersects the line $x = -3$ at $(-3, 0)$ and $(-3, 4)$. Now to find the equations of the asymptotes, let's replace the constant term of $(y - 2)^2 - 4(x + 3)^2 = 4$ with 0 and solve for y.

$$(y - 2)^2 - 4(x + 3)^2 = 0$$
$$[(y - 2) + 2\,(x + 3)][(y - 2) - 2(x + 3)] = 0$$
$$(y - 2 + 2x + 6)(y - 2 - 2x - 6) = 0$$
$$(y + 2x + 4)(y - 2x - 8) = 0$$
$$y + 2x + 4 = 0 \quad \text{or} \quad y - 2x - 8 = 0$$
$$y = -2x - 4 \quad \text{or} \quad y = 2x + 8$$

Therefore, the equations of the asymptotes are $y = -2x - 4$ and $y = 2x + 8$. Drawing the asymptotes and plotting the points $(-3, 0)$ and $(-3, 4)$, we can graph the hyperbola as shown in Figure 12.59.

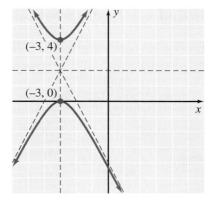

Figure 12.59

As a way of summarizing our work with conic sections, let's focus our attention on the continuity pattern used in this chapter. In Sections 12.4 and 12.5, we studied parabolas by considering variations of the basic quadratic equation $y = ax^2 + bx + c$. Also in Section 12.5, we used the definition of a circle to generate a standard form for the equation of a circle. Then, in Sections 12.6 and 12.7, we discussed ellipses and hyperbolas, not from a definition viewpoint, but by considering variations of the equations $Ax^2 + By^2 = C$ and $A(x - h)^2 + B(y - k)^2 = C$. In a subsequent mathematics course, parabolas, ellipses, and hyperbolas will also be developed from a definition viewpoint. That is, first each conic section will be defined, and then the definition will be used to generate a standard form of its equation.

CONCEPT QUIZ

For Problems 1–6, answer true or false.

1. The graph of an equation of the form $Ax^2 + By^2 = C$, where A, B, and C are nonzero real numbers, is a hyperbola if A and B are of like sign.

2. The graph of a hyperbola always has two branches.

3. Each branch of the graph of a hyperbola approaches one of the asymptotes but never intersects with the asymptote.

4. To find the equations for the asymptotes, we replace the constant term in the equation of the hyperbola with zero and solve for y.

5. The hyperbola $9(x + 1)^2 - 4(y - 3)^2 = 36$ has its center at $(-1, 3)$.

6. The asymptotes of the graph of a hyperbola intersect at the center of the hyperbola.

PROBLEM SET 12.6

For Problems 1–18, graph each hyperbola.

1. $x^2 - y^2 = 1$

2. $x^2 - y^2 = 4$

3. $y^2 - 4x^2 = 9$

4. $4y^2 - x^2 = 16$

5. $5x^2 - 2y^2 = 20$

6. $9x^2 - 4y^2 = 9$

7. $y^2 - 16x^2 = 4$

8. $y^2 - 9x^2 = 16$

9. $-4x^2 + y^2 = -4$

10. $-9x^2 + y^2 = -36$

11. $25y^2 - 3x^2 = 75$

12. $16y^2 - 5x^2 = 80$

13. $-4x^2 + 32x + 9y^2 - 18y - 91 = 0$

14. $x^2 - 4x - y^2 + 6y - 14 = 0$

15. $-4x^2 + 24x + 16y^2 + 64y - 36 = 0$

16. $x^2 + 4x - 9y^2 + 54y - 113 = 0$

17. $4x^2 - 24x - 9y^2 = 0$

18. $16y^2 + 64y - x^2 = 0$

19. The graphs of equations of the form $xy = k$, where k is a nonzero constant, are also hyperbolas, sometimes referred to as rectangular hyperbolas. Graph each of the following.

a. $xy = 3$

b. $xy = 5$

c. $xy = -2$

d. $xy = -4$

20. What is the graph of $xy = 0$? Defend your answer.

21. We have graphed various equations of the form $Ax^2 + By^2 = C$, where C is a nonzero constant. Now graph each of the following.

a. $x^2 + y^2 = 0$

b. $2x^2 + 3y^2 = 0$

c. $x^2 - y^2 = 0$

d. $4y^2 - x^2 = 0$

■ ■ ■ Thoughts into words

22. Explain the concept of an asymptote.

23. Explain how asymptotes can be used to help graph hyperbolas.

24. Are the graphs of $x^2 - y^2 = 0$ and $y^2 - x^2 = 0$ identical? Are the graphs of $x^2 - y^2 = 4$ and $y^2 - x^2 = 4$ identical? Explain your answers.

 ### Graphing calculator activities

25. To graph the hyperbola in Example 1 of this section, we can make the following assignments for the graphing calculator.

$Y_1 = \sqrt{x^2 - 9}$ $Y_2 = -Y_1$

$Y_3 = x$ $Y_4 = -Y_3$

Do this and see whether your graph agrees with Figure 12.51. Also graph the asymptotes and hyperbolas for Examples 2 and 3.

26. Use a graphing calculator to check your graphs for Problems 1– 6.

27. Use a graphing calculator to check your graphs for Problems 13–18.

28. For each of the following equations, (1) predict the type and location of the graph, and (2) use your graphing calculator to check your predictions.

a. $x^2 + y^2 = 100$ **b.** $x^2 - y^2 = 100$ **h.** $x^2 + y^2 - 4y - 2 = 0$

c. $y^2 - x^2 = 100$ **d.** $y = -x^2 + 9$ **i.** $y = x^2 + 16$ **j.** $y^2 = x^2 + 16$

e. $2x^2 + y^2 = 14$ **f.** $x^2 + 2y^2 = 14$ **k.** $9x^2 - 4y^2 = 72$ **l.** $4x^2 - 9y^2 = 72$

g. $x^2 + 2x + y^2 - 4 = 0$ **m.** $y^2 = -x^2 - 4x + 6$

Answers to Concept Quiz

1. False   **2.** True   **3.** True   **4.** True   **5.** True   **6.** True

CHAPTER 12

SUMMARY

(12.1) The **Cartesian** (or **rectangular**) **coordinate system** is used to graph ordered pairs of real numbers. The first number, a, of the ordered pair (a, b) is called the **abscissa**, and the second number, b, is called the **ordinate;** together they are referred to as the **coordinates** of a point.

Two basic kinds of problems exist in coordinate geometry:

1. Given an algebraic equation, find its geometric graph.
2. Given a set of conditions that pertains to a geometric figure, find its algebraic equation.

A **solution** of an equation in two variables is an ordered pair of real numbers that satisfies the equation.

The following suggestions are offered for **graphing an equation** in two variables.

1. Determine what type of symmetry the equation exhibits.
2. Find the intercepts.
3. Solve the equation for y in terms of x or for x in terms of y, if it is not already in such a form.
4. Set up a table of ordered pairs that satisfy the equation. The type of symmetry will affect your choice of values in the table.
5. Plot the points associated with the ordered pairs from the table, and connect them with a smooth curve. Then, if appropriate, reflect this part of the curve according to the symmetry shown by the equation.

The distance between any two points (x_1, y_1) and (x_2, y_2) is given by the **distance formula,**

$$d = \sqrt{(x_2 - x_1)^2 + (y_2 - y_1)^2}$$

The **slope** (denoted by m) of a line determined by the points (x_1, y_1) and (x_2, y_2) is given by the slope formula,

$$m = \frac{y_2 - y_1}{x_2 - x_1}, \qquad x_2 \neq x_1$$

(12.2) The equation $y = mx + b$ is referred to as the **slope-intercept form** of the equation of a straight line. If the equation of a nonvertical line is written in this y form,

then the coefficient of x is the slope of the line, and the constant term is the y intercept.

If two lines have slopes m_1 and m_2, respectively, then

1. The two lines are parallel if and only if $m_1 = m_2$.
2. The two lines are perpendicular if and only if $(m_1)(m_2) = -1$.

To determine the equation of a straight line, given a set of conditions, we can use the point-slope form, $y - y_1 = m(x - x_1)$, or $\dfrac{y - y_1}{x - x_1} = m$. The conditions generally fall into one of the following four categories:

1. Given the slope and a point contained in the line
2. Given two points contained in the line
3. Given a point contained in the line and that the line is parallel to another line
4. Given a point contained in the line and that the line is perpendicular to another line

The result can then be expressed in standard form or slope-intercept form.

(12.3) and (12.4) The graph of any quadratic equation of the form $y = ax^2 + bx + c$, where a, b, and c are real numbers and $a \neq 0$, is a **parabola.**

The following diagram summarizes the graphing of parabolas.

$$y = x^2 \longleftarrow \quad \text{Basic parabola}$$

$$y = ⓐ(x - ⓗ)^2 + ⓚ$$

| Affects the width and which way the parabola opens | Moves the parabola right or left | Moves the parabola up or down |

The **standard form of the equation of a circle** with its center at (h, k) and a radius of length r is

$$(x - h)^2 + (y - k)^2 = r^2$$

The standard form of the equation of a circle with its center at the origin and a radius of length r is

$$x^2 + y^2 = r^2$$

(12.5) The graph of an equation of the form $Ax^2 + By^2 = C$ or of the form $A(x - h)^2 + B(y - k)^2 = C$, where A, B, and C are nonzero real numbers of the same sign and $A \neq B$, is an **ellipse.**

(12.6) The graph of an equation of the form $Ax^2 + By^2 = C$ or of the form $A(x - h)^2 + B(y - k)^2 = C$, where A, B, and C are nonzero real numbers with A and B of unlike signs, is a **hyperbola.**

The equations of the asymptotes of a hyperbola can be found by replacing the constant term of the equation of the hyperbola with zero and solving the resulting equation for y.

Circles, ellipses, parabolas, and hyperbolas are often referred to as **conic sections.**

CHAPTER 12 REVIEW PROBLEM SET

1. Find the slope of the line determined by each pair of points.

   a. $(3, 4), (-2, -2)$ **b.** $(-2, 3), (4, -1)$

2. Find the slope of each of the following lines.

   a. $4x + y = 7$ **b.** $2x - 7y = 3$

3. Find the lengths of the sides of a triangle whose vertices are at $(2, 3), (5, -1)$, and $(-4, -5)$.

For Problems 4–8, write the equation of the line that satisfies the stated conditions. Express final equations in standard form.

4. Containing the points $(-1, 2)$ and $(3, -5)$

5. Having a slope of $-\dfrac{3}{7}$ and a y intercept of 4

6. Containing the point $(-1, -6)$ and having a slope of $\dfrac{2}{3}$

7. Containing the point $(2, 5)$ and parallel to the line $x - 2y = 4$

8. Containing the point $(-2, -6)$ and perpendicular to the line $3x + 2y = 12$

For Problems 9–13, determine the type(s) of symmetry (with respect to the x axis, y axis, and/or origin) exhibited by the graph of each equation. Do not sketch the graphs.

9. $x^2 = y + 1$ 10. $y^2 = x^3$

11. $y = -2x$ 12. $y = \dfrac{8}{x^2 + 6}$

13. $x^2 - 2y^2 = -3$

For Problems 14–19, graph each equation.

14. $2x - y = 6$ 15. $y = -2x^2 - 1$

16. $y = -2x - 1$ 17. $y = -4x$

18. $xy^2 = -1$ 19. $y = \dfrac{-3}{x^2 + 1}$

20. A certain highway has a 6% grade. How many feet does it rise in a horizontal distance of 1 mile?

21. If the ratio of rise to run is to be $\dfrac{2}{3}$ for the steps of a staircase, and the run is 12 inches, find the rise.

For Problems 22–27, find the vertex of each parabola.

22. $y = x^2 + 6$ 23 $y = -x^2 - 8$

24. $y = (x + 3)^2 - 1$ 25. $y = x^2 - 14x + 54$

26. $y = -x^2 + 12x - 44$ 27. $y = 3x^2 + 24x + 39$

For Problems 28–30, write the equation of the circle that satisfies the given conditions. Express your answers in the form $x^2 + y^2 + Dx + Ey + F = 0$.

28. Center at $(2, -6)$ and $r = 5$

29. Center at $(-4, -8)$ and $r = 2\sqrt{3}$

30. Center at $(0, 5)$ and passes through the origin

For Problems 31–34, find the center and the length of a radius for each circle.

31. $x^2 + 14x + y^2 - 8y + 16 = 0$

32. $x^2 + 16x + y^2 + 39 = 0$

33. $x^2 - 12x + y^2 + 16y = 0$

34. $x^2 + y^2 = 24$

For Problems 35–38, find the length of the major axis and the length of the minor axis of each ellipse.

35. $4x^2 + 25y^2 = 100$

36. $2x^2 + 7y^2 = 28$

37. $x^2 - 4x + 9y^2 + 54y + 76 = 0$

38. $9x^2 + 72x + 4y^2 - 8y + 112 = 0$

For Problems 39–42, find the equations of the asymptotes of each hyperbola.

39. $4x^2 - 9y^2 = 16$

40. $16y^2 - 4x^2 = 17$

41. $25x^2 + 100x - 4y^2 + 24y - 36 = 0$

42. $36y^2 - 288y - x^2 + 2x + 539 = 0$

For Problems 43–52, graph each equation.

43. $9x^2 + y^2 = 81$

44. $9x^2 - y^2 = 81$

45. $y = -2x^2 + 3$

46. $y = 4x^2 - 16x + 19$

47. $x^2 + 4x + y^2 + 8y + 11 = 0$

48. $4x^2 - 8x + y^2 + 8y + 4 = 0$

49. $y^2 + 6y - 4x^2 - 24x - 63 = 0$

50. $y = -2x^2 - 4x - 3$

51. $x^2 - y^2 = -9$

52. $4x^2 + 16y^2 + 96y = 0$

CHAPTER 12
TEST

 applies to all problems in this Chapter Test.

1. Find the slope of the line determined by the points $(-2, 4)$ and $(3, -2)$.

2. Find the slope of the line determined by the equation $3x - 7y = 12$.

3. Find the length of the line segment whose endpoints are $(4, 2)$ and $(-3, -1)$. Express the answer in simplest radical form.

4. Find the equation of the line that has a slope of $-\dfrac{3}{2}$ and contains the point $(4, -5)$. Express the equation in standard form.

5. Find the equation of the line that contains the points $(-4, 2)$ and $(2, 1)$. Express the equation in slope-intercept form.

6. Find the equation of the line that is parallel to the line $5x + 2y = 7$ and contains the point $(-2, -4)$. Express the equation in standard form.

7. Find the equation of the line that is perpendicular to the line $x - 6y = 9$ and contains the point $(4, 7)$. Express the equation in standard form.

8. The grade of a highway up a hill is 25%. How much change in horizontal distance is there if the vertical height of the hill is 120 feet?

9. If the ratio of rise to run is to be $\dfrac{3}{4}$ for the steps of a staircase, and the rise is 32 centimeters, find the run to the nearest centimeter.

10. Find the vertex of the parabola $y = x^2 - 6x + 9$.

11. Find the vertex of the parabola $y = 4x^2 + 32x + 62$.

12. Write the equation of the circle with its center at $(2, 8)$ and a radius of length 3 units.

13. Find the center and the length of a radius of the circle $x^2 - 12x + y^2 + 8y + 3 = 0$.

14. Find the length of the major axis of the ellipse $9x^2 + 2y^2 = 32$.

15. Find the length of the minor axis of the ellipse $8x^2 - 32x + 5y^2 + 30y + 45 = 0$.

16. Find the equations of the asymptotes for the hyperbola $y^2 - 16x^2 = 36$.

For Problems 17–25, graph each equation.

17. $\dfrac{1}{3}x + \dfrac{1}{2}y = 2$

18. $y = \dfrac{-x - 1}{4}$

19. $x^2 - 4y^2 = -16$

20. $y = x^2 + 4x$

21. $x^2 + 2x + y^2 + 8y + 8 = 0$

22. $2x^2 + 3y^2 = 12$

23. $y = 2x^2 + 12x + 22$

24. $9x^2 - y^2 = 9$

25. $3x^2 - 12x + 5y^2 + 10y - 10 = 0$

CUMULATIVE PRACTICE TEST *Chapters 1-12*

1. Evaluate each of the following numerical expressions.

   a. $\sqrt[3]{-\dfrac{8}{27}}$ **b.** $-16^{\frac{5}{4}}$

   c. $\left(\dfrac{2^{-1}}{2^{-2}}\right)^{-2}$ **d.** $\left(\dfrac{1}{3}-\dfrac{1}{5}\right)^{-1}$

2. Express each of the following in simplest radical form.

   a. $\sqrt{96}$ **b.** $\sqrt[3]{40}$

   c. $\dfrac{4\sqrt{2}}{6\sqrt{3}}$ **d.** $\dfrac{2}{\sqrt{7}-\sqrt{2}}$

For Problems 3–7, perform the indicated operations and express the answers in simplifed form.

3. $(4xy^2)(-3x^2y^3)(2x)$

4. $(2x+1)(3x^2-4x-2)$

5. $(x^4+5x^3+5x^2-7x-12)\div(x+3)$

6. $\dfrac{2x-1}{4}-\dfrac{3x+2}{3}$

7. $(2\sqrt{3}-\sqrt{2})(3\sqrt{3}+2\sqrt{2})$

8. Identify the type of symmetry that each of the following equations exhibits.

   a. $xy=-6$

   b. $x^2+3x+y^2-4=0$

   c. $y=x^2-14$

   d. $x^2-2y^2-3y-6=0$

9. Evaluate $(3x^2-2x-1)-(3x^2-3x+2)$ for $x=-18$.

10. Evaluate $\dfrac{36x^{-2}y^4}{9x^{-3}y^3}$ for $x=6$ and $y=-8$.

11. Find the least common multiple of 6, 9, and 12.

12. Find the slope of the line determined by the equation $x-4y=-6$.

13. Find the equation of the line that contains the two points $(2,7)$ and $(-1,3)$.

For Problems 14–18, solve each equation.

14. $(x+2)(x-6)=-16$

15. $\dfrac{2}{3}(x-1)-\dfrac{3}{4}(2x+3)=\dfrac{3}{2}$

16. $0.06x+0.07(7000-x)=460$

17. $|3x-5|=10$

18. $3x^2-x-1=0$

For Problems 19–21, solve each inequality and express the solutions using interval notation.

19. $|2x-3|\le 5$

20. $\dfrac{x-2}{x-3}>0$

21. $x^2-5x-6<0$

For Problems 22–25, use an equation or a system of equations to help solve each problem.

22. One of two complementary angles is 6° larger than one-half of the other angle. Find the measure of each angle.

23. Abby has 37 coins, consisting only of dimes and quarters, worth $7.45. How many dimes and how many quarters does she have?

24. Juan started walking at 4 miles per hour. An hour and a half later, Cathy starts jogging along the same route at 6 miles per hour. How long will it take Cathy to catch up with Juan?

25. Sue bought 3 packages of cookies and 2 sacks of potato chips for $11.35. Later, at the same prices, she bought 2 packages of cookies and 5 sacks of potato chips for $18.53. Find the price of a package of cookies.

The price of goods may be decided by using a function to describe the relationship between the price and the demand. Such a function gives us a means of studying the demand when the price is varied.

© Bill Aron /PhotoEdit

Functions

A golf pro-shop operator finds that she can sell 30 sets of golf clubs at $500 per set in a year. Furthermore, she predicts that for each $25 decrease in price, 3 additional sets of golf clubs could be sold. At what price should she sell the clubs to maximize her gross income? We can use the quadratic function $f(x) = (30 + 3x)(500 - 25x)$ to determine that the clubs should be sold at $375 per set.

One of the fundamental concepts of mathematics is the function concept. Functions are used to unify mathematics and also to apply mathematics to many real-world problems. Functions provide a means of studying quantities that vary with one another — that is, a change in one quantity causes a corresponding change in the other.

In this chapter we will (1) introduce the basic ideas that pertain to the function concept, (2) review and extend some concepts from Chapter 12, and (3) discuss some applications of functions.

InfoTrac Project

Do a keyword search on linear relationships. Find an article concerning language skills in cocaine-exposed infants. Write a brief paragraph on the general conclusion of the article. Suppose the relationship between the level (x) of benzoylecgonine (a cocaine by-product) in the newborn's meconium and the age level of auditory comprehension ($f(x)$) at age 1 can be expressed by the following equation: $f(x) = -0.057x + 1.0$. What is the age level of auditory comprehension for a child who was not exposed to cocaine during fetal development? As the level of benzoylecgonine rises, what happens to the age level of auditory comprehension? What would the level of benzoylecgonine need to be for a child to have the age level of auditory comprehension of a 6-month-old child? Graph this relationship on a graphing utility.

13.1   Relations and Functions

Objectives

- Understand the definitions of a function and a relation.
- Use function notation.
- Specify the domain and range.
- Find the difference quotient for a function.

Mathematically, a **function** is a special kind of relation, so we will begin our discussion with a simple definition of a relation.

DEFINITION 13.1

A **relation** is a set of ordered pairs.

Thus a set of ordered pairs such as $\{(1, 2), (3, 7), (8, 14)\}$ is a relation. The set of all first components of the ordered pairs is the **domain** of the relation, and the set of all second components is the **range** of the relation. The relation $\{(1, 2), (3, 7), (8, 14)\}$ has a domain of $\{1, 3, 8\}$ and a range of $\{2, 7, 14\}$.

The ordered pairs we refer to in Definition 13.1 may be generated by various means, such as a graph or a chart. However, one of the most common ways of generating ordered pairs is by using equations. Because the solution set of an equation in two variables is a set of ordered pairs, such an equation describes a relation. Each of the following equations describes a relation between the variables x and y. We have listed *some* of the infinitely many ordered pairs (x, y) of each relation.

1. $x^2 + y^2 = 4$: $(1, \sqrt{3}), (1, -\sqrt{3}), (0, 2), (0, -2)$
2. $y^2 = x^3$: $(0, 0), (1, 1), (1, -1), (4, 8), (4, -8)$
3. $y = x + 2$: $(0, 2), (1, 3), (2, 4), (-1, 1), (5, 7)$
4. $y = \dfrac{1}{x - 1}$: $(0, -1), (2, 1), \left(3, \dfrac{1}{2}\right), \left(-1, -\dfrac{1}{2}\right), \left(-2, -\dfrac{1}{3}\right)$
5. $y = x^2$: $(0, 0), (1, 1), (2, 4), (-1, 1), (-2, 4)$

Now we direct your attention to the ordered pairs associated with equations 3, 4, and 5. Note that in each case, no two ordered pairs have the same first component. Such a set of ordered pairs is called a **function.**

DEFINITION 13.2

A **function** is a relation in which no two ordered pairs have the same first component.

Stated another way, Definition 13.2 means that a function is a relation wherein each member of the domain is assigned *one and only one* member of the range. Thus it is easy to determine that each of the following sets of ordered pairs is a function.

$$f = \{(x, y) \mid y = x + 2\}$$
$$g = \left\{(x, y) \,\middle|\, y = \frac{1}{x - 1}\right\}$$
$$h = \{(x, y) \mid y = x^2\}$$

In each case there is one and only one value of y (an element of the range) associated with each value of x (an element of the domain).

Note that we named the previous functions f, g, and h. It is customary to name functions by means of a single letter, and the letters f, g, and h are often used. We suggest making more meaningful choices when functions are used to portray real-world situations. For example, if a problem involves a profit function, then naming the function p or even P would seem natural.

The symbol for a function can be used along with a variable that represents an element in the domain to represent the associated element in the range. For example, suppose that we have a function f specified in terms of the variable x. The symbol $f(x)$, which is read "f of x" or "the value of f at x," represents the element in the range associated with the element x from the domain. The function $f = \{(x, y) \mid y = x + 2\}$ can be written as $f = \{(x, f(x)) \mid f(x) = x + 2\}$ and is usually shortened to read "f is the function determined by the equation $f(x) = x + 2$."

REMARK: Be careful with the notation $f(x)$. As we stated above, it means the value of the function f at x. It does not mean f times x.

This **function notation** is very convenient when we are computing and expressing various values of the function. For example, the value of the function $f(x) = 3x - 5$ at $x = 1$ is

$$f(1) = 3(1) - 5 = -2$$

Likewise, the functional values for $x = 2$, $x = -1$, and $x = 5$ are

$$f(2) = 3(2) - 5 = 1$$
$$f(-1) = 3(-1) - 5 = -8$$
$$f(5) = 3(5) - 5 = 10$$

Thus, this function f contains the ordered pairs $(1, -2), (2, 1), (-1, -8), (5, 10)$, and in general all ordered pairs of the form $(x, f(x))$, where $f(x) = 3x - 5$ and x is any real number.

It may be helpful for you to picture the concept of a function in terms of a *function machine,* as shown in Figure 13.1. Each time that a value of x is put into the machine, the equation $f(x) = x + 2$ is used to generate one and only one value for $f(x)$ to be ejected from the machine. For example, if 3 is put into this machine, then $f(3) = 3 + 2 = 5$, and 5 is ejected. Thus the ordered pair $(3, 5)$ is one element of the function. Now let's look at some examples to help pull together some of the ideas about functions.

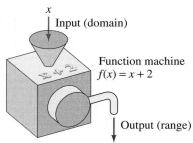

Figure 13.1

EXAMPLE 1

Determine whether the relation $\{(x, y) | y^2 = x\}$ is a function and specify its domain and range.

Solution

Because $y^2 = x$ is equivalent to $y = \pm\sqrt{x}$, to each value of x where $x > 0$, there are assigned *two* values for y. Therefore, this relation is not a function. The expression $\sqrt{x}$ requires that x be nonnegative; therefore, the domain (D) is

$$D = \{x | x \geq 0\}$$

To each nonnegative real number, the relation assigns two real numbers, $\sqrt{x}$ and $-\sqrt{x}$. Thus the range (R) is

$$R = \{y | y \text{ is a real number}\}$$ ■

EXAMPLE 2

Consider the function $f(x) = x^2$.

(a) Specify its domain.

(b) Determine its range.

(c) Evaluate $f(-2), f(0)$, and $f(4)$.

Solution

(a) Any real number can be squared; therefore, the domain (D) is

$$D = \{x | x \text{ is a real number}\}$$

(b) Squaring a real number always produces a nonnegative result. Thus the range (R) is

$$R = \{f(x) | f(x) \geq 0\}$$

(c) $f(-2) = (-2)^2 = 4$

$\quad f(0) = (0)^2 = 0$

$\quad f(4) = (4)^2 = 16$

For our purposes in this text, if the domain of a function is not specifically indicated or determined by a real-world application, then we assume the domain to be all **real number** replacements for the variable, which represents an element in the domain that will produce **real number** functional values. Consider the following examples.

E X A M P L E   3

Specify the domain for each of the following.

(a) $f(x) = \dfrac{1}{x - 1}$ **(b)** $f(t) = \dfrac{1}{t^2 - 4}$ **(c)** $f(s) = \sqrt{s - 3}$

Solution

(a) We can replace x with any real number except 1, because 1 makes the denominator zero. Thus the domain is given by

$$D = \{x \mid x \neq 1\}$$

(b) We need to eliminate any value of t that will make the denominator zero. Thus let's solve the equation $t^2 - 4 = 0$.

$$t^2 - 4 = 0$$
$$t^2 = 4$$
$$t = \pm 2$$

The domain is the set

$$D = \{t \mid t \neq -2 \text{ and } t \neq 2\}$$

(c) The radicand, $s - 3$, must be nonnegative.

$$s - 3 \geq 0$$
$$s \geq 3$$

The domain is the set

$$D = \{s \mid s \geq 3\}$$

REMARK: Certainly interval notation could be used to express the domains of functions, as in Example 3. However, we have chosen to use this section to give you a little more experience with set-builder notation.

E X A M P L E 4

If $f(x) = -2x + 7$ and $g(x) = x^2 - 5x + 6$, find $f(3), f(-4), g(2)$, and $g(-1)$.

Solution

$f(x) = -2x + 7$	$g(x) = x^2 - 5x + 6$
$f(3) = -2(3) + 7 = 1$	$g(2) = 2^2 - 5(2) + 6 = 0$
$f(-4) = -2(-4) + 7 = 15$	$g(-1) = (-1)^2 - 5(-1) + 6 = 12$ ∎

In Example 4, note that we are working with two different functions in the same problem. Thus different names, f and g, are used.

The quotient $\dfrac{f(a + h) - f(a)}{h}$ is often called a **difference quotient,** and we use it extensively with functions when studying the limit concept in calculus. The next two examples show how we found the difference quotient for two specific functions.

E X A M P L E 5

If $f(x) = 3x - 5$, find $\dfrac{f(a + h) - f(a)}{h}$.

Solution

$$f(a + h) = 3(a + h) - 5$$
$$= 3a + 3h - 5$$

and

$$f(a) = 3a - 5$$

Therefore,

$$f(a + h) - f(a) = (3a + 3h - 5) - (3a - 5)$$
$$= 3a + 3h - 5 - 3a + 5$$
$$= 3h$$

and

$$\frac{f(a + h) - f(a)}{h} = \frac{3h}{h} = 3$$ ∎

E X A M P L E 6

If $f(x) = x^2 + 2x - 3$, find $\dfrac{f(a + h) - f(a)}{h}$.

Solution

$$f(a + h) = (a + h)^2 + 2(a + h) - 3$$
$$= a^2 + 2ah + h^2 + 2a + 2h - 3$$

and

$$f(a) = a^2 + 2a - 3$$

Therefore,

$$f(a + h) - f(a) = (a^2 + 2ah + h^2 + 2a + 2h - 3) - (a^2 + 2a - 3)$$
$$= a^2 + 2ah + h^2 + 2a + 2h - 3 - a^2 - 2a + 3$$
$$= 2ah + h^2 + 2h$$

and

$$\frac{f(a + h) - f(a)}{h} = \frac{2ah + h^2 + 2h}{h}$$
$$= \frac{h(2a + h + 2)}{h}$$
$$= 2a + h + 2$$

Functions and functional notation provide the basis for describing many real-world relationships. The next example illustrates this point.

EXAMPLE 7

Suppose a factory determines that the overhead for producing a quantity of a certain item is $500, and the cost for each item is $25. Express the total expenses as a function of the number of items produced, and compute the expenses for producing 12, 25, 50, 75, and 100 items.

Solution

Let n represent the number of items produced. Then $25n + 500$ represents the total expenses. Let's use E to represent the *expense function*, so that we have

$$E(n) = 25n + 500, \quad \text{where } n \text{ is a whole number}$$

from which we obtain

$$E(12) = 25(12) + 500 = 800$$
$$E(25) = 25(25) + 500 = 1125$$
$$E(50) = 25(50) + 500 = 1750$$
$$E(75) = 25(75) + 500 = 2375$$
$$E(100) = 25(100) + 500 = 3000$$

Thus the total expenses for producing 12, 25, 50, 75, and 100 items are $800, $1125, $1750, $2375, and $3000, respectively.

CONCEPT QUIZ

For Problems 1–5, answer true or false.

1. A function is a special type of relation.

2. The relation {(John, Mary), (Mike, Ada), (Kyle, Jenn), (Mike, Sydney)} is a function.

3. Given $f(x) = 3x + 4$, the notation $f(7)$ means to find the value of f when $x = 7$.

4. The set of all first components of the ordered pairs of a relation is called the range.

5. The domain of a function can never be the set of all real numbers.

PROBLEM SET 13.1

For Problems 1–10, specify the domain and the range for each relation. Also state whether or not the relation is a function.

1. $\{(1, 5), (2, 8), (3, 11), (4, 14)\}$

2. $\{(0, 0), (2, 10), (4, 20), (6, 30), (8, 40)\}$

3. $\{(0, 5), (0, -5), (1, 2\sqrt{6}), (1, -2\sqrt{6})\}$

4. $\{(1, 1), (1, 2), (1, -1), (1, -2), (1, 3)\}$

5. $\{(1, 2), (2, 5), (3, 10), (4, 17), (5, 26)\}$

6. $\{(-1, 5), (0, 1), (1, -3), (2, -7)\}$

7. $\{(x, y)|\ 5x - 2y = 6\}$

8. $\{(x, y)|\ y = -3x\}$

9. $\{(x, y)|\ x^2 = y^3\}$

10. $\{(x, y)|\ x^2 - y^2 = 16\}$

For Problems 11–36, specify the domain for each of the functions.

11. $f(x) = 7x - 2$

12. $f(x) = x^2 + 1$

13. $f(x) = \dfrac{1}{x - 1}$

14. $f(x) = \dfrac{-3}{x + 4}$

15. $g(x) = \dfrac{3x}{4x - 3}$

16. $g(x) = \dfrac{5x}{2x + 7}$

17. $h(x) = \dfrac{2}{(x + 1)(x - 4)}$

18. $h(x) = \dfrac{-3}{(x - 6)(2x + 1)}$

19. $f(x) = \dfrac{14}{x^2 + 3x - 40}$

20. $f(x) = \dfrac{7}{x^2 - 8x - 20}$

21. $f(x) = \dfrac{-4}{x^2 + 6x}$

22. $f(x) = \dfrac{9}{x^2 - 12x}$

23. $f(t) = \dfrac{4}{t^2 + 9}$

24. $f(t) = \dfrac{8}{t^2 + 1}$

25. $f(t) = \dfrac{3t}{t^2 - 4}$

26. $f(t) = \dfrac{-2t}{t^2 - 25}$

27. $h(x) = \sqrt{x + 4}$

28. $h(x) = \sqrt{5x - 3}$

29. $f(s) = \sqrt{4s - 5}$

30. $f(s) = \sqrt{s - 2} + 5$

31. $f(x) = \sqrt{x^2 - 16}$

32. $f(x) = \sqrt{x^2 - 49}$

33. $f(x) = \sqrt{x^2 - 3x - 18}$

34. $f(x) = \sqrt{x^2 + 4x - 32}$

35. $f(x) = \sqrt{1 - x^2}$

36. $f(x) = \sqrt{9 - x^2}$

37. If $f(x) = 5x - 2$, find $f(0), f(2), f(-1)$, and $f(-4)$.

38. If $f(x) = -3x - 4$, find $f(-2), f(-1), f(3)$, and $f(5)$.

39. If $f(x) = \dfrac{1}{2}x - \dfrac{3}{4}$, find $f(-2), f(0)$, $f\left(\dfrac{1}{2}\right), f\left(\dfrac{2}{3}\right)$

40. If $g(x) = x^2 + 3x - 1$, find $g(1), g(-1), g(3)$, and $g(-4)$.

41. If $g(x) = 2x^2 - 5x - 7$, find $g(-1), g(2), g(-3)$, and $g(4)$.

42. If $h(x) = -x^2 - 3$, find $h(1), h(-1), h(-3)$, and $h(5)$.

43. If $h(x) = -2x^2 - x + 4$, find $h(-2), h(-3), h(4)$, and $h(5)$.

44. If $f(x) = \sqrt{x - 1}$, find $f(1), f(5), f(13)$, and $f(26)$.

45. If $f(x) = \sqrt{2x + 1}$, find $f(3), f(4), f(10)$, and $f(12)$.

46. If $f(x) = \dfrac{3}{x - 2}$, find $f(3), f(0), f(-1)$, and $f(-5)$.

47. If $f(x) = \dfrac{-4}{x + 3}$, find $f(1), f(-1), f(3)$, and $f(-6)$.

48. If $f(x) = 2x^2 - 7$ and $g(x) = x^2 + x - 1$, find $f(-2), f(3)$, $g(-4)$, and $g(5)$.

49. If $f(x) = 5x^2 - 2x + 3$ and $g(x) = -x^2 + 4x - 5$, find $f(-2), f(3), g(-4)$, and $g(6)$.

50. If $f(x) = |3x - 2|$ and $g(x) = |x| + 2$, find $f(1)$, $f(-1)$, $g(2)$, and $g(-3)$.

51. If $f(x) = 3|x| - 1$ and $g(x) = -|x| + 1$, find $f(-2)$, $f(3)$, $g(-4)$, and $g(5)$.

For Problems 52–59, find $\dfrac{f(a + h) - f(a)}{h}$ for each of the given functions.

52. $f(x) = 5x - 4$ **53.** $f(x) = -3x + 6$

54. $f(x) = x^2 + 5$ **55.** $f(x) = -x^2 - 1$

56. $f(x) = x^2 - 3x + 7$ **57.** $f(x) = 2x^2 - x + 8$

58. $f(x) = -3x^2 + 4x - 1$ **59.** $f(x) = -4x^2 - 7x - 9$

60. Suppose that the cost function for producing a certain item is given by $C(n) = 3n + 5$, where n represents the number of items produced. Compute $C(150)$, $C(500)$, $C(750)$, and $C(1500)$.

61. The height of a projectile fired vertically into the air (neglecting air resistance) at an initial velocity of 64 feet per second is a function of the time (t) and is given by the equation $h(t) = 64t - 16t^2$. Compute $h(1)$, $h(2)$, $h(3)$, and $h(4)$.

62. The profit function for selling n items is given by $P(n) = -n^2 + 500n - 61{,}500$. Compute $P(200)$, $P(230)$, $P(250)$, and $P(260)$.

63. A car rental agency charges \$50 per day plus \$.32 a mile. Therefore, the daily charge for renting a car is a function of the number of miles traveled (m) and can be expressed as $C(m) = 50 + 0.32m$. Compute $C(75)$, $C(150)$, $C(225)$, and $C(650)$.

64. The equation $A(r) = \pi r^2$ expresses the area of a circular region as a function of the length of a radius (r). Use 3.14 as an approximation for π, and compute $A(2)$, $A(3)$, $A(12)$, and $A(17)$.

65. The equation $I(r) = 2500r$ expresses the amount of simple interest earned by an investment of \$2500 for 1 year as a function of the rate of interest (r). Compute $I(0.03)$, $I(0.04)$, $I(0.05)$, and $I(0.065)$.

■ ■ ■ **Thoughts into words**

66. Are all functions also relations? Are all relations also functions? Defend your answers.

67. What does it mean to say that the domain of a function may be restricted if the function represents a real-world situation? Give two or three examples of such situations.

68. Does $f(a + b) = f(a) + f(b)$ for all functions? Defend your answer.

69. Are there any functions for which $f(a + b) = f(a) + f(b)$? Defend your answer.

Answers to Concept Quiz

1. True   **2.** False   **3.** True   **4.** False   **5.** False

13.2   Functions: Their Graphs and Applications

Objectives

■   Graph linear and quadratic functions.

■   Find the vertex and axis of symmetry of the graph of a quadratic function.

■   Apply functions to solve word problems.

In Section 5.1, we used phrases such as *the graph of the solution set of the equation* $y = x - 1$ or simply *the graph of the equation* $y = x - 1$ to indicate a line that contains the points $(0, -1)$ and $(1, 0)$. Because the equation $y = x - 1$ (which can be written as $f(x) = x - 1$) can be used to specify a function, that line we previously referred to is also called the **graph of the function specified by the equation** or simply the **graph of the function.** Generally speaking, the graph of any equation that determines a function is also called the graph of the function. Thus the graphing techniques we discussed earlier will continue to play an important role as we graph functions.

As we use the function concept in our study of mathematics, it is helpful to classify certain types of functions and become familiar with their equations, characteristics, and graphs. In this section we will discuss two special types of functions — **linear** and **quadratic functions.** These functions are merely an outgrowth of our earlier study of linear and quadratic equations.

Linear Functions

Any function defined by an equation that can be written in the form

$$f(x) = ax + b$$

where a and b are real numbers, is called a **linear function.** The following equations are examples of linear functions.

$$f(x) = -3x + 6 \qquad f(x) = 2x + 4 \qquad f(x) = -\frac{1}{2}x - \frac{3}{4}$$

Graphing linear functions is quite easy because the graph of every linear function is a straight line. Therefore, all we need to do is determine two points of the graph and draw the line determined by those two points. You may want to continue using a third point as a check point.

EXAMPLE 1   Graph the function $f(x) = -3x + 6$.

Solution

Because $f(0) = 6$, the point $(0, 6)$ is on the graph. Likewise, because $f(1) = 3$, the point $(1, 3)$ is on the graph. Plot these two points and draw the line determined by the two points to produce Figure 13.2.

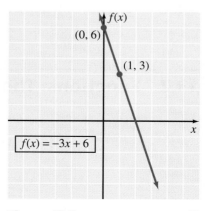

Figure 13.2

REMARK: Note that in Figure 13.2, we labeled the vertical axis $f(x)$. We could also label it y, because $f(x) = -3x + 6$ and $y = -3x + 6$ mean the same thing. We will continue to use the $f(x)$ label in this chapter to help you adjust to the function notation.

E X A M P L E   2   Graph the function $f(x) = x$.

Solution

The equation $f(x) = x$ can be written as $f(x) = 1x + 0$; thus it is a linear function. Because $f(0) = 0$ and $f(2) = 2$, the points $(0, 0)$ and $(2, 2)$ determine the line in Figure 13.3. The function $f(x) = x$ is often called the **identity function.**

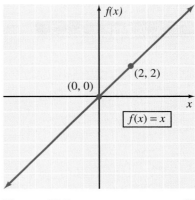

Figure 13.3

As you use function notation to graph functions, it is often helpful to think of the ordinate of every point on the graph as the value of the function at a specific value of x. Geometrically, this functional value is the directed distance of the point from the x axis, as illustrated in Figure 13.4, with the function $f(x) = 2x - 4$. For example, consider the graph of the function $f(x) = 2$. The function $f(x) = 2$ means that every functional value is 2, or, geometrically, that every point on the graph is 2 units above the x axis. Thus the graph is the horizontal line shown in Figure 13.5.

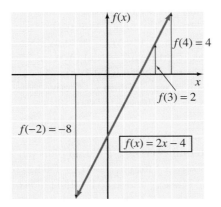

Figure 13.4

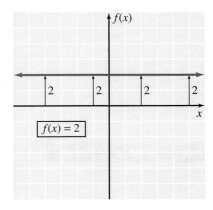

Figure 13.5

Any linear function of the form $f(x) = ax + b$, where $a = 0$, is called a **constant function,** and its graph is a *horizontal line.*

Applications of Linear Functions

We worked with some applications of linear equations in Section 5.2. Let's consider some additional applications at this time and use the concept of a linear function to connect mathematics to the real world.

E X A M P L E   3   The cost of burning a 60-watt light bulb is given by the function $c(h) = 0.0036h$, where h represents the number of hours that the bulb is burning.

(a) How much does it cost to burn a 60-watt bulb for 3 hours per night for a 30-day month?
(b) Graph the function $c(h) = 0.0036h$.
(c) Suppose that a 60-watt light bulb is left burning in a closet for a week before it is discovered and turned off. Use the graph from part (b) to approximate the cost of allowing the bulb to burn for a week. Then use the function to find the exact cost.

Solution

(a) $c(90) = 0.0036(90) = 0.324$ The cost, to the nearest cent, is $.32.

(b) Because $c(0) = 0$ and $c(100) = 0.36$, we can use the points $(0, 0)$ and $(100, 0.36)$ to graph the linear function $c(h) = 0.0036h$ (Figure 13.6).

(c) If the bulb burns for 24 hours per day for a week, it burns for $24(7) = 168$ hours. Reading from the graph, we can approximate 168 on the horizontal axis, read up to the line, and then read across to the vertical axis. It looks as if it will cost approximately 60 cents. Using $c(h) = 0.0036h$, we obtain exactly $c(168) = 0.0036(168) = 0.6048$.

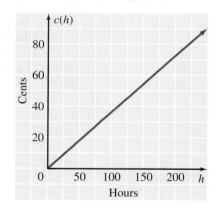

Figure 13.6

E X A M P L E   4   The EZ Car Rental charges a fixed amount per day plus an amount per mile for renting a car. For two different day trips, Ed has rented a car from EZ. He paid $70 for 100 miles on one day and $120 for 350 miles on another day. Determine the linear function that the EZ Car Rental uses to determine its daily rental charges.

Solution

The linear function $f(x) = ax + b$, where x represents the number of miles, models this situation. Ed's two day trips can be represented by the ordered pairs $(100, 70)$ and $(350, 120)$. From these two ordered pairs we can determine a, which is the slope of the line.

$$a = \frac{120 - 70}{350 - 100} = \frac{50}{250} = \frac{1}{5} = 0.2$$

Thus $f(x) = ax + b$ becomes $f(x) = 0.2x + b$. Now either ordered pair can be used to determine the value of b. Using $(100, 70)$ we have $f(100) = 70$; therefore,

$$f(100) = 0.2(100) + b = 70$$
$$b = 50$$

The linear function is $f(x) = 0.2x + 50$. In other words, the EZ Car Rental charges a daily fee of $50 plus $0.20 per mile. ■

EXAMPLE 5

Suppose that Ed (Example 4) also has access to the A-OK Car Rental agency, which charges a daily fee of $25 plus $0.30 per mile. Should Ed use the EZ Car Rental from Example 4 or A-OK Car Rental?

Solution

The linear function $g(x) = 0.3x + 25$, where x represents the number of miles, can be used to determine the daily charges of A-OK Car Rental. Let's graph this function and $f(x) = 0.2x + 50$ from Example 4 on the same set of axes (Figure 13.7).

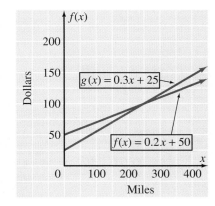

Now we see that the two functions have equal values at the point of intersection of the two lines. To find the coordinates of this point, we can set $0.3x + 25$ equal to $0.2x + 50$ and solve for x.

$$0.3x + 25 = 0.2x + 50$$
$$0.1x = 25$$
$$x = 250$$

Figure 13.7

If $x = 250$, then $0.3(250) + 25 = 100$ and the point of intersection is $(250, 100)$. Again looking at the lines in Figure 13.7, we see that Ed should use A-OK Car Rental for day trips of less than 250 miles, but he should use EZ Car Rental for day trips of more than 250 miles. ■

Quadratic Functions

Any function defined by an equation that can be written in the form

$$f(x) = ax^2 + bx + c$$

where a, b, and c are real numbers with $a \neq 0$, is called a **quadratic function.** The following equations are examples of quadratic functions:

$$f(x) = 3x^2 \qquad f(x) = -2x^2 + 5x \qquad f(x) = 4x^2 - 7x + 1$$

The techniques discussed in Chapter 12 relative to graphing quadratic equations of the form $y = ax^2 + bx + c$ provide the basis for graphing quadratic functions. Let's review some work from Chapter 12 with an example.

EXAMPLE 6

Graph the function $f(x) = 2x^2 - 4x + 5$.

Solution

$$\begin{aligned}
f(x) &= 2x^2 - 4x + 5 \\
&= 2(x^2 - 2x + \underline{\quad}) + 5 \qquad \text{Recall the process of completing the square!} \\
&= 2(x^2 - 2x + 1) + 5 - 2 \\
&= 2(x - 1)^2 + 3
\end{aligned}$$

From this form we can obtain the following information about the parabola.

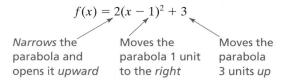

$$f(x) = 2(x - 1)^2 + 3$$

Narrows the parabola and opens it *upward*   Moves the parabola 1 unit to the *right*   Moves the parabola 3 units *up*

Thus the parabola can be drawn as in Figure 13.8.

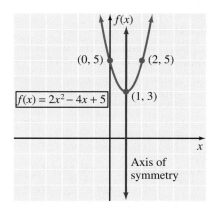

Figure 13.8

In general, if we complete the square on

$$f(x) = ax^2 + bx + c$$

we obtain

$$f(x) = a\left(x^2 + \frac{b}{a}x + \underline{}\right) + c$$

$$= a\left(x^2 + \frac{b}{a}x + \frac{b^2}{4a^2}\right) + c - \frac{b^2}{4a}$$

$$= a\left(x + \frac{b}{2a}\right)^2 + \frac{4ac - b^2}{4a}$$

Therefore, the parabola associated with $f(x) = ax^2 + bx + c$ has its vertex at $\left(-\frac{b}{2a}, \frac{4ac - b^2}{4a}\right)$ and the equation of its axis of symmetry is $x = -\frac{b}{2a}$. These facts are illustrated in Figure 13.9.

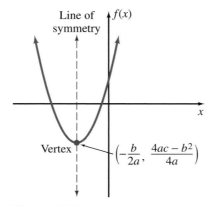

Figure 13.9

By using the information from Figure 13.9, we now have another way of graphing quadratic functions of the form $f(x) = ax^2 + bx + c$. It consists of the following steps.

1. Determine whether the parabola opens upward (if $a > 0$) or downward (if $a < 0$).

2. Find $-\frac{b}{2a}$, which is the x coordinate of the vertex.

3. Find $f\left(-\frac{b}{2a}\right)$, which is the y coordinate of the vertex. $\left(\text{You could also find}\right.$ the y coordinate by evaluating $\left.\frac{4ac - b^2}{4a}.\right)$

4. Locate another point on the parabola, and also locate its image across the line of symmetry, $x = -\frac{b}{2a}$.

The three points in steps 2, 3, and 4 should determine the general shape of the parabola. Let's use these steps in the following two examples.

E X A M P L E   7   Graph $f(x) = 3x^2 - 6x + 5$.

Solution

> **STEP 1**   Because a = 3, the parabola opens upward.
>
> **STEP 2** $-\dfrac{b}{2a} = -\dfrac{-6}{6} = 1$
>
> **STEP 3** $f\left(-\dfrac{b}{2a}\right) = f(1) = 3 - 6 + 5 = 2$. Thus the vertex is at $(1, 2)$.
>
> **STEP 4**   Letting $x = 2$, we obtain $f(2) = 12 - 12 + 5 = 5$. Thus $(2, 5)$ is on the graph, and so is its reflection $(0, 5)$ across the line of symmetry $x = 1$.

The three points $(1, 2)$, $(2, 5)$, and $(0, 5)$ are used to graph the parabola in Figure 13.10.

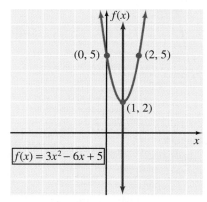

$(0, 5)$ $(2, 5)$

$(1, 2)$

$f(x) = 3x^2 - 6x + 5$

Figure 13.10

E X A M P L E   8   Graph $f(x) = -x^2 - 4x - 7$.

Solution

> **STEP 1**   Because $a = -1$, the parabola opens downward.
>
> **STEP 2** $-\dfrac{b}{2a} = -\dfrac{-4}{-2} = -2$.
>
> **STEP 3** $f\left(-\dfrac{b}{2a}\right) = f(-2) = -(-2)^2 - 4(-2) - 7 = -3$. Thus the vertex is at $(-2, -3)$.
>
> **STEP 4**   Letting $x = 0$, we obtain $f(0) = -7$. Thus $(0, -7)$ is on the graph, and so is its reflection $(-4, -7)$ across the line of symmetry $x = -2$.

The three points $(-2, -3)$, $(0, -7)$, and $(-4, -7)$ are used to draw the parabola in Figure 13.11.

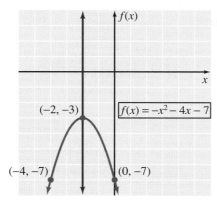

Figure 13.11

In summary, to graph a quadratic function, we have two methods.

1. We can express the equation in the form $f(x) = a(x - h)^2 + k$ and use the values of a, h, and k to determine the parabola.
2. We can express the equation in the form $f(x) = ax^2 + bx + c$ and use the approach demonstrated in Examples 7 and 8.

Problem Solving Using Quadratic Functions

As we have seen, the vertex of the graph of a quadratic function is either the lowest or the highest point on the graph. Thus the term *minimum value* or *maximum value* of a function is often used in applications of the parabola. The x value of the vertex indicates where the minimum or maximum occurs, and $f(x)$ yields the minimum or maximum value of the function. Let's consider some examples that illustrate these ideas.

EXAMPLE 9

A farmer has 120 rods of fencing and wants to enclose a rectangular plot of land that requires fencing on only three sides because it is bounded by a river on one side. Find the length and width of the plot that will maximize the area.

Solution

Let x represent the width; then $120 - 2x$ represents the length as indicated in Figure 13.12.

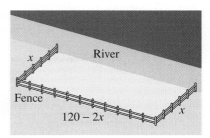

Figure 13.12

The function $A(x) = x(120 - 2x)$ represents the area of the plot in terms of the width x. Because

$$A(x) = x(120 - 2x)$$
$$= 120x - 2x^2$$
$$= -2x^2 + 120x$$

we have a quadratic function with $a = -2$, $b = 120$, and $c = 0$. Therefore, the x value where the maximum value of the function is obtained is

$$-\frac{b}{2a} = -\frac{120}{2(-2)} = 30$$

If $x = 30$, then $120 - 2x = 120 - 2(30) = 60$. Thus the farmer should make the plot 30 rods wide and 60 rods long to maximize the area at $(30)(60) = 1800$ square rods. ■

EXAMPLE 10   Find two numbers whose sum is 30, such that the sum of their squares is a minimum.

Solution
Let x represent one of the numbers; then $30 - x$ represents the other number. By expressing the sum of the squares as a function of x, we obtain

$$f(x) = x^2 + (30 - x)^2$$

which can be simplified to

$$f(x) = x^2 + 900 - 60x + x^2$$
$$= 2x^2 - 60x + 900$$

This is a quadratic function with $a = 2$, $b = -60$, and $c = 900$. Therefore, the x value where the minimum occurs is

$$-\frac{b}{2a} = -\frac{-60}{4} = 15$$

If $x = 15$, then $30 - x = 30 - (15) = 15$. Thus the two numbers should both be 15. ■

EXAMPLE 11   A golf pro-shop operator finds that she can sell 30 sets of golf clubs at $500 per set in a year. Furthermore, she predicts that for each $25 decrease in price, three extra sets of golf clubs could be sold. At what price should she sell the clubs to maximize gross income?

Solution
Sometimes when we are analyzing such a problem, it helps to set up a table.

	Number of sets	×	Price per set	=	Income
3 additional sets can	30	×	$500	=	$15,000
be sold for a $25	33	×	$475	=	$15,675
decrease in price	36	×	$450	=	$16,200

Let x represent the number of \$25 decreases in price. Then we can express the income as a function of x as follows:

$$f(x) = (30 + 3x)(500 - 25x)$$

Number of sets

Price per set

When we simplify, we obtain

$$f(x) = 15{,}000 - 750x + 1500x - 75x^2$$
$$= -75x^2 + 750x + 15{,}000$$

Completing the square yields

$$f(x) = -75x^2 + 750x + 15{,}000$$
$$= -75(x^2 - 10x + \underline{\quad}) + 15{,}000$$
$$= -75(x^2 - 10x + 25) + 15{,}000 + 1875$$
$$= -75(x - 5)^2 + 16{,}875$$

From this form we know that the vertex of the parabola is at (5, 16875). Thus 5 decreases of \$25 each — that is, a \$125 reduction in price — will give a maximum income of \$16,875. The golf clubs should be sold at \$375 per set. ■

What we know about parabolas and the process of completing the square can be helpful when we are using a graphing utility to graph a quadratic function. Consider the following example.

EXAMPLE   12

Use a graphing utility to obtain the graph of the quadratic function

$$f(x) = -x^2 + 37x - 311$$

Solution

First, we know that the parabola opens downward and that its width is the same as that of the basic parabola $f(x) = x^2$. Then we can start the process of completing the square to determine an approximate location of the vertex.

$$f(x) = -x^2 + 37x - 311$$
$$= -(x^2 - 37x + \underline{\quad}) - 311$$
$$= -\left[x^2 - 37x + \left(\frac{37}{2}\right)^2\right] - 311 + \left(\frac{37}{2}\right)^2$$
$$= -[(x^2 - 37x + (18.5)^2] - 311 + 342.25$$
$$= -(x - 18.5)^2 + 31.25$$

Thus the vertex is near $x = 18$ and $y = 31$. Therefore, setting the boundaries of the viewing rectangle so that $-2 \le x \le 25$ and $-10 \le y \le 35$, we obtain the graph shown in Figure 13.13.

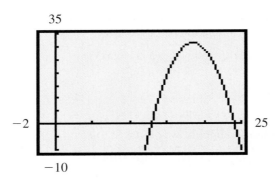

Figure 13.13

REMARK: The graph in Figure 13.13 is sufficient for most purposes, because it shows the vertex and the x intercepts of the parabola. Certainly other boundaries could be used that would also give this information.

CONCEPT QUIZ

For Problems 1–6, answer true or false.

1. The function $f(x) = 3x^2 + 4$ is a linear function.

2. The graph of a linear function of the form $f(x) = b$ is a horizontal line.

3. The graph of a quadratic function is a parabola.

4. The vertex of the graph of a quadratic function is either the lowest or highest point on the graph.

5. The axis of symmetry for a parabola passes through the vertex of the parabola.

6. The parabola for the quadratic function $f(x) = -2x^2 + 3x + 7$ opens upward.

PROBLEM SET 13.2

Graph each of the following linear and quadratic functions (Problems 1–30).

1. $f(x) = 2x - 4$

2. $f(x) = 3x + 3$

3. $f(x) = -2x^2$

4. $f(x) = -4x^2$

5. $f(x) = -3x$

6. $f(x) = -4x$

7. $f(x) = -(x + 1)^2 - 2$

8. $f(x) = -(x - 2)^2 + 4$

9. $f(x) = -x + 3$

10. $f(x) = -2x - 4$

11. $f(x) = x^2 + 2x - 2$

12. $f(x) = x^2 - 4x - 1$

13. $f(x) = -x^2 + 6x - 8$

14. $f(x) = -x^2 - 8x - 15$

15. $f(x) = -3$

16. $f(x) = 1$

17. $f(x) = 2x^2 - 20x + 52$

18. $f(x) = 2x^2 + 12x + 14$

19. $f(x) = -3x^2 + 6x$

20. $f(x) = -4x^2 - 8x$

21. $f(x) = x^2 - x + 2$

22. $f(x) = x^2 + 3x + 2$

23. $f(x) = 2x^2 + 10x + 11$

24. $f(x) = 2x^2 - 10x + 15$

25. $f(x) = -2x^2 - 1$

26. $f(x) = -3x^2 + 2$

27. $f(x) = -3x^2 + 12x - 7$

28. $f(x) = -3x^2 - 18x - 23$

29. $f(x) = -2x^2 + 14x - 25$

30. $f(x) = -2x^2 - 10x - 14$

31. The cost for burning a 75-watt bulb is given by the function $c(h) = 0.0045h$, where h represents the number of hours that the bulb burns.

   a. How much does it cost to burn a 75-watt bulb for 3 hours per night for a 31-day month? Express your answer to the nearest cent.

   b. Graph the function $c(h) = 0.0045h$.

   c. Use the graph in part **b** to approximate the cost of burning a 75-watt bulb for 225 hours.

   d. Use $c(h) = 0.0045h$ to find the exact cost, to the nearest cent, of burning a 75-watt bulb for 225 hours.

32. The Rent-Me Car Rental charges $15 per day plus $.22 per mile to rent a car. Determine a linear function that can be used to calculate daily car rentals. Then use that function to determine the cost of renting a car for a day and driving 175 miles; 220 miles; 300 miles; 460 miles.

33. The ABC Car Rental uses the function $f(x) = 26$ for any daily use of a car up to and including 200 miles. For driving more than 200 miles per day, ABC uses the function $g(x) = 26 + 0.15(x - 200)$ to determine the charges. How much would ABC charge for daily driving of 150 miles? of 230 miles? of 360 miles? of 430 miles?

34. Suppose that a car rental agency charges a fixed amount per day plus an amount per mile for renting a car. Heidi rented a car one day and paid $80 for 200 miles. On another day she rented a car from the same agency and paid $117.50 for 350 miles. Determine the linear function that the agency could use to determine its daily rental charges.

35. A retailer has a number of items that she wants to sell, and she wants to make a profit of 40% of the cost of each item. The function $s(c) = c + 0.4c = 1.4c$, where c represents the cost of an item, can be used to determine the selling price. Find the selling price of items that cost $1.50, $3.25, $14.80, $21, and $24.20.

36. Zack wants to sell five items that cost him $1.20, $2.30, $6.50, $12, and $15.60. He wants to make a profit of 60% of the cost. Create a function that you can use to determine the selling price of each item, and then use the function to calculate each selling price.

37. "All Items 20% Off Marked Price" is a sign at a local golf course. Create a function and then use it to determine how much one has to pay for each of the following marked items: a $9.50 hat, a $15 umbrella, a $75 pair of golf shoes, a $12.50 golf glove, a $750 set of golf clubs.

38. The linear depreciation method assumes that an item depreciates the same amount each year. Suppose that a new piece of machinery costs $32,500 and that it depreciates $1950 each year for t years.

   a. Set up a linear function that yields the value of the machinery after t years.

   b. Find the value of the machinery after 5 years.

   c. Find the value of the machinery after 8 years.

   d. Graph the function from part **a.**

   e. Use the graph from part **d** to approximate how many years it takes for the value of the machinery to become zero.

   f. Use the function to determine how long it takes for the value of the machinery to become zero.

39. Suppose that the cost function for a particular item is given by the equation $C(x) = 2x^2 - 320x + 12{,}920$, where x represents the number of items. How many items should be produced to minimize the cost?

40. Suppose that the equation $p(x) = -2x^2 + 280x - 1000$, where x represents the number of items sold, describes the profit function for a certain business. How many items should be sold to maximize the profit?

41. Find two numbers whose sum is 30, such that the sum of the square of one number plus ten times the other number is a minimum.

42. The height of a projectile fired vertically into the air (neglecting air resistance) at an initial velocity of 96 feet per second is a function of the time and is given by the equation $f(x) = 96x - 16x^2$, where x represents the time. Find the highest point reached by the projectile.

43. Two hundred and forty meters of fencing are available to enclose a rectangular playground. What should be the dimensions of the playground to maximize the area?

44. Find two numbers whose sum is 50 and whose product is a maximum.

45. A cable TV company has 1000 subscribers, and each pays $15 per month. On the basis of a survey, company managers feel that for each decrease of $.25 on the monthly rate, they can obtain 20 additional subscribers. At what rate will maximum revenue be obtained, and how many subscribers will it take at that rate?

46. A motel advertises that it will provide dinner, a dance, and drinks at $50 per couple for a New Year's Eve party. It must have a guarantee of 30 couples. Furthermore, it will agree that for each couple in excess of 30, it will reduce the price per couple for all attending by $.50. How many couples will it take to maximize the motel's revenue?

■ ■ ▢ **Thoughts into words**

47. Give a step-by-step description of how you would use the ideas of this section to graph $f(x) = -4x^2 + 16x - 13$.

48. Is $f(x) = (3x - 2) - (2x + 1)$ a linear function? Explain your answer.

49. Suppose that Bianca walks at a constant rate of 3 miles per hour. Explain what it means to say that the distance Bianca walks is a linear function of the time that she walks.

 Graphing calculator activities

50. Use a graphing calculator to check your graphs for Problems 17–30.

51. Graph each of the following parabolas, and keep in mind that you may need to change the dimensions of the viewing window to obtain a good picture.

  a. $f(x) = x^2 - 2x + 12$ **b.** $f(x) = -x^2 - 4x - 16$

  c. $f(x) = x^2 + 12x + 44$ **d.** $f(x) = x^2 - 30x + 229$

  e. $f(x) = -2x^2 + 8x - 19$

52. Graph each of the following parabolas, and use the TRACE feature to find whole number estimates of the vertex. Then either complete the square or use $\left(-\dfrac{b}{2a}, \dfrac{4ac - b^2}{4a}\right)$ to find the vertex.

  a. $f(x) = x^2 - 6x + 3$

  b. $f(x) = x^2 - 18x + 66$

  c. $f(x) = -x^2 + 8x - 3$

  d. $f(x) = -x^2 + 24x - 129$

  e. $f(x) = 14x^2 - 7x + 1$

  f. $f(x) = -0.5x^2 + 5x - 8.5$

53. a. Graph $f(x) = |x|$, $f(x) = 2|x|$, $f(x) = 4|x|$, and $f(x) = \dfrac{1}{2}|x|$ on the same set of axes.

  b. Graph $f(x) = |x|$, $f(x) = -|x|$, $f(x) = -3|x|$, and $f(x) = -\dfrac{1}{2}|x|$ on the same set of axes.

  c. Use your results from parts **a** and **b** to make a conjecture about the graphs of $f(x) = a|x|$, where a is a nonzero real number.

  d. Graph $f(x) = |x|$, $f(x) = |x| + 3$, $f(x) = |x| - 4$, and $f(x) = |x| + 1$ on the same set of axes. Make a conjecture about the graphs of $f(x) = |x| + k$, where k is a nonzero real number.

  e. Graph $f(x) = |x|$, $f(x) = |x - 3|$, $f(x) = |x - 1|$, and $f(x) = |x + 4|$ on the same set of axes. Make a conjecture about the graphs of $f(x) = |x - h|$, where h is a nonzero real number.

  f. On the basis of your results from parts **a** through **e**, sketch each of the following graphs. Then use a graphing calculator to check your sketches.

    (1) $f(x) = |x - 2| + 3$ (2) $f(x) = |x + 1| - 4$

    (3) $f(x) = 2|x - 4| - 1$ (4) $f(x) = -3|x + 2| + 4$

    (5) $f(x) = \dfrac{1}{2}|x - 3| - 2$

Answers to Concept Quiz

  1. False   **2.** True   **3.** True   **4.** True   **5.** True   **6.** False

13.3 Graphing Made Easy Via Transformations

Objectives

- ◾ Become familiar with horizontal and vertical translations.
- ◾ Recognize and graph reflections across the x axis or y axis.
- ◾ Understand the concepts of vertical stretching and shrinking.
- ◾ Graph functions that have successive transformations.

From our previous work, we know that Figures 13.14–13.16 show the graphs of the functions $f(x) = x^2$, $f(x) = x^3$, and $f(x) = \dfrac{1}{x}$, respectively.

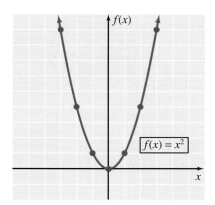

Figure 13.14

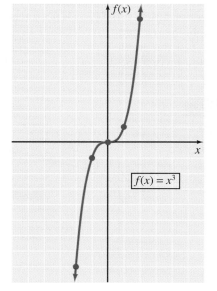

Figure 13.15

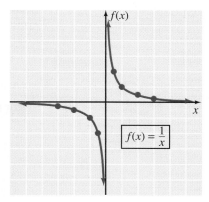

Figure 13.16

To graph a new function — that is, one you are not familiar with — use some of the graphing suggestions we offered in Chapter 12. We will restate those suggestions in terms of function vocabulary and notation. Pay special attention to suggestions 2 and 3, where we have restated the concepts of intercepts and symmetry, using function notation.

1. Determine the domain of the function.
2. Determine any types of symmetry that the equation possesses. If $f(-x) = f(x)$, then the function exhibits y-axis symmetry. If $f(-x) = -f(x)$, then the function exhibits origin symmetry. (Note that the definition of a function rules out the possibility that the graph of a function has x-axis symmetry.)

3. Find the y intercept (we are labeling the y axis with $f(x)$) by evaluating $f(0)$. Find the x intercept by finding the value(s) of x such that $f(x) = 0$.
4. Set up a table of ordered pairs that satisfy the equation. The type of symmetry and the domain will affect your choice of values of x in the table.
5. Plot the points associated with the ordered pairs and connect them with a smooth curve. Then if appropriate, reflect this part of the curve according to any symmetries the graph exhibits.

Let's consider a few examples where we can use some of these suggestions.

E X A M P L E   1   Graph $f(x) = \sqrt{x}$.

Solution

The radicand must be nonnegative, so the domain is the set of nonnegative real numbers. Because $x \geq 0$, $f(-x)$ is not a real number; thus there is no symmetry for this graph. We see that $f(0) = 0$, so both intercepts are 0. That is, the origin $(0, 0)$ is a point of the graph. Now let's set up a table of values, keeping in mind that $x \geq 0$. Plotting these points and connecting them with a smooth curve produces Figure 13.17.

x	f(x)
0	0
1	1
4	2
9	3

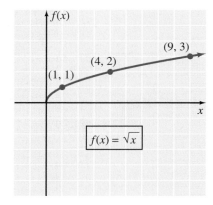

Figure 13.17

Sometimes a new function is defined in terms of old functions. In such cases, the definition plays an important role in the study of the new function. Consider the following example.

E X A M P L E   2   Graph the function $f(x) = |x|$.

Solution

The concept of absolute value is defined for all real numbers as

$$|x| = x \quad \text{if } x \geq 0$$
$$|x| = -x \quad \text{if } x < 0$$

Therefore, we can express the absolute-value function as

$$f(x) = |x| = \begin{cases} x & \text{if } x \geq 0 \\ -x & \text{if } x < 0 \end{cases}$$

The graph of $f(x) = x$ for $x \geq 0$ is the ray in the first quadrant, and the graph of $f(x) = -x$ for $x < 0$ is the half-line in the second quadrant, as indicated in Figure 13.18.

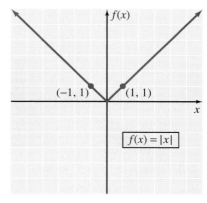

Figure 13.18

REMARK: Note in Example 2 that the equation $f(x) = |x|$ does exhibit y-axis symmetry because $f(-x) = |-x| = |x|$. Even though we did not use the symmetry idea to sketch the curve, you should recognize that the symmetry does exist.

Translations of the Basic Curves

From our work in Chapter 12, we know that the graph of $f(x) = x^2 + 3$ is the graph of $f(x) = x^2$ moved up 3 units. Likewise, the graph of $f(x) = x^2 - 2$ is the graph of $f(x) = x^2$ moved down 2 units. Now we will describe in general the concept of **vertical translation.**

> **Vertical Translation**
>
> The graph of $y = f(x) + k$ is the graph of $y = f(x)$ shifted k units upward if $k > 0$ or shifted $|k|$ units downward if $k < 0$.

In Figure 13.19, we obtain the graph of $f(x) = |x| + 2$ by shifting the graph of $f(x) = |x|$ upward 2 units, and we obtain the graph of $f(x) = |x| - 3$ by shifting the graph of $f(x) = |x|$ downward 3 units. (Remember that we can write $f(x) = |x| - 3$ as $f(x) = |x| + (-3)$.)

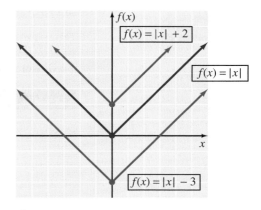

Figure 13.19

We also graphed horizontal translations of the basic parabola in Chapter 12. For example, the graph of $f(x) = (x - 4)^2$ is the graph of $f(x) = x^2$ shifted 4 units to the right, and the graph of $f(x) = (x + 5)^2$ is the graph of $f(x) = x^2$ shifted 5 units to the left. We describe the general concept of a **horizontal translation** as follows:

Horizontal Translation

The graph of $y = f(x - h)$ is the graph of $y = f(x)$ shifted h units to the right if $h > 0$ or shifted $|h|$ units to the left if $h < 0$.

In Figure 13.20, we obtain the graph of $f(x) = (x - 3)^3$ by shifting the graph of $f(x) = x^3$ to the right 3 units. Likewise, we obtain the graph of $f(x) = (x + 2)^3$ by shifting the graph of $f(x) = x^3$ to the left 2 units.

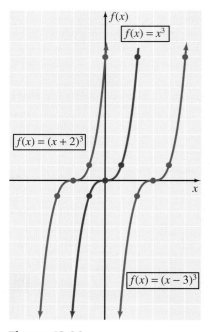

Figure 13.20

Reflections of the Basic Curves

From our work in Chapter 12, we know that the graph of $f(x) = -x^2$ is the graph of $f(x) = x^2$ reflected through the x axis. We describe the general concept of an **x-axis reflection** as follows:

x-Axis Reflection

The graph of $y = -f(x)$ is the graph of $y = f(x)$ reflected through the x axis.

In Figure 13.21, we obtain the graph of $f(x) = -\sqrt{x}$ by reflecting the graph of $f(x) = \sqrt{x}$ through the x axis. Reflections are sometimes referred to as **mirror images.** Thus, in Figure 13.21, if we think of the x axis as a mirror, the graphs of $f(x) = \sqrt{x}$ and $f(x) = -\sqrt{x}$ are mirror images of each other.

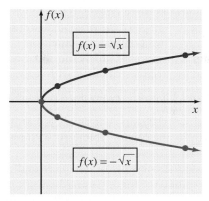

Figure 13.21

In Chapter 12, we did not consider a y-axis reflection of the basic parabola $f(x) = x^2$ because it is symmetric with respect to the y axis. In other words, a y-axis reflection of $f(x) = x^2$ produces the same figure in the same location. At this time we will describe the general concept of a y-axis reflection.

> **y-Axis Reflection**
>
> The graph of $y = f(-x)$ is the graph of $y = f(x)$ reflected through the y axis.

Now suppose that we want to do a y-axis reflection of $f(x) = \sqrt{x}$. Because $f(x) = \sqrt{x}$ is defined for $x \geq 0$, the y-axis reflection $f(x) = \sqrt{-x}$ is defined for $-x \geq 0$, which is equivalent to $x \leq 0$. Figure 13.22 shows the y-axis reflection of $f(x) = \sqrt{x}$.

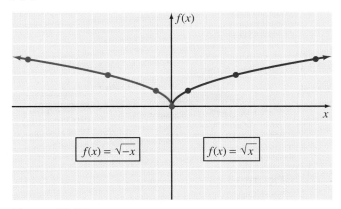

Figure 13.22

Vertical Stretching and Shrinking

Translations and reflections are called **rigid transformations** because the basic shape of the curve being transformed is not changed. In other words, only the positions of the graphs are changed. Now we want to consider some transformations that distort the shape of the original figure somewhat.

In Chapter 12, we graphed the equation $y = 2x^2$ by doubling the y coordinates of the ordered pairs that satisfy the equation $y = x^2$. We obtained a parabola with its vertex at the origin, symmetric with respect to the y axis, but *narrower* than the basic parabola. Likewise, we graphed the equation $y = \frac{1}{2}x^2$ by halving the y coordinates of the ordered pairs that satisfy $y = x^2$. In this case, we obtained a parabola with its vertex at the origin, symmetric with respect to the y axis, but *wider* than the basic parabola.

We can use the concepts of *narrower* and *wider* to describe parabolas, but they cannot accurately be used to describe some other curves. Instead, we use the more general concepts of vertical *stretching* and *shrinking*.

Vertical Stretching and Shrinking

The graph of $y = cf(x)$ is obtained from the graph of $y = f(x)$ by multiplying the y coordinates of $y = f(x)$ by c. If $c > 1$, the graph is said to be *stretched* by a factor of c, and if $0 < c < 1$, the graph is said to be *shrunk* by a factor of c.

In Figure 13.23, the graph of $f(x) = 2\sqrt{x}$ is obtained by doubling the y coordinates of points on the graph of $f(x) = \sqrt{x}$. Likewise, in Figure 13.23, the graph of $f(x) = \frac{1}{2}\sqrt{x}$ is obtained by halving the y coordinates of points on the graph of $f(x) = \sqrt{x}$.

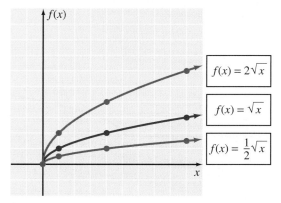

Figure 13.23

Successive Transformations

Some curves are the result of performing more than one transformation on a basic curve. Let's consider the graph of a function that involves a stretching, a reflection, a horizontal translation, and a vertical translation of the basic absolute-value function.

EXAMPLE 3

Graph $f(x) = -2|x - 3| + 1$.

Solution

This is the basic absolute-value curve stretched by a factor of two, reflected through the x axis, shifted 3 units to the right, and shifted 1 unit upward. To sketch the graph, we locate the point $(3, 1)$ and then determine a point on each of the rays. The graph is shown in Figure 13.24.

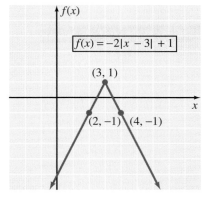

Figure 13.24

REMARK: Note in Example 3 that we did not sketch the original basic curve $f(x) = |x|$ or any of the intermediate transformations. However, it is helpful to picture each transformation mentally. This locates the point $(3, 1)$ and establishes the fact that the two rays point downward. Then a point on each ray determines the final graph.

You also need to realize that changing the order of doing the transformations may produce an incorrect graph. In Example 3, performing the translations first, followed by the stretching and x-axis reflection, would produce an incorrect graph that had its vertex at $(3, -1)$ instead of $(3, 1)$. Unless parentheses indicate otherwise, stretchings, shrinkings, and x-axis reflections should be performed before translations.

EXAMPLE 4

Graph the function $f(x) = \dfrac{1}{x + 2} + 3$.

Solution

This is the basic curve $f(x) = \dfrac{1}{x}$ moved 2 units to the left and 3 units upward. Remember that the x axis is a horizontal asymptote, and the y axis a vertical asymptote for the curve $f(x) = \dfrac{1}{x}$. Thus for this curve, the vertical asymptote is shifted 2 units to the left, and its equation is $x = -2$. Likewise, the horizontal asymptote is shifted 3 units upward, and its equation is $y = 3$. Therefore, in Figure 13.25, we have drawn the asymptotes as dashed lines and then located a few points to help determine each branch of the curve.

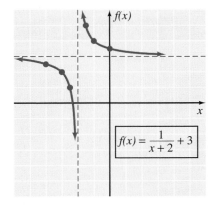

Figure 13.25

Finally, let's use a graphing utility to give another illustration of the concept of stretching and shrinking a curve.

EXAMPLE 5   If $f(x) = \sqrt{25 - x^2}$, sketch a graph of $y = 2(f(x))$ and $y = \dfrac{1}{2}(f(x))$.

Solution

If $y = f(x) = \sqrt{25 - x^2}$, then

$$y = 2(f(x)) = 2\sqrt{25 - x^2} \qquad \text{and} \qquad y = \dfrac{1}{2}(f(x)) = \dfrac{1}{2}\sqrt{25 - x^2}$$

Graphing all three of these functions on the same set of axes produces Figure 13.26.

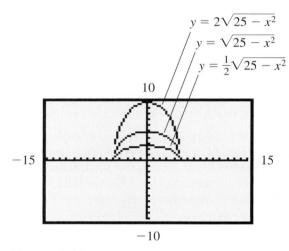

$$y = 2\sqrt{25 - x^2}$$
$$y = \sqrt{25 - x^2}$$
$$y = \tfrac{1}{2}\sqrt{25 - x^2}$$

Figure 13.26

CONCEPT QUIZ

For Problems 1–5, match the function with the description of its graph relative to the graph of $f(x) = \sqrt{x}$.

1. $f(x) = \sqrt{x} + 3$

2. $f(x) = -\sqrt{x}$

3. $f(x) = \sqrt{x + 3}$

4. $f(x) = \sqrt{-x}$

5. $f(x) = 3\sqrt{x}$

A. Stretched by a factor of three

B. Reflected across the y axis

C. Shifted up three units

D. Reflected across the x axis

E. Shifted three units to the left

PROBLEM SET 13.3

For Problems 1–34, graph each of the functions.

1. $f(x) = -x^3$

2. $f(x) = x^3 - 2$

3. $f(x) = -(x - 4)^2 + 2$

4. $f(x) = -2(x + 3)^2 - 4$

5. $f(x) = \dfrac{1}{x} - 2$

6. $f(x) = \dfrac{1}{x - 2}$

7. $f(x) = |x - 1| + 2$

8. $f(x) = -|x + 2|$

9. $f(x) = \dfrac{1}{2}|x|$

10. $f(x) = -2|x|$

11. $f(x) = -2\sqrt{x}$

12. $f(x) = 2\sqrt{x - 1}$

13. $f(x) = \sqrt{x + 2} - 3$

14. $f(x) = -\sqrt{x + 2} + 2$

15. $f(x) = \dfrac{2}{x - 1} + 3$

16. $f(x) = \dfrac{3}{x + 3} - 4$

17. $f(x) = \sqrt{2 - x}$

18. $f(x) = \sqrt{-1 - x}$

19. $f(x) = -3(x - 2)^2 - 1$

20. $f(x) = (x + 5)^2 - 2$

21. $f(x) = 3(x - 2)^3 - 1$

22. $f(x) = -2(x + 1)^3 + 2$

23. $f(x) = 2x^3 + 3$

24. $f(x) = -2x^3 - 1$

25. $f(x) = -2\sqrt{x + 3} + 4$

26. $f(x) = -3\sqrt{x - 1} + 2$

27. $f(x) = \dfrac{-2}{x + 2} + 2$

28. $f(x) = \dfrac{-1}{x - 1} - 1$

29. $f(x) = \dfrac{x - 1}{x}$

30. $f(x) = \dfrac{x + 2}{x}$

31. $f(x) = -3|x + 4| + 3$

32. $f(x) = -2|x - 3| - 4$

33. $f(x) = 4|x| + 2$

34. $f(x) = -3|x| - 4$

35. The graph of $y = f(x)$ with a domain of $-2 \le x \le 2$ is shown in Figure 13.27.

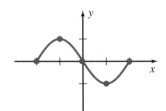

Figure 13.27

Sketch the graph of each of the following transformations of $y = f(x)$.

a. $y = f(x) + 3$

b. $y = f(x - 2)$

c. $y = -f(x)$

d. $y = f(x + 3) - 4$

36. Use the definition of absolute value to help with the following graphs.

a. $f(x) = x + |x|$

b. $f(x) = x - |x|$

c. $f(x) = |x| - x$

d. $f(x) = \dfrac{x}{|x|}$

e. $f(x) = \dfrac{|x|}{x}$

▪▪▫ Thoughts into words

37. Is the graph of $f(x) = x^2 + 2x + 4$ a y-axis reflection of $f(x) = x^2 - 2x + 4$? Defend your answer.

38. Is the graph of $f(x) = x^2 - 4x - 7$ an x-axis reflection of $f(x) = x^2 + 4x + 7$? Defend your answer.

39. Your friend claims that the graph of $f(x) = \dfrac{2x + 1}{x}$ is the graph of $f(x) = \dfrac{1}{x}$ shifted 2 units upward. How could you decide whether she is correct?

 ### Graphing calculator activities

40. Use a graphing calculator to check your graphs for Problems 12–27.

41. Use a graphing calculator to check your graphs for Problem 36.

42. For each of the following, answer the question on the basis of your knowledge of transformations, and then use a graphing calculator to check your answer.

a. Is the graph of $f(x) = 2x^2 + 8x + 13$ a y-axis reflection of $f(x) = 2x^2 - 8x + 13$?

b. Is the graph of $f(x) = 3x^2 - 12x + 16$ an x-axis reflection of $f(x) = -3x^2 + 12x - 16$?

c. Is the graph of $f(x) = \sqrt{4-x}$ a y-axis reflection of $f(x) = \sqrt{x+4}$?

d. Is the graph of $f(x) = \sqrt{3-x}$ a y-axis reflection of $f(x) = \sqrt{x-3}$?

e. Is the graph of $f(x) = -x^3 + x + 1$ a y-axis reflection of $f(x) = x^3 - x + 1$?

f. Is the graph of $f(x) = -(x-2)^3$ an x-axis reflection of $f(x) = (x-2)^3$?

g. Is the graph of $f(x) = -x^3 - x^2 - x + 1$ an x-axis reflection of $f(x) = x^3 + x^2 + x - 1$?

h. Is the graph of $f(x) = \dfrac{3x+1}{x}$ a vertical translation of $f(x) = \dfrac{1}{x}$ upward 3 units?

i. Is the graph of $f(x) = 2 + \dfrac{1}{x}$ a y-axis reflection of $f(x) = \dfrac{2x-1}{x}$?

43. Are the graphs of $f(x) = 2\sqrt{x}$ and $g(x) = \sqrt{2x}$ identical? Defend your answer.

44. Are the graphs of $f(x) = \sqrt{x+4}$ and $g(x) = \sqrt{-x+4}$ y-axis reflections of each other? Defend your answer.

Answers to Concept Quiz

1. C **2.** D **3.** E **4.** B **5.** A

13.4 Composition of Functions

Objectives

■ Given two functions, find the composite function.

■ Evaluate a composite function for a given member of the domain.

■ Determine the domain and range for a composite function.

The basic operations of addition, subtraction, multiplication, and division can be performed on functions. However, for our purposes in this text, there is an additional operation, called **composition**, that we will use in the next chapter. Let's start with the definition and an illustration of this operation.

DEFINITION 13.3

The **composition** of functions f and g is defined by

$$(f \circ g)(x) = f(g(x))$$

for all x in the domain of g such that $g(x)$ is in the domain of f.

The left side, $(f \circ g)(x)$, of the equation in Definition 13.3 can be read as "the composition of f and g," and the right side, $f(g(x))$, can be read as "f of g of x." It may also be helpful for you to picture Definition 13.3 as two function machines *hooked together* to produce another function (often called a **composite function**), as illustrated in Figure 13.28. Note that what comes out of the function g is substituted into the function f. Thus composition is sometimes called the substitution of functions. Figure 13.28 also vividly illustrates the fact that $f \circ g$ is defined *for all x in the domain of g such that g(x) is in the domain of f.* In other words, what comes out of g must be capable of being fed into f. Let's consider some examples.

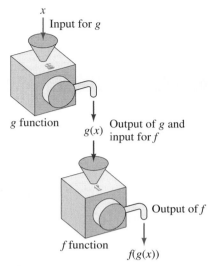

Figure 13.28

E X A M P L E   1

If $f(x) = x^2$ and $g(x) = x - 3$, find $(f \circ g)(x)$ and determine its domain.

Solution

Applying Definition 13.3, we obtain

$$(f \circ g)(x) = f(g(x))$$
$$= f(x - 3)$$
$$= (x - 3)^2$$

Because g and f are both defined for all real numbers, so is $f \circ g$. ∎

E X A M P L E   2

If $f(x) = \sqrt{x}$ and $g(x) = x - 4$, find $(f \circ g)(x)$ and determine its domain.

Solution

Applying Definition 13.3, we obtain

$$(f \circ g)(x) = f(g(x))$$
$$= f(x - 4)$$
$$= \sqrt{x - 4}$$

The domain of g is all real numbers, but the domain of f is only the nonnegative real numbers. Thus $g(x)$, which is $x - 4$, has to be nonnegative. Therefore,

$$x - 4 \geq 0$$
$$x \geq 4$$

and the domain of $f \circ g$ is $D = \{x | x \geq 4\}$. ∎

Definition 13.3, with f and g interchanged, defines the composition of g and f as $(g \circ f)(x) = g(f(x))$.

E X A M P L E   3   If $f(x) = x^2$ and $g(x) = x - 3$, find $(g \circ f)(x)$ and determine its domain.

Solution

$$(g \circ f)(x) = g(f(x))$$
$$= g(x^2)$$
$$= x^2 - 3$$

Because f and g are both defined for all real numbers, the domain of $g \circ f$ is the set of all real numbers.   ■

The results of Examples 1 and 3 demonstrate an important idea: that the composition of functions is *not a commutative operation.* In other words, it is not true that $f \circ g = g \circ f$ for all functions f and g. However, as we will see in the next chapter, there is a special class of functions where $f \circ g = g \circ f$.

E X A M P L E   4   If $f(x) = 2x + 3$ and $g(x) = \sqrt{x - 1}$, determine each of the following.

(a) $(f \circ g)(x)$ **(b)** $(g \circ f)(x)$ **(c)** $(f \circ g)(5)$ **(d)** $(g \circ f)(7)$

Solution

(a) $(f \circ g)(x) = f(g(x))$
$$= f(\sqrt{x - 1})$$
$$= 2\sqrt{x - 1} + 3 \qquad D = \{x | x \geq 1\}$$

(b) $(g \circ f)(x) = g(f(x))$
$$= g(2x + 3)$$
$$= \sqrt{2x + 3 - 1}$$
$$= \sqrt{2x + 2} \qquad D = \{x | x \geq -1\}$$

(c) $(f \circ g)(5) = 2\sqrt{5 - 1} + 3 = 7$

(d) $(g \circ f)(7) = \sqrt{2(7) + 2} = 4$ ■

E X A M P L E   5   If $f(x) = \dfrac{2}{x - 1}$ and $g(x) = \dfrac{1}{x}$, find $(f \circ g)(x)$ and $(g \circ f)(x)$. Determine the domain for each composite function.

Solution

$$(f \circ g)(x) = f(g(x))$$

$$= f\left(\frac{1}{x}\right)$$

$$= \frac{2}{\dfrac{1}{x} - 1} = \frac{2}{\dfrac{1 - x}{x}}$$

$$= \frac{2x}{1 - x}$$

The domain of g is all real numbers except 0, and the domain of f is all real numbers except 1. Because $g(x)$, which is $\dfrac{1}{x}$, cannot equal 1, we have

$$\frac{1}{x} \neq 1$$

$$x \neq 1$$

Therefore, the domain of $f \circ g$ is $D = \{x \mid x \neq 0 \text{ and } x \neq 1\}$.

$$(g \circ f)(x) = g(f(x))$$

$$= g\left(\frac{2}{x - 1}\right)$$

$$= \frac{1}{\dfrac{2}{x - 1}}$$

$$= \frac{x - 1}{2}$$

The domain of f is all real numbers except 1, and the domain of g is all real numbers except 0. Because $f(x)$, which is $\dfrac{2}{x - 1}$, will never equal 0, the domain of $g \circ f$ is $D = \{x \mid x \neq 1\}$. ∎

A graphing utility can be used to find the graph of a composite function without actually forming the function algebraically. Let's see how this works.

EXAMPLE 6   If $f(x) = x^3$ and $g(x) = x - 4$, use a graphing utility to obtain the graph of $y = (f \circ g)(x)$ and of $y = (g \circ f)(x)$.

Solution
To find the graph of $y = (f \circ g)(x)$, we can make the following assignments.

$$Y_1 = x - 4$$
$$Y_2 = (Y_1)^3$$

(Note that we have substituted Y_1 for x in $f(x)$ and assigned this expression to Y_2, much the same way as we would algebraically.) Now, by showing only the graph of Y_2, we obtain Figure 13.29.

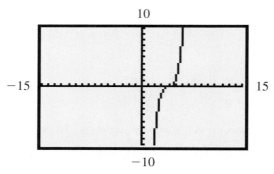

Figure 13.29

To find the graph of $y = (g \circ f)(x)$, we can make the following assignments.

$$Y_1 = x^3$$
$$Y_2 = Y_1 - 4$$

The graph of $y = (g \circ f)(x)$ is the graph of Y_2, as shown in Figure 13.30.

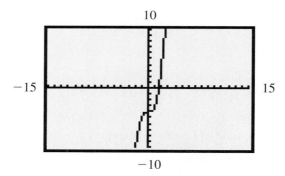

Figure 13.30

Take another look at Figures 13.29 and 13.30. Note that in Figure 13.29, the graph of $y = (f \circ g)(x)$ is the basic cubic curve $f(x) = x^3$ shifted 4 units to the right. Likewise, in Figure 13.30, the graph of $y = (g \circ f)(x)$ is the basic cubic curve shifted 4 units downward.

CONCEPT QUIZ

For Problems 1–5, answer true or false.

1. The composition of functions is a commutative operation.

2. To find $(h \circ k)(x)$ the $k(x)$ will be substituted in the function h.

3. The notation $(f \circ g)(x)$ is read as "the substitution of g and f."

4. The domain for $(f \circ g)(x)$, is always the same as the domain of g.

5. The notation $f(g(x))$ is read "f of g of x."

PROBLEM SET 13.4

For Problems 1–12, determine the indicated functional values.

1. If $f(x) = 9x - 2$ and $g(x) = -4x + 6$, find $(f \circ g)(-2)$ and $(g \circ f)(4)$.

2. If $f(x) = -2x - 6$ and $g(x) = 3x + 10$, find $(f \circ g)(5)$ and $(g \circ f)(-3)$.

3. If $f(x) = 4x^2 - 1$ and $g(x) = 4x + 5$, find $(f \circ g)(1)$ and $(g \circ f)(4)$.

4. If $f(x) = -5x + 2$ and $g(x) = -3x^2 + 4$, find $(f \circ g)(-2)$ and $(g \circ f)(-1)$.

5. If $f(x) = \dfrac{1}{x}$ and $g(x) = \dfrac{2}{x-1}$, find $(f \circ g)(2)$ and $(g \circ f)(-1)$.

6. If $f(x) = \dfrac{2}{x-1}$ and $g(x) = -\dfrac{3}{x}$, find $(f \circ g)(1)$ and $(g \circ f)(-1)$.

7. If $f(x) = \dfrac{1}{x-2}$ and $g(x) = \dfrac{4}{x-1}$, find $(f \circ g)(3)$ and $(g \circ f)(2)$.

8. If $f(x) = \sqrt{x+6}$ and $g(x) = 3x - 1$, find $(f \circ g)(-2)$ and $(g \circ f)(-2)$.

9. If $f(x) = \sqrt{3x-2}$ and $g(x) = -x + 4$, find $(f \circ g)(1)$ and $(g \circ f)(6)$.

10. If $f(x) = -5x + 1$ and $g(x) = \sqrt{4x+1}$, find $(f \circ g)(6)$ and $(g \circ f)(-1)$.

11. If $f(x) = |4x - 5|$ and $g(x) = x^3$, find $(f \circ g)(-2)$ and $(g \circ f)(2)$.

12. If $f(x) = -x^3$ and $g(x) = |2x + 4|$, find $(f \circ g)(-1)$ and $(g \circ f)(-3)$.

For Problems 13–30, determine $(f \circ g)(x)$ and $(g \circ f)(x)$ for each pair of functions. Also specify the domain of $(f \circ g)(x)$ and $(g \circ f)(x)$.

13. $f(x) = 3x$ and $g(x) = 5x - 1$

14. $f(x) = 4x - 3$ and $g(x) = -2x$

15. $f(x) = -2x + 1$ and $g(x) = 7x + 4$

16. $f(x) = 6x - 5$ and $g(x) = -x + 6$

17. $f(x) = 3x + 2$ and $g(x) = x^2 + 3$

18. $f(x) = -2x + 4$ and $g(x) = 2x^2 - 1$

19. $f(x) = 2x^2 - x + 2$ and $g(x) = -x + 3$

20. $f(x) = 3x^2 - 2x - 4$ and $g(x) = -2x + 1$

21. $f(x) = \dfrac{3}{x}$ and $g(x) = 4x - 9$

22. $f(x) = -\dfrac{2}{x}$ and $g(x) = -3x + 6$

23. $f(x) = \sqrt{x+1}$ and $g(x) = 5x + 3$

24. $f(x) = 7x - 2$ and $g(x) = \sqrt{2x-1}$

25. $f(x) = \dfrac{1}{x}$ and $g(x) = \dfrac{1}{x-4}$

26. $f(x) = \dfrac{2}{x+3}$ and $g(x) = -\dfrac{3}{x}$

27. $f(x) = \sqrt{x}$ and $g(x) = \dfrac{4}{x}$

28. $f(x) = \dfrac{2}{x}$ and $g(x) = |x|$

29. $f(x) = \dfrac{3}{2x}$ and $g(x) = \dfrac{1}{x+1}$

30. $f(x) = \dfrac{4}{x-2}$ and $g(x) = \dfrac{3}{4x}$

For Problems 31–38, show that $(f \circ g)(x) = x$ and $(g \circ f)(x) = x$ for each pair of functions.

31. $f(x) = 3x$ and $g(x) = \dfrac{1}{3}x$

32. $f(x) = -2x$ and $g(x) = -\dfrac{1}{2}x$

33. $f(x) = 4x + 2$ and $g(x) = \dfrac{x - 2}{4}$

34. $f(x) = 3x - 7$ and $g(x) = \dfrac{x + 7}{3}$

35. $f(x) = \dfrac{1}{2}x + \dfrac{3}{4}$ and $g(x) = \dfrac{4x - 3}{2}$

36. $f(x) = \dfrac{2}{3}x - \dfrac{1}{5}$ and $g(x) = \dfrac{3}{2}x + \dfrac{3}{10}$

37. $f(x) = -\dfrac{1}{4}x - \dfrac{1}{2}$ and $g(x) = -4x - 2$

38. $f(x) = -\dfrac{3}{4}x + \dfrac{1}{3}$ and $g(x) = -\dfrac{4}{3}x + \dfrac{4}{9}$

■ ■ ▨ Thoughts into words

39. How would you explain the concept of composition of functions to a friend who missed class the day it was discussed?

40. Explain why the composition of functions is not a commutative operation.

 Graphing calculator activities

41. For each of the following, (1) predict the general shape and location of the graph and then (2) use your graphing calculator to graph the function and thus check your prediction. (Your knowledge of the graphs of the basic functions being added or subtracted should be helpful when you make your predictions.)

a. $f(x) = x^3 + x^2$ **b.** $f(x) = x^3 - x^2$

c. $f(x) = x^2 - x^3$ **d.** $f(x) = |x| + \sqrt{x}$

e. $f(x) = |x| - \sqrt{x}$ **f.** $f(x) = \sqrt{x} - |x|$

42. For each of the following, use your graphing calculator to find the graph of $y = (f \circ g)(x)$ and $y = (g \circ f)(x)$. Then find $(f \circ g)(x)$ and $(g \circ f)(x)$ algebraically to see whether your results agree.

a. $f(x) = x^2$ and $g(x) = x - 3$

b. $f(x) = x^3$ and $g(x) = x + 4$

c. $f(x) = x - 2$ and $g(x) = -x^3$

d. $f(x) = x + 6$ and $g(x) = \sqrt{x}$

e. $f(x) = \sqrt{x}$ and $g(x) = x - 5$

13.5   Direct Variation and Inverse Variation

Objectives

■   Translate sentences into equations of variation.

■   Determine the constant of variation.

■   Solve variation word problems.

"The distance a car travels at a fixed rate *varies directly* as the time." "At a constant temperature, the volume of an enclosed gas *varies inversely* as the pressure." Such statements illustrate two basic types of functional relationships, called **direct variation** and **inverse variation,** which are widely used, especially in the physical sciences. These relationships can be expressed by equations that specify functions. The purpose of this section is to investigate these special functions.

The statement "*y* varies directly as *x*" means

$$y = kx$$

where *k* is a nonzero constant called the **constant of variation.** The statement "*y* is directly proportional to *x*" is also used to indicate direct variation; *k* is then referred to as the **constant of proportionality.**

REMARK: Note that the equation $y = kx$ defines a function and could be written as $f(x) = kx$ by using function notation. However, in this section it is more convenient to avoid the function notation and use variables that are meaningful in terms of the physical entities involved in the problem.

Statements that indicate direct variation may also involve powers of *x*. For example, "*y* varies directly as the square of *x*" can be written as

$$y = kx^2$$

In general, "*y* varies directly as the *n*th power of *x* $(n > 0)$" means

$$y = kx^n$$

The three types of problems that deal with direct variation are (1) to translate an English statement into an equation that expresses the direct variation, (2) to find the constant of variation from given values of the variables, and (3) to find additional values of the variables once the constant of variation has been determined. Let's consider an example of each of these types of problems.

E X A M P L E   1

Translate the statement "the tension on a spring varies directly as the distance it is stretched" into an equation and use *k* as the constant of variation.

Solution

If we let *t* represent the tension and *d* the distance, the equation becomes

$$t = kd$$

∎

E X A M P L E   2

If *A* varies directly as the square of *s* and if $A = 28$ when $s = 2$, find the constant of variation.

Solution

Because *A* varies directly as the square of *s*, we have

$$A = ks^2$$

Substituting $A = 28$ and $s = 2$, we obtain

$$28 = k(2)^2$$

Solving this equation for k yields

$$28 = 4k$$
$$7 = k$$

The constant of variation is 7. ∎

EXAMPLE 3   If y is directly proportional to x and if $y = 6$ when $x = 9$, find the value of y when $x = 24$.

Solution

The statement "y is directly proportional to x" translates into

$$y = kx$$

If we let $y = 6$ and $x = 9$, then the constant of variation becomes

$$6 = k(9)$$
$$6 = 9k$$
$$\frac{6}{9} = k$$
$$\frac{2}{3} = k$$

Thus the specific equation is $y = \frac{2}{3}x$. Now, letting $x = 24$, we obtain

$$y = \frac{2}{3}(24) = 16$$

The required value of y is 16. ∎

Inverse Variation

We define the second basic type of variation, called **inverse variation,** as follows: The statement *y varies inversely as x* means

$$y = \frac{k}{x}$$

where k is a nonzero constant; again we refer to it as the constant of variation. The phrase "y is inversely proportional to x" is also used to express inverse variation. As with direct variation, statements that indicate inverse variation may involve powers of x. For example, "y varies inversely as the square of x" can be written as

$$y = \frac{k}{x^2}$$

In general, "y varies inversely as the nth power of x $(n > 0)$" means

$$y = \frac{k}{x^n}$$

The following examples illustrate the three basic kinds of problems we run across that involve inverse variation.

E X A M P L E   4   Translate the statement "the length of a rectangle of a fixed area varies inversely as the width" into an equation that uses k as the constant of variation.

Solution
Let l represent the length and w the width, and the equation is

$$l = \frac{k}{w}$$

■

E X A M P L E   5   If y is inversely proportional to x, and $y = 4$ when $x = 12$, find the constant of variation.

Solution
Because y is inversely proportional to x, we have

$$y = \frac{k}{x}$$

Substituting $y = 4$ and $x = 12$, we obtain

$$4 = \frac{k}{12}$$

Solving this equation for k yields

$$k = 48$$

The constant of variation is 48.

■

E X A M P L E   6   Suppose the number of days it takes to complete a construction job varies inversely as the number of people assigned to the job. If it takes 7 people 8 days to do the job, how long would it take 14 people to complete the job?

Solution
Let d represent the number of days and p the number of people. The phrase "number of days . . . varies inversely as the number of people" translates into

$$d = \frac{k}{p}$$

Let $d = 8$ when $p = 7$, and the constant of variation becomes

$$8 = \frac{k}{7}$$

$$k = 56$$

Thus the specific equation is

$$d = \frac{56}{p}$$

Now, let $p = 14$ to obtain

$$d = \frac{56}{14}$$

$$= 4$$

It should take 14 people 4 days to complete the job. ■

The terms *direct* and *inverse*, as applied to variation, refer to the relative behavior of the variables involved in the equation. That is, in direct variation ($y = kx$), an assignment of *increasing absolute values for x* produces *increasing absolute values for y*, whereas in inverse variation $\left(y = \dfrac{k}{x} \right)$, an assignment of *increasing absolute values for x* produces *decreasing absolute values for y*.

Joint Variation

Variation may involve more than two variables. The following table illustrates some variation statements and their equivalent algebraic equations that use k as the constant of variation.

Variation Statement	Algebraic Equation
1. y varies jointly as x and z	$y = kxz$
2. y varies jointly as x, z, and w	$y = kxzw$
3. V varies jointly as h and the square of r	$V = khr^2$
4. h varies directly as V and inversely as w	$h = \dfrac{kV}{w}$
5. y is directly proportional to x and inversely proportional to the square of z	$y = \dfrac{kx}{z^2}$
6. y varies jointly as w and z and inversely as x	$y = \dfrac{kwz}{x}$

Statements 1, 2, and 3 illustrate the concept of **joint variation.** Statements 4 and 5 show that both direct and inverse variation may occur in the same problem. Statement 6 combines joint variation with inverse variation.

The two final examples of this section illustrate some of these variation situations.

E X A M P L E   7

The length of a rectangular box with a fixed height varies directly as the volume and inversely as the width. If the length is 12 centimeters when the volume is 960 cubic centimeters and the width is 8 centimeters, find the length when the volume is 700 centimeters and the width is 5 centimeters.

Solution

Use l for length, V for volume, and w for width, and the phrase "length varies directly as the volume and inversely as the width" translates into

$$l = \frac{kV}{w}$$

Substitute $l = 12$, $V = 960$, and $w = 8$, and the constant of variation becomes

$$12 = \frac{k(960)}{8}$$

$$12 = 120k$$

$$\frac{1}{10} = k$$

Thus the specific equation is

$$l = \frac{\frac{1}{10}V}{w} = \frac{V}{10w}$$

Now, let $V = 700$ and $w = 5$ to obtain

$$l = \frac{700}{10(5)} = \frac{700}{50} = 14$$

The length is 14 centimeters. ◼

E X A M P L E   8

Suppose that y varies jointly as x and z and inversely as w. If $y = 154$ when $x = 6$, $z = 11$, and $w = 3$, find the constant of variation.

Solution

The statement "y varies jointly as x and z and inversely as w" translates into

$$y = \frac{kxz}{w}$$

Substitute $y = 154$, $x = 6$, $z = 11$, and $w = 3$ to obtain

$$154 = \frac{k(6)(11)}{3}$$

$$154 = 22k$$

$$7 = k$$

The constant of variation is 7. ◼

CONCEPT QUIZ

For Problems 1–5, answer true or false.

1. In the equation $y = kx$, the k is a quantity that varies as y.

2. The equation $y = kx$ defines a function and could be written with functional notation as $f(x) = kx$.

3. When variation involves more than two variables it is called proportional variation.

4. Every equation of variation will have a constant of variation.

5. In joint variation both direct and inverse variation may occur in the same problem.

For Problems 6–10, match the statement of variation with its equation.

6. y varies directly as x	**A.** $y = \dfrac{k}{x}$
7. y varies inversely as x	**B.** $y = kxz$
8. y varies directly as the square of x	**C.** $y = kx^2$
9. y varies directly as the square root of x	**D.** $y = kx$
10. y varies jointly as x and z	**E.** $y = k\sqrt{x}$

PROBLEM SET 13.5

For Problems 1–10, translate each statement of variation into an equation and use k as the constant of variation.

1. y varies inversely as the square of x.

2. y varies directly as the cube of x.

3. C varies directly as g and inversely as the cube of t.

4. V varies jointly as l and w.

5. The volume (V) of a sphere is directly proportional to the cube of its radius (r).

6. At a constant temperature, the volume (V) of a gas varies inversely as the pressure (P).

7. The surface area (S) of a cube varies directly as the square of the length of an edge (e).

8. The intensity of illumination (I) received from a source of light is inversely proportional to the square of the distance (d) from the source.

9. The volume (V) of a cone varies jointly as its height and the square of its radius.

10. The volume (V) of a gas varies directly as the absolute temperature (T) and inversely as the pressure (P).

For Problems 11–24, find the constant of variation for each of the stated conditions.

11. y varies directly as x, and $y = 8$ when $x = 12$.

12. y varies directly as x, and $y = 60$ when $x = 24$.

13. y varies directly as the square of x, and $y = -144$ when $x = 6$.

14. y varies directly as the cube of x, and $y = 48$ when $x = -2$.

15. V varies jointly as B and h, and $V = 96$ when $B = 24$ and $h = 12$.

16. A varies jointly as b and h, and $A = 72$ when $b = 16$ and $h = 9$.

17. y varies inversely as x, and $y = -4$ when $x = \dfrac{1}{2}$.

18. y varies inversely as x, and $y = -6$ when $x = \dfrac{4}{3}$.

19. r varies inversely as the square of t, and $r = \dfrac{1}{8}$ when $t = 4$.

20. r varies inversely as the cube of t, and $r = \dfrac{1}{16}$ when $t = 4$.

21. y varies directly as x and inversely as z, and $y = 45$ when $x = 18$ and $z = 2$.

22. y varies directly as x and inversely as z, and $y = 24$ when $x = 36$ and $z = 18$.

23. y is directly proportional to x and inversely proportional to the square of z, and $y = 81$ when $x = 36$ and $z = 2$.

24. y is directly proportional to the square of x and inversely proportional to the cube of z, and $y = 4\dfrac{1}{2}$ when $x = 6$ and $z = 4$.

Solve each of the following problems.

25. If y is directly proportional to x, and $y = 36$ when $x = 48$, find the value of y when $x = 12$.

26. If y is directly proportional to x, and $y = 42$ when $x = 28$, find the value of y when $x = 38$.

27. If y is inversely proportional to x, and $y = \dfrac{1}{9}$ when $x = 12$, find the value of y when $x = 8$.

28. If y is inversely proportional to x, and $y = \dfrac{1}{35}$ when $x = 14$, find the value of y when $x = 16$.

29. If A varies jointly as b and h, and $A = 60$ when $b = 12$ and $h = 10$, find A when $b = 16$ and $h = 14$.

30. If V varies jointly as B and h, and $V = 51$ when $B = 17$ and $h = 9$, find V when $B = 19$ and $h = 12$.

31. The volume of a gas at a constant temperature varies inversely as the pressure. What is the volume of a gas under pressure of 25 pounds if the gas occupies 15 cubic centimeters under a pressure of 20 pounds?

32. The time required for a car to travel a certain distance varies inversely as the rate at which it travels. If it takes 4 hours at 50 miles per hour to travel the distance, how long will it take at 40 miles per hour?

33. The volume (V) of a gas varies directly as the temperature (T) and inversely as the pressure (P). If $V = 48$ when $T = 320$ and $P = 20$, find V when $T = 280$ and $P = 30$.

34. The distance that a freely falling body falls varies directly as the square of the time it falls. If a body falls 144 feet in 3 seconds, how far will it fall in 5 seconds?

35. The period (the time required for one complete oscillation) of a simple pendulum varies directly as the square root of its length. If a pendulum 12 feet long has a period of 4 seconds, find the period of a pendulum 3 feet long.

36. The simple interest earned by a certain amount of money varies jointly as the rate of interest and the time (in years) that the money is invested. If the money is invested at 12% for 2 years, $120 is earned. How much is earned if the money is invested at 14% for 3 years?

37. The electrical resistance of a wire varies directly as its length and inversely as the square of its diameter. If the resistance of 200 meters of wire that has a diameter of $\dfrac{1}{2}$ centimeter is 1.5 ohms, find the resistance of 400 meters of wire with a diameter of $\dfrac{1}{4}$ centimeter.

38. The volume of a cylinder varies jointly as its altitude and the square of the radius of its base. If the volume of a cylinder is 1386 cubic centimeters when the radius of the base is 7 centimeters and its altitude is 9 centimeters, find the volume of a cylinder that has a base of radius 14 centimeters and an altitude of 5 centimeters.

39. The simple interest earned by a certain amount of money varies jointly as the rate of interest and the time (in years) that the money is invested.

a. If some money invested at 11% for 2 years earns $385, how much would the same amount earn at 12% for 1 year?

b. If some money invested at 12% for 3 years earns $819, how much would the same amount earn at 14% for 2 years?

c. If some money invested at 14% for 4 years earns $1960, how much would the same amount earn at 15% for 2 years?

40. The period (the time required for one complete oscillation) of a simple pendulum varies directly as the square root of its length. If a pendulum 9 inches long has a period of 2.4 seconds, find the period of a pendulum 12 inches long. Express your answer to the nearest tenth of a second.

41. The volume of a cylinder varies jointly as its altitude and the square of the radius of its base. If the volume of a cylinder is 549.5 cubic meters when the radius of the base is 5 meters and its altitude is 7 meters, find the volume of a cylinder that has a base with a radius of 9 meters and an altitude of 14 meters.

42. If y is directly proportional to x and inversely proportional to the square of z, and if $y = 0.336$ when $x = 6$ and $z = 5$, find the constant of variation.

43. If y is inversely proportional to the square root of x, and if $y = 0.08$ when $x = 225$, find y when $x = 625$.

■ ■ ■ **Thoughts into words**

44. How would you explain the difference between direct variation and inverse variation?

45. Suppose that y varies directly as the square of x. Does doubling the value of x also double the value of y? Explain your answer.

46. Suppose that y varies inversely as x. Does doubling the value of x also double the value of y? Explain your answer.

Answers to Concept Quiz

1. False   **2.** True   **3.** False   **4.** True   **5.** True   **6.** D   **7.** A   **8.** C   **9.** E   **10.** B

CHAPTER 13

SUMMARY

(13.1) A **relation** is a set of ordered pairs; a **function** is a relation in which no two ordered pairs have the same first component. The **domain** of a relation (or function) is the set of all first components, and the **range** is the set of all second components of the ordered pairs.

Single symbols such as f, g, and h are commonly used to name functions. The symbol $f(x)$ represents the element in the range associated with x from the domain. Thus if $f(x) = 3x + 7$, then $f(1) = 3(1) + 7 = 10$.

(13.2) Any function defined by an equation that can be written in the form

$$f(x) = ax + b$$

where a and b are real numbers, is a **linear function.** The graph of a linear function is a straight line.

Any function defined by an equation that can be written in the form

$$f(x) = ax^2 + bx + c$$

where a, b, and c are real numbers and $a \neq 0$, is a **quadratic function.** The graph of any quadratic function is a **parabola,** which can be drawn using either of the following methods.

1. Express the function in the form $f(x) = a(x - h)^2 + k$ and use the values of a, h, and k to determine the parabola.
2. Express the function in the form $f(x) = ax^2 + bx + c$ and use the fact that the vertex is at

$$\left(-\frac{b}{2a}, f\left(-\frac{b}{2a} \right) \right)$$

and the axis of symmetry is

$$x = -\frac{b}{2a}$$

We can solve some applications that involve maximum and minimum values with our knowledge of parabolas that are generated by quadratic functions.

(13.3) Another important graphing technique is to be able to recognize equations of the transformations of basic curves. We have worked with the following transformations in this chapter.

Vertical Translation   The graph of $y = f(x) + k$ is the graph of $y = f(x)$ shifted k units upward if $k > 0$ or shifted $|k|$ units downward if $k < 0$.

Horizontal Translation   The graph of $y = f(x - h)$ is the graph of $y = f(x)$ shifted h units to the right if $h > 0$ or shifted $|h|$ units to the left if $h < 0$.

x-axis Reflection   The graph of $y = -f(x)$ is the graph of $y = f(x)$ reflected through the x axis.

y-axis Reflection   The graph of $y = f(-x)$ is the graph of $y = f(x)$ reflected through the y axis.

Vertical Stretching and Shrinking   The graph of $y = cf(x)$ is obtained from the graph of $y = f(x)$ by multiplying the y coordinates of $y = f(x)$ by c. If $c > 1$, the graph is said to be **stretched** by a factor of c, and if $0 < c < 1$, the graph is said to be **shrunk** by a factor of c.

We list the following suggestions for graphing functions you are not familiar with.

1. Determine the domain of the function.
2. Determine any type of symmetry exhibited by the equation.
3. Find the intercepts.
4. Set up a table of values that satisfy the equation.
5. Plot the points associated with the ordered pairs and connect them with a smooth curve. Then, if appropriate, reflect this part of the curve according to any symmetry the graph exhibits.

(13.4) The **composition** of two functions f and g is defined by

$$(f \circ g)(x) = f(g(x))$$

for all x in the domain of g such that $g(x)$ is in the domain of f. Remember that the composition of functions is not a commutative operation.

(13.5) The equation $y = kx$ (k is a nonzero constant) defines a function called **direct variation.** The equation $y = \dfrac{k}{x}$ defines a function called **inverse variation.** In both cases, k is called the **constant of variation.**

CHAPTER 13 REVIEW PROBLEM SET

For Problems 1–4, specify the domain of each function.

1. $f = \{(1, 3), (2, 5), (4, 9)\}$

2. $f(x) = \dfrac{4}{x - 5}$

3. $f(x) = \dfrac{3}{x^2 + 4x}$

4. $f(x) = \sqrt{x^2 - 25}$

5. If $f(x) = x^2 - 2x - 1$, find $f(2), f(-3)$, and $f(a)$.

6. If $f(x) = 2x^2 + x - 7$, find $\dfrac{f(a + h) - f(a)}{h}$.

For Problems 7–16, graph each of the functions.

7. $f(x) = 4$

8. $f(x) = -3x + 2$

9. $f(x) = x^2 + 2x + 2$

10. $f(x) = |x| + 4$

11. $f(x) = -|x - 2|$

12. $f(x) = \sqrt{x - 2} - 3$

13. $f(x) = \dfrac{1}{x^2}$

14. $f(x) = -\dfrac{1}{2}x^2$

15. $f(x) = -3x^2 + 6x - 2$

16. $f(x) = -\sqrt{x + 1} - 2$

17. Find the coordinates of the vertex and the equation of the line of symmetry for each of the following parabolas.

   a. $f(x) = x^2 + 10x - 3$

   b. $f(x) = -2x^2 - 14x + 9$

For Problems 18–20, determine $(f \circ g)(x)$ and $(g \circ f)(x)$ for each pair of functions.

18. $f(x) = 2x - 3$ and $g(x) = 3x - 4$

19. $f(x) = x - 4$ and $g(x) = x^2 - 2x + 3$

20. $f(x) = x^2 - 5$ and $g(x) = -2x + 5$

21. If y varies directly as x and inversely as z, and if $y = 21$ when $x = 14$ and $z = 6$, find the constant of variation.

22. If y varies jointly as x and the square root of z, and if $y = 60$ when $x = 2$ and $z = 9$, find y when $x = 3$ and $z = 16$.

23. The weight of a body above the surface of the earth varies inversely as the square of its distance from the center of the earth. Assume that the radius of the earth is 4000 miles. How much would a man weigh 1000 miles above the earth's surface if he weighs 200 pounds on the surface?

24. Find two numbers whose sum is 40 and whose product is a maximum.

25. Find two numbers whose sum is 50 such that the square of one number plus six times the other number is a minimum.

26. Suppose that 50 students are able to raise $250 for a party when each one contributes $5. Furthermore, they figure that for each additional student they can find to contribute, the cost per student will decrease by a nickel. How many additional students do they need to maximize the amount of money they will have for a party?

27. The surface area of a cube varies directly as the square of the length of an edge. If the surface area of a cube that has edges 8 inches long is 384 square inches, find the surface area of a cube that has edges 10 inches long.

28. The cost for burning a 100-watt bulb is given by the function $c(h) = 0.006h$, where h represents the number of hours that the bulb burns. How much, to the nearest cent, does it cost to burn a 100-watt bulb for 4 hours per night for a 30-day month?

29. "All Items 30% Off Marked Price" is a sign in a local department store. Form a function and then use it to determine how much one has to pay for each of the following marked items: a $65 pair of shoes, a $48 pair of slacks, a $15.50 belt.

CHAPTER 13

TEST

 applies to all problems in this Chapter Test.

1. Determine the domain of the function
$$f(x) = \frac{-3}{2x^2 + 7x - 4}.$$

2. Determine the domain of the function
$$f(x) = \sqrt{5 - 3x}.$$

3. If $f(x) = -\frac{1}{2}x + \frac{1}{3}$, find $f(-3)$.

4. If $f(x) = -x^2 - 6x + 3$, find $f(-2)$.

5. Find the vertex of the parabola
$$f(x) = -2x^2 - 24x - 69.$$

6. If $f(x) = 3x^2 + 2x - 5$, find $\dfrac{f(a + h) - f(a)}{h}$.

7. If $f(x) = -3x + 4$ and $g(x) = 7x + 2$, find $(f \circ g)(x)$.

8. If $f(x) = 2x + 5$ and $g(x) = 2x^2 - x + 3$, find $(g \circ f)(x)$.

9. If $f(x) = \dfrac{3}{x - 2}$ and $g(x) = \dfrac{2}{x}$, find $(f \circ g)(x)$.

10. Determine the domain of the function $f(x) = \sqrt{x^2 + 3x - 10}$.

11. Lola wants to sell three items that cost her $15, $18, and $25. She wants to make a profit of 40% based on the cost. Create a function that you can use to determine the selling price of each item, and then use the function to calculate each selling price.

12. A manufacturer finds that for the first 500 units of its product that are produced and sold, the profit is $50 per unit. The profit on each of the units beyond 500 is decreased by $.10 times the number of additional units sold. What level of output will maximize profit?

13. If y varies inversely as x, and if $y = \frac{1}{2}$ when $x = -8$, find the constant of variation.

14. If y varies jointly as x and z, and if $y = 18$ when $x = 8$ and $z = 9$, find y when $x = 5$ and $z = 12$.

15. Find two numbers whose sum is 60, such that the sum of the square of one number plus twelve times the other number is a minimum.

16. The simple interest earned by a certain amount of money varies jointly as the rate of interest and the time (in years) that the money is invested. If $140 is earned for a certain amount of money invested at 7% for 5 years, how much is earned if the same amount is invested at 8% for 3 years?

For Problems 17–19, use the concepts of translation and/or reflection to describe how the second curve can be obtained from the first curve.

17. $f(x) = x^3$, $f(x) = (x - 6)^3 - 4$

18. $f(x) = |x|$, $f(x) = -|x| + 8$

19. $f(x) = \sqrt{x}$, $f(x) = -\sqrt{x + 5} + 7$

For Problems 20–25, graph each function.

20. $f(x) = -x - 1$

21. $f(x) = -2x^2 - 12x - 14$

22. $f(x) = 2\sqrt{x} - 2$

23. $f(x) = 3|x - 2| - 1$

24. $f(x) = -\dfrac{1}{x} + 3$

25. $f(x) = \sqrt{-x} + 2$

CUMULATIVE PRACTICE TEST *Chapters 1-13*

1. Find the greatest common factor of 48, 60, and 84.

2. Identify the type of symmetry that each of the following equations exhibits.

   a. $2x^2 + y^2 - 3x - 4 = 0$

   b. $f(x) = 3x^2 + 4$

   c. $xy = 18$

   d. $f(x) = x$

3. Find the x intercepts of the graph of the function $f(x) = 2x^2 + 17x - 9$.

4. Find the vertex of the parabola $f(x) = 2x^2 + 4x$.

5. Evaluate $\dfrac{48x^{-3}y^{-2}}{12x^{-4}y^{-3}}$ for $x = -2$ and $y = -4$.

6. Express each of the following in simplest radical form. All variables represent positive real numbers.

   a. $\sqrt{32x^3y^4}$

   b. $\sqrt[3]{32x^3y^4}$

   c. $\dfrac{2\sqrt{6}}{3\sqrt{12}}$

   d. $\dfrac{\sqrt{2}}{3\sqrt{2} + \sqrt{3}}$

7. Evaluate $\dfrac{3x^2 + 5x + 2}{3x^2 - x - 2}$ for $x = 21$.

For Problems 8–11, perform the indicated operations and express the answers in simplest form.

8. $(3x + 4)(2x^2 - x - 5)$

9. $\dfrac{3}{4x} - \dfrac{2}{5x} + \dfrac{7}{10x}$

10. $(2x^4 - 13x^3 + 19x^2 - 25x + 25) \div (x - 5)$

11. $\dfrac{x^3 - 8}{x - 2}$

12. Express $(1.414)(10)^{-3}$ in ordinary decimal notation.

13. 41.4 is what percent of 36?

14. Find the product $(6 + 5i)(-4 - 2i)$ and express the result in the standard form of a complex number.

15. Evaluate each of the following numerical expressions.

   a. $16^{\frac{3}{4}}$

   b. $\left(\dfrac{3}{2} - \dfrac{1}{4}\right)^{-1}$

   c. $-4^{-\frac{3}{2}}$

   d. $\sqrt[3]{\dfrac{64}{27}}$

16. Find the slope of the line determined by the equation $-2x - 3y = 7$.

17. Find the equation of the line that has a slope of $-\dfrac{3}{4}$ and a y intercept of 5.

For Problems 18–21, solve each equation.

18. $3(2x - 1) - (x + 2) = 2(-x + 3)$

19. $\dfrac{3}{2x - 1} = \dfrac{4}{3x - 2}$

20. $3x^2 - 2x + 1 = 0$

21. $\sqrt{x + 1} + \sqrt{x - 4} = 5$

For Problems 22–25, graph each equation.

22. $4x^2 - y^2 = 16$

23. $f(x) = -2x^2 + 8x - 5$

24. $xy^2 = -4$

25. $f(x) = -x - 3$

Because the Richter number for reporting the intensity of an earthquake is calculated from a logarithm, it is referred to as a logarithmic scale. Logarithmic scales are commonly used in science and mathematics to transform very large numbers to a smaller scale.

AP / Wide World Photos

Exponential and Logarithmic Functions

How long will it take $1000 to triple if it is invested at 5% interest compounded annually? We can use the formula $A = P(1 + r)^t$ to generate the equation $3000 = 1000(1 + 0.05)^t$, which can be solved for t by using logarithms. It will take approximately 22.5 years for the money to triple.

This chapter will expand the meaning of an exponent and introduce the concept of a logarithm. We will (1) work with some exponential functions, (2) work with some logarithmic functions, and (3) use the concepts of exponent and logarithm to expand our capabilities for solving problems. Your calculator will be a valuable tool throughout this chapter.

InfoTrac Project

Do a keyword search on compound interest and find the article titled "Tapping the Power of Compound Interest." Write a brief summary of the article. The formula for the amount of money (A) in an account for which interest is compounded continuously is given by $A = Pe^{rt}$, where P is the initial investment, r is the rate of interest, and t is the time (in years) the money is invested. Find the amount of money in an account after 35 years if the initial investment of $2500 was invested at 7% (compounded continuously). At what interest rate must the money be invested for there to be $35,000 in the account at the end of 35 years?

14.1   Exponents and Exponential Functions

Objectives

■   Solve exponential equations.

■   Graph exponential functions.

In Chapter 2, the expression b^n was defined to mean n factors of b, where n is any positive integer and b is any real number. For example,

$$2^3 = 2 \cdot 2 \cdot 2 = 8$$

$$\left(\frac{1}{3}\right)^4 = \left(\frac{1}{3}\right)\left(\frac{1}{3}\right)\left(\frac{1}{3}\right)\left(\frac{1}{3}\right) = \frac{1}{81}$$

$$(-4)^2 = (-4)(-4) = 16$$

$$-(0.5)^3 = -[(0.5)(0.5)(0.5)] = -0.125$$

Then in Chapter 6, by defining $b^0 = 1$ and $b^{-n} = \dfrac{1}{b^n}$, where n is any positive integer and b is any nonzero real number, we extended the concept of an exponent to include all integers. Examples include

$$(0.76)^0 = 1$$

$$2^{-3} = \frac{1}{2^3} = \frac{1}{8}$$

$$\left(\frac{2}{3}\right)^{-2} = \frac{1}{\left(\frac{2}{3}\right)^2} = \frac{1}{\frac{4}{9}} = \frac{9}{4}$$

$$(0.4)^{-1} = \frac{1}{(0.4)^1} = \frac{1}{0.4} = 2.5$$

In Chapter 10 we provided for the use of all rational numbers as exponents by defining

$$b^{m/n} = \sqrt[n]{b^m} = (\sqrt[n]{b})^m$$

where n is a positive integer greater than 1, and b is a real number such that $\sqrt[n]{b}$ exists. Some examples are

$$27^{2/3} = (\sqrt[3]{27})^2 = 9$$

$$16^{1/4} = \sqrt[4]{16^1} = 2$$

$$\left(\frac{1}{9}\right)^{1/2} = \sqrt{\frac{1}{9}} = \frac{1}{3}$$

$$31^{-1/5} = \frac{1}{32^{1/5}} = \frac{1}{\sqrt[5]{32}} = \frac{1}{2}$$

Formally extending the concept of an exponent to include the use of irrational numbers requires some ideas from calculus and is therefore beyond the scope of this text. However, we can take a brief glimpse at the general idea involved. Consider the number $2^{\sqrt{3}}$. By using the nonterminating and nonrepeating decimal representation $1.73205\ldots$ for $\sqrt{3}$, we can form the sequence of numbers 2^1, $2^{1.7}$, $2^{1.73}$, $2^{1.732}$, $2^{1.7320}$, $2^{1.73205}$, $\ldots$. It should seem reasonable that each successive power gets closer to $2^{\sqrt{3}}$. This is precisely what happens if b^n, where n is irrational, is properly defined using the concept of a limit. Furthermore, this will ensure that an expression such as 2^x will yield exactly one value for each value of x.

From now on, then, we can use any real number as an exponent, and the basic properties we stated in Chapter 10 can be extended to include all real numbers as exponents. Let's restate those properties with the restriction that the bases a and b must be positive numbers so that we can avoid expressions such as $(-4)^{1/2}$, which do not represent real numbers.

PROPERTY 14.1

If a and b are positive real numbers and m and n are any real numbers, then

1. $b^n \cdot b^m = b^{n+m}$ Product of two like bases with powers

2. $(b^n)^m = b^{mn}$ Power of a power

3. $(ab)^n = a^n b^n$ Power of a product

4. $\left(\dfrac{a}{b}\right)^n = \dfrac{a^n}{b^n}$ Power of a quotient

5. $\dfrac{b^n}{b^m} = b^{n-m}$ Quotient of two like bases with powers

Another property that can be used to solve certain types of equations that involve exponents can be stated as follows:

PROPERTY 14.2

If $b > 0, b \neq 1$, and m and n are real numbers, then $b^n = b^m$ if and only if $n = m$.

The following examples illustrate the use of Property 14.2.

EXAMPLE 1

Solve $2^x = 32$.

Solution

$$2^x = 32$$
$$2^x = 2^5 \qquad 32 = 2^5$$
$$x = 5 \qquad \text{Apply Property 14.2.}$$

The solution set is $\{5\}$.

EXAMPLE 2   Solve $3^{2x} = \dfrac{1}{9}$.

Solution

$$3^{2x} = \frac{1}{9} = \frac{1}{3^2}$$

$$3^{2x} = 3^{-2}$$

$$2x = -2 \qquad \text{Property 14.2}$$

$$x = -1$$

The solution set is $\{-1\}$. ■

EXAMPLE 3   Solve $\left(\dfrac{1}{5}\right)^{x-4} = \dfrac{1}{125}$.

Solution

$$\left(\frac{1}{5}\right)^{x-4} = \frac{1}{125}$$

$$\left(\frac{1}{5}\right)^{x-4} = \left(\frac{1}{5}\right)^3$$

$$x - 4 = 3 \qquad \text{Property 14.2}$$

$$x = 7$$

The solution set is $\{7\}$. ■

EXAMPLE 4   Solve $8^x = 32$.

Solution

$$8^x = 32$$

$$(2^3)^x = 2^5 \qquad 8 = 2^3$$

$$2^{3x} = 2^5$$

$$3x = 5 \qquad \text{Property 14.2}$$

$$x = \frac{5}{3}$$

The solution set is $\left\{\dfrac{5}{3}\right\}$. ■

EXAMPLE 5   Solve $(3^{x+1})(9^{x-2}) = 27$.

Solution

$$(3^{x+1})(9^{x-2}) = 27$$

$$(3^{x+1})(3^2)^{x-2} = 3^3$$

$$(3^{x+1})(3^{2x-4}) = 3^3$$
$$3^{3x-3} = 3^3$$
$$3x - 3 = 3 \qquad \text{Property 14.2}$$
$$3x = 6$$
$$x = 2$$

The solution set is {2}.

Exponential Functions

If b is any positive number, then the expression b^x designates exactly one real number for every real value of x. Therefore, the equation $f(x) = b^x$ defines a function whose domain is the set of real numbers. Furthermore, if we include the additional restriction $b \neq 1$, then any equation of the form $f(x) = b^x$ describes what we will call later a one-to-one function and is known as an **exponential function.** This leads to the following definition.

DEFINITION 14.1

If $b > 0$ and $b \neq 1$, then the function f defined by

$$f(x) = b^x$$

where x is any real number, is called the **exponential function with base b.**

Now let's consider graphing some exponential functions.

EXAMPLE 6

Graph the function $f(x) = 2^x$.

Solution

Let's set up a table of values; keep in mind that the domain is the set of real numbers and that the equation $f(x) = 2^x$ exhibits no symmetry. Plot these points and connect them with a smooth curve to produce Figure 14.1.

x	2^x
-2	$\dfrac{1}{4}$
-1	$\dfrac{1}{2}$
0	1
1	2
2	4
3	8

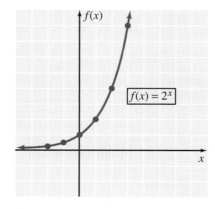

Figure 14.1

In the table for Example 6, we chose integral values for x to keep the computation simple. However, with the use of a calculator, we could easily acquire functional values by using nonintegral exponents. Consider the following additional values for $f(x) = 2^x$.

$$f(0.5) \approx 1.41 \qquad f(1.7) \approx 3.25$$
$$f(-0.5) \approx 0.71 \qquad f(-2.6) \approx 0.16$$

Use your calculator to check these results. Also note that the points generated by these values do fit the graph in Figure 14.1.

E X A M P L E   7   Graph $f(x) = \left(\dfrac{1}{2}\right)^x$.

Solution

Again, let's set up a table of values, plot the points, and connect them with a smooth curve. The graph is shown in Figure 14.2.

x	$\left(\dfrac{1}{2}\right)^x$
-3	8
-2	4
-1	2
0	1
1	$\dfrac{1}{2}$
2	$\dfrac{1}{4}$
3	$\dfrac{1}{8}$

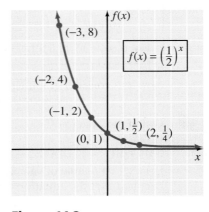

Figure 14.2

REMARK: Because $\left(\dfrac{1}{2}\right)^x = \dfrac{1}{2^x} = 2^{-x}$, the graphs of $f(x) = 2^x$ and $f(x) = \left(\dfrac{1}{2}\right)^x$ are reflections of each other across the y axis. Therefore, Figure 14.2 could have been drawn by reflecting Figure 14.1 across the y axis.

Figures 14.1 and 14.2 illustrate a general behavior pattern of exponential functions. That is to say, if $b > 1$, then the graph of $f(x) = b^x$ goes up to the right, and the function is called an **increasing function.** If $0 < b < 1$, then the graph of $f(x) = b^x$ goes down to the right, and the function is called a **decreasing function.** These facts are illustrated in Figure 14.3. Note that $b^0 = 1$ for any $b > 0$; thus all graphs of $f(x) = b^x$ contain the point $(0, 1)$.

Figure 14.3

As you graph exponential functions, don't forget your previous graphing experiences.

1. The graph of $f(x) = 2^x - 4$ is the graph of $f(x) = 2^x$ *moved down 4 units.*
2. The graph of $f(x) = 2^{x+3}$ is the graph of $f(x) = 2^x$ *moved 3 units to the left.*
3. The graph of $f(x) = -2^x$ is the graph of $f(x) = 2^x$ *reflected across the x axis.*

We used a graphing calculator to graph these four functions on the same set of axes, as shown in Figure 14.4.

If you are faced with an exponential function that is not of the basic form $f(x) = b^x$ or a variation thereof, don't forget the graphing suggestions offered in earlier chapters. Let's consider one such example.

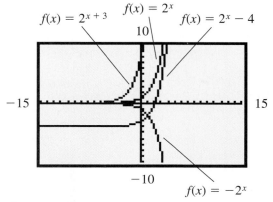

Figure 14.4

E X A M P L E   8

Graph $f(x) = 2^{-x^2}$.

Solution

Because $f(-x) = 2^{-(-x)^2} = 2^{-x^2} = f(x)$, we know that this curve is symmetric with respect to the y axis. Therefore, let's set up a table of values using nonnegative values for x. Plot these points, connect them with a smooth curve, and reflect this portion of the curve across the y axis to produce the graph in Figure 14.5.

x	2^{-x^2}
0	1
$\dfrac{1}{2}$	0.84
1	0.5
$\dfrac{3}{2}$	0.21
2	0.06

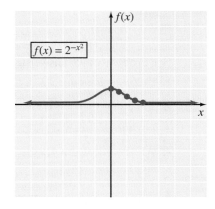

Figure 14.5

Finally, let's consider a problem in which a graphing utility gives us an approximate solution.

Use a graphing utility to obtain a graph of $f(x) = 50(2^x)$ and find an approximate value for x when $f(x) = 15{,}000$.

Solution

First, we must find an appropriate viewing rectangle. Because $50(2^{10}) = 51{,}200$, let's set the boundaries so that $0 \leq x \leq 10$ and $0 \leq y \leq 50{,}000$ with a scale of 10,000 on the y axis. (Certainly other boundaries could be used, but these will give us a graph that we can work with for this problem.) The graph of $f(x) = 50(2^x)$ is shown in Figure 14.6. Now we can use the TRACE and ZOOM features of the graphing utility to find that $x \approx 8.2$ at $y = 15{,}000$.

50,000

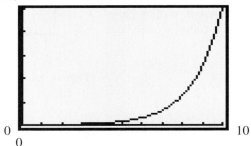

0
0   10

Figure 14.6

REMARK: In Example 9, we used a graphical approach to solve the equation $50(2^x) = 15{,}000$. In Section 14.6, we will use an algebraic approach for solving that same kind of equation.

CONCEPT QUIZ

For Problems 1–5, answer true or false.

1. If $2^{x+1} = 2^{3x}$, then $x + 1 = 3x$.

2. The numerical expression 9^x is equivalent to 3^{2x}.

3. For the exponential function $f(x) = b^x$, the base b can be any positive number.

4. All the graphs of $f(x) = b^x$ for all positive values of b pass through the point $(0,1)$.

5. The graphs of $f(x) = 3^x$ and $f(x) = \left(\dfrac{1}{3}\right)^x$ are reflections of each other across the y axis.

PROBLEM SET 14.1

For Problems 1–32, solve each of the equations.

1. $2^x = 64$

2. $3^x = 81$

3. $3^{2x} = 27$

4. $2^{2x} = 16$

5. $\left(\dfrac{1}{2}\right)^x = \dfrac{1}{128}$

6. $\left(\dfrac{1}{4}\right)^x = \dfrac{1}{256}$

7. $3^{-x} = \dfrac{1}{243}$

8. $3^{x+1} = 9$

9. $6^{3x-1} = 36$

10. $2^{2x+3} = 32$

11. $5^{x+2} = 125$

12. $4^{x-3} = 16$

13. $\left(\dfrac{3}{4}\right)^n = \dfrac{64}{27}$

14. $\left(\dfrac{2}{3}\right)^n = \dfrac{9}{4}$

15. $16^x = 64$

16. $4^x = 8$

17. $\left(\dfrac{1}{2}\right)^{2x} = 64$

18. $\left(\dfrac{1}{3}\right)^{5x} = 243$

19. $6^{2x} + 3 = 39$

20. $5^{2x} - 2 = 123$

21. $27^{4x} = 9^{x+1}$

22. $32^x = 16^{1-x}$

23. $9^{4x-2} = \dfrac{1}{81}$

24. $8^{3x+2} = \dfrac{1}{16}$

25. $10^x = 0.1$

26. $10^x = 0.0001$

27. $(2^{x+1})(2^x) = 64$

28. $(2^{2x-1})(2^{x+2}) = 32$

29. $(27)(3^x) = 9^x$

30. $(3^x)(3^{5x}) = 81$

31. $(4^x)(16^{3x-1}) = 8$

32. $(8^{2x})(4^{2x-1}) = 16$

For Problems 33–52, graph each of the exponential functions.

33. $f(x) = 3^x$

34. $f(x) = 4^x$

35. $f(x) = \left(\dfrac{1}{3}\right)^x$

36. $f(x) = \left(\dfrac{1}{4}\right)^x$

37. $f(x) = \left(\dfrac{3}{2}\right)^x$

38. $f(x) = \left(\dfrac{2}{3}\right)^x$

39. $f(x) = 2^x - 3$

40. $f(x) = 2^x + 1$

41. $f(x) = 2^{x+2}$

42. $f(x) = 2^{x-1}$

43. $f(x) = -2^x$

44. $f(x) = -3^x$

45. $f(x) = 2^{-x-2}$

46. $f(x) = 2^{-x+1}$

47. $f(x) = 2^{x^2}$

48. $f(x) = 2^x + 2^{-x}$

49. $f(x) = 2^{|x|}$

50. $f(x) = 3^{1-x^2}$

51. $f(x) = 2^x - 2^{-x}$

52. $f(x) = 2^{-|x|}$

■ ■ ☐ Thoughts into words

53. Explain how you would solve the equation $(2^{x+1})(8^{2x-3}) = 64$.

54. Why is the base of an exponential function restricted to positive numbers not including 1?

55. Explain how you would graph the function
$$f(x) = -\left(\dfrac{1}{3}\right)^x.$$

 ### Graphing calculator activities

56. Use a graphing calculator to check your graphs for Problems 33–52.

57. Graph $f(x) = 2^x$. Where should the graphs of $f(x) = 2^{x-5}$, $f(x) = 2^{x-7}$, and $f(x) = 2^{x+5}$ be located? Graph all three functions on the same set of axes with $f(x) = 2^x$.

58. Graph $f(x) = 3^x$. Where should the graphs of $f(x) = 3^x + 2$, $f(x) = 3^x - 3$, and $f(x) = 3^x - 7$ be located? Graph all three functions on the same set of axes with $f(x) = 3^x$.

59. Graph $f(x) = \left(\frac{1}{2}\right)^x$. Where should the graphs of

$f(x) = -\left(\frac{1}{2}\right)^x$, $f(x) = \left(\frac{1}{2}\right)^{-x}$, and $f(x) = -\left(\frac{1}{2}\right)^{-x}$ be

located? Graph all three functions on the same set of

axes with $f(x) = \left(\frac{1}{2}\right)^x$.

60. Graph $f(x) = (1.5)^x$, $f(x) = (5.5)^x$, $f(x) = (0.3)^x$, and $f(x) = (0.7)^x$ on the same set of axes. Are these graphs consistent with Figure 14.3?

61. What is the solution for $3^x = 5$? Do you agree that it is between 1 and 2 because $3^1 = 3$ and $3^2 = 9$? Now graph

$f(x) = 3^x - 5$ and use the ZOOM and TRACE features of your graphing calculator to find an approximation, to the nearest hundredth, for the x intercept. You should get an answer of 1.46. Do you see that this is an approximation for the solution of $3^x = 5$? Try it; raise 3 to the 1.46 power.

Find an approximate solution, to the nearest hundredth, for each of the following equations by graphing the appropriate function and finding the x intercept.

a. $2^x = 19$ **b.** $3^x = 50$ **c.** $4^x = 47$

d. $5^x = 120$ **e.** $2^x = 1500$ **f.** $3^{x-1} = 34$

14.2   Applications of Exponential Functions

Objectives

■   Apply exponential functions to growth and decay problems.

■   Calculate compound interest.

■   Use the half-life formula to solve radioactive decay problems.

■   Solve compounding continuously problems.

We can represent many real-world situations exhibiting growth or decay with equations that describe exponential functions. For example, suppose an economist predicts an annual inflation rate of 5% per year for the next 10 years. This means that an item that presently costs $8 will cost $8(105\%) = 8(1.05) = \$8.40$ a year from now. The same item will cost $[8(105\%)](105\%) = 8(1.05)^2 = \$8.82$ in 2 years. In general, the equation

$$P = P_0(1.05)^t$$

yields the predicted price P of an item in t years if the present cost is P_0, and the annual inflation rate is 5%. Using this equation, we can look at some future prices based on the prediction of a 5% inflation rate.

A $1.59 jar of mustard will cost $1.59(1.05)^3 = \$1.84$ in 3 years.

A $2.99 bag of potato chips will cost $2.99(1.05)^5 = \$3.82$ in 5 years.

A $8.59 can of coffee will cost $8.59(1.05)^7 = \$12.09$ in 7 years.

Compound Interest

Compound interest provides another illustration of exponential growth. Suppose that $500, called the **principal,** is invested at an interest rate of 8% *compounded annually.* The interest earned the first year is $500(0.08) = $40, and this amount is added to the original $500 to form a new principal of $540 for the second year. The interest earned during the second year is $540(0.08) = $43.20, and this amount is added to $540 to form a new principal of $583.20 for the third year. Each year a new principal is formed by reinvesting the interest earned during that year.

In general, suppose that a sum of money P (the principal) is invested at an interest rate of r percent compounded annually. The interest earned the first year is Pr, and the new principal for the second year is $P + Pr$, or $P(1 + r)$. Note that the new principal for the second year can be found by multiplying the original principal P by $(1 + r)$. In like fashion, the new principal for the third year can be found by multiplying the previous principal $P(1 + r)$ by $1 + r$, thus obtaining $P(1 + r)^2$. If this process is continued, then *after t years, the total amount of money accumulated, A,* is given by

$$A = P(1 + r)^t$$

Consider the following examples of investments made at a certain rate of interest compounded annually.

1. $750 invested for 5 years at 4% compounded annually produces

$$A = \$750(1.04)^5 = \$912.49$$

2. $1000 invested for 10 years at 6% compounded annually produces

$$A = \$1000(1.06)^{10} = \$1790.85$$

3. $5000 invested for 20 years at 8% compounded annually produces

$$A = \$5000(1.08)^{20} = \$23,304.79$$

We can use the compound interest formula to determine what rate of interest is needed to accumulate a certain amount of money based on a given initial investment. The next example illustrates this idea.

E X A M P L E   1

What rate of interest is needed for an investment of $1000 to yield $2500 in 10 years if the interest is compounded annually?

Solution

Let's substitute $1000 for P, $2500 for A, and 10 years for t in the compound interest formula and solve for r.

$$A = P(1 + r)^t$$
$$2500 = 1000(1 + r)^{10}$$

$$2.5 = (1 + r)^{10}$$

$$(2.5)^{0.1} = [(1 + r)^{10}]^{0.1} \qquad \text{Raise both sides to the 0.1 power.}$$

$$1.095958226 \approx 1 + r$$

$$0.095958226 \approx r$$

$$r = 9.6\% \quad \text{to the nearest tenth of a percent}$$

Therefore, a rate of interest of approximately 9.6% is needed. (Perhaps you should check this answer.) ■

If money invested at a certain rate of interest is compounded more than once a year, then the basic formula $A = P(1 + r)^t$ can be adjusted according to the number of compounding periods in a year. For example, for **semiannual compounding,** the formula becomes $A = P\left(1 + \dfrac{r}{2}\right)^{2t}$; for **quarterly compounding,** the formula becomes $A = P\left(1 + \dfrac{r}{4}\right)^{4t}$. In general, if n represents the number of compounding periods in a year, the formula becomes

$$A = P\left(1 + \frac{r}{n}\right)^{nt}$$

The following examples illustrate the use of the formula.

1. $750 invested for 5 years at 4% compounded semiannually produces

$$A = \$750\left(1 + \frac{0.04}{2}\right)^{2(5)} = \$750(1.02)^{10} = \$914.25$$

2. $1000 invested for 10 years at 6% compounded quarterly produces

$$A = \$1000\left(1 + \frac{0.06}{4}\right)^{4(10)} = \$1000(1.015)^{40} = \$1814.02$$

3. $5000 invested for 20 years at 8% compounded monthly produces

$$A = \$5000\left(1 + \frac{0.08}{12}\right)^{12(20)} = \$5000(1.00667)^{240} = \$24,634.01$$

You may find it interesting to compare these results with those we obtained earlier for annual compounding.

Exponential Decay

Suppose that the value of a car depreciates 15% per year for the first 5 years. Therefore, a car that costs $19,500 will be worth $19,500(100% − 15%) = $19,500(85%) = $19,500(0.85) = $16,575 in 1 year. In 2 years, the value of the car will have declined $19,500(0.85)^2 = $14,089 (to the nearest dollar). The equation

$$V = V_0(0.85)^t$$

yields the value V of a car in t years if the initial cost is V_0, and the value depreciates 15% per year. Therefore, we can estimate some car values to the nearest dollar as follows:

A \$17,000 car will be worth $\$17,000(0.85)^5 = \$7543$ in 5 years.

A \$25,000 car will be worth $\$25,000(0.85)^4 = \$13,050$ in 4 years.

A \$40,000 car will be worth $\$40,000(0.85)^3 = \$24,565$ in 3 years.

Another example of exponential decay is associated with radioactive substances. The rate of decay can be described exponentially and is based on the half-life of a substance. The **half-life** of a radioactive substance is the amount of time that it takes for one-half of an initial amount of the substance to disappear as the result of decay. For example, suppose that we have 200 grams of a certain substance that has a half-life of 5 days. After 5 days, $200\left(\dfrac{1}{2}\right) = 100$ grams remain. After 10 days, $200\left(\dfrac{1}{2}\right)^2 = 50$ grams remain. After 15 days, $200\left(\dfrac{1}{2}\right)^3 = 25$ grams remain. In general, after t days, $200\left(\dfrac{1}{2}\right)^{\frac{t}{5}}$ grams remain.

The previous discussion leads to the following half-life formula. Suppose there is an initial amount, Q_0, of a radioactive substance with a half-life of h. The amount of substance remaining, Q, after a time period of t, is given by the formula

$$Q = Q_0\left(\frac{1}{2}\right)^{\frac{t}{h}}$$

The units of measure for t and h must be the same.

EXAMPLE 2

Barium-140 has a half-life of 13 days. If there are 500 milligrams of barium initially, how many milligrams remain after 26 days? After 100 days?

Solution

When we use $Q_0 = 500$ and $h = 13$, the half-life formula becomes

$$Q = 500\left(\frac{1}{2}\right)^{\frac{t}{13}}$$

If $t = 26$, then

$$Q = 500\left(\frac{1}{2}\right)^{\frac{26}{13}}$$

$$= 500\left(\frac{1}{2}\right)^2$$

$$= 500\left(\frac{1}{4}\right)$$

$$= 125$$

Thus 125 milligrams remain after 26 days. If $t = 100$, then

$$Q = 500\left(\frac{1}{2}\right)^{\frac{100}{13}}$$

$$= 500(0.5)^{\frac{100}{13}}$$

$$= 2.4 \quad \text{to the nearest tenth of a milligram}$$

Approximately 2.4 milligrams remain after 100 days.   ■

REMARK: Example 2 clearly illustrates that a calculator is useful at times but unnecessary at other times. We solved the first part of the problem very easily without a calculator, but it certainly was helpful for the second part of the problem.

Number e

An interesting situation occurs if we consider the compound interest formula for $P = \$1$, $r = 100\%$, and $t = 1$ year. The formula becomes $A = 1\left(1 + \dfrac{1}{n}\right)^n$. The following table shows some values, rounded to eight decimal places, of $\left(1 + \dfrac{1}{n}\right)^n$ for different values of n.

n	$\left(1 + \dfrac{1}{n}\right)^n$
1	2.00000000
10	2.59374246
100	2.70481383
1000	2.71692393
10,000	2.71814593
100,000	2.71826824
1,000,000	2.71828047
10,000,000	2.71828169
100,000,000	2.71828181
1,000,000,000	2.71828183

The table suggests that as n increases, the value of $\left(1 + \dfrac{1}{n}\right)^n$ gets closer and closer to some fixed number. This does happen, and the fixed number is called e. To five decimal places, $e = 2.71828$.

The function defined by the equation $f(x) = e^x$ is the **natural exponential function.** It has a great many real-world applications, some of which we will look at in a moment. First, however, let's get a picture of the natural exponential function. Because $2 < e < 3$, the graph of $f(x) = e^x$ must fall between the graphs of $f(x) = 2^x$

and $f(x) = 3^x$. To be more specific, let's use our calculator to determine a table of values. Use the $\boxed{e^x}$ key, and round the results to the nearest tenth to obtain the following table. Plot the points determined by this table, and connect them with a smooth curve to produce Figure 14.7.

x	$f(x) = e^x$
0	1.0
1	2.7
2	7.4
−1	0.4
−2	0.1

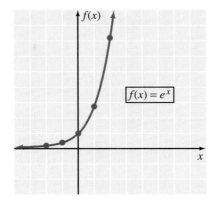

Figure 14.7

Back to Compound Interest

Let's return to the concept of compound interest. If the number of compounding periods in a year is increased indefinitely, we arrive at the concept of **compounding continuously.** Mathematically, we can accomplish this by applying the limit concept to the expression $P\left(1 + \dfrac{r}{n}\right)^{nt}$. We will not show the details here, but the following result is obtained. The formula

$$A = Pe^{rt}$$

yields the accumulated value, A, of a sum of money, P, that has been invested for t years at a rate of r percent compounded continuously. The following examples illustrate the use of the formula.

1. $750 invested for 5 years at 4% compounded continuously produces

   $$A = 750e^{(0.04)(5)} = 750e^{0.2} = \$916.05$$

2. $1000 invested for 10 years at 6% compounded continuously produces

   $$A = 1000e^{(0.06)(10)} = 1000e^{(0.6)} = \$1822.12$$

3. $5000 invested for 20 years at 8% compounded continuously produces

   $$A = 5000e^{(0.08)(20)} = 5000e^{1.6} = \$24765.16$$

Again, you may find it interesting to compare these results with those you obtained earlier when you were using a different number of compounding periods.

Is it better to invest at 6% compounded quarterly or at 5.75% compounded continuously? To answer such a question, we can use the concept of **effective yield** (sometimes called *effective annual rate of interest*). The effective yield of an investment is the simple interest rate that would yield the same amount in 1 year. Thus, for the investment at 6% compounded quarterly, we can calculate the effective yield as follows:

$$P(1 + r) = P\left(1 + \frac{0.06}{4}\right)^4$$

$$1 + r = \left(1 + \frac{0.06}{4}\right)^4 \qquad \text{Multiply both sides by } \frac{1}{P}.$$

$$1 + r = (1.015)^4$$

$$r = (1.015)^4 - 1$$

$$r \approx 0.0613635506$$

$$r = 6.14\% \quad \text{to the nearest hundredth of a percent}$$

Likewise, for the investment at 5.75% compounded continuously, we can calculate the effective yield as follows:

$$P(1 + r) = Pe^{0.0575}$$

$$1 + r = e^{0.0575}$$

$$r = e^{0.0575} - 1$$

$$r \approx 0.0591852707$$

$$r = 5.92\% \quad \text{to the nearest hundredth of a percent}$$

Therefore, comparing the two effective yields, we see that it is better to invest at 6% compounded quarterly than to invest at 5.75% compounded continuously.

Law of Exponential Growth

The ideas behind "compounded continuously" carry over to other growth situations. We use the law of exponential growth,

$$Q(t) = Q_0 e^{kt}$$

as a mathematical model for numerous growth-and-decay applications. In this equation, $Q(t)$ represents the quantity of a given substance at any time t, Q_0 is the initial amount of the substance (when $t = 0$), and k is a constant that depends on the particular application. If $k < 0$, then $Q(t)$ decreases as t increases, and we refer to the model as the **law of decay.**

Let's consider some growth-and-decay applications.

EXAMPLE 3

Suppose that in a certain culture, the equation $Q(t) = 15000e^{0.3t}$ expresses the number of bacteria present as a function of the time t, where t is expressed in hours. Find (a) the initial number of bacteria and (b) the number of bacteria after 3 hours.

Solution

(a) The initial number of bacteria is produced when $t = 0$.

$$Q(0) = 15{,}000e^{0.3(0)}$$
$$= 15{,}000e^0$$
$$= 15{,}000 \qquad e^0 = 1$$

(b) $Q(3) = 15{,}000e^{0.3(3)}$
$$= 15{,}000e^{0.9}$$
$$= 36{,}894 \quad \text{to the nearest whole number}$$

Therefore, there should be approximately 36,894 bacteria present after 3 hours. ■

EXAMPLE 4

Suppose the number of bacteria present in a certain culture after t minutes is given by the equation $Q(t) = Q_0 e^{0.05t}$, where Q_0 represents the initial number of bacteria. If 5000 bacteria are present after 20 minutes, how many bacteria were present initially?

Solution

If 5000 bacteria are present after 20 minutes, then $Q(20) = 5000$.

$$5000 = Q_0 e^{0.05(20)}$$
$$5000 = Q_0 e^1$$
$$\frac{5000}{e} = Q_0$$
$$1839 = Q_0 \quad \text{to the nearest whole number}$$

Therefore, there were approximately 1839 bacteria present initially. ■

EXAMPLE 5

The number of grams of a certain radioactive substance present after t seconds is given by the equation $Q(t) = 200e^{-0.3t}$. How many grams remain after 7 seconds?

Solution

Use $Q(t) = 200e^{-0.3t}$ to obtain

$$Q(7) = 200e^{(-0.3)(7)}$$
$$= 200e^{-2.1}$$
$$= 24.5 \quad \text{to the nearest tenth}$$

Thus approximately 24.5 grams remain after 7 seconds. ■

Finally, let's consider two examples where we use a graphing utility to produce the graph.

E X A M P L E   6

Suppose that $1000 was invested at 6.5% interest compounded continuously. How long would it take for the money to double?

Solution

Substitute $1000 for P and 0.065 for r in the formula $A = Pe^{rt}$ to produce $A = 1000e^{0.065t}$. If we let $y = A$ and $x = t$, we can graph the equation $y = 1000e^{0.065x}$. By letting $x = 20$, we obtain $y = 1000e^{0.065(20)} = 1000e^{1.3} \approx 3670$. Therefore, let's set the boundaries of the viewing rectangle so that $0 \leq x \leq 20$ and $0 \leq y \leq 3700$ with a y scale of 1000. Then we obtain the graph in Figure 14.8. Now we want to find the value of x so that $y = 2000$. (The money is to double.) Using the ZOOM and TRACE features of the graphing utility, we can determine that an x value of approximately 10.7 will produce a y value of 2000. Thus it will take approximately 10.7 years for the $1000 investment to double.

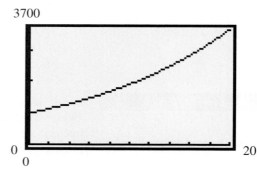

Figure 14.8

E X A M P L E   7

Graph the function $y = \dfrac{1}{\sqrt{2\pi}} e^{-x^2/2}$ and find its maximum value.

Solution

If $x = 0$, then $y = \dfrac{1}{\sqrt{2\pi}} e^0 = \dfrac{1}{\sqrt{2\pi}} \approx 0.4$, so let's set the boundaries of the viewing rectangle such that $-5 \leq x \leq 5$ and $0 \leq y \leq 1$ with a y scale of 0.1; the graph of the function is shown in Figure 14.9. From the graph, we see that the maximum value of the function occurs at $x = 0$, which we have already determined to be approximately 0.4.

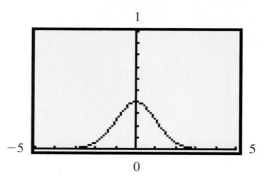

Figure 14.9

REMARK: The curve in Figure 14.9 is called a **normal distribution curve.** You may want to ask your instructor to explain what it means to assign grades on the basis of the normal distribution curve.

CONCEPT QUIZ

For Problems 1–5, match each type of problem with its formula.

1. Compound continuously

2. Exponential growth or decay

3. Interest compounded annually

4. Compound interest

5. Half-life

A. $A = P\left(1 + \dfrac{r}{n}\right)^{nt}$

B. $Q = Q_0\left(\dfrac{1}{2}\right)^{\frac{t}{h}}$

C. $A = P(1 + r)^t$

D. $Q(t) = Q_0 e^{kt}$

E. $A = Pe^{rt}$

PROBLEM SET 14.2

1. Assuming that the rate of inflation is 4% per year, the equation $P = P_0(1.04)^t$ yields the predicted price P of an item in t years that presently costs P_0. Find the predicted price of each of the following items for the indicated years ahead.

   a. $0.89 pack of chewing gum in 3 years

   b. $3.43 hamburger meal in 5 years

   c. $1.99 gallon of gasoline in 4 years

   d. $1.05 soft drink in 10 years

   e. $18,000 car in 5 years (nearest dollar)

   f. $120,000 house in 8 years (nearest dollar)

   g. $500 TV set in 7 years (nearest dollar)

2. Suppose it is estimated that the value of a car depreciates 30% per year for the first 5 years. The equation $A = P_0(0.7)^t$ yields the value (A) of a car after t years if the original price is P_0. Find the value (to the nearest dollar) of each of the following cars after the indicated time.

   a. $16,500 car after 4 years

   b. $22,000 car after 2 years

   c. $27,000 car after 5 years

   d. $40,000 car after 3 years

For Problems 3–6, use the formula $A = P\left(1 + \dfrac{r}{n}\right)^{nt}$ to find the amount for each investment and determine which investment amounts to more.

3. $2200 for 6 years

   a. at 2.5% compounded annually

   b. at 2% compounded quarterly

4. $5850 for 3 years

   a. at 4% compounded annually

   b. at 3.5% compounded quarterly

5. $2000 for 5 years

   a. at 3.5% compounded quarterly

   b. at 3% compounded monthly

6. $13,500 for 4 years

   a. at 5% compounded semiannually

   b. at 4.5% compounded monthly

For Problems 7–10, use the formula $A = Pe^{rt}$ to find the total amount of money accumulated at the end of the indicated time period by compounding continuously.

7. $4000 for 5 years at 3%

8. $5500 for 7 years at 2%

9. $1750 for 3 years at 4%

10. $10,000 for 10 years at 5%

For Problems 11–14, use the formulas $A = P\left(1 + \dfrac{r}{n}\right)^{nt}$ or $A = Pe^{rt}$ to find the amount for each investment and determine which investment amounts to more.

11. $10,500 for 4 years

   a. at 5% compounded continuously

   b. at 5.5% compounded quarterly

12. $1500 for 2 years

   a. at 2.5% compounded continuously

   b. at 3% compounded monthly

13. $4500 for 3 years

   a. at 3% compounded continuously

   b. at 3% compounded monthly

14. $13,750 for 5 years

   a. at 6% compounded continuously

   b. at 6.25% compounded semiannually

15. Rueben has a finance plan with the furniture store where the $4830 he spent accrues finance charges at an annual interest rate of 10.9% compounded monthly for 3 years before he starts to make payments. What will be the balance on the account at the end of those three years?

16. In a certain balloon mortgage loan, the borrower pays the lender all of the principal and interest for the loan at the end of five years. What will be the payoff amount for a loan of $185,000 at 6% annual interest rate where the interest is compounded monthly?

17. Jody took out a $3200 student loan her freshman year of college. The loan was at a 2.5% annual interest rate and accrued interest quarterly. Jody is obligated to begin repaying the loan back in five years. At that time what will be the amount she needs to repay?

18. To pay the tuition for medical school Melissa borrowed $8400 in student loans her first year. The loan is for seven years at an annual interest rate of 3.4%, and interest is compounded semiannually. What will be the amount of principal and interest in seven years?

19. Mark became overextended in his gambling debt and could not pay $500 he owed. The loan person said he could have three weeks to pay off the $500 at 10% interest per week compounded continuously. How much will Mark have to pay at the end of the three weeks?

20. What rate of interest, to the nearest tenth of a percent, compounded annually is needed for an investment of $2000 to grow to $2500 in 5 years?

21. What rate of interest, to the nearest tenth of a percent, compounded quarterly is needed for an investment of $1500 to grow to $2700 in 10 years?

22. Find the effective yield, to the nearest tenth of a percent, of an investment at 4.5% compounded monthly.

23. Find the effective yield, to the nearest hundredth of a percent, of an investment at 4.75% compounded continuously.

24. What investment yields the greater return: 4% compounded monthly or 3.85% compounded continuously?

25. What investment yields the greater return: 5.25% compounded quarterly or 5.3% compounded semiannually?

26. Suppose that a certain radioactive substance has a half-life of 20 years. If there are presently 2500 milligrams of the substance, how much, to the nearest milligram, will remain after 40 years? After 50 years?

27. Strontium-90 has a half-life of 29 years. If there are 400 grams of strontium-90 initially, how much, to the nearest gram, will remain after 87 years? After 100 years?

28. The half-life of radium is approximately 1600 years. If the present amount of radium in a certain location is 500 grams, how much will remain after 800 years? Express your answer to the nearest gram.

29. Suppose that in a certain culture, the equation $Q(t) = 1000e^{0.4t}$ expresses the number of bacteria present as a function of the time t, where t is expressed in hours. How many bacteria are present at the end of 2 hours? 3 hours? 5 hours?

30. The number of bacteria present at a given time under certain conditions is given by the equation $Q = 5000e^{0.05t}$, where t is expressed in minutes. How many bacteria are present at the end of 10 minutes? 30 minutes? 1 hour?

31. The number of bacteria present in a certain culture after t hours is given by the equation $Q = Q_0 e^{0.3t}$, where Q_0 represents the initial number of bacteria. If 6640 bacteria are present after 4 hours, how many bacteria were present initially?

32. The number of grams Q of a certain radioactive substance present after t seconds is given by the equation $Q = 1500e^{-0.4t}$. How many grams remain after 5 seconds? 10 seconds? 20 seconds?

33. The atmospheric pressure, measured in pounds per square inch (psi), is a function of the altitude above sea level. The equation $P(a) = 14.7e^{-0.21a}$, where a is the altitude measured in miles, can be used to approximate atmospheric pressure. Find the atmospheric pressure at each of the following locations.

a. Mount McKinley in Alaska: altitude of 3.85 miles

b. Denver, Colorado: the "mile-high" city

c. Asheville, North Carolina: altitude of 1985 feet

d. Phoenix, Arizona: altitude of 1090 feet

34. Suppose that the present population of a city is 75,000. Using the equation $P(t) = 75,000e^{0.01t}$ to estimate future growth, estimate the population (a) 10 years from now, (b) 15 years from now, and (c) 25 years from now.

For Problems 35–40, graph each of the exponential functions.

35. $f(x) = e^x + 1$

36. $f(x) = e^x - 2$

37. $f(x) = 2e^x$

38. $f(x) = -e^x$

39. $f(x) = e^{2x}$

40. $f(x) = e^{-x}$

■ ■ ■ Thoughts into words

41. Explain the difference between simple interest and compound interest.

42. Would it be better to invest $5000 at 6.25% interest compounded annually for 5 years or to invest $5000 at 6.25% interest compounded continuously for 5 years? Explain your answer.

43. How would you explain the concept of effective yield to someone who missed class when it was discussed?

44. How would you explain the half-life formula to someone who missed class when it was discussed?

■ ■ ■ Further investigations

45. Complete the following chart, which illustrates what happens to $1000 invested at various rates of interest for different lengths of time but always compounded continuously. Round your answers to the nearest dollar.

$1000 Compounded continuously

	2%	3%	4%	5%
5 years				
10 years				
15 years				
20 years				
25 years				

46. Complete the following chart, which illustrates what happens to $1000 invested at 6% for different lengths of time and different numbers of compounding periods. Round all of your answers to the nearest dollar.

$1000 at 6%

	1 year	5 years	10 years	20 years
Compounded annually				
Compounded semiannually				
Compounded quarterly				
Compounded monthly				
Compounded continuously				

47. Complete the following chart, which illustrates what happens to $1000 in 10 years on the basis of different rates of interest and different numbers of compounding periods. Round your answers to the nearest dollar.

$1000 for 10 years

	2%	3%	4%	5%
Compounded annually				
Compounded semiannually				
Compounded quarterly				
Compounded monthly				
Compounded continuously				

For Problems 48–52, graph each of the functions.

48. $f(x) = x(2^x)$

49. $f(x) = \dfrac{e^x + e^{-x}}{2}$

50. $f(x) = \dfrac{2}{e^x + e^{-x}}$

51. $f(x) = \dfrac{e^x - e^{-x}}{2}$

52. $f(x) = \dfrac{2}{e^x - e^{-x}}$

Graphing calculator activities

53. Use a graphing calculator to check your graphs for Problems 48–52.

54. Graph $f(x) = 2^x$, $f(x) = e^x$, and $f(x) = 3^x$ on the same set of axes. Are these graphs consistent with the discussion prior to Figure 14.7?

55. Graph $f(x) = e^x$. Where should the graphs of $f(x) = e^{x-4}$, $f(x) = e^{x-6}$, and $f(x) = e^{x+5}$ be located? Graph all three functions on the same set of axes with $f(x) = e^x$.

56. Graph $f(x) = e^x$. Now predict the graphs for $f(x) = -e^x$, $f(x) = e^{-x}$, and $f(x) = -e^{-x}$. Graph all three functions on the same set of axes with $f(x) = e^x$.

57. How do you think the graphs of $f(x) = e^x$, $f(x) = e^{2x}$, and $f(x) = 2e^x$ will compare? Graph them on the same set of axes to see if you were correct.

58. Find an approximate solution, to the nearest hundredth, for each of the following equations by graphing the appropriate function and finding the x intercept.

   a. $e^x = 7$ **b.** $e^x = 21$

   c. $e^x = 53$ **d.** $2e^x = 60$

   e. $e^{x+1} = 150$ **f.** $e^{x-2} = 300$

59. Use a graphing approach to argue that it is better to invest money at 6% compounded quarterly than at 5.75% compounded continuously.

60. How long will it take $500 to be worth $1500 if it is invested at 7.5% interest compounded semiannually?

61. How long will it take $5000 to triple if it is invested at 6.75% interest compounded quarterly?

14.3 Inverse Functions

Objectives

■ Determine if a function is one-to-one.

■ Verify that two functions are inverse functions.

■ Find inverse functions.

■ Find intervals where a function is increasing or decreasing.

Recall the vertical-line test: If each vertical line intersects a graph in no more than one point, then the graph represents a function. There is also a useful distinction between two basic types of functions. Consider the graphs of the two functions in Figure 14.10(a), $f(x) = 2x - 1$, and Figure 14.10(b), $g(x) = x^2$. In Figure 14.10(a), any *horizontal line* will intersect the graph in no more than one point. Therefore, every value of $f(x)$ has only one value of x associated with it. Any function that has this property of having exactly one value of x associated with each value of $f(x)$ is called a **one-to-one function.** Thus $g(x) = x^2$ is not a one-to-one function, because the horizontal line in Figure 14.10(b) intersects the parabola in two points.

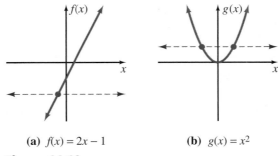

(a) $f(x) = 2x - 1$ (b) $g(x) = x^2$

Figure 14.10

The statement that for a function f to be a one-to-one function, **every value of $f(x)$ has only one value of x associated with it,** can be equivalently stated as "if $f(x_1) = f(x_2)$ for x_1 and x_2 in the domain of f, then $x_1 = x_2$." Let's use this last if-then statement to verify that $f(x) = 2x - 1$ is a one-to-one function. We start with the assumption that $f(x_1) = f(x_2)$.

$$2x_1 - 1 = 2x_2 - 1$$
$$2x_1 = 2x_2$$
$$x_1 = x_2$$

Thus $f(x) = 2x - 1$ is a one-to-one function.

To show that $g(x) = x^2$ is not a one-to-one function, we simply need to find two distinct real numbers in the domain of f that produce the same functional value. For example, $g(-2) = (-2)^2 = 4$, and $g(2) = 2^2 = 4$. Thus $g(x) = x^2$ is not a one-to-one function.

Now let's consider a one-to-one function f that assigns to each x in its domain D the value $f(x)$ in its range R (Figure 14.11a). We can define a new function g that goes from R to D; it assigns $f(x)$ in R back to x in D, as indicated in Figure 14.11(b).

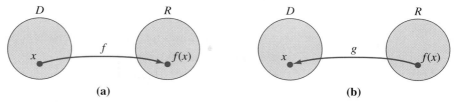

(a)   **(b)**

Figure 14.11

The functions f and g are called **inverse functions** of one another. The following definition precisely states this concept.

DEFINITION 14.2

Let f be a one-to-one function with a domain of X and a range of Y. A function g with a domain of Y and a range of X is called the **inverse function** of f if

$$(f \circ g)(x) = x \qquad \text{for every } x \text{ in } Y$$

and

$$(g \circ f)(x) = x \qquad \text{for every } x \text{ in } X$$

In Definition 14.2, note that for f and g to be inverses of each other, the domain of f must equal the range of g, and the range of f must equal the domain of g. Furthermore, g must reverse the correspondences given by f, and f must reverse the correspondences given by g. In other words, inverse functions *undo* each other. Let's use Definition 14.2 to verify that two specific functions are inverses of each other.

E X A M P L E   1   Verify that $f(x) = 4x - 5$ and $g(x) = \dfrac{x + 5}{4}$ are inverse functions.

Solution

Because the set of real numbers is the domain and range of both functions, we know that the domain of f equals the range of g and that the range of f equals the domain of g. Furthermore,

$$(f \circ g)(x) = f(g(x))$$

$$= f\left(\frac{x + 5}{4}\right)$$

$$= 4\left(\frac{x + 5}{4}\right) - 5 = x$$

and

$$(g \circ f)(x) = g(f(x))$$
$$= g(4x - 5)$$
$$= \frac{4x - 5 + 5}{4} = x$$

Therefore, f and g are inverses of each other. ■

E X A M P L E   2

Verify that $f(x) = x^2 + 1$ for $x \geq 0$ and $g(x) = \sqrt{x - 1}$ for $x \geq 1$ are inverse functions.

Solution

First, note that the domain of f equals the range of g—namely, the set of nonnegative real numbers. Also, the range of f equals the domain of g—namely, the set of real numbers greater than or equal to 1. Furthermore,

$$(f \circ g)(x) = f(g(x))$$
$$= f(\sqrt{x - 1})$$
$$= (\sqrt{x - 1})^2 + 1$$
$$= x - 1 + 1 = x$$

and

$$(g \circ f)(x) = g(f(x))$$
$$= g(x^2 + 1)$$
$$= \sqrt{x^2 + 1 - 1} = \sqrt{x^2} = x \qquad \sqrt{x^2} = x \text{ because } x \geq 1$$

Therefore, f and g are inverses of each other. ■

The inverse of a function f is commonly denoted by f^{-1}, read "f inverse or the inverse of f." Do not confuse the -1 in f^{-1} with a negative exponent. The symbol f^{-1} *does not* mean $1/f^1$ but rather refers to the inverse function of function f.

Remember that a function can also be thought of as a set of ordered pairs no two of which have the same first element. Along those lines, a one-to-one function further requires that no two of the ordered pairs have the same second element. Then, if the components of each ordered pair of a given one-to-one function are interchanged, the resulting function and the given function are inverses of each other. Thus if

$$f = \{(1, 4), (2, 7), (5, 9)\}$$

then

$$f^{-1} = \{(4, 1), (7, 2), (9, 5)\}$$

Graphically, two functions that are inverses of each other are **mirror images with reference to the line $y = x$.** This is due to the fact that ordered pairs (a, b) and (b, a) are reflections of each other with respect to the line $y = x$, as illustrated in Figure 14.12. (You will verify this in the next set of exercises.) Therefore, if the graph of a function f is known, as in Figure 14.13(a), then the graph of f^{-1} can be determined by reflecting f across the line $y = x$ (Figure 14.13b).

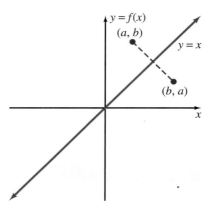

Figure 14.12

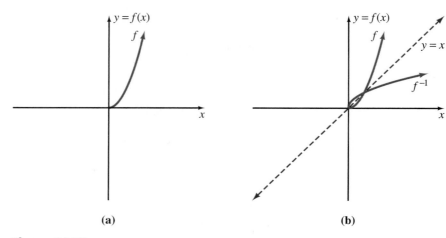

        (a)   (b)

Figure 14.13

Finding Inverse Functions

The idea of inverse functions *undoing each other* provides the basis for an informal approach to finding the inverse of a function. Consider the function

$$f(x) = 2x + 1$$

To each x, this function assigns twice x plus 1. To undo this function, we can subtract 1 and divide by 2. Hence the inverse is

$$f^{-1}(x) = \frac{x - 1}{2}$$

Now let's verify that f and f^{-1} are indeed inverses of each other.

$$(f \circ f^{-1})(x) = f(f^{-1}(x)) \qquad\qquad (f^{-1} \circ f)(x) = f^{-1}(f(x))$$

$$= f\left(\frac{x-1}{2}\right) \qquad\qquad\qquad = f^{-1}(2x + 1)$$

$$= 2\left(\frac{x-1}{2}\right) + 1 \qquad\qquad = \frac{2x + 1 - 1}{2}$$

$$= x - 1 + 1 = x \qquad\qquad\qquad = \frac{2x}{2} = x$$

Thus the inverse of $f(x) = 2x + 1$ is $f^{-1}(x) = \dfrac{x-1}{2}$.

This informal approach may not work very well with more complex functions, but it does emphasize how inverse functions are related to each other. A more formal and systematic technique for finding the inverse of a function can be described as follows:

1. Replace the symbol $f(x)$ with y.
2. Interchange x and y.
3. Solve the equation for y in terms of x.
4. Replace y with the symbol $f^{-1}(x)$.

The following examples illustrate this technique.

E X A M P L E   3   Find the inverse of $f(x) = \dfrac{2}{3}x + \dfrac{3}{5}$.

Solution

When we replace $f(x)$ with y, the equation becomes $y = \dfrac{2}{3}x + \dfrac{3}{5}$. Interchanging x and y produces $x = \dfrac{2}{3}y + \dfrac{3}{5}$.

Now, solving for y, we obtain

$$x = \frac{2}{3}y + \frac{3}{5}$$

$$15(x) = 15\left(\frac{2}{3}y + \frac{3}{5}\right)$$

$$15x = 10y + 9$$

$$15x - 9 = 10y$$

$$\frac{15x - 9}{10} = y$$

Finally, by replacing y with $f^{-1}(x)$, we can express the inverse function as

$$f^{-1}(x) = \frac{15x - 9}{10}$$

The domain of f is equal to the range of f^{-1} (both are the set of real numbers), and the range of f equals the domain of f^{-1} (both are the set of real numbers). Furthermore, we could show that $(f \circ f^{-1})(x) = x$ and $(f^{-1} \circ f)(x) = x$. We leave this for you to complete. ∎

Does the function $f(x) = x^2 - 2$ have an inverse? Sometimes a graph of the function helps answer such a question. In Figure 14.14(a), it should be evident that f is not a one-to-one function and therefore cannot have an inverse. However, it should also be apparent from the graph that if we restrict the domain of f to the nonnegative real numbers, then f is a one-to-one function and should have an inverse (Figure 14.14b). The next example illustrates how to find the inverse function.

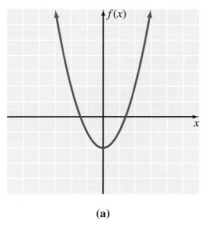
(a)

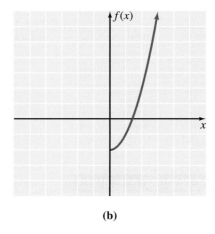
(b)

Figure 14.14

E X A M P L E   4

Find the inverse of $f(x) = x^2 - 2$, where $x \geq 0$.

Solution

When we replace $f(x)$ with y, the equation becomes

$$y = x^2 - 2, \qquad x \geq 0$$

Interchanging x and y produces

$$x = y^2 - 2, \qquad y \geq 0$$

Now let's solve for y; keep in mind that y is to be nonnegative.

$$x = y^2 - 2$$
$$x + 2 = y^2$$
$$\sqrt{x + 2} = y, \qquad x \geq -2$$

Finally, by replacing y with $f^{-1}(x)$, we can express the inverse function as

$$f^{-1}(x) = \sqrt{x + 2}, \qquad x \geq -2$$

The domain of f equals the range of f^{-1} (both are the nonnegative real numbers), and the range of f equals the domain of f^{-1} (both are the real numbers greater than or equal to -2). It can also be shown that $(f \circ f^{-1})(x) = x$ and $(f^{-1} \circ f)(x) = x$. Again, we leave this for you to complete. ■

Increasing and Decreasing Functions

In Section 14.1, we used exponential functions as examples of increasing and decreasing functions. In reality, one function can be both increasing and decreasing over certain intervals. For example, in Figure 14.15, the function f is said to be *increasing* on the intervals $(-\infty, x_1]$ and $[x_2, \infty)$ and is said to be *decreasing* on the interval $[x_1, x_2]$. More specifically, increasing and decreasing functions are defined as follows:

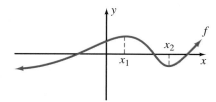

Figure 14.15

DEFINITION 14.3

Let f be a function, with the interval I a subset of the domain of f. Let x_1 and x_2 be in I. Then

    1. f is *increasing on I* if $f(x_1) < f(x_2)$ whenever $x_1 < x_2$.
    2. f is *decreasing on I* if $f(x_1) > f(x_2)$ whenever $x_1 < x_2$.
    3. f is *constant on I* if $f(x_1) = f(x_2)$ for every x_1 and x_2.

Apply Definition 14.3, and you will see that the quadratic function $f(x) = x^2$ shown in Figure 14.16 is decreasing on $(-\infty, 0]$ and increasing on $[0, \infty)$. Likewise,

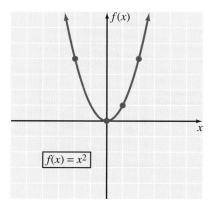

$$f(x) = x^2$$

Figure 14.16

the linear function $f(x) = 2x$ in Figure 14.17 is increasing throughout its domain of real numbers, so we say that it is increasing on $(-\infty, \infty)$. The function $f(x) = -2x$

in Figure 14.18 is decreasing on $(-\infty, \infty)$. For our purposes in this text, we will rely on our knowledge of the graphs of the functions to determine where functions are increasing and decreasing. More formal techniques for determining where functions increase and decrease will be developed in the calculus.

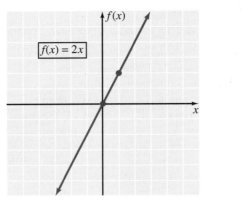

Figure 14.17

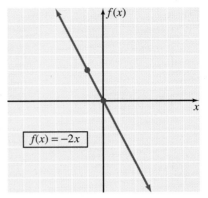

Figure 14.18

A function that is always increasing (or is always decreasing) over its entire domain is one-to-one and so has an inverse. Furthermore, as illustrated by Example 4, even if a function is not one-to-one over its entire domain, it may be so over some subset of the domain. It then has an inverse over this restricted domain.

As functions become more complex, a graphing utility can be used to help with the problems we have discussed in this section. For example, suppose that we want to know whether the function $f(x) = \dfrac{3x + 1}{x - 4}$ is a one-to-one function and therefore has an inverse. Using a graphing utility, we can quickly get a sketch of the graph (see Figure 14.19). Then, by applying the horizontal-line test to the graph, we can be fairly certain that the function is one-to-one.

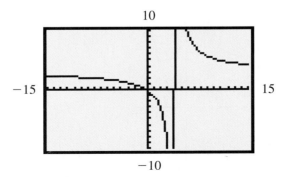

Figure 14.19

A graphing utility can also be used to help determine intervals on which a function is increasing or decreasing. For example, to determine such intervals for the function $f(x) = \sqrt{x^2 + 4}$, let's use a graphing utility to get a sketch of the curve (Figure 14.20). From this graph, we see that the function is decreasing on the interval $(-\infty, 0]$ and is increasing on the interval $[0, \infty)$.

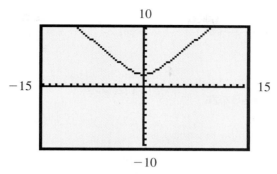

Figure 14.20

CONCEPT QUIZ

For Problems 1–7, answer true or false.

1. If a horizontal line intersects the graph of a function in exactly two points, then the function is said to be one-to-one.

2. The notation f^{-1} refers to the inverse of function f.

3. The graphs of two functions that are inverses of each other are mirror images with reference to the y axis.

4. If $g = \{(1, 3), (5, 9)\}$, then $g^{-1} = \{(3, 1), (9, 5)\}$.

5. Given that f and g are inverse functions, then the range of f is the domain of g.

6. A linear function whose graph has a negative slope is an increasing function.

7. A function that is increasing over its entire domain is a one-to-one function.

PROBLEM SET 14.3

For Problems 1–6, determine whether the graph represents a one-to-one function.

1.

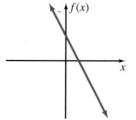

Figure 14.21

2.

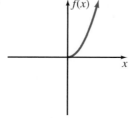

Figure 14.22

3.

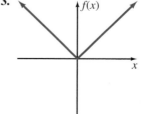

Figure 14.23

4.

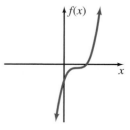

Figure 14.24

5.

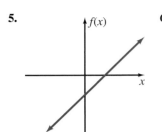

Figure 14.25

6.

$f(x)$

x

Figure 14.26

For Problems 7–14, determine whether the function f is one-to-one.

7. $f(x) = 5x + 4$

8. $f(x) = -3x + 4$

9. $f(x) = x^3$

10. $f(x) = x^5 + 1$

11. $f(x) = |x| + 1$

12. $f(x) = -|x| - 2$

13. $f(x) = -x^4$

14. $f(x) = x^4 + 1$

For Problems 15–18, (a) list the domain and range of the function, (b) form the inverse function f^{-1}, and (c) list the domain and range of f^{-1}.

15. $f = \{(1, 5), (2, 9), (5, 21)\}$

16. $f = \{(1, 1), (4, 2), (9, 3), (16, 4)\}$

17. $f = \{(0, 0), (2, 8), (-1, -1), (-2, -8)\}$

18. $f = \{(-1, 1), (-2, 4), (-3, 9), (-4, 16)\}$

For Problems 19–26, verify that the two given functions are inverses of each other.

19. $f(x) = 5x - 9$ and $g(x) = \dfrac{x + 9}{5}$

20. $f(x) = -3x + 4$ and $g(x) = \dfrac{4 - x}{3}$

21. $f(x) = -\dfrac{1}{2}x + \dfrac{5}{6}$ and $g(x) = -2x + \dfrac{5}{3}$

22. $f(x) = x^3 + 1$ and $g(x) = \sqrt[3]{x - 1}$

23. $f(x) = \dfrac{1}{x - 1}$ for $x > 1$, and

$g(x) = \dfrac{x + 1}{x}$ for $x > 0$

24. $f(x) = x^2 + 2$ for $x \geq 0$, and
$g(x) = \sqrt{x - 2}$ for $x \geq 2$

25. $f(x) = \sqrt{2x - 4}$ for $x \geq 2$, and

$g(x) = \dfrac{x^2 + 4}{2}$ for $x \geq 0$

26. $f(x) = x^2 - 4$ for $x \geq 0$, and
$g(x) = \sqrt{x + 4}$ for $x \geq -4$

For Problems 27–36, determine whether f and g are inverse functions.

27. $f(x) = 3x$ and $g(x) = -\dfrac{1}{3}x$

28. $f(x) = \dfrac{3}{4}x - 2$ and $g(x) = \dfrac{4}{3}x + \dfrac{8}{3}$

29. $f(x) = x^3$ and $g(x) = \sqrt[3]{x}$

30. $f(x) = \dfrac{1}{x + 1}$ and $g(x) = \dfrac{1 - x}{x}$

31. $f(x) = x$ and $g(x) = \dfrac{1}{x}$

32. $f(x) = \dfrac{3}{5}x + \dfrac{1}{3}$ and $g(x) = \dfrac{5}{3}x - 3$

33. $f(x) = x^2 - 3$ for $x \geq 0$, and
$g(x) = \sqrt{x + 3}$ for $x \geq -3$

34. $f(x) = |x - 1|$ for $x \geq 1$, and
$g(x) = |x + 1|$ for $x \geq 0$

35. $f(x) = \sqrt{x + 1}$ and $g(x) = x^2 - 1$ for $x \geq 0$

36. $f(x) = \sqrt{2x - 2}$ and $g(x) = \dfrac{1}{2}x^2 + 1$

For Problems 37–50, (a) find f^{-1} and (b) verify that $(f \circ f^{-1})(x) = x$ and $(f^{-1} \circ f)(x) = x$.

37. $f(x) = x - 4$

38. $f(x) = 2x - 1$

39. $f(x) = -3x - 4$

40. $f(x) = -5x + 6$

41. $f(x) = \dfrac{3}{4}x - \dfrac{5}{6}$

42. $f(x) = \dfrac{2}{3}x - \dfrac{1}{4}$

43. $f(x) = -\dfrac{2}{3}x$

44. $f(x) = \dfrac{4}{3}x$

45. $f(x) = \sqrt{x}$ for $x \geq 0$

46. $f(x) = \dfrac{1}{x}$ for $x \neq 0$

47. $f(x) = x^2 + 4$ for $x \geq 0$

48. $f(x) = x^2 + 1$ for $x \leq 0$

49. $f(x) = 1 + \dfrac{1}{x}$ for $x > 0$

50. $f(x) = \dfrac{x}{x+1}$ for $x > -1$

For Problems 51–58, (a) find f^{-1} and (b) graph f and f^{-1} on the same set of axes.

51. $f(x) = 3x$

52. $f(x) = -x$

53. $f(x) = 2x + 1$

54. $f(x) = -3x - 3$

55. $f(x) = \dfrac{2}{x-1}$ for $x > 1$

56. $f(x) = \dfrac{-1}{x-2}$ for $x > 2$

57. $f(x) = x^2 - 4$ for $x \geq 0$

58. $f(x) = \sqrt{x-3}$ for $x \geq 3$

For Problems 59–66, find the intervals on which the given function is increasing and the intervals on which it is decreasing.

59. $f(x) = x^2 + 1$

60. $f(x) = x^3$

61. $f(x) = -3x + 1$

62. $f(x) = (x-3)^2 + 1$

63. $f(x) = -(x+2)^2 - 1$

64. $f(x) = x^2 - 2x + 6$

65. $f(x) = -2x^2 - 16x - 35$

66. $f(x) = x^2 + 3x - 1$

■ ■ ■ **Thoughts into words**

67. Does the function $f(x) = 4$ have an inverse? Explain your answer.

68. Explain why every nonconstant linear function has an inverse.

69. Are the functions $f(x) = x^4$ and $g(x) = \sqrt[4]{x}$ inverses of each other? Explain your answer.

70. What does it mean to say that 2 and -2 are additive inverses of each other? What does it mean to say that 2 and $\dfrac{1}{2}$ are multiplicative inverses of each other? What does it mean to say that the functions $f(x) = x - 2$ and $f(x) = x + 2$ are inverses of each other? Do you think that the concept of "inverse" is being used in a consistent manner? Explain your answer.

■ ■ ■ **Further investigations**

71. The function notation and the operation of composition can be used to find inverses. Say we want to find the inverse of $f(x) = 5x + 3$. We know that $f(f^{-1}(x))$ must produce x. Therefore,

$$f(f^{-1}(x)) = 5[f^{-1}(x)] + 3 = x$$
$$5[f^{-1}(x)] = x - 3$$
$$f^{-1}(x) = \frac{x-3}{5}$$

Use this approach to find the inverse of each of the following functions.

a. $f(x) = 3x - 9$ **b.** $f(x) = -2x + 6$

c. $f(x) = -x + 1$ **d.** $f(x) = 2x$

e. $f(x) = -5x$ **f.** $f(x) = x^2 + 6$ for $x \geq 0$

72. If $f(x) = 2x + 3$ and $g(x) = 3x - 5$, find

a. $(f \circ g)^{-1}(x)$ **b.** $(f^{-1} \circ g^{-1})(x)$

c. $(g^{-1} \circ f^{-1})(x)$

▦ **Graphing calculator activities**

73. For Problems 37–44, graph the given function, the inverse function that you found, and $f(x) = x$ on the same set of axes. In each case, the given function and its inverse should produce graphs that are reflections of each other through the line $f(x) = x$.

74. There is another way in which we can use the graphing calculator to help show that two functions are inverses of each other. Suppose we want to show that $f(x) = x^2 - 2$ for $x \geq 0$ and $g(x) = \sqrt{x + 2}$ for $x \geq -2$ are inverses of each other. Let's make the following assignments for our graphing calculator.

$$f: \quad Y_1 = x^2 - 2$$
$$g: \quad Y_2 = \sqrt{x + 2}$$
$$f \circ g: \quad Y_3 = (Y_2)^2 - 2$$
$$g \circ f: \quad Y_4 = \sqrt{Y_1 + 2}$$

Now we can proceed as follows:

1. Graph $Y_1 = x^2 - 2$, and note that for $x > 0$, the range is greater than or equal to -2.

2. Graph $Y_2 = \sqrt{x + 2}$, and note that for $x \geq -2$, the range is greater than or equal to 0.

Thus the domain of f equals the range of g, and the range of f equals the domain of g.

3. Graph $Y_3 = (Y_2)^2 - 2$ for $x \geq -2$, and observe the line $y = x$ for $x \geq -2$.

4. Graph $Y_4 = \sqrt{Y_1} + 2$ for $x \geq 0$, and observe the line $y = x$ for $x \geq 0$.

Thus $(f \circ g)(x) = x$ and $(g \circ f)(x) = x$, and the two functions are inverses of each other.

Use this approach to check your answers for Problems 45–50.

75. Use the technique demonstrated in Problem 74 to show that

$$f(x) = \frac{x}{\sqrt{x^2 + 1}}$$

and

$$g(x) = \frac{x}{\sqrt{1 - x^2}} \qquad \text{for } -1 < x < 1$$

are inverses of each other.

Answers to Concept Quiz

1. False   **2.** True   **3.** False   **4.** True   **5.** True   **6.** False   **7.** True

---

14.4   Logarithms

Objectives

▪   Change form between exponential statements and logarithmic statements.

▪   Evaluate a logarithmic expression.

▪   Solve logarithmic equations.

▪   Apply properties of logarithms to change the form of logarithmic expressions.

In Sections 14.1 and 14.2, we discussed exponential expressions of the form b^n, where b is any positive real number and n is any real number; we used exponential expressions of the form b^n to define exponential functions; and we used exponential functions to help solve problems. In the next three sections, we will follow the same basic pattern with respect to a new concept — that of a logarithm. Let's begin with the following definition.

DEFINITION 14.4

If r is any positive real number, then the unique exponent t such that $b^t = r$ is called the **logarithm of r with base b** and is denoted by $\log_b r$.

According to Definition 14.4, the logarithm of 16 base 2 is the exponent t such that $2^t = 16$; thus we can write $\log_2 16 = 4$. Likewise, we can write $\log_{10} 1000 = 3$ because $10^3 = 1000$. In general, we can remember Definition 14.4 by the statement

$$\log_b r = t \quad \text{is equivalent to} \quad b^t = r$$

Therefore, we can easily switch back and forth between exponential and logarithmic forms of equations, as the next examples illustrate.

$$\log_2 8 = 3 \qquad \text{is equivalent to } 2^3 = 8$$
$$\log_{10} 100 = 2 \qquad \text{is equivalent to } 10^2 = 100$$
$$\log_3 81 = 4 \qquad \text{is equivalent to } 3^4 = 81$$
$$\log_{10} 0.001 = -3 \qquad \text{is equivalent to } 10^{-3} = 0.001$$
$$2^7 = 128 \qquad \text{is equivalent to } \log_2 128 = 7$$
$$5^3 = 125 \qquad \text{is equivalent to } \log_5 125 = 3$$
$$\left(\frac{1}{2}\right)^4 = \frac{1}{16} \qquad \text{is equivalent to } \log_{1/2}\left(\frac{1}{16}\right) = 4$$
$$10^{-2} = 0.01 \qquad \text{is equivalent to } \log_{10} 0.01 = -2$$

Some logarithms can be determined by changing to exponential form and using the properties of exponents, as the next two examples illustrate.

EXAMPLE 1

Evaluate $\log_{10} 0.0001$.

Solution

Let $\log_{10} 0.0001 = x$. Then, changing to exponential form yields $10^x = 0.0001$, which can be solved as follows:

$$10^x = 0.0001$$
$$10^x = 10^{-4} \qquad 0.0001 = \frac{1}{10,000} = \frac{1}{10^4} = 10^{-4}$$
$$x = -4$$

Thus we have $\log_{10} 0.0001 = -4$.

EXAMPLE 2 Evaluate $\log_9\left(\dfrac{\sqrt[5]{27}}{3}\right)$.

Solution

Let $\log_9\left(\dfrac{\sqrt[5]{27}}{3}\right) = x$. Then, changing to exponential form yields $9^x = \dfrac{\sqrt[5]{27}}{3}$, which can be solved as follows:

$$9^x = \frac{(27)^{1/5}}{3}$$

$$(3^2)^x = \frac{(3^3)^{1/5}}{3}$$

$$3^{2x} = \frac{3^{3/5}}{3}$$

$$3^{2x} = 3^{-2/5}$$

$$2x = -\frac{2}{5}$$

$$x = -\frac{1}{5}$$

Therefore, we have $\log_9\left(\dfrac{\sqrt[5]{27}}{3}\right) = -\dfrac{1}{5}$. ■

Some equations that involve logarithms can also be solved by changing to exponential form and using our knowledge of exponents.

EXAMPLE 3 Solve $\log_8 x = \dfrac{2}{3}$.

Solution

Changing $\log_8 x = \dfrac{2}{3}$ to exponential form, we obtain

$$8^{2/3} = x$$

Therefore,

$$x = (\sqrt[3]{8})^2$$
$$= 2^2$$
$$= 4$$

The solution set is {4}. ■

EXAMPLE 4   Solve $\log_b\left(\dfrac{27}{64}\right) = 3$.

Solution

Change $\log_b\left(\dfrac{27}{64}\right) = 3$ to exponential form to obtain

$$b^3 = \frac{27}{64}$$

Therefore,

$$b = \sqrt[3]{\frac{27}{64}}$$

$$= \frac{3}{4}$$

The solution set is $\left\{\dfrac{3}{4}\right\}$.   ■

Properties of Logarithms

There are some properties of logarithms that are a direct consequence of Definition 14.2 and the properties of exponents. For example, the following property is obtained by writing the exponential equations $b^1 = b$ and $b^0 = 1$ in logarithmic form.

PROPERTY 14.3

For $b > 0$ and $b \neq 1$,

$$\log_b b = 1 \qquad \text{and} \qquad \log_b 1 = 0$$

Therefore, according to Property 14.3, we can write

$$\log_{10} 10 = 1 \qquad \log_4 4 = 1$$
$$\log_{10} 1 = 0 \qquad \log_5 1 = 0$$

Also from Definition 14.2, we know that $\log_b r$ is the exponent t such that $b^t = r$. Therefore, raising b to the $\log_b r$ power must produce r. This fact is stated in Property 14.4.

PROPERTY 14.4

For $b > 0$, $b \neq 1$, and $r > 0$,

$$b^{\log_b r} = r$$

Therefore, according to Property 14.4, we can write

$$10^{\log_{10} 72} = 72 \qquad 3^{\log_3 85} = 85 \qquad e^{\log_e 7} = 7$$

Because a logarithm is by definition an exponent, it would seem reasonable to predict that some properties of logarithms correspond to the basic exponential properties. This is an accurate prediction; these properties provide a basis for computational work with logarithms. Let's state the first of these properties and show how we can use our knowledge of exponents to verify it.

PROPERTY 14.5

For positive numbers b, r, and s, where $b \neq 1$,

$$\log_b rs = \log_b r + \log_b s$$

To verify Property 14.5, we can proceed as follows: Let $m = \log_b r$ and $n = \log_b s$. Change each of these equations to exponential form.

$$m = \log_b r \quad \text{becomes } r = b^m$$
$$n = \log_b s \quad \text{becomes } s = b^n$$

Thus the product rs becomes

$$rs = b^m \cdot b^n = b^{m+n}$$

Now, changing $rs = b^{m+n}$ back to logarithmic form produces

$$\log_b rs = m + n$$

Replace m with $\log_b r$ and replace n with $\log_b s$ to yield

$$\log_b rs = \log_b r + \log_b s$$

The following two examples illustrate the use of Property 14.5.

EXAMPLE 5   If $\log_2 5 = 2.3222$ and $\log_2 3 = 1.5850$, evaluate $\log_2 15$.

Solution

Because $15 = 5 \cdot 3$, we can apply Property 14.5 as follows:

$$\log_2 15 = \log_2(5 \cdot 3)$$
$$= \log_2 5 + \log_2 3$$
$$= 2.3222 + 1.5850 = 3.9072$$

EXAMPLE 6   Given that $\log_{10} 178 = 2.2504$ and $\log_{10} 89 = 1.9494$, evaluate $\log_{10}(178 \cdot 89)$.

Solution

$$\log_{10}(178 \cdot 89) = \log_{10} 178 + \log_{10} 89$$
$$= 2.2504 + 1.9494 = 4.1998$$

Because $\dfrac{b^m}{b^n} = b^{m-n}$, we would expect a corresponding property that pertains to logarithms. Property 14.6 is that property. We can verify it by using an approach similar to the one we used to verify Property 14.5. This verification is left for you to do as an exercise in the next problem set.

PROPERTY 14.6

For positive numbers b, r, and s, where $b \neq 1$,

$$\log_b\left(\frac{r}{s}\right) = \log_b r - \log_b s$$

We can use Property 14.6 to change a division problem into an equivalent subtraction problem, as the next two examples illustrate.

EXAMPLE 7   If $\log_5 36 = 2.2266$ and $\log_5 4 = 0.8614$, evaluate $\log_5 9$.

Solution

Because $9 = \dfrac{36}{4}$, we can use Property 14.6 as follows:

$$\log_5 9 = \log_5\left(\frac{36}{4}\right)$$
$$= \log_5 36 - \log_5 4$$
$$= 2.2266 - 0.8614 = 1.3652$$

EXAMPLE 8   Evaluate $\log_{10}\left(\dfrac{379}{86}\right)$, given that $\log_{10} 379 = 2.5786$ and $\log_{10} 86 = 1.9345$.

Solution

$$\log_{10}\left(\frac{379}{86}\right) = \log_{10} 379 - \log_{10} 86$$
$$= 2.5786 - 1.9345$$
$$= 0.6441$$

Another property of exponents states that $(b^n)^m = b^{mn}$. The corresponding property of logarithms is stated in Property 14.7. Again, we will leave the verification of this property as an exercise for you to do in the next set of problems.

PROPERTY 14.7

If r is a positive real number, b is a positive real number other than 1, and p is any real number, then

$$\log_b r^p = p(\log_b r)$$

We will use Property 14.7 in the next two examples.

E X A M P L E   9

Evaluate $\log_2 22^{1/3}$ given that $\log_2 22 = 4.4598$.

Solution

$$\log_2 22^{1/3} = \frac{1}{3} \log_2 22 \qquad \text{Property 14.7}$$

$$= \frac{1}{3}(4.4598)$$

$$= 1.4866$$

E X A M P L E   1 0

Evaluate $\log_{10} (8540)^{3/5}$ given that $\log_{10} 8540 = 3.9315$.

Solution

$$\log_{10}(8540)^{3/5} = \frac{3}{5} \log_{10} 8540$$

$$= \frac{3}{5}(3.9315)$$

$$= 2.3589$$

Used together, the properties of logarithms enable us to change the forms of various logarithmic expressions. For example, we can rewrite an expression such as $\log_b \sqrt{\dfrac{xy}{z}}$ in terms of sums and differences of simpler logarithmic quantities as follows:

$$\log_b \sqrt{\frac{xy}{z}} = \log_b \left(\frac{xy}{z}\right)^{1/2}$$

$$= \frac{1}{2} \log_b \left(\frac{xy}{z}\right) \qquad \text{Property 14.7}$$

$$= \frac{1}{2}(\log_b xy - \log_b z) \qquad \text{Property 14.6}$$

$$= \frac{1}{2}(\log_b x + \log_b y - \log_b z) \qquad \text{Property 14.5}$$

Sometimes we need to change from an indicated sum or difference of logarithmic quantities to an indicated product or quotient. This is especially helpful when solving certain kinds of equations that involve logarithms. Note in these next two examples how we can use the properties, along with the process of changing from logarithmic form to exponential form, to solve some equations.

E X A M P L E   1 1   Solve $\log_{10} x + \log_{10}(x + 9) = 1$.

Solution

$$\log_{10} x + \log_{10}(x + 9) = 1$$
$$\log_{10}[x(x + 9)] = 1 \qquad \text{Property 14.5}$$
$$10^1 = x(x + 9) \qquad \text{Change to exponential form.}$$
$$10 = x^2 + 9x$$
$$0 = x^2 + 9x - 10$$
$$0 = (x + 10)(x - 1)$$
$$x + 10 = 0 \qquad \text{or} \qquad x - 1 = 0$$
$$x = -10 \qquad\qquad x = 1$$

Logarithms are defined only for positive numbers, so x and $x + 9$ have to be positive. Therefore, the solution of -10 must be discarded. The solution set is $\{1\}$. ■

E X A M P L E   1 2   Solve $\log_5(x + 4) - \log_5 x = 2$.

Solution

$$\log_5(x + 4) - \log_5 x = 2$$
$$\log_5\left(\frac{x + 4}{x}\right) = 2 \qquad \text{Property 14.6}$$
$$5^2 = \frac{x + 4}{x} \qquad \text{Change to exponential form.}$$
$$25 = \frac{x + 4}{x}$$
$$25x = x + 4$$
$$24x = 4$$
$$x = \frac{4}{24} = \frac{1}{6}$$

The solution set is $\left\{\dfrac{1}{6}\right\}$.   ■

CONCEPT QUIZ

For Problems 1–6, answer true or false.

1. The $\log_m n = q$ is equivalent to $m^q = n$.

2. The $\log_7 7$ equals 0.

3. A logarithm is by definition an exponent.

4. The $\log_5 9^2$ is equivalent to $2 \log_5 9$.

5. For the expression $\log_3 9$ the base of the logarithm is 9.

6. The expression $\log_2 x - \log_2 y + \log_2 z$ is equivalent to $\log_2 xyz$.

PROBLEM SET 14.4

For Problems 1–10, write each exponential statement in logarithmic form. For example, $2^5 = 32$ becomes $\log_2 32 = 5$ in logarithmic form.

1. $2^7 = 128$

2. $3^3 = 27$

3. $5^3 = 125$

4. $2^6 = 64$

5. $10^3 = 1000$

6. $10^1 = 10$

7. $2^{-2} = \dfrac{1}{4}$

8. $3^{-4} = \dfrac{1}{81}$

9. $10^{-1} = 0.1$

10. $10^{-2} = 0.01$

For Problems 11–20, write each logarithmic statement in exponential form. For example, $\log_2 8 = 3$ becomes $2^3 = 8$ in exponential form.

11. $\log_3 81 = 4$

12. $\log_2 256 = 8$

13. $\log_4 64 = 3$

14. $\log_5 25 = 2$

15. $\log_{10} 10{,}000 = 4$

16. $\log_{10} 100{,}000 = 5$

17. $\log_2 \left(\dfrac{1}{16} \right) = -4$

18. $\log_5 \left(\dfrac{1}{125} \right) = -3$

19. $\log_{10} 0.001 = -3$

20. $\log_{10} 0.000001 = -6$

For Problems 21–40, evaluate each logarithmic expression.

21. $\log_2 16$

22. $\log_3 9$

23. $\log_3 81$

24. $\log_2 512$

25. $\log_6 216$

26. $\log_4 256$

27. $\log_7 \sqrt{7}$

28. $\log_2 \sqrt[3]{2}$

29. $\log_{10} 1$

30. $\log_{10} 10$

31. $\log_{10} 0.1$

32. $\log_{10} 0.0001$

33. $10^{\log_{10} 5}$

34. $10^{\log_{10} 14}$

35. $\log_2 \left(\dfrac{1}{32} \right)$

36. $\log_5 \left(\dfrac{1}{25} \right)$

37. $\log_5(\log_2 32)$

38. $\log_2(\log_4 16)$

39. $\log_{10}(\log_7 7)$

40. $\log_2(\log_5 5)$

For Problems 41–50, solve each equation.

41. $\log_7 x = 2$

42. $\log_2 x = 5$

43. $\log_8 x = \dfrac{4}{3}$

44. $\log_{16} x = \dfrac{3}{2}$

45. $\log_9 x = \dfrac{3}{2}$

46. $\log_8 x = -\dfrac{2}{3}$

47. $\log_4 x = -\dfrac{3}{2}$

48. $\log_9 x = -\dfrac{5}{2}$

49. $\log_x 2 = \dfrac{1}{2}$

50. $\log_x 3 = \dfrac{1}{2}$

For Problems 51–59, given that $\log_2 5 = 2.3219$ and $\log_2 7 = 2.8074$, evaluate each expression by using Properties 14.5–14.7.

51. $\log_2 35$

52. $\log_2 \left(\dfrac{7}{5} \right)$

53. $\log_2 125$

54. $\log_2 49$

55. $\log_2 \sqrt{7}$

56. $\log_2 \sqrt[3]{5}$

57. $\log_2 175$

58. $\log_2 56$

59. $\log_2 80$

For Problems 60–68, given that $\log_8 5 = 0.7740$ and $\log_8 11 = 1.1531$, evaluate each expression using Properties 14.5–14.7.

60. $\log_8 55$

61. $\log_8 \left(\dfrac{5}{11} \right)$

62. $\log_8 25$

63. $\log_8 \sqrt{11}$

64. $\log_8 (5)^{2/3}$

65. $\log_8 88$

66. $\log_8 320$

67. $\log_8 \left(\dfrac{25}{11} \right)$

68. $\log_8 \left(\dfrac{121}{25} \right)$

For Problems 69–80, express each of the following as the sum or difference of simpler logarithmic quantities. Assume that all variables represent positive real numbers. For example,

$$\log_b \frac{x^3}{y^2} = \log_b x^3 - \log_b y^2$$

$$= 3 \log_b x - 2 \log_b y$$

69. $\log_b xyz$

70. $\log_b 5x$

71. $\log_b \left(\dfrac{y}{z} \right)$

72. $\log_b \left(\dfrac{x^2}{y} \right)$

73. $\log_b y^3 z^4$

74. $\log_b x^2 y^3$

75. $\log_b \left(\dfrac{x^{1/2} y^{1/3}}{z^4} \right)$

76. $\log_b x^{2/3} y^{3/4}$

77. $\log_b \sqrt[3]{x^2 z}$

78. $\log_b \sqrt{xy}$

79. $\log_b \left(x \sqrt{\dfrac{x}{y}} \right)$

80. $\log_b \sqrt{\dfrac{x}{y}}$

For Problems 81–88, express each of the following as a single logarithm. (Assume that all variables represent positive real numbers.) For example,

$$3 \log_b x + 5 \log_b y = \log_b x^3 y^5$$

81. $2 \log_b x - 4 \log_b y$

82. $\log_b x + \log_b y - \log_b z$

83. $\log_b x - (\log_b y - \log_b z)$

84. $(\log_b x - \log_b y) - \log_b z$

85. $2 \log_b x + 4 \log_b y - 3 \log_b z$

86. $\log_b x + \dfrac{1}{2} \log_b y$

87. $\dfrac{1}{2} \log_b x - \log_b x + 4 \log_b y$

88. $2 \log_b x + \dfrac{1}{2} \log_b (x - 1) - 4 \log_b (2x + 5)$

For Problems 89–100, solve each equation.

89. $\log_3 x + \log_3 4 = 2$

90. $\log_7 5 + \log_7 x = 1$

91. $\log_{10} x + \log_{10} (x - 21) = 2$

92. $\log_{10} x + \log_{10} (x - 3) = 1$

93. $\log_2 x + \log_2 (x - 3) = 2$

94. $\log_3 x + \log_3 (x - 2) = 1$

95. $\log_{10} (2x - 1) - \log_{10} (x - 2) = 1$

96. $\log_{10} (9x - 2) = 1 + \log_{10} (x - 4)$

97. $\log_5 (3x - 2) = 1 + \log_5 (x - 4)$

98. $\log_6 x + \log_6 (x + 5) = 2$

99. $\log_8 (x + 7) + \log_8 x = 1$

100. $\log_6 (x + 1) + \log_6 (x - 4) = 2$

101. Verify Property 14.6.

102. Verify Property 14.7.

■ ■ ■ **Thoughts into words**

103. Explain, without using Property 14.4, why $4^{\log_4 9}$ equals 9.

104. How would you explain the concept of a logarithm to someone who had just completed an elementary algebra course?

105. In the next section, we will show that the logarithmic function $f(x) = \log_2 x$ is the inverse of the exponential function $f(x) = 2^x$. From that information, how could you sketch a graph of $f(x) = \log_2 x$?

Answers to Concept Quiz

1. True   **2.** False   **3.** True   **4.** True   **5.** False   **6.** False

14.5 Logarithmic Functions

Objectives

■ Graph logarithmic functions.

■ Evaluate common and natural logarithms using a calculator.

■ Solve common and natural logarithmic equations using a calculator.

We can now use the concept of a logarithm to define a logarithmic function.

DEFINITION 14.5

If $b > 0$ and $b \neq 1$, then the function defined by

$$f(x) = \log_b x$$

where x is any positive real number, is called the **logarithmic function with base b.**

We can obtain the graph of a specific logarithmic function in various ways. For example, the equation $y = \log_2 x$ can be changed to the exponential equation $2^y = x$, for which we can determine a table of values. (The next set of exercises asks you to use this approach to graph some logarithmic functions.) We can also set up a table of values directly from the logarithmic equation and sketch the graph from the table. Example 1 illustrates this approach.

EXAMPLE 1

Graph $f(x) = \log_2 x$.

Solution

Let's choose some values for x that allow us to easily determine the corresponding values for $\log_2 x$. (Remember that logarithms are defined only for the positive real numbers.)

x	f(x)
$\dfrac{1}{8}$	-3
$\dfrac{1}{4}$	-2
$\dfrac{1}{2}$	-1
1	0
2	1
4	2
8	3

$\mathrm{Log}_2 \dfrac{1}{8} = -3$ because $2^{-3} = \dfrac{1}{2^3} = \dfrac{1}{8}$.

$\mathrm{Log}_2 1 = 0$ because $2^0 = 1$.

Plot these points and connect them with a smooth curve to produce Figure 14.27.

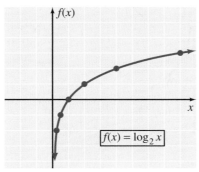

Figure 14.27

Now suppose that we consider two functions f and g as follows:

$f(x) = b^x$ Domain: all real numbers

                Range: positive real numbers

$g(x) = \log_b x$ Domain: positive real numbers

                Range: all real numbers

Furthermore, suppose that we consider the composition of f and g and the composition of g and f:

$$(f \circ g)(x) = f(g(x)) = f(\log_b x) = b^{\log_b x} = x$$
$$(g \circ f)(x) = g(f(x)) = g(b^x) = \log_b b^x = x \log_b b = x(1) = x$$

Because the domain of f is the range of g, the range of f is the domain of g, $f(g(x)) = x$, and $g(f(x)) = x$, the two functions f and g *are inverses of each other.*

Remember that the graph of a function and the graph of its inverse are reflections of each other through the line $y = x$. Thus we can determine the graph of a logarithmic function by reflecting the graph of its inverse exponential function through the line $y = x$. We demonstrate this idea in Figure 14.28, where the graph of $y = 2^x$ has been reflected across the line $y = x$ to produce the graph of $y = \log_2 x$.

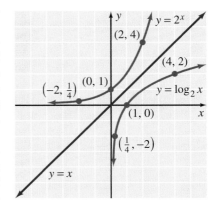

The general behavior patterns of exponential functions were illustrated back in Figure 14.3. We can now reflect each of these graphs through the line $y = x$ and observe the general behavior patterns of logarithmic functions shown in Figure 14.29.

Figure 14.28

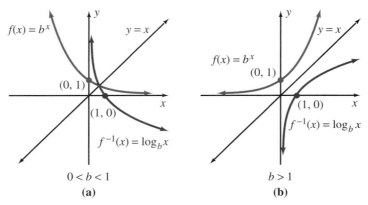

Figure 14.29

As you graph logarithmic functions, don't forget about transformations of basic curves.

1. The graph of $f(x) = 3 + \log_2 x$ is the graph of $f(x) = \log_2 x$ *moved up 3 units.* (Because $\log_2 x + 3$ is likely to be confused with $\log_2(x + 3)$, we commonly write $3 + \log_2 x$.)
2. The graph of $f(x) = \log_2(x - 4)$ is the graph of $f(x) = \log_2 x$ *moved 4 units to the right.*
3. The graph of $f(x) = -\log_2 x$ is the graph of $f(x) = \log_2 x$ *reflected across the x axis.*

Common Logarithms—Base 10

The properties of logarithms we discussed in Section 14.4 are true for any valid base. However, because the Hindu-Arabic numeration system that we use is a base-10 system, logarithms to base 10 have historically been used for computational purposes. Base-10 logarithms are called **common logarithms.**

Originally, common logarithms were developed to aid in complicated numerical calculations that involve products, quotients, and powers of real numbers. Today they are seldom used for that purpose because the calculator and computer can much more effectively handle the messy computational problems. However, common logarithms do still occur in applications, so they deserve our attention.

As we know from earlier work, the definition of a logarithm provides the basis for evaluating $\log_{10} x$ for values of x that are integral powers of 10. Consider the following examples.

$$\log_{10} 1000 = 3 \qquad \text{because } 10^3 = 1000$$
$$\log_{10} 100 = 2 \qquad \text{because } 10^2 = 100$$
$$\log_{10} 10 = 1 \qquad \text{because } 10^1 = 10$$
$$\log_{10} 1 = 0 \qquad \text{because } 10^0 = 1$$

$$\log_{10} 0.1 = -1 \qquad \text{because } 10^{-1} = \frac{1}{10} = 0.1$$

$$\log_{10} 0.01 = -2 \qquad \text{because } 10^{-2} = \frac{1}{10^2} = 0.01$$

$$\log_{10} 0.001 = -3 \qquad \text{because } 10^{-3} = \frac{1}{10^3} = 0.001$$

When working exclusively with base-10 logarithms, it is customary to omit writing the numeral 10 to designate the base. Thus the expression $\log_{10} x$ is written as $\log x$, and a statement such as $\log_{10} 1000 = 3$ becomes $\log 1000 = 3$. We will follow this practice from now on in this chapter, but don't forget that the base is understood to be 10.

$$\log_{10} x = \log x$$

To find the common logarithm of a positive number that is not an integral power of 10, we can use an appropriately equipped calculator. A calculator equipped with a common logarithm function (ordinarily, a key labeled $\boxed{\log}$ is used) gives us the following results rounded to four decimal places.

$\log 1.75 = 0.2430$

$\log 23.8 = 1.3766$ Be sure that you can use a calculator and obtain these
                          results.
$\log 134 = 2.1271$

$\log 0.192 = -0.7167$

$\log 0.0246 = -1.6091$

In order to use logarithms to solve problems, we sometimes need to be able to determine a number when the logarithm of the number is known. That is, we may need to determine x if $\log x$ is known. Let's consider an example.

EXAMPLE 2

Find x if $\log x = 0.2430$.

Solution

If $\log x = 0.2430$, then changing to exponential form yields $10^{0.2430} = x$, so we use the $\boxed{10^x}$ key to find x.

$$x = 10^{0.2430} \approx 1.749846689$$

Therefore, $x = 1.7498$ rounded to five significant digits. ∎

Be sure that you can use your calculator and obtain the following results. We rounded the values for x to five significant digits.

If $\log x = 0.7629$, then $x = 10^{0.7629} = 5.7930$.

If $\log x = 1.4825$, then $x = 10^{1.4825} = 30.374$.

If $\log x = 4.0214$, then $x = 10^{4.0214} = 10,505$.

If $\log x = -1.5162$, then $x = 10^{-1.5162} = 0.030465$.

If $\log x = -3.8921$, then $x = 10^{-3.8921} = 0.00012820$.

The **common logarithmic function** is defined by the equation $f(x) = \log x$. It should now be a simple matter to set up a table of values and sketch the function. You will do this in the next set of exercises. Remember that $f(x) = 10^x$ and $g(x) = \log x$ are inverses of each other. Therefore, we could also get the graph of $g(x) = \log x$ by reflecting the exponential curve $f(x) = 10^x$ across the line $y = x$.

Natural Logarithms—Base e

In many practical applications of logarithms, the number e (remember that $e \approx 2.71828$) is used as a base. Logarithms with a base of e are called **natural logarithms,** and the symbol $\ln x$ is commonly used instead of $\log_e x$.

$$\log_e x = \ln x$$

Natural logarithms can be found with an appropriately equipped calculator. A calculator with a natural logarithm function (ordinarily, a key labeled $\boxed{\ln x}$) gives us the following results rounded to four decimal places.

$\ln 3.21 = 1.1663$

$\ln 47.28 = 3.8561$

$\ln 842 = 6.7358$

$\ln 0.21 = -1.5606$

$\ln 0.0046 = -5.3817$

$\ln 10 = 2.3026$

Be sure that you can use your calculator to obtain these results. Keep in mind the significance of a statement such as $\ln 3.21 = 1.1663$. By changing to exponential form, we are claiming that e raised to the 1.1663 power is approximately 3.21. Using a calculator, we obtain $e^{1.1663} = 3.210093293$.

Let's do a few more problems to find x when given $\ln x$. Be sure that you agree with these results.

If $\ln x = 2.4156$, then $x = e^{2.4156} = 11.196$.

If $\ln x = 0.9847$, then $x = e^{0.9847} = 2.6770$.

If $\ln x = 4.1482$, then $x = e^{4.1482} = 63.320$.

If $\ln x = -1.7654$, then $x = e^{-1.7654} = 0.17112$.

The **natural logarithmic function** is defined by the equation $f(x) = \ln x$. It is the inverse of the natural exponential function $f(x) = e^x$. Thus one way to graph $f(x) = \ln x$ is to reflect the graph of $f(x) = e^x$ across the line $y = x$. We will ask you to do this in the next set of problems.

In Figure 14.30, we used a graphing utility to sketch the graph of $f(x) = e^x$. Now, on the basis of our previous work with transformations, we should be able to make the following statements.

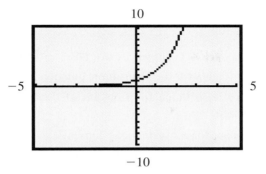

Figure 14.30

1. The graph of $f(x) = -e^x$ is the graph of $f(x) = e^x$ reflected through the x axis.
2. The graph of $f(x) = e^{-x}$ is the graph of $f(x) = e^x$ reflected through the y axis.
3. The graph of $f(x) = e^x + 4$ is the graph of $f(x) = e^x$ shifted upward 4 units.
4. The graph of $f(x) = e^{x+2}$ is the graph of $f(x) = e^x$ shifted 2 units to the left.

These statements are verified in Figure 14.31, which shows the result of graphing these four functions on the same set of axes by using a graphing utility.

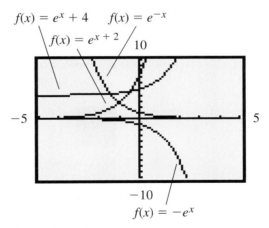

Figure 14.31

REMARK: So far, we have used a graphing utility to graph only common logarithmic and natural logarithmic functions. In the next section, we will see how logarithms with bases other than 10 or e are related to common and natural logarithms. This will provide a way of using a graphing utility to graph a logarithmic function with any valid base.

For Problems 1–6, answer true or false.

1. The domain for the logarithmic function $f(x) = \log_b x$ is all real numbers.

2. Every logarithmic function has an inverse function that is an exponential function.

3. The base for common logarithms is 2.

4. Logarithms with a base of e are called empirical logarithms.

5. The symbol $\ln x$ is usually written instead of $\log_e x$.

6. The graph of $f(x) = \log_4 x$ is the reflection with reference to the line $y = x$ of the graph $f(x) = 4^x$.

PROBLEM SET 14.5

For Problems 1–10, use a calculator to find each **common logarithm.** Express answers to four decimal places.

1. log 7.24

2. log 2.05

3. log 52.23

4. log 825.8

5. log 3214.1

6. log 14,189

7. log 0.729

8. log 0.04376

9. log 0.00034

10. log 0.000069

For Problems 11–20, use your calculator to find x when given $\log x$. Express answers to five significant digits.

11. $\log x = 2.6143$

12. $\log x = 1.5263$

13. $\log x = 4.9547$

14. $\log x = 3.9335$

15. $\log x = 1.9006$

16. $\log x = 0.5517$

17. $\log x = -1.3148$

18. $\log x = -0.1452$

19. $\log x = -2.1928$

20. $\log x = -2.6542$

For Problems 21–30, use your calculator to find each **natural logarithm.** Express answers to four decimal places.

21. ln 5

22. ln 18

23. ln 32.6

24. ln 79.5

25. ln 430

26. ln 371.8

27. ln 0.46

28. ln 0.524

29. ln 0.0314

30. ln 0.008142

For Problems 31–40, use your calculator to find x when given $\ln x$. Express answers to five significant digits.

31. $\ln x = 0.4721$

32. $\ln x = 0.9413$

33. $\ln x = 1.1425$

34. $\ln x = 2.7619$

35. $\ln x = 4.6873$

36. $\ln x = 3.0259$

37. $\ln x = -0.7284$

38. $\ln x = -1.6246$

39. $\ln x = -3.3244$

40. $\ln x = -2.3745$

41. a. Complete the following table, and then graph $f(x) = \log x$. (Express the values for $\log x$ to the nearest tenth.)

x	0.1	0.5	1	2	4	8	10
log x							

b. Complete the following table, expressing values for 10^x to the nearest tenth.

x	−1	−0.3	0	0.3	0.6	0.9	1
10ˣ							

Then graph $f(x) = 10^x$, and reflect it across the line $y = x$ to produce the graph for $f(x) = \log x$.

42. a. Complete the following table, and then graph $f(x) = \ln x$. (Express the values for $\ln x$ to the nearest tenth.)

x	0.1	0.5	1	2	4	8	10
ln x							

b. Complete the following table, expressing values for e^x to the nearest tenth.

x	−2.3	−0.7	0	0.7	1.4	2.1	2.3
e^x							

Then graph $f(x) = e^x$, and reflect it across the line $y = x$ to produce the graph for $f(x) = \ln x$.

43. Graph $y = \log_{\frac{1}{2}} x$ by graphing $\left(\dfrac{1}{2}\right)^y = x$.

44. Graph $y = \log_2 x$ by graphing $2^y = x$.

45. Graph $f(x) = \log_3 x$ by reflecting the graph of $g(x) = 3^x$ across the line $y = x$.

46. Graph $f(x) = \log_4 x$ by reflecting the graph of $g(x) = 4^x$ across the line $y = x$.

For Problems 47–53, graph each of the functions. Remember that the graph of $f(x) = \log_2 x$ is given in Figure 14.27.

47. $f(x) = 3 + \log_2 x$ **48.** $f(x) = -2 + \log_2 x$

49. $f(x) = \log_2(x + 3)$

50. $f(x) = \log_2(x - 2)$

51. $f(x) = \log_2 2x$

52. $f(x) = -\log_2 x$

53. $f(x) = 2 \log_2 x$

For Problems 54–61, perform the following calculations and express answers to the nearest hundredth. (These calculations are assigned in preparation for our work in the next section.)

54. $\dfrac{\log 7}{\log 3}$

55. $\dfrac{\ln 2}{\ln 7}$

56. $\dfrac{2 \ln 3}{\ln 8}$

57. $\dfrac{\ln 5}{2 \ln 3}$

58. $\dfrac{\ln 3}{0.04}$

59. $\dfrac{\ln 2}{0.03}$

60. $\dfrac{\log 2}{5 \log 1.02}$

61. $\dfrac{\log 5}{3 \log 1.07}$

■■■ **Thoughts into words**

62. Why is the number 1 excluded from being a base of a logarithm?

63. How do we know that $\log_2 6$ is between 2 and 3?

 Graphing calculator activities

64. Graph $f(x) = x$, $f(x) = e^x$, and $f(x) = \ln x$ on the same set of axes.

65. Graph $f(x) = x$, $f(x) = 10^x$, and $f(x) = \log x$ on the same set of axes.

66. Graph $f(x) = \ln x$. How should the graphs of $f(x) = 2 \ln x$, $f(x) = 4 \ln x$, and $f(x) = 6 \ln x$ compare to the graph of $f(x) = \ln x$? Graph the three functions on the same set of axes with $f(x) = \ln x$.

67. Graph $f(x) = \log x$. Now predict the graphs for $f(x) = 2 + \log x$, $f(x) = -2 + \log x$, and $f(x) = -6 + \log x$. Graph the three functions on the same set of axes with $f(x) = \log x$.

68. Graph $f(x) = \ln x$. Now predict the graphs for $f(x) = \ln(x - 2)$, $f(x) = \ln(x - 6)$, and $f(x) = \ln(x + 4)$. Graph the three functions on the same set of axes with $f(x) = \ln x$.

69. For each of the following, (a) predict the general shape and location of the graph, and (b) use your graphing calculator to graph the function and thus check your prediction.

   a. $f(x) = \log x + \ln x$ **b.** $f(x) = \log x - \ln x$

   c. $f(x) = \ln x - \log x$ **d.** $f(x) = \ln x^2$

Answers to Concept Quiz

1. False   **2.** True   **3.** False   **4.** False   **5.** True   **6.** True

14.6   Exponential Equations, Logarithmic Equations, and Problem Solving

Objectives

■   Expand the methods of solving logarithmic equations.

■   Solve problems involving compound interest, Richter numbers, and growth and decay.

■   Use a change-of-base formula for logarithms.

In Section 14.1, we solved exponential equations such as $3^x = 81$ by expressing both sides of the equation as a power of 3 and then applying the property "if $b^n = b^m$, then $n = m$." However, if we try this same approach with an equation such as $3^x = 5$, we face the difficulty of expressing 5 as a power of 3. We can solve this type of problem by using the properties of logarithms and the following property of equality.

PROPERTY 14.8

If $x > 0$, $y > 0$, $b > 0$, and $b \neq 1$, then $x = y$ if and only if $\log_b x = \log_b y$.

Property 14.8 is stated in terms of any valid base b; however, for most applications, we use either common logarithms or natural logarithms. Let's consider some examples.

EXAMPLE 1   Solve $3^x = 5$ to the nearest hundredth.

Solution

By using common logarithms, we can proceed as follows:

$$3^x = 5$$
$$\log 3^x = \log 5 \qquad \text{Property 14.8}$$
$$x \log 3 = \log 5 \qquad \log r^p = p \log r$$
$$x = \frac{\log 5}{\log 3}$$
$$x = 1.46 \quad \text{to the nearest hundredth}$$

 Check

Because $3^{1.46} \approx 4.972754647$, we say that, to the nearest hundredth, the solution set for $3^x = 5$ is {1.46}.   ■

EXAMPLE 2   Solve $e^{x+1} = 5$ to the nearest hundredth.

Solution

Because base e is used in the exponential expression, let's use natural logarithms to help solve this equation.

$$e^{x+1} = 5$$

$$\ln e^{x+1} = \ln 5 \qquad \text{Property 14.8}$$

$$(x + 1)\ln e = \ln 5 \qquad \ln r^p = p \ln r$$

$$(x + 1)(1) = \ln 5 \qquad \ln e = 1$$

$$x = \ln 5 - 1$$

$$x = 0.61 \quad \text{to the nearest hundredth}$$

The solution set is {0.61}. Check it!

E X A M P L E   3

Solve $2^{3x-2} = 3^{2x+1}$ to the nearest hundredth.

Solution

$$2^{3x-2} = 3^{2x+1}$$

$$\log 2^{3x-2} = \log 3^{2x+1}$$

$$(3x - 2)\log 2 = (2x + 1)\log 3$$

$$3x \log 2 - 2 \log 2 = 2x \log 3 + \log 3$$

$$3x \log 2 - 2x \log 3 = \log 3 + 2 \log 2$$

$$x(3 \log 2 - 2 \log 3) = \log 3 + 2 \log 2$$

$$x = \frac{\log 3 + 2 \log 2}{3 \log 2 - 2 \log 3}$$

$$x = -21.10 \quad \text{to the nearest hundredth}$$

The solution set is {−21.10}. Check it!

Logarithmic Equations

In Example 11 of Section 14.4, we solved the logarithmic equation

$$\log_{10} x + \log_{10}(x + 9) = 1$$

by simplifying the left side of the equation to $\log_{10}[x(x + 9)]$ and then changing the equation to exponential form to complete the solution. Now, using Property 14.8, we can solve such a logarithmic equation another way and also expand our equation-solving capabilities. Let's consider some examples.

E X A M P L E   4

Solve $\log x + \log(x - 15) = 2$.

Solution

Because $\log 100 = 2$, the given equation becomes

$$\log x + \log(x - 15) = \log 100$$

Now simplify the left side, apply Property 14.8, and proceed as follows:

$$\log(x)(x - 15) = \log 100$$
$$x(x - 15) = 100$$
$$x^2 - 15x - 100 = 0$$
$$(x - 20)(x + 5) = 0$$
$$x - 20 = 0 \quad \text{or} \quad x + 5 = 0$$
$$x = 20 \qquad\qquad x = -5$$

The domain of a logarithmic function must contain only positive numbers, so x and $x - 15$ must be positive in this problem. Therefore, we discard the solution -5; the solution set is $\{20\}$. ◼

E X A M P L E   5   Solve $\ln(x + 2) = \ln(x - 4) + \ln 3$.

Solution

$$\ln(x + 2) = \ln(x - 4) + \ln 3$$
$$\ln(x + 2) = \ln[3(x - 4)]$$
$$x + 2 = 3(x - 4)$$
$$x + 2 = 3x - 12$$
$$14 = 2x$$
$$7 = x$$

The solution set is $\{7\}$. ◼

E X A M P L E   6   Solve $\log_b(x + 2) + \log_b(2x - 1) = \log_b x$.

Solution

$$\log_b(x + 2) + \log_b(2x - 1) = \log_b x$$
$$\log_b[(x + 2)(2x - 1)] = \log_b x$$
$$(x + 2)(2x - 1) = x$$
$$2x^2 + 3x - 2 = x$$
$$2x^2 + 2x - 2 = 0$$
$$x^2 + x - 1 = 0$$

Using the quadratic formula, we obtain

$$x = \frac{-1 \pm \sqrt{1 + 4}}{2}$$
$$= \frac{-1 \pm \sqrt{5}}{2}$$

Because $x + 2$, $2x - 1$, and x have to be positive, we must discard the solution $\dfrac{-1 - \sqrt{5}}{2}$; the solution set is $\left\{\dfrac{-1 + \sqrt{5}}{2}\right\}$.   ■

Problem Solving

In Section 14.2, we used the compound interest formula

$$A = P\left(1 + \frac{r}{n}\right)^{nt}$$

to determine the amount of money (A) accumulated at the end of t years if P dollars is invested at rate of interest r compounded n times per year. Now let's use this formula to solve other types of problems that deal with compound interest.

EXAMPLE 7   How long will it take for \$500 to double if it is invested at 12% compounded quarterly?

Solution

"To double" means that the \$500 must grow into \$1000. Thus

$$1000 = 500\left(1 + \frac{0.12}{4}\right)^{4t}$$
$$= 500(1 + 0.03)^{4t}$$
$$= 500(1.03)^{4t}$$

Multiplying both sides of $1000 = 500(1.03)^{4t}$ by $\dfrac{1}{500}$ yields

$$2 = (1.03)^{4t}$$

Therefore,

$$\log 2 = \log(1.03)^{4t} \qquad \text{Property 14.8}$$
$$= 4t \log 1.03 \qquad \log r^p = p \log r$$

Now let's solve for t.

$$4t \log 1.03 = \log 2$$
$$t = \frac{\log 2}{4 \log 1.03}$$
$$t = 5.9 \quad \text{to the nearest tenth}$$

Therefore, we are claiming that \$500 invested at 12% interest compounded quarterly will double in approximately 5.9 years.

✓ **Check**

$500 invested at 12% compounded quarterly for 5.9 years will produce

$$A = \$500\left(1 + \frac{0.12}{4}\right)^{4(5.9)}$$

$$= \$500(1.03)^{23.6}$$

$$= \$1004.45$$

■

E X A M P L E   8   Suppose that the number of bacteria present in a certain culture after t minutes is given by the equation $Q(t) = Q_0 e^{0.04t}$, where Q_0 represents the initial number of bacteria. How long will it take for the bacteria count to grow from 500 to 2000?

Solution

Substituting into $Q(t) = Q_0 e^{0.04t}$ and solving for t, we obtain

$$2000 = 500e^{0.04t}$$

$$4 = e^{0.04t}$$

$$\ln 4 = \ln e^{0.04t}$$

$$\ln 4 = 0.04t \ln e$$

$$\ln 4 = 0.04t \qquad \ln e = 1$$

$$\frac{\ln 4}{0.04} = t$$

$$34.7 = t \quad \text{to the nearest tenth}$$

It will take approximately 34.7 minutes.

■

Richter Numbers

Seismologists use the Richter scale to measure and report the magnitude of earthquakes. The equation

$$R = \log \frac{I}{I_0} \qquad R \text{ is called a Richter number.}$$

compares the intensity I of an earthquake to a minimal or reference intensity I_0. The reference intensity is the smallest earth movement that can be recorded on a seismograph. Suppose that the intensity of an earthquake was determined to be 50,000 times the reference intensity. In this case, $I = 50{,}000\ I_0$, and the Richter number is calculated as follows:

$$R = \log \frac{50{,}000\, I_0}{I_0}$$
$$= \log 50{,}000$$
$$\approx 4.698970004$$

Thus a Richter number of 4.7 would be reported. Let's consider two more examples that involve Richter numbers.

EXAMPLE 9

An earthquake that occurred in the San Francisco area in 1989 was reported to have a Richter number of 6.9. How did its intensity compare to the reference intensity?

Solution

$$6.9 = \log \frac{I}{I_0}$$
$$10^{6.9} = \frac{I}{I_0}$$
$$I = (10^{6.9})(I_0)$$
$$I \approx 7{,}943{,}282\, I_0$$

Its intensity was a little less than 8 million times the reference intensity.   ■

EXAMPLE 10

An earthquake that occurred in Iran in 1990 had a Richter number of 7.7. Compare the intensity of this earthquake to that of the one in San Francisco referred to in Example 9.

Solution

From Example 9, we have $I = (10^{6.9})(I_0)$ for the earthquake in San Francisco. Then, using a Richter number of 7.7, we obtain $I = (10^{7.7})(I_0)$ for the earthquake in Iran. Therefore, by comparison,

$$\frac{(10^{7.7})(I_0)}{(10^{6.9})(I_0)} = 10^{7.7 - 6.9}$$
$$= 10^{0.8} \approx 6.3$$

The earthquake in Iran was about 6 times as intense as the one in San Francisco.   ■

Logarithms with Base Other Than 10 or e

The basic approach whereby we apply Property 14.8 and use either common or natural logarithms can also be used to evaluate a logarithm to some base other than 10 or e. Consider the following example.

EXAMPLE 11

Evaluate $\log_3 41$.

Solution

Let $x = \log_3 41$. Change to exponential form to obtain

$$3^x = 41$$

Now we can apply Property 14.8 and proceed as follows:

$$\log 3^x = \log 41$$
$$x \log 3 = \log 41$$
$$x = \frac{\log 41}{\log 3}$$
$$x = 3.3802 \quad \text{rounded to four decimal places}$$

Therefore, we are claiming that 3 raised to the 3.3802 power will produce approximately 41. Check it! ■

The method of Example 11 to evaluate $\log_a r$ produces the following formula, which we often refer to as the **change-of-base formula for logarithms.**

PROPERTY 14.9

If a, b, and r are positive numbers, with $a \neq 1$ and $b \neq 1$, then

$$\log_a r = \frac{\log_b r}{\log_b a}$$

By using Property 14.9, we can easily determine a relationship between logarithms of different bases. For example, suppose that in Property 14.9 we let $a = 10$ and $b = e$.

$$\log_a r = \frac{\log_b r}{\log_b a}$$

becomes

$$\log_{10} r = \frac{\log_e r}{\log_e 10}$$
$$\log_e r = (\log_e 10)(\log_{10} r)$$
$$\log_e r = (2.3026)(\log_{10} r)$$

Thus the natural logarithm of any positive number is approximately equal to the common logarithm of the number times 2.3026.

Now we can use a graphing utility to graph logarithmic functions such as $f(x) = \log_2 x$. Using the change-of-base formula, we can express this function as $f(x) = \frac{\log x}{\log 2}$ or as $f(x) = \frac{\ln x}{\ln 2}$. The graph of $f(x) = \log_2 x$ is shown in Figure 14.32.

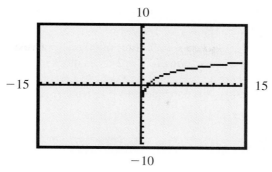

Figure 14.32

Finally, let's use a graphical approach to solve an equation that is cumbersome to solve with an algebraic approach.

EXAMPLE 12 Solve the equation $(5^x - 5^{-x})/2 = 3$.

Solution

First, we need to recognize that the solutions for the equation $(5^x - 5^{-x})/2 = 3$ are the x intercepts of the graph of the equation $y = (5^x - 5^{-x})/2 - 3$. We can use a graphing utility to obtain the graph of this equation as shown in Figure 14.33. Use the ZOOM and TRACE features to determine that the graph crosses the x axis at approximately 1.13. Thus the solution set of the original expression is {1.13}.

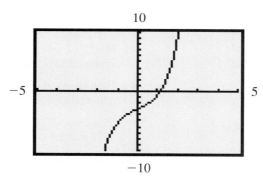

Figure 14.33

CONCEPT QUIZ

For Problems 1–5, answer true or false.

1. The equation $5^{2x+1} = 70$ can be solved by taking the logarithm of both sides of the equation.

2. All solutions of the equation $\log x + \log(x + 2) = 3$ will be positive numbers.

3. The Richter number compares the intensity of an earthquake to the logarithm of 100,000.

4. If the difference in Richter numbers for two earthquakes is 2, then it can be said that one earthquake was 2 times more intense than the other earthquake.

5. The $\log_3 7$ is equivalent to $\dfrac{\ln 7}{\ln 3}$.

PROBLEM SET 14.6

For Problems 1–20, solve each exponential equation and express approximate solutions to the nearest hundredth.

1. $3^x = 13$

2. $2^x = 21$

3. $4^n = 35$

4. $5^n = 75$

5. $2^x + 7 = 50$

6. $3^x - 6 = 25$

7. $3^{x-2} = 11$

8. $2^{x+1} = 7$

9. $5^{3t+1} = 9$

10. $7^{2t-1} = 35$

11. $e^x = 27$

12. $e^x = 86$

13. $e^{x-2} = 13.1$

14. $e^{x-1} = 8.2$

15. $3e^x - 1 = 17$

16. $2e^x = 12.4$

17. $5^{2x+1} = 7^{x+3}$

18. $3^{x-1} = 2^{x+3}$

19. $3^{2x+1} = 2^{3x+2}$

20. $5^{x-1} = 2^{2x+1}$

For Problems 21–32, solve each logarithmic equation and express irrational solutions in lowest radical form.

21. $\log x + \log(x + 21) = 2$

22. $\log x + \log(x + 3) = 1$

23. $\log(3x - 1) = 1 + \log(5x - 2)$

24. $\log(2x - 1) - \log(x - 3) = 1$

25. $\log(x + 1) = \log 3 - \log(2x - 1)$

26. $\log(x - 2) = 1 - \log(x + 3)$

27. $\log(x + 2) - \log(2x + 1) = \log x$

28. $\log(x + 1) - \log(x + 2) = \log \dfrac{1}{x}$

29. $\ln(2t + 5) = \ln 3 + \ln(t - 1)$

30. $\ln(3t - 4) - \ln(t + 1) = \ln 2$

31. $\log \sqrt{x} = \sqrt{\log x}$

32. $\log x^2 = (\log x)^2$

For Problems 33–42, approximate each logarithm to three decimal places. (Example 11 and/or Property 14.9 should be of some help.)

33. $\log_2 40$

34. $\log_2 93$

35. $\log_3 16$

36. $\log_3 37$

37. $\log_4 1.6$

38. $\log_4 3.2$

39. $\log_5 0.26$

40. $\log_5 0.047$

41. $\log_7 500$

42. $\log_8 750$

For Problems 43–55, solve each problem and express answers to the nearest tenth unless stated otherwise.

43. How long will it take $750 to be worth $1000 if it is invested at 12% interest compounded quarterly?

44. How long will it take $1000 to double if it is invested at 9% interest compounded semiannually?

45. How long will it take $2000 to double if it is invested at 13% interest compounded continuously?

46. How long will it take $500 to triple if it is invested at 9% interest compounded continuously?

47. What rate of interest compounded continuously is needed for an investment of $500 to grow to $900 in 10 years?

48. What rate of interest compounded continuously is needed for an investment of $2500 to grow to $10,000 in 20 years?

49. For a certain strain of bacteria, the number of bacteria present after t hours is given by the equation

$Q = Q_0 e^{0.34t}$, where Q_0 represents the initial number of bacteria. How long will it take 400 bacteria to increase to 4000 bacteria?

50. A piece of machinery valued at $30,000 depreciates at a rate of 10% yearly. How long will it take for it to reach a value of $15,000?

51. The equation $P(a) = 14.7e^{-0.21a}$, where a is the altitude above sea level measured in miles, yields the atmospheric pressure in pounds per square inch. If the atmospheric pressure at Cheyenne, Wyoming, is approximately 11.53 pounds per square inch, find that city's altitude above sea level. Express your answer to the nearest hundred feet.

52. The number of grams of a certain radioactive substance present after t hours is given by the equation $Q = Q_0 e^{-0.45t}$, where Q_0 represents the initial number of grams. How long would it take 2500 grams to be reduced to 1250 grams?

53. For a certain culture, the equation $Q(t) = Q_0 e^{0.4t}$, where Q_0 is an initial number of bacteria and t is time measured in hours, yields the number of bacteria as a func-

tion of time. How long will it take 500 bacteria to increase to 2000?

54. Suppose that the equation $P(t) = P_0 e^{0.02t}$, where P_0 represents an initial population and t is the time in years, is used to predict population growth. How long would it take a city of 50,000 to double its population?

55. An earthquake that occurred in Los Angeles in 1971 had an intensity of approximately five million times the reference intensity. What was the Richter number associated with that earthquake?

56. An earthquake that occurred in San Francisco in 1906 was reported to have a Richter number of 8.3. How did its intensity compare to the reference intensity?

57. Calculate how many times more intense an earthquake with a Richter number of 7.3 is than an earthquake with a Richter number of 6.4.

58. Calculate how many times more intense an earthquake with a Richter number of 8.9 is than an earthquake with a Richter number of 6.2.

■ ■ ■ **Thoughts into words**

59. Explain how to determine $\log_4 76$ without using Property 14.9.

60. Explain the concept of a Richter number.

61. Explain how you would solve the equation $2^x = 64$ and also how you would solve the equation $2^x = 53$.

62. How do logarithms with a base of 9 compare to logarithms with a base of 3? Explain how you reached this conclusion.

■ ■ ■ **Further investigations**

63. Use the approach of Example 11 to develop Property 14.9.

64. Let $r = b$ in Property 14.9, and verify that $\log_a b = \dfrac{1}{\log_b a}$.

65. Use an algebraic approach to solve the equation $\dfrac{5^x - 5^{-x}}{2} = 3$. Express your answer to the nearest hundredth.

66. Solve the equation $y = \dfrac{10^x + 10^{-x}}{2}$ for x in terms of y.

67. Solve the equation $y = \dfrac{e^x - e^{-x}}{2}$ for x in terms of y.

Graphing calculator activities

68. Check your answers for Problems 17–20 by graphing the appropriate function and finding the x intercept.

69. Graph $f(x) = x, f(x) = 2^x$, and $f(x) = \log_2 x$ on the same set of axes.

70. Graph $f(x) = x, f(x) = (0.5)^x$, and $f(x) = \log_{0.5} x$ on the same set of axes.

71. Graph $f(x) = \log_2 x$. Now predict the graphs for $f(x) = \log_3 x, f(x) = \log_4 x$, and $f(x) = \log_8 x$. Graph these three functions on the same set of axes with $f(x) = \log_2 x$.

72. Graph $f(x) = \log_5 x$. Now predict the graphs for $f(x) = 2 \log_5 x, f(x) = -4 \log_5 x$, and $f(x) = \log_5(x + 4)$. Graph these three functions on the same set of axes with $f(x) = \log_5 x$.

73. Use both a graphical and an algebraic approach to solve the equation $\dfrac{2^x - 2^{-x}}{3} = 4$.

Answers to Concept Quiz

1. True   **2.** True   **3.** False   **4.** False   **5.** True

CHAPTER 14

SUMMARY

(14.1) If a and b are positive real numbers and m and n are any real numbers, then

1. $b^n \cdot b^m = b^{n+m}$ ⠀ Product of two like bases with powers

2. $(b^n)^m = b^{mn}$ ⠀⠀ Power of a power

3. $(ab)^n = a^n b^n$ ⠀⠀ Power of a product

4. $\left(\dfrac{a}{b}\right)^n = \dfrac{a^n}{b^n}$ ⠀⠀ Power of a quotient

5. $\dfrac{b^n}{b^m} = b^{n-m}$ ⠀⠀ Quotient of two like bases with powers

If $b > 0$, $b \neq 1$, and m and n are real numbers, then $b^n = b^m$ if and only if $n = m$. A function defined by an equation of the form

$$f(x) = b^x, \qquad b > 0 \text{ and } b \neq 1$$

is called an **exponential function.**

(14.2) A general formula for any principal P compounded n times per year for t years, at a rate r, is

$$A = P\left(1 + \frac{r}{n}\right)^{nt}$$

where A represents the total amount of money accumulated at the end of the t years. The value of $\left(1 + \dfrac{1}{n}\right)^n$, as n gets infinitely large, approaches the number e, where e equals 2.71828 to five decimal places.

The formula

$$A = Pe^{rt}$$

yields the accumulated value A of a sum of money P that has been invested for t years at a rate of r percent **compounded continuously.**

The formula

$$Q = Q_0\left(\frac{1}{2}\right)^{\frac{t}{h}}$$

is referred to as the **half-life** formula.

The equation

$$Q(t) = Q_0 e^{kt}$$

is used as a mathematical model for many growth-and-decay applications.

(14.3) A function f is said to be a one-to-one function if every value of $f(x)$ has only one value of x associated with it.

In terms of ordered pairs, a one-to-one function is a function such that no two ordered pairs have the same second component.

Let f be a one-to-one function with a domain of X and a range of Y. A function g, with a domain of Y and a range of X, is called the **inverse function** of f if $(f \circ g)(x) = x$ for every x in Y and $(g \circ f)(x) = x$ for every x in X.

If the components of each ordered pair of a given one-to-one function are interchanged, the resulting function and the given function are **inverses** of each other.

The inverse of a function f is denoted by f^{-1}. Graphically, two functions that are inverses of each other are mirror images with reference to the line $y = x$.

A systematic technique for finding the inverse of a function can be described as follows:

1. Let $y = f(x)$.

2. Interchange x and y.

3. Solve the equation for y in terms of x.

4. The inverse function $f^{-1}(x)$ is determined by the equation in step 3.

Don't forget that the domain of f must equal the range of f^{-1}, and the domain of f^{-1} must equal the range of f.

Let f be a function, with the interval I a subset of the domain of f. Let x_1 and x_2 be in I.

1. f is increasing on I if $f(x_1) < f(x_2)$ whenever $x_1 < x_2$.

2. f is decreasing on I if $f(x_1) > f(x_2)$ whenever $x_1 < x_2$.

3. f is constant on I if $f(x_1) = f(x_2)$ for every x_1 and x_2.

(14.4) If r is any positive real number, then the unique exponent t such that $b^t = r$ is called the **logarithm of r with base b** and is denoted by $\log_b r$. For $b \geq 0$, $b \neq 1$, and $r > 0$,

1. $\log_b b = 1$

2. $\log_b 1 = 0$

3. $r = b^{\log_b r}$

The following properties of logarithms are derived from the definition of a logarithm and the properties of exponents. For positive real numbers b, r, and s, where $b \neq 1$,

1. $\log_b rs = \log_b r + \log_b s$

2. $\log_b\left(\dfrac{r}{s}\right) = \log_b r - \log_b s$

3. $\log_b r^p = p \log_b r$,   where p is any real number

(14.5) A function defined by an equation of the form

$$f(x) = \log_b x, \qquad b > 0 \text{ and } b \neq 1$$

is called a **logarithmic function.** The equation $y = \log_b x$ is equivalent to $x = b^y$. The two functions $f(x) = b^x$ and $g(x) = \log_b x$ are inverses of each other.

Logarithms with a base of 10 are called **common logarithms.** The expression $\log_{10} x$ is commonly written as $\log x$.

Many calculators are equipped with a common logarithm function. Often, a key labeled $\boxed{\log}$ is used to find common logarithms.

Natural logarithms are logarithms that have a base of e, where e is an irrational number whose decimal approximation to eight digits is 2.7182818. Natural logarithms are denoted by $\log_e x$ or $\ln x$.

Many calculators are also equipped with a natural logarithm function. Often, a key labeled $\boxed{\ln x}$ is used for this purpose.

(14.6) Together, the properties of equality and the properties of exponents and logarithms can help us solve a variety of exponential and logarithmic equations. These properties also help us solve problems that deal with various applications, including compound interest and growth problems.

The formula

$$R = \log \frac{I}{I_0}$$

yields the Richter number associated with an earthquake.

The formula

$$\log_a r = \frac{\log_b r}{\log_b a}$$

is often called the **change-of-base formula.**

CHAPTER 14 REVIEW PROBLEM SET

For Problems 1–10, evaluate each of the following:

1. $8^{5/3}$

2. $-25^{3/2}$

3. $(-27)^{4/3}$

4. $\log_6 216$

5. $\log_7\left(\dfrac{1}{49}\right)$

6. $\log_2 \sqrt[3]{2}$

7. $\log_2\left(\dfrac{\sqrt[4]{32}}{2}\right)$

8. $\log_{10} 0.00001$

9. $\ln e$

10. $7^{\log_7 12}$

For Problems 11–24, solve each equation. Express approximate solutions to the nearest hundredth.

11. $\log_{10} 2 + \log_{10} x = 1$ **12.** $\log_3 x = -2$

13. $4^x = 128$

14. $3^t = 42$

15. $\log_2 x = 3$

16. $\left(\dfrac{1}{27}\right)^{3x} = 3^{2x-1}$

17. $2e^x = 14$

18. $2^{2x+1} = 3^{x+1}$

19. $\ln(x + 4) - \ln(x + 2) = \ln x$

20. $\log x + \log(x - 15) = 2$

21. $\log(\log x) = 2$

22. $\log(7x - 4) - \log(x - 1) = 1$

23. $\ln(2t - 1) = \ln 4 + \ln(t - 3)$

24. $64^{2t+1} = 8^{-t+2}$

For Problems 25–28, if log 3 = 0.4771 and log 7 = 0.8451, evaluate each of the following:

25. $\log\left(\dfrac{7}{3}\right)$

26. log 21

27. log 27

28. $\log 7^{2/3}$

29. Express each of the following as the sum or difference of simpler logarithmic quantities. Assume that all variables represent positive real numbers.

a. $\log_b\left(\dfrac{x}{y^2}\right)$

b. $\log_b \sqrt[4]{xy^2}$

c. $\log_b\left(\dfrac{\sqrt{x}}{y^3}\right)$

30. Express each of the following as a single logarithm. Assume that all variables represent positive real numbers.

a. $3\log_b x + 2\log_b y$ **b.** $\dfrac{1}{2}\log_b y - 4\log_b x$

c. $\dfrac{1}{2}(\log_b x + \log_b y) - 2\log_b z$

For Problems 31–34, approximate each of the logarithms to the nearest hundredth.

31. $\log_2 3$

32. $\log_3 2$

33. $\log_4 191$

34. $\log_2 0.23$

For Problems 35–42, graph each of the functions.

35. $f(x) = \left(\dfrac{3}{4}\right)^x$

36. $f(x) = 2^{x+2}$

37. $f(x) = e^{x-1}$

38. $f(x) = -1 + \log x$

39. $f(x) = 3^x - 3^{-x}$

40. $f(x) = e^{-x^2/2}$

41. $f(x) = \log_2(x - 3)$

42. $f(x) = 3\log_3 x$

For Problems 43–45, use the compound interest formula $A = P\left(1 + \dfrac{r}{n}\right)^{nt}$ to find the total amount of money accumulated at the end of the indicated time period for each of the investments.

43. $7250 for 10 years at 7% compounded quarterly

44. $12500 for 15 years at 5% compounded monthly

45. $2500 for 20 years at 4.5% compounded semiannually

For Problems 46–49, determine whether f and g are inverse functions.

46. $f(x) = 7x - 1$ and $g(x) = \dfrac{x + 1}{7}$

47. $f(x) = -\dfrac{2}{3}x$ and $g(x) = \dfrac{3}{2}x$

48. $f(x) = x^2 - 6$ for $x \geq 0$, and $g(x) = \sqrt{x + 6}$ for $x \geq -6$

49. $f(x) = 2 - x^2$ for $x \geq 0$, and $g(x) = \sqrt{2 - x}$ for $x \leq 2$

For Problems 50–53, (a) find f^{-1}, and (b) verify that $(f \circ f^{-1})(x) = x$ and $(f^{-1} \circ f)(x) = x$.

50. $f(x) = 4x + 5$

51. $f(x) = -3x - 7$

52. $f(x) = \dfrac{5}{6}x - \dfrac{1}{3}$

53. $f(x) = -2 - x^2$ for $x \geq 0$

For Problems 54 and 55, find the intervals on which the function is increasing and the intervals on which it is decreasing.

54. $f(x) = -2x^2 + 16x - 35$

55. $f(x) = 2\sqrt{x - 3}$

56. How long will it take $100 to double if it is invested at 8% interest compounded annually?

57. How long will it take $1000 to be worth $3500 if it is invested at 5.5% interest compounded quarterly?

58. What rate of interest (to the nearest tenth of a percent) compounded continuously is needed for an investment of $500 to grow to $1000 in 8 years?

59. Suppose that the present population of a city is 50,000. Use the equation $P(t) = P_0 e^{0.02t}$ (where P_0 represents an initial population) to estimate future populations, and estimate the population of that city in 10 years, 15 years, and 20 years.

60. The number of bacteria present in a certain culture after t hours is given by the equation $Q = Q_0 e^{0.29t}$, where Q_0 represents the initial number of bacteria. How long will it take 500 bacteria to increase to 2000 bacteria?

61. Suppose that a certain radioactive substance has a half-life of 40 days. If there are presently 750 grams of the substance, how much, to the nearest gram, will remain after 100 days?

62. An earthquake that occurred in Mexico City in 1985 had an intensity level about 125,000,000 times the reference intensity. Find the Richter number for that earthquake.

applies to all problems in this Chapter Test.

For Problems 1–4, evaluate each expression.

1. $\log_3 \sqrt{3}$

2. $\log_2(\log_2 4)$

3. $-2 + \ln e^3$

4. $\log_2(0.5)$

For Problems 5–10, solve each equation.

5. $4^x = \dfrac{1}{64}$

6. $9^x = \dfrac{1}{27}$

7. $2^{3x-1} = 128$

8. $\log_9 x = \dfrac{5}{2}$

9. $\log x + \log(x + 48) = 2$

10. $\ln x = \ln 2 + \ln(3x - 1)$

For Problems 11–13, given that $\log_3 4 = 1.2619$ and $\log_3 5 = 1.4650$, evaluate each of the following.

11. $\log_3 100$

12. $\log_3 \dfrac{5}{4}$

13. $\log_3 \sqrt{5}$

14. Find the inverse of the function $f(x) = -3x - 6$.

15. Solve $e^x = 176$ to the nearest hundredth.

16. Solve $2^{x-2} = 314$ to the nearest hundredth.

17. Determine $\log_5 632$ to four decimal places.

18. Find the inverse of the function $f(x) = \dfrac{2}{3}x - \dfrac{3}{5}$.

19. If $3500 was invested at 7.5% interest compounded quarterly, how much money has accumulated at the end of 8 years?

20. How long will it take $5000 to be worth $12,500 if it is invested at 4% compounded annually? Express your answer to the nearest tenth of a year.

21. The number of bacteria present in a certain culture after t hours is given by $Q(t) = Q_0 e^{0.23t}$, where Q_0 represents the initial number of bacteria. How long will it take 400 bacteria to increase to 2400 bacteria? Express your answer to the nearest tenth of an hour.

22. Suppose that a certain radioactive substance has a half-life of 50 years. If there are presently 7500 grams of the substance, how much will remain after 32 years? Express your answer to the nearest gram.

For Problems 23–25, graph each of the functions.

23. $f(x) = e^x - 2$

24. $f(x) = -3^{-x}$

25. $f(x) = \log_2(x - 2)$

CUMULATIVE REVIEW PROBLEM SET *Chapters 1-14*

For Problems 1–5, evaluate each algebraic expression for the given values of the variables.

1. $-5(x - 1) - 3(2x + 4) + 3(3x - 1)$ for $x = -2$

2. $\dfrac{14a^3b^2}{7a^2b}$ for $a = -1$ and $b = 4$

3. $\dfrac{2}{n} - \dfrac{3}{2n} + \dfrac{5}{3n}$ for $n = 4$

4. $4\sqrt{2x - y} + 5\sqrt{3x + y}$ for $x = 16$ and $y = 16$

5. $\dfrac{3}{x - 2} - \dfrac{5}{x + 3}$ for $x = 3$

For Problems 6–15, perform the indicated operations and express answers in simplified form.

6. $(-5\sqrt{6})(3\sqrt{12})$

7. $(2\sqrt{x} - 3)(\sqrt{x} + 4)$

8. $(3\sqrt{2} - \sqrt{6})(\sqrt{2} + 4\sqrt{6})$

9. $(2x - 1)(x^2 + 6x - 4)$

10. $\dfrac{x^2 - x}{x + 5} \cdot \dfrac{x^2 + 5x + 4}{x^4 - x^2}$

11. $\dfrac{16x^2y}{24xy^3} \div \dfrac{9xy}{8x^2y^2}$

12. $\dfrac{x + 3}{10} + \dfrac{2x + 1}{15} - \dfrac{x - 2}{18}$

13. $\dfrac{7}{12ab} - \dfrac{11}{15a^2}$

14. $\dfrac{8}{x^2 - 4x} + \dfrac{2}{x}$

15. $(8x^3 - 6x^2 - 15x + 4) \div (4x - 1)$

For Problems 16–19, simplify each of the complex fractions.

16. $\dfrac{\dfrac{5}{x^2} - \dfrac{3}{x}}{\dfrac{1}{y} + \dfrac{2}{y^2}}$

17. $\dfrac{\dfrac{2}{x} - 3}{\dfrac{3}{y} + 4}$

18. $\dfrac{2 - \dfrac{1}{n - 2}}{3 + \dfrac{4}{n + 3}}$

19. $\dfrac{3a}{2 - \dfrac{1}{a}} - 1$

For Problems 20–25, factor each of the algebraic expressions completely.

20. $20x^2 + 7x - 6$

21. $16x^3 + 54$

22. $4x^4 - 25x^2 + 36$

23. $12x^3 - 52x^2 - 40x$

24. $xy - 6x + 3y - 18$

25. $10 + 9x - 9x^2$

For Problems 26–35, evaluate each of the numerical expressions.

26. $\left(\dfrac{2}{3}\right)^{-4}$

27. $\dfrac{3}{\left(\dfrac{4}{3}\right)^{-1}}$

28. $\sqrt[3]{-\dfrac{27}{64}}$

29. $-\sqrt{0.09}$

30. $(27)^{-4/3}$

31. $4^0 + 4^{-1} + 4^{-2}$

32. $\left(\dfrac{3^{-1}}{2^{-3}}\right)^{-2}$

33. $(2^{-3} - 3^{-2})^{-1}$

34. $\log_2 64$

35. $\log_3\left(\dfrac{1}{9}\right)$

For Problems 36–38, find the indicated products and quotients; express final answers with positive integral exponents only.

36. $(-3x^{-1}y^2)(4x^{-2}y^{-3})$

37. $\dfrac{48x^{-4}y^2}{6xy}$

38. $\left(\dfrac{27a^{-4}b^{-3}}{-3a^{-1}b^{-4}}\right)^{-1}$

For Problems 39–46, express each radical expression in simplest radical form.

39. $\sqrt{80}$

40. $-2\sqrt{54}$

41. $\sqrt{\dfrac{75}{81}}$

42. $\dfrac{4\sqrt{6}}{3\sqrt{8}}$

43. $\sqrt[3]{56}$

44. $\dfrac{\sqrt[3]{3}}{\sqrt[3]{4}}$

45. $4\sqrt{52x^3y^2}$

46. $\sqrt{\dfrac{2x}{3y}}$

For Problems 47–49, use the distributive property to help simplify each of the following:

47. $-3\sqrt{24} + 6\sqrt{54} - \sqrt{6}$

48. $\dfrac{\sqrt{8}}{3} - \dfrac{3\sqrt{18}}{4} - \dfrac{5\sqrt{50}}{2}$

49. $8\sqrt[3]{3} - 6\sqrt[3]{24} - 4\sqrt[3]{81}$

For Problems 50 and 51, rationalize the denominator and simplify.

50. $\dfrac{\sqrt{3}}{\sqrt{6} - 2\sqrt{2}}$

51. $\dfrac{3\sqrt{5} - \sqrt{3}}{2\sqrt{3} + \sqrt{7}}$

For Problems 52–54, use scientific notation to help perform the indicated operations.

52. $\dfrac{(0.00016)(300)(0.028)}{0.064}$

53. $\dfrac{0.00072}{0.0000024}$

54. $\sqrt{0.00000009}$

For Problems 55–58, find each of the indicated products or quotients and express answers in standard form.

55. $(5 - 2i)(4 + 6i)$

56. $(-3 - i)(5 - 2i)$

57. $\dfrac{5}{4i}$

58. $\dfrac{-1 + 6i}{7 - 2i}$

59. Find the slope of the line determined by the points $(2, -3)$ and $(-1, 7)$.

60. Find the slope of the line determined by the equation $4x - 7y = 9$.

61. Find the length of the line segment whose endpoints are $(4, 5)$ and $(-2, 1)$.

62. Write the equation of the line that contains the points $(3, -1)$ and $(7, 4)$.

63. Write the equation of the line that is perpendicular to the line $3x - 4y = 6$ and contains the point $(-3, -2)$.

64. Find the center and the length of a radius of the circle $x^2 + 4x + y^2 - 12y + 31 = 0$.

65. Find the coordinates of the vertex of the parabola $y = x^2 + 10x + 21$.

66. Find the length of the major axis of the ellipse $x^2 + 4y^2 = 16$.

For Problems 67–76, graph each of the functions.

67. $f(x) = -2x - 4$

68. $f(x) = -2x^2 - 2$

69. $f(x) = x^2 - 2x - 2$

70. $f(x) = \sqrt{x + 1} + 2$

71. $f(x) = 2x^2 + 8x + 9$

72. $f(x) = -|x - 2| + 1$

73. $f(x) = 2^x + 2$

74. $f(x) = \log_2(x - 2)$

75. $f(x) = -x(x + 1)(x - 2)$

76. $f(x) = \dfrac{-x}{x + 2}$

77. If $f(x) = x - 3$ and $g(x) = 2x^2 - x - 1$, find $(g \circ f)(x)$ and $(f \circ g)(x)$.

78. Find the inverse (f^{-1}) of $f(x) = 3x - 7$.

79. Find the inverse of $f(x) = -\dfrac{1}{2}x + \dfrac{2}{3}$.

80. Find the constant of variation if y varies directly as x, and $y = 2$ when $x = -\dfrac{2}{3}$.

81. If y is inversely proportional to the square of x, and $y = 4$ when $x = 3$, find y when $x = 6$.

82. The volume of gas at a constant temperature varies inversely as the pressure. What is the volume of a gas under a pressure of 25 pounds if the gas occupies 15 cubic centimeters under a pressure of 20 pounds?

For Problems 83–110, solve each equation.

83. $3(2x - 1) - 2(5x + 1) = 4(3x + 4)$

84. $n + \dfrac{3n - 1}{9} - 4 = \dfrac{3n + 1}{3}$

85. $0.92 + 0.9(x - 0.3) = 2x - 5.95$

86. $|4x - 1| = 11$

87. $3x^2 = 7x$

88. $x^3 - 36x = 0$

89. $30x^2 + 13x - 10 = 0$

90. $8x^3 + 12x^2 - 36x = 0$

91. $x^4 + 8x^2 - 9 = 0$

92. $(n + 4)(n - 6) = 11$

93. $2 - \dfrac{3x}{x - 4} = \dfrac{14}{x + 7}$

94. $\dfrac{2n}{6n^2 + 7n - 3} - \dfrac{n - 3}{3n^2 + 11n - 4} = \dfrac{5}{2n^2 + 11n + 12}$

95. $\sqrt{3y} - y = -6$

96. $\sqrt{x + 19} - \sqrt{x + 28} = -1$

97. $(3x - 1)^2 = 45$

98. $(2x + 5)^2 = -32$

99. $2x^2 - 3x + 4 = 0$

100. $3n^2 - 6n + 2 = 0$

101. $\dfrac{5}{n - 3} - \dfrac{3}{n + 3} = 1$

102. $12x^4 - 19x^2 + 5 = 0$

103. $2x^2 + 5x + 5 = 0$

104. $(3x - 5)(4x + 1) = 0$

105. $|2x + 9| = -4$

106. $16^x = 64$

107. $\log_3 x = 4$

108. $\log_{10} x + \log_{10} 25 = 2$

109. $\ln(3x - 4) - \ln(x + 1) = \ln 2$

110. $27^{4x} = 9^{x+1}$

For Problems 111–120, solve each inequality and express solutions using interval notation.

111. $-5(y - 1) + 3 > 3y - 4 - 4y$

112. $0.06x + 0.08(250 - x) \geq 19$

113. $|5x - 2| > 13$ **114.** $|6x + 2| < 8$

115. $\dfrac{x - 2}{5} - \dfrac{3x - 1}{4} \leq \dfrac{3}{10}$

116. $(x - 2)(x + 4) \leq 0$

117. $(3x - 1)(x - 4) > 0$ **118.** $x(x + 5) < 24$

119. $\dfrac{x - 3}{x - 7} \geq 0$ **120.** $\dfrac{2x}{x + 3} > 4$

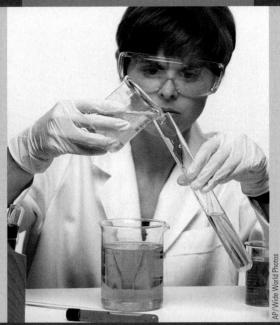

When mixing different solutions, a chemist could use a system of equations to determine how much of each solution is needed to produce a specific concentration.

AP/Wide World Photos

Systems of Equations: Matrices and Determinants

A 10% salt solution is to be mixed with a 20% salt solution to produce 20 gallons of a 17.5% salt solution. How many gallons of the 10% solution and how many gallons of the 20% solution should be mixed? The two equations $x + y = 20$ and $0.10x + 0.20y = 0.175(20)$ algebraically represent the conditions of the problem; x represents the number of gallons of the 10% solution, and y represents the number of gallons of the 20% solution. The two equations considered together form a system of linear equations, and the problem can be solved by solving the system of equations.

Throughout most of this chapter, we consider systems of linear equations and their applications. We will discuss various techniques for solving systems of linear equations. Then, in the last section, we consider systems that involve nonlinear equations.

InfoTrac Project

Do a keyword search on per capita debt and find an article on municipal debt and Arkansas cities. Write a brief summary of the article. The article mentions that one source of debt in cities is revenue bonds. Suppose the equation relating the net revenue (y) (in hundreds of thousands of dollars) received by the cities to the face value (x) of the bonds that are sold could be expressed by $y = x - 600{,}000$. Explain what you think the 600,000 represents in the equation. At some point in time, the bonds will mature and become an expense to the cities. Suppose the equation representing the relationship of the cost (y) (in hundreds of thousands of dollars) to the face value (x) of the bonds could be expressed by $y = 0.12x$. Explain what the 0.12 represents. Using a graphing utility, graph the two equations and find the coordinates of the point where net revenue and cost are the same. For what face value of the bonds does this occur? Now solve this system of equations using algebra. Were your results the same?

15.1   Systems of Two Linear Equations: A Brief Review

Objectives

■   Use substitution or elimination-by-addition methods to solve systems of two linear equations.

■   Solve word problems by using a system of equations.

In Chapter 5 we solved systems of two linear equations in two variables using three different methods: a graphing method, a substitution method, and an elimination-by-addition method. In subsequent chapters you probably have solved some word problems using a system of two linear equations. Even so, a brief review of that material will be helpful before we introduce some additional techniques for solving systems of equations. Let's use this first section to pull together the basic ideas of Chapter 5.

Remember that any equation of the form $Ax + By = C$, where A, B, and C are real numbers (A and B not both zero), is a **linear equation** in the two variables x and y, and its graph is a straight line. Two linear equations in two variables considered together form a **system of linear equations in two variables,** as illustrated by the following examples.

$$\begin{pmatrix} x + y = 6 \\ x - y = 2 \end{pmatrix} \qquad \begin{pmatrix} 3x + 2y = 1 \\ 5x - 2y = 23 \end{pmatrix} \qquad \begin{pmatrix} 4x - 5y = 21 \\ -3x + y = -7 \end{pmatrix}$$

To solve a system (such as any of these three examples) means to find all of the ordered pairs that simultaneously satisfy both equations in the system. For example, if we graph the two equations $x + y = 6$ and $x - y = 2$ on the same set of axes, as in Figure 15.1, then the ordered pair associated with the point of intersection of the two lines is the **solution of the system.** Thus we say that $\{(4, 2)\}$ is the solution set of the system

$$\begin{pmatrix} x + y = 6 \\ x - y = 2 \end{pmatrix}$$

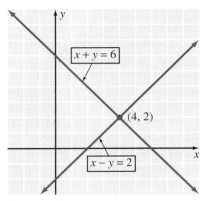

Figure 15.1

To check the solution, we substitute 4 for x and 2 for y in the two equations

$x + y = 6$ becomes $4 + 2 = 6$, a true statement
$x - y = 2$ becomes $4 - 2 = 2$, a true statement

Because the graph of a linear equation in two variables is a straight line, there are three possible situations that can occur when we are solving a system of two linear equations in two variables. These situations are shown in Figure 15.2.

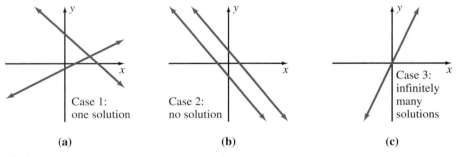

Figure 15.2

> **CASE 1** The graphs of the two equations are two lines intersecting in *one* point. There is exactly one solution, and the system is called a **consistent system.**
>
> **CASE 2** The graphs of the two equations are parallel lines. There is *no solution,* and the system is called an **inconsistent system.**
>
> **CASE 3** The graphs of the two equations are the same line, and there are *infinitely many solutions* of the system. Any pair of real numbers that satisfies one of the equations also satisfies the other equation, and we say that the equations are **dependent.**

Thus as we solve a system of two linear equations in two variables, we can expect one of three outcomes: The system will have *no* solutions, *one* ordered pair as a solution, or *infinitely many* ordered pairs as solutions.

The Substitution Method

Solving specific systems of equations by graphing requires accurate graphs. However, unless the solutions are integers, it is difficult to obtain exact solutions from a graph. Therefore, we will consider some other techniques for solving systems of equations.

The **substitution method,** which works especially well with systems of two equations in two unknowns, can be described as follows:

STEP 1   Solve one of the equations for one variable in terms of the other. (If possible, make a choice that will avoid fractions.)

STEP 2   Substitute the expression obtained in step 1 into the other equation, producing an equation in one variable.

STEP 3   Solve the equation obtained in step 2.

STEP 4   Use the solution obtained in step 3, along with the equation obtained in step 1, to determine the solution of the system.

E X A M P L E   1   Solve the system $\begin{pmatrix} x - 3y = -25 \\ 4x + 5y = 19 \end{pmatrix}$.

Solution

Solve the first equation for x in terms of y to produce

$$x = 3y - 25$$

Substitute $3y - 25$ for x in the second equation and solve for y.

$$4x + 5y = 19$$
$$4(3y - 25) + 5y = 19$$
$$12y - 100 + 5y = 19$$
$$17y = 119$$
$$y = 7$$

Next, substitute 7 for y in the equation $x = 3y - 25$ to obtain

$$x = 3(7) - 25 = -4$$

The solution set of the given system is $\{(-4, 7)\}$. (You should check this solution in both of the original equations.)   ◼

E X A M P L E   2   Solve the system $\begin{pmatrix} 5x + 9y = -2 \\ 2x + 4y = -1 \end{pmatrix}$.

Solution

A glance at the system should tell us that solving either equation for either variable will produce a fractional form, so let's just use the first equation and solve for x in terms of y.

$$5x + 9y = -2$$
$$5x = -9y - 2$$
$$x = \frac{-9y - 2}{5}$$

Now we can substitute this value for x into the second equation and solve for y.

$$2x + 4y = -1$$

$$2\left(\frac{-9y - 2}{5}\right) + 4y = -1$$

$$2(-9y - 2) + 20y = -5 \qquad \text{Multiplied both sides by 5.}$$

$$-18y - 4 + 20y = -5$$

$$2y - 4 = -5$$

$$2y = -1$$

$$y = -\frac{1}{2}$$

Now we can substitute $-\dfrac{1}{2}$ for y in $x = \dfrac{-9y - 2}{5}$.

$$x = \frac{-9\left(-\dfrac{1}{2}\right) - 2}{5} = \frac{\dfrac{9}{2} - 2}{5} = \frac{1}{2}$$

The solution set is $\left\{\left(\dfrac{1}{2}, -\dfrac{1}{2}\right)\right\}$.

E X A M P L E   3   Solve the system

$$\left(\begin{array}{l} 6x - 4y = 18 \\ y = \dfrac{3}{2}x - \dfrac{9}{2} \end{array}\right)$$

Solution

The second equation is given in appropriate form for us to begin the substitution process. Substitute $\dfrac{3}{2}x - \dfrac{9}{2}$ for y in the first equation to yield

$$6x - 4y = 18$$

$$6x - 4\left(\frac{3}{2}x - \frac{9}{2}\right) = 18$$

$$6x - 6x + 18 = 18$$

$$18 = 18$$

Our obtaining a true numerical statement ($18 = 18$) indicates that the system has infinitely many solutions. Any ordered pair that satisfies one of the equations will also satisfy the other equation. Thus in the second equation of the original system, if we let $x = k$, then $y = \dfrac{3}{2}k - \dfrac{9}{2}$. Therefore, the solution set can be expressed $\left\{\left(k, \dfrac{3}{2}k - \dfrac{9}{2}\right) \;\middle|\; k \text{ is a real number}\right\}$. If some specific solutions are needed, they can be generated by the ordered pair $\left(k, \frac{3}{2}k - \frac{9}{2}\right)$. For example, if we let $k = 1$, then we get $\frac{3}{2}(1) - \frac{9}{2} = -\frac{6}{2} = -3$. Thus the ordered pair $(1, -3)$ is a member of the solution set of the given system. ∎

The Elimination-by-Addition Method

Now let's consider the **elimination-by-addition method** for solving a system of equations. This is a very important method because it is the basis for developing other techniques for solving systems that contain many equations and variables. The method involves replacing systems of equations with *simpler equivalent systems* until we obtain a system where the solutions are obvious. **Equivalent systems of equations are systems that have exactly the same solution set.** The following operations or transformations can be applied to a system of equations to produce an equivalent system.

1. Any two equations of the system can be interchanged.
2. Both sides of any equation of the system can be multiplied by any nonzero real number.
3. Any equation of the system can be replaced by the sum of that equation and a nonzero multiple of another equation.

EXAMPLE 4

Solve the system $\begin{pmatrix} 3x + 5y = -9 \\ 2x - 3y = 13 \end{pmatrix}$.

$$(1)$$
$$(2)$$

Solution

We can replace the given system with an equivalent system by multiplying equation (2) by -3.

$$\begin{pmatrix} 3x + 5y = -9 \\ -6x + 9y = -39 \end{pmatrix}$$

$$(3)$$
$$(4)$$

Now let's replace equation (4) with an equation formed by multiplying equation (3) by 2 and adding this result to equation (4).

$$\begin{pmatrix} 3x + 5y = -9 \\ 19y = -57 \end{pmatrix}$$

$$(5)$$
$$(6)$$

From equation (6), we can easily determine that $y = -3$. Then, substituting -3 for y in equation (5) produces

$$3x + 5(-3) = -9$$
$$3x - 15 = -9$$
$$3x = 6$$
$$x = 2$$

The solution set for the given system is $\{(2, -3)\}$. ∎

REMARK: We are using a format for the elimination-by-addition method that highlights the use of equivalent systems. In Section 15.3, this format will lead naturally to an approach using matrices. Thus it is beneficial to stress the use of equivalent systems at this time.

EXAMPLE 5   Solve the system

$$\left(\begin{array}{l} \dfrac{1}{2}x + \dfrac{2}{3}y = -4 \\[2mm] \dfrac{1}{4}x - \dfrac{3}{2}y = 20 \end{array} \right)$$

(7)

(8)

Solution

The given system can be replaced with an equivalent system by multiplying equation (7) by 6 and equation (8) by 4.

$$\left(\begin{array}{l} 3x + 4y = -24 \\ x - 6y = 80 \end{array} \right)$$

(9)

(10)

Now let's exchange equations (9) and (10).

$$\left(\begin{array}{l} x - 6y = 80 \\ 3x + 4y = -24 \end{array} \right)$$

(11)

(12)

We can replace equation (12) with an equation formed by multiplying equation (11) by -3 and adding this result to equation (12).

$$\left(\begin{array}{l} x - 6y = 80 \\ 22y = -264 \end{array} \right)$$

(13)

(14)

From equation (14) we can determine that $y = -12$. Then, substituting -12 for y in equation (13) produces

$$x - 6(-12) = 80$$
$$x + 72 = 80$$
$$x = 8$$

The solution set of the given system is $\{(8, -12)\}$. (Check this!)   ■

EXAMPLE 6   Solve the system $\left(\begin{array}{l} x - 4y = 9 \\ x - 4y = 3 \end{array} \right)$.

(15)

(16)

Solution

We can replace equation (16) with an equation formed by multiplying equation (15) by -1 and adding this result to equation (16).

$$\left(\begin{array}{l} x - 4y = 9 \\ 0 = -6 \end{array} \right)$$

(17)

(18)

The statement $0 = -6$ is a contradiction, and therefore the original system is *inconsistent;* it has no solution. The solution set is $\varnothing$.   ■

Both the elimination-by-addition method and the substitution method can be used to obtain exact solutions for any system of two linear equations in two unknowns. Sometimes it is a matter of deciding which method to use on a particular system. Some systems lend themselves to one or the other of the methods by virtue of the original format of the equations. We will illustrate this idea in a moment when we solve some word problems.

Using Systems to Solve Problems

Many word problems that we solved earlier in this text with one variable and one equation can also be solved by using a system of two linear equations in two variables. In fact, in many of these problems, you may find it more natural to use two variables and two equations.

The two-variable expression $10t + u$ can be used to represent any two-digit whole number. The t represents the tens digit, and the u represents the units digit. For example, if $t = 4$ and $u = 8$, then $10t + u$ becomes $10(4) + 8 = 48$. Now let's use this general representation for a two-digit number to help solve a problem.

PROBLEM 1

The units digit of a two-digit number is 1 more than twice the tens digit. The number with the digits reversed is 45 larger than the original number. Find the original number.

Solution

Let u represent the units digit of the original number, and let t represent the tens digit. Then $10t + u$ represents the original number, and $10u + t$ represents the new number with the digits reversed. The problem translates into the following system.

$$\begin{pmatrix} u = 2t + 1 \\ 10u + t = 10t + u + 45 \end{pmatrix}$$

The units digit is 1 more than twice the tens digit.

The number with the digits reversed is 45 larger than the original number.

Simplify the second equation, and the system becomes

$$\begin{pmatrix} u = 2t + 1 \\ u - t = 5 \end{pmatrix}$$

Because of the form of the first equation, this system lends itself to solution by the substitution method. Substitute $2t + 1$ for u in the second equation to produce

$$(2t + 1) - t = 5$$
$$t + 1 = 5$$
$$t = 4$$

Now substitute 4 for t in the equation $u = 2t + 1$ to get

$$u = 2(4) + 1 = 9$$

The tens digit is 4 and the units digit is 9, so the number is 49.   ∎

PROBLEM 2

Lucinda invested $950, part of it at 11% interest and the remainder at 12%. Her total yearly income from the two investments was $111.50. How much did she invest at each rate?

Solution

Let x represent the amount invested at 11% and y the amount invested at 12%. The problem translates into the following system.

$$\begin{pmatrix} x + y = 950 \\ 0.11x + 0.12y = 111.50 \end{pmatrix}$$ ← The two investments total $950.
← The yearly interest from the two investments totals $111.50.

Multiply the second equation by 100 to produce an equivalent system.

$$\begin{pmatrix} x + y = 950 \\ 11x + 12y = 11150 \end{pmatrix}$$

Because neither equation is solved for one variable in terms of the other, let's use the elimination-by-addition method to solve the system. The second equation can be replaced by an equation formed by multiplying the first equation by -11 and adding this result to the second equation.

$$\begin{pmatrix} x + y = 950 \\ y = 700 \end{pmatrix}$$

Now we substitute 700 for y in the equation $x + y = 950$.

$$x + 700 = 950$$
$$x = 250$$

Therefore, Lucinda must have invested $250 at 11% and $700 at 12%. ∎

In our final example of this section, we will use a graphing utility to help solve a system of equations.

EXAMPLE 7

Solve the system.

$$\begin{pmatrix} 1.14x + 2.35y = -7.12 \\ 3.26x - 5.05y = 26.72 \end{pmatrix}.$$

Solution

First, we need to solve each equation for y in terms of x. Thus the system becomes

$$\begin{pmatrix} y = \dfrac{-7.12 - 1.14x}{2.35} \\ y = \dfrac{3.26x - 26.72}{5.05} \end{pmatrix}$$

Now we can enter both of these equations into a graphing utility and obtain Figure 15.3. From this figure, it appears that the point of intersection is at approximately $x = 2$ and $y = -4$. By direct substitution into the given equations, we can verify that the point of intersection is exactly $(2, -4)$.

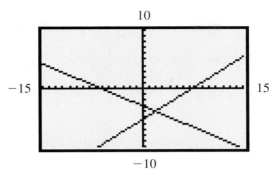

Figure 15.3

CONCEPT QUIZ

For Problems 1–5, answer true or false.

1. If we graph the equations in a system of equations, the ordered pair associated with the point of intersection of the graphs is a solution of the system.

2. When a system of two linear equations has exactly one solution, then the system is said to be consistent.

3. When the graphs for a system of two linear equations are parallel lines, then there are an infinite number of solutions.

4. Equivalent systems of equations are systems that have exactly the same solution set.

5. The substitution method for solving a system of two linear equations can only be used if one of the coefficients of the variables is 1.

PROBLEM SET 15.1

For Problems 1–18, solve each system by using the substitution method.

1. $\begin{pmatrix} x + y = 16 \\ y = x + 2 \end{pmatrix}$

2. $\begin{pmatrix} 2x + 3y = -5 \\ y = 2x + 9 \end{pmatrix}$

3. $\begin{pmatrix} x = 3y - 25 \\ 4x + 5y = 19 \end{pmatrix}$

4. $\begin{pmatrix} 3x - 5y = 25 \\ x = y + 7 \end{pmatrix}$

5. $\begin{pmatrix} y = \dfrac{2}{3}x - 1 \\ 5x - 7y = 9 \end{pmatrix}$

6. $\begin{pmatrix} y = \dfrac{3}{4}x + 5 \\ 4x - 3y = -1 \end{pmatrix}$

7. $\begin{pmatrix} a = 4b + 13 \\ 3a + 6b = -33 \end{pmatrix}$

8. $\begin{pmatrix} 9a - 2b = 28 \\ b = -3a + 1 \end{pmatrix}$

9. $\begin{pmatrix} 2x - 3y = 4 \\ y = \dfrac{2}{3}x - \dfrac{4}{3} \end{pmatrix}$

10. $\begin{pmatrix} t + u = 11 \\ t = u + 7 \end{pmatrix}$

11. $\begin{pmatrix} u = t - 2 \\ t + u = 12 \end{pmatrix}$

12. $\begin{pmatrix} y = 5x - 9 \\ 5x - y = 9 \end{pmatrix}$

13. $\begin{pmatrix} 4x + 3y = -7 \\ 3x - 2y = 16 \end{pmatrix}$

14. $\begin{pmatrix} 5x - 3y = -34 \\ 2x + 7y = -30 \end{pmatrix}$

15. $\begin{pmatrix} 5x - y = 4 \\ y = 5x + 9 \end{pmatrix}$ **16.** $\begin{pmatrix} 2x + 3y = 3 \\ 4x - 9y = -4 \end{pmatrix}$

17. $\begin{pmatrix} 4x - 5y = 3 \\ 8x + 15y = -24 \end{pmatrix}$ **18.** $\begin{pmatrix} 4x + y = 9 \\ y = 15 - 4x \end{pmatrix}$

For Problems 19–34, solve each system by using the elimination-by-addition method.

19. $\begin{pmatrix} 3x + 2y = 1 \\ 5x - 2y = 3 \end{pmatrix}$ **20.** $\begin{pmatrix} 4x + 3y = -22 \\ 4x - 5y = 26 \end{pmatrix}$

21. $\begin{pmatrix} x - 3y = -22 \\ 2x + 7y = 60 \end{pmatrix}$ **22.** $\begin{pmatrix} 6x - y = 3 \\ 5x + 3y = -9 \end{pmatrix}$

23. $\begin{pmatrix} 4x - 5y = 21 \\ 3x + 7y = -38 \end{pmatrix}$ **24.** $\begin{pmatrix} 5x - 3y = -34 \\ 2x + 7y = -30 \end{pmatrix}$

25. $\begin{pmatrix} 5x - 2y = 19 \\ 5x - 2y = 7 \end{pmatrix}$ **26.** $\begin{pmatrix} 4a + 2b = -4 \\ 6a - 5b = 18 \end{pmatrix}$

27. $\begin{pmatrix} 5a + 6b = 8 \\ 2a - 15b = 9 \end{pmatrix}$ **28.** $\begin{pmatrix} 7x + 2y = 11 \\ 7x + 2y = -4 \end{pmatrix}$

29. $\begin{pmatrix} \dfrac{2}{3}s + \dfrac{1}{4}t = -1 \\ \dfrac{1}{2}s - \dfrac{1}{3}t = -7 \end{pmatrix}$ **30.** $\begin{pmatrix} \dfrac{1}{4}s - \dfrac{2}{3}t = -3 \\ \dfrac{1}{3}s + \dfrac{1}{3}t = 7 \end{pmatrix}$

31. $\begin{pmatrix} \dfrac{x}{2} - \dfrac{2y}{5} = \dfrac{-23}{60} \\ \dfrac{2x}{3} + \dfrac{y}{4} = \dfrac{-1}{4} \end{pmatrix}$ **32.** $\begin{pmatrix} \dfrac{2x}{3} - \dfrac{y}{2} = \dfrac{3}{5} \\ \dfrac{x}{4} + \dfrac{y}{2} = \dfrac{7}{80} \end{pmatrix}$

33. $\begin{pmatrix} \dfrac{2x}{3} + \dfrac{y}{2} = \dfrac{1}{6} \\ 4x + 6y = -1 \end{pmatrix}$ **34.** $\begin{pmatrix} \dfrac{1}{2}x + \dfrac{2}{3}y = -\dfrac{3}{10} \\ 5x + 4y = -1 \end{pmatrix}$

For Problems 35–50, solve each system by using either the substitution method or the elimination-by-addition method, whichever seems more appropriate.

35. $\begin{pmatrix} 5x - y = -22 \\ 2x + 3y = -2 \end{pmatrix}$ **36.** $\begin{pmatrix} 4x + 5y = -41 \\ 3x - 2y = 21 \end{pmatrix}$

37. $\begin{pmatrix} x = 3y + 1 \\ x = -2y + 3 \end{pmatrix}$ **38.** $\begin{pmatrix} y = 4x - 24 \\ 7x + y = 42 \end{pmatrix}$

39. $\begin{pmatrix} 3x - 5y = 9 \\ 6x - 10y = -1 \end{pmatrix}$ **40.** $\begin{pmatrix} y = \dfrac{2}{5}x - 3 \\ 4x - 7y = 33 \end{pmatrix}$

41. $\begin{pmatrix} \dfrac{1}{2}x - \dfrac{2}{3}y = 22 \\ \dfrac{1}{2}x + \dfrac{1}{4}y = 0 \end{pmatrix}$ **42.** $\begin{pmatrix} \dfrac{2}{5}x - \dfrac{1}{3}y = -9 \\ \dfrac{3}{4}x + \dfrac{1}{3}y = -14 \end{pmatrix}$

43. $\begin{pmatrix} t = 2u + 2 \\ 9u - 9t = -45 \end{pmatrix}$ **44.** $\begin{pmatrix} 9u - 9t = 36 \\ u = 2t + 1 \end{pmatrix}$

45. $\begin{pmatrix} x + y = 1000 \\ 0.12x + 0.14y = 136 \end{pmatrix}$

46. $\begin{pmatrix} x + y = 10 \\ 0.3x + 0.7y = 4 \end{pmatrix}$

47. $\begin{pmatrix} y = 2x \\ 0.09x + 0.12y = 132 \end{pmatrix}$ **48.** $\begin{pmatrix} y = 3x \\ 0.1x + 0.11y = 64.5 \end{pmatrix}$

49. $\begin{pmatrix} x + y = 10.5 \\ 0.5x + 0.8y = 7.35 \end{pmatrix}$ **50.** $\begin{pmatrix} 2x + y = 7.75 \\ 3x + 2y = 12.5 \end{pmatrix}$

For Problems 51–70, solve each problem by using a system of equations.

51. The sum of two numbers is 53 and their difference is 19. Find the numbers.

52. The sum of two numbers is −3 and their difference is 25. Find the numbers.

53. The measure of the larger of two complementary angles is 15° more than four times the measure of the smaller angle. Find the measures of both angles.

54. Assume that a plane is flying at a constant speed under unvarying wind conditions. Traveling against a head wind, it takes the plane 4 hours to travel 1540 miles. Traveling with a tail wind, the plane flies 1365 miles in 3 hours. Find the speed of the plane and the speed of the wind.

55. The tens digit of a two-digit number is 1 more than three times the units digit. If the sum of the digits is 9, find the number.

56. The units digit of a two-digit number is 1 less than twice the tens digit. The sum of the digits is 8. Find the number.

57. The sum of the digits of a two-digit number is 7. If the digits are reversed, the newly formed number is 9 larger than the original number. Find the original number.

58. The units digit of a two-digit number is 1 less than twice the tens digit. If the digits are reversed, the newly formed number is 27 larger than the original number. Find the original number.

59. A car rental agency rents sedans at $35 a day and convertibles at $48 a day. If 32 cars were rented one day for a total of $1276, how many convertibles were rented?

60. A video store rents new release movies for $5 and favorites for $2.75. One day the number of new release movies rented was twice the number of favorites. If the total income for that day was $956.25, how many movies of each kind were rented?

61. The income from a high school band fundraiser concert was $3360. The price of a student ticket was $6 and parent tickets were sold at $10 each. Four hundred twenty tickets were sold. How many tickets of each kind were sold?

62. Michelle can enter a small business as a full partner and receive a salary of $40,000 a year and 15% of the year's profit, or she can be sales manager for a salary of $65,000 plus 5% of the year's profit. What must the year's profit be for her total earnings to be the same whether she is a full partner or a sales manager?

63. Melinda invested three times as much money at 8% yearly interest as she did at 6%. Her total yearly interest from the two investments was $1500. How much did she invest at each rate?

64. Simon invested $13,500, part of it at 4% and the rest at 7% yearly interest. His yearly income from the 7% investment was $30 less than twice the income from the 4% investment. How much did he invest at each rate?

65. One day last summer, Jim went kayaking on the Little Susitna River in Alaska. Paddling upstream against the current, he traveled 20 miles in 4 hours. Then he turned around and paddled twice as fast downstream, and with the help of the current, traveled 19 miles in 1 hour. Find the rate of the current.

66. One solution contains 30% alcohol and a second solution contains 70% alcohol. How many liters of each solution should be mixed to make 10 liters containing 40% alcohol?

67. Santo bought 4 gallons of green latex paint and 2 gallons of primer for a total of $116. Not having enough paint to finish the project, Santo returned to the same store and bought 3 gallons of green latex paint and 1 gallon of primer for a total of $80. What is the price of a gallon of green latex paint?

68. Six cans of soda and 2 bags of potato chips cost $7.08. At the same prices, 8 cans of soda and 5 bags of potato chips cost $12.45. Find the price per can of soda and the price per bag of potato chips.

69. A cash drawer contains only five- and ten-dollar bills. There are 12 more five-dollar bills than ten-dollar bills. If the drawer contains $330, find the number of each kind of bill.

70. Brad has a collection of dimes and quarters totaling $47.50. The number of quarters is ten more than twice the number of dimes. How many coins of each kind does he have?

■ ■ ■ **Thoughts into words**

71. Give a general description of how to use the substitution method to solve a system of two linear equations in two variables.

72. Give a general description of how to use the elimination-by-addition method to solve a system of two linear equations in two variables.

73. Which method would you use to solve the system $\begin{pmatrix} 9x + 4y = 7 \\ 3x + 2y = 6 \end{pmatrix}$? Why?

74. Which method would you use to solve the system $\begin{pmatrix} 5x + 3y = 12 \\ 3x - y = 10 \end{pmatrix}$? Why?

■ ■ ■ **Further investigations**

A system such as

$$\left(\begin{array}{l} \dfrac{2}{x} + \dfrac{3}{y} = \dfrac{19}{15} \\ -\dfrac{2}{x} + \dfrac{1}{y} = -\dfrac{7}{15} \end{array} \right)$$

is not a linear system, but it can be solved using the elimination-by-addition method as follows. Add the first equation to the second to produce the equivalent system

$$\left(\begin{array}{l} \dfrac{2}{x} + \dfrac{3}{y} = \dfrac{19}{15} \\ \dfrac{4}{y} = \dfrac{12}{15} \end{array} \right)$$

Now solve $\dfrac{4}{y} = \dfrac{12}{15}$ to produce $y = 5$. Substitute 5 for y in the first equation, and solve for x to produce

$$\dfrac{2}{x} + \dfrac{3}{5} = \dfrac{19}{15}$$

$$\dfrac{2}{x} = \dfrac{10}{15}$$

$$10x = 30$$

$$x = 3$$

The solution set of the original system is $\{(3, 5)\}$.

For Problems 75–80, solve each system.

75. $\left(\begin{array}{l} \dfrac{1}{x} + \dfrac{2}{y} = \dfrac{7}{12} \\ \dfrac{3}{x} - \dfrac{2}{y} = \dfrac{5}{12} \end{array} \right)$
76. $\left(\begin{array}{l} \dfrac{3}{x} + \dfrac{2}{y} = 2 \\ \dfrac{2}{x} - \dfrac{3}{y} = \dfrac{1}{4} \end{array} \right)$

77. $\left(\begin{array}{l} \dfrac{3}{x} - \dfrac{2}{y} = \dfrac{13}{6} \\ \dfrac{2}{x} + \dfrac{3}{y} = 0 \end{array} \right)$
78. $\left(\begin{array}{l} \dfrac{4}{x} + \dfrac{1}{y} = 11 \\ \dfrac{3}{x} - \dfrac{5}{y} = -9 \end{array} \right)$

79. $\left(\begin{array}{l} \dfrac{5}{x} - \dfrac{2}{y} = 23 \\ \dfrac{4}{x} + \dfrac{3}{y} = \dfrac{23}{2} \end{array} \right)$
80. $\left(\begin{array}{l} \dfrac{2}{x} - \dfrac{7}{y} = \dfrac{9}{10} \\ \dfrac{5}{x} + \dfrac{4}{y} = -\dfrac{41}{20} \end{array} \right)$

81. Consider the linear system $\left(\begin{array}{l} a_1 x + b_1 y = c_1 \\ a_2 x + b_2 y = c_2 \end{array} \right)$.

   a. Prove that this system has exactly one solution if and only if $\dfrac{a_1}{a_2} \neq \dfrac{b_1}{b_2}$.

   b. Prove that this system has no solution if and only if $\dfrac{a_1}{a_2} = \dfrac{b_1}{b_2} \neq \dfrac{c_1}{c_2}$.

   c. Prove that this system has infinitely many solutions if and only if $\dfrac{a_1}{a_2} = \dfrac{b_1}{b_2} = \dfrac{c_1}{c_2}$.

82. For each of the following systems, use the results from Problem 81 to determine whether the system is consistent or inconsistent or the equations are dependent.

   a. $\left(\begin{array}{l} 5x + y = 9 \\ x - 5y = 4 \end{array} \right)$
   b. $\left(\begin{array}{l} 3x - 2y = 14 \\ 2x + 3y = 9 \end{array} \right)$

   c. $\left(\begin{array}{l} x - 7y = 4 \\ x - 7y = 9 \end{array} \right)$
   d. $\left(\begin{array}{l} 3x - 5y = 10 \\ 6x - 10y = 1 \end{array} \right)$

   e. $\left(\begin{array}{l} 3x + 6y = 2 \\ \dfrac{3}{5}x + \dfrac{6}{5}y = \dfrac{2}{5} \end{array} \right)$
   f. $\left(\begin{array}{l} \dfrac{2}{3}x - \dfrac{3}{4}y = 2 \\ \dfrac{1}{2}x + \dfrac{2}{5}y = 9 \end{array} \right)$

   g. $\left(\begin{array}{l} 7x + 9y = 14 \\ 8x - 3y = 12 \end{array} \right)$
   h. $\left(\begin{array}{l} 4x - 5y = 3 \\ 12x - 15y = 9 \end{array} \right)$

▦ **Graphing calculator activities**

83. For each of the systems of equations in Problem 82, use your graphing calculator to help determine whether the system is consistent or inconsistent or the equations are dependent.

84. Use your graphing calculator to help determine the solution set for each of the following systems. Be sure to check your answers.

a. $\begin{pmatrix} y = 3x - 1 \\ y = 9 - 2x \end{pmatrix}$ **b.** $\begin{pmatrix} 5x + y = -9 \\ 3x - 2y = 5 \end{pmatrix}$ **e.** $\begin{pmatrix} 13x - 12y = 37 \\ 15x + 13y = -11 \end{pmatrix}$

c. $\begin{pmatrix} 4x - 3y = 18 \\ 5x + 6y = 3 \end{pmatrix}$ **d.** $\begin{pmatrix} 2x - y = 20 \\ 7x + y = 79 \end{pmatrix}$ **f.** $\begin{pmatrix} 1.98x + 2.49y = 13.92 \\ 1.19x + 3.45y = 16.18 \end{pmatrix}$

Answers to Concept Quiz

1. True   **2.** True   **3.** False   **4.** True   **5.** False

---

15.2   Systems of Three Linear Equations in Three Variables

Objectives

■   Solve systems of three linear equations in three variables.

■   Use a system of three linear equations to solve word problems.

Consider a linear equation in three variables x, y, and z, such as $3x - 2y + z = 7$. Any **ordered triple** (x, y, z) that makes the equation a true numerical statement is said to be a **solution** of the equation. For example, the ordered triple $(2, 1, 3)$ is a solution because $3(2) - 2(1) + 3 = 7$. However, the ordered triple $(5, 2, 4)$ is not a solution because $3(5) - 2(2) + 4 \neq 7$. There are infinitely many solutions in the solution set.

REMARK: The idea of a linear equation is generalized to include equations of more than two variables. Thus an equation such as $5x - 2y + 9z = 8$ is called a *linear equation in three variables,* the equation $5x - 7y + 2z - 11w = 1$ is called a *linear equation in four variables,* and so on.

To *solve* a system of three linear equations in three variables, such as

$$\begin{pmatrix} 3x - y + 2z = 13 \\ 4x + 2y + 5z = 30 \\ 5x - 3y - z = 3 \end{pmatrix}$$

means to find all of the ordered triples that satisfy all three equations. In other words, the solution set of the system is the intersection of the solution sets of the three equations in the system.

The graph of a linear equation in three variables is a *plane,* not a line. In fact, graphing equations in three variables requires the use of a three-dimensional coordinate system. Thus using a graphing approach to solve systems of three linear equations in three variables is not at all practical. However, a simple graphical analysis does provide us with some indication of what we can expect as we begin solving such systems.

In general, because each linear equation in three variables produces a plane, a system of three such equations produces three planes. There are various ways in which three planes can be related. For example, they may be mutually parallel; or two of the planes may be parallel, with the third intersecting the other two. (You may want to analyze all of the other possibilities for the three planes!) However, for our purposes at this time, we need to realize that from a solution set viewpoint, a system of three linear equations in three variables produces one of the following possibilities.

1. There is *one ordered triple* that satisfies all three equations. The three planes have a common point of intersection, as indicated in Figure 15.4.

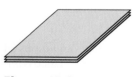

Figure 15.4

2. There are *infinitely many ordered triples* in the solution set, all of which are coordinates of *points on a line* common to the three planes. This happens if the three planes have a common line of intersection (Figure 15.5a) or if two of the planes coincide and the third plane intersects them (Figure 15.5b).

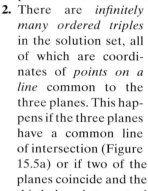
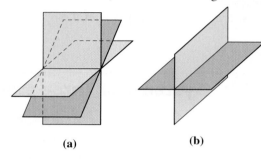

(a)   **(b)**

Figure 15.5

3. There are *infinitely many ordered triples* in the solution set, all of which are coordinates of *points on a plane*. This happens if the three planes coincide, as illustrated in Figure 15.6.

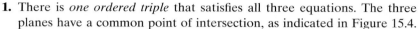

Figure 15.6

4. The solution set is *empty;* thus, we write ∅. This can happen in various ways, as illustrated in Figure 15.7. Note that in each situation there are no points common to all three planes.

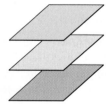

(a) Three parallel planes

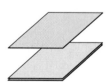

(b) Two planes coincide and the third one is parallel to the coinciding planes.

(c) Two planes are parallel and the third intersects them in parallel lines.

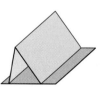

(d) No two planes are parallel, but two of them intersect in a line that is parallel to the third plane.

Figure 15.7

Now that we know what possibilities exist, let's consider finding the solution sets for some systems. Our approach will be the elimination-by-addition method, whereby systems are replaced with equivalent systems until a system is obtained that allows us to easily determine the solution set. The details of this approach will become apparent as we work a few examples.

EXAMPLE 1

Solve the system

$$\begin{pmatrix} 4x - 3y - 2z = 5 \\ 5y + z = -11 \\ 3z = 12 \end{pmatrix}$$

(1)
(2)
(3)

Solution

The form of this system makes it easy to solve. From equation (3), we obtain $z = 4$. Then, substituting 4 for z in equation (2), we get

$$5y + 4 = -11$$
$$5y = -15$$
$$y = -3$$

Finally, substituting 4 for z and -3 for y in equation (1) yields

$$4x - 3(-3) - 2(4) = 5$$
$$4x + 1 = 5$$
$$4x = 4$$
$$x = 1$$

Thus the solution set of the given system is $\{(1, -3, 4)\}$.   ■

EXAMPLE 2

Solve the system

$$\begin{pmatrix} x - 2y + 3z = 22 \\ 2x - 3y - z = 5 \\ 3x + y - 5z = -32 \end{pmatrix}$$

(4)
(5)
(6)

Solution

Equation (5) can be replaced with the equation formed by multiplying equation (4) by -2 and adding this result to equation (5). Equation (6) can be replaced with the equation formed by multiplying equation (4) by -3 and adding this result to equation (6). The following equivalent system is produced, in which equations (8) and (9) contain only the two variables y and z.

$$\begin{pmatrix} x - 2y + 3z = 22 \\ y - 7z = -39 \\ 7y - 14z = -98 \end{pmatrix}$$

(7)
(8)
(9)

Equation (9) can be replaced with the equation formed by multiplying equation (8) by -7 and adding this result to equation (9). This produces the following equivalent system.

$$\left(\begin{array}{l} x - 2y + 3z = 22 \\ y - 7z = -39 \\ 35z = 175 \end{array}\right) \qquad \begin{array}{l}\textbf{(10)}\\\textbf{(11)}\\\textbf{(12)}\end{array}$$

From equation (12) we obtain $z = 5$. Then, substituting 5 for z in equation (11), we obtain

$$y - 7(5) = -39$$
$$y - 35 = -39$$
$$y = -4$$

Finally, substituting -4 for y and 5 for z in equation (10) produces

$$x - 2(-4) + 3(5) = 22$$
$$x + 8 + 15 = 22$$
$$x + 23 = 22$$
$$x = -1$$

The solution set of the original system is $\{(-1, -4, 5)\}$. (Perhaps you should check this ordered triple in all three of the original equations.) ■

EXAMPLE 3

Solve the system

$$\left(\begin{array}{l} 3x - y + 2z = 13 \\ 5x - 3y - z = 3 \\ 4x + 2y + 5z = 30 \end{array}\right) \qquad \begin{array}{l}\textbf{(13)}\\\textbf{(14)}\\\textbf{(15)}\end{array}$$

Solution

Equation (14) can be replaced with the equation formed by multiplying equation (13) by -3 and adding this result to equation (14). Equation (15) can be replaced with the equation formed by multiplying equation (13) by 2 and adding this result to equation (15). Thus we produce the following equivalent system, in which equations (17) and (18) contain only the two variables x and z.

$$\left(\begin{array}{l} 3x - y + 2z = 13 \\ -4x - 7z = -36 \\ 10x + 9z = 56 \end{array}\right) \qquad \begin{array}{l}\textbf{(16)}\\\textbf{(17)}\\\textbf{(18)}\end{array}$$

Now, if we multiply equation (17) by 5 and equation (18) by 2, we get the following equivalent system.

$$\left(\begin{array}{l} 3x - y + 2z = 13 \\ -20x - 35z = -180 \\ 20x + 18z = 112 \end{array}\right) \qquad \begin{array}{l}\textbf{(19)}\\\textbf{(20)}\\\textbf{(21)}\end{array}$$

Equation (21) can be replaced with the equation formed by adding equation (20) to equation (21).

$$\begin{pmatrix} 3x - y + 2z = 13 \\ -20x \qquad - 35z = -180 \\ \qquad - 17z = -68 \end{pmatrix}$$

$$\begin{aligned} &\textbf{(22)} \\ &\textbf{(23)} \\ &\textbf{(24)} \end{aligned}$$

From equation (24), we obtain $z = 4$. Then we can substitute 4 for z in equation (23).

$$-20x - 35(4) = -180$$
$$-20x - 140 = -180$$
$$-20x = -40$$
$$x = 2$$

Now we can substitute 2 for x and 4 for z in equation (22).

$$3(2) - y + 2(4) = 13$$
$$6 - y + 8 = 13$$
$$-y + 14 = 13$$
$$-y = -1$$
$$y = 1$$

The solution set of the original system is $\{(2, 1, 4)\}$.   ■

E X A M P L E   4   Solve the system

$$\begin{pmatrix} x + 2y - z = 4 \\ 2x + 4y - 2z = 7 \\ 3x - y + z = -1 \end{pmatrix}$$

$$\begin{aligned} &\textbf{(25)} \\ &\textbf{(26)} \\ &\textbf{(27)} \end{aligned}$$

Solution
Equation (26) can be replaced with the equation formed by multiplying equation (25) by -2 and adding the result to equation (26).

$$\begin{pmatrix} x + 2y - z = 4 \\ 0 = -1 \\ 3x - y + z = -1 \end{pmatrix}$$

$$\begin{aligned} &\textbf{(28)} \\ &\textbf{(29)} \\ &\textbf{(30)} \end{aligned}$$

Equation (29) is a contradiction; therefore there is no solution for the system of equations. The solution set of the original system is $\varnothing$.   ■

E X A M P L E   5   Solve the system

$$\begin{pmatrix} x - y + 2z = 3 \\ -2x + 3y - z = 1 \\ -x + 2y + z = 4 \end{pmatrix}$$

$$\begin{aligned} &\textbf{(31)} \\ &\textbf{(32)} \\ &\textbf{(33)} \end{aligned}$$

Solution

Equation (32) can be replaced with the equation formed by multiplying equation (31) by 2 and adding the result to equation (32). Equation (33) can be replaced with the equation formed by adding equation (31) to equation (33). The following equivalent system is produced.

$$\begin{pmatrix} x - y + 2z = 3 \\ \quad\quad y + 3z = 7 \\ \quad\quad y + 3z = 7 \end{pmatrix} \qquad\qquad \begin{matrix}(34)\\(35)\\(36)\end{matrix}$$

Equations (35) and (36) are identical. If we continue, equation (36) can be replaced with the equation formed by multiplying equation (35) by -1 and adding the result to equation (36). Then the equivalent system is produced.

$$\begin{pmatrix} x - y + 2z = 3 \\ \quad\quad y + 3z = 7 \\ \quad\quad\quad 0 = 0 \end{pmatrix} \qquad\qquad \begin{matrix}(37)\\(38)\\(39)\end{matrix}$$

Equation (39) is an identity; therefore the original system has an infinite number of solutions. To represent the ordered triples we can solve equation (38) for y obtaining $y = -3z + 7$. We can substitute for y in equation (37) and solve that equation for x obtaining $x = -5z + 10$. Therefore, if we let $z = k$, where k is any real number, the solution set of infinitely many ordered triples can be represented by $\{(-5k + 10, -3k + 7, k)|k$ is a real number$\}$. We can generate a specific solution by replacing k with a number. For example, if $k = 4$, then $-5k + 10$ becomes $-5(4) + 10 = -10$, and $-3k + 7$ becomes $-3(4) + 7 = -5$. Thus the ordered triple $(-10, -5, 4)$ is a solution of the original system. ■

The ability to solve systems of three linear equations in three unknowns enhances our problem-solving capabilities. Let's conclude this section with a problem that we can solve using such a system.

PROBLEM 1

A small company that manufactures sporting equipment produces three different styles of golf shirts. Each style of shirt requires the services of three departments, as indicated by the following table.

	Style A	Style B	Style C
Cutting department	0.1 hour	0.1 hour	0.3 hour
Sewing department	0.3 hour	0.2 hour	0.4 hour
Packaging department	0.1 hour	0.2 hour	0.1 hour

The cutting, sewing, and packaging departments have available a maximum of 340, 580, and 255 work-hours per week, respectively. How many of each style of golf shirt should be produced each week so that the company is operating at full capacity?

Solution

Let a represent the number of shirts of style A produced per week, b the number of style B per week, and c the number of style C per week. Then the problem translates into the following system of equations.

$$\begin{pmatrix} 0.1a + 0.1b + 0.3c = 340 \\ 0.3a + 0.2b + 0.4c = 580 \\ 0.1a + 0.2b + 0.1c = 255 \end{pmatrix} \begin{array}{l} \longleftarrow \text{Cutting department} \\ \longleftarrow \text{Sewing department} \\ \longleftarrow \text{Packaging department} \end{array}$$

Solving this system (we will leave the details for you to carry out) produces $a = 500$, $b = 650$, and $c = 750$. Thus the company should produce 500 golf shirts of style A, 650 of style B, and 750 of style C per week. ■

CONCEPT QUIZ

For Problems 1–6, answer true or false.

1. The graph of a linear equation in three variables is a line.

2. A system of three linear equations in three variables produces three planes when graphed.

3. Three planes can be related by intersecting in exactly two points.

4. One way three planes can be related is if two of the planes are parallel, and the third plane intersects them in parallel lines.

5. A system of three linear equations in three variables always has an infinite number of solutions.

6. A system of three linear equations in three variables can have one ordered triple as a solution.

PROBLEM SET 15.2

For Problems 1–24, solve each system.

1. $\begin{pmatrix} 2x - 3y + 4z = 10 \\ 5y - 2z = -16 \\ 3z = 9 \end{pmatrix}$

2. $\begin{pmatrix} -3x + 2y + z = -9 \\ 4x \quad\quad -3z = 18 \\ 4z = -8 \end{pmatrix}$

3. $\begin{pmatrix} x + 2y - 3z = 2 \\ 3y - z = 13 \\ 3y + 5z = 25 \end{pmatrix}$

4. $\begin{pmatrix} 2x + 3y - 4z = -10 \\ 2y + 3z = 16 \\ 2y - 5z = -16 \end{pmatrix}$

5. $\begin{pmatrix} 3x + 2y - 2z = 14 \\ x \quad\quad - 6z = 16 \\ 2x \quad\quad + 5z = -2 \end{pmatrix}$

6. $\begin{pmatrix} 3x + 2y - z = -11 \\ 2x - 3y \quad = -1 \\ 4x + 5y \quad = -13 \end{pmatrix}$

7. $\begin{pmatrix} x - 2y + 3z = 7 \\ 2x + y + 5z = 17 \\ 3x - 4y - 2z = 1 \end{pmatrix}$

8. $\begin{pmatrix} x + 2y - z = 8 \\ x + 3y + 3z = 10 \\ -x - 3y - 3z = -10 \end{pmatrix}$

9. $\begin{pmatrix} 2x + y + 2z = 12 \\ 4x + y - z = -6 \\ -y - 5z = 10 \end{pmatrix}$

10. $\begin{pmatrix} x - 2y + z = -4 \\ 2x + 4y - 3z = -1 \\ -3x - 6y + 7z = 4 \end{pmatrix}$

11. $\begin{pmatrix} 2x - y + z = 0 \\ 3x - 2y + 4z = 11 \\ 5x + y - 6z = -32 \end{pmatrix}$

12. $\begin{pmatrix} 2x - y + 3z = -14 \\ 4x + 2y - z = 12 \\ 6x - 3y + 4z = -22 \end{pmatrix}$

13. $\begin{pmatrix} 3x + 2y - z = -11 \\ 2x - 3y + 4z = 11 \\ 5x + y - 2z = -17 \end{pmatrix}$

14. $\begin{pmatrix} 9x + 4y - z = 0 \\ 3x - 2y + 4z = 6 \\ 6x - 8y - 3z = 3 \end{pmatrix}$

15. $\begin{pmatrix} 2x + 3y - 4z = -10 \\ 4x - 5y + 3z = 2 \\ 2y + z = 8 \end{pmatrix}$

16. $\begin{pmatrix} x + 2y - 3z = 2 \\ 3x \quad\;\; - z = -8 \\ 2x - 3y + 5z = -9 \end{pmatrix}$

17. $\begin{pmatrix} x + 3y - 2z = -4 \\ 2x + 7y + z = 5 \\ y + 5z = 13 \end{pmatrix}$

18. $\begin{pmatrix} 2x - 3y + z = -1 \\ 4x - 6y + 2z = 5 \\ x + y + 2z = 4 \end{pmatrix}$

19. $\begin{pmatrix} 3x + 2y - 2z = 14 \\ 2x - 5y + 3z = 7 \\ 4x - 3y + 7z = 5 \end{pmatrix}$

20. $\begin{pmatrix} 4x + 3y - 2z = -11 \\ 3x - 7y + 3z = 10 \\ 9x - 8y + 5z = 9 \end{pmatrix}$

21. $\begin{pmatrix} 2x - 3y + 4z = -12 \\ 4x + 2y - 3z = -13 \\ 6x - 5y + 7z = -31 \end{pmatrix}$

22. $\begin{pmatrix} 2x + 5y - 2z = -26 \\ 5x - 2y + 4z = 27 \\ 7x + 3y - 6z = -55 \end{pmatrix}$

23. $\begin{pmatrix} 5x - 3y - 6z = 22 \\ x - y + z = -3 \\ -3x + 7y - 5z = 23 \end{pmatrix}$

24. $\begin{pmatrix} 4x + 3y - 5z = -29 \\ 3x - 7y - z = -19 \\ 2x + 5y + 2z = -10 \end{pmatrix}$

For Problems 25–42, solve each problem by setting up and solving a system of three linear equations in three variables.

25. The sum of three numbers is 20. The sum of the first and third numbers is 2 more than twice the second number. The third number minus the first yields three times the second number. Find the numbers.

26. The sum of three numbers is 40. The third number is 10 less than the sum of the first two numbers. The second number is 1 larger than the first. Find the numbers.

27. Mike bought a motorcycle helmet, jacket, and gloves for $650. The jacket costs $100 more than the helmet. The cost of the helmet and gloves together was $50 less than the cost of the jacket. How much did each item cost?

28. One binder, 2 reams of paper, and 5 spiral notebooks cost $14.82. Three binders, 1 ream of paper, and 4 spiral notebooks cost $14.32. Two binders, 3 reams of paper, and 3 spiral notebooks cost $19.82. Find the cost for each item.

29. In a certain triangle, the measure of $\angle A$ is five times the measure of $\angle B$. The sum of the measures of $\angle B$ and $\angle C$ is 60° less than the measure of $\angle A$. Find the measure of each angle.

30. Shannon purchased a skirt, blouse, and sweater for $72. The cost of the skirt and sweater was $2 more than six times the cost of the blouse. The skirt cost twice the sum of the costs of the blouse and sweater. Find the cost of each item.

31. The wages for a crew consisting of a plumber, an apprentice, and a laborer are $80 an hour. The plumber earns $20 an hour more than the sum of the wages of the apprentice and the laborer. The plumber earns five times as much as the laborer. Find the hourly wage of each.

32. A catering group that has a chef, a salad maker, and a server costs the customer $70 per hour. The salad maker costs $5 per hour more than the server. The chef costs the same as the salad maker and the server cost together. Find the cost per hour of each.

33. A gift store is making a mixture of almonds, pecans, and peanuts, which sell for $3.50 per pound, $4 per pound, and $2 per pound, respectively. The storekeeper wants to make 20 pounds of the mix to sell at $2.70 per pound. The number of pounds of peanuts is to be three times the number of pounds of pecans. Find the number of pounds of each to be used in the mixture.

34. The organizer for a church picnic ordered coleslaw, potato salad, and beans amounting to 50 pounds. There was to be three times as much potato salad as coleslaw. The number of pounds of beans was to be six less than the number of pounds of potato salad. Find the number of pounds of each.

35. A box contains $7.15 in nickels, dimes, and quarters. There are 42 coins in all, and the sum of the numbers of nickels and dimes is two less than the number of quarters. How many coins of each kind are there?

36. A handful of 65 coins consists of pennies, nickels, and dimes. The number of nickels is four less than twice the number of pennies, and there are 13 more dimes than nickels. How many coins of each kind are there?

37. The measure of the largest angle of a triangle is twice the measure of the smallest angle. The sum of the smallest angle and the largest angle is twice the other angle. Find the measure of each angle.

38. The perimeter of a triangle is 45 centimeters. The longest side is 4 centimeters less than twice the shortest side. The sum of the lengths of the shortest and longest sides is 7 centimeters less than three times the length of the remaining side. Find the lengths of all three sides of the triangle.

39. Part of $30,000 is invested at 2%, another part at 3%, and the remainder at 4% yearly interest. The total yearly income from the three investments is $1000. The sum of the amounts invested at 2% and 3% equals the amount invested at 4%. How much is invested at each rate?

40. Different amounts are invested at 4%, 5%, and 6% yearly interest. The amount invested at 5% is $3000 more than what is invested at 4%, and the total yearly income from all three investments is $1500. A total of $29,000 is invested. Find the amount invested at each rate.

41. A small company makes three different types of bird houses. Each type requires the services of three different departments, as indicated by the following table.

	Type A	Type B	Type C
Cutting department	0.1 hour	0.2 hour	0.1 hour
Finishing department	0.4 hour	0.4 hour	0.3 hour
Assembly department	0.2 hour	0.1 hour	0.3 hour

The cutting, finishing, and assembly departments have available a maximum of 35, 95, and 62.5 work-hours per week, respectively. How many bird houses of each type should be made per week so that the company is operating at full capacity?

42. A certain diet consists of dishes A, B, and C. Each serving of A has 1 gram of fat, 2 grams of carbohydrate, and 4 grams of protein. Each serving of B has 2 grams of fat, 1 gram of carbohydrate, and 3 grams of protein. Each serving of C has 2 grams of fat, 4 grams of carbohydrate, and 3 grams of protein. The diet allows 15 grams of fat, 24 grams of carbohydrate, and 30 grams of protein. How many servings of each dish can be eaten?

■ ■ ■ **Thoughts into words**

43. Give a general description of how to solve a system of three linear equations in three variables.

44. Give a step-by-step description of how to solve the system

$$\begin{pmatrix} x - 2y + 3z = -23 \\ 5y - 2z = 32 \\ 4z = -24 \end{pmatrix}$$

45. Give a step-by-step description of how to solve the system

$$\begin{pmatrix} 3x - 2y + 7z = 9 \\ x - 3z = 4 \\ 2x + z = 9 \end{pmatrix}$$

15.3   A Matrix Approach to Solving Systems

■   Represent a system of equations as an augmented matrix.

■   Transform an augmented matrix to the form $\begin{bmatrix} 1 & 0 & 0 & \vdots & a \\ 0 & 1 & 0 & \vdots & b \\ 0 & 0 & 1 & \vdots & c \end{bmatrix}$.

■   Solve a system of equations by using an augmented matrix.

■   Use reduced echelon form to solve a system of equations.

In the first two sections of this chapter, we found that the substitution and elimination- by-addition techniques worked effectively with two equations and two unknowns, but they started to get a bit cumbersome with three equations and three unknowns. Therefore, we shall now begin to analyze some techniques that lend themselves to use with larger systems of equations. Furthermore, some of these techniques form the basis for using a computer to solve systems. Even though these techniques are primarily designed for large systems of equations, we shall study them in the context of small systems so that we won't get bogged down with the computational aspects of the techniques.

Matrices

A **matrix** is an array of numbers arranged in horizontal rows and vertical columns and enclosed in brackets. For example, the matrix

2 rows $\longrightarrow$ $\begin{bmatrix} 2 & 3 & -1 \\ -4 & 7 & 12 \end{bmatrix}$

$\uparrow$ $\uparrow$ $\uparrow$

3 columns

has 2 rows and 3 columns and is called a 2×3 (this is read "two by three") matrix. Each number in a matrix is called an **element** of the matrix. Some additional examples of matrices (*matrices* is the plural of *matrix*) follow.

3×2 2×2 1×2 4×1

$\begin{bmatrix} 2 & 1 \\ 1 & -4 \\ \dfrac{1}{2} & \dfrac{2}{3} \end{bmatrix}$ $\begin{bmatrix} 17 & 18 \\ -14 & 16 \end{bmatrix}$ $\begin{bmatrix} 7 & 14 \end{bmatrix}$ $\begin{bmatrix} 3 \\ -2 \\ 1 \\ 19 \end{bmatrix}$

In general, a matrix of m rows and n columns is called a matrix of **dimension $m \times n$** or **order $m \times n$.**

With every system of linear equations, we can associate a matrix that consists of the coefficients and constant terms. For example, with the system

$$\begin{pmatrix} a_1x + b_1y + c_1z = d_1 \\ a_2x + b_2y + c_2z = d_2 \\ a_3x + b_3y + c_3z = d_3 \end{pmatrix}$$

we can associate the matrix

$$\begin{bmatrix} a_1 & b_1 & c_1 & \vdots & d_1 \\ a_2 & b_2 & c_2 & \vdots & d_2 \\ a_3 & b_3 & c_3 & \vdots & d_3 \end{bmatrix}$$

which is commonly called the **augmented matrix** of the system of equations. The dashed line simply separates the coefficients from the constant terms and reminds us that we are working with an augmented matrix.

In Section 15.1, we considered the operations or transformations that can be applied to a system of equations to produce an equivalent system. Because augmented matrices are essentially abbreviated forms of systems of linear equations, there are analogous transformations that can be applied to augmented matrices. These transformations are usually referred to as **elementary row operations** and can be stated as follows:

For any augmented matrix of a system of linear equations, the following elementary row operations will produce a matrix of an equivalent system.

1. Any two rows of the matrix can be interchanged.
2. Any row of the matrix can be multiplied by a nonzero real number.
3. Any row of the matrix can be replaced by the sum of a nonzero multiple of another row plus that row.

Let's illustrate the use of augmented matrices and elementary row operations to solve a system of two linear equations in two variables.

EXAMPLE 1

Solve the system

$$\begin{pmatrix} x - 3y = -17 \\ 2x + 7y = 31 \end{pmatrix}$$

Solution

The augmented matrix of the system is

$$\begin{bmatrix} 1 & -3 & \vdots & -17 \\ 2 & 7 & \vdots & 31 \end{bmatrix}$$

We would like to change this matrix to one of the form

$$\begin{bmatrix} 1 & 0 & \vdots & a \\ 0 & 1 & \vdots & b \end{bmatrix}$$

where we can easily determine that the solution is $x = a$ and $y = b$. Let's begin by adding -2 times row 1 to row 2 to produce a new row 2.

$$\begin{bmatrix} 1 & -3 & \vdots & -17 \\ 0 & 13 & \vdots & 65 \end{bmatrix}$$

Now we can multiply row 2 by $\dfrac{1}{13}$.

$$\begin{bmatrix} 1 & -3 & \vdots & -17 \\ 0 & 1 & \vdots & 5 \end{bmatrix}$$

Finally, we can add 3 times row 2 to row 1 to produce a new row 1.

$$\begin{bmatrix} 1 & 0 & \vdots & -2 \\ 0 & 1 & \vdots & 5 \end{bmatrix}$$

From this last matrix, we see that $x = -2$ and $y = 5$. In other words, the solution set of the original system is $\{(-2, 5)\}$. ■

It may seem that the matrix approach does not provide us with much extra power for solving systems of two linear equations in two unknowns. However, as the systems get larger, the compactness of the matrix approach becomes more convenient. Let's consider a system of three equations in three variables.

EXAMPLE 2   Solve the system

$$\left(\begin{array}{c} x + 2y - 3z = 15 \\ -2x - 3y + z = -15 \\ 4x + 9y - 4z = 49 \end{array} \right)$$

Solution

The augmented matrix of this system is

$$\begin{bmatrix} 1 & 2 & -3 & \vdots & 15 \\ -2 & -3 & 1 & \vdots & -15 \\ 4 & 9 & -4 & \vdots & 49 \end{bmatrix}$$

If the system has a unique solution, then we will be able to change the augmented matrix to the form

$$\begin{bmatrix} 1 & 0 & 0 & \vdots & a \\ 0 & 1 & 0 & \vdots & b \\ 0 & 0 & 1 & \vdots & c \end{bmatrix}$$

where we will be able to read the solution $x = a$, $y = b$, and $z = c$.

Add 2 times row 1 to row 2 to produce a new row 2. Likewise, add -4 times row 1 to row 3 to produce a new row 3.

$$\begin{bmatrix} 1 & 2 & -3 & \vdots & 15 \\ 0 & 1 & -5 & \vdots & 15 \\ 0 & 1 & 8 & \vdots & -11 \end{bmatrix}$$

Now add -2 times row 2 to row 1 to produce a new row 1. Also, add -1 times row 2 to row 3 to produce a new row 3.

$$\begin{bmatrix} 1 & 0 & 7 & \vdots & -15 \\ 0 & 1 & -5 & \vdots & 15 \\ 0 & 0 & 13 & \vdots & -26 \end{bmatrix}$$

Now let's multiply row 3 by $\dfrac{1}{13}$.

$$\begin{bmatrix} 1 & 0 & 7 & \vdots & -15 \\ 0 & 1 & -5 & \vdots & 15 \\ 0 & 0 & 1 & \vdots & -2 \end{bmatrix}$$

Finally, we can add -7 times row 3 to row 1 to produce a new row 1, and we can add 5 times row 3 to row 2 for a new row 2.

$$\begin{bmatrix} 1 & 0 & 0 & \vdots & -1 \\ 0 & 1 & 0 & \vdots & 5 \\ 0 & 0 & 1 & \vdots & -2 \end{bmatrix}$$

From this last matrix, we can see that the solution set of the original system is $\{(-1, 5, -2)\}$. ∎

The final matrices of Examples 1 and 2,

$$\begin{bmatrix} 1 & 0 & \vdots & -2 \\ 0 & 1 & \vdots & 5 \end{bmatrix} \quad \text{and} \quad \begin{bmatrix} 1 & 0 & 0 & \vdots & -1 \\ 0 & 1 & 0 & \vdots & 5 \\ 0 & 0 & 1 & \vdots & -2 \end{bmatrix}$$

are said to be in **reduced echelon form.** In general, a matrix is in reduced echelon form if the following conditions are satisfied:

1. Reading from left to right, the first nonzero entry of each row is 1.
2. In the *column* containing the leftmost 1 of a row, all the remaining entries are zeros.
3. The leftmost 1 of any row is to the right of the leftmost 1 of the preceding row.
4. Rows containing only zeros are below all the rows containing nonzero entries.

Like the final matrices of Examples 1 and 2, the following are in reduced echelon form.

$$\begin{bmatrix} 1 & 2 & \vdots & -3 \\ 0 & 0 & \vdots & 0 \end{bmatrix} \qquad \begin{bmatrix} 1 & 0 & -2 & \vdots & 5 \\ 0 & 1 & 4 & \vdots & 7 \\ 0 & 0 & 0 & \vdots & 0 \end{bmatrix} \qquad \begin{bmatrix} 1 & 0 & 0 & 0 & \vdots & 8 \\ 0 & 1 & 0 & 0 & \vdots & -9 \\ 0 & 0 & 1 & 0 & \vdots & -2 \\ 0 & 0 & 0 & 1 & \vdots & 12 \end{bmatrix}$$

In contrast, the following matrices are *not* in reduced echelon form for the reason indicated below each matrix.

$$\begin{bmatrix} 1 & 0 & 0 & \vdots & 11 \\ 0 & 3 & 0 & \vdots & -1 \\ 0 & 0 & 1 & \vdots & -2 \end{bmatrix} \qquad \begin{bmatrix} 1 & 2 & -3 & \vdots & 5 \\ 0 & 1 & 7 & \vdots & 9 \\ 0 & 0 & 1 & \vdots & -6 \end{bmatrix}$$

Violates condition 1   Violates condition 2

$$\begin{bmatrix} 1 & 0 & 0 & \vdots & 7 \\ 0 & 0 & 1 & \vdots & -8 \\ 0 & 1 & 0 & \vdots & 14 \end{bmatrix} \qquad \begin{bmatrix} 1 & 0 & 0 & 0 & \vdots & -1 \\ 0 & 0 & 0 & 0 & \vdots & 0 \\ 0 & 0 & 1 & 0 & \vdots & 7 \\ 0 & 0 & 0 & 0 & \vdots & 0 \end{bmatrix}$$

Violates condition 3   Violates condition 4

Once we have an augmented matrix in reduced echelon form, it is easy to determine the solution set of the system. Furthermore, the procedure for changing a given augmented matrix to reduced echelon form can be described in a very systematic way. For example, if an augmented matrix of a system of three linear equations in three unknowns has a unique solution, then it can be changed to reduced echelon form as follows:

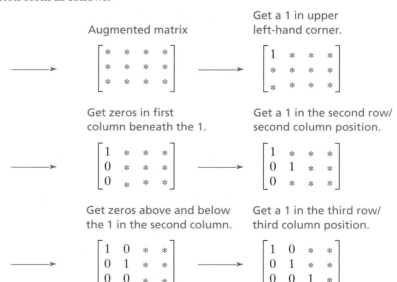

Get zeros above the 1
in the third column.

$$\longrightarrow \begin{bmatrix} 1 & 0 & 0 & * \\ 0 & 1 & 0 & * \\ 0 & 0 & 1 & * \end{bmatrix}$$

We can identify inconsistent and dependent systems while we are changing a matrix to reduced echelon form. We will show some examples of such cases in a moment, but first let's consider another example of a system of three linear equations in three unknowns where there is a unique solution.

EXAMPLE   3

Solve the system

$$\begin{pmatrix} 2x + 4y - 5z = 37 \\ x + 3y - 4z = 29 \\ 5x - y + 3z = -20 \end{pmatrix}$$

Solution

The augmented matrix

$$\begin{bmatrix} 2 & 4 & -5 & \vdots & 37 \\ 1 & 3 & -4 & \vdots & 29 \\ 5 & -1 & 3 & \vdots & -20 \end{bmatrix}$$

does not have a one in the upper left-hand corner, but this can be remedied by exchanging rows 1 and 2.

$$\begin{bmatrix} 1 & 3 & -4 & \vdots & 29 \\ 2 & 4 & -5 & \vdots & 37 \\ 5 & -1 & 3 & \vdots & -20 \end{bmatrix}$$

Now we can get zeros in the first column beneath the one by adding -2 times row 1 to row 2 and by adding -5 times row 1 to row 3.

$$\begin{bmatrix} 1 & 3 & -4 & \vdots & 29 \\ 0 & -2 & 3 & \vdots & -21 \\ 0 & -16 & 23 & \vdots & -165 \end{bmatrix}$$

Next, we can get a one for the first nonzero entry of the second row by multiplying the second row by $-\dfrac{1}{2}$.

$$\begin{bmatrix} 1 & 3 & -4 & \vdots & 29 \\ 0 & 1 & -\dfrac{3}{2} & \vdots & \dfrac{21}{2} \\ 0 & -16 & 23 & \vdots & -165 \end{bmatrix}$$

Now we can get zeros above and below the one in the second column by adding -3 times row 2 to row 1 and by adding 16 times row 2 to row 3.

$$\begin{bmatrix} 1 & 0 & \frac{1}{2} & \vdots & -\frac{5}{2} \\ 0 & 1 & -\frac{3}{2} & \vdots & \frac{21}{2} \\ 0 & 0 & -1 & \vdots & 3 \end{bmatrix}$$

Next, we can get a one in the first nonzero entry of the third row by multiplying the third row by -1.

$$\begin{bmatrix} 1 & 0 & \frac{1}{2} & \vdots & -\frac{5}{2} \\ 0 & 1 & -\frac{3}{2} & \vdots & \frac{21}{2} \\ 0 & 0 & 1 & \vdots & -3 \end{bmatrix}$$

Finally, we can get zeros above the one in the third column by adding $-\frac{1}{2}$ times row 3 to row 1 and by adding $\frac{3}{2}$ times row 3 to row 2.

$$\begin{bmatrix} 1 & 0 & 0 & \vdots & -1 \\ 0 & 1 & 0 & \vdots & 6 \\ 0 & 0 & 1 & \vdots & -3 \end{bmatrix}$$

From this last matrix, we see that the solution set of the original system is $\{(-1, 6, -3)\}$. ∎

Example 3 illustrates that even though the process of changing to reduced echelon form can be systematically described, it can involve some rather messy calculations. However, with the aid of a computer, such calculations are not troublesome. For our purposes in this text, the examples and problems involve systems that minimize messy calculations. This will allow us to concentrate on the procedures.

We want to call your attention to another issue in the solution of Example 3. Consider the matrix

$$\begin{bmatrix} 1 & 3 & -4 & \vdots & 29 \\ 0 & 1 & -\frac{3}{2} & \vdots & \frac{21}{2} \\ 0 & -16 & 23 & \vdots & -165 \end{bmatrix}$$

which is obtained about halfway through the solution. At this step, it seems evident that the calculations are getting a little messy. Therefore, instead of continuing toward the reduced echelon form, let's add 16 times row 2 to row 3 to produce a new row 3.

$$\begin{bmatrix} 1 & 3 & -4 & \vdots & 29 \\ 0 & 1 & -\dfrac{3}{2} & \vdots & \dfrac{21}{2} \\ 0 & 0 & -1 & \vdots & 3 \end{bmatrix}$$

The system represented by this matrix is

$$\left(\begin{array}{l} x + 3y - 4z = 29 \\ \quad\quad y - \dfrac{3}{2}z = \dfrac{21}{2} \\ \quad\quad\quad -z = 3 \end{array} \right)$$

and it is said to be in **triangular form.** The last equation determines the value for z; then we can use the process of back-substitution to determine the values for y and x.

Finally, let's consider two examples to illustrate what happens when we use the matrix approach on inconsistent and dependent systems.

EXAMPLE 4

Solve the system

$$\left(\begin{array}{l} x - 2y + 3z = 3 \\ 5x - 9y + 4z = 2 \\ 2x - 4y + 6z = -1 \end{array} \right)$$

Solution

The augmented matrix of the system is

$$\begin{bmatrix} 1 & -2 & 3 & \vdots & 3 \\ 5 & -9 & 4 & \vdots & 2 \\ 2 & -4 & 6 & \vdots & -1 \end{bmatrix}$$

We can get zeros below the one in the first column by adding -5 times row 1 to row 2 and by adding -2 times row 1 to row 3.

$$\begin{bmatrix} 1 & -2 & 3 & \vdots & 3 \\ 0 & 1 & -11 & \vdots & -13 \\ 0 & 0 & 0 & \vdots & -7 \end{bmatrix}$$

At this step we can stop, because the bottom row of the matrix represents the statement $0(x) + 0(y) + 0(z) = -7$, which is obviously false for all values of x, y, and z. Thus the original system is inconsistent; its solution set is $\varnothing$. ∎

EXAMPLE 5

Solve the system

$$\left(\begin{array}{l} x + 2y + 2z = 9 \\ x + 3y - 4z = 5 \\ 2x + 5y - 2z = 14 \end{array} \right)$$

Solution

The augmented matrix of the system is

$$\left[\begin{array}{ccc:c} 1 & 2 & 2 & 9 \\ 1 & 3 & -4 & 5 \\ 2 & 5 & -2 & 14 \end{array}\right]$$

We can get zeros in the first column below the one in the upper left-hand corner by adding -1 times row 1 to row 2 and adding -2 times row 1 to row 3.

$$\left[\begin{array}{ccc:c} 1 & 2 & 2 & 9 \\ 0 & 1 & -6 & -4 \\ 0 & 1 & -6 & -4 \end{array}\right]$$

Now we can get zeros in the second column above and below the one in the second row by adding -2 times row 2 to row 1 and adding -1 times row 2 to row 3.

$$\left[\begin{array}{ccc:c} 1 & 0 & 14 & 17 \\ 0 & 1 & -6 & -4 \\ 0 & 0 & 0 & 0 \end{array}\right]$$

The bottom row of zeros represents the statement $0(x) + 0(y) + 0(z) = 0$, which is true for all values of x, y, and z. The second row represents the statement $y - 6z = -4$, which can be rewritten $y = 6z - 4$. The top row represents the statement $x + 14z = 17$, which can be rewritten $x = -14z + 17$. Therefore, if we let $z = k$, where k is any real number, the solution set of infinitely many ordered triples can be represented by $\{(-14k + 17, 6k - 4, k) | k \text{ is a real number}\}$. Specific solutions can be generated by letting k take on a value. For example, if $k = 2$, then $6k - 4$ becomes $6(2) - 4 = 8$ and $-14k + 17$ becomes $-14(2) + 17 = -11$. Thus the ordered triple $(-11, 8, 2)$ is a member of the solution set. ■

CONCEPT QUIZ

For Problems 1–8, answer true or false.

1. A matrix of dimension 2×6, has 6 rows and 2 columns.

2. The augmented matrix of a system of equations is a matrix of the coefficients and constant terms of the equations.

3. Transformations that are applied to augmented matrices are called elementary column operations.

4. For any augmented matrix, two rows can be interchanged to produce an equivalent matrix.

5. For any augmented matrix, any column may be multiplied by a nonzero real number to produce an equivalent matrix.

6. The matrix $\begin{bmatrix} 1 & 0 & 0 & \vdots & 6 \\ 0 & 1 & 1 & \vdots & -3 \\ 0 & 0 & 1 & \vdots & 2 \end{bmatrix}$ is in reduced echelon form.

7. The system of equations $\begin{pmatrix} x + y + 2z = 7 \\ y - 3z = 4 \\ z = -5 \end{pmatrix}$ is in triangular form.

8. Given that the matrix $\begin{bmatrix} 1 & 0 & 0 & \vdots & 2 \\ 0 & 1 & 0 & \vdots & -3 \\ 0 & 0 & 1 & \vdots & 4 \end{bmatrix}$ represents a system of equations, the solution of the system is the ordered triple $(2, -3, 4)$.

PROBLEM SET 15.3

For Problems 1–10, indicate whether each matrix is in reduced echelon form.

1. $\begin{bmatrix} 1 & 0 & \vdots & -4 \\ 0 & 1 & \vdots & 14 \end{bmatrix}$

2. $\begin{bmatrix} 1 & 2 & \vdots & 8 \\ 0 & 0 & \vdots & 0 \end{bmatrix}$

3. $\begin{bmatrix} 1 & 0 & 2 & \vdots & 5 \\ 0 & 1 & 3 & \vdots & 7 \\ 0 & 0 & 0 & \vdots & 0 \end{bmatrix}$

4. $\begin{bmatrix} 1 & 0 & 0 & \vdots & 5 \\ 0 & 3 & 0 & \vdots & 8 \\ 0 & 0 & 1 & \vdots & -11 \end{bmatrix}$

5. $\begin{bmatrix} 1 & 0 & 0 & \vdots & 17 \\ 0 & 0 & 0 & \vdots & 0 \\ 0 & 1 & 0 & \vdots & -14 \end{bmatrix}$

6. $\begin{bmatrix} 1 & 0 & 0 & \vdots & -7 \\ 0 & 1 & 0 & \vdots & 0 \\ 0 & 0 & 1 & \vdots & 9 \end{bmatrix}$

7. $\begin{bmatrix} 1 & 1 & 0 & \vdots & -3 \\ 0 & 1 & 2 & \vdots & 5 \\ 0 & 0 & 1 & \vdots & 7 \end{bmatrix}$

8. $\begin{bmatrix} 1 & 0 & 3 & \vdots & 8 \\ 0 & 1 & 2 & \vdots & -6 \\ 0 & 0 & 0 & \vdots & 0 \end{bmatrix}$

9. $\begin{bmatrix} 1 & 0 & 0 & 3 & \vdots & 4 \\ 0 & 1 & 0 & 5 & \vdots & -3 \\ 0 & 0 & 1 & -1 & \vdots & 7 \\ 0 & 0 & 0 & 0 & \vdots & 0 \end{bmatrix}$

10. $\begin{bmatrix} 1 & 0 & 0 & 0 & \vdots & 2 \\ 0 & 0 & 1 & 0 & \vdots & 4 \\ 0 & 1 & 0 & 0 & \vdots & -3 \\ 0 & 0 & 0 & 1 & \vdots & 9 \end{bmatrix}$

For Problems 11–30, use a matrix approach to solve each system.

11. $\begin{pmatrix} x - 3y = 14 \\ 3x + 2y = -13 \end{pmatrix}$

12. $\begin{pmatrix} x + 5y = -18 \\ -2x + 3y = -16 \end{pmatrix}$

13. $\begin{pmatrix} 3x - 4y = 33 \\ x + 7y = -39 \end{pmatrix}$

14. $\begin{pmatrix} 2x + 7y = -55 \\ x - 4y = 25 \end{pmatrix}$

15. $\begin{pmatrix} x - 6y = -2 \\ 2x - 12y = 5 \end{pmatrix}$

16. $\begin{pmatrix} 2x - 3y = -12 \\ 3x + 2y = 8 \end{pmatrix}$

17. $\begin{pmatrix} 3x - 5y = 39 \\ 2x + 7y = -67 \end{pmatrix}$

18. $\begin{pmatrix} 3x + 9y = -1 \\ x + 3y = 10 \end{pmatrix}$

19. $\begin{pmatrix} x - 2y - 3z = -6 \\ 3x - 5y - z = 4 \\ 2x + y + 2z = 2 \end{pmatrix}$

20. $\begin{pmatrix} x + 3y - 4z = 13 \\ 2x + 7y - 3z = 11 \\ -2x - y + 2z = -8 \end{pmatrix}$

21. $\begin{pmatrix} -2x - 5y + 3z = 11 \\ x + 3y - 3z = -12 \\ 3x - 2y + 5z = 31 \end{pmatrix}$

22. $\begin{pmatrix} -3x + 2y + z = 17 \\ x - y + 5z = -2 \\ 4x - 5y - 3z = -36 \end{pmatrix}$

23. $\begin{pmatrix} x - 3y - z = 2 \\ 3x + y - 4z = -18 \\ -2x + 5y + 3z = 2 \end{pmatrix}$

24. $\begin{pmatrix} x - 4y + 3z = 16 \\ 2x + 3y - 4z = -22 \\ -3x + 11y - z = -36 \end{pmatrix}$

25. $\begin{pmatrix} x - y + 2z = 1 \\ -3x + 4y - z = 4 \\ -x + 2y + 3z = 6 \end{pmatrix}$ **26.** $\begin{pmatrix} x + 2y - 5z = -1 \\ 2x + 3y - 2z = 2 \\ 3x + 5y - 7z = 4 \end{pmatrix}$

34. $\begin{pmatrix} x_1 + 2x_2 - 3x_3 + x_4 = -2 \\ -2x_1 - 3x_2 + x_3 - x_4 = 5 \\ 4x_1 + 9x_2 - 2x_3 - 2x_4 = -28 \\ -5x_1 - 9x_2 + 2x_3 - 3x_4 = 14 \end{pmatrix}$

27. $\begin{pmatrix} -2x + y + 5z = -5 \\ 3x + 8y - z = -34 \\ x + 2y + z = -12 \end{pmatrix}$

28. $\begin{pmatrix} 4x - 10y + 3z = -19 \\ 2x + 5y - z = -7 \\ x - 3y - 2z = -2 \end{pmatrix}$

29. $\begin{pmatrix} 2x + 3y - z = 7 \\ 3x + 4y + 5z = -2 \\ 5x + y + 3z = 13 \end{pmatrix}$ **30.** $\begin{pmatrix} 4x + 3y - z = 0 \\ 3x + 2y + 5z = 6 \\ 5x - y - 3z = 3 \end{pmatrix}$

Subscript notation is frequently used for working with larger systems of equations. For Problems 31–34, use a matrix approach to solve each system. Express the solutions as 4-tuples of the form (x_1, x_2, x_3, x_4).

31. $\begin{pmatrix} x_1 - 3x_2 - 2x_3 + x_4 = -3 \\ -2x_1 + 7x_2 + x_3 - 2x_4 = -1 \\ 3x_1 - 7x_2 - 3x_3 + 3x_4 = -5 \\ 5x_1 + x_2 + 4x_3 - 2x_4 = 18 \end{pmatrix}$

32. $\begin{pmatrix} x_1 - 2x_2 + 2x_3 - x_4 = -2 \\ -3x_1 + 5x_2 - x_3 - 3x_4 = 2 \\ 2x_1 + 3x_2 + 3x_3 + 5x_4 = -9 \\ 4x_1 - x_2 - x_3 - 2x_4 = 8 \end{pmatrix}$

33. $\begin{pmatrix} x_1 + 3x_2 - x_3 + 2x_4 = -2 \\ 2x_1 + 7x_2 + 2x_3 - x_4 = 19 \\ -3x_1 - 8x_2 + 3x_3 + x_4 = -7 \\ 4x_1 + 11x_2 - 2x_3 - 3x_4 = 19 \end{pmatrix}$

In Problems 35–42, each matrix is the reduced echelon matrix for a system with variables x_1, x_2, x_3, and x_4. Find the solution set of each system.

35. $\begin{bmatrix} 1 & 0 & 0 & 0 & \vdots & -2 \\ 0 & 1 & 0 & 0 & \vdots & 4 \\ 0 & 0 & 1 & 0 & \vdots & -3 \\ 0 & 0 & 0 & 1 & \vdots & 0 \end{bmatrix}$ **36.** $\begin{bmatrix} 1 & 0 & 0 & 0 & \vdots & 0 \\ 0 & 1 & 0 & 0 & \vdots & -5 \\ 0 & 0 & 1 & 0 & \vdots & 0 \\ 0 & 0 & 0 & 1 & \vdots & 4 \end{bmatrix}$

37. $\begin{bmatrix} 1 & 0 & 0 & 0 & \vdots & -8 \\ 0 & 1 & 0 & 0 & \vdots & 5 \\ 0 & 0 & 1 & 0 & \vdots & -2 \\ 0 & 0 & 0 & 0 & \vdots & 1 \end{bmatrix}$ **38.** $\begin{bmatrix} 1 & 0 & 0 & 0 & \vdots & 2 \\ 0 & 1 & 0 & 2 & \vdots & -3 \\ 0 & 0 & 1 & 3 & \vdots & 4 \\ 0 & 0 & 0 & 0 & \vdots & 0 \end{bmatrix}$

39. $\begin{bmatrix} 1 & 0 & 0 & 3 & \vdots & 5 \\ 0 & 1 & 0 & 0 & \vdots & -1 \\ 0 & 0 & 1 & 4 & \vdots & 2 \\ 0 & 0 & 0 & 0 & \vdots & 0 \end{bmatrix}$ **40.** $\begin{bmatrix} 1 & 3 & 0 & 2 & \vdots & 0 \\ 0 & 0 & 1 & 0 & \vdots & 0 \\ 0 & 0 & 0 & 0 & \vdots & 1 \\ 0 & 0 & 0 & 0 & \vdots & 0 \end{bmatrix}$

41. $\begin{bmatrix} 1 & 3 & 0 & 0 & \vdots & 9 \\ 0 & 0 & 1 & 0 & \vdots & 2 \\ 0 & 0 & 0 & 1 & \vdots & -3 \\ 0 & 0 & 0 & 0 & \vdots & 0 \end{bmatrix}$

42. $\begin{bmatrix} 1 & 0 & 0 & 0 & \vdots & 7 \\ 0 & 1 & 0 & 0 & \vdots & -3 \\ 0 & 0 & 1 & -2 & \vdots & 5 \\ 0 & 0 & 0 & 0 & \vdots & 0 \end{bmatrix}$

■ ■ ■ Thoughts into words

43. What is a matrix? What is an augmented matrix of a system of linear equations?

44. Describe how to use matrices to solve the system $\begin{pmatrix} x - 2y = 5 \\ 2x + 7y = 9 \end{pmatrix}$.

■ ■ ■ Further investigations

For Problems 45–50, change each augmented matrix of the system to reduced echelon form and then indicate the solutions of the system.

45. $\begin{pmatrix} x - 2y + 3z = 4 \\ 3x - 5y - z = 7 \end{pmatrix}$

46. $\begin{pmatrix} x + 3y - 2z = -1 \\ 2x - 5y + 7z = 4 \end{pmatrix}$

47. $\begin{pmatrix} 2x - 4y + 3z = 8 \\ 3x + 5y - z = 7 \end{pmatrix}$

48. $\begin{pmatrix} 3x + 6y - z = 9 \\ 2x - 3y + 4z = 1 \end{pmatrix}$

50. $\begin{pmatrix} x + y - 2z = -1 \\ 3x + 3y - 6z = -3 \end{pmatrix}$

49. $\begin{pmatrix} x - 2y + 4z = 9 \\ 2x - 4y + 8z = 3 \end{pmatrix}$

 Graphing calculator activities

51. If your graphing calculator has the capability of manipulating matrices, this is a good time to become familiar with those operations. You may need to refer to your user's manual for the key-punching instructions. To be- gin the familiarization process, load your calculator with the three augmented matrices in Examples 1, 2, and 3. Then, for each one, carry out the row operations as described in the text.

Answers to Concept Quiz

1. False   **2.** True   **3.** False   **4.** True   **5.** False   **6.** False   **7.** True   **8.** True

15.4   Determinants

Objectives

■ Evaluate the determinant of a matrix.

■ Compute the cofactor of an element in a matrix.

■ Expand a determinant about a row or column.

■ Apply the properties of determinants to simplify the evaluation of a determinant.

Before we introduce the concept of a determinant, let's agree on some convenient new notation. A **general $m \times n$ (m-by-n) matrix** can be represented by

$$A = \begin{bmatrix} a_{11} & a_{12} & a_{13} & \cdots & a_{1n} \\ a_{21} & a_{22} & a_{23} & \cdots & a_{2n} \\ \cdot & \cdot & \cdot & & \cdot \\ \cdot & \cdot & \cdot & & \cdot \\ \cdot & \cdot & \cdot & & \cdot \\ a_{m1} & a_{m2} & a_{m3} & \cdots & a_{mn} \end{bmatrix}$$

where the double subscripts are used to identify the number of the row and the number of the column, in that order. For example, a_{23} is the entry at the intersection of the second row and the third column. In general, the entry at the intersection of row i and column j is denoted by a_{ij}.

A **square matrix** is one that has the same number of rows as columns. Each square matrix A with real number entries can be associated with a real number called the **determinant** of the matrix, denoted by $|A|$. We will first define $|A|$ for a 2×2 matrix.

DEFINITION 15.1

If $A = \begin{bmatrix} a_{11} & a_{12} \\ a_{21} & a_{22} \end{bmatrix}$, then

$$|A| = \begin{vmatrix} a_{11} & a_{12} \\ a_{21} & a_{22} \end{vmatrix} = a_{11}a_{22} - a_{12}a_{21}$$

EXAMPLE 1 If $A = \begin{bmatrix} 3 & -2 \\ 5 & 8 \end{bmatrix}$, find $|A|$.

Solution
Use Definition 15.1 to obtain

$$|A| = \begin{vmatrix} 3 & -2 \\ 5 & 8 \end{vmatrix} = 3(8) - (-2)(5)$$
$$= 24 + 10$$
$$= 34$$

Finding the determinant of a square matrix is commonly called **evaluating the determinant,** and the matrix notation is often omitted.

EXAMPLE 2 Evaluate $\begin{vmatrix} -3 & 6 \\ 2 & 8 \end{vmatrix}$.

Solution

$$\begin{vmatrix} -3 & 6 \\ 2 & 8 \end{vmatrix} = (-3)(8) - (6)(2)$$
$$= -24 - 12$$
$$= -36$$

To find the determinants of 3×3 and larger square matrices, it is convenient to introduce some additional terminology.

DEFINITION 15.2

If A is a 3×3 matrix, then the **minor** (denoted by M_{ij}) of the a_{ij} element is the determinant of the 2×2 matrix obtained by deleting row i and column j of A.

EXAMPLE 3   If $A = \begin{bmatrix} 2 & 1 & 4 \\ -6 & 3 & -2 \\ 4 & 2 & 5 \end{bmatrix}$, find (a) M_{11} and (b) M_{23}.

Solution

(a) To find M_{11}, we first delete row 1 and column 1 of matrix A.

$$\begin{bmatrix} 2 & 1 & 4 \\ -6 & 3 & -2 \\ 4 & 2 & 5 \end{bmatrix}$$

Thus

$$M_{11} = \begin{vmatrix} 3 & -2 \\ 2 & 5 \end{vmatrix} = 3(5) - (-2)(2) = 19$$

(b) To find M_{23}, we first delete row 2 and column 3 of matrix A.

$$\begin{bmatrix} 2 & 1 & 4 \\ -6 & 3 & -2 \\ 4 & 2 & 5 \end{bmatrix}$$

Thus

$$M_{23} = \begin{vmatrix} 2 & 1 \\ 4 & 2 \end{vmatrix} = 2(2) - (1)(4) = 0$$

The following definition will also be used.

DEFINITION 15.3

If A is a 3×3 matrix, then the **cofactor** (denoted by C_{ij}) of the element a_{ij} is defined by

$$C_{ij} = (-1)^{i+j} M_{ij}$$

According to Definition 15.3, to find the cofactor of any element a_{ij} of a square matrix A, we find the minor of a_{ij} and multiply it by 1 if $i + j$ is even, or multiply it by -1 if $i + j$ is odd.

EXAMPLE 4   If $A = \begin{bmatrix} 3 & 2 & -4 \\ 1 & 5 & 4 \\ 2 & -3 & 1 \end{bmatrix}$, find C_{32}.

Solution

First, let's find M_{32} by deleting row 3 and column 2 of matrix A.

$$\begin{bmatrix} 3 & 2 & -4 \\ 1 & 5 & 4 \\ 2 & -3 & 1 \end{bmatrix}$$

Thus

$$M_{32} = \begin{vmatrix} 3 & -4 \\ 1 & 4 \end{vmatrix} = 3(4) - (-4)(1) = 16$$

Therefore,

$$C_{32} = (-1)^{3+2}M_{32} = (-1)^5(16) = -16 \quad \blacksquare$$

The concept of a cofactor can be used to define the determinant of a 3×3 matrix as follows:

DEFINITION 15.4

If $A = \begin{bmatrix} a_{11} & a_{12} & a_{13} \\ a_{21} & a_{22} & a_{23} \\ a_{31} & a_{32} & a_{33} \end{bmatrix}$, then

$$|A| = a_{11}C_{11} + a_{21}C_{21} + a_{31}C_{31}$$

Definition 15.4 simply states that the determinant of a 3×3 matrix can be found by multiplying each element of the first column by its corresponding cofactor and then adding the three results. Let's illustrate this procedure.

EXAMPLE 5   Find $|A|$ if $A = \begin{bmatrix} -2 & 1 & 4 \\ 3 & 0 & 5 \\ 1 & -4 & -6 \end{bmatrix}$.

Solution

$$|A| = a_{11}C_{11} + a_{21}C_{21} + a_{31}C_{31}$$

$$= (-2)(-1)^{1+1}\begin{vmatrix} 0 & 5 \\ -4 & -6 \end{vmatrix} + (3)(-1)^{2+1}\begin{vmatrix} 1 & 4 \\ -4 & -6 \end{vmatrix} + (1)(-1)^{3+1}\begin{vmatrix} 1 & 4 \\ 0 & 5 \end{vmatrix}$$

$$= (-2)(1)(20) + (3)(-1)(10) + (1)(1)(5)$$

$$= -40 - 30 + 5$$

$$= -65 \quad \blacksquare$$

When we use Definition 15.4, we often say that "the determinant is being expanded about the first column." It can also be shown that **any row or column can be used to expand a determinant.** For example, for matrix A in Example 5, the expansion of the determinant about the *second row* is as follows:

$$\begin{vmatrix} -2 & 1 & 4 \\ 3 & 0 & 5 \\ 1 & -4 & -6 \end{vmatrix} = (3)(-1)^{2+1}\begin{vmatrix} 1 & 4 \\ -4 & -6 \end{vmatrix} + (0)(-1)^{2+2}\begin{vmatrix} -2 & 4 \\ 1 & -6 \end{vmatrix} + (5)(-1)^{2+3}\begin{vmatrix} -2 & 1 \\ 1 & -4 \end{vmatrix}$$

$$= (3)(-1)(10) + (0)(1)(8) + (5)(-1)(7)$$

$$= -30 + 0 - 35$$

$$= -65$$

Note that when we expanded about the second row, the computation was simplified by the presence of a zero. In general, it is helpful to expand about the row or column that contains the most zeros.

The concepts of minor and cofactor have been defined in terms of 3×3 matrices. Analogous definitions can be given for any square matrix (that is, any $n \times n$ matrix with $n \geq 2$), and the determinant can then be expanded about any row or column. Certainly, as the matrices become larger than 3×3, the computations get more tedious. We will concentrate most of our efforts in this text on 2×2 and 3×3 matrices.

Properties of Determinants

Determinants have several interesting properties, some of which are important primarily from a theoretical standpoint. But some of the properties are also very useful when evaluating determinants. We will state these properties for square matrices in general, but we will use 2×2 or 3×3 matrices as examples. We can demonstrate some of the proofs of these properties by evaluating the determinants involved; some of the proofs for 3×3 matrices will be left for you to verify in the next problem set.

> **PROPERTY 15.1**
>
> If any row (or column) of a square matrix A contains only zeros, then $|A| = 0$.

If every element of a row (or column) of a square matrix A is 0, then it should be evident that expanding the determinant about that row (or column) of zeros will produce 0.

> **PROPERTY 15.2**
>
> If square matrix B is obtained from square matrix A by interchanging two rows (or two columns), then $|B| = -|A|$.

Property 15.2 states that **interchanging two rows (or columns) changes the sign of the determinant.** As an example of this property, suppose that

$$A = \begin{bmatrix} 2 & 5 \\ -1 & 6 \end{bmatrix}$$

and that rows 1 and 2 are interchanged to form

$$B = \begin{bmatrix} -1 & 6 \\ 2 & 5 \end{bmatrix}$$

Calculating $|A|$ and $|B|$ yields

$$|A| = \begin{vmatrix} 2 & 5 \\ -1 & 6 \end{vmatrix} = 2(6) - (5)(-1) = 17$$

and

$$|B| = \begin{vmatrix} -1 & 6 \\ 2 & 5 \end{vmatrix} = (-1)(5) - (6)(2) = -17$$

PROPERTY 15.3

If square matrix B is obtained from square matrix A by multiplying each element of any row (or column) of A by some real number k, then $|B| = k|A|$.

Property 15.3 states that **multiplying any row (or column) by a factor of k affects the value of the determinant by a factor of k.** As an example of this property, suppose that

$$A = \begin{bmatrix} 1 & -2 & 8 \\ 2 & 1 & 12 \\ 3 & 2 & -16 \end{bmatrix}$$

and that B is formed by multiplying each element of the third column by $\dfrac{1}{4}$.

$$B = \begin{bmatrix} 1 & -2 & 2 \\ 2 & 1 & 3 \\ 3 & 2 & -4 \end{bmatrix}$$

Now let's calculate $|A|$ and $|B|$ by expanding about the third column in each case.

$$|A| = \begin{vmatrix} 1 & -2 & 8 \\ 2 & 1 & 12 \\ 3 & 2 & -16 \end{vmatrix} = (8)(-1)^{1+3}\begin{vmatrix} 2 & 1 \\ 3 & 2 \end{vmatrix} + (12)(-1)^{2+3}\begin{vmatrix} 1 & -2 \\ 3 & 2 \end{vmatrix} + (-16)(-1)^{3+3}\begin{vmatrix} 1 & -2 \\ 2 & 1 \end{vmatrix}$$

$$= (8)(1)(1) + (12)(-1)(8) + (-16)(1)(5)$$

$$= -168$$

$$|B| = \begin{vmatrix} 1 & -2 & 2 \\ 2 & 1 & 3 \\ 3 & 2 & -4 \end{vmatrix} = (2)(-1)^{1+3}\begin{vmatrix} 2 & 1 \\ 3 & 2 \end{vmatrix} + (3)(-1)^{2+3}\begin{vmatrix} 1 & -2 \\ 3 & 2 \end{vmatrix} + (-4)(-1)^{3+3}\begin{vmatrix} 1 & -2 \\ 2 & 1 \end{vmatrix}$$

$$= (2)(1)(1) + (3)(-1)(8) + (-4)(1)(5)$$

$$= -42$$

We see that $|B| = \dfrac{1}{4}|A|$. This example also illustrates the computational use of Property 15.3: We can factor out a common factor from a row or column, and then adjust the value of the determinant by that factor. For example,

$$\begin{vmatrix} 2 & 6 & 8 \\ -1 & 2 & 7 \\ 5 & 2 & 1 \end{vmatrix} = 2\begin{vmatrix} 1 & 3 & 4 \\ -1 & 2 & 7 \\ 5 & 2 & 1 \end{vmatrix}$$

Factor a 2 from the top row.

PROPERTY 15.4

If square matrix B is obtained from square matrix A by adding k times a row (or column) of A to another row (or column) of A, then $|B| = |A|$.

Property 15.4 states that **adding the product of k times a row (or column) to another row (or column) does not affect the value of the determinant.** As an example of this property, suppose that

$$A = \begin{bmatrix} 1 & 2 & 4 \\ 2 & 4 & 7 \\ -1 & 3 & 5 \end{bmatrix}$$

Now let's form B by replacing row 2 with the result of adding -2 times row 1 to row 2.

$$B = \begin{bmatrix} 1 & 2 & 4 \\ 0 & 0 & -1 \\ -1 & 3 & 5 \end{bmatrix}$$

Next, let's evaluate $|A|$ and $|B|$ by expanding about the second row in each case.

$$|A| = \begin{vmatrix} 1 & 2 & 4 \\ 2 & 4 & 7 \\ -1 & 3 & 5 \end{vmatrix} = (2)(-1)^{2+1}\begin{vmatrix} 2 & 4 \\ 3 & 5 \end{vmatrix} + (4)(-1)^{2+2}\begin{vmatrix} 1 & 4 \\ -1 & 5 \end{vmatrix} + (7)(-1)^{2+3}\begin{vmatrix} 1 & 2 \\ -1 & 3 \end{vmatrix}$$

$$= 2(-1)(-2) + (4)(1)(9) + (7)(-1)(5)$$

$$= 5$$

$$|B| = \begin{vmatrix} 1 & 2 & 4 \\ 0 & 0 & -1 \\ -1 & 3 & 5 \end{vmatrix} = (0)(-1)^{2+1}\begin{vmatrix} 2 & 4 \\ 3 & 5 \end{vmatrix} + (0)(-1)^{2+2}\begin{vmatrix} 1 & 4 \\ -1 & 5 \end{vmatrix} + (-1)(-1)^{2+3}\begin{vmatrix} 1 & 2 \\ -1 & 3 \end{vmatrix}$$

$$= 0 + 0 + (-1)(-1)(5)$$

$$= 5$$

Note that $|B| = |A|$. Furthermore, note that because of the zeros in the second row, evaluating $|B|$ is much easier than evaluating $|A|$. Property 15.4 can often be used to obtain some zeros before evaluating a determinant.

A word of caution is in order at this time. Be careful not to confuse Properties 15.2, 15.3, and 15.4 with the three elementary row transformations of augmented matrices that were used in Section 15.3. The statements of the two sets of properties do resemble each other, but the properties pertain to *two different concepts,* so be sure you understand the distinction between them.

One final property of determinants should be mentioned.

PROPERTY 15.5

If two rows (or columns) of a square matrix A are identical, then $|A| = 0$.

Property 15.5 is a direct consequence of Property 15.2. Suppose that A is a square matrix (any size) with two identical rows. Square matrix B can be formed from A by interchanging the two identical rows. Because identical rows were interchanged, $|B| = |A|$. *But* by Property 15.2, $|B| = -|A|$. For both of these statements to hold, $|A| = 0$.

Let's conclude this section by evaluating a 4×4 determinant, using Properties 15.3 and 15.4 to facilitate the computation.

E X A M P L E   6   Evaluate $\begin{vmatrix} 6 & 2 & 1 & -2 \\ 9 & -1 & 4 & 1 \\ 12 & -2 & 3 & -1 \\ 0 & 0 & 9 & 3 \end{vmatrix}$.

Solution

First, let's add -3 times the fourth column to the third column.

$$\begin{vmatrix} 6 & 2 & 7 & -2 \\ 9 & -1 & 1 & 1 \\ 12 & -2 & 6 & -1 \\ 0 & 0 & 0 & 3 \end{vmatrix}$$

Now, if we expand about the fourth row, we get only one nonzero product.

$$(3)(-1)^{4+4} \begin{vmatrix} 6 & 2 & 7 \\ 9 & -1 & 1 \\ 12 & -2 & 6 \end{vmatrix}$$

Factoring a 3 out of the first column of the 3×3 determinant yields

$$(3)(-1)^8(3) \begin{vmatrix} 2 & 2 & 7 \\ 3 & -1 & 1 \\ 4 & -2 & 6 \end{vmatrix}$$

Next, working with the 3×3 determinant, we can first add column 3 to column 2 and then add -3 times column 3 to column 1.

$$(3)(-1)^8(3) \begin{vmatrix} -19 & 9 & 7 \\ 0 & 0 & 1 \\ -14 & 4 & 6 \end{vmatrix}$$

Finally, by expanding this 3×3 determinant about the second row, we obtain

$$(3)(-1)^8(3)(1)(-1)^{2+3} \begin{vmatrix} -19 & 9 \\ -14 & 4 \end{vmatrix}$$

Our final result is

$$(3)(-1)^8(3)(1)(-1)^5(50) = -450$$

CONCEPT QUIZ

For Problems 1–8, answer true or false.

1. A square matrix has the same number of rows and columns.

2. A determinant can be calculated for any matrix.

3. The determinant of a matrix can be zero.

4. The a_{14} element of a matrix is in the 4th row and 1st column.

5. The minor of the element a_{31} is the determinant of the matrix obtained by deleting the 3rd row and 1st column.

6. Given $A = \begin{bmatrix} 1 & -2 & 3 \\ 0 & 4 & -1 \\ 5 & -3 & 2 \end{bmatrix}$, the cofactor $C_{23} = \begin{vmatrix} 1 & -2 \\ 5 & -3 \end{vmatrix}$.

7. Interchanging two columns of a matrix has no effect on the determinant.

8. Multiplying a row of a matrix by 2 affects the determinant of that matrix by a factor of $\dfrac{1}{2}$.

PROBLEM SET 15.4

For Problems 1–12, evaluate each 2×2 determinant by using Definition 15.1.

For Problems 13–28, evaluate each 3×3 determinant. Use the properties of determinants to your advantage.

1. $\begin{vmatrix} 4 & 3 \\ 2 & 7 \end{vmatrix}$

2. $\begin{vmatrix} 3 & 5 \\ 6 & 4 \end{vmatrix}$

3. $\begin{vmatrix} -3 & 2 \\ 7 & 5 \end{vmatrix}$

13. $\begin{vmatrix} 1 & 2 & -1 \\ 3 & 1 & 2 \\ 2 & 4 & 3 \end{vmatrix}$

14. $\begin{vmatrix} 1 & -2 & 1 \\ 2 & 1 & -1 \\ 3 & 2 & 4 \end{vmatrix}$

4. $\begin{vmatrix} 5 & 3 \\ 6 & -1 \end{vmatrix}$

5. $\begin{vmatrix} 2 & -3 \\ 8 & -2 \end{vmatrix}$

6. $\begin{vmatrix} -5 & 5 \\ -6 & 2 \end{vmatrix}$

15. $\begin{vmatrix} 1 & -4 & 1 \\ 2 & 5 & -1 \\ 3 & 3 & 4 \end{vmatrix}$

16. $\begin{vmatrix} 3 & -2 & 1 \\ 2 & 1 & 4 \\ -1 & 3 & 5 \end{vmatrix}$

7. $\begin{vmatrix} -2 & -3 \\ -1 & -4 \end{vmatrix}$

8. $\begin{vmatrix} -4 & -3 \\ -5 & -7 \end{vmatrix}$

9. $\begin{vmatrix} \dfrac{1}{2} & \dfrac{1}{3} \\ -3 & -6 \end{vmatrix}$

17. $\begin{vmatrix} 6 & 12 & 3 \\ -1 & 5 & 1 \\ -3 & 6 & 2 \end{vmatrix}$

18. $\begin{vmatrix} 2 & 35 & 5 \\ 1 & -5 & 1 \\ -4 & 15 & 2 \end{vmatrix}$

10. $\begin{vmatrix} \dfrac{2}{3} & \dfrac{3}{4} \\ 8 & 6 \end{vmatrix}$

11. $\begin{vmatrix} \dfrac{1}{2} & \dfrac{2}{3} \\ \dfrac{3}{4} & -\dfrac{1}{3} \end{vmatrix}$

12. $\begin{vmatrix} \dfrac{2}{3} & \dfrac{1}{5} \\ -\dfrac{1}{4} & \dfrac{3}{2} \end{vmatrix}$

19. $\begin{vmatrix} 2 & -1 & 3 \\ 0 & 3 & 1 \\ 1 & -2 & -1 \end{vmatrix}$

20. $\begin{vmatrix} 2 & -17 & 3 \\ 0 & 5 & 1 \\ 1 & -3 & -1 \end{vmatrix}$

21. $\begin{vmatrix} -3 & -2 & 1 \\ 5 & 0 & 6 \\ 2 & 1 & -4 \end{vmatrix}$

22. $\begin{vmatrix} -5 & 1 & -1 \\ 3 & 4 & 2 \\ 0 & 2 & -3 \end{vmatrix}$

23. $\begin{vmatrix} 3 & -4 & -2 \\ 5 & -2 & 1 \\ 1 & 0 & 0 \end{vmatrix}$

24. $\begin{vmatrix} -6 & 5 & 3 \\ 2 & 0 & -1 \\ 4 & 0 & 7 \end{vmatrix}$

25. $\begin{vmatrix} 24 & -1 & 4 \\ 40 & 2 & 0 \\ -16 & 6 & 0 \end{vmatrix}$

26. $\begin{vmatrix} 2 & -1 & 3 \\ 0 & 3 & 1 \\ 4 & -8 & -4 \end{vmatrix}$

27. $\begin{vmatrix} 2 & 3 & -4 \\ 4 & 6 & -1 \\ -6 & 1 & -2 \end{vmatrix}$

28. $\begin{vmatrix} 1 & 2 & -3 \\ -3 & -1 & 1 \\ 4 & 5 & 4 \end{vmatrix}$

For Problems 29–32, evaluate each 4×4 determinant. Use the properties of determinants to your advantage.

29. $\begin{vmatrix} 1 & -2 & 3 & 2 \\ 2 & -1 & 0 & 4 \\ -3 & 4 & 0 & -2 \\ -1 & 1 & 1 & 5 \end{vmatrix}$

30. $\begin{vmatrix} 1 & 2 & 5 & 7 \\ -6 & 3 & 0 & 9 \\ -3 & 5 & 2 & 7 \\ 2 & 1 & 4 & 3 \end{vmatrix}$

31. $\begin{vmatrix} 3 & -1 & 2 & 3 \\ 1 & 0 & 2 & 1 \\ 2 & 3 & 0 & 1 \\ 5 & 2 & 4 & -5 \end{vmatrix}$

32. $\begin{vmatrix} 1 & 2 & 0 & 0 \\ 3 & -1 & 4 & 5 \\ -2 & 4 & 1 & 6 \\ 2 & -1 & -2 & -3 \end{vmatrix}$

For Problems 33–42, use the appropriate property of determinants from this section to justify each true statement. *Do not* evaluate the determinants.

33. $(-4)\begin{vmatrix} 2 & 1 & -1 \\ 3 & 2 & 1 \\ 2 & 1 & 3 \end{vmatrix} = \begin{vmatrix} 2 & -4 & -1 \\ 3 & -8 & 1 \\ 2 & -4 & 3 \end{vmatrix}$

34. $\begin{vmatrix} 1 & -2 & 3 \\ 4 & -6 & -8 \\ 0 & 2 & 7 \end{vmatrix} = (-2)\begin{vmatrix} 1 & -2 & 3 \\ -2 & 3 & 4 \\ 0 & 2 & 7 \end{vmatrix}$

35. $\begin{vmatrix} 4 & 7 & 9 \\ 6 & -8 & 2 \\ 4 & 3 & -1 \end{vmatrix} = -\begin{vmatrix} 4 & 9 & 7 \\ 6 & 2 & -8 \\ 4 & -1 & 3 \end{vmatrix}$

36. $\begin{vmatrix} 3 & -1 & 4 \\ 5 & 2 & 7 \\ 3 & -1 & 4 \end{vmatrix} = 0$

37. $\begin{vmatrix} 1 & 3 & 4 \\ -2 & 5 & 7 \\ -3 & -1 & 2 \end{vmatrix} = \begin{vmatrix} 1 & 3 & 4 \\ -2 & 5 & 7 \\ 0 & 8 & 14 \end{vmatrix}$

38. $\begin{vmatrix} 3 & 2 & 0 \\ 1 & 4 & 1 \\ -4 & 9 & 2 \end{vmatrix} = \begin{vmatrix} 3 & 2 & -3 \\ 1 & 4 & 0 \\ -4 & 9 & 6 \end{vmatrix}$

39. $\begin{vmatrix} 6 & 2 & 2 \\ 3 & -1 & 4 \\ 9 & -3 & 6 \end{vmatrix} = 6\begin{vmatrix} 2 & 2 & 1 \\ 1 & -1 & 2 \\ 3 & -3 & 3 \end{vmatrix} = 18\begin{vmatrix} 2 & 2 & 1 \\ 1 & -1 & 2 \\ 1 & -1 & 1 \end{vmatrix}$

40. $\begin{vmatrix} 2 & 1 & -3 \\ 0 & 2 & -4 \\ -5 & 1 & 3 \end{vmatrix} = -\begin{vmatrix} 2 & 1 & -3 \\ -5 & 1 & 3 \\ 0 & 2 & -4 \end{vmatrix}$

41. $\begin{vmatrix} 2 & -3 & 2 \\ 1 & -4 & 1 \\ 7 & 8 & 7 \end{vmatrix} = 0$

42. $\begin{vmatrix} 3 & 1 & 2 \\ -4 & 5 & -1 \\ 2 & -2 & -4 \end{vmatrix} = \begin{vmatrix} 3 & 1 & 0 \\ -4 & 5 & -11 \\ 2 & -2 & 0 \end{vmatrix}$

■■■ **Thoughts into words**

43. Explain the difference between a matrix and a determinant.

44. Explain the concept of a cofactor and how it is used to help expand a determinant.

45. What does it mean to say that any row or column can be used to expand a determinant?

46. Give a step-by-step explanation of how to evaluate the determinant

$\begin{vmatrix} 3 & 0 & 2 \\ 1 & -2 & 5 \\ 6 & 0 & 9 \end{vmatrix}$

■ ■ ■ **Further investigations**

For Problems 47–50, use

$$A = \begin{bmatrix} a_{11} & a_{12} & a_{13} \\ a_{21} & a_{22} & a_{23} \\ a_{31} & a_{32} & a_{33} \end{bmatrix}$$

as a general representation for any 3×3 matrix.

47. Verify Property 15.2 for 3×3 matrices.

48. Verify Property 15.3 for 3×3 matrices.

49. Verify Property 15.4 for 3×3 matrices.

50. Show that $|A| = a_{11}a_{22}a_{33}a_{44}$ if

$$A = \begin{bmatrix} a_{11} & a_{12} & a_{13} & a_{14} \\ 0 & a_{22} & a_{23} & a_{24} \\ 0 & 0 & a_{33} & a_{34} \\ 0 & 0 & 0 & a_{44} \end{bmatrix}$$

 Graphing calculator activities

51. Use a calculator to check your answers for Problems 29–32.

52. Consider the following matrix:

$$A = \begin{bmatrix} 2 & 5 & 7 & 9 \\ -4 & 6 & 2 & 4 \\ 6 & 9 & 12 & 3 \\ 5 & 4 & -2 & 8 \end{bmatrix}$$

Form matrix B by interchanging rows 1 and 3 of matrix A. Now use your calculator to show that $|B| = -|A|$.

53. Consider the following matrix:

$$A = \begin{bmatrix} 2 & 1 & 7 & 6 & 8 \\ 3 & -2 & 4 & 5 & -1 \\ 6 & 7 & 9 & 12 & 13 \\ -4 & -7 & 6 & 2 & 1 \\ 9 & 8 & 12 & 14 & 17 \end{bmatrix}$$

Form matrix B by multiplying each element of the second row of matrix A by 3. Now use your calculator to show that $|B| = 3|A|$.

54. Consider the following matrix:

$$A = \begin{bmatrix} 4 & 3 & 2 & 1 & 5 & -3 \\ 5 & 2 & 7 & 8 & 6 & 3 \\ 0 & 9 & 1 & 4 & 7 & 2 \\ 4 & 3 & 2 & 1 & 5 & -3 \\ -4 & -6 & 7 & 12 & 11 & 9 \\ 5 & 8 & 6 & -3 & 2 & -1 \end{bmatrix}$$

Use your calculator to show that $|A| = 0$.

Answers to Concept Quiz

1. True   **2.** False   **3.** True   **4.** False   **5.** True   **6.** False   **7.** False   **8.** False

15.5   Cramer's Rule

Objectives

■   Set up and evaluate the determinants necessary to use Cramer's rule.

■   Use Cramer's rule to solve a system of equations.

Determinants provide the basis for another method of solving linear systems. Consider the following linear system of two equations and two unknowns.

$$\begin{pmatrix} a_1x + b_1y = c_1 \\ a_2x + b_2y = c_2 \end{pmatrix}$$

The augmented matrix of this system is

$$\left[\begin{array}{cc:c} a_1 & b_1 & c_1 \\ a_2 & b_2 & c_2 \end{array}\right]$$

Using the elementary row transformation of augmented matrices, we can change this matrix to the following reduced echelon form. (The details of this are left for you to do as an exercise.)

$$\left[\begin{array}{cc:c} 1 & 0 & \dfrac{c_1 b_2 - c_2 b_1}{a_1 b_2 - a_2 b_1} \\[2ex] 0 & 1 & \dfrac{a_1 c_2 - a_2 c_1}{a_1 b_2 - a_2 b_1} \end{array}\right], \qquad a_1 b_2 - a_2 b_1 \neq 0$$

The solution for x and y can be expressed in determinant form as follows:

$$x = \frac{c_1 b_2 - c_2 b_1}{a_1 b_2 - a_2 b_1} = \frac{\begin{vmatrix} c_1 & b_1 \\ c_2 & b_2 \end{vmatrix}}{\begin{vmatrix} a_1 & b_1 \\ a_2 & b_2 \end{vmatrix}}$$

$$y = \frac{a_1 c_2 - a_2 c_1}{a_1 b_2 - a_2 b_1} = \frac{\begin{vmatrix} a_1 & c_1 \\ a_2 & c_2 \end{vmatrix}}{\begin{vmatrix} a_1 & b_1 \\ a_2 & b_2 \end{vmatrix}}$$

This method of using determinants to solve a system of two linear equations in two variables is called **Cramer's rule** and can be stated as follows:

Cramer's Rule (2 × 2 case)

Given the system

$$\begin{pmatrix} a_1 x + b_1 y = c_1 \\ a_2 x + b_2 y = c_2 \end{pmatrix}$$

with

$$D = \begin{vmatrix} a_1 & b_1 \\ a_2 & b_2 \end{vmatrix} \neq 0$$

$$D_x = \begin{vmatrix} c_1 & b_1 \\ c_2 & b_2 \end{vmatrix}, \quad \text{and} \quad D_y = \begin{vmatrix} a_1 & c_1 \\ a_2 & c_2 \end{vmatrix}$$

the solution for this system is given by

$$x = \frac{D_x}{D} \quad \text{and} \quad y = \frac{D_y}{D}$$

Note that the elements of D are the coefficients of the variables in the given system. In D_x the coefficients of x are replaced by the corresponding constants, and in D_y

the coefficients of y are replaced by the corresponding constants. Let's illustrate the use of Cramer's rule to solve some systems.

EXAMPLE 1

Solve the system $\begin{pmatrix} 6x + 3y = 2 \\ 3x + 2y = -4 \end{pmatrix}$.

Solution

The system is in the proper form for us to apply Cramer's rule, so let's determine D, D_x, and D_y.

$$D = \begin{vmatrix} 6 & 3 \\ 3 & 2 \end{vmatrix} = 12 - 9 = 3$$

$$D_x = \begin{vmatrix} 2 & 3 \\ -4 & 2 \end{vmatrix} = 4 + 12 = 16$$

$$D_y = \begin{vmatrix} 6 & 2 \\ 3 & -4 \end{vmatrix} = -24 - 6 = -30$$

Therefore,

$$x = \frac{D_x}{D} = \frac{16}{3}$$

and

$$y = \frac{D_y}{D} = \frac{-30}{3} = -10$$

The solution set is $\left\{ \left(\frac{16}{3}, -10 \right) \right\}$. ∎

EXAMPLE 2

Solve the system $\begin{pmatrix} y = -2x - 2 \\ 4x - 5y = 17 \end{pmatrix}$.

Solution

To begin, we must change the form of the first equation so that the system fits the form given in Cramer's rule. The equation $y = -2x - 2$ can be rewritten $2x + y = -2$. The system now becomes

$$\begin{pmatrix} 2x + y = -2 \\ 4x - 5y = 17 \end{pmatrix}$$

and we can proceed to determine D, D_x, and D_y.

$$D = \begin{vmatrix} 2 & 1 \\ 4 & -5 \end{vmatrix} = -10 - 4 = -14$$

$$D_x = \begin{vmatrix} -2 & 1 \\ 17 & -5 \end{vmatrix} = 10 - 17 = -7$$

$$D_y = \begin{vmatrix} 2 & -2 \\ 4 & 17 \end{vmatrix} = 34 - (-8) = 42$$

Thus,

$$x = \frac{D_x}{D} = \frac{-7}{-14} = \frac{1}{2} \quad \text{and} \quad y = \frac{D_y}{D} = \frac{42}{-14} = -3$$

The solution set is $\left\{ \left(\frac{1}{2}, -3 \right) \right\}$, which can be verified, as always, by substituting back into the original equations. ■

EXAMPLE 3   Solve the system

$$\begin{pmatrix} \dfrac{1}{2}x + \dfrac{2}{3}y = -4 \\ \dfrac{1}{4}x - \dfrac{3}{2}y = 20 \end{pmatrix}$$

Solution

With such a system, either we can first produce an equivalent system with integral coefficients and then apply Cramer's rule, or we can apply the rule immediately. Let's avoid some work with fractions by multiplying the first equation by 6 and the second equation by 4 to produce the following equivalent system.

$$\begin{pmatrix} 3x + 4y = -24 \\ x - 6y = 80 \end{pmatrix}$$

Now we can proceed as before.

$$D = \begin{vmatrix} 3 & 4 \\ 1 & -6 \end{vmatrix} = -18 - 4 = -22$$

$$D_x = \begin{vmatrix} -24 & 4 \\ 80 & -6 \end{vmatrix} = 144 - 320 = -176$$

$$D_y = \begin{vmatrix} 3 & -24 \\ 1 & 80 \end{vmatrix} = 240 - (-24) = 264$$

Therefore,

$$x = \frac{D_x}{D} = \frac{-176}{-22} = 8 \quad \text{and} \quad y = \frac{D_y}{D} = \frac{264}{-22} = -12$$

The solution set is $\{(8, -12)\}$. ■

In the statement of Cramer's rule, the condition that $D \neq 0$ was imposed. If $D = 0$ and either D_x or D_y (or both) is nonzero, then the system is inconsistent and has no solution. If $D = 0$, $D_x = 0$, and $D_y = 0$, then the equations are dependent and there are infinitely many solutions.

Cramer's Rule Extended

Without showing the details, we will simply state that Cramer's rule also applies to solving systems of three linear equations in three variables. It can be stated as follows:

Cramer's Rule (3×3 case)

Given the system

$$\begin{pmatrix} a_1x + b_1y + c_1z = d_1 \\ a_2x + b_2y + c_2z = d_2 \\ a_3x + b_3y + c_3z = d_3 \end{pmatrix}$$

with

$$D = \begin{vmatrix} a_1 & b_1 & c_1 \\ a_2 & b_2 & c_2 \\ a_3 & b_3 & c_3 \end{vmatrix} \neq 0 \qquad D_x = \begin{vmatrix} d_1 & b_1 & c_1 \\ d_2 & b_2 & c_2 \\ d_3 & b_3 & c_3 \end{vmatrix}$$

$$D_y = \begin{vmatrix} a_1 & d_1 & c_1 \\ a_2 & d_2 & c_2 \\ a_3 & d_3 & c_3 \end{vmatrix} \qquad D_z = \begin{vmatrix} a_1 & b_1 & d_1 \\ a_2 & b_2 & d_2 \\ a_3 & b_3 & d_3 \end{vmatrix}$$

we have

$$x = \frac{D_x}{D}, \qquad y = \frac{D_y}{D}, \qquad \text{and} \qquad z = \frac{D_z}{D}$$

Again, note the restriction that $D \neq 0$. If $D = 0$ and at least one of D_x, D_y, and D_z is not zero, then the system is inconsistent. If D, D_x, D_y, and D_z are all zero, then the equations are dependent, and there are infinitely many solutions.

EXAMPLE 4

Solve the system

$$\begin{pmatrix} x - 2y + z = -4 \\ 2x + y - z = 5 \\ 3x + 2y + 4z = 3 \end{pmatrix}$$

Solution

We will simply indicate the values of D, D_x, D_y, and D_z and leave the computations for you to check.

$$D = \begin{vmatrix} 1 & -2 & 1 \\ 2 & 1 & -1 \\ 3 & 2 & 4 \end{vmatrix} = 29 \qquad D_x = \begin{vmatrix} -4 & -2 & 1 \\ 5 & 1 & -1 \\ 3 & 2 & 4 \end{vmatrix} = 29$$

$$D_y = \begin{vmatrix} 1 & -4 & 1 \\ 2 & 5 & -1 \\ 3 & 3 & 4 \end{vmatrix} = 58 \qquad D_z = \begin{vmatrix} 1 & -2 & -4 \\ 2 & 1 & 5 \\ 3 & 2 & 3 \end{vmatrix} = -29$$

Therefore,

$$x = \frac{D_x}{D} = \frac{29}{29} = 1$$

$$y = \frac{D_y}{D} = \frac{58}{29} = 2$$

and

$$z = \frac{D_z}{D} = \frac{-29}{29} = -1$$

The solution set is $\{(1, 2, -1)\}$. (Be sure to check it!) ■

E X A M P L E   5   Solve the system

$$\begin{pmatrix} x + 3y - z = 4 \\ 3x - 2y + z = 7 \\ 2x + 6y - 2z = 1 \end{pmatrix}$$

Solution

$$D = \begin{vmatrix} 1 & 3 & -1 \\ 3 & -2 & 1 \\ 2 & 6 & -2 \end{vmatrix} = 2\begin{vmatrix} 1 & 3 & -1 \\ 3 & -2 & 1 \\ 1 & 3 & -1 \end{vmatrix} = 2(0) = 0$$

$$D_x = \begin{vmatrix} 4 & 3 & -1 \\ 7 & -2 & 1 \\ 1 & 6 & -2 \end{vmatrix} = -7$$

Therefore, because $D = 0$ and at least one of D_x, D_y, and D_z is not zero, the system is inconsistent. The solution set is $\varnothing$. ■

Example 5 illustrates why D should be determined first. Once we found that $D = 0$ and $D_x \neq 0$, we knew that the system was inconsistent, and there was no need to find D_y and D_z.

Finally, it should be noted that Cramer's rule can be extended to systems of n linear equations in n variables; however, that method is not considered to be a very efficient way of solving a large system of linear equations.

CONCEPT QUIZ

For Problems 1–5, answer true or false.

1. Cramer's rule is a method of solving a system of equations by using matrices.

2. If $D = 0$, then the system of equations is either inconsistent or the equations are dependent.

3. If $D \neq 0$, then the system of equations has either one solution or infinitely many solutions.

4. If $D = 0$ and $D_z = 4$ then the system of equations is inconsistent.

5. Cramer's rule can be extended to systems of n linear equations in n variables.

PROBLEM SET 15.5

For Problems 1–32, use Cramer's rule to find the solution set for each system. If the equations are dependent, simply indicate that there are infinitely many solutions.

1. $\begin{pmatrix} 2x - y = -2 \\ 3x + 2y = 11 \end{pmatrix}$

2. $\begin{pmatrix} 3x + y = -9 \\ 4x - 3y = 1 \end{pmatrix}$

3. $\begin{pmatrix} 5x + 2y = 5 \\ 3x - 4y = 29 \end{pmatrix}$

4. $\begin{pmatrix} 4x - 7y = -23 \\ 2x + 5y = -3 \end{pmatrix}$

5. $\begin{pmatrix} 5x - 4y = 14 \\ -x + 2y = -4 \end{pmatrix}$

6. $\begin{pmatrix} -x + 2y = 10 \\ 3x - y = -10 \end{pmatrix}$

7. $\begin{pmatrix} y = 2x - 4 \\ 6x - 3y = 1 \end{pmatrix}$

8. $\begin{pmatrix} -3x - 4y = 14 \\ -2x + 3y = -19 \end{pmatrix}$

9. $\begin{pmatrix} -4x + 3y = 3 \\ 4x - 6y = -5 \end{pmatrix}$

10. $\begin{pmatrix} x = 4y - 1 \\ 2x - 8y = -2 \end{pmatrix}$

11. $\begin{pmatrix} 9x - y = -2 \\ 8x + y = 4 \end{pmatrix}$

12. $\begin{pmatrix} 6x - 5y = 1 \\ 4x - 7y = 2 \end{pmatrix}$

13. $\begin{pmatrix} -\dfrac{2}{3}x + \dfrac{1}{2}y = -7 \\ \dfrac{1}{3}x - \dfrac{3}{2}y = 6 \end{pmatrix}$

14. $\begin{pmatrix} \dfrac{1}{2}x + \dfrac{2}{3}y = -6 \\ \dfrac{1}{4}x - \dfrac{1}{3}y = -1 \end{pmatrix}$

15. $\begin{pmatrix} 2x + 7y = -1 \\ x = 2 \end{pmatrix}$

16. $\begin{pmatrix} 5x - 3y = 2 \\ y = 4 \end{pmatrix}$

17. $\begin{pmatrix} x - y + 2z = -8 \\ 2x + 3y - 4z = 18 \\ -x + 2y - z = 7 \end{pmatrix}$

18. $\begin{pmatrix} x - 2y + z = 3 \\ 3x + 2y + z = -3 \\ 2x - 3y - 3z = -5 \end{pmatrix}$

19. $\begin{pmatrix} 2x - 3y + z = -7 \\ -3x + y - z = -7 \\ x - 2y - 5z = -45 \end{pmatrix}$

20. $\begin{pmatrix} 3x - y - z = 18 \\ 4x + 3y - 2z = 10 \\ -5x - 2y + 3z = -22 \end{pmatrix}$

21. $\begin{pmatrix} 4x + 5y - 2z = -14 \\ 7x - y + 2z = 42 \\ 3x + y + 4z = 28 \end{pmatrix}$

22. $\begin{pmatrix} -5x + 6y + 4z = -4 \\ -7x - 8y + 2z = -2 \\ 2x + 9y - z = 1 \end{pmatrix}$

23. $\begin{pmatrix} 2x - y + 3z = -17 \\ 3y + z = 5 \\ x - 2y - z = -3 \end{pmatrix}$

24. $\begin{pmatrix} 2x - y + 3z = -5 \\ 3x + 4y - 2z = -25 \\ -x \quad\quad + z = 6 \end{pmatrix}$

25. $\begin{pmatrix} x + 3y - 4z = -1 \\ 2x - y + z = 2 \\ 4x + 5y - 7z = 0 \end{pmatrix}$

26. $\begin{pmatrix} x - 2y + z = 1 \\ 3x + y - z = 2 \\ 2x - 4y + 2z = -1 \end{pmatrix}$

27. $\begin{pmatrix} 3x - 2y - 3z = -5 \\ x + 2y + 3z = -3 \\ -x + 4y - 6z = 8 \end{pmatrix}$

28. $\begin{pmatrix} 3x - 2y + z = 11 \\ 5x + 3y \quad\quad = 17 \\ x + y - 2z = 6 \end{pmatrix}$

29. $\begin{pmatrix} x - 2y + 3z = 1 \\ -2x + 4y - 3z = -3 \\ 5x - 6y + 6z = 10 \end{pmatrix}$

30. $\begin{pmatrix} 2x - y + 2z = -1 \\ 4x + 3y - 4z = 2 \\ x + 5y - z = 9 \end{pmatrix}$

31. $\begin{pmatrix} -x - y + 3z = -2 \\ -2x + y + 7z = 14 \\ 3x + 4y - 5z = 12 \end{pmatrix}$

32. $\begin{pmatrix} -2x + y - 3z = -4 \\ x + 5y - 4z = 13 \\ 7x - 2y - z = 37 \end{pmatrix}$

■ ■ ■ Thoughts into words

33. Give a step-by-step description of how you would solve the system

$$\begin{pmatrix} 2x - y + 3z = 31 \\ x - 2y - z = 8 \\ 3x + 5y + 8z = 35 \end{pmatrix}$$

34. Give a step-by-step description of how you would find the value of x in the solution for the system

$$\begin{pmatrix} x + 5y - z = -9 \\ 2x - y + z = 11 \\ -3x - 2y + 4z = 20 \end{pmatrix}$$

■ ■ ■ Further investigations

35. A linear system in which the constant terms are all zero is called a **homogeneous system.**

   a. Verify that for a 3×3 homogeneous system, if $D \neq 0$, then $(0, 0, 0)$ is the only solution for the system.

   b. Verify that for a 3×3 homogeneous system, if $D = 0$, then the equations are dependent.

For Problems 36–39, solve each of the homogeneous systems (see Problem 35). If the equations are dependent, indicate that the system has infinitely many solutions.

36. $\begin{pmatrix} x - 2y + 5z = 0 \\ 3x + y - 2z = 0 \\ 4x - y + 3z = 0 \end{pmatrix}$ **37.** $\begin{pmatrix} 2x - y + z = 0 \\ 3x + 2y + 5z = 0 \\ 4x - 7y + z = 0 \end{pmatrix}$

38. $\begin{pmatrix} 3x + y - z = 0 \\ x - y + 2z = 0 \\ 4x - 5y - 2z = 0 \end{pmatrix}$ **39.** $\begin{pmatrix} 2x - y + 2z = 0 \\ x + 2y + z = 0 \\ x - 3y + z = 0 \end{pmatrix}$

 ### Graphing calculator activities

40. Use determinants and your calculator to solve each of the following systems.

   a. $\begin{pmatrix} 4x - 3y + z = 10 \\ 8x + 5y - 2z = -6 \\ -12x - 2y + 3z = -2 \end{pmatrix}$

   b. $\begin{pmatrix} 2x + y - z + w = -4 \\ x + 2y + 2z - 3w = 6 \\ 3x - y - z + 2w = 0 \\ 2x + 3y + z + 4w = -5 \end{pmatrix}$

   c. $\begin{pmatrix} x - 2y + z - 3w = 4 \\ 2x + 3y - z - 2w = -4 \\ 3x - 4y + 2z - 4w = 12 \\ 2x - y - 3z + 2w = -2 \end{pmatrix}$

   d. $\begin{pmatrix} 1.98x + 2.49y + 3.45z = 80.10 \\ 2.15x + 3.20y + 4.19z = 97.16 \\ 1.49x + 4.49y + 2.79z = 83.92 \end{pmatrix}$

Answers to Concept Quiz

1. False   **2.** True   **3.** False   **4.** True   **5.** True

15.6   Systems Involving Nonlinear Equations

Objectives

■   Graph systems of nonlinear equations.

■   Solve systems of nonlinear equations.

In Section 15.1 we reviewed the use of the substitution method and the elimination-by-addition method to solve a system of two linear equations. We will use both of those techniques in this section to solve systems that contain at least one nonlinear equation. Furthermore, we will use our knowledge of graphing lines, circles, parabolas, ellipses, and hyperbolas to get a visual read on the systems. This will give us a basis for predicting approximate real number solutions if there are any. In other words, we have once again arrived at a topic that vividly illustrates the merging of mathematical ideas. Let's begin by considering a system that contains one linear and one nonlinear equation.

E X A M P L E   1

Solve the system $\left(\begin{array}{c} x^2 + y^2 = 13 \\ 3x + 2y = 0 \end{array} \right)$.

Solution

From our previous graphing experiences, we should recognize that $x^2 + y^2 = 13$ is a circle and $3x + 2y = 0$ is a straight line. Thus the system can be pictured as in Figure 15.8. The graph indicates that the solution set of this system should consist of two ordered pairs of real numbers, which represent the points of intersection in the second and fourth quadrants.

Now let's solve the system analytically by using the *substitution method.* Change the form of $3x + 2y = 0$ to $y = -3x/2$, and then substitute $-3x/2$ for y in the other equation to produce

$$x^2 + \left(-\frac{3x}{2}\right)^2 = 13$$

This equation can now be solved for x.

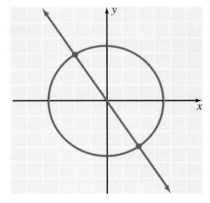

Figure 15.8

$$x^2 + \frac{9x^2}{4} = 13$$

$$4x^2 + 9x^2 = 52$$

$$13x^2 = 52$$

$$x^2 = 4$$

$$x = \pm 2$$

Substitute 2 for x and then -2 for x in the second equation of the system to produce two values for y.

$$3x + 2y = 0 \qquad\qquad 3x + 2y = 0$$
$$3(2) + 2y = 0 \qquad\quad 3(-2) + 2y = 0$$
$$2y = -6 \qquad\qquad\quad 2y = 6$$
$$y = -3 \qquad\qquad\quad\; y = 3$$

Therefore, the solution set of the system is $\{(2, -3), (-2, 3)\}$. ■

REMARK: Don't forget that, as always, you can check the solutions by substituting them back into the original equations. Graphing the system permits you to approximate any possible real number solutions before solving the system. Then, after solving the system, you can use the graph again to check that the answers are reasonable.

E X A M P L E   2

Solve the system $\left(\begin{array}{l} x^2 + y^2 = 16 \\ y^2 - x^2 = 4 \end{array} \right)$.

Solution

Graphing the system produces Figure 15.9. This figure indicates that there should be four ordered pairs of real numbers in the solution set of the system. Solving the system by using the *elimination method* works nicely. We can simply add the two equations, which eliminates the x's.

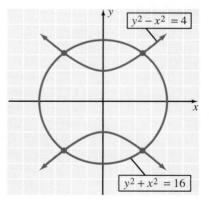

$y^2 - x^2 = 4$

$y^2 + x^2 = 16$

Figure 15.9

$$x^2 + y^2 = 16$$
$$\underline{-x^2 + y^2 = 4}$$
$$2y^2 = 20$$
$$y^2 = 10$$
$$y = \pm\sqrt{10}$$

Substituting $\sqrt{10}$ for y in the first equation yields

$$x^2 + y^2 = 16$$
$$x^2 + (\sqrt{10})^2 = 16$$
$$x^2 + 10 = 16$$
$$x^2 = 6$$
$$x = \pm\sqrt{6}$$

Thus $(\sqrt{6}, \sqrt{10})$ and $(-\sqrt{6}, \sqrt{10})$ are solutions. Substituting $-\sqrt{10}$ for y in the first equation yields

$$x^2 + y^2 = 16$$
$$x^2 + (-\sqrt{10})^2 = 16$$

$$x^2 + 10 = 16$$
$$x^2 = 6$$
$$x = \pm\sqrt{6}$$

Thus $(\sqrt{6}, -\sqrt{10})$ and $(-\sqrt{6}, -\sqrt{10})$ are also solutions. The solution set is $\{(-\sqrt{6}, \sqrt{10}), (-\sqrt{6}, -\sqrt{10}), (\sqrt{6}, \sqrt{10}), (\sqrt{6}, -\sqrt{10})\}$. ∎

Sometimes a sketch of the graph of a system may not clearly indicate whether the system contains any real number solutions. The next example illustrates such a situation.

E X A M P L E   3   Solve the system $\left(\begin{array}{c} y = x^2 + 2 \\ 6x - 4y = -5 \end{array} \right)$.

Solution

From our previous graphing experience, we recognize that $y = x^2 + 2$ is the basic parabola shifted upward 2 units, and that $6x - 4y = -5$ is a straight line (see Figure 15.10). Because of the close proximity of the curves, it is difficult to tell whether they intersect. In other words, the graph does not definitely indicate any real number solutions for the system.

Let's solve the system by using the substitution method. We can substitute $x^2 + 2$ for y in the second equation, which produces two values for x.

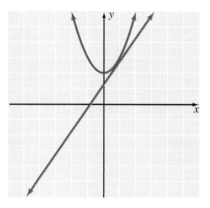

Figure 15.10

$$6x - 4(x^2 + 2) = -5$$
$$6x - 4x^2 - 8 = -5$$
$$-4x^2 + 6x - 3 = 0$$
$$4x^2 - 6x + 3 = 0$$
$$x = \frac{6 \pm \sqrt{36 - 48}}{8}$$
$$= \frac{6 \pm \sqrt{-12}}{8}$$
$$= \frac{6 \pm 2i\sqrt{3}}{8}$$
$$= \frac{3 \pm i\sqrt{3}}{4}$$

It is now obvious that the system has no real number solutions. That is, the line and the parabola do not intersect in the real number plane. However, there will be two pairs of complex numbers in the solution set. We can substitute $(3 + i\sqrt{3})/4$ for x in the first equation.

$$y = \left(\frac{3 + i\sqrt{3}}{4}\right)^2 + 2$$

$$= \frac{6 + 6i\sqrt{3}}{16} + 2$$

$$= \frac{6 + 6i\sqrt{3} + 32}{16}$$

$$= \frac{38 + 6i\sqrt{3}}{16}$$

$$= \frac{19 + 3i\sqrt{3}}{8}$$

Likewise, we can substitute $(3 - i\sqrt{3})/4$ for x in the first equation.

$$y = \left(\frac{3 - i\sqrt{3}}{4}\right)^2 + 2$$

$$= \frac{6 - 6i\sqrt{3}}{16} + 2$$

$$= \frac{6 - 6i\sqrt{3} + 32}{16}$$

$$= \frac{38 - 6i\sqrt{3}}{16}$$

$$= \frac{19 - 3i\sqrt{3}}{8}$$

The solution set is $\left\{\left(\dfrac{3 + i\sqrt{3}}{4}, \dfrac{19 + 3i\sqrt{3}}{8}\right), \left(\dfrac{3 - i\sqrt{3}}{4}, \dfrac{19 - 3i\sqrt{3}}{8}\right)\right\}$. ■

In Example 3, the use of a graphing utility may not, at first, indicate whether the system has any real number solutions. Suppose that we graph the system using a viewing rectangle such that $-15 \le x \le 15$ and $-10 \le y \le 10$. As shown in the display in Figure 15.11, we cannot tell whether the line and the parabola intersect. However, if we change the viewing rectangle so that

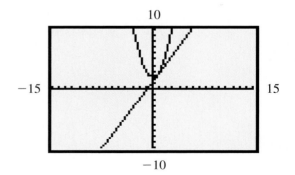

Figure 15.11

$0 \le x \le 2$ and $0 \le y \le 4$, as shown in Figure 15.12, it becomes apparent that the two graphs do not intersect.

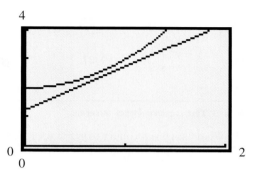

Figure 15.12

CONCEPT QUIZ

For Problems 1–5, answer true or false.

1. Graphing a system of equations is a method of approximating the solutions.

2. Every system of nonlinear equations has a real number solution.

3. Every nonlinear system of equations can be solved by substitution.

4. Every nonlinear system of equations can be solved by the elimination method.

5. Graphs of a circle and a line will have one, two, or no points of intersection.

PROBLEM SET 15.6

For Problems 1–30, (a) graph the system so that approximate real number solutions (if there are any) can be predicted, and (b) solve the system by the substitution or elimination method.

1. $\begin{pmatrix} x^2 + y^2 = 5 \\ x + 2y = 5 \end{pmatrix}$

2. $\begin{pmatrix} x^2 + y^2 = 13 \\ 2x + 3y = 13 \end{pmatrix}$

3. $\begin{pmatrix} x^2 + y^2 = 26 \\ x + y = -4 \end{pmatrix}$

4. $\begin{pmatrix} x^2 + y^2 = 10 \\ x + y = -2 \end{pmatrix}$

5. $\begin{pmatrix} x^2 + y^2 = 2 \\ x - y = 4 \end{pmatrix}$

6. $\begin{pmatrix} x^2 + y^2 = 3 \\ x - y = -5 \end{pmatrix}$

7. $\begin{pmatrix} y = x^2 + 6x + 7 \\ 2x + y = -5 \end{pmatrix}$

8. $\begin{pmatrix} y = x^2 - 4x + 5 \\ y - x = 1 \end{pmatrix}$

9. $\begin{pmatrix} 2x + y = -2 \\ y = x^2 + 4x + 7 \end{pmatrix}$

10. $\begin{pmatrix} 2x + y = 0 \\ y = -x^2 + 2x - 4 \end{pmatrix}$

11. $\begin{pmatrix} y = x^2 - 3 \\ x + y = -4 \end{pmatrix}$

12. $\begin{pmatrix} y = -x^2 + 1 \\ x + y = 2 \end{pmatrix}$

13. $\begin{pmatrix} x^2 + 2y^2 = 9 \\ x - 4y = -9 \end{pmatrix}$

14. $\begin{pmatrix} 2x - y = 7 \\ 3x^2 + y^2 = 21 \end{pmatrix}$

15. $\begin{pmatrix} x + y = -3 \\ x^2 + 2y^2 - 12y - 18 = 0 \end{pmatrix}$

16. $\begin{pmatrix} 4x^2 + 9y^2 = 25 \\ 2x + 3y = 7 \end{pmatrix}$

17. $\begin{pmatrix} x - y = 2 \\ x^2 - y^2 = 16 \end{pmatrix}$

18. $\begin{pmatrix} x^2 - 4y^2 = 16 \\ 2y - x = 2 \end{pmatrix}$

19. $\begin{pmatrix} y = -x^2 + 3 \\ y = x^2 + 1 \end{pmatrix}$

20. $\begin{pmatrix} y = x^2 \\ y = x^2 - 4x + 4 \end{pmatrix}$

21. $\begin{pmatrix} y = x^2 + 2x - 1 \\ y = x^2 + 4x + 5 \end{pmatrix}$

22. $\begin{pmatrix} y = -x^2 + 1 \\ y = x^2 - 2 \end{pmatrix}$

23. $\begin{pmatrix} x^2 - y^2 = 4 \\ x^2 + y^2 = 4 \end{pmatrix}$

24. $\left(\begin{array}{l} 2x^2 + y^2 = 8 \\ x^2 + y^2 = 4 \end{array} \right)$ **25.** $\left(\begin{array}{l} 8y^2 - 9x^2 = 6 \\ 8x^2 - 3y^2 = 7 \end{array} \right)$ **28.** $\left(\begin{array}{l} 4x^2 + 3y^2 = 9 \\ y^2 - 4x^2 = 7 \end{array} \right)$ **29.** $\left(\begin{array}{l} xy = 3 \\ 2x + 2y = 7 \end{array} \right)$

26. $\left(\begin{array}{l} 2x^2 + y^2 = 11 \\ x^2 - y^2 = 4 \end{array} \right)$ **27.** $\left(\begin{array}{l} 2x^2 - 3y^2 = -1 \\ 2x^2 + 3y^2 = 5 \end{array} \right)$ **30.** $\left(\begin{array}{l} x^2 + 4y^2 = 25 \\ xy = 6 \end{array} \right)$

■ ■ ■ Thoughts into words

31. What happens if you try to graph the following system?

$$\left(\begin{array}{l} 7x^2 + 8y^2 = 36 \\ 11x^2 + 5y^2 = -4 \end{array} \right)$$

32. For what value(s) of k will the line $x + y = k$ touch the ellipse $x^2 + 2y^2 = 6$ in one and only one point? Defend your answer.

33. The system

$$\left(\begin{array}{l} x^2 - 6x + y^2 - 4y + 4 = 0 \\ x^2 - 4x + y^2 + 8y - 5 = 0 \end{array} \right)$$

represents two circles that intersect in two points. An equivalent system can be formed by replacing the second equation with the result of adding -1 times the first equation to the second equation. Thus we obtain the system

$$\left(\begin{array}{l} x^2 - 6x + y^2 - 4y + 4 = 0 \\ 2x + 12y - 9 = 0 \end{array} \right)$$

Explain why the linear equation in this system is the equation of the common chord of the original two intersecting circles.

 ### Graphing calculator activities

34. Graph the system of equations $\left(\begin{array}{l} y = x^2 + 2 \\ 6x - 4y = -5 \end{array} \right)$, and use the TRACE and ZOOM features of your calculator to show that this system has no real number solutions.

For Problems 35–40, use a graphing calculator to approximate, to the nearest tenth, the real number solutions for each system of equations.

35. $\left(\begin{array}{l} y = e^x + 1 \\ y = x^3 + x^2 - 2x - 1 \end{array} \right)$

36. $\left(\begin{array}{l} y = x^3 + 2x^2 - 3x + 2 \\ y = -x^3 - x^2 + 1 \end{array} \right)$

37. $\left(\begin{array}{l} y = 2^x + 1 \\ y = 2^{-x} + 2 \end{array} \right)$ **38.** $\left(\begin{array}{l} y = \ln(x - 1) \\ y = x^2 - 16x + 64 \end{array} \right)$

39. $\left(\begin{array}{l} x = y^2 - 2y + 3 \\ x^2 + y^2 = 25 \end{array} \right)$ **40.** $\left(\begin{array}{l} y^2 - x^2 = 16 \\ 2y^2 - x^2 = 8 \end{array} \right)$

Answers to Concept Quiz

1. True   **2.** False   **3.** True   **4.** False   **5.** True

CHAPTER 15

SUMMARY

(15.1 and 15.2) The primary focus of this entire chapter is the development of different techniques for solving systems of linear equations.

Substitution Method

With the aid of an example, we can describe the substitution method as follows. Suppose we want to solve the system

$$\begin{pmatrix} x - 2y = 22 \\ 3x + 4y = -24 \end{pmatrix}$$

STEP 1 Solve the first equation for x in terms of y.

$$x - 2y = 22$$
$$x = 2y + 22$$

STEP 2 Substitute $2y + 22$ for x in the second equation.

$$3(2y + 22) + 4y = -24$$

STEP 3 Solve the equation obtained in step 2.

$$6y + 66 + 4y = -24$$
$$10y + 66 = -24$$
$$10y = -90$$
$$y = -9$$

STEP 4 Substitute -9 for y in the equation of step 1.

$$x = 2(-9) + 22 = 4$$

The solution set is $\{(4, -9)\}$.

Elimination-by-Addition Method

This method enables us to replace systems of equations with *simpler equivalent systems* until we obtain a system for which we can easily determine the solution. The following operations produce equivalent systems:

1. Any two equations of a system can be interchanged.
2. Both sides of any equation of the system can be multiplied by any nonzero real number.

3. Any equation of the system can be replaced by the sum of a nonzero multiple of another equation plus that equation.

For example, through a sequence of operations, we can transform the system

$$\begin{pmatrix} 5x + 3y = -28 \\ \frac{1}{2}x - y = -8 \end{pmatrix}$$

to the equivalent system

$$\begin{pmatrix} x - 2y = -16 \\ 13y = 52 \end{pmatrix}$$

for which we can easily determine the solution set $\{(-8, 4)\}$.

Matrix Approach

(15.3) We can change the augmented matrix of a system to reduced echelon form by applying the following elementary row operations.

1. Any two rows of the matrix can be interchanged.
2. Any row of the matrix can be multiplied by a nonzero real number.
3. Any row of the matrix can be replaced by the sum of a nonzero multiple of another row plus that row.

For example, the augmented matrix of the system

$$\begin{pmatrix} x - 2y + 3z = 4 \\ 2x + y - 4z = 3 \\ -3x + 4y - z = -2 \end{pmatrix}$$

is

$$\begin{bmatrix} 1 & -2 & 3 & \vdots & 4 \\ 2 & 1 & -4 & \vdots & 3 \\ -3 & 4 & -1 & \vdots & -2 \end{bmatrix}$$

We can change this matrix to the reduced echelon form

$$\begin{bmatrix} 1 & 0 & 0 & \vdots & 4 \\ 0 & 1 & 0 & \vdots & 3 \\ 0 & 0 & 1 & \vdots & 2 \end{bmatrix}$$

where the solution set $\{(4, 3, 2)\}$ is obvious.

(15.4) A rectangular array of numbers is called a **matrix**. A **square matrix** has the same number of rows as columns. For a 2×2 matrix

$$\begin{bmatrix} a_1 & b_1 \\ a_2 & b_2 \end{bmatrix}$$

the **determinant** of the matrix is written as

$$\begin{vmatrix} a_1 & b_1 \\ a_2 & b_2 \end{vmatrix}$$

and is defined by

$$\begin{vmatrix} a_1 & b_1 \\ a_2 & b_2 \end{vmatrix} = a_1 b_2 - a_2 b_1$$

The determinant of a 3×3 (or larger) square matrix can be evaluated by expansion of minors of the elements of any row or any column. The concepts of minor and cofactor are needed for this purpose; these terms are defined in Definitions 15.2 and 15.3.

The following properties are helpful when we are evaluating determinants:

1. If any row (or column) of a square matrix A contains only zeros, then $|A| = 0$.
2. If square matrix B is obtained from square matrix A by interchanging two rows (or two columns), then $|B| = -|A|$.
3. If square matrix B is obtained from square matrix A by multiplying each element of any row (or column) of A by some real number k, then $|B| = k|A|$.
4. If square matrix B is obtained from square matrix A by adding k times a row (or column) of A to another row (or column) of A, then $|B| = |A|$.
5. If two rows (or columns) of a square matrix A are identical, then $|A| = 0$.

(15.5) Cramer's rule for solving a system of two linear equations in two variables is stated as follows: Given the system

$$\begin{pmatrix} a_1 x + b_1 y = c_1 \\ a_2 x + b_2 y = c_2 \end{pmatrix}$$

with

$$D = \begin{vmatrix} a_1 & b_1 \\ a_2 & b_2 \end{vmatrix} \neq 0$$

$$D_x = \begin{vmatrix} c_1 & b_1 \\ c_2 & b_2 \end{vmatrix} \qquad D_y = \begin{vmatrix} a_1 & c_1 \\ a_2 & c_2 \end{vmatrix}$$

then

$$x = \frac{D_x}{D} \qquad \text{and} \qquad y = \frac{D_y}{D}$$

Cramer's rule for solving a system of three linear equations in three variables is stated as follows: Given the system

$$\begin{pmatrix} a_1 x + b_1 y + c_1 z = d_1 \\ a_2 x + b_2 y + c_2 z = d_2 \\ a_3 x + b_3 y + c_3 z = d_3 \end{pmatrix}$$

with

$$D = \begin{vmatrix} a_1 & b_1 & c_1 \\ a_2 & b_2 & c_2 \\ a_3 & b_3 & c_3 \end{vmatrix} \neq 0 \qquad D_x = \begin{vmatrix} d_1 & b_1 & c_1 \\ d_2 & b_2 & c_2 \\ d_3 & b_3 & c_3 \end{vmatrix}$$

$$D_y = \begin{vmatrix} a_1 & d_1 & c_1 \\ a_2 & d_2 & c_2 \\ a_3 & d_3 & c_3 \end{vmatrix} \qquad D_z = \begin{vmatrix} a_1 & b_1 & d_1 \\ a_2 & b_2 & d_2 \\ a_3 & b_3 & d_3 \end{vmatrix}$$

then

$$x = \frac{D_x}{D}, \qquad y = \frac{D_y}{D}, \qquad \text{and} \qquad z = \frac{D_z}{D}$$

(15.6) Systems that contain at least one nonlinear equation can often be solved by substitution or by the elimination method. Graphing the system will often provide a basis for predicting approximate real number solutions if there are any.

CHAPTER 15 REVIEW PROBLEM SET

For Problems 1–4, solve each system by using the *substitution* method.

1. $\begin{pmatrix} 3x - y = 16 \\ 5x + 7y = -34 \end{pmatrix}$ **2.** $\begin{pmatrix} 6x + 5y = -21 \\ x - 4y = 11 \end{pmatrix}$

3. $\begin{pmatrix} 2x - 3y = 12 \\ 3x + 5y = -20 \end{pmatrix}$ **4.** $\begin{pmatrix} 5x + 8y = 1 \\ 4x + 7y = -2 \end{pmatrix}$

For Problems 5–8, solve each system by using the *elimination-by-addition* method.

5. $\begin{pmatrix} 4x - 3y = 34 \\ 3x + 2y = 0 \end{pmatrix}$ **6.** $\begin{pmatrix} \frac{1}{2}x - \frac{2}{3}y = 1 \\ \frac{3}{4}x + \frac{1}{6}y = -1 \end{pmatrix}$

7. $\begin{pmatrix} 2x - y + 3z = -19 \\ 3x + 2y - 4z = 21 \\ 5x - 4y - z = -8 \end{pmatrix}$ **8.** $\begin{pmatrix} 3x + 2y - 4z = 4 \\ 5x + 3y - z = 2 \\ 4x - 2y + 3z = 11 \end{pmatrix}$

For Problems 9–12, solve each system by *changing the augmented matrix to reduced echelon form.*

9. $\begin{pmatrix} x - 3y = 17 \\ -3x + 2y = -23 \end{pmatrix}$ **10.** $\begin{pmatrix} 2x + 3y = 25 \\ 3x - 5y = -29 \end{pmatrix}$

11. $\begin{pmatrix} x - 2y + z = -7 \\ 2x - 3y + 4z = -14 \\ -3x + y - 2z = 10 \end{pmatrix}$ **12.** $\begin{pmatrix} -2x - 7y + z = 9 \\ x + 3y - 4z = -11 \\ 4x + 5y - 3z = -11 \end{pmatrix}$

For Problems 13–16, solve each system by using *Cramer's rule.*

13. $\begin{pmatrix} 5x + 3y = -18 \\ 4x - 9y = -3 \end{pmatrix}$ **14.** $\begin{pmatrix} 0.2x + 0.3y = 2.6 \\ 0.5x - 0.1y = 1.4 \end{pmatrix}$

15. $\begin{pmatrix} 2x - 3y - 3z = 25 \\ 3x + y + 2z = -5 \\ 5x - 2y - 4z = 32 \end{pmatrix}$ **16.** $\begin{pmatrix} 3x - y + z = -10 \\ 6x - 2y + 5z = -35 \\ 7x + 3y - 4z = 19 \end{pmatrix}$

For Problems 17–26, solve each system by using the method you think is most appropriate.

17. $\begin{pmatrix} 4x + 7y = -15 \\ 3x - 2y = 25 \end{pmatrix}$ **18.** $\begin{pmatrix} \frac{3}{4}x - \frac{1}{2}y = -15 \\ \frac{2}{3}x + \frac{1}{4}y = -5 \end{pmatrix}$

19. $\begin{pmatrix} x + 4y = 3 \\ 3x - 2y = 1 \end{pmatrix}$ **20.** $\begin{pmatrix} 7x - 3y = -49 \\ y = \frac{3}{5}x - 1 \end{pmatrix}$

21. $\begin{pmatrix} 2x + 4y = -1 \\ 3x + 6y = 5 \end{pmatrix}$

22. $\begin{pmatrix} x - y - z = 4 \\ -3x + 2y + 5z = -21 \\ 5x - 3y - 7z = 30 \end{pmatrix}$

23. $\begin{pmatrix} 2x - y + z = -7 \\ -5x + 2y - 3z = 17 \\ 3x + y + 7z = -5 \end{pmatrix}$

24. $\begin{pmatrix} 3x - 2y - 5z = 2 \\ -4x + 3y + 11z = 3 \\ 2x - y + z = -1 \end{pmatrix}$

25. $\begin{pmatrix} x - y + z = 7 \\ 2x - 3y - z = 2 \\ 3x - 4y = 9 \end{pmatrix}$

26. $\begin{pmatrix} 7x - y + z = -4 \\ -2x + 9y - 3z = -50 \\ x - 5y + 4z = 42 \end{pmatrix}$

For Problems 27–32, evaluate each determinant.

27. $\begin{vmatrix} -2 & 6 \\ 3 & 8 \end{vmatrix}$ **28.** $\begin{vmatrix} 5 & -4 \\ 7 & -3 \end{vmatrix}$

29. $\begin{vmatrix} 2 & 3 & -1 \\ 3 & 4 & -5 \\ 6 & 4 & 2 \end{vmatrix}$ **30.** $\begin{vmatrix} 3 & -2 & 4 \\ 1 & 0 & 6 \\ 3 & -3 & 5 \end{vmatrix}$

31. $\begin{vmatrix} 5 & 4 & 3 \\ 2 & -7 & 0 \\ 3 & -2 & 0 \end{vmatrix}$ **32.** $\begin{vmatrix} 5 & -4 & 2 & 1 \\ 3 & 7 & 6 & -2 \\ 2 & 1 & -5 & 0 \\ 3 & -2 & 4 & 0 \end{vmatrix}$

For Problems 33–38, (a) graph the system, and (b) solve the system by using the substitution or elimination method.

33. $\begin{pmatrix} x^2 + y^2 = 17 \\ x - 4y = -17 \end{pmatrix}$ **34.** $\begin{pmatrix} x^2 - y^2 = 8 \\ 3x - y = 8 \end{pmatrix}$

35. $\begin{pmatrix} x - y = 1 \\ y = x^2 + 4x + 1 \end{pmatrix}$ **36.** $\begin{pmatrix} 4x^2 - y^2 = 16 \\ 9x^2 + 9y^2 = 16 \end{pmatrix}$

37. $\begin{pmatrix} x^2 + 2y^2 = 8 \\ 2x^2 + 3y^2 = 12 \end{pmatrix}$ **38.** $\begin{pmatrix} y^2 - x^2 = 1 \\ 4x^2 + y^2 = 4 \end{pmatrix}$

For Problems 39–42, solve each problem by setting up and solving a system of linear equations.

39. The sum of the digits of a two-digit number is 9. If the digits are reversed, the newly formed number is 45 less than the original number. Find the original number.

40. Sara invested $2500, part of it at 10% and the rest at 12% yearly interest. The yearly income on the 12% investment was $102 more than the income on the 10% investment. How much money did she invest at each rate?

41. A box contains $17.70 in nickels, dimes, and quarters. The number of dimes is 8 less than twice the number of nickels. The number of quarters is 2 more than the sum of the nickels and dimes. How many coins of each kind are there in the box?

42. The measure of the largest angle of a triangle is 10° more than four times the smallest angle. The sum of the measures of the smallest and largest angles is three times the measure of the other angle. Find the measure of each angle of the triangle.

TEST

 applies to all problems in this Chapter Test.

For Problems 1–4, refer to the following systems of equations.

$$\text{I. } \begin{pmatrix} 3x - 2y = 4 \\ 9x - 6y = 12 \end{pmatrix} \qquad \text{II. } \begin{pmatrix} 5x - y = 4 \\ 3x + 7y = 9 \end{pmatrix}$$

$$\text{III. } \begin{pmatrix} 2x - y = 4 \\ 2x - y = -6 \end{pmatrix}$$

1. For which system are the graphs parallel lines?

2. For which system are the equations dependent?

3. For which system is the solution set $\varnothing$?

4. Which system is consistent?

For Problems 5–8, evaluate each determinant.

5. $\begin{vmatrix} -2 & 4 \\ -5 & 6 \end{vmatrix}$

6. $\begin{vmatrix} \dfrac{1}{2} & \dfrac{1}{3} \\[2mm] \dfrac{3}{4} & -\dfrac{2}{3} \end{vmatrix}$

7. $\begin{vmatrix} -1 & 2 & 1 \\ 3 & 1 & -2 \\ 2 & -1 & 1 \end{vmatrix}$

8. $\begin{vmatrix} 2 & 4 & -5 \\ -4 & 3 & 0 \\ -2 & 6 & 1 \end{vmatrix}$

9. How many ordered pairs of real numbers are in the solution set for the system $\begin{pmatrix} y = 3x - 4 \\ 9x - 3y = 12 \end{pmatrix}$?

10. Solve the system $\begin{pmatrix} 3x - 2y = -14 \\ 7x + 2y = -6 \end{pmatrix}$.

11. Solve the system $\begin{pmatrix} 4x - 5y = 17 \\ y = -3x + 8 \end{pmatrix}$.

12. Find the value of x in the solution for the following system.

$$\begin{pmatrix} \dfrac{3}{4}x - \dfrac{1}{2}y = -21 \\[3mm] \dfrac{2}{3}x + \dfrac{1}{6}y = -4 \end{pmatrix}$$

13. Find the value of y in the solution for the system $\begin{pmatrix} 4x - y = 7 \\ 3x + 2y = 2 \end{pmatrix}$.

14. How many real number solutions are there for the system

$$\begin{pmatrix} x^2 + y^2 = 16 \\ x^2 - 4y = 8 \end{pmatrix}?$$

15. Suppose that the augmented matrix of a system of three linear equations in the three variables x, y, and z can be changed to the matrix

$$\begin{bmatrix} 1 & 1 & -4 & \vdots & 3 \\ 0 & 1 & 4 & \vdots & 5 \\ 0 & 0 & 3 & \vdots & 6 \end{bmatrix}$$

Find the value of x in the solution for the system.

16. Suppose that the augmented matrix of a system of three linear equations in the three variables x, y, and z can be changed to the matrix

$$\begin{bmatrix} 1 & 2 & -3 & \vdots & 4 \\ 0 & 1 & 2 & \vdots & 5 \\ 0 & 0 & 2 & \vdots & -8 \end{bmatrix}$$

Find the value of y in the solution for the system.

17. How many ordered triples are there in the solution set for the following system?

$$\begin{pmatrix} x + 3y - z = 5 \\ 2x - y - z = 7 \\ 5x + 8y - 4z = 22 \end{pmatrix}$$

18. How many ordered triples are there in the solution set for the following system?

$$\begin{pmatrix} 3x - y - 2z = 1 \\ 4x + 2y + z = 5 \\ 6x - 2y - 4z = 9 \end{pmatrix}$$

19. Solve the following system.

$$\begin{pmatrix} 5x - 3y - 2z = -1 \\ 4y + 7z = 3 \\ 4z = -12 \end{pmatrix}$$

20. Solve the following system.

$$\begin{pmatrix} x - 2y + z = 0 \\ y - 3z = -1 \\ 2y + 5z = -2 \end{pmatrix}$$

21. Find the value of x in the solution for the system

$$\begin{pmatrix} x - 4y + z = 12 \\ -2x + 3y - z = -11 \\ 5x - 3y + 2z = 17 \end{pmatrix}$$

22. Find the value of y in the solution for the system

$$\begin{pmatrix} x - 3y + z = -13 \\ 3x + 5y - z = 17 \\ 5x - 2y + 2z = -13 \end{pmatrix}$$

23. Solve the system $\begin{pmatrix} x^2 + 4y^2 = 25 \\ xy = 6 \end{pmatrix}$.

24. One solution is 30% alcohol and another solution is 70% alcohol. Some of each of the two solutions is mixed to produce 8 liters of a 40% solution. How many liters of the 70% solution should be used?

25. A box contains $7.25 in nickels, dimes, and quarters. There are 43 coins, and the number of quarters is 1 more than three times the number of nickels. Find the number of quarters in the box.

CUMULATIVE REVIEW PROBLEM SET *Word Problems*

Set up an equation, an inequality, or a system of equations to help solve each of the following problems.

1. A car repair bill without tax was $340. This included $145 for parts and 3 hours of labor. Find the cost per hour for the labor.

2. Find three consecutive odd integers whose sum is 57.

3. The supplement of an angle is 10° more than five times its complement. Find the measure of the angle.

4. Eric has a collection of 63 coins consisting of nickels, dimes, and quarters. The number of dimes is 6 more than the number of nickels, and the number of quarters is 1 more than twice the number of nickels. How many coins of each kind are in the collection?

5. The largest angle of a triangle is 10° more than three times the smallest angle. The other angle is 20° more than the smallest angle. Find the measure of each angle of the triangle.

6. Kaya is paid "time and a half" for each hour over 40 hours in a week. Last week she worked 46 hours and earned $455.70. What is her normal hourly rate?

7. If a DVD player costs an audio shop $300, at what price should the shop sell it to make a profit of 50% on the selling price?

8. Beth invested a certain amount of money at 8% interest and $300 more than that amount at 9%. Her total yearly interest was $316. How much did she invest at each rate?

9. Sam shot rounds of 70, 73, and 76 on the first 3 days of a golf tournament. What must he shoot on the fourth day to average 72 or less for the 4 days?

10. Eric bought a number of shares of stock for $300. A month later he sold all but 10 shares at a profit of $5 per share and regained his original investment of $300. How many shares did he originally buy and at what price per share?

11. The perimeter of a rectangle is 44 inches and its area is 112 square inches. Find the dimensions of the rectangle.

12. The cube of a number equals nine times the same number. Find the number.

13. Two motorcycles leave Daytona Beach at the same time, one traveling north and the other traveling south. At the end of 4.5 hours, they are 639 miles apart. If the rate of the motorcycle traveling north is 10 miles per hour greater than that of the other motorcycle, find their rates.

14. A 10-quart radiator contains a 50% solution of antifreeze. How much needs to be drained out and replaced with pure antifreeze to obtain a 70% antifreeze solution?

15. How long will it take $750 to double itself if it is invested at 6% simple interest?

16. How long will it take $750 to double itself if it is invested at 6% interest compounded quarterly?

17. How long will it take $750 to double itself if it is invested at 6% interest compounded continuously?

18. The perimeter of a square is 4 centimeters less than twice the perimeter of an equilateral triangle. The length of a side of the square is 2 centimeters more than the length of a side of the equilateral triangle. Find the length of a side of the equilateral triangle.

19. Heidi starts jogging at 4 miles per hour. One-half hour later Ed starts jogging on the same route at 6 miles per hour. How long will it take Ed to catch Heidi?

20. A strip of uniform width is to be cut off both sides and both ends of a sheet of paper that is 8 inches by 14 inches to reduce the size of the paper to an area of 72 square inches. Find the width of the strip.

21. A sum of $2450 is to be divided between two people in the ratio of 3 to 4. How much does each person receive?

22. Working together, Sue and Dean can complete a task in $1\frac{1}{5}$ hours. Dean can do the task by himself in 2 hours. How long would it take Sue to complete the task by herself?

23. The units digit of a two-digit number is 1 more than twice the tens digit. The sum of the digits is 10. Find the number.

24. Suppose the number of days it takes to complete a job varies inversely as the number of people assigned to the job. If it takes 12 people 8 days to do the job, how long would it take 20 people to do the job?

25. The cost of labor varies jointly as the number of workers and the number of days that they work. If it costs $3750 to have 15 people work for 5 days, how much will it cost to have 20 people work for 4 days?

26. It takes a freight train 2 hours longer to travel 300 miles than it takes an express train to travel 280 miles. The rate of the express train is 20 miles per hour greater than the rate of the freight train. Find the times and rates of both trains.

27. Suppose that we want the temperature in a room to be between 19° and 22° Celsius. What Fahrenheit temperatures should be maintained? Remember the formula

$$C = \frac{5}{9}(F - 32).$$

28. A country fair BBQ booth sells pork dinners for $12 and rib dinners for $15. If 270 dinners were sold for a total of $3690, how many rib dinners were sold?

29. Find two numbers such that their sum is 2 and their product is −1.

30. Larry drove 156 miles in 1 hour more than it took Nita to drive 108 miles. Nita drove at an average rate of 2 miles per hour faster than Larry. How fast did each one travel?

31. An auditorium in a local high school contains 300 seats. There are 5 fewer rows than the number of seats per row. Find the number of rows and the number of seats per row.

32. The area of a certain circle is numerically equal to twice the circumference of the circle. Find the length of a radius of the circle.

33. A class trip was to cost a total of $3000. If there had been 10 more students, it would have cost each student $25 less. How many students took the trip?

34. The difference in the lengths of the two legs of a right triangle is 2 yards. If the length of the hypotenuse is $2\sqrt{13}$ yards, find the length of each leg.

35. The length of the hypotenuse of an isosceles right triangle is 12 inches. Find the length of each leg.

36. The rental charge for 3 movies and 2 video games is $20.75. At the same prices, 5 movies and 3 video games cost $32.92. Find the price of renting a movie and the price of renting a video game.

37. Find three consecutive whole numbers such that the sum of the first plus twice the second plus three times the third is 134.

38. Ike has some nickels and dimes amounting to $2.90. The number of dimes is 1 less than twice the number of nickels. How many coins of each kind does he have?

39. Fourteen increased by twice a number is less than or equal to three times the number. Find the numbers that satisfy this relationship.

40. How many milliliters of pure acid must be added to 150 milliliters of a 30% solution of acid to obtain a 40% solution?

When objects are arranged in a sequence, the total number of objects is the sum of the terms of the sequence.

Royalty-Free /CORBIS

Miscellaneous Topics: Problem Solving

S uppose that an auditorium has 35 seats in the first row, 40 seats in the second row, 45 seats in the third row, and so on, for ten rows. The numbers 35, 40, 45, 50, . . . , 80 represent the number of seats per row from row 1 through row 10. The list of numbers has a constant difference of 5 between any two successive numbers in the list; such a list is called an **arithmetic sequence.** (Used in this sense, the word "arithmetic" is pronounced with the accent on the syllable "met.")

Suppose that a fungus culture growing under controlled conditions doubles in size each day. If today the size of the culture is 6 units, then the numbers 12, 24, 48, 96, 192 represent the size of the culture for the next 5 days. In this list of numbers, each number after the first is twice the previous number; such a list is called a **geometric sequence.** Arithmetic and geometric sequences will be used as problem-solving tools in this chapter.

With an ordinary deck of 52 playing cards, there is *1 chance out of 54,145* that you will be dealt four aces in a five-card hand.

The radio is predicting a 40% chance of locally severe thunderstorms by late afternoon. The *odds in favor of* the Cubs winning the pennant are 2 to 3. Suppose that in a box containing 50 light bulbs, 45 are good ones and 5 are burned out. If 2 bulbs are chosen at random, the probability of getting at least 1 good bulb is $\frac{243}{245}$. Historically, many basic probability concepts have been developed as a result of studying various games of chance. In recent years, however, applications of probability have been surfacing at a phenomenal rate in a wide variety of fields, such as physics, biology, psychology, economics, insurance, military science, manufacturing, and even politics. It is our purpose in this chapter first to introduce some counting techniques and then to use those techniques to solve basic probability problems.

InfoTrac Project

Do a keyword search on bacteria growth and find the article titled "Logarithms and Life." Write a brief summary of the article. Using the information in the article, find the total number of bacteria (assuming none die) 24 hours after the fly landed on the food.

16.1   Arithmetic Sequences

Objectives

■ Find specified terms of an arithmetic sequence.

■ Determine the general term for an arithmetic sequence.

■ Find the sum of a specific number of terms of an arithmetic sequence.

■ Solve word problems involving arithmetic sequences.

An **infinite sequence** is a function whose domain is the set of positive integers. For example, consider the function defined by the equation

$$f(n) = 5n + 1$$

where the domain is the set of positive integers. If we substitute the numbers of the domain in order, starting with 1, we can list the resulting ordered pairs:

$$(1, 6) \qquad (2, 11) \qquad (3, 16) \qquad (4, 21) \qquad (5, 26)$$

and so on. However, because we know we are using the domain of positive integers in order, starting with 1, there is no need to use ordered pairs. We can simply express the infinite sequence as

$$6, 11, 16, 21, 26, \ldots$$

Often, the letter a is used to represent sequential functions, and the functional value of a at n is written a_n (this is read "a sub n") instead of $a(n)$. The sequence is then expressed

$$a_1, a_2, a_3, a_4, \ldots$$

where a_1 is the **first term,** a_2 is the **second term,** a_3 is the **third term,** and so on. The expression a_n, which defines the sequence, is called the **general term** of the sequence. Knowing the general term of a sequence enables us to find as many terms of the sequence as necessary and also to find any specific terms. Consider the following example.

E X A M P L E   1   Find the first five terms of the sequence where $a_n = 2n^2 - 3$; find the 20th term.

Solution

The first five terms are generated by replacing n with 1, 2, 3, 4, and 5.

$$a_1 = 2(1)^2 - 3 = -1 \qquad\qquad a_2 = 2(2)^2 - 3 = 5$$
$$a_3 = 2(3)^2 - 3 = 15 \qquad\qquad a_4 = 2(4)^2 - 3 = 29$$
$$a_5 = 2(5)^2 - 3 = 47$$

The first five terms are thus $-1, 5, 15, 29$, and 47. The 20th term is

$$a_{20} = 2(20)^2 - 3 = 797$$

Arithmetic Sequences

An **arithmetic sequence** (also called an arithmetic progression) is a sequence that has a common difference between successive terms. The following are examples of arithmetic sequences:

$$1, 8, 15, 22, 29, \ldots$$
$$4, 7, 10, 13, 16, \ldots$$
$$4, 1, -2, -5, -8, \ldots$$
$$-1, -6, -11, -16, -21, \ldots$$

The common difference in the first sequence is 7. That is, $8 - 1 = 7$, $15 - 8 = 7$, $22 - 15 = 7$, $29 - 22 = 7$, and so on. The common differences for the next three sequences are 3, -3, and -5, respectively.

In a more general setting, we say that the sequence

$$a_1, a_2, a_3, a_4, \ldots, a_n, \ldots$$

is an arithmetic sequence if and only if there is a real number d such that

$$a_{k+1} - a_k = d$$

for every positive integer k. The number d is called the **common difference.**

From the definition, we see that $a_{k+1} = a_k + d$. In other words, we can generate an arithmetic sequence that has a common difference of d by starting with a first term a_1 and then simply adding d to each successive term.

First term: a_1

Second term: $a_1 + d$

Third term: $a_1 + 2d$ $(a_1 + d) + d = a_1 + 2d$

Fourth term: $a_1 + 3d$

.

.

.

nth term: $a_1 + (n - 1)d$

Thus the **general term** of an arithmetic sequence is given by

$$a_n = a_1 + (n - 1)d$$

where a_1 is the first term and d is the common difference. This formula for the general term can be used to solve a variety of problems involving arithmetic sequences.

EXAMPLE 2

Find the general-term expression for the arithmetic sequence $6, 2, -2, -6, \ldots$.

Solution

The common difference, d, is $2 - 6 = -4$, and the first term, a_1, is 6. Substitute these values into $a_n = a_1 + (n - 1)d$ and simplify to obtain

$$a_n = a_1 + (n - 1)d$$
$$= 6 + (n - 1)(-4)$$
$$= 6 - 4n + 4$$
$$= -4n + 10$$

EXAMPLE 3

Find the 40th term of the arithmetic sequence $1, 5, 9, 13, \ldots$.

Solution

Using $a_n = a_1 + (n - 1)d$, we obtain

$$a_{40} = 1 + (40 - 1)4$$
$$= 1 + (39)(4)$$
$$= 157$$

EXAMPLE 4

Find the first term of the arithmetic sequence where the fourth term is 26 and the ninth term is 61.

Solution

Using $a_n = a_1 + (n-1)d$ with $a_4 = 26$ (the fourth term is 26) and $a_9 = 61$ (the ninth term is 61), we have

$$26 = a_1 + (4-1)d = a_1 + 3d$$
$$61 = a_1 + (9-1)d = a_1 + 8d$$

Solving the system of equations

$$\begin{pmatrix} a_1 + 3d = 26 \\ a_1 + 8d = 61 \end{pmatrix}$$

yields $a_1 = 5$ and $d = 7$. Thus the first term is 5.   ■

Sums of Arithmetic Sequences

We often use sequences to solve problems, so we need to be able to find the sum of a certain number of terms of the sequence. Before we develop a general-sum formula for arithmetic sequences, let's consider an approach to a specific problem that we can then use in a general setting.

EXAMPLE 5

Find the sum of the first 100 positive integers.

Solution

We are being asked to find the sum of $1 + 2 + 3 + 4 + \cdots + 100$. Rather than adding in the usual way, let's find the sum in the following manner.

$$\begin{array}{ccccccccc} 1 & + & 2 & + & 3 & + & 4 & + \cdots + & 100 \\ 100 & + & 99 & + & 98 & + & 97 & + \cdots + & 1 \\ \hline 101 & + & 101 & + & 101 & + & 101 & + \cdots + & 101 \end{array}$$

$$\frac{\overset{50}{\cancel{100}}(101)}{\cancel{2}} = 5050$$

Note that we simply wrote the indicated sum forward and backward, and then we added the results. In so doing, we produced 100 sums of 101, but half of them are repeats. For example, $100 + 1$ and $1 + 100$ are both counted in this process. Thus we divide the product $(100)(101)$ by 2, which yields the final result of 5050.   ■

The *forward-backward* approach we used in Example 5 can be used to develop a formula for finding the sum of the first n terms of any arithmetic sequence. Consider an arithmetic sequence $a_1, a_2, a_3, a_4, \ldots, a_n$ with a common difference of d. Use S_n to represent the sum of the first n terms, and proceed as follows:

$$S_n = a_1 + (a_1 + d) + (a_1 + 2d) + \cdots + (a_n - 2d) + (a_n - d) + a_n$$

Now write this sum in reverse.

$$S_n = a_n + (a_n - d) + (a_n - 2d) + \cdots + (a_1 + 2d) + (a_1 + d) + a_1$$

Add the two equations to produce

$$2S_n = (a_1 + a_n) + (a_1 + a_n) + (a_1 + a_n) + \cdots + (a_1 + a_n) + (a_1 + a_n) + (a_1 + a_n)$$

That is, we have n sums $a_1 + a_n$, so

$$2S_n = n(a_1 + a_n)$$

from which we obtain a **sum formula:**

$$S_n = \frac{n(a_1 + a_n)}{2}$$

Using the nth-term formula and/or the sum formula, we can solve a variety of problems involving arithmetic sequences.

EXAMPLE 6   Find the sum of the first 30 terms of the arithmetic sequence 3, 7, 11, 15, ...

Solution

Using $a_n = a_1 + (n - 1)d$, we can find the 30th term.

$$a_{30} = 3 + (30 - 1)4 = 3 + 29(4) = 119$$

Now we can use the sum formula.

$$S_{30} = \frac{30(3 + 119)}{2} = 1830$$

EXAMPLE 7   Find the sum $7 + 10 + 13 + \cdots + 157$.

Solution

To use the sum formula, we need to know the number of terms. Applying the nth-term formula will give us that information.

$$a_n = a_1 + (n - 1)d$$
$$157 = 7 + (n - 1)3$$
$$157 = 7 + 3n - 3$$
$$157 = 3n + 4$$
$$153 = 3n$$
$$51 = n$$

Now we can use the sum formula.

$$S_{51} = \frac{51(7 + 157)}{2} = 4182$$

Keep in mind that we developed the sum formula for an arithmetic sequence by using the forward-backward technique, which we had previously used on a specific problem. Now that we have the sum formula, we have two choices when solving problems. We can either memorize the formula and use it, or simply use the forward-backward technique. If you choose to use the formula and some day you forget it, don't panic. Just use the forward-backward technique. In other words, understanding the development of a formula often enables you to do problems even when you forget the formula itself.

Problem Solving

Now let's use arithmetic sequences as another problem-solving tool. First, let's restate some previous problem-solving suggestions that continue to apply, and consider some new suggestions in light of our work with sequences. (We have indicated the new suggestions with an asterisk.)

Suggestions for Solving Word Problems

1. Read the problem carefully and make certain that you understand the meanings of all the words. Be especially alert for any technical terms used in the statement of the problem.

2. Read the problem a second time (perhaps even a third time) to get an overview of the situation being described and to determine the known facts, as well as what you are to find.

3. Sketch a figure, diagram, or chart that might be helpful in analyzing the problem.

*4. Write down the first few terms of the sequence to describe what is taking place in the problem. Be sure that you understand, term by term, what the sequence represents in the problem.

*5. Determine whether the problem is asking for a specific term of the sequence or for the sum of a certain number of terms.

*6. Carry out the necessary calculations and check your answer for reasonableness.

As we solve some problems, these suggestions will become more meaningful.

PROBLEM 1

Domenica started to work in 1995 at an annual salary of $24,500. She received a $1350 raise each year. What was her annual salary in 2004?

Solution

The following sequence represents her annual salary beginning in 1995.

$$24{,}500, \ 25{,}850, \ 27{,}200, \ 28{,}550, \ \ldots$$

This is an arithmetic sequence, with $a_1 = 24{,}500$ and $d = 1350$. Because each term of the sequence represents her annual salary, we are looking for the tenth term.

$$a_{10} = 24{,}500 + (10 - 1)1350 = 24{,}500 + 9(1350) = 36{,}650$$

Her annual salary in 2004 was $36,650. ∎

PROBLEM 2   An auditorium has 20 seats in the front row, 24 seats in the second row, 28 seats in the third row, and so on, for 15 rows. How many seats are there in the auditorium?

Solution

The following sequence represents the number of seats per row, starting with the first row.

$$20, 24, 28, 32, \ldots$$

This is an arithmetic sequence, with $a_1 = 20$ and $d = 4$. Therefore, the 15th term, which represents the number of seats in the 15th row, is given by

$$a_{15} = 20 + (15 - 1)4 = 20 + 14(4) = 76$$

The total number of seats in the auditorium is represented by

$$20 + 24 + 28 + \cdots + 76$$

Use the sum formula for an arithmetic sequence to obtain

$$S_{15} = \frac{15}{2}(20 + 76) = 720$$

There are 720 seats in the auditorium.

CONCEPT QUIZ

For Problems 1–5, answer true or false.

1. The domain of an infinite sequence is the set of integers.

2. The notation a_6 denotes the sixth term of a sequence.

3. Every arithmetic sequence has a common difference between successive terms.

4. To find the 8th term of an arithmetic sequence, we add 8 times the common difference to the first term.

5. The sum formula for n terms of an arithmetic sequence is n times the average of the first and last term in the sum.

PROBLEM SET 16.1

For Problems 1–10, write the first five terms of the sequence that has the indicated general term.

1. $a_n = 3n - 7$

2. $a_n = 5n - 2$

3. $a_n = -2n + 4$

4. $a_n = -4n + 7$

5. $a_n = 3n^2 - 1$

6. $a_n = 2n^2 - 6$

7. $a_n = n(n - 1)$

8. $a_n = (n + 1)(n + 2)$

9. $a_n = 2^{n+1}$

10. $a_n = 3^{n-1}$

11. Find the 15th and 30th terms of the sequence where $a_n = -5n - 4$.

12. Find the 20th and 50th terms of the sequence where $a_n = -n - 3$.

13. Find the 25th and 50th terms of the sequence where $a_n = (-1)^{n+1}$.

14. Find the 10th and 15th terms of the sequence where $a_n = -n^2 - 10$.

For Problems 15–24, find the general term (the nth term) for each arithmetic sequence.

15. $11, 13, 15, 17, 19, \ldots$

16. $7, 10, 13, 16, 19, \ldots$

17. $2, -1, -4, -7, -10, \ldots$

18. $4, 2, 0, -2, -4, \ldots$

19. $\dfrac{3}{2}, 2, \dfrac{5}{2}, 3, \dfrac{7}{2}, \ldots$

20. $0, \dfrac{1}{2}, 1, \dfrac{3}{2}, 2, \ldots$

21. $2, 6, 10, 14, 18, \ldots$

22. $2, 7, 12, 17, 22, \ldots$

23. $-3, -6, -9, -12, -15, \ldots$

24. $-4, -8, -12, -16, -20, \ldots$

For Problems 25–30, find the required term for each arithmetic sequence.

25. The 15th term of $3, 8, 13, 18, \ldots$

26. The 20th term of $4, 11, 18, 25, \ldots$

27. The 30th term of $15, 26, 37, 48, \ldots$

28. The 35th term of $9, 17, 25, 33, \ldots$

29. The 52nd term of $1, \dfrac{5}{3}, \dfrac{7}{3}, 3, \ldots$

30. The 47th term of $\dfrac{1}{2}, \dfrac{5}{4}, 2, \dfrac{11}{4}, \ldots$

For Problems 31–42, solve each problem.

31. If the 6th term of an arithmetic sequence is 12 and the 10th term is 16, find the first term.

32. If the 5th term of an arithmetic sequence is 14 and the 12th term is 42, find the first term.

33. If the 3rd term of an arithmetic sequence is 20 and the 7th term is 32, find the 25th term.

34. If the 5th term of an arithmetic sequence is -5 and the 15th term is -25, find the 50th term.

35. Find the sum of the first 50 terms of the arithmetic sequence $5, 7, 9, 11, 13, \ldots$.

36. Find the sum of the first 30 terms of the arithmetic sequence $0, 2, 4, 6, 8, \ldots$.

37. Find the sum of the first 40 terms of the arithmetic sequence $2, 6, 10, 14, 18, \ldots$.

38. Find the sum of the first 60 terms of the arithmetic sequence $-2, 3, 8, 13, 18, \ldots$.

39. Find the sum of the first 75 terms of the arithmetic sequence $5, 2, -1, -4, -7, \ldots$.

40. Find the sum of the first 80 terms of the arithmetic sequence $7, 3, -1, -5, -9, \ldots$.

41. Find the sum of the first 50 terms of the arithmetic sequence $\dfrac{1}{2}, 1, \dfrac{3}{2}, 2, \dfrac{5}{2}, \ldots$.

42. Find the sum of the first 100 terms of the arithmetic sequence $-\dfrac{1}{3}, \dfrac{1}{3}, 1, \dfrac{5}{3}, \dfrac{7}{3}, \ldots$.

For Problems 43–50, find the indicated sum.

43. $1 + 5 + 9 + 13 + \cdots + 197$

44. $3 + 8 + 13 + 18 + \cdots + 398$

45. $2 + 8 + 14 + 20 + \cdots + 146$

46. $6 + 9 + 12 + 15 + \cdots + 93$

47. $(-7) + (-10) + (-13) + (-16) + \cdots + (-109)$

48. $(-5) + (-9) + (-13) + (-17) + \cdots + (-169)$

49. $(-5) + (-3) + (-1) + 1 + \cdots + 119$

50. $(-7) + (-4) + (-1) + 2 + \cdots + 131$

For Problems 51–68, solve each problem.

51. Find the sum of the first 200 odd whole numbers.

52. Find the sum of the first 175 positive even whole numbers.

53. Find the sum of all even numbers between 18 and 482, inclusive.

54. Find the sum of all odd numbers between 17 and 379, inclusive.

55. Find the sum of the first 30 terms of the arithmetic sequence with the general term $a_n = 5n - 4$.

56. Find the sum of the first 40 terms of the arithmetic sequence with the general term $a_n = 4n - 7$.

57. Find the sum of the first 25 terms of the arithmetic sequence with the general term $a_n = -4n - 1$.

58. Find the sum of the first 35 terms of the arithmetic sequence with the general term $a_n = -5n - 3$.

59. A man started to work in 1980 at an annual salary of $19,500. He received a $1170 raise each year. How much was his annual salary in 2001?

60. A woman started to work in 1990 at an annual salary of $23,400. She received a $1500 raise each year. How much was her annual salary in 2005?

61. Online University had an enrollment of 600 students in 2000. Each year the enrollment increased by 2150 students. What was the enrollment in 2005?

62. Math University had an enrollment of 12,800 students in 1997. Each year the enrollment decreased by 75 students. What was the enrollment in 2004?

63. Sue is saving quarters. She saves 1 quarter the first day, 2 quarters the second day, 3 quarters the third day, and so on for 30 days. How much money will she have saved in 30 days?

64. Suppose you save a penny the first day of a month, 2 cents the second day, 3 cents the third day, and so on for 31 days. What will be your total savings for the 31 days?

65. An auditorium has 40 seats in the front row, 44 seats in the second row, 48 seats in the third row, and so on, for 25 rows. How many seats are there in the last row? How many seats are there in the auditorium?

66. A display in a grocery store has cans stacked with 25 cans in the bottom row, 23 cans in the second row from the bottom, 21 cans in the third row from the bottom, and so on until there is only 1 can in the top row. How many cans are there in the display?

67. Ms. Bryan invested $1500 at 12% simple interest at the beginning of each year for a period of 10 years. Find the total accumulated value of all the investments at the end of the 10-year period.

68. A well driller charges $9.00 per foot for the first 10 feet, $9.10 per foot for the next 10 feet, $9.20 per foot for the next 10 feet, and so on, at a price increase of $.10 per foot for succeeding intervals of 10 feet. How much does it cost to drill a well to a depth of 150 feet?

■ ■ ■ **Thoughts into words**

69. Before developing the formula $a_n = a_1 + (n - 1)d$, we stated the equation $a_{k+1} - a_k = d$. In your own words, explain what this equation says.

70. Explain how to find the sum $1 + 2 + 3 + 4 + \cdots + 175$ without using the sum formula.

71. Explain in words how to find the sum of the first n terms of an arithmetic sequence.

72. Explain how one can tell that a particular sequence is an arithmetic sequence.

Answers to Concept Quiz

1. False   **2.** True   **3.** True   **4.** False   **5.** True

16.2   Geometric Sequences

Objectives

- Find the general term for a geometric sequence.

- Determine specified terms of a geometric sequence.

- Find the sum of a specific number of terms of a geometric sequence.

- Solve word problems involving geometric sequences.

A **geometric sequence** or **geometric progression** is a sequence in which we obtain each term after the first by multiplying the preceding term by a common multiplier called the **common ratio** of the sequence. We can find the common ratio of a geometric sequence by dividing any term (other than the first) by the preceding term.

The following geometric sequences have common ratios of 3, 2, $\frac{1}{2}$, and -4, respectively.

$$1, 3, 9, 27, 81, \ldots$$
$$3, 6, 12, 24, 48, \ldots$$
$$16, 8, 4, 2, 1, \ldots$$
$$-1, 4, -16, 64, -256, \ldots$$

In a more general setting, we say that the sequence $a_1, a_2, a_3, \ldots, a_n, \ldots$ is a geometric sequence if and only if there is a nonzero real number r such that

$$a_{k+1} = ra_k$$

for every positive integer k. The nonzero real number r is called the common ratio of the sequence.

The previous equation can be used to generate a general geometric sequence that has a_1 as the first term and r as the common ratio. We can proceed as follows.

First term: a_1

Second term: $a_1 r$

Third term: $a_1 r^2$ $\qquad$ $(a_1 r)(r) = a_1 r^2$

Fourth term: $a_1 r^3$

$.$

$.$

$.$

$.$

nth term: $a_1 r^{n-1}$

Thus the **general term** of a geometric sequence is given by

$$a_n = a_1 r^{n-1}$$

where a_1 is the first term and r is the common ratio.

E X A M P L E   1   Find the general term for the geometric sequence $8, 16, 32, 64, \ldots$.

Solution

Using $a_n = a_1 r^{n-1}$, we obtain

$$a_n = 8(2)^{n-1} = (2^3)(2)^{n-1} = 2^{n+2}$$

EXAMPLE 2   Find the ninth term of the geometric sequence 27, 9, 3, 1, ...

Solution

Using $a_n = a_1 r^{n-1}$, we can find the ninth term as follows:

$$a_9 = 27\left(\frac{1}{3}\right)^{9-1} = 27\left(\frac{1}{3}\right)^8 = \frac{3^3}{3^8} = \frac{1}{3^5} = \frac{1}{243}$$

Sums of Geometric Sequences

As with arithmetic sequences, we often need to find the sum of a certain number of terms of a geometric sequence. Before we develop a general-sum formula for geometric sequences, let's consider an approach to a specific problem that we can then use in a general setting.

EXAMPLE 3   Find the sum of $1 + 3 + 9 + 27 + \cdots + 6561$.

Solution

Let S represent the sum and proceed as follows:

$$S = 1 + 3 + 9 + 27 + \cdots + 6561 \tag{1}$$

$$3S = \quad\ \ 3 + 9 + 27 + \cdots + 6561 + 19683 \tag{2}$$

Equation (2) is the result of multiplying equation (1) by the common ratio 3. Subtracting equation (1) from equation (2) produces

$$2S = 19683 - 1 = 19682$$

$$S = 9841$$

Now let's consider a general geometric sequence $a_1, a_1 r, a_1 r^2, \ldots, a_1 r^{n-1}$. By applying a procedure similar to the one we used in Example 3, we can develop a formula for finding the sum of the first n terms of any geometric sequence. We let S_n represent the sum of the first n terms.

$$S_n = a_1 + a_1 r + a_1 r^2 + \cdots + a_1 r^{n-1} \tag{3}$$

Next, we multiply both sides of equation (3) by the common ratio r.

$$rS_n = a_1 r + a_1 r^2 + a_1 r^3 + \cdots + a_1 r^n \tag{4}$$

We then subtract equation (3) from equation (4).

$$rS_n - S_n = a_1 r^n - a_1$$

When we apply the distributive property to the left side and then solve for S_n, we obtain

$$S_n(r - 1) = a_1 r^n - a_1$$

$$S_n = \frac{a_1 r^n - a_1}{r - 1}, \qquad r \neq 1$$

Therefore, the sum of the first n terms of a geometric sequence with a first term a_1 and a common ratio r is given by

$$S_n = \frac{a_1 r^n - a_1}{r - 1}, \qquad r \neq 1$$

EXAMPLE 4 Find the sum of the first eight terms of the geometric sequence $1, 2, 4, 8, \ldots$.

Solution
Use the sum formula to obtain

$$S_8 = \frac{1(2)^8 - 1}{2 - 1} = \frac{2^8 - 1}{1} = 255$$ ∎

If the common ratio of a geometric sequence is less than 1, it may be more convenient to change the form of the sum formula. That is, the fraction

$$\frac{a_1 r^n - a_1}{r - 1}$$

can be changed to

$$\frac{a_1 - a_1 r^n}{1 - r}$$

by multiplying both the numerator and the denominator by -1. Thus, by using

$$S_n = \frac{a_1 - a_1 r^n}{1 - r}$$

we can sometimes avoid unnecessary work with negative numbers when $r < 1$, as the next example illustrates.

EXAMPLE 5 Find the sum $1 + \dfrac{1}{2} + \dfrac{1}{4} + \cdots + \dfrac{1}{256}$.

Solution A
To use the sum formula, we need to know the number of terms, which can be found by counting them or by applying the nth-term formula, as follows.

$$a_n = a_1 r^{n-1}$$
$$\frac{1}{256} = 1\left(\frac{1}{2}\right)^{n-1}$$
$$\left(\frac{1}{2}\right)^8 = \left(\frac{1}{2}\right)^{n-1}$$
$$8 = n - 1 \qquad \text{If } b^n = b^m, \text{ then } n = m.$$
$$9 = n$$

Now we use $n = 9$, $a_1 = 1$, and $r = \dfrac{1}{2}$ in the sum formula of the form

$$S_n = \frac{a_1 - a_1 r^n}{1 - r}$$

$$S_9 = \frac{1 - 1\left(\dfrac{1}{2}\right)^9}{1 - \dfrac{1}{2}} = \frac{1 - \dfrac{1}{512}}{\dfrac{1}{2}} = \frac{\dfrac{511}{512}}{\dfrac{1}{2}} = 1\frac{255}{256}$$

We can also do a problem like Example 5 without finding the number of terms; we use the general approach illustrated in Example 3. Solution B demonstrates this idea.

Solution B

Let S represent the desired sum.

$$S = 1 + \frac{1}{2} + \frac{1}{4} + \cdots + \frac{1}{256}$$

Multiply both sides by the common ratio $\dfrac{1}{2}$.

$$\frac{1}{2}S = \frac{1}{2} + \frac{1}{4} + \frac{1}{8} + \cdots + \frac{1}{256} + \frac{1}{512}$$

Subtract the second equation from the first, and solve for S.

$$\frac{1}{2}S = 1 - \frac{1}{512} = \frac{511}{512}$$

$$S = \frac{511}{256} = 1\frac{255}{256}$$

The Sum of an Infinite Geometric Sequence

Let's take the formula

$$S_n = \frac{a_1 - a_1 r^n}{1 - r}$$

and rewrite the right-hand side by applying the property

$$\frac{a - b}{c} = \frac{a}{c} - \frac{b}{c}$$

Thus we obtain

$$S_n = \frac{a_1}{1 - r} - \frac{a_1 r^n}{1 - r}$$

Now let's examine the behavior of r^n for $|r| < 1$ — that is, for $-1 < r < 1$. For example, suppose that $r = \dfrac{1}{2}$. Then

$$r^2 = \left(\frac{1}{2}\right)^2 = \frac{1}{4} \qquad r^3 = \left(\frac{1}{2}\right)^3 = \frac{1}{8}$$

$$r^4 = \left(\frac{1}{2}\right)^4 = \frac{1}{16} \qquad r^5 = \left(\frac{1}{2}\right)^5 = \frac{1}{32}$$

and so on. We can make $\left(\dfrac{1}{2}\right)^n$ as close to zero as we please by choosing sufficiently large values for n. In general, for values of r such that $|r| < 1$, the expression r^n approaches zero as n gets larger and larger. Therefore, the fraction $a_1 r^n / (1 - r)$ in equation (1) approaches zero as n increases. We say that **the sum of the infinite geometric sequence** is given by

$$S_\infty = \frac{a_1}{1 - r}, \qquad |r| < 1$$

EXAMPLE 6

Find the sum of the infinite geometric sequence

$$1, \frac{1}{2}, \frac{1}{4}, \frac{1}{8}, \dots$$

Solution

Because $a_1 = 1$ and $r = \dfrac{1}{2}$, we obtain

$$S_\infty = \frac{1}{1 - \dfrac{1}{2}} = \frac{1}{\dfrac{1}{2}} = 2$$

When we state that $S_\infty = 2$ in Example 6, we mean that as we add more and more terms, the sum approaches 2. Observe what happens when we calculate the sum up to five terms.

First term: $\qquad\qquad\qquad\qquad 1$

Sum of first two terms: $\qquad\qquad 1 + \dfrac{1}{2} = 1\dfrac{1}{2}$

Sum of first three terms: $\qquad\quad 1 + \dfrac{1}{2} + \dfrac{1}{4} = 1\dfrac{3}{4}$

Sum of first four terms: $\qquad\quad 1 + \dfrac{1}{2} + \dfrac{1}{4} + \dfrac{1}{8} = 1\dfrac{7}{8}$

Sum of first five terms: $\qquad\quad 1 + \dfrac{1}{2} + \dfrac{1}{4} + \dfrac{1}{8} + \dfrac{1}{16} = 1\dfrac{15}{16}$

If $|r| > 1$, the absolute value of r^n increases without bound as n increases. Consider the following two examples, and note the unbounded growth of the absolute value of r^n.

Let $r = 3$	Let $r = -2$	
$r^2 = 3^2 = 9$	$r^2 = (-2)^2 = 4$	
$r^3 = 3^3 = 27$	$r^3 = (-2)^3 = -8$	$|-8| = 8$
$r^4 = 3^4 = 81$	$r^4 = (-2)^4 = 16$	
$r^5 = 3^5 = 243$	$r^5 = (-2)^5 = -32$	$|-32| = 32$

If $r = 1$, then $S_n = na_1$, and as n increases without bound, $|S_n|$ also increases without bound. If $r = -1$, then S_n will be either a_1 or 0. Therefore, we say that the sum of any infinite geometric sequence where $|r| \geq 1$ *does not exist.*

Repeating Decimals as Sums of Infinite Geometric Sequences

In Section 2.3, we defined rational numbers to be numbers that have either a terminating or a repeating decimal representation. For example,

$$2.23 \qquad 0.147 \qquad 0.\overline{3} \qquad 0.\overline{14} \qquad \text{and} \qquad 0.5\overline{6}$$

are rational numbers. (Remember that $0.\overline{3}$ means $0.3333\ldots$.) Place value provides the basis for changing terminating decimals such as 2.23 and 0.147 to a/b form, where a and b are integers and $b \neq 0$.

$$2.23 = \frac{223}{100} \qquad \text{and} \qquad 0.147 = \frac{147}{1000}$$

However, changing repeating decimals to a/b form requires a different technique, and our work with sums of infinite geometric sequences provides the basis for one such approach. Consider the following examples.

EXAMPLE 7

Change $0.\overline{14}$ to a/b form, where a and b are integers and $b \neq 0$.

Solution

The repeating decimal $0.\overline{14}$ can be written as the indicated sum of an infinite geometric sequence with first term 0.14 and common ratio 0.01.

$$0.14 + 0.0014 + 0.000014 + \cdots$$

Using $S_\infty = a_1/(1 - r)$, we obtain

$$S_\infty = \frac{0.14}{1 - 0.01} = \frac{0.14}{0.99} = \frac{14}{99}$$

Thus $0.\overline{14} = \frac{14}{99}$.

If the repeating block of digits does not begin immediately after the decimal point, as in $0.5\overline{6}$, we can make an adjustment in the technique we used in Example 7.

EXAMPLE 8

Change $0.5\overline{6}$ to a/b form, where a and b are integers and $b \neq 0$.

Solution

The repeating decimal $0.5\overline{6}$ can be written

$$(0.5) + (0.06 + 0.006 + 0.0006 + \cdots)$$

where

$$0.06 + 0.006 + 0.0006 + \cdots$$

is the indicated sum of the infinite geometric sequence with $a_1 = 0.06$ and $r = 0.1$. Therefore,

$$S_\infty = \frac{0.06}{1 - 0.1} = \frac{0.06}{0.9} = \frac{6}{90} = \frac{1}{15}$$

Now we can add 0.5 and $\dfrac{1}{15}$.

$$0.5\overline{6} = 0.5 + \frac{1}{15} = \frac{1}{2} + \frac{1}{15} = \frac{15}{30} + \frac{2}{30} = \frac{17}{30}$$ ■

Back to Problem Solving

Suggestions 4 and 5 from the list of problem-solving suggestions in Section 16.1 continue to apply here as we use geometric sequences to solve problems. It is important to write down the first few terms of the sequence, making sure that you understand, term by term, what the sequence represents relative to the problem. Then you must determine whether the problem is asking for a specific term of the sequence or for the sum of a certain number of terms. Let's consider two examples to emphasize these points.

PROBLEM 1

Suppose that you save 25 cents the first day of a week, 50 cents the second day, and one dollar the third day and that you continue to double your savings each day. How much will you save on the seventh day? What will be your total savings for the week?

Solution

The following sequence represents your savings per day, expressed in cents.

$$25, 50, 100, \ldots$$

This is a geometric sequence, with $a_1 = 25$ and $r = 2$. Your savings on the seventh day is the seventh term of this sequence. Therefore, using $a_n = a_1 r^{n-1}$, we obtain

$$a_7 = 25(2)^6 = 1600$$

You will save $16 on the seventh day. Your total savings for the 7 days is given by

$$25 + 50 + 100 + \cdots + 1600$$

Use the sum formula for a geometric sequence to obtain

$$S_7 = \frac{25(2)^7 - 25}{2 - 1} = \frac{25(2^7 - 1)}{1} = 3175$$

Thus your savings for the entire week is $31.75.   ■

P R O B L E M   2   A pump is attached to a container for the purpose of creating a vacuum. For each stroke of the pump, $\frac{1}{4}$ of the air that remains in the container is removed. To the nearest tenth of a percent, how much of the air remains in the container after six strokes?

Solution
Let's draw a diagram to help with the analysis of this problem.

First stroke: $\frac{1}{4}$ of the air is removed | $1 - \frac{1}{4} = \frac{3}{4}$ of the air remains

Second stroke: $\frac{1}{4}\left(\frac{3}{4}\right) = \frac{3}{16}$ of the air is removed | $\frac{3}{4} - \frac{3}{16} = \frac{9}{16}$ of the air remains

Third stroke: $\frac{1}{4}\left(\frac{9}{16}\right) = \frac{9}{64}$ of the air is removed | $\frac{9}{16} - \frac{9}{64} = \frac{27}{64}$ of the air remains

The diagram suggests two approaches to the problem.

Approach A   The sequence $\frac{1}{4}, \frac{3}{16}, \frac{9}{64}, \ldots$ represents, term by term, the fractional amount of air that is removed with each successive stroke. Therefore, we can find the total amount removed and subtract it from 100%. The sequence is geometric with $a_1 = \frac{1}{4}$ and $r = \frac{3}{4}$.

$$S_6 = \frac{\frac{1}{4} - \frac{1}{4}\left(\frac{3}{4}\right)^6}{1 - \frac{3}{4}} = \frac{\frac{1}{4}\left[1 - \left(\frac{3}{4}\right)^6\right]}{\frac{1}{4}}$$

$$= 1 - \frac{729}{4096} = \frac{3367}{4096} = 82.2\%$$

Therefore, $100\% - 82.2\% = 17.8\%$ of the air remains after six strokes.   ■

Approach B   The sequence

$$\frac{3}{4}, \frac{9}{16}, \frac{27}{64}, \dots$$

represents, term by term, the amount of air that remains in the container after each stroke. Therefore, when we find the sixth term of this geometric sequence, we will have the answer to the problem. Because $a_1 = \dfrac{3}{4}$ and $r = \dfrac{3}{4}$, we obtain

$$a_6 = \frac{3}{4}\left(\frac{3}{4}\right)^5 = \left(\frac{3}{4}\right)^6 = \frac{729}{4096} = 17.8\%$$

Therefore, 17.8% of the air remains after six strokes.   ■

It will be helpful for you to take another look at the two approaches we used to solve Problem 2. Note that in approach B, finding the sixth term of the sequence produced the answer to the problem without any further calculations. In approach A, we had to find the sum of six terms of the sequence and then subtract that amount from 100%. As we solve problems that involve sequences, we *must* understand what each particular sequence represents on a term-by-term basis.

CONCEPT QUIZ

For Problems 1–5, answer true or false.

1. The common ratio of a geometric sequence is found by dividing any term by the following term.

2. The common ratio is always a positive number.

3. For an infinite geometric sequence, the sum of the terms exists if the common ratio is greater than 1.

4. A repeating or terminating decimal can be written as the sum of an infinite geometric sequence.

5. The nth term of a geometric sequence is the first term multiplied by the common ratio raised to the $(n - 1)$ power.

PROBLEM SET 16.2

For Problems 1–12, find the general term (the nth term) for each geometric sequence.

1. $3, 6, 12, 24, \dots$

2. $2, 6, 18, 54, \dots$

3. $3, 9, 27, 81, \dots$

4. $2, 4, 8, 16, \dots$

5. $\dfrac{1}{4}, \dfrac{1}{8}, \dfrac{1}{16}, \dfrac{1}{32}, \dots$

6. $8, 4, 2, 1, \dots$

7. $4, 16, 64, 256, \dots$

8. $6, 2, \dfrac{2}{3}, \dfrac{2}{9}, \dots$

9. $1, 0.3, 0.09, 0.027, \dots$

10. $0.2, 0.04, 0.008, 0.0016, \dots$

11. $1, -2, 4, -8, \dots$

12. $-3, 9, -27, 81, \dots$

For Problems 13–20, find the required term for each geometric sequence.

13. The 8th term of $\frac{1}{2}$, 1, 2, 4, . . .

14. The 7th term of 2, 6, 18, 54, . . .

15. The 9th term of 729, 243, 81, 27, . . .

16. The 11th term of 768, 384, 192, 96, . . .

17. The 10th term of 1, −2, 4, −8, . . .

18. The 8th term of $-1, -\frac{3}{2}, -\frac{9}{4}, -\frac{27}{8}, \ldots$

19. The 8th term of $\frac{1}{2}, \frac{1}{6}, \frac{1}{18}, \frac{1}{54}, \ldots$

20. The 9th term of $\frac{16}{81}, \frac{8}{27}, \frac{4}{9}, \frac{2}{3}, \ldots$

For Problems 21–32, solve each problem.

21. Find the first term of the geometric sequence with 5th term $\frac{32}{3}$ and common ratio 2.

22. Find the first term of the geometric sequence with 4th term $\frac{27}{128}$ and common ratio $\frac{3}{4}$.

23. Find the common ratio of the geometric sequence with 3rd term 12 and 6th term 96.

24. Find the common ratio of the geometric sequence with 2nd term $\frac{8}{3}$ and 5th term $\frac{64}{81}$.

25. Find the sum of the first ten terms of the geometric sequence 1, 2, 4, 8, ...

26. Find the sum of the first seven terms of the geometric sequence 3, 9, 27, 81, ...

27. Find the sum of the first nine terms of the geometric sequence 2, 6, 18, 54, ...

28. Find the sum of the first ten terms of the geometric sequence 5, 10, 20, 40, ...

29. Find the sum of the first eight terms of the geometric sequence 8, 12, 18, 27, ...

30. Find the sum of the first eight terms of the geometric sequence 9, 12, 16, $\frac{64}{3}$, ...

31. Find the sum of the first ten terms of the geometric sequence −4, 8, −16, 32, ...

32. Find the sum of the first nine terms of the geometric sequence −2, 6, −18, 54, ...

For Problems 33–38, find each indicated sum.

33. $9 + 27 + 81 + \cdots + 729$

34. $2 + 8 + 32 + \cdots + 8192$

35. $4 + 2 + 1 + \cdots + \frac{1}{512}$

36. $1 + (-2) + 4 + \cdots + 256$

37. $(-1) + 3 + (-9) + \cdots + (-729)$

38. $16 + 8 + 4 + \cdots + \frac{1}{32}$

For Problems 39–50, find the sum of each infinite geometric sequence. If the sequence has no sum, so state.

39. $2, 1, \frac{1}{2}, \frac{1}{4}, \ldots$

40. $9, 3, 1, \frac{1}{3}, \ldots$

41. $1, \frac{2}{3}, \frac{4}{9}, \frac{8}{27}, \ldots$

42. $5, 3, \frac{9}{5}, \frac{27}{25}, \ldots$

43. $4, 8, 16, 32, \ldots$

44. $32, 16, 8, 4, \ldots$

45. $9, -3, 1, -\frac{1}{3}, \ldots$

46. $2, -6, 18, -54, \ldots$

47. $\frac{1}{2}, \frac{3}{8}, \frac{9}{32}, \frac{27}{128}, \ldots$

48. $4, -\frac{4}{3}, \frac{4}{9}, -\frac{4}{27}, \ldots$

49. $8, -4, 2, -1, \ldots$

50. $7, \frac{14}{5}, \frac{28}{25}, \frac{56}{125}, \ldots$

For Problems 51–62, change each repeating decimal to a/b form, where a and b are integers and $b \neq 0$. Express a/b in reduced form.

51. $0.\overline{3}$

52. $0.\overline{4}$

53. $0.\overline{26}$

54. $0.\overline{18}$

55. $0.\overline{123}$

56. $0.\overline{273}$

57. $0.2\overline{6}$

58. $0.4\overline{3}$

59. $0.2\overline{14}$

60. $0.3\overline{71}$

61. $2.\overline{3}$ –

62. $3.\overline{7}$

For Problems 63–82, use either a geometric or arithmetic sequence to help solve the problem. (Note that some of these problems do require an arithmetic sequence approach!)

63. The enrollment at University X is predicted to increase at the rate of 10% per year. If the enrollment for 1999 was 5000 students, find the predicted enrollment for 2003. Express your answer to the nearest whole number.

64. If you pay $18,000 for a car and it depreciates 20% per year, how much will it be worth in 5 years? Express your answer to the nearest dollar.

65. A tank contains 16,000 liters of water. Each day one-half of the water in the tank is removed and not replaced. How much water remains in the tank at the end of 7 days?

66. If the price of a pound of coffee is $8.20 and the projected rate of inflation is 5% per year, how much per pound should we expect coffee to cost in 5 years? Express your answer to the nearest cent.

67. A tank contains 5832 gallons of water. Each day one-third of the water in the tank is removed and not replaced. How much water remains in the tank at the end of 6 days?

68. A fungus culture growing under controlled conditions doubles in size each day. How many units will the culture contain after 7 days if it originally contained 4 units?

69. Suppose you save a penny the first day of a month, 2 cents the second day, 4 cents the third day, and continue to double your savings each day. How much will you save on the 15th day of the month? How much will your total savings be for the 15 days?

70. Eric saved a nickel the first day of a month, a dime the second day, and 20 cents the third day and then continued to double this daily savings each day for 14 days. What was his daily savings on the 14th day? What was his total savings for the 14 days?

71. Mr. Woodley invested $12,000 at 4% simple interest at the beginning of each year for a period of 8 years. Find the total accumulated value of all the investments at the end of the 8-year period.

72. An object falling from rest in a vacuum falls approximately 16 feet the first second, 48 feet the second second, 80 feet the third second, 112 feet the fourth second, and so on. How far will it fall in 11 seconds?

73. A raffle is organized so that the amount paid for each ticket is determined by the number on the ticket. The tickets are numbered with the consecutive odd whole numbers 1, 3, 5, 7, ... Each contestant pays as many cents as the number on the ticket drawn. How much money will the raffle take in if 1000 tickets are sold?

74. Suppose an element has a half-life of 4 hours. This means that if n grams of it exist at a specific time, then only $\frac{1}{2}n$ grams remain 4 hours later. If at a particular moment we have 60 grams of the element, how many grams of it will remain 24 hours later?

75. Suppose an element has a half-life of 3 hours. (See Problem 74 for a definition of half-life.) If at a particular moment we have 768 grams of the element, how many grams of it will remain 24 hours later?

76. A rubber ball is dropped from a height of 1458 feet, and at each bounce it rebounds one-third of the height from which it last fell. How far has the ball traveled by the time it strikes the ground for the sixth time?

77. A rubber ball is dropped from a height of 100 feet, and at each bounce it rebounds one-half of the height from which it last fell. What distance has the ball traveled up to the instant it hits the ground for the eighth time?

78. A pile of logs has 25 logs in the bottom layer, 24 logs in the next layer, 23 logs in the next layer, and so on, until the top layer has 1 log. How many logs are in the pile?

79. A pump is attached to a container for the purpose of creating a vacuum. For each stroke of the pump, $\frac{1}{3}$ of the air remaining in the container is removed. To the nearest tenth of a percent, how much of the air remains in the container after seven strokes?

80. Suppose that in Problem 79, each stroke of the pump removes $\frac{1}{2}$ of the air remaining in the container. What fractional part of the air has been removed after six strokes?

81. A tank contains 20 gallons of water. One-half of the water is removed and replaced with antifreeze. Then one-half of this mixture is removed and replaced with antifreeze. This process is repeated eight times. How much water remains in the tank after the eighth replacement process?

82. The radiator of a truck contains 10 gallons of water. Suppose we remove 1 gallon of water and replace it with antifreeze. Then we remove 1 gallon of this mixture and replace it with antifreeze. This process is carried out seven times. To the nearest tenth of a gallon, how much antifreeze is in the final mixture?

■ ■ ▫ **Thoughts into words**

83. Your friend solves Problem 64 as follows: If the car depreciates 20% per year, then at the end of 5 years it will have depreciated 100% and be worth zero dollars. How would you convince him that his reasoning is incorrect?

84. A contractor wants you to clear some land for a housing project. He anticipates that it will take 20 working days to do the job. He offers to pay you one of two ways: (1) a fixed amount of $3000 or (2) a penny the first day, 2 cents the second day, 4 cents the third day, and so on, doubling your daily wages each day for the 20 days. Which offer should you take and why?

85. Explain the difference between an arithmetic sequence and a geometric sequence.

86. What does it mean to say that the sum of the infinite geometric sequence $1, \frac{1}{2}, \frac{1}{4}, \frac{1}{8}, \ldots$ is 2?

87. What do we mean when we say that the infinite geometric sequence $1, 2, 4, 8, \ldots$ has no sum?

88. Why don't we discuss the sum of an infinite arithmetic sequence?

Answers to Concept Quiz

1. False   **2.** False   **3.** False   **4.** False   **5.** True

16.3   Fundamental Principle of Counting

Objective

■   Count the number of outcomes possible for accomplishing a task.

One very useful counting principle is referred to as the **fundamental principle of counting.** We will offer some examples, state the property, and then use it to solve a variety of counting problems. Let's consider two examples to lead up to the statement of the property.

PROBLEM 1

A woman has four skirts and five blouses. Assuming that each blouse can be worn with each skirt, how many different skirt-blouse outfits does she have?

Solution

For *each* of the four skirts, she has a choice of five blouses. Therefore, she has $4(5) = 20$ different skirt-blouse outfits from which to choose.   ■

PROBLEM 2

Eric is shopping for a new bicycle and has two different models (5-speed or 10-speed) and four different colors (red, white, blue, or silver) from which to choose. How many different choices does he have?

Solution

His different choices can be counted with the help of a **tree diagram.**

Models	Colors	Choices
5-speed •	Red	5-speed red
	White	5-speed white
	Blue	5-speed blue
	Silver	5-speed silver
10-speed •	Red	10-speed red
	White	10-speed white
	Blue	10-speed blue
	Silver	10-speed silver

For each of the two model choices, there are four choices of color. Altogether, then, Eric has $2(4) = 8$ choices. ■

These two problems exemplify the following general principle.

Fundamental Principle of Counting

If one task can be accomplished in x different ways and, following this task, a second task can be accomplished in y different ways, then the first task followed by the second task can be accomplished in $x \cdot y$ different ways. (This counting principle can be extended to any finite number of tasks.)

As you apply the fundamental principle of counting, it is often helpful to analyze a problem systematically in terms of the tasks to be accomplished. Let's consider some examples.

PROBLEM 3

How many numbers of three different digits each can be formed by choosing from the digits 1, 2, 3, 4, 5, and 6?

Solution

Let's analyze this problem in terms of three tasks.

TASK 1   Choose the hundreds digit, for which there are six choices.

TASK 2   Now choose the tens digit, for which there are only five choices, because one digit was used in the hundreds place.

TASK 3   Now choose the units digit, for which there are only four choices, because two digits have been used for the other places.

Therefore, task 1 followed by task 2 followed by task 3 can be accomplished in $(6)(5)(4) = 120$ ways. In other words, there are 120 numbers of three different digits that can be formed by choosing from the six given digits. ■

Now look back over the solution for Problem 3 and think about each of the following questions.

1. Can we solve the problem by choosing the units digit first, then the tens digit, and finally the hundreds digit?
2. How many three-digit numbers can be formed from 1, 2, 3, 4, 5, and 6 if we do not require each number to have three *different* digits? (Your answer should be 216.)
3. Suppose that the digits from which to choose are 0, 1, 2, 3, 4, and 5. Now how many numbers of three different digits each can be formed, assuming that we do not want zero in the hundreds place? (Your answer should be 100.)
4. Suppose that we want to know the number of *even* numbers with three different digits each that can be formed by choosing from 1, 2, 3, 4, 5, and 6. How many are there? (Your answer should be 60.)

PROBLEM 4

Employee ID numbers at a certain factory consist of one capital letter followed by a three-digit number that contains no repeat digits. For example, A-014 is an ID number. How many such ID numbers can be formed? How many can be formed if repeated digits *are* allowed?

Solution

Again, let's analyze the problem in terms of tasks to be completed.

TASK 1 Choose the letter part of the ID number: there are 26 choices.

TASK 2 Choose the first digit of the three-digit number: there are ten choices.

TASK 3 Choose the second digit: there are nine choices.

TASK 4 Choose the third digit: there are eight choices.

Therefore, applying the fundamental principle, we obtain $(26)(10)(9)(8) = 18{,}720$ possible ID numbers.

If repeat digits were allowed, then there would be $(26)(10)(10)(10) = 26{,}000$ possible ID numbers. ■

PROBLEM 5

In how many ways can Al, Barb, Chad, Dan, and Edna be seated in a row of five seats so that Al and Barb are seated side by side?

Solution

This problem can be analyzed in terms of three tasks.

TASK 1 Choose the two adjacent seats to be occupied by Al and Barb. An illustration such as Figure 16.1 helps us to see that there are four choices for the two adjacent seats.

Figure 16.1

TASK 2 Determine the number of ways in which Al and Barb can be seated. Because Al can be seated on the left and Barb on the right, or vice versa, there are two ways to seat Al and Barb for each pair of adjacent seats.

TASK 3 The remaining three people must be seated in the remaining three seats. This can be done in $(3)(2)(1) = 6$ different ways.

Therefore, by the fundamental principle, task 1 followed by task 2 followed by task 3 can be done in $(4)(2)(6) = 48$ ways. ■

Suppose that in Problem 5, we wanted instead the number of ways in which the five people can sit so that Al and Barb are *not* side by side. We can determine this number by using either of two basically different techniques: (1) analyze and count the number of nonadjacent positions for Al and Barb, or (2) subtract the number of seating arrangements determined in Problem 5 from the total number of ways in which five people can be seated in five seats. Try doing this problem both ways, and see whether you agree with the answer of 72 ways.

As you apply the fundamental principle of counting, you may find that for certain problems, simply thinking about an appropriate tree diagram is helpful, even though the size of the problem may make it inappropriate to write out the diagram in detail. Consider the following problem.

P R O B L E M   6

Suppose that the undergraduate students in three departments — geography, history, and psychology — are to be classified according to sex and year in school. How many categories are needed?

Solution

Let's represent the various classifications symbolically as follows:

M: Male	1. Freshman	G: Geography
F: Female	2. Sophomore	H: History
	3. Junior	P: Psychology
	4. Senior	

We can mentally picture a tree diagram such that each of the two sex classifications branches into four school-year classifications, which in turn branch into three department classifications. Thus we have $(2)(4)(3) = 24$ different categories. ■

Another technique that works on certain problems involves what some people call the *back door* approach. For example, suppose we know that the classroom contains 50 seats. On some days, it may be easier to determine the number of students present by counting the number of empty seats and subtracting from 50 than by counting the number of students in attendance. (We suggested this back door approach as one way to count the nonadjacent seating arrangements in the discussion following Problem 5.) The next example further illustrates this approach.

PROBLEM 7   When rolling a pair of dice, in how many ways can we obtain a sum greater than 4?

Solution

For clarification purposes, let's use a red die and a white die. (It is not necessary to use different-colored dice, but it does help us analyze the different possible outcomes.) With a moment of thought, you will see that there are more ways to get a sum greater than 4 than there are ways to get a sum of 4 or less. Therefore, let's determine the number of possibilities for getting a sum of 4 or less; then we'll subtract that number from the total number of possible outcomes when rolling a pair of dice.

First, we can simply list and count the ways of getting a sum of 4 or less.

Red Die	White Die
1	1
1	2
1	3
2	1
2	2
3	1

There are six ways of getting a sum of 4 or less.

Second, because there are six possible outcomes on the red die and six possible outcomes on the white die, there is a total of $(6)(6) = 36$ possible outcomes when rolling a pair of dice.

Therefore, subtracting the number of ways of getting 4 or less from the total number of possible outcomes, we obtain $36 - 6 = 30$ ways of getting a sum greater than 4. ■

CONCEPT QUIZ

For Problems 1–6, match the problem with the correct expression for the solution.

1. How many different two-digit numbers can be formed by choosing from the digits 1, 2, 3, 4, 5, 6, 7, 8, and 9 if repetition is allowed?

2. How many different even two-digit numbers can be formed by choosing from the digits 1, 2, 3, 4, 5, 6, 7, 8, and 9 if repetition is allowed?

3. How many different odd two-digit numbers can be formed by choosing from the digits 1, 2, 3, 4, 5, 6, 7, 8, and 9 if repetition is allowed?

4. How many different two-digit numbers can be formed by choosing from the digits 1, 2, 3, 4, 5, 6, 7, 8, and 9 if repetition is not allowed?

5. How many different two-digit numbers greater than 30 can be formed by choosing from the digits 1, 2, 3, 4, 5, 6, 7, 8, and 9 if repetition is allowed?

6. How many different two-digit numbers less than 30 can be formed by choosing from the digits 1, 2, 3, 4, 5, 6, 7, 8, and 9 if repetition is allowed?

A. $9 \cdot 8$ **B.** $2 \cdot 9$ **C.** $9 \cdot 4$ **D.** $9 \cdot 9$ **E.** $9 \cdot 5$ **F.** $7 \cdot 9$

PROBLEM SET 16.3

Solve Problems 1–37.

1. If a woman has two skirts and ten blouses, how many different skirt-blouse combinations does she have?

2. If a man has eight shirts, five pairs of slacks, and three pairs of shoes, how many different shirt-slacks-shoe combinations does he have?

3. In how many ways can four people be seated in a row of four seats?

4. How many numbers of two different digits can be formed by choosing from the digits 1, 2, 3, 4, 5, 6, and 7?

5. How many *even* numbers of three different digits can be formed by choosing from the digits 2, 3, 4, 5, 6, 7, 8, and 9?

6. How many *odd* numbers of four different digits can be formed by choosing from the digits 1, 2, 3, 4, 5, 6, 7, and 8?

7. Suppose that the students at a certain university are to be classified according to their college (College of Applied Science, College of Arts and Sciences, College of Business, College of Education, College of Fine Arts, College of Health and Physical Education), sex (female, male), and year in school (1, 2, 3, 4). How many categories are possible?

8. A medical researcher classifies subjects according to sex (female, male), smoking habits (smoker, nonsmoker), and weight (below average, average, above average). How many different combined classifications are used?

9. A pollster classifies voters according to sex (female, male), party affiliation (Democrat, Republican, Independent), and family income (below $10,000, $10,000–$19,999, $20,000–$29,999, $30,000–$39,999, $40,000–$49,999, $50,000 and above). How many combined classifications does the pollster use?

10. A couple is planning to have four children. How many ways can this happen in terms of boy-girl classification? (For example, BBBG indicates that the first three children are boys and the last is a girl.)

11. In how many ways can three officers — president, secretary, and treasurer — be selected from a club that has 20 members?

12. In how many ways can three officers — president, secretary, and treasurer — be selected from a club with 15 female and 10 male members so that the president is female and the secretary and treasurer are male?

13. A disc jockey wants to play six songs once each in a half-hour program. How many different ways can these songs be ordered?

14. A state has agreed to have its automobile license plates consist of two letters followed by four digits. State officials do not want to repeat any letters or digits in any license numbers. How many different license plates will be available?

15. In how many ways can six people be seated in a row of six seats?

16. In how many ways can Al, Bob, Carlos, Don, Ed, and Fern be seated in a row of six seats if Al and Bob want to sit side by side?

17. In how many ways can Amy, Bob, Cindy, Dan, and Elmer be seated in a row of five seats so that neither Amy nor Bob occupies an end seat?

18. In how many ways can Al, Bob, Carlos, Don, Ed, and Fern be seated in a row of six seats if Al and Bob are not to be seated side by side? [*Hint:* Either Al and Bob will be seated side by side or they will not be seated side by side.]

19. In how many ways can Al, Bob, Carol, Dawn, and Ed be seated in a row of five chairs if Al is to be seated in the middle chair?

20. In how many ways can three letters be dropped in five mailboxes?

21. In how many ways can five letters be dropped in three mailboxes?

22. In how many ways can four letters be dropped in six mailboxes so that no two letters go in the same box?

23. In how many ways can six letters be dropped in four mailboxes so that no two letters go in the same box?

24. If five coins are tossed, in how many ways can they fall?

25. If three dice are tossed, in how many ways can they fall?

26. In how many ways can a sum less than ten be obtained when tossing a pair of dice?

27. In how many ways can a sum greater than five be obtained when tossing a pair of dice?

28. In how many ways can a sum greater than four be obtained when tossing three dice?

29. If no number contains repeated digits, how many numbers greater than 400 can be formed by choosing from

the digits 2, 3, 4, and 5? [*Hint:* Consider both three-digit and four-digit numbers.]

30. If no number contains repeated digits, how many numbers greater than 5000 can be formed by choosing from the digits 1, 2, 3, 4, 5, and 6?

31. In how many ways can four boys and three girls be seated in a row of seven seats so that boys and girls occupy alternating seats?

32. In how many ways can three different mathematics books and four different history books be exhibited on a shelf so that all of the books in a subject area are side by side?

33. In how many ways can a true-false test of ten questions be answered?

34. If no number contains repeated digits, how many even numbers greater than 3000 can be formed by choosing from the digits 1, 2, 3, and 4?

35. If no number contains repeated digits, how many odd numbers greater than 40,000 can be formed by choosing from the digits 1, 2, 3, 4, and 5?

36. In how many ways can Al, Bob, Carol, Don, Ed, Faye, and George be seated in a row of seven seats so that Al, Bob, and Carol occupy consecutive seats in some order?

37. The license plates for a certain state consist of two letters followed by a four-digit number such that the first digit of the number is not zero. An example is PK-2446.

   a. How many different license plates can be produced?

   b. How many different plates do not have a repeated letter?

   c. How many plates do not have any repeated digits in the number part of the plate?

   d. How many plates do not have a repeated letter and also do not have any repeated digits?

■ ■ ■ **Thoughts into words**

38. How would you explain the fundamental principle of counting to a friend who missed class the day it was discussed?

39. Give two or three simple illustrations of the fundamental principle of counting.

40. Explain how you solved Problem 29.

16.4   Permutations and Combinations

Objectives

■   Evaluate factorial expressions.

■   Determine the number of permutations or combinations of n objects taken r at a time.

As we develop the material in this section, **factorial notation** becomes very useful. The notation $n!$ (which is read "n factorial") is used with positive integers as follows:

$$1! = 1$$
$$2! = 2 \cdot 1 = 2$$
$$3! = 3 \cdot 2 \cdot 1 = 6$$
$$4! = 4 \cdot 3 \cdot 2 \cdot 1 = 24$$

Note that the factorial notation refers to an *indicated product.* In general, we write

$$n! = n(n - 1)(n - 2) \cdots 3 \cdot 2 \cdot 1$$

We also define $0! = 1$ so that certain formulas will be true for all nonnegative integers.

Now, as an introduction to the first concept of this section, let's consider a counting problem that closely resembles problems from the previous section.

PROBLEM 1   In how many ways can the three letters A, B, and C be arranged in a row?

Solution A

Certainly one approach to the problem is simply to list and count the arrangements.

ABC   ACB   BAC   BCA   CAB   CBA

There are six arrangements of the three letters.

Solution B

Another approach, one that can be generalized for more difficult problems, uses the fundamental principle of counting. Because there are three choices for the first letter of an arrangement, two choices for the second letter, and one choice for the third letter, there are $(3)(2)(1) = 6$ arrangements.   ■

Ordered arrangements are called **permutations.** In general, a permutation of a set of n elements is an ordered arrangement of the n elements; we will use the symbol $P(n, n)$ to denote the number of such permutations. For example, from Problem 1, we know that $P(3, 3) = 6$. Furthermore, by using the same basic approach as in solution B of Problem 1, we can obtain

$$P(1, 1) = 1 = 1!$$
$$P(2, 2) = 2 \cdot 1 = 2!$$
$$P(4, 4) = 4 \cdot 3 \cdot 2 \cdot 1 = 4!$$
$$P(5, 5) = 5 \cdot 4 \cdot 3 \cdot 2 \cdot 1 = 5!$$

In general, the following formula becomes evident.

$$P(n, n) = n!$$

Now suppose that we are interested in the number of two-letter permutations that can be formed by choosing from the four letters A, B, C, and D. (Some examples of such permutations are AB, BA, AC, BC, and CB.) In other words, we want to find the number of two-element permutations that can be formed from a set of four elements. We denote this number by $P(4, 2)$. To find $P(4, 2)$, we can reason as follows. First, we can choose any one of the four letters to occupy the first position in the permutation, and then we can choose any one of the three remaining letters for the second position. Therefore, by the fundamental principle of counting, we have $(4)(3) = 12$ different two-letter permutations; that is, $P(4, 2) = 12$. By using a similar line of reasoning, we can determine the following numbers. (Make sure that you agree with each of these.)

$$P(4, 3) = 4 \cdot 3 \cdot 2 = 24$$
$$P(5, 2) = 5 \cdot 4 = 20$$
$$P(6, 4) = 6 \cdot 5 \cdot 4 \cdot 3 = 360$$
$$P(7, 3) = 7 \cdot 6 \cdot 5 = 210$$

In general, we say that **the number of r-element permutations that can be formed from a set of n elements is given by**

$$P(n, r) = \underbrace{n(n - 1)(n - 2) \cdots}_{r \text{ factors}}$$

Note that the indicated product for $P(n, r)$ begins with n. Thereafter, each factor is 1 less than the previous one, and there is a total of r factors. For example,

$$P(6, 2) = 6 \cdot 5 = 30$$
$$P(8, 3) = 8 \cdot 7 \cdot 6 = 336$$
$$P(9, 4) = 9 \cdot 8 \cdot 7 \cdot 6 = 3024$$

Let's consider two problems that illustrate the use of $P(n, n)$ and $P(n, r)$.

| PROBLEM 2 | In how many ways can five students be seated in a row of five seats? |

Solution

The problem is asking for the number of five-element permutations that can be formed from a set of five elements. Thus we can apply $P(n, n) = n!$.

$$P(5, 5) = 5! = 5 \cdot 4 \cdot 3 \cdot 2 \cdot 1 = 120$$

| PROBLEM 3 | Suppose that seven people enter a swimming race. In how many ways can first, second, and third prizes be awarded? |

Solution

This problem is asking for the number of three-element permutations that can be formed from a set of seven elements. Therefore, using the formula for $P(n, r)$, we obtain

$$P(7, 3) = 7 \cdot 6 \cdot 5 = 210$$

It should be evident that both Problem 2 and Problem 3 could have been solved by applying the fundamental principle of counting. In fact, the formulas for $P(n, n)$ and $P(n, r)$ do not really give us much additional problem-solving power. However, as we will see in a moment, they do provide the basis for developing a formula that is very useful as a problem-solving tool.

Permutations Involving Nondistinguishable Objects

Suppose we have two identical Hs and one T in an arrangement such as HTH. If we switch the two identical Hs, the newly formed arrangement, HTH, will not be distinguishable from the original. In other words, there are fewer distinguishable permutations of n elements when some of those elements are identical than when the n elements are distinctly different.

To see the effect of identical elements on the number of distinguishable permutations, let's look at some specific examples.

2 identical Hs   1 permutation (HH)

2 different letters   2! permutations (HT, TH)

Therefore, having two different letters affects the number of permutations by a *factor* of 2!.

3 identical Hs   1 permutation (HHH)

3 different letters   3! permutations

Therefore, having three different letters affects the number of permutations by a *factor* of 3!.

4 identical Hs   1 permutation (HHHH)

4 different letters   4! permutations

Therefore, having four different letters affects the number of permutations by a *factor* of 4!.

Now let's solve a specific problem.

PROBLEM 4

How many distinguishable permutations can be formed from three identical Hs and two identical Ts?

Solution

If we had five distinctly different letters, we could form 5! permutations. But the three identical Hs affect the number of distinguishable permutations by a factor of 3!, and the two identical Ts affect the number of permutations by a factor of 2!. Therefore, we must divide 5! by 3! and 2!. Thus we obtain

$$\frac{5!}{(3!)(2!)} = \frac{5 \cdot \overset{2}{\cancel{4}} \cdot \cancel{3} \cdot \cancel{2} \cdot 1}{\cancel{3} \cdot \cancel{2} \cdot 1 \cdot \cancel{2} \cdot 1} = 10$$

distinguishable permutations of three Hs and two Ts.   ▪

The type of reasoning used in Problem 4 leads us to the following general counting technique. If there are n elements to be arranged, where there are r_1 of one kind, r_2 of another kind, r_3 of another kind, . . . , r_k of a kth kind, then the total number of distinguishable permutations is given by the expression

$$\frac{n!}{(r_1!)(r_2!)(r_3!) \cdots (r_k!)}$$

PROBLEM 5

How many different 11-letter permutations can be formed from the 11 letters of the word MISSISSIPPI?

Solution

Because there are 4 Is, 4 Ss, and 2 Ps, we can form

$$\frac{11!}{(4!)(4!)(2!)} = \frac{11 \cdot 10 \cdot 9 \cdot 8 \cdot 7 \cdot 6 \cdot 5 \cdot 4 \cdot 3 \cdot 2 \cdot 1}{4 \cdot 3 \cdot 2 \cdot 1 \cdot 4 \cdot 3 \cdot 2 \cdot 1 \cdot 2 \cdot 1} = 34{,}650$$

distinguishable permutations.   ▪

Combinations (Subsets)

Permutations are *ordered* arrangements; however, *order* is often not a consideration. For example, suppose that we want to determine the number of three-person committees that can be formed from the five people Al, Barb, Carol, Dawn, and Eric. Certainly the committee consisting of Al, Barb, and Eric is the same as the committee consisting of Barb, Eric, and Al. In other words, the order in which we choose or list the members is not important. Therefore, we are really dealing with subsets; that is, we are looking for the number of three-element subsets that can be formed from a set of five elements. Traditionally in this context, subsets have been called **combinations.** Stated another way, then, we are looking for the number of combinations of five things taken three at a time. In general, r-element subsets taken from a set of n elements are called **combinations of n things taken r at a time.** The symbol $C(n, r)$ denotes the number of these combinations.

Now let's restate that committee problem and show a detailed solution that can be generalized to handle a variety of problems dealing with combinations.

P R O B L E M   6

How many three-person committees can be formed from the five people Al, Barb, Carol, Dawn, and Eric?

Solution

Let's use the set {A, B, C, D, E} to represent the five people. Consider one possible three-person committee (subset), such as {A, B, C}; there are 3! permutations of these three letters. Now take another committee, such as {A, B, D}; there are also 3! permutations of these three letters. If we were to continue this process with all of the three-letter subsets that can be formed from the five letters, we would be counting all possible three-letter permutations of the five letters. That is, we would obtain $P(5, 3)$. Therefore, if we let $C(5, 3)$ represent the number of three-element subsets, then

$$(3!) \cdot C(5, 3) = P(5, 3)$$

Solving this equation for $C(5, 3)$ yields

$$C(5, 3) = \frac{P(5, 3)}{3!} = \frac{5 \cdot 4 \cdot 3}{3 \cdot 2 \cdot 1} = 10$$

Thus 10 three-person committees can be formed from the five people.   ■

In general, $C(n, r)$ times $r!$ yields $P(n, r)$. Thus

$$(r!) \cdot C(n, r) = P(n, r)$$

and solving this equation for $C(n, r)$ produces

$$C(n, r) = \frac{P(n, r)}{r!}$$

In other words, we can find the number of *combinations* of n things taken r at a time by dividing by $r!$ the number of permutations of n things taken r at a time. The following examples illustrate this idea.

$$C(7, 3) = \frac{P(7, 3)}{3!} = \frac{7 \cdot 6 \cdot 5}{3 \cdot 2 \cdot 1} = 35$$

$$C(9, 2) = \frac{P(9, 2)}{2!} = \frac{9 \cdot 8}{2 \cdot 1} = 36$$

$$C(10, 4) = \frac{P(10, 4)}{4!} = \frac{10 \cdot 9 \cdot 8 \cdot 7}{4 \cdot 3 \cdot 2 \cdot 1} = 210$$

P R O B L E M   7

How many different five-card hands can be dealt from a deck of 52 playing cards?

Solution

Because the order in which the cards are dealt is not an issue, we are working with a combination (subset) problem. Thus, using the formula for $C(n, r)$, we obtain

$$C(52, 5) = \frac{P(52, 5)}{5!} = \frac{52 \cdot 51 \cdot 50 \cdot 49 \cdot 48}{5 \cdot 4 \cdot 3 \cdot 2 \cdot 1} = 2{,}598{,}960$$

There are 2,598,960 different five-card hands that can be dealt from a deck of 52 playing cards.   ■

Some counting problems, such as Problem 8, can be solved by using the fundamental principle of counting along with the combination formula.

P R O B L E M   8

How many committees that consist of three women and two men can be formed from a group of five women and four men?

Solution

Let's think of this problem in terms of two tasks.

> **TASK 1**   Choose a subset of three women from the five women. This can be done in
>
> $$C(5, 3) = \frac{P(5, 3)}{3!} = \frac{5 \cdot 4 \cdot 3}{3 \cdot 2 \cdot 1} = 10 \text{ ways}$$
>
> **TASK 2**   Choose a subset of two men from the four men. This can be done in
>
> $$C(4, 2) = \frac{P(4, 2)}{2!} = \frac{4 \cdot 3}{2 \cdot 1} = 6 \text{ ways}$$

Task 1 followed by task 2 can be done in $(10)(6) = 60$ ways. Therefore, there are 60 committees consisting of three women and two men that can be formed.   ■

Sometimes it takes a little thought to decide whether permutations or combinations should be used. Remember that **if order is to be considered, permutations should be used, but if order does not matter, then use combinations.** It is helpful to think of combinations as subsets.

PROBLEM 9

A small accounting firm has 12 computer programmers. Three of these people are to be promoted to systems analysts. In how many ways can the firm select the three people to be promoted?

Solution

Let's call the people A, B, C, D, E, F, G, H, I, J, K, and L. Suppose A, B, and C are chosen for promotion. Is this any different from choosing B, C, and A? Obviously not, so order does not matter, and we are being asked a question about combinations. More specifically, we need to find the number of combinations of 12 people taken three at a time. Thus there are

$$C(12, 3) = \frac{P(12, 3)}{3!} = \frac{12 \cdot 11 \cdot 10}{3 \cdot 2 \cdot 1} = 220$$

different ways to choose the three people to be promoted.

PROBLEM 10

A club is to elect three officers — president, secretary, and treasurer — from a group of six people, all of whom are willing to serve in any office. In how many different ways can the officers be chosen?

Solution

Let's call the candidates A, B, C, D, E, and F. Is electing A as president, B as secretary, and C as treasurer different from electing B as president, C as secretary, and A as treasurer? Obviously it is, so we are working with permutations. Thus there are

$$P(6, 3) = 6 \cdot 5 \cdot 4 = 120$$

different ways of filling the offices.

CONCEPT QUIZ

For Problems 1–6, answer true or false.

1. The notation $n!$ is read as "n factorial."

2. By definition, $0! = 0$.

3. Ordered arrangements are called permutations.

4. The number of distinguishable permutations when some of the elements are identical is the same as when the elements are distinctly different.

5. The number of two-element subsets that can be formed from a set of six elements is called a combination.

6. In a counting problem when order does matter, permutations should be used.

PROBLEM SET 16.4

In Problems 1–12, evaluate each permutation or combination.

1. $P(5, 3)$

2. $P(8, 2)$

3. $P(6, 4)$

4. $P(9, 3)$

5. $C(7, 2)$

6. $C(8, 5)$

7. $C(10, 5)$

8. $C(12, 4)$

9. $C(15, 2)$

10. $P(5, 5)$

11. $C(5, 5)$

12. $C(11, 1)$

For Problems 13–44, solve each problem.

13. How many permutations of the four letters A, B, C, and D can be formed by using all the letters in each permutation?

14. In how many ways can six students be seated in a row of six seats?

15. How many three-person committees can be formed from a group of nine people?

16. How many two-card hands can be dealt from a deck of 52 playing cards?

17. a. How many three-letter permutations can be formed from the first eight letters of the alphabet if repetitions are not allowed?

   b. How many can be formed if repetitions are allowed?

18. In a seven-team baseball league, in how many ways can the top three positions in the final standings be filled?

19. In how many ways can the manager of a baseball team arrange his batting order of nine starters if he wants his best hitters in the top four positions?

20. In a baseball league of nine teams, how many games are needed to complete the schedule if each team plays 12 games with each other team?

21. How many committees consisting of four women and four men can be chosen from a group of seven women and eight men?

22. How many three-element subsets containing one vowel and two consonants can be formed from the set {a, b, c, d, e, f, g, h, i}?

23. Five associate professors are being considered for promotion to the rank of full professor, but only three will be promoted. How many different combinations of three could be promoted?

24. How many numbers of four different digits can be formed from the digits 1, 2, 3, 4, 5, 6, 7, 8, and 9 if each number must consist of two odd and two even digits?

25. How many three-element subsets containing the letter A can be formed from the set {A, B, C, D, E, F}?

26. How many four-person committees can be chosen from five women and three men if each committee must contain at least one man?

27. How many different seven-letter permutations can be formed from four identical Hs and three identical Ts?

28. How many different eight-letter permutations can be formed from six identical Hs and two identical Ts?

29. How many different nine-letter permutations can be formed from three identical As, four identical Bs, and two identical Cs?

30. How many different ten-letter permutations can be formed from five identical As, four identical Bs, and one C?

31. How many different seven-letter permutations can be formed from the seven letters of the word ALGEBRA?

32. How many different 11-letter permutations can be formed from the 11 letters of the word MATHEMATICS?

33. In how many ways can x^4y^2 be written without using exponents? [*Hint:* One way is *xxxxyy.*]

34. In how many ways can $x^3y^4z^3$ be written without using exponents?

35. Ten basketball players are going to be divided into two teams of five players each for a game. In how many ways can this be done?

36. Ten basketball players are going to be divided into two teams of five in such a way that the two best players are on opposite teams. In how many ways can this be done?

37. A box contains nine good light bulbs and four defective bulbs. How many samples of three bulbs contain one defective bulb? How many samples of three bulbs contain *at least* one defective bulb?

38. How many five-person committees consisting of two juniors and three seniors can be formed from a group of six juniors and eight seniors?

39. In how many ways can six people be divided into two groups so that there are four in one group and two in the other? In how many ways can six people be divided into two groups of three each?

40. How many five-element subsets containing A and B can be formed from the set {A, B, C, D, E, F, G, H}?

41. How many four-element subsets containing A or B but not both A and B can be formed from the set {A, B, C, D, E, F, G}?

42. How many different five-person committees can be selected from nine people if two of those people refuse to serve together on a committee?

43. How many different line segments are determined by five points? By six points? By seven points? By *n* points?

44. a. How many five-card hands consisting of two kings and three aces can be dealt from a deck of 52 playing cards?

b. How many five-card hands consisting of three kings and two aces can be dealt from a deck of 52 playing cards?

c. How many five-card hands consisting of three cards of one face value and two cards of another face value can be dealt from a deck of 52 playing cards?

■ ■ ▨ **Thoughts into words**

45. Explain the difference between a permutation and a combination. Give an example of each one to illustrate your explanation.

46. Your friend is having difficulty distinguishing between permutations and combinations in problem-solving situations. What can you do to help her?

■ ▨ ▨ **Further investigations**

47. In how many ways can six people be seated at a circular table? [*Hint:* Moving each person one place to the right (or left) does not create a new seating.]

48. The quantity $P(8, 3)$ can be expressed completely in factorial notation as follows:

$$P(8, 3) = \frac{P(8, 3) \cdot 5!}{5!} = \frac{(8 \cdot 7 \cdot 6)(5 \cdot 4 \cdot 3 \cdot 2 \cdot 1)}{5!} = \frac{8!}{5!}$$

Express each of the following in terms of factorial notation.

a. $P(7, 3)$

b. $P(9, 2)$

c. $P(10, 7)$

d. $P(n, r)$, $r \le n$, and 0! is defined to be 1

49. Sometimes the formula

$$C(n, r) = \frac{n!}{r!(n - r)!}$$

is used to find the number of combinations of *n* things taken *r* at a time. Use the result from part d of Problem 48 and develop this formula.

50. Compute $C(7, 3)$ and $C(7, 4)$. Compute $C(8, 2)$ and $C(8, 6)$. Compute $C(9, 8)$ and $C(9, 1)$. Now argue that $C(n, r) = C(n, n - r)$ for $r \le n$.

 Graphing calculator activities

Before doing Problems 51–56, be sure that you can use your calculator to compute the number of permutations and combinations. Your calculator may possess a special sequence of keys for such computations. You may need to refer to your user's manual for this information.

51. Use your calculator to check your answers for Problems 1–12.

52. How many different five-card hands can be dealt from a deck of 52 playing cards?

53. How many different seven-card hands can be dealt from a deck of 52 playing cards?

54. How many different five-person committees can be formed from a group of 50 people?

55. How many different juries consisting of 11 people can be chosen from a group of 30 people?

56. How many seven-person committees consisting of three juniors and four seniors can be formed from 45 juniors and 53 seniors?

Answers to Concept Quiz

1. True   **2.** False   **3.** True   **4.** False   **5.** True   **6.** True

---

16.5   Some Basic Probability Ideas

Objectives

■   Determine the number of outcomes in an event or sample space.

■   Apply the definition of probability to calculate probabilities of events.

In order to introduce some terminology and notation, let's consider a simple experiment of tossing a regular six-sided die. There are six possible outcomes to this experiment: The 1, the 2, the 3, the 4, the 5, or the 6 will land up. This set of possible outcomes is called a *sample space,* and the individual elements of the sample space are called *sample points.* We will use S (sometimes with subscripts for identification purposes) to refer to a particular sample space of an experiment; then we will denote the number of sample points by $n(S)$. Thus for the experiment of tossing a die, $S = \{1, 2, 3, 4, 5, 6\}$ and $n(S) = 6$.

In general, the set of all possible outcomes of a given experiment is called the **sample space,** and the individual elements of the sample space are called **sample points.** (In this text, we will be working only with sample spaces that are finite.)

Now suppose we are interested in some of the various possible outcomes in the die-tossing experiment. For example, we might be interested in the event *An even number comes up.* In this case we are satisfied if a 2, 4, or 6 appears on the top face of the die, and therefore the event *An even number comes up* is the subset $E = \{2, 4, 6\}$, where $n(E) = 3$. Perhaps, instead, we might be interested in the event *A multiple of 3 comes up.* This event determines the subset $F = \{3, 6\}$, where $n(F) = 2$.

In general, any subset of a sample space is called an **event** or an **event space.** If the event consists of exactly one element of the sample space, then it is called a **simple event.** Any nonempty event that is not simple is called a **compound event.** A compound event can be represented as the union of simple events.

It is now possible to give a very simple definition for *probability* as we want to use the term in this text.

DEFINITION 16.1

In an experiment where all possible outcomes in the sample space S are equally likely to occur, the **probability** of an event E is defined by

$$P(E) = \frac{n(E)}{n(S)}$$

where $n(E)$ denotes the number of elements in the event E and $n(S)$ denotes the number of elements in the sample space S.

Many probability problems can be solved by applying Definition 16.1. Such an approach requires that we be able to determine the number of elements in the sample space and the number of elements in the event space. For example, returning to the die-tossing experiment, the probability of getting an even number with one toss of the die is given by

$$P(E) = \frac{n(E)}{n(S)} = \frac{3}{6} = \frac{1}{2}$$

Let's consider two examples where both the number of elements in the sample space and the number in the event space are quite easy to determine.

PROBLEM 1

A coin is tossed. Find the probability that a head turns up.

Solution

Let the sample space be $S = \{H, T\}$; then $n(S) = 2$. The event of a head turning up is the subset $E = \{H\}$, so $n(E) = 1$. Therefore, the probability of getting a head with one flip of a coin is given by

$$P(E) = \frac{n(E)}{n(S)} = \frac{1}{2}$$ ∎

PROBLEM 2

Two coins are tossed. What is the probability that *at least* one head will turn up?

Solution

For clarification purposes, let the coins be a penny and a nickel. The possible outcomes of this experiment are (1) a head on both coins, (2) a head on the penny and a tail on the nickel, (3) a tail on the penny and a head on the nickel, and (4) a tail on both coins. Using ordered-pair notation, where the first entry of a pair represents the penny and the second entry the nickel, we can write the sample space

$$S = \{(H, H), (H, T), (T, H), (T, T)\}$$

and $n(S) = 4$.

Let E be the event of getting at least one head. Thus $E = \{(H, H), (H, T), (T, H)\}$ and $n(E) = 3$. Therefore, the probability of getting at least one head with one toss of two coins is

$$P(E) = \frac{n(E)}{n(S)} = \frac{3}{4}$$ ■

As you might expect, the counting techniques discussed in Sections 16.3 and 16.4 can often be used to solve probability problems.

PROBLEM 3   Four coins are tossed. Find the probability of getting three heads and one tail.

Solution

The sample space consists of the possible outcomes for tossing four coins. Because there are two things that can happen on each coin, by the fundamental principle of counting there are $2 \cdot 2 \cdot 2 \cdot 2 = 16$ possible outcomes for tossing four coins. Thus we know that $n(S) = 16$ without taking the time to list all of the elements. The event of getting three heads and one tail is the subset $E = \{(H, H, H, T), (H, H, T, H), (H, T, H, H), (T, H, H, H)\}$, where $n(E) = 4$. Therefore, the requested probability is

$$P(E) = \frac{n(E)}{n(S)} = \frac{4}{16} = \frac{1}{4}$$ ■

PROBLEM 4   Al, Bob, Chad, Dorcas, Eve, and Françoise are randomly seated in a row of six chairs. What is the probability that Al and Bob are seated in the end seats?

Solution

The sample space consists of all possible ways of seating six people in six chairs; in other words, the permutations of six things taken six at a time. Thus $n(S) = P(6, 6) = 6! = 6 \cdot 5 \cdot 4 \cdot 3 \cdot 2 \cdot 1 = 720$.

The event space consists of all possible ways of seating the six people so that Al and Bob both occupy end seats. The number of these possibilities can be determined as follows:

TASK 1   Put Al and Bob in the end seats. This can be done in two ways because Al can be on the left end and Bob on the right end, or vice versa.

TASK 2   Put the other four people in the remaining four seats. This can be done in $4! = 4 \cdot 3 \cdot 2 \cdot 1 = 24$ different ways.

Therefore, task 1 followed by task 2 can be done in $(2)(24) = 48$ different ways, so $n(E) = 48$. Thus the requested probability is

$$P(E) = \frac{n(E)}{n(S)} = \frac{48}{720} = \frac{1}{15}$$

■

Note that in Problem 3, by using the fundamental principle of counting to determine the number of elements in the sample space, we did not actually have to list all of the elements. For the event space, we listed the elements and counted them in the usual way. In Problem 4, we used the permutation formula $P(n, n) = n!$ to determine the number of elements in the sample space, and then we used the fundamental principle of counting to determine the number of elements in the event space. There are no definite rules about when to list the elements and when to apply some sort of counting technique. In general, we suggest that if you do not immediately see a counting pattern for a particular problem, you should begin the listing process. If a counting pattern then emerges as you are listing the elements, use the pattern at that time.

The combination (subset) formula we developed in Section 16.4, $C(n, r) = P(n, r)/r!$, is also a very useful tool for solving certain kinds of probability problems. The next three examples illustrate some problems of this type.

PROBLEM 5

A committee of three people is randomly selected from Alice, Bjorn, Chad, Dee, and Eric. What is the probability that Alice is on the committee?

Solution

The sample space, S, consists of all possible three-person committees that can be formed from the five people. Therefore,

$$n(S) = C(5, 3) = \frac{P(5, 3)}{3!} = \frac{5 \cdot 4 \cdot 3}{3 \cdot 2 \cdot 1} = 10$$

The event space, E, consists of all the three-person committees that have Alice as a member. Each of those committees contains Alice and two other people chosen from the four remaining people. Thus the number of such committees is $C(4, 2)$, so we obtain

$$n(E) = C(4, 2) = \frac{P(4, 2)}{2!} = \frac{4 \cdot 3}{2 \cdot 1} = 6$$

The requested probability is

$$P(E) = \frac{n(E)}{n(S)} = \frac{6}{10} = \frac{3}{5}$$

■

PROBLEM 6

A committee of four is chosen at random from a group of five seniors and four juniors. Find the probability that the committee will contain two seniors and two juniors.

Solution

The sample space, S, consists of all possible four-person committees that can be formed from the nine people. Thus

$$n(S) = C(9, 4)$$
$$= \frac{P(9, 4)}{4!}$$
$$= \frac{9 \cdot 8 \cdot 7 \cdot 6}{4 \cdot 3 \cdot 2 \cdot 1} = 126$$

The event space, E, consists of all four-person committees that contain two seniors and two juniors. They can be counted as follows:

TASK 1 Choose two seniors from the five available seniors in $C(5, 2) = 10$ ways.

TASK 2 Choose two juniors from the four available juniors in $C(4, 2) = 6$ ways.

Therefore, there are $10 \cdot 6 = 60$ committees consisting of two seniors and two juniors. The requested probability is

$$P(E) = \frac{n(E)}{n(S)} = \frac{60}{126} = \frac{10}{21}$$

PROBLEM 7 Eight coins are tossed. Find the probability of getting two heads and six tails.

Solution

Because either of two things can happen on each coin, the total number of possible outcomes, $n(S)$, is $2^8 = 256$.

We can select two coins, which are to fall heads, in $C(8, 2) = 28$ ways. For each of these ways, there is only one way to select the other six coins that are to fall tails. Therefore, there are $28 \cdot 1 = 28$ ways of getting two heads and six tails, so $n(E) = 28$. The requested probability is

$$P(E) = \frac{n(E)}{n(S)} = \frac{28}{256} = \frac{7}{64}$$

CONCEPT QUIZ

For Problems 1–5, answer true or false.

1. The set of all possible outcomes of a given experiment is called the sample space.

2. If two dice are rolled, the event of rolling a number greater than 4 is a simple event.

3. A compound event is the union of simple events.

4. The probability of an event can never be equal to 1.

5. The probability of an event can be equal to 0.

PROBLEM SET 16.5

For Problems 1–4, *two* coins are tossed. Find the probability of tossing each of the following events.

1. One head and one tail   **2.** Two tails

3. At least one tail   **4.** No tails

For Problems 5–8, *three* coins are tossed. Find the probability of tossing each of the following events.

5. Three heads   **6.** Two heads and a tail

7. At least one head   **8.** Exactly one tail

For Problems 9–12, *four* coins are tossed. Find the probability of tossing each of the following events.

9. Four heads   **10.** Three heads and a tail

11. Two heads and two tails

12. At least one head

For Problems 13–16, *one* die is rolled. Find the probability of rolling each of the following events.

13. A multiple of 3   **14.** A prime number

15. An even number   **16.** A multiple of 7

For Problems 17–22, *two* dice are rolled. Find the probability of rolling each of the following events.

17. A sum of 6   **18.** A sum of 11

19. A sum less than 5   **20.** A 5 on exactly one die

21. A 4 on at least one die   **22.** A sum greater than 4

For Problems 23–26, *one* card is drawn from a standard deck of 52 playing cards. Find the probability of each of the following events.

23. A heart is drawn.   **24.** A king is drawn.

25. A spade or a diamond is drawn.

26. A red jack is drawn.

For Problems 27–30, suppose that 25 slips of paper numbered 1 to 25, inclusive, are put in a hat and then one is drawn out at random. Find the probability of each of the following events.

27. The slip with the 5 on it is drawn.

28. A slip with an even number on it is drawn.

29. A slip with a prime number on it is drawn.

30. A slip with a multiple of 6 on it is drawn.

For Problems 31–34, suppose that a committee of two boys is to be chosen at random from the five boys Al, Bill, Carl, Dan, and Eli. Find the probability of each of the following events.

31. Dan is on the committee.

32. Dan and Eli are both on the committee.

33. Bill and Carl are not both on the committee.

34. Dan or Eli, but not both of them, is on the committee.

For Problems 35–38, suppose that a five-person committee is selected at random from the eight people Al, Barb, Chad, Dominique, Eric, Fern, George, and Harriet. Find the probability of each of the following events.

35. Al and Barb are both on the committee.

36. George is not on the committee.

37. Either Chad or Dominique, but not both, is on the committee.

38. Neither Al nor Barb is on the committee.

For Problems 39–41, suppose that a box of ten items from a manufacturing company is known to contain two defective and eight nondefective items. A sample of three items is selected at random. Find the probability of each of the following events.

39. The sample contains all nondefective items.

40. The sample contains one defective and two nondefective items.

41. The sample contains two defective and one nondefective item.

For Problems 42–60, solve each problem.

42. A building has five doors. Find the probability that two people, entering the building at random, will choose the same door.

43. Bill, Carol, and Alice are to be seated at random in a row of three seats. Find the probability that Bill and Carol will be seated side by side.

44. April, Bill, Carl, and Denise are to be seated at random in a row of four chairs. What is the probability that April and Bill will occupy the end seats?

45. A committee of four girls is to be chosen at random from the five girls Alice, Becky, Candy, Dee, and Elaine. Find the probability that Elaine is not on the committee.

46. Three boys and two girls are to be seated at random in a row of five seats. What is the probability that the boys and girls will be in alternating seats?

47. Four different mathematics books and five different history books are randomly placed on a shelf. What is the probability that all of the books on a subject are side by side?

48. Each of three letters is to be mailed in any one of five different mailboxes. What is the probability that all will be mailed in the same mailbox?

49. Randomly form a four-digit number by using the digits 2, 3, 4, and 6 once each. What is the probability that the number formed is greater than 4000?

50. Randomly select one of the 120 permutations of the letters a, b, c, d, and e. Find the probability that in the chosen permutation, the letter a precedes the b (the a is to the left of the b).

51. A committee of four is chosen at random from a group of six women and five men. Find the probability that the committee contains two women and two men.

52. A committee of three is chosen at random from a group of four women and five men. Find the probability that the committee contains at least one man.

53. Ahmed, Bob, Carl, Dan, Ed, Frank, Gino, Harry, Julio, and Mike are randomly divided into two five-man teams for a basketball game. What is the probability that Ahmed, Bob, and Carl are on the same team?

54. Seven coins are tossed. Find the probability of getting four heads and three tails.

55. Nine coins are tossed. Find the probability of getting three heads and six tails.

56. Six coins are tossed. Find the probability of getting at least four heads.

57. Five coins are tossed. Find the probability of getting no more than three heads.

58. Each arrangement of the 11 letters of the word MISSISSIPPI is put on a slip of paper and placed in a hat. One slip is drawn at random from the hat. Find the probability that the slip contains an arrangement of the letters with the four Ss at the beginning.

59. Each arrangement of the seven letters of the word OSMOSIS is put on a slip of paper and placed in a hat. One slip is drawn at random from the hat. Find the probability that the slip contains an arrangement of the letters with an O at the beginning and an O at the end.

60. Consider all possible arrangements of three identical Hs and three identical Ts. Suppose that one of these arrangements is selected at random. What is the probability that the selected arrangement has the three Hs in consecutive positions?

■ ■ ■ **Thoughts into words**

61. Explain the concepts of sample space and event space.

62. Why must probability answers fall between 0 and 1, inclusive? Give an example of a situation for which the probability is zero. Also give an example for which the probability is one.

■ ■ ■ **Further investigations**

In Section 16.4, we found that there are 2,598,960 different five-card hands that can be dealt from a deck of 52 playing cards. Therefore, probabilities for certain kinds of five-card poker hands can be calculated by using 2,598,960 as the number of elements in the sample space. For Problems 63–71, determine how many different five-card poker hands of the indicated type can be obtained.

63. A straight flush (five cards in sequence and of the same suit; aces are both low and high, so A2345 and 10JQKA are both acceptable)

64. Four of a kind (four of the same face value, such as four kings)

65. A full house (three cards of one face value and two cards of another face value)

66. A flush (five cards of the same suit but not in sequence)

67. A straight (five cards in sequence but not all of the same suit)

68. Three of a kind (three cards of one face value and two cards of two different face values)

69. Two pairs

70. Exactly one pair

71. No pairs

Answers to Concept Quiz

1. True   **2.** False   **3.** True   **4.** False   **5.** True

---

16.6   **Binomial Expansions Revisited**

Objectives

■   Use the binomial theorem to expand a binomial.

■   Find specific terms of a binomial expansion.

In Chapter 6, when multiplying polynomials, we used the patterns $(x + y)^2 = x^2 + 2xy + y^2$ and $(x - y)^2 = x^2 - 2xy + y^2$ to square binomials as the following examples demonstrate.

$$(2x + 3y)^2 = (2x)^2 + 2(2x)(3y) + (3y)^2$$
$$= 4x^2 + 12xy + 9y^2$$

$$(a - 4b)^2 = (a)^2 - 2(a)(4b) + (4b)^2$$
$$= a^2 - 8ab + 16b^2$$

Then in Chapter 8 we used the patterns $(x + y)^3 = x^3 + 3x^2y + 3xy^2 + y^3$ and $(x - y)^3 = x^3 - 3x^2y + 3xy^2 - y^3$ to cube binomials.

$$(3x + 4y)^3 = (3x)^3 + 3(3x)^2(4y) + 3(3x)(4y)^2 + (4y)^3$$
$$= 27x^3 + 108x^2y + 144xy^2 + 64y^3$$

$$(2a - b)^3 = (2a)^3 - 3(2a)^2(b) + 3(2a)(b)^2 - (b)^3$$
$$= 8a^3 - 12a^2b + 6ab^2 - b^3$$

Finally, in Chapter 8, we tackled the general binomial expansion pattern for $(x + y)^n$, where n is any positive integer. We did this by first looking at some specific expansions that can be verified by direct multiplication. (Note that the patterns for squaring and cubing a binomial are part of this list.)

$$(x + y)^1 = x + y$$
$$(x + y)^2 = x^2 + 2xy + y^2$$
$$(x + y)^3 = x^3 + 3x^2y + 3xy^2 + y^3$$
$$(x + y)^4 = x^4 + 4x^3y + 6x^2y^2 + 4xy^3 + y^4$$
$$(x + y)^5 = x^5 + 5x^4y + 10x^3y^2 + 10x^2y^3 + 5xy^4 + y^5$$

First, note the pattern of the exponents for x and y on a term-by-term basis. The exponents of x begin with the exponent of the binomial and decrease by 1, term by term, until the last term has x^0, which is 1. The exponents of y begin with zero ($y^0 = 1$) and increase by 1, term by term, until the last term contains y to the power of the binomial. In other words, the variables in the expansion of $(x + y)^n$ have the following pattern.

$$x^n, \qquad x^{n-1}y, \qquad x^{n-2}y^2, \qquad x^{n-3}y^3, \qquad \ldots, \qquad xy^{n-1}, \qquad y^n$$

Note that for each term, the sum of the exponents of x and y is n.

Next, let's arrange the *coefficients* in a triangular formation; this yields an easy-to-remember pattern (Pascal's triangle).

```
                    1           1
              1           2           1
        1           3           3           1
  1           4           6           4           1
1           5          10          10           5           1
```

Row number n in the formation contains the coefficients of the expansion of $(x + y)^n$. For example, the fifth row contains 1   5   10   10   5   1, and these numbers are the coefficients of the terms in the expansion of $(x + y)^5$. Furthermore, each can be formed from the previous row as follows:

1. Start and end each row with 1.
2. All other entries result from adding the two numbers in the row immediately above, one number to the left and one number to the right.

Thus from row 5, we can form row 6.

Row 5:   1   5   10   10   5   1
                   Add   Add   Add   Add   Add
                   ↘ ↙   ↘ ↙   ↘ ↙   ↘ ↙   ↘ ↙

Row 6:   1   6   15   20   15   6   1

Now we can use these seven coefficients and our discussion about the exponents to write out the expansion for $(x + y)^6$.

$$(x + y)^6 = x^6 + 6x^5y + 15x^4y^2 + 20x^3y^3 + 15x^2y^4 + 6xy^5 + y^6$$

The Seemingly Unrelated Become Related

In Sections 16.3 and 16.4 we introduced the fundamental principle of counting and then used that idea to develop some formulas for counting permutations and combinations. On the surface, it may seem as if those topics and binomial expansions in this section have nothing in common. This is not the case; there is an important link between the concepts. Let's take another look at the coefficients in the expansion of $(x + y)^5$.

$$(x + y)^5 = x^5 + 5x^4y^1 + 10x^3y^2 + 10x^2y^3 + 5x^1y^4 + 1y^5$$

$$\uparrow \qquad \uparrow \qquad \uparrow \qquad \uparrow \qquad \uparrow$$

$$C(5, 1) \quad C(5, 2) \quad C(5, 3) \quad C(5, 4) \quad C(5, 5)$$

As indicated by the arrows, the coefficients are numbers that arise as different-sized combinations of five things. To see why this happens, consider the coefficient for the term containing x^3y^2. The two y's (for y^2) come from two of the factors of $(x + y)$, and therefore the three x's (for x^3) must come from the other three factors of $(x + y)$. In other words, the coefficient is $C(5, 2)$.

We can now state a general expansion formula for $(x + y)^n$; this formula is often called the **binomial theorem.** But before stating it, let's make a small switch in notation. Instead of $C(n, r)$, we shall write $\binom{n}{r}$, which will prove to be a little more convenient at this time. The symbol $\binom{n}{r}$ still refers to the number of combinations of n things taken r at a time, but in this context, it is often called a **binomial coefficient.**

Binomial Theorem

For any binomial $(x + y)$ and any natural number n,

$$(x + y)^n = x^n + \binom{n}{1}x^{n-1}y + \binom{n}{2}x^{n-2}y^2 + \cdots + \binom{n}{n}y^n$$

EXAMPLE 1   Expand $(x + y)^7$.

Solution

$$(x + y)^7 = x^7 + \binom{7}{1}x^6y + \binom{7}{2}x^5y^2 + \binom{7}{3}x^4y^3 + \binom{7}{4}x^3y^4 + \binom{7}{5}x^2y^5 + \binom{7}{6}xy^6 + \binom{7}{7}y^7$$

$$= x^7 + 7x^6y + 21x^5y^2 + 35x^4y^3 + 35x^3y^4 + 21x^2y^5 + 7xy^6 + y^7 \qquad \blacksquare$$

E X A M P L E   2   Expand $(x - y)^5$.

Solution
We shall treat $(x - y)^5$ as $[x + (-y)]^5$.

$$[x + (-y)]^5 = x^5 + \binom{5}{1}x^4(-y) + \binom{5}{2}x^3(-y)^2 + \binom{5}{3}x^2(-y)^3 + \binom{5}{4}x(-y)^4 + \binom{5}{5}(-y)^5$$

$$= x^5 - 5x^4y + 10x^3y^2 - 10x^2y^3 + 5xy^4 - y^5$$ ∎

E X A M P L E   3   Expand $(2a + 3b)^4$.

Solution
Let $x = 2a$ and $y = 3b$ in the binomial theorem.

$$(2a + 3b)^4 = (2a)^4 + \binom{4}{1}(2a)^3(3b) + \binom{4}{2}(2a)^2(3b)^2 + \binom{4}{3}(2a)(3b)^3 + \binom{4}{4}(3b)^4$$

$$= 16a^4 + 96a^3b + 216a^2b^2 + 216ab^3 + 81b^4$$ ∎

E X A M P L E   4   Expand $\left(a + \dfrac{1}{n}\right)^5$.

Solution

$$\left(a + \frac{1}{n}\right)^5 = a^5 + \binom{5}{1}a^4\left(\frac{1}{n}\right) + \binom{5}{2}a^3\left(\frac{1}{n}\right)^2 + \binom{5}{3}a^2\left(\frac{1}{n}\right)^3 + \binom{5}{4}a\left(\frac{1}{n}\right)^4 + \binom{5}{5}\left(\frac{1}{n}\right)^5$$

$$= a^5 + \frac{5a^4}{n} + \frac{10a^3}{n^2} + \frac{10a^2}{n^3} + \frac{5a}{n^4} + \frac{1}{n^5}$$ ∎

E X A M P L E   5   Expand $(x^2 - 2y^3)^6$.

Solution

$$[x^2 + (-2y^3)]^6 = (x^2)^6 + \binom{6}{1}(x^2)^5(-2y^3) + \binom{6}{2}(x^2)^4(-2y^3)^2$$

$$+ \binom{6}{3}(x^2)^3(-2y^3)^3 + \binom{6}{4}(x^2)^2(-2y^3)^4$$

$$+ \binom{6}{5}(x^2)(-2y^3)^5 + \binom{6}{6}(-2y^3)^6$$

$$= x^{12} - 12x^{10}y^3 + 60x^8y^6 - 160x^6y^9 + 240x^4y^{12} - 192x^2y^{15} + 64y^{18}$$ ∎

Finding Specific Terms

Sometimes it is convenient to be able to write down the specific term of a binomial expansion without writing out the entire expansion. For example, suppose that we want the sixth term of the expansion $(x + y)^{12}$. We can proceed as follows: The sixth term will contain y^5. (Note in the binomial theorem that the **exponent of y is always one less than the number of the term.**) Because the sum of the exponents for x and y must be 12 (the exponent of the binomial), the sixth term will also contain x^7. The coefficient is $\binom{12}{5}$, where the 5 agrees with the exponent of y^5. Therefore, the sixth term of $(x + y)^{12}$ is

$$\binom{12}{5} x^7 y^5 = 792 x^7 y^5$$

EXAMPLE 6

Find the fourth term of $(3a + 2b)^7$.

Solution

The fourth term will contain $(2b)^3$, and therefore it will also contain $(3a)^4$. The coefficient is $\binom{7}{3}$. Thus the fourth term is

$$\binom{7}{3}(3a)^4(2b)^3 = (35)(81a^4)(8b^3) = 22{,}680a^4b^3 \qquad \blacksquare$$

EXAMPLE 7

Find the sixth term of $(4x - y)^9$.

Solution

The sixth term will contain $(-y)^5$, and therefore it will also contain $(4x)^4$. The coefficient is $\binom{9}{5}$. Thus the sixth term is

$$\binom{9}{5}(4x)^4(-y)^5 = (126)(256x^4)(-y^5)$$
$$= -32{,}256x^4y^5 \qquad \blacksquare$$

CONCEPT QUIZ

For Problems 1–5, answer true or false.

1. For the expansion of $(x + y)^n$, the exponents on x increase term by term.

2. For the expansion of $(x + y)^n$, the exponent for y in the last term is n.

3. The symbol $\binom{n}{r}$ is called the binomial coefficient.

4. The symbol $\binom{n}{r}$ refers to the number of permutations of n things taken r at a time.

5. For the expansion of $(x + y)^n$, the exponent for y is always one less than the number of the term.

PROBLEM SET 16.6

For Problems 1–26, expand and simplify each binomial.

1. $(x + y)^8$ **2.** $(x + y)^9$ **3.** $(x - y)^6$

4. $(x - y)^4$ **5.** $(a + 2b)^4$ **6.** $(3a + b)^4$

7. $(x - 3y)^5$ **8.** $(2x - y)^6$ **9.** $(2a - 3b)^4$

10. $(3a - 2b)^5$ **11.** $(x^2 + y)^5$ **12.** $(x + y^3)^6$

13. $(2x^2 - y^2)^4$ **14.** $(3x^2 - 2y^2)^5$ **15.** $(x + 3)^6$

16. $(x + 2)^7$ **17.** $(x - 1)^9$ **18.** $(x - 3)^4$

19. $\left(1 + \dfrac{1}{n}\right)^4$ **20.** $\left(2 + \dfrac{1}{n}\right)^5$

21. $\left(a - \dfrac{1}{n}\right)^6$ **22.** $\left(2a - \dfrac{1}{n}\right)^5$

23. $(1 + \sqrt{2})^4$ **24.** $(2 + \sqrt{3})^3$

25. $(3 - \sqrt{2})^5$ **26.** $(1 - \sqrt{3})^4$

For Problems 27–36, write the first four terms of each expansion.

27. $(x + y)^{12}$ **28.** $(x + y)^{15}$ **29.** $(x - y)^{20}$

30. $(a - 2b)^{13}$ **31.** $(x^2 - 2y^3)^{14}$ **32.** $(x^3 - 3y^2)^{11}$

33. $\left(a + \dfrac{1}{n}\right)^9$ **34.** $\left(2 - \dfrac{1}{n}\right)^6$

35. $(-x + 2y)^{10}$ **36.** $(-a - b)^{14}$

For Problems 37–46, find the specified term for each binomial expansion.

37. The fourth term of $(x + y)^8$

38. The seventh term of $(x + y)^{11}$

39. The fifth term of $(x - y)^9$

40. The fourth term of $(x - 2y)^6$

41. The sixth term of $(3a + b)^7$

42. The third term of $(2x - 5y)^5$

43. The eighth term of $(x^2 + y^3)^{10}$

44. The ninth term of $(a + b^3)^{12}$

45. The seventh term of $\left(1 - \dfrac{1}{n}\right)^{15}$

46. The eighth term of $\left(1 - \dfrac{1}{n}\right)^{13}$

■■□ Thoughts into words

47. How would you explain binomial expansions to an elementary algebra student?

48. Explain how to find the fifth term of the expansion of $(2x + 3y)^9$ without writing out the entire expansion.

49. Is the tenth term of the expansion $(x - 2)^{15}$ positive or negative? Explain how you determined the answer to this question.

■■□ Further investigations

For Problems 50–53, expand and simplify each complex number.

50. $(1 + 2i)^5$ **51.** $(2 + i)^6$

52. $(2 - i)^6$ **53.** $(3 - 2i)^5$

Answers to Concept Quiz

1. False   **2.** True   **3.** True   **4.** False   **5.** True

CHAPTER 16

SUMMARY

(16.1) The sequence $a_1, a_2, a_3, a_4, \ldots$ is called **arithmetic** if and only if

$$a_{k+1} - a_k = d$$

for every positive integer k. In other words, there is a **common difference,** d, between successive terms.

The **general term** of an arithmetic sequence is given by the formula

$$a_n = a_1 + (n - 1)d$$

where a_1 is the first term, n is the number of terms, and d is the common difference.

The **sum** of the first n terms of an arithmetic sequence is given by the formula

$$S_n = \frac{n(a_1 + a_n)}{2}$$

Many of the problem-solving suggestions offered earlier in this text are still appropriate when we are solving problems that deal with sequences. However, there are also some special suggestions pertaining to sequence problems.

1. Write down the first few terms of the sequence to describe what is taking place in the problem. Drawing a picture or diagram may help with this step.

2. Be sure that you understand, term by term, what the sequence represents in the problem.

3. Determine whether the sequence is arithmetic or geometric. (Those are the only kinds of sequences we are working with in this text.)

4. Determine whether the problem is asking for a specific term or for the sum of a certain number of terms.

(16.2) The sequence $a_1, a_2, a_3, a_4, \ldots$ is called **geometric** if and only if

$$a_{k+1} = ra_k$$

for every positive integer k. There is a **common ratio, r,** between successive terms.

The **general term** of a geometric sequence is given by the formula

$$a_n = a_1 r^{n-1}$$

where a_1 is the first term, n is the number of terms, and r is the common ratio.

The **sum** of the first n terms of a geometric sequence is given by the formula

$$S_n = \frac{a_1 r^n - a_1}{r - 1} \qquad r \neq 1$$

The **sum of an infinite geometric sequence** is given by the formula

$$S_\infty = \frac{a_1}{1 - r} \qquad \text{for } |r| < 1$$

If $|r| \geq 1$, then the sequence has no sum.

Repeating decimals (such as $0.\overline{4}$) can be changed to a/b form, where a and b are integers and $b \neq 0$, by treating them as the sum of an infinite geometric sequence. For example, the repeating decimal $0.\overline{4}$ can be written $0.4 + 0.04 + 0.004 + 0.0004 + \cdots$.

(16.3) The **fundamental principle of counting** states that if a first task can be accomplished in x ways and, following this task, a second task can be accomplished in y ways, then task 1 followed by task 2 can be accomplished in $x \cdot y$ ways. The principle extends to any finite number of tasks. As you solve problems involving the fundamental principle of counting, it is often helpful to analyze the problem in terms of the tasks to be completed.

(16.4) Ordered arrangements are called **permutations.** The number of permutations of n things taken n at a time is given by

$$P(n, n) = n!$$

The number of r-element permutations that can be formed from a set of n elements is given by

$$P(n, r) = \underbrace{n(n - 1)(n - 2) \cdots}_{r \text{ factors}}$$

If there are n elements to be arranged, where there are r_1 of one kind, r_2 of another kind, r_3 of another kind, $\ldots$ r_k of a kth kind, then the number of distinguishable permutations is given by

$$\frac{n!}{(r_1!)(r_2!)(r_3!) \ldots (r_k!)}$$

Combinations are subsets; the order in which the elements appear does not make a difference. The number of r-element combinations (subsets) that can be formed from a set of n elements is given by

$$C(n, r) = \frac{P(n, r)}{r!}$$

Does the order in which the elements appear make any difference? This is a key question to consider when trying to decide whether a particular problem involves permutations or combinations. If the answer to the question is yes, then it is a permutation problem; if the answer is no, then it is a combination problem. Don't forget that combinations are subsets.

(16.5) In an experiment where all possible outcomes in the sample space S are equally likely to occur, the **probability** of an event E is defined by

$$P(E) = \frac{n(E)}{n(S)}$$

where $n(E)$ denotes the number of elements in the event E, and $n(S)$ denotes the number of elements in the sample space S. The numbers $n(E)$ and $n(S)$ can often be determined by using one or more of the previously listed counting techniques. For all events E, it is always

true that $0 \le P(E) \le 1$. That is, all probabilities fall in the range from 0 to 1, inclusive.

(16.6) For any binomial $(x + y)$ and any natural number n,

$$(x + y)^n =$$
$$x^n + \binom{n}{1}x^{n-1}y + \binom{n}{2}x^{n-2}y^2 + \cdots + \binom{n}{n}y^n$$

Note the following patterns in a binomial expansion.

1. In each term, the sum of the exponents of x and y is n.
2. The exponents of x begin with the exponent of the binomial and decrease by 1, term by term, until the last term has x^0, which is 1. The exponents of y begin with zero ($y^0 = 1$) and increase by 1, term by term, until the last term contains y to the power of the binomial.
3. The coefficient of any term is given by $\binom{n}{r}$, where the value of r agrees with the exponent of y for that term. For example, if the term contains y^3, then the coefficient of that term is $\binom{n}{3}$.
4. The expansion of $(x + y)^n$ contains $n + 1$ terms.

CHAPTER 16 REVIEW PROBLEM SET

For Problems 1–10, find the general term (the nth term) for each sequence. These problems include both arithmetic sequences and geometric sequences.

1. $3, 9, 15, 21, \ldots$

2. $\frac{1}{3}, 1, 3, 9, \ldots$

3. $10, 20, 40, 80, \ldots$

4. $5, 2, -1, -4, \ldots$

5. $-5, -3, -1, 1, \ldots$

6. $9, 3, 1, \frac{1}{3}, \ldots$

7. $-1, 2, -4, 8, \ldots$

8. $12, 15, 18, 21, \ldots$

9. $\frac{2}{3}, 1, \frac{4}{3}, \frac{5}{3}, \ldots$

10. $1, 4, 16, 64, \ldots$

For Problems 11–16, find the required term of each of the sequences.

11. The 19th term of $1, 5, 9, 13, \ldots$

12. The 28th term of $-2, 2, 6, 10, \ldots$

13. The 9th term of $8, 4, 2, 1, \ldots$

14. The 8th term of $\frac{243}{32}, \frac{81}{16}, \frac{27}{8}, \frac{9}{4}, \ldots$

15. The 34th term of $7, 4, 1, -2, \ldots$

16. The 10th term of $-32, 16, -8, 4, \ldots$

For Problems 17–36, solve each problem.

17. If the fifth term of an arithmetic sequence is -19 and the eighth term is -34, find the common difference of the sequence.

18. If the 8th term of an arithmetic sequence is 37 and the 13th term is 57, find the 20th term.

19. Find the first term of a geometric sequence if the third term is 5 and the sixth term is 135.

20. Find the common ratio of a geometric sequence if the second term is $\frac{1}{2}$ and the sixth term is 8.

21. Find the sum of the first nine terms of the sequence 81, 27, 9, 3, ...

22. Find the sum of the first 70 terms of the sequence -3, 0, 3, 6, ...

23. Find the sum of the first 75 terms of the sequence 5, 1, $-3, -7, \ldots$.

24. Find the sum of the first 10 terms of the sequence where $a_n = 2^{5-n}$.

25. Find the sum of the first 95 terms of the sequence where $a_n = 7n + 1$.

26. Find the sum $5 + 7 + 9 + \cdots + 137$.

27. Find the sum $64 + 16 + 4 + \cdots + \frac{1}{64}$.

28. Find the sum of all even numbers between 8 and 384, inclusive.

29. Find the sum of all multiples of 3 between 27 and 276, inclusive.

30. Find the sum of the infinite geometric sequence 64, 16, 4, 1, ...

31. Change $0.\overline{36}$ to reduced a/b form, where a and b are integers and $b \neq 0$.

32. Change $0.4\overline{5}$ to reduced a/b form, where a and b are integers and $b \neq 0$.

33. Suppose that your savings account contains $3750 at the beginning of a year. If you withdraw $250 per month from the account, how much will it contain at the end of the year?

34. Sonya decides to start saving dimes. She plans to save 1 dime the first day of April, 2 dimes the second day, 3 dimes the third day, 4 dimes the fourth day, and so on for the 30 days of April. How much money will she save in April?

35. Nancy decides to start saving dimes. She plans to save 1 dime the first day of April, 2 dimes the second day, 4 dimes the third day, 8 dimes the fourth day, and so on for the first 15 days of April. How much will she save in 15 days?

36. A tank contains 61,440 gallons of water. Each day one-fourth of the water is drained out. How much water remains in the tank at the end of 6 days?

Problems 37–50 are counting problems.

37. How many different arrangements of the letters A, B, C, D, E, and F can be made?

38. How many different nine-letter arrangements can be formed from the nine letters of the word APPARATUS?

39. How many odd numbers of three different digits each can be formed by choosing from the digits 1, 2, 3, 5, 7, 8, and 9?

40. In how many ways can Arlene, Brent, Carlos, Dave, Ernie, Frank, and Gladys be seated in a row of seven seats so that Arlene and Carlos are side by side?

41. In how many ways can a committee of three people be chosen from six people?

42. How many committees consisting of three men and two women can be formed from seven men and six women?

43. How many different five-card hands consisting of all hearts can be formed from a deck of 52 playing cards?

44. If no number contains repeated digits, how many numbers greater than 500 can be formed by choosing from the digits 2, 3, 4, 5, and 6?

45. How many three-person committees can be formed from four men and five women so that each committee contains at least one man?

46. How many different four-person committees can be formed from eight people if two particular people refuse to serve together on a committee?

47. How many four-element subsets containing A or B, but not both A and B, can be formed from the set {A, B, C, D, E, F, G, H}?

48. How many different six-letter permutations can be formed from four identical Hs and two identical Ts?

49. How many four-person committees consisting of two seniors, one sophomore, and one junior can be formed from three seniors, four juniors, and five sophomores?

50. In a baseball league of six teams, how many games are needed to complete a schedule if each team plays eight games with each other team?

Problems 51–63 pose some probability questions.

51. If three coins are tossed, find the probability of getting two heads and one tail.

52. If five coins are tossed, find the probability of getting three heads and two tails.

53. What is the probability of getting a sum of 8 with one roll of a pair of dice?

54. What is the probability of getting a sum more than 5 with one roll of a pair of dice?

55. Aimée, Brenda, Chuck, Dave, and Eli are randomly seated in a row of five seats. Find the probability that Aimée and Chuck are not seated side by side.

56. Four girls and three boys are to be randomly seated in a row of seven seats. Find the probability that the girls and boys will be seated in alternating seats.

57. Six coins are tossed. Find the probability of getting at least two heads.

58. Two cards are randomly chosen from a deck of 52 playing cards. What is the probability that two jacks are drawn?

59. Each arrangement of the six letters of the word CYCLIC is put on a slip of paper and placed in a hat. One slip is drawn at random. Find the probability that the slip contains an arrangement with the Y at the beginning.

60. A committee of three is randomly chosen from one man and six women. What is the probability that the man is not on the committee?

61. A four-person committee is selected at random from the eight people Alice, Bob, Carl, Dee, Enrique, Fred, Gina, and Hilda. Find the probability that Alice or Bob, but not both, is on the committee.

62. A committee of three is chosen at random from a group of five men and four women. Find the probability that the committee contains two men and one woman.

63. A committee of four is chosen at random from a group of six men and seven women. Find the probability that the committee contains at least one woman.

For Problems 64–69, expand each binomial and simplify.

64. $(x + 2y)^5$

65. $(x - y)^8$

66. $(a^2 - 3b^3)^4$

67. $\left(x + \dfrac{1}{n}\right)^6$

68. $(1 - \sqrt{2})^5$

69. $(-a + b)^3$

70. Find the fourth term of the expansion of $(x - 2y)^{12}$.

71. Find the tenth term of the expansion of $(3a + b^2)^{13}$.

TEST

applies to all problems in this Chapter Test.

1. Find the general term of the sequence $5, \dfrac{5}{2}, \dfrac{5}{4}, \dfrac{5}{8}, \ldots$

2. Find the general term of the sequence 10, 16, 22, 28, ...

3. Find the 75th term of the sequence 1, 4, 7, 10, ...

4. Find the sum of the first eight terms of the sequence 3, 6, 12, 24, ...

5. Find the sum of the first 45 terms of the sequence for which $a_n = 7n - 2$.

6. Find the sum of the first ten terms of the sequence for which $a_n = 3(2)^n$.

7. Find the sum of the first 150 positive even whole numbers.

8. Find the sum of the infinite geometric sequence $3, \dfrac{3}{2}, \dfrac{3}{4}, \dfrac{3}{8}, \ldots$

9. Change $0.\overline{18}$ to reduced a/b form, where a and b are integers and $b \neq 0$.

10. A tank contains 49,152 liters of gasoline. Each day, three-fourths of the gasoline remaining in the tank is pumped out and not replaced. How much gasoline remains in the tank at the end of 7 days?

11. Suppose that you save a dime the first day of a month, $.20 the second day, and $.40 the third day, and that you continue to double your savings each day for 15 days. Find the total amount that you will save at the end of 15 days.

12. A marching band lines up with 8 members in the front row, 12 members in the second row, 16 members in the third row, and so on, for 10 rows. How many band members are there in the last row? How many members are there in the band?

13. In a baseball league of ten teams, how many games are needed to complete the schedule if each team plays six games against each other team?

14. How many four-element subsets containing A or B, but not both A and B, can be formed from the set {A, B, C, D, E, F, G}?

15. How many five-card hands consisting of two aces, two kings, and one queen can be dealt from a deck of 52 playing cards?

16. How many different nine-letter arrangements can be formed from the nine letters of the word SASSAFRAS?

17. How many committees consisting of four men and three women can be formed from a group of seven men and five women?

18. What is the probability of rolling a sum less than 9 with a pair of dice?

19. Six coins are tossed. Find the probability of getting three heads and three tails.

20. All possible numbers of three different digits each are formed from the digits 1, 2, 3, 4, 5, and 6. If one number is then chosen at random, find the probability that it is greater than 200.

21. A four-person committee is selected at random from Anwar, Barb, Chad, Dick, Edna, Fern, and Giraldo. What is the probability that neither Anwar nor Barb is on the committee?

22. From a group of three men and five women, a three-person committee is selected at random. Find the probability that the committee contains at least one man.

23. Expand and simplify $\left(2 - \dfrac{1}{n}\right)^6$.

24. Expand and simplify $(3x + 2y)^5$.

25. Find the ninth term of the expansion of $\left(x - \dfrac{1}{2}\right)^{12}$.

Answers to Odd-Numbered Problems and All Chapter Review, Chapter Test, and Cumulative Review Problems

CHAPTER 1

Problem Set 1.1 (page 8)
1. 16 **3.** 35 **5.** 51 **7.** 72 **9.** 82
11. 55 **13.** 60 **15.** 66 **17.** 26 **19.** 2
21. 47 **23.** 21 **25.** 11 **27.** 15 **29.** 14
31. 79 **33.** 6 **35.** 74 **37.** 12 **39.** 187
41. 884 **43.** 9 **45.** 18 **47.** 55 **49.** 99
51. 72 **53.** 11 **55.** 48 **57.** 21
59. 40 **61.** 170 **63.** 164 **65.** 153
71. $36 + 12 \div (3 + 3) + 6 \cdot 2$
73. $36 + (12 \div 3 + 3) + 6 \cdot 2$

Problem Set 1.2 (page 14)
1. True **3.** False **5.** True **7.** True
9. True **11.** False **13.** True **15.** False
17. True **19.** False **21.** Prime **23.** Prime
25. Composite **27.** Prime **29.** Composite
31. $2 \cdot 13$ **33.** $2 \cdot 2 \cdot 3 \cdot 3$ **35.** $7 \cdot 7$
37. $2 \cdot 2 \cdot 2 \cdot 7$ **39.** $2 \cdot 2 \cdot 2 \cdot 3 \cdot 5$
41. $3 \cdot 3 \cdot 3 \cdot 5$ **43.** 4 **45.** 8 **47.** 9
49. 12 **51.** 18 **53.** 12 **55.** 24
57. 48 **59.** 140 **61.** 392 **63.** 168
65. 90 **69.** All other even numbers are divisible by 2.
71. 61 **73.** x **75.** xy **77.** $2 \cdot 2 \cdot 19$
79. $3 \cdot 41$ **81.** $5 \cdot 23$ **83.** $3 \cdot 3 \cdot 7 \cdot 7$
85. $3 \cdot 3 \cdot 17$

Problem Set 1.3 (page 21)
1. 2 **3.** -4 **5.** -7 **7.** 6 **9.** -6
11. 8 **13.** -11 **15.** -15 **17.** -7
19. -31 **21.** -19 **23.** 9 **25.** -61
27. -18 **29.** -92 **31.** -5 **33.** -13
35. 12 **37.** 6 **39.** -1 **41.** -45

43. -29 **45.** 27 **47.** -65 **49.** -29
51. -11 **53.** -1 **55.** -8 **57.** -13
59. -35 **61.** -15 **63.** -32 **65.** 2
67. -4 **69.** -31 **71.** -9 **73.** 18
75. 8 **77.** -29 **79.** -7 **81.** 15
83. 1 **85.** 36 **87.** -39 **89.** -24
91. 7 **93.** -1 **95.** 10 **97.** 9
99. -17 **101.** -3 **103.** -10 **105.** -3
107. 11 **109.** 5 **111.** -65 **113.** -100
115. -25 **117.** 130 **119.** 80
121. $-17 + 14 = -3(-3°F)$
123. $3 + (-2) + (-3) + (-5) = -7$ (7 under par)
125. $-2 + 1 + 3 + 1 + (-2) = 1$

Problem Set 1.4 (page 28)
1. -30 **3.** -9 **5.** 7 **7.** -56 **9.** 60
11. -12 **13.** -126 **15.** 154 **17.** -9
19. 11 **21.** 225 **23.** -14 **25.** 0
27. 23 **29.** -19 **31.** 90 **33.** 14
35. Undefined **37.** -4 **39.** -972
41. -47 **43.** 18 **45.** 69 **47.** 4
49. 4 **51.** -6 **53.** 31 **55.** 4 **57.** 28
59. -7 **61.** 10 **63.** -59 **65.** 66
67. 7 **69.** 69 **71.** -7 **73.** 126
75. -70 **77.** 15 **79.** -10 **81.** -25
83. 77 **85.** 104 **87.** 14
89. $800(19) + 800(2) + 800(4)(-1) = 13,600$
91. $5 + 4(-3) = -7$

Problem Set 1.5 (page 37)
1. Distributive property
3. Associative property for addition
5. Commutative property for multiplication
7. Additive inverse property

877

9. Identity property for addition
11. Associative property for multiplication
13. 56   **15.** 7   **17.** 1800   **19.** $-14{,}400$
21. -3700 **23.** 5900   **25.** -338 **27.** -38
29. 7   **31.** $-5x$ **33.** $-3m$ **35.** $-11y$
37. $-3x - 2y$ **39.** $-16a - 4b$ **41.** $-7xy + 3x$
43. $10x + 5$ **45.** $6xy - 4$ **47.** $-6a - 5b$
49. $5ab - 11a$ **51.** $8x + 36$ **53.** $11x + 28$
55. $8x + 44$ **57.** $5a + 29$ **59.** $3m + 29$
61. $-8y + 6$ **63.** -5 **65.** -40 **67.** 72
69. -18 **71.** 37   **73.** -74 **75.** 180
77. 34   **79.** -65

Chapter 1 Review Problem Set (page 39)

1. -3 **2.** -25 **3.** -5 **4.** -15
5. -1 **6.** 2   **7.** -156 **8.** 252   **9.** 6
10. -13 **11.** Prime   **12.** Composite
13. Composite   **14.** Composite   **15.** Composite
16. $2 \cdot 2 \cdot 2 \cdot 3$ **17.** $3 \cdot 3 \cdot 7$ **18.** $3 \cdot 19$
19. $2 \cdot 2 \cdot 2 \cdot 2 \cdot 2 \cdot 2$ **20.** $2 \cdot 2 \cdot 3 \cdot 7$ **21.** 18
22. 12   **23.** 180   **24.** 945   **25.** 66
26. -7 **27.** -2 **28.** 4   **29.** -18
30. 12   **31.** -34 **32.** -27 **33.** -38
34. -93 **35.** 2   **36.** 3   **37.** 35   **38.** 27
39. $175°F$ **40.** 20,602 feet
41. $2(6) - 4 + 3(8) - 1 = 31$ **42.** $3444
43. $8x$ **44.** $-5y - 9$ **45.** $-5x + 4y$
46. $13a - 6b$ **47.** $-ab - 2a$ **48.** $-3xy - y$
49. $10x + 74$ **50.** $2x + 7$ **51.** $-7x - 18$
52. $-3x + 12$ **53.** $-2a + 4$ **54.** $-2a - 4$
55. -59 **56.** -57 **57.** 2   **58.** 1
59. 12   **60.** 13   **61.** 22   **62.** 32
63. -9 **64.** 37   **65.** -39 **66.** -32
67. 9   **68.** -44

Chapter 1 Test (page 42)

1. 7   **2.** 45   **3.** 38   **4.** -11 **5.** -58
6. -58 **7.** 4   **8.** -1 **9.** -20
10. -7 **11.** $-6°F$ **12.** 26   **13.** -36
14. 9   **15.** -57 **16.** -47 **17.** 5
18. Prime   **19.** $2 \cdot 2 \cdot 2 \cdot 3 \cdot 3 \cdot 5$ **20.** 12
21. 72   **22.** Associative property of addition
23. Distributive property   **24.** $-13x + 6y$
25. $-13x - 21$

CHAPTER 2

Problem Set 2.1 (page 51)

1. $\dfrac{2}{3}$ **3.** $\dfrac{2}{3}$ **5.** $\dfrac{5}{3}$ **7.** $-\dfrac{1}{6}$ **9.** $-\dfrac{3}{4}$

11. $\dfrac{27}{28}$ **13.** $\dfrac{6}{11}$ **15.** $\dfrac{3x}{7y}$ **17.** $\dfrac{2x}{5}$

19. $-\dfrac{5a}{13c}$ **21.** $\dfrac{8z}{7x}$ **23.** $\dfrac{5b}{7}$ **25.** $\dfrac{15}{28}$

27. $\dfrac{10}{21}$ **29.** $\dfrac{3}{10}$ **31.** $-\dfrac{4}{3}$ **33.** $\dfrac{7}{5}$

35. $-\dfrac{3}{10}$ **37.** $\dfrac{1}{4}$ **39.** -27 **41.** $\dfrac{35}{27}$

43. $\dfrac{8}{21}$ **45.** $-\dfrac{5}{6y}$ **47.** $2a$ **49.** $\dfrac{2}{5}$

51. $\dfrac{y}{2x}$ **53.** $\dfrac{20}{13}$ **55.** $-\dfrac{7}{9}$ **57.** $\dfrac{2}{9}$ **59.** $\dfrac{2}{5}$

61. $\dfrac{13}{28}$ **63.** $\dfrac{8}{5}$ **65.** -4 **67.** $\dfrac{36}{49}$ **69.** 1

71. $\dfrac{2}{3}$ **73.** $\dfrac{20}{9}$ **75.** $\dfrac{1}{4}$ **77.** $2\dfrac{1}{4}$ cups

79. $1\dfrac{3}{4}$ cups   **81.** $16\dfrac{1}{4}$ yards

85. a. 8   **c.** 40   **e.** 5

87. a. $\dfrac{11}{13}$ **c.** $-\dfrac{37}{41}$ **e.** $\dfrac{6}{11}$ **g.** $\dfrac{7}{11}$

Problem Set 2.2 (page 60)

1. $\dfrac{5}{7}$ **3.** $\dfrac{5}{9}$ **5.** 3   **7.** $\dfrac{2}{3}$ **9.** $-\dfrac{1}{2}$

11. $\dfrac{2}{3}$ **13.** $\dfrac{15}{x}$ **15.** $\dfrac{2}{y}$ **17.** $\dfrac{8}{15}$ **19.** $\dfrac{9}{16}$

21. $\dfrac{37}{30}$ **23.** $\dfrac{59}{96}$ **25.** $-\dfrac{19}{72}$ **27.** $-\dfrac{1}{24}$

29. $-\dfrac{1}{3}$ **31.** $-\dfrac{1}{6}$ **33.** $-\dfrac{31}{7}$ **35.** $-\dfrac{21}{4}$

37. $\dfrac{3y + 4x}{xy}$ **39.** $\dfrac{7b - 2a}{ab}$ **41.** $\dfrac{11}{2x}$ **43.** $\dfrac{4}{3x}$

45. $-\dfrac{2}{5x}$ **47.** $\dfrac{19}{6y}$ **49.** $\dfrac{1}{24y}$ **51.** $-\dfrac{17}{24n}$

53. $\dfrac{5y + 7x}{3xy}$ **55.** $\dfrac{32y + 15x}{20xy}$ **57.** $\dfrac{63y - 20x}{36xy}$

59. $\dfrac{-6y - 5x}{4xy}$ **61.** $\dfrac{3x + 2}{x}$ **63.** $\dfrac{4x - 3}{2x}$

65. $\dfrac{1}{4}$ **67.** $\dfrac{37}{30}$ **69.** $\dfrac{1}{3}$ **71.** $-\dfrac{12}{5}$

73. $-\dfrac{1}{30}$ **75.** 14   **77.** 68   **79.** $\dfrac{7}{26}$

81. $\dfrac{11}{15}x$ **83.** $\dfrac{5}{24}a$ **85.** $\dfrac{4}{3}x$ **87.** $\dfrac{13}{20}n$

89. $\dfrac{20}{9}n$ **91.** $-\dfrac{79}{36}n$ **93.** $\dfrac{13}{14}x + \dfrac{9}{8}y$

95. $-\dfrac{11}{45}x - \dfrac{9}{20}y$ **97.** $2\dfrac{1}{8}$ yards   **99.** $1\dfrac{3}{4}$ miles

101. $36\dfrac{2}{3}$ yards

Problem Set 2.3 (page 71)

1. Real, rational, integer, and negative
3. Real, irrational, and positive
5. Real, rational, noninteger, and positive
7. Real, rational, noninteger, and negative
9. 0.62 **11.** 1.45 **13.** -3.3 **15.** 7.5
17. 7.8 **19.** -0.9 **21.** 1.16 **23.** -0.272
25. -24.3 **27.** 44.8 **29.** 1.2 **31.** -7.4
33. 0.38 **35.** 7.2 **37.** -0.42 **39.** 0.76
41. 4.7 **43.** 4.3 **45.** -14.8 **47.** 1.3
49. $-1.2x$ **51.** $3n$ **53.** $0.5t$
55. $-5.8x + 2.8y$ **57.** $0.1x + 1.2$
59. $-3x - 2.3$ **61.** $\dfrac{11}{12}$ **63.** $\dfrac{4}{3}$
65. 17.3 **67.** -97.8 **69.** 2.2 **71.** 13.75
73. 0.6 **75.** \$9910.00 **77.** 19.1 centimeters
79. 4.7 centimeters **81.** \$6.55
89. a. The denominator has only factors of 2.
 c. $\dfrac{7}{8}, \dfrac{11}{16}, \dfrac{13}{32}, \dfrac{17}{40}, \dfrac{9}{20}$, and $\dfrac{3}{64}$

Problem Set 2.4 (page 78)

1. 64 **3.** 81 **5.** -8 **7.** -9 **9.** 16
11. $\dfrac{16}{81}$ **13.** $-\dfrac{1}{8}$ **15.** $\dfrac{9}{4}$ **17.** 0.027
19. -1.44 **21.** -47 **23.** -33 **25.** 11
27. -75 **29.** -60 **31.** 31 **33.** -13
35. $9x^2$ **37.** $12xy^2$ **39.** $-18x^4y$ **41.** $15xy$
43. $12x^4$ **45.** $8a^5$ **47.** $-8x^2$ **49.** $4y^3$
51. $-2x^2 + 6y^2$ **53.** $-\dfrac{11}{60}n^2$ **55.** $-2x^2 - 6x$
57. $7x^2 - 3x + 8$ **59.** $\dfrac{3y}{5}$ **61.** $\dfrac{11}{3y}$ **63.** $\dfrac{7b^2}{17a}$
65. $-\dfrac{3ac}{4}$ **67.** $\dfrac{x^2y^2}{4}$ **69.** $\dfrac{4x}{9}$ **71.** $\dfrac{5}{12ab}$
73. $\dfrac{6y^2 + 5x}{xy^2}$ **75.** $\dfrac{5 - 7x^2}{x^4}$ **77.** $\dfrac{3 + 12x^2}{2x^3}$
79. $\dfrac{13}{12x^2}$ **81.** $\dfrac{11b^2 - 14a^2}{a^2b^2}$ **83.** $\dfrac{3 - 8x}{6x^3}$
85. $\dfrac{3y - 4x - 5}{xy}$ **87.** 79 **89.** $\dfrac{23}{36}$ **91.** $\dfrac{25}{4}$
93. -64 **95.** -25 **97.** -33 **99.** 0.45

Problem Set 2.5 (page 86)

Answers may vary somewhat for Problems 1–11.
1. The difference of a and b
3. One-third of the product of B and h
5. Two times the quantity, l plus w

7. The quotient of A divided by w
9. The quantity, a plus b, divided by 2
11. Two more than three times y **13.** $l + w$
15. ab **17.** $\dfrac{d}{t}$ **19.** lwh **21.** $y - x$
23. $xy + 2$ **25.** $7 - y^2$ **27.** $\dfrac{x - y}{4}$
29. $10 - x$ **31.** $10(n + 2)$ **33.** $xy - 7$
35. $xy - 12$ **37.** $35 - n$ **39.** $n + 45$
41. $y + 10$ **43.** $2x - 3$ **45.** $10d + 25q$
47. $\dfrac{d}{t}$ **49.** $\dfrac{d}{p}$ **51.** $\dfrac{d}{12}$ **53.** $n + 1$ **55.** $n + 2$
57. $3y - 2$ **59.** $36y + 12f$ **61.** $\dfrac{f}{3}$ **63.** $8w$
65. $3l - 4$ **67.** $48f + 72$ **69.** $2w^2$ **71.** $9s^2$

Chapter 2 Review Problem Set (page 88)

1. 64 **2.** -27 **3.** -16 **4.** 125 **5.** $\dfrac{9}{16}$ **6.** $-\dfrac{1}{4}$
7. $\dfrac{49}{36}$ **8.** 0.216 **9.** 0.0144 **10.** 0.0036
11. $-\dfrac{8}{27}$ **12.** $\dfrac{1}{16}$ **13.** $-\dfrac{1}{64}$ **14.** 0.25 **15.** $\dfrac{4}{9}$
16. $\dfrac{19}{24}$ **17.** $\dfrac{39}{70}$ **18.** $\dfrac{1}{15}$ **19.** $\dfrac{14y + 9x}{2xy}$
20. $\dfrac{5x - 8y}{x^2y}$ **21.** $\dfrac{7y}{20}$ **22.** $\dfrac{4x^3}{5y^2}$ **23.** $\dfrac{2}{7}$
24. 1 **25.** $\dfrac{27n^2}{28}$ **26.** $\dfrac{1}{24}$ **27.** $-\dfrac{13}{8}$
28. $\dfrac{7}{9}$ **29.** $\dfrac{29}{12}$ **30.** $\dfrac{1}{2}$ **31.** 0.67
32. 0.49 **33.** 2.4 **34.** -0.11 **35.** 1.76
36. 36 **37.** 1.92 **38.** $\dfrac{5}{56}x^2 + \dfrac{7}{20}y^2$
39. $-0.58ab + 0.36bc$ **40.** $\dfrac{11x}{24}$ **41.** $2.2a + 1.7b$
42. $-\dfrac{1}{10}n$ **43.** $\dfrac{41}{20}n$ **44.** $\dfrac{19}{42}$ **45.** $-\dfrac{1}{72}$
46. -0.75 **47.** -0.35 **48.** $\dfrac{1}{17}$ **49.** -8
50. $72 - n$ **51.** $p + 10d$ **52.** $\dfrac{x}{60}$ **53.** $2y - 3$
54. $5n + 3$ **55.** $36y + 12f$ **56.** $100m$
57. $5n + 10d + 25q$ **58.** $n - 5$ **59.** $5 - n$
60. $10(x - 2)$ **61.** $10x - 2$ **62.** $x - 3$
63. $\dfrac{d}{r}$ **64.** $x^2 + 9$ **65.** $(x + 9)^2$ **66.** $x^3 + y^3$
67. $xy - 4$

Chapter 2 Test (page 90)

1. a. 81   **b.** -64 **c.** 0.008   **2.** $\dfrac{7}{9}$

3. $\dfrac{9xy}{16}$ **4.** -2.6 **5.** 3.04   **6.** -0.56

7. $\dfrac{1}{256}$ **8.** $\dfrac{2}{9}$ **9.** $-\dfrac{5}{24}$ **10.** $\dfrac{187}{60}$ or $3\dfrac{7}{60}$

11. $-\dfrac{13}{48}$ **12.** $\dfrac{4y}{5}$ **13.** $2x^2$ **14.** $\dfrac{4y^2 - 5x}{xy^2}$

15. $\dfrac{8}{3x}$ **16.** $\dfrac{35y + 27}{21y^2}$ **17.** $\dfrac{10a^2b}{9}$

18. $-x + 5xy$ **19.** $-3a^2 - 2b^2$ **20.** $\dfrac{37}{36}$

21. -0.48 **22.** $-\dfrac{31}{40}$ **23.** 2.85

24. $5n + 10d + 25q$ **25.** $4n - 3$

Cumulative Review Problem Set (Chapters 1 and 2) (page 91)

1. 10   **2.** -30 **3.** 1   **4.** -26 **5.** -29

6. 17   **7.** $\dfrac{1}{2}$ **8.** $-\dfrac{7}{6}$ **9.** $\dfrac{1}{36}$

10. -64 **11.** 200   **12.** 0.173   **13.** -142

14. 136   **15.** $\dfrac{19}{9}$ **16.** -0.01 **17.** -2.4

18. $\dfrac{79}{40}$ **19.** $\dfrac{1}{4}$ **20.** $\dfrac{3}{5}$ **21.** $2 \cdot 3 \cdot 3 \cdot 3$

22. $2 \cdot 3 \cdot 13$ **23.** $7 \cdot 13$ **24.** $3 \cdot 3 \cdot 17$

25. 14   **26.** 9   **27.** 4   **28.** 6   **29.** 140

30. 200   **31.** 108   **32.** 80   **33.** $-\dfrac{1}{12}x - \dfrac{11}{12}y$

34. $-\dfrac{1}{15}n$ **35.** $-3a + 1.9b$ **36.** $-2n + 6$

37. $-x - 15$ **38.** $-9a - 13$ **39.** $\dfrac{11}{48}$

40. $-\dfrac{31}{36}$ **41.** $\dfrac{5 - 2y + 3x}{xy}$ **42.** $\dfrac{-7y + 9x}{x^2y}$

43. $\dfrac{2x}{3}$ **44.** $\dfrac{8a^2}{21b}$ **45.** $\dfrac{4x^2}{3y}$ **46.** $-\dfrac{27}{16}$

47. $p + 5n + 10d$ **48.** $4n - 5$

49. $36y + 12f + i$ **50.** $200x + 200y$ or $200(x + y)$

CHAPTER 3

Problem Set 3.1 (page 100)

1. {8}   **3.** $\{-6\}$ **5.** $\{-9\}$ **7.** $\{-6\}$

9. {13}   **11.** {48}   **13.** {23}   **15.** $\{-7\}$

17. $\left\{\dfrac{17}{12}\right\}$ **19.** $\left\{-\dfrac{4}{15}\right\}$ **21.** {0.27}   **23.** $\{-3.5\}$

25. $\{-17\}$ **27.** $\{-35\}$ **29.** $\{-8\}$ **31.** $\{-17\}$

33. $\left\{\dfrac{37}{5}\right\}$ **35.** $\{-3\}$ **37.** $\left\{\dfrac{13}{2}\right\}$ **39.** {144}

41. {24}   **43.** $\{-15\}$ **45.** {24}   **47.** $\{-35\}$

49. $\left\{\dfrac{3}{10}\right\}$ **51.** $\left\{-\dfrac{9}{10}\right\}$ **53.** $\left\{\dfrac{1}{2}\right\}$ **55.** $\left\{-\dfrac{1}{3}\right\}$

57. $\left\{\dfrac{27}{32}\right\}$ **59.** $\left\{-\dfrac{5}{14}\right\}$ **61.** $\left\{-\dfrac{7}{5}\right\}$ **63.** $\left\{-\dfrac{1}{12}\right\}$

65. $\left\{-\dfrac{3}{20}\right\}$ **67.** {0.3}   **69.** {9}   **71.** $\{-5\}$

Problem Set 3.2 (page 106)

1. {4}   **3.** {6}   **5.** {8}   **7.** {11}   **9.** $\left\{\dfrac{17}{6}\right\}$

11. $\left\{\dfrac{19}{2}\right\}$ **13.** {6}   **15.** $\{-1\}$ **17.** $\{-5\}$

19. $\{-6\}$ **21.** $\left\{\dfrac{11}{2}\right\}$ **23.** $\{-2\}$ **25.** $\left\{\dfrac{10}{7}\right\}$

27. {18}   **29.** $\left\{-\dfrac{25}{4}\right\}$ **31.** $\{-7\}$ **33.** $\left\{-\dfrac{24}{7}\right\}$

35. $\left\{\dfrac{5}{2}\right\}$ **37.** $\left\{\dfrac{4}{17}\right\}$ **39.** $\left\{-\dfrac{12}{5}\right\}$ **41.** 9

43. 22   **45.** \$18   **47.** 35 years old

49. \$7.25   **51.** 6   **53.** 5   **55.** 11

57. 8   **59.** 3   **61.** \$300   **63.** 4 meters

65. 341 million   **67.** 1.25 hours

Problem Set 3.3 (page 112)

1. {5}   **3.** $\{-8\}$ **5.** $\left\{\dfrac{8}{5}\right\}$ **7.** $\{-11\}$

9. $\left\{-\dfrac{5}{2}\right\}$ **11.** $\{-9\}$ **13.** {2}   **15.** $\{-3\}$

17. $\left\{\dfrac{13}{2}\right\}$ **19.** $\left\{\dfrac{5}{3}\right\}$ **21.** {17}   **23.** $\left\{-\dfrac{13}{2}\right\}$

25. $\left\{\dfrac{16}{3}\right\}$ **27.** {2}   **29.** $\left\{-\dfrac{1}{3}\right\}$ **31.** $\left\{-\dfrac{19}{10}\right\}$

33. 17   **35.** 35 and 37   **37.** 36, 38, and 40

39. $\dfrac{3}{2}$ **41.** -6 **43.** 32° and 58°

45. 50° and 130°   **47.** 65° and 75°   **49.** \$42

51. \$9 per hour   **53.** 150 men and 450 women

55. \$91   **57.** \$145

Problem Set 3.4 (page 121)

1. {1}   **3.** {10}   **5.** $\{-9\}$ **7.** $\left\{\dfrac{29}{4}\right\}$

9. $\left\{-\dfrac{17}{3}\right\}$ **11.** {10} **13.** {44} **15.** {26}

17. {All reals} **19.** $\varnothing$ **21.** {3} **23.** {-1}

25. {-2} **27.** {16} **29.** $\left\{\dfrac{22}{3}\right\}$ **31.** {All reals}

33. $\left\{-\dfrac{1}{6}\right\}$ **35.** {-57} **37.** $\left\{-\dfrac{7}{5}\right\}$ **39.** {2}

41. {-3} **43.** $\left\{\dfrac{27}{10}\right\}$ **45.** $\left\{\dfrac{3}{28}\right\}$ **47.** $\left\{\dfrac{18}{5}\right\}$

49. $\left\{\dfrac{24}{7}\right\}$ **51.** {5} **53.** {0} **55.** $\left\{-\dfrac{51}{10}\right\}$

57. {-12} **59.** {15} **61.** 7 and 8
63. 14, 15, and 16 **65.** 6 and 11 **67.** 48
69. 12 and 18 **71.** 8 feet and 12 feet
73. 15 nickels and 20 quarters
75. 40 nickels, 80 dimes, and 90 quarters
77. 8 dimes and 10 quarters
79. 4 crabs, 12 fish, and 6 plants **81.** 30°
83. 20°, 50°, and 110° **85.** 40°
93. Any three consecutive integers

Problem Set 3.5 (page 131)
1. True **3.** False **5.** False **7.** True **9.** True

11. {$x \mid x > -2$} or $(-2, \infty)$

13. {$x \mid x \le 3$} or $(-\infty, 3]$

15. {$x \mid x > 2$} or $(2, \infty)$

17. {$x \mid x \le -2$} or $(-\infty, -2]$

19. {$x \mid x < -1$} or $(-\infty, -1)$

21. {$x \mid x < 2$} or $(-\infty, 2)$

23. {$x \mid x < -20$} or $(-\infty, -20)$
25. {$x \mid x \ge -9$} or $[-9, \infty)$ **27.** {$x \mid x > 9$} or $(9, \infty)$

29. $\left\{x \mid x < \dfrac{10}{3}\right\}$ or $\left(-\infty, \dfrac{10}{3}\right)$

31. {$x \mid x < -8$} or $(-\infty, -8)$ **33.** {$n \mid n \ge 8$} or $[8, \infty)$

35. $\left\{n \mid n > -\dfrac{24}{7}\right\}$ or $\left(-\dfrac{24}{7}, \infty\right)$

37. {$n \mid n > 7$} or $(7, \infty)$ **39.** {$x \mid x > 5$} or $(5, \infty)$
41. {$x \mid x \le 6$} or $(-\infty, 6]$
43. {$x \mid x \le -21$} or $(-\infty, -21]$

45. $\left\{x \mid x < \dfrac{8}{3}\right\}$ or $\left(-\infty, \dfrac{8}{3}\right)$

47. $\left\{x \mid x < \dfrac{5}{4}\right\}$ or $\left(-\infty, \dfrac{5}{4}\right)$

49. {$x \mid x < 1$} or $(-\infty, 1)$ **51.** {$t \mid t \ge 4$} or $[4, \infty)$
53. {$x \mid x > 14$} or $(14, \infty)$

55. $\left\{x \mid x > \dfrac{3}{2}\right\}$ or $\left(\dfrac{3}{2}, \infty\right)$ **57.** $\left\{t \mid t \ge \dfrac{1}{4}\right\}$ or $\left[\dfrac{1}{4}, \infty\right)$

59. $\left\{x \mid x < -\dfrac{9}{4}\right\}$ or $\left(-\infty, -\dfrac{9}{4}\right)$

65. All real numbers **67.** $\varnothing$
69. All real numbers **71.** $\varnothing$

Problem Set 3.6 (page 138)
1. {$x \mid x > 2$} or $(2, \infty)$ **3.** {$x \mid x < -1$} or $(-\infty, -1)$

5. $\left\{x \mid x > -\dfrac{10}{3}\right\}$ or $\left(-\dfrac{10}{3}, \infty\right)$

7. {$n \mid n \ge -11$} or $[-11, \infty)$ **9.** {$t \mid t \le 11$} or $(-\infty, 11]$

11. $\left\{x \mid x > -\dfrac{11}{5}\right\}$ or $\left(-\dfrac{11}{5}, \infty\right)$

13. $\left\{x \mid x < \dfrac{5}{2}\right\}$ or $\left(-\infty, \dfrac{5}{2}\right)$ **15.** {$x \mid x \le 8$} or $(-\infty, 8]$

17. $\left\{n \mid n > \dfrac{3}{2}\right\}$ or $\left(\dfrac{3}{2}, \infty\right)$ **19.** {$y \mid y > -3$} or $(-3, \infty)$

21. $\left\{x \mid x < \dfrac{5}{2}\right\}$ or $\left(-\infty, \dfrac{5}{2}\right)$

23. {$x \mid x < 8$} or $(-\infty, 8)$ **25.** {$x \mid x < 21$} or $(-\infty, 21)$
27. {$x \mid x < 6$} or $(-\infty, 6)$

29. $\left\{n \mid n > -\dfrac{17}{2}\right\}$ or $\left(-\dfrac{17}{2}, \infty\right)$

31. {$n \mid n \le 42$} or $(-\infty, 42]$

33. $\left\{n \mid n > -\dfrac{9}{2}\right\}$ or $\left(-\dfrac{9}{2}, \infty\right)$

35. $\left\{x \mid x > \dfrac{4}{3}\right\}$ or $\left(\dfrac{4}{3}, \infty\right)$ **37.** {$n \mid n \ge 4$} or $[4, \infty)$

39. {$t \mid t > 300$} or $(300, \infty)$ **41.** {$x \mid x \le 50$} or $(-\infty, 50]$
43. {$x \mid x > 0$} or $(0, \infty)$ **45.** {$x \mid x > 64$} or $(64, \infty)$

47. $\left\{n \mid n > \dfrac{33}{5}\right\}$ or $\left(\dfrac{33}{5}, \infty\right)$

49. $\left\{x \mid x \ge -\dfrac{16}{3}\right\}$ or $\left[-\dfrac{16}{3}, \infty\right)$

51. **53.**

55. **57.**

59. **61.** $\varnothing$

63. **65.** all reals

67. All numbers greater than 7 **69.** 15 inches
71. 158 or higher **73.** Greater than 90
75. More than 250 sales **77.** 77 or less

Chapter 3 Review Problem Set (page 141)

1. $\{-3\}$ **2.** $\{1\}$ **3.** $\left\{-\dfrac{3}{4}\right\}$ **4.** $\{9\}$

5. $\{-4\}$ **6.** $\left\{\dfrac{40}{3}\right\}$ **7.** $\left\{\dfrac{9}{4}\right\}$ **8.** $\left\{-\dfrac{15}{8}\right\}$

9. $\{-7\}$ **10.** $\left\{\dfrac{2}{41}\right\}$ **11.** $\left\{\dfrac{19}{7}\right\}$ **12.** $\left\{\dfrac{1}{2}\right\}$

13. $\{-32\}$ **14.** $\{-12\}$ **15.** $\{21\}$ **16.** $\{-60\}$

17. $\{10\}$ **18.** $\left\{-\dfrac{11}{4}\right\}$ **19.** $\left\{-\dfrac{8}{5}\right\}$ **20.** $\left\{\dfrac{5}{21}\right\}$

21. $\{x \mid x > 4\}$ or $(4, \infty)$ **22.** $\{x \mid x > -4\}$ or $(-4, \infty)$

23. $\{x \mid x \geq 13\}$ or $[13, \infty)$

24. $\left\{x \mid x \geq \dfrac{11}{2}\right\}$ or $\left[\dfrac{11}{2}, \infty\right)$ **25.** $\{x \mid x > 35\}$ or $(35, \infty)$

26. $\left\{x \mid x < \dfrac{26}{5}\right\}$ or $\left(-\infty, \dfrac{26}{5}\right)$

27. $\{n \mid n < 2\}$ or $(-\infty, 2)$

28. $\left\{n \mid n > \dfrac{5}{11}\right\}$ or $\left(\dfrac{5}{11}, \infty\right)$

29. $\{y \mid y < 24\}$ or $(-\infty, 24)$

30. $\{x \mid x > 10\}$ or $(10, \infty)$

31. $\left\{n \mid n < \dfrac{2}{11}\right\}$ or $\left(-\infty, \dfrac{2}{11}\right)$

32. $\{n \mid n > 33\}$ or $(33, \infty)$ **33.** $\{n \mid n \leq 120\}$ or $(-\infty, 120]$

34. $\left\{n \mid n \leq -\dfrac{180}{13}\right\}$ or $\left(-\infty, -\dfrac{180}{13}\right]$

35. $\left\{x \mid x > \dfrac{9}{2}\right\}$ or $\left(\dfrac{9}{2}, \infty\right)$

36. $\left\{x \mid x < -\dfrac{43}{3}\right\}$ or $\left(-\infty, -\dfrac{43}{3}\right)$

37.   **38.**

39. all reals   **40.**

41. 24   **42.** 7   **43.** 33   **44.** 8

45. 89 or higher   **46.** 16 and 24   **47.** 18

48. 88 or higher   **49.** 8 nickels and 22 dimes

50. 8 nickels, 25 dimes, and 50 quarters   **51.** $52°$

52. 700 miles

Chapter 3 Test (page 144)

1. $\{2\}$ **2.** $\{3\}$ **3.** $\{-9\}$ **4.** $\{-5\}$

5. $\{-53\}$ **6.** $\{-18\}$ **7.** $\left\{-\dfrac{5}{2}\right\}$ **8.** $\left\{\dfrac{35}{18}\right\}$

9. $\{12\}$ **10.** $\left\{\dfrac{11}{5}\right\}$ **11.** $\{22\}$ **12.** $\left\{\dfrac{31}{2}\right\}$

13. $\{x \mid x < 5\}$ or $(-\infty, 5)$ **14.** $\{x \mid x \leq 1\}$ or $(-\infty, 1]$

15. $\{x \mid x \geq -9\}$ or $[-9, \infty)$ **16.** $\{x \mid x < 0\}$ or $(-\infty, 0)$

17. $\left\{x \mid x > -\dfrac{23}{2}\right\}$ or $\left(-\dfrac{23}{2}, \infty\right)$

18. $\{n \mid n \geq 12\}$ or $[12, \infty)$

19.   **20.**

21. \$72.00 per hour

22. 15 meters, 25 meters, and 30 meters

23. 96 or higher

24. 17 nickels, 33 dimes, and 53 quarters

25. $60°, 30°$, and $90°$

Cumulative Practice Test (Chapters 1–3) (page 145)

1. 3   **2.** -128 **3.** -2.4 **4.** $-\dfrac{5}{12}$

5. 20   **6.** $-\dfrac{19}{90}$ **7.** 0.09   **8.** 12   **9.** 36

10. $-\dfrac{1}{24}x - \dfrac{9}{28}y$ **11.** $3x + 40$ **12.** $\dfrac{2}{9}x^3$

13. $\dfrac{5y^2 - 6x}{xy^2}$ **14.** $\dfrac{3b^4}{a^2}$ **15.** $\dfrac{25y - 6}{15y^2}$

16. $\{16\}$ **17.** $\{35\}$ **18.** $\left\{-\dfrac{17}{11}\right\}$

19. $\{x \mid x \leq 16\}$ or $(-\infty, 16]$

20. $\{x \mid x < -1\}$ or $(-\infty, -1)$ **21.** $2 \cdot 2 \cdot 3 \cdot 5 \cdot 5$

22.

23. 9 on Friday and 33 on Saturday   **24.** 67 or fewer

25. 11 and 13

CHAPTER 4

Problem Set 4.1 (page 153)

1. $\{9\}$ **3.** $\{10\}$ **5.** $\left\{\dfrac{15}{2}\right\}$ **7.** $\{-22\}$

9. $\{-4\}$ **11.** $\{6\}$ **13.** $\{-28\}$ **15.** $\{34\}$

17. $\{6\}$ **19.** $\left\{-\dfrac{8}{5}\right\}$ **21.** $\{7\}$ **23.** $\left\{\dfrac{9}{2}\right\}$

25. $\left\{-\dfrac{53}{2}\right\}$ **27.** $\{50\}$ **29.** $\{120\}$ **31.** $\left\{\dfrac{9}{7}\right\}$

33. 55%   **35.** 60%   **37.** $16\dfrac{2}{3}$%   **39.** $37\dfrac{1}{2}$%

41. 150%   **43.** 240%   **45.** 2.66   **47.** 42

49. 80%   **51.** 60   **53.** 115%   **55.** 90

57. 15 feet by $19\frac{1}{2}$ feet   **59.** 330 miles

61. 60 centimeters   **63.** 7.5 pounds

65. $33\frac{1}{3}$ pounds   **67.** 90,000

69. 137.5 grams   **71.** $300

73. $150,000   **77.** All real numbers except 2

79. {0}   **81.** All real numbers

Problem Set 4.2 (page 160)

1. {1.11}   **3.** {6.6}   **5.** {0.48}   **7.** {80}

9. {3}   **11.** {50}   **13.** {70}   **15.** {200}

17. {450}   **19.** {150}   **21.** {2200}

23. $50   **25.** $3600   **27.** $20.80   **29.** 30%

31. $8.50   **33.** $12.40   **35.** $1000

37. 40%   **39.** 8%   **41.** $4166.67

43. $3400   **45.** $633.33

49. Yes, if the profit is figured as a percent of the selling price

51. Yes   **53.** {1.625}   **55.** {350}

57. {0.06}   **59.** {15.4}

Problem Set 4.3 (page 169)

1. 7   **3.** 500   **5.** 20   **7.** 48   **9.** 9

11. 46 centimeters   **13.** 15 inches

15. 504 square feet   **17.** $6   **19.** 7 inches

21. 150π square centimeters   **23.** $\frac{1}{4}\pi$ square yards

25. $S = 324\pi$ square inches and $V = 972\pi$ cubic inches

27. $V = 1152\pi$ cubic feet and $S = 416\pi$ square feet

29. 12 inches   **31.** 8 feet   **33.** $h = \dfrac{V}{B}$

35. $B = \dfrac{3V}{h}$ **37.** $w = \dfrac{P - 2\ell}{2}$ **39.** $h = \dfrac{3V}{\pi r^2}$

41. $C = \dfrac{5}{9}(F - 32)$ **43.** $h = \dfrac{A - 2\pi r^2}{2\pi r}$

45. $x = \dfrac{9 - 7y}{3}$ **47.** $y = \dfrac{9x - 13}{6}$

49. $x = \dfrac{11y - 14}{2}$ **51.** $x = \dfrac{-y - 4}{3}$

53. $y = \dfrac{3}{2}x$ **55.** $y = \dfrac{ax - c}{b}$ **57.** $x = \dfrac{2y - 22}{5}$

59. $y = mx + b$ **65.** 125.6 square centimeters

67. 245 square centimeters   **69.** 65 cubic inches

Problem Set 4.4 (page 177)

1. $\left\{8\frac{1}{3}\right\}$ **3.** {16}   **5.** {25}   **7.** {7}   **9.** {24}

11. {4}   **13.** $12\frac{1}{2}$ years   **15.** $33\frac{1}{3}$ years

17. The width is 14 inches and the length is 42 inches.

19. The width is 12 centimeters and the length is 34 centimeters.

21. 80 square inches   **23.** 24 feet, 31 feet, and 45 feet

25. 6 centimeters, 19 centimeters, and 21 centimeters

27. 12 centimeters   **29.** 7 centimeters   **31.** 9 hours

33. $2\frac{1}{2}$ hours   **35.** 55 miles per hour

37. 64 and 72 miles per hour   **39.** 60 miles

Problem Set 4.5 (page 184)

1. {15}   **3.** $\left\{\dfrac{20}{7}\right\}$ **5.** $\left\{\dfrac{15}{4}\right\}$ **7.** $\left\{\dfrac{5}{3}\right\}$ **9.** {2}

11. $\left\{\dfrac{33}{10}\right\}$ **13.** 12.5 milliliters   **15.** 15 centiliters

17. $7\frac{1}{2}$ quarts of the 30% solution and $2\frac{1}{2}$ quarts of the 50% solution

19. 5 gallons   **21.** 3 quarts   **23.** 12 gallons

25. 16.25%

27. The square is 6 inches by 6 inches, and the rectangle is 9 inches long and 3 inches wide.

29. 40 minutes   **31.** Pam is 9 and Bill is 18.

33. $1500 at 3% and $2250 at 5%

35. $500 at 4% and $700 at 6%

37. $4000   **39.** $600 at 6% and $1700 at 8%

41. $3000 at 8% and $2400 at 10%

Chapter 4 Review Problem Set (page 186)

1. $\left\{\dfrac{17}{12}\right\}$ **2.** {5}   **3.** {800}   **4.** {16}

5. {73}   **6.** $w = 6$ **7.** $C = 25$

8. $t = \dfrac{A - P}{Pr}$ **9.** $x = \dfrac{13 + 3y}{2}$

10. 77 square inches   **11.** 6 centimeters

12. 15 feet   **13.** 60%   **14.** 40 and 56   **15.** 40

16. The length is 17 meters and the width is 6 meters.

17. $1\frac{1}{2}$ hours   **18.** 20 liters

19. 15 centimeters by 40 centimeters

20. The length is 29 yards and the width is 10 yards.

21. 20°   **22.** 30 gallons

23. $675 at 3% and $1425 at 5%   **24.** $40

25. 35%   **26.** 34° and 99°   **27.** 5 hours

28. 18 gallons   **29.** 26%   **30.** $367.50

Chapter 4 Test (page 188)

1. $\{-22\}$ **2.** $\left\{-\dfrac{17}{18}\right\}$ **3.** $\{-77\}$ **4.** $\left\{\dfrac{4}{3}\right\}$

5. $\{14\}$ **6.** $\left\{\dfrac{12}{5}\right\}$ **7.** $\{100\}$ **8.** $\{70\}$

9. $\{250\}$ **10.** $\left\{\dfrac{11}{2}\right\}$ **11.** $C = \dfrac{5F - 160}{9}$

12. $x = \dfrac{y + 8}{2}$ **13.** $y = \dfrac{9x + 47}{4}$

14. 64π square centimeters
15. 576 square inches **16.** 14 yards
17. 125% **18.** 70 **19.** $189
20. $52 **21.** 40% **22.** 875 women
23. 10 hours **24.** 4 centiliters **25.** 11.1 years

Cumulative Review Problem Set (Chapters 1–4) (page 189)

1. $-16x$ **2.** $4a - 6$ **3.** $12x + 27$
4. $-5x + 1$ **5.** $9n - 8$ **6.** $14n - 5$

7. $\dfrac{1}{4}x$ **8.** $-\dfrac{1}{10}n$ **9.** $-0.1x$ **10.** $0.7x + 0.2$

11. -65 **12.** -51 **13.** 20 **14.** 32

15. 25 **16.** $-\dfrac{5}{6}$ **17.** -0.28 **18.** $-\dfrac{1}{4}$

19. 5 **20.** $-\dfrac{1}{4}$ **21.** 81 **22.** -64

23. 0.064 **24.** $-\dfrac{1}{32}$ **25.** $\dfrac{25}{36}$ **26.** $-\dfrac{1}{512}$

27. $\{-4\}$ **28.** $\{-2\}$ **29.** $\{28\}$ **30.** $\{-8\}$

31. $\left\{\dfrac{25}{2}\right\}$ **32.** $\left\{-\dfrac{4}{7}\right\}$ **33.** $\left\{\dfrac{34}{3}\right\}$ **34.** $\{200\}$

35. $\left\{\dfrac{10}{7}\right\}$ **36.** $\{11\}$ **37.** $\left\{\dfrac{3}{2}\right\}$ **38.** $\{0\}$

39. $\{x \mid x > 7\}$ or $(7, \infty)$ **40.** $\{x \mid x > -6\}$ or $(-6, \infty)$

41. $\left\{n \mid n \geq \dfrac{7}{5}\right\}$ or $\left[\dfrac{7}{5}, \infty\right)$ **42.** $\{x \mid x \geq 21\}$ or $[21, \infty)$

43. $\{t \mid t < 100\}$ or $(-\infty, 100)$
44. $\{x \mid x < -1\}$ or $(-\infty, -1)$

45. $\{n \mid n \geq 18\}$ or $[18, \infty)$ **46.** $\left\{x \mid x < \dfrac{5}{3}\right\}$ or $\left(-\infty, \dfrac{5}{3}\right)$

47. $15,000 **48.** 45° and 135°
49. 8 nickels and 17 dimes **50.** 130 or higher
51. 12 feet and 18 feet **52.** $40
53. 45 miles per hour and 50 miles per hour
54. 5 liters

Problem Set 5.1 (page 200)

1.

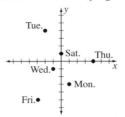

3. $y = \dfrac{13 - 3x}{7}$ **5.** $x = 3y + 9$ **7.** $y = \dfrac{x + 14}{5}$

9. $x = \dfrac{y - 7}{3}$ **11.** $y = \dfrac{2x - 5}{3}$

13.

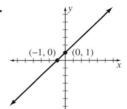

15.

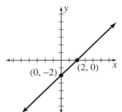

17.

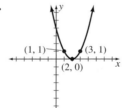

19.

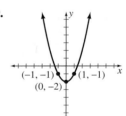

21.

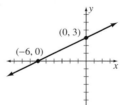

23.

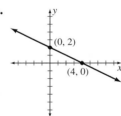

25.

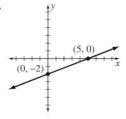

27.

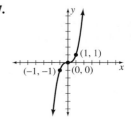

29.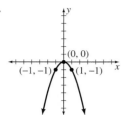

31.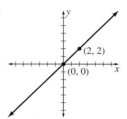

13.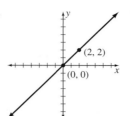

15.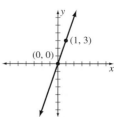

33.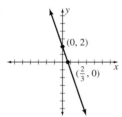

35.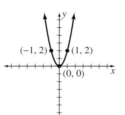

17.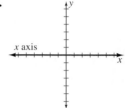

19.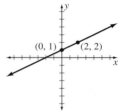

Problem Set 5.2 (page 208)

21.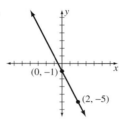

23.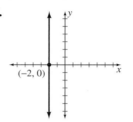

1.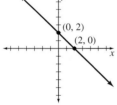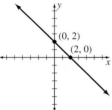

3.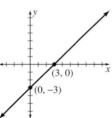

25.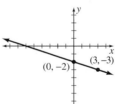

27.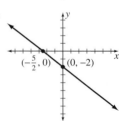

5.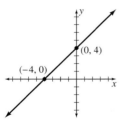

7.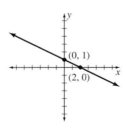

29.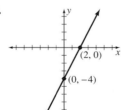

31.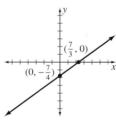

9.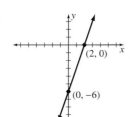

11.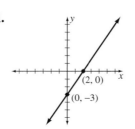

33.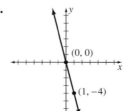

35.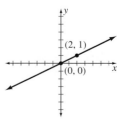

37.
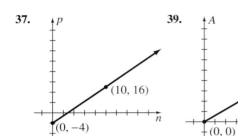
$(10, 16)$
$(0, -4)$

39.
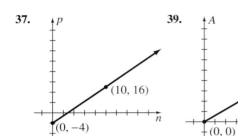
$(5, 15)$
$(0, 0)$

45.
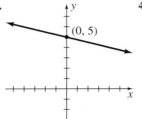
$(0, 5)$

47.
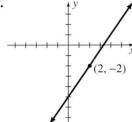
$(2, -2)$

41. $41.76; $43.20; $44.40; $46.50; $46.92

43. a.

C	0	5	10	15	20	−5	−10	−15	−20	−25
F	32	41	50	59	68	23	14	5	−4	−13

49. $-\dfrac{3}{2}$ **51.** $\dfrac{5}{4}$ **53.** $-\dfrac{1}{5}$ **55.** 2   **57.** 0

59. $\dfrac{2}{5}$ **61.** $\dfrac{6}{5}$ **63.** -3 **65.** 4   **67.** $\dfrac{2}{3}$

69. 5.1%   **71.** 32 cm   **73.** 1.0 feet

49.
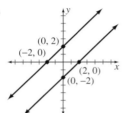
$(0, 2)$
$(-2, 0)$
$(2, 0)$
$(0, -2)$

51.
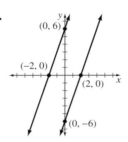
$(0, 6)$
$(-2, 0)$
$(2, 0)$
$(0, -6)$

Problem Set 5.4 (page 226)

1. $\{(2, -1)\}$ **3.** $\{(2, 1)\}$ **5.** $\varnothing$ **7.** $\{(0, 0)\}$
9. $\{(1, -1)\}$ **11.** $\{(x, y)\,|\,y = -2x + 3\}$ **13.** $\{(1, 3)\}$
15. $\{(3, -2)\}$ **17.** $\{(2, 4)\}$ **19.** $\{(-2, -3)\}$
21. $\{(8, 12)\}$ **23.** $\{(-4, -6)\}$ **25.** $\{(-9, 3)\}$

27. $\varnothing$ **29.** $\left\{\left(5, \dfrac{3}{2}\right)\right\}$ **31.** $\left\{\left(\dfrac{9}{5}, -\dfrac{7}{25}\right)\right\}$

33. $\{(-2, -4)\}$ **35.** $\{(5, 2)\}$ **37.** $\{(-4, -8)\}$

39. $\left\{\left(\dfrac{11}{20}, \dfrac{7}{20}\right)\right\}$ **41.** $\{(x, y)\,|\,x - 2y = -3\}$

43. $\left\{\left(-\dfrac{3}{4}, -\dfrac{6}{5}\right)\right\}$ **45.** $\left\{\left(\dfrac{5}{27}, -\dfrac{26}{27}\right)\right\}$

47. $2000 at 7% and $8000 at 8%
49. 34 and 97   **51.** 42 women
53. 20 inches by 27 inches
55. 60 five-dollar bills and 40 ten-dollar bills
57. 2500 student tickets and 500 nonstudent tickets

Problem Set 5.3 (page 217)

1. $\dfrac{3}{4}$ **3.** $\dfrac{7}{5}$ **5.** $-\dfrac{6}{5}$

7. $-\dfrac{10}{3}$ **9.** $\dfrac{3}{4}$

11. 0   **13.** $-\dfrac{3}{2}$ **15.** Undefined

17. 1

19. $\dfrac{b - d}{a - c}$ or $\dfrac{d - b}{c - a}$ **21.** 4   **23.** -6

25–31. Answers will vary.   **33.** Negative
35. Positive   **37.** Zero   **39.** Negative

41.
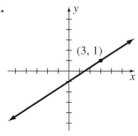
$(3, 1)$

43.
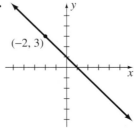
$(-2, 3)$

Problem Set 5.5 (page 236)

1. $\{(4, -3)\}$ **3.** $\{(-1, -3)\}$ **5.** $\{(-8, 2)\}$
7. $\{(-4, 0)\}$ **9.** $\{(1, -1)\}$ **11.** $\varnothing$

13. $\left\{\left(-\dfrac{1}{11}, \dfrac{4}{11}\right)\right\}$ **15.** $\left\{\left(\dfrac{3}{2}, -\dfrac{1}{3}\right)\right\}$

17. $\{(4, -9)\}$ **19.** $\{(7, 0)\}$ **21.** $\{(7, 12)\}$

23. $\left\{\left(\dfrac{7}{11}, \dfrac{2}{11}\right)\right\}$ **25.** $\varnothing$ **27.** $\left\{\left(\dfrac{51}{31}, -\dfrac{32}{31}\right)\right\}$

29. $\{(-2, -4)\}$ **31.** $\left\{\left(-1, -\dfrac{14}{3}\right)\right\}$

33. $\{(-6, 12)\}$ **35.** $\{(2, 8)\}$ **37.** $\{(-1, 3)\}$

39. $\{(16, -12)\}$ **41.** $\left\{\left(-\dfrac{3}{4}, \dfrac{3}{2}\right)\right\}$ **43.** $\{(5, -5)\}$

45. 5 gallons of 10% solution and 15 gallons of 20% solution
47. $2 for a tennis ball and $3 for a golf ball

49. 40 double rooms and 15 single rooms   **51.** $\dfrac{3}{4}$

53. 18 centimeters by 24 centimeters   **55.** 8 feet

59. a. Consistent   **c.** Consistent
e. Dependent   **g.** Inconsistent

61. $x = \dfrac{b_1c_2 - b_2c_1}{a_2b_1 - a_1b_2}$ and $y = \dfrac{a_2c_1 - a_1c_2}{a_2b_1 - a_1b_2}$

Problem Set 5.6 (page 243)

1.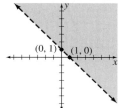

3.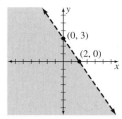

5.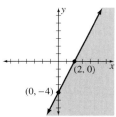

7.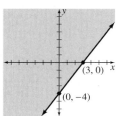

9.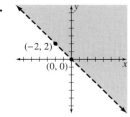

11.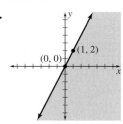

13.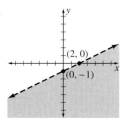

15.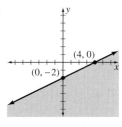

17.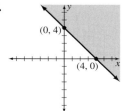

19.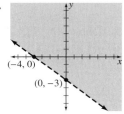

21.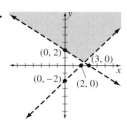

23.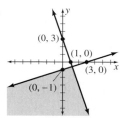

25.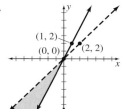

27.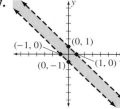

29.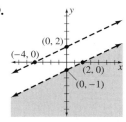

33. $\varnothing$
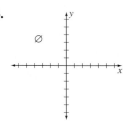

35.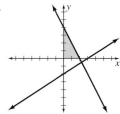

Chapter 5 Review Problem Set (page 246)

1.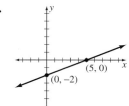

2.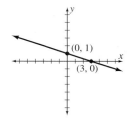

3.

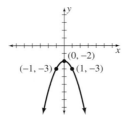

4.

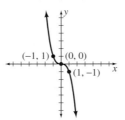

5.

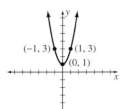

6.

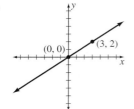

7. x intercept of -1, y intercept of -4
8. x intercept of 2, y intercept of -1
9. x intercept of -2, y intercept of 6
10. x intercept of $\dfrac{1}{4}$, y intercept of 1

11. -2 **12.** Undefined   **13.** $\{(3, -2)\}$

14. $\{(7, 13)\}$ **15.** $\{(16, -5)\}$ **16.** $\left\{\left(\dfrac{41}{23}, \dfrac{19}{23}\right)\right\}$

17. $\{(10, 25)\}$ **18.** $\{(-6, -8)\}$ **19.** $\{(400, 600)\}$

20. $\varnothing$ **21.** $\left\{\left(\dfrac{5}{16}, -\dfrac{17}{16}\right)\right\}$ **22.** $t = 4$ and $u = 8$

23. $t = 8$ and $u = 4$ **24.** $t = 3$ and $u = 7$ **25.** $\{(-9, 6)\}$
26. 38 and 75   **27.** $250 at 6% and $300 at 8%
28. 18 nickels and 25 dimes
29. Length of 19 inches and width of 6 inches
30. $32°$ and $58°$ **31.** $50°$ and $130°$
32. $3.25 for a cheeseburger and $2.50 for a milkshake
33. $1.59 for orange juice and $0.99 for water

34.

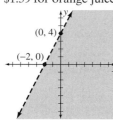

35.

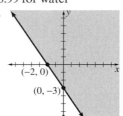

36.

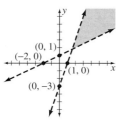

Chapter 5 Test (page 248)
1. Yes   **2.** Yes   **3.** -4 and 4

4. 4   **5.** $-\dfrac{1}{2}$

6.

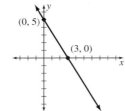

7.

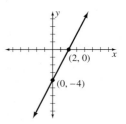

8.

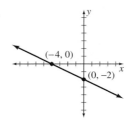

9.

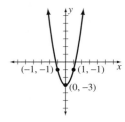

10. No   **11.** Yes   **12.** $\{(2, 5)\}$
13. $\{(3, 4)\}$ **14.** $\{(-4, 6)\}$ **15.** $\{(-1, -4)\}$
16. $\{(3, -6)\}$ **17.** Yes   **18.** No   **19.** Yes

20.

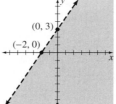

21.

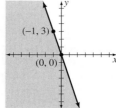

22.

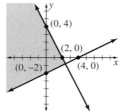

23. 16 dimes and 24 quarters
24. 13 inches   **25.** 3 liters

Cumulative Practice Test (Chapters 1–5) (page 249)

1. -1.5 **2.** $\dfrac{1}{16}$ **3.** 1.384 **4.** $-\dfrac{16}{5}$

5. 0.95 **6.** -80 **7.** $\dfrac{3y + 2x - 4}{xy}$ **8.** $\dfrac{2x}{3y^2}$

9. $\dfrac{3ab}{2}$ **10.** $\{-20\}$ **11.** $\left\{-\dfrac{15}{2}\right\}$

12. $\left\{-\dfrac{5}{2}\right\}$ **13.** $\{-62\}$ **14.** $\{600\}$

15. $y = \dfrac{2x - 13}{3}$ **16.** $\{(-2, 5)\}$

17. $\{(5, -3)\}$ **18.** $\{x \mid x > -30\}$ or $(-30, \infty)$

19. $\left\{x \mid x \le -\dfrac{36}{5}\right\}$ or $\left(-\infty, -\dfrac{36}{5}\right]$

20.   **21.**

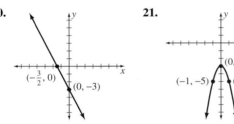

22.

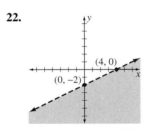

23. 36 and 99   **24.** 77 or less
25. 4 gallons of the 10% salt solution and 6 gallons of the 15% salt solution

CHAPTER 6

Problem Set 6.1 (page 256)
1. 3   **3.** 2   **5.** 3   **7.** 2   **9.** $8x + 11$
11. $4y + 10$ **13.** $2x^2 - 2x - 23$ **15.** $17x - 19$
17. $6x^2 - 5x - 4$ **19.** $5n - 6$
21. $-7x^2 - 13x - 7$ **23.** $5x + 5$ **25.** $-2x - 5$
27. $-3x + 7$ **29.** $2x^2 + 15x - 6$
31. $5n^2 + 2n + 3$ **33.** $-3x^3 + 2x^2 - 11$
35. $9x - 2$ **37.** $2a + 15$ **39.** $-2x^2 + 5$
41. $6x^3 + 12x^2 - 5$ **43.** $4x^3 - 8x^2 + 13x + 12$
45. $x + 11$ **47.** $-3x - 14$
49. $-x^2 - 13x - 12$ **51.** $x^2 - 11x - 8$
53. $-10a - 3b$ **55.** $-n^2 + 2n - 17$
57. $8x + 1$ **59.** $-5n - 1$ **61.** $-2a + 6$
63. $11x + 6$ **65.** $6x + 7$ **67.** $-5n + 7$
69. $8x + 6$ **71.** $20x^2$

Problem Set 6.2 (page 263)
1. $45x^2$ **3.** $21x^3$ **5.** $-6x^2y^2$ **7.** $14x^3y$
9. $-48a^3b^3$ **11.** $5x^4y$ **13.** $104a^3b^2c^2$
15. $30x^6$ **17.** $-56x^2y^3$ **19.** $-6a^2b^3$
21. $72c^3d^3$ **23.** $\dfrac{2}{5}x^3y^5$ **25.** $-\dfrac{2}{9}a^2b^5$
27. $0.28x^8$ **29.** $-6.4a^4b^2$ **31.** $4x^8$
33. $9a^4b^6$ **35.** $27x^6$ **37.** $-64x^{12}$
39. $81x^8y^{10}$ **41.** $16x^8y^4$ **43.** $81a^{12}b^8$
45. $x^{12}y^6$ **47.** $15x^2 + 10x$ **49.** $18x^3 - 6x^2$
51. $-28x^3 + 16x$ **53.** $2x^3 - 8x^2 + 12x$
55. $-18a^3 + 30a^2 + 42a$ **57.** $28x^3y - 7x^2y + 35xy$
59. $-9x^3y + 2x^2y + 6xy$ **61.** $13x + 22y$
63. $-2x - 9y$ **65.** $4x^3 - 3x^2 - 14x$
67. $-x + 14$ **69.** $-7x + 12$ **71.** $18x^5$
73. $-432x^5$ **75.** $25x^7y^8$ **77.** $-a^{12}b^5c^9$
79. $-16x^{11}y^{17}$ **81.** $7x + 5$ **83.** $3\pi x^2$
89. x^{7n} **91.** x^{6n+1} **93.** x^{6n+3}
95. $-20x^{10n}$ **97.** $12x^{7n}$

Problem Set 6.3 (page 270)
1. $xy + 3x + 2y + 6$ **3.** $xy + x - 4y - 4$
5. $xy - 6x - 5y + 30$ **7.** $xy + xz + x + 2y + 2z + 2$
9. $6xy + 2x + 9y + 3$ **11.** $x^2 + 10x + 21$
13. $x^2 + 5x - 24$ **15.** $x^2 - 6x - 7$
17. $n^2 - 10n + 24$ **19.** $3n^2 + 19n + 6$
21. $15x^2 + 29x - 14$ **23.** $x^3 + 7x^2 + 21x + 27$
25. $x^3 + 3x^2 - 10x - 24$ **27.** $2x^3 - 7x^2 - 22x + 35$
29. $8a^3 - 14a^2 + 23a - 9$ **31.** $3a^3 + 2a^2 - 8a - 5$
33. $x^4 + 7x^3 + 17x^2 + 23x + 12$
35. $x^4 - 3x^3 - 34x^2 + 33x + 63$ **37.** $x^2 + 11x + 18$
39. $x^2 + 4x - 12$ **41.** $x^2 - 8x - 33$
43. $n^2 - 7n + 12$ **45.** $n^2 + 18n + 72$
47. $y^2 - 4y - 21$ **49.** $y^2 - 19y + 84$
51. $x^2 + 2x - 35$ **53.** $x^2 - 6x - 112$
55. $a^2 + a - 90$ **57.** $2a^2 + 13a + 6$
59. $5x^2 + 33x - 14$ **61.** $6x^2 - 11x - 7$
63. $12a^2 - 7a - 12$ **65.** $12n^2 - 28n + 15$
67. $14x^2 + 13x - 12$ **69.** $45 - 19x + 2x^2$
71. $-8x^2 + 22x - 15$ **73.** $-9x^2 + 9x + 4$
75. $72n^2 - 5n - 12$ **77.** $27 - 21x + 2x^2$
79. $20x^2 - 7x - 6$ **81.** $x^2 + 14x + 49$
83. $25x^2 - 4$ **85.** $x^2 - 2x + 1$
87. $9x^2 + 42x + 49$ **89.** $4x^2 - 12x + 9$
91. $4x^2 - 9y^2$ **93.** $1 - 10n + 25n^2$
95. $9x^2 + 24xy + 16y^2$ **97.** $9 + 24y + 16y^2$
99. $1 - 49n^2$ **101.** $16a^2 - 56ab + 49b^2$
103. $x^2 + 16xy + 64y^2$ **105.** $25x^2 - 121y^2$
107. $64x^3 - x$ **109.** $-32x^3 + 2xy^2$

111. $x^3 + 6x^2 + 12x + 8$
113. $x^3 - 9x^2 + 27x - 27$
115. $8n^3 + 12n^2 + 6n + 1$
117. $27n^3 - 54n^2 + 36n - 8$
121. $V = 4x^3 - 56x^2 + 196x$; $S = 196 - 4x^2$

Problem Set 6.4 (page 276)

1. x^8 **3.** $2x^2$ **5.** $-8n^4$ **7.** -8
9. $13xy^2$ **11.** $7ab^2$ **13.** $18xy^4$
15. $-32x^5y^2$ **17.** $-8x^5y^4$ **19.** -1
21. $14ab^2c^4$ **23.** $16yz^4$ **25.** $4x^2 + 6x^3$
27. $3x^3 - 8x$ **29.** $-7n^3 + 9$
31. $5x^4 - 8x^3 - 12x$ **33.** $4n^5 - 8n^2 + 13$
35. $5a^6 + 8a^2$ **37.** $-3xy + 5y$
39. $-8ab - 10a^2b^3$ **41.** $-3bc + 13b^2c^4$
43. $-9xy^2 + 12x^2y^3$ **45.** $-3x^4 - 5x^2 + 7$
47. $-3a^2 + 7a + 13b$ **49.** $-1 + 5xy^2 - 7xy^5$

Problem Set 6.5 (page 281)

1. $x + 12$ **3.** $x + 2$
5. $x + 8$ with a remainder of 4
7. $x + 4$ with a remainder of -7 **9.** $5n + 4$
11. $8y - 3$ with a remainder of 2   **13.** $4x - 7$
15. $3x + 2$ with a remainder of -6
17. $2x^2 + 3x + 4$ **19.** $5n^2 - 4n - 3$
21. $n^2 + 6n - 4$ **23.** $x^2 + 3x + 9$
25. $9x^2 + 12x + 16$
27. $3n - 8$ with a remainder of 17
29. $3t + 2$ with a remainder of 6   **31.** $3n^2 - n - 4$
33. $4x^2 - 5x + 5$ with a remainder of -3
35. $x + 4$ with a remainder of $5x - 1$
37. $2x - 12$ with a remainder of $49x - 5$
39. $x^3 - 2x^2 + 4x - 8$

Problem Set 6.6 (page 288)

1. $\dfrac{1}{9}$ **3.** $\dfrac{1}{64}$ **5.** $\dfrac{2}{3}$ **7.** 16 **9.** 1

11. $-\dfrac{27}{8}$ **13.** $\dfrac{1}{4}$ **15.** $-\dfrac{1}{9}$ **17.** $\dfrac{27}{64}$

19. $\dfrac{1}{8}$ **21.** 27 **23.** 1000 **25.** $\dfrac{1}{1000}$ or 0.001

27. 18 **29.** 144 **31.** x^5 **33.** $\dfrac{1}{n^2}$

35. $\dfrac{1}{a^5}$ **37.** $8x$ **39.** $\dfrac{27}{x^4}$ **41.** $-\dfrac{15}{y^3}$

43. 96 **45.** x^{10} **47.** $\dfrac{1}{n^4}$ **49.** $2n^2$

51. $-\dfrac{3}{x^4}$ **53.** 4 **55.** x^6 **57.** $\dfrac{1}{x^4}$

59. $\dfrac{1}{x^3y^4}$ **61.** $\dfrac{1}{x^6y^3}$ **63.** $\dfrac{8}{n^6}$ **65.** $\dfrac{1}{16n^6}$

67. $\dfrac{81}{a^8}$ **69.** $\dfrac{x^2}{25}$ **71.** $\dfrac{x^2y}{2}$ **73.** $\dfrac{y}{x^2}$

75. a^4b^8 **77.** $\dfrac{x^2}{y^6}$ **79.** x **81.** $\dfrac{1}{8x^3}$

83. $\dfrac{x^4}{4}$ **85.** $(3.21)(10^2)$ **87.** $(8)(10^3)$

89. $(2.46)(10^{-3})$ **91.** $(1.79)(10^{-5})$
93. $(8.7)(10^7)$ **95.** 8000 **97.** $52{,}100$
99. $11{,}400{,}000$ **101.** 0.07 **103.** 0.000987
105. 0.00000864 **107.** 0.84 **109.** 450
111. $4{,}000{,}000$ **113.** 0.0000002 **115.** 0.3
117. 0.000007

Chapter 6 Review Problem Set (page 290)

1. $8x^2 - 13x + 2$ **2.** $3y^2 + 11y - 9$
3. $3x^2 + 2x - 9$ **4.** $-8x^2 + 18$ **5.** $11x + 8$
6. $-9x^2 + 8x - 20$ **7.** $2y^2 - 54y + 18$
8. $-13a - 30$ **9.** $-27a - 7$ **10.** $n - 2$
11. $-5n^2 - 2n$ **12.** $17n^2 - 14n - 16$ **13.** $35x^6$
14. $-54x^8$ **15.** $24x^3y^5$ **16.** $-6a^4b^9$
17. $8a^6b^9$ **18.** $9x^2y^4$ **19.** $35x^2 + 15x$
20. $-24x^3 + 3x^2$ **21.** $x^2 + 17x + 72$
22. $3x^2 + 10x + 7$ **23.** $x^2 - 3x - 10$
24. $y^2 - 13y + 36$ **25.** $14x^2 - x - 3$
26. $20a^2 - 3a - 56$ **27.** $9a^2 - 30a + 25$
28. $2x^3 + 17x^2 + 26x - 24$ **29.** $30n^2 + 19n - 5$
30. $12n^2 + 13n - 4$ **31.** $4n^2 - 1$
32. $16n^2 - 25$ **33.** $4a^2 + 28a + 49$
34. $9a^2 + 30a + 25$ **35.** $x^3 - 3x^2 + 8x - 12$
36. $2x^3 + 7x^2 + 10x - 7$ **37.** $a^3 + 15a^2 + 75a + 125$
38. $a^3 - 18a^2 + 108a - 216$
39. $x^4 + x^3 + 2x^2 - 7x - 5$
40. $n^4 - 5n^3 - 11n^2 - 30n - 4$ **41.** $-12x^3y^3$
42. $7a^3b^4$ **43.** $-3x^2y - 9x^4$
44. $10a^4b^9 - 13a^3b^7$ **45.** $14x^2 - 10x - 8$
46. $x + 4$, R $= -21$ **47.** $7x - 6$
48. $2x^2 + x + 4$, R $= 4$ **49.** 13 **50.** 25
51. $\dfrac{1}{16}$ **52.** 1 **53.** -1 **54.** 9 **55.** $\dfrac{16}{9}$

56. $\dfrac{1}{4}$ **57.** -8 **58.** $\dfrac{11}{18}$ **59.** $\dfrac{5}{4}$

60. $\dfrac{1}{25}$ **61.** $\dfrac{1}{x^3}$ **62.** $12x^3$ **63.** x^2

64. $\dfrac{1}{x^2}$ **65.** $8a^6$ **66.** $\dfrac{4}{n}$ **67.** $\dfrac{x^2}{y}$

68. $\dfrac{b^6}{a^4}$ **69.** $\dfrac{1}{2x}$ **70.** $\dfrac{1}{9n^4}$ **71.** $\dfrac{n^3}{8}$

72. $-12b$　**73.** 610　**74.** 56,000　**75.** 0.08
76. 0.00092　**77.** $(9)(10^3)$　**78.** $(4.7)(10)$
79. $(4.7)(10^{-2})$　**80.** $(2.1)(10^{-4})$　**81.** 0.48
82. 4.2　**83.** 2000　**84.** 0.00000002

Chapter 6 Test (page 293)

1. $-2x^2 - 2x + 5$　**2.** $-3x^2 - 6x + 20$
3. $-13x + 2$　**4.** $-28x^3y^5$　**5.** $12x^5y^5$
6. $x^2 - 7x - 18$　**7.** $n^2 + 7n - 98$
8. $40a^2 + 59a + 21$　**9.** $9x^2 - 42xy + 49y^2$
10. $2x^3 + 2x^2 - 19x - 21$　**11.** $81x^2 - 25y^2$
12. $15x^2 - 68x + 77$　**13.** $8x^2y^4$
14. $-7x + 9y$　**15.** $x^2 + 4x - 5$

16. $4x^2 - x + 6$　**17.** $\dfrac{27}{8}$　**18.** $1\dfrac{5}{16}$　**19.** 16

20. $-\dfrac{24}{x^2}$　**21.** $\dfrac{x^3}{4}$　**22.** $\dfrac{x^6}{y^{10}}$　**23.** $(2.7)(10^{-4})$
24. 9,200,000　**25.** 0.006

Cumulative Review Problem Set
(Chapters 1–6) (page 294)

1. 130　**2.** -1　**3.** 27　**4.** -16　**5.** 81

6. -32　**7.** $\dfrac{3}{2}$　**8.** 16　**9.** 36　**10.** $1\dfrac{3}{4}$

11. 0　**12.** $\dfrac{13}{40}$　**13.** $-\dfrac{2}{13}$　**14.** 5

15. -1　**16.** -33　**17.** $-15x^3y^7$　**18.** $12ab^7$
19. $-8x^6y^{15}$　**20.** $-6x^2y + 15xy^2$
21. $15x^2 - 11x + 2$　**22.** $21x^2 + 25x - 4$
23. $-2x^2 - 7x - 6$　**24.** $49 - 4y^2$
25. $3x^3 - 7x^2 - 2x + 8$　**26.** $2x^3 - 3x^2 - 13x + 20$
27. $8n^3 + 36n^2 + 54n + 27$　**28.** $1 - 6n + 12n^2 - 8n^3$
29. $2x^4 + x^3 - 4x^2 + 42x - 36$　**30.** $-4x^2y^2$
31. $14ab^2$　**32.** $7y - 8x^2 - 9x^3y^3$

33. $2x^2 - 4x - 7$　**34.** $x^2 + 6x + 4$　**35.** $-\dfrac{6}{x}$

36. $\dfrac{2}{x}$　**37.** $\dfrac{xy^2}{3}$　**38.** $\dfrac{z^2}{x^2y^4}$　**39.** 0.12

40. 0.0000000018　**41.** 200　**42.** $\{11\}$

43. $\{-1\}$　**44.** $\{48\}$　**45.** $\left\{-\dfrac{3}{7}\right\}$　**46.** $\{9\}$

47. $\{13\}$　**48.** $\left\{\dfrac{9}{14}\right\}$　**49.** $\{500\}$

50. $\{x \mid x \le -1\}$ or $(-\infty, -1]$
51. $\{x \mid x > 0\}$ or $(0, \infty)$

52. $\left\{x \mid x < \dfrac{4}{5}\right\}$ or $\left(-\infty, \dfrac{4}{5}\right)$

53. $\{x \mid x < -2\}$ or $(-\infty, -2)$ 891

54. $\left\{x \mid x \ge \dfrac{12}{7}\right\}$ or $\left[\dfrac{12}{7}, \infty\right)$

55. $\{x \mid x \ge 300\}$ or $[300, \infty)$

56.
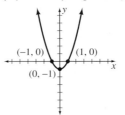
(−1, 0)　(1, 0)
(0, −1)

57.
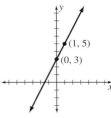
(1, 5)
(0, 3)

58.
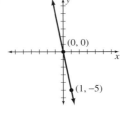
(0, 0)
(1, −5)

59.
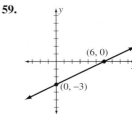
(6, 0)
(0, −3)

60.
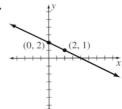
(0, 2)　(2, 1)

61. 3　**62.** 40　**63.** 8 dimes and 10 quarters
64. $700 at 8% and $800 at 9%　**65.** 3 gallons

66. $3\dfrac{1}{2}$ hours

67. The length is 15 meters and the width is 7 meters.

CHAPTER 7

Problem Set 7.1 (page 302)

1. $6y$　**3.** $12xy$　**5.** $14ab^2$　**7.** $2x$
9. $8a^2b^2$　**11.** $4(2x + 3y)$　**13.** $7y(2x - 3)$
15. $9x(2x + 5)$　**17.** $6xy(2y - 5x)$
19. $12a^2b(3 - 5ab^3)$　**21.** $xy^2(16y + 25x)$
23. $8(8ab - 9cd)$　**25.** $9a^2b(b^3 - 3)$
27. $4x^4y(13y + 15x^2)$　**29.** $8x^2y(5y + 1)$
31. $3x(4 + 5y + 7x)$　**33.** $x(2x^2 - 3x + 4)$
35. $4y^2(11y^3 - 6y - 5)$　**37.** $7ab(2ab^2 + 5b - 7a^2)$
39. $(y + 1)(x + z)$　**41.** $(b - 4)(a - c)$
43. $(x + 3)(x + 6)$　**45.** $(x + 1)(2x - 3)$

47. $(x + y)(5 + b)$ **49.** $(x - y)(b - c)$
51. $(a + b)(c + 1)$ **53.** $(x + 5)(x + 12)$
55. $(x - 2)(x - 8)$ **57.** $(2x + 1)(x - 5)$
59. $(2n - 1)(3n - 4)$ **61.** $\{0, 8\}$ **63.** $\{-1, 0\}$

65. $\{0, 5\}$ **67.** $\left\{0, \dfrac{3}{2}\right\}$ **69.** $\left\{-\dfrac{3}{7}, 0\right\}$

71. $\{-5, 0\}$ **73.** $\left\{0, \dfrac{3}{2}\right\}$ **75.** $\{0, 7\}$

77. $\{0, 13\}$ **79.** $\left\{-\dfrac{5}{2}, 0\right\}$ **81.** $\{-5, 4\}$

83. $\{4, 6\}$ **85.** 0 or 9   **87.** 20 units   **89.** $\dfrac{4}{\pi}$

91. The square is 3 inches by 3 inches and the rectangle is 3 inches by 6 inches.

95. a. $116   **c.** $750   **97.** $x = 0$ or $x = \dfrac{c}{b^2}$

99. $y = \dfrac{c}{1 + a - b}$

Problem Set 7.2 (page 309)
1. $(x - 1)(x + 1)$ **3.** $(x - 10)(x + 10)$
5. $(x - 2y)(x + 2y)$ **7.** $(3x - y)(3x + y)$
9. $(6a - 5b)(6a + 5b)$ **11.** $(1 - 2n)(1 + 2n)$
13. $5(x - 2)(x + 2)$ **15.** $8(x^2 + 4)$
17. $2(x - 3y)(x + 3y)$ **19.** $x(x - 5)(x + 5)$
21. Not factorable   **23.** $9x(5x - 4y)$
25. $4(3 - x)(3 + x)$ **27.** $4a^2(a^2 + 4)$
29. $(x - 3)(x + 3)(x^2 + 9)$ **31.** $x^2(x^2 + 1)$
33. $3x(x^2 + 16)$ **35.** $5x(1 - 2x)(1 + 2x)$
37. $4(x - 4)(x + 4)$ **39.** $3xy(5x - 2y)(5x + 2y)$

41. $\{-3, 3\}$ **43.** $\{-2, 2\}$ **45.** $\left\{-\dfrac{4}{3}, \dfrac{4}{3}\right\}$

47. $\{-11, 11\}$ **49.** $\left\{-\dfrac{2}{5}, \dfrac{2}{5}\right\}$ **51.** $\{-5, 5\}$

53. $\{-4, 0, 4\}$ **55.** $\{-4, 0, 4\}$ **57.** $\left\{-\dfrac{1}{3}, \dfrac{1}{3}\right\}$

59. $\{-10, 0, 10\}$ **61.** $\left\{-\dfrac{9}{8}, \dfrac{9}{8}\right\}$

63. $\left\{-\dfrac{1}{2}, 0, \dfrac{1}{2}\right\}$ **65.** -7 or 7   **67.** $-4, 0,$ or 4

69. 3 inches and 15 inches
71. The length is 20 centimeters and the width is 8 centimeters.
73. 4 meters and 8 meters   **75.** 5 centimeters
81. $(x - 2)(x^2 + 2x + 4)$ **83.** $(n + 4)(n^2 - 4n + 16)$
85. $(3a - 4b)(9a^2 + 12ab + 16b^2)$
87. $(1 + 3a)(1 - 3a + 9a^2)$
89. $(2x - y)(4x^2 + 2xy + y^2)$

91. $(3x - 2y)(9x^2 + 6xy + 4y^2)$
93. $(5x + 2y)(25x^2 - 10xy + 4y^2)$
95. $(4 + x)(16 - 4x + x^2)$

Problem Set 7.3 (page 317)
1. $(x + 4)(x + 6)$ **3.** $(x + 5)(x + 8)$
5. $(x - 2)(x - 9)$ **7.** $(n - 7)(n - 4)$
9. $(n + 9)(n - 3)$ **11.** $(n - 10)(n + 4)$
13. Not factorable   **15.** $(x - 6)(x - 12)$
17. $(x + 11)(x - 6)$ **19.** $(y - 9)(y + 8)$
21. $(x + 5)(x + 16)$ **23.** $(x + 12)(x - 6)$
25. Not factorable   **27.** $(x - 2y)(x + 5y)$
29. $(a - 8b)(a + 4b)$ **31.** $\{-7, -3\}$ **33.** $\{3, 6\}$
35. $\{-2, 5\}$ **37.** $\{-9, 4\}$ **39.** $\{-4, 10\}$
41. $\{-8, 7\}$ **43.** $\{2, 14\}$ **45.** $\{-12, 1\}$
47. $\{2, 8\}$ **49.** $\{-6, 4\}$
51. 7 and 8 or -7 and -8 **53.** 12 and 14
55. $-4, -3, -2,$ and -1 or 7, 8, 9, and 10
57. 4 and 7 or 0 and 3
59. The length is 9 inches and the width is 6 inches.
61. 9 centimeters by 6 centimeters
63. 7 rows
65. 8 feet, 15 feet, and 17 feet
67. 6 inches and 8 inches   **73.** $(x^a + 8)(x^a + 5)$
75. $(x^a + 9)(x^a - 3)$

Problem Set 7.4 (page 322)
1. $(3x + 1)(x + 2)$ **3.** $(2x + 5)(3x + 2)$
5. $(4x - 1)(x - 6)$ **7.** $(4x - 5)(3x - 4)$
9. $(5y + 2)(y - 7)$ **11.** $2(2n - 3)(n + 8)$
13. Not factorable   **15.** $(3x + 7)(6x + 1)$
17. $3(7x - 2)(x - 4)$ **19.** $(4x + 7)(2x - 3)$
21. $(3t + 2)(3t - 7)$ **23.** $(12y - 5)(y + 7)$
25. Not factorable   **27.** $(7x + 3)(2x + 7)$

29. $(4x - 3)(5x - 4)$ **31.** $\left\{-6, -\dfrac{1}{2}\right\}$

33. $\left\{-\dfrac{2}{3}, -\dfrac{1}{4}\right\}$ **35.** $\left\{\dfrac{1}{3}, 8\right\}$ **37.** $\left\{\dfrac{2}{5}, \dfrac{7}{3}\right\}$

39. $\left\{-7, \dfrac{5}{6}\right\}$ **41.** $\left\{-\dfrac{3}{8}, \dfrac{3}{2}\right\}$ **43.** $\left\{-\dfrac{2}{3}, \dfrac{4}{3}\right\}$

45. $\left\{\dfrac{2}{5}, \dfrac{5}{2}\right\}$ **47.** $\left\{-\dfrac{5}{2}, -\dfrac{2}{3}\right\}$ **49.** $\left\{-\dfrac{5}{4}, \dfrac{1}{4}\right\}$

Problem Set 7.5 (page 330)
1. $(x + 2)^2$ **3.** $(x - 5)^2$ **5.** $(3n + 2)^2$
7. $(4a - 1)^2$ **9.** $(2 + 9x)^2$ **11.** $(4x - 3y)^2$
13. $(2x + 1)(x + 8)$ **15.** $2x(x - 6)(x + 6)$
17. $(n - 12)(n + 5)$ **19.** Not factorable
21. $8(x^2 + 9)$ **23.** $(3x + 5)^2$ **25.** $5(x + 2)(3x + 7)$
27. $(4x - 3)(6x + 5)$ **29.** $(x + 5)(y - 8)$

31. $(5x - y)(4x + 7y)$　　**33.** $3(2x - 3)(4x + 9)$
35. $6(2x^2 + x + 5)$　　**37.** $5(x - 2)(x + 2)(x^2 + 4)$
39. $(x + 6y)^2$　　**41.** $\{0, 5\}$　　**43.** $\{-3, 12\}$
45. $\{-2, 0, 2\}$　　**47.** $\left\{-\dfrac{2}{3}, \dfrac{11}{2}\right\}$　　**49.** $\left\{\dfrac{1}{3}, \dfrac{3}{4}\right\}$
51. $\{-3, -1\}$　　**53.** $\{0, 6\}$　　**55.** $\{-4, 0, 6\}$
57. $\left\{\dfrac{2}{5}, \dfrac{6}{5}\right\}$　　**59.** $\{12, 16\}$　　**61.** $\left\{-\dfrac{10}{3}, 1\right\}$
63. $\{0, 6\}$　　**65.** $\left\{\dfrac{4}{3}\right\}$　　**67.** $\{-5, 0\}$
69. $\left\{-\dfrac{4}{3}, \dfrac{5}{8}\right\}$　　**71.** $\dfrac{5}{4}$ and 12 or -5 and -3
73. -1 and 1 or $-\dfrac{1}{2}$ and 2
75. 4 and 9 or $-\dfrac{24}{5}$ and $-\dfrac{43}{5}$
77. 6 rows and 9 chairs per row
79. One square is 6 feet by 6 feet and the other one is 18 feet by 18 feet.
81. 11 centimeters long and 5 centimeters wide
83. The side is 17 inches long and the altitude to that side is 6 inches long.
85. $1\dfrac{1}{2}$ inches　　**87.** 6 inches and 12 inches

Chapter 7 Review Problem Set (page 333)
1. $(x - 2)(x - 7)$　　**2.** $3x(x + 7)$
3. $(3x + 2)(3x - 2)$　　**4.** $(2x - 1)(2x + 5)$
5. $(5x - 6)^2$　　**6.** $n(n + 5)(n + 8)$
7. $(y + 12)(y - 1)$　　**8.** $3xy(y + 2x)$
9. $(x + 1)(x - 1)(x^2 + 1)$　　**10.** $(6n + 5)(3n - 1)$
11. Not factorable　　**12.** $(4x - 7)(x + 1)$
13. $3(n + 6)(n - 5)$　　**14.** $x(x + y)(x - y)$
15. $(2x - y)(x + 2y)$　　**16.** $2(n - 4)(2n + 5)$
17. $(x + y)(5 + a)$　　**18.** $(7t - 4)(3t + 1)$
19. $2x(x + 1)(x - 1)$　　**20.** $3x(x + 6)(x - 6)$
21. $(4x + 5)^2$　　**22.** $(y - 3)(x - 2)$
23. $(5x + y)(3x - 2y)$　　**24.** $n^2(2n - 1)(3n - 1)$
25. $\{-6, 2\}$　　**26.** $\{0, 11\}$　　**27.** $\left\{-4, \dfrac{5}{2}\right\}$
28. $\left\{-\dfrac{8}{3}, \dfrac{1}{3}\right\}$　　**29.** $\{-2, 2\}$　　**30.** $\left\{-\dfrac{5}{4}\right\}$
31. $\{-1, 0, 1\}$　　**32.** $\left\{-\dfrac{9}{4}, -\dfrac{2}{7}\right\}$　　**33.** $\{-7, 4\}$
34. $\{-5, 5\}$　　**35.** $\left\{-6, \dfrac{3}{5}\right\}$　　**36.** $\left\{-\dfrac{7}{2}, 1\right\}$
37. $\{-2, 0, 2\}$　　**38.** $\{8, 12\}$　　**39.** $\left\{-5, \dfrac{3}{4}\right\}$

40. $\{-2, 3\}$　　**41.** $\left\{-\dfrac{7}{3}, \dfrac{5}{2}\right\}$　　**42.** $\{-9, 6\}$
43. $\left\{-5, \dfrac{3}{2}\right\}$　　**44.** $\left\{\dfrac{4}{3}, \dfrac{5}{2}\right\}$
45. $-\dfrac{8}{3}$ and $-\dfrac{19}{3}$ or 4 and 7
46. The length is 8 centimeters and the width is 2 centimeters.
47. A 2-by-2-inch square and a 10-by-10-inch square
48. 8 by 15 by 17　　**49.** $-\dfrac{13}{6}$ and -12 or 2 and 13
50. 7, 9, and 11　　**51.** 4 shelves
52. A 5-by-5-yard square and a 5-by-40-yard rectangle
53. -18 and -17 or 17 and 18　　**54.** 6 units
55. 2 meters and 7 meters　　**56.** 9 and 11
57. 2 centimeters　　**58.** 15 feet

Chapter 7 Test (page 335)
1. $(x + 5)(x - 2)$　　**2.** $(x + 3)(x - 8)$
3. $2x(x + 1)(x - 1)$　　**4.** $(x + 9)(x + 12)$
5. $3(2n + 1)(3n + 2)$　　**6.** $(x + y)(a + 2b)$
7. $(4x - 3)(x + 5)$　　**8.** $6(x^2 + 4)$
9. $2x(5x - 6)(3x - 4)$　　**10.** $(7 - 2x)(4 + 3x)$
11. $\{-3, 3\}$　　**12.** $\{-6, 1\}$　　**13.** $\{0, 8\}$
14. $\left\{-\dfrac{5}{2}, \dfrac{2}{3}\right\}$　　**15.** $\{-6, 2\}$　　**16.** $\{-12, -4, 0\}$
17. $\{5, 9\}$　　**18.** $\left\{-12, \dfrac{1}{3}\right\}$　　**19.** $\left\{-4, \dfrac{2}{3}\right\}$
20. $\{-5, 0, 5\}$　　**21.** $\left\{\dfrac{7}{5}\right\}$　　**22.** 14 inches
23. 12 centimeters　　**24.** 16 chairs per row
25. 12 units

Cumulative Review Problem Set (Chapters 1–7) (page 336)
1. $\dfrac{5}{2}$　　**2.** 6　　**3.** 0.6　　**4.** 20　　**5.** 18
6. -21　　**7.** $\dfrac{1}{27}$　　**8.** $\dfrac{3}{2}$　　**9.** 1　　**10.** $\dfrac{12}{7}$
11. $-\dfrac{1}{16}$　　**12.** -16　　**13.** $\dfrac{25}{4}$　　**14.** $-\dfrac{1}{27}$
15. $\dfrac{19}{10x}$　　**16.** $\dfrac{2y}{3x}$　　**17.** $-35x^5y^5$　　**18.** $81a^2b^6$
19. $-15n^4 - 18n^3 + 6n^2$　　**20.** $15x^2 + 17x - 4$
21. $4x^2 + 20x + 25$　　**22.** $2x^3 + x^2 - 7x - 2$
23. $x^4 + x^3 - 6x^2 + x + 3$　　**24.** $-6x^2 + 11x + 7$
25. $3xy - 6x^3y^3$　　**26.** $7x + 4$　　**27.** $3x(x^2 + 5x + 9)$
28. $(x + 10)(x - 10)$　　**29.** $(5x - 2)(x - 4)$

30. $(4x + 7)(2x - 9)$ **31.** $(n + 16)(n + 9)$
32. $(x + y)(n - 2)$ **33.** $3x(x + 1)(x - 1)$
34. $2x(x - 9)(x + 6)$ **35.** $(6x - 5)^2$

36. $(3x + y)(x - 2y)$ **37.** $\left\{\dfrac{16}{3}\right\}$ **38.** $\{-11, 0\}$

39. $\left\{\dfrac{1}{14}\right\}$ **40.** $\{-1, 1\}$ **41.** $\{-6, 1\}$

42. $\left\{\dfrac{11}{12}\right\}$ **43.** $\{1, 2\}$ **44.** $\left\{-\dfrac{7}{2}, \dfrac{1}{3}\right\}$

45. $\left\{\dfrac{1}{2}, 8\right\}$ **46.** $\{-9, 2\}$

47. $\{x \mid x \le -1\}$ or $(-\infty, -1]$
48. $\{x \mid x > 13\}$ or $(13, \infty)$

49. $\left\{x \mid x \le \dfrac{3}{2}\right\}$ or $\left(-\infty, \dfrac{3}{2}\right]$

50. $\left\{x \mid x > \dfrac{48}{5}\right\}$ or $\left(\dfrac{48}{5}, \infty\right)$

51. $\{x \mid x \le 500\}$ or $(-\infty, 500]$

52. 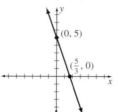 **53.**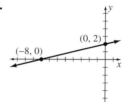

54. 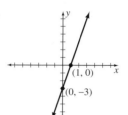 **55.**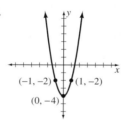

56. $\{(-6, -4)\}$ **57.** $\{(0, -3)\}$ **58.** $\left\{\left(-\dfrac{1}{2}, 5\right)\right\}$

59. $\{(-10, 6)\}$ **60.** Composite **61.** 6
62. 72 **63.** 175% **64.** $(2.4)(10^{-3})$ **65.** 3140
66.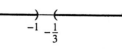

67. 16π square centimeters
68. 18 **69.**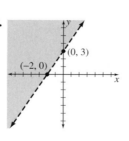

70. 6 inches, 8 inches, and 10 inches
71. 100 milliliters **72.** 6 feet by 8 feet
73. 2.5 hours **74.** 7.5 centimeters
75. 27 gallons **76.** 50° and 130°
77. 40°, 50°, and 90°
78. 40 pennies, 50 nickels, and 85 dimes
79. 14 dimes and 25 quarters
80. 97 or better **81.** 900 girls and 750 boys
82. $2500 **83.** $1275 **84.** 1 quart
85. $1.60 for a tennis ball and $2.25 for a golf ball

CHAPTER 8

Problem Set 8.1 (page 346)

1. $\{4\}$ **3.** $\{-1\}$ **5.** $\{6\}$ **7.** $\left\{-\dfrac{7}{3}\right\}$

9. $\left\{\dfrac{50}{3}\right\}$ **11.** $\left\{-\dfrac{53}{2}\right\}$ **13.** $\left\{\dfrac{27}{22}\right\}$

15. $\{-19\}$ **17.** $\left\{-\dfrac{19}{16}\right\}$ **19.** $\varnothing$ **21.** $\{275\}$

23. $\{x \mid x \text{ is a real number}\}$ **25.** $\left\{-\dfrac{64}{79}\right\}$

27. $\{(2, -4)\}$ **29.** $\{(-3, -6)\}$ **31.** $\varnothing$
33. $\{(-6, 0)\}$ **35.** $\{(-8, 6)\}$
37. $-16, -15,$ and -14
39. 14 males and 37 females
41. $42 skirt, $48 sweater, $34 shoes
43. 14 centimeters **45.** $25
47. 125 lb of sodium and 75 lb of chlorine
49. 8 meters and 15 meters
51. 16 miles per hour for Javier and 19 miles per hour for Domenica

Problem Set 8.2 (page 358)

1. $(-6, \infty)$ **3.** $(-5, \infty)$ **5.** $\left(-\infty, \dfrac{5}{3}\right]$

7. $(-36, \infty)$ **9.** $\left(-\infty, -\dfrac{8}{17}\right]$ **11.** $\left(-\dfrac{11}{2}, \infty\right)$

13. $(23, \infty)$ **15.** $\left(-\infty, -\dfrac{37}{3}\right]$ **17.** $\left(-\infty, -\dfrac{19}{6}\right)$

19. $(-\infty, 50]$ **21.** $(300, \infty)$ **23.** $[4, \infty)$

25. $[3, \infty)$

27. $\left(\dfrac{1}{3}, \dfrac{2}{5}\right)$

29. $(-\infty, -1) \cup \left(-\dfrac{1}{3}, \infty\right)$

31. $\left(-\dfrac{1}{4}, \dfrac{11}{4}\right)$ **33.** $[-11, 13]$ **35.** $(-1, 5)$

37. More than \$200   **39.** 96 or better

41. Between $-20°C$ and $-5°C$, inclusive

43. 8.8 to 15.4, inclusive

Problem Set 8.3 (page 365)

1. $(-5, 5)$

3. $[-2, 2]$

5. $(-\infty, -2) \cup (2, \infty)$

7. $(-1, 3)$

9. $[-6, 2]$

11. $(-\infty, -3) \cup (-1, \infty)$

13. $(-\infty, 1] \cup [5, \infty)$

15. $\{-7, 9\}$ **17.** $(-\infty, -4) \cup (8, \infty)$

19. $(-8, 2)$ **21.** $\{-1, 5\}$ **23.** $[-4, 5]$

25. $\left(-\infty, -\dfrac{7}{2}\right] \cup \left[\dfrac{5}{2}, \infty\right)$ **27.** $\left\{-5, \dfrac{7}{3}\right\}$

29. $\{-1, 5\}$ **31.** $(-\infty, -2) \cup (6, \infty)$

33. $\left(-\dfrac{1}{2}, \dfrac{3}{2}\right)$ **35.** $\left[-5, \dfrac{7}{5}\right]$ **37.** $\left\{\dfrac{1}{12}, \dfrac{17}{12}\right\}$

39. $[-3, 10]$ **41.** $(-5, 11)$

43. $\left(-\infty, -\dfrac{3}{2}\right) \cup \left(\dfrac{1}{2}, \infty\right)$ **45.** $\{0, 3\}$

47. $(-\infty, -14] \cup [0, \infty)$ **49.** $[-2, 3]$

51. $\varnothing$ **53.** $(-\infty, \infty)$ **55.** $\left\{\dfrac{2}{5}\right\}$ **57.** $\varnothing$

59. $\varnothing$ **65.** $\left\{-2, -\dfrac{4}{3}\right\}$ **67.** $\{-2\}$ **69.** $\{0\}$

Problem Set 8.4 (page 374)

1. $-\dfrac{1}{6} x^3 y^4$ **3.** $30x^6$ **5.** $-18x^9$

7. $-3x^6 y^6$ **9.** $81a^4 b^{12}$ **11.** $-16a^4 b^4$

13. $4x^2 + 33x + 35$ **15.** $9y^2 - 1$

17. $14x^2 + 3x - 2$ **19.** $5 + 3t - 2t^2$

21. $9t^2 + 42t + 49$ **23.** $4 - 25x^2$

25. $x^3 - 4x^2 + x + 6$ **27.** $x^3 - x^2 - 9x + 9$

29. $x^3 + x^2 - 24x + 16$ **31.** $2x^3 + 9x^2 + 2x - 30$

33. $12x^3 - 7x^2 + 25x - 6$

35. $x^4 + 5x^3 + 11x^2 + 11x + 4$

37. $64x^3 - 48x^2 + 12x - 1$

39. $125x^3 + 150x^2 + 60x + 8$

41. $x^{2n} - 16$ **43.** $x^{2a} + 4x^a - 12$

45. $6x^{2n} + x^n - 35$ **47.** $x^{4a} - 10x^{2a} + 21$

49. $4x^{2n} + 20x^n + 25$ **51.** $3x^3 y^3$ **53.** $-5x^3 y^2$

55. $9bc^2$ **57.** $-18xyz^4$ **59.** $-a^2 b^3 c^2$

61. $a^7 + 7a^6 b + 21a^5 b^2 + 35a^4 b^3 + 35a^3 b^4 + 21a^2 b^5 + 7ab^6 + b^7$

63. $x^5 - 5x^4 y + 10x^3 y^2 - 10x^2 y^3 + 5xy^4 - y^5$

65. $x^4 + 8x^3 y + 24x^2 y^2 + 32xy^3 + 16y^4$

67. $64a^6 - 192a^5 b + 240a^4 b^2 - 160a^3 b^3 + 60a^2 b^4 - 12ab^5 + b^6$

69. $x^{14} + 7x^{12}y + 21x^{10}y^2 + 35x^8 y^3 + 35x^6 y^4 + 21x^4 y^5 + 7x^2 y^6 + y^7$

71. $32a^5 - 240a^4 b + 720a^3 b^2 - 1080a^2 b^3 + 810ab^4 - 243b^5$

Problem Set 8.5 (page 381)

1. $3x^3 + 6x^2$ **3.** $-6x^4 + 9x^6$ **5.** $3a^2 - 5a - 8$

7. $-13x^2 + 17x - 28$ **9.** $-3xy + 4x^2 y - 8xy^2$

11. Q: $4x + 5$ **13.** Q: $t^2 + 2t - 4$ **15.** Q: $2x + 5$
     R: 0   R: 0   R: 1

17. Q: $3x - 4$ **19.** Q: $5y - 1$ **21.** Q: $4a + 6$
     R: $3x - 1$ R: $-8y - 2$ R: $7a - 19$

23. Q: $3x + 4$ **25.** Q: $x + 6$ **27.** Q: $4x - 3$
     R: 0   R: 14   R: 2

29. Q: $x^2 - 1$ **31.** Q: $3x^3 - 4x^2 + 6x - 13$
     R: 0   R: 12

33. Q: $x^2 - 2x - 3$ **35.** Q: $x^3 + 7x^2 + 21x + 56$
     R: 0   R: 167

37. Q: $x^2 + 3x + 2$ **39.** Q: $x^4 + x^3 + x^2 + x + 1$
     R: 0   R: 0

41. Q: $x^4 + x^3 + x^2 + x + 1$ **43.** Q: $2x^2 + 2x - 3$
     R: 2   R: $\dfrac{9}{2}$

45. Q: $4x^3 + 2x^2 - 4x - 2$
     R: 0

Problem Set 8.6 (page 394)

1. $2xy(3 - 4y)$ **3.** $(z + 3)(x + y)$

5. $(x + y)(3 + a)$ **7.** $(x - y)(a - b)$

9. $(3x + 5)(3x - 5)$ **11.** $(1 + 9n)(1 - 9n)$

13. $(x + 4 + y)(x + 4 - y)$

15. $(3s + 2t - 1)(3s - 2t + 1)$ **17.** $(x - 7)(x + 2)$

19. $(5 + x)(3 - x)$ **21.** Not factorable

23. $(3x - 5)(x - 2)$ **25.** $(5x + 1)(2x - 7)$

27. $(x - 2)(x^2 + 2x + 4)$

29. $(4x + 3y)(16x^2 - 12xy + 9y^2)$ **31.** $4(x^2 + 4)$

33. $x(x + 3)(x - 3)$ **35.** $(3a - 7)^2$

37. $2n(n^2 + 3n + 5)$ **39.** $(5x - 3)(2x + 9)$
41. $(6a - 1)^2$ **43.** $(4x - y)(2x + y)$
45. Not factorable   **47.** $2n(n^2 + 7n - 10)$
49. $4(x + 2)(x^2 - 2x + 4)$ **51.** $\{-3, -1\}$
53. $\{-12, -6\}$ **55.** $\{4, 9\}$ **57.** $\{-6, 2\}$
59. $\{-1, 5\}$ **61.** $\{-13, -12\}$ **63.** $\left\{-5, \dfrac{1}{3}\right\}$
65. $\left\{-\dfrac{7}{2}, -\dfrac{2}{3}\right\}$ **67.** $\{0, 4\}$ **69.** $\left\{\dfrac{1}{6}, 2\right\}$
71. $\{-6, 0, 6\}$ **73.** $\{-4, 6\}$ **75.** $\left\{\dfrac{1}{3}\right\}$
77. $\{-11, 4\}$ **79.** $\{-5, 5\}$ **81.** 6 units
83. 8 and 13   **85.** 6, 8, 10   **87.** 6 inches
89. 4 centimeters by 4 centimeters and 6 centimeters by 8 centimeters

Chapter 8 Review Problem Set (page 400)

1. $\{18\}$ **2.** $\{-14\}$ **3.** $\{0\}$ **4.** $\left\{\dfrac{1}{2}\right\}$
5. $\{10\}$ **6.** $\left\{\dfrac{7}{3}\right\}$ **7.** $\left\{\dfrac{28}{17}\right\}$ **8.** $\left\{-\dfrac{1}{38}\right\}$
9. $\left\{\dfrac{27}{17}\right\}$ **10.** $\left\{-\dfrac{10}{3}, 4\right\}$ **11.** $\{50\}$
12. $\left\{-\dfrac{39}{2}\right\}$ **13.** $\{200\}$ **14.** $\{-8\}$
15. $\{-3, 3\}$ **16.** $\{-6, 1\}$ **17.** $\left\{\dfrac{2}{7}\right\}$
18. $\left\{-\dfrac{2}{5}, \dfrac{1}{3}\right\}$ **19.** $\left\{-\dfrac{1}{3}, 3\right\}$ **20.** $\{-3, 0, 3\}$
21. $\left\{-\dfrac{4}{7}, \dfrac{2}{7}\right\}$ **22.** $\left\{-\dfrac{4}{5}, \dfrac{5}{6}\right\}$ **23.** $\{0, 1, 8\}$
24. $\left\{-10, \dfrac{1}{4}\right\}$ **25.** $x = \dfrac{2b + 2}{a}$ **26.** $x = \dfrac{c}{a - b}$
27. $x = \dfrac{pb - ma}{m - p}$ **28.** $x = \dfrac{11 + 7y}{5}$
29. $x = \dfrac{by + b + ac}{c}$ **30.** $[-5, \infty)$ **31.** $(4, \infty)$
32. $\left(-\dfrac{7}{3}, \infty\right)$ **33.** $\left[\dfrac{17}{2}, \infty\right)$ **34.** $\left(-\infty, \dfrac{1}{3}\right)$
35. $\left(\dfrac{53}{11}, \infty\right)$ **36.** $[6, \infty)$ **37.** $(-\infty, 100]$
38. $(-5, 6)$ **39.** $\left(-\infty, -\dfrac{11}{3}\right) \cup (3, \infty)$
40.
41.

42.
43.

44. $5x - 3$ **45.** $3x^2 + 12x - 2$ **46.** $12x^2 - x + 5$
47. $-20x^5y^7$ **48.** $-6a^5b^5$ **49.** $15a^4 - 10a^3 - 5a^2$
50. $24x^2 + 2xy - 15y^2$ **51.** $3x^3 + 7x^2 - 21x - 4$
52. $256x^8y^{12}$ **53.** $9x^2 - 12xy + 4y^2$
54. $-8x^6y^9z^3$ **55.** $-13x^2y$ **56.** $2x + y - 2$
57. $x^4 + x^3 - 18x^2 - x + 35$ **58.** $21 + 26x - 15x^2$
59. $-12a^5b^7$ **60.** $-8a^7b^3$ **61.** $7x^2 + 19x - 36$
62. $6x^3 - 11x^2 - 7x + 2$ **63.** $6x^{4n}$
64. $4x^2 + 20xy + 25y^2$ **65.** $x^3 - 6x^2 + 12x - 8$
66. Q: $3x^2 - 4x - 6$ **67.** Q: $5x^2 + 3x - 2$
 R: 29   R: 8
68. Q: $x^3 - 4x^2 - 7x - 1$ **69.** Q: $2x^3 + 3x^2 - 4x - 2$
 R: 0   R: 0
70. $(x + 7)(x - 4)$ **71.** $2(t + 3)(t - 3)$
72. Not factorable   **73.** $(4n - 1)(3n - 1)$
74. $x^2(x^2 + 1)(x + 1)(x - 1)$ **75.** $x(x - 12)(x + 6)$
76. $2a^2b(3a + 2b - c)$ **77.** $(x - y + 1)(x + y - 1)$
78. $4(2x^2 + 3)$ **79.** $(4x + 7)(3x - 5)$
80. $(4n - 5)^2$ **81.** $4n(n - 2)$
82. $3w(w^2 + 6w - 8)$ **83.** $(5x + 2y)(4x - y)$
84. $16a(a - 4)$ **85.** $3x(x + 1)(x - 6)$
86. $(n + 8)(n - 16)$ **87.** $(t + 5)(t - 5)(t^2 + 3)$
88. $(5x - 3)(7x + 2)$ **89.** $(3 - x)(5 - 3x)$
90. $(4n - 3)(16n^2 + 12n + 9)$
91. $2(2x + 5)(4x^2 - 10x + 25)$
92. The length is 15 meters and the width is 7 meters.
93. $2000 at 7% and $3000 at 8%
94. 88 or better   **95.** 4, 5, and 6
96. $10.50 per hour
97. 20 nickels, 50 dimes, and 75 quarters
98. 80°   **99.** $45.60   **100.** 30 or more
101. 55 miles per hour
102. Sonya for $3\dfrac{1}{4}$ hours and Rita for $4\dfrac{1}{2}$ hours
103. $6\dfrac{1}{4}$ cups   **104.** 12 miles and 16 miles
105. 4 meters by 12 meters
106. 9 rows and 16 chairs per row
107. The side is 13 feet long and the altitude is 6 feet.

Chapter 8 Test (page 402)

1. $2x - 11$ **2.** $20x^2 + 17x - 63$
3. $2x^3 + 11x^2 - 11x - 30$
4. $x^3 - 12x^2y + 48xy^2 - 64y^3$
5. Q: $2x^2 - 3x - 4$ **6.** Q: $3x^3 - x^2 - 2x - 6$
 R: 0   R: 3

7. $(x - y)(x + 4)$ **8.** $3(2x + 1)(2x - 1)$

9. $\left\{\dfrac{16}{5}\right\}$ **10.** $\left\{-\dfrac{14}{5}\right\}$ **11.** $\left\{-\dfrac{3}{2}, 3\right\}$

12. $\{3\}$ **13.** $\{650\}$ **14.** $\left\{0, \dfrac{1}{4}\right\}$

15. $\left\{\dfrac{3}{2}\right\}$ **16.** $\{-9, 0, 2\}$ **17.** $\left\{-\dfrac{3}{7}, \dfrac{4}{5}\right\}$

18. $\left\{-\dfrac{1}{3}, 2\right\}$ **19.** $\left(-1, \dfrac{7}{3}\right)$ **20.** $(3, \infty)$

21. $(-\infty, -35]$ **22.** $\dfrac{2}{3}$ of a cup

23. 97 or better   **24.** $70°$ **25.** 8 feet

CHAPTER 9

Problem Set 9.1 (page 409)

1. $\dfrac{3}{4}$ **3.** $\dfrac{5}{6}$ **5.** $-\dfrac{2}{5}$ **7.** $\dfrac{2}{7}$ **9.** $\dfrac{2x}{7}$

11. $\dfrac{2a}{5b}$ **13.** $-\dfrac{y}{4x}$ **15.** $-\dfrac{9c}{13d}$ **17.** $\dfrac{5x^2}{3y^3}$

19. $\dfrac{x - 2}{x}$ **21.** $\dfrac{3x + 2}{2x - 1}$ **23.** $\dfrac{a + 5}{a - 9}$

25. $\dfrac{n - 3}{5n - 1}$ **27.** $\dfrac{5x^2 + 7}{10x}$ **29.** $\dfrac{3x + 5}{4x + 1}$

31. $\dfrac{3x}{x^2 + 4x + 16}$ **33.** $\dfrac{x + 6}{3x - 1}$ **35.** $\dfrac{x(2x + 7)}{y(x + 9)}$

37. $\dfrac{y + 4}{5y - 2}$ **39.** $\dfrac{3x(x - 1)}{x^2 + 1}$ **41.** $\dfrac{2(x + 3y)}{3x(3x + y)}$

43. $\dfrac{3n - 4}{7n + 2}$ **45.** $\dfrac{4 - x}{5 + 3x}$ **47.** $\dfrac{9x^2 + 3x + 1}{2(x + 2)}$

49. $\dfrac{-2(x - 1)}{x + 1}$ **51.** $\dfrac{y + b}{y + c}$ **53.** $\dfrac{x + 2y}{2x + y}$

55. $\dfrac{x + 1}{x - 6}$ **57.** $\dfrac{2s + 5}{3s + 1}$ **59.** -1 **61.** $-n - 7$

63. $-\dfrac{2}{x + 1}$ **65.** -2 **67.** $-\dfrac{n + 3}{n + 5}$

Problem Set 9.2 (page 416)

1. $\dfrac{1}{10}$ **3.** $-\dfrac{4}{15}$ **5.** $\dfrac{3}{16}$ **7.** $-\dfrac{5}{6}$ **9.** $-\dfrac{2}{3}$

11. $\dfrac{10}{11}$ **13.** $-\dfrac{5x^3}{12y^2}$ **15.** $\dfrac{2a^3}{3b}$ **17.** $\dfrac{3x^3}{4}$

19. $\dfrac{25x^3}{108y^2}$ **21.** $\dfrac{ac^2}{2b^2}$ **23.** $\dfrac{3x}{4y}$ **25.** $\dfrac{3(x^2 + 4)}{5y(x + 8)}$

27. $\dfrac{5(a + 3)}{a(a - 2)}$ **29.** $\dfrac{3}{2}$ **31.** $\dfrac{3xy}{4(x + 6)}$

33. $\dfrac{5(x - 2y)}{7y}$ **35.** $\dfrac{5 + n}{3 - n}$ **37.** $\dfrac{x^2 + 1}{x^2 - 10}$

39. $\dfrac{6x + 5}{3x + 4}$ **41.** $\dfrac{2t^2 + 5}{2(t^2 + 1)(t + 1)}$ **43.** $\dfrac{t(t + 6)}{4t + 5}$

45. $\dfrac{n + 3}{n(n - 2)}$ **47.** $\dfrac{25x^3y^3}{4(x + 1)}$ **49.** $\dfrac{2(a - 2b)}{a(3a - 2b)}$

Problem Set 9.3 (page 425)

1. $\dfrac{13}{12}$ **3.** $\dfrac{11}{40}$ **5.** $\dfrac{19}{20}$ **7.** $\dfrac{49}{75}$ **9.** $\dfrac{17}{30}$

11. $-\dfrac{11}{84}$ **13.** $\dfrac{2x + 4}{x - 1}$ **15.** 4 **17.** $\dfrac{7y - 10}{7y}$

19. $\dfrac{5x + 3}{6}$ **21.** $\dfrac{12a + 1}{12}$ **23.** $\dfrac{n + 14}{18}$

25. $-\dfrac{11}{15}$ **27.** $\dfrac{3x - 25}{30}$ **29.** $\dfrac{43}{40x}$

31. $\dfrac{20y - 77x}{28xy}$ **33.** $\dfrac{16y + 15x - 12xy}{12xy}$

35. $\dfrac{21 + 22x}{30x^2}$ **37.** $\dfrac{10n - 21}{7n^2}$ **39.** $\dfrac{45 - 6n + 20n^2}{15n^2}$

41. $\dfrac{11x - 10}{6x^2}$ **43.** $\dfrac{42t + 43}{35t^3}$ **45.** $\dfrac{20b^2 - 33a^3}{96a^2b}$

47. $\dfrac{14 - 24y^3 + 45xy}{18xy^3}$ **49.** $\dfrac{2x^2 + 3x - 3}{x(x - 1)}$

51. $\dfrac{a^2 - a - 8}{a(a + 4)}$ **53.** $\dfrac{-41n - 55}{(4n + 5)(3n + 5)}$

55. $\dfrac{-3x + 17}{(x + 4)(7x - 1)}$ **57.** $\dfrac{-x + 74}{(3x - 5)(2x + 7)}$

59. $\dfrac{38x + 13}{(3x - 2)(4x + 5)}$ **61.** $\dfrac{5x + 5}{2x + 5}$ **63.** $\dfrac{x + 15}{x - 5}$

65. $\dfrac{-2x - 4}{2x + 1}$ **67. a.** -1 **c.** 0

Problem Set 9.4 (page 436)

1. $\dfrac{7x + 20}{x(x + 4)}$ **3.** $\dfrac{-x - 3}{x(x + 7)}$ **5.** $\dfrac{6x - 5}{(x + 1)(x - 1)}$

7. $\dfrac{1}{a + 1}$ **9.** $\dfrac{5n + 15}{4(n + 5)(n - 5)}$ **11.** $\dfrac{x^2 + 60}{x(x + 6)}$

13. $\dfrac{11x + 13}{(x + 2)(x + 7)(2x + 1)}$

15. $\dfrac{-3a + 1}{(a - 5)(a + 2)(a + 9)}$

17. $\dfrac{9a^2 + 17a + 1}{(5a + 1)(4a - 3)(3a + 4)}$

19. $\dfrac{3x^2 + 20x - 111}{(x^2 + 3)(x + 7)(x - 3)}$

21. $\dfrac{-7y - 14}{(y + 8)(y - 2)}$ **23.** $\dfrac{-2x^2 - 4x + 3}{(x + 2)(x - 2)}$

25. $\dfrac{2x^2 + 14x - 19}{(x + 10)(x - 2)}$ **27.** $\dfrac{2n + 1}{n - 6}$

29. $\dfrac{2x^2 - 32x + 16}{(x + 1)(2x - 1)(3x - 2)}$ **31.** $\dfrac{1}{(n^2 + 1)(n + 1)}$

33. $\dfrac{-16x}{(5x - 2)(x - 1)}$ **35.** $\dfrac{t + 1}{t - 2}$ **37.** $\dfrac{2}{11}$

39. $-\dfrac{7}{27}$ **41.** $\dfrac{x}{4}$ **43.** $\dfrac{3y - 2x}{4x - 7}$

45. $\dfrac{6ab^2 - 5a^2}{12b^2 + 2a^2b}$ **47.** $\dfrac{2y - 3xy}{3x + 4xy}$ **49.** $\dfrac{3n + 14}{5n + 19}$

51. $\dfrac{5n - 17}{4n - 13}$ **53.** $\dfrac{-x + 5y - 10}{3y - 10}$ **55.** $\dfrac{-x + 15}{-2x - 1}$

57. $\dfrac{3a^2 - 2a + 1}{2a - 1}$ **59.** $\dfrac{-x^2 + 6x - 4}{3x - 2}$

Problem Set 9.5 (page 444)

1. $\{2\}$ **3.** $\{-3\}$ **5.** $\{6\}$ **7.** $\left\{-\dfrac{85}{18}\right\}$

9. $\left\{\dfrac{7}{10}\right\}$ **11.** $\{5\}$ **13.** $\{58\}$

15. $\left\{\dfrac{1}{4}, 4\right\}$ **17.** $\left\{-\dfrac{2}{5}, 5\right\}$ **19.** $\{-16\}$

21. $\left\{-\dfrac{13}{3}\right\}$ **23.** $\{-3, 1\}$ **25.** $\left\{-\dfrac{5}{2}\right\}$

27. $\{-51\}$ **29.** $\left\{-\dfrac{5}{3}, 4\right\}$ **31.** $\varnothing$

33. $\left\{-\dfrac{11}{8}, 2\right\}$ **35.** $\{-29, 0\}$ **37.** $\{-9, 3\}$

39. $\left\{-2, \dfrac{23}{8}\right\}$ **41.** $\left\{\dfrac{11}{23}\right\}$

43. $750 and $1000   **45.** 48° and 72°   **47.** 60°
49. $2280   **51.** $120 for Tammy and $90 for Laura
53. 8 and 82   **55.** 14 feet and 6 feet
57. 12,690 females and 8460 males

Problem Set 9.6 (page 453)

1. $\{-21\}$ **3.** $\{-1, 2\}$ **5.** $\{2\}$ **7.** $\left\{\dfrac{37}{15}\right\}$

9. $\{-1\}$ **11.** $\{-1\}$ **13.** $\left\{0, \dfrac{13}{2}\right\}$

15. $\left\{-2, \dfrac{19}{2}\right\}$ **17.** $\{-2\}$ **19.** $\left\{-\dfrac{1}{5}\right\}$ **21.** $\varnothing$

23. $\left\{\dfrac{7}{2}\right\}$ **25.** $\{-3\}$ **27.** $\left\{-\dfrac{7}{9}\right\}$ **29.** $\left\{-\dfrac{7}{6}\right\}$

31. $x = \dfrac{18y - 4}{15}$ **33.** $y = \dfrac{-5x + 22}{2}$

35. $M = \dfrac{IC}{100}$ **37.** $R = \dfrac{ST}{S + T}$

39. $y = \dfrac{bx - x - 3b + a}{a - 3}$ **41.** $y = \dfrac{ab - bx}{a}$

43. $y = \dfrac{-2x - 9}{3}$

45. 50 miles per hour for Dave and 54 miles per hour for Kent
47. 60 minutes
49. 60 words per minute for Connie and 40 words per minute for Katie
51. Plane B could travel at 400 miles per hour for 5 hours and plane A at 350 miles per hour for 4 hours, or plane B could travel at 250 miles per hour for 8 hours and plane A at 200 miles per hour for 7 hours.
53. 60 minutes for Nancy and 120 minutes for Amy
55. 3 hours
57. 16 miles per hour on the way out and 12 miles per hour on the way back, or 12 miles per hour out and 8 miles per hour back

Chapter 9 Review Problem Set (page 457)

1. $\dfrac{2y}{3x^2}$ **2.** $\dfrac{a - 3}{a}$ **3.** $\dfrac{n - 5}{n - 1}$ **4.** $\dfrac{x^2 + 1}{x}$

5. $\dfrac{2x + 1}{3}$ **6.** $\dfrac{x^2 - 10}{2x^2 + 1}$ **7.** $\dfrac{3}{22}$ **8.** $\dfrac{18y + 20x}{48y - 9x}$

9. $\dfrac{3x + 2}{3x - 2}$ **10.** $\dfrac{x - 1}{2x - 1}$ **11.** $\dfrac{2x}{7y^2}$ **12.** $3b$

13. $\dfrac{n(n + 5)}{n - 1}$ **14.** $\dfrac{x(x - 3y)}{x^2 + 9y^2}$ **15.** $\dfrac{23x - 6}{20}$

16. $\dfrac{57 - 2n}{18n}$ **17.** $\dfrac{3x^2 - 2x - 14}{x(x + 7)}$ **18.** $\dfrac{2}{x - 5}$

19. $\dfrac{5n - 21}{(n - 9)(n + 4)(n - 1)}$ **20.** $\dfrac{6y - 23}{(2y + 3)(y - 6)}$

21. $\dfrac{1}{x + 1}$ **22.** $\dfrac{3n^2 - 20n + 20}{(n + 5)(n + 1)(n - 8)}$ **23.** $\left\{\dfrac{4}{13}\right\}$

24. $\left\{\dfrac{3}{16}\right\}$ **25.** $\varnothing$ **26.** $\{-17\}$ **27.** $\left\{\dfrac{2}{7}, \dfrac{7}{2}\right\}$

28. $\{22\}$ **29.** $\left\{-\dfrac{6}{7}, 3\right\}$ **30.** $\left\{\dfrac{3}{4}, \dfrac{5}{2}\right\}$ **31.** $\left\{\dfrac{9}{7}\right\}$

32. $\left\{-\dfrac{5}{4}\right\}$ **33.** $y = \dfrac{3x + 27}{4}$ **34.** $y = \dfrac{bx - ab}{a}$

35. $525 and $875
36. 20 minutes for Julio and 30 minutes for Dan

37. 50 miles per hour and 55 miles per hour or $8\frac{1}{3}$ miles per hour for A and $13\frac{1}{3}$ miles for B

38. 9 hours   **39.** 80 hours

40. 13 miles per hour

Chapter 9 Test (page 459)

1. $\dfrac{13y^2}{24x}$ **2.** $\dfrac{3x-1}{x(x-6)}$ **3.** $\dfrac{2n-3}{n+4}$ **4.** $-\dfrac{2x}{x+1}$

5. $\dfrac{3y^2}{8}$ **6.** $\dfrac{a-b}{4(2a+b)}$ **7.** $\dfrac{x+4}{5x-1}$ **8.** $\dfrac{13x+7}{12}$

9. $\dfrac{3x}{2}$ **10.** $\dfrac{10n-26}{15n}$ **11.** $\dfrac{3x^2+2x-12}{x(x-6)}$

12. $\dfrac{11-2x}{x(x-1)}$ **13.** $\dfrac{13n+46}{(2n+5)(n-2)(n+7)}$

14. $\dfrac{23x+6}{6x(x-3)(x+2)}$ **15.** $\dfrac{18-2x}{8+9x}$

16. $y=\dfrac{4x+20}{3}$ **17.** $\{1\}$ **18.** $\left\{\dfrac{1}{10}\right\}$

19. $\{-35\}$ **20.** $\{-1,5\}$ **21.** $\left\{\dfrac{5}{3}\right\}$

22. $\left\{-\dfrac{9}{13}\right\}$ **23.** $\dfrac{27}{72}$ **24.** 1 hour

25. 15 miles per hour

Cumulative Practice Test (Chapters 1–9) (page 460)

1. -287 **2.** $\dfrac{8}{7}$ **3.** 48

4. Q: $2x^3+5x^2-7x-6$ R: 3

5. $\{(-3,4)\}$ **6.** $\{(5,-6)\}$

7.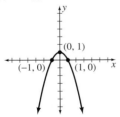

8.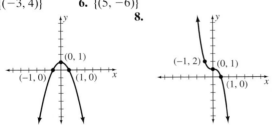

9.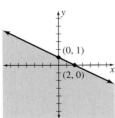

10. 36   **11.** 225%   **12.** $(1.3)(10)^{-4}$

13. $\{-8\}$ **14.** $\left\{\dfrac{1}{2},6\right\}$ **15.** $\left\{-\dfrac{2}{5},0,\dfrac{1}{3}\right\}$

16. $\left\{-\dfrac{6}{5},\dfrac{8}{5}\right\}$ **17.** $\left\{-1,\dfrac{5}{3}\right\}$ **18.** $[-2,1]$

19. $\left(-\infty,-\dfrac{9}{2}\right)$ **20.** $(-\infty,0)\cup(1,\infty)$ **21.** $20

22. 20 gallons   **23.** $26\frac{2}{3}$ minutes

24. 15 centimeters and 20 centimeters

25. 95 or better

CHAPTER 10

Problem Set 10.1 (page 468)

1. $\dfrac{1}{27}$ **3.** $-\dfrac{1}{100}$ **5.** 81   **7.** -27 **9.** -8

11. 1   **13.** $\dfrac{9}{49}$ **15.** 16   **17.** $\dfrac{1}{1000}$

19. $\dfrac{1}{1000}$ **21.** 27   **23.** $\dfrac{1}{125}$ **25.** $\dfrac{9}{8}$

27. $\dfrac{256}{25}$ **29.** $\dfrac{2}{25}$ **31.** $\dfrac{81}{4}$ **33.** 81

35. $\dfrac{1}{10,000}$ **37.** $\dfrac{13}{36}$ **39.** $\dfrac{1}{2}$ **41.** $\dfrac{72}{17}$

43. $\dfrac{1}{x^6}$ **45.** $\dfrac{1}{a^3}$ **47.** $\dfrac{1}{a^8}$ **49.** $\dfrac{y^6}{x^2}$ **51.** $\dfrac{c^8}{a^4b^{12}}$

53. $\dfrac{y^{12}}{8x^9}$ **55.** $\dfrac{x^3}{y^{12}}$ **57.** $\dfrac{4a^4}{9b^2}$ **59.** $\dfrac{1}{x^2}$

61. a^5b^2 **63.** $\dfrac{6y^3}{x}$ **65.** $7b^2$ **67.** $\dfrac{7x}{y^2}$

69. $-\dfrac{12b^3}{a}$ **71.** $\dfrac{x^5y^5}{5}$ **73.** $\dfrac{b^{20}}{81}$

75. $\dfrac{x+1}{x^3}$ **77.** $\dfrac{y-x^3}{x^3y}$ **79.** $\dfrac{3b+4a^2}{a^2b}$

81. $\dfrac{1-x^2y}{xy^2}$ **83.** $\dfrac{2x-3}{x^2}$ **85.** $(4)(10)^7$

87. $(3.764)(10)^2$ **89.** $(3.47)(10)^{-1}$

91. $(2.14)(10)^{-2}$ **93.** 31,400,000,000

95. 0.43   **97.** 0.000914   **99.** 0.00000005123

101. 1000   **103.** 1000

Problem Set 10.2 (page 480)

1. 8   **3.** -10 **5.** 3   **7.** -4 **9.** 3

11. $\dfrac{4}{5}$ **13.** $-\dfrac{6}{7}$ **15.** $\dfrac{1}{2}$ **17.** $\dfrac{3}{4}$ **19.** 8

21. $3\sqrt{3}$ **23.** $4\sqrt{2}$ **25.** $4\sqrt{5}$ **27.** $4\sqrt{10}$

29. $12\sqrt{2}$ **31.** $-12\sqrt{5}$ **33.** $2\sqrt{3}$ **35.** $3\sqrt{6}$

37. $-\dfrac{5}{3}\sqrt{7}$ **39.** $\dfrac{\sqrt{19}}{2}$ **41.** $\dfrac{3\sqrt{3}}{4}$ **43.** $\dfrac{5\sqrt{3}}{9}$

45. $\dfrac{\sqrt{14}}{7}$ **47.** $\dfrac{\sqrt{6}}{3}$ **49.** $\dfrac{\sqrt{15}}{6}$ **51.** $\dfrac{\sqrt{66}}{12}$

53. $\dfrac{\sqrt{6}}{3}$ **55.** $\sqrt{5}$ **57.** $\dfrac{2\sqrt{21}}{7}$ **59.** $-\dfrac{8\sqrt{15}}{5}$

61. $\dfrac{\sqrt{6}}{4}$ **63.** $-\dfrac{12}{25}$ **65.** $2\sqrt[3]{2}$ **67.** $6\sqrt[3]{3}$

69. $\dfrac{2\sqrt[3]{3}}{3}$ **71.** $\dfrac{3\sqrt[3]{2}}{2}$ **73.** $\dfrac{\sqrt[3]{12}}{2}$

75. 42 miles per hour; 49 miles per hour; 65 miles per hour
77. 107 square centimeters   **79.** 140 square inches
85. a. 1.414   **c.** 12.490   **e.** 57.000
    g. 0.374   **i.** 0.930

Problem Set 10.3 (page 486)

1. $13\sqrt{2}$ **3.** $54\sqrt{3}$ **5.** $-30\sqrt{2}$ **7.** $-\sqrt{5}$

9. $-21\sqrt{6}$ **11.** $-\dfrac{7\sqrt{7}}{12}$ **13.** $\dfrac{37\sqrt{10}}{10}$

15. $\dfrac{41\sqrt{2}}{20}$ **17.** $-9\sqrt[3]{3}$ **19.** $10\sqrt[3]{2}$ **21.** $4\sqrt{2x}$

23. $5x\sqrt{3}$ **25.** $2x\sqrt{5y}$ **27.** $8xy^3\sqrt{xy}$

29. $3a^2b\sqrt{6b}$ **31.** $3x^3y^4\sqrt{7}$ **33.** $4a\sqrt{10a}$

35. $\dfrac{8y}{3}\sqrt{6xy}$ **37.** $\dfrac{\sqrt{10xy}}{5y}$ **39.** $\dfrac{\sqrt{15}}{6x^2}$

41. $\dfrac{5\sqrt{2y}}{6y}$ **43.** $\dfrac{\sqrt{14xy}}{4y^3}$ **45.** $\dfrac{3y\sqrt{2xy}}{4x}$

47. $\dfrac{2\sqrt{42ab}}{7b^2}$ **49.** $2\sqrt[3]{3y}$ **51.** $2x\sqrt[3]{2x}$

53. $2x^2y^2\sqrt[3]{7y^2}$ **55.** $\dfrac{\sqrt[3]{21x}}{3x}$ **57.** $\dfrac{\sqrt[3]{12x^2y}}{4x^2}$

59. $\dfrac{\sqrt[3]{4x^2y^2}}{xy^2}$ **61.** $2\sqrt{2x}+3y$ **63.** $4\sqrt{x}+3y$

65. $33\sqrt{x}$ **67.** $-30\sqrt{2x}$ **69.** $7\sqrt{3n}$

71. $-40\sqrt{ab}$ **73.** $-7x\sqrt{2x}$

Problem Set 10.4 (page 493)

1. $6\sqrt{2}$ **3.** $18\sqrt{2}$ **5.** $-24\sqrt{10}$ **7.** $24\sqrt{6}$

9. 120   **11.** 24   **13.** $56\sqrt[3]{3}$ **15.** $\sqrt{6}+\sqrt{10}$

17. $6\sqrt{10}-3\sqrt{35}$ **19.** $24\sqrt{3}-60\sqrt{2}$

21. $-40-32\sqrt{15}$ **23.** $15\sqrt{2x}+3\sqrt{xy}$

25. $5xy-6x\sqrt{y}$ **27.** $2\sqrt{10xy}+2y\sqrt{15y}$

29. $-25\sqrt{6}$ **31.** $-25-3\sqrt{3}$ **33.** $23-9\sqrt{5}$

35. $6\sqrt{35}+3\sqrt{10}-4\sqrt{21}-2\sqrt{6}$

37. $8\sqrt{3}-36\sqrt{2}+6\sqrt{10}-18\sqrt{15}$

39. $11+13\sqrt{30}$ **41.** $141-51\sqrt{6}$ **43.** -10

45. -8 **47.** $2x-3y$ **49.** $10\sqrt[3]{12}+2\sqrt[3]{18}$

51. $12-36\sqrt[3]{2}$ **53.** $\dfrac{\sqrt{7}-1}{3}$ **55.** $\dfrac{-3\sqrt{2}-15}{23}$

57. $\dfrac{\sqrt{7}-\sqrt{2}}{5}$ **59.** $\dfrac{2\sqrt{5}+\sqrt{6}}{7}$

61. $\dfrac{\sqrt{15}-2\sqrt{3}}{2}$ **63.** $\dfrac{6\sqrt{7}+4\sqrt{6}}{13}$

65. $\sqrt{3}-\sqrt{2}$ **67.** $\dfrac{2\sqrt{x}-8}{x-16}$ **69.** $\dfrac{x+5\sqrt{x}}{x-25}$

71. $\dfrac{x-8\sqrt{x}+12}{x-36}$ **73.** $\dfrac{x-2\sqrt{xy}}{x-4y}$

75. $\dfrac{6\sqrt{xy}+9y}{4x-9y}$

Problem Set 10.5 (page 499)

1. $\{20\}$ **3.** $\varnothing$ **5.** $\left\{\dfrac{25}{4}\right\}$ **7.** $\left\{\dfrac{4}{9}\right\}$ **9.** $\{5\}$

11. $\left\{\dfrac{39}{4}\right\}$ **13.** $\varnothing$ **15.** $\{1\}$ **17.** $\left\{\dfrac{3}{2}\right\}$

19. $\{3\}$ **21.** $\left\{\dfrac{61}{25}\right\}$ **23.** $\{-3, 3\}$ **25.** $\{-9, -4\}$

27. $\{0\}$ **29.** $\{3\}$ **31.** $\{4\}$ **33.** $\{-4, -3\}$

35. $\{12\}$ **37.** $\{25\}$ **39.** $\{29\}$ **41.** $\{-15\}$

43. $\left\{-\dfrac{1}{3}\right\}$ **45.** $\{-3\}$ **47.** $\{0\}$ **49.** $\{5\}$

51. $\{2, 6\}$ **53.** 56 feet; 106 feet; 148 feet
55. 3.2 feet; 5.1 feet; 7.3 feet

Problem Set 10.6 (page 505)

1. 9   **3.** 3   **5.** -2 **7.** -5 **9.** $\dfrac{1}{6}$

11. 3   **13.** 8   **15.** 81   **17.** -1

19. -32 **21.** $\dfrac{81}{16}$ **23.** 4   **25.** $\dfrac{1}{128}$

27. -125 **29.** 625   **31.** $\sqrt[3]{x^4}$ **33.** $3\sqrt{x}$

35. $\sqrt[3]{2y}$ **37.** $\sqrt{2x-3y}$ **39.** $\sqrt[3]{(2a-3b)^2}$

41. $\sqrt[3]{x^2y}$ **43.** $-3\sqrt[5]{xy^2}$ **45.** $5^{\frac{1}{2}}y^{\frac{1}{2}}$ **47.** $3y^{\frac{1}{2}}$

49. $x^{\frac{1}{3}}y^{\frac{1}{3}}$ **51.** $a^{\frac{1}{2}}b^{\frac{3}{4}}$ **53.** $(2x-y)^{\frac{3}{5}}$ **55.** $5xy^{\frac{1}{2}}$

57. $-(x+y)^{\frac{1}{3}}$ **59.** $12x^{\frac{13}{20}}$ **61.** $y^{\frac{1}{12}}$ **63.** $\dfrac{4}{x^{\frac{1}{10}}}$

65. $16xy^2$ **67.** $2x^2y$ **69.** $4x^{\frac{4}{15}}$

71. $\dfrac{4}{b^{\frac{5}{12}}}$ **73.** $\dfrac{36x^{\frac{4}{5}}}{49y^{\frac{4}{3}}}$ **75.** $\dfrac{y^{\frac{3}{2}}}{x}$ **77.** $4x^{\frac{1}{6}}$

79. $\dfrac{16}{a^{\frac{11}{10}}}$ **81.** $\sqrt[6]{243}$ **83.** $\sqrt[4]{216}$ **85.** $\sqrt[12]{3}$

87. $\sqrt{2}$ **89.** $\sqrt[4]{3}$

93. a. 12   **c.** 7   **e.** 11

95. a. 1024   **c.** 512   **e.** 49

Chapter 10 Review Problem Set (page 508)

1. $\dfrac{1}{64}$ **2.** $\dfrac{9}{4}$ **3.** 3   **4.** -2 **5.** $\dfrac{2}{3}$ **6.** 32

7. 1   **8.** $\dfrac{4}{9}$ **9.** -64 **10.** 32   **11.** 1

12. 27   **13.** $3\sqrt{6}$ **14.** $4x\sqrt{3xy}$ **15.** $2\sqrt{2}$

16. $\dfrac{\sqrt{15x}}{6x^2}$ **17.** $2\sqrt[3]{7}$ **18.** $\dfrac{\sqrt[3]{6}}{3}$ **19.** $\dfrac{3\sqrt{5}}{5}$

20. $\dfrac{x\sqrt{21x}}{7}$ **21.** $3xy^2\sqrt[3]{4xy^2}$ **22.** $\dfrac{15\sqrt{6}}{4}$

23. $2y\sqrt{5xy}$ **24.** $2\sqrt{x}$ **25.** $24\sqrt{10}$ **26.** 60

27. $24\sqrt{3} - 6\sqrt{14}$ **28.** $x - 2\sqrt{x} - 15$ **29.** 17

30. $12 - 8\sqrt{3}$ **31.** $6a - 5\sqrt{ab} - 4b$ **32.** 70

33. $\dfrac{2(\sqrt{7} + 1)}{3}$ **34.** $\dfrac{2\sqrt{6} - \sqrt{15}}{3}$

35. $\dfrac{3\sqrt{5} - 2\sqrt{3}}{11}$ **36.** $\dfrac{6\sqrt{3} + 3\sqrt{5}}{7}$ **37.** $\dfrac{x^6}{y^8}$

38. $\dfrac{27a^3b^{12}}{8}$ **39.** $20x^{\frac{7}{10}}$ **40.** $7a^{\frac{5}{12}}$ **41.** $\dfrac{y^{\frac{4}{3}}}{x}$

42. $\dfrac{x^{12}}{9}$ **43.** $\sqrt{5}$ **44.** $5\sqrt[3]{3}$ **45.** $\dfrac{29\sqrt{6}}{5}$

46. $-15\sqrt{3x}$ **47.** $\dfrac{y + x^2}{x^2y}$ **48.** $\dfrac{b - 2a}{a^2b}$

49. $\left\{\dfrac{19}{7}\right\}$ **50.** $\{4\}$ **51.** $\{8\}$ **52.** $\varnothing$

53. $\{14\}$ **54.** $\{-10, 1\}$ **55.** $\{2\}$ **56.** $\{8\}$

57. 0.000000006   **58.** 36,000,000,000   **59.** 6

60. 0.15   **61.** 0.000028   **62.** 0.002

63. 0.002   **64.** 8,000,000,000

Chapter 10 Test (page 509)

1. $\dfrac{1}{32}$ **2.** -32 **3.** $\dfrac{81}{16}$ **4.** $\dfrac{1}{4}$ **5.** $3\sqrt{7}$

6. $3\sqrt[3]{4}$ **7.** $2x^2y\sqrt{13y}$ **8.** $\dfrac{5\sqrt{6}}{6}$ **9.** $\dfrac{\sqrt{42x}}{12x^2}$

10. $72\sqrt{2}$ **11.** $-5\sqrt{6}$ **12.** $-38\sqrt{2}$

13. $\dfrac{3\sqrt{6} + 3}{10}$ **14.** $\dfrac{9x^2y^2}{4}$ **15.** $-\dfrac{12}{a^{\frac{3}{10}}}$

16. $\dfrac{y^3 + x}{xy^3}$ **17.** $-12x^{\frac{1}{4}}$ **18.** 33   **19.** 600

20. 0.003   **21.** $\left\{\dfrac{8}{3}\right\}$ **22.** $\{2\}$ **23.** $\{4\}$

24. $\{5\}$ **25.** $\{4, 6\}$

Cumulative Practice Test (Chapters 1–10) (page 510)

1. a. $\dfrac{12}{5}$ **b.** $\dfrac{324}{121}$ **c.** $-\dfrac{1}{2}$ **d.** 81

2. $-\dfrac{4y^5}{x^4}$ **3.** $-\dfrac{12}{y^4}$ **4.** $6x^3 + 16x^2 - 18x + 4$

5. $4x^2 - 7x - 2$ **6.** $\{(-2, 7)\}$

7. a. $6\sqrt{14}$ **b.** $4\sqrt[3]{7}$ **c.** $\dfrac{\sqrt{3}}{4}$ **d.** $\dfrac{\sqrt{3}}{2}$

8. 31.11   **9.** 72   **10.** 429   **11.** $-\dfrac{7}{3}$

12. a. $2 \cdot 2 \cdot 13$ **b.** $2 \cdot 2 \cdot 2 \cdot 2 \cdot 5$ **c.** $7 \cdot 13$

d. $2 \cdot 3 \cdot 13$ **13.** $\dfrac{5y - 6x}{9y + 5x}$ **14.** $\dfrac{2x - 33}{20}$

15. $\dfrac{2x^2}{3y}$ **16.** $\dfrac{2}{3x - 4}$ **17.** $\{3, 4\}$

18. $\{-1, 4\}$ **19.** $\left\{-\dfrac{7}{2}, \dfrac{5}{3}\right\}$ **20.** $\varnothing$ **21.** $\{2\}$

22. The altitude is 6 inches and the side is 17 inches.

23. Pedro is 23 years old and Brad is 29 years old.

24. $75   **25.** 7 inches by 11 inches

CHAPTER 11

Problem Set 11.1 (page 518)

1. False   **3.** True   **5.** True   **7.** True

9. $10 + 8i$ **11.** $-6 + 10i$ **13.** $-2 - 5i$

15. $-12 + 5i$ **17.** $-1 - 23i$ **19.** $-4 - 5i$

21. $1 + 3i$ **23.** $\dfrac{5}{3} - \dfrac{5}{12}i$ **25.** $-\dfrac{17}{9} + \dfrac{23}{30}i$

27. $9i$ **29.** $i\sqrt{14}$ **31.** $\dfrac{4}{5}i$ **33.** $3i\sqrt{2}$

35. $5i\sqrt{3}$ **37.** $6i\sqrt{7}$ **39.** $-8i\sqrt{5}$

41. $36i\sqrt{10}$ **43.** -8 **45.** $-\sqrt{15}$ **47.** $-3\sqrt{6}$

49. $-5\sqrt{3}$ **51.** $-3\sqrt{6}$ **53.** $4i\sqrt{3}$ **55.** $\dfrac{5}{2}$

57. $2\sqrt{2}$ **59.** $2i$ **61.** $-20 + 0i$ **63.** $42 + 0i$

65. $15 + 6i$ **67.** $-42 + 12i$ **69.** $7 + 22i$

71. $40 - 20i$ **73.** $-3 - 28i$ **75.** $-3 - 15i$
77. $-9 + 40i$ **79.** $-12 + 16i$ **81.** $85 + 0i$

83. $5 + 0i$ **85.** $\dfrac{3}{5} + \dfrac{3}{10}i$ **87.** $\dfrac{5}{17} - \dfrac{3}{17}i$

89. $2 + \dfrac{2}{3}i$ **91.** $0 - \dfrac{2}{7}i$ **93.** $\dfrac{22}{25} - \dfrac{4}{25}i$

95. $-\dfrac{18}{41} + \dfrac{39}{41}i$ **97.** $\dfrac{9}{2} - \dfrac{5}{2}i$ **99.** $\dfrac{4}{13} - \dfrac{1}{26}i$

Problem Set 11.2 (page 526)

1. $\{0, 9\}$ **3.** $\{-3, 0\}$ **5.** $\{-4, 0\}$ **7.** $\left\{0, \dfrac{9}{5}\right\}$

9. $\{-6, 5\}$ **11.** $\{7, 12\}$ **13.** $\left\{-8, -\dfrac{3}{2}\right\}$

15. $\left\{-\dfrac{7}{3}, \dfrac{2}{5}\right\}$ **17.** $\left\{\dfrac{3}{5}\right\}$ **19.** $\left\{-\dfrac{3}{2}, \dfrac{7}{3}\right\}$ **21.** $\{1, 4\}$

23. $\{8\}$ **25.** $\{12\}$ **27.** $\{0, 5k\}$ **29.** $\{0, 16k^2\}$

31. $\{5k, 7k\}$ **33.** $\left\{\dfrac{k}{2}, -3k\right\}$ **35.** $\{\pm 1\}$ **37.** $\{\pm 6i\}$

39. $\{\pm\sqrt{14}\}$ **41.** $\{\pm 2\sqrt{7}\}$ **43.** $\{\pm 3\sqrt{2}\}$

45. $\left\{\pm\dfrac{\sqrt{14}}{2}\right\}$ **47.** $\left\{\pm\dfrac{2\sqrt{3}}{3}\right\}$ **49.** $\left\{\pm\dfrac{2i\sqrt{30}}{5}\right\}$

51. $\left\{\pm\dfrac{\sqrt{6}}{2}\right\}$ **53.** $\{-1, 5\}$ **55.** $\{-8, 2\}$

57. $\{-6 \pm 2i\}$ **59.** $\{1, 2\}$ **61.** $\{4 \pm \sqrt{5}\}$

63. $\{-5 \pm 2\sqrt{3}\}$ **65.** $\left\{\dfrac{2 \pm 3i\sqrt{3}}{3}\right\}$ **67.** $\{-12, -2\}$

69. $\left\{\dfrac{2 \pm \sqrt{10}}{5}\right\}$ **71.** $2\sqrt{13}$ centimeters

73. $4\sqrt{5}$ inches   **75.** 8 yards   **77.** $6\sqrt{2}$ inches
79. $a = b = 4\sqrt{2}$ meters
81. $b = 3\sqrt{3}$ inches and $c = 6$ inches
83. $a = 7$ centimeters and $b = 7\sqrt{3}$ centimeters

85. $a = \dfrac{10\sqrt{3}}{3}$ feet and $c = \dfrac{20\sqrt{3}}{3}$ feet

87. 17.9 feet   **89.** 38 meters
91. 53 meters   **95.** 10.8 centimeters   **97.** $h = s\sqrt{2}$

Problem Set 11.3 (page 533)
1. $\{-6, 10\}$ **3.** $\{4, 10\}$ **5.** $\{-5, 10\}$ **7.** $\{-8, 1\}$

9. $\left\{-\dfrac{5}{2}, 3\right\}$ **11.** $\left\{-3, \dfrac{2}{3}\right\}$ **13.** $\{-16, 10\}$

15. $\{-2 \pm \sqrt{6}\}$ **17.** $\{-3 \pm 2\sqrt{3}\}$
19. $\{5 \pm \sqrt{26}\}$ **21.** $\{4 \pm i\}$ **23.** $\{-6 \pm 3\sqrt{3}\}$

25. $\{-1 \pm i\sqrt{5}\}$ **27.** $\left\{\dfrac{-3 \pm \sqrt{17}}{2}\right\}$

29. $\left\{\dfrac{-5 \pm \sqrt{21}}{2}\right\}$ **31.** $\left\{\dfrac{7 \pm \sqrt{37}}{2}\right\}$

33. $\left\{\dfrac{-2 \pm \sqrt{10}}{2}\right\}$ **35.** $\left\{\dfrac{3 \pm i\sqrt{6}}{3}\right\}$

37. $\left\{\dfrac{-5 \pm \sqrt{37}}{6}\right\}$ **39.** $\{-12, 4\}$

41. $\left\{\dfrac{4 \pm \sqrt{10}}{2}\right\}$ **43.** $\left\{-\dfrac{9}{2}, \dfrac{1}{3}\right\}$ **45.** $\{-3, 8\}$

47. $\{3 \pm 2\sqrt{3}\}$ **49.** $\left\{\dfrac{3 \pm i\sqrt{3}}{3}\right\}$

51. $\{-20, 12\}$ **53.** $\left\{-6, -\dfrac{11}{3}\right\}$ **55.** $\left\{-\dfrac{7}{3}, -\dfrac{3}{2}\right\}$

57. $\{-6 \pm 2\sqrt{10}\}$ **59.** $\left\{\dfrac{1}{4}, \dfrac{1}{3}\right\}$

61. $\left\{\dfrac{-b \pm \sqrt{b^2 - 4ac}}{2a}\right\}$ **65.** $x = \dfrac{a\sqrt{b^2 - y^2}}{b}$

67. $r = \dfrac{\sqrt{A\pi}}{\pi}$ **69.** $\{2a, 3a\}$ **71.** $\left\{\dfrac{a}{2}, -\dfrac{2a}{3}\right\}$

73. $\left\{\dfrac{2b}{3}\right\}$

Problem Set 11.4 (page 544)
1. Two real solutions; $\{-7, 3\}$

3. One real solution; $\left\{\dfrac{1}{3}\right\}$

5. Two complex solutions; $\left\{\dfrac{7 \pm i\sqrt{3}}{2}\right\}$

7. Two real solutions; $\left\{-\dfrac{4}{3}, \dfrac{1}{5}\right\}$

9. Two real solutions; $\left\{\dfrac{-2 \pm \sqrt{10}}{3}\right\}$

11. $\{-1 \pm \sqrt{2}\}$ **13.** $\left\{\dfrac{-5 \pm \sqrt{37}}{2}\right\}$

15. $\{4 \pm 2\sqrt{5}\}$ **17.** $\left\{\dfrac{-5 \pm i\sqrt{7}}{2}\right\}$

19. $\{8, 10\}$ **21.** $\left\{\dfrac{9 \pm \sqrt{61}}{2}\right\}$

23. $\left\{\dfrac{-1 \pm \sqrt{33}}{4}\right\}$ **25.** $\left\{\dfrac{-1 \pm i\sqrt{3}}{4}\right\}$

27. $\left\{\dfrac{4 \pm \sqrt{10}}{3}\right\}$ **29.** $\left\{-1, \dfrac{5}{2}\right\}$ **31.** $\left\{-5, -\dfrac{4}{3}\right\}$

33. $\left\{\dfrac{5}{6}\right\}$ **35.** $\left\{\dfrac{1 \pm \sqrt{13}}{4}\right\}$ **37.** $\left\{0, \dfrac{13}{5}\right\}$

39. $\left\{\pm\dfrac{\sqrt{15}}{3}\right\}$ **41.** $\left\{\dfrac{-1 \pm \sqrt{73}}{12}\right\}$ **43.** $\{-18, -14\}$

45. $\left\{\dfrac{11}{4}, \dfrac{10}{3}\right\}$ **47.** $\left\{\dfrac{2 \pm i\sqrt{2}}{2}\right\}$ **49.** $\left\{\dfrac{1 \pm \sqrt{7}}{6}\right\}$

55. $\{-1.381, 17.381\}$ **57.** $\{-13.426, 3.426\}$
59. $\{-8.653, -0.347\}$ **61.** $\{0.119, 1.681\}$
63. $\{-0.708, 4.708\}$ **65.** $k = 4$ or $k = -4$

Problem Set 11.5 (page 554)

1. $\{2 \pm \sqrt{10}\}$ **3.** $\left\{-9, \dfrac{4}{3}\right\}$ **5.** $\{9 \pm 3\sqrt{10}\}$

7. $\left\{\dfrac{3 \pm i\sqrt{23}}{4}\right\}$ **9.** $\{-15, -9\}$ **11.** $\{-8, 1\}$

13. $\left\{\dfrac{2 \pm i\sqrt{10}}{2}\right\}$ **15.** $\{9 \pm \sqrt{66}\}$ **17.** $\left\{-\dfrac{5}{4}, \dfrac{2}{5}\right\}$

19. $\left\{\dfrac{-1 \pm \sqrt{2}}{2}\right\}$ **21.** $\left\{\dfrac{3}{4}, 4\right\}$

23. $\left\{\dfrac{11 \pm \sqrt{109}}{2}\right\}$ **25.** $\left\{\dfrac{3}{7}, 4\right\}$

27. $\left\{\dfrac{7 \pm \sqrt{129}}{10}\right\}$ **29.** $\left\{-\dfrac{10}{7}, 3\right\}$

31. $\{1 \pm \sqrt{34}\}$ **33.** $\{\pm\sqrt{6}, \pm2\sqrt{3}\}$

35. $\left\{\pm3, \pm\dfrac{2\sqrt{6}}{3}\right\}$ **37.** $\left\{\pm\dfrac{i\sqrt{15}}{3}, \pm2i\right\}$

39. $\left\{\pm\dfrac{\sqrt{14}}{2}, \pm\dfrac{2\sqrt{3}}{3}\right\}$ **41.** 8 and 9   **43.** 9 and 12

45. $5 + \sqrt{3}$ and $5 - \sqrt{3}$ **47.** 3 and 6
49. 9 inches and 12 inches   **51.** 1 meter
53. 8 inches by 14 inches
55. 20 miles per hour for Lorraine and 25 miles per hour
for Charlotte, or 45 miles per hour for Lorraine and
50 miles per hour for Charlotte
57. 55 miles per hour
59. 6 hours for Tom and 8 hours for Terry
61. 30 hours   **63.** 8 people
65. 40 shares at $20 per share
67. 50 numbers   **69.** 7%   **71.** 6%

77. $\{9, 36\}$ **79.** $\{1\}$ **81.** $\left\{-\dfrac{8}{27}, \dfrac{27}{8}\right\}$

83. $\left\{-4, \dfrac{3}{5}\right\}$

85. a. $\{2.6, 5.4\}$ **c.** $\{-6.5, 2.5\}$ **e.** $\{-1.7, 1.7\}$
g. $\{-0.8, 1.3\}$

Problem Set 11.6 (page 561)
1. $(-\infty, -2) \cup (1, \infty)$

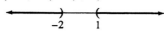

3. $(-4, -1)$

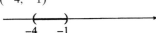

5. $\left(-\infty, -\dfrac{7}{3}\right] \cup \left[\dfrac{1}{2}, \infty\right)$

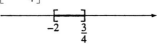

7. $\left[-2, \dfrac{3}{4}\right]$

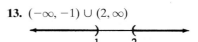

9. $(-1, 1) \cup (3, \infty)$

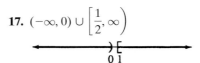

11. $(-\infty, -2] \cup [0, 4]$

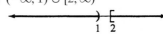

13. $(-\infty, -1) \cup (2, \infty)$

15. $(-2, 3)$

17. $(-\infty, 0) \cup \left[\dfrac{1}{2}, \infty\right)$

19. $(-\infty, 1) \cup [2, \infty)$

21. $(-7, 5)$ **23.** $(-\infty, 4) \cup (7, \infty)$ **25.** $\left[-5, \dfrac{2}{3}\right]$

27. $\left(-\infty, -\dfrac{5}{2}\right] \cup \left[-\dfrac{1}{4}, \infty\right)$ **29.** $\left(-\infty, -\dfrac{4}{5}\right) \cup (8, \infty)$

31. $(-\infty, \infty)$ **33.** $\left\{-\dfrac{5}{2}\right\}$ **35.** $(-1, 3) \cup (3, \infty)$

37. $(-6, -3)$ **39.** $(-\infty, 5) \cup [9, \infty)$

41. $\left(-\infty, \dfrac{4}{3}\right) \cup (3, \infty)$ **43.** $(-4, 6]$ **45.** $(-\infty, 2)$

55. a. $(-\infty, -2] \cup [1, \infty)$ **c.** $(-1, 1) \cup (3, \infty)$
e. $(-\infty, -4) \cup (-1, 5)$

Chapter 11 Review Problem Set (page 565)
1. $2 - 2i$ **2.** $-3 - i$ **3.** $30 + 15i$ **4.** $86 - 2i$

5. $-32 + 4i$ **6.** $25 + 0i$ **7.** $\dfrac{9}{20} + \dfrac{13}{20}i$

8. $-\dfrac{3}{29} + \dfrac{7}{29}i$ **9.** Two equal real solutions

10. Two nonreal complex solutions

11. Two unequal real solutions

12. Two unequal real solutions   **13.** $\{0, 17\}$

14. $\{-4, 8\}$ **15.** $\left\{\dfrac{1 \pm 8i}{2}\right\}$ **16.** $\{-3, 7\}$

17. $\{-1 \pm \sqrt{10}\}$ **18.** $\{3 \pm 5i\}$ **19.** $\{25\}$

20. $\left\{-4, \dfrac{2}{3}\right\}$ **21.** $\{-10, 20\}$ **22.** $\left\{\dfrac{-1 \pm \sqrt{61}}{6}\right\}$

23. $\left\{\dfrac{1 \pm i\sqrt{11}}{2}\right\}$ **24.** $\left\{\dfrac{5 \pm i\sqrt{23}}{4}\right\}$

25. $\left\{\dfrac{-2 \pm \sqrt{14}}{2}\right\}$ **26.** $\{-9, 4\}$

27. $\{-2 \pm i\sqrt{5}\}$ **28.** $\{-6, 12\}$ **29.** $\{1 \pm \sqrt{10}\}$

30. $\left\{\pm\dfrac{\sqrt{14}}{2}, \pm 2\sqrt{2}\right\}$ **31.** $\left\{\dfrac{-3 \pm \sqrt{97}}{2}\right\}$

32. $(-\infty, -5) \cup (2, \infty)$ **33.** $\left[-\dfrac{7}{2}, 3\right]$

34. $(-\infty, -6) \cup [4, \infty)$ **35.** $\left(-\dfrac{5}{2}, -1\right)$

36. $3 + \sqrt{7}$ and $3 - \sqrt{7}$

37. 20 shares at \$15 per share

38. 45 miles per hour and 52 miles per hour

39. 8 units   **40.** 8 and 10

41. 7 inches by 12 inches

42. 4 hours for Reena and 6 hours for Billy

43. 10 meters

Chapter 11 Test (page 567)

1. $39 - 2i$ **2.** $-\dfrac{6}{25} - \dfrac{17}{25}i$ **3.** $\{0, 7\}$

4. $\{-1, 7\}$ **5.** $\{-6, 3\}$ **6.** $\{1 - \sqrt{2}, 1 + \sqrt{2}\}$

7. $\left\{\dfrac{1 - 2i}{5}, \dfrac{1 + 2i}{5}\right\}$ **8.** $\{-16, -14\}$

9. $\left\{\dfrac{1 - 6i}{3}, \dfrac{1 + 6i}{3}\right\}$ **10.** $\left\{-\dfrac{7}{4}, \dfrac{6}{5}\right\}$

11. $\left\{-3, \dfrac{19}{6}\right\}$ **12.** $\left\{-\dfrac{10}{3}, 4\right\}$

13. $\{-2, 2, -4i, 4i\}$ **14.** $\left\{-\dfrac{3}{4}, 1\right\}$

15. $\left\{\dfrac{1 - \sqrt{10}}{3}, \dfrac{1 + \sqrt{10}}{3}\right\}$

16. Two equal real solutions

17. Two nonreal complex solutions   **18.** $[-6, 9]$

19. $(-\infty, -2) \cup \left(\dfrac{1}{3}, \infty\right)$ **20.** $[-10, -6)$

21. 20.8 feet   **22.** 29 meters   **23.** 150 shares

24. $6\dfrac{1}{2}$ inches   **25.** $3 + \sqrt{5}$

Cumulative Review Problem Set (Chapters 1–11) (page 568)

1. $\dfrac{64}{15}$ **2.** $\dfrac{11}{3}$ **3.** $\dfrac{1}{6}$ **4.** $-\dfrac{44}{5}$ **5.** -7

6. $-24a^4b^5$ **7.** $2x^3 + 5x^2 - 7x - 12$ **8.** $\dfrac{3x^2y^2}{8}$

9. $\dfrac{a(a + 1)}{2a - 1}$ **10.** $\dfrac{-x + 14}{18}$ **11.** $\dfrac{5x + 19}{x(x + 3)}$

12. $\dfrac{2}{n + 8}$ **13.** $\dfrac{x - 14}{(5x - 2)(x + 1)(x - 4)}$

14. $y^2 - 5y + 6$ **15.** $x^2 - 3x - 2$ **16.** $20 + 7\sqrt{10}$

17. $2x - 2\sqrt{xy} - 12y$ **18.** $-\dfrac{3}{8}$ **19.** $-\dfrac{2}{3}$

20. 0.2 **21.** $\dfrac{1}{2}$ **22.** $\dfrac{13}{9}$ **23.** -27

24. $\dfrac{16}{9}$ **25.** $\dfrac{8}{27}$ **26.** $3x(x + 3)(x^2 - 3x + 9)$

27. $(6x - 5)(x + 4)$ **28.** $(4 + 7x)(3 - 2x)$

29. $(3x + 2)(3x - 2)(x^2 + 8)$ **30.** $(2x - y)(a - b)$

31. $(3x - 2y)(9x^2 + 6xy + 4y^2)$ **32.** $\left\{-\dfrac{12}{7}\right\}$

33. $\{150\}$ **34.** $\{25\}$ **35.** $\{0\}$ **36.** $\{-2, 2\}$

37. $\{-7\}$ **38.** $\left\{-6, \dfrac{4}{3}\right\}$ **39.** $\left\{\dfrac{5}{4}\right\}$ **40.** $\{3\}$

41. $\left\{\dfrac{4}{5}, 1\right\}$ **42.** $\left\{-\dfrac{10}{3}, 4\right\}$ **43.** $\left\{\dfrac{1}{4}, \dfrac{2}{3}\right\}$

44. $\left\{-\dfrac{3}{2}, 3\right\}$ **45.** $\left\{\dfrac{1}{5}\right\}$ **46.** $\left\{\dfrac{5}{7}\right\}$

47. $\left\{-2, 2, \pm\dfrac{i\sqrt{6}}{6}\right\}$ **48.** $\{0, \pm i\}$ **49.** $\{-6, 19\}$

50. $\left\{-\dfrac{3}{4}, \dfrac{2}{3}\right\}$ **51.** $\{1 \pm 5i\}$ **52.** $\{1, 3\}$

53. $\left\{-4, \dfrac{1}{3}\right\}$ **54.** $\{-2 \pm 4i\}$ **55.** $\left\{\dfrac{1 \pm \sqrt{33}}{4}\right\}$

56. $\{-\infty, -2]$ **57.** $\left(-\infty, \dfrac{19}{5}\right)$ **58.** $\left(\dfrac{1}{4}, \infty\right)$

59. $(-2, 3)$ **60.** $\left(-\infty, -\dfrac{13}{3}\right) \cup (3, \infty)$

61. $(-\infty, 29]$ **62.** $[-2, 4]$

63. $(-\infty, -5) \cup \left(\dfrac{1}{3}, \infty\right)$ **64.** $(-\infty, -2] \cup (7, \infty)$

65. $(-3, 4)$ **66.** $\{(-1, -2)\}$ **67.** $\{(8, -9)\}$

68. $\{(4, 6)\}$ **69.** $-\dfrac{30y^2}{x^2}$

70. a. $12\sqrt{6}$ **b.** $6\sqrt[3]{2}$

c. $\dfrac{3\sqrt{15}}{20}$ **d.** $2xy^2\sqrt{6xy}$

71. a. $(7.652)(10)^3$ **b.** $(2.6)(10)^{-5}$
c. $(1.414)(10)^0$ **d.** $(1)(10)^3$

72. 6 liters **73.** $900 and $1350

74. 12 inches by 17 inches **75.** 5 hours

76. 7 golf balls **77.** 12 minutes **78.** 7%

79. 15 chairs per row **80.** 140 shares

81. 60° and 80° **82.** 70° **83.** More than 8.5%

84. $0.25 per lemon and $0.33 per orange **85.** $90

CHAPTER 12

Problem Set 12.1 (page 586)

1. 15 **3.** $\sqrt{13}$ **5.** $3\sqrt{2}$ **7.** $3\sqrt{5}$
9. 6 **11.** $3\sqrt{10}$

13. The lengths of the sides are $10, 5\sqrt{5}$, and 5. Because $10^2 + 5^2 = (5\sqrt{5})^2$, it is a right triangle.

15. The distances between $(3, 6)$ and $(7, 12)$, between $(7, 12)$ and $(11, 18)$, and between $(11, 18)$ and $(15, 24)$ are all $2\sqrt{13}$ units.

17. $\dfrac{4}{3}$ **19.** Undefined **21.** -2 **23.** $\dfrac{3}{5}$

25. 0 **27.** $\dfrac{1}{2}$ **29.** 7 **31.** -2

33. $(-3, -1); (3, 1); (3, -1)$
35. $(7, 2); (-7, -2); (-7, 2)$
37. $(5, 0); (-5, 0); (-5, 0)$ **39.** x axis
41. y axis **43.** x axis, y axis, and origin
45. x axis **47.** None **49.** Origin **51.** y axis

53. 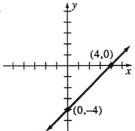 **55.**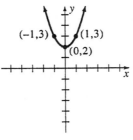

57. 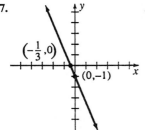 **59.**

61. 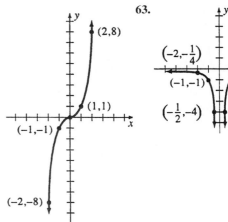 **63.**

65. 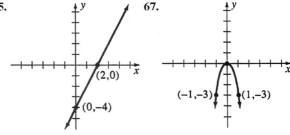 **67.**

69. 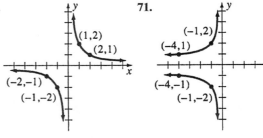 **71.**

73. 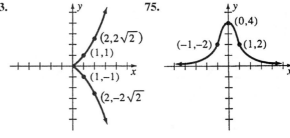 **75.**

77.

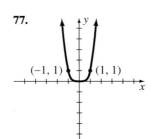

(−1, 1)   (1, 1)

79.

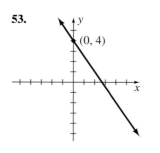

(−6, 2)   (1, 1)
(2, 0)
−6
(3, −1) (10, −2)
−6

57. $y = \dfrac{1}{1000}x + 2$ **59.** $y = \dfrac{9}{5}x + 32$

Problem Set 12.3 (page 607)

1.

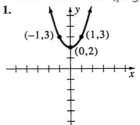

(−1,3)   (1,3)
(0,2)

3.

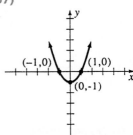

(−1,0)   (1,0)
(0,−1)

85. a. 7   **b.** 10   **c.** $\dfrac{1}{3}$ **d.** $-\dfrac{3}{5}$ **e.** −7   **f.** $-\dfrac{16}{3}$

Problem Set 12.2 (page 597)

1. $x - 2y = -7$ **3.** $3x - y = -10$
5. $3x + 4y = -15$ **7.** $5x - 4y = 28$ **9.** $x - y = 1$
11. $5x - 2y = -4$ **13.** $x + 7y = 11$
15. $x + 2y = -9$ **17.** $7x - 5y = 0$

19. $y = \dfrac{3}{7}x + 4$ **21.** $y = 2x - 3$

23. $y = -\dfrac{2}{5}x + 1$ **25.** $y = 0(x) - 4$

27. $2x - y = 4$ **29.** $5x + 8y = -15$
31. $x + 0(y) = 2$ **33.** $0(x) + y = 6$
35. $x + 5y = 16$ **37.** $4x - 7y = 0$
39. $x + 2y = 5$ **41.** $3x + 2y = 0$

43. $m = -3$ and $b = 7$ **45.** $m = -\dfrac{3}{2}$ and $b = \dfrac{9}{2}$

47. $m = \dfrac{1}{5}$ and $b = -\dfrac{12}{5}$

49.

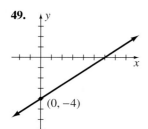

(0, −4)

51.

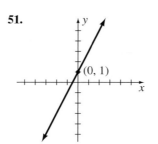

(0, 1)

53.

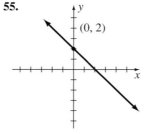

(0, 4)

55.

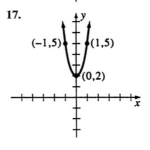

(0, 2)

5.

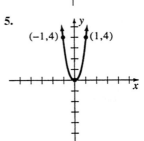

(−1,4)   (1,4)

7.

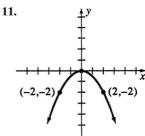

(−1,−3)   (1,−3)

9.

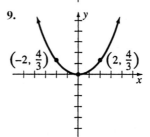

$\left(-2, \dfrac{4}{3}\right)$ $\left(2, \dfrac{4}{3}\right)$

11.

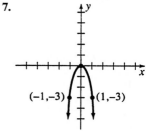

(−2,−2)   (2,−2)

13.

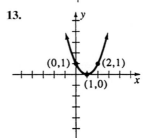

(0,1)   (2,1)
(1,0)

15.

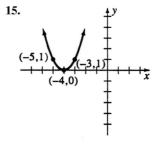

(−5,1)   (−3,1)
(−4,0)

17.

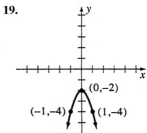

(−1,5)   (1,5)
(0,2)

19.

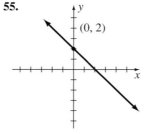

(0,−2)
(−1,−4)   (1,−4)

21.

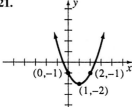

23.

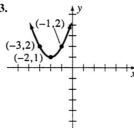

9.

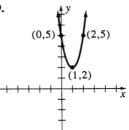

11.

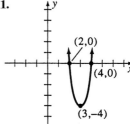

25.

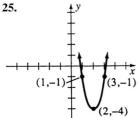

27.

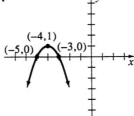

13.

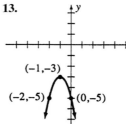

15.

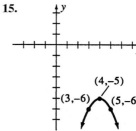

29.

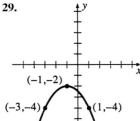

17.

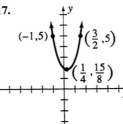

19.

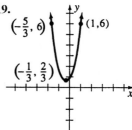

Problem Set 12.4 (page 615)

1.

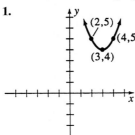

3.

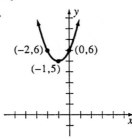

21.

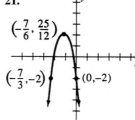
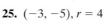

23. $(1, 3), r = 4$

25. $(-3, -5), r = 4$ **27.** $(0, 0), r = \sqrt{10}$
29. $(8, -3), r = \sqrt{2}$ **31.** $(-3, 4), r = 5$
33. $\left(-\dfrac{1}{2}, 4\right), r = 2\sqrt{2}$

5.

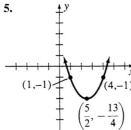

7.

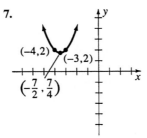

35.

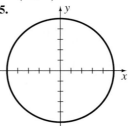

37.

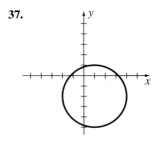

39.
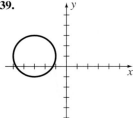

41.
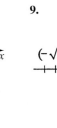

9.
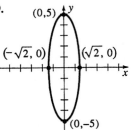

11.
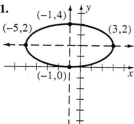

43.
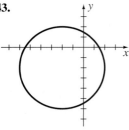

13.
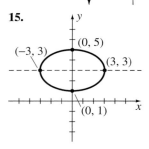

15.
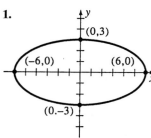

45. $x^2 + y^2 - 6x - 10y + 9 = 0$
47. $x^2 + y^2 + 8x - 2y - 47 = 0$
49. $x^2 + y^2 + 4x + 12y + 22 = 0$
51. $x^2 + y^2 - 20 = 0$ **53.** $x^2 + y^2 - 10x + 16y - 7 = 0$
55. $x^2 + y^2 - 8y = 0$ **57.** $x^2 + y^2 + 8x - 6y = 0$
63. a. $(1, 4), r = 3$ **c.** $(-6, -4), r = 8$
    e. $(0, 6), r = 9$

Problem Set 12.5 (page 622)

1.
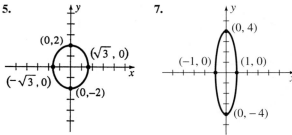

3.
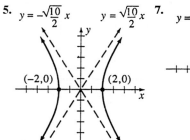

5.

7.
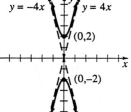

Problem Set 12.6 (page 630)

1.
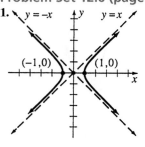

3.
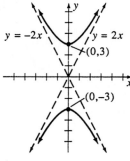

5.
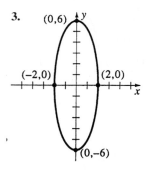

7.

9.

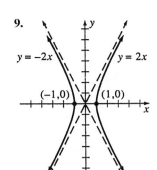

$y = -2x$ $y = 2x$

$(-1,0)$ $(1,0)$

11. $y = -\sqrt{\dfrac{3}{5}}\,x$ $y = \sqrt{\dfrac{3}{5}}\,x$

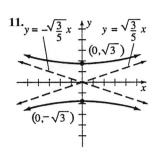

$(0,\sqrt{3}\,)$

$(0,-\sqrt{3}\,)$

21. a. Origin

c.

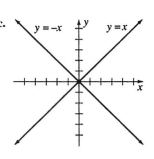

$y = -x$ $y = x$

13.

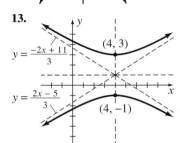

$y = \dfrac{-2x+11}{3}$

$(4, 3)$

$y = \dfrac{2x-5}{3}$

$(4, -1)$

Chapter 12 Review Problem Set (page 633)

1. a. $\dfrac{6}{5}$ **b.** $-\dfrac{2}{3}$ **2. a.** -4 **b.** $\dfrac{2}{7}$

3. 5, 10, and $\sqrt{97}$ **4.** $7x + 4y = 1$

5. $3x + 7y = 28$ **6.** $2x - 3y = 16$

7. $x - 2y = -8$ **8.** $2x - 3y = 14$

9. y axis   **10.** x axis   **11.** Origin

12. y axis   **13.** x axis, y axis, and origin

15.

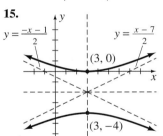

$y = \dfrac{-x-1}{2}$ $y = \dfrac{x-7}{2}$

$(3, 0)$

$(3, -4)$

14.

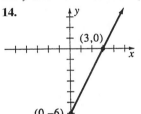

$(3,0)$

$(0,-6)$

15.

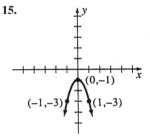

$(0,-1)$

$(-1,-3)$ $(1,-3)$

17.

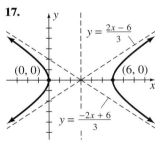

$y = \dfrac{2x-6}{3}$

$(0, 0)$ $(6, 0)$

$y = \dfrac{-2x+6}{3}$

16.

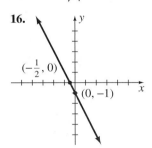

$\left(-\dfrac{1}{2}, 0\right)$

$(0, -1)$

17.

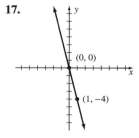

$(0, 0)$

$(1, -4)$

19a.

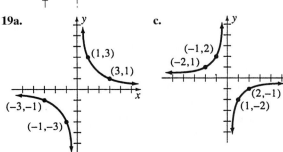

$(1,3)$

$(3,1)$

$(-3,-1)$

$(-1,-3)$

c.

$(-1,2)$

$(-2,1)$

$(2,-1)$

$(1,-2)$

18.

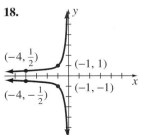

$\left(-4,\dfrac{1}{2}\right)$ $(-1, 1)$

$\left(-4,-\dfrac{1}{2}\right)$ $(-1, -1)$

19.

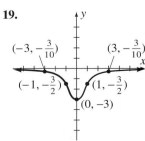

$\left(-3, -\dfrac{3}{10}\right)$ $\left(3, -\dfrac{3}{10}\right)$

$\left(-1, -\dfrac{3}{2}\right)$ $\left(1, -\dfrac{3}{2}\right)$

$(0, -3)$

20. 316.8 feet   **21.** 8 inches

22. $(0, 6)$ **23.** $(0, -8)$ **24.** $(-3, -1)$

25. $(7, 5)$ **26.** $(6, -8)$ **27.** $(-4, -9)$

28. $x^2 - 4x + y^2 + 12y + 15 = 0$

29. $x^2 + 8x + y^2 + 16y + 68 = 0$

30. $x^2 + y^2 - 10y = 0$ **31.** $(-7, 4)$ and $r = 7$

32. $(-8, 0)$ and $r = 5$ **33.** $(6, -8)$ and $r = 10$

34. $(0, 0)$ and $r = 2\sqrt{6}$ **35.** 10 and 4

36. $2\sqrt{14}$ and 4   **37.** 6 and 2   **38.** 6 and 4

39. $y = \pm\dfrac{2}{3}x$ **40.** $y = \pm\dfrac{1}{2}x$

41. $y = -\dfrac{5}{2}x - 2$ and $y = \dfrac{5}{2}x + 8$

42. $y = -\dfrac{1}{6}x + \dfrac{25}{6}$ and $y = \dfrac{1}{6}x + \dfrac{23}{6}$

43.

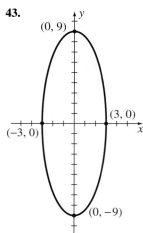

44.

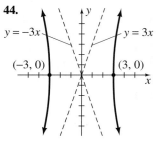

45.

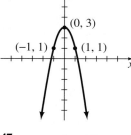

46.

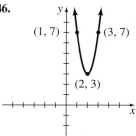

47.

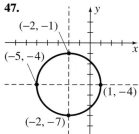

48.

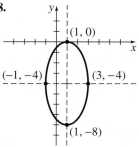

49.

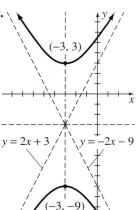

50.

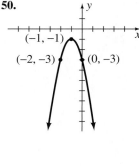

51.

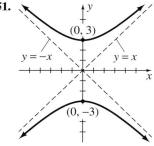

52.

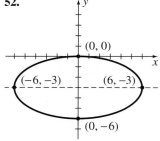

Chapter 12 Test (page 635)

1. $-\dfrac{6}{5}$ **2.** $\dfrac{3}{7}$ **3.** $\sqrt{58}$ **4.** $3x + 2y = 2$

5. $y = -\dfrac{1}{6}x + \dfrac{4}{3}$ **6.** $5x + 2y = -18$

7. $6x + y = 31$ **8.** 480 feet

9. 43 centimeters   **10.** $(3, 0)$

11. $(-4, -2)$ **12.** $x^2 - 4x + y^2 - 16y + 59 = 0$

13. $(6, -4)$ and $r = 7$ **14.** 8 units

15. 4 units   **16.** $y = \pm 4x$

17.

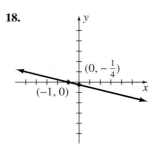

18.

19.

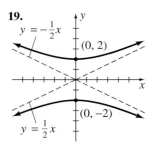

20.

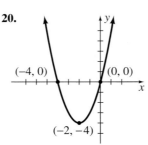

21.

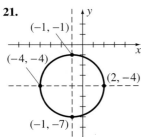

22.

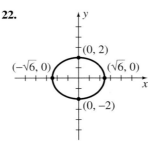

23.

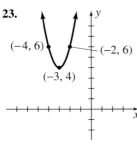

24.

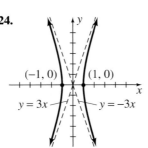

25.

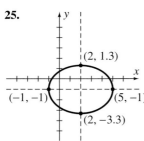

Cumulative Practice Test (Chapters 1–12) (page 636)

1. a. $-\dfrac{2}{3}$ **b.** -32 **c.** $\dfrac{1}{4}$ **d.** $\dfrac{15}{2}$

2. a. $4\sqrt{6}$ **b.** $2\sqrt[3]{5}$ **c.** $\dfrac{2\sqrt{6}}{9}$

  d. $\dfrac{2(\sqrt{7}+\sqrt{2})}{5}$ **3.** $-24x^4y^5$

4. $6x^3 - 5x^2 - 8x - 2$ **5.** $x^3 + 2x^2 - x - 4$

6. $\dfrac{-6x - 11}{12}$ **7.** $14 + \sqrt{6}$

8. a. Origin **b.** x axis **c.** y axis **d.** y axis

9. -21 **10.** -192 **11.** 36 **12.** $\dfrac{1}{4}$

13. $4x - 3y = -13$ **14.** $\{2\}$ **15.** $\left\{-\dfrac{53}{10}\right\}$

16. $\{3000\}$ **17.** $\left\{-\dfrac{5}{3}, 5\right\}$

18. $\left\{\dfrac{1 \pm \sqrt{13}}{6}\right\}$ **19.** $[-1, 4]$

20. $(-\infty, 2) \cup (3, \infty)$ **21.** $(-1, 6)$

22. $34°$ and $56°$ **23.** 12 dimes and 25 quarters

24. 3 hours **25.** $1.79

CHAPTER 13

Problem Set 13.1 (page 644)

1. $D = \{1, 2, 3, 4\}$, $R = \{5, 8, 11, 14\}$ It is a function.
3. $D = \{0, 1\}$, $R = \{-2\sqrt{6}, -5, 5, 2\sqrt{6}\}$ It is not a function.
5. $D = \{1, 2, 3, 4, 5\}$, $R = \{2, 5, 10, 17, 26\}$ It is a function.
7. $D = \{$All reals$\}$, $R = \{$All reals$\}$ It is a function.
9. $D = \{$All reals$\}$, $R = \{y \,|\, y \geq 0\}$ It is a function.

11. $\{$All reals$\}$ **13.** $\{x \,|\, x \neq 1\}$ **15.** $\left\{x \,\middle|\, x \neq \dfrac{3}{4}\right\}$

17. $\{x \,|\, x \neq -1$ and $x \neq 4\}$ **19.** $\{x \,|\, x \neq -8$ and $x \neq 5\}$
21. $\{x \,|\, x \neq -6$ and $x \neq 0\}$ **23.** $\{$All reals$\}$
25. $\{t \,|\, t \neq -2$ and $t \neq 2\}$ **27.** $\{x \,|\, x \geq -4\}$

29. $\left\{s \,\middle|\, s \geq \dfrac{5}{4}\right\}$ **31.** $\{x \,|\, x \leq -4$ or $x \geq 4\}$

33. $\{x \,|\, x \leq -3$ or $x \geq 6\}$ **35.** $\{x \,|\, -1 \leq x \leq 1\}$
37. $f(0) = -2, f(2) = 8, f(-1) = -7, f(-4) = -22$

39. $f(-2) = -\dfrac{7}{4}, f(0) = -\dfrac{3}{4}, f\left(\dfrac{1}{2}\right) = -\dfrac{1}{2}, f\left(\dfrac{2}{3}\right) = -\dfrac{5}{12}$

41. $g(-1) = 0$; $g(2) = -9$; $g(-3) = 26$; $g(4) = 5$
43. $h(-2) = -2$; $h(-3) = -11$; $h(4) = -32$; $h(5) = -51$
45. $f(3) = \sqrt{7}$; $f(4) = 3$; $f(10) = \sqrt{21}$; $f(12) = 5$

47. $f(1) = -1$; $f(-1) = -2$; $f(3) = -\dfrac{2}{3}$; $f(-6) = \dfrac{4}{3}$

49. $f(-2) = 27; f(3) = 42; g(-4) = -37; g(6) = -17$
51. $f(-2) = 5; f(3) = 8; g(-4) = -3; g(5) = -4$
53. -3 **55.** $-2a - h$ **57.** $4a - 1 + 2h$
59. $-8a - 7 - 4h$
61. $h(1) = 48; h(2) = 64; h(3) = 48; h(4) = 0$
63. $C(75) = \$74; C(150) = \$98; C(225) = \$122;$
      $C(650) = \$258$
65. $I(0.03) = 75; I(0.04) = 100; I(0.05) = 125;$
      $I(0.065) = 162.50$

17.
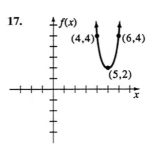

19.
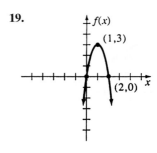

Problem Set 13.2 (page 656)

1.
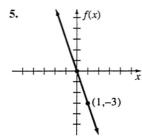

3.
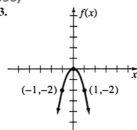

21.
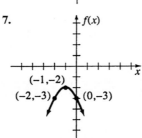

23.
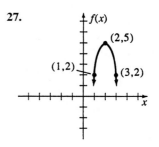

5.
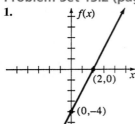

7.
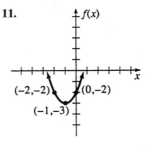

25.
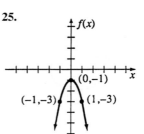

27.

9.
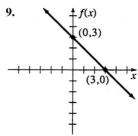

11.

29.
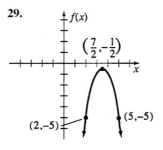

13.
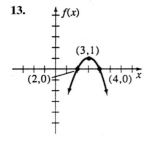

15.
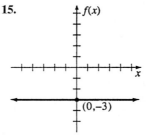

31. a. $0.42   **c.** Answers will vary.
33. $26; $30.50; $50; $60.50
35. $2.10; $4.55; $20.72; $29.40; $33.88
37. $f(p) = 0.8p;$ $7.60; $12; $60; $10; $600
39. 80 items   **41.** 5 and 25
43. 60 meters by 60 meters
45. 1100 subscribers at $13.75 per month

Problem Set 13.3 (page 667)

1.
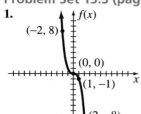

3.
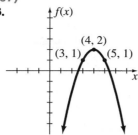

21.
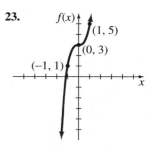

23.
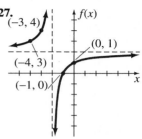

5.
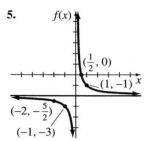

7.
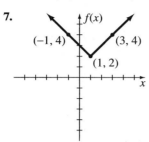

25.
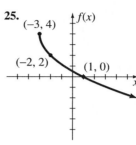

27.
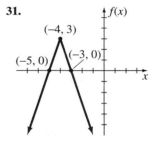

9.
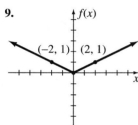

11.
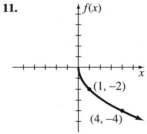

29.
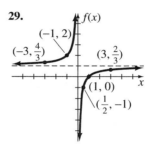

31.
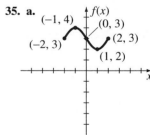

13.
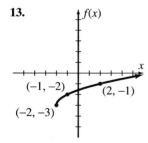

15.
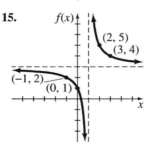

33.
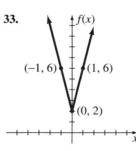

35. a.
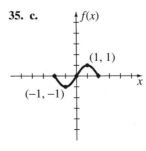

17.
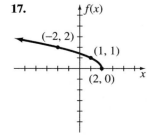

19.
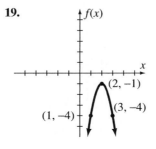

35. c.

Problem Set 13.4 (page 673)

1. 124 and −130   **3.** 323 and 257   **5.** $\dfrac{1}{2}$ and −1

7. Undefined and undefined   **9.** $\sqrt{7}$ and 0

11. 37 and 27

13. $(f \circ g)(x) = 15x - 3, D = \{\text{all reals}\}$
$(g \circ f)(x) = 15x - 1, D = \{\text{all reals}\}$

15. $(f \circ g)(x) = -14x - 7, D = \{\text{all reals}\}$
$(g \circ f)(x) = -14x + 11, D = \{\text{all reals}\}$

17. $(f \circ g)(x) = 3x^2 + 11, D = \{\text{all reals}\}$
$(g \circ f)(x) = 9x^2 + 12x + 7, D = \{\text{all reals}\}$

19. $(f \circ g)(x) = 2x^2 - 11x + 17, D = \{\text{all reals}\}$
$(g \circ f)(x) = -2x^2 + x + 1, D = \{\text{all reals}\}$

21. $(f \circ g)(x) = \dfrac{3}{4x - 9}, D = \left\{ x \mid x \neq \dfrac{9}{4} \right\}$

$(g \circ f)(x) = \dfrac{12 - 9x}{x}, D = \left\{ x \mid x \neq 0 \right\}$

23. $(f \circ g)(x) = \sqrt{5x + 4}, D = \left\{ x \mid x \geq -\dfrac{4}{5} \right\}$

$(g \circ f)(x) = 5\sqrt{x + 1} + 3, D = \left\{ x \mid x \geq -1 \right\}$

25. $(f \circ g)(x) = x - 4, D = \left\{ x \mid x \neq 4 \right\}$

$(g \circ f)(x) = \dfrac{x}{1 - 4x}, D = \left\{ x \mid x \neq 0 \text{ and } x \neq \dfrac{1}{4} \right\}$

27. $(f \circ g)(x) = \dfrac{2\sqrt{x}}{x}, D = \left\{ x \mid x > 0 \right\}$

$(g \circ f)(x) = \dfrac{4\sqrt{x}}{x}, D = \left\{ x \mid x > 0 \right\}$

29. $(f \circ g)(x) = \dfrac{3x + 3}{2}, D = \left\{ x \mid x \neq -1 \right\}$

$(g \circ f)(x) = \dfrac{2x}{2x + 3}, D = \left\{ x \mid x \neq 0 \text{ and } x \neq -\dfrac{3}{2} \right\}$

Problem Set 13.5 (page 680)

1. $y = \dfrac{k}{x^2}$ **3.** $C = \dfrac{kg}{t^3}$

5. $V = kr^3$ **7.** $S = ke^2$

9. $V = khr^2$ **11.** $\dfrac{2}{3}$ **13.** −4   **15.** $\dfrac{1}{3}$

17. −2   **19.** 2   **21.** 5   **23.** 9   **25.** 9

27. $\dfrac{1}{6}$ **29.** 112   **31.** 12 cubic centimeters

33. 28   **35.** 2 seconds   **37.** 12 ohms

39. a. $210   **c.** $1050

41. 3560.76 cubic meters   **43.** 0.048

Chapter 13 Review Problem Set (page 684)

1. $D = \{1, 2, 4\}$ **2.** $D = \{x \mid x \neq 5\}$

3. $D = \{x \mid x \neq 0 \text{ and } x \neq -4\}$

4. $D = \{x \mid x \geq 5 \text{ or } x \leq -5\}$ **5.** $f(2) = -1, f(-3) = 14;$
$f(a) = a^2 - 2a - 1$

6. $4a + 2h + 1$

7.

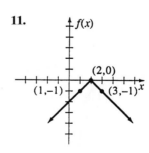

8.

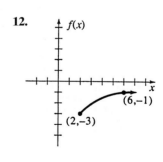

9.

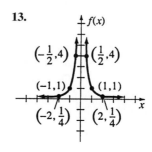

10.

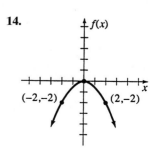

11.

12.

13.

14.

15.

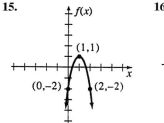

(1,1)

(0,−2)   (2,−2)

16.

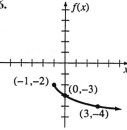

(−1,−2)   (0,−3)

(3,−4)

22.

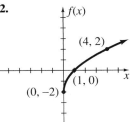

(4, 2)

(1, 0)

(0, −2)

23

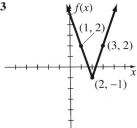

(1, 2)

(3, 2)

(2, −1)

17. a. $(-5, -28); x = -5$ **b.** $\left(-\dfrac{7}{2}, \dfrac{67}{2}\right); x = -\dfrac{7}{2}$

18. $(f \circ g)(x) = 6x - 11$ and $(g \circ f)(x) = 6x - 13$

19. $(f \circ g)(x) = x^2 - 2x - 1$ and $(g \circ f)(x) = x^2 - 10x + 27$

20. $(f \circ g)(x) = 4x^2 - 20x + 20$ and $(g \circ f)(x) = -2x^2 + 15$

21. $k = 9$ **22.** $y = 120$

23. 128 pounds   **24.** 20 and 20   **25.** 3 and 47

26. 25 students   **27.** 600 square inches

28. $0.72   **29.** $f(x) = 0.7x$; $45.50; $33.60; $10.85

24.

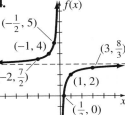

$\left(-\dfrac{1}{2}, 5\right)$

$(-1, 4)$

$\left(3, \dfrac{8}{3}\right)$

$\left(-2, \dfrac{7}{2}\right)$ $(1, 2)$

$\left(\dfrac{1}{3}, 0\right)$

25.

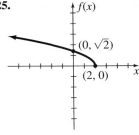

$(0, \sqrt{2})$

$(2, 0)$

Chapter 13 Test (page 685)

1. $\left\{x \mid x \neq -4 \text{ and } x \neq \dfrac{1}{2}\right\}$ **2.** $\left\{x \mid x \leq \dfrac{5}{3}\right\}$ **3.** $\dfrac{11}{6}$

4. 11   **5.** $(-6, 3)$ **6.** $6a + 3h + 2$

7. $(f \circ g)(x) = -21x - 2$

8. $(g \circ f)(x) = 8x^2 + 38x + 48$

9. $(f \circ g)(x) = \dfrac{3x}{2 - 2x}$

10. $\{x \mid x \leq -5 \text{ or } x \geq 2\}$

11. $f(c) = 1.4c$, $21, $25.20, $35

12. 750 units   **13.** -4 **14.** 15

15. 6 and 54   **16.** $96

17. The graph of $f(x) = (x - 6)^3 - 4$ is the graph of $f(x) = x^3$ translated 6 units to the right and 4 units downward.

18. The graph of $f(x) = -|x| + 8$ is the graph of $f(x) = |x|$ reflected across the x axis and translated 8 units upward.

19. The graph of $f(x) = -\sqrt{x + 5} + 7$ is the graph of $f(x) = \sqrt{x}$ reflected across the x axis and translated 5 units to the left and 7 units upward.

20.

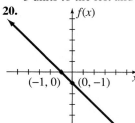

(−1, 0)   (0, −1)

21.

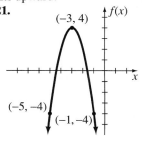

(−3, 4)

(−5, −4)   (−1, −4)

Cumulative Practice Test (Chapters 1–13) (page 686)

1. 12   **2. a.** x axis   **b.** $f(x)$ axis, or y axis

c. Origin   **d.** Origin

3. -9 and $\dfrac{1}{2}$ **4.** $(-1, -2)$ **5.** 32

6. a. $4xy^2\sqrt{2x}$ **b.** $2xy\sqrt[3]{4y}$ **c.** $\dfrac{\sqrt{2}}{3}$

d. $\dfrac{6 - \sqrt{6}}{15}$ **7.** $\dfrac{11}{10}$ **8.** $6x^3 + 5x^2 - 19x - 20$

9. $\dfrac{21}{20x}$ **10.** $2x^3 - 3x^2 + 4x - 5$ **11.** $x^2 + 2x + 4$

12. 0.001414   **13.** 115%   **14.** $-14 - 32i$

15. a. 8   **b.** $\dfrac{4}{5}$ **c.** $-\dfrac{1}{8}$ **d.** $\dfrac{4}{3}$

16. $-\dfrac{2}{3}$ **17.** $3x + 4y = 20$ **18.** $\left\{\dfrac{11}{7}\right\}$

19. $\{2\}$ **20.** $\left\{\dfrac{1 \pm i\sqrt{2}}{3}\right\}$ **21.** $\{8\}$

22.

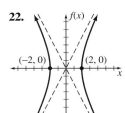

(−2, 0)   (2, 0)

23.

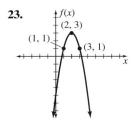

(2, 3)

(1, 1)   (3, 1)

24.
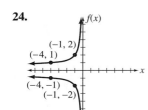

25.
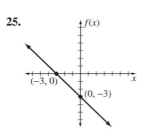

45.
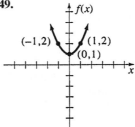

47.
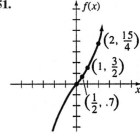

CHAPTER 14

Problem Set 14.1 (page 695)

1. {6}   **3.** $\left\{\dfrac{3}{2}\right\}$ **5.** {7}   **7.** {5}   **9.** {1}

11. {1}   **13.** {−3}   **15.** $\left\{\dfrac{3}{2}\right\}$ **17.** {−3}

19. {1}   **21.** $\left\{\dfrac{1}{5}\right\}$ **23.** {0}

25. {−1}   **27.** $\left\{\dfrac{5}{2}\right\}$ **29.** {3}   **31.** $\left\{\dfrac{1}{2}\right\}$

33.

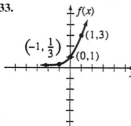

35.
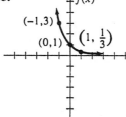

49.
f(x) graph

51.
f(x) graph

37.
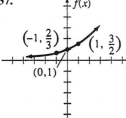

39.
f(x) graph

Problem Set 14.2 (page 705)
1. a. $1.00   **c.** $2.33   **e.** $21,900   **g.** $658
3. a. $2551.33
  b. $2479.75; 2.5% compounded annually
5. a. $2380.68
  b. $2323.23; 3.5% compounded quarterly
7. $4,647.34   **9.** $1973.12
11. a. $12824.73
  b. $13064.21; 5.5% compounded quarterly
13. a. $4923.78
  b. $4923.23; 3% compounded continuously
15. $6688.37   **17.** $3624.66   **19.** 674.93
21. 5.9%   **23.** 4.87%
25. 5.3% compounded semiannually
27. 50 grams; 37 grams
29. 2226; 3320; 7389
31. 2000   **33. a.** 6.5 psi   **c.** 13.6 psi

41.
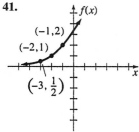

43.
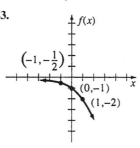
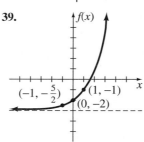

35.
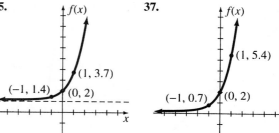

37.
f(x) graph

39.

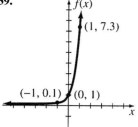

45.

	2%	3%	4%	5%
5 years	$1105	1162	1221	1284
10 years	1221	1350	1492	1649
15 years	1350	1568	1822	2117
20 years	1492	1822	2226	2718
25 years	1649	2117	2718	3490

47.

	2%	3%	4%	5%
Compounded annually	$1219	1344	1480	1629
Compounded semiannually	1220	1347	1486	1639
Compounded quarterly	1221	1348	1489	1644
Compounded monthly	1221	1349	1491	1647
Compounded continuously	1221	1350	1492	1649

49.

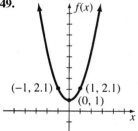

51.

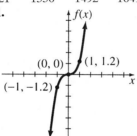

Problem Set 14.3 (page 717)

1. Yes   **3.** No   **5.** Yes   **7.** Yes

9. Yes   **11.** No   **13.** No

15. Domain of f: $\{1, 2, 5\}$
Range of f: $\{5, 9, 21\}$
$f^{-1} = \{(5, 1), (9, 2), (21, 5)\}$
Domain of f^{-1}: $\{5, 9, 21\}$
Range of f^{-1}: $\{1, 2, 5\}$

17. Domain of f: $\{0, 2, -1, -2\}$
Range of f: $\{0, 8, -1, -8\}$
f^{-1}: $\{(0, 0), (8, 2), (-1, -1), (-8, -2)\}$
Domain of f^{-1}: $\{0, 8, -1, -8\}$
Range of f^{-1}: $\{0, 2, -1, -2\}$

27. No   **29.** Yes   **31.** No   **33.** Yes

35. Yes   **37.** $f^{-1}(x) = x + 4$

39. $f^{-1}(x) = \dfrac{-x - 4}{3}$ **41.** $f^{-1}(x) = \dfrac{12x + 10}{9}$

43. $f^{-1}(x) = -\dfrac{3}{2}x$ **45.** $f^{-1}(x) = x^2$ for $x \geq 0$

47. $f^{-1}(x) = \sqrt{x - 4}$ for $x \geq 4$

49. $f^{-1}(x) = \dfrac{1}{x - 1}$ for $x > 1$

51. $f^{-1}(x) = \dfrac{1}{3}x$ **53.** $f^{-1}(x) = \dfrac{x - 1}{2}$

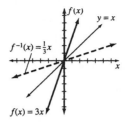
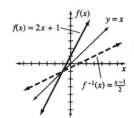

55. $f^{-1}(x) = \dfrac{x + 2}{x}$ for $x > 0$

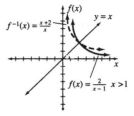

57. $f^{-1}(x) = \sqrt{x + 4}$ for $x \geq -4$

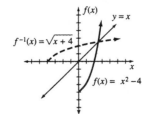

59. Increasing on $[0, \infty)$ and decreasing on $(-\infty, 0]$
61. Decreasing on $(-\infty, \infty)$
63. Increasing on $(-\infty, -2]$ and decreasing on $[-2, \infty)$
65. Increasing on $(-\infty, -4]$ and decreasing on $[-4, \infty)$

71. a. $f^{-1}(x) = \dfrac{x + 9}{3}$ **c.** $f^{-1}(x) = -x + 1$

e. $f^{-1}(x) = -\dfrac{1}{5}x$

Problem Set 14.4 (page 728)

1. $\log_2 128 = 7$ **3.** $\log_5 125 = 3$

5. $\log_{10} 1000 = 3$ **7.** $\log_2 \left(\dfrac{1}{4}\right) = -2$

9. $\log_{10} 0.1 = -1$ **11.** $3^4 = 81$ **13.** $4^3 = 64$

15. $10^4 = 10,000$ **17.** $2^{-4} = \dfrac{1}{16}$ **19.** $10^{-3} = 0.001$

21. 4 **23.** 4 **25.** 3 **27.** $\dfrac{1}{2}$ **29.** 0

31. -1 **33.** 5 **35.** -5 **37.** 1 **39.** 0

41. {49} **43.** {16} **45.** {27} **47.** $\left\{\dfrac{1}{8}\right\}$

49. {4} **51.** 5.1293 **53.** 6.9657 **55.** 1.4037

57. 7.4512 **59.** 6.3219 **61.** -0.3791

63. 0.5766 **65.** 2.1531 **67.** 0.3949

69. $\log_b x + \log_b y + \log_b z$ **71.** $\log_b y - \log_b z$

73. $3 \log_b y + 4 \log_b z$ **75.** $\dfrac{1}{2} \log_b x + \dfrac{1}{3} \log_b y - 4 \log_b z$

77. $\dfrac{2}{3} \log_b x + \dfrac{1}{3} \log_b z$ **79.** $\dfrac{3}{2} \log_b x - \dfrac{1}{2} \log_b y$

81. $\log_b \left(\dfrac{x^2}{y^4} \right)$ **83.** $\log_b \left(\dfrac{xz}{y} \right)$ **85.** $\log_b \left(\dfrac{x^2 y^4}{z^3} \right)$

87. $\log_b \left(\dfrac{y^4 \sqrt{x}}{x} \right)$ **89.** $\left\{\dfrac{9}{4}\right\}$ **91.** {25}

93. {4} **95.** $\left\{\dfrac{19}{8}\right\}$ **97.** {9} **99.** {1}

Problem Set 14.5 (page 736)

1. 0.8597 **3.** 1.7179 **5.** 3.5071 **7.** -0.1373

9. -3.4685 **11.** 411.43 **13.** 90,095

15. 79.543 **17.** 0.048440 **19.** 0.0064150

21. 1.6094 **23.** 3.4843 **25.** 6.0638

27. -0.7765 **29.** -3.4609 **31.** 1.6034

33. 3.1346 **35.** 108.56 **37.** 0.48268

39. 0.035994

41.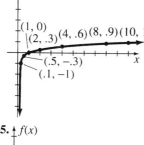

43.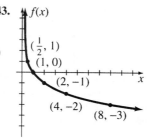

45.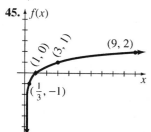

47.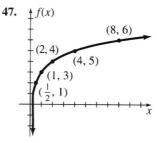

49.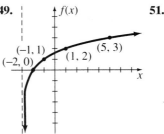

51.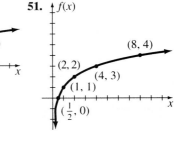

53.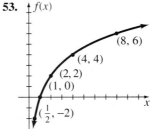

55. 0.36 **57.** 0.73 **59.** 23.10 **61.** 7.93

Problem Set 14.6 (page 746)

1. {2.33} **3.** {2.56} **5.** {5.43} **7.** {4.18}

9. {0.12} **11.** {3.30} **13.** {4.57} **15.** {1.79}

17. {3.32} **19.** {2.44} **21.** {4} **23.** $\left\{\dfrac{19}{47}\right\}$

25. $\left\{\dfrac{-1 + \sqrt{33}}{4}\right\}$ **27.** {1} **29.** {8}

31. {1,10000} **33.** 5.322 **35.** 2.524 **37.** 0.339

39. -0.837 **41.** 3.194 **43.** 2.4 years

45. 5.3 years **47.** 5.9% **49.** 6.8 hours

51. 6100 feet **53.** 3.5 hours **55.** 6.7

57. Approximately 8 times **65.** {1.13}

67. $x = \ln(y + \sqrt{y^2 + 1})$

Chapter 14 Review Problem Set (page 750)

1. 32 **2.** -125 **3.** 81 **4.** 3 **5.** -2

6. $\dfrac{1}{3}$ **7.** $\dfrac{1}{4}$ **8.** -5 **9.** 1 **10.** 12

11. {5} **12.** $\left\{\dfrac{1}{9}\right\}$ **13.** $\left\{\dfrac{7}{2}\right\}$ **14.** {3.40}

15. {8} **16.** $\left\{\dfrac{1}{11}\right\}$ **17.** {1.95} **18.** {1.41}

19. {1.56} **20.** {20} **21.** $\{10^{100}\}$ **22.** {2}

23. $\left\{\dfrac{11}{2}\right\}$ **24.** {0} **25.** 0.3680 **26.** 1.3222

27. 1.4313 **28.** 0.5634

29. a. $\log_b x - 2 \log_b y$ **b.** $\dfrac{1}{4} \log_b x + \dfrac{1}{2} \log_b y$

c. $\dfrac{1}{2} \log_b x - 3 \log_b y$

30. a. $\log_b x^3 y^2$ **b.** $\log_b\left(\dfrac{\sqrt{y}}{x^4}\right)$ **c.** $\log_b\left(\dfrac{\sqrt{xy}}{z^2}\right)$

31. 1.58   **32.** 0.63   **33.** 3.79   **34.** -2.12

35.

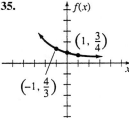

36.

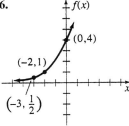

37.

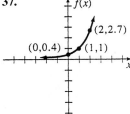

38.

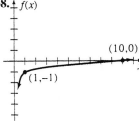

39.

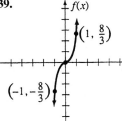

40.

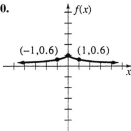

41.

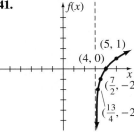

42.

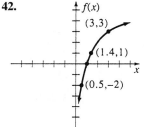

43. \$14,511.58   **44.** \$26,421.30   **45.** \$6087.97

46. Yes   **47.** No   **48.** Yes   **49.** Yes

50. $f^{-1}(x) = \dfrac{x - 5}{4}$ **51.** $f^{-1}(x) = \dfrac{-x - 7}{3}$

52. $f^{-1}(x) = \dfrac{6x + 2}{5}$

53. $f^{-1}(x) = \sqrt{-2 - x}$ for $x \le 2$

54. Increasing on $(-\infty, 4]$ and decreasing on $[4, \infty)$

55. Increasing on $[3, \infty)$

56. Approximately 9.0 years

57. Approximately 22.9 years

58. Approximately 8.7%

59. 61,070; 67,493; 74,591

60. Approximately 4.8 hours   **61.** 133 grams

62. 8.1

Chapter 14 Test (page 752)

1. $\dfrac{1}{2}$ **2.** 1   **3.** 1   **4.** -1 **5.** $\{-3\}$

6. $\left\{-\dfrac{3}{2}\right\}$ **7.** $\left\{\dfrac{8}{3}\right\}$ **8.** $\{243\}$ **9.** $\{2\}$ **10.** $\left\{\dfrac{2}{5}\right\}$

11. 4.1919   **12.** 0.2031   **13.** 0.7325

14. $f^{-1}(x) = \dfrac{-6 - x}{3}$ **15.** $\{5.17\}$

16. $\{10.29\}$ **17.** 4.0069

18. $f^{-1}(x) = \dfrac{3}{2}x + \dfrac{9}{10}$

19. \$6342.08   **20.** 23.4 years

21. 7.8 hours   **22.** 4813 grams

23.

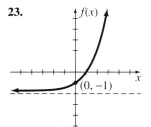

24.

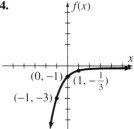

25.

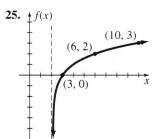

**Cumulative Review Problem Set
(Chapters 1–14) (page 753)**

1. -6 **2.** -8 **3.** $\dfrac{13}{24}$ **4.** 56 **5.** $\dfrac{13}{6}$

6. $-90\sqrt{2}$ **7.** $2x + 5\sqrt{x} - 12$

8. $-18 + 22\sqrt{3}$

9. $2x^3 + 11x^2 - 14x + 4$ **10.** $\dfrac{x + 4}{x(x + 5)}$

11. $\dfrac{16x^2}{27y}$ **12.** $\dfrac{16x + 43}{90}$ **13.** $\dfrac{35a - 44b}{60a^2b}$

14. $\dfrac{2}{x - 4}$ **15.** $2x^2 - x - 4$ **16.** $\dfrac{5y^2 - 3xy^2}{x^2y + 2x^2}$

17. $\dfrac{2y - 3xy}{3x + 4xy}$ **18.** $\dfrac{(2n - 5)(n + 3)}{(n - 2)(3n + 13)}$

19. $\dfrac{3a^2 - 2a + 1}{2a - 1}$ **20.** $(5x - 2)(4x + 3)$

21. $2(2x + 3)(4x^2 - 6x + 9)$
22. $(2x + 3)(2x - 3)(x + 2)(x - 2)$
23. $4x(3x + 2)(x - 5)$ **24.** $(y - 6)(x + 3)$

25. $(5 - 3x)(2 + 3x)$ **26.** $\dfrac{81}{16}$ **27.** 4 **28.** $-\dfrac{3}{4}$

29. -0.3 **30.** $\dfrac{1}{81}$ **31.** $\dfrac{21}{16}$ **32.** $\dfrac{9}{64}$

33. 72 **34.** 6 **35.** -2 **36.** $\dfrac{-12}{x^3y}$

37. $\dfrac{8y}{x^5}$ **38.** $-\dfrac{a^3}{9b}$ **39.** $4\sqrt{5}$ **40.** $-6\sqrt{6}$

41. $\dfrac{5\sqrt{3}}{9}$ **42.** $\dfrac{2\sqrt{3}}{3}$ **43.** $2\sqrt[3]{7}$ **44.** $\dfrac{\sqrt[3]{6}}{2}$

45. $8xy\sqrt{13x}$ **46.** $\dfrac{\sqrt{6xy}}{3y}$ **47.** $11\sqrt{6}$

48. $-\dfrac{169\sqrt{2}}{12}$ **49.** $-16\sqrt[3]{3}$ **50.** $\dfrac{-3\sqrt{2} - 2\sqrt{6}}{2}$

51. $\dfrac{6\sqrt{15} - 3\sqrt{35} - 6 + \sqrt{21}}{5}$ **52.** 0.021

53. 300 **54.** 0.0003 **55.** $32 + 22i$

56. $-17 + i$ **57.** $0 - \dfrac{5}{4}i$ **58.** $-\dfrac{19}{53} + \dfrac{40}{53}i$

59. $-\dfrac{10}{3}$ **60.** $\dfrac{4}{7}$ **61.** $2\sqrt{13}$

62. $5x - 4y = 19$ **63.** $4x + 3y = -18$
64. $(-2, 6)$ and $r = 3$ **65.** $(-5, -4)$ **66.** 8 units

67.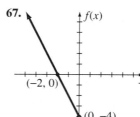

68.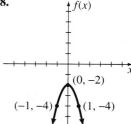

69.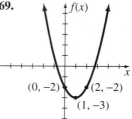

70.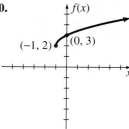

71.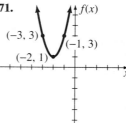

72.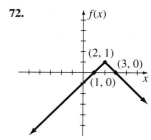

73.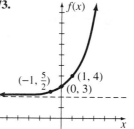

74.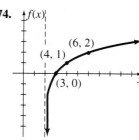

75.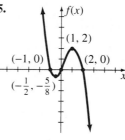

76.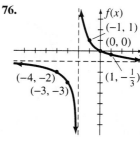

77. $(g \circ f)(x) = 2x^2 - 13x + 20$; $(f \circ g)(x) = 2x^2 - x - 4$

78. $f^{-1}(x) = \dfrac{x + 7}{3}$ **79.** $f^{-1}(x) = -2x + \dfrac{4}{3}$

80. $k = -3$ **81.** $y = 1$ **82.** 12 cubic centimeters

83. $\left\{-\dfrac{21}{16}\right\}$ **84.** $\left\{\dfrac{40}{3}\right\}$ **85.** $\{6\}$

86. $\left\{-\dfrac{5}{2}, 3\right\}$ **87.** $\left\{0, \dfrac{7}{3}\right\}$ **88.** $\{-6, 0, 6\}$

89. $\left\{-\dfrac{5}{6}, \dfrac{2}{5}\right\}$ **90.** $\left\{-3, 0, \dfrac{3}{2}\right\}$ **91.** $\{\pm 1, \pm 3i\}$

92. $\{-5, 7\}$ **93.** $\{-29, 0\}$ **94.** $\left\{\dfrac{7}{2}\right\}$

95. $\{12\}$ **96.** $\{-3\}$ **97.** $\left\{\dfrac{1 \pm 3\sqrt{5}}{3}\right\}$

98. $\left\{\dfrac{-5 \pm 4i\sqrt{2}}{2}\right\}$ **99.** $\left\{\dfrac{3 \pm i\sqrt{23}}{4}\right\}$

100. $\left\{\dfrac{3 \pm \sqrt{3}}{3}\right\}$ **101.** $\{1 \pm \sqrt{34}\}$

102. $\left\{\pm\dfrac{\sqrt{5}}{2}, \pm\dfrac{\sqrt{3}}{3}\right\}$ **103.** $\left\{\dfrac{-5 \pm i\sqrt{15}}{4}\right\}$

104. $\left\{-\dfrac{1}{4}, \dfrac{5}{3}\right\}$ **105.** $\varnothing$ **106.** $\left\{\dfrac{3}{2}\right\}$

107. $\{81\}$ **108.** $\{4\}$ **109.** $\{6\}$ **110.** $\left\{\dfrac{1}{5}\right\}$

111. $(-\infty, 3)$ **112.** $(-\infty, 50]$

113. $\left(-\infty, -\dfrac{11}{5}\right) \cup (3, \infty)$ **114.** $\left(-\dfrac{5}{3}, 1\right)$

115. $\left[-\dfrac{9}{11}, \infty\right)$ **116.** $[-4, 2]$

117. $\left(-\infty, \dfrac{1}{3}\right) \cup (4, \infty)$ **118.** $(-8, 3)$

119. $(-\infty, 3] \cup (7, \infty)$ **120.** $(-6, -3)$

CHAPTER 15

Problem Set 15.1 (page 765)
1. $\{(7, 9)\}$ **3.** $\{(-4, 7)\}$ **5.** $\{(6, 3)\}$

7. $a = -3$ and $b = -4$

9. $\left\{\left(k, \dfrac{2}{3}k - \dfrac{4}{3}\right)\right\}$, a dependent system

11. $u = 5$ and $t = 7$ **13.** $\{(2, -5)\}$

15. $\varnothing$, an inconsistent system   **17.** $\left\{\left(-\dfrac{3}{4}, -\dfrac{6}{5}\right)\right\}$

19. $\left\{\left(\dfrac{1}{2}, -\dfrac{1}{4}\right)\right\}$ **21.** $\{(2, 8)\}$ **23.** $\{(-1, -5)\}$

25. $\varnothing$, an inconsistent system

27. $a = 2$ and $b = -\dfrac{1}{3}$ **29.** $s = -6$ and $t = 12$

31. $\left\{\left(-\dfrac{1}{2}, \dfrac{1}{3}\right)\right\}$ **33.** $\left\{\left(\dfrac{3}{4}, -\dfrac{2}{3}\right)\right\}$ **35.** $\{(-4, 2)\}$

37. $\left\{\left(\dfrac{11}{5}, \dfrac{2}{5}\right)\right\}$ **39.** $\varnothing$, an inconsistent system

41. $\{(12, -24)\}$ **43.** $t = 8$ and $u = 3$

45. $\{(200, 800)\}$ **47.** $\{(400, 800)\}$ **49.** $\{(3.5, 7)\}$

51. 17 and 36   **53.** $15°, 75°$ **55.** 72

57. 34   **59.** 12

61. 210 student tickets and 210 parent tickets

63. \$5000 at 6% and \$15,000 at 8%

65. 3 miles per hour

67. \$22

69. 30 five-dollar bills and 18 ten-dollar bills

75. $\{(4, 6)\}$ **77.** $\{(2, -3)\}$ **79.** $\left\{\left(\dfrac{1}{4}, -\dfrac{2}{3}\right)\right\}$

Problem Set 15.2 (page 775)
1. $\{(-4, -2, 3)\}$ **3.** $\{(-2, 5, 2)\}$

5. $\{(4, -1, -2)\}$ **7.** $\{(3, 1, 2)\}$ **9.** $\varnothing$

11. $\{(-1, 3, 5)\}$ **13.** $\{(-2, -1, 3)\}$

15. $\{(0, 2, 4)\}$

17. $\{(17k - 43, -5k + 13, k)|k$ is a real number$\}$

19. $\{(4, -1, -2)\}$

21. $\{(-4, 0, -1)\}$ **23.** $\{(2, 2, -3)\}$ **25.** $-2, 6, 16$

27. Helmet = \$250; jacket = \$350; gloves = \$50

29. $\angle A = 120°$; $\angle B = 24°$; $\angle C = 36°$

31. Plumber = \$50 per hour; apprentice = \$20 per hour; laborer = \$10 per hour

33. 4 pounds of pecans, 4 pounds of almonds, and 12 pounds of peanuts

35. 7 nickels, 13 dimes, and 22 quarters

37. $40°, 60°,$ and $80°$

39. \$5000 at 2%, \$10,000 at 3%, \$15,000 at 4%

41. 50 of type A, 75 of type B, and 150 of type C

Problem Set 15.3 (page 787)
1. Yes   **3.** Yes   **5.** No   **7.** No

9. Yes   **11.** $\{(-1, -5)\}$ **13.** $\{(3, -6)\}$

15. $\varnothing$ **17.** $\{(-2, -9)\}$ **19.** $\{(-1, -2, 3)\}$

21. $\{(3, -1, 4)\}$ **23.** $\{(0, -2, 4)\}$

25. $\{(-7k + 8, -5k + 7, k)\}$ **27.** $\{(-4, -3, -2)\}$

29. $\{(4, -1, -2)\}$ **31.** $\{(1, -1, 2, -3)\}$

33. $\{(2, 1, 3, -2)\}$ **35.** $\{(-2, 4, -3, 0)\}$

37. $\varnothing$ **39.** $\{(-3k + 5, -1, -4k + 2, k)\}$

41. $\{(-3k + 9, k, 2, -3)\}$

45. $\{(17k - 6, 10k - 5, k)\}$

47. $\left\{\left(-\dfrac{1}{2}k + \dfrac{34}{11}, \dfrac{1}{2}k - \dfrac{5}{11}, k\right)\right\}$ **49.** $\varnothing$

Problem Set 15.4 (page 797)

1. 22   **3.** -29 **5.** 20   **7.** 5   **9.** -2

11. $-\dfrac{2}{3}$ **13.** -25 **15.** 58   **17.** 39

19. -12 **21.** -41 **23.** -8 **25.** 1088

27. -140 **29.** 81   **31.** 146

33. Property 15.3   **35.** Property 15.2

37. Property 15.4   **39.** Property 15.3

41. Property 15.5

Problem Set 15.5 (page 805)

1. $\{(1, 4)\}$ **3.** $\{(3, -5)\}$ **5.** $\{(2, -1)\}$

7. $\varnothing$ **9.** $\left\{\left(-\dfrac{1}{4}, \dfrac{2}{3}\right)\right\}$ **11.** $\left\{\left(\dfrac{2}{17}, \dfrac{52}{17}\right)\right\}$

13. $\{(9, -2)\}$ **15.** $\left\{\left(2, -\dfrac{5}{7}\right)\right\}$ **17.** $\{(0, 2, -3)\}$

19. $\{(2, 6, 7)\}$ **21.** $\{(4, -4, 5)\}$

23. $\{(-1, 3, -4)\}$ **25.** Infinitely many solutions

27. $\left\{\left(-2, \dfrac{1}{2}, -\dfrac{2}{3}\right)\right\}$ **29.** $\left\{\left(3, \dfrac{1}{2}, -\dfrac{1}{3}\right)\right\}$

31. $(-4, 6, 0)$ **37.** $(0, 0, 0)$

39. Infinitely many solutions

Problem Set 15.6 (page 811)

1. $\{(1, 2)\}$ **3.** $\{(1, -5), (-5, 1)\}$

5. $\{(2 + i\sqrt{3}, -2 + i\sqrt{3}), (2 - i\sqrt{3}, -2 - i\sqrt{3})\}$

7. $\{(-6, 7), (-2, -1)\}$ **9.** $\{(-3, 4)\}$

11. $\left\{\left(\dfrac{-1 + i\sqrt{3}}{2}, \dfrac{-7 - i\sqrt{3}}{2}\right),\right.$ $\left.\left(\dfrac{-1 - i\sqrt{3}}{2}, \dfrac{-7 + i\sqrt{3}}{2}\right)\right\}$

13. $\{(-1, 2)\}$ **15.** $\{(-6, 3), (-2, -1)\}$

17. $\{(5, 3)\}$ **19.** $\{(1, 2), (-1, 2)\}$ **21.** $\{(-3, 2)\}$

23. $\{(2, 0), (-2, 0)\}$

25. $\{(\sqrt{2}, \sqrt{3}), (\sqrt{2}, -\sqrt{3}), (-\sqrt{2}, \sqrt{3}), (-\sqrt{2}, -\sqrt{3})\}$

27. $\{(1, 1), (1, -1), (-1, 1), (-1, -1)\}$

29. $\left\{\left(2, \dfrac{3}{2}\right), \left(\dfrac{3}{2}, 2\right)\right\}$

Chapter 15 Review Problem Set (page 815)

1. $\{(3, -7)\}$ **2.** $\{(-1, -3)\}$ **3.** $\{(0, -4)\}$

4. $\left\{\left(\dfrac{23}{3}, -\dfrac{14}{3}\right)\right\}$ **5.** $\{(4, -6)\}$

6. $\left\{\left(-\dfrac{6}{7}, -\dfrac{15}{7}\right)\right\}$ **7.** $\{(-1, 2, -5)\}$

8. $\{(2, -3, -1)\}$ **9.** $\{(5, -4)\}$ **10.** $\{(2, 7)\}$

11. $\{(-2, 2, -1)\}$ **12.** $\{(0, -1, 2)\}$

13. $\{(-3, -1)\}$ **14.** $\{(4, 6)\}$ **15.** $\{(2, -3, -4)\}$

16. $\{(-1, 2, -5)\}$ **17.** $\{(5, -5)\}$

18. $\{(-12, 12)\}$ **19.** $\left\{\left(\dfrac{5}{7}, \dfrac{4}{7}\right)\right\}$

20. $\{(-10, -7)\}$ **21.** $\varnothing$ **22.** $\{(1, 1, -4)\}$

23. $\{(-4, 0, 1)\}$ **24.** $\varnothing$

25. $\{(-4k + 19, -3k + 12, k)|k$ is a real number$\}$

26. $\{(-2, -4, 6)\}$

27. -34 **28.** 13   **29.** -40 **30.** 16

31. 51   **32.** 125   **33.** $\{(-1, 4)\}$

34. $\{(3, 1)\}$ **35.** $\{(-1, -2), (-2, -3)\}$

36. $\left\{\left(\dfrac{4\sqrt{2}}{3}, \dfrac{4}{3}i\right), \left(\dfrac{4\sqrt{2}}{3}, -\dfrac{4}{3}i\right), \left(-\dfrac{4\sqrt{2}}{3}, \dfrac{4}{3}i\right),\right.$ $\left.\left(-\dfrac{4\sqrt{2}}{3}, -\dfrac{4}{3}i\right)\right\}$

37. $\{(0, 2), (0, -2)\}$

38. $\left\{\left(\dfrac{\sqrt{15}}{5}, \dfrac{2\sqrt{10}}{5}\right), \left(\dfrac{\sqrt{15}}{5}, -\dfrac{2\sqrt{10}}{5}\right), \left(-\dfrac{\sqrt{15}}{5}, \dfrac{2\sqrt{10}}{5}\right),\right.$ $\left.\left(-\dfrac{\sqrt{15}}{5}, -\dfrac{2\sqrt{10}}{5}\right)\right\}$

39. 72   **40.** $900 at 10% and $1600 at 12%

41. 20 nickels, 32 dimes, and 54 quarters

42. $25°$, $45°$, and $110°$

Chapter 15 Test (page 817)

1. III   **2.** I   **3.** III   **4.** II

5. 8   **6.** $-\dfrac{7}{12}$ **7.** -18 **8.** 112

9. Infinitely many   **10.** $\{(-2, 4)\}$

11. $\{(3, -1)\}$ **12.** $x = -12$ **13.** $y = -\dfrac{13}{11}$

14. Two   **15.** $x = 14$ **16.** $y = 13$

17. Infinitely many   **18.** None

19. $\left\{\left(\dfrac{11}{5}, 6, -3\right)\right\}$ **20.** $\{(-2, -1, 0)\}$

21. $x = 1$ **22.** $y = 4$

23. $\{(3, 2), (-3, -2), \left(4, \dfrac{3}{2}\right), \left(-4, -\dfrac{3}{2}\right)\}$

24. 2 liters   **25.** 22 quarters

Cumulative Review Problem Set (Word Problems) (page 819)

1. $65 per hour   **2.** 17, 19, and 21   **3.** $70°$

4. 14 nickels, 20 dimes, and 29 quarters

5. $30°$, $50°$, and $100°$ **6.** $9.30 per hour   **7.** $600

8. $1700 at 8% and $2000 at 9%   **9.** 69 or less

10. 30 shares at $10 per share

11. 8 inches by 14 inches   **12.** -3, 0, or 3

13. 66 miles per hour and 76 miles per hour

14. 4 quarts   **15.** $16\frac{2}{3}$ years

16. Approximately 11.64 years
17. Approximately 11.55 years   **18.** 6 centimeters
19. 1 hour   **20.** 1 inch
21. $1050 and $1400   **22.** 3 hours   **23.** 37

24. $4\frac{4}{5}$ days   **25.** $4000

26. Freight train: 6 hours at 50 miles per hour; express
train: 4 hours at 70 miles per hour
27. $66.2° < F < 71.6°$
28. 150   **29.** $1 + \sqrt{2}$ and $1 - \sqrt{2}$
30. Larry: 52 miles per hour; Nita: 54 miles per hour
31. 15 rows and 20 seats per row
32. 4 units   **33.** 30 students
34. 4 yards and 6 yards   **35.** $6\sqrt{2}$ inches
36. $3.59 per movie and $4.99 per game
37. 21, 22, and 23   **38.** 12 nickels and 23 dimes
39. $[14, \infty)$ **40.** 25 milliliters

CHAPTER 16

Problem Set 16.1 (page 828)
1. $-4, -1, 2, 5, 8$ **3.** $2, 0, -2, -4, -6$
5. $2, 11, 26, 47, 74$ **7.** $0, 2, 6, 12, 20$
9. $4, 8, 16, 32, 64$ **11.** $a_{15} = -79; a_{30} = -154$
13. $a_{25} = 1; a_{50} = -1$ **15.** $a_n = 2n + 9$

17. $a_n = -3n + 5$ **19.** $a_n = \dfrac{n + 2}{2}$

21. $a_n = 4n - 2$ **23.** $a_n = -3n$ **25.** 73
27. 334   **29.** 35   **31.** 7   **33.** 86
35. 2700   **37.** 3200   **39.** -7950 **41.** 637.5
43. 4950   **45.** 1850   **47.** -2030 **49.** 3591
51. 40,000   **53.** 58,250   **55.** 2205
57. -1325 **59.** $44,070   **61.** 11,350 students
63. $116.25   **65.** 136 seats, 2200 seats
67. $24,900

Problem Set 16.2 (page 839)

1. $a_n = 3(2)^{n-1}$ **3.** $a_n = 3^n$ **5.** $a_n = \left(\dfrac{1}{2}\right)^{n+1}$

7. $a_n = 4^n$ **9.** $a_n = (0.3)^{n-1}$

11. $a_n = (-2)^{n-1}$ **13.** 64   **15.** $\dfrac{1}{9}$

17. -512 **19.** $\dfrac{1}{4374}$ **21.** $\dfrac{2}{3}$ **23.** 2

25. 1023   **27.** 19,682   **29.** $394\dfrac{1}{16}$

31. 1364   **33.** 1089   **35.** $7\dfrac{511}{512}$ **37.** -547

39. 4   **41.** 3   **43.** No sum   **45.** $\dfrac{27}{4}$

47. 2   **49.** $\dfrac{16}{3}$ **51.** $\dfrac{1}{3}$ **53.** $\dfrac{26}{99}$

55. $\dfrac{41}{333}$ **57.** $\dfrac{4}{15}$ **59.** $\dfrac{106}{495}$ **61.** $\dfrac{7}{3}$

63. 7320   **65.** 125 liters   **67.** 512 gallons
69. $163.84; $327.67   **71.** $113,280   **73.** $10,000

75. 3 grams   **77.** $298\dfrac{7}{16}$ feet   **79.** 5.9%

81. $\dfrac{5}{64}$ of a gallon

Problem Set 16.3 (page 847)
1. 20   **3.** 24   **5.** 168   **7.** 48   **9.** 36
11. 6840   **13.** 720   **15.** 720   **17.** 36
19. 24   **21.** 243   **23.** Impossible   **25.** 216
27. 26   **29.** 36   **31.** 144   **33.** 1024
35. 30   **37. a.** 6,084,000   **c.** 3,066,336

Problem Set 16.4 (page 856)
1. 60   **3.** 360   **5.** 21   **7.** 252   **9.** 105
11. 1   **13.** 24   **15.** 84   **17. a.** 336
19. 2880   **21.** 2450   **23.** 10   **25.** 10
27. 35   **29.** 1260   **31.** 2520   **33.** 15
35. 126   **37.** 144; 202   **39.** 15; 10   **41.** 20

43. 10; 15; 21; $\dfrac{n(n-1)}{2}$ **47.** 120

53. 133,784,560   **55.** 54,627,300

Problem Set 16.5 (page 863)

1. $\dfrac{1}{2}$ **3.** $\dfrac{3}{4}$ **5.** $\dfrac{1}{8}$ **7.** $\dfrac{7}{8}$ **9.** $\dfrac{1}{16}$

11. $\dfrac{3}{8}$ **13.** $\dfrac{1}{3}$ **15.** $\dfrac{1}{2}$ **17.** $\dfrac{5}{36}$ **19.** $\dfrac{1}{6}$

21. $\dfrac{11}{36}$ **23.** $\dfrac{1}{4}$ **25.** $\dfrac{1}{2}$ **27.** $\dfrac{1}{25}$ **29.** $\dfrac{9}{25}$

31. $\dfrac{2}{5}$ **33.** $\dfrac{9}{10}$ **35.** $\dfrac{5}{14}$ **37.** $\dfrac{15}{28}$

39. $\dfrac{7}{15}$ **41.** $\dfrac{1}{15}$ **43.** $\dfrac{2}{3}$ **45.** $\dfrac{1}{5}$ **47.** $\dfrac{1}{63}$

49. $\dfrac{1}{2}$ **51.** $\dfrac{5}{11}$ **53.** $\dfrac{1}{6}$ **55.** $\dfrac{21}{128}$

57. $\dfrac{13}{16}$ **59.** $\dfrac{1}{21}$ **63.** 40   **65.** 3744

67. 10,200   **69.** 123,552   **71.** 1,302,540

Problem Set 16.6 (page 870)

1. $x^8 + 8x^7y + 28x^6y^2 + 56x^5y^3 + 70x^4y^4 + 56x^3y^5 + 28x^2y^6 + 8xy^7 + y^8$

3. $x^6 - 6x^5y + 15x^4y^2 - 20x^3y^3 + 15x^2y^4 - 6xy^5 + y^6$

5. $a^4 + 8a^3b + 24a^2b^2 + 32ab^3 + 16b^4$

7. $x^5 - 15x^4y + 90x^3y^2 - 270x^2y^3 + 405xy^4 - 243y^5$

9. $16a^4 - 96a^3b + 216a^2b^2 - 216ab^3 + 81b^4$

11. $x^{10} + 5x^8y + 10x^6y^2 + 10x^4y^3 + 5x^2y^4 + y^5$

13. $16x^8 - 32x^6y^2 + 24x^4y^4 - 8x^2y^6 + y^8$

15. $x^6 + 18x^5 + 135x^4 + 540x^3 + 1215x^2 + 1458x + 729$

17. $x^9 - 9x^8 + 36x^7 - 84x^6 + 126x^5 - 126x^4 + 84x^3 - 36x^2 + 9x - 1$

19. $1 + \dfrac{4}{n} + \dfrac{6}{n^2} + \dfrac{4}{n^3} + \dfrac{1}{n^4}$

21. $a^6 - \dfrac{6a^5}{n} + \dfrac{15a^4}{n^2} - \dfrac{20a^3}{n^3} + \dfrac{15a^2}{n^4} - \dfrac{6a}{n^5} + \dfrac{1}{n^6}$

23. $17 + 12\sqrt{2}$ **25.** $843 - 589\sqrt{2}$

27. $x^{12} + 12x^{11}y + 66x^{10}y^2 + 220x^9y^3$

29. $x^{20} - 20x^{19}y + 190x^{18}y^2 - 1140x^{17}y^3$

31. $x^{28} - 28x^{26}y^3 + 364x^{24}y^6 - 2912x^{22}y^9$

33. $a^9 + \dfrac{9a^8}{n} + \dfrac{36a^7}{n^2} + \dfrac{84a^6}{n^3}$

35. $x^{10} - 20x^9y + 180x^8y^2 - 960x^7y^3$ **37.** $56x^5y^3$

39. $126x^5y^4$ **41.** $189a^2b^5$ **43.** $120x^6y^{21}$

45. $\dfrac{5005}{n^6}$ **51.** $-117 + 44i$ **53.** $-597 - 122i$

Chapter 16 Review Problem Set (page 872)

1. $a_n = 6n - 3$ **2.** $a_n = 3^{n-2}$ **3.** $a_n = 5 \cdot 2^n$

4. $a_n = -3n + 8$ **5.** $a_n = 2n - 7$

6. $a_n = 3^{3-n}$ **7.** $a_n = -(-2)^{n-1}$

8. $a_n = 3n + 9$ **9.** $a_n = \dfrac{n+1}{3}$

10. $a_n = 4^{n-1}$ **11.** 73   **12.** 106   **13.** $\dfrac{1}{32}$

14. $\dfrac{4}{9}$ **15.** -92 **16.** $\dfrac{1}{16}$ **17.** -5

18. 85   **19.** $\dfrac{5}{9}$ **20.** 2 or -2 **21.** $121\dfrac{40}{81}$

22. 7035   **23.** $-10,725$ **24.** $31\dfrac{31}{32}$

25. 32,015   **26.** 4757   **27.** $85\dfrac{21}{64}$

28. 37,044   **29.** 12,726   **30.** $85\dfrac{1}{3}$

31. $\dfrac{4}{11}$ **32.** $\dfrac{41}{90}$ **33.** $750   **34.** $46.50

35. $3276.70   **36.** 10,935 gallons

37. 720   **38.** 30,240   **39.** 150   **40.** 1440

41. 20   **42.** 525   **43.** 1287   **44.** 264

45. 74   **46.** 55   **47.** 40   **48.** 15

49. 60   **50.** 120   **51.** $\dfrac{3}{8}$ **52.** $\dfrac{5}{16}$

53. $\dfrac{5}{36}$ **54.** $\dfrac{13}{18}$ **55.** $\dfrac{3}{5}$ **56.** $\dfrac{1}{35}$

57. $\dfrac{57}{64}$ **58.** $\dfrac{1}{221}$ **59.** $\dfrac{1}{6}$ **60.** $\dfrac{4}{7}$

61. $\dfrac{4}{7}$ **62.** $\dfrac{10}{21}$ **63.** $\dfrac{140}{143}$

64. $x^5 + 10x^4y + 40x^3y^2 + 80x^2y^3 + 80xy^4 + 32y^5$

65. $x^8 - 8x^7y + 28x^6y^2 - 56x^5y^3 + 70x^4y^4 - 56x^3y^5 + 28x^2y^6 - 8xy^7 + y^8$

66. $a^8 - 12a^6b^3 + 54a^4b^6 - 108a^2b^9 + 81b^{12}$

67. $x^6 + \dfrac{6x^5}{n} + \dfrac{15x^4}{n^2} + \dfrac{20x^3}{n^3} + \dfrac{15x^2}{n^4} + \dfrac{6x}{n^5} + \dfrac{1}{n^6}$

68. $41 - 29\sqrt{2}$ **69.** $-a^3 + 3a^2b - 3ab^2 + b^3$

70. $-1760x^9y^3$ **71.** $57915a^4b^{18}$

Chapter 16 Test (page 875)

1. $a_n = 5(2)^{1-n}$ **2.** $a_n = 6n + 4$ **3.** 223

4. 765   **5.** 7155   **6.** 6138   **7.** 22,650

8. 6   **9.** $\dfrac{2}{11}$ **10.** 3 liters   **11.** $3276.70

12. 44 members; 260 members   **13.** 270   **14.** 20

15. 144   **16.** 2520   **17.** 350   **18.** $\dfrac{13}{18}$ **19.** $\dfrac{5}{16}$

20. $\dfrac{5}{6}$ **21.** $\dfrac{1}{7}$ **22.** $\dfrac{23}{28}$

23. $64 - \dfrac{192}{n} + \dfrac{240}{n^2} - \dfrac{160}{n^3} + \dfrac{60}{n^4} - \dfrac{12}{n^5} + \dfrac{1}{n^6}$

24. $243x^5 + 810x^4y + 1080x^3y^2 + 720x^2y^3 + 240xy^4 + 32y^5$

25. $\dfrac{495}{256}x^4$

Index

Metric System

Basic Prefixes

kilo-	10^3	= 1000	**kilo** means 1000 times basic unit
hecto-	10^2	= 100	**hecto** means 100 times basic unit
deka-	10^1	= 10	**deka** means 10 times basic unit
deci-	$10^{-1} = \dfrac{1}{10} = .1$		**deci** means .1 times basic unit
centi-	$10^{-2} = \dfrac{1}{100} = .01$		**centi** means .01 times basic unit
milli-	$10^{-3} = \dfrac{1}{1000} = .001$		**milli** means .001 times basic unit

Length

The **meter** is the basic unit.

1 **kilo**meter	= 1000 meters	1 km	= 1000 m
1 **hecto**meter	= 100 meters	1 hm	= 100 m
1 **deka**meter	= 10 meters	1 dam	= 10 m
1 **deci**meter	= $\dfrac{1}{10}$ of a meter	1 dm	= .1 m
1 **centi**meter	= $\dfrac{1}{100}$ of a meter	1 cm	= .01 m
1 **milli**meter	= $\dfrac{1}{1000}$ of a meter	1 mm	= .001 m

Volume

The **liter** is the basic unit.

1 **kilo**liter	= 1000 liters	1 kL	= 1000 L
1 **hecto**liter	= 100 liters	1 hL	= 100 L
1 **deka**liter	= 10 liters	1 daL	= 10 L
1 **deci**liter	= $\dfrac{1}{10}$ of a liter	1 dL	= .1 L
1 **centi**liter	= $\dfrac{1}{100}$ of a liter	1 cL	= .01 L
1 **milli**liter	= $\dfrac{1}{1000}$ of a liter	1 mL	= .001 L